Earth Environments

Cyclone Nargis just before it hits land in Mayanmar (Burma). Source: NASA

Earth Environments

Past, Present and Future

David Huddart and **Tim Stott**

Liverpool John Moores University, UK

A John Wiley & Sons, Ltd., Publication

This edition first published 2010
©2010 by John Wiley & Sons, Ltd

Wiley-Blackwell is an imprint of John Wiley & Sons, formed by the merger of Wiley's global Scientific, Technical and Medical business with Blackwell Publishing.

Registered office: John Wiley & Sons Ltd, The Atrium, Southern Gate, Chichester, West Sussex, PO19 8SQ, UK

Other Editorial Offices:
9600 Garsington Road, Oxford, OX4 2DQ, UK
111 River Street, Hoboken, NJ 07030-5774, USA

For details of our global editorial offices, for customer services and for information about how to apply for permission to reuse the copyright material in this book please see our website at www.wiley.com/wiley-blackwell

Library of Congress Cataloguing-in-Publication Data

Huddart, David.
    Earth environments : past, present, and future / David Huddart and Tim Stott.
        p. cm.
    ISBN 978-0-471-48532-2 (cloth)
    1. Earth sciences–Textbooks. I. Stott, Tim. II. Title.
    QE28.H84 2010
    550–dc22
                                                        2009049245

ISBN: 978-0-471-4853-2 (HB)   978-0-471-48533-9 (PB)

A catalogue record for this book is available from the British Library.

Set in 10/12 pt Times by Laserwords Private Limited, Chennai, India
Printed and bound in Malaysia by Vivar Printing Sdn Bhd

First printing   2010

Contents

Introduction   xi

SECTION I   Introduction to Earth Systems   1

1   Introduction to Earth Systems   3

1.1   Introduction to the Earth's formation   4
1.2   Introduction to Earth spheres   5
1.3   Scales in space and time   6
1.4   Systems and feedback   7
1.5   Open and closed flow systems   8
1.6   Equilibrium in systems   9
1.7   Time cycles in systems   11
     Exercises   13
     References   13

SECTION II   Atmospheric and Ocean Systems   15

2   Structure and Composition of the Atmosphere   17

2.1   Structure of the atmosphere   18
2.2   Composition of the atmosphere   18
     Exercises   22
     References   23
     Further reading   23

3   Energy in the Atmosphere and the Earth Heat Budget   25

3.1   Introduction   26
3.2   Solar radiation   26
     Exercises   36
     References   36
     Further reading   36

4   Moisture in the Atmosphere   37

4.1   Introduction   38
4.2   The global hydrological cycle   38
4.3   Air stability and instability   42
4.4   Clouds   42
4.5   Precipitation   44
     Exercises   48
     References   48
     Further reading   48

5   Atmospheric Motion   49

5.1   Introduction   50
5.2   Atmospheric pressure   50
5.3   Winds and pressure gradients   51
5.4   The global pattern of atmospheric circulation   55
     Exercises   58
     References   59
     Further reading   59

6   Weather Systems   61

6.1   Introduction   62
6.2   Macroscale synoptic systems   62
6.3   Meso-scale: Local winds   71
6.4   Microclimates   74
6.5   Weather observation and forecasting   79
     Exercises   86
     References   87
     Further reading   87

7   World Climates   89

7.1   Introduction   90

7.2   Classification of climate   90

      Exercises   100

      References   100

      Further reading   100

8 **Ocean Structure and Circulation Patterns**   **101**

   8.1   Introduction   102

   8.2   Physical structure of the oceans   102

   8.3   Temperature structure of the oceans   105

   8.4   Ocean circulation   105

   8.5   Sea-level change   109

         Exercises   110

         References   110

         Further reading   110

9 **Atmospheric Evolution and Climate Change**   **111**

   9.1   Evolution of the Earth's atmosphere   112

         Exercises   115

         References   115

         Further reading   115

10 **Principles of Climate Change**   **117**

    10.1   Introduction   118

    10.2   Evidence for climate change   118

    10.3   Causes of climate change   125

           Exercises   136

           References   136

           Further reading   136

SECTION III   Endogenic Geological
**   Systems   137**

11 **Earth Materials: Mineralogy, Rocks and**
    the Rock Cycle   **139**

    11.1   What is a mineral?   140

    11.2   Rocks and the rock cycle   148

    11.3   Vulcanicity and igneous rocks   151

    11.4   Sedimentary rocks, fossils and
           sedimentary structures   153

    11.5   Metamorphic rocks   159

      Exercises   162

      References   163

      Further reading   163

12 **The Internal Structure of the Earth**   **165**

    12.1   Introduction   166

    12.2   Evidence of the Earth's composition
           from drilling   166

    12.3   Evidence of the Earth's composition
           from volcanoes   167

    12.4   Evidence of the Earth's composition
           from meteorites   167

    12.5   Using earthquake seismic waves as
           Earth probes   168

           Exercises   171

           References   171

           Further reading   171

13 **Plate Tectonics and Volcanism: Processes,**
    Products and Landforms   **173**

    13.1   Introduction   174

    13.2   Global tectonics: how plates, basins
           and mountains are created   174

    13.3   Volcanic processes and the global
           tectonic model   177

    13.4   Magma eruption   186

    13.5   Explosive volcanism   194

    13.6   Petrographic features of volcaniclastic
           sediments   200

    13.7   Transport and deposition of
           pyroclastic materials   200

    13.8   The relationship between volcanic
           processes and the Earth's atmosphere
           and climate   210

    13.9   Relationships between volcanic
           eruptions and biotic evolution   216

    13.10 Plate tectonics, uniformitarianism and
           Earth history   217

           Exercises   223

           References   223

           Further reading   225

14 Geotectonics: Processes, Structures and Landforms 227
  14.1 Introduction 228
  14.2 Tectonic structures 228
  14.3 Tectonic structures as lines of weakness in landscape evolution 234
    Exercises 235
    References 235
    Further reading 235

SECTION IV Exogenic Geological Systems 237

15 Weathering Processes and Products 239
  15.1 Introduction 240
  15.2 Physical or mechanical weathering 242
  15.3 Chemical weathering 251
  15.4 Measuring weathering rates 262
  15.5 Weathering landforms 262
    Exercises 267
    References 267
    Further reading 268

16 Slope Processes and Morphology 269
  16.1 Introduction 270
  16.2 Slopes: mass movement 270
  16.3 Hillslope hydrology and slope processes 297
  16.4 Slope morphology and its evolution 305
    Exercises 318
    References 318
    Further reading 319

17 Fluvial Processes and Landform– Sediment Assemblages 321
  17.1 Introduction 322
  17.2 Loose boundary hydraulics 322
  17.3 The energy of a river and its ability to do work 323
  17.4 Transport of the sediment load 325

  17.5 Types of sediment load 327
  17.6 River hydrology 328
  17.7 The drainage basin 329
  17.8 Drainage patterns and their interpretation 332
  17.9 Fluvial channel geomorphology 332
    Exercises 376
    References 376
    Further reading 378

18 Carbonate Sedimentary Environments and Karst Processes and Landforms 379
  18.1 Introduction 380
  18.2 Carbonate sedimentary environments and the creation of carbonate rock characteristics 380
  18.3 Evaporites 394
  18.4 Carbonate facies models 395
  18.5 Karst processes 401
    Exercises 427
    References 428
    Further reading 429

19 Coastal Processes, Landforms and Sediments 431
  19.1 Introduction to the coastal zone 432
  19.2 Sea waves, tides and tsunamis 433
  19.3 Tides 439
  19.4 Tsunamis 445
  19.5 Coastal landsystems 445
  19.6 Distribution of coastal landsystems 489
  19.7 The impact of climatic change on coastal landsystems: What lies in the future? 492
    Exercises 498
    References 498
    Further reading 499

20 Glacial Processes and Landsystems 501
  20.1 Introduction 502
  20.2 Mass balance and glacier formation 504

20.3 Mass balance and glacier flow 510
20.4 Surging or galloping glaciers 512
20.5 Processes of glacial erosion and deposition 515
20.6 Glacial landsystems 536
Exercises 562
References 562
Further reading 563

21 Periglacial Processes and Landform–Sediment Assemblages 565
21.1 Introduction to the term 'periglacial' 566
21.2 Permafrost 566
21.3 Periglacial processes and landforms 569
21.4 Frost heaving and frost thrusting 571
21.5 Landforms associated with frost sorting 573
21.6 Needle ice development 574
21.7 Frost cracking and the development of ice wedges 574
21.8 Growth of ground ice and its decay, and the development of pingos, thufurs and palsas 578
21.9 Processes associated with snowbanks (nivation processes) 583
21.10 Cryoplanation or altiplanation processes and their resultant landforms 585
21.11 The development of tors 588
21.12 Slope processes associated with the short summer melt season 593
21.13 Cambering and associated structures 596
21.14 Wind action in a periglacial climate 597
21.15 Fluvial processes in a periglacial environment 600
21.16 Alluvial fans in a periglacial region 602
21.17 An overview of the importance of periglacial processes in shaping the landscape of upland Britain 603
21.18 The periglaciation of lowland Britain 607
Exercises 607
References 607

22 Aeolian (Wind) Processes and Landform–Sediment Assemblages 611
22.1 Introduction 612
22.2 Current controls on wind systems 613
22.3 Sediment entrainment and processes of sand movement 613
22.4 Processes of wind transport 614
22.5 Aeolian bedforms 616
22.6 Dune and aeolian sediments 629
22.7 Dust and loess deposition 631
22.8 Wind erosion landforms 633
Exercises 637
References 637
Further reading 639

SECTION V Principles of Ecology and Biogeography 641
23 Principles of Ecology and Biogeography 643
23.1 Introduction 644
23.2 Why do organisms live where they do? 644
23.3 Components of ecosystems 647
23.4 Energy flow in ecosystems 651
23.5 Food chains and webs 657
23.6 Pathways of mineral matter (biogeochemical cycling) 660
23.7 Vegetation succession and climaxes 665
23.8 Concluding remarks 681
Exercises 682
References 683
Further reading 683

24 Soil-forming Processes and Products 685
24.1 Introduction 686
24.2 Controls on soil formation 686
24.3 Soils as systems 689
24.4 Soil profile development 690
24.5 Soil properties 696
24.6 Soil description in the field 704
24.7 Key soil types, with a description and typical profile 707
24.8 Podsolization: theories 711

24.9   Soil classification   712
24.10   Regional and local soil distribution   720
24.11   The development of dune soils: an
        example from the Sefton coast   729
24.12   The development of woodland soils in
        Delamere Forest   730
24.13   Intrazonal soils caused by topographic
        change   731
24.14   Palaeosols   731
        Exercises   732
        References   732
        Further reading   733

25 World Ecosystems   735
25.1   Introduction   736
25.2   The tundra ecozone   737
25.3   The tropical (equatorial) rain forest, or
        humid tropics *sensu stricto*, ecozone   744
25.4   The seasonal tropics or savanna
        ecozone   750
25.5   Potential effects of global warming on
        the world's ecozones   757
        Exercises   759
        References   759
        Further reading   759

SECTION VI   Global Environmental Change:
             Past, Present and Future   761

26 The Earth as a Planet: Geological Evolution
    and Change   763
26.1   Introduction   764
26.2   How unique is the Earth as a planet?   764
26.3   What do we really know about the early
        Earth?   765
26.4   The early geological record   765
26.5   The first Earth system   768
26.6   How did the Earth's core form?   768
26.7   Evolution of the Earth's mantle   769
26.8   Evolution of the continental crust   777
        Exercises   778
        References   779

        Further reading   779

27 Atmospheric Evolution and Climate Change   781
27.1   Evolution of the Earth's atmosphere   782
27.2   Future climate change   783
        Exercises   788
        References   788
        Further reading   788

28 Change in Ocean Circulation and the
    Hydrosphere   789
28.1   Introduction   790
28.2   Sea-level change and the
        supercontinental cycle   790
28.3   Ocean circulation in a warming climate 794
        Exercises   795
        References   795
        Further reading   795

29 Biosphere Evolution and Change in the
    Biosphere   797
29.1   Introduction   798
29.2   Mechanisms of evolution in the fossil
        record   798
29.3   The origins of life   801
29.4   An outline history of the Earth's
        biospheric evolution   803
29.5   Mass extinctions and catastrophes in
        the history of life on Earth   817
        Exercises   824
        References   825
        Further reading   825

30 Environmental Change: Greenhouse and
    Icehouse Earth Phases and Climates Prior
    to Recent Changes   827
30.1   Introduction   828
30.2   Early glaciations in the Proterozoic
        phase of the Pre-Cambrian (the
        Snowball Earth hypothesis)   828
30.3   Examples of changes from greenhouse
        to icehouse climates in the Earth's past 833

30.4 Late Cenozoic ice ages: rapid climate
      change in the Quaternary   845
30.5 Late Glacial climates and evidence for
      rapid change   852
30.6 The Medieval Warm Period or Medieval
      Climate Optimum and the Little Ice
      Age   860
      Exercises   864
      References   864
      Further reading   866

31 Global Environmental Change in the Future   867
31.1 Introduction   868

31.2 Future climate change   868
31.3 Change in the geosphere   869
31.4 Change in the oceans and hydrosphere 872
31.5 Change in the biosphere   873
31.6 A timeline for future Earth   873
31.7 Causes for future optimism?   874
31.8 Concluding remarks   876
      Exercises   877
      References   877
      Further reading   878

Index   879

Introduction

In the year that the major part of this book has been written there has never been a greater need for an understanding of modern Earth processes, how these have changed over geological time and how they may impact on the planet's future. As we write this preface in early May 2008 a natural disaster has just occurred in the Irrawaddy delta in Mayanmar (the former Burma), when a tropical cyclone in the Bay of Bengal (see Frontispiece) produced a storm surge in the delta area, resulting in catastrophic loss of life, initially from extremely strong winds and flooding (Figure 0.1). Media reports on such natural disasters, whether floods, tsunamis, earthquakes, volcanoes or mass movements, rightly stress the human impact, but there is much to understand too about the physical processes behind them.

We might ask ourselves whether global warming made Cyclone Nargis worse, as some scientists have argued that storms are more likely in a warming world. Similar debates followed Hurricane Katrina in the Gulf of Mexico in 2005. Warmer seas could make such storms more intense, though not more frequent. In 2007 reports from the Intergovernmental Panel on Climate Change suggested that it was likely that future cyclones would be more intense, while other research has suggested that future storm strength might increase in some places but decrease in others.

The fact that global warming is now having acknowledged repercussions, which are reported in the press and on television on a regular basis, has to have a central place in a book on Earth environments. Currently there is a preoccupation with climate change brought about by the activities of mankind since the Industrial Revolution, and rightly so, because global warming is playing a crucial role in present-day global processes, whether in the atmosphere, the oceans, geomorphological processes or the ecology. There are effects throughout all the Earth's systems and many feedback loops are occurring because of the warming. We are rightly concerned with what might happen to the Earth in the future as a result of these climatic changes, but it has to be realized that the planet has had many climatic oscillations, both warmer and cooler, on a bigger scale than that seen today, though these occurred prior to human evolution. The difference now is that humans are the dominant controller of certain atmospheric processes that are contributing in a major way to climatic change. We hope that we can learn from similar climate changes in the geological record in order to show that we need to alter our current wilful disregard for our environment.

The 'anthropocene', as the current period of man-induced change has been called, is hopefully not the next phase of global mass extinction, where the causes of that extinction are man-imposed. Whilst global mass extinctions have similarly been common in the history of life on Earth, they have never before been directly caused by us. Currently, however, this is what is happening, and the World Wildlife Fund Living Planet Index illustrates this very well. It monitors the 302 species of mammal, 811 birds, 83 amphibians and 40 reptiles on the planet and has

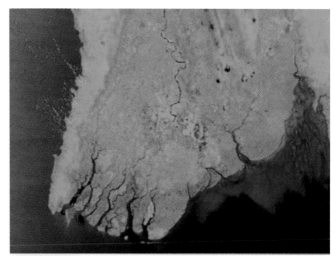

Figure 0.1 Before and after satellite images – taken on 15 April 2008 (top) and 5 May 2008 (bottom) – showing the extent of flooding along the Mayanmar (Burma) coast as a result of Cyclone Nargis. (Source: NASA/MODIS Rapid Response Team)

found from figures published this year that populations decreased by an average of 27% between 1970 and 2005. Land-based species fell by 25% over this period, whilst marine species were particularly hard-hit, falling by 28%. Seabirds have suffered a rapid decline of around 30% since the mid 1990s. Most of the problem is due to development, overfishing, intensive farming, habitat loss, wildlife trade, pollution and man-made climate change. The latter will be an increasingly important factor affecting species in the next 30 years. Overall we are consuming some 25% more natural resources than the Earth can replace.

In the past, mass extinctions have usually been caused by the repercussions of impact from extraterrestrial bodies such as comets; by major volcanic processes, the build-up of huge plateau basalt provinces and supervolcanoes; by major phases of glaciation; and by the effects all of these have had on the Earth's atmospheric processes. In fact, it appears that human populations may have been decimated by the Toba supervolcano 70 000 years ago and that the current human domination of the planet started after the repercussions of that event from a very small population base, estimated to have been as low as 2000 people.

Another environmental natural disaster has just occurred literally in the middle of the writing of this preface: a large earthquake, 7.9 on the Richter scale, in Wenchuan County in the Sichuan Province of China, 92 km north-west of Chengdu, on 12 May 2008 (Figure 0.2). The death toll stands already at over 50 000 people after only a few days. A further strong

USGS ShakeMap: EASTERN SICHUAN, CHINA
Mon May 12, 2008 06:28:01 GMT M 7.9 N31.02 E103.37 Depth: 19.0 km ID:2008ryan

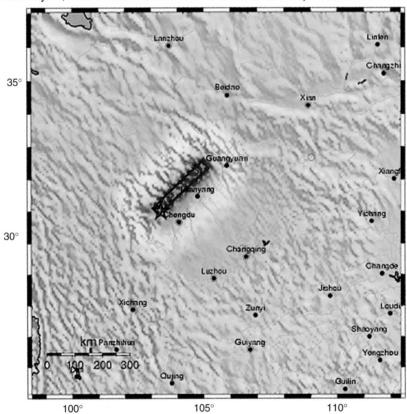

Map Version 7 Processed Tue May 13, 2005 09:12:25 AM MDT–NDT REVIEWED BY HUMAN

PERCEIVED SHAKING	Not felt	Weak	Light	Moderate	Strong	Very Strong	Severe	Violent	Extreme
POTENTIAL DAMAGE	none	none	none	Very light	Light	Moderate	Moderate Heavy	Heavy	Very Heavy
PEAK ACC(%g)	<.17	.17-1.4	1.4-3.9	3.9-9.2	9.2-18	18-34	34-65	65-124	>124
PEAK VEL(cm/s)	<0.1	0.1-1.1	1.1-3.4	3.4-8.1	8.1-16	16-31	31-60	60-116	>116
INSTRUMENTAL INTENSITY	I	II-III	IV	V	VI	VII	VIII	IX	X+

Figure 0.2 Earthquake shake map. (Source: Wikimedia Commons. USGS provided data on 12 May 2008, http://www.earthquake.usgs.gov/eqcenter/shakeup/global/shake/2008ryan)

aftershock at 5.9 occurred on 16 May and there have been many others since, with associated landslides, mudflows and threats of floods. This emphasises that it is not just atmospheric processes that humans have to worry about, but uncontrollable natural tectonic and volcanic processes on a variety of scales, inflicting damage wherever humans happen to live close by. We can do little to prevent such disasters, except perhaps to use increasingly sophisticated technology to, for example, build earthquake-proof buildings. Unfortunately such building do not appear to have existed in China, where many schools and poorly-constructed flats failed (Figure 0.3). It may be possible to accelerate small-scale movement along known fault lines, rather than to let the pressure build up (so that when the energy is liberated it is on a devastating scale), or to divert lava flows away from settlements. The search for predictors of natural disasters is another way of moving people quickly away from the likely impact zones.

However, when it comes to the megascale tsunami, super-volcanic eruption or superearthquake, humans can do nothing much in the face of such potentially devastating natural disasters. Some of these megascale events have not been witnessed by human populations in historical time but they have occurred in the geological past and will do so again as part of the Earth's set of processes, which operate both internally within the Earth and externally from the solar system.

For these reasons we feel that there is a need for more education about the past, current and future processes that occur on and in our planet. This is especially so when the United Nations Educational, Scientific and Cultural Organization (UNESCO) and the International Union of Geological Sciences have jointly declared 2008 as the International Year of Planet Earth (IYPE), which actually lasts from late 2007 to early 2009. The aim of this year is to raise 'worldwide public and political awareness

of the vast (but often under-used) potential of Earth sciences for improving the quality of life and safeguarding the planet.' It is hoped that the public's imagination can be captured so that information related to the Earth can be used to ensure that it is a safer, healthier and wealthier place for our future children. The science programme consists of 10 broad, socially relevant and multidisciplinary themes: health, climate, ground water, ocean, soils, deep Earth, megacities, resources, hazards and life. Further information related to each theme can be obtained at http://www.yearofplanetearth.org.

2007–2009 is also the 4th International Polar Year, established by the International Council for Science and the World Meteorological Organization, where the biggest challenge for scientists is to understand the relationships between changing climatic conditions and the dynamics of polar ice.

We consider that there is also the need for everyone on the planet to become more knowledgeable about the processes that currently operate on and in the Earth. Humans need to be educated about the fascinating, yet potentially lethal, set of atmospheric, surface and internal processes that interact to produce our living environment. Without education, decisions cannot be made sensibly by individuals, by politicians, or by professionals with direct involvement with environmental organizations of various kinds. Everyone can benefit from better education about the planet on which we live. We hope that reading this book will foster in the reader not just a greater awareness but a greater enthusiasm for Earth processes. However, our main aim is to provide a textbook for introductory university courses in Earth Systems Science, Environmental Sciences, Ecology, Geology, Earth Sciences and Physical Geography, and thus this book has been designed for use in a wide range of courses which discuss environmental and Earth processes, both those that currently operate and those that have operated in the geological past.

In order to foster a greater understanding of Earth systems science we need to build an understanding of the whole Earth system, and to do that we have to increase our knowledge of the component parts and the ways in which these interact. So we need to know how the Earth works as a planet today and how some components of the systems have evolved over geological time in response to changes in others. We must try to be in a position to predict future changes and how the Earth's processes will operate following them.

At the heart of environmental sciences today, in the past and in the future is the climate system. Government organizations are trying to develop risk-based predictions of the future state of the climate on all kinds of scales, both spatially and in terms of time. These predictions are extremely important as they form the basis on which society can build adaptation and mitigation strategies, although there is little in this book that covers these strategies and the accuracy of such predictions is difficult to establish. However, we do try to predict wherever possible the likely future impacts of climate change on whichever part of the Earth system we are studying. One of the major effects already noticeable is on the biodiversity of the planet, the huge variety of plants and animals which forms a key aspect of global ecosystems. It is apparent

Figure 0.3 Building collapse and destruction: here a single door frame bearing a portrait of Chairman Mao remained standing in a pile of debris, on the road heading to Wenyuan, the epicentre. (Source: Wikimedia Commons, http://zh.wikipedia.org/wiki/User:Miniwiki)

that environmental change has already led to loss of biodiversity, which plays a key role in the resilience of ecosystems.

We have already mentioned some of the natural disasters that can occur on the planet and we must better understand earthquakes, volcanoes, tsunamis, coastal erosion, sea-level rise, landslides, flooding and storm processes before we can improve our forecasts of natural hazards, which will hopefully bring tangible humanitarian and economic benefits in the future. Of course, technological solutions to some of the most pressing environmental problems and challenges will continue to evolve, for example in observing and monitoring the environment and providing sophisticated models of environmental processes in order to predict the future state of aspects of the environment and develop mitigation solutions, such as carbon capture and storage.

However, at the same time as these technological solutions are being developed the world's population continues to grow: the global demand for natural resources increases continually, so that it is necessary to develop ways of creating sustainable economic systems and managing resource use within the Earth's environmental limits. To help achieve this we need to be able to provide ways of growing more food, limiting the spread of diseases and providing solutions to water contamination, soil erosion and the pollution of the atmosphere. In fact, a sustainable use of resources on the planet Earth is a necessity for the maintenance of mankind's current dominant position in the world's ecosystems. Whether we remain in this position in the future remains debateable.

Whilst all these issues are fundamental to mankind's use of the Earth, a detailed consideration of many related issues is outside the scope of this book and we only consider some of the scientific aspects of the Earth systems, whilst acknowledging that there are political, technological, philosophical, ethical and moral aspects to humanity's use of the Earth's resources. The WWF believes there is a need for all of us to move to a one-planet future and that 'Biodiversity underpins the health of the planet and has a great impact on all our lives, so that it is alarming that despite an increased awareness of environmental issues, we continue to see a downward trend' (Colin Butfield, WWF-UK campaign chief). We hope that this book can contribute to the education necessary to reverse this trend.

The book is divided into six major sections, with the overall aim being to produce a holistic approach to understanding the Earth system and to give insights into the complex planet on which we live. A major issue is providing a sufficient depth of information while still covering a breadth that integrates the components of the system adequately. This is always a difficult task yet we feel that it is essential to have this holistic approach. There is cross-referencing between chapters to make sure that the synthesis is apparent. We hope that we have produced a book that gives the reader an appreciation of the breadth of the Earth system but also provides solid foundations in the core disciplines, though in a way that does not compartmentalize these disciplines. The aim is to give an understanding of how the world works, not only currently but throughout its history. It is a basic tenet of geology and geomorphology that the present is the key to the past, but we must realize that this fundamental principle does not apply to the Earth's very early history. There is likely to have been a completely different set of processes operating during planetary development. Here we need to ask questions such as: When did plate tectonics develop and why? How did the atmosphere develop and what was it like initially? How and why did the first life on Earth develop?

The authors are physical geographers and geologists by academic and research training and it is inevitable that there is an emphasis on these subject areas. There are bound to be gaps in the coverage in some sections, whilst in others there will be a more in-depth treatment because of our inherent interests and knowledge base. Nevertheless, we feel that we have achieved a broad, integrated coverage of how Earth environments have developed and how they currently operate.

Section 1 is an introduction to Earth systems, explaining that there are complex couplings and feedback mechanisms linking the atmosphere, biosphere, hydrosphere and cryosphere with the Earth's internal system, the geosphere. We need to understand that the Earth is a single integrated system but that there are interrelationships between all of its components. There is a concentration on the cycling of energy and matter and an examination of the dynamics and relationships between the processes on time scales that range from the present day back through the Earth's history until its creation.

Section 2 concentrates on atmospheric and ocean systems, including the Earth's heat budget, moisture in the atmosphere, atmospheric motion, weather systems and climates, with a section showing the evidence for current global warming and its effects.

Section 3 establishes the major geospheric processes that operate internally within the planet, how these manifest themselves on the Earth's surface and crust as plate tectonics, volcanic processes and landforms and how the rock cycle links the internal and external processes.

Section 4 shows how these crustal rocks are weathered and moved around the Earth's surface in various geomorphic systems, such as the fluvial, coastal and glacial systems. In this section there are geological examples of how the present is the key to the past in the interpretation of fossil sedimentary environments.

In Section 5 these newly-created surface environments are then colonized by a diversity of plants and animals, which form ecosystems ranging from the soil system to the tropical rain forest and the Arctic tundra. The ecological principles behind the world's ecosystems on all scales are covered as part of this section.

Section 6 links the other five sections into the conclusions, covering global environmental change in the past, present and future. For example, we cover how the Earth's core, mantle and crust have developed and how the atmosphere has evolved through geological time. We see that there have been many icehouse and greenhouse Earth phases throughout history. Detailed case studies are evaluated, such as the Proterozoic Snowball Earth, the Mesozoic greenhouse world, the oscillating climates between greenhouse and icehouse in the Eocene and Oligocene, and the move into the Cenozoic ice ages, which include the rapid climatic and environmental changes of the Quaternary Late-Glacial, the Medieval Warm Period and the Little Ice Age. Accompanying these atmospheric, ocean circulation and climatic changes there have been radical changes in the biosphere, including major changes in biodiversity during several mass extinctions, and the controls on the evolution of this biosphere are evaluated.

Throughout the book there is extensive use of recent research examples into particular aspects of the Earth environments and processes under consideration. These are separated from the main text in Research Boxes, or as Case Studies on the associated Web site, but they should be considered an integral part of the book as they illustrate some significant new research ideas and examples of particular topics.

Each chapter has a set of review exercises through which readers can establish the depth of their knowledge base. There is also a bibliography and list of further reading, which can be used for more in-depth study. Every Research Box also has a relevant set of references for a similar purpose. Each figure in the associated website and the Research Boxes are highlighted in colour so that readers can gain additional important information.

On the companion Web site (www.wileyeurope.com/college/huddart) you can find a glossary, virtual field guides and additional figures for which there was insufficient room in the text. These can be used as extra teaching and learning resources and as an aid to increasing understanding of some of the environments discussed in the book. A basic training in science has been assumed throughout, although technical terms are defined in either the text or the glossary.

There is also a brief concluding section which attempts to look forward into the likely future changes on the Earth. All the areas that we cover in the book have become more quantitative in nature, more analytical and more rigorously scientific than in the past. Subject areas have become increasingly specialized and research information has escalated in the light of technical advances, computing capacity and numerical modelling, with the result that all research areas are highly technical and complex. Scientists are now specialists and experts in ever-narrowing research fields. We feel that, whilst this is excellent for the furtherance of scientific knowledge and debate related to the Earth, it is inevitably more difficult for one person to have a sufficient depth of knowledge to fully understand all these specialist areas. There is a greater need therefore to acknowledge these research developments in a book like this, whilst stressing the breadth of understanding that is required to have a holistic overview of our planet. This is necessary for undergraduates and interested members of the general public. A broad knowledge base is crucial before specialization takes place at postgraduate level. This book provides a single source of information for the first two years for most undergraduates requiring information on basic geology, geomorphology, biogeography, oceanography, meteorology and climatology; it is not a replacement for the specialist texts that support each of these subjects. It should facilitate further study and we hope it will help launch students towards primary, journal-based literature, towards specialist texts and eventually towards a career in some aspect of the Earth's systems.

We have always been excited and enthusiastic in our teaching and research, particularly in our specialist areas of fluvial geomorphology and hydrology, glacial and volcanic processes, and sedimentology. We hope that through our efforts in writing this book many more future students will be stimulated and motivated to study aspects of the Earth. Whilst we have emphasized the academic aspects of Earth environments, we hope that readers will realize that there are also many important practical applications, for example in natural-hazard mitigation, geomorphology of slopes, coasts and rivers, conservation and management of wildlife, and mitigating the effects of global warming in the atmosphere and the oceans. There is no space to develop these areas in this volume and the reader will have to go to other sources for information on them. Usually at least one source for the applied aspects of a subject is provided in the further reading for each chapter so that the text will be even more comprehensive.

David Huddart and Tim Stott
Liverpool, 2009

SECTION I
Introduction to Earth Systems

1 Introduction to Earth Systems

Flying over the Murrumbidgee River, New South Wales, Australia. Photo: T. Stott

Earth Environments   David Huddart and Tim Stott
© 2010 John Wiley & Sons, Ltd

Learning Outcomes

After reading this chapter and completing the exercises you will be able to:

➤ Describe the position of the Earth within the universe and solar system.

➤ Describe the four spheres of the Earth and how they interact.

➤ Explain what is meant by a system and a systems approach.

➤ Explain about scales in time and space.

➤ Explain the difference between open and closed systems.

➤ Describe what is meant by feedback, equilibrium and cycles in relation to Earth systems.

1.1 Introduction to the Earth's formation

Scientists now believe that the Earth formed some 4.6 billion years ago as matter thrown into space following the Big Bang some 15 billion years ago, which slowly came together in nebulae (dust and gas clouds). These cooled and condensed, undergoing a process now known as planetary accretion to form our solar system. The Big Bang explosion had scattered atoms of various elements throughout space. Most of the atoms were hydrogen and helium, though small percentages of other chemical elements were also present. The atoms began to form a swirling cloud from this cosmic gas, which, due to gravity, gradually thickened over a very long time and became hotter and denser. Near the centre of the cloud, hydrogen and helium began to fuse together to form heavier elements and our Sun was born some 6 billion years ago.

Later, the cool outer portions of the gas cloud also became compacted enough to allow solid objects to condense (rather like ice condenses from water vapour to form snow). These solid objects later began colliding into each other, and eventually became the planets, Moons and asteroids of our solar system (Figure 1.1).

It is theorized that the Large Hadron Collider (LHC) (Box 1.1, see in the companion website) will produce the elusive Higgs boson, the last unobserved particle among those predicted by the Standard Model of particle physics. The verification of the existence of the Higgs boson would shed light on the mechanism of electroweak symmetry breaking, through which the particles of the Standard Model are thought to acquire their mass. In addition to the Higgs boson, new particles predicted by possible extensions of the Standard Model might be produced at the LHC.

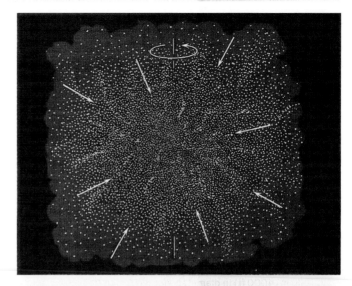

Figure 1.1 Atoms in space remaining from the Big Bang gather to form a swirling gas cloud or nebula. The centre of the gas cloud eventually becomes the Sun, the planets and the Moons, formed by condensation of the outer portions of the cloud. (Source: Skinner and Porter, 1992, p. 19)

The planets and Moons nearest the Sun, where temperatures are highest, contain compounds that condense at high temperatures, such as silicon, aluminum, iron and magnesium. Planets and Moons distant from the Sun, where temperatures are lower, contain larger proportions of volatile elements such as hydrogen and sulphur, which do not condense at high temperatures and form compounds that are solid at low temperatures. These rocky fragments formed from condensation are called planetesimals, which were drawn together by gravitational attraction. The largest ones, sweeping up more of these fragments, grew larger to become planets. Meteorites still fall to Earth, proving that these ancient rocky fragments still exist in space, and there are plenty of examples of meteorite impact craters still visible on the Earth's surface, such as Meteor Crater in Arizona, USA (Figure 1.2).

Meteorites and the scars of ancient impacts provide evidence of the way our terrestrial planets grew to their present size. The planets of our solar system can be divided into two groups based on their densities and distance from the Sun (Table 1.1). The inner planets (Mercury, Venus, Earth and Mars) are small, rocky

Figure 1.2 Meteor Crater is located about 35 miles (55 km) east of Flagstaff, AZ, USA. It lies at an elevation of about 1740 m (5709 ft) above sea level and was created around 50 000 years ago. It is about 1200 m (4000 ft) in diameter, some 170 m (570 ft) deep, and is surrounded by a rim that rises 45 m (150 ft) above the surrounding plains. The crater is filled with 210–240 m (700–800 ft) of rubble lying above crater bedrock. (Source: T. Stott collection)

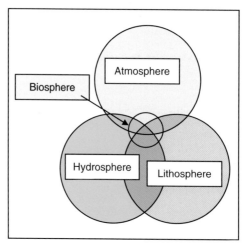

Figure 1.3 The Earth spheres, shown as intersecting circles

and dense and are called the terrestrial planets (Table 1.1). The outer or Jovian (or gas) planets (Jupiter, Saturn, Uranus and Neptune) are much larger than the terrestrial planets, yet much less dense and separated by far greater distances (Table 1.1). Pluto was originally considered a planet, but is now known as a dwarf-planet and recognized as the largest member of a distinct region called the Kuiper belt.

1.2 Introduction to Earth spheres

We will now explore the components of the Earth system, how they interact and how they are changing. We can simplify the Earth system by breaking it down into four key components or spheres: the atmosphere, the hydrosphere, the lithosphere and the biosphere (Figure 1.3). Section 3 will examine the Earth's internal structure and composition in more detail.

The atmosphere is a gaseous layer surrounding the planet that receives heat from the Sun, which is absorbed into the Earth's surface and then reemitted. The atmosphere therefore receives heat and moisture from the Earth's surface and redistributes them. Vital elements needed to sustain life are contained within the atmosphere: carbon, nitrogen, oxygen and hydrogen.

The lithosphere is the solid part of the planet. We are most familiar with the crust, which is the hard rocky surface surrounding the planet like the skin of an orange. Below is the region where molten magma is located – the products of which we occasionally encounter at volcanoes – on which the crust 'floats'. In many places the lithosphere develops a layer of soil in which nutrient elements become available to living organisms and thereby provide a basis for life. Weathering and erosion of the lithosphere result in landforms such as mountains, hills, plains and river valleys (discussed in Section 4), which provide varied habitats for plants, animals and humans.

Water in its liquid, gaseous (vapour) and solid (ice) forms makes up the hydrosphere. The main part of the hydrosphere lies in the oceans, where some 97.5% of the planet's water resides (Figure 1.4). This leaves only around 2.5% as fresh water that can

Table 1.1 Data on the orbits and properties of the planets, showing the Earth to be the third planet from the Sun, with the highest density, and the fifth highest mass and diameter. (Source: Skinner and Porter, 1992)

	Mercury	Venus	Earth	Mars	Jupiter	Saturn	Uranus	Neptune
Diameter (km)	4880	12 104	12 756	6787	142 800	120 000	51 800	49 500
Mass (Earth = 1)	0.055	0.815	1	0.108	317.8	95.2	14.4	17.2
Density, g/cm^3 (water = 1)	5.44	5.2	5.52	3.93	1.3	0.69	1.28	1.64
No. Moons	0	0	1	2	16	18	15	8
Length of day (in Earth hours)	1416	5832	24	24.6	9.8	10.2	17.2	16.1
Period of one revolution around the Sun (in Earth years)	0.24	0.62	1.00	1.88	11.86	29.5	84.0	164.9
Average distance from the Sun (millions km)	58	108	150	228	778	1427	2870	4497

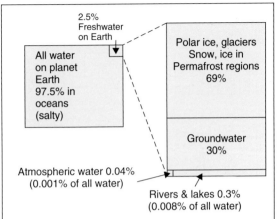

Figure 1.4 The distribution of water on Earth

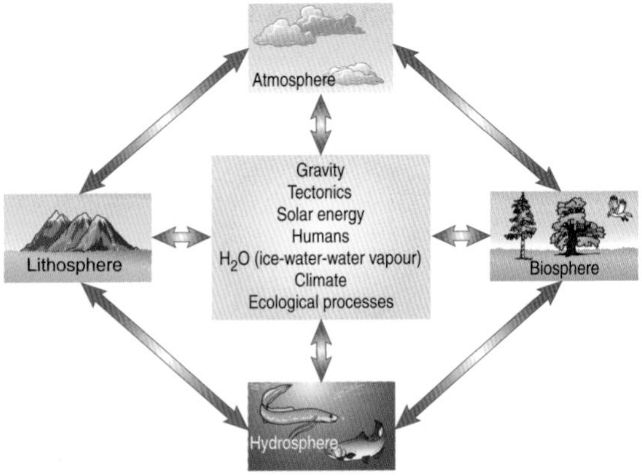

Figure 1.5 Components of the Earth system. The atmosphere, hydrosphere, biosphere and lithosphere all interact with one another. (Source: Holden, 2005, p. 4, adapted by TAS)

be used by humans. Of this, around two-thirds is frozen in polar ice caps and glaciers. Water in the atmosphere exists as gaseous vapour, liquid droplets and solid ice crystals, making up around 0.04% of the fresh water on the planet. Water is also found on top of, and within the lithosphere in soil (as soil water) and in rocks (as ground water, which represents 30% of fresh water). The remaining 0.3% of liquid fresh water, and the most accessible form to humans, is found in streams, rivers and lakes. Water is essential to life.

The biosphere contains all living organisms, which utilize gases from the atmosphere, water from the hydrosphere and nutrients from the lithosphere, in that order of priority. For example, humans would die in a few minutes if deprived of the vital gases they breathe from the atmosphere (namely oxygen), a few days if deprived of water, and a few weeks if deprived of nutrients (food). The biosphere is dependent on all three of the other spheres, and so appears as a central circle overlapping and linking with the other three spheres in Figure 1.3. Most of the biosphere is contained in a shallow surface zone, sometimes called the life layer, which includes the lower part of the atmosphere, the surface of the lithosphere and the upper 100 m or so of the oceans.

These four spheres interact with one another in many ways (Figure 1.5).

Take, for example, water. While water is clearly the main constituent of the hydrosphere, it can exist as ice on land, as ground water within rocks, as liquid within plants and animals and as vapour within the atmosphere, both invisible and visible as clouds. Water also transports nutrients between the rock, soil and plants, and from soils to rivers and the ocean. The lithosphere is worn down by weathering and erosion, which are driven by gravity and water (as ice in glaciers, as water in rivers and as rain from clouds in the atmosphere). This creates the landscape. So water is constantly moving within and between these four spheres or Earth systems.

We as humans provide another example of an interaction between these systems. We eat plant or animal material, which we might grow in the soil of the lithosphere. We also breathe gases from the atmosphere, which help us digest our food. Our excreted

products, which have therefore come from the atmosphere and biosphere, then return to the soil or ocean, where they are broken down by microbes and the nutrients are returned to the lithosphere or hydrosphere.

Throughout this book we will often refer to the Earth's spheres and the life layer and their interactions. The themes and subjects of this book concern systems and cycles that involve the atmosphere, hydrosphere, lithosphere and biosphere, and usually more than one of these spheres simultaneously. Because natural processes are always active, Earth's environments are constantly changing. At times these changes are so slow and subtle that you may not even see or notice them. For example, the movement of crustal plates (more in Section 3) over geologic time to create continents and ocean basins is very slow. At other times changes are rapid, as when a hurricane flattens large areas of forests or human settlements.

It is now becoming widely accepted that environmental change is not only produced by the natural processes that have acted on the planet for millions of years, but also by human activity. Humans have now populated the planet so widely that few places remain free of some kind of human impact. Global change therefore involves both natural processes and the human processes that interact with them.

1.3 Scales in space and time

The processes within, and interactions between, the four spheres of the Earth operate on various scales. On a global scale the Earth is an almost spherical body which rotates about an axis once each day and orbits the Sun once per year. The Sun is the energy source that powers most of the processes which take place on Earth, and from which living things derive their energy. The Earth–Sun relationship is therefore extremely important when considering climate, climate change and past and future cycles and trends. In Section 2 we will consider the Earth and its global energy balance system as a whole.

Energy from the Sun is not absorbed evenly by the Earth's water and land surfaces, which heat up at different rates. This unequal solar heating produces winds in the atmosphere and currents in the oceans, which give rise to the global atmospheric and ocean circulation systems. To study these systems we need to move to the continental scale, so that we can distinguish continents and oceans and learn about the speed and direction of winds and ocean currents. At the regional scale we can begin to identify the cloud patterns associated with weather systems and to track their regular movement over time. Regional climates help determine the soil types, and in turn the types of plants and animals which adapt to different regions of the Earth. However, factors at the local scale ultimately determine the precise patterns of soils and vegetation. For example, the top of an upland crag may offer a cold, harsh and windswept environment for plants and animals to live in, whereas a few metres away on the Sun-facing, leeward side of the crag it may be calm, warm and sheltered, providing a very different set of local factors to which plants and animals would need to adapt.

Time scales, like spatial scales, vary greatly too. It may take millions of years for the Earth's tectonic forces to produce mountain ranges like the Himalayas, Rockies or Alps. On the other hand, an earthquake creating a fault in the Earth's crust, or a tornado destroying a crop in a farmer's field, may last less than one minute. Likewise, the processes of soil formation are too slow to see, but if you watch soil being washed down a gully during an intense rain storm, you begin to understand how years of rainfall can slowly help to carve river valleys into the landscape.

So, systems and cycles operating in the Earth's environments act over a wide range of scales in time and space. Sometimes it is easiest to understand systems by considering the whole Earth (e.g. the global circulation of the atmosphere), while at other times we need to concern ourselves with very local phenomena, such as sand grains moving along a river bed.

This book is organized so that as we move through the sections we generally go from the large to the small scale. Understanding this organization should help in learning about the systems and processes operating within the Earth's environments.

1.4 Systems and feedback

The processes that interact within the four spheres and result in the different Earth environments are complex, and a helpful way to start to understand the relationships between these processes is to study them as systems. We are going to use the concept of systems a great deal throughout this book so in this chapter we will look at systems in detail.

The problem of trying to understand all the processes on Earth, and between Earth and space, is immense. Thinking in terms of systems involves breaking down reality into elements and identifying the linkages between them. A simple example of this is a car (Figure 1.6). The engine is driven by fuel, which produces changes in other elements of the system, and eventually the wheels in contact with the road turn and the car moves. The

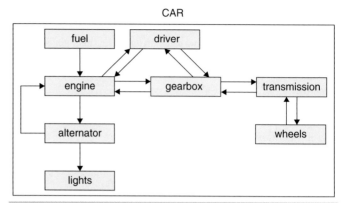

Figure 1.6 The car as a system. (Source: Clowes and Comfort, 1982, p. 10)

arrows in the figure show that most of the links in the system operate in both directions. For example, the alternator is driven by the engine, but it produces electricity that is returned to the spark plugs in the engine, as well as charging the battery and working the lights.

Flow in one pathway that acts to increase or decrease the flow in another pathway is known as feedback and we come across this in many of Earth's systems. In a car, the noise of the engine gives the driver feedback so he or she knows when to change gear. If fuel were fed constantly into the system, the engine would go faster and faster and draw more fuel, until in the end it would break down. This is an example of positive feedback, which, unless regulated in some way, leads to a breakdown in the system. You can also hear positive feedback as a squealing noise in public address systems when the volume is turned up too high. The person speaking into the microphone has their voice amplified through loudspeakers. The sound from the loudspeakers is picked up as well as the speaker's voice, making the sound through the loudspeakers even louder. As this sequence repeats itself, the sound gets louder still, producing the unpleasant squeal. It reinforces the sound energy in the pathway between the microphone and speakers.

Negative feedback occurs when an increase in the activity of one element of the system leads to a decrease in the activity of another. The driver taking pressure off the accelerator of the car restricts the fuel flow and the engine slows down. A thermostat which controls temperature in a greenhouse provides another example of negative feedback. When the heating system is on, the greenhouse warms, but eventually the warming trips a temperature-sensitive switch that turns the heating system off. So the heat energy from the heating system produces a negative feedback that reduces the amount of heat produced. It is this kind of feedback which climate scientists are concerned about in connection with global warming (more in Section 6). Feedback normally regulates a system and tends to maintain a steady state.

Another example of a system with which most of us will be familiar is a country's road and motorway system. In this engineered system, the roads serve as pathways for the flow of vehicles. The roads are connected by roundabouts, junctions and interchanges which allow vehicles to move from one road to another. The vehicles on the road are part of this special type of system, called a flow system, in which traffic moves from one place to another. Almost all of Earth's systems are flow systems.

In the road network system, it is the vehicles containing people and goods that 'flow'. The vehicles are real objects or pieces of matter that move around the road network and so we can call this type of system a matter flow system. In Section 2 we will concern ourselves with a system that describes the flow of solar energy from the Sun to the Earth and its atmosphere. Such a system is called an energy flow system.

Flow systems have a structure in which pathways of flow are connected (the pathways are the roads in the road network system). Sometimes, pathways are not so obvious, such as when solar radiation reaches the top of our atmosphere: some is reflected back into space, some is absorbed and some is scattered (deflected away from the Earth's surface). It is quite a difficult task, therefore, to identify the energy pathways. We refer to the pattern of pathways and their interconnections as the structure of the system. It is also convenient to use the term components to refer to the parts of the system, as in the car system we saw earlier. In Figure 1.5 the boxes tell us nothing about the complexity of the internal combustion engine, which is regarded as a black box, with inputs and outputs. We do not need to know how the engine works to be able to drive the car, but a car mechanic needs to know how it works if it breaks down. For the mechanic, the box becomes transparent as he or she comes to understand the operation of the systems within it.

Inputs and outputs are important characteristics of many flow systems, and in the road network system the input is represented by new cars being either manufactured or imported, and the output by automobile scrap yards, where cars are broken and their components recycled, and by vehicle export. An essential feature of flow systems is the need for some kind of power source. In the road system, vehicles are powered by fuel. Natural systems on Earth are powered generally by natural sources, which include: energy from the Sun to the Earth; the power stored in the inertia of the Earth's rotation; and the outflow of heat from the Earth's interior (geothermal energy), which eventually produces movements in the Earth's crust.

An example of a natural Earth system might be a river system, which is a matter flow system of water, mineral and organic matter in a set of connected stream channels. The channels are pathways, connected in a structure called the channel network, which organizes the flow of water, mineral and organic matter from high ground to the ocean. Another example of a natural system is the food chain of an ecosystem, in which energy in the form of food flows between the plant and animal components of the system.

In summary, a flow system is made up of a structure of connected pathways within which matter and/or energy flows. In the Earth environment this could consist of light, air, water, heat, ozone, pollutants, organic matter or sediment. Flow systems have inputs and outputs and all require a power source to run them.

1.5 Open and closed flow systems

A road network is an example of an open flow system, into which there are inputs of matter (new vehicles) and from which there are outputs of matter (broken or damaged vehicles). A river system is also an open flow system. Precipitation (rain, snow, hail, etc.) provides the input, and water is output to the ocean.

However, there is another type of matter flow system in which there are no inputs or outputs. In a closed system, matter flows endlessly in a series of interconnected paths or loops, also called a cycle. An example might be a motorcycle race around a track, where the bikes are powered by the flow of fuel from their tanks, and remain on the track throughout the race, so there is no input or output of bikes as there is on a national road network. Natural closed flow systems are usually much more complicated than this example. They frequently have complicated structures with many looping pathways that are interconnected in many places. For example, we can take the open system example of a river, but move the boundary to enclose the Earth and atmosphere (Figure 1.7) so that the system, now called the hydrological cycle, becomes a closed system in which the loops are flow paths of water in solid, liquid and gaseous forms. Water moves as ice in glaciers, as liquid in streams, rivers and soil water, and as vapour in flows of moist air. A second example of a natural closed system could be the carbon cycle. The element carbon travels or loops in different chemical forms between rocks, ocean water, the atmosphere and the matter of plants and animals in very complicated and intricate pathways.

As we can see from Figure 1.7, whether a matter flow system is open or closed depends on where we draw the boundary. If we draw the boundary around the river system itself (Figure 1.7(a)) then precipitation becomes an input to the system (crossing from the boundary from the atmosphere system into the river system) and water flow to the ocean (crossing the boundary from the river system into the ocean system) becomes an output. If, however, we extend the system boundary to include the whole planet (lithosphere, atmosphere, biosphere and hydrosphere) then since water does not escape into space, nor does it arrive from space, the system becomes a closed system (Figure 1.7(b)). In this system, a new pathway has been added: the return flow of water from the ocean to the atmosphere by evaporation and transpiration. As there is no input or output, the system is now a closed system.

If we extend this idea of global systems, we can see that all global systems are more or less closed, apart from some gas molecules moving from the edge of the atmosphere into space, and the input of occasional meteorites. So, as well as the global carbon cycle, the oxygen, nitrogen and sulphur cycles are also closed matter flow systems. However, this is not true for global energy systems. Since the Earth is warmer than the space around it, heat escapes from the Earth as radiant energy. Radiant energy is also received by the Earth from the Sun, so there is always an energy input and output. Therefore, energy flow systems are open systems.

In summary then, matter flow systems can be open (having inputs and outputs) or closed (lacking inputs and outputs). Closed systems may be referred to as cycles. Global matter flow

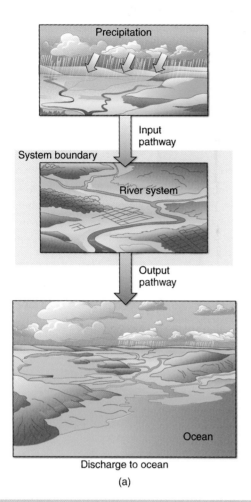

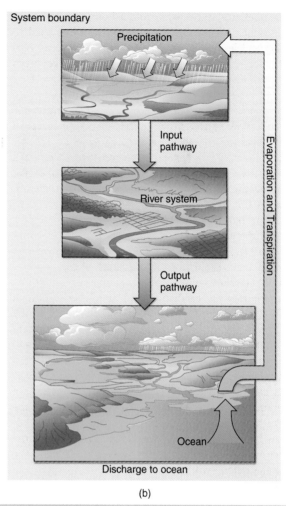

Figure 1.7 (a) A river system as an open matter flow system. (b) When the boundary is moved to enclose the Earth and atmosphere, the system becomes a closed, global flow system (the hydrological cycle). (Source: Strahler and Strahler, 1997, p. 9)

systems, such as the nitrogen cycle, are always closed, while energy flow systems are always open.

A discussion of the Gaia Hypothesis can be found in the associated companion website Box 1.2.

1.6 Equilibrium in systems

When a system is in equilibrium it is in a steady state in which the flow rates in its various pathways stay about the same. A good example of this is a lake within a closed basin system in an arid climate (Figure 1.8(a)), such as the Great Salt Lake in Utah, USA, which has no stream outlets (besides evaporation) and would dry up completely if it were not fed by three rivers that rise in the Uinta mountain range in north-eastern Utah. The amount of evaporation depends on the surface area of the lake. If the climate becomes wetter, more water enters the lake (Figure 1.8(b)), the water level rises and the surface area of the lake expands, resulting in more evaporation. When the lake level rises to the point where the increased evaporation rate equals the increased inflow rate,

the lake is said to have reached equilibrium. If the climate changes again and the input is reduced, the lake level will fall, the surface area will decrease and evaporation will decrease. Eventually the lake will move to a new, lower, equilibrium level. In this example, the link between input, lake surface area and evaporation is a negative feedback. Systems which come into equilibrium like this are normally stabilized by negative feedback loops or pathways.

The global climate system has numerous pathways of energy flow, which include important negative feedbacks that draw the global mean temperature towards a mean value. Climate scientists now agree that human activity over the past three centuries is responsible for the rises in the global mean temperature that we are now experiencing. However, there are many uncertainties in the climate prediction models being used to forecast global temperatures over the next 40–100 years, one of which is the role of clouds. The effect of clouds on the global climate system serves as a good example to demonstrate the role of positive and negative feedback on an equilibrium system. Low white clouds reflect sunlight back to space more effectively than dark ocean or land surfaces. In terms of energy flows, low clouds create a

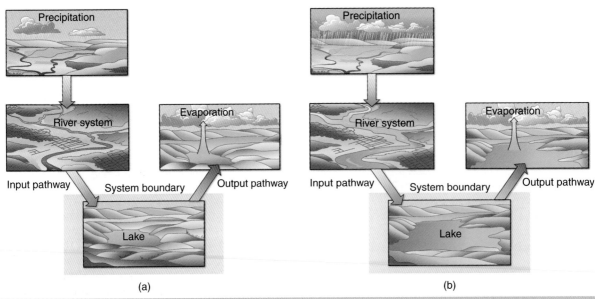

Figure 1.8 (a) Lake in a closed basin as a matter flow system in equilibrium. (b) Precipitation increases, the lake level rises, lake surface area increases and the result is more evaporation, which helps to balance the greater input and tends the lake towards equilibrium. (Source: Strahler and Strahler, 1997, p. 10)

pathway in which some solar energy is redirected back into space. This tends to cool the surface, as it is deprived of this energy, and so acts as a negative feedback. On the other hand, high clouds tend to absorb the outgoing heat flow from Earth to space, redirecting it back towards the Earth. So, high clouds provide a positive feedback that warms the surface. What would happen if there were a small increase in global surface temperature? It is generally believed that if the Earth were warmer, more water would evaporate from the oceans, lakes and wet soil and vegetation, and so more clouds would form. But would this increase in cloud cover cause global cooling or warming? The best estimates currently suggest that in this scenario, more high clouds than low clouds would form, so the effect would be a positive feedback that would tend to make the surface even warmer. However, as you can imagine, clouds systems are very complex and their effect on climate is not well understood. This topic will be revisited in Section 6.

Another example of equilibrium was put forward by Hack (1960) when he revived Gilbert's late-nineteenth-century concept of 'dynamic equilibrium'. Hack suggested that every slope in every channel in an erosional system is adjusted to all the others, and that relief and form can be explained in spatial terms rather than historic ones. Rather than being exactly concave, he suggested that river profiles were determined by sediment building up from hillslopes. When sediment builds up in a river, the river has to steepen itself in order to move that sediment. Once removed, the river may become less steep in profile. Figure 1.9 is based on Gilbert's concepts and Hack's work and expands on how the equilibrium between a river's energy and the sediment load it carries is maintained by its carrying out erosion and deposition.

The nature of the equilibrium in a natural system, however, depends on the time scale under which it is being investigated. Figure 1.10 shows forms of equilibrium over three time scales.

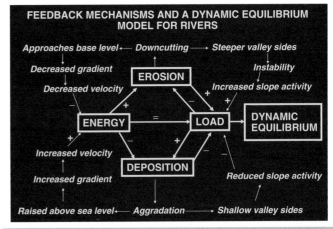

Figure 1.9 Feedback mechanisms and a dynamic equilibrium model for rivers. (Source: Tim Stott's lecture slide, adapted from Hack, 1960)

Over short time scales of a year or so, where there is no change over time, there may be a static equilibrium (Figure 1.10(a)) or a steady-state equilibrium (Figure 1.10(b)), where there are short-term fluctuations about a longer-term mean over say 100 years. Over longer periods of 10 000 years or so, there might be dynamic equilibrium (Figure 1.10(c)), where there are shorter-term fluctuations about a changing longer-term mean value. So the concept of equilibrium is clearly very time-dependent, and it may depend on where and when you measure something as to whether it will demonstrate equilibrium. For example, in Figure 1.10 measurements were taken at two times (t_1 and t_2) for each case. However, because of the timing of the measurements, the nature of the long-term change may have been incorrectly

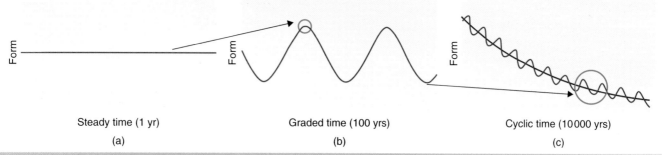

Figure 1.10 Equilibrium over three time scales: (a) dynamic equilibrium; (b) steady-state equilibrium; (c) static equilibrium. (Source: after Schumm and Lichty, 1965, adapted from Holden, 2005, p. 9)

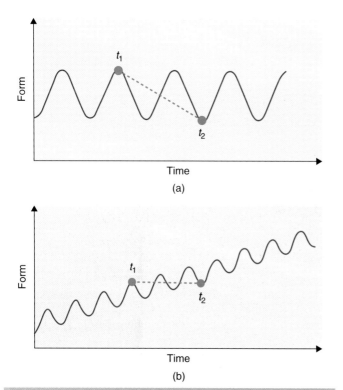

Figure 1.11 Two systems showing short-term measurements which, because of their timing, draw incorrect conclusions about the long-term trend. (Source: Holden, 2005, p. 9)

identified in each case. In Figure 1.11(a) a downward trend has been established, when really there is no long-term trend. In Figure 1.11(b) no change has been identified, when really there is a long-term upward trend.

1.7 Time cycles in systems

Changes in rates of flow of energy and matter in open and closed flow systems are common. Flow rates can speed up or slow down and these changes in activity can even be reversed at certain intervals of time. The rates can alternate between fast and slow in what are called time cycles.

There is a rhythm of increasing and decreasing flow rates in many natural systems. The rotation of the Earth on its own axis once every day causes a cycle of day and night, known as a diurnal cycle, which has a great impact on the energy flow in natural systems on the Earth. The revolution of the Earth around the Sun once every year generates a time cycle of energy flow in natural systems which results in a seasonal rhythm. The Moon orbits the Earth once a month, which sets up its own time cycle. This lunar cycle affects the tidal range, with higher high tides and lower low tides (called spring tides) coinciding with full and new Moons. Other time cycles include ice ages, with time scales of thousands to tens of thousands of years, during which great ice sheets spread over and modify the Earth's surface, and the drifting of the supercontinents, over time scales of millions of years, which form, break apart and reform.

These concepts of systems and cycles help us to understand and organize the Earth's processes and phenomena. Scientists today devote much time to systems thinking. Whole systems, as well as the pathways, interconnections, energy sources and other components of these systems are put under observation. Theories are developed to help our understanding of how systems are organized, and experiments are devised to test these theories. Increasingly today, mathematics and computers are used to model system behaviour and try to predict behaviour in the future. There is now a whole subject called Systems Theory that attempts to explain and understand how systems work. Clearly, systems thinking is very important to students who study natural processes – geographers, geologists, ecologists, oceanographers, atmospheric scientists and others. These disciplines study the Earth and life phenomena, but because of the scales of most natural systems, it can be difficult to conduct experiments. For example, it would not be possible to increase the Sun's output by say 20% to see what effect it had on the Earth's climate. While scientists have attempted to build models of rivers in their laboratories, they encounter so many problems in working at a smaller scale that it becomes very difficult to say whether the results of their experiments in flumes can be transferred out into the real world with any accuracy. So natural scientists tend to do much of their work by treating the Earth as their natural laboratory. They observe the Earth and its processes at different times and places and on different time scales. Then,

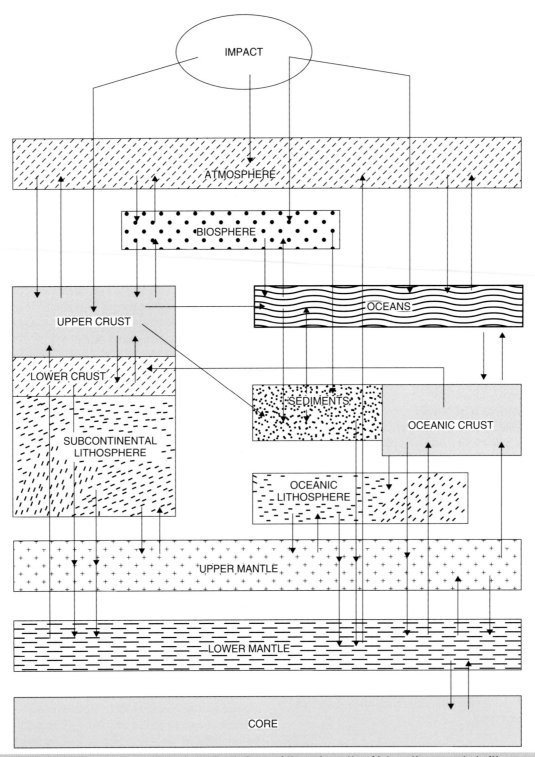

Figure 1.12 Diagram showing the major subsystems in the Earth. Some of the major paths of interaction are noted with arrows. (Source: Condie, 2005, p. 10)

using a systems approach to guide these observations, they come to understand and predict the behaviour of these processes.

Whether in systems thinking or elsewhere, critical thinking about how things work and how they are related to one another is a skill that everyone relies upon. This book will help you to develop this skill and to apply it to a better understanding of the Earth's environments.

Some of the major pathways of interaction among Earth systems and between Earth and extraterrestrial systems are summarized in Figure 1.12.

As an example, the crystallization of metal onto the surface of the inner core might give off enough heat to create mantle plumes just above the core–mantle interface. As such plumes rose into the upper layers of the mantle, spread beneath the lithosphere and began to melt, large volumes of basalt might underplate the crust and erupt at the Earth's surface. These eruptions might pump large quantities of CO_2 into the atmosphere, and since CO_2 is a greenhouse gas, this would warm the atmosphere, giving rise to warmer climates. This, in turn, might affect the continents by increasing weathering and erosion rates; the oceans by increasing the rate of limestone deposition; and life by changing the climate, leading to the extinction of some species that are unable to adapt. Thus, through such a linked sequence of events, processes occurring in the Earth's core could lead to the extinction of life forms on the surface. Before such changes can affect life, however, negative feedback processes might return the atmosphere–ocean system back to an equilibrium level, or even reverse the changes altogether. For example, increased weathering rates caused by increased CO_2 levels in the atmosphere might drain the atmosphere of its excess CO_2, which would then be transported in streams and rivers to the oceans and deposited as limestone. If enough cooling occurred, this could lead to widespread glaciations, in turn causing the extinction of many life forms.

In order to prepare for the continued survival of living systems, it is important that we understand the nature and causes of these interactions among Earth systems and between Earth and extraterrestrial systems.

Exercises

1. Describe the Big Bang theory and its relationship with planetary accretion.

2. For what purpose was the Large Hadron Collider built by the European Organization for Nuclear Research (CERN)?

3. What are planetesimals?

4. Define the terms (a) atmosphere, (b) lithosphere, (c) hydrosphere, (d) biosphere.

5. Describe the distribution of water on the planet Earth.

6. Using terms like *continental, local, global* and *regional*, discuss the concept of scales in space.

7. With reference to the processes that interact within the four spheres on Earth and result in the different Earth environments, define what is meant by the term *system*.

8. Explain the difference between positive and negative feedback in one Earth system.

9. What is the fundamental difference between an open and a closed system?

References

Condie, K.C. 2005. Earth as an Evolving Planetary System. London, Elsevier Academic Press.

Holden, J. (Ed.). 2005. An Introduction to Physical Geography and the Environment. Harlow, Pearson Education.

Skinner, B.J. and Porter, S.C. 1992. The Dynamic Earth: An Introduction to Physical Geology. Chichester, John Wiley & Sons.

Strahler, A.H. and Strahler, A. 1997. Physical Geography: Science and Systems of the Human Environment. Chichester, John Wiley & Sons, Inc. Introduction, pp. 2–13.

SECTION II
Atmospheric and Ocean Systems

2 Structure and Composition of the Atmosphere

The Ecrins National Park, SE France. Photo: T. Stott

Earth Environments   David Huddart and Tim Stott
© 2010 John Wiley & Sons, Ltd

Learning Outcomes

After reading this chapter and completing the exercises you will be able to:

➤ Describe the structure of the Earth's atmosphere and name the four layers and boundaries between them.

➤ Describe and name the gaseous and nongaseous components of the atmosphere.

➤ Understand how temperature varies with height in the atmosphere.

➤ Describe the environmental lapse rate in the lower atmosphere.

➤ Understand the location, processes and function of stratospheric ozone and the threats to its existence.

➤ Understand the importance of variable greenhouse gases like carbon dioxide, methane, nitrous oxide, ozone and other pollutants and how they vary over space and time.

➤ Describe the sources of aerosols and understand the causes and effects of acid rain.

2.1 Structure of the atmosphere

The word 'atmosphere' is derived from the ancient Greek words *atmos* (meaning vapour) and *sphaira* (meaning sphere). The atmosphere is an envelope of transparent, odourless gases held to the Earth by gravitational attraction. Almost all of the atmosphere (97%) lies within 30 km of the Earth's surface. The upper limit of the atmosphere is at a height of approximately 1000 km above the Earth's surface, but most of the atmosphere is concentrated within 16 km of the Earth's surface at the equator and within 8 km at the poles. Recordings made by radiosondes, weather balloons, satellites and rockets show that atmospheric pressure decreases rapidly with height, but that the change in temperature with height is more complex. Based on changes in temperature, the atmosphere is divided into four layers (Figure 2.1).

Moving outwards from the Earth's surface, the four layers are:

1. **Troposphere:** This is the zone of our weather and extends from 0 to 12 km on average, but is thinner over the poles (5–6 km) and thicker over the equator (15–16 km). Temperatures in the troposphere fall from around 15 °C at the Earth's surface to −57 °C at the tropopause (the upper boundary of the troposphere). This decrease of around 6.4 °C per km is called the environmental lapse rate (ELR), more of which in Chapter 3. This happens because the Earth's surface is warmed by incoming solar radiation, which it absorbs and then re-radiates, heating the layer of atmosphere next

to it by conduction, convection and radiation. Air pressure decreases with height as the effect of gravity weakens, although wind speeds usually increase with height. The troposphere is inherently unstable. It contains most of the atmosphere's water vapour, dust, cloud and a mixture of natural and anthropogenically-produced gases. At the tropopause is a layer extending from around 12 to 20 km at which temperatures remain constant despite increasing height. This layer is known as an isothermal layer.

2. **Stratosphere:** This layer extends from around 12 to 48 km, and moving up through it there is a steady increase in temperature (called a temperature inversion), which is caused by a concentration of ozone (O_3) which absorbs incoming short-wave ultraviolet (UV) radiation from the Sun. Pressure continues to fall and the air is dry. Winds are generally light in the lower part of the stratosphere, but increase with height. The stratopause marks the upper boundary of the stratosphere, and is another isothermal layer, like the tropopause.

3. **Mesosphere:** There is no water vapour, cloud, dust or ozone in this layer, which extends from 48 to 80 km, so temperatures fall rapidly because there is little to absorb the Sun's incoming radiation. Within this layer the atmosphere's lowest temperatures (−90 °C) and strongest winds (around 3000 mph) are reached. The mesopause marks the upper boundary of the mesosphere at around 80 km from the Earth's surface, and like the tropopause and stratopause, is isothermal.

4. **Thermosphere:** In this outer layer of the atmosphere, which extends outwards from 80 km, temperatures rise rapidly up to around 1500 °C due to an increasing proportion of free oxygen in the atmosphere, which, like ozone, absorbs incoming extreme UV radiation (0.125–0.205 um) from the Sun. However, since temperature is really a measure of the mean kinetic energy of the molecules of an object, and there are no air molecules here, these temperatures are just 'theoretical'.

2.2 Composition of the atmosphere

Air is a mixture of a number of different gases, some of which are 'constant' constituents, while others (e.g. water vapour and carbon dioxide) vary over space and time. If we consider these gases in terms of dry air, Table 2.1 shows the percentages by weight and volume.

Nitrogen gas exists as a molecule consisting of two nitrogen atoms (N_2). It is not very chemically reactive and can be thought of as a neutral substance.

Oxygen gas (O_2), on the other hand, is highly chemically reactive and combines readily with other elements in the process of oxidation. Combustion of fuels is a rapid form of oxidation, while certain forms of rock decay (weathering) are very slow forms of oxidation.

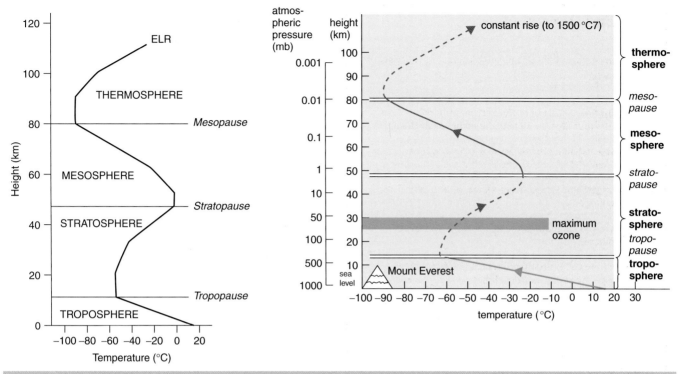

Figure 2.1 The vertical structure of the atmosphere. (Source: Thompson, 2002, p. 5)

Table 2.1 Average composition of dry atmosphere (below 80 km). (Source: after Thompson, 2002)

Component	Symbol	% Weight	% Volume
Nitrogen	N_2	5.51	78.08
Oxygen	O_2	23.15	20.95
Argon	Ar	1.28	0.93

Argon (Ar) is an inactive gas of little importance in natural processes.

These three gases make up 99.96% of dry air; the remaining 0.04% consists of minute quantities of various inert gases (mainly neon, helium, krypton, xenon), hydrogen (H_2), ozone (O_3), carbon dioxide (CO_2), methane (CH_4), halogen derivatives of organochloride compounds like chlorofluorocarbons (CFCs), nitrous oxide (NO_2) and aerosols. While O_3, CO_2, CH_4 and CFCs are variable, the other 'fixed' gases do not vary up to 80 km. This means, therefore, that their production at the surface is balanced by their destruction. For example, nitrogen is removed from the atmosphere by nitrogen-fixing bacteria in plant roots, but is returned when plants and animals die and decay. Oxygen is removed from the atmosphere when plants and animals die and decompose, when oxidation takes place with a wide range of substances (like iron oxide when it rusts), or when we take in oxygen from the air we breathe. Oxygen is returned to the atmosphere during photosynthesis, when carbon dioxide is combined with water to produce sugars and oxygen is given off as a byproduct. However, in terms of their interactions with other components of the climate system, these fixed gases have little significance in the atmosphere. Far more important are the variable and trace gases like carbon dioxide, nitrous oxide, methane, aerosols, CFCs, ozone and water vapour. Table 2.2 shows their volumes and concentrations in the atmosphere but hides the fact that although their proportion in the total composition of the atmosphere is extremely small, they are very important in influencing large-scale global warming and cooling events and air quality.

2.2.1 Ozone in the upper atmosphere

Ozone is mainly found in the stratosphere, where it is produced by gaseous chemical reactions. An oxygen molecule (O_2) absorbs UV light energy and splits into two oxygen atoms (O + O). A free oxygen atom (O) then combines with an O_2 molecule to form ozone (O_3). Once formed, ozone can also be destroyed by absorbing UV light energy and splitting into O_2 + O. Two oxygen atoms (O + O) can join to form an oxygen molecule (O_2). The net effect of all these changes is that O_3, O_2 and O are constantly formed, destroyed and reformed in the ozone layer, absorbing UV radiation with each transformation. This absorption of harmful UV radiation from the Sun helps to shield life below from its harmful effects. If the concentration of ozone is reduced, fewer transformations among O_3, O_2 and O occur, and so UV absorption is reduced. Early in the Earth's history, the lack of ozone in the atmosphere confined living things to the depths of the oceans. If solar UV radiation were to reach the Earth's surface

Table 2.2 Variable gases in the atmosphere. (Source: Thompson, 2002, p. 4)

Gas (and particles)	Symbol	% Volume	Concentration parts per million (ppm)
Water vapour	H_2O	0–4	–
Carbon dioxide	CO_2	0.035	355
Methane	CH_4	0.000 17	1.7
Nitrous oxide	N_2O	0.000 03	0.3
Ozone	O_3	0.000 004	0.04
Particles/ aerosols	(dust, sulphate, etc.)	0.000 001	0.01
Chlorofluoro-carbons	CFCs	0.000 000 01	0.0001

at full intensity again, all bacteria exposed at the surface would die and animal and human tissue would be severely damaged. So, although ozone only represents 0.000 004% by volume of the gases in the atmosphere (Table 2.2), its presence is essential if life on the Earth's surface is to be maintained.

The release of chlorofluorocarbons (CFCs) into the atmosphere is a serious threat to the ozone layer. CFCs have the smallest concentration in Table 2.2, at 0.0001 ppm. They are synthetic industrial compounds containing chlorine, fluorine and carbon atoms, and the molecules of CFCs are very stable close to the Earth's surface. However, they do move upwards by diffusion, and on reaching the stratosphere, like ozone, they absorb UV light energy and start to break up. Chlorine oxide molecules are formed, which in turn attack ozone molecules, setting up a chain reaction which converts them to oxygen molecules. This reduces the ozone concentration in the stratosphere, meaning that there are fewer ozone molecules to absorb UV radiation, so that the intensity reaching the Earth's surface is increased. Other human-produced gases like nitrogen oxides, bromine oxides and hydrogen oxides can act in the same way as chlorine oxides to reduce stratospheric ozone. However, not all threats to the ozone layer are caused by humans. The June 1991 eruption of Mount Pinatubo in the Philippines injected volcanic dust into the stratosphere and is estimated to have reduced ozone there by 4% during the following year, with reductions over the mid-latitudes of up to 9%. Such reductions are predicted to increase the incidence of skin cancer in humans, may reduce crop yields and could cause the death of some forms of aquatic life. As in all systems, the links and feedback mechanisms may not be fully understood and so it is difficult to accurately predict the full effects of such changes.

2.2.2 Ozone holes

Each year for the past few decades during the southern hemisphere spring, chemical reactions involving chlorine and bromine have caused ozone in the southern polar region to be destroyed

rapidly and severely. This depleted region is known as the 'ozone hole'. Ozone levels in the atmosphere over Antarctica have been monitored by NASA scientists since 1979. At the Ozone Hole Watch Web site, http://ozonewatch.gsfc.nasa.gov/, you can check on the latest status of the ozone layer over the South Pole. A hole in the ozone layer over Antarctica was first discovered in 1984, since which time a similar one has been reported over the Arctic. By the late 1980s scientists had agreed that the decrease in the ozone layer was escalating much faster than had been predicted. In the Antarctic, scientists noted a seasonal thinning of the ozone layer during the early spring, with ozone reaching a minimum during October.

The area of the ozone hole is determined from a map of total column ozone. It is calculated from the area on the Earth that is enclosed by a line with a constant value of 220 Dobson units. The value of 220 Dobson units is chosen because total ozone values of less than 220 Dobson units were not found in the historic observations over Antarctica prior to 1979.

As the global ozone layer thins we would expect the rate of incoming UV radiation to increase. Scientists have estimated that for each 1% decrease in global ozone, UV radiation should increase by 2%. However, measurements of UV radiation reaching the Earth's surface have not always shown increases in recent years, and several other factors may affect year-to-year changes, including cloud cover and particulates in the atmosphere (from pollution). Nevertheless, 23 nations endorsed a United Nations plan in 1987 to cut global CFC production by 50% by 1999. Since then, international agreements on CFC control have been strengthened, leading to a great reduction, or even a halting, of CFC manufacture and consumption. Even so, CFCs already in the lower atmosphere will continue to work their way upwards to the stratosphere for several decades yet. Figure 2.2 shows that the size of the hole was greatest in 2006.

Still, the CFC treaty is seen as something of an environmental success story because of the overwhelming public support for the ban, and implementation was relatively quick. In this example, global cooperation has resulted in a global solution to a global problem.

See the companion website for Low-level ozone stops plants absorbing carbon dioxide.

2.2.3 Carbon dioxide and methane

The contribution of other variable trace gases, such as carbon dioxide (CO_2) and methane (CH_4), to the greenhouse effect has received considerable media attention in recent years. CO_2 represents 0.035% of the volume of the atmosphere and is produced from combustion (e.g. burning fossil fuels), respiration, soil processes and evaporation from oceans. It is consumed by the process of photosynthesis, which all plants carry out. These processes are not always balanced, but the oceans help to regulate the balance by absorbing CO_2. It has been estimated that during the Upper Carboniferous period of coal deposition 300 million years ago, 10^{14} tons of CO_2 were withdrawn from the atmosphere and 'locked up'. Some of this 'locked up' CO_2 has been released back into the atmosphere by burning fossil fuels,

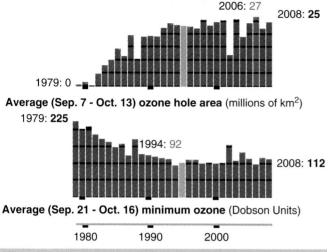

2006: 27
2008: **25**

1979: 0

Average (Sep. 7 - Oct. 13) ozone hole area (millions of km^2)

1979: **225**

1994: 92

2008: **112**

Average (Sep. 21 - Oct. 16) minimum ozone (Dobson Units)

1980   1990   2000

Figure 2.2 Annual South Pole ozone record since 1979. The severity of the ozone hole varies somewhat from year to year. These fluctuations are superimposed on a trend extending over the last three decades. The red bars indicate the largest area and the lowest minimum value. (Source: http://ozonewatch.gsfc.nasa.gov/)

Table 2.3 Global aerosol production estimates (10^9 kg per year). (Source: Thompson, 2002, p. 7)

Sources	All sizes	<5 μm radius
Natural		
Primary production		
Sea salt	30–100	500
Windblown dust	70–500	250
Volcanic emissions	4–150	25
Forest fires	3–150	5
Secondary production (gas → particle)		
Sulphate from H_2S	37–420	335
Nitrates from NO_2	75–700	60
Converted plant hydrocarbons	75–1095	75
Total natural	**501–4025**	**1250**
Anthropogenic		
Primary production		
Transportation	2.2	1.8
Stationary combustion	43.3	9.6
Industrial processes	56.4	12.4
Solid waste disposal	2.4	0.4
Miscellaneous	28.8	5.4
Secondary production (gas → particle)		
Sulphate from SO_2	110–220	200
Nitrates from NO_x	23–40	35
Converted	15–90	15
Total anthropogenic	**185–483**	**280**
Overall total	**686–4508**	**1530**

over the past three centuries in particular. It is estimated that the CO_2 concentration of the atmosphere has increased by around 20% since 1900. This gas is of great importance in atmospheric processes because of its ability to absorb radiant heat passing through the atmosphere from the Earth. CO_2 thus adds to the warming of the lower atmosphere.

2.2.4 Water vapour

Another important component of the atmosphere is water vapour, the gaseous form of water, the molecules of which mix freely throughout the atmosphere like the other gases. However, water vapour can vary greatly in its concentration, and the ability of the atmosphere to hold water vapour and the condensed form of water droplets depends on its temperature. On average, water vapour makes up less than 1% of the atmosphere, but under very warm conditions this can rise to 2–4% and since, like CO_2, it is a good absorber of heat radiation, it also plays a major role in warming the lower atmosphere. Most of the moisture in the atmosphere is held in the lowest few km since this is near the source of water from the Earth's ocean and land surface. Uncertainties about how rises in global mean temperature will affect evaporation rates and the water vapour content of the atmosphere make the job of climate modellers a difficult one. Water vapour is vital in the atmosphere, since it initiates the water cycles and maintains the global water balance.

2.2.5 Particulate matter

In addition to the various gases discussed so far, the atmosphere also contains a considerable amount of particulate matter or aerosols, which are held in suspension (Table 2.3).

Aerosols are too small to be visible to the naked eye, but collectively they can be seen as haze. They can help to scatter solar radiation back to space, and some theories suggest that global cooling episodes leading to ice ages may have been triggered by increases in particulates and aerosols, perhaps from massive volcanic eruptions. Aerosols may be picked up from the Earth's surface by winds, injected into the atmosphere by volcanoes and forest fires, or originate from anthropogenic industrial/manufacturing processes or car exhausts. Table 2.3 confirms the importance of the naturally-supplied sources, particularly sea salt and windblown dust.

2.2.6 Acid rain

Sulphates from hydrogen sulphide (H_2S) contribute up to 98% of the total atmospheric aerosol production. The input of sulphates

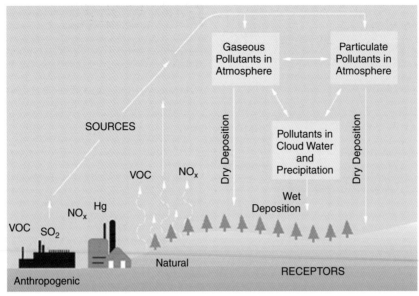

Figure 2.4 Natural and anthropogenic sources and receptors of acid rain. (Source: Wikimedia Commons, http://en.wikipedia.org/wiki/File:Origins.gif)

from SO_2 resulting from the burning of coal and oil represents the only significant anthropogenic contribution. These sulphates contribute to acid deposition, more generally known as acid rain, which consists of rain water, fog droplets or dry matter contaminated by natural and anthropogenic aerosols (especially sulphates and nitrates) (Figure 2.4).

Dry fallout tends to occur at, or close to, the source, but acid deposition in rain has become a serious environmental problem in North America, northern Europe and, more recently, southeast China, particularly since the pollutants can be transported several thousand kilometres from their source. Acidification of lakes in southern Scandinavia in the 1980s has been attributed to pollution from British power stations and road traffic, which was transported by the south-westerly prevailing winds. Acidity in rain water is measured on the logarithmic pH scale, with a value of 1 being very acid, through 7 being neutral and 14 being very alkaline. Due to water droplets falling through the atmosphere picking up CO_2, rain is naturally a weak carbonic acid with a pH of between 4.8 and 5.6. However, when anthropogenic combustion of coal and oil introduces pollutants such as SO_2 and NO_2 into the lower/middle troposphere, the pH of rain water can be lower. Following the great smogs of London in the early 1950s, the 1956 Clean Air Act resulted in power plants using taller chimney stacks to disperse the pollutants higher up into the troposphere, where strong winds transported them huge distances. The increased residence times of the pollutants in the troposphere allowed them to react with water droplets to create sulphuric acid (H_2SO_4) and nitric acid (HNO_3), with pH values below 3. On one occasion at Pitlochry in Scotland, a rain water pH value of 2.4 was recorded.

The consequences of acid rain deposition have been the decimation of fish populations in streams and lakes, particularly in areas with granite bedrock, which is incapable of buffering the acidity. Acid deposition can also deplete the soil of basic cations, like magnesium and potassium, leading to nutrient deficiency in plants and trees, which can cause dieback at growing points, as well as leaving plants more susceptible to disease. In severely stressed ecosystems, where acid-tolerant species thrive, biodiversity may be reduced. Acid deposition can also increase the release of toxic cations, such as aluminum, which can be detrimental to fish.

Exercises

1. What is the 'atmosphere system'? Describe its components and their interactions in your answer.

2. What is the difference between 'weather' and 'climate'?

3. Draw a simple diagram to show the structure of the atmosphere.

4. While nitrogen and oxygen make up around 99% of the Earth's atmosphere by volume, list five other variable gases that you might expect to find in a sample of air.

5. Ozone is a gas found in its highest concentrations in the stratosphere. Discuss why holes in the ozone layer have been discovered and what has been done to limit its destruction.

6. Describe the main controls on the concentrations of carbon dioxide and methane in the Earth's atmosphere.

7. What is the relationship between the percentage of water vapour and temperature in the troposphere?

8. Outline the main sources of particulate matter in the Earth's atmosphere.

9. What causes 'acid rain'?

10. Give some examples of the effects of acid rain on ecosystems.

References

Thompson, R.D. 2002. Atmospheric Processes and Systems, Routledge Introductions to Environment Series. London, Routledge.

Further reading

Barry, R.G. and Chorley, R.J. 2003. Atmosphere, Weather and Climate. 8th edition. London, Routledge. Ch. 1.

Hamblin, W.K. and Christiansen, E.H. 2004. Earth's Dynamic Systems. 10th edition. Upper Saddle River, NJ, Prentice-Hall. Ch. 9.

Holden, J. (Ed.). 2005. An Introduction to Physical Geography and the Environment. Harlow, Pearson Education. Ch. 2.

Ozone Hole Watch Web site: http://ozonewatch.gsfc.nasa.gov/.

Robinson, P.J. and Henderson-Sellers, A. 1999. Contemporary Climatology. Harlow, Pearson Education.

Smithson, P., Addison, K. and Atkinson, K. 2002. Fundamentals of the Physical Environment. 3rd edition. London, Routledge. Ch. 3.

3 Energy in the Atmosphere and the Earth Heat Budget

Sunset over Vancouver Island, western Canada. Photo: T. Stott

Earth Environments   David Huddart and Tim Stott
© 2010 John Wiley & Sons, Ltd

Learning Outcomes

After reading this chapter and completing the exercises you will be able to:

➤ Understand the concept of solar radiation and how it is intercepted by the Earth.

➤ Understand about the total energy flux to Earth and the solar constant.

➤ Understand the difference between sensible heat and latent heat.

➤ Know that electromagnetic radiation is emitted by all objects and that the emission of radiation by an object is described by the radiation laws: the Stefan–Boltzmann law and Wien's law.

➤ Describe the electromagnetic spectrum.

➤ Understand the term insolation and how it is used to describe the flow of energy intercepted by an exposed surface and how it varies with season and latitude.

➤ Define the term albedo and describe typical albedo values for a range of surfaces.

➤ Understand the term net radiation and explain how it varies over the Earth's surface.

3.1 Introduction

This chapter explains how solar radiation is intercepted by our planet, flows through the Earth's atmosphere and interacts with the Earth's land and ocean surfaces. Solar radiation is the driving power source for wind, waves, weather, rivers and ocean currents. Most natural processes that occur at the Earth's surface are directly or indirectly solar powered, for example the growth of a field of corn, the flow of a river, the breaking of waves onto a beach. The Earth can be considered as a flow system with an energy budget that includes all gains of incoming energy and all losses of outgoing energy. The planet is approximately in equilibrium, so the sum of the gains should be approximately equal to the sum of the losses. The total amount, or flux, of energy entering the Earth's atmosphere is measured in watts (W). Watts are a unit of power, which means energy per unit time. The total energy flux to Earth is estimated at 174 petawatts (1 petawatt $= 10^{15}$ watts). The vast majority (99.978%) comes from solar radiation (this is equivalent to around 340 W m^{-2} when averaged over the whole of the Earth's surface). A small amount of energy (0.013%) comes from geothermal energy (about 0.045 W m^{-2}), which is produced by stored heat and heat produced by radioactive decay leaking out of the Earth's interior. Tidal energy contributes 0.002% (or

about 0.0059 W m^{-2}), which is produced by the interaction of the Earth's mass with the gravitational fields of other bodies, such as the Moon and Sun. Finally, waste heat from fossil fuel burning contributes about 0.007% (or around 0.025 W m^{-2}).

The flow of energy from the Sun to the Earth's atmosphere and surface and then back out into space is a complex flow system. Energy exists in a variety of forms, including heat, radiation, potential energy, kinetic energy, chemical energy and electromagnetic energy. Energy is neither created nor destroyed, and from this it follows that all forms of energy are convertible to all other forms. It is important here to distinguish between temperature and heat. Temperature is a measure of the mean kinetic energy (speed) per molecule of an object, while heat is a measure of the total kinetic energy of all the molecules in that object. Thus, the temperature of the air is simply a measure of the 'internal energy' of the air, which is associated with the random motion of the air molecules.

As the temperature of a parcel of air decreases, it eventually reaches a state in which the molecules are at complete rest and there is no internal energy, a point on the temperature scale known as absolute zero. This is 273.15 °C below the melting point of pure ice. Physicists use the Kelvin temperature scale, which is measured upwards from absolute zero in Celsius units, making 0 °C equal to 273.15 K.

Heat naturally transfers from high- to low-temperature objects, which alters either the temperature or the state of the substance, or both. If a heated body acquires a higher temperature, this is known as sensible heat. On the other hand, a heated body might change to a higher state (solid to liquid or liquid to gas) and therefore acquire latent heat. So if ice at 0 °C was heated it could melt to become water also at 0 °C. The extra heat is used to change the state of the solid ice to liquid water and is known as latent heat. The transfer of heat to or from a substance is affected by the processes of conduction, convection and radiation. Heat conduction, or thermal conduction, is the spontaneous transfer of thermal energy through matter, from a region of higher temperature to a region of lower temperature, without the transfer of matter itself, and hence acts to even out temperature differences. It is the process by which heat passes through solids like soil or rock, though liquids and gases are poor conductors of heat in comparison. This is why we wear layers of clothes to trap air, which is a poor heat conductor, so we stay warmer. Convection in the most general terms refers to the movement of currents within fluids (i.e. liquids and gases) and is a mode of heat transfer in the oceans and atmosphere involving the movement of substantial volumes of the substance itself. Radiation is the final form of heat transfer and is transmitted by any object that is not at a temperature of absolute zero (0 K).

3.2 Solar radiation

Any study of global energy systems needs to begin with the subject of radiation – that is, electromagnetic radiation – which is

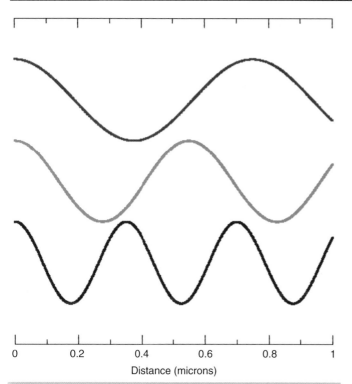

Distance (microns)

Figure 3.1 Electromagnetic radiation is a collection of energy waves with different wavelengths. Wavelength is the crest-to-crest distance between successive wave crests. This figure shows three electromagnetic modes – blue (450 nm), green (550 nm) and red (750 nm) – with a distance scale in microns (micrometres) along the x-axis. (Source: Wikimedia Commons)

emitted by all objects that have a temperature above absolute zero. Electromagnetic radiation can be best considered as a collection, or spectrum, of waves of a wide range of wavelengths travelling at high speed away from an object. The distance between one wave crest and the next is called the wavelength (Figure 3.1). The unit used to measure wavelength is called the micrometre or μm (formerly called the micron), which is one millionth of a metre. So, one centimetre is 10 000 μm. The electromagnetic spectrum (Figure 3.2) extends from very short X-rays through ultraviolet and visible light to infrared, microwaves and long radio waves.

3.2.1 Radiation laws

The emission of radiation by an object is described by the radiation laws. Note that a perfect black body is one which absorbs all the radiation falling on it and which emits, at any temperature, the maximum amount of radiant energy. A black body does not have to be black in colour. Indeed, snow is an excellent black body in the infrared part of the spectrum.

In summary, the radiation laws include:

1. The Stefan–Boltzmann law, also known as Stefan's law, a power law which states that the total energy radiated per unit surface area of a black body in unit time (known also as the black body irradiance, energy flux density or radiant flux), F, is directly proportional to the fourth power of the black body's thermodynamic temperature T (also called absolute temperature):

$$F = \sigma T^4$$

where F is the flux of radiation (watts per square metre or $W\,m^{-2}$), T is the absolute temperature (kelvin) and σ is Stefan's constant ($10^{-8}\,W\,m^{-2}\,K^{-4}$). Because of this relationship, a small increase in temperature can mean a large increase in the rate at which radiation is given off by an object. For example, if the absolute temperature of an object is doubled, the flow of radiant energy from it will be 16 times larger.

2. Wien's law, also known as the Wien displacement law, which states that a black body emits radiation with a range of frequencies, but with a maximum frequency of λ_{max}. It can be shown by the Wein displacement law that the wavelength of maximum energy, λ_{max}, is inversely proportional to the absolute temperature:

$$\lambda_{max} = \alpha/T$$

where α is a constant ($1.05 \times 10^{11}\,K^{-1}\,s^{-1}$). So there is an inverse relationship between the wavelength of radiation that an object emits and the temperature of the object. For example, the Sun, a very hot object, emits radiation with short wavelengths. In contrast, the Earth, a much cooler object, emits radiation with longer wavelengths.

Radiation is emitted from the Sun, 150 million km away, which has a surface temperature of around 6000 °C, and travels in all directions in straight lines, or rays, at the speed of light ($3 \times 10^8\,ms^{-1}$ or $300\,000\,kms^{-1}$). It therefore takes just over eight minutes to travel from the Sun to the Earth. The Earth intercepts about half of one-billionth of the Sun's total energy output. The portion of the electromagnetic spectrum (Figure 3.2) that contains radiation emitted by the Sun and the Earth can be further divided into four major portions, which are presented in Figure 3.3. Gamma rays, X-rays and ultraviolet (UV) radiation, on the left of Figure 3.3 (wavelength 0.2–0.4 μm), are short waves high in energy, and are nearly all absorbed by the gas molecules in the atmosphere (particularly ozone in the stratosphere), but if they penetrate to the Earth's surface they can damage living tissue. Visible light (towards the centre of Figure 3.3) has wavelengths of 0.4 to 0.7 μm, which our eyes can see as the colours of the rainbow. Shorter wavelengths we see as violet, and longer ones as red (with blue, green, yellow and orange in between). Incoming radiation peaks in the blue wavelengths, hence we see the sky as blue. Only a small amount of visible light is absorbed by gases in the atmosphere, so nearly all reaches the Earth's surface. Short-wave infrared radiation (0.3–0.7 μm) acts like visible light and penetrates the atmosphere easily, but our eyes are not sensitive

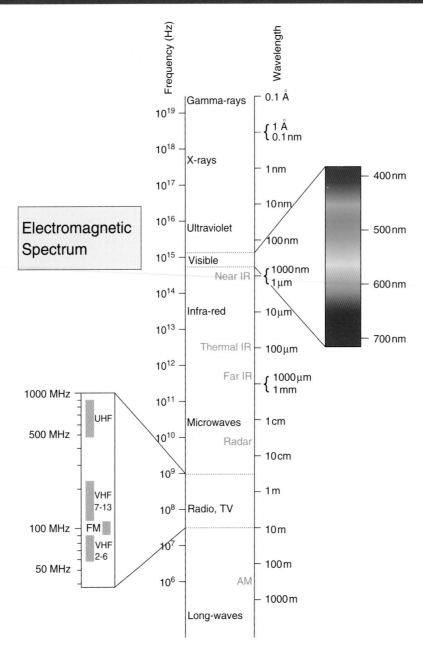

Figure 3.2 The electromagnetic spectrum. (Originally taken from en.wikipedia en:Electromagnetic-Spectrum.svg and en:Electromagnetic-Spectrum.png)

to it. Together, UV, visible light and short-wave infrared are collectively known as short-wave radiation. At the right-hand side of Figure 3.3 are the thermal infrared wavelengths, longer than 3 μm. These are emitted by cooler objects such as the Earth. We perceive this type of radiation as heat and it is collectively referred to as long-wave radiation.

The Sun does not emit all wavelengths of radiation equally. Inside the Sun, hydrogen is converted to helium at very high temperature and pressure in a process known as nuclear fusion. The rate of production of these vast amounts of energy is fairly constant, and the amount the Earth receives (around half of one billionth) is known as the solar constant, with a value estimated at around $1360\,\mathrm{W\,m^{-2}}$.

The Earth is much cooler than the Sun, so according to Wein's law the Earth emits long-wave radiation (3–30 μm), and according to Stefan's law the flux of its radiation is much lower

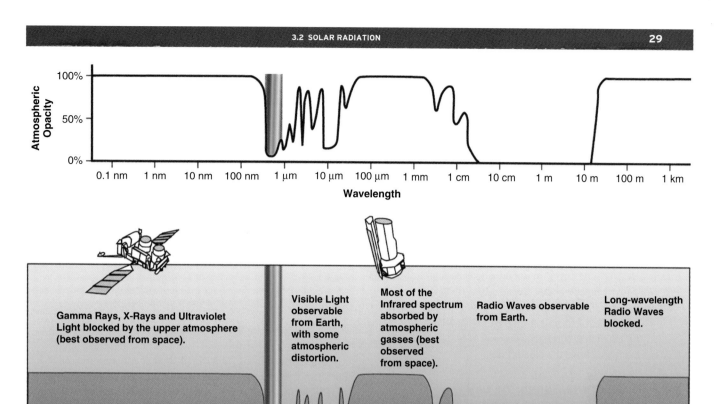

Figure 3.3 Atmospheric opacity in relation to the electromagnetic spectrum. (Source: Vectorized by User:Mysid in Inkscape, original NASA image from File:Atmospheric electromagnetic transmittance or opacity.jpg, from Wikimedia Commons)

than that of the Sun. Water vapour and carbon dioxide in the atmosphere absorb much of the radiation leaving the surface.

3.2.2 Global radiation

Global radiation is the sum of all short-wave radiation received, both directly from the Sun (direct solar radiation) and indirectly from the sky and clouds (diffuse radiation). Part of the Sun's radiation is scattered back into space by clouds and dust in the atmosphere, and some is scattered back from the Earth's surface without being absorbed. The remaining short-wave radiation from the Sun is absorbed by either the atmosphere or land and ocean surfaces. Land and ocean surfaces in turn reemit energy as long-wave radiation, which tends to lower the temperature and thus cool the planet as it leaves, heading for outer space. Over large time scales, these flows balance, so that incoming radiation absorbed equals outgoing radiation emitted.

The flow of solar radiation to the Earth as a whole remains constant, but it varies from place to place and from time to time. The term insolation (incoming solar radiation) is often used to describe the flow of energy intercepted by an exposed surface for the case of a uniformly spherical Earth with no atmosphere.

The amount of incoming radiation received by the Earth is determined by four astronomical factors.

The solar constant

This is the amount of incoming solar electromagnetic radiation per unit area, measured on the outer surface of the Earth's atmosphere, in a plane perpendicular to the rays. The solar constant includes all types of solar radiation, not just visible light. It is measured by satellites to be roughly 1366 watts per square metre, though it fluctuates by about 6.9% during a year – from 1412 W m^{-2} in early January to 1321 W m^{-2} in early July – due to the Earth's varying distance from the Sun, and by a few parts per thousand from day to day. Solar variations are changes in the amount of radiant energy emitted by the Sun. There are periodic components to these variations, the principal one being the 11-year solar cycle (or sunspot cycle), as well as aperiodic fluctuations. Any changes in the amount of radiant energy emitted by the Sun are referred to as 'solar forcing' and recent results suggest that there had been around a 0.1% variation in the solar constant over the last 2000 years. Solar activity has been measured via satellites during recent decades, and was measured through 'proxy' variables before satellites were available. Climate scientists are interested in understanding what, if any, effect variations in solar activity have on the Earth.

The distance from the Sun

The Earth completes a revolution around the Sun in 365.256366 days – just over a quarter of a day more than the calendar year of 365 days. Every four years, or every 'leap year', therefore, the extra quarter day is added on as the 29th of February to correct for this effect. The Earth's orbit around the Sun is shaped like an ellipse, or oval, so the distance between Sun and Earth varies throughout the year. At perihelion, which occurs around 3 January, the Earth is 147 098 074 km from the Sun, whereas at aphelion, on or around 4 July, it is 152 097 701 km from the Sun. So the distance between Earth and Sun varies by about 3.3% in one revolution, and for most purposes we can disregard this as an important influence on the Earth's radiation budget. Figure 3.4 shows the direction in which the Earth and Moon revolve and rotate – anticlockwise when viewed from above the North Pole.

The altitude of the Sun in the sky

The spherical shape of the Earth results in large temperature differences, because, as Figure 3.5 shows, the same amount of solar radiation is spread out over a larger area near the poles, where, because of the angle at which the bundle of solar radiation approaches, it has to penetrate a greater depth of atmosphere.

Compared to that at the equatorial regions Figure 3.5(b), incoming solar radiation at the polar regions Figure 3.5(a) is less intense for two reasons: first, the solar radiation arrives at an oblique angle nearer the poles, so that the energy spreads over a larger surface area, lessening its intensity; and second, the solar radiation travels a longer distance through the atmosphere, which absorbs, scatters and reflects it. Tropical areas (i.e. lower latitudes, nearer the equator) receive solar radiation that is closer to vertical. The angle of incidence of the rays, combined with the albedo of the surface, also has a strong influence on the amount of energy absorbed (or reflected) at the surface. In the ice-covered polar zones, almost all direct energy from the Sun is reflected, because it is white and the angle is small. In short, the angle of incidence affects the heating of the surface in three different ways:

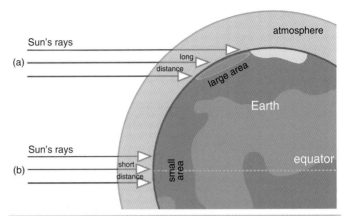

Figure 3.5 Why the polar regions are colder: the effect of the Earth's shape and atmosphere on incoming solar radiation. (Source: Oblique_rays_02_Pengo.svgy, Wikimedia Commons)

length of atmospheric track, variable flux and variable reflection. For simplicity, Figure 3.5 ignores the axial tilt of the Earth, which causes each pole to slip into darkness for around six months of the year and means that the equator's ground is generally not perpendicular to the Sun's light.

Direct solar radiation casts shadows and its intensity on a horizontal surface varies with the angle of the Sun (solar angle) above the horizon. Diffuse radiation is the result of scattering and absorption in clouds and its intensity depends on cloud structure and thickness, not just on solar angle. It does not cast shadows, it only varies slightly during the day and it is much less intense than direct insolation. It is therefore quite possible that people may get sunburned on cloudy days, especially at high altitude, from diffuse radiation.

Global radiation with a large direct component shows a strong diurnal variation in intensity, with maximum values when the Sun is at its greatest angle above the horizon at midday. On cloudy days, this diurnal variation becomes almost nonexistent, when the diffuse component dominates.

The length of day and night

The length of day and night varies with the seasons, which are related to the orientation of the Earth's axis of rotation to the Sun. In other words, the Earth's axis is tilted by 23.5°. This means that when the North Pole is tilted towards the Sun (Figure 3.6(a)) all latitudes north of the Arctic circle (66.5° N) are continuously in the Sun and that part of the Earth receives continuous daylight in what is known as the Arctic summer. The latitudes to the south of the Antarctic circle (66.5° S) are in total darkness during this time (Figure 3.6(b)), and this is called the Antarctic winter.

The direction in which the axis tilts does not change as the Earth revolves, so the North Pole is tilted towards the Sun during the summer (summer solstice, 21 June, Figure 3.7) and is tilted away from the Sun during the winter (winter solstice, 21 December, Figure 3.7). Both hemispheres are illuminated equally during the spring (vernal) and autumn equinoxes.

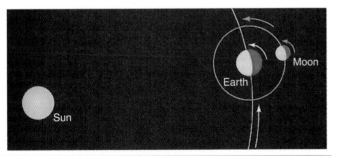

Figure 3.4 The Earth's direction of revolution around the Sun and of rotation on its own axis, and the directions of revolution and rotation of the Moon. (Source: Strahler and Strahler, 1997, p. 28)

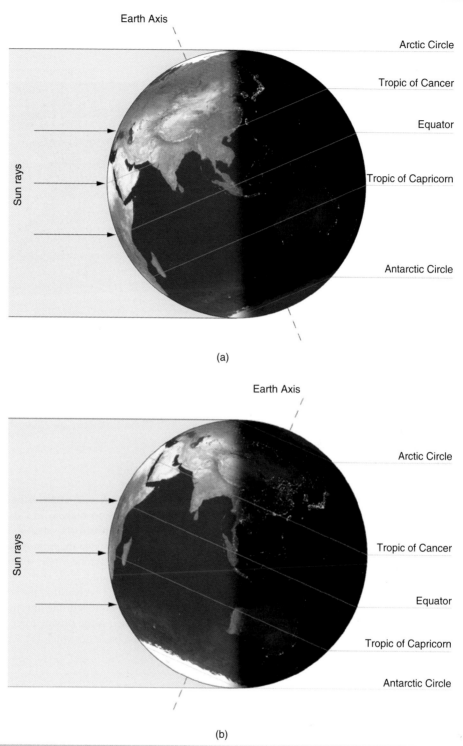

(a)

(b)

Figure 3.6 (a) The North Pole is tilted towards the Sun, giving the northern hemisphere summer. (b) The South Pole is tilted towards the Sun, giving the southern hemisphere summer. (Source: (a) File:Earth-lighting-summer-solstice EN.png, Wikimedia Commons; (b) File:Earth-lighting-winter-solstice EN.png, Wikimedia Commons)

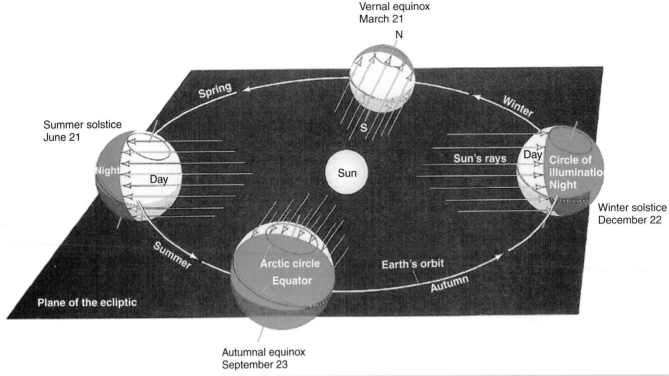

Figure 3.7 The Earth's tilted axis, which keeps a constant orientation in space as it revolves around the Sun, is the cause of the four seasons. This tips the northern hemisphere towards the Sun for the summer solstice, and away from the Sun for the winter solstice. Both hemispheres are illuminated equally during the spring (vernal) and autumn equinoxes. (Source: Strahler and Strahler, 1997, p. 29). Reprinted from Strahler and Strahler, Physical Geography, 1997, p. 29 with permission from John Wiley & Sons Ltd.

The effect of the Earth's tilted axis combined with its revolution around the Sun is that at any particular place (or latitude), the Sun's path in the sky changes greatly in position and height above the horizon from summer to winter. Figure 3.8 shows the Sun's path, giving the times of sunrise and sunset for a location 40° N (e.g. New York). This is the way it would appear to an observer standing at 40° N. The Earth's surface appears flat and the Sun seems to travel inside a vast dome in the sky.

At the winter solstice the Sun is low in the sky, sunrise is at 7.30 am and sunset at 4.30 pm, so the period of daylight is short (9 hours). At the summer solstice, the Sun is high in the sky, it rises at 4.30 am and sets at 7.30 pm, so the daylight period is long (15 hours). Clearly, the solar radiation reaching the surface will be much greater during the summer than during the winter solstice. In summer the angle of incidence of the Sun's rays is greatest, and the length of time the Sun's radiation is exposed to the Earth's surface is also greatest.

3.2.3 Insolation through the year and by latitude

When daily insolation values are plotted for a full year, they form a wavelike curve on a graph (Figure 3.9). The curve for latitude 40° N exhibits greater daily insolation at the summer solstice (June) and lower daily insolation at the winter solstice

(December). However, not every latitude shows this simple progression from low to high to low daily insolation through the year. Latitudes between the equator (0°) and the Tropic of Cancer (23.5° N) show two maximum values; others show only one. Poleward of the Arctic circle (between 66.5 and 90° N), insolation is zero for some of the year.

From this section, it is clear that the seasonal pattern of insolation is directly related to latitude. It is important to understand this pattern because it is the driving force for the annual cycle of climate, and is the basis on which the Earth's climatic belts are determined.

3.2.4 Insolation losses in the atmosphere: The solar energy cascade

As short-wave radiation from the Sun penetrates the atmosphere, its energy is absorbed or diverted in various ways. Figure 3.10 gives typical values for losses of incoming solar energy.

Figure 3.10 describes the relative amounts (based on 100 units available at the top of the atmosphere) of short-wave radiation lost to various atmospheric processes as it passes through the atmosphere. The diagram indicates that 19 units of insolation are absorbed (and therefore transferred into heat energy and long-wave radiation) in the atmosphere by stratospheric absorption of UV radiation by ozone (2 units) and tropospheric absorption

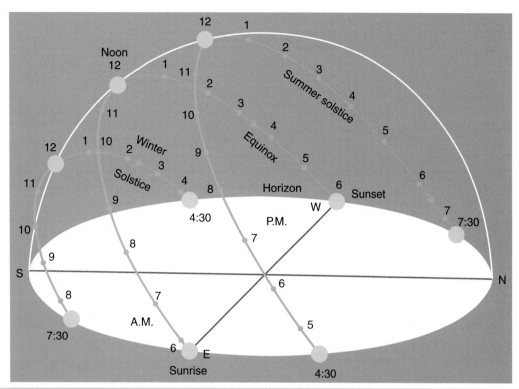

Figure 3.8 The Sun's path in the sky for a location 40° N. At the winter solstice the Sun is low in the sky, and the period of daylight is short. At the summer solstice the Sun is high, and the daylight period is long. (Source: Strahler and Strahler, 1997, p. 43). Reprinted from Strahler and Strahler, Physical Geography, 1997, p. 43 with permission from John Wiley & Sons Ltd.

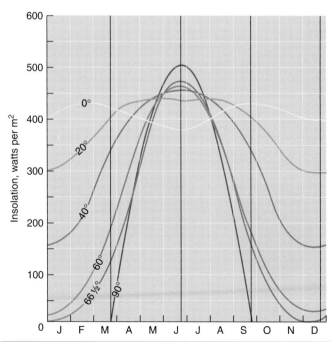

Figure 3.9 Insolation curves at various latitudes in the northern hemisphere as measured at the top of the atmosphere. Black lines mark the equinoxes and solstices. (Source: Strahler and Strahler, 1997, p. 44). Reprinted from Strahler and Strahler, Physical Geography, 1997, p. 44 with permission from John Wiley & Sons Ltd.

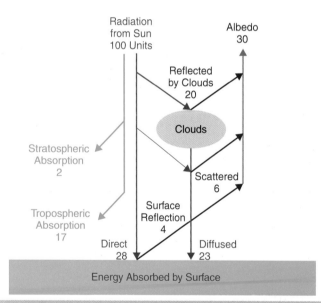

Figure 3.10 The global short-wave radiation cascade: absorption, reflection and scattering of incoming short-wave radiation by the Earth's atmosphere. (Source: http://www.physicalgeography.net/fundamentals/images/incoming.gif, accessed 28 October 2007)

of insolation by clouds and aerosols (17 units). 23 units of solar radiation scattered in the atmosphere are later absorbed at the surface as diffuse insolation. 28 units of the incoming solar radiation are absorbed at the surface as direct insolation. The total amount of solar insolation absorbed at the surface is 51 units. The total amount of shortwave radiation absorbed at the surface and in the atmosphere is 70 units.

Three main losses of solar radiation back to space occur in the Earth's short-wave radiation cascade. Four units of sunlight are returned to space from surface reflection. Cloud reflection returns another 20 units of solar radiation. Back-scattering of sunlight returns 6 units to space. The total loss of shortwave radiation from these processes is 30 units.

So, as it approaches the top of the atmosphere at around 150 km, the solar radiation spectrum possesses almost 100% of its original energy. By the time it penetrates to 88 km, most of the X-rays and some of the UV rays have been absorbed. As it moves into the lower, denser layers of the atmosphere, gas molecules, dust and other particles cause visible light rays to be scattered. Scattered radiation is not changed, but it follows a new path, in which it becomes known as diffuse radiation. Some goes back out to space, while some carries on down to the Earth's surface. Clouds can increase the amount of incoming solar radiation reflected back to space, and they can absorb radiation too.

Table 3.1 Values of albedo for various surfaces. (Sources: Seller, 1965; List, 1966; Paterson, 1969; Monteith, 1973 (adapted from Oke, 1987, p. 12); Holden, 2005)

Surface	Albedo (percentage of incoming short-wave radiation that is reflected)
Fresh, dry snow	80-95
Old snow	40-50
Sea ice	30-40
Glacier ice	20-40
Dry light sandy soil	35-45
Desert	20-45
Tundra	18-25
Grass (long, 1.0 m)	16-20
Grass (short, 0.02 m)	20-26
Dry steppe	20-30
Orchard	15-20
Coniferous forest	10-15
Deciduous forest	15-20
Dark pavement	2-3
Water surface with low Sun angle	3-10
Water surface with high Sun angle	10-100

3.2.5 Albedo

The percentage of short-wave radiation reflected or scattered upwards from a surface is called its albedo. The albedo of a surface is important because it determines how quickly the surface heats up. Table 3.1 gives some albedo values for various surfaces.

The albedo of a water surface depends a lot on the angle at which the solar radiation approaches it, and whether the surface is calm or choppy. If the rays are nearly vertical, and the surface disturbed, the absorption is high and the albedo is very low (e.g. 2%), but if the angle is low and the surface is calm, it can act like a mirror and the albedo can be very high (e.g. 95%).

See the companion website for Box 3.1 Earth's Albedo Tells and interesting story.

3.2.6 Re-radiation and the greenhouse effect

The global long-wave radiation cascade indicates that energy leaves the Earth's surface through three different processes (Figure 3.12).

Seven units leave the surface as sensible heat. This heat is transferred into the atmosphere by conduction and convection. The melting and evaporation of water at the Earth's surface incorporates 23 units of energy into the atmosphere as latent heat. This latent heat is released into the atmosphere when the water condenses, or becomes solid. Both of these processes become part of the emission of long-wave radiation by the atmosphere and clouds.

The surface of the Earth emits 117 units of long-wave radiation. Of this emission, only six units are directly lost to space. The other 111 units are absorbed by greenhouse gases in the atmosphere and converted into heat energy and then into atmospheric emissions of long-wave radiation (the greenhouse effect).

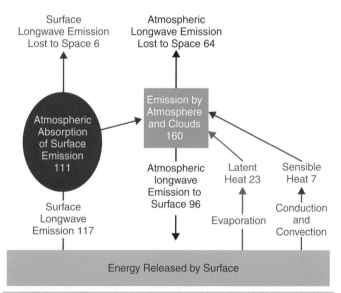

Figure 3.12 Global long-wave radiation cascade. (Source: www.physicalgeography.net/fundamentals/7i.html, accessed 28 October 2007)

The atmosphere emits 160 units of long-wave energy. Contributions to these 160 units come from surface emissions of long-wave radiation (111 units), latent heat transfer (23 units), sensible heat transfer (7 units) and the absorption of short-wave radiation by atmospheric gases and clouds (19 units; see Figure 3.11). Atmospheric emissions travel in two directions. 64 units of atmospheric emission are lost directly to space. 96 units travel to the Earth's surface, where they are absorbed and transferred into heat energy. The total amount of energy lost to space in the global long-wave radiation cascade is 70 units (surface emission, 6 units + atmospheric emission, 64 units). This is the same amount of energy that was added to the Earth's atmosphere and surface by the global short-wave radiation cascade, which we examined in Figure 3.10.

Finally, to balance the surface energy exchanges in this cascade we have to account for 51 units of missing energy (atmosphere and cloud long-wave emission (96 units) minus surface long-wave emission (117 units) minus latent heat transfer (23 units) minus sensible heat transfer (7 units) = −51 units). This missing component to the radiation balance is the 51 units of energy absorbed at the Earth's surface as direct and diffuse short-wave radiation.

This analysis helps us to understand the very important role of the atmosphere in trapping heat through the greenhouse effect. Without this effect, the Earth would be a very cold, forbidding place on which we would be unable to survive.

3.2.7 Net radiation and the energy balance

The Earth's geological record shows that the Earth is neither cooling nor warming to any appreciable extent. The radiation received from the Sun over the long term must therefore be balanced by the outgoing losses from the Earth. However, for any given surface, incoming and outgoing radiation flows do not have to be in balance. At night, even though there is no incoming radiation, the Earth's surface and atmosphere still emit outgoing radiation. So, in any one place at any one time it is very unlikely that incoming radiation will exactly equal that outgoing.

Net radiation is the difference between all incoming and all outgoing radiation. So where the incoming radiation is greater than the outgoing radiation we see that net radiation will be positive, and where incoming radiation is less than the outgoing

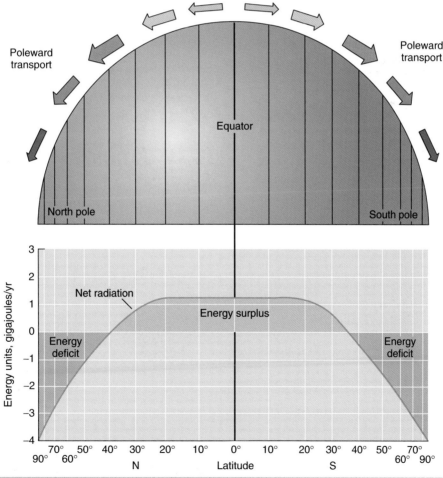

Figure 3.13 Annual surface net radiation variation with latitude, showing regions of surplus and deficit. (Source: Strahler and Strahler, 1997, p. 54)

radiation, net radiation will be negative. In Figure 3.9 we saw that solar energy input varies with latitude. The effect of this difference on net radiation is seen in the lower part of Figure 3.13, which shows the global net radiation balance from pole to pole.

Between latitudes 40° N and 40° S there is a net radiation gain or 'energy surplus'. However, poleward of 40° N and 40° S the net radiation is negative, or in 'deficit'. The orange area of the graph labelled 'surplus' exactly equals the two blue areas of the graph labelled 'deficit'. However, these areas of surplus and deficit cannot continue to build in the long term. There has to be an equalization of this energy. Two processes are responsible for this: the circulation pattern of the oceans, moving heat energy from tropical areas polewards in a sensible heat transfer, and the movement of moist air from equatorial regions poleward, carrying latent heat energy to the higher latitudes. These two energy transport processes are summarized in the upper part of Figure 3.13.

Exercises

1. The total amount, or flux, of energy entering the Earth's atmosphere is measured in watts. Define 'watts'.

2. Outline what is meant by the 'electromagnetic spectrum'.

3. Define the Stefan–Boltzmann law.

4. Explain Wien's law.

5. The amount of incoming radiation received by the Earth is determined by which four astronomical factors?

6. Draw a diagram to show how Earth's tilted axis is the cause of the four seasons.

7. The term 'solar energy cascade' is sometimes used to describe the losses of incoming solar radiation between the top of the atmosphere and the Earth's surface. Outline where these losses occur.

8. With reference to the Earth's surface, define the term 'albedo'.

9. How might the Earth's cloud cover and albedo be related?

10. With reference to the Earth's incoming and outgoing radiation, explain what is meant by the term 'net radiation'.

References

Holden, J. (Ed.). 2005. An Introduction to Physical Geography and the Environment. Harlow, Pearson Education.

List, R.J. 1966. Smithsonian Meteorological Tables. 6th edition. Smithsonian Institute, Washington, DC.

Monteith, J.L. 1973. Principles of Environmental Physics. London, Edward Arnold.

Oke, T.R. 1987. Boundary Layer Climates. 2nd edition. New York, Methuen.

Paterson, W.S.B. 1969. The Physics of Glaciers. Oxford, Pergamon.

Sellers, W.D. 1965. Physical Climatology. Chicago, University of Chicago Press.

Strahler, A.H. and Strahler, A. 1997. Physical Geography: Science and Systems of the Human Environment. Chichester, John Wiley & Sons, Inc.

Further reading

Barry, R.G. and Chorley, R.J. 2003. Atmosphere, Weather and Climate. 8th edition. London, Routledge. Ch. 1.

Hamblin, W.K. and Christiansen, E.H. 2004. Earth's Dynamic Systems. 10th edition. Upper Saddle River, NJ, Prentice-Hall. Ch. 9.

Robinson, P.J. and Henderson-Sellers, A. 1999. Contemporary Climatology. Harlow, Pearson Education.

Smithson, P., Addison, K. and Atkinson, K. 2002. Fundamentals of the Physical Environment. 3rd edition. London, Routledge. Ch. 3

Thompson, R.D. 2002. Atmospheric Processes and Systems. Routledge Introductions to Environment Series. London, Routledge. Chs 3–5.

4 Moisture in the Atmosphere

Cloud forming over Castle Creek Glacier, Cariboo Mountains, northern British Columbia, Canada.
Photo: J. Claxton

Earth Environments   David Huddart and Tim Stott
© 2010 John Wiley & Sons, Ltd

Learning Outcomes

After reading this chapter and completing the exercises you will be able to:

➤ Understand the role of water in the atmosphere in absorbing radiation and in transporting energy from the Earth's surface into the atmosphere and from the tropics to the poles.

➤ Describe how water exists in three states: liquid, solid and gas, and how it can be converted from one state to another through processes which involve gain or loss of latent heat.

➤ Understand the distribution of water on the Earth, the proportion which is in the atmosphere and the global hydrological cycle.

➤ Understand the process of evaporation, the factors which affect it and the differences between absolute, specific and relative humidity.

➤ Describe the process of condensation, the role of hygroscopic nuclei and the conditions in the natural environment which favour condensation.

➤ Understand the differences between dry and saturated adiabatic lapse rates, and how rising parcels of air can be stable, conditionally stable or unstable.

➤ Describe how clouds develop and how they are classified.

➤ Understand the mechanisms which explain how water droplets in clouds become large and heavy enough to fall as precipitation, and describe the three main types of rainfall.

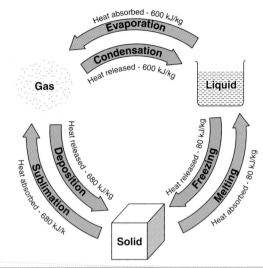

Figure 4.1 Schematic diagram to show the three states of water in the atmosphere. Arrows name the processes by which one state can change to another, indicating the release or absorption of latent heat, which depends on the direction of the change. Evaporation: heat absorbed = 2260 kJ kg^{-1}; condensation: heat released = 2260 kJ kg^{-1}; melting: heat absorbed = 335 kJ kg^{-1}; freezing: heat released = 335 kJ kg^{-1}; sublimation: heat absorbed = 3014 kJ kg^{-1}; deposition: heat released = 3014 kJ kg^{-1}. (Source: http://earth.usc.edu/~stott/Catalina/images/latent.gif, accessed 30 October 2007)

4.1 Introduction

Water is a vital component of the atmosphere as it is essential for life on Earth, as we have seen in Chapter 3. Atmospheric moisture plays a significant role in absorbing radiation and in transporting energy from the surface of the Earth into the atmosphere. It is of vital significance in the transfer of surplus energy from the tropics to the poles, as we saw in the last chapter. Water occurs in the atmosphere and above and below the Earth's surface in three states: liquid, solid and gas, all of which can be converted from one to another with the gain or loss of latent heat (Figure 4.1).

Melting, freezing, evaporation and condensation should all be familiar terms, but sublimation describes the direct transition from solid to vapour. This can happen to ice cubes in the dry air of a freezer, where they lose mass over time, but never melt. Deposition is the reverse process, when water vapour crystallizes directly as ice, as when frost forms as ice crystals on a surface at night, even though it has not rained. The amount of heat per kg taken up and released in each process is indicated next to the arrows in Figure 4.1.

The distribution of water on the Earth was illustrated in Figure 1.4. Around 97.5% of all water on the planet is held in the oceans, where it may reside for up to 4000 years. The remaining 2.5% is fresh water. Of this, some 69% is frozen in the planet's ice caps and glaciers, a figure which is decreasing all the time as global warming continues. Water can be trapped as ice for up to 15 000 years, or for even longer in the permafrost regions of the planet. Around 30% of the planet's fresh water is held underground as ground water, where it can reside for up to 10 000 years. Free fresh water on the Earth's surface makes up around 0.3% of total fresh water. Water in this part of the hydrosphere usually resides in rivers for up to two weeks, as soil moisture for up to a year and in lakes for up to ten years. The 0.04% of the planet's fresh water which exists as atmospheric moisture may reside in the atmosphere for a matter of just days. Although this atmospheric moisture reservoir seems very small, its importance should not be overlooked. It provides the supply of precipitation that purifies and replenishes all fresh water (including that in ice caps and glaciers) on the land. The movement of water among the great global reservoirs just described constitutes the water or hydrological cycle.

4.2 The global hydrological cycle

The simplified global hydrological or water cycle in Figure 4.2 shows the general cycling of water in an anticlockwise direction.

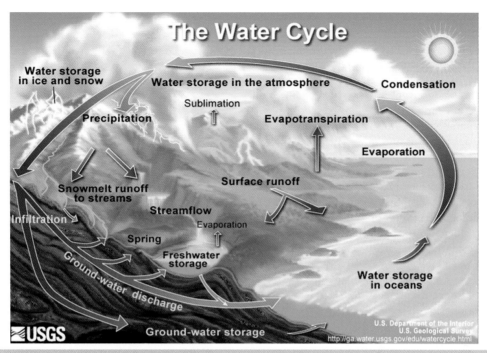

Figure 4.2 The hydrological cycle. (Source: Wikimedia Commons, http://ga.water.usgs.gov/edu/watercycleprint.html)

Most descriptions of the hydrological cycle start at the oceans, where solar radiation causes evaporation. This evaporated water rises, expands, cools and condenses to form clouds, which eventually cause precipitation. When precipitation falls over land it makes its way back to the ocean via streams and rivers, soil or ground water to repeat the cycle.

> With 97% of the world's water in the sea and 2% in deep freeze, the world evidently is a fine place for whales and penguins, but it has its shortcomings for man. In addition, 17% of the land area is under ice or frozen, and 32% is arid or semi-arid. Small wonder that man, throughout history, has sought to interfere with the water cycle.
>
> Chorley and Barry, 1969, p. 40

4.2.1 Evaporation, humidity and condensation

Evaporation

Evaporation occurs whenever energy is transported to a suitable evaporating surface and the vapour pressure of the air next to it is below the saturated value. As we saw in Figure 4.1, the change in state from liquid to vapour requires energy to be expended to overcome the intermolecular attractions of the liquid water molecules (Figure 4.3).

The evaporation process thus causes an apparent heat loss (latent heat), which results in a drop in temperature of the remaining liquid water. The latent heat of vaporization to evaporate 1 kg of water at 0 °C is 2260 kJ. As the faster molecules will generally be the first to escape, so the average energy (and therefore the temperature) of those left in the liquid will decrease and the amount of energy needed to release them will get greater. At the same time, of course, water vapour molecules in the lower layers of air are also in continual motion, and some of them will collide with the water surface and rejoin the underlying mass of water. The rate of evaporation at any given time will therefore depend on the number of molecules leaving the water surface, less the number returning.

The rate of evaporation depends on a number of factors, the most important of which are:

1. **Radiation:** Solar radiation is now generally regarded as the most important single factor governing evaporation.

2. **Temperature:** Since air and water temperature are largely dependent on solar radiation, it is no surprise that there is a fairly close relationship between them all. The temperature of the water surface is important as it governs the rate at which the water molecules leave the surface and enter the overlying air, as we saw in Figure 4.3.

3. **Saturation vapour pressure of the air next to the evaporating surface:** If the air next to the evaporation surface is already saturated with water molecules then while escape of new molecules is still possible at 100% relative humidity, there is a much higher chance of molecules from the air colliding with the water surface, so reversing the evaporation process.

4. **Wind speed:** Evaporation from a water surface into a still layer of air will be continually slowed down as the layer of

air approaches saturation point, and as more water vapour molecules in the air re-enter the water surface. Wind, which causes air movement, is therefore necessary to stir up the air and remove the lowest-saturated air layers and mix them with upper, drier layers. However, the relationship between wind speed and evaporation only holds up to a certain point; beyond that point any further increase in wind speed will not increase the evaporation rate. In this sense, wind does not actually *cause* evaporation, it just 'clears the air' to allow a given rate of evaporation to be maintained.

Other geographical factors which influence actual evaporation rates include: water quality, the depth of a water body and the size of a water surface. Additional factors affecting evaporation from soil surfaces include: soil moisture content and soil characteristics such as particle size, colour, the presence or absence of vegetation and the depth of the water table.

The simplest and most reliable way to obtain direct measurements of evaporation from a free water surface is to measure the loss from an evaporation pan (Figure 4.4), where:

$$Ev = WL_2 - WL_1 + R$$

where Ev = daily evaporation (mm), WL_1 = water level at 0900 on Day 1 (mm); WL_2 = water level at 0900 on Day 2 (mm); and R = rainfall between 0900 on Day 1 and 0900 on Day 2.

The pan in Figure 4.4 has a small vertical pipe within which the water surface is still and can be accurately determined by drawing up a brass hook on a screw thread with a Vernier scale to just break the water surface. For various reasons, pan evaporation rates are normally greater than those from a water surface exposed to the same meteorological conditions and so correction coefficients are often needed.

Figure 4.4 Evaporation pan as part of a weather station at Malham Tarn, Yorkshire, England. (Photo: T. Stott)

Humidity

The term 'humidity' refers to the amount of water vapour present in the air. Absolute humidity is the mass of water vapour in a given volume of air, measured in grams per cubic metre ($g\,m^{-3}$). The amount of water vapour air can hold depends on its temperature (Figure 4.5). Specific humidity is similar, but is expressed in grams of water per kilogram of air ($g\,kg^{-1}$). The measure of humidity

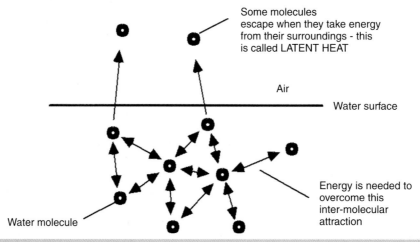

Figure 4.3 Diagram to illustrate how water molecules overcome intermolecular attraction to escape from a water surface by taking latent heat energy from their surroundings. (Drawn by T. Stott)

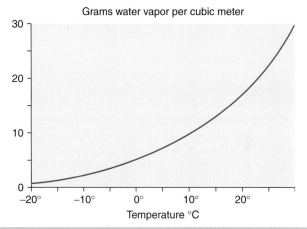

Figure 4.5 The air temperatures and absolute humidity for saturated air. This curve indicates the maximum absolute humidity. (Source: http://www.eduspace.esa.int/Background/images/grafenb.jpg, accessed 13 November 2007)

that is most often encountered, however, is relative humidity (RH). This compares the amount of water vapour present to the amount that the air can hold at the same temperature, expressed as a percentage. At any given temperature, there is a limit to how much moisture the air can hold, and when this is reached the air becomes saturated. The cold air of the Arctic, for example, can hold almost nothing, whereas the warm wet regions of the tropics can hold up to 4–5% of a given volume of air.

For example, if the air holds half the moisture possible at its present temperature, the relative humidity is 50%. When the humidity reaches 100%, the air holds the maximum amount possible and is saturated. Air is generally termed as 'moist' when RH is over 80% and 'dry' when RH is less than 50%, and figures as low as 10% have been recorded over hot deserts.

When unsaturated air is cooled and atmospheric pressure remains constant, a critical temperature will be reached at which the air becomes saturated (RH 100%), and this is known as the dew point. Any further cooling beyond this will result in the condensation of excess vapour, either into water droplets containing condensation nuclei, or into ice crystals if the air is below 0 °C.

See the companion website for Box 4.1 Rise in surface humidity attributed to greenhouse gas emissions.

Condensation

Condensation is the process by which water vapour in the atmosphere is changed into a liquid or, if the temperature is below 0 °C, a solid. Condensation does not occur readily in clean air. If air is perfectly clean, it can actually be cooled below its dew point to become supersaturated, with an RH in excess of 100%. Clean, supersaturated air can be cooled in a laboratory to −40 °C before condensation occurs, or in this case sublimation. However, in natural situations air is hardly ever pure and almost always contains large numbers of condensation nuclei. These microscopic particles, referred to as hygroscopic nuclei, attract water. They

include volcanic dust, dust from windblown soil, smoke particles and sulphuric acid originating from industry and salt from sea spray. As you might expect, they are most numerous over cities, where there might be 1 million per cm^3, and are found in much lower concentrations over oceans (e.g. 10 per cm^3).

Condensation in the natural environment usually results from the air being cooled until it is saturated, which may be achieved by:

1. **Radiation (contact) cooling:** Imagine a night with clear skies. The net loss of radiation from the ground causes its temperature to fall. Air close to the ground surface is cooled by conduction and by radiation of heat from the warmer air to the cooling ground. The air becomes cooler and, as it approaches its saturation point, water droplets begin to form around condensation nuclei. This type of cooling, with clear skies and light winds to help to mix the lower layers of air, forms a layer of water droplets near ground level which results in radiation fog or mist. When the temperature drops below 0 °C at ground level, a hoar frost may occur, and ground surfaces will appear frosty and white.

2. **Orographic and frontal uplift:** Far more widespread methods of cooling air are those connected with upslope motion, which occurs when an air stream meets a mountain barrier and the air is forced to rise. An air parcel will expand as it rises, since the air pressure decreases on moving upwards through the atmosphere. The individual air molecules become more widely spaced, so do not strike each other as frequently, and this gives the air a lower sensible temperature. In this rising and cooling air, humidity increases, dew point is approached and condensation begins, which results in clouds forming on upper mountain slopes. A similar form of cooling resulting from rising air occurs at a weather front, which is a sloping surface at which two differing air masses meet. Air masses are bodies of air which extend horizontally over the Earth's surface for hundreds of kilometres and which have similar temperature and humidity characteristics. When a warmer and less dense air mass meets a cooler, denser one, the warm air tends to glide upwards over the colder air. This results in cooling and condensation, which leads to cloud formation (Chapter 6 gives more details).

3. **Convective or adiabatic cooling:** This type of cooling occurs when air rises in thermals during the daytime. Solar radiation striking a varied surface will cause differential heating. Pockets or 'bubbles' of air heat up and rise (Figure 4.6). As the air parcels expand, they cool sufficiently for condensation to occur. Because air is cooled by the reduction of pressure with height, rather than by loss of heat to the surrounding air, it is said to be adiabatically cooled. Glider and parapente pilots try to gain altitude by searching for thermals.

4. **Advection cooling:** This occurs when there is a horizontal transfer of warm moist air over an adjacent cold surface. In this case, direct contact cooling occurs and condensation may take place in the lower layers, resulting in advection fog. As

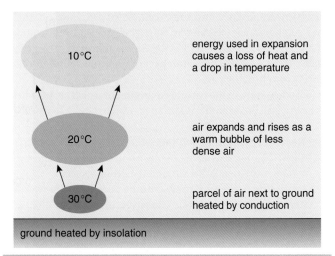

Figure 4.6 Convective cooling. (Source: after Waugh, 2000)

with radiation fog, this is usually of restricted occurrence and limited vertical extent. Advection fogs occur where air above a warm ocean current crosses over onto a cooler current. In the case of Newfoundland, the Gulf Stream at 14 °C and the Labrador Current at −1 °C make an ideal place for the formation of these fogs.

4.3 Air stability and instability

If a balloonist made an ascent through the troposphere, he or she would find that the air temperature fell. This change is known as the environmental lapse rate (ELR). The ELR varies from place to place and from time to time. The principle of convective uplift in a thermal and its associated cooling and condensation was noted in Figure 4.6. This fall in temperature is known as an adiabatic lapse rate, 'adiabatic' indicating that it is the reduced pressure which causes the temperature change of the air parcel, not the loss of heat to the surrounding air. This is an important principle of physics, that if a gas is allowed to expand it cools, and if compressed it warms. If you use a bicycle pump to pump up a tyre, it gets hot because the air inside it is being compressed.

If a rising parcel of air is unsaturated it will cool at the dry adiabatic lapse rate of about 10 °C per 1000 m. In many cases, however, the rising and cooling air will approach its dew point temperature and condensation will begin. If the air continues to rise it will continue to cool, but at a lower rate resulting from the release of latent heat as vapour is condensed into droplets. As water vapour molecules come together, the space between them decreases and they have energy to give up to their surroundings. This means that the parcel of air cools more slowly at what is known as the saturated adiabatic lapse rate, which ranges between 4 and 9 °C per 1000 m at high and low temperatures, respectively.

These three lapse rates can be represented on temperature height diagrams (Figure 4.7).

Figure 4.7(a) shows a position of stability, where the ELR lies to the right of the DALR. If a parcel of unsaturated air is forced to rise it cools at the DALR (10 °C per 1000 m), but the temperature of the air surrounding it is indicated by the ELR line in Figure 4.7(a). In this example, at a height of 500 m the rising parcel of air will have cooled to around 5 °C, but it will be surrounded by air at 8 °C. If the force uplifting the parcel of air were removed, it would sink back down to its starting point, because the parcel is cooler and denser than the surrounding air.

In Figure 4.7(b) the rising air reaches its dew point at 800 m, after which it continues to cool at the saturated ALR (SALR), which in this example has been plotted as 6 °C per 1000 m. Even though at the SALR the air parcel cools more slowly than it would at the DALR, at an altitude of 1100 m the rising parcel of air is still cooler than the surrounding air and would still sink back without its uplifting force, so it is still a stable position.

Since the ELR is variable, however, Figure 4.7(c) shows the situation with a different ELR (this may be a different time or different air mass). Here the rising air parcel cools at the DALR to begin with, reaching its dew point at 6 °C, at which point it is cooler than its surroundings. However, as it continues to rise above the condensation level it cools at the SALR and its temperature slowly approaches that of the surrounding air as it continues to rise. At 1200 m it reaches the same temperature as the surrounding air. Above this point the air becomes warmer and less dense. It will then continue to rise freely, even if the uplifting force is removed, and is now said to be unstable, but it will only remain in this state as long as its saturation level remains at that level or greater. This state is known as conditional instability.

Figure 4.7(a)–(c) has shown constant ELRs, whereas Figure 4.7(d) gives an example of a more realistic situation, in which the ELR varies with height. The cloud base is controlled by the condensation level, at which the dew point is reached. The upper surface of the cloud may lie at the top of the zone of instability, although rising thermals in the cloud can push this upwards.

As we have seen in Figure 4.7(a)–(d), there is a fall in temperature with height. However, there are occasions when a temperature inversion can develop, and a rise in temperature occurs with height (Figure 4.7(e)). Inversions can occur near ground level or high in the troposphere. High-level inversions are found in depressions, where warm air overrides cold air at the warm front or is undercut by colder air at the cold front (see Chapter 6). Low-level or ground inversions usually occur under calm, clear conditions, where there is a rapid loss of heat from the ground due to radiation at night. Under these conditions, fog or frost may form in valleys and hollows (Figure 4.8).

4.4 Clouds

Clouds in the atmosphere are made up of water droplets, or ice crystals suspended in the air. The particles have a diameter in the range 20–50 μm (1 μm is one thousandth of a millimetre). At the centre of each droplet is a hygroscopic nucleus, which has a diameter in the range 0.1–1 μm. Air which contains abundant hygroscopic nuclei will enhance cloud formation,

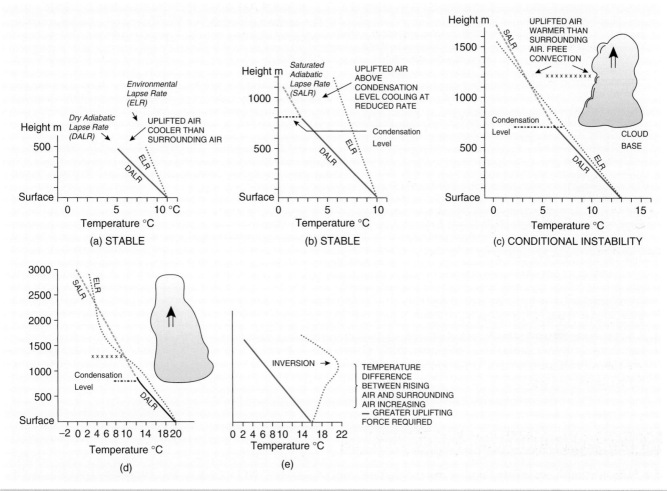

Figure 4.7 Lapse rates, stability and instability. (Source: after Hilton, 1985)

Figure 4.8 (a) Temperature inversion and valley cloud in Skeldal Valley, north-east Greenland. (b) Temperature inversion and valley fog in the Cairngorm Mountains, UK. (Photos: T. Stott)

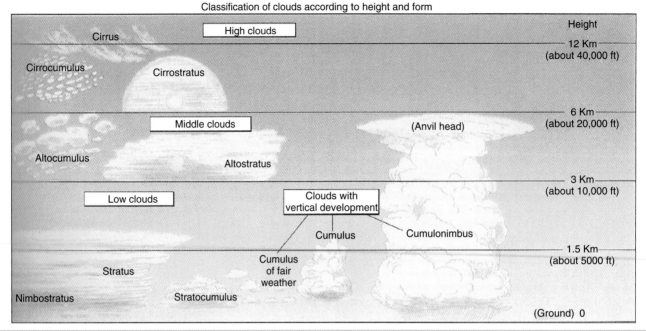

Figure 4.9 Classification of clouds based on their altitude and general shape. (Source: Strahler and Strahler, 1997, p. 97)

and sources include: salt crystals from the sea, dust from volcanoes, wind-eroded soil/dust from deserts and dust and chemicals from industry and transport (urban areas). As cloud temperature becomes cooler, with height, the proportion of ice crystals becomes greater. The highest clouds, between 6 and 12 km, will normally consist entirely of ice crystals.

Clouds come in many shapes and sizes and the great variety necessitates a classification system for the purposes of description and weather forecasting. The internationally adopted system is based upon:

1. **General shape, structure and vertical extent:** On the basis of shape, clouds are grouped into two major classes: stratiform, or layered clouds, and cumuliform, or globular (fluffy) clouds. Stratiform clouds are blanket-like and cover large areas (e.g. stratus). Cumuliform clouds are globular masses of cloud that are associated with small to large parcels of rising air, which rise like bubbles in a fluid (e.g. cumulus).

2. **Altitude:** Most meteorologists classify clouds into four groups based on their height: high clouds, middle-level clouds, low clouds and clouds with vertical development. These, along with the names of individual cloud types, are shown in Figure 4.9.

In 1803 Luke Howard proposed a general descriptive classification of clouds based on shape and height. He used Latin terms, the meanings of which are given in Table 4.1.

Howard also combined terms, for example cirrostratus, cumulonimbus, and added the prefix 'alto' for middle-level clouds, for example altocumulus. For some images of different forms of clouds, see Figure 4.10.

Table 4.1 The meanings of Latin terms for clouds proposed in 1803 by Luke Howard

Latin name	Meaning	Description
Cirrus	A lock of curly hair	High clouds consisting of ice crystals
Cumulus	A heap or pile	Fluffy or globular clouds which usually show vertical development
Stratus	A layer	Layered clouds
Nimbus	Rain-bearing	Grey or black clouds

4.5 Precipitation

Precipitation is the all-encompassing term for all forms of moisture transfer from the atmosphere to the Earth's surface. In most parts of the world it takes the form of rain, or snow in colder regions. However, sleet, hail, dew, hoar frost, fog and rime, drizzle, mist and even mizzle are all terms which have been used to describe particular forms of precipitation.

4.5.1 How clouds make precipitation

Not all clouds produce precipitation. Cloud droplets are small, 1–50 μm in radius, and held in the atmosphere by updrafts, without which they might fall though the atmosphere slowly, perhaps taking five days to descend 100 m. The average raindrop is a million times bigger, 1000–2000 μm in radius, and would

Figure 4.10 (a) High-level cirrus and cirrostratus clouds at Ethabuka Reserve, Simpson Desert, Australia. (b) Mid-level stratus clouds with low-level cumulus below, Kilimanjaro, East Africa. (c) Isolated lenticular cloud forming on a warm afternoon in the Swiss Alps, low-level cumulus in distance. (d) Cumulonimbus (rain-bearing) clouds in the English Lake District, UK. (Photos: (a)–(c): T. Stott, photo (d): D. Hardy)

take around two minutes to fall 100 m through the atmosphere. So how do cloud droplets become raindrops?

There are two favoured theories: in the first, cloud droplets collide and coalesce into larger and larger water droplets that eventually fall as rain. This is known as the collision and coalescence theory, proposed by Longmuir. Warm clouds which contain no ice crystals contain water droplets of differing sizes that are swept upwards at different velocities, and as they do so they collide with other droplets. The larger the droplet, the greater the chance that it will collide with another droplet. When coalescing droplets reach an optimum size of 3 mm, their motion causes them to disintegrate to make a fresh supply of droplets. In thicker clouds that have vertical development (Figure 4.9), such as cumulo-nimbus, the droplets have more time in which to grow, and subsequently they will fall faster as thundery showers.

The second theory of the process by which cloud droplets form rain is known as the ice crystal mechanism, or Bergeron–Findeisen theory. When air temperature is between −5 °C and

Figure 4.11 Snow crystals magnified. (Source: Strahler and Strahler, 1997, p. 99)

−25 °C, supercooled water droplets and ice crystals exist in the same cloud. Supercooling occurs when there is a lack of condensation nuclei in the air. When an ice crystal collides with a droplet of supercooled water, it causes the droplet to freeze. The ice crystals then join together to become snowflakes, which can become heavy enough to fall from the cloud. Some ice crystals can grow directly from the process of deposition, which was outlined at the beginning of this chapter. Snowflakes formed entirely by deposition (Figure 4.11) can have intricate crystal structures.

When the layer of air below these clouds is freezing, the snowflakes reach the ground as snow, but when the air is above freezing the snow melts and falls as rain. In some cases a mixture of ice crystals and rain falls, called sleet, or rain may be lifted back up by rising air currents, where it may freeze to form a hail pellet. This can happen a number of times, so that the hail pellet gets coated with several layers of ice until it becomes so heavy that it falls down to the Earth's surface (Figure 4.12).

These two theories may complement each other.

4.5.2 Types of precipitation

As we saw earlier, precipitation includes sleet, hail, dew, hoar frost, fog and rime, but it is only rain and snow which provide any significant totals in the hydrological cycle. There are three main types of rainfall, all of which depend on the uplift and adiabatic cooling of air until it reaches saturation point:

1. Orographic or relief rainfall results when warm, moist air meets a mountain barrier and is forced to rise. Providing the air is moist enough, and the mountain barrier high enough, orographic precipitation will occur (Figure 4.13).
   Moist air arriving at the mountain barrier (1) has usually travelled over an ocean, where it has picked up its moisture. As the air rises on the windward side of the mountain range

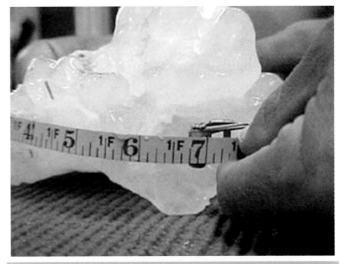

Figure 4.12 The largest hailstone ever recovered in the United States fell in Aurora, Nebraska on 22 June 2003, with a record 17.5 cm (7 inch) diameter and a circumference of 47 cm (18.75 inches). (Source: http://www.noaanews.noaa.gov/stories/s2008, accessed 16 March 2004, courtesy NOAA)

it is cooled at the dry adiabatic lapse rate, and when the dew point is reached, moisture condenses to form clouds (2). Cooling now continues at the saturated adiabatic lapse rate and eventually precipitation begins. After passing over the mountain summit (3) the air begins to descend, and as it does so it is compressed and warmed. Cloud droplets and ice crystals evaporate, or sublimate, and the air clears. Far less rain falls in this area, which is known as the rain shadow (4).

2. Convectional rainfall occurs when the ground surface is locally overheated and the air next to it gets heat from the

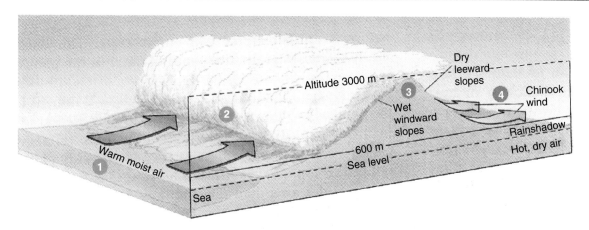

Figure 4.13 Warm, moist air forced to ascend over a mountain barrier rises, cools and condenses. Precipitation is heaviest on the windward slopes and summit, but usually lighter on the lee slopes and beyond, where there is a rain shadow. (Source: Strahler and Strahler, 1997, p. 100). Reprinted from Strahler and Strahler, Physical Geography, 1997, p. 97 with permission from John Wiley & Sons Ltd.

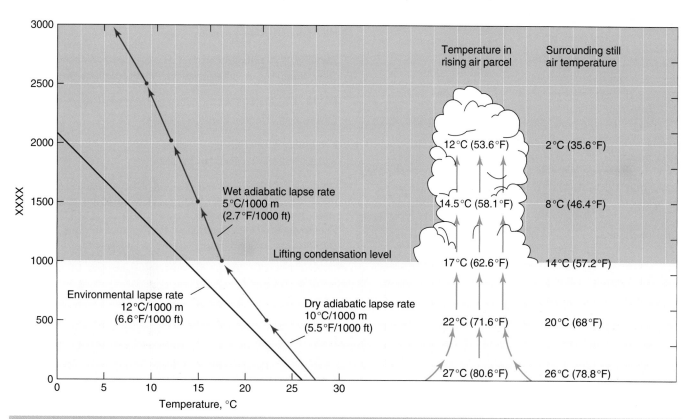

Figure 4.14 Convection in unstable air, where a parcel of air that is heated from below (by the ground surface) rises to great heights. (Source: Strahler and Strahler, 1997, p. 103)

ground by conduction. This warm air rises and expands, and as it does so it remains warmer than the surrounding air, making it conditionally unstable, as we saw in Fig 4.7(c). Towering cumulo-nimbus clouds form and the air rises in a 'chimney' (Figure 4.14), the updraft being maintained by energy released by latent heat as the water vapour molecules reach their dew point and condense.

3. Frontal or cyclonic rainfall occurs at a weather front where two air masses with different characteristics meet. When air in a warm air mass meets air in a cold air mass, the warm air tends to rise as it is less stable. The rising air expands and cools, and moisture in it condenses to form clouds and rain. This type of rain is associated with fronts which form from low-pressure systems called depressions (see Chapter 6).

Exercises

1. Draw a simplified diagram to show the main pathways and storage points in the global hydrological cycle.

2. List at least four factors which affect evaporation rate.

3. Explain the difference between absolute, specific and relative humidity.

4. What is 'dew point'?

5. Explain the importance of hygroscopic nuclei in the process of cloud formation.

6. Give four ways in which moisture in air condenses in the natural environment.

7. Explain briefly the difference between dry and saturated adiabatic lapse rates, making reference to air stability, conditional stability and temperature inversions.

8. On what basis are clouds classified? Name four cloud types.

9. Describe two theories which explain how clouds make rain.

10. Explain the difference between orographic, convectional and frontal rainfall.

References

Chorley, R.J. and Barry, R.G. 1969. Water, Earth and Man: A Synthesis of Hydrology, Geomorphology and Socio-economic Geography. London, Methuen.

Hilton, K. 1985. Process and Pattern in Physical Geography. 2nd edition. London, Unwin Hyman.

Strahler, A.H. and Strahler, A. 1997. Physical Geography: Science and Systems of the Human Environment. Chichester, John Wiley & Sons, Inc.

Waugh, D. 2000. Geography: An Integrated Approach. 3rd edition. London, Nelson.

Further reading

Barry, R.G. and Chorley, R.J. 2003. Atmosphere, Weather and Climate. 8th edition. London, Routledge. Ch. 2.

Hamblin, W.K. and Christiansen, E.H. 2004. Earth's Dynamic Systems. 10th edition. Upper Saddle River, NJ. Prentice-Hall. Ch. 9.

Holden, J. (Ed.). 2005. An Introduction to Physical Geography and the Environment. Harlow, Pearson Education. Chs 2–4.

Robinson, P.J. and Henderson-Sellers, A. 1999. Contemporary Climatology. Harlow, Pearson Education.

Smithson, P., Addison, K. and Atkinson, K. 2002. Fundamentals of the Physical Environment. 3rd edition. London, Routledge. Chs 4–5.

Thompson, R.D. 2002. Atmospheric Processes and Systems. Routledge Introductions to Environment Series. London, Routledge. Chs 6–8.

5 Atmospheric Motion

Winds gusting over Cwm Idwal, North Wales, UK. Photo: T. Stott

Earth Environments   David Huddart and Tim Stott
© 2010 John Wiley & Sons, Ltd

Learning Outcomes

After reading this chapter and completing the exercises you will be able to:

➤ Understand the causes of winds and vertical air movements within the atmosphere.

➤ Understand the concept of atmospheric pressure, how it is measured and how it varies with altitude.

➤ Describe the various forces which affect air movement and explain how they interact to determine wind direction with respect to isobars on weather charts.

➤ Understand the differences between depressions and anticyclones.

➤ Describe the tricellular model of atmospheric circulation and understand how this model gives rise to prevailing winds at a global scale.

➤ Describe Rossby waves and jet streams and explain their role in affecting global weather patterns.

5.1 Introduction

The Earth's atmosphere is in perpetual motion. Movement of the air may be vertical (rising or subsiding) or horizontal. We know horizontal air movements as wind. Winds result from differences in air pressure, which are caused by differential heating of the Earth's surface. In Chapter 3 we saw how a number of factors contribute to the fact that not all places on the Earth are heated equally. In fact, equatorial regions receive two to three times more heat from the Sun than polar regions. It is this imbalance that drives the atmospheric circulation system, which strives to redistribute this heat.

Consider the situation in Figure 5.1, which illustrates the basic principle of winds.

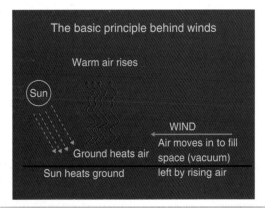

Figure 5.1 The basic principle of winds

The Sun's radiation heats the Earth's surface. Air in contact with the surface gains heat by conduction, expands and rises. As air molecules rise, they leave below them a lower concentration of air molecules, into which surrounding air is drawn. The horizontal movement of air molecules into the space left by the rising air parcel is wind. This principle explains many types of local wind, such as the sea breeze you experience on an afternoon at the beach. It is also the cause of global wind motions, due to the global imbalance of the Sun's radiation received by the poles and equatorial regions.

5.2 Atmospheric pressure

Atmospheric pressure exists because air has mass and is being pulled towards the Earth by the force of gravity. Pressure increases downward through the atmosphere, as the mass of air above each layer increases. Atmospheric pressure is greatest at the Earth's

Column of atmosphere one cm in cross section

1 kg

(a)

Figure 5.2 Atmospheric pressure explained by imagining the weight of a column of air. A column of air 1 cm × 1 cm is balanced by a 1 kg weight. (Source: Strahler and Strahler, 1997, p. 118). Reprinted from Strahler and Strahler, Physical Geography, 1997, p. 118 with permission from John Wiley & Sons Ltd.

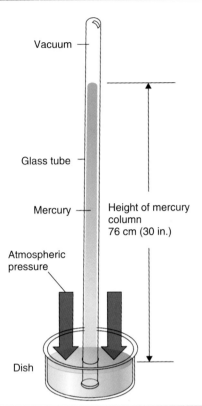

Figure 5.4 Pocket-size aneroid barometer used by climbers and hill-walkers as a portable means of measuring their altitude based on air pressure. (Photo: T. Stott)

Figure 5.3 A simple mercury barometer. At sea level, the average height of the mercury level is about 76 cm. (Source: Strahler and Strahler, 1997, p. 118). Reprinted from Strahler and Strahler, Physical Geography, 1997, p. 118 with permission from John Wiley & Sons Ltd.

surface, where the atmosphere is at its greatest density. Pressure is a force per unit area. The unit of pressure in the metric system is the pascal (Pa), defined as one newton per square metre. A newton is the force produced when a 1 kg mass is accelerated by $1 \, \mathrm{m \, s^{-2}}$. In the atmosphere, the force is produced by the acceleration of gravity acting on a column of air. Figure 5.2 illustrates this.

Although the pascal is the correct metric unit for pressure, most atmospheric units in common use today are based on the bar and millibar (mb), where 1 bar = 1000 mb and 1 mb = 100 Pa. Standard sea-level atmospheric pressure is 1013.2 mb. Atmospheric pressure is measured by a barometer, a simple type being a mercury barometer (shown in Figure 5.3). This is a 1 m glass column, sealed at one end, and completely filled with mercury. The column is inverted into a dish of mercury until the downward force exerted by atmospheric pressure on the surface of the mercury in the dish equals that of the column of the mercury which has flowed out of the glass column and created a vacuum in the top of the tube. When this equalizes, the height of the column of mercury at sea level will be 76 cm. So, you may sometimes see sea-level pressure expressed as 76 cm Hg (Hg being the chemical symbol for mercury).

Mercury is now a banned substance, so you are unlikely to see this type of barometer in use in the future. You are more likely to come across an aneroid barometer (Figure 5.4), which uses a sealed canister of air, with a slight partial vacuum tensioning a diaphragm on the side of the can against a spring. The diaphragm flexes as the air pressure changes. This moves a spring and changes a needle, which moves along a scale to indicate the pressure.

The pocket-size aneroid barometer (Figure 5.4), used by climbers and hill-walkers (also now widely available in wrist watches), is a portable means of measuring altitude based on air pressure. As a rule of thumb, air pressure decreases by 10 mb for every 10 m of vertical ascent within the troposphere, although at higher altitudes the decrease is much slower (Figure 5.5).

This decrease of air pressure with altitude affects human physiology. With the decreased air pressure of higher altitudes, oxygen moves into the lung tissues more slowly, with the result that high-altitude climbers feel a shortness of breath and fatigue, which can be accompanied by headaches and nausea, referred to as 'mountain sickness' (the onset of which can occur at any height above 3000 m). Most people can acclimatize to the reduced oxygen after a few days.

Another effect of the reduction in air pressure with altitude is that the boiling point of water becomes lower. For example, water boils at 100 °C at sea level, but at 3000 m the boiling point is 90 °C, and at 5000 m it is reduced to 84 °C. This makes cooking food at altitude more difficult, and some climbers use pressure cookers to speed up the cooking process.

5.3 Winds and pressure gradients

Wind is defined as air motion with respect to the Earth's surface. Wind movement is mainly horizontal. Vertical winds are generally

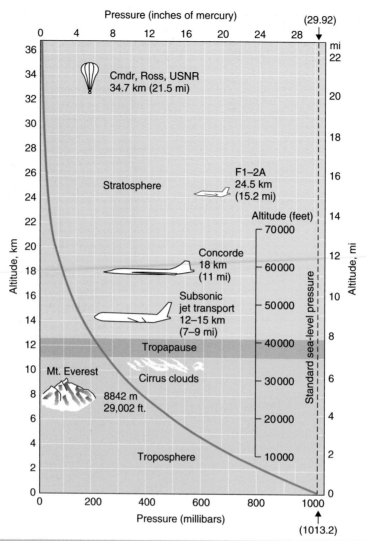

Figure 5.5 Pressure change with altitude. (Source: Strahler and Strahler, 1997). Reprinted from Strahler and Strahler, Physical Geography, 1997, p. 119 with permission from John Wiley & Sons Ltd.

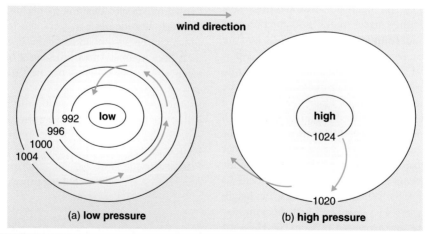

Figure 5.6 Isobars are lines joining places with equal pressure. The units of measurement shown here are millibars (mb). The lines are plotted at 4 mb intervals. Since pressure falls with altitude, pressure readings are normally reduced to represent pressure at sea level. (Source: Waugh, 2000, p. 224)

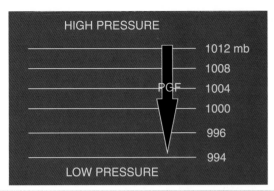

Figure 5.7 The pressure gradient force shows the tendency for the horizontal movement of air from high to low pressure

called updrafts or downdrafts and are much less common. As we saw at the start of this chapter, wind is caused by differences in atmospheric pressure from place to place, caused by differential heating of the surface beneath the air. Just as heated air rises to cause low pressure (Figure 5.6(a)), cooling air can descend to cause high pressure (Figure 5.6(b)). Differences in pressure are shown on maps by isobars, which are lines joining places of equal pressure (Figure 5.6).

Air tends to move from high to low pressure, across the isobars, until the air pressures are equal. The black arrow in Figure 5.7 illustrates this tendency, which is known as the pressure gradient force.

5.3.1 Land and sea breezes

If you have ever spent time on a beach at the coast, you will probably be familiar with the diurnal, horizontal movement of air from high- to low-pressure areas between the sea and the land, known as a land and sea breeze (Figure 5.8). This circulation

system is common in coastal areas, particularly during good weather (high-pressure anticyclones). It develops each day due to the different albedos and heat absorption capacities of land and sea. As the Sun rises during the morning, the land heats up faster than the sea, and air over the land expands and rises, drawing in cooler, more stable air from over the sea to equalize the pressure difference created. Cooler air aloft over the sea is drawn downwards to complete a simple circulatory or convection cell.

Later in the evening, clear skies cause heat to radiate from the land faster than from the sea and a cooler layer of air develops over the land. The pressure gradient is reversed and air is drawn from the land out to sea, and the pressure distribution at night reverses.

5.3.2 The Coriolis effect and surface winds

For local winds that move relatively short distances, the simple convective cell which operates in land and sea breezes holds true, but on a larger global scale the direction of air movement is influenced by the Coriolis effect, which results from the Earth's rotation. This force results from both the spin and the curvature of the Earth. Since the Earth rotates once in 24 hours, a point on the equator has to travel a far greater distance, and therefore much faster, than one near either of the poles. If the Earth's diameter is 12 750 km, its circumference is just over 40 000 km. For a point to travel this distance in 24 hours, its speed will be around 1670 km per hour. We can represent the Earth by using a rotating turntable, as shown in Figure 5.9. A person with a ball stands at the centre of the clockwise rotating turntable. The person at the centre throws the ball towards a catcher on the edge. However, during the time it takes the ball to travel from the centre to the edge, the catcher has moved to a new position and the ball appears to have travelled in the path of the curved (dotted) line.

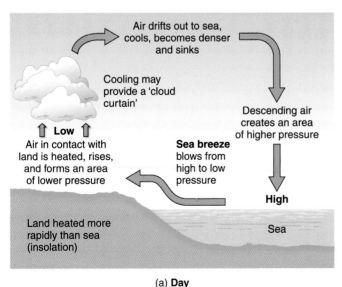

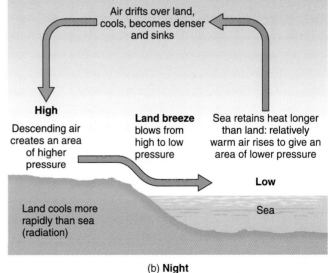

(a) **Day**   (b) **Night**

Figure 5.8 Diurnal land and sea breezes. (Source: Waugh, 2000, p. 240)

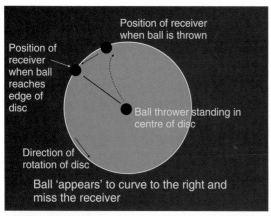

Figure 5.9 What happens to a ball thrown from the centre of a rotating turntable to a catcher on its edge

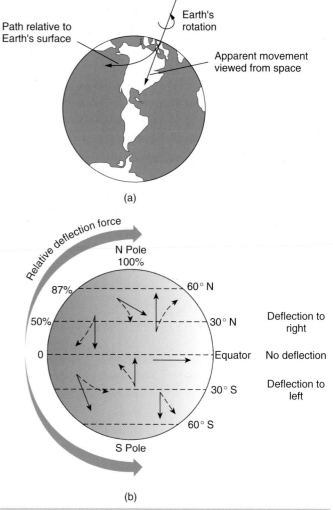

Figure 5.11 (a) The effect of the Earth's rotation on air movement. (b) The changing magnitude of the Coriolis force with latitude. (Source: Smithson *et al.*, 2002. p. 104). Reprinted from *Fundamentals of the Physical Environment*, Smithson, P., Addison, K., Atkinson, K, 2002, p. 104, with permission from Taylor and Francis

Scaling up this model, if a rocket were fired from London due south towards the equator along the Greenwich Meridian, the Earth would be rotating beneath its trajectory from west to east (Figure 5.10). Because the rocket would be travelling towards the equator, which is travelling at a greater speed than London, it would appear to curve to the right with respect to the Earth beneath. Likewise, the same rocket fired towards the North Pole along the Greenwich Meridian would also appear to be deflected to the east (right).

The Coriolis force results in an apparent deflection to the right in the northern hemisphere, and to the left in the southern hemisphere. The effect is strongest near the poles and decreases to no apparent deflection at the equator (Figure 5.11).

The Coriolis force can be seen to act in the opposite direction to the pressure gradient force (Figure 5.12). The result of these two opposing forces gives rise to a geostrophic wind, which blows parallel to the isobars above about 500 m (Figure 5.12).

The final force that has an important effect on air movement is that due to friction with the Earth's surface. Below about 500 m, where the Earth's surface is flat, friction begins to decrease the wind velocity below its geostrophic value. This has an effect on the deflective force (which depends on velocity), causing it also to decrease. The result is summarized in the Ekman spiral shown in Figure 5.13.

Winds at the surface therefore cut obliquely across the isobars, blowing towards low pressure. Buys Ballot's law states that in the northern hemisphere, if you stand with your back to the wind,

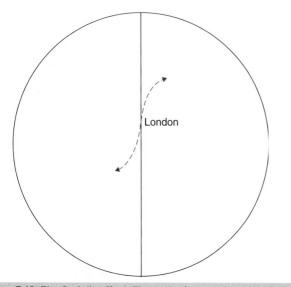

Figure 5.10 The Coriolis effect. The path of a rocket (red dotted arrows) launched from London due south along the Greenwich Meridian towards the equator will appear to be deflected to the right by the Earth's rotation. (Source: Google Earth; adapted by T. Stott)

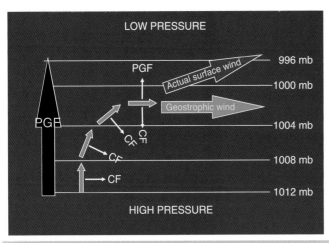

Figure 5.12 The pressure gradient force (PFG), Coriolis force (CF), geostrophic wind and actual surface wind, which is the geostrophic wind modified by the effect of friction

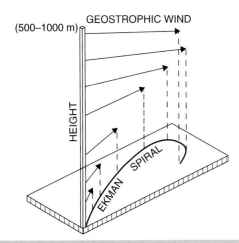

Figure 5.13 The Ekman spiral of wind with height in the northern hemisphere. The wind reaches the geostrophic velocity between 500 and 1000 m. (Source: Barry and Chorley, 1987, p. 121). Reprinted from Atmosphere, weather & climate, Barry & Chorley, 1987, p. 121, with permission from Taylor and Francis

the low-pressure area will be on your left. In other words, wind travels anticlockwise around low-pressure zones (Figure 5.14), and clockwise around high-pressure areas (anticyclones), in the northern hemisphere.

Viewed in cross-section, rather than in plan form as in Figure 5.14, low-pressure centres, called cyclones or depressions, are associated with rising air, while high-pressure areas, or anticyclones, are associated with descending air (Figure 5.15).

When air rises, it tends to expand and cool, and moisture in it condenses, causing clouds. For this reason, low-pressure conditions are generally associated with cloudy, changeable weather and rain. In anticyclones, descending air warms and any moisture

within it turns to vapour, so high pressure is generally associated with clear, stable and dry weather conditions.

5.4 The global pattern of atmospheric circulation

5.4.1 The tricellular model

As we saw above, there is a surplus of heat at the equator, and a deficit at the poles. In theory heat should be transferred from areas of surplus to areas of deficit by means of a single convective cell, as seen in Figure 5.16. This idea was first advanced by Halley in 1686, and would perhaps hold true if the planet were not rotating.

However, the discovery of three different cells was made by Ferrel (1856) and refined by Rossby (1941), and this tricellular model still forms the basis of our understanding of the general circulation of the atmosphere today (Figure 5.17).

The driving force for this mechanism begins at the inter-tropical convergence zone (ITCZ), where the trade winds driven by the Hadley cell meet and pick up latent heat as they cross the warm, tropical oceans. They begin to rise when they meet violent convection currents at the equator, forming towering cumulo-nimbus clouds, frequent afternoon thunderstorms and large areas of low pressure. It is these strong upward currents that drive the general circulation model. The heated air rises in the tropics, drifting north, before sinking around 30° N and S in the Hadley cell circulation. At the surface, air flow is easterly and creates the NE and SE trade winds (Figure 5.17). Remember, the direction in which these winds blow is deflected by the Coriolis force, discussed earlier.

The area at which the cold air sinks between the Hadley and Ferrel cells creates high pressure. The descending air is compressed and warmed, and gives cloudless, stable conditions. In the mid-latitudes (about 30–60° N and S) the circulation is dominated by the Ferrell cell, which, at the Earth's surface, causes the polar-front westerlies, strong prevailing winds which dominate mid-latitude weather systems, and again are deflected by the Coriolis force. The area around 60° N and S is an area in which air from the Ferrel and weak polar cells converges and is uplifted, creating an area of low pressure (mid-latitude depressions) – the reason European weather is so unsettled! At high latitudes the simple, convection-driven circulation is called the polar cell. The surface winds created by this weak circulation are known as the returning polar easterlies. Regions of high and low surface pressure are marked 'H' and 'L', respectively, on Figure 5.17.

5.4.2 Rossby waves and jet streams

The first evidence of strong winds in the upper part of the troposphere came when First World War Zeppelin airships were blown off course, and later balloons were observed travelling

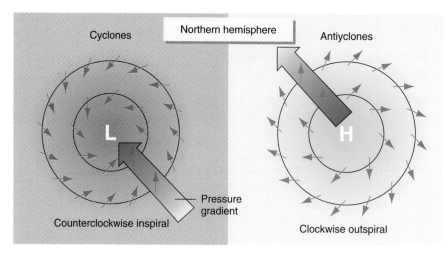

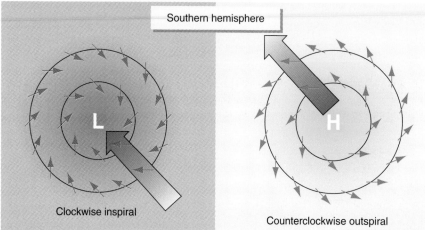

Figure 5.14 Surface winds spiral inward toward the centre of a low-pressure area (depression or cyclone), but outward and away from the centre of an anticyclone. Note the different directions in the northern and southern hemispheres. (Source: Strahler & Strahler, 1997, p. 124). Reprinted from Physical Geography, Strahler and Strahler, 1997, p. 124, with permission from John Wiley & Sons Ltd.

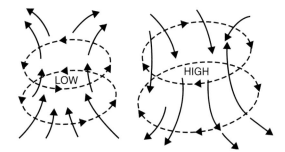

Figure 5.15 Low-pressure systems are associated with rising air (left), while high-pressure areas are associated with descending air (right)

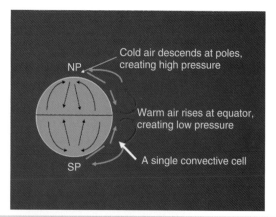

Figure 5.16 Air movement on a non-rotating planet (drawn by T. Stott)

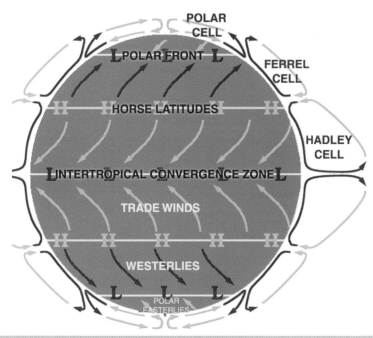

Figure 5.17 The tricellular model, showing the general circulation of the atmosphere in the northern hemisphere. The general direction of trade winds is driven by the Hadley cell, the westerlies are driven by the Ferrel cell and returning polar easterlies are driven by the weak polar cell. (Source: http://commons.wikimedia.org/wiki/File:AtmosphCirc2.png#file, Wikimedia Commons)

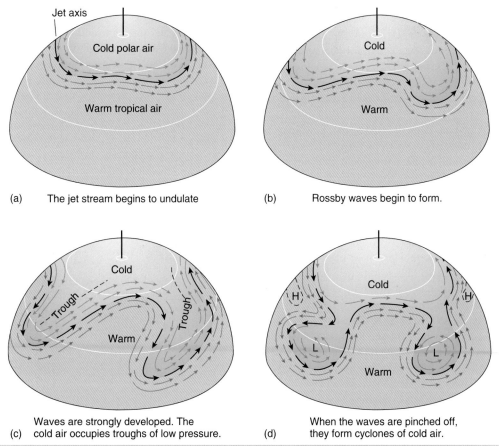

(a)   The jet stream begins to undulate

(b)   Rossby waves begin to form.

(c)   Waves are strongly developed. The cold air occupies troughs of low pressure.

(d)   When the waves are pinched off, they form cyclones of cold air.

Figure 5.18 Development of upper-air Rossby waves in the westerlies of the northern hemisphere. (Source: Strahler and Strahler, 1997, p. 134). Reprinted from Physical Geography, Strahler and Strahler, 1997, p. 134, with permission from John Wiley and Sons Ltd.

(a)

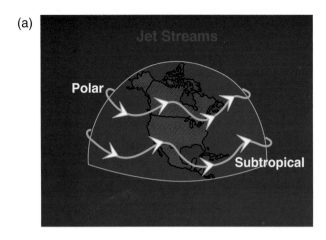

(b)

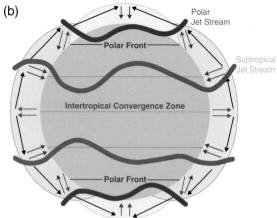

Figure 5.19 (a) Jet streams typically flow at heights of 10–12 km in broadly curving tracks around the Earth. (Source: http://nevis.k12.mn.us/academics/science/2jetstreams.gif, accessed 27 November 2007). (b) Locations of jet streams in relation to the tricellular global circulation model. (Source: www.physicalgeography.net/fundamentals/7q.html, accessed 27 November 2007)

at speeds of 200 km per hour. Pilots flying at heights above 8 km soon noticed that eastward flights were much faster than westward ones, whilst those flying north–south often got blown off course. A belt of fast-moving, upper-air westerlies called Rossby waves was the explanation. These follow a meandering path, with the number of meanders, or waves, varying with the season. There are usually four to six in summer, and three in winter. Figure 5.18 shows how these waves develop and grow in the northern hemisphere, the amplitude of the waves increasing from (a) to (d), before type (a) becomes reestablished. The waves arise in the zone of contact between the cold polar air and the warm tropical air.

For a period of several days or weeks, the flow may be smooth, but then an undulation develops (Figure 5.18(b)) and warm air pushes poleward, while a tongue of cold air moves south (Figure 5.18(c)). Eventually, the tongue is pinched off (Figure 5.18(d)), leaving a pool of cold air at a latitude far south of its normal location, but after some weeks this warms and the situation shown in Figure 5.18(a) returns.

Jet streams are important features of upper-air circulation. They are narrow zones of fast-moving air at high altitudes, in which wind speeds can reach 350–450 km per hour. The jet streams follow broadly curving tracks (Figure 5.19), with wind speeds highest in the centre. The poleward jet stream is located along the polar front (40–60° N or S) at the moving boundary between the Ferrel and Polar cells (Figure 5.19(a)) and is called either the polar-front jet stream or simply 'polar jet', and is present in both hemispheres. Where, in the northern hemisphere, this jet stream moves south, it brings with it cold air, which descends in a clockwise direction to give dry, stable conditions associated with areas of high pressure (anticyclones). When the now-warmed jet stream moves back north, it takes warm air with it, which rises to give areas of low pressure (depressions), associated with windy and wet weather. Sometimes the path of the polar-front jet stream is altered by a stationary 'blocking anticyclone', which

can produce extremes of climate, such as the hot, dry summers of 1976 and 1989 or the cold January of 1987 in the UK.

See Companion website for Box 5.1 Warmer winters in the UK and the role of the North Atlantic Oscillation.

The second type of jet stream forms at the tropopause between the Ferrel and Hadley cells (Figure 5.19(a)) between 25 and 30° N or S and is called the subtropical jet stream. It has lower wind velocities than the polar jet, but follows a similar west–east path.

A third type of jet stream found at even lower latitudes is known as the tropical easterly jet stream. It runs in the opposite direction to the two previously mentioned, from east to west, and only occurs in the summer season. It is limited to the northern hemisphere and is located over South-East Asia, India and Africa.

The global atmospheric circulation system is complex and dynamic, and therefore challenging to model and predict. We will see more of the attempts by climatologists to model the effect of future climate change on our atmosphere in Section 6.

Exercises

1. Explain what causes winds to blow on the Earth's surface.

2. Explain what is meant by air pressure and give the average sea-level pressure in mb.

3. What is the relationship between air pressure and altitude in the lower part of the troposphere?

4. Draw a simple diagram showing isobars and mark the following with arrows: (i) pressure gradient force, (ii) Coriolis force, (iii) geostrophic wind, (iv) surface wind.

5. What does Buys Ballot's law state?

6. Draw a quarter section of the Earth to illustrate the tricellular model of global atmospheric circulation.

7. What are: (i) jet streams, (ii) Rossby waves?

8. Explain the potential link between UK winter temperatures and the North Atlantic Oscillation.

9. Explain how the polar-front jet stream can influence UK weather systems.

10. What is the difference between the subtropical and tropical easterly jet streams?

References

Barry, R.G. and Chorley, R.J. 2003. Atmosphere, Weather and Climate. 8th edition. London, Routledge.

Smithson, P., Addison, K. and Atkinson, K. 2002. Fundamentals of the Physical Environment. 3rd edition. London, Routledge.

Strahler, A.H. and Strahler, A. 1997. Physical Geography: Science and Systems of the Human Environment. Chichester, John Wiley & Sons, Inc.

Further reading

Hamblin, W.K. and Christiansen, E.H. 2004. Earth's Dynamic Systems. 10th edition. Upper Saddle River, NJ. Prentice-Hall. Ch. 9.

Holden, J. (Ed.). 2005. An Introduction to Physical Geography and the Environment. Harlow, Pearson Education. Chs 2–5.

Robinson, P.J. and Henderson-Sellers, A. 1999. Contemporary Climatology. Harlow, Pearson Education.

Thompson, R.D. 2002. Atmospheric Processes and Systems. Routledge Introductions to Environment Series. London, Routledge. Chs 9–11.

Union City Tornado, Source: http://www.photolib.noaa.gov/nssl/nssl0123.htm; Image ID: nssl0123, National Severe Storms Laboratory (NSSL) Collection; Location: Union City, Oklahoma; Photo Date: 24 May 1973; Credit: NOAA Photo Library, NOAA Central Library; OAR/ERL/National Severe Storms Laboratory (NSSL)

Earth Environments   David Huddart and Tim Stott
© 2010 John Wiley & Sons, Ltd

Learning Outcomes

After reading this chapter and completing the exercises you will be able to:

➤ Understand the scale and three-dimensional dynamic nature of weather systems such as depressions, anticyclones, hurricanes and tornados.

➤ Understand about air masses and describe the typical weather pattern associated with the passage of a north Atlantic depression.

➤ Describe the formation, characteristics and classification systems for cyclones, hurricanes, typhoons and tornados.

➤ Understand the conditions for land and sea breezes, anabatic and katabatic winds in mountains and föhn winds.

➤ Describe the microclimates of cities, forests, lakes and coasts.

➤ Understand how weather observations are recorded in a standardized way and how these data are used to prepare weather forecasts.

➤ Interpret weather systems using weather charts and satellite imagery.

Figure 6.1 Hurricane Katrina was a category 5 hurricane on 28 August 2005, seen here one day before it made landfall on the Gulf Coast. (Source: http://web.mit.edu/12.000/www/m2010/images/katrina-08-28-2005.jpg, accessed 28 November 2007, NOAA Photograph)

6.1 Introduction

A weather system is an organized state of the atmosphere associated with a characteristic weather pattern. Cold fronts, warm fronts, depressions, anticyclones, hurricanes and tornados are all examples of weather systems that will be discussed in this chapter.

On 29 August 2005, Hurricane Katrina (Figure 6.1) devastated New Orleans and surrounding areas. There were more than a thousand fatalities and billions of dollars in property losses, and tens of thousands of people were stranded and/or lost their homes.

6.2 Macroscale synoptic systems

Cyclones and anticyclones are weather systems that involve masses of air moving in a spiralling motion. Cyclones are also called depressions in Europe, a term which we will use hereafter. In a depression or low-pressure system, air rises and spirals inwards, whereas in an anticyclone air descends and spirals outwards. The direction of the spiral depends on which hemisphere the weather system is in. In the northern hemisphere, air moves in an anticlockwise direction around a depression, and clockwise around an anticyclone. The reverse holds for the southern hemisphere. Depressions and anticyclones are normally associated with the motion of air masses.

6.2.1 Air masses

An air mass is a large body of air with fairly uniform temperature and humidity characteristics. If air remains stationary in an area for several days, it tends to assume the temperature and humidity properties of the surface over which it has been lying, be it ocean or land. An air mass can be several thousand kilometres across, and can extend upwards to the top of the troposphere. The areas in which air masses develop are called source regions and are classified according to: (i) the latitude in which they lie, which determines their general temperature, e.g. arctic (A), polar (P), tropical (T) or equatorial (E); and (ii) the nature of the surface, which affects the moisture content of the air mass, e.g. a source region over an ocean is termed maritime (m) while one over land is termed continental (c). The five major air masses which affect the British Isles are shown in Figure 6.2.

Air masses are named by combining the latitude and humidity terms (Am, Pm, Tm, Pc, Tc). When an air mass moves from one region to another under the influence of barometric pressure gradients, it is modified by the surface over which it passes, and this changes its temperature, humidity and stability. So tropical air moving northwards gets cooled and becomes more stable, whereas polar air moving south gets warmed and becomes more unstable. The general conditions associated with each of the five air masses affecting Britain are given in Figure 6.2.

6.2.2 Formation of weather fronts

When two air masses with contrasting characteristics meet, they do not mix well due to differences in temperature and density. The point at which they meet is called a front. Figure 6.3 shows a warm front in cross-section. The warm Tm air mass is advancing and being forced to rise over the cold Pm air. A cold front occurs when advancing cold air undercuts a warm air mass.

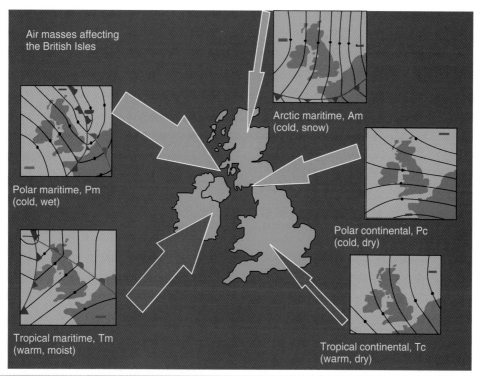

Figure 6.2 Five air masses affecting the British Isles. The width of the arrows is in approximate proportion to the frequency with which each air mass crosses the country. Weather map inserts show the isobar patterns which give rise to these types of air flow. (Adapted from Langmuir, 1984, p. 161)

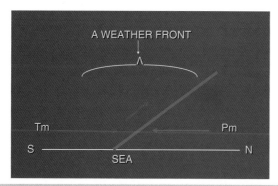

Figure 6.3 Cross-section through a warm front where a tropical maritime (Tm) meets a polar maritime (Pm) air mass. Note: the gradient of the front is exaggerated

When maritime (moist) air rises at a front, it expands and cools, and moisture vapour within it condenses to form clouds. These clouds often generate precipitation. The most notable type of front occurs when warm, moist tropical maritime (Tm) air meets cold, moist polar maritime (Pm) air, forming a polar front. The clouds associated with a polar front moving west to east over Britain on 28 November 2007 can be seen in Figure 6.4.

Fronts can extend for hundreds of kilometres and have low gradients of around 1 : 50 in the case of cold fronts and 1 : 150 in warm fronts. The symbols used to denote different types of front on weather charts are shown in Figure 6.5.

The concept of fronts was put forward in the 1920s by Norwegian meteorologists, who realized that there was often a sharp dividing line between warm and cold air, with cloud along the division. The term 'front' was derived from the First World War expression meaning a division between opposing forces. It is at the polar front that depressions form.

6.2.3 Depressions

Depressions are areas of low pressure, which form frequently over the oceans in mid-latitudes and move eastwards, bringing cloud and rain to the western margins of continents. Where cold polar air meets warmer subtropical air, the tendency is for the cold, dense air to undercut the warm, less-dense air, forcing it to rise. The result is cooling, condensation and the formation of cloud. The effect of the front on weather, in particular precipitation, depends largely on the temperature contrast. Fronts can vary from just a line of cloud with little depth, to a mass of vertically-developed cloud producing persistent heavy rain. If we examine a polar front over the Atlantic, where Pm air from the north meets Tm air from the south (Figure 6.6(a)), we see that small waves often run along the front in the flow (Figure 6.6(b)), and if the pressure at the tip of the wave continues to fall, a depression known as an open-wave depression begins to form. It is important to understand that a depression is not simply a surface phenomenon, and Figure 6.6(c)–(e) attempts to show the development of a depression in three dimensions.

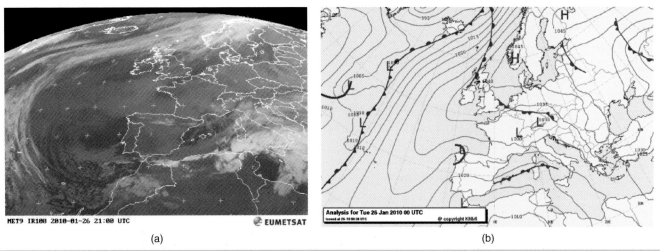

(a)   (b)

Figure 6.4 (a) Meteosat satellite image taken over north-west Europe at 1500 hours on 28 November 2007, showing bands of cloud associated with a passing weather front seen on Meteorological Office weather charts (b). (Source: (a) http://www.eumetsat.int/; (b) http://www.metoffice.gov.uk/, accessed 28 November 2007)

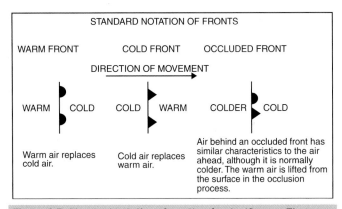

Figure 6.5 Standard notation of weather fronts. (Source: Thomas, 1995, p. 50)

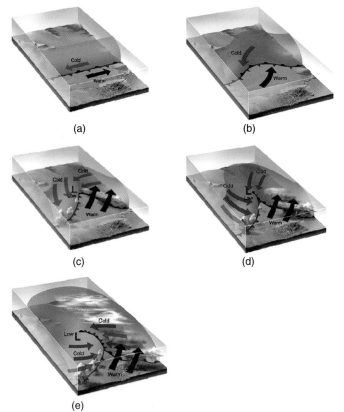

(a)   (b)

(c)   (d)

(e)

Figure 6.6 Stages in the development of a mid-Atlantic depression. (Source: http://rst.gsfc.nasa.gov/Sect14/Sect14_1d.html, accessed 1 December 2007. Courtesy of NASA)

There are three kinds of movement involved: the cyclonic movement of air in an anticlockwise direction round the centre of the low-pressure area, giving rise to Buys Ballot's law that 'in the northern hemisphere, if you stand with your back to the wind, the low-pressure area will be on your left'. The second type of movement is the upward movement of warm air drawn in at low levels (convergence) and spreading out at high levels (divergence). The third movement is the generally eastward tracking of the whole system at speeds which vary between 20 and 80 km per hour. Figure 6.6(c) shows it in its embryo stage. The polar front becomes increasingly distorted (Figure 6.6(d)) as it matures, with distinct warm and cold fronts being formed. The warm tropical air, known as the warm sector, lies between the two, and gradually gets squeezed and lifted off the surface as the cold front catches up the warm front (Figure 6.6(e)). When this happens, an occluded front forms; the weather map notation for this is shown in Figure 6.5. The depression will continue to deepen after the occlusion has started to form, perhaps for

up to 18 hours, after which the decaying process begins. The movement of the depression becomes slower as it starts to fill (pressure rises), and the tendency is for it to turn to the left before becoming stationary.

The clouds associated with a frontal system can extend several hundred miles ahead of the surface front and have a characteristic pattern that is easily recognized. It should be remembered that the frontal line marked on weather charts shows the point at which the air masses meet on the Earth's surface. The division between the cold air ahead and the warm air behind is a forward-sloping rather than a vertical gradient. A band of cloud is associated with the warm and cold front.

Figure 6.7 shows a weather chart plan view with cross-section (a)–(b) through a typical north Atlantic depression below. The weather map shows the positions of the fronts in relation to the isobars. Note the change in direction of both the isobars and the wind direction at the fronts. The wind direction in the cold air mass immediately behind the cold front is north-westerly, whereas in the warm sector immediately in front of the cold front it has become south-westerly. The wind direction is said to have backed (moved anticlockwise, as opposed to veered, which is a

clockwise shift). Wind speeds are also considerably reduced in the warm sector.

The first signs of an approaching depression available to an observer standing at point X in Figure 6.7(a) would be high cirrus clouds on the leading edge of the warm front, accompanied by a slight and gradual rise in temperature. The clouds would later begin to change to cirro-stratus, followed by thickening alto-stratus and the onset of rain as the darker nimbo-stratus clouds pass overhead. The pressure would consistently drop and the temperature would rise suddenly as the warm sector passed over. This would be seen by the observer as a breaking up of the cloud with some clear sky, possibly with showers. The approach of the cold front would be accompanied by tall or towering cumulo-nimbus clouds, which would bring heavy rain but for a shorter duration than at the warm front. Pressure would suddenly start to rise after the cold front had passed, and temperature would plummet again as the cold air mass replaced the tropical air of

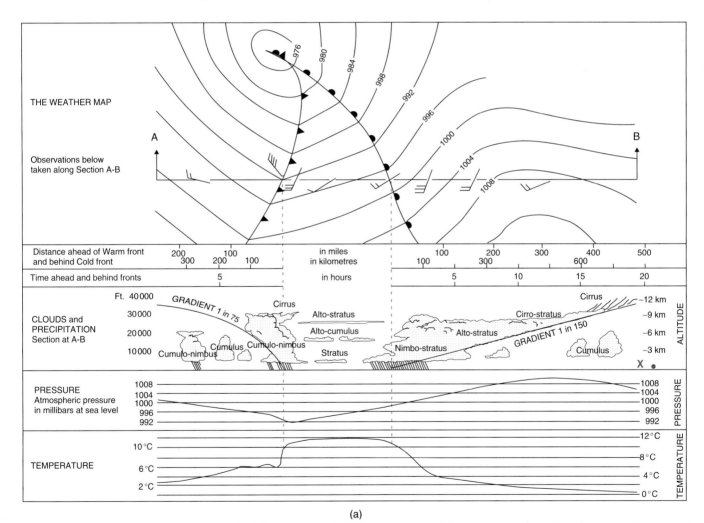

(a)

Figure 6.7 (a) Plan view and cross-section through a typical north Atlantic depression showing the weather chart appearance, gradient and cloud patterns associated with the warm and cold front, and changes in pressure and temperature as the system passes. (b) Cross-section through a typical north Atlantic depression showing changes in weather variables as the system passes. (Source: after (a) Langmuir, 1984 and (b) Waugh, 2000)

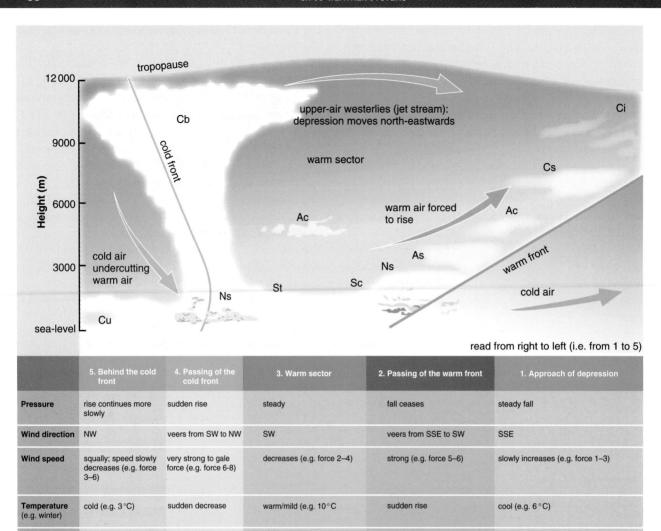

read from right to left (i.e. from 1 to 5)

	5. Behind the cold front	4. Passing of the cold front	3. Warm sector	2. Passing of the warm front	1. Approach of depression
Pressure	rise continues more slowly	sudden rise	steady	fall ceases	steady fall
Wind direction	NW	veers from SW to NW	SW	veers from SSE to SW	SSE
Wind speed	squally; speed slowly decreases (e.g. force 3–6)	very strong to gale force (e.g. force 6-8)	decreases (e.g. force 2–4)	strong (e.g. force 5–6)	slowly increases (e.g. force 1–3)
Temperature (e.g. winter)	cold (e.g. 3 °C)	sudden decrease	warm/mild (e.g. 10 °C	sudden rise	cool (e.g. 6 °C)
Relative humidity	rapid fall	high during precipitation	steady and high	high during precipitation	slow rise
Cloud (Figure 9.20)	decreasing; in succession, Cb and Cu	very thick and towering Cb	low or may clear; St, Sc, Ac	low and thick Ns	high and thin; in succession, Ci, Cs, Ac, A
Precipitation	heavy showers	short period of heavy rain or hail	drizzle or stops raining	continuous rainfall, steady and quite heavy	none
Visibility	very good; poor in showers	poor	often poor	decreases rapidly	good but beginning to decrease

(b)

Figure 6.7 (*continued*)

the warm sector. In winter, when temperatures are low enough, clouds at a cold front might produce snow.

As can be seen in Figure 6.8, depressions are not unique to Britain and the north Atlantic. They form wherever warm tropical air encounters cold polar air. In North America they tend to be called wave cyclones rather than depressions. The western coast of North America commonly receives wave cyclones that develop over the North Pacific Ocean. They track in an easterly direction, mainly in the middle latitudes 30–70° N or S, but can stray outside of this zone at times. In certain areas they tend to track in common paths until they dissolve. You may notice that the tracking arrows in the southern hemisphere in Figure 6.8 appear to be more uniform than those in the northern hemisphere. This is because there is a more uniform pattern of ocean surface circling the globe at these latitudes, only broken by the southern tip of South America.

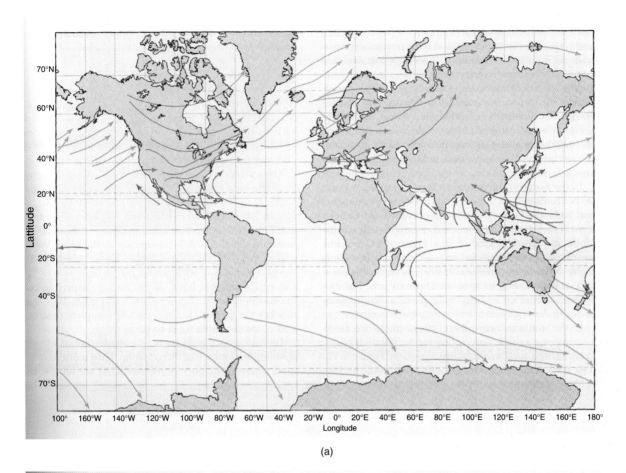

(a)

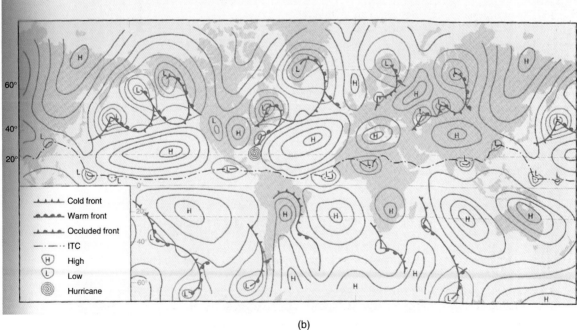

(b)

Figure 6.8 (a) Typical paths of depressions in the mid-latitudes (blue lines) and tropical cyclones (red lines). (Based on data from S. Petterson, B. Haurwitz and N.M. Austin, J. Namias, M.J. Rubin and J.-H. Chang.) (b) A typical daily weather map of the world for a given day during July or August; this is a composite of typical weather conditions. (After M.A. Garbell; Source: Strahler and Strahler, 1997, p. 155)

6.2.4 Tropical cyclones

The most powerful and destructive type of cyclonic storm is the tropical cyclone, known as a hurricane in the Atlantic Ocean or as a typhoon in the Pacific and Indian Oceans. A cyclone is an atmospheric low-pressure system that gives rise to a roughly circular, inward-spiralling wind motion, called vorticity. It can be 600 km across, regional in extent and can exist for several weeks. This type of storm develops over oceans 8–15° N or S of the equator, where average ocean temperatures remain above 27 °C for at least one month of the year. Figure 6.9 shows the most common tracks and the frequencies of tropical cyclones.

As can be seen from the tracks in Figure 6.9, tropical cyclones tend to track westward through the trade-wind belt, often intensifying as they travel. They can then curve northward and penetrate into mid-latitudes.

Tropical cyclones are characterized by winds of extreme velocity and are accompanied by torrential rainfall: two factors which can cause widespread damage and loss of life. The system consists of an almost circular storm centre of extremely low pressure. Winds spiral inward at high speed, accompanied by torrential rain (Figure 6.10).

The storm gains its energy through the release of latent heat as vapour turns to rain. The diameter could be 150–500 km, height 10–15 km and wind speeds up to 300 km per hour. Barometric pressure in the storm centre can be as low as 950 mb, the Atlantic basin record of 882 mb being recorded for Hurricane Wilma in 2005 (www.noaa.gov). The central eye of the storm is a cloud-free vortex, and as the eye passes over a specific location, calm prevails for up to 30 minutes before the spiralling bands of wind and rain return, but with winds now in the opposite direction.

Hurricanes are the tropical cyclones of the Atlantic. Unlike depressions, hurricanes need large areas of uniform temperature, humidity and pressure to form intense low-pressure areas. There must also be a continuous source of heat to maintain the rising air currents, plus a supply of moisture to provide the latent

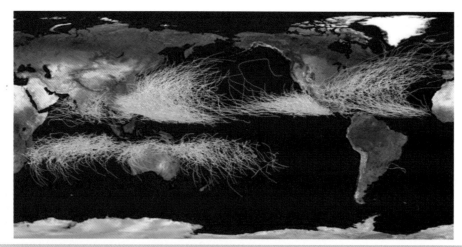

Figure 6.9 The tracks of all the tropical cyclones that formed worldwide from 1985 to 2005. The points show the locations of the storms at six-hourly intervals and use the colour scheme from the Saffir–Simpson Hurricane Scale. (Source: http://commons.wikimedia.org/wiki/File:Global_tropical_cyclone_tracks-edit.jpg, Wikimedia Commons)

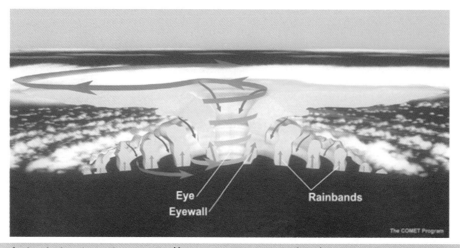

Figure 6.10 Structure of a tropical cyclone. (Source: http://commons.wikimedia.org/wiki/File:Eye_structure.jpg, Wikimedia Commons)

heat released by condensation to drive the storm and deliver the precipitation. On reaching cooler oceans or land, the energy of these storms is rapidly dissipated. They are another mechanism by which surplus heat is transferred away from the equator. The average lifespan is 7–14 days. Tropical cyclones are major natural hazards due to high winds, which can damage crops, buildings and electricity supplies; ocean storm (tidal) surges resulting from the low pressure can inundate coastal areas, which is a major problem in flat, densely-populated countries like Bangladesh; flooding can result from the torrential precipitation (e.g. in 1974 in Honduras 800 000 people died when homes were washed away by floods; in 2005 more than 1800 people died as a result of flooding caused by Hurricane Katrina, mainly in and around New Orleans (Figure 6.1)); landslides can result from heavy rainfall and can damage buildings and people in their path.

Tropical cyclones are classified according to their maximum wind speeds, using several scales. These scales are provided by various bodies, including the World Meteorological Organization, the National Hurricane Center and the Bureau of Meteorology. The National Hurricane Center uses the Saffir–Simpson Hurricane Scale (Table 6.1) for hurricanes in the eastern Pacific and Atlantic basins. It is a 1–5 rating based on a hurricane's present intensity, used to give an estimate of the potential property damage and flooding expected along the coast from a hurricane landfall. Wind

Table 6.1 The Saffir–Simpson scale for hurricanes in the eastern Pacific and Atlantic basins. (Based on data from the National Hurricane Center http://www.nhc.noaa.gov/)

Category	Description
One	Winds 74–95 mph (64–82 kt or 119–153 km per hour). No significant damage to building structures. Damage primarily to unanchored mobile homes, shrubbery and trees. Some damage to poorly constructed signs. Also, some coastal road flooding and minor pier damage. Hurricane Lili of 2002 made landfall on the Louisiana coast as a Category One hurricane. Hurricane Gaston of 2004 was a Category One hurricane that made landfall along the central South Carolina coast.
Two	Winds 96–110 mph (83–95 kt or 154–177 km per hour). Some roofing material, door and window damage of buildings. Considerable damage to shrubbery and trees, with some trees blown down. Considerable damage to mobile homes, poorly constructed signs and piers. Coastal and low-lying escape routes flood two to four hours before arrival of the hurricane centre. Small craft in unprotected anchorages break moorings. Hurricane Frances of 2004 made landfall over the southern end of Hutchinson Island, Florida as a Category Two hurricane. Hurricane Isabel of 2003 made landfall near Drum Inlet on the Outer Banks of North Carolina as a Category Two hurricane.
Three	Winds 111–130 mph (96–113 kt or 178–209 km per hour). Some structural damage to small residences and utility buildings, with a minor amount of curtainwall failures. Damage to shrubbery and trees, with foliage blown off trees and large trees blown down. Mobile homes and poorly constructed signs destroyed. Low-lying escape routes cut by rising water three to five hours before arrival of the centre of the hurricane. Flooding near the coast destroys smaller structures, with larger structures damaged by battering from floating debris. Terrain continuously lower than 5 ft above mean sea level may be flooded inland 8 miles (13 km) or more. Evacuation of low-lying residences with several blocks of the shoreline may be required. Hurricanes Jeanne and Ivan of 2004 were Category Three hurricanes when they made landfall in Florida and in Alabama, respectively.
Four	Winds 131–155 mph (114–135 kt or 210–249 km per hour). More extensive curtainwall failures, with some complete roof structure failures on small residences. Shrubs, trees and all signs blown down. Complete destruction of mobile homes. Extensive damage to doors and windows. Low-lying escape routes may be cut by rising water three to five hours before arrival of the centre of the hurricane. Major damage to lower floors of structures near the shore. Terrain lower than 10 ft above sea level may be flooded, requiring massive evacuation of residential areas as far inland as 6 miles (10 km). Hurricane Charley of 2004 was a Category Four hurricane when it made landfall in Charlotte County, Florida, with winds of 150 mph. Hurricane Dennis of 2005 struck the island of Cuba as a Category Four hurricane.
Five	Winds greater than 155 mph (135 kt or 249 km per hour). Complete roof failure on many residences and industrial buildings. Some complete building failures, with small utility buildings blown over or away. All shrubs, trees and signs blown down. Complete destruction of mobile homes. Severe and extensive window and door damage. Low-lying escape routes cut by rising water three to five hours before arrival of the centre of the hurricane. Major damage to lower floors of all structures located less than 15 ft above sea level and within 500 yards of the shoreline. Massive evacuation of residential areas on low ground within 5–10 miles (8–16 km) of the shoreline may be required. Only three Category Five hurricanes have made landfall in the United States since records began: the Labour Day Hurricane of 1935, Hurricane Camille in 1969 and Hurricane Andrew in August 1992. The 1935 Labour Day Hurricane struck the Florida Keys with a minimum pressure of 892 mb – the lowest pressure ever observed in the United States. Hurricane Camille struck the Mississippi Gulf Coast, causing a 25-foot storm surge, which inundated Pass Christian. Hurricane Katrina, a Category Five storm over the Gulf of Mexico, was still responsible for at least 81 billion dollars of property damage when it struck the US Gulf Coast as a Category Three. It is by far the costliest hurricane to ever strike the United States. In addition, Hurricane Wilma of 2005 was a Category Five hurricane at peak intensity and is the strongest Atlantic tropical cyclone on record, with a minimum pressure of 882 mb.

speed is the determining factor in the scale, as storm surge values are highly dependent on the slope of the continental shelf and the shape of the coastline in the landfall region. Note that all winds use the US 1 minute average.

Australia uses a different set of tropical cyclone categories for its region. Many basins have different names for storms of hurricane/typhoon/cyclone strength. The use of different definitions for maximum sustained wind creates additional confusion in the definitions of cyclone categories worldwide.

See companion website for Box 6.1 Stormy Debates continue between scientists over the evidence linking hurricanes to global warming.

6.2.5 Tornados

A tornado is a cyclonic storm with a very intense low-pressure centre. Tornados are short-lived and local in extent, but they can be extremely violent. They typically follow very narrow paths in the range of 300–400 m wide. US National Weather Service records show that tornados have the power to drive 2 × 4″ wood boards through brick walls, lift an 83 ton railroad car, and carry a home freezer over 2 km. A tornado appears as a dark funnel cloud hanging from the base of a dense cumulo-nimbus cloud (Figure 6.12).

When over water, the resulting funnel cloud is known as a 'waterspout'; if developed over a desert, it is said to be a 'dust devil' (usually small and nondestructive). The base of the funnel appears dark because of the density of condensing moisture, dust and debris being swept up by the wind. Estimates of wind speeds run as high as 400 km per hour and exceed any other known type of storm. In 1971 Professor Theodore Fujita of the University of Chicago set out to develop a scale to categorize tornados by their intensity and area and to estimate a wind speed associated with the damage caused by a tornado. He developed the Fujita

Table 6.2 The enhanced F scale for tornado intensity

FUJITA SCALE-DERIVED EF SCALE			OPERATIONAL EF SCALE	
F number	Fastest 1/4 mile (mph)	Three-second gust (mph)	EF number	Three-second gust (mph)
0	40–72	45–78	0	65–85
1	73–112	79–117	1	86–110
2	113–157	118–161	2	111–135
3	158–207	162–209	3	136–165
4	208–260	210–261	4	166–200
5	261–318	262–317	5	Over 200

or F Scale. This original F Scale has been revised and the new scale used to categorize tornados is called the 'F Enhanced Scale', 'Enhanced F Scale' or 'EF Scale'. The new scale was implemented on 1 February 2007. Information on the new scale is available at http://www.spc.noaa.gov/efscale/. It incorporates 28 damage indicators and three-second gusts to determine an EF rating. Table 6.2 shows how tornados are now classified in miles per hour.

While they can occur anywhere in the world, tornados are most common in the central and south-eastern United States. Tornados typically form along the cold front of a fast-moving mid-latitude cyclonic storm system. When a moving mass of cold air overtakes and traps an underlying layer of warm air, the warm air is drawn up into the core of the storm in an upward spiralling motion. At the same time, the cold air spirals down, creating a vortex or twisting funnel cloud. The air pressure within the vortex

Figure 6.12 Approaching ground level, a tornado appears dark because it draws up dust, soil and debris into the vortex, which is a partial vacuum. (Source: http://rst.gsfc.nasa.gov/Sect14/tornado.jpg, accessesed 1 December 2007)

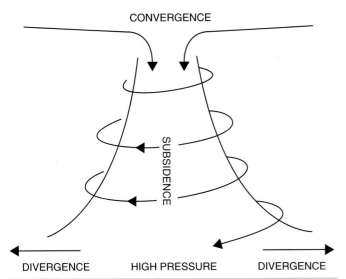

Figure 6.13 A model illustrating the main characteristics of an anticyclone (Source: after Langmuir, 1984)

may be as little as 60% of that outside, so this partial vacuum causes the tornado to suck up soil and debris, which give the funnel cloud its dark and ominous appearance.

6.2.6 Anticyclones

An anticyclone is a mass of relatively high-pressure air within which the air is subsiding (Figure 6.13), and light winds blow outward and in a clockwise direction (in the northern hemisphere).

Figure 6.14 compares the air flow in a depression with an anticyclone and makes the point that rate of air inflow compared to outflow determines the stability of the system. In a depression (left), as long as the rate of outflow (upward) is greater than the rate of inflow (at its base) the depression will continue to deepen. In the anticyclone (right), as long as the inflow (downward) is greater than the outflow at the base, the anticyclone will continue to stabilize and persist.

The source of the air for an anticyclone is the upper atmosphere, where the amount of water vapour is low. As it descends, air warms at the dry adiabatic lapse rate, bringing dry weather conditions. Pressure gradients are gentle, giving only weak winds or calms.

Anticyclones (Figure 6.15) are much larger than depressions, and may be up to 3000 km in diameter. Once established, they can give several days (or even weeks) of settled dry, calm, clear weather. In summer this may take the form of hot, sunny days with an absence of rain. Rapid radiation at night, however, can lead to temperature inversions and dew or mist can form overnight, though this usually burns off quickly the next morning. After several days in summer, thermals may develop, which can lead to thunderstorms. In winter, although the skies are clear, the low angle of the Sun means that temperatures do not climb very much. At night, the lack of clouds leads to radiation cooling and the formation of fog and frost, which may take a long time to disperse the following day; sometimes they can persist all day.

Blocking anticyclones (Figure 6.15) occur when cells of high pressure detach themselves from major high-pressure areas of the subtropics or poles. Once formed, they can last for several days or even weeks, and 'block' the eastward moving depressions, creating unusual weather conditions such as the dry hot summer of 1995 and the cold winter of 1987 in Britain.

6.3 Meso-scale: Local winds

Three meso-scale circulation systems are introduced in this section. Land and sea breezes (already introduced in Chapter 5, Section 5.3.1) and mountain and valley winds are caused by local temperature differences. Föhn winds result from pressure differences on either side of a mountain range.

6.3.1 Land and sea breezes

As we saw in Chapter 5, the land and sea breeze mechanism is an example, on a diurnal timescale, of a circulation system that

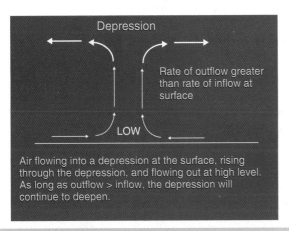

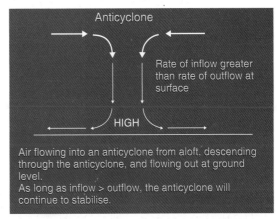

Figure 6.14 Air flow in a depression and an anticyclone

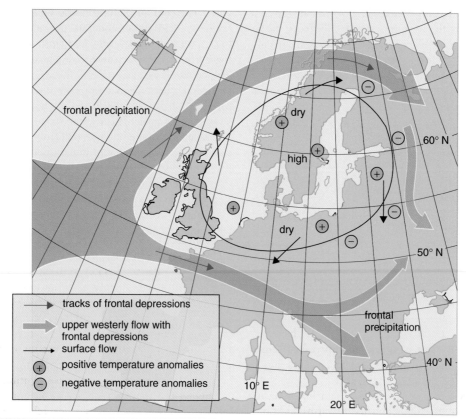

Figure 6.15 High pressure over northern Europe gives light clockwise winds. If the anticyclone stabilizes and persists, it can become a 'blocking anticyclone', which deflects depressions around it and has been the cause of droughts in the UK (e.g. in 1976). (Source: Waugh, 2000, p. 234)

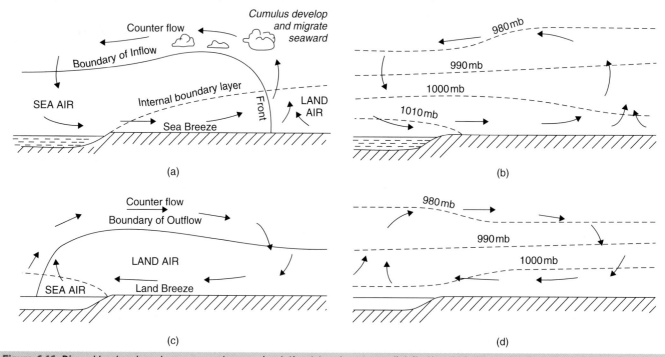

Figure 6.16 Diurnal land and sea breezes: sea breeze circulation (a) and pressure distribution (b) in the early afternoon; land breeze circulation (c) and pressure distribution at night (d). (Source: Barry and Chorley, 1987, p. 130). Reprinted from Atmosphere, weather & climate, Barry & Chorley, 1987, p. 130, with permission from Taylor and Francis

is caused by differential heating and cooling of coastal land and the adjacent sea. During the morning, the land heats up faster than the sea, and air over the land expands and rises, drawing in cooler, more-stable air from over the sea (Figure 6.16(a)) to equalize the pressure difference (Figure 6.16(b)). Cooler air above the sea is drawn downwards to complete a simple circulatory or convection cell.

Later in the evening, clear skies cause heat to radiate from the land faster than from the sea and a cooler layer of air develops over the land. The pressure gradient is reversed and air is drawn from the land out to sea (Figure 6.16(c)) and the pressure distribution at night reverses (Figure 6.16(d)).

6.3.2 Mountain and valley winds

In many mountain areas anabatic and katabatic winds develop, particularly during periods of calm, clear, settled weather. The anabatic wind is an upslope flow which develops during the day. The valley sides, particularly those with an aspect facing the Sun which are steep and vegetation free, heat up more than air at the same height in the centre of the valley. Air rises above the ridges and feeds an upper return current along the line of the valley (Figure 6.17). Hang-gliders and para-gliders seek out these rising thermals.

The rising, conditionally unstable air can, on hot summer afternoons, produce cumulus clouds, and in very warm conditions cumulo-nimbus clouds may develop, giving rise to thunderstorms on mountain ridges.

During a clear evening, the valley loses heat by radiation and the air on the ridges and valley sides cools, becomes denser and drains downslope under gravity towards the lowest point in the valley bottom. This down-valley air flow is known as a katabatic

wind. The cold air pooling in the valley bottom gives rise to a temperature inversion, and if the air is moist enough it may create fog. In locations where this frequently happens, and where the fog freezes in winter to cause frost, the term 'frost hollow' is used. These conditions in winter can be hazardous for drivers. Where crop cultivation takes place in frost hollows, large 'anti-frost' fans may be placed next to crops to keep the air moving and prevent frost damage. Katabatic winds are strongest where they blow over glaciers or permanently snow-covered slopes as temperature contrasts can be greater in these environments. In Antarctica, katabatic winds can reach hurricane force.

6.3.3 Föhn winds

As a parcel of air rises or is forced up a mountain side, it expands due to the reduction in atmospheric pressure. The energy required to cause this expansion is lost from the air as latent heat and the air parcel cools as it ascends. If the air is moist, it will cool at the saturated adiabatic lapse rate (Chapter 4, Section 4.3) of around 0.6 °C per 100 m. A moist parcel of air rising up the windward side of a mountain will cool until it reaches its dew point, the temperature at which the water vapour in the air condenses into droplets, and clouds form. From this point upwards the air is saturated and cools at the slower saturated adiabatic lapse rate, possibly shedding moisture as it goes in the form of rain (see Section 4.5.2, 'Orographic or relief rainfall'). As the air passes over the mountain summit, having deposited its moisture (Figure 6.18), the air is dry and warms up as it expands at the dry adiabatic lapse rate of around 1 °C per 100 m.

The effect of this passage of air over the mountain is that the air is both drier and warmer than it was on the windward side. In the Alps the difference may be substantial, but even in the

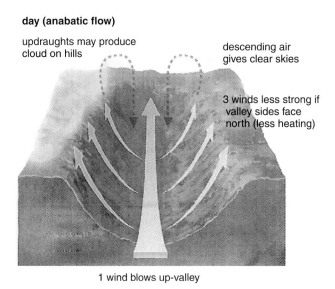

day (anabatic flow)

updraughts may produce cloud on hills

descending air gives clear skies

3 winds less strong if valley sides face north (less heating)

1 wind blows up-valley

(a)

night (katabatic flow)

2 under clear skies, cold dense air sinks under gravity: can form fog and frost hollows in valley (temperature inversion)

1 wind blows down-valley

(b)

Figure 6.17 Mountain and valley winds: (a) anabatic (upslope) flow; (b) katabatic (down-valley) flow. (Source: Waugh, 2000, p. 241)

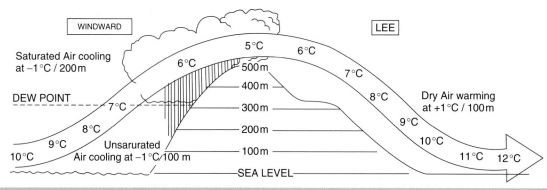

Figure 6.18 The föhn effect. Moist air rising on the windward side passes the dew point and cools at the saturated adiabatic lapse rate, but on descending on the lee side it warms at the dry adiabatic lapse rate. (Source: after Langmuir, 1984). Reprinted from Geography: An Integrated Approach, David Waugh, 2000, p. 241, with permission from Nelson Thornes

British Isles the föhn effect can result in the air temperature being 1–2 °C warmer on the east side of the Scottish Highlands when a moist south-west wind is blowing. This wind is also known as the Chinook on the American prairies, where it is most likely to blow in spring. The name chinook is Native American for 'snow-eater' and this wind can melt vast amounts of snow quickly, enabling the spring wheat to be sown. In Switzerland, the föhn wind clears the alpine meadows of snow quickly, but the rapid thaw sometimes causes avalanches, forest fires and even the premature budding of trees and shrubs.

6.4 Microclimates

Microclimatology is the study of local climate over a small area. Since we live in the lowest few metres of the atmosphere, we need to take a special interest in the climate of this zone as it can be very complicated and diverse, and climatic differences equivalent to a change in latitude of several degrees can occur over a space of just a few metres. For example, take a climber completing a climb on a south-facing crag (in the northern hemisphere) in winter with a cold northerly wind blowing. During the climb the crag is sheltered from the wind and the Sun is shining directly onto the rock face, so air temperatures may reach 10–15 °C. However, as soon as the climber reaches the top of the climb, he or she steps into a very cold northerly wind, the temperature drops to 3–4 °C and there is a high wind-chill factor. This dramatic change in the space of less than a metre is what makes the study of microclimatology both interesting and important. Take another example: you walk barefoot onto a Mediterranean beach in the middle of the day in summer and the sand has become too hot to stand on. It may have reached 50–65 °C. The air temperature near your head may be 30 °C, but in the shade (where official temperature observations are made) of a beach umbrella it may be 20 °C. Similar variations can be found in humidity. So, what is it about this layer near the ground which produces such major gradients – gradients that are not found anywhere else in the atmosphere?

Such variability occurs because the ground–atmosphere interface, or the boundary layer, is the main energy-exchange zone

between the ground surface and the atmosphere. Energy, mainly from the Sun but also from the atmosphere itself, reaches the ground and is absorbed, returned to the atmosphere in a different form, or stored in the ground or water body as heat. The absorption process is very dependent on the nature of the surface, the albedo (as we saw in Chapter 3, Section 3.2.5), which varies hugely from dark surfaces like roads and roofs in urban areas and dark soils and coniferous forests to light-coloured surfaces like a sandy beach, a glacier or fresh snow. In this section we will examine microclimates in urban, forest and water-influenced localities.

6.4.1 Urban microclimate

Large cities and conurbations experience climatic conditions which differ from those of the surrounding countryside. Instead of a mixture of soil and vegetation, much of an urban area is covered with a mosaic of tarmac, concrete, brick, glass, slate and stone

Table 6.3 Effects of urbanization on climate: average urban climatic differences expressed as a percentage of rural conditions (except temperature, in °C). (Based on data from Barry and Chorley, 1987; WMO, 1970; and Smithson et al., 2002)

Measure	Warm season	Cold season	Annual
Pollution	+250	+1000	+500
Solar radiation	−5	−15	−10
Temperature	+1	+3	+2
Humidity	−10	−2	−5
Visibility	−10	−20	−15
Fog	+5	+15	+10
Wind speed	−30	−20	−25
Cloudiness	+10	+5	+8
Rainfall	+10	0	+5
Thunderstorms	+30	+5	+15

surfaces, which may extend vertically for hundreds of metres. However, throughout urban areas there may still be parks with lakes, avenues of trees and areas of grass scattered throughout. Table 6.3 summarizes the average effects of urbanization on climate.

The composition of the atmosphere in conurbations is influenced by both industry and transport. Pollution levels can be much higher, for example, on average: carbon dioxide ($\times 2$); sulphur dioxide ($\times 200$); nitrogen oxides ($\times 10$); carbon monoxide ($\times 200+$); total hydrocarbons ($\times 20$); and particulate matter ($\times 3$–7). In calm anticyclonic conditions, or where urban areas are in hollows, as in the case of Los Angeles, the pollution levels can persist for days, weeks or even months and become a hazard to human health. A typical pollution dome is shown in Figure 6.19(a), with the modification caused by a light wind shown in Figure 6.19(b). During such conditions the air quality is poor and people with respiratory problems may be advised to stay indoors. The pollution may be completely dispersed with stronger winds.

The increased level of particulates in the atmosphere above urban areas has further implications for the microclimate. These particulates increase the number of hygroscopic nuclei drastically. Over an ocean thousands of km from land there may be 10 particles per cm^3 of the size suitable for hygroscopic nuclei, whereas over a city centre there may be 1 million per cm^3. This means that water vapour in the air may condense more readily over urban areas, which may lead to more fog ($+10\%$), clouds ($+8\%$) and associated rainfall ($+5\%$). The combined effect of pollution, fog and cloud on the receipt of bright sunshine in London is illustrated in Figure 6.20.

Figure 6.20(a) shows clearly the effects of winter atmospheric pollution in the city, which may be a combination of the increased burning of coal fires emitting particulates and the development of fogs and smogs (a combination of smoke and fog) for which London was so famous around this period. Figure 6.20(b), however, shows the effect of the 1956 Clean Air Act, which effectively banned the burning of all but smokeless fuels in the centre of London, on increasing the receipt of winter sunshine. The figure shows that the improvement was greatest in the winter months.

Concrete, brick, asphalt and glass respond differently to radiative exchange than do the vegetation surfaces of the rural surrounding areas. Concrete and brick have high heat capacities, so large quantities of heat added to the material while the Sun is shining are slowly released at night, adding heat to the urban atmosphere. In addition, heat from the bodies of the people in the city, heat from vehicle engines and exhausts, plus heat from buildings all contribute to what has been termed the 'urban heat island effect'. In May 1959, Chandler (1965) (Figure 6.21) identified this effect over London.

Figure 6.22 shows the temperature along a cross-section through the city of Chester in north-west England; the highest temperature along the transect corresponds to the city centre.

Early blooming of flowers and decreased incidence of frosts and the time when snow lies are indicators of this effect, which is greatest in winter during calm high-pressure conditions.

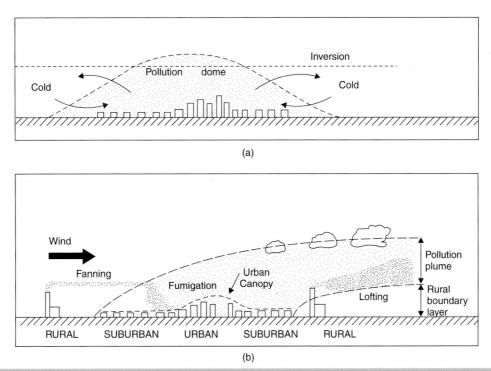

Figure 6.19 (a) Urban pollution dome, which can develop in calm weather when there is little or no wind to mix and disperse it. (b) Urban pollution dome with light wind. Fanning is indicative of vertical atmospheric stability. (After Oke, 1978; Source: Barry and Chorley, 1987, p. 356)

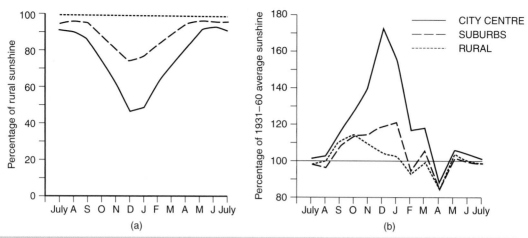

Figure 6.20 Sunshine in and around London. (a) Mean monthly bright sunshine recorded in the city and suburbs for 1921–1950, expressed as a percentage of that recorded in the adjacent rural areas. (After Chandler, 1965.) (b) Mean monthly bright sunshine recorded in the city, suburbs and surrounding rural areas during the period 1958–1967, expressed as a percentage of the averages for the period 1931–1960. (After Jenkins, 1969; Source: Barry and Chorley, 1987, p. 352). Reprinted from Atmosphere, Weather and Climate, Barry & Chorley, 1987, p. 356, with permission from Taylor and Francis

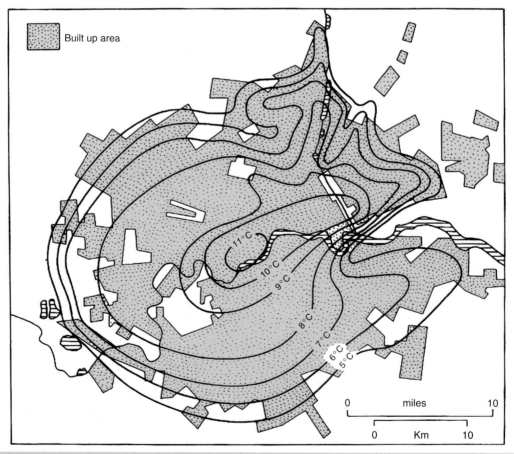

Figure 6.21 Distribution of minimum temperatures (°C) in London on 14 May 1959, showing the urban heat island directly over the built-up area (after Chandler, 1959; Source: Barry and Chorley, 1987, p. 360). Reprinted from Atmosphere, Weather and Climate, Barry & Chorley, 1987, p. 352, with permission from Taylor and Francis

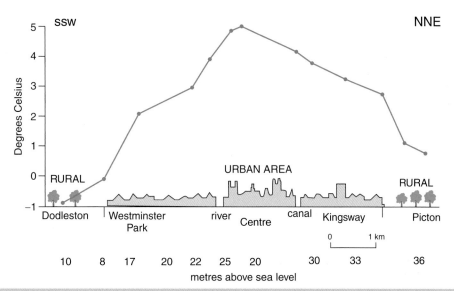

Figure 6.22 Transect through the city centre of Chester in north-west England, showing the urban heat island effect. (After Nelder, 1985; Source: Smithson *et al.*, 2002, p. 149). Reprinted from Atmosphere, Weather and Climate, Barry & Chorley, 1987, p. 360, with permission from Taylor and Francis

Table 6.4 Fog frequencies in London and south-east England (hours per year). (Source: Waugh, 2000, p. 243)

Location	Very dense fog: visibility less than 40 m	Less dense fog: visibility less than 1000 m
Kingsway (central London)	19	940
Kew (middle suburbs)	79	633
London Airport (outer suburbs)	46	562
South-east England (mean of 7 stations)	20	494

Humidity in urban areas is on average slightly lower (−5%) than that above surrounding rural areas because concrete and asphalt road surfaces do not store water in the same way as soil. Urban surfaces are generally impermeable and designed so that rainfall runs off quickly into underground drains and sewer systems. Therefore, little water is available for evaporation or transpiration into the atmosphere and humidity levels are generally lower.

Visibility in urban areas is on average 15% poorer than in surrounding rural areas. Table 6.4 demonstrates this with fog frequency data for London and south-east England.

With the exception of the very dense fog frequency in central London, which may be related to the 1956 Clean Air Act, there is a decrease in fog frequency in both the very dense and less dense categories with distance from the urban centre. Fog frequency is far lower at the weather stations in rural areas of south-east England.

The presence of the urban structure tends to slow air movement down on average (Figure 6.23), although the degree of gustiness may be higher.

As the air flows over the very irregular urban surface, friction with the buildings retards the wind in the lowest layers in particular. The presence of tall buildings can produce eddies (Figure 6.24) which can cause strong local gusts. Architects planning shopping precincts have had to take these problems into account in their designs.

The increase in cloud and precipitation in cities reported in Table 6.3 was demonstrated in some American work on St Louis, which showed that additional heating of the air crossing the city, increases in pollutants (hygroscopic nuclei in particular), the frictional and turbulent effects on air flow and altered moisture all appear to play a role in the preferential development of clouds and rain, particularly in summer. Figure 6.25 illustrates this phenomenon in south-east England, where the highest rainfall totals in the region are centred over London.

So urban areas modify the microclimate in a range of ways. They exhibit an urban heat island effect, which raises daytime temperatures by an average of 0.6 °C and night-time temperatures by 3–4 °C, as dust and cloud act like a blanket to reduce radiation heat losses, while buildings and vehicles give out heat like storage radiators. Air pollution can be much higher in urban areas compared to surrounding rural areas. Urban areas receive less sunlight and attract more fog. While humidity in urban areas is lower, they tend to have more hygroscopic nuclei, and so attract more cloud and rain. Average wind speeds in urban areas are lower, but locally may be considerably higher where air is squeezed between buildings.

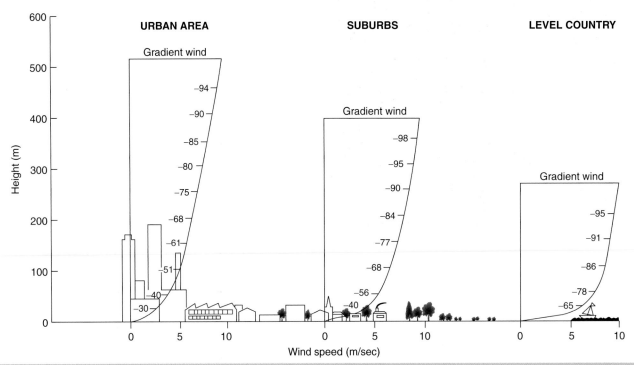

Figure 6.23 The effect of an urban area on the wind-speed profile. With decreasing roughness the depth of the modified layer becomes shallower and the profile steeper. (Source: Smithson *et al.*, 2002, p. 152). Reprinted from Fundamentals of the Physical Environment, Smithson, P., Addison, K., Atkinson, K, 2002, p. 149, with permission from Taylor and Francis

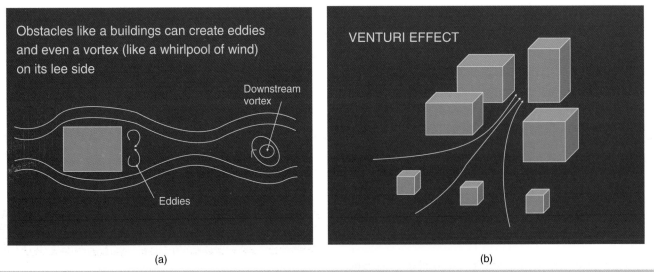

Figure 6.24 Details of urban airflow. (a) Formation of eddies on lee side of buildings. (b) Air being squeezed through gaps causes it to accelerate (the Venturi effect). (c) Transverse currents form when buildings are offset. (d) Flow around two buildings of different sizes and shapes; the numbers give relative wind speeds. Stippled areas are those of high wind velocity and turbulence at street level. SP, stagnation point; CS, cornerstream; VF, vortex flow; L, lee eddy. (Source: after Plate, 1972; Oke, 1978; and Barry and Chorley, 1987).

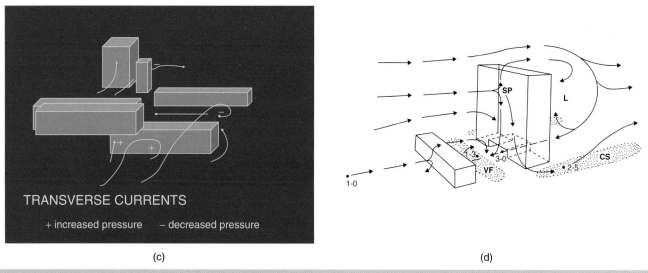

(c)

(d)

Figure 6.24 (*continued*)

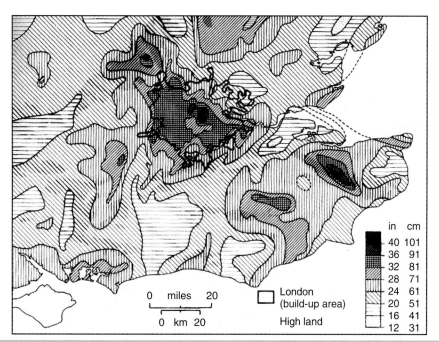

Figure 6.25 The distribution of total thunderstorm rain in south-east England during the period 1951–1960. (After Atkinson, 1968; Source: Barry and Chorley, 1987, p. 363). Reprinted from Atmosphere, weather & climate, Barry & Chorley, 1987, p. 361, with permission from Taylor and Francis

6.4.2 Forest, lake and coastal microclimates

Different land surfaces produce distinctive local microclimates. Table 6.5 summarizes and compares the key features of forest, lake and coastal microclimates.

The time and space scales of the various atmospheric phenomena discussed so far in this chapter are related in Figure 6.26, the shaded area of which delimits the 'zone of weather' or troposphere.

6.5 Weather observation and forecasting

Today, thousands of weather observations are gathered around the world every minute. Most weather data are now collected by electronic instruments, the measurements taken being recorded into data loggers. Measurements are then transferred to computer databases for analysis, either to generate a forecast or to learn more about day-to-day weather fluctuations or longer-term changes in climate.

Table 6.5 Microclimates of forest, lake and coast. (Source: after Waugh, 2000, p. 243)

	Forest (coniferous and deciduous)	Lake	Coast/Sea
Incoming radiation and albedo	General dark colour (green/brown) means that a high proportion of incoming radiation is absorbed and trapped. Albedo for coniferous forest is 15%; deciduous 25% in summer and 35% in winter.	Less insolation absorbed and trapped. Albedo may be over 60% and is highest on calm days when water surface it still.	Water surface rarely still, albedo generally low (5-10%) but can be very high when Sun is at low angle (>90%). Beaches generally light colours, so albedo 35-45%. Not much heat absorbed.
Temperature	Small diurnal range due to blanket effect of the canopy. Forest floors are protected from direct sunlight. Evapotranspiration important and may cause some heat to be lost.	Small diurnal range because water has a higher specific heat capacity, leading to cooler summers and milder winters. Lakesides have longer growing season.	Small diurnal range because sea has a higher specific heat capacity, giving cooler summers and milder winters. Coastal areas have longer growing season. Land and sea breezes can keep coastal areas cool on hot summer days.
Relative humidity	Higher during daytime and summer, especially in deciduous forest due to high evapotranspiration rates, which depend on length of day, leaf surface area, wind speed, etc.	Very high, especially when radiation is strong as evaporation takes place, transferring water molecules from the water to the atmosphere above.	High, especially when radiation is strong as evaporation takes place. Wind caused by land and sea breeze systems can increase evaporation rates.
Cloud and precipitation	Heavy rain can result from high evapotranspiration rates, e.g. tropical rain forests, where cloud builds up during the day and heavy rain falls each afternoon. On average 30-40% of rain is intercepted, though this can be as high at 80% in dense coniferous forest plantations.	Air is humid and if forced to rise can be unstable and produce cloud and rain. Fewer condensation nuclei mean rainfall amounts are usually lower. Fogs can form in calm conditions.	Air is humid and if forced to rise can be unstable and produce cloud and rain. Fewer condensation nuclei mean rainfall amounts are usually lower. Advection fogs can form.
Wind speed and direction	Trees reduce wind speeds, particularly at ground level, by as much as 50%. Trees are often planted as wind breaks. Trees can produce eddies.	Wind may be strong due to reduced friction. Large lakes like Lake Victoria in Africa, can create their own land and sea breezes.	Wind may be strong due to reduced friction, though large waves can increase friction, reducing wind speeds to 70% of geostrophic wind. Coasts develop land and sea breezes, which can bring strong winds on hot days, with predictable diurnal reversal of wind direction.

Weather observations are made on board ships, aircraft, moored and drifting buoys, oil rigs, balloons and at manned weather stations on land. Additional data are recorded by satellites and radar (Figure 6.27).

6.5.1 Measuring weather elements at a land-based weather station

Land-based weather stations are generally located in open positions, not too close to buildings, forests or water bodies that might affect the local climate. Instruments need to be checked and calibrated regularly. At manually-read weather stations, normally attended by amateur observers, daily recordings (usually at 0900 GMT) are made. At professionally-manned weather stations (e.g. coastguards, airports), readings may be taken more frequently (e.g. every three hours).

Temperature and humidity

Air temperature is widely measured using the Celsius temperature scale (0–100 °C), though some parts of the world (e.g. the USA) still use the Fahrenheit scale (32–212 °F). Traditionally temperature has been measured using a thermometer – a glass tube inside which liquid expands and contracts with temperature changes. Many observations today are also made using thermistors. A

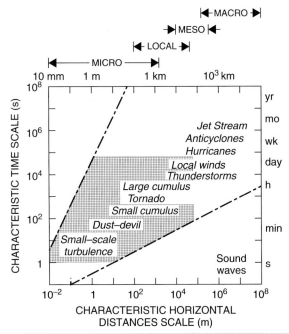

Figure 6.26 Time and space scales of various atmospheric phenomena. (Adpated from Smagorinski, 1974 and Oke, 1978). Reprinted from Atmosphere, Weather and Climate, Barry & Chorley, 1987, p. 363, with permisson from Taylor and Francis

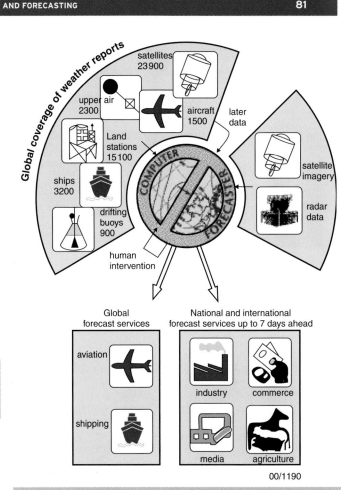

Figure 6.27 The flow of information from observations to forecasts. The numbers give an indication of how many observations are used in each 24 hour period as a basis for the global forecasts. (Source: UK Meteorological Office, http://www.metoffice.gov.uk, accessed 25 May 2001)

thermistor is a type of resistor used to measure temperature changes, relying on the change in electrical resistance with changing temperature. Since temperature can vary with height, it is measured at a standard level 1.2 m above the ground surface. A thermometer shelter or Stevenson screen (named after the designer) is a white louvered box that holds thermometers or other weather instruments at the correct height, while sheltering them from the direct rays of the Sun (Figure 6.28).

Many weather stations are now equipped with temperature measurement systems that use thermistors. Data are recorded on data loggers and may be transmitted via wireless communication to a console and/or computer (Figure 6.29).

A Stevenson screen normally contains a wet-bulb and a dry-bulb thermometer. The wet-bulb thermometer is kept moist by a wick connected to a bottle of water (Figure 6.30(a)). Water evaporating from the wick in contact with the thermometer takes heat energy from the glass bulb of the thermometer, causing it to cool or depress the temperature reading. Comparison of the wet-bulb reading with the dry-bulb one allows the calculation of relative humidity. The whirling hygrometer shown in Figure 6.30(b) is a portable instrument containing both wet- and dry-bulb thermometers and is capable of measuring air temperature and humidity.

Wind speed and direction

Wind speed is usually measured by a cup anemometer (Figure 6.29(a)), which rotates and has the advantage of being able to measure speed whatever the direction. A wind vane points to the direction from which the wind is coming (Figure 6.29(a)). The direction given for the wind refers to the direction from which it comes. For example, a westerly wind is blowing from the west towards the east. Measurements of wind strength are made at 10 metres (33 feet) above the ground. A specified height has to be used because the wind speed decreases towards the ground. Wind speeds are normally measured in knots (nautical miles per hour). However, forecast winds are often given in miles per hour (where 1 knot is equivalent to 1.15 mph) or in terms of the Beaufort scale (Table 6.6), though sometimes km per hour and $m\,s^{-1}$ may also be used.

Precipitation

Precipitation is measured by collecting rain or snowfall in a cylindrical container, which normally has a funnelled opening to allow rain and snow to enter, but to reduce the risk of evaporation. Figure 6.31 shows a range of types of rain and snow gauges.

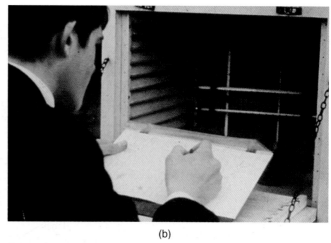

(a)

(b)

Figure 6.28 (a) Thermometer shelter or Stevenson screen. (b) Recording observations from thermometers inside a Stevenson screen. (Photos: T. Stott)

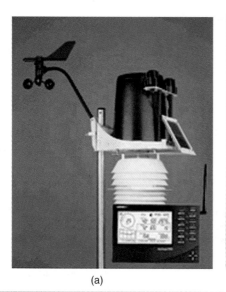

(a)

(b)

Figure 6.29 (a) Davis Vantage Pro wireless weather station with thermistor shelter at base, rain gauge, wind vane and anemometer. (b) Mountain-summit automatic weather station on Nuolja (1169 m), near Abisko, Arctic Sweden. (Photos: T. Stott)

The size or area of the collecting funnel is related to the volume of water collected:

$$\text{Rainfall (mm)} = \text{volume (ml)}/\text{area of gauge (cm}^2)$$

Snow is melted; the amount of water it contains, the water equivalent, is of most interest to hydrologists. The standard rain gauge (Figure 6.31(a)) has the water it has collected measured daily, but the autographic or tipping bucket type (Figure 6.31(b) and Figure 6.29(a)) tips when the upper side of the seesaw mechanism fills with rain, so the other side begins to fill. The rate at which the seesaw tips is a measure of the rainfall rate or rainfall intensity, which is clearly very useful for flood forecasting. The ground-level rain gauge (Figure 6.31(c)) was developed for use in windy mountainous environments. Air turbulence around the top of standard rain gauges in such environments was found to result in rain total estimates some 8–10% lower than those of the ground-level gauge with antisplash grid (shown in lower photograph, Figure 6.31 (c)).

6.5.2 Weather maps and forecasting

A weather map or synoptic chart (Figure 6.32(a)) shows the weather for an area at one particular time. Meteorologists collate weather data for numerous weather stations and the data are plotted using internationally-accepted weather symbols (Figure 6.32(b)).

Weather elements, with the exception of wind direction, are plotted in a fixed position around the circle so that individual elements can easily be identified.

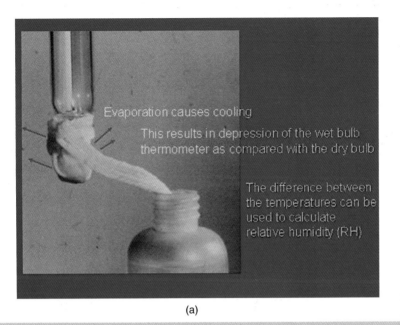

(a)   (b)

Figure 6.30 (a) Wet-bulb thermometer. (b) Whirling hygrometer with wet- and dry-bulb thermometers

Table 6.6 The Beaufort Wind Scale. (Source: http://www.bbc.co.uk, accessed 19 December 2007)

No.	Knots	Mph	Description	Effects at sea	Effects on land
0	0	0	Calm	Sea like a mirror	Smoke rises vertically
1	1-3	1-3	Light air	Ripples, but no foam crests	Smoke drifts in the wind
2	4-6	4-7	Light breeze	Small wavelets	Leaves rustle; wind felt on face
3	7-10	8-12	Gentle breeze	Large wavelets; crests, not breaking	Small twigs in constant motion; light flags extended
4	11-16	13-18	Moderate wind	Numerous whitecaps	Dust, leaves and loose paper raised; small branches moved
5	17-21	19-24	Fresh wind	Many whitecaps; some spray	Small trees sway
6	22-27	25-31	Strong wind	Larger waves form; whitecaps everywhere; more spray	Large branches move; whistling in phone wires; difficult to use umbrellas
7	28-33	32-38	Very strong wind	White foam from breaking waves begins to be blown in streaks	Whole trees in motion
8	34-40	39-46	Gale	Edges of wave crests begin to break into spindrift	Twigs break off trees; difficult to walk
9	41-47	47-54	Severe gale	High waves; sea begins to roll; spray may reduce visibility	Chimney pots and slates removed
10	48-55	55-63	Storm	Very high waves with overhanging crests; blowing foam gives sea a white appearance	Trees uprooted; structural damage
11	56-63	64-72	Severe storm	Exceptionally high waves	Widespread damage
12	63+	73+	Hurricane force	Air filled with foam; sea completely white; visibility greatly reduced	Widespread damage; very rarely experienced on land

Figure 6.31 (a) Standard British Meteorological Office rain gauge with 100 mm (4 inch) diameter opening. (b) Autographic or tipping bucket rain gauge, which gives a 'continuous' record of rainfall. (c) Snow gauge used in Scottish Highlands, which is tall (to avoid burial) and has a heated funnel to melt snow as it enters, and a ground-level rain gauge. (d) Diagram of types of rain gauge. (Source: (a)–(c): photos by T. Stott; (d): Smithson *et al.*, 2002, p. 80)

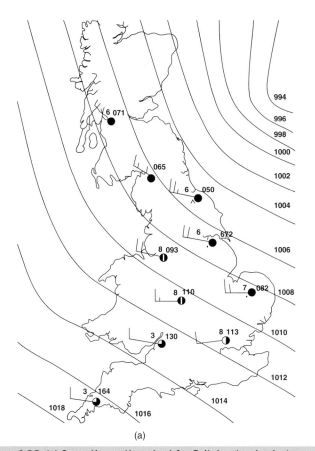

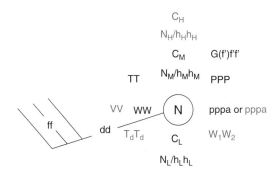

Decode of elements plotted on a land station circle (note the colour coding)

Identifier	Description
N	Total amount of cloud (in oktas)
C_L	Type of low cloud
N_L	Amount of low cloud (in oktas)
h_Lh_L	Height of low cloud (in feet)
C_M	Type of medium cloud
N_M	Amount of medium cloud (in oktas)
h_Mh_M	Height of medium cloud (in feet)
C_H	Type of high cloud
N_H	Amount of high cloud (in oktas)
h_Hh_H	Height of high cloud (in feet)
TT	Dry-bulb air temperature (in degrees Celsius)
ww	Present weather
dd	Wind direction (in degrees)
ff	Wind speed (in knots)
VV	Visibility (in metres or kilometres)
T_dT_d	Dew point temperature (in degrees Celsius)
W_1W_2	Past weather
pppa or pppa	Pressure tendency and trend (black: rising, red: falling) (in millibars)
PPP	(Atmospheric pressure (in millibars)
G(f')f'f'	Wind gust (in knots)

Table 1. Decode of elements plotted on a land station circle.

(a)

(b)

Figure 6.32 (a) Synoptic weather chart for Britain, showing isobars and station models for selected weather stations. (b) International land station circle plot with decode of elements. (Source: National Meteorological Library and Archive, UK Meteorological Office). Reprinted from *Fundamentals of the Physical Environment*, Smithson, P., Addison, K., Atkinson, K, 2002, p. 80, with permission from Taylor and Francis

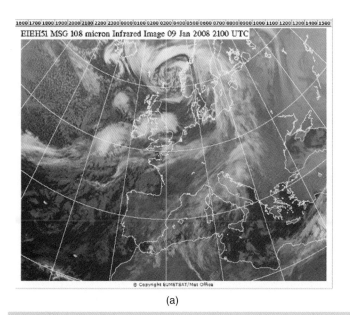

(a)

(b)

Figure 6.33 Time sequence of infrared satellite imagery over Europe on 9–10 January 2008. (a) 2100 hrs on 09 Jan. (b) 0300 hrs on 10 Jan. (c) 0900 hrs on 10 Jan. (d) 1500 hrs on 10 Jan. (Source: UK Meteorological Office, www.metoffice.gov.uk, accessed 10 January 2008)

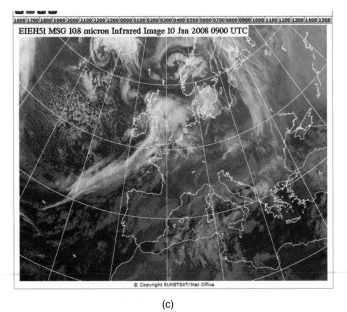

EIEH51 MSG 10.8 micron Infrared Image 10 Jan 2008 0900 UTC

© Copyright EUMETSAT/Met Office

(c)

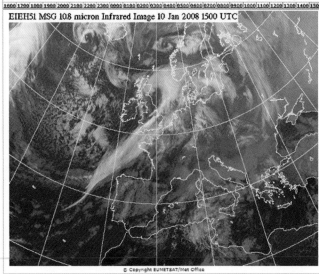

EIEH51 MSG 10.8 micron Infrared Image 10 Jan 2008 1500 UTC

© Copyright EUMETSAT/Met Office

(d)

Figure 6.33 (*continued*)

The role of the weather forecaster is to try to predict the speed and direction in which various air masses and associated weather fronts will move. Satellite images (Figure 6.33) are now an integral part of a weather forecaster's toolkit.

Even though weather forecasting is assisted by satellite images, radar (RAdio Detection And Ranging) and computers, which show upper air as well as surface conditions in a three-dimensional way, the complexity and unpredictability of the atmosphere can still catch the forecaster by surprise. This is because the data used in the forecaster's models still remains a sample of the total atmospheric conditions, and there is always a risk that an unmeasured part of the model will become reality. This happened on 16 October 1987 when an unforeseen storm with hurricane-force winds travelled up from the east of Florida overnight and unexpectedly hit the south coast of England in the early hours of the morning.

Exercises

1. Define the term 'air mass' and give examples of typical source regions.

2. Draw a sequence of simple diagrams to show the formation of a weather front in plan view.

3. Draw a cross-section of a typical mid-Atlantic depression, showing the location of: (i) warm front, (ii) cold front, (iii) rainfall, (iv) cirrus cloud, (v) stratus cloud, (vi) cumulus and cumulo-nimbus cloud.

4. As a depression passes overhead, explain what happens to (i) temperature, (ii) pressure, (iii) wind direction.

5. What are the conditions that favour the development of tropical cyclones?

6. Name the scale used to classify hurricanes.

7. What is the difference between a hurricane and a tornado?

8. Describe the weather patterns associated with an anti-cyclone.

9. Draw a simple diagram to show the operation of land and sea breezes.

10. Explain how anabatic and katabatic winds blow.

11. What is the föhn effect?

12. Explain how an urban microclimate differs from a rural one.

13. How do forests affect the microclimate of an area?

14. Describe the instruments used in a standard ground-based weather station.

References

Atkinson, B.W. 1968. A preliminary examination of the possible effect of London's urban area on the distribution of thunder rainfall 1951–60. *Transactions of the Institute of British Geographers*. **44**, 97–118.

Barry, R.G. and Chorley, R.J. 1987. Atmosphere, Weather and Climate. 5th edition. New York, Methuen.

Chandler, T.J. 1965. The Climate of London. London, Hutchinson.

Jenkins, I. 1969. Increases in averages of sunshine in Greater London. *Weather*. **24**, 52–54.

Langmuir, E. 1984. Mountaincraft and Leadership. Edinburgh, Scottish Sports Council and The Mountainwalking Leader Training Board.

Oke, T.R. 1978. Boundary Layer Climates, London, Methuen.

Plate, E. 1972. Berücksichtigung von Windströmungen in der Bauleitplanung. In: Seminarberichte Rahmenthema Umweltschutz. Institut für Stätebau und Landesplanung, Selbstverlag, Karlsruhe. pp. 201–229.

Smagorinski, J. 1974. Global atmospheric modelling and the numerical simulation of climate. In: Hess, W.N. (Ed.). Weather and Climate Modification. New York, Wiley.

Smithson, P., Addison, K. and Atkinson, K. 2002. Fundamentals of the Physical Environment. 3rd edition. London, Routledge.

Strahler, A.H. and Strahler, A. 1997. Physical Geography: Science and Systems of the Human Environment. Chichester, John Wiley & Sons, Inc.

Thomas, M. 1995. Weather for Hillwalkers and Climbers. Stroud, Alan Sutton Publishing.

Waugh, D. 2000. Geography: An Integrated Approach. 3rd edition. London, Nelson.

WMO. 1970. Urban climates. WMO Technical Note 108.

Further reading

Barry, R.G. and Chorley, R.J. 2003. Atmosphere, Weather and Climate. 8th edition. London, Routledge. Chs 4–7.

Holden, J. (Ed.). 2005. An Introduction to Physical Geography and the Environment. Harlow, Pearson Education. Ch. 4

Robinson, P.J. and Henderson-Sellers, A. 1999. Contemporary Climatology. Harlow, Pearson Education.

Thompson, R.D. 2002. Atmospheric Processes and Systems. Routledge Introductions to Environment Series. London, Routledge. Chs 12–16.

Dust sampling equipment at Diamantina National Park, western Queensland, Australia. Photo: T. Stott

Learning Outcomes

After reading this chapter and completing the exercises you will be able to:

➤ Understand the difference between weather and climate.

➤ Understand the terms 'temperature' and 'precipitation regime' and how these vary with latitude and maritime/continental location.

➤ Describe the Köppen climate classification system and explain the differences between the different climate groups and sub-groups.

➤ Be aware of alternative climate classification systems.

7.1 Introduction

The climate of a location is the weather averaged over a period of time (usually at least 30 years). This includes the region's general pattern of weather conditions, seasons and weather extremes, like hurricanes, droughts and excessive rainy periods. The Earth's climate varies from place to place, creating a variety of environments. Thus, in various parts of the Earth, we find deserts, tropical rain forests, tundra (frozen, treeless plains), coniferous forests (which consist of cone-bearing trees and bushes), prairies and expanses of glacial ice. As we have seen in earlier chapters, there are a number of weather elements, such as temperature, wind speed and direction, barometric pressure, solar radiation, cloud cover and precipitation, which could be used to calculate an area's climate. However, since detailed observations of all these elements are not made regularly at most weather stations around the world, the two which climatologists focus on to characterize worldwide climate are temperature and precipitation. These two elements strongly influence the natural vegetation of a region. For example, forests only tend to grow in areas where there is sufficient precipitation. In regions with low precipitation, grassland, scrub or desert will more likely be found. The natural vegetation in an area will determine the animals which inhabit it. Temperature and precipitation are also important elements for the cultivation of crops and development of soils, so regional classification on the basis of climate (temperature and precipitation) helps our understanding of the potential of certain regions to grow food, produce timber or to harvest rain water.

7.2 Classification of climate

Recall from earlier chapters that the Sun's rays arrive at the equator at a direct angle between 23° N and 23° S latitude, where

radiation is at its most intense. At higher latitudes the rays arrive at an angle to the surface and are less intense. The closer a place is to one of the poles, the smaller the angle and therefore the less intense the radiation. As we saw in Chapter 3, the Earth's climate system is based on the location of these hot and cold air mass regions and the atmospheric circulation created by trade winds and westerlies. Trade winds north of the equator blow from the north-east, whereas south of the equator, they blow from the south-east. The trade winds of the two hemispheres meet near the equator and cause the air to rise. As the rising air expands and cools, clouds and rain develop. The resulting bands of cloudy and rainy weather near the equator create tropical conditions.

Westerlies blow from the south-west in the northern hemisphere and from the north-west in the southern hemisphere. Westerlies steer storms from west to east across middle latitudes. Both westerlies and trade winds blow away from the 30° latitude belt. Over large areas centred at 30° latitude, surface winds are light. Air slowly descends (between the Hadley and Ferrel cells) to replace the air that blows away. Any moisture that the air contains evaporates in the intense heat. The tropical deserts, such as the Sahara of Africa and the Sonoran of Mexico, have developed in these regions.

7.2.1 Global temperature regimes

As you will recall from Chapter 5, the Earth rotates about its axis, which is tilted at 23.5°, once every 24 hours. The daily rotation causes day and night and a temperature signal like that in Figure 7.1(a), with temperature peaking sometime after midday and falling to a minimum a few hours after midnight. The tilt in the Earth's axis causes the seasons and a temperature signal like that in Figure 7.1(b), where, in the northern hemisphere, the highest temperatures occur in the months of June or July (northern hemisphere summer) and the lowest in December and January (northern hemisphere winter). However, in consecutive years the seasonal pattern might differ (Figure 7.1(c)), with warmer and cooler years.

Temperature regimes, as shown for various places in the northern and southern hemisphere in Figure 7.2, give the mean monthly air temperature (shown by red dots joined by a red curve in the figure) for each month, but each value may be the average of 30+ years.

The patterns in Figure 7.2 are known as temperature regimes – distinctive types of annual temperature cycle related to latitude and location. The effect of latitude on continental and maritime location is demonstrated in Figure 7.2(a). In Figure 7.2(b) each station has its latitude given and is labelled according to its latitude zone: equatorial, tropical, midlatitude and subarctic. The four stations at the top of Figure 7.2(a) and on the left (west) in Figure 7.2(b) are closer to the sea (known as maritime locations) and show flatter temperature curves with a lower annual range. The four at the bottom of Figure 7.2(a) and to the right (north and west) in Figure 7.2(b) are inland (or continental locations) and show

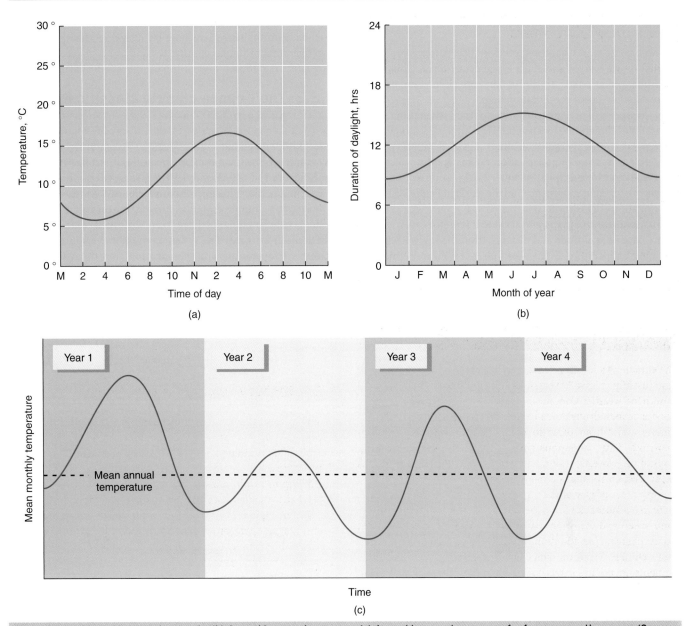

Figure 7.1 (a) Diurnal temperature cycle. (b) Annual temperature curve. (c) Annual temperature curves for four consecutive years. (Source: Strahler and Strahler, 1997, p. 171). Reprinted from Physical Geography, Strahler and Strahler, 1997, p. 171, with permission from John Wiley & Sons Ltd.

greater extremes of temperature and a higher annual temperature range. The equatorial regime at Douala, Cameroon, in West Africa, has a high monthly mean temperature of around 27 °C all year, with very little variation because insolation is almost uniform throughout the year. In contrast, In Salah, Algeria, where there is a very strong temperature cycle, there are mean monthly temperatures in the summer months in the mid 30 °Cs, due to its continental location. Compare In Salah with Walvis Bay in south-west Africa, which is at nearly the same latitude but has only a weak annual temperature cycle with no extreme heat. This is because of the moderating effects of the sea at Walvis,

which we saw at Douala, and can also be seen at Monterey, California and Sitka, Alaska. The other continental regimes show strong annual temperature cycles and the effect of latitude is to increase the temperature range and to depress the mean annual temperatures.

To summarize, the annual variation in insolation is determined by latitude and this gives the basic control on temperature regime, but the effect of location, such as continental interior or maritime, moderates this variation. The influences of latitude and the continents and oceans on the January and July mean sea-level temperatures can be seen in Figure 7.3.

7.2.2 Global precipitation regimes

As we have seen in Chapter 4, precipitation patterns are closely related to air masses and their movements. On a global scale, precipitation is linked to global air and circulation patterns. Figure 7.4 is a global precipitation map.

Two aspects of precipitation are important when using it as an indicator of climate: the mean annual total and the seasonal distribution. Like the temperature regime graphs earlier in this chapter, climatologists also plot precipitation regime graphs, which show the mean monthly precipitation total. Precipitation regime graphs are shown for eight stations around the world in Figure 7.5. Where precipitation falls as snow, it is treated as if it were water.

In looking for patterns and seeking a sense of order, geographers have used latitude and distance from the sea as key factors affecting temperature. Since warm air can hold more moisture than cold air, cold regions are generally drier than warmer ones.

In addition, precipitation totals will tend to be higher during the warmer months. These facts have been used to attempt to group together areas of similar climate.

7.2.3 The Köppen climate classification system

The Köppen climate classification system, developed by Wladimir Köppen, a German climatologist, around 1900 (with modifications in 1918 and 1936), is one of the most widely used climate classification systems (Figure 7.6). Based on the concept that natural vegetation is the best expression of climate, climate zone boundaries have been delimited with vegetation distribution in mind. The system combines average annual and monthly temperatures and precipitation, and the seasonality of precipitation.

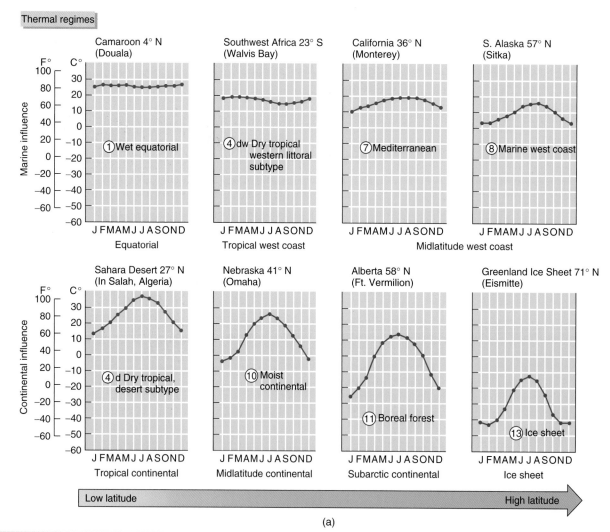

(a)

Figure 7.2 (a) Climate graphs showing the influence of latitude and continental/maritime location. (b) Mean annual temperature curves at various places in the northern and southern hemispheres. (Source: Strahler and Strahler, 1997, pp. 172–173). Reprinted from Physical Geography, Strahler and Strahler, 1997, p. 172-3, with permission from John Wiley & Sons Ltd.

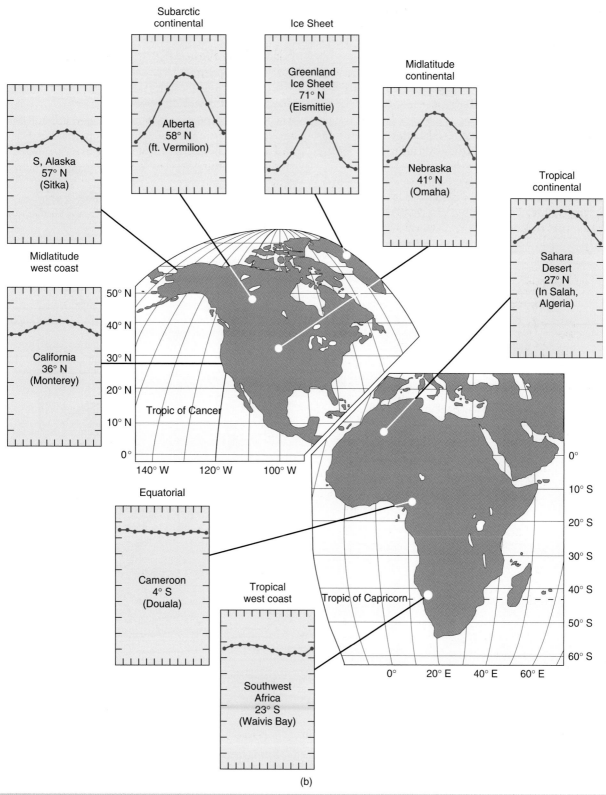

(b)

Figure 7.2 (continued)

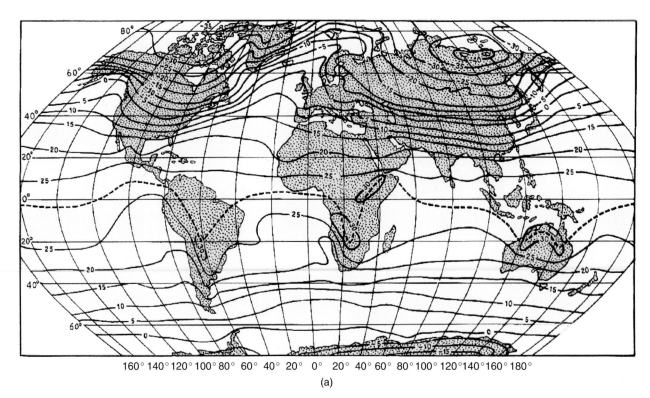

(a)

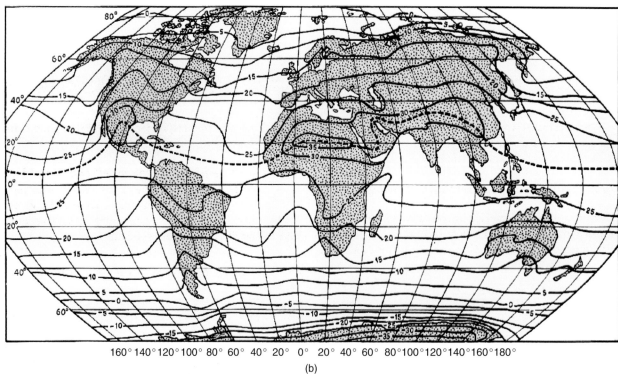

(b)

Figure 7.3 (a) January mean sea-level temperatures (°C). (b) July mean sea-level temperatures (°C). The approximate position of the thermal equator (the zone of maximum temperature) is shown by the dashed line. (Source: Barry and Chorley, 1987, pp. 24–25). Reprinted from Atmosphere, Weather and Climate, Barry & Chorley, 1987, p. 24–5, with permission from Taylor and Francis

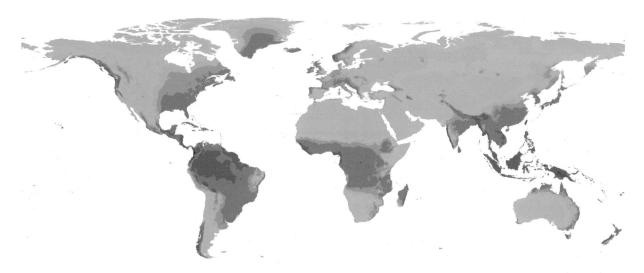

Total precipitation per annum, over land

0–300 mm
300–500 mm
500–700 mm
700–1000 mm
1000–2100 mm
2100–4200 mm
4200–6301 mm
6301–8401 mm
8401–10 501 mm
n/a

Figure 7.4 Global precipitation map. (Source: http://commons.wikimedia.org/wiki/File:World_precip_annual.png, Wikimedia Commons)

The Köppen climate classification scheme divides climates into five main groups, with several types and subtypes. Each particular climate type is represented by a two- to four-letter symbol.

Group A: Tropical/megathermal climates

Tropical climates are characterized by constant high temperature (at sea level and low elevations), where all 12 months of the year have average temperatures of 18 °C or higher. They are subdivided as follows:

Tropical rainforest climates (Af)
All 12 months have average precipitation of at least 60 mm. These climates usually occur within 5–10° latitude of the equator and are dominated by the equatorial low-pressure system all year round, and therefore have no natural seasons. The climate of Singapore is a good example of this group.

Tropical monsoon climates (Am)
Found mainly in southern Asia and West Africa, this type of climate results from the monsoon winds, which change direction with the seasons. This climate has a driest month (which nearly always occurs at or soon after the 'winter' solstice for that side of the equator) with rainfall less than 60 mm, but more than

(100 - [total annual precipitation {mm}/25]). A good example is found at Chittagong, Bangladesh (Figure 7.5(a)).

Tropical wet and dry or savannah climates (Aw)
These climates have a pronounced dry season, with the driest month having precipitation less than 60 mm and also less than (100 - [total annual precipitation {mm}/25]). Bangalore, India and Townsville, Australia are good examples.

Group B: Dry (arid and semiarid) climates

In general terms, the climate of a locale or region is said to be arid when it is characterized by a severe lack of available water, to the extent of hindering or even preventing the growth and development of plant and animal life. A semiarid climate or steppe climate generally describes climatic regions that receive low annual rainfall (250–500 mm or 10–20 inches). In these climates precipitation is less than potential evapotranspiration (PET). Potential evapotranspiration is a representation of the environmental demand for evapotranspiration and represents the evapotranspiration rate of a short green crop, completely shading the ground, of uniform height and with adequate water status in the soil profile. It is a reflection of the energy available

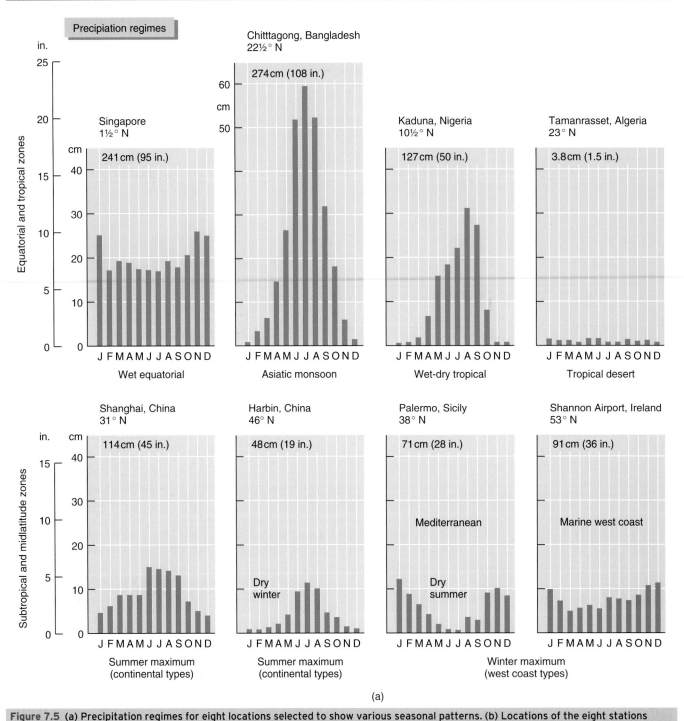

Figure 7.5 (a) Precipitation regimes for eight locations selected to show various seasonal patterns. (b) Locations of the eight stations around the world. (Source: Strahler and Strahler, 1997, pp. 180-181). Reprinted from Physical Geography, Strahler and Strahler, 1997, p. 180-81, with permission from John Wiley & Sons Ltd.

to evaporate water, and of the wind available to transport the water vapour from the ground up into the lower atmosphere. The precipitation threshold (in mm) is found as follows:

• Multiply the average annual temperature in °C by 20, then add 280 if 70% or more of the total precipitation is in the high-Sun half of the year (April through September in the

northern hemisphere; October through March in the southern hemisphere), 140 if 30–70% of the total precipitation is received during the aforementioned period, or 0 if less than 30% of the total precipitation is so received.

• If the annual precipitation is less than half the threshold for Group B, it is classified as BW (desert climate); if it is less than

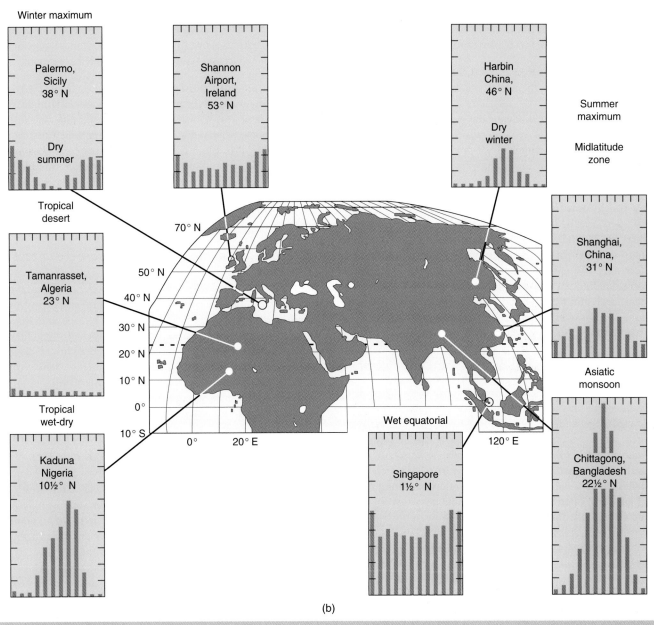

(b)

Figure 7.5 (continued)

the threshold but more than half the threshold, it is classified as BS (steppe climate).

- A third letter can be included to indicate temperature. Originally, *h* signified low-latitude climate (average annual temperature above 18 °C), while *k* signified middle-latitude climate (average annual temperature below 18 °C), but the more common practice today (especially in the United States) is to use *h* to mean that the coldest month has an average temperature that is above 0 °C, with *k* denoting that at least one month averages below 0 °C.

Places with examples of this type of climate include Cobar, Australia (*BSh*) and Murcia, Spain (*BSh*).

Group C: Temperate/mesothermal climates

These climates have an average temperature above 10 °C in their warmest months, and a coldest-month average between −3 and 18 °C. The second letter indicates the precipitation pattern: *w* indicates dry winters (driest winter month average precipitation less than 10% of the wettest summer month average precipitation); *s* indicates dry summers (driest summer month less than 30 mm average precipitation and less than 33% of the wettest winter month precipitation); and *f* means significant precipitation in all seasons (neither above-mentioned set of conditions fulfilled).

The third letter indicates the degree of summer heat: *a* indicates warmest-month average temperature above 22 °C; *b* indicates warmest month average temperature below 22 °C, with at least four months averaging above 10 °C; while *c* means three or fewer months with mean temperatures above 10 °C.

Group C climates are subdivided as follows.

Mediterranean climates (Csa, Csb)

These usually occur on the western sides of continents between the latitudes of 30° and 45°. They are in the polar-front region in winter, and thus have moderate temperatures and changeable, rainy weather. Summers are generally hot and dry, due to the domination of the subtropical high-pressure systems. Palermo, Italy (*Csa*) and San Francisco, California (*Csb*) are examples.

Humid subtropical climates (Cfa, Cwa)

These usually occur in the interiors of continents, or on their east coasts, between the latitudes of 25° and 40° (46° N in Europe). Unlike the Mediterranean climates, the summers are humid, due to unstable tropical air masses or onshore trade winds. In eastern Asia, winters can be dry (and colder than other places at a corresponding latitude) because of the Siberian high-pressure system, and summers very wet due to monsoonal influence. Houston, Texas (*Cfa*) and Brisbane, Australia (*Cfa*) are examples, with both having uniform precipitation distribution.

Maritime temperate climates or oceanic climates (Cfb, Cwb)

These usually occur on the western sides of continents between the latitudes of 45° and 55°. They are typically situated immediately poleward of the Mediterranean climates, although in Australia this climate is found immediately poleward of the humid subtropical climate, and at a somewhat lower latitude. In Western Europe, this climate occurs in coastal areas up to 63° latitude. Such climates are dominated all year round by the polar front, leading to changeable, often overcast weather. Summers are cool due to cloud cover, but winters are milder than those of other climates in similar latitudes. Limoges, France (*Cfb*) and Bergen, Norway (*Cfb*) are examples, with both having uniform precipitation distribution.

Maritime subarctic climates or subpolar oceanic climates (Cfc)

These occur poleward of the maritime temperate climates, and are confined either to narrow coastal strips on the western poleward margins of the continents, or, especially in the northern hemisphere, to islands off such coasts. Punta Arenas, Chile (*Cfc*) and Reykjavík, Iceland (*Cfc*) are examples, both having uniform precipitation distribution.

Group D: Continental/microthermal climates

These climates have an average temperature above 10 °C in their warmest months, and a coldest-month average below −3 °C. They usually occur in the interiors of continents, or on their east coasts, north of latitude 40° N. In the southern hemisphere, Group D climates are extremely rare, due to the smaller land masses in the middle latitudes and the almost complete absence of land south of latitude 40° S, existing only in some highland locations in New Zealand, which have heavy winter snows.

The second and third letters are used as for Group C climates, while a third letter, *d*, indicates three or fewer months with mean temperatures above 10 °C and a coldest-month temperature below −38 °C. Group D climates are subdivided as follows:

Hot summer continental climates (Dfa, Dwa, Dsa)

These usually occur in the high 30s and low 40s in latitude, and in eastern Asia *Dwa* climates extend further south due to the influence of the Siberian high-pressure system, which also causes winters to be dry, while summers can be very wet because of monsoon circulation. Boston, Massachusetts is an example of a *Dfa* climate, with uniform precipitation distribution, while Seoul, South Korea is a *Dwa* example.

Dsa exists only at higher elevations adjacent to areas with Mediterranean climates.

Warm summer continental or hemiboreal climates (Dfb, Dwb, Dsb)

These occur immediately north of hot summer continental climates, generally in the high 40s and low 50s in latitude in North America, and also in central and eastern Europe and Russia, between the maritime temperate and continental subarctic climates, where they extend up to high 50s and even low 60 degrees latitude. Examples include Moncton, Canada (*Dfb* – uniform precipitation distribution) and Vladivostok, Russia (*Dwb*). *Dsb* arises from the same scenario as *Dsa*, but at even higher altitudes, and mainly in North America, since there the Mediterranean climates extend further poleward than in Eurasia.

Continental subarctic or boreal (taiga) climates (Dfc, Dwc, Dsc)

These occur poleward of the other Group D climates, mostly in the 50s N latitude, although they might occur as far north as 70° latitude. Anchorage, Alaska (*Dfc* – summer wetter than winter) and Mount Robson, Canada (*Dfc* – summer drier than winter) are examples. *Dsc*, like *Dsa* and *Dsb*, is confined exclusively to highland locations near areas that have Mediterranean climates, and is the rarest of the three as a still-higher altitude is needed to produce it.

Continental subarctic climates with extremely severe winters (Dfd, Dwd)

These ccur only in eastern Siberia. The names of some of the places that have such a climate, most notably Verkhoyansk and Oymyakon, have become veritable synonyms for extreme, severe winter cold.

Group E: Polar climates

Polar climates are characterized by average temperatures below 10 °C in all twelve months of the year.

Tundra climates (ET)

The warmest month has an average temperature between 0 °C and 10 °C. These climates occur on the northern edges of the North American and Eurasian landmasses, and on nearby islands. They also exist along the outer fringes of Antarctica. Iqaluit, Canada,

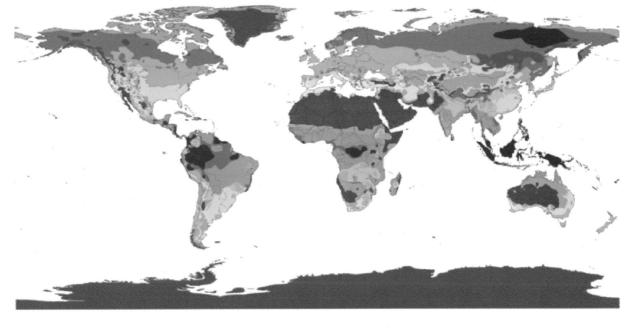

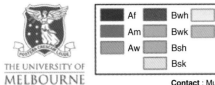

Af		Bwh		Csa		Cwa		Cfa		Dsa		Dwa		Dfa		ET
Am		Bwk		Csb		Cwb		Cfb		Dsb		Dwb		Dfb		EF
Aw		Bsh				Cwc		Cfc		Dsc		Dwc		Dfc	
		Bsk								Dsd		Dwd		Dfd	

DATA SOURCE : GHCN v2.0 station data Temperature (N = 4,844) and Precipitation (N = 12396)

PERIOD OF RECORD : All available

MIN LENGTH : ≥30 for each month

RESOLUTION : 0.1 degree lat/long

THE UNIVERSITY OF MELBOURNE

Contact : Murray C. Peel (mpeel@unimelb.edu.au) for further information

Figure 7.6 World map of the Köppen climate classification system. (Source: Dr Murray Peel, University of Melbourne). Reprinted with permission from Peel MC, Finlayson BL & McMahon TA (2007), Updated world map of the Köppen-Geiger climate classification, Hydrol. Earth Syst. Sci., 11, 1633–1644

Longyearbyen, Svalbard and Deception Island, Antarctica are examples.

ET is also found at high elevations outside the polar regions, above the tree line, as at Mount Washington, New Hampshire.

Ice cap climates (EF)

All twelve months of the year have average temperatures below 0 °C. These climates are dominant in Antarctica (e.g. Scott Base) and inner Greenland (e.g. Eismitte or North Ice).

See companion website for Box 7.1 Widening of the tropical belt in a changing climate.

7.2.4 Other climate classification systems

Some climatologists have argued that Köppen's system could be improved. The most common modification of the Köppen system today is that of the late University of Wisconsin geographer Glen Trewartha. Trewartha's system is based upon 23 climatic regions and attempts to redefine the broad climatic groups in such a way as to be closer to vegetational zoning, especially in the United States. Under the standard Köppen system, western Washington and Oregon are classed into the same climate as southern California, even though the regions have strikingly different vegetation. It also classes southern New England into

the same climate as the Gulf Coast. Trewartha's modifications seek to reclassify the Pacific Northwest seaboard as a different climate from California, and New England as different from the Gulf Coast. It adds a sixth highland group and provides an option to add a universal thermal scale.

In the 1930s and 1940s C.W. Thornthwaite proposed a classification system with a more quantitative approach, which was based upon 'effectiveness of precipitation', or the P/E Index. This was obtained by dividing the mean monthly precipitation of a place by the mean monthly evapotranspiration, and taking the sum of the 12 months. This resulted in a classification with 32 climatic regions. However, there were and still are difficulties measuring transpiration losses from vegetation.

In Britain, Austin Miller proposed a relatively simple classification system in the 1930s based on five latitudinal temperature zones, which he determined by using temperature thresholds: 21 °C (limit for growth of coconut palms), 10 °C (minimum for tree growth) and 6 °C (minimum for grass growth). He then subdivided these zones longitudinally based on the seasonal distribution of precipitation.

All these systems have advantages and disadvantages. They do not allow for local microclimate variations within each zone; some are too complex while others are too simplistic; they do not show transition zones, just arbitrarily placed lines; and they don't account for human influences like urbanization (e.g. urban heat islands), global warming and climate change. Indeed, as we shall see in Section 6, global warming is already causing climate zones to shift.

Exercises

1. Over what period of time must weather conditions be measured in order to determine a region's climate?

2. Explain how bands of cloudy and rainy weather near the equator are formed to create tropical conditions.

3. What is meant by the term 'global temperature regime'?

4. Complete the blanks: Locations closer to the sea (known as maritime locations) show _____ temperature curves, which have a _____ annual range. Locations inland (or continental locations) show _____ extremes of temperature and a _____ annual temperature range.

5. Which two aspects of precipitation are important when using it as an indicator of climate?

6. On what concept is the Köppen climate classification system based?

7. The Köppen climate classification scheme divides the climates into five main groups: name them.

8. By what techniques have scientists determined that the width of the tropical belt has increased since 1979?

9. On how many climatic regions is the Trewartha climatic classification system based?

10. On what measure is the Thornthwaite climatic classification system based?

References

Barry, R.G. and Chorley, R.J. 1987. Atmosphere, Weather and Climate. 5th edition. New York, Methuen.

Strahler, A.H. and Strahler, A. 1997. Physical Geography: Science and Systems of the Human Environment. Chichester, John Wiley & Sons, Inc.

Further reading

Barry, R.G., and Chorley, R.J. 2003. Atmosphere, Weather and Climate. 8th edition. London, Routledge. Chs 5–6.

Goudie, A. 2001. Nature of the Environment. 4th edition. London, Blackwell. Chs 4–7.

Holden, J. (Ed.). 2005. An Introduction to Physical Geography and the Environment. Harlow, Pearson Education. Ch. 3.

Smithson, P., Addison, K. and Atkinson, K. 2002. Fundamentals of the Physical Environment. 3rd edition. London, Routledge. Chs 24–28.

8 Ocean Structure and Circulation Patterns

Coastline of South Australia, near Adelaide. Photo: T. Stott

Earth Environments   David Huddart and Tim Stott
© 2010 John Wiley & Sons, Ltd

Learning Outcomes

After reading this chapter and completing the exercises you will be able to:

➤ Understand the relative importance of the ocean and atmospheric circulation systems in the transport of energy from the tropics to the poles.

➤ Understand the nature of stratification in oceans, the term thermocline, and how this varies with latitude and season.

➤ Describe the processes of upwelling and downwelling in ocean circulatory systems and explain the link between winds and ocean currents.

➤ Know what is meant by an ocean gyre and be able to locate examples on a map of the world's oceans.

➤ Understand the nature of deep water or thermohaline circulation and the term 'global ocean conveyor'.

8.1 Introduction

Oceans are a major influence on the climate of the Earth. In terms of surface area, they cover 71% of the Earth's surface and store 97% of the planet's water by volume. The circulation of the oceans is responsible for about half of the transport of energy from the tropics to the poles, the atmosphere transferring the other half. As we saw in Chapter 3, atmospheric circulation is driven by the heating of surface waters in the tropics and cooling at the poles. Cold surface currents travel equatorwards and warm surface currents travel polewards. It is therefore important that we have some understanding of the patterns of ocean circulation, particularly as these patterns could change in a warmer future, with important implications for the Earth's climate.

8.2 Physical structure of the oceans

In 1979 a submersible named *Alvin* descended to the Galapagos Rift, where scientists observed mounds of metal-rich sediments and black smokers (chimneylike vents discharging plumes of hot, black water). These vents form near spreading ridges (joins between two crustal tectonic plates beneath the oceans), where sea water seeps through cracks and fissures into the oceanic crust, gets heated by hot rocks and then rises and is discharged into the sea floor. Near these hydrothermal vents live communities of organisms which include bacteria, mussels, starfish, crabs and tubeworms, many never before identified. At such depths no sunlight is available, so rather than photosynthesis, bacteria practice chemosynthesis, in which they oxidize sulphur compounds from the hot vent waters, providing their own nutrients as well as a base for other members of the food chain.

8.2.1 Gathering information

The Deep Sea Drilling Project, an international programme in which the research vessel the *Glomar Challenger* was used to drill in water more than 6000 m deep, began in 1969. By 1983 the *Glomar Challenger* had drilled over 1000 holes into the sea floor, retrieving valuable information about geological structures and deep-sea sediments. The first measurements of ocean depths were made by lowering a weighted line to the sea floor, but nowadays echo sounders are used. Sound waves are reflected from the sea floor and detected by instruments on a ship, giving a continuous profile of the floor. By knowing the velocity of the sound waves in water, and their travel time, depth can be determined. Seismic profiling is similar, but is even more detailed than echo sounding. Strong waves are generated at an energy source, the waves penetrate the layers beneath the sea floor, and some of the energy is reflected from the various horizons back to the surface, thus giving information about the oceanic crust below the sea-floor sediments, as well as about the structure of the deep-sea sediments themselves.

8.2.2 Structure and composition of the oceanic crust

Most of the oceanic crust melts and is consumed back into the mantle at subduction zones (zones where one tectonic plate collides with another – see Chapter 13), but a small amount is not subducted. Small slices of oceanic crust along with deep-sea sediments and pieces of the upper mantle, called ophiolites, become emplaced in mountain ranges on land, usually by moving along large fractures known as thrust faults. An ophiolite consists of a layer of deep-sea sedimentary rocks underlain successively by a layer of pillow lavas and lava flows, and a complex of vertical basaltic dykes. Further down is gabbro and then rock representing the upper mantle. Thus, a complete ophiolite sequence consists of deep-sea sedimentary rock, oceanic crust and upper mantle rocks (Figure 8.1).

8.2.3 Sea-floor and continental margin topography

The sea-floor is as varied as the land surface, not flat as was once believed. Figure 8.2 is a map showing the topography of the ocean floors.

The abyssal plains are not completely flat plains but exhibit deep-sea mountains and long mid-oceanic ridges. The global topography of the sea floor only became known after special techniques using satellite measurements were made possible in the late 1990s. The Mariana Trench (or Mariana's Trench) is the deepest part of the world's oceans, and the deepest location on the surface of the Earth's crust. It has a maximum depth of about 11 km, and is located in the western North Pacific Ocean, to the east and south of the Mariana Islands, near Guam. The trench forms the boundary between two tectonic plates, where the Pacific Plate is subducted beneath the Philippine Plate. The

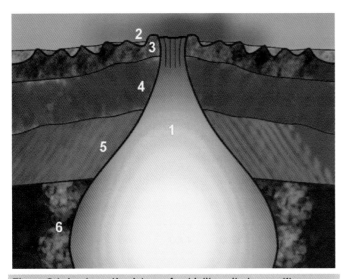

Figure 8.1 A schematic picture of ophiolite suite layers: (1) magma chamber; (2) sediments; (3) pillow basalts; (4) basaltic sheeted dyke swarms; (5) layered gabbro; (6) fractionation cumulates. New oceanic crust consisting of the layers shown here forms as magma rises beneath oceanic ridges. (Source: http://commons.wikimedia. org/wiki/File:Ophiolite_suite_scheme.jpg, Wikimedia Commons)

bottom of the trench is farther below sea level than Mount Everest is above it (8850 m).

Continental margins are zones separating a continent above sea-level from the deep-sea floor. They consist of a gently sloping continental shelf, a more steeply sloping continental slope, and in some cases a deeper, gently-sloping continental rise (Figure 8.3).

While continents are generally thought of as the land above present sea level, in fact the true geologic margin of continents (i.e. where continental crust changes to oceanic crust) is well below sea level, generally somewhere beneath the continental slope, so that marginal parts of continents are submerged.

The continental shelf is an area in which the sea floor slopes less than 1° seaward, whereas the sea floor slope is more like 0.1° and the continental slope is several degrees. On average, the oceans are more than 3800 m deep, though in some places the abyssal plains may be interrupted by peaks rising more than 1000 m. In general, however, they are the flattest, most featureless places on Earth, so deep that light does not penetrate and little is known about them other than data retrieved by echo-sounding and drilling. Abyssal plains are covered by fine-grained sediment derived from continental erosion. The sediment was deposited far from land by the settling of small particles in sufficient quantities to bury the otherwise rugged sea floor.

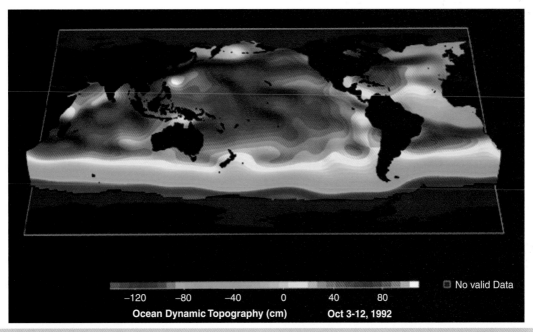

Figure 8.2 TOPEX/Poseidon was the first space mission that allowed scientists to map ocean topography with sufficient accuracy to study the large-scale current systems of the world's oceans. The total relief of ocean topography shown in this image is about 2 m. The colour scale corresponds to the grades of the relief in centimetres. The vertical scale is greatly exaggerated to illustrate the three-dimensional perspective of the topography. In this image, the maximum sea level (shown in white) is located in the western Pacific Ocean and the minimum sea level (indicated by magenta and dark blue) is shown around Antarctica. In the northern hemisphere, ocean currents flow clockwise around the highs of ocean topography and anticlockwise around the lows; this process is reversed in the southern hemisphere. These highs and lows are the oceanic counterparts of atmospheric circulation systems. While the basic structure of these ocean systems is constant, the details of the systems are constantly changing. Although this image was constructed from only 10 days of TOPEX/Poseidon data (3–12 October 1992), it reveals most of the current systems than have been identified by shipboard observations collected over the last 100 years. (Source: http://commons.wikimedia.org/wiki/File:Ocean_dynamic_topography.jpg, Wikimedia Commons)

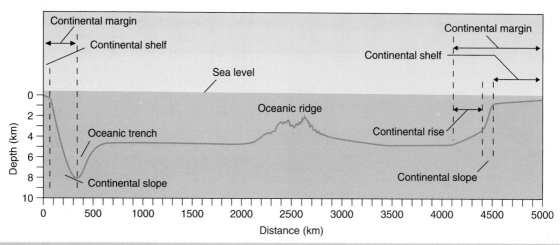

Figure 8.3 Profile of the sea floor with vertical exaggeration. (Source: after Dutch *et al.*, 1998)

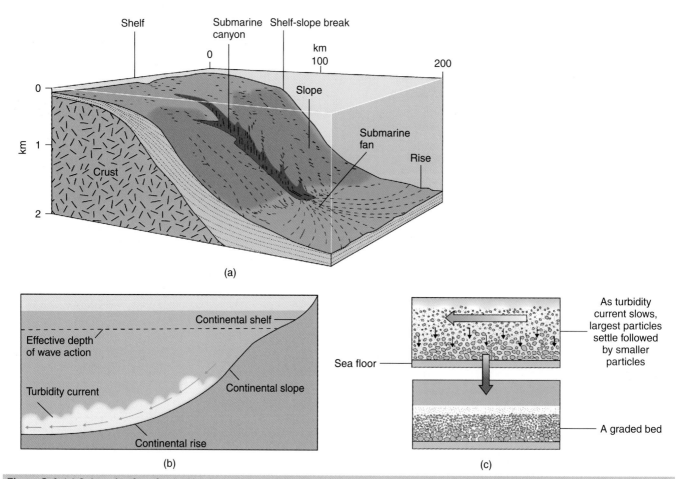

Figure 8.4 (a) Submarine fans formed by deposition of sediments carried down submarine canyons by turbidity currents. Much of the continental shelf rise is composed of overlapping submarine fans. (b) Turbidity currents flow downslope along the sea floor (or lake bottom), because of their density. (c) Graded bedding formed by deposition from a turbidity current. (Adapted from Dutch *et al.*, 1998)

8.2.4 Continental shelf break and deep-sea sediments

Between the shoreline and continental slope of all continents lies the continental shelf. Deep, steep-sided submarine canyons are frequently found on the continental slope (Figure 8.4), but some of them extend well up onto the continental shelf. Some of these canyons lie offshore from the mouths of large rivers. At times during the Pleistocene, when the sea level was up to 120 m lower than at present. Rivers flowed across these exposed shelves and eroded deep canyons, which were later flooded when the sea level rose.

Turbidity currents are sediment–water mixtures denser than normal sea water that flow downslope to the deep-sea floor, usually through submarine canyons. An individual turbidity current flows onto the relatively flat deep-ocean floor, where it slows and begins depositing sediment: the coarsest particles are deposited first, followed by progressively smaller ones, so forming graded bedding (Figure 8.4(c)). These deposits accumulate as a series of overlapping submarine fans, which constitute a large part of the continental rise (Figure 8.3 and Figure 8.4(a)).

In northern Europe and North America, glaciers extended onto the exposed shelves, creating glacially-formed troughs and depositing gravels, sands and mud (Figure 8.5).

That the continental shelf sediments were, in fact, deposited on land is indicated by the presence in them of human settlements and the fossils of mammoths and other land-dwelling animals.

Deep-sea sediments consist mainly of fine-grained deposition because there are few mechanisms by which coarser sediments can be transported far from land. One notable exception is the coarse grains that can be transported by icebergs, and a broad range of glacimarine sediments that have been deposited next to Antarctica and Greenland (see Figure 8.5).

Most of the fine-grained sediment in the deep sea is windblown (and cosmic) dust, volcanic ash and the shells of microscopic organisms that live in the upper ocean layers. Most of the deeper parts of the ocean basins are covered by brown or reddish clay derived from continents and ocean islands. Deposits resulting from chemical reactions in sea water also form an important constituent of ocean-floor sediments. These include nodules of manganese and iron oxides, and also contain copper, cobalt and nickel.

8.3 Temperature structure of the oceans

Oceans are layered or stratified based on temperature and salinity. Solar radiation flux at the ocean surface strongly affects the upper $100–200 \, m$, which has a temperature range of $0–30 \, °C$, with a global mean of $17 \, °C$ (Figure 8.6).

The term thermocline is given to a layer within a body of water in which the temperature changes rapidly with depth (see Figure 8.6). Since ocean water is not perfectly transparent, most sunlight is absorbed in the surface layer, which heats up. Wind and waves circulate the water in the surface layer, distributing heat within it so that the temperature may be quite uniform for the first $100–200 \, m$. Below this mixed layer, however, the temperature drops rapidly, perhaps as much as $20 \, °C$ with an additional 150 m of depth. Below this area of rapid transition (the thermocline) the temperature continues to drop with depth, but much more gradually. 90% of the water in the Earth's oceans is below the thermocline, and since the surface layer above is warmer and therefore more buoyant, mixing with the deeper layer is prevented. Deep oceans consist of layers of equal density, being poorly mixed, and may be as cold as $0–3 \, °C$. Temperature and salinity together control the density of ocean water. If the density increases with depth, the column of water will be stable, whereas if a dense layer overlies a less dense one, the situation will be unstable and vertical mixing will occur until the density is evened out. It is this vertical mixing that drives the three-dimensional circulation of the oceans.

The depth of the thermocline varies with latitude and season. In the tropics it is permanent, in temperate climates it is more variable (strongest during the summer) and in polar regions, where the water column is cold from the surface to the bottom, it is weak to nonexistent. Figure 8.7 shows how the warm surface layer disappears in Arctic and Antarctic latitudes.

In comparison to surface solar heating, geothermal heat flow through oceanic crust, even at mid-ocean ridges, contributes little to the thermal balance of the oceans.

8.4 Ocean circulation

The atmosphere and ocean are both fluids in which thermally-driven, circulatory systems develop. In a similar way to the atmosphere, we can envisage tropically-heated surface water moving poleward, cooling and then downwelling (sinking) to form lower, return currents which upwell (rise) at the equator. Since ocean currents move cold water towards the equator and warm water towards the poles, they are important regulators of the temperature of overlying air masses, which in turn affect our weather and climate. Warm ocean currents like the Gulf Stream keep coastal winter temperatures in the British Isles from falling below freezing for any length of time. The cool Peru or Humboldt Current keeps the west coast of South America cool in summer.

Friction between the wind and ocean surface is a transfer of energy from wind to water. Like air, the motion of water in the oceans is influenced by the Coriolis effect (see Chapter 5), which draws water to the right of its path in the northern hemisphere and to the left south of the equator. However, the shape of the world's ocean basins constrains the movement of ocean currents far more than the continental topography affects air movement. Also, ocean water can hold heat far longer than the atmosphere due to its higher specific heat capacity and slower-moving mass.

8.4.1 Surface ocean currents

An ocean current is a generally horizontal persistent flow of ocean water. Surface currents are classed as either warm (when they flow within the warm layer above the thermocline) or cold (when they

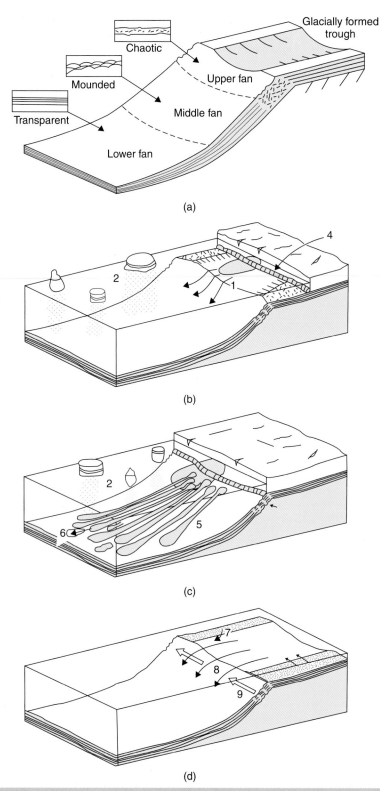

Key

1. Meltwater plumes from the ice sheet
2. Ice rafting from icebergs calved off the ice sheet
3. Sliding/slumping of material, driven by gravity
4. Sediment push and squeeze at the base of the ice sheet
5. Debris flows
6. Turbidity currents
7. Shallow bank winnowing
8. Outflow of cold, dense shelf water
9. Upper slope winnowing

Figure 8.5 Morphology and processes within glacigenic trough mouth fan systems. (a) Generalized diagram showing the seismic signature of sediments. Fan sedimentary processes operating under: (b) glacial advance; (c) glacial maximum; (d) glacial retreat. (Source: Siegert, 2001, p. 57). Reprinted from Ice sheets and Late Quaternary environmental change, Martin J. Siegert, 2001, p. 57, with permission from John Wiley & Sons Ltd

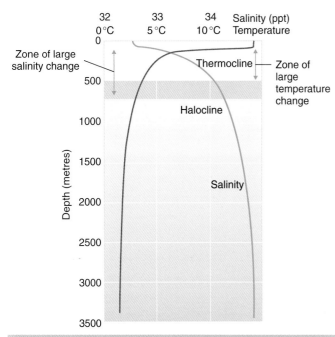

Figure 8.6 Vertical temperature and salinity profile in the ocean based on the NE Pacific. Note the fastest changes in the first 1 km. (Source: after Sverdrup *et al.*, 2003, reproduced with permission of McGraw Hill Companies. Source: Holden, 2005, p. 122). Reprinted from An introduction to Physical Geography and the Environment, Joe Holden, 2005, p. 122, with permission from Pearson Education

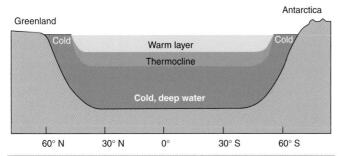

Figure 8.7 A schematic north-south cross-section of the world ocean, showing how the warm surface layer (scale exaggerated) disappears in Arctic and Antarctic latitudes, where very cold water lies at the surface. (Source: Strahler and Strahler, 1997, p. 136). Reprinted from Physical Geography, Strahler and Strahler, 1997, p. 136, with permission from John Wiley & Sons Ltd

are found in the cold surface layer in the Arctic or Antarctic; see Figure 8.7). Almost all large surface ocean currents are started by the prevailing surface winds, which impart a frictional force or wind stress on the ocean that is proportional to the square of the wind speed. This creates a film of surface waves, which overlies a more persistent, slower current some 50–100 m below the surface, moving at 3–5% of the wind speed, and deflected about 45° from the direction of the prevailing wind by the Coriolis force.

The seasonally-mobile, circulatory cells in the atmosphere, which were discussed in Chapter 5, are known as gyres in the

ocean. Figure 8.8 shows the general features of the world's ocean circulation. The circulation shows the movements of the large circular gyres that are centred around latitudes 30° N and S. The motions follow the movement of air around the subtropical high-pressure cells which we saw in Chapter 5. In the equatorial region, west-flowing, equatorial currents approach land and are turned poleward along the west sides of the oceans, giving warm currents which flow parallel to the coast. The Gulf Stream is one example. It flows up the east coast of North America, then joins the North Atlantic Drift to flow past the British Isles up into the Arctic, keeping the south and west coasts of Svalbard ice-free in summer, even though it is 78–80° N.

At 30–40° N in the Pacific, westerly currents approach the western shores of the continents and are deflected equatorward to give cool currents along the coast, which are often accompanied by upwelling along the continental margins. Here, colder water from greater depths rises to the surface. Examples include the Peru or Humboldt Current off the coast of Chile and Peru, which gives rise to one of the major upwelling systems of the world, supporting an extraordinary abundance of marine life. This is one of the world's largest fisheries and 18–20% of the world's fish catch comes from the Humboldt Current. The California Current, the Benguela Current off the west coast of South Africa and the Canaries Current off the north-west coast of Africa are further examples.

Other westerly currents also move poleward to join Arctic and Antarctic circulations such as the North Atlantic Drift. In the northern hemisphere, where the polar sea is more or less landlocked, cold currents flow equatorward, such as the Kamchatka Current, which flows along the Asian coast from Alaska, and the Labrador Current, which flows between Labrador and Greenland to reach the coasts of Newfoundland, Nova Scotia and New England. Another major feature of the global ocean circulation is the circumpolar currents which flow around Antarctica, with a westerly current below 60° S and an easterly current above 60° S.

See the companion website for Box 8.1 A new arrival through the Northwest Passage marks a change in the Arctic Ocean Circulation.

8.4.2 Deep-ocean currents

The large-scale deep circulation of the world's oceans is not driven by wind but by density variations, which are controlled by both salinity and temperature. Combining these two factors, the resulting circulation is called the thermohaline circulation. The principle is that when a body of water becomes denser than the water around it, it will sink. When water evaporates or freezes, it leaves behind its salt, making the remaining water more saline and therefore denser. At present the most important regions in which deep-water currents form are the North Atlantic/Arctic Ocean and the Antarctic Ocean (Figure 8.10), where water annually freezes into the ice caps, leaving behind a slightly saltier ocean, which is both very cold and very salty, and therefore sinks and spreads equatorwards.

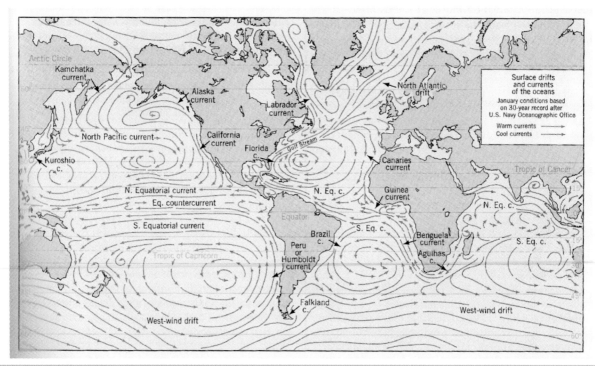

Figure 8.8 Surface drifts and currents of shallow ocean waters in January. (Based on data from US Navy Oceanographic Office, redrawn and revised by A.N. Strahler; source: Strahler and Strahler, 1997, p. 137). Reprinted from Physical Geography, Strahler and Strahler, 1997, p. 137, with permission from John Wiley & Sons Ltd

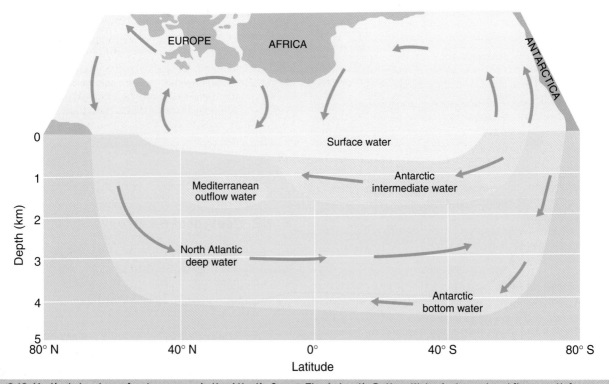

Figure 8.10 Vertical structure of water masses in the Atlantic Ocean. The Antarctic Bottom Water is densest and flows north from Antarctica. The North Atlantic Deep Water flows south from Greenland above the Antarctic Bottom Water. Above these are the intermediate water masses that are formed nearer to the equator. (Adapted from Holden, 2005). Reprinted from Earth's dynamic systems, W. Kenneth Hamblin, Eric H. Christiansen. 10th ed. 2004, p. 239, with permission from Pearson Education Ltd

The thermohaline circulation moves at speeds of 10–50 km per year and is isolated from the surface except where it forms in the North Atlantic/Arctic Ocean and the Antarctic Ocean. It has an average cycling time of 500–2000 years. As the warm currents approach the polar oceans, they start to cool rapidly. The North Atlantic Deep Water (NADW) flows south from Greenland above the Antarctic Bottom Water, penetrating to 60° S, between even denser, north-moving Antarctic Bottom Water (AABW) and less-dense Antarctic Intermediate Water (AAIW). AABW undercuts the NADW to about 40° N. However, this same movement cannot occur in the Indian Ocean, and although AABW also flows into the Pacific Ocean, there is no equivalent northern Pacific cold outflow. Less-dense Pacific and Indian Ocean Common Water (PICW) therefore completes the circulation into the Atlantic Ocean, rejoining the Gulf Stream. Subsidence of the NADW therefore contributes to a convective, density-driven system which draws a return surface flow of warm water into the North Atlantic. Since 1.4 billion km³ of ocean water is involved in this thermohaline circulation, it is a very slow process. The term Global Ocean Conveyor (Figure 8.11) is now used to describe it, particularly with reference to the implications that changes in this circulation might have for the Earth's climate.

8.5 Sea-level change

Geologists have observed an obvious cyclicity of sedimentary deposits and popular theories suggest that this cyclicity represents the response of depositional processes to the rise and fall of sea level. In the rock record, geologists see periods in which sea level was extremely low, alternating with periods in which it was much higher than today. These anomalies often appear worldwide. For example, in the middle of the last ice age 18 000 years ago, when hundreds of thousands of cubic kilometres of ice were lying on the continents as glaciers, sea level was 120 m lower than today. Places that supported coral reefs were left high and dry, and coastlines were many kilometres further seaward than the present-day coastline. Submerged forests are evidence for this (Figure 8.12).

During this time of very low sea level there was a dry land connection between Asia and Alaska, over which humans are believed to have migrated to North America (via the Bering Land Bridge), and there was no water in the English Channel, so that the British Isles were joined to continental Europe.

Over the past 6000 years the sea level has gradually risen to the level we see today. During the previous interglacial about 120 000 years ago, the sea level was for a short time about 6 m higher than today, as evidenced by wave-cut notches along cliffs and coral reefs.

The sea level has risen about 120 m since the peak of the last ice age about 18 000 years ago. From 3000 years ago to the start of the nineteenth century, the sea level was almost constant, rising at 0.1–0.2 mm per year. Since 1900 the level has risen at 1–2 mm per year; since 1993 satellite altimetry indicates a rate of rise of 3.1 ± 0.7 mm per year. Between 1870 and 2004 there

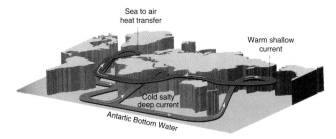

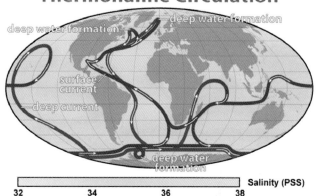

Thermohaline Circulation

Figure 8.11 The pattern of thermohaline circulation, also known as 'meridional overturning circulation' or the 'Global Ocean Conveyor'. This collection of currents is responsible for the large-scale exchange of water masses in the ocean, including providing oxygen to the deep ocean. The entire circulation pattern takes ~2000 years. Deep-water circulation (blue arrows) originates in the North Atlantic from the sinking of cold surface waters north of Iceland. This water flows southward at depth along the western side of the ocean basin and into the South Atlantic ocean. Along the shores of Antarctica, it is joined by more cold, sinking water and then flows eastward into the deep basins of the Indian and Pacific oceans. Upwelling returns some of this water to the surface, which warms as it crosses the equator. A warm surface current from the Pacific (red arrows) returns water to the North Atlantic. (Source: http://commons.wikimedia.org/wiki/File:Thermohaline_Circulation_2.png, Wikimedia Common)

Figure 8.12 Buried forest seen at low water at Dove Point, on the Cheshire coast, 1913 (England). Photo Clement Reid. (Source: Wikimedia Commons)

was a sea-level rise of 195 mm, giving a twentieth century rate of sea-level rise of 1.7 ± 0.3 mm per year; a significant acceleration of sea-level rise of 0.013 ± 0.006 mm per year. If this acceleration remains constant, the 1990–2100 rise will range from 280 to 340 mm. Sea-level rise might be a product of global warming through two main processes: thermal expansion of sea water and widespread melting of land ice. Global warming is predicted to cause significant rises in sea level over the course of the twenty-first century.

Exercises

1. Why is the Earth's ocean circulation important for climate and future climate change?

2. Describe how scientists have been able to map the structure and topography of the oceans.

3. What are ophiolites?

4. What is the average slope angle of the continental shelf compared to the continental slope?

5. Describe what is meant by the term 'turbidity current'.

6. With reference to the temperature structure of oceans, what is a thermocline?

7. What are ocean gyres?

8. Describe what is meant by the term 'thermohaline circulation'.

9. Describe the changes in sea level which have taken place since the middle of the last ice age 18 000 years ago.

10. Sea levels rise due to two main causes. What are they?

References

Dutch, S.I., Monroe, J.S. and Moran, J.M. 1998. Earth Science. Boston, International Thomson Publishing.

Holden, J. 2005. An Introduction to Physical Geography and the Environment. Harlow, Pearson Education.

Siegert, M.J. 2001. Ice Sheets and Late Quaternary Environmental Change. Chichester, John Wiley & Sons.

Strahler, A.H. and Strahler, A. 1997. Physical Geography: Science and Systems of the Human Environment. Chichester, John Wiley & Sons, Inc.

Sverdrup, K.A., Duxbury, A.C. and Duxbury, A.B. 2003. An Introduction to the World's Oceans. London, McGraw-Hill.

Further reading

Hamblin, W.K. and Christiansen, E.H. 2004. Earth's Dynamic Systems. 10th edition. Upper Saddle River, NJ, Prentice-Hall. Ch. 9.

Smithson, P., Addison, K. and Atkinson, K. 2008. Fundamentals of the Physical Environment. 4th edition. London, Routledge. Ch. 11.

Thurman, H.V. 2001. Introductory Oceanography. Upper Saddle River, NJ, Prentice-Hall.

9 Atmospheric Evolution and Climate Change

Ethabuka Reserve, Simpson Desert, SE Queensland, Australia. Photo: T. Stott

Earth Environments   David Huddart and Tim Stott
© 2010 John Wiley & Sons, Ltd

Table 9.1 Inventory of carbon near the Earth's surface (normalized). (Source: Henderson-Sellars, 1985)

Biosphere marine	1
Biosphere nonmarine	1
Atmosphere (in CO_2)	70
Ocean (in dissolved CO_2)	4000
Fossil fuels	800*
Shales	8 000 000
Carbonate rocks	2 000 000

*The release of CO_2 from fossil fuels back into the atmosphere over the 200–300 years since the start of the Industrial Revolution is causing concern today as it has been unequivocally linked with global warming

9.1 Evolution of the Earth's atmosphere

In the 4.6 billion years since the Earth's formation there have been some traumatic upheavals affecting the evolution of the atmosphere. Scientists believe that the Earth's atmosphere contained only hydrogen and helium at the time of its formation 4.6 billion years ago. Some of the influences on the evolution of the Earth's atmosphere include: changes in solar luminosity, volcanic activity, the formation of the oceans, mountain building and continental drift, and the origin of life on the planet.

9.1.1 The early atmosphere and the effect of volcanic eruptions

It is generally agreed that the Earth and other planets near the Sun (the terrestrial planets – Mercury, Venus and Mars) lost any primordial atmosphere through solar heating early in their lives. The present atmospheres are thought to be a result of the loss of gas from the mantle, including both volcanic outgassing and the vaporization caused by the impact of meteorites. Basic or shield volcanoes, which produce fluid lava, also emit gases into the atmosphere when they erupt. After millions of years of such outgassing, the atmosphere's composition began to change. The composition of volcanic gaseous emissions and their percentage by volume included: water vapour (79%), carbon dioxide (12%), sulphur dioxide (7%), nitrogen (1%) and hydrogen (0.5%). Water vapour and carbon dioxide were abundant, but at this early stage oxygen was absent. Scientists believe that photosynthesis and photodissociation processes helped to achieve today's oxygen levels. However, the most important factor, by a long way, in differentiating the Earth's atmosphere from those of our neighbouring planets, was the mean global surface temperature at the time when the atmosphere began to form. This determined where the water went. On Venus the temperature was high enough for water to be kept in its vapour state – this resulted in a runaway greenhouse effect, which has made Venus a hot desert today. On

Mars, the much lower temperatures meant that the water couldn't even melt, so the atmosphere remained thin, with water trapped in frozen reservoirs below the surface. On Earth, however, the intermediate position resulted in temperatures which ensured condensation of water vapour released into the atmosphere, forming large oceans, permitting carbon dioxide solution and leading to the formation of sedimentary rocks as rain carried eroded material into the oceans. With carbon dioxide being removed from the atmosphere (see Table 9.1) the future surface-temperature evolution was determined.

Once the atmosphere was established, the global surface temperature was increased by the greenhouse effect (see Chapter 3), whereby the atmosphere acts as a blanket, trapping heat that would otherwise be radiated back into space. The amount of heat the greenhouse gases in the Earth's atmosphere retain depends on the types of gas and their proportions. Water is critically important because in its vapour form it is one of the most powerful greenhouse gases. It also controls temperature on the Earth, both by the greenhouse effect and by the extent to which snow and ice cover or clouds change the Earth's reflectivity (its albedo). These feedback mechanisms (see Chapter 1) have ensured that mean global temperature has stayed close to the present 15 °C, permitting liquid water to dominate the planet's surface and providing a suitable environment for life.

9.1.2 The rise of oxygen and the biological era

With increasing amounts of water vapour being emitted from volcanic eruptions, the vapour would probably have condensed and formed clouds. These in turn would have precipitated and formed bodies of water.

Oxygen increased in stages, first through photolysis (Figure 9.1) of water vapour and carbon dioxide by ultraviolet energy and perhaps lightning. This process produced oxygen for the early atmosphere before photosynthesis became dominant.

The photolysis of water vapour and carbon dioxide includes these reactions: Water molecules (H_2O) split to produce hydrogen

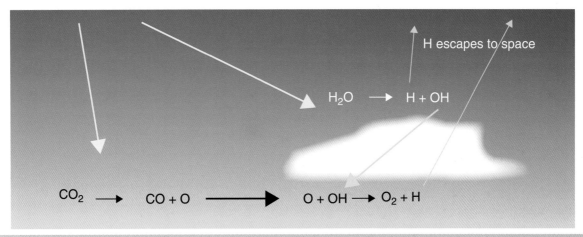

Figure 9.1 Photolysis of water vapour and carbon dioxide produces hydroxyl and atomic oxygen, respectively, which in turn produce oxygen in small concentrations. (Source: http://www.globalchange.umich.edu/globalchange1/current/lectures/samson/evolution_atm/#evolution, accessed 28 December 2007)

(H) and a hydroxyl radical (OH):

$$H_2O \rightarrow H + OH$$

Carbon dioxide molecules (CO_2) split to form carbon monoxide (CO) and an atomic oxygen (O):

$$CO_2 \rightarrow CO + O$$

The OH is very reactive and combines with the O:

$$O + OH \rightarrow O_2 + H$$

The hydrogen atoms formed in these reactions are light and a small proportion escape to space (Figure 9.1), allowing the O_2 to build up to a very low concentration, probably less than 1% of the oxygen in today's atmosphere.

The composition of the atmosphere as we know it now, however, needed oxygen to be present at much higher levels: levels sufficient to sustain life. Life was in fact needed to create these levels of oxygen. This era of evolution of the atmosphere is called the 'biological era'. The biological era was marked by a decrease in atmospheric carbon dioxide (CO_2) and at the same time an increase in oxygen (O_2) due to life processes starting to become established. Once life became established it played a big part in determining the chemical composition of the atmosphere. Primitive life originated, according to the most widely-accepted theories, in the subsurface layers of the early oceans, where it was protected from harmful ultraviolet radiation. To the first organisms, oxygen was a dangerous poison, a waste product produced from their life processes which dissolved in the ocean waters, slowly diffusing into the atmosphere above.

Photosynthesis has led to maintenance of the ~20% present-day level of O_2 and was first carried out by prokaryotes, which are molecules surrounded by a membrane and a cell wall that may have photosynthetic pigments, such as are found in cyanobacteria (also known as cyanophyta or blue-green algae). Such ancient

prokaryotes would have been the first life forms responsible for introducing the oxygen into the atmosphere.

At the same time that photosynthetic life began to decrease the carbon dioxide content of the atmosphere, it also started to produce oxygen. The oxygen did not build up in the atmosphere for a long time, since it was bound up by chemical reactions into the surface rocks, which could be easily oxidized (rusted). To the present day, most of the oxygen produced over time is locked up in the ancient 'banded rock' and 'red bed' rock formations found in ancient sedimentary rock. It was not until around one billion years ago that the reservoirs of oxidizable rock became saturated and the free oxygen stayed in the air. Once the level of oxygen had built up to about 0.2%, or ~1% of its present abundance, eukaryotic metabolism could begin. This must have occurred by approximately two billion years ago, according to the fossil record. Thus, the eukaryotes came about as a consequence of the long, steady, but less efficient earlier photosynthesis carried out by prokaryotes.

Once enough oxygen had accumulated in the atmosphere, it changed the rules for living organisms in two ways. First, it provided a new source of energy for life forms that were able to adapt to the process of respiration. Second, once enough oxygen had accumulated in the stratosphere, it was acted on by sunlight to form ozone (O_3), which filtered out the Sun's harmful ultraviolet radiation, making it feasible for life to move out of the oceans and colonize the land. The first evidence for vascular plant colonization of the land dates back to ~400 million years ago.

The availability of oxygen enabled a diversification of metabolic pathways, leading to a great increase in efficiency. Most oxygen formed once life began on the planet, principally through the process of photosynthesis:

$$6CO_2 + 6H_2O \leftrightarrow C_6H_{12}O_6 + 6O_2$$

where carbon dioxide and water vapour, in the presence of light, produce sugars and oxygen. The reaction can go either way, so in the case of respiration or decay the organic matter combines

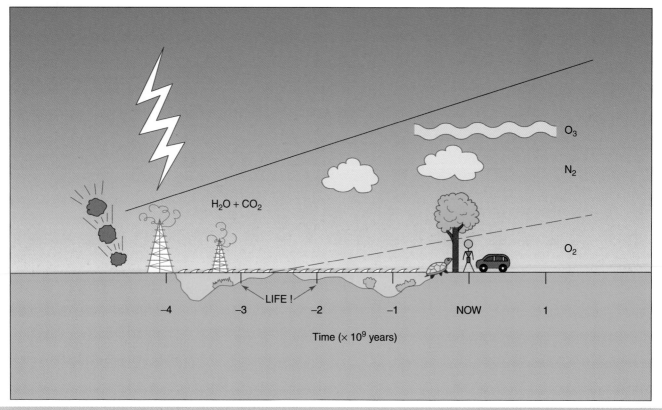

Figure 9.2 Schematic diagram of the history of the Earth's atmosphere. (Source: Henderson-Sellars, 1985)

with oxygen to form carbon dioxide and water vapour. Life started to have a major impact on the Earth's atmosphere once photosynthetic organisms had evolved. These organisms fed off atmospheric carbon dioxide and converted much of it into marine sediments, consisting of the innumerable shells and decomposed remnants of sea creatures.

The oxygen built up to today's value only after the colonization of land by green plants, which led to efficient and ubiquitous photosynthesis. The current level of 21% seems stable since significantly lower or higher levels would be damaging to life. If we had <15% oxygen, fires would not burn, yet at >25% oxygen even wet organic matter would burn freely. Figure 9.2 is a schematic diagram of the history of the Earth's atmosphere.

See companion website Box 9.1. The habitat and nature of early life.

9.1.3 Fluctuations in oxygen

The evolution of macroscopic life on Earth is divided into three great eras: the Palaeozoic, Mesozoic and Cenozoic. Each era is then divided into periods. The latter half of the Palaeozoic era includes the Devonian period, which ended around 360 million years ago, the Carboniferous period, which ended around 280 million years ago, and the Permian period, which ended around 250 million years ago. According to geochemical models, oxygen

levels are thought to have risen to a maximum of 35% and then declined to a low of 15% during a 120 million year period that ended in a mass extinction 250 million years ago at the end of the Permian. Such a dramatic change in oxygen levels would have had huge biological consequences, increasing diffusion-dependent processes such as respiration and thus allowing insects such as spiders, scorpions, centipedes and dragonflies to grow to very large sizes. Fossil records indicate, for example, that one species of dragonfly had a wingspan of 0.8 m!

9.1.4 The effect of humans on the evolution of the Earth's atmosphere

With oxygen levels at 21%, life could spread on a grand scale, which had a dramatic effect on the carbon dioxide content of the atmosphere. Sedimentary deposits buried dead plants and animals, the remains of which were rich in carbon. They solidified to form sedimentary rocks, particularly coal, limestone and shale, and oil reservoirs, resulting in the semi-permanent removal of vast quantities of carbon from the atmosphere. The reason this removal was only semi-permanent is because during the last few millennia, and the last few hundred years in particular, humans have become involved in extracting some of the carbon-rich material in the form of fossil fuels and building materials. The implications of this will be discussed in the following chapters.

Exercises

1. At the time of its formation 4.6 billion years ago, which two gases did the Earth's early atmosphere contain?

2. After millions of years of volcanic outgassing, the atmosphere's composition began to reflect the composition of volcanic gaseous emissions. Which two processes do scientists believe helped to achieve today's oxygen levels?

3. Until the recent industrial revolution in the last three centuries, CO_2 was systematically removed from the atmosphere. Where was this CO_2 being stored?

4. Describe the process of photolysis of water vapour and carbon dioxide.

5. What are prokaryotes? What is their importance in terms of present-day oxygen levels?

6. What is the significance of the ancient 'banded rock' and 'red bed' rock formations?

7. Once enough oxygen had accumulated in the atmosphere, it changed the rules for living organisms in which two ways?

8. What is the main effect that humans have had on the balance of atmospheric gases in the past millennium?

References

Henderson-Sellars, A. 1985. The evolution of the Earth's atmosphere. In: Fifield, R. (Ed.). The Making of the Earth. Blackwell and New Scientist.

Further reading

Henderson-Sellars, A. and Robinson, P. J. 1999. Contemporary Climatology. London, Longman.

10 Principles of Climate Change

Winter scene in north Wales, UK. Photo: T. Stott

Earth Environments   David Huddart and Tim Stott
© 2010 John Wiley & Sons, Ltd

Learning Outcomes

After reading this chapter and completing the exercises you will be able to:

➤ Understand the importance and limitations of instrumental records for understanding past climatic change.

➤ Understand what is meant by palaeoclimate and describe the range of proxy climate indicators which have been used to interpret past climate.

➤ Describe past climate fluctuations at a range of scales.

➤ Explain the external and internal factors that are generally agreed to force climate change.

➤ Understand about feedback mechanisms and how these are used to explain rapid climate change.

➤ Describe the role of the oceans in climate feedback and climate change.

10.1 Introduction

Our discussions in Section 2 have tended to focus largely on the Earth's present weather and climate. However, there is abundant evidence that the Earth's climate has rarely, if ever, been the same as it is today. Indeed, the day-to-day and seasonal variations in our climate make it quite difficult to even characterize it at present.

Given all the attention on climate change in the media, it is important that we are well informed about past climatic variations, since we base our knowledge on two things:

• Knowledge of how the climate has behaved in the past.

• Our ability to explain past climate changes.

We can find evidence of the Earth's past climates from the distant geological past to the most recent millennium. For example, coal deposits tell us of a tropical climate that existed in the Carboniferous period, while we can find signs that ice once existed in certain areas in the way it has shaped the landscape and the deposits it has left behind.

If we plot temperature values at one place through time (Figure 10.1), it is clear that values vary from year to year. The pattern may be entirely random, or values may oscillate between warmer and cooler periods, with no obvious long-term trend. Direct instrumental records really began about 200 years ago with the advent of modern instrument like thermometers, rain gauges and anemometers (see Chapter 6). While they provide a fairly accurate record, they have been around for too short a period to sample the whole range of climate changes that have taken place in the past, and that could occur again in the future.

10.2 Evidence for climate change

We will now examine the types of evidence that have been used to determine whether there have been changes in the Earth's climate over time and the extent of these changes, concentrating on the recent geological past. In the past, the positions of the continents on the Earth changed (see Section 3), and changes in climate may have been caused by this process rather than by changes in the Earth's atmosphere.

10.2.1 Proxy indicators for climate change

There is a range of evidence which can help to give scientists an insight into what climatic conditions prevailed in the past. The best type of evidence is where we have instrumental records at a particular site, so that we know the date, what weather elements were observed and how the observations were taken. However, when instruments were first used back in the seventeenth century, neither they nor their use was standardized, so it is difficult to compare values between locations, or over time.

Climate that predates the instrumental period of direct weather observations is known as palaeoclimate. Prior to the invention of instruments, many weather observations were recorded in documents such as weather diaries, which were often kept in monasteries, for example. Such diaries would normally note extreme weather events such as storms, floods and droughts, and other general observations, such as a good summer for the harvest or a wet winter. Other examples of documentary evidence include farming reports, ships' logs, grain price records and wine harvest times, but the information is often only qualitative, so for example we may know that a particular summer was wet, but not be able to tell how wet.

Prior to these written records, which span just a tiny fraction of the Earth's 4.6 billion year history, we have to depend upon proxy records or indirect data. There are many types of proxy indicator used by climatologists, as summarized in Table 10.1. They are often used in combination to build up a reliable picture of the past. Each has its merits and limitations.

Geological and geomorphological evidence is based on the assumption that specific lithologies and landforms can be identified that form under a known climatic type. In most cases these will be fossil landforms that no longer form under the present climatic conditions but may still provide evidence of a different climate in the past. Dramatic evidence comes from landscape features such as U-shaped valleys (Figure 10.2).

Landforms of glacial erosion and deposition are some of the best known indicators of environmental change. The parallel roads of Glen Roy, Scotland (Figure 10.3) indicate the former levels of an ice-dammed lake which occupied the glen at the end of the last ice age, 10 000–12 000 years ago.

When carefully mapped, glacial deposition features can reveal information about the thickness, areal extent and direction of flow of the former ice masses that produced these landforms.

However, the interpretations of past climate made from such landforms depend on increasingly complex models of what caused

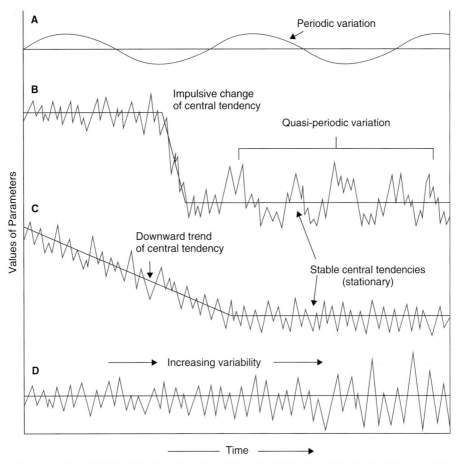

Figure 10.1 Idealized examples of changes in climatic records. (Source: Smithson *et al.*, 2002, p. 160). Reprinted from Fundamentals of the Physical Environment, Smithson, P., Addison, K., Atkinson, K, 2002, p. 160, with permission from Taylor and Francis

the proxy data and under what conditions. For example, glaciers form when snow accumulates at higher elevations, compacts to form ice, then flows downhill to eventually melt and/or evaporate. Present-day observations of this process are needed in order to calibrate a model that can predict how it happened in the past (this can sometimes be done using a mathematical model based on first principles). Therefore, proxy data must be calibrated against observations from instrumental data in order for a meaningful palaeoclimate interpretation to be made.

In addition to glacial landforms, which are associated with full ice cover, there may also be fossil features that have developed in periglacial conditions, such as the fossil ice wedge cast seen in Figure 10.4. The presence of sediments affected by cold climate is dramatically demonstrated by occasional fossil ice wedge casts. Repeated freezing and thawing at the surface of permafrost creates polygonal patterns of ice wedges. The soil is forced apart by the expansion of ice, and when it melts the sediments fall in from above to form the fossil ice wedge cast.

Dating of glacial landforms can be achieved by using plant fragments retrieved from terminal moraines, for example. The plant fragments are dated from radioactive carbon concentrations. Short-lived isotopes such as ^{14}C, which is formed in the atmosphere when cosmic rays strike air molecules and absorbed

in small quantities by all living things, gradually disappears when the organic material dies. Radiocarbon dating is useful for dating material up to 50 000 years old. Dating for much older material has to rely on other isotopes with longer decay times, such as uranium-238, which decays to uranium-234 and then thorium-230. Ratios of ^{230}Th to ^{234}U can be used to date back millions of years.

Evidence for annual or even seasonal variations in past climate can be gained by analysing isotopes from coral growth layers. The width or density of annual growth rings in trees is also a useful source of proxy data. Some records from species like the bristlecone pine of the western Unites States can go back thousands of years. The isotopic composition, chemistry, thickness of annual layers and gas composition trapped in air bubbles in ice cores can be used to reconstruct past temperatures. In 2006 a ten-country consortium extracted a 3.2 km long ice core from a region of Antarctica known as Dome Concordia (Dome C). It was drilled out by the European Project for Ice Coring in Antarctica (EPICA) and reveals the Earth's natural climate rhythm over the last 800 000 years, the longest time span yet analysed. Scientists concluded that whenever carbon dioxide levels changed there was always an accompanying climate change. Over the last 200 years human activity has increased carbon dioxide to well outside

Table 10.1 Proxy indicators of climate-related variables

Indicator	Property measured	Time resolution	Time span	Climate-related information obtained
Geological and geomorphological				
Sedimentary rocks	Layers, grain size, fossils	Centuries	Centuries to millennia	Prevailing wind direction, wind speed, aridity
Relict soils	Soil type, thickness of horizons, chemical analysis, fossil remains	Decades	Centuries to millennia	Temperature, rainfall, fire
Lake and bog sediments	Deposition rates from varves, species assemblages from shells and pollen, macrofossils, charcoal, artefacts, human remains	Annual	Millennia	Rainfall (floods and droughts), vegetation type, fire,
Aeolian deposits: loess, desert dust, sand dunes	Orientation, grain size	Centuries	Millennia	Prevailing wind direction, wind speed, aridity
Evaporites, tufas	Isotope dating	Decades	Millennia	Temperature, salinity
Speleothems	Isotope dating of stalactites and stalagmites	Decades	Millennia	Temperature
Ocean sediment cores	Diatoms, foraminifera, chironomids, ostracods, cladocera	Usually multiple decades or centuries	Millennia	Sea temperatures, salinity, acidity, ice volumes and sea levels, river outflows, aridity, land vegetation
Coastal landforms	Marine shorelines, raised beaches, submerged forests	Decades to centuries	Decades to millennia	Former sea levels
Glaciological				
Mountain glaciers and ice sheets	Glacial moraines, mapping location, dating (e.g. lichenometric dating)	Decades	Decades to millennia	Ice mass, volume and extent gives indication of temperature, precipitation
Periglacial features	Ice wedges	Decades to centuries	Decades to millennia	Temperature and precipitation
Ice cores	Oxygen isotopes, fractional melting, annual layer thickness, dust grain size/properties, trapped gas bubbles, insect remains	Annual	Millennia	Temperature, snow accumulation rate, wind, trapped gas concentrations
Old groundwater	Isotopes, noble gases	Centuries	Millennia	Temperature
Biological				
Tree rings	Width, density, isotopic ratios, trace elements	Annual	Centuries to millennia	Temperature, rainfall, fire
Lichens	Mean size (growth rate)	Decades	Centuries	Temperature, rainfall,
Coral growth rings	Density, isotope ratios, fluorescence	Annual	Centuries	Temperatures, gas concentrations
Plant and animal fossils	Species, relative abundance or absolute concentration	Decades	Millennia	Mass extinctions
Archaeological				
Written records	Reports of extreme harvests, floods, fires, droughts	Annual	Centuries to millennia	Temperature, precipitation, fire, drought

Table 10.1 *(continued)*

Indicator	Property measured	Time resolution	Time span	Climate-related information obtained
Plant and animal remains, including humans	Bone analysis, skin, hair, teeth, DNA, radiocarbon dating	Decades	Centuries to millennia	Various
Rock art	Cave paintings, tombs	Centuries	Centuries to millennia	Various
Ancient dwellings	Ground penetrating radar, electro-resistivity survey	Centuries	Centuries to millennia	Air temperature, precipitation, windiness
Artefacts	Bone, stone, wood, metal, shell, leather	Centuries	Centuries to millennia	Various

Figure 10.2 The U-shaped glacial valley above Lauterbrunnen, Switzerland in 2004. This valley was carved out by a large glacier during the last glacial period some 20 000 years ago. The glacier, visible in the distance, has now receded many kilometres up the valley in response to a rise in global mean temperatures of around just 4–5 °C in the last 10 000–20 000 years. The glacier is retreating faster now due to warming in the past few decades. This is dramatic evidence of past natural climate change, and of the potential impacts of future climate change due to human activities. (Photo: T. Stott)

Figure 10.3 The parallel roads of Glen Roy indicate the shore lines of a former ice-dammed lake, which existed at the end of the last ice age. (Source: http://commons.wikimedia.org/wiki/File: Parallel_Roads.JPG, Wikimedia Commons)

the natural range. Figure 10.5 shows an 84 million-year-old air bubble trapped in amber (fossilized tree sap). Using a quadrupole mass spectrometer, scientists can learn what the atmosphere was like when the dinosaurs roamed the Earth.

As well as the surface landforms, subsurface sediments can sometimes allow an interpretation of the environmental conditions under which they formed, indirectly giving an insight into the climate at that time. The thickness of layers and particle sizes in varves in lake- or ocean-bottom sediments can indicate past climate variations. Deep-ocean cores can span thousands

or even millions of years. On land, even where there are no geomorphological features, there are many parts of temperate latitudes covered by wind-blown sediments such as cover sands and loess. Analysis of thick sequences of loess indicates that they contain signs of cyclical climatic change, with phases of loess accumulation interbedded with phases of soil formation during warmer stages. The aeolian sediment may also give an indication of wind direction.

Much of the palaeoclimatic value of preserved sediments comes from the biological material they contain. Interpretation of this material is based on the principle of uniformitarianism. This principle assumes that the present is the key to the past, or in other words, if evidence for certain plant or animal species (normally fossils) is found preserved in ancient sediments, that the conditions under which they live today must have prevailed at that time and location in the past.

Figure 10.4 A fossil ice wedge cast developed in permafrost. (Source: http://commons.wikimedia.org/wiki/File:Permafrost_-_ice_wedge.jpg, NASA)

Figure 10.6 Pollen from a variety of common plants: sunflower (*Helianthus annuus*), morning glory (*Ipomoea purpurea*), hollyhock (*Sildalcea malviflora*), lily (*Lilium auratum*), primrose (*Oenothera fruticosa*) and castor bean (*Ricinus communis*). The image is magnified some 500 times, so the bean-shaped grain in the bottom-left corner is about 50 μm long. (Source: Dartmouth Electron Microscope Facility, Dartmouth College; available at http://commons.wikimedia.org/wiki/File:Misc_pollen.jpg, Wikimedia Commons)

Figure 10.5 84 million-year-old air bubble trapped in amber. (Source: USGS http://commons.wikimedia.org/wiki/File:Treesmed.gif, Wikimedia Commons)

of what is known about the patterns of woodland clearance and early farming practices has been obtained from pollen analytical evidence.

A range of other plants and animals also provide proxy climate information. These include insects such as beetles, diatoms (a member of the algae family, often found preserved in lakes) and ostracods, foraminifera and molluscs (found in marine sediments).

Reconstructing former climate on the basis of proxy evidence like this is complex and prone to error. If plants are used, the communities need to be established from pollen evidence, then climatic interpretations have to be made, errors being involved at both stages as climate is only one of a range of factors involved in a lifeform's existence.

10.2.2 Glacial periods

By studying many different proxy data sources from places around the world, scientists have found evidence of global-scale climate changes, from 'ice ages' or 'glacial periods', when huge ice sheets covered most of the Earth, to periods like the present, when ice is largely confined to the polar regions. Climate changes that have taken place over the past 2 million years, the Quaternary geological period, have a fairly regular cyclic manner, with glacial periods lasting around 100 000 years, and warmer interglacials of around 10 000 years occurring between them (Figure 10.7). These climatic fluctuations were, not surprisingly, accompanied

Although some organic material is visible to the naked eye, most is not. This means it has to be extracted by laboratory techniques and examined by microscope or more advanced techniques. The most widely used evidence comes from pollen from wind-pollinated plants. Because of their large size, the pollen of trees is usually the most abundant. Each species can be identified from its unique pollen grain (see Figure 10.6).

Pollen grains are produced by seed-producing plants and are spread over wide areas by wind, water, animals and insects. Pollen grains from the lower plants are all wind-blown. The grains become incorporated into peats and soils, and are often preserved in water-laid sediments, such as are found in lakes and bogs, wherever anaerobic conditions exist. Fossil pollen obtained from layered sequences of sediment provides a record of vegetational (and hence environmental) change through time. Much

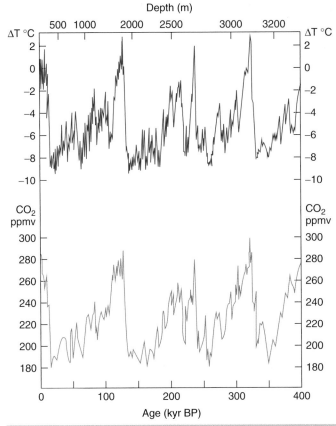

Figure 10.7 Temperature and CO_2 variations derived from the Vostok deep ice core in Antarctica show a close relationship over the past 400 000 years. (Adapted from Petit *et al.*, 1999; source: Smithson *et al.*, 2002, p. 174). Reprinted from Late Quaternary environmental change : physical and human perspectives, M. Bell, M.J.C. Walker. 2nd ed., 2005, p. 28, with permission from Pearson Education Ltd.

by large variations in global average sea level of up to 120 m (Figure 10.8).

Figure 10.8 indicates that about 18 000 years ago large parts of the northern hemisphere, especially Europe and North America, were ice-covered, with huge areas of continental shelf exposed. Notice how Alaska is joined to Siberia, mainland Australia is joined to New Guinea and Tasmania, and Britain is joined to Ireland and Europe. Sea surface temperatures estimated from foraminiferal remains and oxygen isotope analysis show major decreases in some areas, such as the north-east Atlantic, where the warm ocean current changed its position and temperatures dropped 10 °C. 18 000 years ago was the maximum of the last Devensian glaciation; after that time the ice sheets melted and contracted and the continental shelves were again slowly flooded over thousands of years.

10.2.3 The postglacial period

Climate amelioration after the last glacial maximum was rapid, though it was not without fluctuations (Figure 10.9). By 14 500 BP

(before present) ice in the Lake District, UK, was beginning to melt rapidly, the rate of warming being so sudden that in many cases the vegetation seems to have been out of phase with the climate, as indicated by comparing vegetation and insect evidence. By around 12 200 BP the climate of Britain is believed to have been similar to that of the present day, but soon there was another short phase of rapid cooling in north-west Europe and Russia. In the British uplands, corrie glaciers became reestablished, mean July temperatures fell below 10 °C and trees temporarily disappeared.

This brief period of around 1200 years, recently determined from Greenland ice cores to have been 12 900–11 700 BP, is known as the Loch Lomond stadial. It was too short for the extensive glacier growth of the previous glacial, but the piles of angular debris which can still be found in many British upland corries (termed cwm in Wales) show the deposition that took place when these small corrie glaciers melted (Figure 10.10).

Following the last retreat of the continental ice sheets from Europe and North America between 10 000 and 7000 BP, the climate warmed rapidly again until the climatic optimum of the present interglacial (known as the Hypsithermal in North America) occurred between 7500 and 4500 BP, when mean temperatures are believed to have been 1–2 °C warmer than at present. It was during this phase that Britain was cut off from continental Europe by the flooding of the English Channel. This thermal maximum was followed by general cooling and fluctuations in precipitation from around 6000 BP onwards. Interpretation of the terrestrial proxy records, however, becomes more problematic after this time since it is often difficult to disentangle the climatic signal from anthropogenic effects. By this time Neolithic nomadic hunter-gatherers were spreading north, and changes in forest composition, for example, which had previously been deemed an indicator of climate change, may now equally well reflect the activities of these early human groups. Within the general cooling there were cooler and wetter phases around 2000 BP, which began in the Iron Age, and warmer spells such as the Medieval Warm Phase around AD 800–1200 allowed Viking colonies to be established in Greenland and northern Newfoundland and raids to take place further south into Britain.

10.2.4 The Little Ice Age

Evidence from around the North Atlantic suggests a cool phase began in the fourteenth century, and it was around this time that some of the first instrumental records start to become available. For example, the mean annual temperature in central England in the 1690s was measured to be 8.1 °C, some 2.1 °C below the current mean. There is evidence that many farms in upland Britain were abandoned around this time due to the shorter growing season, while glaciers in the Alps advanced. Globally, the extent of ice and snow on land seems to have reached its peak in the early seventeenth century, the coolest since the Loch Lomond stadial around 12 000 BP. The timing of this seems to have been earlier in the United States than in Europe and the southern hemisphere.

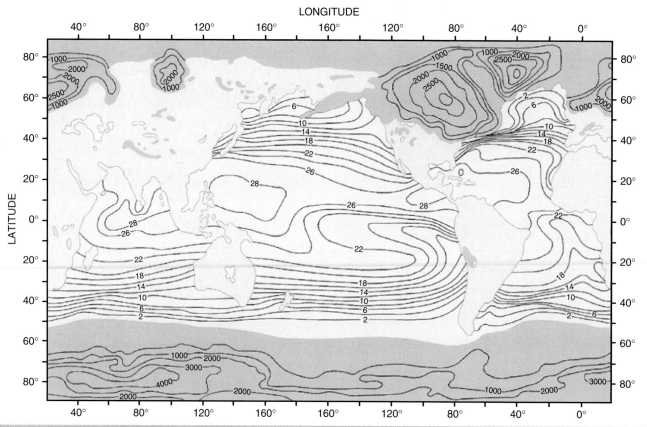

Figure 10.8 Sea surface temperatures, ice extent (blue shading) and ice elevation for southern winter 18 000 years ago, with sea level at −85 m. (After CLIMAP Project, 1976; source: Smithson *et al.*, 2002, p. 167)

The trend was not only towards colder and wetter summers, but also colder and wetter winters with increased storminess. During these severe winters the River Thames sometimes froze solid, allowing frost fairs to be held on the river.

10.2.5 The present climate

The effects of the Little Ice Age began to ease in most parts of the world by the middle of the nineteenth century, from which time there has been a steady warming, particularly in the twentieth century (Figure 10.11). The first phase of warming peaked in the 1940s, followed by a slight decline in global mean temperatures, when climatologists were predicting another 'Little Ice Age'. From the mid-1970s, however, the cooling trend reversed and mean temperatures rose rapidly through the 1980s, 1990s and into the twenty-first century. Concern is now firmly focussed on the effects of global warming and the extent to which anthropogenically-produced greenhouse gases are contributing to it.

However, while the global mean is rising, this can mask that fact that in some regions of the world temperatures are rising above this rate, while in others they are cooling. Rainfalls patterns vary too, and may or may not be linked to global mean temperatures. The most well-known example of a significant recent change in

rainfall is that of north Africa, where the political and social upheaval in countries like Ethiopia, Somalia, Niger and Sudan in the 1980s coincided with the driest year (1984) in almost a hundred years, with devastating consequences. We can see from the rainfall record (Figure 10.12) that rainfall in this region is variable, but since the late 1960s it has consistently been lower than the 1961–1990 average.

According to the Fourth Assessment Report of the Intergovernmental Panel on Climate Change (IPCC), which reported in 2007, warming of the climate system is unequivocal. This is now evident from observations of increases in global average air and ocean temperatures, widespread melting of snow and ice, and rising global average sea level (Figure 10.13).

Eleven of the twelve years (1995–2006) rank among the twelve warmest years in the instrumental record of global surface temperature (since 1850). The 100 year linear trend (1906–2005) of 0.74 (0.56–0.92) °C is larger than the corresponding trend of 0.6 (0.4–0.8) °C (1901–2000) given in the IPCC's Third Assessment Report in 2001. The temperature increase is widespread over the globe, and is greater at higher northern latitudes.

Land regions have generally warmed faster than the oceans. Rising sea level is consistent with warming (Figure 10.13(b)). Global average sea level has risen since 1961 at an average rate of

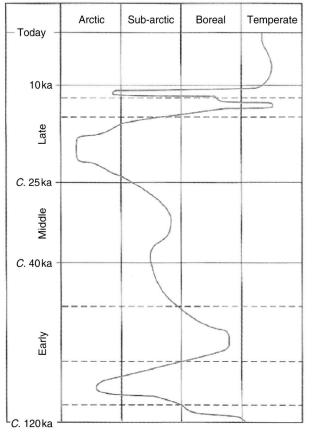

Figure 10.9 Estimated mean summer temperatures over the British Isles for the last 120 000 years. (After Jones and Keen, 1993; source: Smithson *et al.*

Figure 10.10 Mounds of glacial moraine and debris deposited during the Loch Lomond stadial at Cwm Silyn, Nantlle in western Snowdonia, Wales. (Photo: T. Stott).

1.8 (1.3–2.3) mm per year, and since 1993 at 3.1 (2.4–3.8) mm per year, with thermal expansion, melting glaciers and ice caps, and the polar ice sheets contributing.

Observed decreases in snow and ice extent are also consistent with warming (Figure 10.13(c)). Satellite data since 1978 show that annual average Arctic sea ice extent has shrunk by 2.7 (2.1–3.3)% per decade, with larger decreases in the summers of 7.4 (5.0–9.8)% per decade. Mountain glaciers and snow cover on average have declined in both hemispheres. From 1900 to 2005, precipitation increased significantly in eastern parts of North and South America, northern Europe and northern and central Asia, but declined in the Sahel, the Mediterranean, southern Africa and parts of southern Asia. Globally, the area affected by drought has likely increased since the 1970s. It is very likely that over the past 50 years, cold days, cold nights and frosts have become less frequent over most land areas, and hot days and hot nights have become more frequent. It is likely that heat waves have become more frequent over most land areas, the frequency of heavy precipitation events has increased over most areas, and since 1975 the incidence of extreme high sea level has increased worldwide.

There is observational evidence of an increase in intense tropical cyclone activity in the North Atlantic since about 1970,

with some evidence of increases elsewhere. There is no clear trend in the annual numbers of tropical cyclones and it is difficult to ascertain longer-term trends in cyclone activity, particularly prior to 1970. Average northern hemisphere temperatures during the second half of the twentieth century were very likely higher than during any other 50 year period in the last 500 years, and were likely the highest in at least the past 1300 years.

Facts like these demonstrate that our climate is still changing, and possibly at a rate that is speeding up. The debate seems to have shifted from questions like 'How much of this observed change is natural and how much is human-induced?' at the start of the last decade to 'How fast will future changes occur and what can humankind do about it?' now.

10.3 Causes of climate change

From the previous sections it is clear that climate has varied a great deal over the 4.6 billion years of the Earth's history. Humans have only been present on the Earth for a very small fraction of this time. If we were to compress the Earth's history into one calendar year, humans would have been around for about the

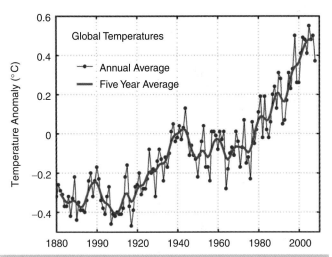

Figure 10.11 Instrumental record of global average temperatures as compiled by NASA's Goddard Institute for Space Studies. The data set used follows the methodology outlined by Hansen (2006). Following the common practice of the IPCC, the zero on this figure is the mean temperature from 1961 to 1990. This figure was originally prepared by Robert A. Rohde from publicly-available data and is incorporated into the Global Warming Art project. (Source: http://commons.wikimedia.org/wiki/File:Instrumental_Temperature _Record.png, Wikimedia Commons)

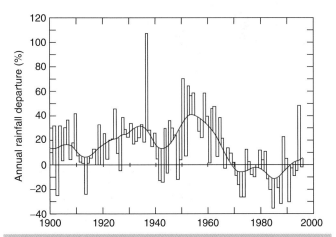

Figure 10.12 Annual rainfall departures (%) from the 1961-1990 average for the African Sahel (1900-1996) the smooth curve is the ten-year running mean. (After Mike Hulme, Climatic Research Unit, University of East Anglia; source: Smithson et al., 2002, p. 170)

last minute of that year. It is reasonable to conclude, therefore, that the climatic changes identified by the proxy data must be the result of natural processes acting on the Earth–atmosphere system, as, apart from the last few thousand years, they have occurred well before human activity was sufficient to have an impact on climate.

As we saw in Section 10.2, there are at least four different time scales over which climatic changes have taken place: glacial–interglacial (e.g. Devensian glaciation); stadial–interstadial (e.g.

Loch Lomond interstadial); postglacial fluctuations (e.g. Medieval Warm Phase, Little Ice Age); and the past few hundred years since instrumental records began (e.g. eleven of the twelve years (1995–2006) rank among the twelve warmest years in the instrumental record of global surface temperature (since 1850)). In looking for causes for these changes, we need to take into account that fact that the Earth's climate is the product of a complex interrelated set of systems: the atmosphere, lithosphere, hydrosphere and biosphere. Changes may be forced upon these systems due to alterations in the Sun's radiation or in the position of the Earth in relation to the Sun. There may be internal or external changes and there will be feedback mechanisms (see Chapter 1) which interact with the atmosphere to exacerbate or ameliorate change (Figure 10.14).

10.3.1 External factors

Variations in solar output

As we saw in Chapter 3, the basic source of all energy for driving the Earth's climate system is radiation from the Sun. So, it would be logical to assume that any variations in solar output would drive changes in the climate. Satellite observations of the solar beam intensity suggest small variations in the output which are only loosely related to the well-known eleven-year sunspot cycle. Estimates made over long periods suggest that variations of less than 1% in solar output due to sunspot activity or other natural changes are likely. A 1% increase in solar output could lead to a 0.6 °C increase in mean annual temperature, insufficient to explain the kinds of changes observed in the proxy climate record. All reconstructions of solar radiation changes over the twentieth century due to solar variability indicate that it was only about 20–25% of the total change in radiation at the Earth's surface. The bulk of the changes in radiation were due to increased greenhouse gases, which we shall examine later.

Changes in the Earth's orbit

Changes in solar radiation reaching the Earth's surface as a result of orbital variations were first established by James Croll, a British scientist, in 1867. His ideas were developed in 1920 by Milutin Milankovitch, a climatologist from Yugoslavia, whose name is usually linked with the theory. Three interacting variations are known to occur, involving regular changes in: (i) the shape of the Earth's orbit around the Sun; (ii) the tilt of the Earth's axis of rotation; (3) the time of year at which the Earth is closest to the Sun. At present the Earth's orbit around the Sun is approximately elliptical. The nearest point of this orbit to the Sun is known as the perihelion and is about 147.1 million km from the Sun, while the furthest point, the aphelion, is about 152 million km from the Sun. At present the perihelion occurs around 3 January, when a maximum of 1400 W m^{-2} of solar radiation is received at the outer limits of the Earth's atmosphere. The aphelion occurs around 5 July, when 1311 W m^{-2} of solar radiation is received. This difference, about 7%, means that if the Earth is at perihelion

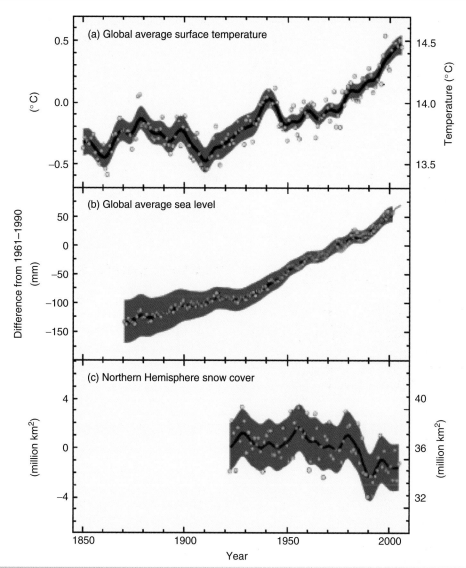

Figure 10.13 Observed changes in: (a) global average surface temperature; (b) global average sea level from tide gauge (blue) and satellite (red) data; (c) northern hemisphere snow cover for March–April. All differences are relative to corresponding averages for the period 1961–1990. Smoothed curves represent decadal averaged values while circles show yearly values. The shaded areas are the uncertainty intervals estimated from a comprehensive analysis of known uncertainties ((a) and (b)) and from the time series c). (Source: IPCC). Reprinted from Fundamentals of the Physical Environment, Smithson, P., Addison, K., Atkinson, K, 2002, p. 170, with permission from Taylor and Francis

during the northern hemisphere winter then it will receive more energy and therefore be warmer than if it was at aphelion at that time. The time of year of the perihelion does change over time. A complete cycle takes 22 000 years and is termed the precession of the equinoxes (Figure 10.15).

The degree of ellipticity of the Earth's orbit also changes over time, with a cycle of about 96 000 years, and this is known as the eccentricity of the orbit. At times the Earth's orbit is more circular, with little difference between perihelion and aphelion, while at other times it stretches to become more elliptical.

The final source of variation in the solar inputs is due to changes in the tilt of the Earth's axis of rotation. The axis of spin is tilted with respect to the axis of the Earth's orbit, the angle of

tilt varying between 21.8 and 24.2° (currently it is 23.5°), with a period of 41 000 years. One obvious question which arises is: when, on the Milankovitch theory, is the next ice age due? It so happens that we are currently in a period of relatively small solar radiation variation and the best projections for the long term are of a longer than normal interglacial period, leading to the beginning of another ice age in around 50 000 years' time.

As the Earth's orbit changes its relationship to the Sun, although the total quantity of solar radiation reaching the Earth varies very little, the distribution of that radiation with latitude and season over the Earth's surface changes considerably. More careful study of the relationship between the ice ages and the Earth's orbital variations shows that the size of the climate

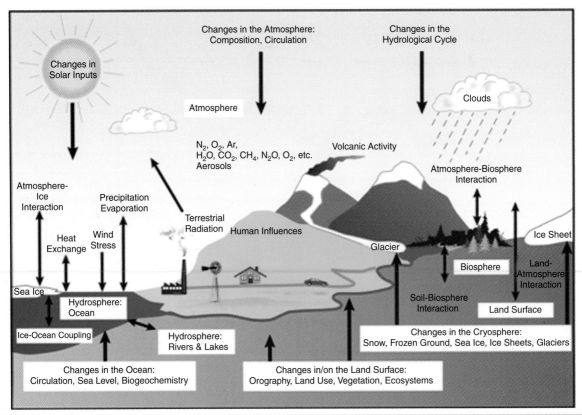

Figure 10.14 Physical processes and properties that control global climate changes. (Source: IPCC AR4, 2007 http://www.ipcc.ch/graphics/graphics/ar4-wg1/ppt/figure01.ppt#262,7,FAQ%201.2,%20Figure%201, accessed 4 January 2008)

change is larger than might be expected from forcing by the radiation changes alone. Other processes that enhance the effect of radiation changes, such as positive feedback processes (see Chapter 1), have to be introduced to explain the climate variations observed in ocean and ice cores, many of which are far too rapid to have solar output or orbital changes as their only driving cause. For example, ice cores from Greenland have confirmed that a temperature increase of 7 °C at the end of the Younger Dryas stadial approximately 12 700–11 500 BP could have taken only 10–20 years, with an accompanying increase in precipitation of 50%. Twenty-three such rapid climate fluctuations have been identified between 110 000 and 23 000 BP, now known as Dansgaard-Oeschger events. It seems highly unlikely that the orbital variations could have been responsible for such events, particularly as in the case described here the sudden warming preceded the orbit-induced event that was supposed to have caused it. To explain such changes, we must look to other mechanisms.

10.3.2 Internal forcing

Surface changes

Figure 10.14 showed the internal mechanisms that can also drive climate change. Some, such as the change in the distribution of land and sea (called the Earth's orography) brought about by

plate tectonics, are likely to operate only very slowly. The Earth's climate is likely to be very different if there were no mountain ranges like the Himalayas and Rockies, which interrupt and mix airflows. It has only been appreciated recently that the shape of ocean basins may affect ocean circulation, salinity levels and the interactions between surface and deep waters (see Chapter 8).

Changes in surface features other than the distribution of continents and oceans can have large effects on climate. For example, deforestation, brought about by humans or natural causes, can cause a sudden change to the Earth's surface, changing surface albedo and heat budget, which could affect climate. Clearance of the Earth's tropical rainforests, sometimes called 'the lungs of the Earth' is believed to have the potential to modify climate. Some model estimates predict that a change of Brazilian rainforests to savannah would cause a decrease in evapotranspiration of up to 40%, with runoff increasing from 14% of the rainfall to 43% and an average increase in soil temperature from 27 °C to 32 °C. Some of the effects of clearfelling are summarized in Figure 10.16.

Degradation of vegetation has also been cited as a factor contributing to climate change in the Sahel region of northern Africa.

When snow or ice melts at the Earth's surface, its albedo (see Chapter 3, Section 3.2) will change from a relatively high figure such as 90% to a much lower one such as 20%; as a result, less solar radiation will be reflected from the surface and

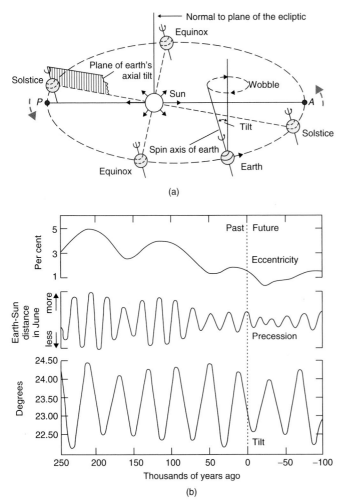

Figure 10.15 (a) Geometry of the Sun–Earth system, showing the factors causing variation in solar radiation receipt by the Earth. (b) Changes in eccentricity, tilt and precession for the last 250 000 years and the next 100 000 years. (After Imbrie and Imbrie, 1979; source: Smithson *et al.*, 2002, p. 172)

more will be absorbed, warming the surface and the surrounding atmosphere. Where previously any heat available would have been used to melt the snow or ice, now it can warm the atmosphere instead, producing a marked increase in surface air temperatures. Conversely, a cold year could result in a lot more snow and ice lying on the surface, which would trigger a positive feedback to enhance cooling.

Changes in atmospheric composition: Volcanoes, cosmic collisions and aerosols

When Mount Tamboro in Indonesia erupted in 1815 it was the largest eruption on the Earth since that of Lake Taupo in AD 181. The explosion was heard on Sumatra (more than 2000 km away). Heavy volcanic ash falls occurred as far away as Borneo, Sulawesi, Java and Maluku. The death toll was at least 71 000 people, of which 11 000–12 000 were killed directly by the eruption. It created global climate anomalies: the following year, 1816, became known as the 'Year without a Summer' because of the effect on North American and European weather systems. Agricultural crops failed and livestock died in much of the northern hemisphere, resulting in the worst famine of the nineteenth century.

The most recent major volcanic eruption was Mount Pinatubo in the Philippines in 1991, which, like Tamboro, Krakatoa (1883) and many others, injected large quantities of dust particles and reactive gases, such as sulphur dioxide and hydrogen sulphide, into the lower and upper atmosphere. The particles in the lower atmosphere, the troposphere, are generally washed out by rain within a few weeks or months, but those which enter the upper atmosphere, the stratosphere, can remain for many years, resulting in warming of a few degrees centigrade in the stratosphere and cooling of up to 1 °C at the Earth's surface. These effects, however, might only last a few years, so unless there were a series of major volcanic eruptions there would be unlikely to be a major effect on global climate. This effect may have contributed to the severity of the Little Ice Age, with major volcanic eruptions in 1750–1770 and 1810–1835.

As well as volcanic dust, the atmosphere contains dust blown up from arid parts of the Earth's surface. Occasionally falls of Saharan red dust blown from north Africa reach southern England, for example. In the 1930s, drought in the Great Plains of the central United States led to major soil losses through wind erosion, with the finer soil particles entering the atmosphere and being carried west as far as Washington and New York, with local impacts on climate.

Geological evidence such as a thin stratum of iridium-rich clay found at the Cretaceous–Tertiary geological boundary of 65 million years ago may have had an extraterrestrial origin, attributed to an asteroid or comet impact. This theory is widely accepted to explain the demise of the dinosaurs. Dust injected into the atmosphere following such a collision might have blocked solar radiation, killing plant life and the food source of dinosaurs. There are, however, at least a dozen other competing theories that have been offered to explain the decline of the dinosaurs. In the 1980s, the threat of a nuclear war inspired climatologists to examine the potential effects on climate. Again, the release of large clouds of smoke from burning cities was projected to have had an effect termed a 'nuclear winter'. However, this and major cosmic collisions are generally deemed highly unlikely over the next century, so are largely ignored in current climatic projections.

More relevant to current projections is the presence of large quantities of small particles in the lower atmosphere, commonly referred to as 'aerosols'. The main source of these is from the emission of sulphurous gases and soot particles from the burning of fossil fuels in industry and transport, though some come from natural dust blown up from arid parts of the Earth's surface. The continued injection of such aerosols into the atmosphere has led climatologists to coin the term global dimming. This is the gradual reduction in the amount of global direct radiation at the Earth's surface that was observed for several decades after the start of systematic measurements in 1950s. The effect varies by location, but worldwide it has been estimated to be of the order

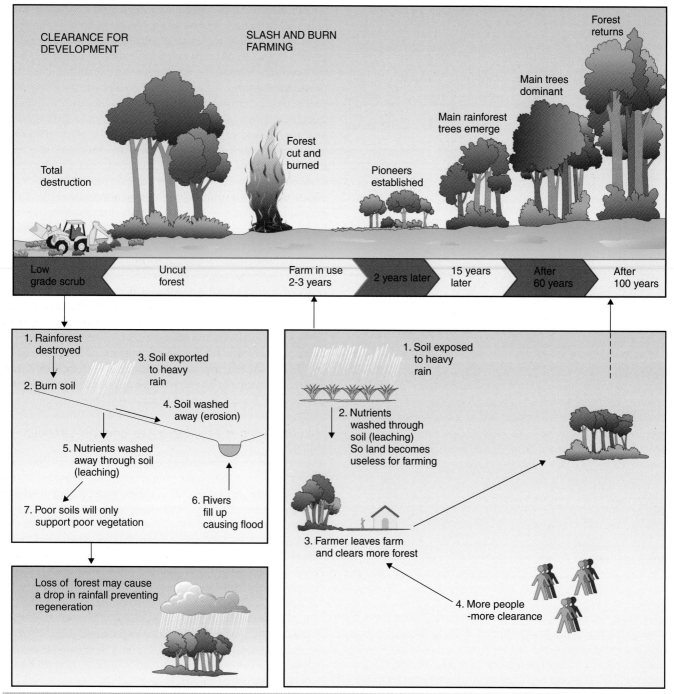

Figure 10.16 Some effects of the clearance of rainforest. (Source: after Dady, 1990). Reprinted from Fundamentals of the Physical Environment, Smithson, P., Addison, K., Atkinson, K, 2002, p. 172, with permission from Taylor and Francis

of a 4% reduction over the three decades from 1960 to 1990. The incomplete combustion of fossil fuels (such as diesel) and wood releases black carbon into the air. Though black carbon, most of which is soot, is an extremely small component of air pollution at land surface levels, the phenomenon has a significant heating effect on the atmosphere at altitudes above two kilometres. There, aerosols and other particulates are believed to absorb solar energy and reflect sunlight back into space. The pollutants can also become hygroscopic nuclei for cloud droplets (see Chapter 4). Increased pollution causes more particulates and thereby creates clouds consisting of a greater number of smaller droplets (i.e. the same amount of water is spread over more droplets). The smaller droplets make clouds more reflective, so more incoming sunlight is reflected back into space and less reaches the Earth's surface.

Figure 10.17 A NASA photograph of aircraft contrails, taken from I-95 in northern Virginia, 26 January 2001 by NASA scientist Louis Ngyyen. (Source: http://en.wikipedia.org/wiki/Image:Sfc.contrail.1.26.01.JPG, Wikimedia Commons, accessed 5 January 2008)

Clouds intercept heat from the Sun and heat radiated from the Earth. Their effects are complex and vary over time, location and altitude. During daytime, clouds mainly intercept sunlight, giving a cooling effect. However, at night the re-radiation of heat to the Earth slows the Earth's heat loss. Global dimming has interfered with the hydrological cycle. In the 1990s in Europe, Israel and North America, scientists noticed something that at the time was considered very strange: the rate of pan evaporation (evaporation rates measured using evaporation pans, see Chapter 4, Figure 4.8) was falling, where they had expected it to increase due to global warming. Many climate scientists believe the pan evaporation data to be the most convincing evidence of solar dimming. However, pan evaporation depends on other factors like vapour pressure deficit and wind speed as well as net radiation from the Sun. Ambient air temperature turns out to be a negligible factor. The pan evaporation data support the data gathered by radiometers. With adjustments to these factors, pan evaporation data have been compared to the results of climate simulations.

Some climate scientists have theorized that aircraft contrails (also called vapour trails) (Figure 10.17) are implicated in global dimming, but the constant flow of air traffic previously meant that this could not be tested.

Following the 11 September 2001 attacks in New York, all civil air traffic was grounded for three days. This unique occurrence afforded a rare opportunity for scientists to observe the climate of the United States without the constant addition of new contrails. During this three-day period, an increase in diurnal temperature variation of over 1 °C was observed in some parts of the USA, suggesting that aircraft contrails may have been raising night temperatures and/or lowering daytime temperatures by much more than previously thought.

A NASA-sponsored satellite-based study in 2007 shed light on scientists' observations that while the amount of sunlight reaching the Earth's surface had been steadily declining in previous decades, it had suddenly started to rebound around 1990. This switch from a 'global dimming' trend to a 'brightening' one seems to have coincided with the time at which global aerosol levels began to decline (Figure 10.18).

Most governments of developed nations have introduced Clean Air Acts since the 1950s, often to improve human health, particularly in cities. Global dimming, it seems, may have been creating a cooling effect that partially masked the effect of greenhouse gases and global warming. Solving the effects of global dimming may therefore lead to increases in predictions of future temperature rise.

Some scientists have suggested using aerosols to stave off the effects of global warming as an emergency measure. In 1974, Russian expert Mikhail Budyko suggested that if global warming became a problem, we could cool down the planet by burning sulphur in the stratosphere, which would create a haze. However, sulfates cause other environmental problems such as acid rain; carbon black causes human health problems; and dimming causes ecological problems such as changes in evaporation and rainfall patterns, with droughts and increased rainfall both causing problems for agriculture. Artificially increasing aerosol emissions to counteract global warming implies an ever-increasing amount of emissions in order to match the accumulated greenhouse gases in the atmosphere, with ever-increasing monetary and health costs. In essence, the sources of *both* greenhouse gases *and* air particulates must be addressed simultaneously.

Variations in greenhouse gas concentrations

Another aspect of atmospheric composition known to vary over time is the proportion of greenhouse gases (see Chapter 3). The most powerful of these include carbon dioxide (CO_2), methane (CH_4), nitrous oxide (N_2O), chlorofluorocarbons (CFCs) and water vapour (H_2O).

Carbon dioxide is one of the natural atmospheric gases that contribute to the greenhouse effect (see Section 3.2.6). We would expect an increase in the proportion of carbon dioxide to trap more long-wave radiation from the Earth's surface and increase the mean temperature of the planet. Temperature and CO_2 variations derived from the Vostok deep ice core in Antarctica (Figure 10.7) show a close relationship over the past 400 000 years, though it is not clear which increased first, or whether the changes are synchronous.

Since 1958, precise measurements of CO_2 levels have been taken at the Mauna Loa observatory on Hawaii (Figure 10.19). They show an increase from 315 ppm by volume in 1958 to about 379 ppm in 2005, which is largely the effect of fossil fuel burning, though deforestation and other land-use changes have had an impact (Figure 10.20).

Global atmospheric concentrations of CO_2, methane (CH_4) and nitrous oxide (N_2O) have increased markedly as a result of human activities since 1750 and now far exceed pre-industrial values determined from ice cores spanning many thousands of years. Atmospheric concentrations of CO_2 (379 ppm) and CH_4 (1774 ppb) in 2005 exceed the natural range over the last 650 000 years.

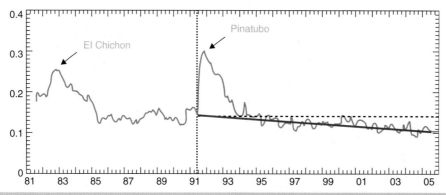

Figure 10.18 Sun-blocking aerosols around the world steadily declined (red line) following the 1991 eruption of Mount Pinatubo, according to satellite estimates. The decline appears to have brought an end to the 'global dimming' earlier in the century. (Source: Michael Mishchenko, NASA; available at: http://www.nasa.gov/centers/goddard/news/topstory/2007/aerosol_dimming.html, accessed 5 January 2008)

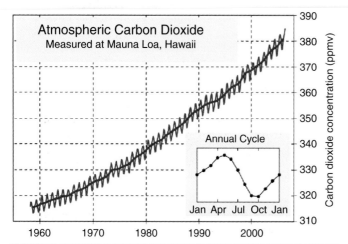

Figure 10.19 The history of atmospheric carbon dioxide concentrations as directly measured at Mauna Loa, Hawaii. This curve is known as the Keeling curve, and is an essential piece of evidence of the manmade increases in greenhouse gases that are believed to be the cause of global warming. The longest such record exists at Mauna Loa, but these measurements have been independently confirmed at many other sites around the world. (Source: http://www.globalwarmingart.com/wiki/Image:Mauna_Loa_Carbon_Dioxide.png, accessed 5 January 2008)

Although less publicized, methane is causing concern as it is a byproduct of both energy consumption and agricultural activity. The observed increase in CH_4 concentration is believed to be a result of anaerobic decomposition of organic matter from rice paddy cultivation and the digestive processes of ruminants such as cattle. As both rice area and ruminant numbers have increased with the rising human population over the last 200 years, the present annual increase is around 1% per year, compared with 0.5% for CO_2.

The remarkable similarity between late Quaternary atmospheric methane and temperature variations recorded in ice cores suggests that methane has played a significant role in past climate change. Abrupt global warming observed in ocean and ice cores suggest that changes in mean global temperature of several degrees centigrade in less than one human lifespan were possible during and at the end of the last ice age. Recent research indicates that a likely culprit is methane hydrate (clathrate), a vast reservoir of frozen methane in ocean sediments. Destabilization of the reservoir through changes in temperature and pressure changes brought about through sea-level changes may release the powerful greenhouse gas methane into the atmosphere, with drastic climatic consequences. The process has become known as the clathrate gun hypothesis.

The permafrost in the bogs of subarctic Sweden is undergoing dramatic changes due to recent global warming. The part of the soil that thaws in the summer, called the active layer, is becoming thicker in many areas, and in some places the permafrost has disappeared altogether. This has led to significant changes in the vegetation and to a subsequent increase in emissions of methane, which is released from the breakdown of plant material under wet soil conditions. At Stordalen Mire in northern Sweden, researchers from Lund University have been able to estimate an increase in methane emissions of at least 20%, but maybe as much as 60%, from 1970 to 2000. As a greenhouse gas, methane is deemed to be 25 times more powerful than carbon dioxide, so this finding could be important.

Nitrous oxide (N_2O) is used as a common anaesthetic, known as laughing gas, and is another greenhouse gas. Its atmospheric concentration is around 0.3 ppm but is rising at around 0.25% per year and is already 16% higher than in pre-industrial times. It resides in the atmosphere for up to 115 years and comes from natural and agricultural ecosystems, with increased use of fertilizers being the main cause of the increase.

Chlorofluorocarbons (CFCs) are compounds containing chlorine, fluorine and carbon. They were formerly used as refrigerants, propellants and cleaning solvents but their use was banned by the 1989 Montreal Protocol, because of their effects on the ozone layer (see Chapter 2). CFCs are extremely stable molecules which gradually disperse throughout the atmosphere and are destroyed by ultraviolet radiation in the stratosphere, yielding free chlorine atoms. The highly-reactive chlorine reacts with ozone in a chain reaction that will continue to destroy ozone for several decades

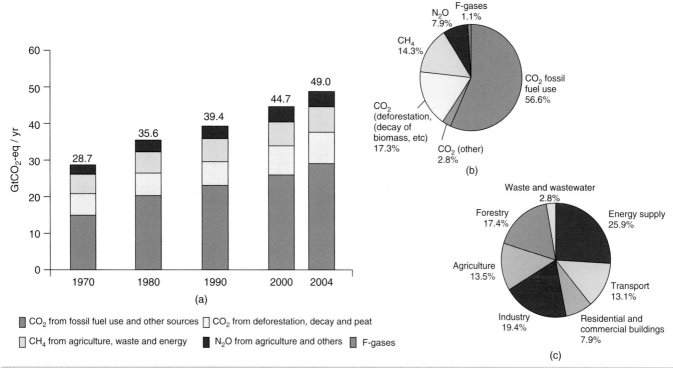

Figure 10.20 (a) Global annual emissions of anthropogenic greenhouse gases from 1970 to 2004. (b) Share of different anthropogenic greenhouse gases in total emissions in 2004 in terms of CO_2 equivalent. (c) Share of different sectors in total anthropogenic greenhouse gas emissions in 2004 in terms of CO_2 equivalent ('forestry' includes deforestation). (Source: IPCC, 2007, Figure 2.1). Reprinted from *Fundamentals of the Physical Environment*, Smithson, P., Addison, K., Atkinson, K, 2002, p. 174, with permission from Taylor and Francis

after the ban on CFCs. CFCs are also very effective absorbers of long-wave radiation, making them a powerful greenhouse gas.

Finally, direct warming of the atmosphere by waste heat also affects the temperature of the atmosphere, particularly in cities (see Chapter 6, the urban heat island effect). Estimates of global energy production have suggested that 8×106 MW are generated annually, most in densely-populated urban and industrial areas.

The Fourth (2007) Assessment Report of the Intergovernmental Panel on Climate Change concludes that most of the observed increase in globally-averaged temperatures since the mid-twentieth century is very likely due to the observed increase in anthropogenic greenhouse gas concentrations. It is likely there has been significant anthropogenic warming over the past 50 years averaged over each continent (except Antarctica). The model simulations presented in Figure 10.21 show that models which use natural forcings (due to solar activity and volcanoes) only (blue-shaded lines in the figure) are unable to predict temperatures observed in the 1906–2005 100 year period, whereas those incorporating anthropogenic forcings (greenhouse gas production) do predict the observed temperature trend quite well.

10.3.3 Feedback effects

It is clear that the external factors and internal forcing effects discussed above all have the potential to cause climate change. However, due to complex interactions with the other Earth systems – the biosphere, the hydrosphere and the lithosphere – the results are not always straightforward. Indeed, feedback mechanisms are likely to have been responsible for the more rapid changes in climate that we have seen earlier in this section.

Positive feedback leads to bigger and more dramatic changes. Small changes in the environment might lead to large adjustments in the system, which could explain why climate can change so abruptly without any corresponding evidence of changes of the same magnitude in external factors. One mechanism proposed for how an ice age could start is given in Figure 10.22. A small lowering of temperature at the poles of say $1–2\,°C$ would delay the normal summer melting of the Arctic ice cap. Ice survives for longer and so the albedo of the surface stays higher, in turn reflecting more incoming short-wave radiation. The surface is not heated as much, so the ice survives longer, reflecting even more heat back to space. The cycle is self-perpetuating and once initiated, this positive feedback process would magnify the small initial decrease in summer temperature, which might have been brought about by an external factor.

In terms of the climate system, one of the most important feedbacks is concerned with water vapour. With a warmer atmosphere, more evaporation will occur from the ocean and wet surfaces on land, so a warmer atmosphere will be a wetter one with a higher vapour content. Since water vapour is a powerful greenhouse gas, on average a positive feedback result that is large enough, models predict, to more or less double the increase in global average

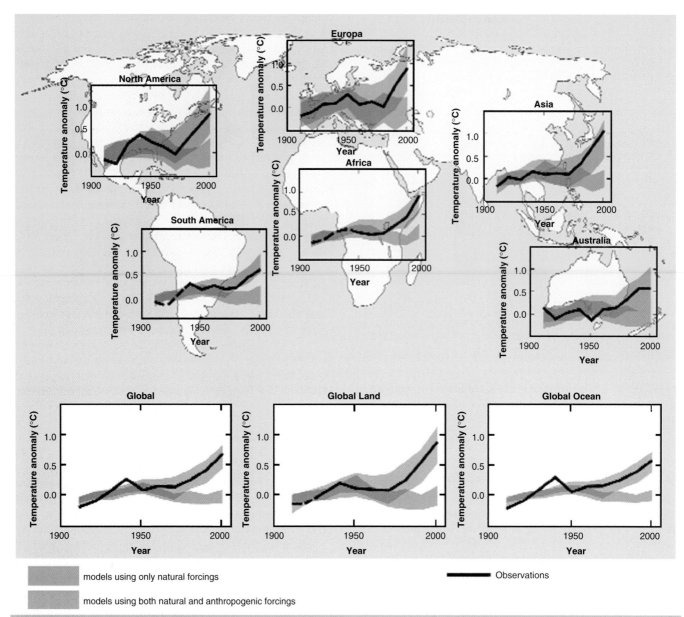

models using only natural forcings

models using both natural and anthropogenic forcings

Observations

Figure 10.21 Comparison of observed continental- and global-scale changes in surface temperature, with results simulated by climate models using either natural or both natural and anthropogenic forcings. Decadal averages of observations are shown for the period 1906–2005 (black line), plotted against the centre of the decade and relative to the corresponding average for the period 1901–1950. Lines are dashed where spatial coverage is less than 50%. Blue-shaded bands show the 5–95% range for 19 simulations from 5 climate models using only the natural forcings due to solar activity and volcanoes. Red-shaded bands show the 5–95% range for 58 simulations from 14 climate models using both natural and anthropogenic forcings. (Source: IPCC, 2007, Figure 2.5)

temperature that would arise if the water vapour remained fixed. This is a very important positive feedback in the climate system.

Climate is very sensitive to possible changes in cloud amount and structure. Clouds interfere with the transfer of radiation from the atmosphere in two ways (Figure 10.23). First, clouds reflect a certain proportion of solar radiation back to space, the amount depending on their albedo (colour, thickness). Second, they act as blankets to thermal radiation from the Earth's surface by absorbing thermal radiation passing from the Earth's surface

to space. They also re-emit some of this absorbed radiation, so reducing heat loss to space and acting in the same way as other greenhouse gases. The process that is most important depends on the cloud's temperature (high clouds are generally colder; low ones are warmer) and on its albedo (which will depend on whether it contains water droplets or ice particles, their size and density). For low clouds, the reflectivity effect wins, so they tend to cool the Earth–atmosphere system by reflecting away solar radiation. For high clouds, the blanketing effect is more dominant, and they

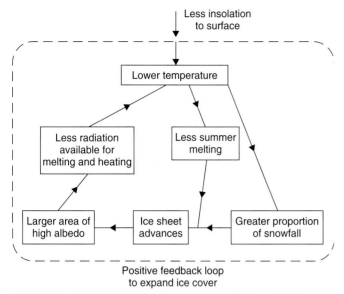

Figure 10.22 A positive feedback loop shows how a decrease in insolation and lower surface temperatures may result in further cooling and even lead to an ice age. (Source: Smithson *et al.*, 2002, p. 176)

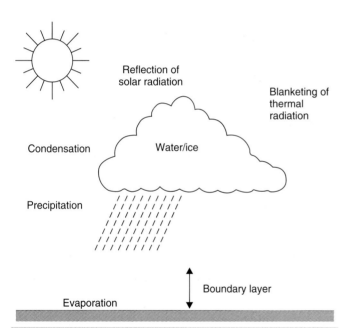

Figure 10.23 The physical processes associated with cloud radiation feedback. (Source: Haughton, 2004, p. 91)

tend to cause warming. The overall feedback effect of clouds can therefore be positive or negative.

The oceans also play a large part in determining the Earth's climate and cause climate change in three different ways. First, since the oceans cover 71% of the planet's surface, evaporation from the oceans provides the main source of atmospheric water vapour, which, through its latent heat of condensation in clouds, provides the largest single source of heat to the atmosphere. The

atmosphere causes wind stress on the ocean surface, which, in turn, drives the ocean circulation. Second, oceans have a large heat capacity compared with the atmosphere and continents. The entire heat capacity of the atmosphere is equivalent to only 3 m of the oceans. So oceans can absorb a vast amount of heat, and in a warming planet the oceans will warm up much more slowly than the atmosphere. They will have an ameliorating effect and exert a dominant control on the rate at which atmospheric changes occur. Third, the ocean circulatory system redistributes heat throughout the climate system, and it transports as much heat from the equator to the polar regions as the atmosphere. However, more heat is transferred to some regions than to others. For example, 20–30 times more heat is transferred by the Gulf Stream northwards between Greenland and Europe than is moved north by the North Pacific Current through the Bering Strait between Alaska and Siberia. The Gulf Stream brings as much heat to the British Isles and northern Europe as that region receives from solar radiation. If this ocean transfer of heat were to reduce, it would have huge implications for the climate in that part of the Earth. Therefore, any accurate simulation of future climate change would need to take ocean circulation changes into account.

Three feedbacks have been identified here, all of which play an important part in the determination of climate and future climate change. Not all parts of the climate system respond to change at the same rate. In general, the atmosphere will respond more rapidly to a forced change than will ice sheets or ocean currents. There may be a considerable lag time between a rise in air temperature and a melting response in glaciers and ice sheets. There is then likely to be a further lag before a response in the ocean circulation system, which may be triggered by additional fresh water changing the salinity and density of the oceans. On the other hand, while melting of ice sheets appears to be slow, there can be sudden breakups, as happened in 2002 when the northern section of the Larsen B ice shelf, a large floating ice mass on the eastern side of the Antarctic Peninsula, shattered and separated from the continent (Box 10.1, see the companion website). A total of about 3250 km^2 of shelf area disintegrated in a 35 day period in February and March 2002. This kind of observation raises the question of thresholds. However, a recent article in the *Journal of Glaciology* suggests that this breakup may not have been entirely the fault of global warming: it was weakened by cracks and had been teetering on the brink of breakup for decades. It was undermined not just by raised air temperatures but by changes in the atmosphere, sea and nearby glaciers.

The examples of the methane clathrate gun hypothesis and the increased rate of release of methane from permafrost regions may help to explain the abrupt global warming of several °C observed in ocean and ice cores in less than one human lifespan during and at the end of the last ice age. As we shall see in the next section, climatologists develop models to try to predict future climate, but given the concept of thresholds, which, once exceeded, give rise to extremely rapid change over relatively short time scales, the task of modelling future climate is a complex and challenging one. Some of the tipping points that could cause runaway warming or catastrophic sea level rise are discussed in Box 10.2 (see the companion website).

See also the companion web site for Box 10.1 Larsen Ice Shelf collapse : more than global warming? and Box 10.3. The current position of the recent Antarctic ice mass loss.

Exercises

1. Define the term 'palaeoclimate'.

2. What is the significance of the parallel roads of Glen Roy in Scotland?

3. Give examples of evidence for former ice ages that may still be visible in the landscape.

4. In 2006 a ten-country consortium extracted a 3.2 km long ice core from a region of Antarctica known as Dome Concordia. What was its significance in terms of past climate?

5. Interpretation of preserved sediments and the biological material they contain is based on the principle of uniformitarianism. What does this principle assume?

6. How can pollen grains be used to aid our interpretation of past climate?

7. When was the last Devensian glaciation maximum?

8. To what does the term 'Loch Lomond stadial' refer?

9. Approximately when did the 'climatic optimum' of the present interglacial occur?

10. What was the Little Ice Age?

11. Explain how the two external factors may be forcing climate change.

12. Explain how changes in the distribution of land and sea (called Earth's orography) could affect the Earth's climate.

13. How might changes in surface features like the distribution of forests, deserts and ice sheets have large effects on climate?

14. Discuss how volcanoes, cosmic collisions and aerosols can influence climate through a process called global dimming.

15. How can methane hydrate (clathrate), a vast reservoir of frozen methane in ocean sediments, explain abrupt changes in climate in the past?

16. Give an example of how a positive feedback process could magnify the small initial decrease in summer temperature (brought about by an external factor) to start an ice age.

References

CLIMAP Project. 1976. The surface of the ice-age Earth. *Science.* **191**, 1131–1137.

Dady, B.J. 1990. Environmental Issues. London, Hodder and Stoughton.

Hansen, J. *et al.* 2006. Global temperature change, *Proceedings of the National Academy of Sciences.* **103**, 14 288–14 293.

Houghton, J.T. 2004. Global Warming: The Complete Briefing. 3rd edition. Cambridge, Cambridge University Press.

Imbrie, J. and Imbrie, J.Z. 1979. Modeling the climatic response to orbital variations. *Science.* **29**(207:4434) 943–953. DOI: 10.1126/science.207.4434.943.

IPCC AR4. 2007. Climate Change 2007: Synthesis Report.

Jones, R. and Keen, D.H. 1993. Pleistocene Environments in the British Isles. London, Kluwer Academic Publishers.

Petit, R., Jouzel, J., Raynaud, D., Barkov, N.I., Barnola, J.M., Basile, I., Bender, M., Chappellaz, J., Davis, J., Delaygue, G., Delmotte, M., Kotlyakov, V.M., Legrand, M., Lipenkov, V., Lorius, C., Pépin, L., Ritz, C., Saltzman, E. and Stievenard, M. 1999. Climate and atmospheric history of the past 420 000 years from the Vostok ice core, Antarctica. *Nature.* **399**, 429–436.

Smithson, P., Addison, K. and Atkinson, K. 2008. Fundamentals of the Physical Environment. 4th edition. London, Routledge.

Further reading

Houghton, J. 2004. Global Warming: The Complete Briefing. 3rd edition. Cambridge, Cambridge University Press.

Kennett, J.P., Cannariato, K.G., Hendy, I.L. and Behl, R.J. 2003. Methane Hydrates in Quaternary Climate Change: The Clathrate Gun Hypothesis. Washington, DC, American Geophysical Union Special Publications Series Volume 54.

Pittock, A.B. 2007. Climate Change: Turning Up the Heat. Collingwood, Victoria. CSIRO Publishing.

SECTION III
Endogenic Geological Systems

11 Earth Materials: Mineralogy, Rocks and the Rock Cycle

Conglomerate bed, The Olgas near Uluru, Australia. Photo: T. Stott

Earth Environments   David Huddart and Tim Stott
© 2010 John Wiley & Sons, Ltd

Learning Outcomes

After reading this chapter and completing the exercises you will be able to:

➤ Understand the chemical makeup of the Earth.

➤ Understand about atoms, their structure and common types of bonds.

➤ Identify common minerals.

➤ Know the difference between a rock and a mineral.

➤ Understand about the major groups of rocks and how they are related through the rock cycle.

➤ Know what is meant by intrusive and extrusive vulcanicity and about the birth of igneous rocks.

➤ Understand the formation of sedimentary rocks and the characteristics by which they are classified.

➤ Know how fossils and sedimentary structures are formed.

➤ Understand the term 'metamorphic rock' and name some common examples.

11.1 What is a mineral?

In geology, a mineral is defined as a naturally-occurring crystalline chemical compound. Crystalline means that the compound has an ordered internal arrangement of atoms. All materials are made up of atoms, which when joined to other atoms in definite patterns and proportions make up chemical compounds. Minerals are simply the chemical compounds that make up rock, which is the main component of the Earth's crust on which we live.

A mineral may be practically anything of value extracted from the Earth's crust. Minerals are natural, homogeneous inorganic solids with crystalline atomic structures that have a definite and limited range of composition, expressible as a chemical formula. Minerals are characterized by certain physical properties such as hardness, density and colour. About 3500 minerals have been described and named, and more are being found all the time.

Minerals are named in a variety of ways, usually by their discoverer, who will most often name a new mineral after the place where it occurs, its chemical makeup or properties, or some person whom they wish to honour.

Minerals and rocks are essential resources upon which the human race depends. Iron, copper, gold, coal, oil, natural gas, building stone and many other resources are found in rocks, for example. Many gemstones such as diamonds, topaz and rubies are minerals that for various reasons we find attractive and valuable.

11.1.1 From what is the Earth made?

From studies of meteorites and materials brought to the surface by crustal movements and volcanoes, scientists have been able to estimate the chemical composition of the Earth. The Earth's interior consists of a mantle rich in silicon, oxygen, magnesium and iron, and a core of iron and nickel. As a whole, 35% of the Earth is oxygen, 24% iron, 17% silicon, 14% magnesium and 6% sulphur, with aluminum and calcium both around 1% (Figure 11.1(a)).

The Earth's crust has a different composition to its interior. For example, in Figure 11.1(b) iron makes up just 5.5%, but it is the second most abundant element in Figure 11.1(a), at 24%. Most of the iron is in the Earth's core. Oxygen (47%) and silicon (27%) make up almost three quarters of the continental crust, and together with aluminum, iron, calcium, sodium, potassium and magnesium they account for 99% of the continental crust (Figure 11.1(b)). Titanium, hydrogen, phosphorus and manganese account for most of the remaining 1%.

Some important resources such as aluminum and iron are listed in Figure 11.1(b), but others like copper, lead, zinc, gold, silver, platinum and uranium are not. Along with many other metals, these make up only a tiny proportion of the crust. They are mined in some localities because geologic processes have concentrated them to form ore deposits.

11.1.2 Atoms and their structure

The fundamental units that make up all familiar materials are small particles known as atoms (Figure 11.2). At the centre of an atom is a tiny nucleus, which is about 1/100 000 the diameter of the atom, but contains virtually all the atom's mass. The nucleus is made from protons (positive electrical charges) and neutrons (no charge, but almost the same mass as protons). Negatively-charged electrons orbit the nucleus. They are grouped into distinct electron shells and it is the electrons which control the atom's interaction with other atoms. The nucleus determines how many electrons an atom has, because the positive protons attract and hold the negative electrons in their orbit.

The number of protons in an atom's nucleus gives its atomic number, for example H = 1, C = 6, Fe = 26. Atoms are arranged according to their atomic number in the periodic table (Figure 11.3). Each chemical element, or simply element, is composed of a specific type of atom.

All the atoms of an element have the same atomic number, but can have varying numbers of neutrons. For example, oxygen always has 8 protons, but can have 8, 9 or even 10 neutrons. The element's atomic weight is found by adding together all its protons and neutrons. Atoms of an element with different numbers of neutrons are called isotopes. Isotopes of an element are chemically identical but may differ in other ways – some may be radioactive while others are stable. Radioactive isotopes are important in determining the age of rocks. Each element has a name and a one- or two-letter symbol, for example N = nitrogen, Si = silicon, Fe = iron.

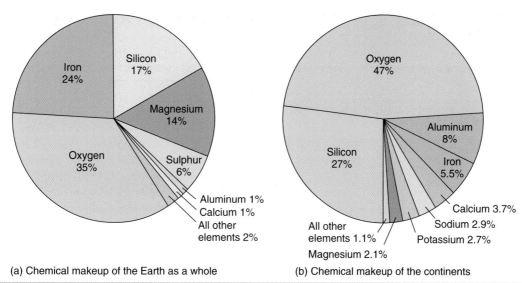

(a) Chemical makeup of the Earth as a whole

(b) Chemical makeup of the continents

Figure 11.1 (a) Chemical composition of the Earth as a whole. (b) Chemical composition of the continental crust. Note that iron comprises 24% of the Earth as a whole but only 5.5% of the continental crust. Most iron is contained in the Earth's core. The other elements that make up the remaining 1.1% of the continents are mainly titanium, hydrogen, phosphorus and manganese. Carbon, an essential element in all living organisms, makes up just 0.03%. (Source: Dutch *et al.*, 1998, p. 19)

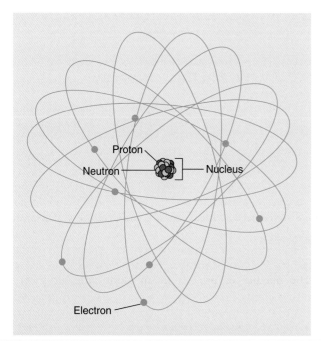

Figure 11.2 The structure of an atom. The dense nucleus of protons and neutrons is surrounded by a cloud of electrons arranged in electron shells. (Source: Dutch *et al.*, 1998, p. 19)

11.1.3 Atoms and bonding

While large bodies in the universe such as planets and stars are held together by gravity, smaller objects like atoms are held together by bonds. Bonds result from the interactions of the electron shells around atoms (Figure 11.4). Atoms can also bond by sharing outer electrons. The shared electrons orbit both atoms in what is called covalent bonding (Figure 11.5).

Most carbon-bearing compounds (including most in the human body) have covalent bonds. So does the hardest mineral known: diamond (Figure 11.6).

In metals, electrons wander freely in the space between the atoms, and the material is held together by metallic or van der Waals bonds. Because electrons move about freely, metals make good conductors of electricity and heat. The layers of atoms in metals are not bonded tightly together so can slip over each other easily, making metals malleable (easily deformed).

11.1.4 Identifying minerals

To identify minerals we need to be familiar with certain distinctive mineral properties.

Colour

Colour is the most obvious property, but some minerals have distinctive colours while others are white or transparent. Copper minerals are commonly bright green or blue. Quartz, on the other hand, is normally colourless but can be tinted any colour by small amounts of impurities. Often, the apparent colour is a surface coating, and the true colour is only revealed when the mineral is broken or scratched. This is called the streak on a streak plate.

The most important colouring material in minerals and rocks is iron. Ferric iron (Fe^{+++}) has lost three electrons and has a +3 charge. It produces Earth tones: red, brown or yellow. Ferrous iron (Fe^{++}) has lost two electrons and has a +2 charge. It gives the dark green or black colours of many silicate minerals and of rocks composed of these minerals. Manganese (pink), copper

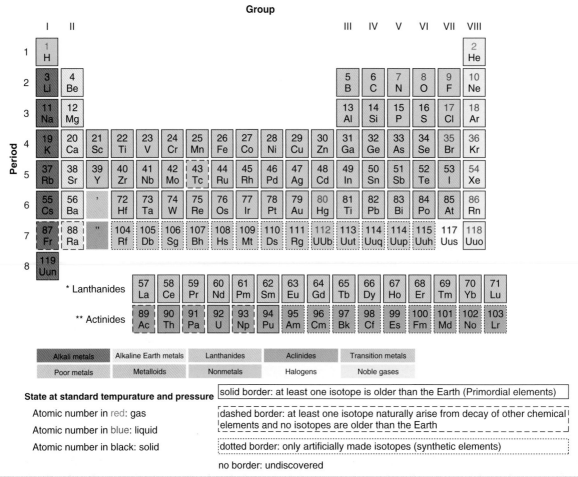

Figure 11.3 The periodic table of the elements. (Source: http://commons.wikimedia.org/wiki/File:Periodic_Table_Armtuk3.svg, Wikimedia Commons)

(blue, green) and chromium (pink, green) also act as colouring agents in minerals.

Lustre

Lustre refers to the way a mineral reflects light. Most are either metallic (shiny) or nonmetallic (dull). A variety of terms are used to describe the lustre of nonmetallic minerals: glassy/vitreous, waxy, resinous and silky.

Cleavage

Cleavage is the tendency for minerals to break along smooth planes. Cleavage occurs along planes of weakness between atoms and is one of the most consistent and distinctive properties of minerals (Figure 11.7).

Minerals known as micas (biotite and muscovite) cleave into thin sheets. Amphiboles and pyroxenes cleave into long splinters while halite (salt) and galena (lead ore) cleave into cubes. Fluorite and diamond exhibit cleavage in four directions, forming a double pyramid crystal. Although diamond is the hardest mineral, it still has excellent cleavage, and diamond cutters exploit this by cutting along diamonds' planes of weakness. Quartz is a common mineral that lacks cleavage, however.

Hardness

Hardness is another reliable property used to identify minerals. It is a direct measure of how tightly the atoms in a mineral are bonded together. The German mineralogist Friedrich Mohs devised a hardness scale by arranging 10 common minerals in order of relative hardness (Table 11.1).

For geologists, hardness is a measure of a mineral's resistance to scratching. They use everyday objects such as a fingernail (hardness 2.5), a copper coin (3), glass (5.5–6), a steel file and the head of a geology hammer (6.5) to help determine a mineral's hardness.

Density

Density is a further reliable property of minerals (Figure 11.8) because it is directly related to the weight of atoms in a mineral and how closely they are arranged (Figure 11.8). When we describe lead as 'heavier' than aluminum, we really mean it is denser. Density is mass per unit volume expressed as $g\,cm^{-3}$, or in terms of

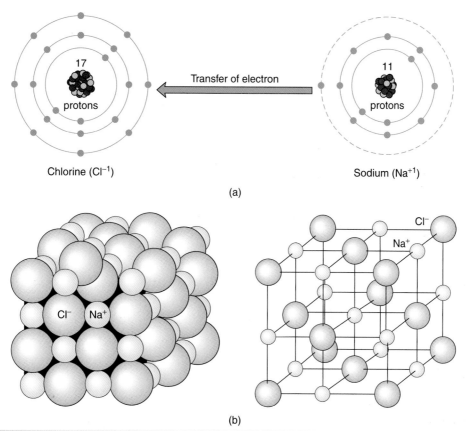

Figure 11.4 Atoms gain or lose electrons to achieve a more stable arrangement of eight electrons in their outer shell. These atoms become electrically-charged ions. (a) A sodium atom has 11 protons and 11 electrons. It easily loses an electron to become a positive ion (Na^+). A chlorine atom has 17 neutrons and 17 electrons. It gains an electron to fill its outer shell and becomes a negatively-charged ion (Cl^-). These opposite charges attract each other to become an ionic bond. The compound formed by this bond is sodium chloride (NaCl), or the mineral halite (salt). (b) The crystal structure of halite shows the relative sizes (left) and locations (right) of its ions. (Source: Dutch *et al.*, 1998, p. 21)

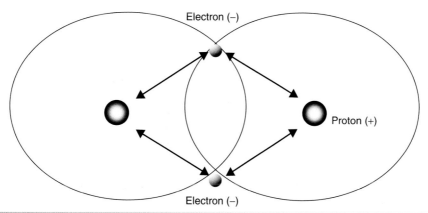

Figure 11.5 In covalent bonds, nuclei repel, but are also attracted by, the pair of negative electrons. Sometimes one atom provides both electrons for a covalent bond (called a dative bond). A covalent bond involves a balance between attraction and repulsion. The atoms are continuously vibrating as if they are on springs. Covalent bonds will break if enough energy is provided. (Source: http://www.webchem.net/images/bonds/covale2.gif, accessed 2 February 2008)

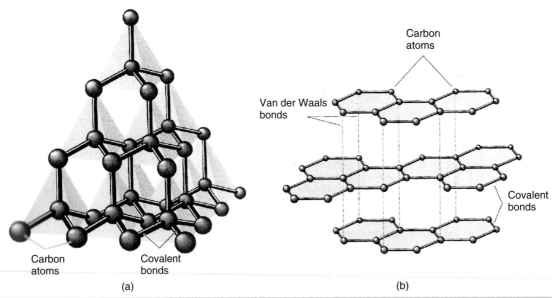

Figure 11.6 Crystal structure of: (a) diamond, where each carbon atom shares its four outer-shell electrons with four other carbon atoms, forming covalent bonds and a three-dimensional geometric arrangement. (Source: Skinner and Porter, 1992, p. 47); (b) graphite, where bonds within the sheets of atoms are strong covalent bonds, but bonds between the sheets are weaker van der Waals bonds. (Source: Skinner and Porter, 1992, p. 48)

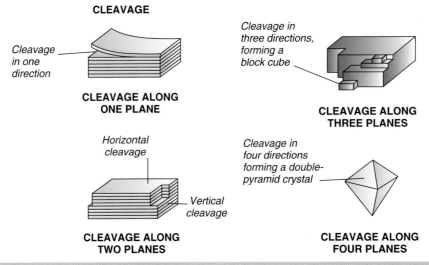

Figure 11.7 Cleavage in minerals. (Source: Stalker, 1994, p. 24)

specific gravity, which is the density of a material relative to water. Most common minerals have specific gravities 2.5–3.5, but lead is 7.6 and pure gold is 19.3.

11.1.5 Common groups of minerals

Minerals can be divided into two broad groups: silicates, which contain silicon (Si), and nonsilicates, which do not. The most common nonsilicate minerals have simpler chemical formulas and crystal structures than most silicate minerals. The important nonsilicate minerals and their uses are shown in Table 11.2.

A few of the most chemically-inert elements, like gold and silver, occur uncombined with other elements and form minerals known as native elements. Diamond and graphite are also native elements, both being composed of carbon (C), but each mineral has a unique crystal structure.

Diamond and graphite are polymorphs because they are the same compound occuring in more than one crystal structure. While diamond has a tetrahedral structure, with atoms held together strongly by covalent bonds, graphite (which we know as pencil lead) has sheets of carbon atoms that are themselves held together strongly, but the bonds holding the sheets together are weak van der Waals bonds. Thus, when a pencil is used to draw or write on paper, the sheets of carbon slide off and stick to the paper (held by electrostatic forces) to leave a grey line. This is

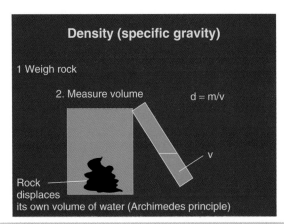

Figure 11.8 Measuring the specific gravity of a mineral. d, density; m, mass; v, volume

Table 11.1 Mohs scale of hardness for minerals. (Source: Skinner and Porter, 1992, p. 56)

	Relative number in the scale	Mineral	Hardness of some common objects
	10	Diamond	
	9	Corundum	
Decreasing	8	Topaz	
	7	Quartz	
	6	Potassium feldspar	Pocketknife; glass
	5	Apatite	
	4	Fluorite	Copper penny
	3	Calcite	Fingernail
	2	Gypsum	
	1	Talc	

By far the most abundant minerals are the silicates, composed of silicon and oxygen, which together make up almost 75% of the Earth's crust by weight. The important combination of the elements oxygen and silicon is usually referred to as silica, which is abundant, and is capable of forming numerous different atomic structures. In silicates, a silicon atom bonds with four oxygen atoms to form a three-sided pyramid, or silica tetrahedron, with an oxygen atom at each corner and a silicon atom in the centre (Figure 11.11).

Each silicate tetrahedron has four negative charges. Silicon has four electrons in its outer shell and oxygen has six. The silicon atom in the silicate tetrahedron shares one of its electrons with each of the oxygen atoms. The four oxygen atoms, in turn, each share one of their electrons with the silicon. This leaves the silicon with a stable outer shell of eight electrons, but each of the four oxygen atoms still requires an additional electron for a stable outer shell of eight. They can attain these in two ways:

1. They can accept electrons from, or share electrons with, cations. For example in olivine (Mg_2SiO_4), two magnesium (Mg) atoms give up their outer-shell electrons to the oxygen atoms, forming ionic bonds.

2. They can bond with two Si atoms at the same time. The shared oxygen is covalently bonded to each of the two silicon atoms. The outer-shell electron of the shared oxygen is now filled because it accepts an electron from each silicon and the two tetrahedral-shaped silicate anions become joined $(Si_2O_7)^{6-}$. This process of linking silicate tetrahedra by oxygen-sharing can be extended to form huge anions and is called polymerization.

By sharing oxygen atoms, silica tetrahedra can link together to form rings, sheets, chains and three-dimensional structures (Figure 11.12).

Most common silicate minerals contain very large anions as a result of polymerization. The main mineral groups are listed below.

The olivine group

A glassy-looking group of minerals, usually pale green in colour (Figure 11.13(a)). Fe^{2+} can substitute readily for Mg^{2+} in olivine, giving rise to the group formula $(Mg, Fe)_2SiO_4$. It is the Fe^{2+} that gives olivine its green colour. Olivine is one of the most abundant mineral groups in the Earth, being a very common constituent of igneous rocks in the oceanic crust and upper mantle.

The garnet group

This is the second most important mineral group, with isolated silicate tetrahedra. As with olivine, ionic substitution gives garnet an even wider range of compositions. The garnet group has the complex formula $A_3B_2(SiO_4)_3$, where A can be any of the cations Mg^{2+}, Fe^{2+}, Ca^{2+} or Mn^{2+}, or any mixture of them, while B can be either Al^{3+}, Fe^{3+} or Cr^{3+}, or a mixture of them. Garnet is

easily removed by an eraser, which breaks the electrostatic forces and releases the carbon sheets from the paper's surface.

Minerals known as sulphides have a positively-charged ion combined with sulphur and are important in the environment because the refining of sulphide ore minerals and the burning of coal (which contains the sulphide mineral pyrite (FeS_2)) are two of the main causes of acid rain (as we saw in Chapter 2).

Other important minerals include the halides, in which positively-charged ions combine with chlorine or fluorine, and the oxides, in which positively-charged ions combine with oxygen. The sulphates contain the complex SO_4 radical, while the carbonates contain the CO_3 radical. Calcite ($CaCO_3$) is the most common carbonate mineral and is the main constituent of the sedimentary rock limestone. Calcite also has a polymorphic form called aragonite. Although most of these minerals are not as common as the silicates, several of them are nevertheless very important resources.

Table 11.2 Important nonsilicate minerals and their uses. (Source: Dutch *et al.*, 1998, p. 25)

Name	Formula	Uses
Native elements: chemical elements not combined with other elements		
Gold	Au	Coins; jewellery; dentistry
Silver	Ag	Coins; photography; dentistry
Sulphur	S	Sulphuric acid
Diamond	C	Jewellery; abrasives
Graphite	C	Pencil leads; lubricants
Sulphides: elements combined with sulphur (−2)		
Pyrite	FeS_2	'Fool's gold'; minor source of iron; principal contributor to acid runoff and precipitation
Chalcopyrite	$CuFeS_2$	Principal ore of copper
Sphalerite	ZnS	Principal ore of zinc
Galena	CaF_2	Principal ore of lead
Halides: elements combined with chloride or fluorine (−1)		
Halite	$NaCl$	Table salt
Fluorite	CaF_2	Metallurgy; source of fluorine
Oxides: elements combined with oxygen (−2)		
Hematite	Fe_2O_3	Principal ore of iron; common rust has the same chemical composition
Corundum	Al_2O_3	Abrasives; ruby and sapphire are gem varieties
Hydroxides: elements combined with OH (−1)		
Limonite		Minor ore of iron; principal colouring agent in soils
Bauxite		Principal ore of aluminum
Sulphates: elements combined with SO$_4$ (−2)		
Gypsum	$CaSO_4 \bullet 2H_2O$	Principal ingredient of plaster
Barite	$BaSO_4$	Drilling oil wells
Carbonates: elements combined with CO$_3$ (−2)		
Calcite	$CaCO_3$	Main component of limestone; principal ingredient of cement
Dolomite	$CaMg(CO_3)_2$	Main component of dolostone

usually found in metamorphic rocks of the continental crust but is also found in some igneous rocks. Garnet can form beautiful crystals of almost any colour (Figure 11.13(b)). An important property is its hardness, which makes it useful as an abrasive for grinding and polishing.

The pyroxene and amphibole groups

These contain long, chainlike anions. Pyroxenes (Figure 11.13(c)) are built from a polymerized chain of tetrahedra, each of which shares two oxygens, so the anion has the general formula $(SiO_3)_n^{2-}$, where n is a large number. Amphiboles are built from double chains of tetrahedra equivalent to two pyroxene chains, in which half the tetrahedra share two oxygens and the other half share three, giving the general anion formula of $(Si_4O_{11})_n^{6-}$. Both pyroxene and amphibole chains are bonded together by cations such as Ca^{2+}, Mg^{2+} and Fe^{2+}, which transfer electrons and form bonds with oxygens in adjacent chains that have unfilled electron shells. The general formula for a pyroxene is $AB(SiO_3)_2$, where A and B can be any number of cations

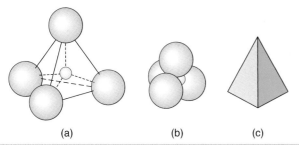

Figure 11.11 The silica tetrahedron: (a) expanded view showing oxygen atoms at the corners of a tetrahedron and a small silicon atom at the centre; (b) view of the silica tetrahedron as it really exists with the oxygen atoms touching; (c) diagrammatic representation of the silica tetrahedron. (Source: Dutch *et al.*, 1998, p. 27). Reprinted from The Dynamic Earth, BJ Skinner & SC Porter, 2nd ed., 1992, p. 47 and 48, with permission from John Wiley & Sons Ltd.

(e.g. Mg^{2+}, Fe^{2+}, Ca^{2+}, Mn^{2+}, Na^{1+}, Al^{3+}). The pyroxenes are most commonly found in igneous rocks of the oceanic crust and mantle, but also occur in igneous and metamorphic rocks of the continental crust. The most common pyroxene is a shiny black variety called augite (Figure 11.13(d)), with the approximate formula $Ca(Mg,Fe,Al)[(Si,Al)O_3]_2$. Amphiboles probably have the most complicated formulas of all the common minerals. An example is hornblende (Figure 11.13(e)).

The clays, micas and chlorites

Clays, micas and chlorites all contain polymerized sheets of silicate tetrahedra and because of this they are related mineral groups. Each tetrahedron shares three of its oxygen molecules with adjacent tetrahedra to give the general anion formula $(Si_4O_{10})_n{}^{4-}$. This leaves a single oxygen molecule in each tetrahedron with an unfilled electron shell. Cations like Al^{3+}, Mg^{2+} and Fe^{2+} bond with the spare oxygen to hold the polymerized sheets together in the crystal structure. The Si–O bonds are much stronger than the bonds between oxygen and the other cations, so these minerals all have very pronounced cleavage parallel to the sheets. Kaolinite (Figure 11.13(f)) is a clay mineral, very common in soil, in which the oxygens with unfilled shells bond with Al^{3+} cations, giving $Al_4Si_4O_{10}(OH)_8$. There are two common types of mica: muscovite (Figure 11.13(g)), which has the formula $KAl_2(Si_3Al)O_{10}(OH)_2$ and is a clear, almost colourless mineral, famous for producing large cleavage sheets which have been used in the windows of welding masks and central heating boilers, its high melting point giving it heat-resistant properties; and biotite (Figure 11.13(h)), which has the formula $K(Mg,Fe)_3(Si_3Al)O_{10}(OH)2$ and is a dark-brown colour, caused by the iron it contains. The micas are common minerals found in both igneous and metamorphic rocks. The chlorite group has the formula $(Mg,Fe,Al)_{6-}(Si,Al)_4O_{10}(OH)_8$, is usually green in colour (Figure 11.13(i)) and is a common alteration product from other minerals like biotite, augite, hornblende and olivine, since it contains iron and magnesium. Oceanic rocks commonly contain olivine and augite, and when they come into contact with sea water they alter to chlorite, the OH in chlorite

being one of the chemicals that enhances melting of oceanic crust during subduction.

Quartz

This is the only common mineral composed entirely of silicon and oxygen (SiO_2). It provides an example of polymerization in that it fills all unfilled oxygen shells, forming a three-dimensional network of tetrahedra. It characteristically forms six-sided crystals (Figure 11.13(j)) and although usually colourless, it can be found in a variety of colours given small amounts of iron, aluminum, titanium and other elements provided through ionic substitution. Quartz occurs in igneous, metamorphic and sedimentary rocks and is the most important constituent of sand.

The feldspar group

The name 'feldspar' comes from two Swedish words, *feld* (field) and *spar* (mineral). It is a common mineral which had to be cleared from fields before they could be farmed. Feldspar is the most common mineral in the Earth's crust, accounting for 60% of all minerals in the continental crust, and is also abundant in minerals of the seafloor. Like quartz, feldspar has a structure formed by polymerization of all the oxygen atoms in the silicate tetrahedra. Unlike quartz though, some of the tetrahedra contain Al^{3+} substituting for Si^{4+}, so another cation must be added to the structure to donate an electron to balance the charges – usually K^{1+}, Na^{1+} or Ca^{2+}. Feldspars have a range of compositions, the most common of which are potassium feldspar, $K(Si_3Al)O_8$, and plagioclase, $(Na,Ca)(Si,Al)_3O_8$ (Figure 11.13(k)).

Carbonate, phosphate and sulphate minerals

The carbonate anion $(CO_3)^{2-}$ forms three important common minerals: calcite (Figure 11.13(l)), aragonite and dolomite. Calcite and aragonite have the same chemical composition, $CaCO_3$, and are polymorphs. Dolomite has the formula $CaMg(CO_3)_2$. Calcite and dolomite are much more common than aragonite. Both look similar, have the same vitreous lustre and distinctive cleavage, and are relatively soft and common in sedimentary rocks (limestone and dolostone). Calcite reacts vigorously with dilute hydrochloric acid (HCl), whereas dolomite reacts very slowly. Apatite (Figure 11.13(m)) is one of the most important phosphate minerals. It contains the complex anion $(PO_4)^{3-}$, has the general formula $Ca_5(PO_4)_3(F,OH)$ and is the substance from which our bones and teeth are made. Sulphate minerals contain the sulphate anion $(SO_4)^{2-}$. The two most common are calcium sulphate minerals: anhydrite, $CaSO_4$, and gypsum, $CaSO_42H_2O$ (Figure 11.13(n)), both formed when sea water evaporates.

The ore minerals

The term 'ore mineral' is used for minerals that are mined for their valuable metal contents. These minerals tend to be sulphides or oxides. Sulphide minerals have metallic lustre (shiny appearance) and high specific gravity (they feel heavy for their size). The

			Formula of negatively charged ion group	Silicon to oxygen ratio	Example
(a)	Isolated tetrahedra (nesosilicates)		$(SiO_4)^{-4}$	1:4	Olivine
(b)	Continuous chains of tetrahedra (inosilicates)	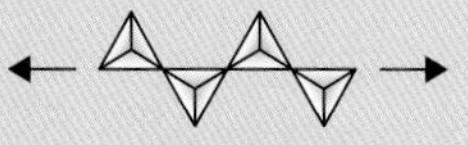 Single chain	$(SiO_3)^{-2}$	1:3	Pyroxene group
		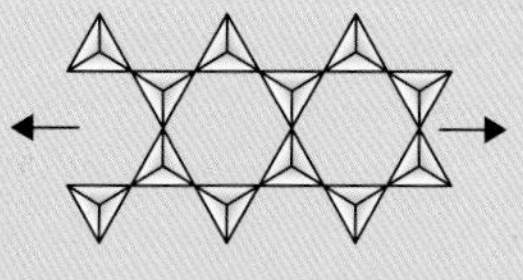 Double chain	$(Si_4O_{11})^{-6}$	4:11	Amphibole group
(c)	Continuous sheets (phyllosilicates)	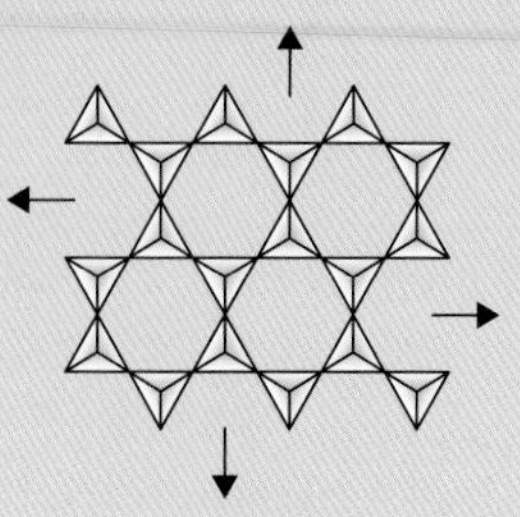	$(Si_4O_{10})^{-4}$	2:5	Micas
(d)	Three-dimensional networks (tectosilicates)	Too complex to be shown by a simple two-dimensional drawing	$(SiO_2)^0$	1:2	Quartz

Figure 11.12 Structures of some of the common silicate minerals shown by various arrangements of silica tetrahedra: (a) isolated tetrahedra; (b) continuous chains; (c) continuous sheets; (d) networks. (Source: Dutch *et al.*, 1998, p. 28)

two most common are pyrite (FeS_2) (Figure 11.13(o)), otherwise known as 'fool's gold', and pyrrhotite (FeS), though these are not actually mined for their iron content. Most of the world's lead comes from galena (PbS) (Figure 11.13(p)), most of the zinc from sphalerite (ZnS) and most of the copper from chalcopyrite ($CuFeS_2$), while other metals like cobalt, mercury, molybdenum and silver also come from sulphide ore minerals. Since iron is one of the most abundant elements in the crust, the iron oxides magnetite (Fe_3O_4) and hematite (Fe_2O_3) (Figure 11.13(q)) are the two most common oxide minerals. Other oxide ore minerals are uraninite (U_3O_8) (Figure 11.13(r)), the main source of uranium; cassiterite (SnO_2), the main ore mineral for tin; and rutile (TiO_2), the main source of titanium. (See Box 3.1 Diamond sources and their discovery in the companion website.)

11.1.6 Environments in which minerals form

As well as being economically valuable, minerals can give an insight into the chemical and physical conditions in parts of the Earth that we cannot observe and measure directly, through examination of their structures and compositions. By studying

them in the laboratory, scientists have begun to understand the temperatures and pressures at which diamond will form rather than its polymorph, graphite, for example (Figure 11.14). Rock samples that contain diamonds are likely to have come from at least 150 km below the Earth's surface. Minerals can tell us a great deal about past environments from the way they have formed and subsequently weathered.

11.2 Rocks and the rock cycle

Rocks are defined as consolidated or poorly-consolidated aggregates of one or more minerals, glass or solidified organic matter (such as coal) that cover a significant part of the Earth's crust. Some geologists refer to minerals as the 'building blocks' of rocks.

An important feature of rock is texture, which refers to the appearance of the rock due to the size, shape and arrangement of its constituent mineral grains. Another important feature is the kind of minerals that are present. Some rocks, such as limestone, are composed of one mineral, in this case calcite, but most rocks

(a) Olivine

(b) Garnet mineral specimen containing large crystals of the garnet mineral spessartine (red)

(c) Pyroxene. A ferro-magnesian mineral. Dark green to black

(d) Augite crystal set in rock matrix

1cm

(e) Hornblende, an amphibole. A dark-green-to-black mineral that looks rather like augite but can be distinguished from augite because the angles between cleavage surfaces differ

(f) Kaolinite, a clay mineral commonly found in soils

(g) Muscovite. Clear or colourless mica

#3

Biotite

(h) Biotite. Dark-brown mica

#33

Chlorite

(i) Chlorite. Identified by its greenish colour. Similarities to the micas in appearance

Figure 11.13 Some common minerals. (Sources: (a) http://skywalker.cochise.edu/wellerr/mineral/olivine/6olivine64.jpg; (b) http://pubs. usgs.gov/fs/2006/3149/images/coverphoto.jpg; (c) http://csmres.jmu.edu/geollab/fichter/RockMin/Pyrox-70.JPG; (d) http://www. dkimages.com/discover/previews/892/75002046.JPG; (e) http://www.soes.soton.ac.uk/resources/collection/minerals/minerals/ images/M34-A-Hornblende.jpg; (f) http://www.mii.org/Minerals/Minpics1/Kaolinite%202.jpg; (g) http://www.cropsoil.uga.edu/soils andhydrology/images/Muscovite.jpg; (h) http://cmsc.minotstateu.edu/Labs/web%20minerals/BIOTITE.jpg; (i) http://cmsc.minotstateu. edu/Labs/web%20minerals/Chlorite.jpg; (j) http://earthnet-geonet.ca/images/glossary/quartz.jpg; (k) http://www.mii.org/Minerals/ Minpics1/Plagioclase%20feldspar.jpg; (l) http://www.windows.ucar.edu/earth/geology/images/calcite_med.jpg; (m) http://gwydir. demon.co.uk/jo/minerals/pix/apatite2.jpg; (n) http://www.pitt.edu/~cejones/GeoImages/1Minerals/2SedimentaryMineralz/Gypsum_ Halite/GypsumSelenite.JPG; (o) http://www.3dchem.com/imagesofmolecules/pyrite2.jpg; (p) http://csm.jmu.edu/minerals/minerals% 5C+Galena.jpg; (q) T. Stott; (r) http://www.ecolo.org/photos/uranium/uraninite-NorthCarolina.jpg)

(j) Quartz. Characteristically forms six-sided crystals

(k) Plagioclase feldspar

(l) Calcite

(m) Apatite, one of the most important phosphate minerals used in making phosphate fertilisers

(n) Gypsum. With a hardness of 2 on the Mhos scale, gypsum is easily scratched, as seen here. Used in making plaster and plasterboard

(o) Pyrite or 'fool's gold'. Iron sulphide

(p) Galena, lead ore

(q) Hematite, iron ore

(r) Uraninite, uranium ore used in nuclear-power generation

Figure 11.13 (continued)

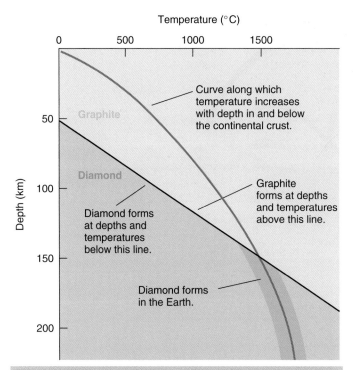

Figure 11.14 Line separating the conditions of temperature and pressure (plotted as depth) at which the two polymorphs of carbon, diamond and graphite, grow. At pressure equal to that at a depth of 150 km, the diamond–graphite line intersects the curve showing the way the Earth's temperature changes with depth. Diamond can therefore only form at depths of greater than 150 km below the surface – one reason why it is so difficult to find. (Source: Skinner and Porter, 1992, p. 66)

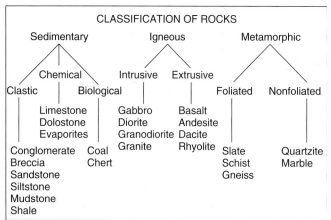

Figure 11.15 Classification of rocks. (Source: http://volcano.und. edu/vwdocs/vwlessons/volcanic_rocks.html, accessed 13 June 2008). Reprinted from The Dynamic Earth, BJ Skinner & SC Porter, 2nd ed., 1992, p. 66, with permission from John Wiley & Sons Ltd.

contain two or more minerals. Granite, for example, contains feldspar, quartz and usually mica. The variety and abundance of different minerals in rocks is known as the mineral assemblage and is important information for interpreting how a rock was formed.

The mineral grains in some rocks are quite loosely packed and these rocks can easily be broken apart. In other rocks, the grains are held together very strongly and are intricately interlocked.

There are three classes of rock, based upon their origin:

1. Igneous rocks (Latin *ignis* = fire) are crystallized from molten or partly-molten material that cools within the crust or at the surface, or they may result from consolidation of fragments ejected from volcanoes during explosive eruptions.

2. Sedimentary rocks are made up of the particles and chemicals yielded by the breakdown of older rocks. They include both lithified (that is, turned to stone) fragments of preexisting rock, called sediments, and rocks that were formed through chemical or biological action.

3. Metamorphic rocks are those that have been changed by heat and/or pressure within the Earth's crust.

Figure 11.15 shows a broad classification scheme for rocks, with some named examples.

The rock cycle is one of many natural cycles on the Earth and is a useful way to summarize the relationships between the three families of rocks. Figure 11.16 shows the interactions of energy, Earth materials and the geological processes that form and destroy rocks and minerals.

Starting at the bottom of Figure 11.16, molten rock, called magma, rises from the Earth's mantle, invading the crust and cooling and hardening to form igneous rocks. These rocks eventually become exposed at the surface by erosion, and are attacked by weathering. The weathered fragments are transported by erosion processes (water, wind, ice and gravity) to a lake or sea, where they are deposited in layers, which harden to become sedimentary rocks. If these sedimentary rocks are buried deeply enough they may be changed by heat and pressure into metamorphic rocks. The heating can actually melt metamorphic rock, turning it back into magma and starting the cycle again. The arrows within the circle in Figure 11.16 show that any possible change can occur within the rock cycle. For example, sedimentary rocks may not be metamorphosed, but can be uplifted and exposed at the surface instead. Igneous, sedimentary and metamorphic rocks can all be attacked by weathering and used as the raw material for new sedimentary rock. Any of the three rock types can be changed into metamorphic rock or melted to form new igneous rock. Thus, any of the three types can be transformed into any other type.

11.3 Vulcanicity and igneous rocks

Molten material or magma from beneath the Earth's crust that escapes from the Earth's surface is called lava. Lava cools rapidly, forming igneous rocks with small crystals. Figure 11.17 shows how volcanic activity which extrudes magma on to and above the

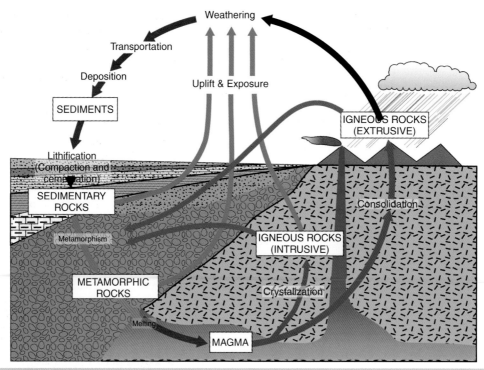

Figure 11.16 The rock cycle. (Source: http://commons.wikimedia.org/wiki/File:Rock_cycle_NASA.jpg)

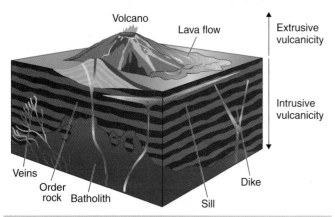

Figure 11.17 Extrusive and intrusive vulcanicity. (Source: Strahler and Strahler, 1997, p. 260)

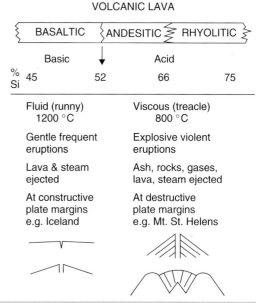

Figure 11.18 The relationship between lava composition, type of eruption and landforms produced by volcanoes

Earth's surface is known as extrusive vulcanicity, while magma which cools within the crust is known as intrusive vulcanicity.

The nature of the magmatic material has a large influence on the type of volcano formed and the nature of its eruptions. Figure 11.18 shows how the proportion of silica in the lava determines whether the lava is fluid (runny like paint) or viscous (like treacle).

Vulcanicity gives rise to certain types of igneous rock, as seen in Table 11.3. Igneous rocks are classified according to their texture and mineral composition. A rock's texture is a function of the size and shape of its mineral grains, and for igneous rocks

this is determined by how quickly or slowly a melted mass cools (Figure 11.19).

Figure 11.20 shows a schematic overview of igneous and volcanic structures/bodies. Deep magma cools slowly, while magma extruded on to the Earth's surface, particularly beneath the oceans, cools rapidly.

Rapid cooling

Slow cooling

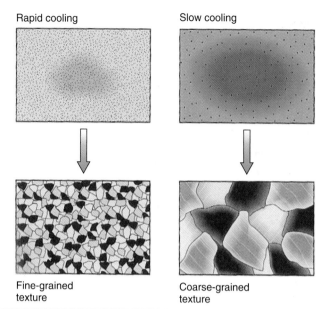

Fine-grained
texture

Coarse-grained
texture

Figure 11.19 How rate of cooling affects igneous rock texture.
(Source: Dutch *et al.*, 1998, p. 35)

Table 11.3 shows how igneous rocks can be classified based on the location (rate of cooling) and proportion of silica (SiO_2) present. The percentage of silica in an igneous rock is a measure of how 'distilled' or fractionated its original magma was. The higher the percentage of SiO_2 present, the more steps of fractionation it has gone through.

Three important and relatively common glassy (composed of finely crystalline SiO_2) volcanic rocks are obsidian, which looks like black glass; pumice, which has the composition of glass but is not really glassy-looking; and tuff, which is consolidated volcanic ash or cinders. Tuff has a pyroclastic texture, resulting from fragmentation during violent eruptions.

Other examples from Table 11.3 can be seen in Figure 11.21.

The geologist N.L. Bowen observed that minerals in most igneous rocks tend to form in a specific sequence, which has become known as Bowen's series, given in Figure 11.22. Bowen's series summarizes the processes that occur when magma melts or solidifies. As minerals solidify in magma, some chemical components leave the liquid magma and form into the solid minerals. The composition of the remaining liquid magma therefore changes, and other minerals begin to form in turn, until the magma has solidified.

11.4 Sedimentary rocks, fossils and sedimentary structures

A range of processes collectively known as weathering constantly attacks rocks found at or near the Earth's surface, dissolving some materials and disintegrating solid rock into fragments. This *in situ* breakdown results in the formation of sediments and dissolved materials, which are the raw materials of sedimentary rocks.

Sediment is particulate matter derived from the physical or chemical weathering of materials of the Earth's crust and by

Table 11.3 Classification of igneous rocks. (Source: after Gresswell, 1963, p. 17 and Dutch *et al.*, 1998, p. 36)

Also described as	VOLCANIC Extrusive	HYPABYSSAL Intermediate	PLUTONIC Intrusive
Solidifies	On surface	below surface	at depth
Cooling rate	Rapid	medium	slow
Crystal size	Small	medium	large
Forming	Lava	dykes & sills	great masses
ACID or Felsic >66% SiO_2	Rhyolite	Granophyre	Granite
Key minerals	amphibole, biotite, sodium plagioclase, potassium feldspar, quartz		
INTERMEDIATE 50–66% SiO_2	Andesite	Porphyries	Diorite
Key minerals	pyroxene, amphibole, calcium plagioclase, sodium plagioclase		
BASIC or Mafic <50% SiO_2	Basalt	Dolorite	Gabbro
Key minerals	olivine, pyroxene, calcium plagioclase		

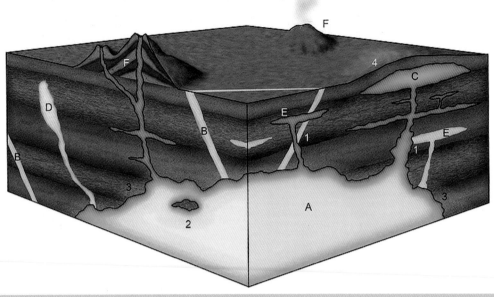

Figure 11.20 Schematic overview of igneous and volcanic structures/bodies. (A) Magmachamber (batholith). (B) Dyke/dike. (C) Laccolith. (D) Pegmatite. (E) Sill. (F) Stratovolcano. Processes: 1, newer intrusion cutting through older one; 2, xenolith or roof pendant; 3, contact metamorphism; 4, uplift due to laccolith emplacement

(a) Obsidian　　(b) Pumice　　(c) Tuff

(d) Rhyolite　　(e) Andesite　　(f) Basalt

(g) Granite　　(h) Diorite　　(i) Gabbro

Figure 11.21 Some common igneous rocks. (Sources: (a) http://www.oregongeology.com/sub/learnmore/jpegs/orig/obsidian.jpg; (b) http://www.swisseduc.ch/stromboli/glossary/icons/pumice_2.jpg; (c) http://z.about.com/d/geology/1/0/u/W/tuff_grnvly2.jpg; (d) http://volcano.und.edu/vwdocs/vwlessons/rocks_pics/rhyolite.jpg; (e) http://www.gc.maricopa.edu/earthsci/imagearchive/ANDESITE_1_big.jpg; (f) http://itc.gsw.edu/faculty/tweiland/basalt.jpg; (g) http://library.thinkquest.org/05aug/00461/images/granite.jpg; (h) http://www.pitt.edu/~cejones/GeoImages/2IgneousRocks/IgneousCompositions/4Diorite/Diorite.jpg; (i) http://www.es.ucl.ac.uk/schools/Glossary/gabbro.jpg)

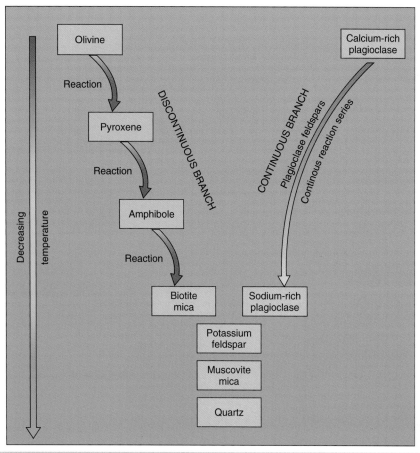

Figure 11.22 Bowen's series, showing the order of crystallization of minerals common in igneous rocks. (Source: Dutch *et al.*, 1998, p. 38)

certain organic processes. It can be transported and redeposited by streams, glaciers, wind, waves or gravity. The process of removal and transport of weathered particles by streams, glaciers, wind, waves or gravity is known as erosion. Running water is a particularly important agent of sediment transport, in which dissolved substances are transported to lakes or the ocean, where they are deposited. Sediment deposited in layers will form sedimentary rock, which is sediment that has been lithified (turned to stone) by pressure from deep burial, by cementation or by both of these processes. Dissolved substances may become concentrated and eventually precipitate as minerals. An area of sediment deposition is known as a depositional environment and three major depositional settings are recognized: continental (glacial deposits, alluvial fans, sediment in lakes); marginal marine (deltas, beaches, estuaries, continental shelf); and marine (submarine fans, deep sea sediments).

11.4.1 Classifying sedimentary rocks

Sedimentary rocks can be classified into two broad categories: detrital sedimentary rocks, which are composed of detritus (fragments of preexisting rocks and minerals with clastic texture), and clastic sedimentary rocks, which are classified according to their grain size. Thus, gravel-sized sediments lithify to conglomerate, sand-sized sediments to sandstone and clay-sized sediments to shale (Figure 11.23).

Figure 11.24 shows examples of detrital sedimentary rocks named in Figure 11.23. Particles measuring more than 2 mm in diameter are classed as gravel (Table 11.4) and cannot normally be transported by wind or slow-moving water. These sediments therefore indicate a high-energy depositional environment. Such rocks are called conglomerates (rounded pebbles) (Figure 11.24(a)) or breccias (angular pebbles) (Figure 11.24(b)). 'Sand' refers to detrital particles measuring 1/16–2 mm in diameter, so a rock composed of sand is called a sandstone (Figure 11.24(c) and (d)). Most sand is made of quartz (SiO_2), which is common and durable, but a number of other minerals are usually present in small quantities. Mudrocks include all detrital sedimentary rocks composed of silt-sized (1/256–1/16 mm) and clay-sized (<1/256 mm) particles. Examples include siltstone (Figure 11.24(e)) and shale (Figure 11.24(f)).

Other sediments result from chemical and biological activity. Chemical sedimentary rocks, which may be clastic or nonclastic,

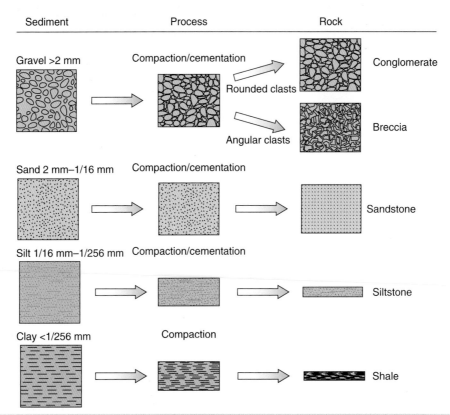

Figure 11.23 How sediments are transformed into detrital sedimentary rocks. (Source: Dutch *et al.*, 1998, p. 41)

Figure 11.24 Some common detrital sedimentary rocks. (Sources: (a) http://www.dkimages.com/discover/previews/1051/25002102.JPG; (b) http://www.gc.maricopa.edu/earthsci/imagearchive/BRECCIA_big.jpg; (c) http://www.geomore.com/images/sandstone_very_coarse.jpg; (d) http://www.answersincreation.org/curriculum/geology/images/Sandstone.jpg; (e) http://z.about.com/d/geology/1/0/j/E/siltstone.jpg; (f) http://www.usoe.k12.ut.us/curr/Science/core/8thgrd/sciber8/geology/images/shale4.jpg)

(a) Limestone   (b) Rock salt   (c) Gypsum

(d) Coal   (e) Chalk   (f) Chert

Figure 11.25 Some common chemical sedimentary rocks. (Sources: (a) http://www.mii.org/Minerals/Minpics1/Limestone%202.jpg; (b) http://www.dkimages.com/discover/previews/973/90006784.JPG; (c) http://www.mii.org/Minerals/Minpics1/Satin%20spar%20gypsum.jpg; (d) http://www.mii.org/Minerals/Minpics1/CoalAnthracite.jpg; (e) http://www.dkimages.com/discover/previews/1004/795272.JPG; (f) http://itc.gsw.edu/faculty/tweiland/chert.jpg)

Table 11.4 Classification of sediments

Phi units*	Size	Wentworth size class	Sediment/ rock name
−8	256 mm	Boulders	Sediment: GRAVEL
−6	64 mm	Cobbles	
−2	4 mm	Pebbles	Rocks: RUDITES (conglomerates, breccias)
−1	2 mm	Granules	
0	1 mm	Very coarse sand	Sediment: SAND
1	1/2 mm	Coarse sand	
2	1/4 mm	Medium sand	Rocks: SANDSTONES (arenites, wackes)
3	1/8 mm	Fine sand	
4	1/16 mm	Very fine sand	
8	1/256 mm	Silt	Sediment: MUD
		Clay	Rocks: LUTITES (mudrocks)

*Udden–Wentworth scale

include chemically-precipitated limestone, rock salt and gypsum (Figure 11.25(a)–(c)). Biogenic sedimentary rocks are produced directly by biological activity, such as coal (lithified plant debris), some limestone and chalk (calcium carbonate, shell material), and chert (siliceous shells) (Figure 11.25(d)–(f)). Table 11.5 shows this classification of sedimentary rocks.

11.4.2 Fossils and sedimentary structures

Sedimentary rocks harbour fossils. Fossils are the remains of plants and animals that have been preserved in rock. A fossil may be the preserved remains of an organism itself, an impression of an organism in the rock, or the preserved traces (known as trace fossils) left by an organism when it was alive, such as organic carbon outlines, fossilized footprints or droppings (e.g. fossilized dinosaur droppings are known as coprolites).

Most dead organisms soon rot away or are eaten by scavengers. Therefore, for fossilization to occur, rapid burial by sediment is needed. The organism decays, but the harder parts like bones, teeth and shells are preserved and hardened by minerals from the surrounding sediment. Even if the hard parts of the organism are dissolved, an impression or mould may be left, which can be filled by minerals, creating a cast of the organism.

The study of fossils (called palaeontology) can show how living things have evolved and helps to reveal the Earth's geological history by aiding in the dating of rock strata, for example.

Table 11.5 Classification of sedimentary rocks. (Source: Pipkin and Trent, 2001, p. 40)

DETRITAL SEDIMENTARY ROCKS (CLASSIC TEXTURE)

Sediment	Description	Rock name
Gravel (>2.0 mm)	Rounded rock fragments	Conglomerate
	Angular rock fragments	Breccia
Sand (0.062–2.0 mm)	Quartz predominant	Quartz sandstone
	>25% feldspars	Arkose
Silt (0.004–0.062 mm)	Quartz predominant; gritty feel	Siltstone
Clay, mud (<0.004 mm)	Laminated, splits into thin sheets	Shale
	Thick beds, blocky	Mudstone

CHEMICAL SEDIMENTARY ROCKS

Texture	Composition	Rock name
Clastic	Calcite ($CaCO_3$)	Limestone
	Dolomite ($CaMg(CO_3)_2$)	Dolostone
Crystalline	Halite (NaCl)	Rock salt
	Gypsum ($CaSO_4 \bullet 2H_2O$)	Rock gypsum

BIOGENIC SEDIMENTARY ROCKS

Texture	Composition	Rock name
Clastic	Shell calcite, skeletons, broken shells	Limestone, coquina
	Microscopic shells ($CaCO_3$)	Chalk
Nonclastic (altered)	Microscopic shells (SiO_2), recrystallized silica	Chert
	Consolidated plant remains (largely carbon)	Coal

environment. For example, among the most common features are the distinct layers known as strata or beds. Such beds can be seen in Figure 11.28 and Figure 11.29.

PROCESS OF FOSSILIZATION

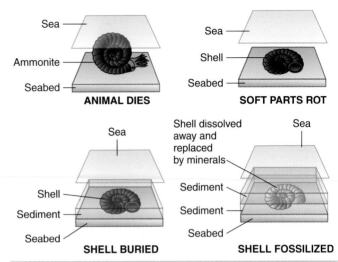

Figure 11.26 The process of fossilization. (Source: Dorling Kindersley Visual Dictionary of the Earth, 1994, p. 30)

(a) Trilobite fossil (Permian period)

(b) Giant fossil ammonite from Perryfield Quarry, Dorset, UK (Jurassic period)

Figure 11.27 Examples of fossils in sedimentary rock. (Sources: (a) http://science.nationalgeographic.com/science/photos/permian-period/trilobite-fossil.html; (b) http://www.perryfield.ukfossils.co.uk/location-photos/9.jpg; (c) http://users.aristotle.net/~russjohn/nature/roadcut03.jpg)

Figure 11.26 shows the process of fossilization, while Figure 11.27 shows examples of fossils in sedimentary rock.

As well as fossils, sedimentary structures are frequently preserved in sedimentary rocks, which form as a result of the physical, chemical and biological processes operating in the depositional

(c) Fossil crinoid (sea lily) stems (Carboniferous period)

Figure 11.27 *(continued)*

Figure 11.29 Sedimentary strata in Diamantina National Park, Queensland, Australia. (Photo: T. Stott)

Figure 11.28 Sedimentary beds in sea cliffs of Isfjord, Svalbard. (Photo: T. Stott)

Most sedimentary rocks have some kind of layering, with individual layers showing differences in colour, particle size or mineral composition. Layers are not always horizontal, as can be seen in Figure 11.30.

Many sedimentary rocks are cross-bedded, which means that layers within a layer are inclined at a different angle. Excellent examples can be seen in the Vermillion Cliffs area in north-central Arizona and south-central Utah, where a vast area of ancient sand dunes has been shaped and moulded by the elements of nature. Cross-bedding of sand dunes from 190 million years ago has long been solidified (see Figure 11.31).

Ripple marks are another sedimentary structure commonly seen in sedimentary rocks (Figure 11.32). They result from wave action, so typically form on lake shores or beaches.

Another sedimentary structure occurs when clay-rich sediments dry out and form mud cracks which are later infilled by other, perhaps wind-blown, sediments (Figure 11.33).

Both fossils and sedimentary structures are useful in determining the depositional environment of sedimentary rocks, which in turn provide much of our knowledge of Earth history. From an economic viewpoint, an understanding of how sediments were deposited is useful in exploration for petroleum, natural gas and coal, which are all found in sedimentary deposits.

11.5 Metamorphic rocks

Rocks that have been changed from preexisting rocks by heat, pressure or chemical processes are classified as metamorphic rocks. The process of metamorphism results in new structures, textures and minerals. Metamorphic rocks form in one of two ways.

Contact metamorphism takes place when rocks are heated by an adjacent mass of hot rock, such as an intrusion (as seen in Figure 11.20). When, for example, the sedimentary rock limestone comes into contact with a hot igneous intrusion it becomes heated and metamorphosed into marble. Contact metamorphic effects rarely extend more than about a kilometre from the source of the heat.

Most of the Earth's metamorphic rocks are formed by the second process, called regional metamorphism (Figure 11.34), which results from intense pressure usually related to plate tectonic processes (Chapter 13). Regional metamorphism extends over vast areas of the Earth's surface and is most often accompanied by deformation of rocks.

Figure 11.30 Vertical beds exposed in the sea cliffs at Marloes Sands, Pembrokeshire, South Wales, UK, showing differential erosion

Figure 11.31 This is a panoramic view of 'the wave' in the Vermilion Cliffs, clearly showing the diverse cross-bedding of the compacted ancient sand dunes. (Source: http://commons.wikimedia.org/wiki/File:Verm_wave.jpg, Wikimedia Commons)

Figure 11.32 Ripple marks in Moenkopi formation rock, Capitol Reef National Park, Utah, USA. (Photo: Daniel Mayer; source: http://commons.wikimedia.org/wiki/File:Ripple_marks_in_Moenkopi_Formation_rock_off_of_Capitol_Reef_Scenic_Drive2.jpeg, Wikimedia Commons)

Figure 11.33 Mud cracking in dried-out clays of the Fraser River estuary, Rockhampton, Queensland, Australia. (Photo: T. Stott)

REGIONAL METAMORPHISM

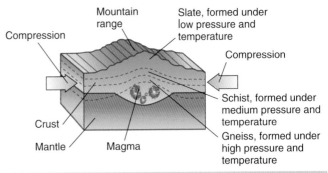

Compression

Mountain range

Slate, formed under low pressure and temperature

Compression

Crust

Mantle

Magma

Schist, formed under medium pressure and temperature

Gneiss, formed under high pressure and temperature

Figure 11.34 Regional metamorphism results from intense pressure, usually resulting from plate tectonic movements. (Source: The Visual Dictionary of the Earth, Dorling Kindersley, 1993, p. 26)

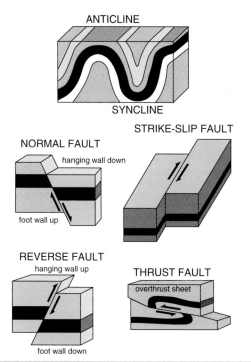

ANTICLINE

SYNCLINE

NORMAL FAULT
hanging wall down

foot wall up

STRIKE-SLIP FAULT

REVERSE FAULT
hanging wall up

foot wall down

THRUST FAULT
overthrust sheet

Figure 11.35 A schematic diagram showing folds (anticlines and synclines) and faults (normal, reverse, thrust and strike-slip). (Source: http://upload.wikimedia.org/wikipedia/commons/e/e4/Faults_and_folds.jpg, Wikimedia Commons)

Table 11.6 Common metamorphic rocks. (Source: Pipkin and Trent, 2001, p. 44)

Rock	Parent rock	Characteristics
FOLIATED OR LAYERED		
Slate	Shale and mudstone	Splits into thin sheets
Schist	Fine-grained rocks, siltstone, shale, tuff	Mica minerals often crinkled
Gneiss	Coarse-grained rocks	Dark and light layers of aligned minerals
NONFOLIATED OR RECRYSTALLIZED		
Marble	Limestone	Interlocking crystals
Quartzite	Sandstone	Interlocking, almost fused quartz grains

Rock deformation resulting from differential stress on rocks on the Earth's crust can cause them to bend and break, known by geologists as folding and faulting. (Figure 11.35). Faulting is favoured in cold, brittle rocks, where plate tectonic movement is rapid. Some folded strata are shown in Figure 11.36.

(a)

(b)

Figure 11.36 Folding exposed in sea cliffs. (a) Folding in limestones, Castlemartin, Pembrokeshire, South Wales, UK. (b) Recumbent fold in sea cliffs at Amroth, South Wales, UK. (Photos: T. Stott)

Another type of metamorphism, sometimes called metasomatism, occurs when water rich in minerals permeates rocks and adds or removes chemical compounds.

Metamorphic rocks may also form by recrystallization. This occurs when a rock is heated and strained by uniform stresses so that larger, more perfect grains result, or new minerals form. In this way a limestone may recrystallize to marble (Figure 11.37(d)), or a quartz sandstone to quartzite (one of the most resistant rocks) (Figure 11.37(e)). Foliation is the flattening and layering of minerals by nonuniform stresses. Foliated metamorphic rocks such as slate, schist and gneiss (Figure 11.37(a)–(c)) are identified by the thickness and crudeness of their foliation.

Table 11.6 summarizes the characteristics of some common metamorphic rocks, which are shown in Figure 11.37.

The changes which take place during metamorphism depend on both the agents of metamorphism described here and the type of rock that is affected.

(a) Slate

(b) Schist

(c) Gneiss

(d) Marble

(e) Quartzite

Figure 11.37 Common metamorphic rocks. (Sources: (a) http://www.enonhall.com/images/01232006/slate.jpg; (b) http://skywalker.cochise.edu/wellerr/rocks/mtrx/6mrx-schist6.jpg; (c) http://www.gc.maricopa.edu/earthsci/imagearchive/GNEISS_1_big.jpg; (d) http://www.mii.org/Minerals/Minpics1/Marble.jpg; (e) http://www.dkimages.com/discover/previews/884/50002337.JPG)

Exercises

1. What is a mineral?

2. Name the five most abundant elements in the Earth's crust in order, starting with the most abundant.

3. Briefly describe the general structure of atoms.

4. What is an atom's atomic number? What is its atomic weight?

5. Give three ways in which atoms bond with other atoms.

6. What simple tests are used by geologists to identify minerals?

7. What are polymorphs?

8. List five of the most common groups of silicate minerals.

9. Draw a simple diagram to illustrate the rock cycle.

10. Summarize how rocks are classified, giving at least two examples of each class.

11. How does the rate of cooling affect igneous rock texture?

12. What does Bowen's series show?

13. What is the difference between the processes of rock weathering and erosion?

14. What is 'lithification'?

15. What effect does transport of clastic sediments have on their shape?

16. What is the main difference between clay (mudstone) and shale?

17. Explain what is meant by cross-bedding in sedimentary strata.

18. How are limestones formed? Name a simple test which can be used to identify limestone.

19. Explain the difference between 'foliated' and 'unfoliated' metamorphic rocks. Name one example of each.

20. Fill in the blank spaces: Rocks baked by the heat given off by igneous intrusions are metamorphosed by _____ or _____ metamorphism. When mountains are formed by tectonic movements the rocks inside them are metamorphosed by the combined effects of heat and pressure – this is called _____ metamorphism.

References

Dutch, S.I., Monroe, J.S. and Moran, J.M. 1998. Earth Science, Boston, MA, International Thomson Publishing.

Gresswell, R.K. 1963. Geology for geographers. London, Hulton Educational Publishers.

Pipkin, B.W. and Trent, D.D. 2001. Geology and the Environment, Pacific Grove, CA, Brooks/Cole.

Skinner, B.J. and Porter, S.C. 2003. The Dynamic Earth: An Introduction to Physical Geology. 2nd edition. Chichester, John Wiley & Sons, Inc.

Stalker, G. 1994. The Visual Dictionary of the Earth. London, Dorling Kindersley.

Strahler, A.H. and Strahler, A. 1997. Physical Geography: Science and Systems of the Human Environment. Chichester, John Wiley & Sons, Inc.

Further reading

Fry, N. (1984) Field Description of Metamorphic Rocks. Chichester, John Wiley & Sons, Inc.

Gillen, C. (1982) Metamorphic Geology. Allen and Unwin.

Hamblin, W.K. and Christiansen, E.H. 2004. Earth's Dynamic Systems. 10th edition. Upper Saddle River, NJ, Pearson Education, Inc. Chs 3–6.

Thorpe, R and Brown, G. (1985) Field Description of Igneous Rocks. Chichester, John Wiley & Sons, Inc.

Tucker, M.E. (1982) The Field Description of Sedimentary Rocks. Chichester, John Wiley & Sons, Inc.

12 The Internal Structure of the Earth

Lady Knox Geyser, South Island, New Zealand. Photo: T. Stott

Earth Environments   David Huddart and Tim Stott
© 2010 John Wiley & Sons, Ltd

Learning Outcomes

After reading this chapter and completing the exercises you will be able to:

➤ Discuss the evidence for our understanding of the internal structure of the Earth.

➤ Evaluate drilling, volcanoes and meteorites as lines of evidence of the internal structure of the Earth.

➤ Understand the nature of seismic waves and their role in our understanding of the internal structure of the Earth.

12.1 Introduction

Until around two centuries ago, the Earth's interior was perceived as an underground world of vast caverns, heat, sulphurous gases and demons. However, by the 1860s, scientists knew that the Earth's average density was $5.5\,t\,m^{-3}$ and that pressure and temperature increased with depth. Since the average density of crustal rocks was $2-3\,t\,m^{-3}$, it was estimated that the density at the core would need to be around $13\,t\,m^{-3}$ in order for the average of $5.5\,t\,m^{-3}$ to prevail.

Scientists now believe that as terrestrial planets grew larger through the process of planetary accretion, their temperatures must have risen, because energy can be changed from one form to another (e.g. electricity to heat) but it cannot be destroyed. A moving object such as a planetary fragment or meteorite has the energy of motion (kinetic energy). On impact with the Earth, this energy is transformed to heat. In addition, heat was added from the naturally radioactive elements in the Earth's interior. Eventually the Earth began to melt and lighter elements (e.g. Si, Al, Na, K) rose towards the surface, while denser material (molten Fe, Ni) sank to the centre. The melting released volatile elements which escaped as gases through volcanoes. These included water vapour, carbon dioxide, methane and ammonia, which formed the early atmosphere. It was from this source that the water that we now find in the Earth's oceans came. Partial melting changed the Earth from an originally homogeneous planet to a compositionally layered one (Figure 12.1):

- **The core:** At the centre is the densest of the layers, called the core, which is divided into the inner core and the outer core. The core is a sphere made mainly from metallic iron, along with nickel and some other elements. In the inner core, pressures are so great that the iron is solid despite the high temperature. In the outer core, temperature and pressure are so balanced that the iron is molten and exists as a liquid, even though the compositions of the two cores are believed to be the same.

- **The mesosphere:** This is the region at the edge of the outer core, known as the core–mantle or C–M boundary (at 2883 km depth), to a depth of about 350 km. The strength of this material is controlled by temperature and pressure, both of which increase with depth. When a solid is heated, it loses strength. When it is compressed, it gains strength. Differences in temperature and pressure divide the mantle and crust into three distinct strength regions. In the lower part of the mantle, the rock is highly compressed, so that it has high strength even though the temperature is high – this region is solid.

- **The asthenosphere:** Within the upper mantle, from 350 to 100 km below the surface, there is a balance between temperature and pressure so that the material has little strength. It is weak and easily deformed, like warm tar or butter, even though its composition seems to be the same as that of the mesosphere. There are thought to be convection currents in this layer which facilitate the movement of the lithospheric plates of the crust above, which rest on the asthenosphere.

- **The lithosphere:** This is the uppermost 100 km of the mantle and contains all of the hard crust. It is a region where the rocks are cooler, stronger and more rigid than those in the plastic asthenosphere. However, the composition remains the same as those of the asthenosphere and mesosphere; only the temperature and pressure differ. As you will see in the next section, the lithosphere is not one continuous layer, but is made up of several large and numerous smaller plates, which 'float' on the asthenosphere below it.

At this stage you may be wondering how the model in Figure 12.1 has been developed. The fact is that we cannot see into the Earth's interior and so this model is really just an informed estimate. Some of the techniques scientists have used to inform the development of this model are presented in the following sections.

12.2 Evidence of the Earth's composition from drilling

We have a good knowledge of the rocks of the lithosphere because we live on its surface and people have collected, examined and classified these rocks for millennia. However, we have been able to extend our knowledge of the rocks of the lithosphere downwards by drilling. A number of projects have been undertaken, the most successful of which was the Kola Super Deep Borehole in Russia (see Box 12.1, see the companion website).

Some have described such drilling attempts as 'barely penetrating the peel of an orange'. Nevertheless, a great deal of very important information, which could not otherwise have been learned, has come from drilling.

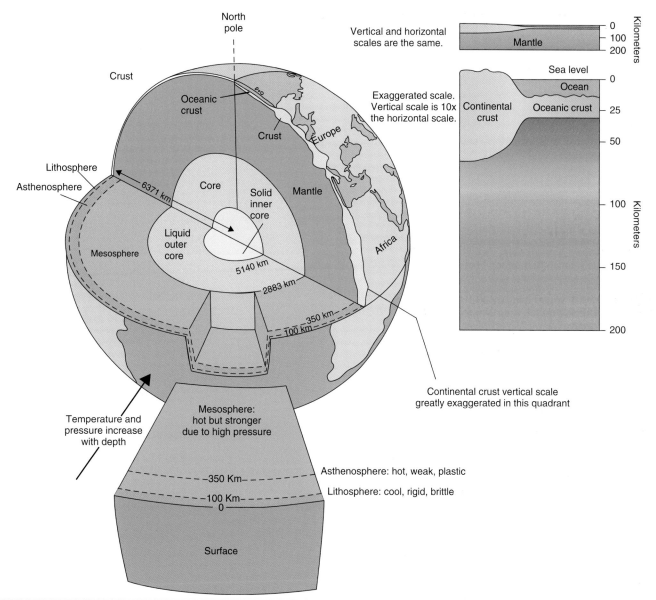

Figure 12.1 A sliced view of the Earth reveals layers of different composition and zones of differing rock strength. Note that the crust is thicker beneath the continents than under the oceans. (Source: Skinner and Porter, 2003, p. 22). Reprinted from The Dynamic Earth, BJ Skinner & SC Porter, 2nd ed., 1992, p. 22, with permission from John Wiley & Sons Ltd.

12.3 Evidence of the Earth's composition from volcanoes

Volcanoes eject a variety of different lava types on to the Earth's surface. Are they from the same source? Are they changed while being transported out through the crust, or do they come from different layers?

The longer material takes to reach the surface, the more time there is for it to change to different temperature and pressure conditions. Rapid ascent results in little change. Rocks gathered after a violent eruption in Hawaii proved to be peridotite,

suggesting that this is the composition of the layer below the crust. However, as we saw in Section 11.3, the silica composition of lava varies considerably and gives rise to a variety of volcanic landforms.

12.4 Evidence of the Earth's composition from meteorites

From time to time fragments of debris from space (e.g. the asteroid belt) become trapped in the Earth's gravitational field.

Most burn up with the friction of passing through the atmosphere, but some survive. Two categories of meteorite are known: stony and metallic. Stony meteorites are like peridotite (which has been recovered from volcanoes). Metallic meteorites are made of iron/nickel alloy. It is thought that they may be fragments from the core of a small terrestrial planet shattered by a gigantic impact early in the history of the solar system. Evidence for meteorite impacts is plentiful on the Earth and our planetary neighbours. See Chapter 1 for a further discussion and illustration of meteorite impact craters.

12.5 Using earthquake seismic waves as Earth probes

One of nature's most frightening and destructive phenomena is the earthquake. The origin, world distribution, measurement, risks and results of earthquakes are discussed in Chapter 13.

As we saw in Section 12.1, the lithosphere is not one continuous layer, but is made up of several large and numerous smaller tectonic plates, which 'float' on the asthenosphere below. An earthquake is defined as the vibration of the Earth caused by the sudden release of energy beneath the Earth's surface, usually as a result of the displacement of rocks along fractures (usually, but not always, along plate boundaries) known as faults.

Any movements (from footsteps to explosions to earthquakes) transmit energy through the ground in the form of wavelike vibrations which can be measured by seismographs. The shaking and destruction resulting from earthquakes are caused by two different types of seismic wave (Figure 12.3): body waves, which travel through the Earth and are somewhat like sound waves; and surface waves, which travel only along the ground surface and are more like ocean waves.

An earthquake generates two types of body wave, P-waves and S-waves:

- P-waves (primary or compressional waves) are the fastest seismic waves and travel quickest in dense rock and slower in fluids (i.e. their speed is related to the density of the material they pass through). They can travel through solids, liquids and gases. They are push–pull waves which move material forward and backward along a line in the same direction that the waves themselves are travelling. The material P-waves travel through is alternately expanded and compressed as the wave moves through it.

- S-waves (secondary, shear or shake waves) are slower than P-waves and can only travel through solids. They are shear waves because they move material perpendicular to their direction of travel, so produce shear stresses in the material they travel through at half the speed of p-waves.

P- and S-waves travel through rock at different speeds. They also respond differently to changing rock properties. Seismographs located around the world provide information on the arrival

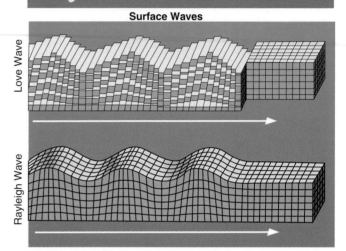

Figure 12.3 Seismic waves. (Source: United Sates Geological Survey). Reprinted from The Dynamic Earth, BJ Skinner & SC Porter, 2nd ed., 1992, p. 423, with permission from John Wiley & Sons Ltd.

times of P- and S-waves that have travelled through the Earth from many different locations. Using these records, seismologists have been able to calculate how the rock properties inside the Earth change and where there are distinct boundaries between layers with different properties. Thus, seismic waves are the most sensitive probes available for us to measure the internal properties of the Earth.

If the Earth's composition were uniform, the velocities of P- and S-waves would increase with depth, since higher pressure leads to an increase in density and rigidity, both of which control wave velocities. Early in the twentieth century a scientist named Mohorovičić noticed that for shallow earthquakes (focus <40 km deep), seismographs about 800 km from the epicentre recorded two distinct sets of P- and S-waves. He concluded that one pair must have travelled directly through the crust (solid red line in Figure 12.4) and that the other pair had arrived slightly earlier because they had been refracted by a boundary at some depth. The refracted waves (dashed red line in Figure 12.4), he deduced, must have penetrated a zone with higher velocity below the crust.

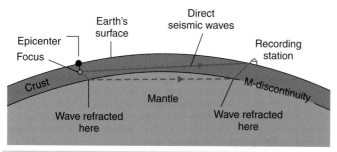

Figure 12.4 Travel paths of direct and refracted seismic waves from a shallow-focus earthquake to a nearby seismograph station. (Source: Skinner and Porter, 1992, p. 423). Reprinted from The Dynamic Earth, BJ Skinner & SC Porter, 2nd ed., 1992, p. 425, with permission from John Wiley & Sons Ltd.

Scientists now refer to this boundary as the Mohorovičić discontinuity (or Moho for short), which marks the base of the crust. Seismic wave velocities increase with depth in the continental crust from 6.0–6.2 km s^{-1} at depths of <10 km to 6.6 km s^{-1} at 25 km depth. Lower crustal velocities range from 6.8 to 7.2 km s^{-1} and in some parts of the continental crust there is

evidence of a small discontinuity at midcrustal depths, called the Conrad discontinuity. This varies in depth and character from region to region, suggesting that, unlike the Moho, it is not a fundamental property of the crust. P-wave velocities in the crust thus range between about 6 and 7 km s^{-1}, which laboratory tests show is consistent with rocks like granite, gabbro and basalt. Using seismic methods, it has been found that the crust varies in thickness from 20 to 60 km, and is thickest beneath major mountain masses.

Beneath the Moho, P-wave velocities are greater than 8 km s^{-1}, suggesting that these rocks are rich in dense minerals such as olivine and pyroxene. There is little direct evidence for what the upper mantle is composed of. However, rare samples of mantle rocks are found in kimberlite pipes, narrow pipelike masses of igneous rock, sometimes containing diamonds, that intrude the crust but originate deep in the mantle.

Both P- and S-waves are strongly refracted by the core–mantle boundary (C–M boundary) at a depth of 2900 km. P-waves reaching this boundary are so strongly reflected and refracted that the boundary casts a P-wave shadow from 103 to 143° in which no P-waves are observed (Figure 12.5, red lines).

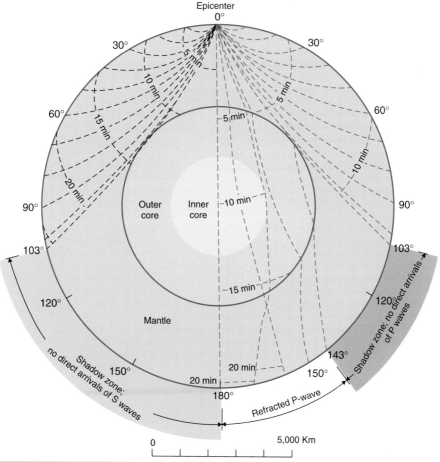

Figure 12.5 Paths of P-waves from an earthquake focus with epicentre at 0°, shown in red on the right. S-wave paths shown in black (left side). (Source: Skinner and Porter, 1992, p. 425). Reprinted from The Dynamic Earth, BJ Skinner & SC Porter, 2nd ed., 1992, p. 426, with permission from John Wiley & Sons Ltd.

The same boundary casts an even more pronounced shadow for S-waves (Figure 12.5, black lines), not because they are reflected and refracted like P-waves, but simply because S-waves cannot travel through liquids. The large S-wave shadow, therefore, tells us that the outer core must be made of liquid.

Seismic-wave velocities in the core suggest that the rock density is around $3.3 \, \text{t m}^{-3}$ at the top of the mantle, increasing to about $5.5 \, \text{t m}^{-3}$ at the base. Given that the average density of the Earth is $5.5 \, \text{t m}^{-3}$, in order to balance the less-dense crust and mantle, the core must have a density of $10-11 \, \text{t m}^{-3}$, such as is found in iron meteorites, which must have come from the cores of ancient planets.

P-wave reflections indicate the presence of a solid, dense inner core with identical composition to the outer core, but under higher pressure. Near the centre of the Earth, pressure rises to a value millions of times greater than atmospheric pressure. In the outer core, temperature and pressure are in balance and iron is molten, but in the inner core pressure increases more than temperature, so iron is solid.

While seismologists cannot determine any compositional boundaries within the mantle, seismic velocities do seem to be variable. As Figure 12.6 shows, seismic-wave velocities do not increase smoothly from the base of the crust to the core–mantle

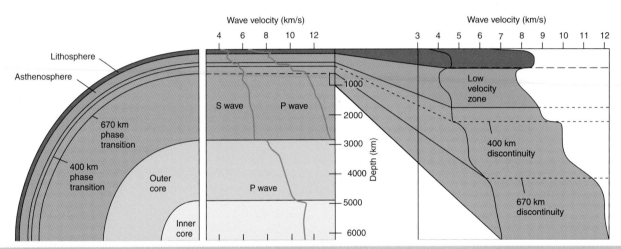

Figure 12.6 Variation in seismic-wave velocity within the Earth. Changes are obvious at the crust–mantle (Moho) boundary and the core–mantle boundary due to changes in composition. However, other changes occur at depths of 100 km at the lithosphere–asthenosphere boundary, and at 400 and 670 km. (Source: Skinner and Porter, 1992, p. 426)

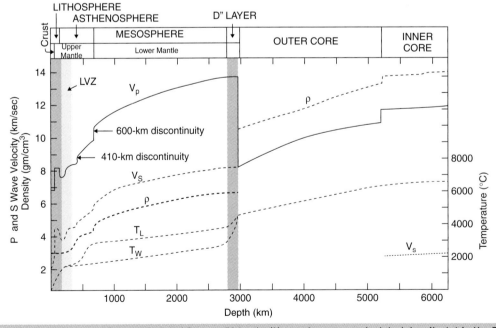

Figure 12.7 The distribution of average P-wave (V_p) and S-wave (V_s) velocities and average calculated density (σ) in the Earth. Also shown are the temperature distributions for whole-mantle convection (T_w) and layered mantle convection (T_L). LVZ = low-velocity zone. (Source: Adapted from Siegert, 2000)

boundary. There are sudden changes due to the physical properties of the mantle, with some zones being less rigid, less elastic and more ductile than adjacent ones.

Figure 12.7 gives even greater detail and shows that there are only two layers in the Earth with anomalously low seismic-velocity gradients: the lithosphere and the D'' layer, just above the core.

These layers coincide with steep temperature gradients – they are thermal boundary layers within the Earth. Both layers play an important role in the cooling of the Earth. Over 90% of cooling occurs through plate tectonics as plates are subducted deep into the mantle. The D'' layer is important in that steep thermal gradients in this layer may generate mantle plumes, many of which rise to the base of the lithosphere, thus bringing heat to the surface (<10% of total Earth cooling).

Exercises

1. Why do scientists believe that the Earth's temperature rose as it grew larger?

2. What might be the underlying reason for the Earth's 'layered' structure, with denser elements (e.g. Fe) at the centre and lighter ones (e.g. Si Al) at the surface?

3. What are the main differences between the inner and outer cores of the Earth?

4. Name the two categories of meteorites that have penetrated the Earth's atmosphere.

5. On what principle do seismographs work?

6. With reference to earthquakes, what are P- and S-waves?

7. What happens to seismic waves as they are transmitted through the Earth?

8. From what type of rock do we believe the mantle to be made?

9. What can explain the existence of P- and S-wave shadow zones?

10. Draw a simple annotated diagram to show the internal structure of the Earth.

References

Skinner, B.J. and Porter, S.C. 2003. *The Dynamic Earth: An Introduction to Physical Geology.* Chichester, John Wiley & Sons, Inc.

Further reading

Condie, K.C. 2005. *Earth as an Evolving Planetary System.* London, Elsevier Academic Press. Ch. 1.

Dutch, S.I., Monroe, J.S. and Moran, J.M. 1998. *Earth Science.* Boston, International Thomson Publishing. Chs 2–3.

Hamblin, W.K. and Christiansen, E.H. 2004. *Earth's Dynamic Systems.* 10th edition. Upper Saddle River, NJ, Pearson Education, Inc. Ch. 1.

Pipkin, B.W. and Trent, D.D. 2001. *Geology and the Environment,* Pacific Grove, Brooks/Cole. Ch. 3.

13 Plate Tectonics and Volcanism: Processes, Products and Landforms

Mount St Helens and Spirit Lake. Photo: Lyn Topinka, 19 May 1982, from USGS

Earth Environments   David Huddart and Tim Stott
© 2010 John Wiley & Sons, Ltd

Learning Outcomes

After reading this chapter and completing the exercises you will be able to:

➤ Understand how plates, basins and mountains are created.

➤ Understand the relationships between volcanic processes and the global tectonic model.

➤ Understand how and where magma is generated and the types of volcanic process associated with different types of eruption.

➤ Understand the relationships between volcanic processes and their potential global atmospheric effects and effects on biotic evolution.

13.1 Introduction

Our understanding of the Earth's history is a function of the quality of the record preserved in the Earth's changing pattern of environments, and as we see in this book, it is an enthralling story of climate change, mountain building, continental accretion and disruption, volcanism and many other processes which can radically affect the evolution of life. It is a story written, though not exclusively, in the sediments produced and in the fossils preserved within those sediments. Hence sedimentology, which is a lasting record of the Earth's changing geomorphology, is important because if no sediment were deposited, no long-term record would survive. Whether this sediment survives is a function of the availability of sedimentary basins, preferably those that are subsiding and can accommodate great sediment thicknesses. The long-term history of these basins in terms of such processes as closure, uplift and subsequent erosion is also key in determining preservation. We will see that the formation and evolution of these subsiding basins is mainly a product of plate tectonic processes, which is why plate tectonics is crucial to understanding the Earth's history and to explaining the large-scale patterns in the Earth's landscape.

Volcanoes have been extremely important as part of the Earth's system throughout the geological record. They have interacted with the atmospheric system by providing gases during eruptions which are likely to have controlled how life has evolved and they have influenced the Earth's climate over time. Volcanic processes have distinctively shaped the landscape across the planet, and since the human species has evolved, volcanoes have provided both natural resources, especially fertile soils, and problems to overcome in terms of recurrent, and different types of, volcanic disasters. However, our understanding of the processes has been greatly helped by the development of plate tectonics and the types of volcanic process and their effects are closely linked to which part of the global tectonic model a particular volcano is associated with.

In this chapter we will first discuss how these global tectonic processes have influenced the volcanic products and landforms, then where magma comes from and how it is transported to the Earth's surface; how volcanoes are linked to the Earth's atmosphere; how volcanic processes may be linked with the evolution of life on the planet; and finally in what ways the human species has suffered from volcanic hazards and disasters. Hence in this we will develop an understanding of the key plate tectonic processes, how these ideas developed into a model that explains much of the Earth's history and how volcanic processes fit into this model.

13.2 Global tectonics: how plates, basins and mountains are created

Ever since James Hutton first discovered that there were cycles of mountain building, erosion and deposition preserved within the stratigraphic record, there has been a mission to find out what the mechanism was for this mountain building. Throughout the nineteenth century the early geologists gathered information on the rock successions over increasingly large areas, which allowed the first global patterns to emerge. One of the key discoveries was that the thickest sedimentary rock sequences tended to be concentrated in long, linear belts. These belts, which have been heavily folded and faulted, occur along the margins of cratons (ancient continental crust). These belts of deformed rock form the Earth's main tectonic belts and consist of rocks that were originally deposited in deep, sedimentary basins, and which have subsequently been folded, faulted, thrust, intruded by igneous rocks and metamorphosed. The uplift of these deformed sediments and associated rocks has subsequently produced long mountain chains. It was the formation of these sedimentary basins and their subsequent compression that provided one of the most enduring problems for the nineteenth century geologists. However, these geologists began to interpret these tectonic belts as geosynclines, which were seen as regional troughs or basins that had gradually subsided as they were filled with sediments. This subsidence ultimately led to the intrusion of igneous rocks and the closure of the basin, causing intense folding and the formation of mountains. This type of hypothesis, relating geosynclines and mountain building, was the paradigm – with minor refinements and modifications – until the mid-1960s, when the concept of plate tectonics emerged and was applied to the formation and closure of the sedimentary basins, which resulted in the formation of tectonic belts. The theory of plate tectonics has unlocked the mystery of mountain building, explaining the mountain belts in terms of sedimentary basins that were opened and closed by the movement of giant lithospheric plates on the Earth's surface.

However, this idea of moving plates or continents was not a new concept in the 1960s and it had been occasionally canvassed throughout the nineteenth century. It found its major advocate in Alfred Wegener, who first articulated the hypothesis of continental drift in 1912. At the heart of this idea was the remarkable jigsaw-like fit between the Atlantic coasts of Africa and South America. More critically, he presented three lines of evidence based around palaeontology, palaeoclimatology and lithostratigraphy. Wegener argued that the leaf remains of the plant *Glossopteris* were common in the Permian successions of the southern hemisphere and that therefore the currently widely spaced continents must at some time have been joined. He also used the distribution of Late Palaeozoic tillites (glacial deposits), which occurred across southern Africa, India, Australia, South America and Antarctica, to argue that these continents must have been linked at that time and glaciated by the same ice sheets (see Chapter 30). Using these types of lithostratigraphy he was able to show that the tectonic belts on either side of the North Atlantic Ocean were remarkably similar, again demonstrating that the continents had once been juxtaposed. Wegener presented his findings in his book *The Origin of Continents and Oceans* (Wegener, 1915). However, the evidence presented was largely stratigraphic and he could not explain adequately the underlying causes of continental drift, and he was largely ignored by the contemporary geological community.

The idea of static continents remained the paradigm until the 1950s, when a range of geophysical observations began to accumulate, until the plate tectonic theory finally emerged in the mid–late 1960s. The key geophysical evidence that emerged was related to palaeomagnetism and global seismicity:

- **Palaeomagnetic data:** Iron-rich minerals within volcanic rocks contain a remanent magnetism, which reflects the position of the rock relative to the poles at the time it cooled below the Curie point. In molten rock the ferrous minerals are free to align in the prevailing magnetic field, but once cooled and crystallized their magnetic properties become set in stone. The location and polarity of the poles have changed through geological time and initially the palaeomagnetic data were used to define polar wandering curves. However, discrepancies began to appear between the polar wandering curves produced from different parts of the Earth's surface for equivalent time periods and these discrepancies could only be accounted for if the continents were in motion around the Earth's surface. Palaeomagnetic data was also to play a role in the study of the mid-ocean ridges, which had been identified in all the main ocean basins as linear mountain chains. Geophysical surveys in the 1950s had produced patterns of magnetic reversals on either side of these ridges. Throughout geological time the polarity of the poles had changed, with the north pole reversing to the south pole and vice versa at various periods. Each side of the mid-ocean ridges was a record of these polarity changes in the seafloor lavas.

- **Global seismicity:** The distribution of seismic events and both active and dormant volcanoes show a distinct pattern, which define a series of linear seismic and volcanic belts. These seismic and volcanic events appear not to have been random but were closely controlled, so that for example beneath ocean trenches there was a zone of seismic and volcanic activity which descended at an angle of around $45°$. This has now been called the Wadati–Benioff zone.

On the basis of this type of evidence the concept of plate tectonics emerged and was refined in the late 1960s. The basic premise was that the Earth's surface is broken up into a series of plates, which move about and interact with each other. The distinctive distribution of seismic events reflects the edges of these plates. For example, where two plates collide, one descends into the mantle beneath the other, generating a distinctive zone of earthquakes along the subduction or Wadati–Benioff zone. Where plates diverge, sea-floor spreading and mid-ocean ridges occur. In these locations new ocean crust is created; as it moves laterally away from the ridge zone, the rock cools, and as it cools the ferrous minerals take on the characteristics of the prevailing magnetic polarity. Hence changes in the polarity of the poles are recorded as magnetic anomalies on each side of the ridge system. The polar wandering curves and palaeontological, palaeoclimatic and lithostratigraphic evidence therefore are all explained, as though geological time plates have moved across the Earth's surface, bringing together and eventually splitting large continents, such as Gondwanaland.

13.2.1 Plate tectonics

The plate tectonic model proposes that the Earth's surface consists of a series of relatively thin but rigid plates that are in constant motion. The surface layer of each plate is composed of either oceanic crust, continental crust or a combination of both. The lower part consists of the rigid, upper layer of the Earth's mantle (the lithospheric mantle). On the surface of the current Earth, these plates vary in area from 10^6 to 10^8 km^2 and they can be up to 70 km thick if composed of oceanic crust, or 150 km thick if incorporating continental crust. However, the number of plates has varied, as plates are divided by new rift-spreading systems or are closed by collision. Currently, plates move at up to 100 mm per year, although the average rate of movement is about 70 mm per year.

There are two fundamental assumptions underlying the plate tectonic model: first, that the surface area of the Earth has not changed and consequently that there is a balance between the crust that is destroyed and the crust that is created at plate boundaries; second, that there is little internal deformation within plates relative to the motion between them. There are three types of plate boundary recognized: divergent, convergent and transform boundaries.

The geomorphology of divergent plate boundaries and associated sedimentary basins

Here plates move apart by sea-floor spreading and new oceanic crust is formed in the gap between two diverging plates as a constructive plate boundary. This boundary is usually marked by a mid-ocean ridge, or rise, and may be offset by strike-slip faults, which are known as transform faults (see Chapter 14). Plate area increases in this location and most divergent plate boundaries occur along the central zone of the Earth's major ocean basins. The Mid-Atlantic Ridge and East Pacific Rise are good examples of this type of plate margin; they provide a mechanism to calculate the rate of spreading from the pattern and age of the magnetic anomalies either side of a spreading ridge. Sections of crust have formed during periods of normal polarity, with a palaeomagnetic remanence orientated towards today's magnetic north; some sections of crust formed during periods of reversed polarity do not have such a remanence. The linear strips of magnetic anomalies that form a symmetrical pattern from a spreading centre and the boundaries of the reversal have been dated so that the spreading rate and direction of spreading can be calculated.

The geomorphology of convergent plate boundaries and associated sedimentary basins

At a convergent plate boundary, two plates are in relative motion towards one another and one of the two plates slides down below the other at an angle of around 45° to be incorporated back into the mantle along a subduction zone. The path of this descending plate can be determined from an analysis of deep foci earthquakes along the Wadati–Benioff zone, whilst the initial point of descent is marked on the surface by a deep ocean trench. Plate area is reduced along the subduction zone and hence these are sometimes known as destructive plate boundaries. The detailed morphology of convergent boundaries depends on whether the converging plates are both composed of oceanic crust, or whether one or both of the plates consists of continental crust.

When two plates of oceanic crust collide, a volcanic island arc forms. One of the two plates, the cooler and denser of the two, is subducted beneath the other and begins to melt at a depth of between 90 and 150 km. The frictional heat generated by subduction and the hydration of the mantle by water and sediment subducted with the ocean plate generate large amounts of steam and silica-rich (acidic) magmas, which rise through the over-riding plate above the subduction zone to form a chain or arc of volcanic islands (see Section 13.3.2). Traversing an island arc from the direction of the descending plate would show the following geomorphological zones:

1. An ocean trench, which marks the point at which the subducting plate descends into the mantle beneath the over-riding plate. These trenches form the lowest points on the Earth's surface. An example is the Marianas Trench.

2. An accretionary prism forms because most of the ocean floor sediment does not descend with the subducting plate but is sliced from it by the leading edge of the over-riding plate. A complex prism of thrust slabs of sediment and sometimes slices of seafloor builds up on the leading edge of the over-riding plate. The size of the prism is a function of the availability of seafloor sediment.

3. A fore-arc basin is a linear basin which may form between the leading edge of the over-riding plate and the chain of volcanic islands behind it. The width of this basin is a function of the descent angle in the subduction zone and if the descent is steep the fore-arc basin will be narrow. This angle is controlled by a range of factors, including the rate of plate movement, the age and therefore temperature and density of the subducting crust, the surface roughness of the incoming plate, the structural integrity of the incoming plate – for example if it breaks up or tears during descent then the angle and spread of subduction may increase – and finally the angle of convergence between the two plates.

4. A series or chain of volcanic islands is formed by rising magma generated by the subducting plate.

5. A back-arc basin may form to the rear of the chain of volcanic islands. At its simplest, this may be a marginal sea between a series of volcanic islands and a continent within the plate interior. Alternatively, it may form as a result of back-arc spreading. This is where thermal thinning of the crust leads to sea-floor spreading behind the island arc. The cause of this is not clear but the heat flux from the volcanism may soften and thin the crust. The development of a tensile stress regime is harder to explain, but is probably due to a function of the broader plate interactions.

This general picture of arc environments is complicated by the interaction of multiple subduction zones at complex margins, such as the interaction between the Pacific, Philippine and Eurasian plates in the Pacific in the vicinity of Japan, where a stacked series of subduction zones occurs. When the geological record of volcanic island arcs is noted, it provides a number of important observations: (a) the average life span of an island arc is between 20 and 80 million years, producing between 20 and 40 km^3 of material per million years of new ocean crust; (b) the structure of most volcanic island arc settings is unstable – volcanic arcs often split and subductions may abandon an arc or switch polarity, with the over-riding plate switching to become the subducted plate; (c) island arcs rarely survive intact and are usually accreted via arc–continent collisions on to continental margin orogens.

Where a plate composed of oceanic crust meets one composed of continental crust, the former is subducted. The molten rock produced in the subduction zone rises to form a volcanic chain along the edge of the continent, which is often associated with mountain building. The basic environments are similar to that of an island arc-trench, accretionary prism, fore-arc, volcanoes and back-arc, but their stability is greater since ocean crust will always descend below continental crust. In some rare cases continent–ocean crust subduction zones appear to be associated with volcanic island arcs as a result of back-arc spreading. This back-arc spreading may rift the edge of the continent, creating a chain of volcanic islands separated by ocean crust from the true continent. In contrast to volcanic island arcs, however, the chain of islands is not composed of new volcanic rock but of ancient continental crust. This type of scenario may be applicable to Japan, which is separated from the Eurasian plate by the South China Sea. In other areas, the back-arc basin consists of continental crust that has been thinned by heating and extensional tectonics to produce a region of horst and graben (see Chapter 14). The Basin and Range Province of North America is a good illustration. In other cases, sediments originally deposited within a back-arc basin may be compressed to produce back-arc fold-thrust belts, the foreland of which is sometimes referred to as a retro-arc basin.

One of the key features of many continent–ocean subduction zones is a chain of mountains and volcanoes, such as the Andes along the western seaboard of South America, or the North American Cordillera along the western seaboard of North America. The origin of such mountains is complex and is discussed later, in Section 13.10.4. Not all such margins need be compressive.

The geomorphology of transform margins and associated sedimentary basins

At a transform or conservative margin, two plates move laterally past each other and oceanic crust is neither created nor destroyed. Transform faults are common along divergent plate boundaries, where they occur at right angles to the diverging plates. The most famous example is the San Andreas Fault in southern California (see Figure 14.13), where two moving plates meet in western California. The Pacific Plate to the west moves north-westward relative to the North American Plate to the east, causing earthquakes along the fault. The lateral displacement between two plates is not always smooth and if the plate edges are irregular, zones of local compression (transpression) and extension (transtension) are created as the plates move past one another. Basins formed by transtension are characteristically deep in relation to their size, and their margins are associated with steep faults. Examples of this type of basin include the Gulf of California and the Dead Sea rift. An example of transpression is the Alpine fault in the South Island of New Zealand, which is responsible for the formation of the Southern Alps (see Figure 14.14.)

13.2.2 Plate motion mechanisms

Although there is general acceptance of the plate tectonic theory, the question 'What causes the plates to move?' has not been fully resolved. Several hyphotheses have been proposed to try and answer it.

The first suggests that flow in the mantle is induced by convection currents that drag and move the lithospheric plates above the asthenosphere. Convection currents rise and spread below divergent plate boundaries, and converge and descend along convergent margins. The convection currents result from three heat sources: the cooling of the Earth's core, radioactivity within the mantle and crust, and the cooling of the mantle. This type of hypothesis has been proposed in several forms over the last 60 years, but in each case the size of the convection currents and the amount of mantle that is in motion has been different. More recent work has questioned the convection hypothesis.

There is no doubt that mantle convection does occur at a range of different scales, but it is suggested that the pattern of divergent plate boundaries – the points of upwelling, according to the convection hypothesis – on the Earth's surface is not consistent with the pattern of heat convection within the upper mantle. Moreover, the presence of transform faults and the geophysical properties of divergent boundaries are also inconsistent with convection.

The second hypothesis invokes the injection of magma at a spreading centre, which pushes plates apart and so causes plate movement.

A third possible mechanism is gravity. It is suggested that oceanic lithosphere thickens as it moves away from a spreading centre and cools: a configuration that might tend to induce plates to slide, under gravity, from a divergent margin towards a convergent one.

Another hypothesis suggests that a cold, dense plate descending into the mantle at a subduction zone might pull the rest of the plate with it and thereby cause plate motion. This last mechanism has gained support in recent years but there is still widespread disagreement over the precise mechanism involved.

13.3 Volcanic processes and the global tectonic model

The Earth can be viewed as a thermal engine in which the heat produced is lost into the atmosphere by either conduction, or principally by convection, and plate tectonics is the simple consequence of the Earth's loss of this heat. Conduction occurs throughout the Earth's surface, but convection is effective only along restricted zones or at isolated spots. This is where the volcanoes occur, seismic activity is concentrated and the various types of plate boundary are located. The Earth's surface is divided into about 12 major, moving plates, pushed by convection currents in the asthenosphere and pulled by the weight of the diving plates. Magma is generated and accreted at divergent plate boundaries and generated in a complex fashion at convergent boundaries. The third major type of plate boundary, the transform fault, is not important for volcanic processes as there is neither destruction nor generation of material in this tectonic setting.

13.3.1 Magma generation

The Earth's interior is solid, with the exception of the outer core. At temperatures that would cause melting at the surface the mantle is mostly solid, as its melting point increases with increasing pressure. The mantle becomes close to its melting point at a mean depth of about 100 km and at a temperature of around 1350 °C. Overlying this thermal boundary is the lithosphere, composed of rigid plates of either oceanic or continental crust underlain by rigid mantle. Below this boundary is the asthenosphere, which is also solid but is able to convectively flow at a speed of between

1 and 10 cm per year. Partial melting of the mantle peridotite can occur in three ways:

1. Pressure release without heat loss (adiabatic compression melting) by upwelling rock along mid-ocean ridge systems or hot spots. At the divergent plate boundaries the mantle moves upwards, in part due to convection but also as a response to the removal of the lithospheric lid above it, which spreads laterally. The upwelling mantle partially melts in response to this process.

2. Increased heat supply, for example by conductive heating through underlying magma intrusions.

3. The addition of water and the lowering of the solidus by water and other volatile components brought into the mantle by subduction, despite the low temperatures and downward flow.

The magmatism at the subduction zones is due to water in the upper layer of the lithosphere and oceanic crust, both of which become hydrated as a result of sea-floor alteration and hydrothermal processes that occur at spreading centres. As this oceanic lithosphere is subducted to greater pressures and temperatures it undergoes metamorphism, which drives off fluids into the overlying mantle wedge. The subduction also causes counterconvection in the wedge-shaped region between the slab and the upper plate. This draws hotter mantle into the region above the slab, into which the volatiles are released. The overall effect is to cause melting, generating water-rich magmas that are more explosive on eruption than those in other tectonic zones. Depending on the global tectonic setting, either one or two of these factors can lead to partial melting of the mantle, melt segregation and the production of basaltic magma. The energy sources that heat the Earth's interior and cause mantle convection and volcanism are a combination of the primordial heat left over from the planet's accumulation and heat that has been produced by the breakdown of naturally-occurring radioactive elements such as potassium, uranium and thorium as they decay through time. However, once a melt is formed at depth it can move relatively quickly, at a rate of metres to kilometres per year, which separates the liquid from its source rock. These melt migrations occur by porous flow along grain-sized channels and by flow through larger conduits (Daines, 2000). As the melts typically differ in bulk composition from their source rocks, the combination of melt formation and rapid migration is the essential process of geochemical differentiation, whereby portions of the large-scale areas of the planet are chemically distinct.

13.3.2 Relationship of the plate boundaries to volcanic processes

It is obvious that there is a close relationship between volcanism and plate tectonics: volcanoes are situated at plate boundaries in narrow belts and are not random. The magma flux from the mantle is concentrated along these boundaries and is broadly

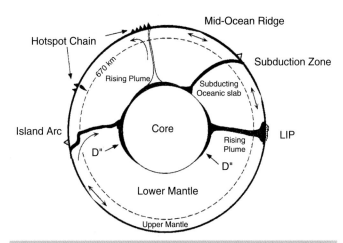

Figure 13.1 Hypothesized recycling of oceanic lithosphere into the mantle and the origin of mantle plumes from the D" seismic layer at the core–mantle boundary. (Source: after Perfit and Davidson, 2000)

correlated with heat flow. However, although this is the current state of affairs, there is no guarantee that it is a constant, steady state. In fact the flux of plume-related volcanism seems highly variable and at times in the Earth's history there have been flood basalt events in which enormous quantities of intraplate basalts were erupted. These are thought to have been the result of a large mantle plume head impinging on the base of the lithosphere (See Figure 13.1 and 13.2).

Mid-ocean ridge systems (MORs)

Around 60% of the Earth's surface is oceanic and divergent plate margins are responsible for about two-thirds of the Earth's yearly magma production (about 3 km^3 per year). Most of this ocean crust has been formed by magmatic processes that occur at mid-ocean ridges (MORs). These represent the largest, most continuous volcanic system on the planet, with a total length of over 60 000 km. In effect, they form a submarine mountain range, although it is broken into segments, or ridge axis discontinuities, which reflect breaks in the volcanic plumbing systems that feed the axial magmatism zone. The unusually elevated segments of some MORs, like Iceland, whose central graben is the extension of the Mid-Atlantic Ridge, or near the Azores, are directly caused by a nearby thermal plume. Where there are slowly spreading boundaries there is a low volcanic output, faulting and tectonism, whereas fast-spreading MORs are controlled more by volcanism. There is also a significant amount of crustal construction by intrusion, with a ratio of extrusion to intrusion at 1:7. The graben, or extensional rift valley, often forms above intruding dykes in volcanic rift zones because a dyke creates two zones of maximum tensional strain at the surface. Dyke intrusion releases horizontal tensional stress that has built up by divergent plate motion since the last rifting. When the dyke-induced graben forms above a dyke, the faulting occurs before the dyke reaches the surface and once an eruption begins lava can then fill, or partially fill, the graben. In the neovolcanic zone (where there

(a)   (b)

Figure 13.2 Flood basalts: (a) Seydisfjordur, Iceland. (b) East Greenland (Source: (b) Greenland Geological Survey, www.geus.dk/departments/geol-mapping/projects/dlc94-51-24c.jpg)

is volcanism and high-temperature hydrothermal activity) there can be a spectrum of morphological forms along the ridge crest axis, depending on the influence of volcanism and faulting/rifting, and this also depends on the rate of spreading (Table 13.1). The ridges are associated with vents (black smokers) which deposit sulphide compounds, and where percolating sea water reacts with hot rocks there is leaching of metals and hot hydrothermal jets rise through the sea floor. The black smoke is composed of metallic and other minerals, which precipitate as the smoker cools.

Table 13.1 Characteristics of MORs where the spreading rates are different. (Source: after Perfit and Davidson, 2000)

Type of MOR	Properties
Slow-spreading ridges (10–40 mm per year), e.g. the Mid-Atlantic Ridge	Large axial valleys, 8–20 km wide, 1–2 km deep; magmatism relatively unfocussed, both across and along the axis; dominantly bulbous, pillow lava constructing hummocks <50 m high and <500 m diameter, hummocky ridges between 1 and 2 km long, small circular seamounts that often coalesce to form axial volcanic ridges along the inner valley floor; high density of faulting; larger earthquakes common reflect dominance of tectonism over volcanism; volcanic events relatively infrequent due to low magma supply, thicker crust and deeper extent of hydrothermal cooling
Fast-spreading ridges (80–160 mm per year), e.g. the South-East Pacific Rise	No axial valley but along most of an axial crest there is a narrow, linear trough (5–40 m deep, 40–250 m wide) marking the locus of the neovolcanic zone, which is very narrow (<2 km); direct consequences of the fast spreading rate are greater magma supply, shallower and more steady-state magma reservoir and more frequent intrusive events; lavas predominantly thin, fluid sheet flows

The volcanic rocks that make up the ocean crust are dominated by low K_2O, olivine, tholeiitic basalt, with very low percentages of volatile elements and a narrow range of major element compositions. It is this, coupled with the great hydrostatic pressures on the sea floor, that explains why explosive eruptions on MORs are very rare. These basalts result from high degrees of partial melting (20–30%) at low pressures, corresponding to a depth of a few tens of kilometres.

Volcanism at convergent plate boundaries

Volcanism here is destructive and marks the subduction of the ocean lithosphere into the mantle. The characteristics of destructive plate boundaries in island arcs are as follows:

- Arcuate chains of islands or linear volcanic belts that are long (100s to 1000s of km) and relatively narrow (200–300 km).

- A deep oceanic trench (6000–11 000 m deep) on the oceanic side.

- Active volcanism where there is an abrupt oceanward boundary to the volcanic zone called the volcanic front, usually parallel to and 100–200 km from the oceanic trench.

- A dipping zone of seismicity (Wadati–Benioff zone) that includes shallow, intermediate and deep-focus earthquakes, marking the descent plane of the oceanic lithosphere into the mantle.

- A characteristic volcanic association, with a range of varied igneous activity, making it a complex set of magma-generating environments.

Potential sources for magmas are varied but include: the mantle wedge above the subducted slab, which consists of (a) a 40–70 km thick section of oceanic lithosphere and (b) a zone of asthenospheric upper mantle of varying thickness; the oceanic crust, consisting of (a) variably metamorphosed ocean-floor basalt, dolerite and gabbro, (b) oceanic sediments; sea water, which is incorporated during hydrothermal alteration of the oceanic crustal layer during ocean-floor metamorphism and by direct circulation of sea water in the island arc crust.

The role of aqueous, slab-derived fluids and partial melts is crucial and appears to distinguish magma generation in the subduction zone from other zones of magma generation. The partial melts of the subducted oceanic crust rise into the mantle wedge, where they react and lose their chemical identity (Figure 13.3). These slab-derived fluids lower the mantle solidus, promoting partial melting, and can be regarded as a catalyst for the bulk of island arc magmatism rather than as a prime source.

Magmatism at these boundaries creates the linear or curved volcanic chain known as a volcanic arc, which can be constructed on oceanic (to create an island arc, as found around the western Pacific Ocean Basin, e.g. the Aleutians or Kuriles) or continental (like the volcanic chain of the Andes) lithosphere. This chain is broadly parallel to the trench on the sea floor that reflects the surface trace of the plate boundary. Behind some volcanic arcs, secondary sea-floor spreading occurs, which results in the development of back-arc or marginal basins. The processes are similar to those on MORs, although the end products may be more geochemically complex.

Subduction is a transient state: the subducted lithosphere is always oceanic, as continental lithosphere is thicker and more buoyant, and cannot be effectively subducted. When two fragments of continental lithosphere collide, the interaction produces deformation and uplift to produce fold mountain belts. Such collisions stop subduction at that location, although new subduction zones will be developed as a response to the new plate organization and stress change. This destructive process is a necessary counterbalance to sea-floor spreading at mid-ocean ridges as the Earth is not expanding. Oceanic lithosphere must be destroyed as fast as it is produced and the scale of such subduction is illustrated by the fact that the Atlantic and Pacific Oceans have been created over the past 200 million years by sea-floor spreading, and hence a comparative amount of lithosphere must have been subducted.

There is a consistent relationship between the position of the volcanic arc front and the depth to the Wadati–Benioff zone (the inclined zone of seismicity which corresponds roughly to the slab top). This depth of 100–150 km implies that the trigger for magmatism is pressure-sensitive. There are compositional variations in the magmas produced both in time and spatially from basalts to rhyolites: early magmatism at oceanic arcs produces basaltic and tholeiitic magmatism, while later magmatism tends to produce more calcalkaline and andesitic magma, with the latter type predominant overall. Classically there is a spatial variation of increasing alkalinity away from the trench, which led to the development of a K–h relationship, whereby the K_2O content of the magmas at a fixed SiO_2 value was apparently correlated with the depth to the Wadati–Benioff zone. Many arc systems do not follow this simple pattern, however.

At greater distance beyond the main volcanic chain there is an extensional zone due to the oceanward migration of the over-riding plate. At island arcs a small spreading centre may develop in a back-arc basin with the accretion of tholeiites. The volcanoes that form are diverse in type and distribution, but the main type is the composite cone, composed of blocky lavas and pyroclastics. Indeed, explosive volcanism is more typical than the effusive eruptions of intraplate and divergent plate margins. This is because of the higher water content in the primary magmas. The relatively protracted differentiation of the primary magmas to produce andesites through rhyolites that are erupted at arc volcanoes exacerbates the explosivity by concentrating the volatiles into progressively more viscous magma. An interesting discussion has taken place since the Sumatran or Indian Ocean earthquake of 26 December 2004, relating these large megathrust earthquakes to volcanic processes (see Box 13.1, see in the companion website).

Destructive plate boundaries where one plate is continental

Magmas generated in this tectonic environment occur along the western coasts of the Americas and in Japan, New Zealand and the Mediterranean. The andesite association of island arcs

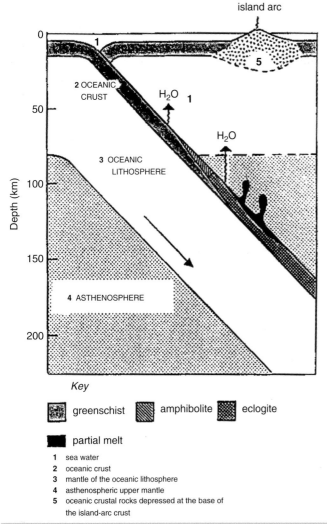

Key

greenschist   amphibolite   eclogite

partial melt

1 sea water
2 oceanic crust
3 mantle of the oceanic lithosphere
4 asthenospheric upper mantle
5 oceanic crustal rocks depressed at the base of the island-arc crust

Figure 13.3 Potential sources involved in island arc magma genesis. (Source: after Wylie, 1988)

also typifies this volcanic suite, although the passage of magmas through thick continental crust creates extra complexities. There is, for example, the greater abundance of more silica-rich magma, such as dacites and rhyolites, much occurring as pyroclastic flows, particularly associated with zones of thickened continental crust, which may have been partially melted.

Figure 13.4 shows a simplified cross-section of an active continental margin. Upon subduction, the cold oceanic lithosphere is heated by the combined effects of friction and conduction, and a series of metamorphic transformations take place in the rocks, from greenschist through amphibolite to ecologite facies. These metamorphic reactions involve dehydration, and the resultant hydrous fluids are released into the mantle wedge, where they lower the solidus and activate partial melting. Any mantle-derived magma passing through a thick section of continental crust must interact with it and therefore assimilation and fractional crystallization processes must be important. Melts are generated at depths of 80–200 km, producing andesitic-predominant volcanic products and massive granite intrusions. These igneous intrusions contribute to a crustal thickening, which in the Andes is around 70 km and is heavily faulted and folded. Such thick crusts along continental margins may well have important implications for our understanding of andesitic melts (Davidson *et al.*, 1990).

Thermal plumes or hot spots

Although plate tectonics and the thermal-boundary-layer model explain much of the surface global volcanism, an exception is provided by thermal plumes or hot spots, with about 12% of magma extruded there. Each hot spot has a chain of volcanoes, showing a systematic age dependence along the chain, making it appear that the volcanism is the result of a stationary source of mantle heat underlying the moving tectonic plate (Figure 13.5). The Hawaiian–Emperor volcanic chain is the largest in terms of heat flux and lava volume, but the Galapagos chain is also a good example; both are found in the Pacific Ocean. The plumes are thought to originate from the lower mantle (as in Figure 13.1), although it is possible that hot spots indicate the surface markings of heat coming from the Earth's core (Jeanloz, 2000). Volcanic activity also indicates that mantle convection is the process by which the Earth has geologically evolved over its 4.5 billion-year history. Geochemistry indicates that the magma melts are partial melts from multicomponent sources, involving original mantle, ancient recycled oceanic crust with basalt and sediments, depleted asthenosphere, depleted oceanic lithosphere and recycled continental lithosphere. Studies of the Hawaiian Islands illustrate that the volcanoes show a characteristic evolutionary sequence from a voluminous, tholeiitic, shield-building phase to a late-stage, alkali basalt phase. Traditionally such sequences have been explained by models where the early eruptives represent moderately large

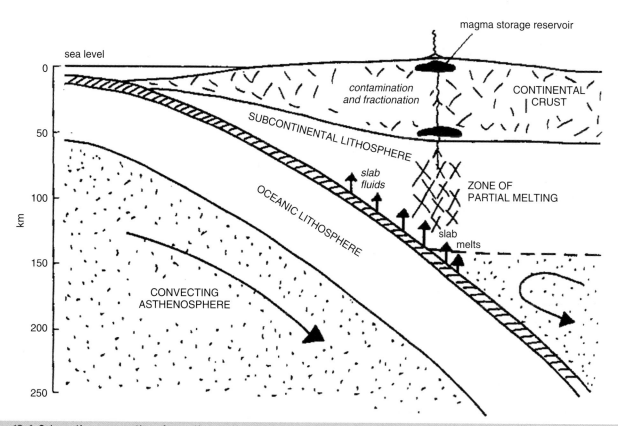

Figure 13.4 Schematic cross-section of an active continental margin. (Source: after Wilson, 1993)

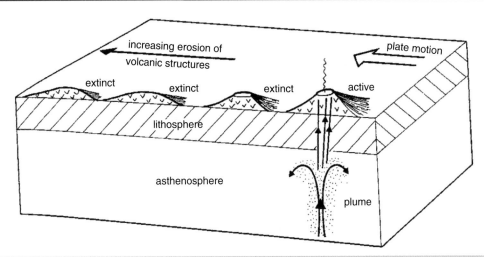

Figure 13.5 Hot-spot model for the generation of linear volcanic chains. (Source: after Wilson, 1993)

degrees of partial melting, at relatively shallow depths, while the late stages represent smaller degrees of partial melting at greater depths. However, in Hawaii the geochemistry precludes derivation from a compositionally homogeneous source and the explanation for Hawaii seems to be that partial melts from the rising plume interact with partial melts of the wall rock, presumed to be oceanic lithosphere. The voluminous tholeiites have isotopic characteristics close to the original mantle and may represent relatively uncontaminated partial melts of the plume source. After the shield-building stage, the volcano gradually moves away from the hot spot and the supply of plume material decreases. The last stage is dominated by partial melting of the lithosphere (Figure 13.6(a),(b)). Evidence in Box 13.2 (see the companion website) suggests that recycling can be recognized from distinctive isotopic signatures.

Tholeiitic flood basalts and continental rifting

When there is a burst of magmatism accompanying the rise of a big plume head to shallow depths, this may have a major effect on the adjacent lithosphere, depending on whether it is oceanic or continental. If the lithosphere is continental it may be uplifted, domed and eventually rifted apart (Figure 13.7). Voluminous magma outpourings form a flood basalt province, flooding the landscape and creating a new subdued-relief landscape – up to the order of 100 000 km³, erupted in a few million years. Some are up to 550 000 km², with repeated lava flows comprising piles several thousand metres thick. They are fed by fissure swarms that migrate with time. A good example is the Deccan Traps in India, with a present area of 500 000 km² and an initial volume estimated at 1 500 000 km³. Eruptions lasted for 5 million years but most of the volume was created in only 0.5 million years, at the site of a new hot spot when India was several thousand kilometres to the south of its current location, near the present position of Reunion Island. The long chain of volcanic islands and aseismic ridges from Deccan to Reunion is the waning trace of that hot spot on the moving Indian plate. Many individual flows in flood basalts have volumes exceeding 10 km³, such as

the 1783 Laki flow in Iceland (Figure 13.8), and giant fields, such as the Columbia River in Washington State, USA, contain flows over 1000 km³ in individual volume. Vents are marked by crater rows of spatter and cinders.

If rifting of the continental lithosphere continues to completion, new oceanic lithosphere is formed and a new divergent plate boundary develops. At this stage the plume and the divergent boundary are coincident, and they will remain so until the plate boundary migrates. This can be seen in the mid-Atlantic, where many hot spots are found close to the Mid-Atlantic Ridge, reflecting their earlier role in the rifting apart of the continental landmasses now located on either side of the basin, in Iceland and east Greenland, for example. Here the Tertiary lavas are thought to reflect the earliest stages of rifting in the North Atlantic region 60 million years ago. If the lithosphere above the plume is oceanic, the end product is an oceanic large igneous province. Plateaus such as the Ontong (Java) and Manihiki, in the western Pacific Ocean, represent some of the largest outpourings on the planet, up to 20 km thick, with a total volume estimated at 100 million km³. In such cases the stronger ocean lithosphere does not rift apart; the increased thickness from the lavas causes isostatic compensation and increased elevation as an extensive submarine plateau. The relative thinness of the oceanic crust as compared to the continental crust allows a larger melt volume to form at the same temperature.

Fundamental chemical and isotopic differences exist between continental flood basalts and oceanic tholeiitic magmas. In general the continental basalts show greater elemental and isotopic diversity, which has been attributed to a variety of processes, including crustal contamination, melting of enriched subcontinental mantle, mixing between depleted and enriched mantle sources, and combinations of melting of enriched mantle and crustal contamination.

Continental rift zone magmatism

This is a dominantly more alkaline – but with a very wide magmatic spectrum – volcanic province, associated with rift/graben

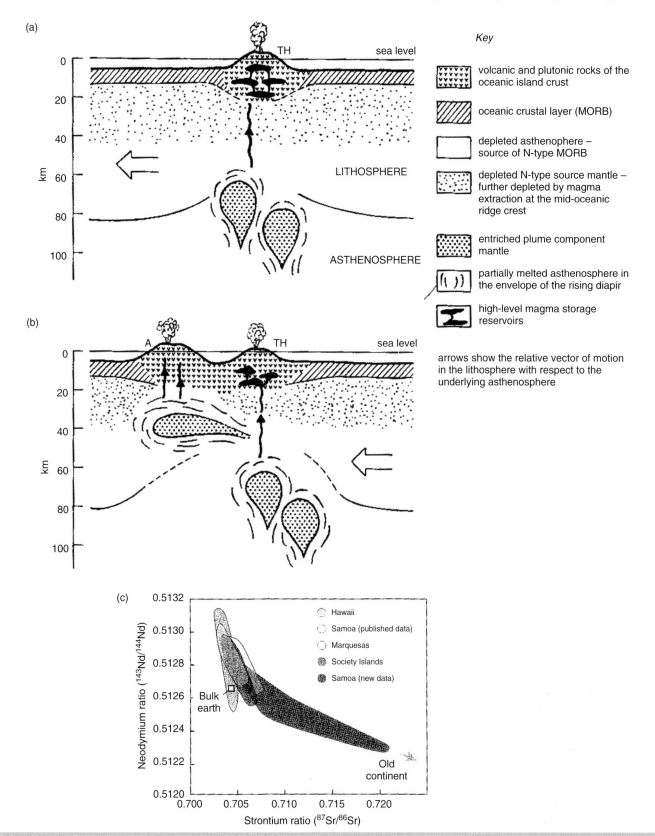

Figure 13.6 (a) and (b) Models for the evolution of a Hawaiian volcano. (c) Adopting continental values. (Sources: (a) after Wilson, 1993; (b) after Hofmann, 2007). Reproduced, with permission, from www.geus.dk/departments/geol-mapping/projects/lavaplateau-k.htm, Copyright: Geological Survey of Denmark and Greenland (GEUS)

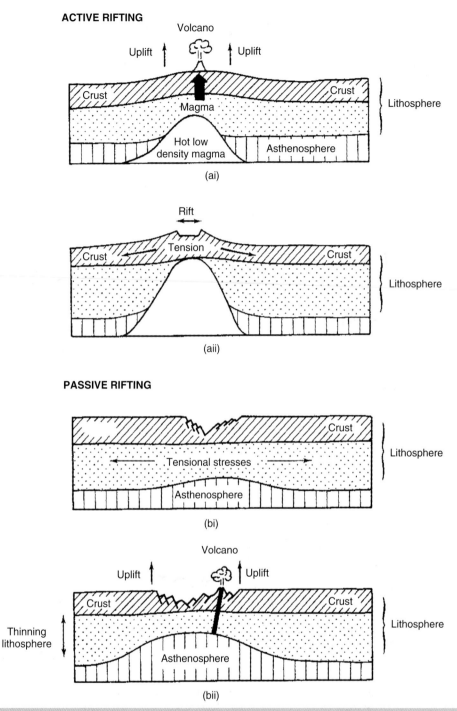

Figure 13.7 Active and passive rift formation. (Source: after Summerfield, 1991)

structures, fed from fissures or central vent volcanoes. It forms areas of localized lithospheric extension, characterized by a central depression, uplifted flanks and a thinning of the underlying crust. High heat flow, broad zones of regional uplift and magmatism are often associated with such structures. They are usually a few tens of kilometres wide and tens to a few hundred kilometres long, and they are usually thought of as areas of incipient continental fragmentation which precedes the creation of new ocean basins, like

the classic example of the East African rift system. There has been discussion as to whether rift zones are produced by upwelling mantle, splitting the continent along preweakened zones (active rifting), or whether the mantle is forced to rise as the continents are pulled apart during lithospheric stretching (passive rifting). However, available geophysical data suggest that regardless of the original mechanism, asthenosphere upwelling is a fundamental feature of all continental rift zones (Figure 13.7). In active rifting

Figure 13.8 Laki, Iceland, 1783 eruption. (a) Linear eruption. Photo looks south-west along the fissure. (b) Areal extent of the eruption. (c) Explosion crater and cone. (d) Linear rift from which lava poured. (Sources: (a) and (b): http://www.killerinourmidst.com/permian%20world.html)

the expected event sequence is volcanic activity, uplift and rifting, whereas in passive rifting the sequence is rifting, volcanism and uplift. In East Africa it is thought that the uplift preceded rifting and so the whole system is active, but the evidence is equivocal and there has been dispute about whether the rifting might have occurred before uplift.

The basaltic magmas erupted are predominantly alkalic, and in general more explosive and pyroclastic rocks may locally dominate the volcanic sequence. This suggests an enrichment of volatiles in the magma source region. The volcanic products can vary across a rift, with flank lavas tending to be more alkaline than those erupted in the axial graben. It has long been held that the main reason magmas produced at continental rifts are so varied in comparison with MOR basalts is their differing depths of origin. However, it has also been argued that the wide variety of volcanic products is due to the selective removal of some minerals, such as olivine, pyroxene and garnet, from the melts, although contamination by crustal rocks may be important too. There is a detailed discussion of the controls on such magmatic processes in Wilson (1993), Chapter 11. In contrast to most continental rifts, which are narrow, the Basin and Range Province in the western United States and northern Mexico is several hundred kilometres across and volcanically diverse. It is characterized by:

• A high mean elevation.

• A marked geothermal heat flux.

• A distinctive topography of basins and dissected uplands.

• Strong seismic as well as volcanic activity.

• A varied rock sequence, ranging from tholeiitic and alkalic basalts to silica-rich, calc-alkaline rocks, the latter associated with ignimbrites and large calderas.

There has been much discussion as to the geological interpretation. Henyey and Lee (1976) and Eaton (1984) have argued that the province is the terrestrial equivalent of an oceanic back-arc basin. Figure 13.9 summarizes one such model for its formation.

Summary of magma generation

We can say that basalt magmas are generated by partial melting of the peridotite mantle by adiabatic decompression beneath MORs to give theoleiitic basalts, and by adiabatic decompression in mantle plumes and depression of the melting point by the addition of volatiles in subduction zones (tholeiitic and calcalkaline basalts). In continental regions, basalts can be produced by adiabatic melting within mantle plumes in regions of extension, although they can also be generated as a result of conductive

heating of lithospheric mantle that contains a small number of volatile elements. The generation of magma at subduction zones is a consequence of the presence of water in the magma source region. The water is supplied to the mantle by the dehydration of water-bearing minerals as they are carried down in the subducted lithosphere. These change and produce water-rich melts at depths between 115 and 75 km, which rise and create more melts as they go. The end products are a range of basaltic, andesitic and rhyolitic magmas.

So partial melting of the mantle rock produces the primary magmas of basic or ultrabasic composition in most tectonic settings, but subsequent differentiation processes, including fractional crystallization, magma mixing and crustal contamination, are responsible for the wide spectrum of surface igneous rocks. The geochemical characteristics of the primary magmas depend upon such factors as the source composition and mineralogy, and the depth and degree of partial melting, all of which can vary depending on the tectonic setting. Most primary magmas seem to be generated within a restricted depth range within the upper 100–200 km of the mantle. Kimberlites, which appear not to be forming as a present-day process, seem to be derived from the deepest terrestrial magmas, originating at depths greater than 200–250 km in the mantle. An excellent discussion of these kinds of igneous process and the geochemistry of magmas is given in Wilson (1993). What is apparent in general is that the plate tectonic model explains the spatial pattern of volcanism.

13.4 Magma eruption

13.4.1 Introduction

Most of the processes operating within magma chambers lead to volume increases, stresses on the crustal rocks and sometimes movement of magma, which is normally less dense, from the

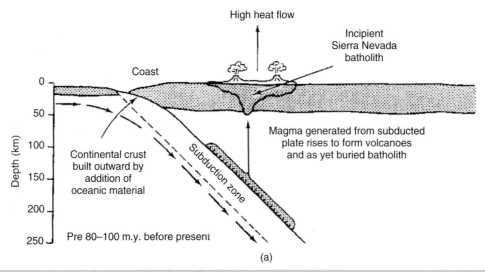

Figure 13.9 The origin of the Basin and Range Province. (Source: after Henyey and Lee, 1976)

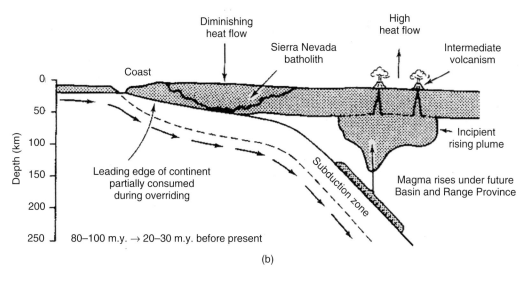

(b)

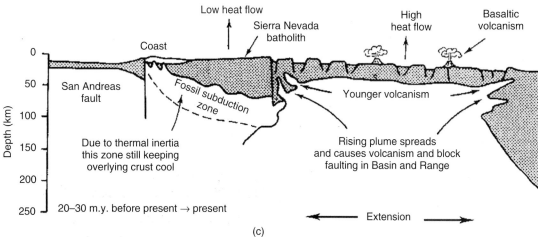

(c)

Figure 13.9 (*continued*)

chamber to the surface as a volcanic eruption through conduits. The processes here include the expansion of volatiles (especially water, CO_2 and SO_2) and the solution of volatiles in magma, which is controlled by pressure. As the pressure is reduced towards the surface, the gases exsolve. However, silica-rich melts are viscous and the higher the viscosity, the more restricted the migration of bubbles. Here eruptions are typified by dome-building, rhyolitic lavas and the development of pyroclastic flows. In silica-poor basalts, though gases are free to exsolve as pressure is reduced and eruptions are characterized by the quiet release of gas, lava flows dominate and pyroclastic sediments are spatially confined. Between these two eruption types are the magmas of more intermediate composition, usually andesite, which typically produce high eruption columns that spread gases and pyroclastic fall deposits over a large area. During the waning of the columns there are often pyroclastic flows and surges. The result is that there are three basic eruption types (Figure 13.10), based on the viscosity and temperature of the magma composition, which give rise to various volcanic styles

and major, subaerial, volcanic landforms (Table 13.2). Volcanoes show huge variations in eruption volumes and rates, and vary widely in their destructiveness. Hence a logarithmic scale is necessary to categorize the sizes of the eruptions. One such scale is the Volcanic Explosivity Index (Table 13.3), which is based on both magnitude (erupted volume) and intensity (eruption column height).

13.4.2 Physical properties of lava

The principal physical attributes for various lavas are summarized in Table 13.4.

The viscosity, or internal resistance to flow, of a lava is controlled by many factors (see Table 13.5) and has a major influence on lava mobility, geometry and geomorphology, but the subject still generates much debate. Due to their relatively high viscosities, most lavas flow in a laminar fashion, which can give a distinctive texture caused by the crystal orientation, vesicles, bubbles or inclusions parallel to the flow direction.

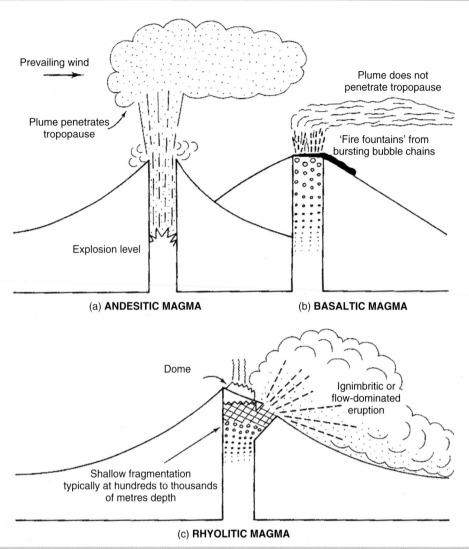

Figure 13.10 Styles of volcanic eruption. (Source: after Clapperton, 1977)

Table 13.2 Major landform types at sites of basaltic, andesitic and rhyolitic volcanism

Magma type	Typical landform
Basalt	Lava shields created from successive flows; low slope angles, usually <6° (Figure 13.12); sometimes a summit crater or lava erupted from flank, parasitic cones; more silica-rich basalts can produce strombolian activity and build strato-shield cones, like Etna
Andesite	Plinian activity, with calc-alkaline magma generation, produces strato-volcanoes composed of both lavas and pyroclastic material (Figure 13.13). Slope angles about 30°
Rhyolites	Silica-rich, viscous magma often produces domes (Figure 13.28(b.ii))

13.4.3 Effusive volcanism

The main requirement for effusive volcanism is the eruption of a magma that is not explosive, or that only has small fire fountains caused by minimal gas escape. This means that there must be a low volatile content in the magma to prevent the build up of gas pressure, which might cause explosive fragmentation.

Effusive basalts

These lavas make up over half of all volcanoes. They form lava shields, strato-volcanoes, monogenetic volcano fields, flood basalts and central volcanoes (which also erupt silicic magma), and can also form under the sea floor and under ice. Flood basalts are mainly pahoehoe flows and are typified by well-developed columnar jointing; they erupted as vast lava lakes which took many years to cool and solidify, during which period the jointing developed. Their characteristics are illustrated in Figure 13.11 and Table 13.6.

Table 13.3 The Volcanic Explosivity Index

VEI Index	0	1	2	3	4	5	6	7	8
General description	Non-explosive		Small		Moderate		Moderate–large		Very large
Qualitative description	Gentle		Effusive		Explosive		Cataclysmic		Paroxysmal
Maximum erupted volume of tephra (m^3)	10^4	10^6	10^7	10^8	10^9	10^{10}	10^{11}	10^{12}	10^{13}
Eruption cloud column height (km)	<0.1	0.1–1	1–5		3–15		10–25		>25
Classification	Hawaiian		Strombolian		Plinian		Ultra-Plinian		Vulcanian

Table 13.4 Principal physical attributes of various lavas

Attribute	Description
Composition	Volume and length of a lava flow is correlated with silica content of parent magma; rhyolitic, andesitic and basaltic are in a sequence of increasing volume and length
Volume	~90% of subaerial flows are basaltic in composition; siliceous lavas >1 km^3 are very rare
Temperature	In Hawaiian volcanoes, around 1200 °C; most others are lower, especially silica-rich lava
Dimension	Largest are continental flood and mid-ocean basalts; largest historical flow was 40 km Laki fissure flow, but Miocene Rosa flow (Columbia River, USA) is 300 m; central basaltic cones have lavas much less and rhyolitic flows are <4 km; thicknesses vary greatly: Columbia River basalts average 25 m, Hawaiian from a few cm to 15 m; andesitic up to 30 m; and rhyolitic short and thick
Discharge rate	Effusion rates depend on lava viscosity and the efficiency of the eruptive conduit. Most eruptions usually decline with time or fluctuate: high initial rates due to the high gas content; basalts $0.3 \geq 45 \ m^3 \ s^{-3}$, more silica-rich <1$m^3 \ s^{-3}$; many prehistoric flood basalts an order of magnitude higher
Texture	Depends on a number of factors, including cooling rate, gas content and temperature
Cooling, velocity	High thermal conductivity and high heat capacity; cool slowly, mainly from surface radiation; flow velocity depends on such factors as effusion rate, viscosity, volume and ground slope: basalts are fastest, up to 64 km per hour but usually much slower; reduced velocity with distance from source

Table 13.5 Controls on lava viscosity

Attribute	Effect
Temperature	Viscosity increases as temperature falls, partly due to crystallization but also due to composition
Chemical composition	Silica and aluminium form bonds with oxygen which are far stronger than other oxygen bonds and therefore silica-rich magmas are generally more viscous than silica-poor magmas
Crystals	The greater the crystal content, the higher the viscosity
Volatiles	At a given temperature the magma viscosity will decrease with increasing water content; but it is more complex than that as solubility depends on other factors, including pressure, temperature and the presence of other volatiles
Pressure	Viscosity reduced as pressure increased in laboratory experiments
Bubbles	In basalts, exsolution of water and other volatiles has little effect on producing low viscosity because viscosity is already low due to temperature and composition, but bubbles will lower the viscosity further; in more acidic magma, the high viscosity may be influenced by the presence of bubbles

Basalts from central volcanoes are well known from Iceland, Hawaii and Etna. They can erupt from coherent flows from openings, from overspill or from breaching of a lava lake that is ponded in the crater, or of fire fountains that reconstitute around the vent and then flow away. Repeated lava eruptions give rise to the classic, low-angle, shield volcanoes (Figure 13.12) and the much higher and steeper, conical, composite strato-volcanoes, which also incorporate pyroclastic materials (cinders and ash). Many have parasitic cinder cones and craters on their flanks and several have been truncated by caldera collapse, such as Vesuvius. A complicated example is given in Figures 13.13(c)

Figure 13.11 Pahoehoe lava. (Source: Wikimedia Commons)

and 13.14, showing the evolution of the Nevado de Toluca in Mexico, where there are lava flows, pyroclastic stages and dome collapse. Monogenetic volcanoes erupt only once, are usually cinder cones – commonly with lava flows – and normally occur in fields (clusters); there are many examples in Mexico, with the Michoacan–Guanajuato field containing over 1000 volcanoes. The fields can be active for 50 000–5 000 000 years. Classic examples are Paricutin (Figure 13.13), which started to erupt in 1943 in a farmer's field, and the Chichinautzen field south of Mexico City.

13.4.4 Lava flows

Lava flows are the most common volcanic landform on this planet, with melting temperatures in the range 800–1200 °C, although rare lavas of carbonatite composition occur (for example, Ol Doinyo Lengai volcano in Tanzania) with temperatures as low as 600 °C, and sulphur lavas down to 150 °C (for example, Lastarria volcano in Chile). Flows are differentiated from lava domes by their extreme elongation downslope. Whether pahoehoe (smooth surfaced), aa (with a thick upper zone of loose rubbly debris) or blocky lava forms occur seems to depend on the rate of shear and viscosity. The change from pahoehoe to aa occurs as viscosity and/or rate of shear increases and a threshold is crossed. This may be caused by factors like an increased viscosity due to cooling and increased shear rates as a flow encounters a steeper slope. If a flow is aa throughout then it is produced by high viscosities and/or high shear rates, usually involving a high eruption rate

whilst the converse applies to flows which are solely pahoehoe. The change from pahoehoe to aa also occurs because, as lava advances downstream, a greater proportion of the gravitational energy flux for the movement of the whole flow is used in deforming a thickening crust. Eventually the energy flux exceeds the critical value, after which the surface must break as aa if the flow is to advance. Blocky surfaces are associated with stronger and more viscous lavas that break before cooling is significant. The aa–blocky transition occurs therefore when the lava interior becomes too crystalline to move forward by flowing alone, so that the surface fracturing is no longer controlled by crustal cooling. The transitions between surface morphologies are irreversible. Once erupted, the lava can be described as simple, meaning it is composed of a single flow unit, or compound, meaning it is made up of several units or overlapping lobes.

Lavas derived from rhyolitic magmas erupt violently but infrequently and are usually associated with the formation of pyroclastic materials. Domes, however, are mounds of rock of silica-rich lavas, as the high viscosity stops flow close to the eruptive vent. There are various classifications of such landforms; these are shown in Figure 13.15. It is known that rhyolitic lavas can flow as far as 6 km, as from the Long Valley caldera, California, and they can be 50–60 m thick. The danger comes not from their growth but during their collapse, as at Unzen (Japan), Soufriere Hills (Montserrat) and Mount St Helens (USA), as we will see later.

Lavas derived from dacitic and andesitic magmas are intermediate in composition and have characteristics in common with both basaltic and rhyolitic flows. The classic landform is the block lava flow, which is of small volume, short in length, slow-moving and has a surface of partially solidified, angular detached blocks (Decker and Decker, 1991). After cooling, the flows pass down into massive flows, although sometimes they are columnar-jointed. Flow fronts are steep and, although their movement mechanisms are similar to those of aa flows, because of the higher viscosity there is greater internal shearing along planes subparallel to the surface over which they passes (Figure 13.16). The largest known example is the Chao lava in Northern Chile, which is a 14.5 km long dacite flow, with flow fronts up to 400 m high.

Subaqueous basaltic lavas form extensively on the ocean floor and occasionally in lakes, where they often produce pillow lavas. The individual pillows in three dimensions are interconnected and represent cross-sections through lava tubes. Their characteristics are illustrated in Figure 13.17(a),(b). They also form subglacially and englacially and are known in Iceland, British Columbia and Antarctica, where they are associated with subglacial and englacial lakes (Figure 13.18). Occasionally basalts cool directly against ice and produce structures like those shown in Figure 13.19.

Subaqueous silica-rich lavas are known to have been erupted under ice in Antarctica and Iceland (Furness et al., 1980; Tuffen et al., 2001) and, although contemporary observations are impossible, the structures exposed later are clear (Figure 13.20). They form lava lobes, chilled within conically-shaped, subglacial cavities, set in poorly-sorted breccias with an ash-grade matrix. A gradational lava–breccia contact at the base of lava lobes represents

Table 13.6 Principal differences between pahoehoe and aa types of lava. (Source: McDonald, 1972; Williams and McBirney, 1979)

Type of lava flow	Surface morphology	Internal composition	Nature of flow
Pahoehoe	Smooth, and sometimes rope-like. The ropy morphology reflects dragging of the partially solidified plastic crust by the still-liquid lava beneath. When the lava flow is a narrow stream the 'ropes' are bent into curves that are convex; pointing towards the distal margins. Cracks in the surface are often sites where liquid lava is 'squeezed up' to form ridges. Sometimes mounds are formed (*turnuli*) by both squeezing up under hydrostatic pressure and differential sagging of the surface following the draubage of lava tubes.	Lava is fed to be flow front by neans of 'tubes' and channels. At the close of an erotion lava often drais from the tube, to produce 'cave' systems.	Following eruption the upper tens of centimetres cool quickly and become highly viscous. The flow remains liquid beneath this crust and, while the eruption continues, frontal flow is maintained. Some less active parts of the interior solidify and movement is often confined to distinct 'tubes' within the flow, which feed the flow front.
Aa	Lava has a sharp, rubbly surface, made up of angular jagged fragments (i.e. clinker). Aa flows do not normally flow in the same way as pahoehoe, but by means of open 'streams' or channels of lava which build their own embankments or 'levées'. At one time it was often stated that aa flows did not contain lava tubes. It is now known that, although relatively uncommon, this is not the case.	Aa flows usually comprise: an upper fragmental layer; a massive interior; and a rubble-rich base.	The sharp, broken surface is produced by fragmentation of the upper viscous material by the movement of the flow beneath. Overall the flow advances 'like a caterpillar tread dumping talus over its front and then overriding the fragments' so accounting for three-fold internal composition (Williams and McBirney 1979, p. 110).

a fossilized fragmentation interface, driven by magma–water interaction as the lava flowed over poorly-consolidated, water-logged debris. Ice-contact structures are represented by columnar jointing on the upper surfaces of lobes.

Extensive descriptions of submarine silica-rich lavas and hyaloclastites are not common either, although Pichler (1965) describes rhyolites extruded into shallow water as dykes that formed domes up to 200 m high (Figure 13.21), which

(a) (b)

Figure 13.12 Shield volcano forms: (a) Mauna Loa from www.fas.org/...does/rst/Sect17/Sect17_3.html (b) Lambahraun, south of Langjökull, central Iceland

Figure 13.13 Strato-volcano forms: (a) Fujiyama from Lake Kawaguchi; (b) Fujiyama from Mount Tano 13.13.c Nevado de Toluca (Source: http://en.wikipedia.org/wki/File:Nevado_%26_Toluca.jpg ;13.13d/13.13e Paricutin); (c) Nevado de Toluca, Central Mexico; (d) Paricutin, Michoacan, Mexico; (e) Paricutin's cone, lava flow and vegetation recolonization since 1943; (f) Popocatepetl

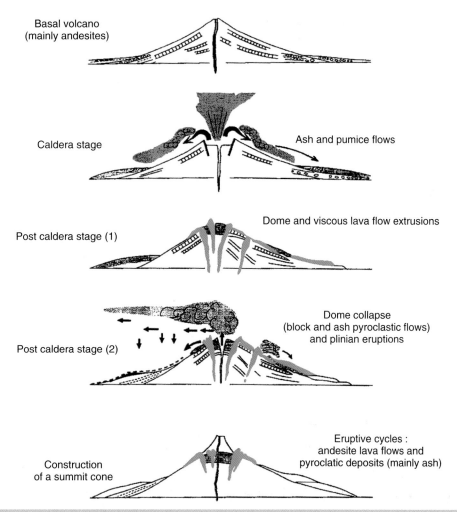

Figure 13.14 Evolution of Nevado de Toluca volcano, Mexico. (Source: after Cantagrel *et al.*, 1981)

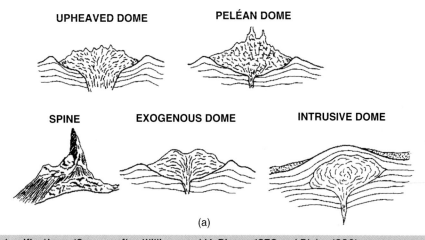

(a)

Figure 13.15 Lava dome classifications. (Source: after Williams and McBirney, 1979 and Blake, 1990)

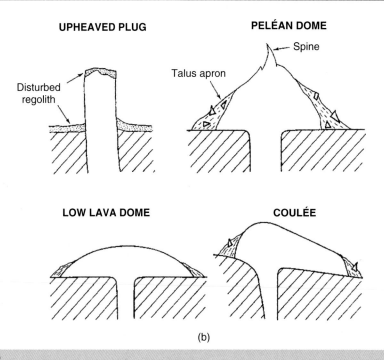

(b)

Figure 13.15 *(continued)*

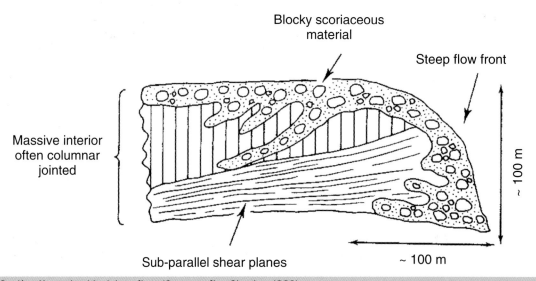

Figure 13.16 Section through a block lava flow. (Source: after Chester, 1993)

chilled quickly to produce outer glassy margins and extensive breaking and jointing. These domes are covered by an extensive hyaloclastite layer. 'Hyaloclastites' denotes a glass-fragment rock and include all rocks produced by eruptions in which water is involved and fragmentation of the magma is the dominant process. They form in a wide variety of shallow-eruption-depth, subaqueous and subglacial environments.

The major differences between subaqueous and subglacial eruptions and the differences between subglacial rhyolite and basalt eruptions are summarized in Table 13.7.

13.5 Explosive volcanism

Explosive volcanoes are the most powerful and destructive volcanic type and usually occur only at destructive plate boundaries. These eruptions produce large quantities of fragmental material (pyroclasts), which are carried vertically upward in convecting columns as plumes, or laterally in high-velocity pyroclastic flows. Explosive volcanism is confined to andesitic and rhyolitic eruptions ((a) and (c) in Figure 13.10). These produce pyroclastic

(a)   (b)

Figure 13.17 Characteristics of pillow lavas: (a) cyclindrical pillows, Jarlhettur subglacial lake facies; (b) linear pillows after slow movement down a palaeoslope, Kalfstindar, central Iceland

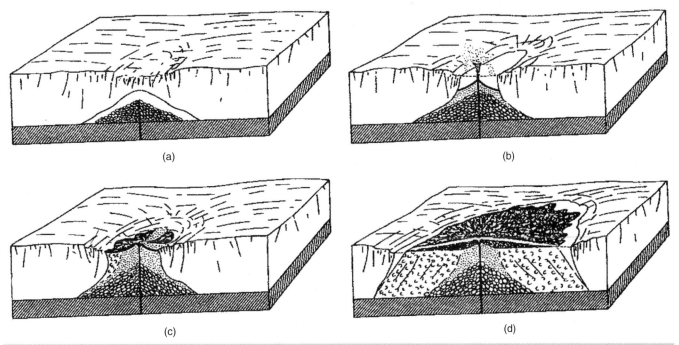

(a)   (b)

(c)   (d)

Figure 13.18 Subglacial lakes and the formation of pillow lavas. (Source: after Jones, 1969)

(a)

(b)

Figure 13.19 Direct cooling of basalt against ice: (a) summit of Fagradalsfjäll. Jarlhettur; (b) cooling joints in basalt cooling injection structure, Laugarvatn area, central Iceland

sedimentary sequences on land or in water, ranging in grain size from volcanic bombs to fine ash, in four model situations (Figure 13.22).

The processes causing the disruption of the magma are varied and result in various styles of volcanic eruption, named after characteristic volcanoes (Figure 13.23(a)). At an open vent, exsolution of volatiles and bubble growth begins during the magma's ascent to the surface. This is because the solubility of gas decreases with decreasing pressure. Bubbles can burst, which causes fragmentation of the magma, and pyroclasts can be ejected in a rapid gaseous flux as a sustained explosive eruption. The pressure drop can be enhanced in some cases, for example where the partial collapse of a lava dome results in accelerated ascent velocity and the extrusion of a lava lobe richer in volatiles, or spectacularly where flank collapses occur, such as at Mount St Helens. In

the superficial magma chamber of basaltic volcanoes, magma is already oversaturated in carbon dioxide and the continuous gas bubble ascent results in accumulation of a foam layer at the top of the chamber. When it reaches a critical thickness, this foam is expelled, producing intermittent explosive eruptions of Strombolian or Hawaiian styles. However, relatively little ash is produced and this falls within a few kilometres of the vent. In these types of eruption the column height is restricted and the umbrella region is missing (Figure 13.24), and falls are generated from the gas-thrust region. Magma mixing can also cause explosions when a chamber of differentiated (different composition) magma is intruded by a hotter basic magma, and the solubility decreases with the rising temperature. It is thought that the 1875 eruption of Askja in Iceland was caused by magma mixing (Sparks et al., 1977).

Hydromagmatic processes through the action of external water can be important too, as they cool the magma and fragment the melt. This can produce maars, tuff rings (Figure 13.25) or littoral cones where lavas flow into the sea. These hydromagmatic processes have been compared to industrial explosions between fuels and coolants, where sometimes the vaporization occurs so rapidly that an explosion occurs. It has been suggested that hydromagmatic fragmentation takes place as a cyclic, five-stage system that involves positive feedback. In stage 1 initial contact of water and magma creates a film of water vapour at the interface, especially where magma rises into a zone of surface water, or water-saturated sediments. In stage 2 this water film expands, but at the outer contact condensation occurs and the film may collapse in places. This releases energy. Expansion and collapse continue and increase so that eventually there is sufficient energy to allow the melt to be penetrated and for fragmentation to occur. In stage 3 the penetration of the water into the melt, or mixing of the collapse film with the melt, increases the surface area of the magma–water interface. In stage 4 increasingly rapid transfer of heat, as water encloses fragments of melt, takes place. In stage 5 a new film of vapour is formed as water is vaporized by superheating and the reversion of the system to stage 2. This further disperses the magma, positive feedback occurs and an exponential increase in surface area leads to an exponential increase in total heat transfer and explosions.

Additional processes of hydromagmatic fragmentation have been described for the submarine environment by Kokelaar (1986). Bulk-interaction steam explosivity occurs where water is either trapped close to the magma or engulfed by it. Here steam expands explosively to tear apart the magma, which is shattered by pressure waves. Then cooling-contraction granulation occurs in deep water, where magma cools by conducting heat. This causes chilling of the surface layers and solidification. When it cools, the interior sets up stresses that cannot be accommodated by the solid surface and cracking occurs. All processes of fragmentation increase in effectiveness as water becomes shallow, volcanic plumes increase in height and fall materials are distributed over greater distances, as described for the Surtsey eruption of 1963, where materials were erupted above the sea surface as black debris-rich clouds.

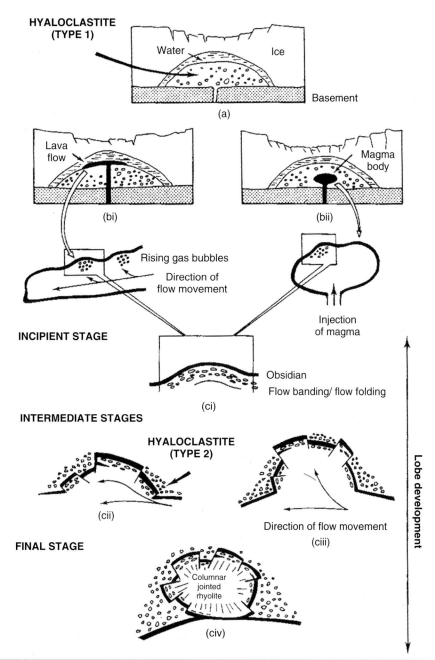

HYALOCLASTITE (TYPE 1)

Water   Ice

Basement

(a)

Lava flow

Magma body

(bi)   (bii)

Rising gas bubbles

Direction of flow movement

INCIPIENT STAGE

Injection of magma

Obsidian

Flow banding/ flow folding

(ci)

INTERMEDIATE STAGES

HYALOCLASTITE (TYPE 2)

(cii)

Direction of flow movement

(ciii)

FINAL STAGE

Columnar jointed rhyolite

(civ)

Lobe development

Figure 13.20 Formation of two types of silica-rich hyaloclastite and lava lobes in subglacial environments. (Source: after Furness *et al.*, 1980)

The Plinian eruptions are highly explosive and involve the injection of large volumes of pyroclasts into the eruption columns. As magma saturated with volatiles rises to the surface, decompression takes place, and because of high melt viscosity, migration of bubbles is inhibited and a relatively uniform foam is formed. When the volume occupied by bubbles exceeds around 75%, the melt is disrupted into highly vesiculated pumice and scoria. The roof of the magma chamber is fractured by the melt buoyancy and velocity and an eruption occurs, in which there can be two critical pressure levels. The exsolution level separates that part of the reservoir in which all volatiles are still dissolved from an upper portion in which volatiles are being exsolved. The fragmentation level is the level at which exsolved bubbles exceed about 75% of the total volume. Above this, rapid acceleration occurs and there is dispersion of gas and pyroclasts. Eruption velocities are typically $200-600 \text{ m s}^{-1}$. Erosion of the vent occurs to give a flared shape, which may lead to a transition from sub- to supersonic flow at the point at which the conduit is narrowest. The eruption columns can go up to 60 km. As the eruption wanes, gas content reduces and/or erosion causes the eruptive vent to increase in radius, so an eruption column cannot be sustained. Collapse occurs, which can generate both pyroclastic flows and surges.

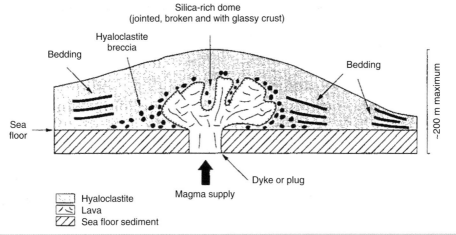

Figure 13.21 Silica-rich lava dome extruded into shallow water at the island of Ponza, Italy. (Source: after Pichler, 1965)

Table 13.7 (a) Major differences between subaqueous and subglacial eruptions. (b) Differences between rhyolitic and basaltic subglacial eruptions. (Source: after Tuffen *et al.*, 2001)

(a) Major differences between subaqueous and subglacial eruptions

Property of eruption	Subglacial environment	Subaqueous environment	Implications
Edifice constraint	Constraint by ice walls during subglacial stage[a]	No constraint	Subglacial edifices likely to be steeper with greater risk of flank collapse if ice recedes
Effective pressure	Varies from glaciostatic to atmospheric, depending on meltwater discharge rate[b]	Variable but hydrostatic	Style of subglacial eruptions likely to be more variable: effusive and explosive
Water/magma ratio	Variable depending on drainage patterns	Consistently high, although steam envelopes may "shield" magma[c]	Subglacial sequences may include "dry" units with little evidence for magma-water interaction

[a]Skilling (1994); [b]Hooke (1984); [c]Kokelaar (1982)

(b) Major differences between subglacial rhyolite and basalt eruptions

Property of eruption	Subglacial rhyolite	Subglacial basalt	Implications
Magma temperature[a]	800–900 °C	1100–1200 °C	Less energy released as rhyolite is quenched
Melting potential[a]	≤ 8 times own volume of ice	≤ 14 times own volume of ice	Positive pressure changes during rhyolite eruption, negative during basalt
Pressure changes during eruption[a]	Positive	Negative	Rhyolite: meltwater tends to drain away: basalt: meltwater tends to collect at vent
Magma viscosity[a]	10^6–10^7 Pa s	10^3–10^4 Pa s	Rhyolite eruptions tend to be more explosive, larger aspect ratio lava flows
Effusion rate[a]	10^1–10^2 m^3 s^{-1}	10^1–10^4 m^3 s^{-1}	Inward ice creep[b] more significant during rhyolite eruptions because edifice growth is slower[a]
Distribution[c]	Iceland, mostly at central volcanoes	Antarctica, Iceland, British Columbia	Basalt much better studied than rhyolite
Recent eruptions	None observed	Gjálp 1996, Iceland[c]	Insight gained on basaltic eruptions, not on rhyolitic

(a)

SUBAERIAL PROCESSES: MAGMATIC ERUPTIONS

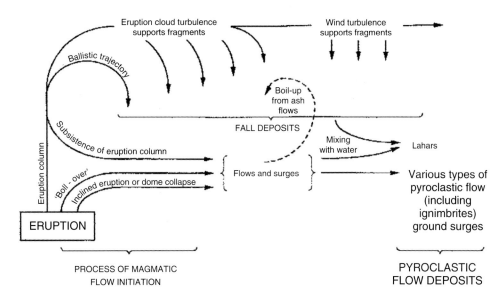

(b)

SUBAERIAL PROCESSES: HYDROVOLCANIC

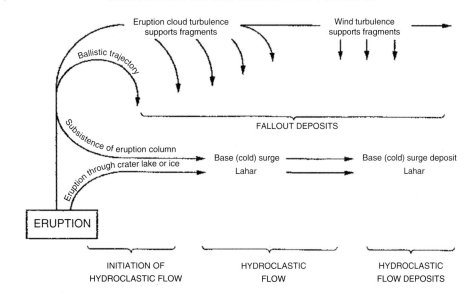

(c)

SUBAQUEOUS DEPOSITS FROM SUBAERIAL VOLCANOES

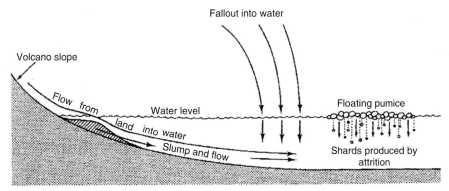

Figure 13.22 Four model situations in which pyroclastic materials are produced and deposited. (Source: after Fisher and Schminke, 1984)

(d)

TEPHRA FROM SUBAQUEOUS ERUPTIONS

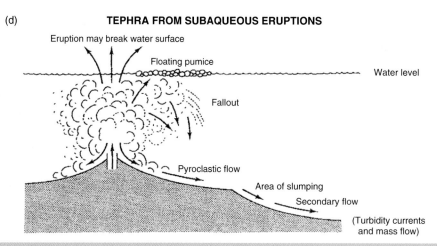

Figure 13.22 *(continued)*

Pelean eruptions are usually associated with silicic lava dome growth and involve viscous andesitic, dacitic and rhyolitic magmas. They are characterized by explosions of moderate-to-extreme violence in which pyroclastic flows are common. Whilst this descriptive terminology for the various types of explosive vulcanicity is generally useful, there are overlaps in the processes and it is much better to try and understand the physical processes of transport and deposition of the pyroclasts.

13.6 Petrographic features of volcaniclastic sediments

Volcaniclastic sediments are immature and have certain mineral characteristics in common. These are that they are composed of glass, pumice (where the magma has been charged with gas bubbles), scoria, well-formed euhedral crystals (created whilst the magma was still fluid) and basaltic hornblende; they have diagenetic mineral alteration (for example, glass alters to clay minerals, zeolites and silica); and in base surges granulated sideromelane is a common occurrence, forming from the drastic chilling of basaltic magma or its palagonite decomposition products. Most pyroclastic sediments tend to be rhyolitic in composition, where the magma is rich in silica and has greater viscosity and gas content, and so favours explosive rather than effusive activity.

13.7 Transport and deposition of pyroclastic materials

Two major types of transport system are known from explosive eruptions. The vertical plumes generate fall deposits from wind-driven clouds at high elevations. Laterally-moving systems generate surge and flow deposits from gravity-controlled, ground-hugging, density currents.

13.7.1 Pyroclastic fall or tephra deposits

These are the simplest pyroclastic deposits and are caused by rain-out of clasts through the atmosphere from an eruption jet and/or plume, forming concave-upwards cones of tephra. Tephra consists of pyroclastic fragments of differing sizes (Figure 13.23(b)) and origins, but ranges in size from ash (<2 mm), through lapilli (2–64 mm), to blocks and bombs (>64 mm) (Figure 13.25(c)). Densities can vary from pumice to solid lava fragments. Isopachs of ash thickness usually form ellipses that are elongated in the downwind direction (Figure 13.23(c)) because the particles are segregated by fall velocity as they are transported downwind. Both the thickness and the maximum grain size decay (according to their settling velocities) exponentially with more or less straight-line relationships, with the coarsest particles nearest to the volcano and the finest furthest away. Wind direction, velocity, turbulence and the height to which the particles were ejected control the fall out pattern. Fall deposits can show variations in grain sizes, often with sharp bedding planes which imply spasmodic, nonsustained eruptions, and there can be grading. Fall deposits usually drape and follow preexisting relief. During an eruption the fall sequences generally become finer vertically in any one section as the transporting power normally decreases during time. The characteristics of ash falls are described in Table 13.8.

13.7.2 Pyroclastic density flows

Pyroclastic density currents produce relatively thin, bedded, cross-bedded or massive deposits. Their characteristics are that they are turbulent, low-density clouds of rock debris, air and/or other gases which move over the ground at high speeds, hugging the ground and, depending on their density and speed, possibly being controlled by the underlying topography. They are of two types, although they may be part of a gradational series. One is called a hot pyroclastic flow, which consists of dry clouds of rock debris and gases that have temperatures well above 100 °C. The other is

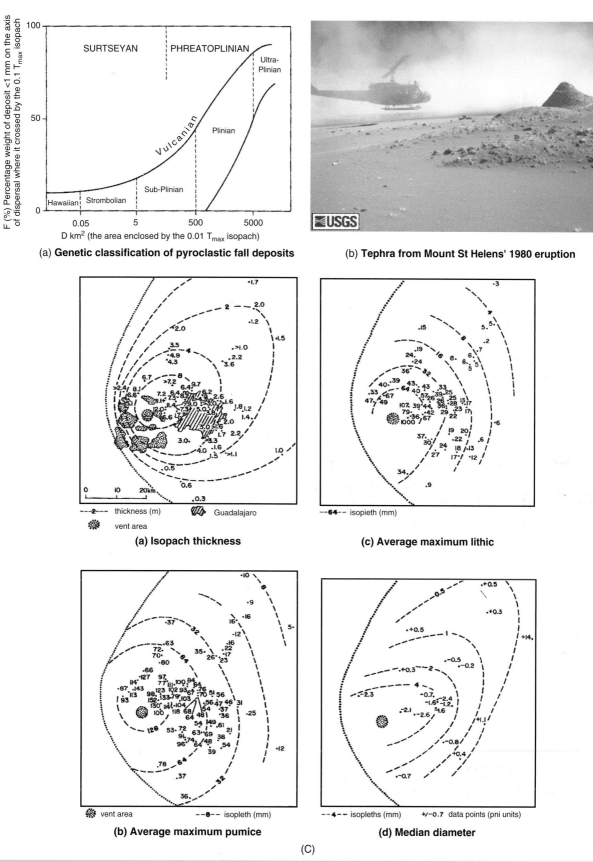

(a) **Genetic classification of pyroclastic fall deposits**

(b) **Tephra from Mount St Helens' 1980 eruption**

(a) **Isopach thickness**

(c) **Average maximum lithic**

(b) **Average maximum pumice**

(d) **Median diameter**

(C)

Figure 13.23 (a) Genetic classification of pyroclastic fall deposits. (b) Tephra from Mount St Helens' 1980 eruption. (c) Isopach thickness and grain-size characteristics of La Primavera rhyolite volcano. (Sources: (a) after Walker, 1973; (b) from USGS; (c) after Walker *et al.*, 1981)

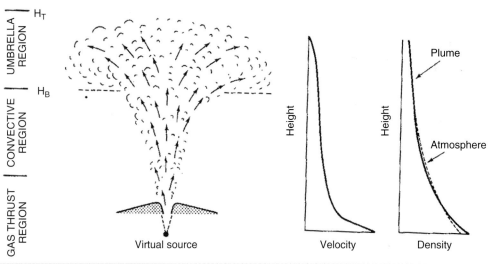

Figure 13.24 Main features of a convecting eruption column. (Source: after Carey and Sparks, 1986)

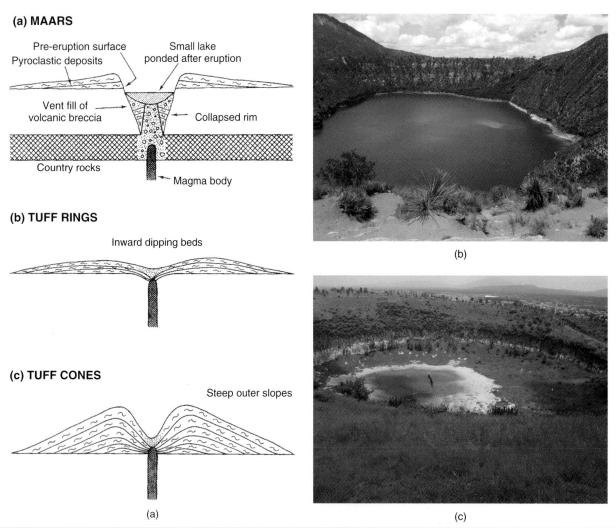

Figure 13.25 (a) Hydromagmatic tuff rings and cones. (b) Maar, Central Mexico. (c) Maar, Central Mexico. (d) Bombs and bomb sags, Alchicheca base surge deposits, Central Mexico. (e),(f) Accretionary lapilli, Toluquilla tuff cone, Valsequillo Basin, Puebla, Central Mexico. (Sources: (a) after Cas and Wright, 1987)

(d)

(e)

(f)

Figure 13.25 *(continued)*

a cold surge, called a base surge, which consists of rock debris and steam, or water, at or below 100 °C.

Base surges

These are high-velocity, density currents that spread out from the base of a rising, vertical ejecta column. They can originate from shallow nuclear and chemical explosions or high-energy impact events, but they are more commonly formed as phreatomagmatic or hydromagmatic surges, as at Taal (1965, Philippines) and Capelinhos (1927–9, Azores). They were first observed from nuclear blasts, for example the Sedan nuclear test (1962), where the initial velocity of the surge was over 50 m s^{-1}, and where it carried debris about twice as far as the ballistic fragments and deposited dune-like bedforms with 3–4 m wavelengths concentrically around the explosion centre. They consist of dilute, turbulent mixtures of particles and steam that expand rapidly in all directions from the eruption site. The deposits have an overall wedge-shape in cross-section, with a nearly logarithmic decrease in thickness away from a vent. In proximal locations, sandwaves, with subsidiary massive beds, are common, whilst in distal locations planar beds predominate over massive beds. Different flow regimes in the flows control the bedforms. Sorting values are intermediate between air fall and pyroclastic flows, but there is wide overlap. They characteristically lack the very-fine and coarse fractions and therefore have a more restricted grain size than pyroclastic flows, so are better sorted (see Figure 13.25(d)).

Wohletz and Sheridan (1979) suggest that the differences in these bedforms could be explained by differences in fluidization. They argue that when erupted, surges are partly fluidized by steam, volatiles and the collapsing column, and that the materials move away from the vent by gravity and/or the velocity imparted by the fluidized particles. Close to the vent, the surges are viscous

Table 13.8 Characteristics of air-fall deposits

Attribute	Description
Composition	May be of any composition, but andesitic and rhyolitic are more common than basaltic
Texture	Generally well sorted; often better sorted further away from vent; the greater the distance away, the finer the grain size
Structure	Drape existing topography; grading often reflects decreasing transport power with time as an eruption wanes
Distribution	Either circular round a vent or in distinct sectors because of the prevailing wind; fan-shaped with apices often coinciding with vents; these fans become wider with distance and materials spread over greater areas
Thickness	Thinner with distance
Associated facies	In proximal locations often associated with lavas, pyroclastic flows, debris avalanches, surges and lahars. In distal locations falls dominate and are often interbedded with nonvolcanic sediments and soils
Components	Crystals, lithics and pumice (including glass)

Table 13.9 Characteristics of cold base surges

Attribute	Description
Field relationships	Narrow and thickest in valleys and thin or absent on interfluves; overbank deposits on either side of valleys that have filled up; currents can also move uphill on inner sides of crater rims; stratigraphic horizons can be traced and correlated for many kilometres
U-shaped channels and antidune cross-stratification	Range from <1 m to >10 m across and may be up to 3+ m in thickness; scour marks and small local unconformities attributed to current erosion; trough-type beds with a marked lenticularity of the deposits; thought to represent lobes within the surge, which act as semi-independent entities and are able to erode deposits already deposited; low-angle cross-stratification (<15°), with a regular sinuosity of the upper surfaces; abundant preservation of stoss-side laminae; lee-side laminae with angles under the angle of repose; some evidence of upcurrent migration of internal crestal axes within individual dunes
Bedding sags and convolute lamination	Produced by the impact of large air-fall blocks on wet, still plastic, surge beds, which causes the beds to be penetrated, distorted and thinned; this can also cause convolution of the beds (Figure 13.25(c))
Accretionary lapilli	Spherical masses of concentrically-layered, cemented ash, formed by accretion of fine material by moisture; can also occur in rain-flushed fine-ash falls (Figure 13.25(e))

and act as Newtonian fluids. The surges are turbulent and are able to build up complex sequences of ripples, dunes and antidunes, which together form the sandwave units. This turbulent flow causes a high shear stress on the surface beneath it, but energy is quickly lost on flat surfaces, or where movement is uphill. The hot cloud of steam, solid fragments and rapidly cooling fluid particles must have a relatively high-volume concentration of solids, especially towards its base, and a relatively high viscosity and density. The deposits must be cohesive due to the presence of water carried as steam droplets within the cloud. This causes adhesion between particles, due to surface tension, and therefore the particles cannot freely interact with a current passing over them. At high velocities the surge can erode channels, but when the velocity decreases, deposition begins. In distal exposures, most gas has escaped and surges are nonfluidized, but zones of densely-fluidized material may remain. Here flow is laminar, producing planar beds, whilst the still-fluidized portions of the flow create the subsidiary massive beds.

Vertical variation can change in detail, but all become finer-grained and thinner-bedded upwards as the surge wanes and sedimentation rates decrease with time. A common vertical sequence from a blast-driven pyroclastic surge, as at Mount St Helens, is a coarse, fines-depleted, massive basal layer overlain by a finer-grained, massive layer with cross beds at the top and an ash-fall layer capping the sequence. The presence of water in wet basaltic surges and their fresh deposits facilitates the alteration of glass to palagonite (a combination of hydrous glass, zeolites and clay). Other characteristics of cold base surges are illustrated in Table 13.9. Other examples of these types of deposit are found at

Ubehebe Crater (Death Valley, California) and the Laacher See (Eifel district of Germany). At Ubehebe there is evidence of only a single base surge, whereas usually there are successive surges, each of which forms a set of deposits and modifies the underlying structures. The notion of one bed equals one surge may be false, however, as it is likely that a single surge includes many turbulent cells of varying strengths, which would cause reworking and erosion within deposits from the same surge.

Pyroclastic flows

When the steep sides of lava domes collapse, or steep-sided Plinian columns collapse, there results high-speed, turbulent flows of hot, incandescent ash, rock fragments and gas down the volcano. The speeds are probably up to 200 km per hour, with temperature estimates between 550 and 950 °C. Such flows tend to follow valleys, where they can typically pond to great thicknesses, but are capable of overtopping the topography and travelling >100 km from the source. Most consist of two parts: a basal flow of coarse fragments and gas that moves along the ground, and a turbulent cloud of finer particles that rise above the basal flow. Small pyroclastic flows can confine themselves by forming lateral block levees and terminating in lobes with steep, levee fronts. Gas seems to have a crucial role in the driving mechanism: exsolution of gas from juvenile particles, which buoys up these particles and reduces friction between them, and breakage of hot lithic fragments with the release of gas and the heating and expansion of air engulfed at the flow front both help in the fluidization of the flow, explaining the flow mobility and its emplacement far from the eruption site. Fluidization involves gas flowing through the solid particle bed; as the speed increases, there comes a point at which the drag force exerted across the bed by the gas is equal to the buoyant weight of the bed. At that point, the bed acts as a fluid. However, pyroclastic flows are not fully fluidized as the size of grains is variable and the smaller ones reach their terminal velocities before the larger ones are fluidized, which means that the smaller ones are removed, or elutriated, from the flow. Pyroclastic flows show great vertical, longitudinal and lateral variations and can produce very complex sequences. A summary of their characteristics is given in Table 13.10 and in Figure 13.26.

Block-and-ash flows

These can form by the collapse of Vulcanian eruption columns, but most historic examples have resulted from the collapse of viscous lava domes as they oversteepen their fronts during growth (e.g. Mount Unzen in Japan or Las Derrumbadas, Central Mexico). They can extend up to 10 km from their source, at speeds up to 100 km per hour, and are accompanied by ash-cloud surges, which can detach from their associated channelized flows. They differ from pyroclastic flows in containing little fine ash, but with a large fraction of dense-to-moderately-vesicular, rarely pumiceous, blocks several metres in diameter. These are derived from the source dome, or from older dome remnants. The deposits are poorly sorted and bedding is usually massive, although vague layering locally may occur. Lapilli pipes are rare, and a fine-grained, or block-free, basal layer may be present. The blocks are either uniformly distributed through the bed or show normal or reverse coarse-tail grading. Block-and-ash flows are usually between 1 and 10 m thick and between 1 and 2 m at their lobate, blunt, termini. Ash-cloud surge deposits are typically <1 m thick, consist of sand and finer ash, and are relatively well-sorted.

Lahars

'Lahars' refer to all water-transported volcanic debris, regardless of origin and sedimentological properties, usually flowing as hyperconcentrated mudflows or debris flows down the volcano flanks. They can travel up to 300 km or more, at speeds locally of up to 60 km per hour, and some contain so much debris (between 60 and 90% by weight) that they look like fast-moving rivers of wet concrete. This abundance of solid debris carries the water – unlike watery floods, in which water carries the fragments – but there is a transition between the two, referred

Table 13.10 Characteristics of pyroclastic flow deposits

Attribute	Description
Flow units	Units deposited as distinct flow lobes, varying from a few centimetres to tens of metres; boundaries marked by differences in grain size, composition, pumice concentration; each flow unit deposited from a single pyroclastic current, although flow units have been observed to split into two or more thinner flows laterally, due to splitting into lobes or to pulsatory motion of the flow
Welding and the effects of high temperatures	During flow, temperatures may be close to those of the parent magma; as the flows are such good insulators they can retain heat for a long time
Texture	More poorly sorted than falls; sorting improves, median grain size decreases and fine ash content increases from proximal to distal locations; usually polymodal because of fragmentation in the vent, segregation during transport and emplacement, and additions to certain components by fragmentation and abrasion during transport or by entrainment of bed material (striations and percussion marks on the bed); commonly pumice is reversely graded and lithics normally graded; elutriation of particles due to both fluidization and segregation within the eruptive column means that flows are enriched with pumice and lithics but depleted of the fine-grained glassy materials; hence the whole rock geochemistry does not reflect the parent magma
Components and chemical composition	Flows consist of varying amounts of crystals, glass fragments, pumice and lithics; usually glassy ash and pumice are the main components of the matrix, followed by crystals and fine lithic fragments; larger flows commonly show vertical compositional zoning due to eruption from zoned magma chambers
Associated materials	Three zones commonly make up the flow: (a) the dense flow itself, (b) hot surges, which are dilute and occur at the head and on the lateral margins of the flow, (c) a dilute ash cloud consisting of material winnowed and elutriated from the flow and lying above it; this can form a buoyant plume many kilometres high and fall material can be deposited over a considerable distance; because flows are generated from silica-rich magma, lava flows associated with them are short and domes often mark the eruption sites; carbonized wood incorporated into the flows and gas lapilli pipes through the flows mark places of vertical gas escape

to as hyperconcentrated flows. These flows have higher bulk densities, are plastic in behaviour and flow as non-Newtonian fluids. They can transform into normal floods as they become

Figure 13.26 Pyroclastic flow processes and deposits. (a) Montserrat pyroclastic flow. (b) Mount St Helens pyroclastic flow, 7 August 1980. At least 17 separate flows occurred during the 1980 eruption. (c) Mount St Helens pyroclastic flow deposits. Pumice blocks at the edge of a flow, 18 May 1980 eruption. (Source: USGS, photos taken by Donald A. Swanson)

increasingly diluted with water downstream. The formation of lahars depends on the right mix of readily-available sediment and water discharge. There are a variety of ways in which they can be formed and they may be primary (syneruptive), secondary (post-eruptive) or not related to an actual eruption:

- **Directly related to eruptions:** Mobilization of materials during eruption through a crater-lake; explosions cause the movement of old crater materials into rivers draining a volcano; pyroclastic flow movement into streams draining a volcano and the mobilization of debris; broken lava flows into streams draining a volcano; lava or pyroclastic flows over snow/ice and mobilization of loose material.

- **Indirectly related to eruptions:** Release of water from a crater lake by the breaching of its walls; rapid melting of snow and ice on the volcano slopes; heavy rainfall or seismic activity causing mobilization of debris on the steep volcano slopes; mobilization of debris by other slope instability processes.

Lahars follow valleys and are absent on interfluves (Figure 13.27(a)). They can vary in thickness from <1 to >200 m. They cause erosion by undercutting steep slopes and valley terraces and by scouring their beds. The final waning stages are erosive, as channels are incised into the freshly-deposited lahar. The surface morphology is usually flat, with gentle slopes towards the flow front, but in places there may be local relief of low-amplitude ridges and valleys, reflecting an irregular, underlying surface topography. Where they are derived from pyroclastic flows they may be very difficult to distinguish from these deposits. They usually have a wide range of grain sizes, are often massive and unstratified, are poorly sorted and often have large blocks that are matrix-supported, reflecting the strength of the sediment. These large particles may be rafted close to or at the surface of the flow, and then migrate forward toward the flow margins, because velocities are greatest near the surface, especially as friction causes retardation at the bed. The gradual incorporation of water at the front of a lahar flowing down an active stream channel causes progressive loss of capacity to carry large gravel, and this progressively lags behind the flow front. Deposition seems to occur incrementally. The deposits can be graded or ungraded, depending on the amount of water that the flows contained and the degree of their downstream evolution. Fabrics are generally weak in lahars, although there is much variation documented, depending on the movement and deposition mechanisms (Fisher and Schminke, 1984). The deposits commonly contain matrix vesicles, which result from the entrapment of air bubbles.

Hyperconcentrated flows

These have characteristics which are between those of debris flows and fluvial deposits. They generally have medium sorting and grain size, they can be massive, but commonly they have weak stratification, defined by thin horizontal beds and low-angle cross-beds. Floodplain facies have grain sizes in the granule gravel to silt size, with occasionally floating coarser clasts. If pumice is an important constituent it is found at the top of overbank deposits. It forms from pumice rafts stranded during falling stage. Channel facies commonly show bimodality and clast support, with concentrations of coarse clasts in the finer matrix, and fabrics are stronger than in debris flow deposits.

Debris avalanches

A debris avalanche is a sudden, very rapid flow of an incoherent, unsorted mixture of rock and soil in response to gravity. These coarse, granular flows can occur where there is a collapse of part of a volcanic mountain as a landslide which is transformed progressively into a flow. They are common features of the entire growth history of composite volcanoes, but the frequency is estimated to be under one per 10 000 years. This collapse can be caused by intrusion of new magma, which is the most frequent type (the Bezymianny type, named after an eruption in Kamtchatka in 1956); a phreatic explosion, with superheated steam explosions (the Bandai type, named after a Japanese volcano in 1988 where there was no new magma or full-blown eruption); or an earthquake, where there is no volcanic activity at all, like at Unzen and Ontake volcanoes in Japan. Slumping of a caldera wall during caldera collapse and major slope failures of oceanic islands can also create submarine debris avalanches, as offshore from Hawaii (Moore *et al.*, 1994) and the Canary Islands (Carracedo, 1996). They can be recognized by various characteristic features, although it is only since observations at Mount St Helens in 1980 that their true origins have become better understood. They have recognizable geomorphic characteristics and internal structures after they have occurred, but the mechanism of flow and emplacement is still controversial. The hummocky topography (Figure 13.27(b.i),(b.ii),(b.vi)) is the main recognizable morphology but the individual hummocks are variable and irregular, although there is proximal–distal transition in volume and height. There can be thousands of small hills and closed depressions covering large areas at the base of the volcano, as at Mount

Shasta, California, the largest known in the world, where the collapse height has been estimated at 3500 m, with a runout distance of at least 45 000 m. Long radial ridges, or local parallelism of the hummock-long axes, have been observed. Natural levees and a distinct marginal and distal cliff are other characteristic features

March 25, 2007

Feburary 9, 2002

(a)

Figure 13.27 (a) Lahar on Mount Ruapheu, New Zealand. Lahars are an ongoing threat from this volcano and on 18 March 2007 one such flow occurred from the caldera. Nine days later the Advanced Spaceborne Thermal Emission and Reflection Radiometer (ASTER) on NASA's *Terra* satellite captured the top image. An earlier one from 9 February 2002 is given as a comparison. (b) Debris avalanche landforms and deposits: (i) Mount Shasta debris flow landforms; (ii) Las Derrumbadas rhyolite dome, Central Mexico, showing the hydrothermally-altered summit rocks, which were the source of the avalanche deposits, and the morphology close to the failure, which included steep secondary fans; (iii) hummock topography, Las Derrumbadas; (iv) fan deposits from Las Derrumbadas, showing several phases, interbedded with rhyolitic ash; (v) striations and percussion marks from pyroclastic flow/block-and-ash flow deposits, Las Derrumbadas; (vi) Popocatepetl south flank, where the debris avalanches occurred; (vii) Popocatepetl proximal debris avalanche morphology; (viii) Popocatepetl debris flow deposits, showing the incorporation of several types of sediment and the deformation involved. (Source: (a) http://www.earthobservatory.nasa.gov/NaturalHazards/Archive/Mar2007/ruapehu_ast_jpg; (b)(i) USGS photo, 22 September 1982, taken by Harry Glicken)

(b)i

(b)ii

(b)iii

(b)iv

(b)v

(b)vi

Figure 13.27 (*continued*)

(b)vii

(b)vii

(b)viii

Figure 13.27 *(continued)*

which marked the limits of an individual avalanche, were caused by the rapid cessation of the avalanche movement. The remnants of temporary river channels can be left on the surface of the valley-filling avalanche deposits and there can be small lakes formed in the depressions. There is usually a well-marked horseshoe-shaped amphitheatre marking the rupture zone, although the shape can vary widely and can be partially or completely hidden by later lava flows or domes. This morphology is marked when the collapse is accompanied by a lateral blast. For example, Mount St Helens was one-third filled by a new dome after 10 years, and at Parinacota, Chile, there are no avalanche scar remains after 13 000 years.

The internal structure of debris avalanche deposits is composed of large homogenous megablocks, surrounded by a more heterogeneous matrix. The deposits are very poorly sorted and the smaller clasts resemble angular breccias (Figure 13.27(b.vi)) Most of the blocks are derived from the source volcano but, although they are fractured and deformed, they preserve many of the primary textures and structures. Some blocks are derived from erosion during transportation and there are often percussion marks, chipping and striations because of contact between large blocks. Fracture patterns called jigsaw cracks are common where there are close fits between blocks, and some blocks show minor faulting. The matrix is composed of a mixture of smaller volcanic rocks derived from the source volcano but the abundance of eroded matrix increases with distance away from the source. Deformed soft sediments are common near the basal contacts, which must reflect higher shear strains at the flow bases. It is clear that volcanic debris avalanches can be transformed into lahars, sometimes within 1–5 km of the source (Scott *et al.*, 2001). The stages can be seen in Figure 13.28.

The transportation process for debris avalanches is not fully understood. There is much controversy over their emplacement, especially their high mobility, and a combination of the many models suggested so far probably applies for individual avalanches. It appears that some kind of lubrication is necessary at the interface between the debris avalanche and the ground to explain the high speeds and relative lack of erosion, and that some

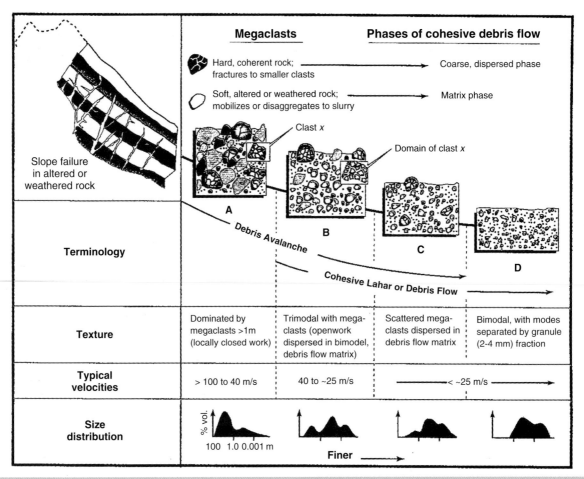

Figure 13.28 Stages in the formation of a lahar from debris avalanches. (Source: after Scott *et al.*, 2001)

form of fluidization process is needed. Several of the models are given in Table 13.11.

13.8 The relationship between volcanic processes and the Earth's atmosphere and climate

13.8.1 Global atmospheric effects of volcanic processes

Volcanic eruptions are part of a complex set of processes which link the Earth's interior to its atmosphere and climate and it seems likely too that certain climatic effects could cause volcanic activity. The processes involve interactions between gaseous, liquid and solid volcanic products, the atmosphere and radiation. The links in this chain are often difficult to study; they require knowledge of a wide range of evidence, from ice core records, through tree rings to written records, and also require an understanding of the structure and dynamics of the atmosphere. However, Figure 13.29 summarizes many of the atmospheric effects caused by mainly explosive volcanic eruptions.

Volcanic aerosols

These eruptions release large quantities of ash and magmatic gases into the atmosphere, but of the gases only sulphur dioxide (SO_2) creates major effects, although volcanic ash provides effective condensation nuclei for water droplets and ice. In the cold lower atmosphere the SO_2 is converted to sulphuric acid by the reaction of the Sun's rays with the water vapour, in turn forming sulphuric acid aerosols, which can stay in the atmosphere for many years. This is important because the amount of sulphur produced by several historic eruptions shows a good correlation with the average decrease in temperature in subsequent years (Figure 13.30(a)) and there is an excellent record of sulphuric acid peaks in Greenland ice cores (Figure 13.30(b)). The 1991 Mount Pinatubo eruption produced the largest sulphur dioxide aerosol for a century (about 20 million tons of SO_2 into the stratosphere) and data show that the stratosphere warmed up and the troposphere cooled down because of the replenishment of this aerosol.

Satellite data have shown that volcanic chlorine can help cause ozone depletion. It is emitted as hydrochloric acid but breaks down and forms chlorine and chlorine monoxide molecules. The sulphate aerosols furnish the locations for the chemical reactions

Table 13.11 Some suggested dynamic emplacement models for debris avalanches

Model	Description
Air fluidization	Flowing on a basal air layer, like a hovercraft, explains long runout and sedimentological characteristics; cannot be the case for submarine avalanches and those of Mars and the Moon
Basal gaseous pore-pressure and self-lubrication	Used to explain fast flows and low apparent angle of friction where pore fluid vaporization took place due to frictional heating during sliding; alternatively, a basal self-lubrication mechanism where rock can be 'smeared-out' beneath the avalanche to provide a zone of high fluidity beneath the front
Mechanical fluidization based on grain contact, linked with granular flow theory	Presence of an intergranular fluid and highly-energetic, interstitial dust could reduce the effective normal pressure on grains and hence frictional resistance; for wet avalanches, mud could form the fluid between clasts, while for dry avalanches fine dust might act as the interstitial fluid
Acoustic fluidization	Produced indirectly by the released fall energy through the propagation of strong sound waves of just the correct frequency through a flowing breccia stream; this requires much less energy than mechanical fluidization
CO_2 gas lubrication	Where there is limestone it is believed that high temperatures developing along a discrete slip surface would continuously dissociate the carbonate rock into a mixture of lime and CO_2 gas during travel, and this gas would provide the lubricant along the basal slip surface
Bingham flow	The natural levees and marginal and distal cliffs suggest non-Newtonian (Bingham) laminar flow, like plug flow, with high yield strengths; alternatively, initial sliding and then plug flow
Granular flow	By liquid nonturbulent granular flows; emplacement from progressive aggradation of the particles related to a loss of the nonturbulent liquid stage; the mass of the avalanche that exerts strong friction on the basement
Basal low-density layers	Based on computer simulations of interacting two-dimensional discs; the basal low-density is a zone in which the particles are very active, with large random velocity components, and this zone supports an over-riding, high-density, low-mobility plug during flow
Mass loss	Suggests that the low apparent angle of friction occurs because avalanches selectively deposit low-velocity debris during movement and the runout length depends on the rate and timing of this mass loss; in a depositional area, debris at the trailing end of the moving avalanche begins to deposit, transferring its energy forward to the moving part; the toe continues to move downslope by the input of momentum from decelerating trailing clasts until the stopping wave catches up with the toe
Seismic energy fluidization and basal pressure wave	Seismic tremors maintain the mobility, combined perhaps with a basal pressure wave, which is a wave propagated along the basal layer at the phase velocity, which is initially greater than the avalanche velocity; with increasing avalanche velocity, the avalanche catches up with the guided wave; over a threshold avalanche velocity, a sonic boom is generated around the basal layer, and thick shock contributes to a loosening of the avalanche into a fluidized state

that then release the chlorine atoms. These are added to the manmade chlorine already present in the stratosphere. All this chlorine then proceeds to destroy ozone, each chlorine atom being recycled many times. Satellite data showed a 15–25% ozone loss at high latitudes after Pinatubo's eruption and a reduction by nearly 50% over the Antarctic after the eruption of Pinatubo and Mount Hudson, Chile, in 1991. However, many of the other gases in the stratosphere, such as carbon dioxide, are in much larger quantities than could be released by volcanic eruptions. The background water vapour is comparable to that released by only the largest eruptions, such as Tambora (1815), and volcanic water vapour and the highly-soluble species (HCl and HBr) rain from the rising volcanic plume as it expands and cools.

Volcanic effects on stratospheric dynamics

The inputs of volcanic aerosols can affect the dynamics in the atmosphere and thereby affect its own distribution in the stratosphere. By absorbing infrared energy rising from the Earth's surface, as well as sunlight in the near infrared wavelengths, aerosols can heat the atmosphere and alter wind patterns. However, this effect depends on how much aerosol is produced. For example, the aerosol eruption from El Chichón, Mexico, in 1982 and other subtropical eruptions appears to remain in its hemisphere of origin, but that from Pinatubo spread within several weeks to a latitude band covering virtually all of the tropics. In the lower tropical stratosphere, temperatures increased by about 3.5 °C within four months of the Pinatubo eruption, and because the warm air rose more vigorously to greater altitudes, this had a marked effect on the ozone layer, where the ozone decreased by 6–8%. This was only a temporary effect and lasted less than a year.

Radiative effects

Historical records have shown us that large volcanic eruptions are associated with colder-than-normal Earth surface temperatures. Aerosols absorb the Earth's infrared radiation, which heats the

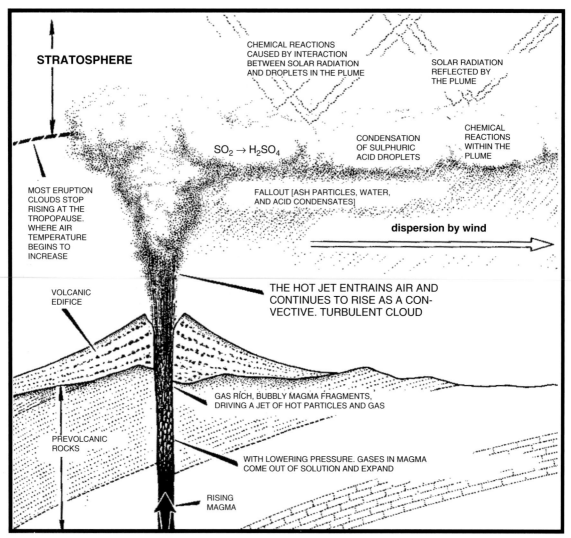

Figure 13.29 Volcanic eruptions and their effects on the Earth's atmospheric system. (Source: after McCormick *et al.*, 1995 and Fisher *et al.*, 1997)

tropical stratosphere, and significant temperature anomalies (up to +2 °C, but little or no effects at higher latitudes) were noted within two seasons of the eruptions of Agung, Bali, in 1963, El Chichón and Pinatubo. These warming events lasted for up to two years, and up to 30 km in the case of El Chichón. However, two years after Pinatubo, stratospheric temperatures had moved from higher than average to new lows. This is likely to have been the result of an observed decrease in ozone after the Pinatubo eruption.

Tropospheric cooling

Whether the effect of volcanic emissions is to cool or heat the Earth depends on the size of the aerosols. For a warming effect to overcome the cooling effect, particles must be larger than about 2 μm in radius. The Pinatubo eruption is thought to have caused the mass-averaged aerosol radius to increase from 0.3 μm to about 1.0 μm, which resulted in net cooling. Particles of radius greater than 2.0 μm fall at rates of over 30 km per year in the lower stratosphere, meaning that they are removed within months, and hence volcanic aerosols probably could not warm the Earth for long periods. Recent major volcanic eruptions though have produced cooling anomalies for between one and three years, but this can be masked by natural temperature variations and global warming.

Possible links between the El Niño – Southern Oscillation and volcanic events

Every few years the appearance of the warmer current (1–3 °C warmer) off the west coast of South America around Christmas causes major disruption of the marine ecosystem and brings torrential rain to the Peruvian coastal deserts. It is part of a worldwide pattern of pressure and temperature change in the

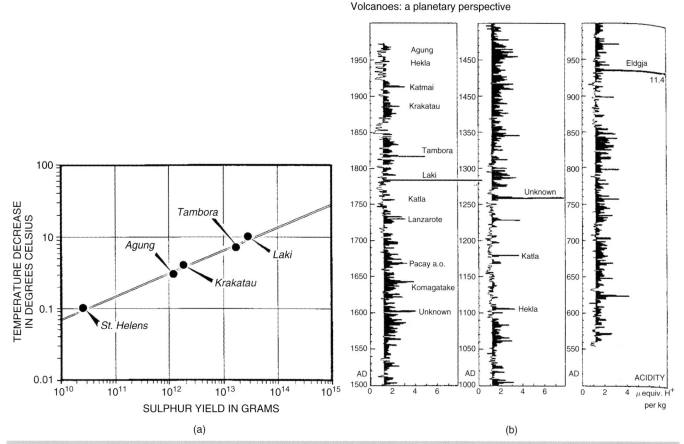

Figure 13.30 (a) Global temperature decrease with an increase in SO₂ emission from volcanoes. (Source: after Fisher *et al.*, 1997.) (b) Acidity profile through a Greenland ice core showing H_2SO_4 peaks due to acid aerosols from historic eruptions. (Source: Hammer *et al.*, 1980)

equatorial regions, the Southern Oscillation, and is a complex part of the interrelated ocean–atmospheric system. It was suggested by Shaw and Moore (1988) that the unusually warm sea temperatures were caused by large-volume, submarine, volcanic eruptions, but that the effects would only be big enough if the eruptions took place over a number of sites along a ridge over several years. Small inputs of volcanic heat might tip the balance, but a major problem with this idea is that the warm oceanic current occurs periodically, every three to seven years, while volcanic eruptions are irregular. Handler (1989) suggested that eruptions taking place at the right time and place could reinforce the effects of the El Niño–Southern Oscillation (ENSO). Temperature changes due to the stratospheric aerosols would occur more quickly over landmasses than the oceans. The land cools faster, causing pressures to become slightly higher, and as a result pressures over the ocean would fall. Due to the fact that most of the Earth's landmasses are in the northern hemisphere, air would be transferred from high-pressure systems over the oceans in both the southern and northern hemispheres to the continents. In the right circumstances pressure gradients in the tropics, ranging from relatively high over southern Eurasia to relatively low over the Pacific, would resemble those of ENSO events. If an eruption happened

to take place in the tropics during such an event the effects would be enhanced, as in 1982 after the El Chichón eruption, so it is possible that the effects of volcanic eruptions may be seen in this subtle manner (See Table 13.12 for a summary of Handler's points).

Adams *et al.* (2003) give further substance to the controversial idea (see Handler and Andsager, 1990) that there might be a statistical correlation between explosive volcanic events and subsequent El Niño events. They use two different palaeoclimatic reconstructions of El Niño activity and two independent, proxy-based chronologies of explosive volcanic activity (the Volcanic Explosivity Index, a measure of the eruption size, and the Ice Core Volcanic Index, a measure of potential climatic impact) from 1649 AD to the present to show the response of the ENSO phenomenon. The results imply roughly a doubling of the probability of an El Niño event occurring in the winter following a volcanic eruption. The dust from an eruption will cool the Earth's surface, but Adams *et al.*'s data suggest that it triggers the warm phase of the cycle, not the cold phase. The volcanic eruptions, it is concluded, do not trigger all El Niño events, but they do help to drive the ocean and atmosphere towards a state in which warming is favoured. These results shed light on how the tropical Pacific ocean–atmosphere system may respond to natural and

Table 13.12 Handler's (1989) model linking aerosol production by volcanoes in different parts of the world to global climatic variations

Summary of Handler's model linking aerosol production by volcanoes in different parts of the world, to variations in global climate (based on Handler 1989)	
Assumptions	
1)	Surface temperature decreases have now been established following major eruptions.
2)	Stratospheric aerosols are restricted initially to a narrow range of latitudes and induced temperature decreases are non-uniform over the surface of the Earth.
3)	Aerosols not only produce gradients of surface temperature, but also induce pressure differences between land masses and oceans.
Argument	
Stage 1	Following the injection of aerosols into the stratosphere, cooling of the land will occur before cooling of the sea. There will be an initial increase in sea-level pressure on land. Due to the conservation of air mass at the global scale, this means a reduction in sea-level pressure over the oceans.
Stage 2	Because most of the land on Earth is in the Northern Hemisphere, the cooling induced by volcanic activity will transfer air mass from oceanic anticyclones to the continents. This will be particularly apparent in the Northern Hemisphere and especially over southern Europe.
Stage 3	Opacity, the month of eruption, its latitude, subsequent transport and lifespan of aerosols are important factors in controlling the direction of redistribution, and the magnitude of anomalous air-mass transfer.
The model	
El Chichón is an example of an eruption which produced a large aerosol injection at the right time of the year and in the right latitude (low latitude – Northern Hemisphere), to produce a strong ENSO in 1982/83.	
The Agung and Krakatau eruptions occurred in the Southern Hemisphere and are examples of eruptions where aerosols were concentrated over the southern oceans, so that they produced minor ENSO events.	
High-latitude eruptions, produce cooling of the regions poleward of the subtropics and this provides an additional sink for air mass leaving southern Eurasia in summer. One result is stronger monsoons. If aerosol are injected by a volcano which lies outside these regions (i.e. neither high or low latitude), then the effects on weather will be mixed.	

anthropogenic radiative forcing. It seems likely that explosive volcanism is a vital catalyst in global climatic interconnections and a major player in the Earth's climatic system (De Silva, 2003).

Possible links between volcanic activity and climatic change

Rampino *et al.* (1979) looked at the timings of large volcanic eruptions and glacial advances, which suggested that rather than preceding global cooling and glacial advances, the volcanic events in many cases appeared to follow them. However, there are all sorts of problems with relating the chronologies of glacial and volcanic events, which make the connections conjectural. Because of the large volume of ice involved in a Pleistocene glacial phase, the geophysical effects may have triggered volcanic events. The ice loading mass is so great that the global spin axis is perturbed, causing subtle changes to the three-dimensional shape of the Earth's surface (the geoid). Stresses set up during

worldwide realignment may trigger the volcanic eruptions. Similarly, during glaciations large water volumes are locked up in the ice sheets and the ocean basins are unloaded slightly, and as they readjust after the cold phase possible eruptive activity may be caused. Both primary and secondary trigger mechanisms whereby climate could increase volcanic activity are given in Table 13.13. In Iceland, several authors have proposed that mechanical effects due to glacial loading and unloading might have caused some variation in magma production in the mantle and/or magma release towards the surface, so affecting volcanism in the Icelandic axial rift zone during glacial times (Sigvaldson *et al.*, 1992; Jull and Mackenzie, 1996). It has also been suggested that the rift zone in northern Iceland jumped some 50 km westwards at the end of the last glaciation and that this jump and other changes in the rifting mode may be due to the glacial load modifying the state of stress of the lithosphere (e.g. Cohen, 1993). A further link between volcanic activity and sea-level change is illustrated in Box 13.3 (see the companion website).

Table 13.13 Possible mechanisms by which climate may initiate or trigger explosive volcanic activity. (Source: after Rampino *et al.*, 1979)

Possible cause of increased volcanism	Processes involved
A) *Primary cause*	
1) Realignment of the geoid due to redistribution of water.	Redistribution of water during successive glaciations and deglaciations causes realignment of the geoid and leads to worldwide changes in crustal stress. This will be most acutely felt at plate margins and major fault intersections. The majority of Pleistocene volcanism was related to these zones (see Anderson 1974). Chappel (1975), argued that the stress gradients beneath continental margins in response to this process are some 10^5 times greater than the stress gradients associated with Earth tides, which are known to affect the timing of some earthquakes and volcanic eruptions.
2) Hydrostatic unloading of ocean basins and its effect on magma movement.	This has been proposed by Matthews (1969) and suggests that the unloading of ocean basins during glacial episodes favours the upward movement of basaltic magmas.
B) *Trigger mechanism*	
3) Change in Earth orbit.	Some authors propose that the timing of glaciations is dependent upon variations in Earth orbit (see Hays *et al.* 1976). If this is the case then Earth tides will be created and may trigger eruptions. Hence, it is possible that both climatic change and volcanic activity in the Pleistocene is linked to extraterrestrial causes.
4) Magma mixing.	The isostatic processes noted above may not initiate volcanism, but may trigger it through magma mixing and greater activity at faults. Injection of basaltic magma into more silica-rich magma chambers would favour explosive volcanism. Mixed magma origin implied for many Quaternary tephra sheets (see Sparks *et al.* 1977).

13.8.2 Examples of individual volcanic eruption climatic disturbances

Laki fissure eruption of 1783

This eruption (Figures 13.8 and 13.31) caused severe climatic disturbances, leading to local- and regional-scale mortality of biotas.

It produced the largest lava flow known in historic time ($565\,km^2$) but little ash and although settlements were overrun by lava, the main damage was from toxic gases. An estimated $1.3-1.6 \times 10^7$ tons of sulphur dioxide was released, which resulted in the stunting of pasture and the death of 50% of cattle, 79% of sheep and 76% of horses. The human population was ravaged by famine and over 10 000 people or 24% of the population died. The climatic effects were more widespread too, and Benjamin Franklin noted in 1784 that there was a constant fog over all Europe and part of North America. Sigurdsson (1982) summarized the palaeoclimatic evidence from historical temperature records and Greenland ice cores. This evidence indicates unusually cold mean temperatures in the winter of 1783–4 and acidity levels in the 1783 ice layer are higher than any recorded in the last 1000 years. Taking the northern hemisphere as a whole, a temperature decrease of about $1\,°C$ seems to have occurred. This shows that an effusive eruption can have climatic effects comparable with those of explosive eruptions, like Tambora and Krakatau.

Toba supervolcano

Toba is the greatest eruption known since humans evolved. It occurred about $73\,500 \pm 3500$ BP on Sumatra, when an estimated $2800\,km^3$ was erupted in a VEI 8 eruption, with eruption clouds up to $32 \pm 5\,km$, to create the $100 \times 40\,km$ caldera known today as Toba. The eruption caused air fall 10 cm thick up to 2000 km away, and although the eruption was relatively sulphur-poor, the enormous volume meant that something like 2000 million tonnes of sulphur dioxide was released. It has been predicted that this would have caused a 'volcanic winter' with possible abrupt regional cooling of up to $15\,°C$ and estimates that the global temperatures would have fallen by at least $3-5\,°C$ for several years. This average cooling would be disastrous for vegetation and would translate into about $10\,°C$ cooling during the growing season in northern temperate latitudes, killing 50% of the trees in every forest. Global climate models suggest that there would have been severe drought in the tropical rainforest belt and in monsoonal regions. In ice cores, it can be seen that ash was still settling as long as six years afterwards. Mankind clearly survived the effects of this eruption but it must have been a global ecological disaster and had major effects. This can be seen from genetic studies, which suggest that sometime prior to about 60 000 years ago humans suffered a severe population bottleneck, with possibly only 3000–10 000 individuals surviving for a period of up to 20 000 years, followed by a rapid population increase, technological innovations and migrations out of Africa of modern human populations. The climatic effect of this Toba eruption could have caused this bottleneck and other similar bottlenecks amongst other organisms (e.g. some chimpanzee populations) at this time (Rampino and Ambrose, 2000). Although this genetic clock date is uncertain by 5000 years either way, no other environmental disaster has yet been identified in the relevant period, so it seems likely that the human population today are all descendents of the survivors of the global-scale environmental disaster which was triggered by the Toba eruption.

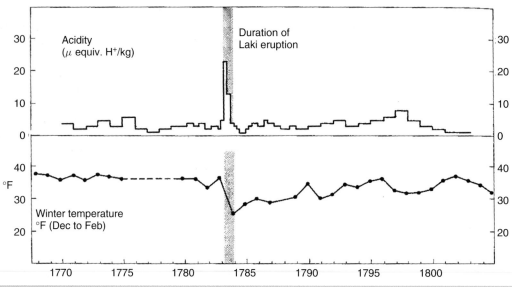

Figure 13.31 Some effects of the Laki 1783 eruption. (Source: after Sigurdsson, 1982)

There are other known supervolcanoes on Earth, such as Cerro Galan in the Argentinian Andes, La Pacana in the Chilean Andes, Yellowstone in Wyoming, Long Valley, California and Valles, New Mexico. All have major calderas and have undergone VEI 8 eruptions in the geological past. Rothery (2003) speculates that the Earth probably suffers the effects of a supervolcanic eruption every 50 000 years, with an associated volcanic winter. All similar supervolcanic eruptions would have major climatic and environmental consequences on the planet and would affect biotic evolution. However, supervolcanic eruptions might not be so environmentally devastating as we thought (see Box 13.4, in the companion website).

13.9 Relationships between volcanic eruptions and biotic evolution

(See further discussion in Chapter 29.)

Mass extinctions occur in the geological record where there are relatively sudden, near-simultaneous disappearances of a large percentage of the living species, affecting a great variety of organisms. They have been recognized at several well-documented periods of geological time: at the end of the Ordovician period (439 million years ago); the Late Devonian (365 million years ago); the end of the Permian (250 million years ago); the end of the Triassic (201 million years ago); and the end of the Cretaceous (65 million years ago). In these major extinctions at least 75% of marine life and many nonmarine species died out. There are also another 20 smaller-scale extinctions, like at the end of the Pleistocene, in which many mega-mammals died out. The latter has been caused either by climatic change or by the effects of hunting by humans, or a combination, but it is likely that many mass extinctions have been caused by catastrophic events like comet or asteroid impact. Mega-eruptions from volcanoes have

also been implicated and both would have had major climatic effects, with both greenhouse effects and volcanic/nuclear winters postulated. These mass extinctions have caused great controversy in the geological literature and often polarized discussion (see for example Sharpton and Ward, 1990). The end-Cretaceous (65 million years ago) mass extinction has usually been correlated with the 10 km wide Chicxulub crater off the Yucatan coast of Mexico, caused by a comet or asteroid, but the near-coincidence in time of the Deccan Flood basalts may have implications for this event too. Despite the fact that some workers have found a coincidence of flood basalt events with a number of mass extinctions (e.g. the Siberian flood basalt province and the end-Permian extinctions 250 million years ago) and a possible cause-and-effect relationship, it is also possible that both types of extreme event could be implicated. Flood basalt activity could represent only one of many interrelated geological factors and it has been suggested that the coincidence of these events with large impacts may be necessary in causing major mass extinction. It seems possible that large impacts could trigger or enhance the volcanism.

13.9.1 Flood basalts and mass extinctions

In at least three cases (Deccan, Newark and Siberia) there seems to be a direct correlation between the age of flood basalts and mass extinctions. The Deccan Flood basalts, where there are over 50 flows in 1200 m of section, are thought to have erupted in about 500 000 years, bracketing the Cretaceous/Tertiary boundary. The impact fallout layer has also been found in sediments interbedded within the lavas, which support the radiometric dating and palaeo-magnetic dating, suggesting that the eruptions began prior to the boundary. In the Newark Basalts of eastern Northern America the pollen changes that mark the Triassic/Jurassic boundary have been located just below the first lava flows. For the Siberian Basalts the palaeontological evidence suggests that Permian/Triassic mass extinctions were in the lowest lava sequence. There seems therefore to be a major probability that the correlations are correct, especially as there are other examples of flood basalt and lesser faunal and floral changes occurring in the geological record (e.g. the Karoo Basalts might correlate with the boundary between the Early and Middle Jurassic, which is marked by a faunal turnover event).

The processes that have been suggested as the environmental effects of flood basalts and therefore linked to mass extinctions (summarized in greater detail in Rampino and Self (2000)) are as follows:

- A cooling caused by the formation and spread of stratospheric aerosols produced by high fire fountaining and associated convective plumes above them. This would be associated with rich dissolved sulphur taken into the stratosphere. Volcanic aerosols at all altitudes should cause short-term climatic cooling but a flood basalt eruption could release such large quantities of sulphuric acid aerosols that a regional tropospheric aerosol cloud could be maintained despite aerosol washout. This could be aided by a suppression of convection in the atmosphere by the cooling, which would mean that rainout would be less effective at removing the aerosols.

- Greenhouse warming caused by large carbon dioxide emissions, although recent estimates suggest that the total increase in CO_2 would have been under 200 ppm, leading to a predicted global warming of only $2\,^{\circ}C$ over the eruption period of the lavas, and it is not likely that such a small and gradual warming would have been an important factor in mass extinctions. Another source of warming could have been from the SO_2 concentrations, which could have led to major severe regional warming. In the case of the Siberian Basalts, it has been suggested that the volcanism caused the release of large quantities of methane from hydrates that had accumulated on adjacent high-latitude continental shelves, leading to greenhouse warming.

13.9.2 The relationship between impacts and flood basalts

Large impacts by a comet or asteroid might excavate initial holes deep enough to result in decompression and upper-mantle melting but there is no firm evidence that impacts can directly cause volcanism, or of local, large impacts at flood basalt locations. However, an indirect connection may be possible as impact energy would cause major lithospheric stresses which might lead to rapid decompression in the mantle above rising mantle plumes and help induce or increase ongoing volcanism. It does seem though that the Deccan lavas began erupting prior to the Chicxulub impact and therefore that the impact could not have triggered volcanism. Moreover, other evidence suggests that the mass extinctions at the Cretaceous/Tertiary boundary were probably caused by the

impact and not by volcanism. For example, there is low iridium (a mineral thought originally to have been brought in from an extraterrestrial source, although volcanic sources have also been found) in the Deccan lavas, in soils and weathering profiles between flows and at the boundary sections, and the shocked quartz and stishovite cannot be produced at pressures characteristic of volcanic eruptions (although there has been dispute here too about whether these shocked minerals can be formed by volcanic processes). Moreover, the tektite glass (impact melt glass) found is not compatible with a volcanic origin. However, there remains much dispute still about the role of volcanism in the Cretaceous/Tertiary extinctions, and Rice (1990) suggested that worldwide volcanism coincident with the end of the Cretaceous provides the source of iridium, shocked minerals, the carbon (CO_2) excursion and the attendant greenhouse effect.

Whatever the causes of mass extinctions, it is nevertheless clear that major volcanic events can have an impact on the climate, can cause major local changes in the ecology and must cause local death and species migration. Ecosystems are changed, populations may be split and isolated from other populations, and hence speciation may be encouraged by volcanism. Despite the hazards caused by volcanoes and the problems for human populations living close to volcanoes from ash fall, pyroclastic flows and lahars in particular, people still live in close proximity to them. This is because the ash undergoes weathering, which turns it into a fertile soil with a rejuvenated mineral content and a favourable capacity to prevent the loss of phosphorous by leaching. Hence volcanic areas, although only covering about 1% of the Earth's surface, have over 10% of the world's population.

13.10 Plate tectonics, uniformitarianism and Earth history

13.10.1 Superplumes, supercontinents, plume tectonics and plate tectonics

There is no doubt that plate tectonics has revolutionized our understanding of the Earth's history, providing an interpretative framework with which to explain and model the movement of plates across the surface of the planet, the changing spatial and temporal patterns of volcanism on the Earth's surface and the evolution of new ocean and sedimentary basins and their closure to produce mountain belts. These are mechanisms that are essential to understanding the morphology of the Earth's surface and to the interpretation of its stratigraphic record. One of the fundamental series of questions however is: has plate tectonics always operated, has it always been in the same way and at what rate has it operated? These are difficult questions to answer because of the paucity of reliable data from the early part of the Earth's history. As regards the uniformity of rate, there is growing evidence to suggest that plate tectonic activity may be subject to periodic accelerations of activity, referred to as pulsation tectonics. These episodes of accelerated activity have been explained by the interaction of surface plates with the movement of plumes of molten rock within the mantle. These plumes, referred to as superplumes to distinguish them from hot spots and other thermal anomalies, appear to originate at the junction between the core and the mantle at a depth of over 2900 km. They are hundreds of kilometres in diameter and may be 250–300 °C hotter than the surrounding mantle. The origin of these superplumes is uncertain but they appear to be generated by instability at the core–mantle boundary. Superplumes rise through the mantle before encountering the lithosphere, where they mushroom out below it, causing it to dome, rift and melt. This process may be associated with the production of large amounts of flood basalt, accelerated rifting and subsequent plate divergence.

Early models of these superplumes suggested that, like the current hot spots in plate interiors, superplumes were simply superimposed and independent of plate tectonics. What is now emerging is the idea of a 'whole Earth' tectonic model in which plate tectonics both influence and is influenced by plume tectonics. The details of such a model remain unclear, but a recurrent theme is the idea that subduction of cold slabs of ocean crust may generate superplumes. Slabs of relatively cold oceanic crust appear to accumulate below subduction zones in the upper mantle at a depth of 600 km. As this cold ocean crust accumulates in the upper mantle it becomes subject to catastrophic collapse, known as avalanching, into the lower mantle, towards the mantle–core boundary. A consequence of this collapse is the rise of a complimentary superplume. In this way material is recycled through the mantle and heat is convected from the core to the surface. Plume tectonics is therefore responsible for periodically accelerating plate tectonic processes, although the detailed processes are somewhat sketchy.

Plate tectonics has operated throughout the Phanerozoic, and the last 500 million years have been dominated by the creation, assembly and dispersal of large supercontinents. This is called the supercontinent cycle. This cycle involves the accretion of continents and microcontinents to form a single supercontinent, and the supercontinent's subsequent rifting and dispersal. In the late Proterozoic the continents were assembled as a single continent, which then rifted in the early Phanerozoic to give a dispersed pattern of continents, which then began to coalesce in the late Palaeozoic through a series of mountain-building episodes to again form a single supercontinent, Pangaea. This cycle has had a profound effect on the distribution of Earth environments, climate and the evolving biosphere. In fact, during the Precambrian this cycle of continental agglomeration and dispersal may have occurred several times.

The cause of supercontinental cycles is another problem. Some geologists suggest that they may result from the contrast in the thermal properties of continental and oceanic crust. Continental crust is only about one-half as efficient at conducting heat as oceanic crust, so if a supercontinent such as Pangaea covers a significant part of the Earth's surface, heat will build up in the mantle below it. Given time the build up of heat beneath the supercontinent may lead to uplift, rifting and continental dispersal. As ocean basins form during continental dispersal, mature subduction zones will develop and basin contraction may start. This will lead to a transition from continental dispersal to continental aggradation and the development of another supercontinent. The problem with this type of model is that the thermal contrast between ocean and continental crust is, according to recent modelling, unlikely to be sufficient to cause supercontinents to rift. Moreover, it is not clear what causes the continents to converge to form a single supercontinent. In view of this, the supercontinent cycle has been linked to plume tectonics.

It has been widely noted that there is an overlap between the dispersal and assembly phase of supercontinents. As one supercontinent is being dispersed, another is being assembled. In part, this has been explained by the fact that mantle upwelling, a prerequisite for continental rifting and dispersal, causes thermal updoming of the Earth's surface to form a geoid high. The geoid is the overall shape of the Earth itself, which is actually quite variable. As individual continents rift from supercontinents they will tend to move towards geoid lows, or towards cool, and therefore lower, areas of the Earth's surface. As continents converge in these geoid lows a concentration of subduction zones is likely. This will further cool the mantle as cold slabs of ocean crust descend. If these slabs become concentrated at the base of the upper mantle along the 660 km seismic discontinuity the process of catastrophic avalanching already described may occur, thereby generating mantle plumes. A likely result of this downwelling and avalanching is the pulling of plates towards subduction zones at accelerated rates, attracting and concentrating continents into a large supercontinent. In this type of model, mantle plumes are created as a consequence of slab avalanches during the construction of the supercontinent, particularly during the Archaean and Proterozoic, when slab avalanching may have been more pronounced.

Once created, the supercontinent may act to shield the mantle from cooling by subduction, which is concentrated around the margins of the supercontinent. As the mantle beneath the supercontinent begins to warm, a mantle plume may form, causing the supercontinent to rift and thereby restart the cycle. Slab avalanching need not therefore be implicated in continental dispersal, although other variations of this model suggest that subduction and slab avalanching around the margins of a supercontinent may actually generate the mantle plume that ultimately leads to its demise.

Although the geological record suggests that several supercontinental cycles may have operated during the Precambrian, conventional plate tectonics might not have operated early in the Earth's history. The Earth's interior has been cooling since its formation some 4600 million years ago and, with it, geotectonic processes have probably changed. The presence of subduction zones within the mantle are recorded in the geological record by the occurrence of high-grade, glaucophane schist belts, which are first recorded during the Mid-Proterozoic and provide clear evidence that from this point on conventional plate tectonic processes have dominated the Earth's tectonic system. But what was it like before this time? This question is difficult to answer and there is currently active debate and discussion of geotectonic models for the Early Archaean and Proterozoic. Most of the recent models and ideas that have been proposed involve plume tectonics (see Chapter 26 for further discussion).

The Earth's first crust probably formed on an ocean of magma some time after 4500 million years ago. Rapid convection in the upper mantle led to recycling of this early crust, but the exact nature of the plate boundaries and in particular the geometry of the associated subduction zones is uncertain. Before 3000 million years ago there were probably only a few small microcontinents, perhaps originating as volcanic island arcs at early subduction zones. Most of these microcontinents were then captured by Late Archaean mountain-building phases to form larger continents or cratons. The mantle may have had a strong thermal stratification, with the upper and lower mantle convecting heat independently, at this time, causing a barrier to the descent of subducted crust, which may have accumulated at the thermal boundary between the upper and lower mantle. As the mantle cooled and this boundary became less pronounced, subducted crust may have periodically avalanched as catastrophic events towards the core, initiating a series of superplumes and causing the rapid accretion continents to form early supercontinents. The process of continental accretion probably involved the formation of numerous island arcs and their accretion on to the growing supercontinent, producing combinations of granites and gneisses formed from the early crust, and greenstone belts associated with the deposition of volcanics and volcaniclastic sediments derived from the arcs. As the mantle continued to cool, the thermal stratification within it decreased such that deeper subduction could start, and the accumulation of crustal slabs at the interface between the upper and lower mantle was less marked. Consequently, conventional plate tectonics began to dominate over plume tectonics. It must be emphasized that this model is one of many and the question of the true nature of tectonic processes on the early Earth awaits further research.

13.10.2 The opening of sedimentary Basins: Continental rifting and the evolution of passive margins

Continental rifting occurs when divergent plate margins develop in continental masses. They are defined as elongated depressions in which the entire lithospheric thickness has deformed because of extensional forces. The crust has been arched upwards, extended and pulled apart and the structure is dominated by parallel normal faults, with large vertical displacements. This forms the rift valleys or grabens and associated uplifted blocks or horsts. The regions are characterized by thin crust, normal faulting, shallow earthquakes and a magmatic assemblage dominated by basalts and lesser rhyolites. The continental rifting creates new continental margins that are marked by normal faults and volcanic rocks which are interdigitated with thick sequences of continental sedimentary rocks. As the region subsides it becomes covered by thick shallow marine sediments.

The passive continental margins along mature rifts are important because they contain the largest accumulations of sediment on the planet. They are also economically important because of their oilfields and reserves. The best known examples are the East African and Red Sea rifts, discussed further in Hamblin and Christiansen (2004, pp. 564–566). It appears that progressive stages in the development of a new passive margin from a continental rift can be recognized (Figure 13.32):

1. The initial event is an upwarp or dome in a continent. As the lithosphere expands, it arches, thins the crust and fractures the brittle outer part. Extension and thinning of the continental crust creates a fault-bounded rift valley. As the crust thins, decompression occurs, basaltic magmas are erupted and rhyolites can be produced by partial melting of the granitic crust by heating from the rising basaltic magma or by differentiation of the basaltic magma. These volcanic rocks are interbedded with the rift-basin sediments.

2. After the continent moves away from the hot, uparched centre, the rifted margins start to subside due to cooling. This subsidence allows thicker sediment sequences to accumulate in shallow oceans at the new passive continental margin.

3. With continued subsidence, oceans fill the depression to form long, narrow, shallow seas, which have restricted circulation. Shallow marine sediments are deposited on the older continental sediments deposited in the original rift valley. Thick salt sequences can form from evaporating sea water where the climate is arid and hot.

4. Finally the rift opens sufficiently for circulating marine water, with marginal organic reefs, lagoons and beaches at the margins, which are covered by river sediments deposited in the oceans as the basin subsides. In deeper water towards the centre there are turbidites and slumps from submarine landslides and deep-marine organic oozes.

Further discussion of the characteristics of rifts, their classification, petrology, structure and origins can be found in Chapter 6 of Kearney and Vine (1990, pp. 107–114).

13.10.3 The closure of sedimentary Basins: Basins to mountains

As illustrated above, the three main types of plate boundary are associated with a range of different sedimentary basins, which on closure may lead to mountain building, or orogenesis. The relationship between mountain building and sedimentary basins can be seen at its simplest through the concept of the Wilson Cycle. Ocean basins are formed by the rifting and break-up of continents and grow through the process of sea-floor spreading. Within the lifetime of an ocean, subduction will begin, and the rate at which the ocean floor is subducted may gradually exceed the rate of sea-floor spreading, causing the basin to close. As a consequence, the ocean will shrink as more sea floor is lost through subduction than is created by sea-floor spreading. Eventually the ocean will

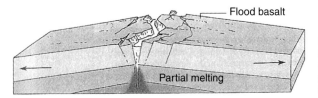
Flood basalt

(A) Continental rifting begins when the crust is uparched and stretched, so that normal faults (red) develop. Continental sediment (yellow) accumulates in the depressions of the downfaulted blocks, and basaltic magma is injected into the rift system. Flood basalt (gray) can be extruded over large areas of the rift zone during this phase.

Partial melting

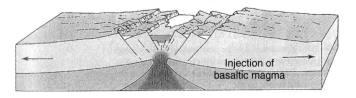
Injection of basaltic magma

(B) Rifting continues, and the continents separate enough for a narrow arm of the ocean to invade the rift zone. The injection of basaltic magma continues and begins to develop new oceanic crust (green).

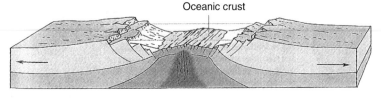
Oceanic crust

(C) As the continents separate, new oceanic crust and new lithosphere are formed in the rift zone, and the ocean basin becomes wider. Remnants of continental sediment can be preserved in the down-dropped blocks of the new continental margins.

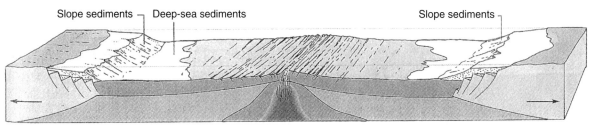
Slope sediments — Deep-sea sediments   Slope sediments

Stages of continental rifting are shown in this series of diagrams. The major geologic processes at divergent plate boundaries are tensional stress, block faulting, and basaltic volcanism.

Figure 13.32 Stages of continental rifting. The major geological processes at divergent plate boundaries are tensional stress, block faulting and basaltic volcanism

close and the continents along its margins will collide and fold the sediments that accumulated in the former ocean to form a chain of mountains. Ocean basins are just one form of sedimentary basin formed by plate tectonics.

Before the main processes of mountain building can be reviewed, some of the basic controls on the evolution of a mountain belts must be considered: the nature of the building blocks available for mountain building, the concept of an accretionary wedge and the role of exogenic processes in mountain building. Most mountain belts are composed of a mixture of different geological terranes; that is, blocks of rock that are distinct, in terms of their stratigraphy, igneous and metamorphic history, and tectonic development, from adjacent units from which they are separated by major tectonic structures. The term allochthonous is often added as a prefix, indicating that the terrane was not formed in its present position. Terranes can consist of continental blocks of various sizes, island arcs, fragments of ocean crust (ophiolite complexes) or blocks of thrust and folded sediment. Even the simplest mountain belts formed by the collision of two continents as an ocean basin closes will consist of two terranes, one from

either side of the closing ocean. However, most mountain belts appear to consist of a series of far-travelled terranes that have become juxtaposed during orogenesis. For example, the Western Cordillera of North America is built up of over 50 major terranes formed by the accretion of different crustal fragments to the North American continent during subduction of the Pacific plate beneath it.

Although the processes of compression within a mountain range are extremely complex, they can be modelled conceptually as an accretionary wedge. In understanding this concept it is instructive to consider the analogy of a bulldozer whose frontal plough is driven into a sheet of sand spread over a melted road on which the plough rests. As the bulldozer moves into the sand it will form a wedge in front of the plough, which tapers into the foreland in front of the advancing bulldozer. The thickness of the wedge depends on the internal strength of the sand and the friction between it and the rigid road surface below, which acts as a slip or décollement surface. The thickness of the wedge expressed in terms of the angle of taper will increase until the gravitational gradient is such that it just balances the frictional forces on the décollement surface over which the wedge is moving. If we add material to the wedge, it will tend to reduce the angle of taper by extension into the foreland. In contrast, if the angle of taper is too shallow then the wedge will thicken by internal shortening until the force balance is restored. The idea is that a wedge with a critical taper will move over the basal décollement surface without internal deformation. This concept was first applied to the evolution of accretionary wedges at subduction zones and to simple fold–thrust systems, but its application had broadened to embrace orogenic episodes in general. It is used as a basis for numerical modelling, although this use is subject to some debate, but its value here is as a conceptual model with which to consider the evolution of orogenic belts. The geometrical configuration of fold–thrust belts depends on the strength of the material, the fluid pressures present and the level of internal friction. A deforming body of rock will tend to deform internally until it arrives at a shape that is characterized by a critical taper, whereupon the mass will slide over a rigid décollement surface without further deformation. The critical taper required depends on the strength of the deforming rock mass relative to the frictional drag along the décollement surface. The stronger the deforming material, for example, the thinner the deforming wedge and the lower the angle of critical taper. Equally, if the friction along the basal décollement is reduced by the presence, for example, of very high fluid pressures, then the wedge will also be thinner and the transport distances for the thrust sheets or nappes will be correspondingly greater. The geometry of the décollement surface can also affect the shape of the deforming wedge. At subducting margins, deformation of the fore-arc and accretionary prism occurs above a décollement defined by the upper surface of the subducting plate. The angle may change through time. For example, if the angle of subduction increases as a subduction zone steepens, the wedge will no longer be critical and will need to thicken. Equally, if the subduction zone were to flatten then the wedge might extend in order to flatten the angle of taper.

The shape and form of a growing wedge is also influenced by the processes of denudation as the mountains develop and are eroded. It is helpful to distinguish two main types of uplift. Surface uplift is the elevation of the land surface with respect to a fixed datum, usually sea level, and results from the balance between crustal uplift (the movement of the rock column below the land surface) and the rate of denudation. If the rate of denudation exceeds the rate of crustal uplift, for example, then no surface uplift and mountain building will occur. As a mountain range develops, denudation will begin to erode the mountain range, and by doing so it will affect the distribution of mass within the orogen and may therefore influence its structural evolution. The distribution of denudation processes within a mountain range is controlled by the prevailing climate and altitude. For example, above a certain altitude, glaciation may be initiated; more importantly, mountains often impact on precipitation patterns: orographic rainfall will tend to fall on the windward side of mountains, leading to enhanced denudation, whilst on the leeside precipitation and consequently denudation may be reduced by a rain shadow effect. We can see this in terms of the accretionary-wedge-type model

introduced above. If denudation rates are high then they will tend to reduce the surface taper of the wedge and as a consequence the wedge will tend to undergo internal deformation in order to increase the angle of taper. Such a wedge is unlikely to propagate into the foreland and will result in a narrow tectonic zone. In contrast, if denudation rates are low the wedge will tend to oversteepen and propagate outwards, via a series of large imbricate nappes and thrust sheets, into the foreland, resulting in a broad deformation zone.

In dry mountain belts the distribution of topography is controlled by the tectonic strain distribution, whilst in wet mountain belts, drainage organization and rock strength properties determine the distribution of relief and to some extent the style of deformation.

13.10.4 Four types of mountain-building scenario

Continental margin orogenesis

Not all mountain chains form by the closure of ocean basins: orogenesis may occur along subducting plate margins. This type of mountain building is referred to as continental margin orogenesis and involves the production of a mountain chain along a convergent margin at which an oceanic plate is descending beneath a plate composed of continental crust. Compression and deformation will occur at the plate margin if the rate at which plate convergence is occurring exceeds the rate of subduction. The subduction rate is controlled by the rate and angle at which the plate of ocean crust descends beneath the plate of continental crust. This is controlled by the buoyancy of the ocean plate or by its temperature, as a cold plate will descend easily. This in turn is determined by the distance between the centre of sea-floor spreading and the subduction zone: the greater the distance, the greater the opportunity for the ocean crust to have cooled. Compression tends to be concentrated in one or both of the following areas: at the leading edge of the continental plate or in the back-arc zone. Compression of the leading edge involves the evolution of an accretionary prism and the deformation of sediments within the fore-arc. Shortening of fore-arc sediments and those sliced from the subducting plate builds up large fold–thrust belts above a basal décollement formed by the surface of the subducting plate. The second zone of possible compression occurs within the back-arc zone (retro-arc basin). Here shortening of foreland sediments, eroded from the volcanic mountains and within the crust, may give rise to a back-arc, thrust–fold system. This is common where there is movement of the continental plate towards the subduction zone. This may occur if, for example, there is active spreading along one of the other margins. In addition to crustal shortening, continental margin orogens are also built of significant percentages of volcanic rocks, rising through the continental crust, causing it to thicken. The development of the orogen may be assisted further by the accretion of terranes. Allocthonous terranes (volcanic island arcs, fragments of ocean crust or accretionary prisms) may be transported on the subducting plate to the continental margin, where their accretion to the continental margin may add to the growing mountain range. The angle of closure between a descending plate and the over-riding continent may not be exactly parallel and oblique angles of closure may lead to lateral or transpressive movement along the orogen. In such scenarios terranes may move laterally along an orogen after they have docked with it.

It is also important to realize that subduction zones are not always zones of compression and that during the history of a single orogen a margin may experience both compressive and extensional phases. For example, thermal upwelling to the rear of the volcanic zone may cause continental extension, which if it were to continue to the rear of the volcanic zone might cause continental extension, which could lead to the development of a new ocean basin. In short, the evolution of a continental margin orogen may be complex and involve several mountain-building phases.

Recently some interesting ideas have been put forward on the subject of the relationship between climate and tectonics in active mountain ranges and it has been suggested that there is a causal relationship between Himalayan erosion and monsoon intensification. These relationships are described in Box 13.5 (see the companion website).

Continental and collage collisions

These involve the production of a range of mountains by the closure of an ocean basin and the collision of two continental terranes along its margins. This is the classic mountain-building scenario, in which sediment is compressed, vice-like, between two advancing continental margins. As the two plates dock, both the ocean crust and the overlying sediments deposited in the former ocean are subject to intense shortening, usually along a series of large thrusts, which form a stacked sequence incorporating both ocean crust and sediment. As the two continental plates continue to dock, subduction ceases, although one of the plates may be partially subducted below the other, and intense shortening of sediment and crust occurs to produce a mountain range along the suture between the two plates. In practice, the interaction between the continental plates may be more complex, as both are buoyant and therefore resistant to subduction. In some cases, as just described, one of the continental slabs may be pulled slightly beneath the other before subduction ceases, but in other cases a continent may produce a flake or a thin sliver of continental crust, shaved from the surface of the underthrust plate and riding over the surface of the over-riding plate. Consequently, the interaction between the two continental terranes may be much more complex. As subduction ceases, volcanism eventually stops in the over-riding plate. Shortening of the crust and sediments adds to the growing tectonic pile and large nappes of rock may begin to move either side of the growing mountain belt, perhaps assisted by gravity. Two foreland basins develop either side of the range and will begin to receive sediment from the eroding mountains. Of these two basins, the peripheral foreland basin is formed by the down-dip of the continental crust, which

is underthrust beneath the over-riding plate, while the retro-arc foreland basin forms in what was once the back-arc zone. The picture of a continental collision and orogenesis can be complicated by the addition of accreted or trapped terranes, which may move considerable distances laterally, particularly if the angle of collision is oblique. Equally, the style of structural deformation within the mountain belt can be very varied and can range from intensely-faulted systems, which can contain a series of stacked thrusts, to giant recumbent folds that become detached along thrust planes to form nappes, which can travel over hundreds of kilometres. The style of deformation depends on the materials involved and the stress and temperature regimes generated by the colliding plates, but may also be partly controlled by the rate of denudation, as we will see later.

In practice, most mountain ranges incorporate a collage of terranes, which may have been accreted to the two advancing plates prior to their collision, or may simply have been trapped within the closing ocean.

Transpression

This involves deformation along a transform margin and the closure of any transform basins present. It forms an important part of many orogenic episodes, since there is usually a strong lateral component within most types of continental collision.

13.10.5 Tectonic basins within plate interiors

A variety of different types of sedimentary basin may form within a plate composed of continental crust. Of these, the intercratonic basin is perhaps the most noteworthy. This type of basin is a broad regional downwarp in continental crust, well removed from the tectonic influence of a plate margin. It does not usually involve faulting, although some – but not all – of the basins may be centred on rift valleys. Intercratonic basins are commonly oval in shape and enclosed, with a central point much lower than the surrounding area. The processes by which these broad downwarps form is not entirely clear, although it has been suggested that they result from the activity of thermal plumes in the underlying mantle. Most volcanic activity is confined to plate boundaries, but local hot spots or thermal plumes do occur within the interior of plates. If a thermal plume rises beneath a continent, it will cause the overlying crust to thin and stretch to produce a rift valley. If this process continues the crust will separate to form two new plates and a new ocean. However, if the thermal event comes to an end before crustal separation occurs, a broad crustal down-sagging may result when the heat is removed, which will give rise to an intercratonic basin. Alternatively, thermal plumes may elevate, but not cause rifting. At the end of a thermal episode the crust will subside back to its original position. However, if it has been thinned by erosion while elevated, the crust will subside below its original level to form a basin. In this way, regional downwarps may form that are not centred on local rift valleys.

Exercises

1. What are the geological characteristics and consequences of two oceanic plates colliding?

2. Describe the characteristics of mid-ocean-ridge-system volcanic processes.

3. What factors control the viscosity of a lava?

4. Explain why there are several types of pyroclastic density flow.

5. Describe the mechanisms that have been suggested for volcanic debris flows. What are the characteristics of sediments formed by such events?

6. What are the differences between lahars and hyper-concentrated flows?

7. What are the characteristics of volcanic air falls?

8. What are the likely global atmospheric effects of a major volcanic eruption?

9. What are the characteristic geological structures and rock types found in the Earth's major mountain belts and how can these be explained?

10. Why do continental rift zone magmatism and hot spot magmatism occur? Describe the rock type differences that may be found in such areas.

References

Adams, J.B., Mann, M.E. and Ammann, C.M. 2003. Proxy evidence for an El Niño-like response to volcanic forcing. *Nature.* **426**, 274–278.

Blake, S. 1990. Viscoplastic models of lava domes. In: Fink, J.H. (Ed.). Lava Flows and Domes. IAVCE Proceedings in Volcanology. Berlin, Springer-Verlag. pp. 88–126.

Cantagrel, J.-M., Robin, C. and Vincent, P.M. 1981. Les grandes étapes d'évolution d'un volcan andésitique: Exemple du Nevado de Toluca. *Bulletin Volcanologique.* **44**, 177–188.

Carey, S.N. and Sparks, R.S.J. 1986. Quantitative models of the fallout and dispersal of tephra from volcanic columns. *Bulletin of Volcanology.* **48**, 109–125.

Carracedo, J.C. 1996. A simple model for the genesis of large gravitational landslide hazards in the Canary Islands. In: McGuire, W.J., Jones, A.P. and Neuberg, J. (Eds). Volcano Instability on the Earth and other Planets. Geological Society Special Publication. **110**, 125–135.

Cas, R.A.F. and Wright, J.V. 1996. Volcanic Successions. London, Chapman and Hall.

Chester, D. 1993. Volcanoes and Society. London, Arnold.

Clapperton, C.M. 1977. Volcanoes in space and time. *Progress in Physical Geography.* **1**, 375–411.

Cohen, S.C. 1993. Does rapid change in ice loading modulate strain accumulation and release in glaciated, tectonically active regions? *Geophysical Research Letters.* **20**, 2123–2126.

Daines, M.J. 2000. Migration of melt. In: Sigurdsson, H. (Ed.) Encyclopedia of Volcanoes. San Diego and London, Academic Press. pp. 69–88.

Davidson, J.P., McMillan, N.J., Moorbath, S, Wörner, G., Harmon, R.S. and Lopez-Escobar, L. 1990. The Nevados de Payachata volcanic region 11: Evidence for widespread crustal involvement in Andean magmatism. *Contributions to Mineralogy and Petrology.* **105**, 412–432.

De Silva, S. 2003. Eruptions linked to El Niño. *Nature.* **426**, 239–241.

Decker, R.W. and Decker, B. 1991. Mountains of Fire. Cambridge, Cambridge University Press.

Eaton, G.P. 1984. The Miocene Great Basin of Western North America as an extending back-arc region. *Tectonophysics.* **102**, 275–295.

Fisher, R.V., Heiken, G. and Hulen, J.B. 1997. Volcanoes: Crucibles of Change. Princeton, NJ, Princeton University Press.

Fisher, R.V. and Schminke, H-U. 1984. Pyroclastic Rocks. Berlin, Springer-Verlag.

Furness, H., Friedliefsson, I.B. and Atkins, F.B. 1980. Subglacial volcanics: On the formation of acid hyaloclastites. *Journal of Volcanology and Geothermal Research.* **8**, 95–110.

Hamblin, W.H. and Christiansen, E.H. 2004. Earth's Dynamic Systems. 10th edition. Upple Saddle River, NJ, Prentice Hall.

Hammer, C.U., Clausen, H.B. and Dansgaard, W. 1980. Greenland ice sheet evidence of postglacial volcanism and its climatic impact. *Nature.* **288**, 230–235.

Handler, P. 1989. The effect of volcanic aerosols on global climate. *Journal of Volcanology and Geothermal Research.* **37**, 233–249.

Handler, P. and Andsager, K. 1990. Possible association between the climatic effects of stratospheric aerosols and sea surface temperatures in the eastern tropical Pacific Ocean. *International Journal of Climatology.* **10**, 413–424.

Henyey, T.L. and Lee, T.C. 1976. Heat flow in Lake Tahoe, California–Nevada and the Sierra Nevada–Basin and Range transition. *Geological Society of America Bulletin.* **87**, 1179–1189.

Hofmann, A.W. 2007. The lost continents. *Nature News and Views.* **448**, 655–656.

Jeanloz, R. 2000. Mantle of the Earth. In: Sigurdsson, H. (Ed.). Encyclopedia of Volcanoes. San Diego and London, Academic Press. pp. 41–54.

Jones, J.J. 1969 Intraglacial volcanoes of south-west Iceland 1. *Quatertly Journal of the Geological Society of London.* **124**, 197–211.

Jull, M. and Mackenzie, D. 1996. The effects of deglaciation on mantle melting beneath Iceland. *Journal of Geophysical Research.* **101**, 21 815–21 828.

Kearney, P. and Vine, F.J. 1990. Global Tectonics. Oxford, Blackwell Scientific.

Kilburn, C.R.J. 2000. Lava flows and flow fields. Sigurdsson, H. (Ed.). Encyclopedia of Volcanoes. San Diego and London, Academic Press. pp. 291–306.

Kokelaar, B.P. 1986. Magma–water interactions in subaqueous and emergent basaltic volcanism. *Bulletin of Volcanology.* **48**, 275–289.

McCormick, M.P., Thomason, L.W. and Trepte, C.R. 1995. Atmospheric effects of the Mt Pinatubo eruption. *Nature.* **373**, 399–404.

McDonald, G.A. 1972. Volcanoes. New Jersey, Prentice Hall.

Moore, J.G., Normark, W.R. and Holcomb, R.T. 1994. Giant Hawaiian landslides. *Annual Reviews of Earth and Planetary Science.* **22**, 119–144.

Olson, P., Silver, P.G. and Carlson, R.W. 1990. The large-scale structure of convection in the Earth's mantle. *Nature.* **344**, 209–215.

Perfit, M.R. and Davidson, J.P. 2000. Plate tectonics and volcanism. In: Sigurdsson, H. (Ed.). Encyclopedia of Volcanoes. San Diego and London, Academic Press. pp. 89–114.

Peterson, D.W. and Tilling, R.I. 1980. Transition of a basaltic lava from pahoehoe to aa, Kilauea volcano, Hawaii. *Journal of Volcanology and Geothermal Research.* **7**, 271–293.

Pichler, H. 1965. Acid hayaloclastites. *Bulletin Volcanologique.* **28**, 293–310.

Rampino, M.R. and Ambrose, S.H. 2000. Volcanic winter in the Garden of Eden: The Toba supereruption and the late Pleistocene human population crash. In: McKoy, F.W. and Heiken, G. (Eds). Volcanic Hazards and Disasters in Human Antiquity. United States Geological Survey Special Paper. **345**, 71–82.

Rampino, M.R. and Self, S. 2000. Volcanism and biotic extinctions. In: Sigurdsson, H. (Ed.). Encyclopedia of Volcanoes. San Diego and London, Academic Press. pp. 1083–1091.

Rampino, M.R., Self, S. and Fairbridge, R.W. 1979. Can rapid climatic change cause volcanic eruptions? *Science.* **206**, 826–829.

Rice, A. 1990. The role of volcanism in K/T extinctions. In: Lockley, M.G. and Rice, A. (Eds). Volcanism and Fossil Biotas. Geological Society of America Special Paper. **244**, 39–56.

Rogers, N. and Hawkesworth, C. 2000. Composition of Magmas. In: Sigurdsson, H. (Ed.). Encyclopedia of Volcanoes. San Diego and London, Academic Press. pp. 115–132.

Rothery, D.A. 2003. Volcanoes. Teach Yourself Series. London, Hodder and Stoughton.

Scott, K.M., Macias, J.L., Naranyo, J.A., Rodríguez, S. and McGeehin, J.P. 2001. Catastrophic Debris Flows transformed from landslides in Volcanic Terrains: Mobility, Hazard Assessment and Mitigation Strategies. United States Geological Survey Professional Paper 1630.

Sharpton, V.L. and Ward, P.D. 1990 Global Catastrophism in Earth History: An Interdisciplinary Conference on Impacts, Volcanism and Mass Mortality. Geological Society of America Special Paper 247.

Shaw, H.R. and Moore, J.G. 1988. Magmatic heat and the El Niňo cycle, Eos. *Transaction of the American Geophysical Union.* **69**, 1553, 1564–1565.

Sigurdsson, H. 1982. Volcanic pollution and climate: The 1783 Laki eruption. *Eos.* **63**, 601–602.

Sigvaldson, G.E., Annertz, K. and Nilsson, M. 1992. Effects of glacier loading/deloading on volcanism: Postglacial volcanic production rate of the Dyngjufjöll area, central Iceland. *Bulletin Volcanologique.* **54**, 385–392.

Sparks, R.S.J., Sigurdsson, H. and Wilson, L. 1977. Magma mixing: A mechanism for triggering acid explosive eruptions. *Nature.* **267**, 315–318.

Summerfield, M.A. 1991. Global Geomorphology. Longman, Harlow.

Tuffen, H., Gilbert, J. and McGarvie, D. 2001. Products of an effusive, subglacial, rhyolite eruption, Bláhnúkur, Torfajökull, Iceland. *Bulletin Volcanologique.* **63**, 179–190.

Walker, G.P.L. 1973. Explosive volcanic eruptions: A new classification scheme. *Geologische Rundschau.* **62**, 431–446.

Walker, G.P.L., Wright, J.V., Clough, B.J. and Booth, B. 1981. Pyroclastic geology of the rhyolitic volcano of La Primavera, Mexico. *Geologische Rundschau.* **70**, 1100–1118.

Wegener, A. 1915. Die Entstehung der Kontinents und Ozeane. Berlin Bornträger. Trans. Skerl, J.G.A. 1924. The Origin of Continents and Oceans. London, Methuen.

Williams, H. and McBirney, A.R. 1979. Volcanology. San Francisco, Freeman Cooper.

Wilson, M. 1993. Igneous Petrogenesis: A Global Tectonic Approach. London, Chapman and Hall.

Wohletz, K.H. and Sheridan, M.F. 1979. A model of pyroclastic surge. Geological Society of America Special Paper. **180**, 177–194.

Wylie, P.J. 1988. Magma genesis, plate tectonics and the chemical differentiation of the Earth. *Review of Geophysics.* **26**, 370–404.

Further reading

Branney, M.J. and Kokelaar, P. 2002. Pyroclastic Density Currents and the Sedimentation of Ignimbrites. Geological Society of London, Memoir 27. Bath, Geological Society Publishing House.

Chester, D.K. 2005. Volcanoes, society and culture. In: Marti, J. and Ernst, G.J. (Eds). Volcanoes and the Environment. Cambridge, Cambridge University Press. pp. 404–439.

Dobran, F. 2001. Volcanic Processes: Mechanisms in Material Transport. New York, Plenum Press.

Parfitt, E.A. and Wilson, L. 2008. Fundamentals of Physical Volcanology. Oxford, Blackwell Scientific.

Schminke, H.-U. 2004. Volcanism. Berlin, Springer Verlag.

Sigurdsson, H. 2000. Encyclopedia of Volcanoes. San Diego and London, Academic Press.

14 Geotectonics: Processes, Structures and Landforms

Sulaiman fold belt. Source: www.fas.org/...docs/rst/Sect17/Sect17_3.html

Learning Outcomes

After reading this chapter and completing the exercises you will be able to:

➤ Understand many types of deformation structures occur and that the tectonic processes that form them have a major influence on the Earth's surface landscape and link the endogenetic and exogenic systems.

➤ Understand that there are many controls on rock deformation and new rock types can be created by the deformation.

(a)

14.1 Introduction

Many rocks and sediments contain evidence of deformation, defined as structures produced as a result of an applied stress. These can vary from small folds within individual laminae to major folded rock masses in the heart of the world's mountain ranges (Figure 14.1(a) and (b)). In some of these mountainous terrains, fold blocks may have been transported hundreds of kilometres and exotic terranes can be juxtaposed in a complex mélange of deformed rock. The causes of this deformation can be varied, for example sediments may deform in response to compression and compaction during lithification, or within mass movements. On the other hand, rocks and sediments can be compressed by the movement of the tectonic plates (see Chapter 13). These tectonic processes have a major influence on the Earth's surface landscape and form a link between the Earth's endogenic and exogenic systems. Tectonics also play an important role in determining the preservation of the geological record and are vital in any reading and understanding of the Earth's history.

Inactive or relict tectonic structures provide a structural grain that can control the landscape's evolution, as joints, faults and folds provide lines of weakness along which exogenic processes can operate. Active tectonics also have a strong impact on surface processes, by controlling both the sources and sinks for sedimentation, and can also generate their own suite of landforms, from small fault-line scarps to continental-scale mountain ranges. In turn, macroscale tectonic processes, such as mountain building, can have a major impact on climate and the regional distribution of exogenic processes. An example of this is that a mountain range may generate orographic rainfall, which will tend to fall on the windward slope of the mountain range, with a rainshadow in the lee. This in turn will impact on the discharge and erosive potential of rivers on either side of the mountain range. The formation, subsidence and closure of sedimentary basins is also critical to the preservation of the rock record.

Initially in this chapter the basic geometry and surface expression of some common geological structures will be reviewed,

(b)

Figure 14.1 (a) Large-scale Anteojo fold in Sierra de la Madera, Cuatrocienegas, Coahuila, Northern Mexico. (b) Small-scale folds in schists from the Republic of Ireland

before their influence as passive landscape elements in a landscape's evolution is considered.

14.2 Tectonic structures

Tectonic structures such as joints, faults and folds result from deformation of sediment or rock under an applied stress. This can be defined as a force applied over a given unit area. Opposing this stress is the strength of the sediment or rock, which is determined by its mineral composition or the lines of weakness within it. Stress can be either positive (compressive), negative (extensional) or neutral, as in the case of a strike-slip fault system where two rock blocks laterally move past one another. Under a positive stress regime, strata are compressed and shortened, via faulting and folding, while in contrast, under a negative stress or tensile regime, strata will tend to extend and thin.

The response to an applied stress depends on the rock properties and the rate and magnitude of the applied stress. As a stress is applied to a rock, it first behaves in an elastic manner, without undergoing a permanent change in shape or volume, before reaching a yield strength at which permanent deformation or strain occurs. This deformation can occur in several ways. It may result in the complete failure of a rock, leading to a fracture or joint, along which displacement may occur to create a fault. Alternatively, a rock may deform in a ductile manner to give a fold, with bent and buckled strata. In the case of brittle failure, the chemical bonds within the rock break along a failure plane and usually stay broken, while in ductile failure, bonds are broken in zones of high stress and material moves to areas of lower stress, where the bonds reform.

14.2.1 Controls on rock deformation

The nature of rock deformation, whether brittle or ductile, is controlled by a range of factors, which include:

- **Lithostatic pressure:** Here the rock which overlies a particular layer imposes a weight or lithostatic pressure, which increases with depth. The greater the pressure, the more ductile a rock is.

- **Heat:** The geothermal heat flux increases towards the Earth's centre and the high temperatures at depth may help promote ductile deformation, while cold surface rocks are more likely to break through brittle failure.

- **Composition and fluid pressure:** The lithology of the rock under stress is critical to the deformation style, as is the fluid pressure within it. Mineral composition, texture and the presence of internal discontinuities all play a role in determining a rock's strength and its response to an applied stress. However, of particular importance is the fluid pressure within a rock or sediment, since high fluid pressures tend to help lubricate movement between grains, or along slip faces. The geological history may also be important as its cumulative strain history may play an important part in determining its behaviour.

- **Magnitude and rate of applied stress:** The rate at which stress is applied is critical to the rock response and the same rock may respond differently to two different stress regimes. For example, if a rock is exposed to a low-stress regime over a

prolonged period it may have time to respond and adjust in a ductile fashion, but if the stress is applied suddenly, the rock may not be able to respond quickly enough and brittle failure will result.

14.2.2 Deformation structures

Folds

These show bent or buckled strata. Some of the basic terminology to describe these structures is shown in Figure 14.2. Folds tend to be classified on the basis of four main features: direction of closing, attitude of the axial surface, size of the interlimb angle and nature of the fold profile. Folds that close upwards (folds with limbs that dip away from the hinge) are called antiforms, while folds that close downwards are called synforms. Folds that close sideways are sometimes referred to as neutral folds. Where the direction of younging has not been inverted (the age of the strata within a rock pile declines upwards), the bedding in antiform folds becomes younger upwards and is referred to as an anticline, while in a synform, younger rock occurs in the core of the fold and is referred to as a syncline (Figure 14.3). In complexly folded regions, however, the direction of younging may be inverted,

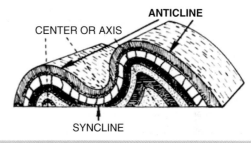

Figure 14.2 Terminology of folds

Figure 14.3 Anticline and syncline in layered pre-Cambrian gneiss, along New Jersey Route 23, west of Butler, NJ. (Source: USGS public domain image)

in which case there may be synformal anticlines and antiformal synclines. The key to this is the direction in which strata get younger. In a syncline the strata are always younger within the core, while in an anticline the strata are older within the fold core.

The next variable with which to classify a fold is the attitude of the axial plane. As the axial plane changes from the vertical to the horizontal and beyond, the terms upright, inclined, recumbent and overfold can be applied. Folds can also be described by the terms gentle, open, closed, tight and isoclinal on the basis of the degree of closure between the two fold limbs. Finally, a fold may be described in relation to its cross-sectional profile, which is the shape of the fold when seen from a plane perpendicular to the fold axis. Typical descriptors are: parallel, similar, concentric and chevron. The complexity of folds may be complicated by episodes of multiphase deformation, in which a fold may be refolded several times to give superimposed folds, or interference may occur between two simultaneous fold systems, to give interference patterns.

Folds can show themselves on the Earth's surface as broad anticlinal swells or synclinal basins (Figures 14.4 and 14.5).

In practice, however, given that folds are not usually associated with near-surface tectonics, fold systems tend to be exhumed, or modified by erosion. The process of folding involves extension and compression of beds. Take the analogy of a sandwich which is gently folded to form a broad, open anticline: the outer layer of bread will undergo extension, perhaps showing itself in a series of tensional joints or gashes, while the lower layer, at the core fold, will undergo compression. The consequence of this is that it is easier for exogenic processes to erode anticlines than synclines. Thus the anticlinal crest of a fold is usually eroded fastest. Then the limbs may form cuestas, which are asymmetrical landforms with a gentle slope following the dip of the folded bed and a steep scarp slope cutting the bed and facing inwards towards the former fold axis. Such landforms are common in southern

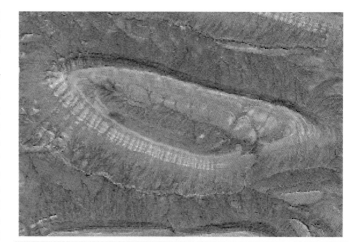

Figure 14.5 Synclinal basin showing erosion of the centre. (Source: TM Satellite image, Alaska, public domain image from landcover.org)

England and form the North and South Downs of Kent and Sussex, for example.

Faults

These are the results of brittle failure and the displacement of rock on either side of a failure plane (Figure 14.6). The basic nomenclature of faults and the main types of fault are shown in Figure 14.7. Faults can be marked by a zone of broken and crushed rock, known as a fault breccia or gouge, produced by the displacement of two fault blocks along the fault plane. The following types of rock can be produced by faulting:

- **Cataclastite:** A fault rock that is cohesive, with a poorly developed or absent planar fabric, or is incohesive, characterized by

Figure 14.4 Synclinal basin, Parc National Torres del Paine, Chile. January 2007. (Photo: Michael Lejeune, CC-BY-SA2.5, Wikimedia Commons)

Figure 14.6 Fault in shale, near Adelaide, Australia

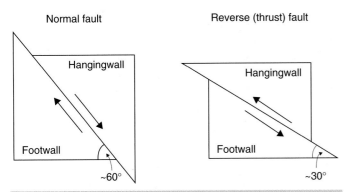

Figure 14.7 Types of fault and fault nomenclature

generally angular clasts and rock fragments in a finer-grained matrix of similar composition.

- **Mylonite:** A fault rock that is cohesive and characterized by a well-developed planar fabric resulting from tectonic reduction of grain size and commonly containing rounded porphyroclasts and rock fragments of similar composition to minerals in the matrix.

- **Tectonic or fault breccia:** A medium- to coarse-grained cataclasite containing <30% visible fragments.

- **Fault gouge:** An incohesive, clay-rich fine- to ultrafine-grained cataclastite, which may possess a planar fabric and contains >30% visible fragments. Rock clasts may be present.

- **Pseudotachylite:** Ultrafine-grained vitreous-looking material, usually black and flinty in appearance, occurring as thin planar veins, injection veins or as a matrix to pseudoconglomerates or breccias, which infills dilation fractures in the host rock.

- **Clay smear:** Clay-rich fault gouge formed in sedimentary sequences containing clay-rich layers that are strongly deformed and sheared into the fault gouge.

The surfaces at which failure takes place may show striations or slickensides (Figure 14.8), which provide evidence for the movement direction along a fault.

Faults are usually found in groups, such as the Craven Faults in Yorkshire, and may collectively form large structures like rift valleys (grabens), as in Iceland and East Africa (Figure 14.9), and horst blocks. A graben (German for 'ditch') is a depressed block of land bordered by parallel faults and is the result of a block of land being downthrown, producing a valley with a distinct scarp on each side. Grabens occur side by side with horsts, and horst and graben structures are indicative of tensional forces and crustal stretching (Figure 14.10). Grabens are produced from parallel normal faults, where the hanging wall is downthrown and the footwall is upthrown. The faults typically dip toward the centre of the graben from both sides. Horsts are parallel blocks that remain between grabens; the bounding faults of a horst typically

Figure 14.8 Slickensides along a fault, Peloponnese, Greece

Figure 14.9 Infrared enhanced satellite image of a graben in the Afar Depression, Ethiopia. (Source: NASA)

dip away from the centre line of the horst. One or more grabens can produce a rift valley.

Where faults occur at the Earth's surface they have an expression dependant on the balance between the rate of fault movement and the rate of surface erosion or denudation. If the rate of denudation exceeds the rate of fault movement then there will be no landform expression of the fault. It is useful therefore to

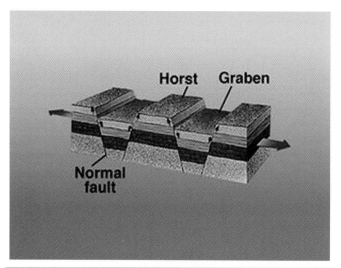

Figure 14.10 Horst and graben structure. (Source: USGS)

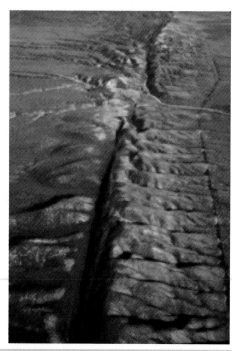

Figure 14.11 San Andreas strike-slip fault. (Source: http://pubs.usgs.gov/gip/earth1/how.html)

Figure 14.12 Landsat image of the San Andreas Fault in southern California. (Source: USGS)

Figure 14.13 The Southern Alps and the Transform Alpine Fault on New Zealand's west coast. Image is almost 500 km wide, with north-west towards the top. (Source: Jet Propulsion Laboratory of NASA, photo ID: PAA06661)

distinguish between two types of uplift: (i) crustal uplift, which is movement of the rock column, and (ii) surface uplift, which is movement of the land surface relative to a fixed datum, usually sea level. If the rate of denudation exceeds the rate of crustal uplift then there will be no fault landform, but if the rate of denudation is less than the rate of fault movement then some form of surface expression can be expected. Normal and reverse faults may result in the formation of a faultline scarp. This landform is most clearly developed in the case of normal faulting, where the scarp face corresponds to the fault surface. Once formed, a faultline scarp is subject to erosion and degradation by a range of slope processes and over time its morphology will become degraded. It is therefore possible to obtain relative age estimates of movement along a series of faults by contrasting the degrees of surface degradation. The importance of such observations is that they provide a relative time element. One key feature of any faulting is that it can juxtapose very different lithologies on either side of the faultline. These lithologies may be more or less resistant to erosion and so differential erosion may give rise to a range of faultline scarps, such as in the Craven Faults in north-west Yorkshire, where Giggleswick Scar is a good example. These are the result of continued erosion rather than fault movement. Strike-slip faults, such as the San Andreas fault system (Figure 14.11) – a major transform fault that runs between the Mendocino Triple Junction in the north and the northern end of the East Pacific Rise somewhere beneath the Imperial Valley in the south – may also have a strong landform expression (Figures 14.12 and another example from New Zealand Figure 14.13). The typical geomorphic landforms are illustrated in Figure 14.14.

The San Jacinto fault zone is a series of dextral strike-slip faults located in southern California. It accommodates a portion of the deformation caused by the opposing movements between the North American and Pacific plates. The San Jacinto fault zone branches westward from the San Andreas fault zone in the Transverse Ranges and is characterized by a series of strike-slip faults that cross through the pre-Cretaceous metasedimentary and

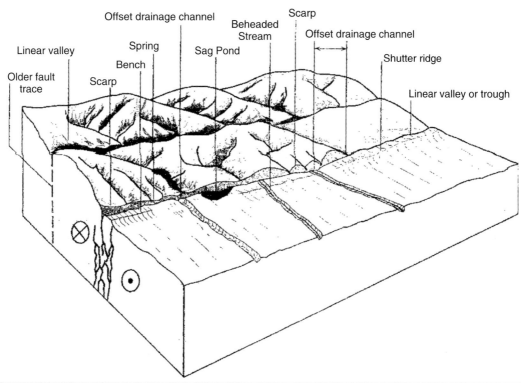

Figure 14.14 Strike-slip geomorphology: linear trough, sag ponds, offset ridges and drainage, springs, scarps and beheaded streams are typical morphological features. (Source: after Wesson *et al.*, 1975)

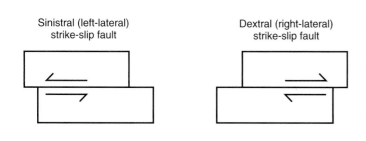

NB: This is a plan view of the Earth's surface

Figure 14.15 Schematic illustration of the two types of strike-slip fault

Cretaceous plutonic rocks of the Peninsular Ranges, as well as late Cenozoic sedimentary rocks of the Salton Trough. Throughout the fault zone there are minor compressional, tensional and oblique zones, but most of the plate motion is taken up by a series of dextral strike-slip faults.

In strike-slip faults the fault surface is usually near vertical and the footwall moves either left, right or laterally, with very little vertical motion. Strike-slip faults with left-lateral motion are also known as sinistral faults, while those with right-lateral motion are also known as dextral faults (Figure 14.15). A special class of strike-slip fault is the transform fault, which in plate tectonics is a fault related to spreading centres, such as mid-ocean ridges.

Joints

These are planes of failure along which no displacement has occurred, and they can result from a wide range of processes. They may be infilled with a mineral, deposited from hydrothermal water (so forming a vein) and annealed if the rock is heated and placed under pressure. Explaining the formation of joints is not always possible. Some are formed as a result of tensile stress from a contracting rock mass, as when lava cools, it shrinks, developing columnar joints (Figure 14.16). Equally, a sediment, as it dries and is compacted as part of the lithification process, may contract, like the desiccation cracks that form on a muddy pool floor as it dries. Other joints may result from unloading, via erosion or from rapid melting of ice sheets, a process which tends to generate large sheet joints that parallel the eroding surface. The processes of deformation also generate a range of fractures associated with the release of tension.

Basin and range

This is a geological term for a type of topography characterized by a series of separate and parallel mountain ranges with broad valleys interposed, extending over a wide area. It is typified by the topography found in the Great Basin in the western United States, which is part of a larger regional topography known as the Basin and Range Province. Basin and range provinces exist in other regions of the world as well, on dry land and on the sea floor.

Figure 14.16 Columnar joints, Svartifoss, Skaftafellsheidi, Southern Iceland

Basin and range topography results from crustal extension. As the crust stretches, faults develop to accommodate the extension. Most of these faults are vertically-oriented normal faults. The basins are downfallen blocks of crust and the ranges are relatively uplifted blocks, many of which tilt slightly in one direction at their tops. The normal arrangement in the basin and range system is that each valley (i.e. basin) is bounded on each side by one or more normal faults, which are orientated along or sub-parallel to the range front. Basin and range faulting is characteristic of incipient rift zones in continental crust, and also of back-arc basins.

Folding and faulting processes are not scale-dependent as they operate in the same way, with the same characteristics, on a scale of just a few millimetres as on a regional, continental scale. It is therefore important to emphasize that a range of different mechanisms may be responsible for compression or extension within different geological systems and at different scales. For example, compressive and extensional deformation regimes may develop within large submarine slumps driven by gravity. Cross-sections through a large push moraine formed in the marginal zone of a Pleistocene ice sheet may show thrusting and the cross-sections of large mountain thrust belts may be similar but formed by the convergence of lithospheric plates. So we must realise that

not all tectonics within the geological record need be associated with plate tectonic stress regimes.

14.3 Tectonic structures as lines of weakness in landscape evolution

In tectonically-deformed terrain, erosion of the land surface will often reveal a range of faults, folds and joints. These structures may have been active just a few thousand years ago, or could have formed hundreds of millions of years in the past. In landscape evolution, such structures have two roles: as discontinuities within the rock mass and as lines of weakness that may be exploited by exogenic processes. As we will see in Chapter 16, the rock mass strength is largely controlled by the frequency, width and continuity of joints and other rock fractures. The orientation of these fractures or discontinuities with respect to the prevailing slope is also critical to the strength of the rock mass and its resistance to erosion. Steep rock slopes occur where the rock mass contains few joints or faults. On a more specific level, prominent joints or faults may act as lines of weakness that can be exploited by exogenic processes. For example, there is often a striking river erosion pattern developed where a river exploits a regional joint pattern and the tectonic influence on drainage patterns is such that an analysis of drainage systems on satellite images or aerial photos often allows important structural data to be obtained. Section 17.8 and Figure 17.10 summarize some common drainage patterns and the tectonic structures that are often associated with them. In cave systems too, the influence of joint orientations and faults can be important in controlling both the passage shape and where the water can erode, and often river or glacial erosion processes pick out weaknesses in the rock caused by tectonic processes.

On a much bigger scale, rivers can continue to downcut across the tectonic belt as fold mountains are created; there are good examples in the Himalayas. This type of drainage is called antecedent drainage. An example from South Africa is shown in Figure 14.17. Here rivers cut across the mountains of the Cape

Figure 14.17 Antecedent drainage in the Cape Fold Belt, South Africa. (Source: Google Earth)

Fold Belt, which comprise resistant rock types (note the tight meander bend (left) and the regular spacing of the rivers). This indicates that the drainage pattern is in part antecedent, but across vast areas of southern Africa the main driver(s) for this inherited style of drainage remains unclear. In the present day, the area is generally considered to be tectonically quiescent, and an arid to semi-arid climate prevails in much of the upper catchment. This suggests that bedrock incision occurs during punctuated high-magnitude flood events, rather than through steady-state erosion.

Exercises

1. Describe several ways in which rock may deform.

2. What factors control the nature of rock deformation?

3. How can folds be classified?

4. Describe the characteristics of rock types created by faulting.

5. Why do faultline scarps occur?

6. What are characteristics of strike-slip faulting?

7. How do joints in rock form?

8. Describe the relationship between tectonic structures and surface landform development.

9. How could you differentiate between horsts, grabens and rift valleys?

10. Explain how slickensides form and what they can indicate.

References

Wesson, R.L., Helley, E.J., Lajoie, K.R. and Wentworth, C.M. 1975. Faults and future earthquakes. In: Borchardt, R.D. (Ed.). Studies for Seismic Zonation of the San Francisco Bay Region. United States Geological Survey Professional Paper 941A, pp. 5–30.

Further reading

Bull, W.B. 2007. Tectonic Geomorphology of Mountains. Oxford, Blackwell Publishing.

Burbank, D.W. and Anderson, R.S. 2001. Tectonic Geomorphology. Oxford, Blackwell Science.

Crustal Deformation and Structural Geology Web site: http://iws.ccccd.edu/rgrayson/Physical%20Geology.

Keller, E.A. 2001. Active Tectonics: Earthquakes, Uplift and Landscape. New Jersey, Prentice Hall.

Keller, E.A. and Pinter, N. 2002. Active Tectonics, Earthquakes and Landscape. 2nd edition. New Jersey, Prentice Hall.

Moores, E.M. and Twiss, R.J. 1995. Tectonics. W.H. Freeman.

Schumm, S.A., Dumont, J.F. and Holbrook, J.M. 2000. Active Tectonics and Alluvial Rivers. Cambridge, Cambridge University Press.

Turcotte, D.L. and Schubert, G. 2001. Geodynamics. Cambridge, Cambridge University Press.

SECTION IV
Exogenic Geological Systems

15 Weathering Processes and Products

Granite domes in Yosemite

Earth Environments   David Huddart and Tim Stott
© 2010 John Wiley & Sons, Ltd

Learning Outcomes

After reading this chapter and completing the exercises you will be able to:

➤ Define weathering and its characteristics.

➤ Understand the processes of mechanical and chemical weathering and be able to describe weathering products.

➤ Understand how small-scale weathering landforms such as rinds and cavernous weathering are formed.

➤ Understand how weathering rates can be measured.

15.1 Introduction

Weathering is one of the most fundamental processes operating on the Earth's surface. If rocks did not weather when subject to attack then very little sediment movement would take place. It not only causes rock disintegration and facilitates erosion, but it is central to the formation of soils and contributes to the solute content of water taken up by plants. As a process, it involves the transformation of unweathered rock or sediment into a range of different products, from rock fragments to minerals to solutes. When rocks are exposed at the Earth's surface to air and water, they are out of equilibrium with the surface natural environment, as they will often have been formed deep within the Earth's crust, perhaps at extremes of temperature and elevated pressure. Minerals and rocks therefore are unstable at the Earth's surface. On exposure, a rock undergoes a transformation and weathering slowly brings it into equilibrium with the prevailing environment, and the prevailing atmospheric, hydrospheric and biospheric conditions (subsequent change in that environment may lead to further weathering as a new equilibrium is sought). Gradually the visible effects of weathering appear in successive stages (Figure 15.1(a)) and a percentage change in volume of minerals over time (Figure 15.1(b)). So weathering is the preparation of bedrock for removal by transporting agencies, but it is continuous throughout the geomorphic system and does not just affect the originally-exposed bedrock; it affects the initial bedrock, the sediment as it is transported on hillslopes and in rivers, and often it continues to influence the sediment when it has been deposited in the sea. For example, glauconite only forms in the sea. Changes take place as the less-resistant minerals disappear. As we will see, it is no surprise that quartz dominates the geological record as it is a resistate.

Weathering is confined to a relatively narrow zone of the Earth's crust, since most of the forces involved are derived from a rock's exposure to the atmosphere. As a consequence, the weathering zone tends to extend from the soil or rock surface to the maximum depth to which gas and energy from the atmosphere can penetrate. This zone may vary from a few centimetres to several metres and is stylized in Figure 15.2, which shows the features occurring in a fully-developed weathering profile on granitic rocks. It is important to emphasize, however, that water may penetrate to considerable depth within the Earth's crust. For example, ground-water circulation to depths of 3 km or more may occur within the Appalachian province of the eastern USA, while enhanced permeability due to fractures may allow water to circulate at depths of 10 km in fault zones. This water circulation has the potential to cause weathering, but despite this, weathering is usually considered to be a near-surface set of processes.

Traditional weathering is distinguished from other exogenic processes by an emphasis on the *in situ* decay of the rock mass, in contrast to erosion, which involves the transport of rock debris. Although this is a useful distinction, it is not strictly true since most weathering involves movement, whether by gravity or through the removal of soluble products. Weathering is not a precursor to erosion, since erosion of unweathered rock also occurs, but it does facilitate erosion. Perhaps the key difference between erosion and weathering is that weathering, at least in the initial stages, does not remove the basic fabric or structure of the rock mass. In a weathered rock mass it is still possible to see sedimentary structures, bedding and mineral texture, features that are only removed in the most advanced stages of weathering. However, weathering is the preparation of rock for removal of sediment and as such it requires a loss in material strength. This could be a loss in physical strength, breaking it down and making it more easily moved by mechanical processes, or it could be a loss in chemical strength, making the chemical constituents more mobile or more easily carried away in solution. In contrast, erosion does destroy the basic fabric of a rock through its piecemeal removal.

It is also traditional to distinguish between mechanical, chemical and biological weathering processes. In practice, weathering operates through a combination of all three processes. For example, mechanical fracturing of a rock is essential to allow the penetration of water and air into a rock mass for chemical weathering, and the rock's susceptibility to chemical weathering must be increased as there is more surface area exposed for reaction. Moreover, advances in weathering study have shown that chemical weathering may be involved in some processes of mechanical rock disintegration, and mechanical stress within minerals is a feature of many chemical processes. There is usually a change in grain size, a change in volume, or both. In a similar way, most processes of biological weathering are either mechanical or chemical. In practice it is an artificial distinction, since most processes of chemical weathering are enhanced by the presence of organic matter.

Despite these reservations, the traditional classification of weathering processes is perpetuated in this chapter, but the definition of weathering that will be adopted is that it is the loss of chemical or physical strength from a material as a result of the action of subaerial or subaqueous processes at the Earth's surface.

STAGES OF WEATHERING OF A ROCK MASS

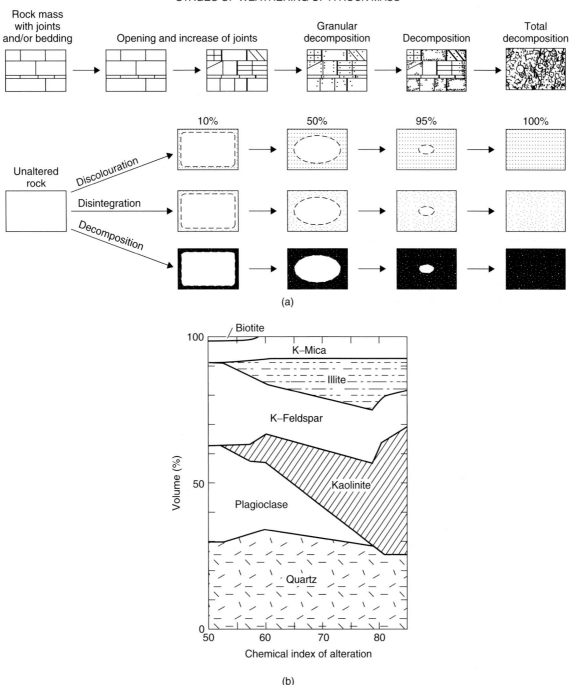

Figure 15.1 (a) Stages of weathering of a rock mass. (b) Mineralogy and the chemical index of alteration. Reprinted courtesy of Alan Colville from www.alancolville.com/weathering.html

As such, only *in situ* changes in the parent material are involved and no net movement away from the site of weathering is implied.

In sections 15.2 and 15.3 the processes of mechanical, chemical and biological weathering are examined. The controls on the rate of weathering and the measurement strategies available for determining that rate are discussed in sections 15.4. In sections 15.5 some of the products and landforms that result from weathering will be explored.

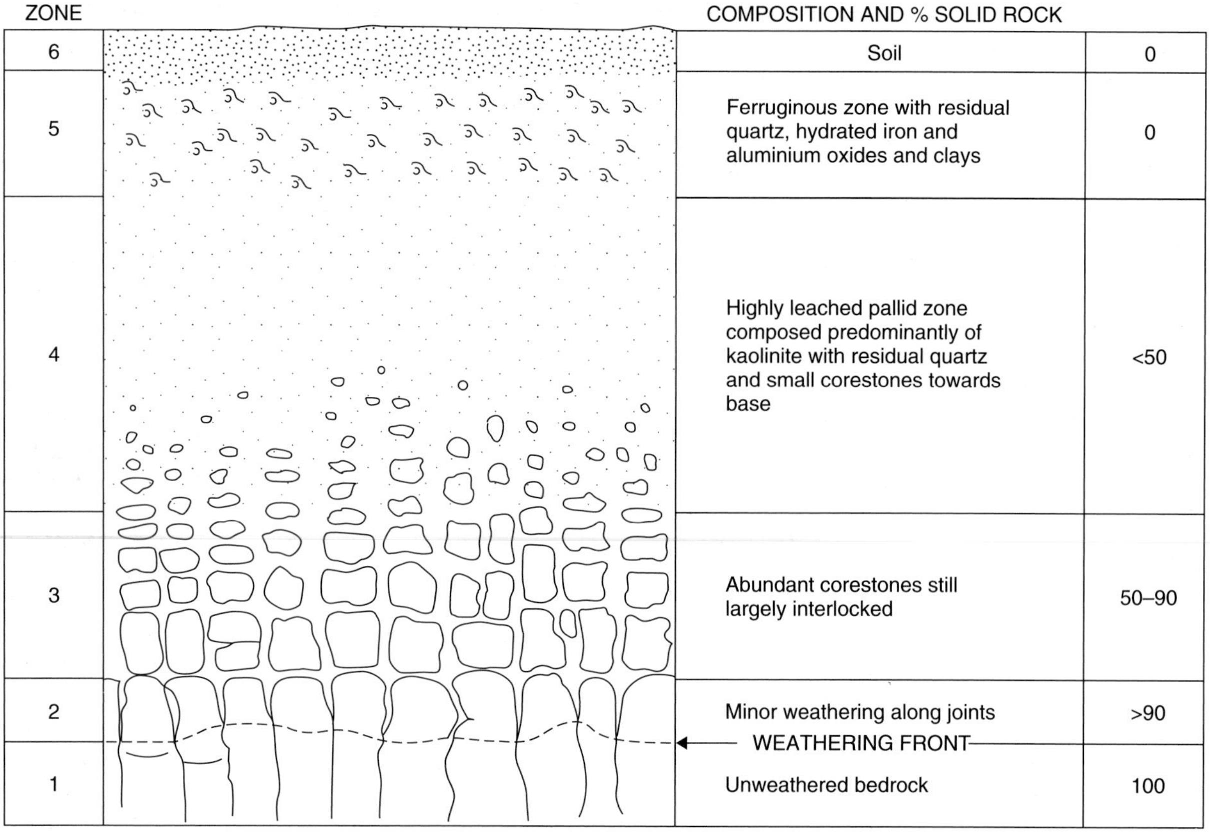

ZONE   COMPOSITION AND % SOLID ROCK

Figure 15.2 Characteristics of a fully-developed weathering profile on granitic rock

15.2 Physical or mechanical weathering

15.2.1 Forces and rock strength

Before the processes of mechanical weathering are reviewed it is necessary to reflect on the rock characteristics that oppose their mechanical disintegration: effectively the strength of the rock. The strength of any rock is a function not only of its intact strength, but also of the nature of its discontinuities, such as joints, faults, fracture and cleavage, within both individual minerals and the rock mass as a whole.

Two elements of intact strength are relevant to weathering: first, the modulus of elasticity, or Young's modulus. When a stress is applied to a rock, it responds by showing deformation, or strain. Up to a point known as the elastic limit, a rock will respond in an elastic fashion; that is, the strain is recoverable once the stress is removed. This property is measured by Young's modulus, determined by the ratio of stress to strain. If the modulus is high, less deformation results from a given stress and the rock is consequently more resistant to deformation. When a stress builds up beyond the elastic limit, a rock will show a sudden and catastrophic failure. Rocks are more susceptible to tensile stress. Common rock types display a wide range of elastic limits, depending on their mineralogy and internal organization. Failure

Table 15.1 Strength (M Nm^{-2}) at failure for different rock types and stresses. (Source: after Bland and Rolls, 1998)

Rock	Compressive stress	Tensile stress	Shear stress
Granite	100-250	7-25	14-50
Dolerite	100-350	15-35	25-60
Basalt	100-300	10-30	20-60
Quartzite	150-300	10-30	20-60
Sandstone	20-170	4-25	8-40
Shale	5-100	2-10	3-30
Limestone	30-250	5-25	10-50

tends to be concentrated along joints or fractures, since maximum stress is usually achieved at the tip of a fracture. The strengths at failure of different common rock types and stresses are shown in Table 15.1.

Of greater importance to the overall strength and weathering behaviour of a rock is the nature of the discontinuities within the rock mass as a whole, since they provide the lines of weakness along which strain may propagate and also allow the penetration of air and water. For example, a fracture network may define the pattern of water flow through a rock mass and therefore the pattern of weathering. Discontinuities are present at a range of scales, from those that cut a rock mass as a whole (such as faults, joints and rock cleavage) to those that are formed of individual minerals (such as mineral cleavage, multi-grain or single-grain fractures controlled by crystal structure, and larger stress fractures, which tend to be independent of crystal structure). Collectively these discontinuities define the likely three-dimensional response of a rock to strain. Response may be either isotropic (equal in all directions) or anisotropic (unequal). Due to the discontinuities present within most rocks, they are normally anisotropic. For example, the compressive strength of schist may be 70% weaker along its layers than normal to them. The width, spacing and continuity of discontinuities within a rock are also important in determining the size of blocks produced by mechanical weathering and the ease with which such blocks may be detached. For example, if a fracture network is already more or less continuous then little fracture propagation is required to generate a discrete block. Similarly, if the fractures are closely spaced then the blocks produced by their expansion are easily removed from the rock mass by the processes of erosion.

The hydraulic properties of a rock are also very important in determining weathering since many of the processes require water saturation. There are two types of porosity, primary porosity, which is a function of the proportion of the voids or pores between the grains or minerals within a rock, and secondary porosity, which is a function of the fracture network within a rock. Primary porosity is determined by factors such as crystal texture, mineral composition, grain size, grain sorting and the presence of cementing minerals. A rock may have a very high porosity, but if the pores do not link up, water cannot pass through the rock; that is, it will be impermeable. Clay provides a good illustration of this. The pore space per unit volume tends to increase as particle size decreases, but the permeability will decrease. Small grains have numerous small pores, which give a high porosity, but small pores tend not to connect easily, which results in low permeability. The spatial variation in grain size and therefore permeability within a rock mass, for example in sandstone with graded beds, may cause variation in the rock's susceptibility to weathering. Pore size is also important to the movement of water within a rock or soil under the influence of capillarity, where the tendency is for water to rise up a narrow tube of pores.

15.2.2 Mechanisms of mechanical weathering

The processes of mechanical weathering can be resolved into two basic groups, those which involve a volumetric change within the rock mass itself, and those which involve a volumetric change within the pores, voids and fissures of a rock mass. They can also be looked at as a response to stresses resulting from an externally-applied force or as a response to stresses resulting from internally-generated forces. The net effects of mechanical

disintegration are a reduction of strength of material along more-or-less well-defined fractures and a decrease in grain size.

Volumetric change of a rock mass

Pressure release associated with the unloading of buried rock may lead to significant expansion and dilatancy of a rock mass. Both tectonic denudation and subaerial erosion may lead to the unroofing of deeply-buried rock, which will expand once the confining pressure is removed. For example, the European Alps have been denuded of a vertical thickness of at least 30 km of rock in the last 30 million years, unroofing the core of this mountain belt. Such unroofing releases confining pressure, allowing rocks to expand by anything from 0.1 to 0.8%; expansion which is directed towards the unconfined surface and may lead to fracture propagation and the development of sheet joints that follow the surface contours of the topography. These joints may cross-cut primary structures and intersect with other fractures to give broad tabular rock bodies, where the spalling (exfoliation) may lead to the development of exfoliation domes in which the surface topography is defined by the geometry of the sheet joints (Figure 15.3(a),(b),(c)). The spacing of joints gets closer towards the ground surface, and they are generally more continuous, too. They can also develop vertically, parallel to linear erosion, like in a river gorge or glaciated valley. The erosional removal of overburden usually happens slowly, between 0.01 and 1 m every 1000 years, but it can happen much more quickly when a valley glacier or ice sheet melts away and a considerable overburden pressure is removed relatively quickly. However, it has been noted that the rate at which sheet jointing develops is very variable and in Greenland it has been found that sheet joints closely parallel the glaciated land surface, which means the joints have developed in the current post-glacial period. In corrie walls, sheeting occurs along steep planes characteristically concave to the sky, whereas in peaks and horns it is nearly vertical. On corrie floors and in broad valleys the sheeting is subhorizontal. On gently-sloping rounded hills and mountains the sheet structures show dome-shaped patterns. Where these hills are later affected by corrie erosion, a typical intersecting pattern of concave and convex sheeting is developed. Examples from Arran are illustrated in Figure 15.3(d) and (e). However, in Maine, USA, the sheet jointing was parallel to the preglacial land surface and was truncated by glaciated valleys. This means that no joints have developed in the post-glacial time and the development of sheeting is slow. The most dramatic development of sheeting is seen in modern quarries, where freshly-exposed faces sometimes literally explode and collapse simply due to stress relief as the rock is blasted from the surface and the jointing develops in the rock, and sheets continually peel away from the surface as a result of the rapid excavation. The sheet jointing is often well-developed in granite, massive sandstone, arkose, conglomerate and limestone. The pattern of loading, particularly within a collection of boulders, may generate patterns of stress which can either lead directly to failure or be exploited by other processes. The rate of sheeting depends on various factors, such as the bedrock type (extremely brittle rocks will accommodate less stress relief and elastic strain

Figure 15.3 (a) Exfoliation of granite, Enchanted Rock State Natural Area, Texas, USA. (b) Royal Arches and North Dome, Yosemite National Park, with sheet joints in the foreground. (c) Sheet jointing in granite, half dome, Yosemite National Park. (d) Sheet joints in granite, Arran, Scotland. (e) Sheet joints in granite, Arran, Scotland. In both (d) and (e), note the subhorizontal joints paralleling the former topography and the vertical joints associated with the backwall of a corrie. (Sources: (a) by Wing-Chi Poon, Wikimedia Commons; (b) and (c) http://www.alancolville.com/weathering.html)

before splitting begins). The rate of removal of overburden is important, so the faster the rate and the greater the amount of rock removed, the faster the sheeting will develop. Finally, the presence of internal stresses due to tectonic activity will cause sheeting to develop more quickly than in situations in which simple overburden is the sole cause.

The thermal expansion of a rock, or its component minerals, may lead to insolation weathering because the expansion may not be even throughout the rock. Ground temperatures in low-latitude deserts can exceed $85\,^\circ$C and diurnal temperature fluctuations greater than $50\,^\circ$C are not unusual. For example, in Death Valley, California, temperature fluctuated between $73\,^\circ$C during the day and $28\,^\circ$C at night in August 1992. Extremes of temperature, however, are not just a feature of warm deserts but are also typical of cold environments. For example, in Victoria Land, Antarctica, a diurnal temperature range of $60\,^\circ$C has been recorded, and due to the very low air temperature, rapid rates of temperature change are possible, because as soon as the Sun stops shining, rapid cooling occurs. Even something as simple as the passing of a cloud may cause rapid fluctuations in temperature, with rates of change as high as $0.8\,^\circ$C per minute. Bush fires also generate extreme temperatures for short periods, with rock temperatures of $500\,^\circ$C achieved in as little as ten minutes. Here the potential for thermal shock is considerable. A rock may respond to such temperature changes in several ways, leading to insolation weathering. First, different minerals expand by different amounts in response to heating. For example, volumetric expansion of quartz is three times that of feldspar. In addition, minerals such as quartz and feldspar expand under heat in an anisotropic fashion, with expansion occurring preferentially along specific crystallographic axes. Mineral expansion along these axes may be 20 times more than along other axes. The result of this differential expansion between minerals, and by different amounts in different directions, is to set up mechanical stress between mineral grains, with the potential to cause granular disintegration. Second, since rocks are poor thermal conductors, steep temperature gradients are set up within the outer layers of a rock exposed to surface heating. Under intense heat, the outer layers of a rock will expand, while a few centimetres below the surface, the rise in temperature and therefore the resultant expansion is minimal. This generates tensile stress between the outer and inner layers of a rock. The inner layers do heat up, but very slowly, and as a result are also slow to cool down, while the outer surface layers will cool rapidly. This generates compressive stress as the surface layers contract around a warmer, and therefore more expanded, inner. Regular diurnal temperature changes lead to repetitive cycles of tensile and compressive stress within a rock surface, resulting in the propagation of microfractures and rock spalling. Experiments show that the rate of temperature change is greatest at the rock surface and decreases with depth; the rate of temperature change is extremely high for short periods of time; and the interior of the rock sample continues to warm as the surface cools following removal of the heat source (see Figure 15.4(b),(c)).

Two main factors apart from heat supply, air temperature, distribution of Sun and shade, and local wind regimes influence the surface temperature of a rock. These are albedo and the thermal conductivity of the rock. Albedo is the amount of heat energy reflected; a dark-coloured rock absorbs more heat energy than a light-coloured one, which reflects a higher proportion of the incident energy. A basalt, for example, may have an albedo of about 12%, a granite 18% and a chalk 25%, as when fresh chalk is brilliant white in colour. A rock's thermal conductivity is also important, as a rock with a high thermal conductivity will have lower surface temperatures, since the heat is conducted away from the surface into the rock's interior more effectively. The thermal gradient and therefore the potential for insolation weathering will also be reduced. The importance of albedo and thermal conductivity has been illustrated by comparing the surface temperature at a depth of 50 mm for four different lithologies on exposure to sunlight (Figure 15.5). The thermal conductivity of a rock depends on its mineral composition and on the proportion of air-filled pores or voids.

Extreme temperature gradients may also be generated by bush or forest fires, with surface spalling occurring on most boulders or exposed rock surfaces after just one event. For example, it was reported that between 70 and 90% of boulders on the Pinedale moraines at Freemont Lake, Wyoming, showed evidence of spalling following a bush fire. Rapid surface heating and cooling as the fire front passes the boulder result in extremely high temperature gradients through the outermost layers.

It is worth noting that although the evidence of insolation weathering can be seen in most regions subject to extreme temperature change, the effectiveness of the process has been questioned by experimental data. Small blocks in the laboratory have been subject to numerous heating and cooling cycles, but failed to show evidence of weathering unless moisture or salt solutions were present, suggesting that the process may involve an element of chemical weathering too. These types of experiment were later repeated and there were some observations of microfracturing under dry conditions, but the rate of weathering was enhanced by the presence of water and it is likely that the expansion and contraction of this water within rock capillaries assists rock disintegration. However, in general these laboratory studies are now considered to be artificial in that they use small blocks of rock, which do not replicate the tensile and compressive stresses that develop in large, more confined rock masses with a range of preexisting discontinuities.

Repetitive cycles of wetting and drying, perhaps under alternating periods of rainfall and evaporation, have potential to cause weathering in rocks with a high proportion of clay minerals, strong structural weaknesses (such as cleavage or schistosity) and a low tensile strength. A fractured mineral outcropping on the face of a fine crack, or along a cleavage plane, may have unsatisfied electrostatic charges. Water molecules are naturally polar; that is, although overall the chemical charges on the molecule are balanced, there is a tendency for positive and negative charge to concentrate at different ends of the molecule (i.e. one end is negative and the other is positive). As a result, water molecules may be adsorbed on to the surface of a crack to balance unsatisfied electrostatic charges. The build-up of such water leads to expansion of the fracture. During periods of evaporation, water is lost, the adsorbed water film on the sides of the crack becomes

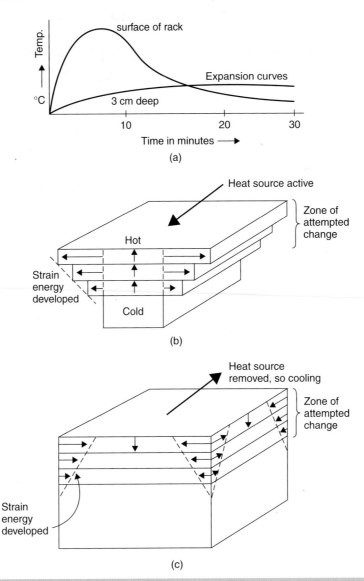

Figure 15.4 (a) Temperature data from a rock block showing how internal temperatures continue to rise even as the surface cools after the removal of the heat source. (b) Development of tensile stresses during heating phase. (c) Development of compressive stresses during the cooling phase. (Sources: (b) and (c) Hall and Hall, 1991)

thinner and in some cases the sides of the crack may be pulled together as the opposing faces try to share the limited amount of water. The repetitive stress of repeated cycles of wetting and drying may lead to rock failure and is thought to be particularly important in shales, slates and some schists.

Volumetric change within pores, voids and fissures

Biological processes, such as the growth of tree roots, are traditionally considered to have a weathering potential through their ability to prise a rock apart. In practice though, roots frequently take the line of least resistance offered by a joint or fracture and the pressures that are exerted are small, in the range 1–3 MPa, and are therefore unlikely to cause fracture propagation. Lichens are known to have mechanical weathering potential. They are composite organisms whose thallus (body) consists of a fungus and alga growing symbiotically. The fungus has a web of fine threads, or hyphae, which can penetrate a rock surface via its pores, or microfractures, up to depths of 1–3 mm below the surface and may result in substantial tensile stress within a rock surface. The lichen thallus and hyphae are also subject to expansion on saturation and may increase their water content between 150 and 300% under certain conditions. Again, this may generate tensile stress within the surface rock layers. In a similar way, certain algae may live within rock (lithophytic algae) and may penetrate fractures, causing mechanical disintegration on expansion when saturated. The suggested effects of lichens in weathering are shown in Table 15.2. Much more effective in fracture propagation and rock disintegration, however, is the role of ice and salt crystal growth.

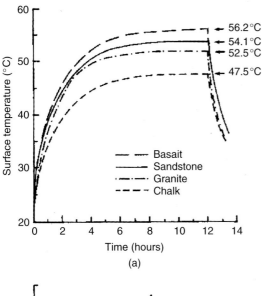

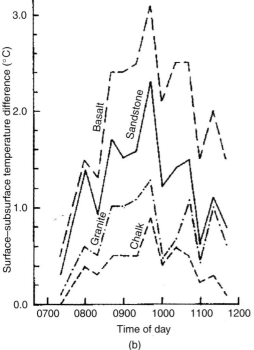

Figure 15.5 Experiments using four rock types: (a) change in rock-surface temperature under closely controlled conditions; (b) differences in temperature between the rock surface and a depth of 50 mm, with the four specimens receiving the same insolation. (Source: after McGreevey, 1985)

Table 15.2 Lichen weathering mechanisms and forms produced (after Robinson *et al*, 1994)

Mechanisms	Results of weathering
Chemical mechanisms	*Results of chemical action*
Chelation of extracellular, soluble compounds	Grooves at endolithic thalli interfaces
Attack by oxalic acid	Etching minerals
Attack by water acidified by respired carbon dioxide	Precipitation of alteration products, e.g. calcium oxalate, which may or may not play a further role in weathering
Physical mechanisms	*Results of physical action*
Rhizoid penetration	Exfoliation of rock surface layer
Thallus expansion and contraction on wetting and drying	Cracking of rock
	Increase in pore volume

Frost weathering, freeze–thaw, frost shattering and frost wedging are all terms referring to the mechanical breakdown of rock by the freezing of water within it. A contrast is often made between macrogelivation, which involves the enlargement and opening of existing fractures in order to isolate a block, and microgelivation, which involves the formation and growth of microfractures and the enlargement of rock pores. The process of frost weathering has been subject to extensive investigations both in the field and in the laboratory in order to try and understand the range of thermal conditions that favour its operation and to elucidate its mechanisms. There are two basic models of frost weathering: (a) volumetric expansion of water on freezing and (b) volumetric expansion of water due to the growth of segregation ice. Traditional models involve the ideas that on freezing water expands by 9% and that in a closed system cooling to a temperature of $-22\,°C$ induces a maximum pressure of 200 MPa, which compares favourably with the average strength of rock, which is around 25 MPa. In practice, water within a rock fracture, pore space or other void does not form a closed system and such pressures are unlikely to develop since (a) in order for a closed system to occur there must be very rapid freezing from the surface downwards to seal a crack or fissure and (b) air bubbles in the ice and pore spaces within rock provide a pressure buffer. It has been argued that although the stress generated by the expansion of ice may not reach its theoretical maximum, the repetitive stress of freeze–thaw cycles can induce failure.

An alternative model has been proposed based on the development of segregation ice within soils, which has also been applied to rocks. Ice segregation involves the migration of moisture to a freezing front, where it forms a body of segregated ice. The growth of such ice bodies within fractures or rock pores has the potential to place a rock under tensile pressure. Ice segregation occurs because ice and water coexist within a sediment or rock, as illustrated in Figure 15.6(a). Normally two phases of a substance can only coexist when the free energies of the phases are equal, which is not the case for water, since ice has less free energy than liquid water. One has only to think of the latent heat released when water freezes to appreciate this point. The coexistence

(a)   **Segregation**

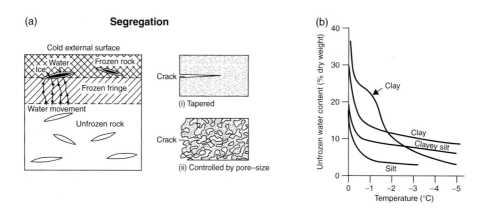

(b)

(c)   **Volumetric expansion**

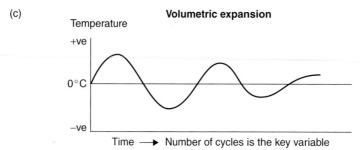

(d)

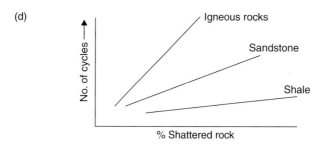

(e)

Figure 15.6 (a) Ice and water can coexist within sediment or rock. (b) The relationship between soil moisture and grain size in frozen soils. (c) Number of cycles across the freezing point, giving volumetric expansion. (d) Relationship between lithology, number of freezing cycles and the percentage shattered rock. (e) Macrogelivated slope in limestone near to Kap Petersen, East Greenland. The rock is fractured by closely-spaced vertical and horizontal joints, which allow water to enter the rock and then freeze

of water and ice is therefore an unstable situation and over time ice, with the lower-energy state, must grow at the expense of water. An energy gradient therefore exists between the water and ice (a gradient of high to low energy), along which the water moves, migrating to the freezing front. The movement of water within a rock or soil towards a freezing front may generate a body of segregated ice. The freezing front is fixed within a soil or rock, either by the limit of frost penetration or more commonly by the balance between the latent heat liberated by the freezing of water and the frost penetration. The key problem is to understand why soil and ice may coexist within a freezing soil or rock. This can be explained by three factors: (a) the presence of salts within soil water may reduce free energy and depress the freezing point; (b) adsorption of water on small soil particles such as clay platelets reduces free energy and the freezing point; hence the greater amount of unfrozen water in fine-grained soils; (c) strong capillary forces may reduce free energy; as a soil freezes, water is confined to successively smaller pores in which capillarity increases. Again, this helps explain the relationship between soil moisture and grain size in frozen soils (Figure 15.6(b)).

It is important to emphasize that each model requires very different conditions to operate, although both models need a wet environment. For example, the expansion model is favoured by a climatic regime which is wet, allowing water penetration of fractures and pores, with rapid rates of freezing and numerous fluctuations in air temperature around $0\,^{\circ}C$. In contrast, ice segregation is favoured by wet conditions, with slow cooling $0.1-0.5\,^{\circ}C$ per hour, sustained subzero temperatures in the range of $-4\,^{\circ}C$ to $-15\,^{\circ}C$ and high porosity (i.e. a large internal surface area).

The two mechanisms also attack a rock in different ways. Ice segregation tends to occur at the base of the frozen zone: at the limit of frost penetration. In permanently-frozen ground, the surface layer, known as the active layer, thaws each summer and the formation of ice segregation reaches a peak at the base of the active layer. Although segregated ice lenses will occupy any available void or fissure, they will tend to develop subparallel fractures (microgelivation) along the frost limit, which is often effectively parallel to the rock surface. This favours spalling of surface rock layers and may also make the process relatively deepseated, depending on the depth to which the freezing front penetrates, or upon the depth of the active layer in the case of permanently frozen rocks. It is interesting to note that many rock profiles in formerly periglaciated areas, such as Southern Britain, show a distinct brecciated surfaced layer, which has been explained in terms of ice segregation and intense frost weathering at the base of the active layer in an area of former permafrost. In contrast, the volumetric expansion of ice in rock fissures tends to be a more near-surface process (perhaps 20 cm or so) and favours macrogelivation, which often leads to a more blocky disintegration. The process is also very dependent on the presence of pre-existing fractures within the rock mass along which water can penetrate.

Some workers believe that these two models of frost weathering can co-exist and that in practice frost weathering is multiprocess;

others have argued that ice segregation is the only possible mechanism. Several lines of evidence have been deployed in favour of the ice segregation model; in particular, acoustic emission was observed in which rock was cooled from $-3\,^{\circ}C$ to $-6\,^{\circ}C$ and it was thought that these emissions were the product of fracture propagation during the growth of segregated ice. Segregated ice lenses have also been observed within the rocks of Arctic regions and much has been made of brecciated weathering profiles in support of this process. In contrast, a large body of experimental data emphasizes the importance of the number of temperature oscillations across the freezing point (Figure 15.6(c)) and supports the idea of conventional ice expansion. In addition, field observations in the Japanese Alps provide support for the enlargement of fractures by the expansion of water. Fractures enlarge during a spring and autumn period of rapid freeze–thaw cycles that are unlikely to be associated with ice segregation.

What does emerge from the available evidence is that frost weathering operates over a very wide range of thermal and hydraulic conditions and that lithology is an important variable in determining frost susceptibility (see Figure 15.6(d),(e)). It is likely therefore given this range of environmental conditions that both processes may have a role to play in frost weathering, in which case the relative dominance of the two processes will depend on the complex interaction of three sets of variables, namely: (a) the thermal regime, in particular the amplitude of a freezing cycle and the rate of freezing; (b) moisture conditions such as the degree of rock saturation, distribution of pore water, solute content and availability of unfrozen water; (c) lithology, essentially the porosity and permeability of the pore or microfracture networks present.

Given appropriate environmental conditions, salt weathering is one of the most effective processes by which the mechanical rock disintegration can be achieved. It is discussed in detail in Goudie and Viles (1997). Most ground or soil water contains dissolved salts derived from rock weathering, saline precipitation near coasts or saline ground water. The common salts include calcium sulphate, sodium chloride, sodium sulphate, sodium carbonate, potassium sulphate and magnesium sulphate. These salts may become concentrated either by evaporation or by the abundance of the supply. As a consequence, saline ground water or pore water tends to be a feature, although not exclusively, of arid and semi-arid regions, or the coastal zone. Saline fluid within a rock may cause weathering in one of three ways:

1. The growth of salt crystals precipitated from pore fluids may exert pressure as a pore or fissure fills, known as crystallization pressure. Precipitation occurs whenever pore water becomes supersaturated with a particular salt and may result either from evaporation or as a result of a temperature change with a near-saturated solution. Depending on the degree of supersaturation and the ambient temperature, crystallization pressures for sodium chloride can range from 54 to 366 MPa, exceeding the strength of most rocks.

2. The hydration and associated expansion of salts precipitated within rock pores or fissures may exert pressure, leading to

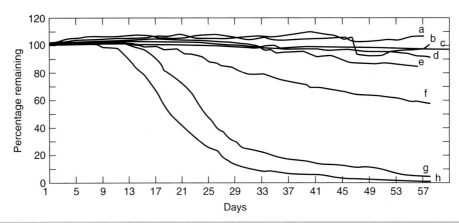

Figure 15.7 The progress of breakdown of a silica-cemented sandstone under the action of different treatments: (a) sodium nitrate crystallization; (b) sea salt crystallization; (c) frost; (d) sodium carbonate crystallization; (e) sodium chloride crystallization; (f) magnesium sulphate; (h) sodium sulphate crystallization. (Source: after Goudie, 1974)

rock disintegration. For example, the hydration of magnesium sulphate involves an expansion by volume of 170%. Hydration is very sensitive to air humidity, which is also linked to temperature, and may therefore vary over a diurnal cycle, leading to repetitive stress in a rock.

3. The thermal expansion of precipitated salts may lead to rock disintegration. For example, most salts have a coefficient of thermal expansion which exceeds that of a typical host rock. A salt such as sodium chloride expands by 1% during a temperature rise of 50 °C, which is greater than most rocks.

An example of a chemical alteration causing a volume change and the weathered product having a greater volume than the original mineral is the chemical alteration of the mineral olivine to serpentine. It involves a great increase in volume and consequently a swelling occurs in the rock, which causes cracking in the surrounding rock. Again, diurnal temperature change within a rock may lead to repetitive episodes of expansion and contraction and therefore stress.

Other examples occur in urban environments where there is, or has been, sulphur dioxide in the atmosphere. Here a weak sulphuric acid may be formed where the sulphur dioxide combines with water and oxygen before combining with limestone building stones to form gypsum. These gypsum crystals grow within the surface layers of the limestone and cause an expansion, since gypsum takes up more space than the original calcite. Pieces then can flake off the stone. This has been especially noticeable on limestones, such as the soft Jurassic limestones like Bath Stone. These have a heavily pock-marked appearance after over a hundred years of exposure, and in the Houses of Parliament in London, which are built of Magnesian limestone, this rock is converted to magnesium sulphate. This is readily soluble and generally easily removed but when it crystallizes out in a dry spell there are disruptive effects in the surface layers.

As a process, salt weathering has been subject to extensive laboratory simulation, in which rock samples are inundated with saline water and then dried in some form of controlled climate cabinet. In this way, experiments have been conducted to assess the different potentials of various common salts (Figure 15.7) and the susceptibility of different rock types (Figure 15.8). These experimental results clearly illustrate that pore-water chemistry is important in determining the weathering potential and that the more porous a given lithology, the greater its potential for weathering.

15.2.3 Controls on the intensity of mechanical weathering

The intensity of mechanical weathering is a function of two variables: the prevailing climatic regime and lithology. Different lithologies have different susceptibilities to weathering, with porosity and permeability two critical variables. For example, as the data in Figure 15.8 show, it is the more porous lithologies that are subject to rapid salt weathering. Equally, frost action appears to be more effective in porous rocks, which can quickly be saturated with water. In addition, most processes of mechanical weathering require some form of discontinuity that can be exploited and propagated, and consequently the properties of the rock mass as a whole, in terms of the spacing, width and continuity of fractures, are critical. Climate is also an important variable: without subzero temperatures, frost weathering cannot occur. Equally, semi-arid conditions favour salt weathering. Of all the mechanical weathering processes, most studies suggest that salt and frost weathering are the most aggressive, and given the range of conditions in which salt weathering can occur, it is probably the primary mechanical agent. Both these processes are favoured by high levels of saturation and therefore environmental conditions and lithologies that facilitate these are likely to be ones that are most affected by mechanical weathering processes. It has also been suggested that global warming could play an important role in mechanical weathering in the European Alps (see Box 15.1, see in the companion website).

It is worth emphasizing that mechanical weathering plays an essential role in facilitating further weathering, whether it

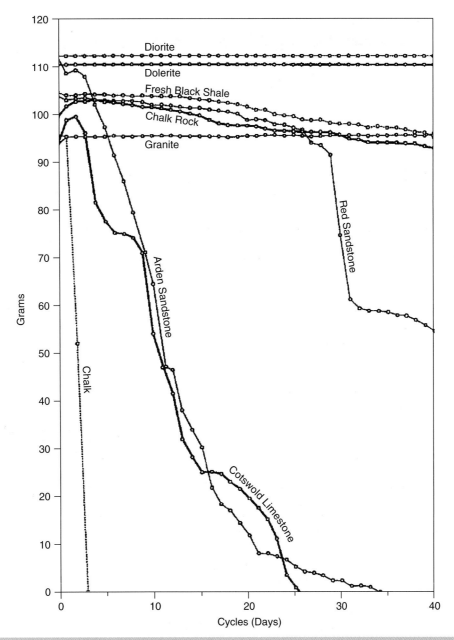

Figure 15.8 Changes in the weight of selected rock types on treatment with sodium sulphate. (Source: after Goudie, 1974)

is mechanical or chemical. The early stages of rock break-up by mechanical processes result in an exponential increase in the internal surface area of a rock, which favours accelerated chemical and mechanical weathering. The physical effects of burrowing animals must be mentioned, because a whole range of organisms including earthworms, moles, rabbits, prairie dogs and meerkats collectively move huge quantities of soil around and physically break it up. The mixing of soils, bringing fresh soil to the surface, can result in accelerated chemical weathering. Roots and stems also disturb soil and there are purported examples in which boulders have been split in two by roots, breaking the rock up much faster than it normally would.

15.3 Chemical weathering

15.3.1 Rock-forming minerals and weathering

In order to appreciate the role of chemical weathering it is necessary to reflect on the nature of rock-forming minerals. The chemical composition of the crust is dominated by 8 elements, despite the fact that 87 or more have been recorded within natural minerals (Table 15.3). The two most important are oxygen and silica. Similarly, although over 2000 rock-forming minerals have

Table 15.3 Average composition of crustal rocks

In terms of elements			In terms of oxides		
Name	Symbol and valency	Percentage	Name	Formula	Percentage
Oxygen	O^{2-}	46.60			
Silicon	Si^{4+}	27.72	Silica	SiO_2	59.26
Aluminum	Al^{3+}	8.13	Alumina	Al_2O_3	15.35
	Fe^{3+}			Fe_2O_3	3.14
Iron		5.00	Iron oxides		3.14
	Fe^{2+}			FeO	3.74
Calcium	Ca^{2+}		Lime	CaO	5.08
Sodium	Na^+	2.83	Soda	Na_2O	3.81
Potassium	K^+	2.59	Potash	K_2O	3.12
Magnesium	Mg^{2+}	2.09	Magnesia	MgO	3.46
Titanium	Ti^{4+}	0.44	Titania	TiO_2	0.73
Hydrogen	H^+	0.14	Water	H_2O	1.26
Phosphorus	P^{5+}	0.12	Phosphorus pentoxide	P_2O_5	0.28
		99.29			99.23

been identified, most rocks are dominated by a dozen or so minerals (Table 15.4). Individual elements combine to form minerals as a result of chemical bonding. However, it is the architecture of the complex lattices of tetrahedra in different silicate minerals (Figure 15.9) that provides a conceptual starting point from which to consider chemical weathering. The removal of metal ions by solution or hydrolysis, the addition or removal of water, and the substitution of one ion by another all have the potential not only to weaken or stress a crystal lattice but to change its properties, such as its solubility. They can also transform a mineral by removal and substitution to create a new, secondary mineral, often a clay mineral. The products of chemical weathering are therefore: (a) a residual framework of minerals resistant to weathering; (b) secondary clay minerals formed by transformation of primary silicate minerals; (c) the loss of minerals and both cations and ions in solution.

15.3.2 Processes of chemical weathering

There are a range of chemical weathering processes by which rock-forming minerals may be attacked, altered or transformed, which include solution, hydrolysis, hydration, isomorphous replacement and oxidation/reduction reactions. It is not the intention to examine the detailed chemistry of these reactions but rather to focus on their general characteristics and emphasize some of the variables that control their operation.

Solution

This is rather a loose term, describing a range of slightly different reactions. Some substances, such as salt, dissolve by migration of ions from the mineral to a dispersed form within the solvent, usually water. In other cases, substances enter into solution largely through reaction with compounds dissolved within water. For example, organic compounds dissolved within water may react with ions within a mineral to form new organic compounds that are soluble in water and are removed with it, a process known as chelation. Equally, solution may be complete (congruent), where the components of a solid are equally soluble, or incongruent, where only selected components are soluble. Solution will occur until the solvent becomes saturated, at which point dissolution will normally cease. The rate at which this saturation occurs is governed by the solution velocity of particular reactions.

In Figure 15.10 the solubility of some common chemical components involved in chemical weathering is shown in relation to pH, which is the availability of hydrogen ions. As the pH value falls, more free hydrogen ions become available to combine with minerals, thereby causing their dissolution. It is worth noting that only a few mineral constituents are soluble over the range of natural pH found within soils. The rate at which solution occurs is governed by the solubility of the mineral involved and the flux of solvent past the mineral (the flushing rate). The rate of flushing determines the supply of reactants to mineral surfaces and the removal of the weathered products. The contact time between the mineral and solvent is also governed by the flow

Table 15.4 Average mineral composition of common rocks

Minerals	Granite	Basalt	Sandstone	Shale	Limestone
Quartz	31.3	-	69.8	31.9	3.7
Feldspars	52.3	46.2	8.4	17.6	2.2
Micas	11.5	-	1.2	18.4	-
Clay minerals	-	-	6.9	10.0	1.0
Chlorite	-	-	1.1	6.4	-
Amphiboles (mainly hornblende)	2.4	-	-	-	-
Pyroxenes (mainly augite)	rare	36.9	-	-	-
Olivine	-	7.6	-	-	-
Calcite and dolomite	-	-	10.6	7.9	92.8
Iron ores	2.0	6.5	1.7	5.4	0.1
Other	0.5	2.8	0.3	2.4	0.3

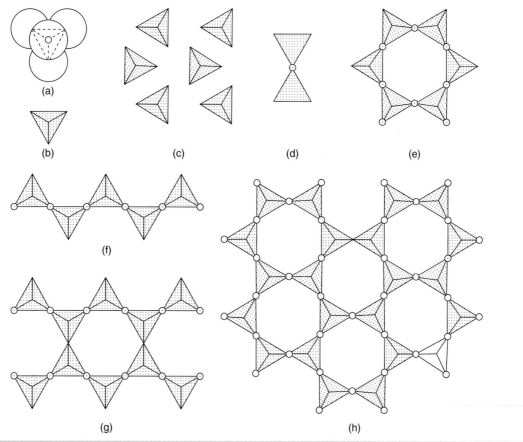

Figure 15.9 Some of the chief structural arrangements of the SiO_4 tetrahedron in crystals: (a) the SiO_4 tetrahedron; (b) conventional representation of the SiO_4 tetrahedron, shown by broken lines in (a); (c) no oxygens shared (e.g. olivine); (d) a pair of tetrahedra, each sharing an oxygen (e.g. melilite); (e) a ring of six tetrahedra, each sharing two oxygens (e.g. beryl); (f) a single chain of tetrahedra, each sharing two oxygens (e.g. pyroxenes); (g) a double chain of tetrahedra (e.g. amphiboles); (h) a sheet of tetrahedra, each sharing three oxygens and forming a continuous network with hexagonal holes as in (g) (e.g. mica). In (d)-(h) shared oxygens are shown by open circles

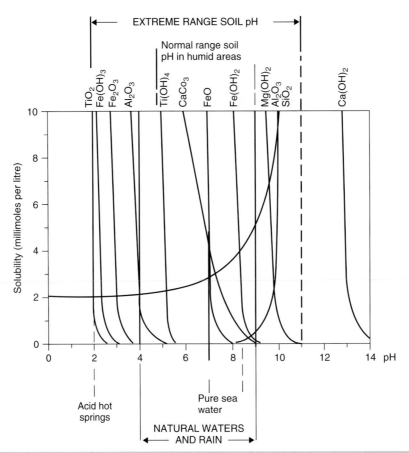

Figure 15.10 Relationship between solubility and pH for some products of chemical weathering, and the limits of pH for some natural environments. (Source: after Loughnan, 1969)

rate. Rapid water flow leads to a short contact time and little opportunity for dissolution, while slow rates of flow prolong the contact time, allowing greater saturation to be achieved. Ideal conditions therefore occur when the flow rate allows sufficient contact time for dissolution to proceed, but is sufficiently rapid to prevent the build-up of saturated solvent in contact with the mineral surface, which would inhibit dissolution. On a rock surface, water flow is rapid and anything that slows it down, such as rock texture, topography or vegetation, will favour enhanced dissolution. In contrast, within a soil, water flow is usually slow and dissolution is therefore encouraged, but may become limited by the slow rate at which the dissolved products are removed. In this case any factor which increases the rate of water flow will aid dissolution. It is possible to identify two situations that provide limiting conditions for dissolution. They are:

1. **Transport-limited:** If the products of dissolution are not removed effectively from the mineral surfaces, the solvent in contact with them will become saturated and the weathering rate will fall. This may occur because of a slow rate of water flow across the mineral faces, or alternatively due to a very high reaction rate, perhaps associated with very acid ground waters. The weathering rate is controlled by the solubility of the minerals involved and weathering tends to be uniform across a rock surface irrespective of the different solution

velocities of the component minerals (i.e. there is time for everything to dissolve that will).

2. **Reaction-limited:** If the rate of water flow is rapid, the contact time is consequently limited, and any dissolution products that result are removed quickly. In this case the weathering rate is determined by the reaction rate or solution velocity of a mineral–solvent interaction. Weathering will tend to occur more rapidly for those minerals that have high solution velocities and differential weathering of a rock surface is likely (i.e. those minerals which attain saturation quickly will tend to be preferentially removed).

Reaction-limited and transport-limited situations during dissolution are illustrated in Figure 15.11.

Solution is assisted by the presence of free hydrogen or hydroxyl ions, in a process referred to as hydrolysis. Water naturally dissociates to give free hydrogen and hydroxyl ions. This dissociation is greatly accelerated by the presence of carbon dioxide, which combines with water to form carbonic acid (H_2CO_3):

$$CO_2 + H_2O \rightarrow H_2CO_3^- + H^+$$

The ions produced by this reaction are free to take part in mineral weathering. For example, in limestone, which is composed

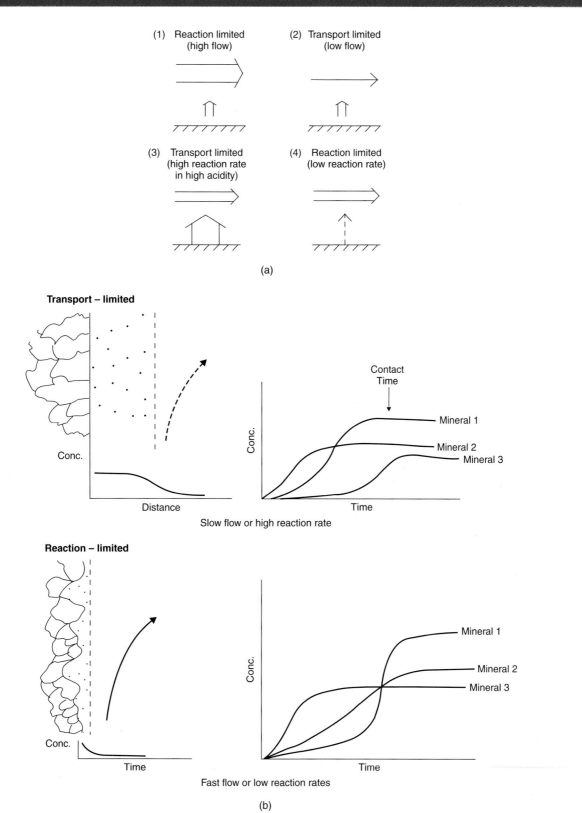

Figure 15.11 Reaction-limited and transport-limited situations during dissolution. (a) Transport-limited situations occur where the products of dissolution are not removed effectively from the mineral surfaces (2 and 3) and thus the weathering rate will fall (2), but there could be a high reaction rate (3). Reaction-limited situations occur where the water flow is rapid (1 and 4), contact time is limited and dissolution products are removed quickly. (b) Weathering depends on the reaction rate or solution velocity of a mineral–solvent interaction. Differences are illustrated in minerals 1–3

of calcium carbonate ($CaCO_3$), the following reaction occurs, in which the hydrogen carbonate (HCO_3^-) and calcium ions are removed in solution:

$$CaCO_3 + HCO_3^- + H^+ \rightarrow Ca^{++} + 2HCO_3^-$$

A similar reaction occurs with orthoclase feldspar ($KAlSi_3O_8$), which not only produces soluble products such as potassium carbonate (K_2CO_3) but also the clay mineral kaolinite ($Al_2Si_2O_5(OH)_4$):

$$2KAl\,Si_3O_8 + 2H_2O + CO_2 \rightarrow Al_2Si_2O_5(OH)_4 + K_2CO_3 + 4SiO_2$$

These reactions occur as a result of the combination of water and carbon dioxide and although rainfall may dissolve some atmospheric carbon dioxide, the primary source is from the soil. Carbon dioxide may build up within a soil as a consequence of respiration by plant roots, the bacterial oxidation of carbon and other more general processes in which soil biomass is reduced through decomposition. The build-up of carbon dioxide is greatest in the most impermeable soils, since these do not connect as freely to the ground surface as a coarse-grained soil. However, these soils are unlikely to be associated with subsurface water flow. Carbon dioxide levels also increase with depth as the void space and its connectivity decline away from the surface. However, this picture is complicated by the concentration of carbon dioxide production in the root zone of the soil. Temperature and moisture are also critical to the rate of carbon dioxide production, as shown by the experimental data in Figure 15.12. Production rates are favoured by warm, wet soils, which facilitate bacterial attack of soil organic matter. As a consequence, there is often a distinct seasonality to carbon dioxide levels within a soil (Figure 15.13). In general, carbon dioxide levels in a soil are between 10 and 10 000 times that in the atmosphere and therefore the production of carbonic acid by percolating ground water is favoured by soil development. It is also interesting to note that the solubility of carbon dioxide is a function of temperature, with the highest solubility occurring in the coldest environments.

Other sources of acidic ground water (which aids the hydrolysis process) include a range of organic acids secreted during the life cycle of organisms such as lichens, bacteria, filamentous algae and cyanobacteria. Bacteria play a major role in the nitrogen cycle and nitrification (the conversion of ammonium ions to nitrates) is facilitated by bacterial processing of dead plant and animal matter, which results in the release of free hydrogen ions and therefore the acidification of soil and associated soil waters. Desert varnish has been attributed to blue-green algae colonies that mobilize iron ions and produce a concentration of oxides on rock surfaces. It is also important to recognize the impact of humans in producing acidic rainfall from air pollution in urban areas or downwind of them.

Hydration

This involves the uptake of the whole water molecule by the mineral and its subsequent expansion, which is a process that can

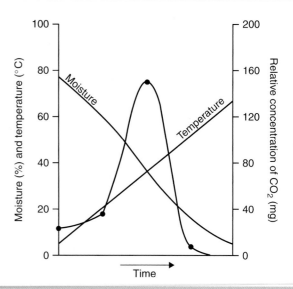

Figure 15.12 Carbon dioxide evolution during gradual heating and drying of soil, showing optimum productivity at intermediate levels of both temperature and moisture

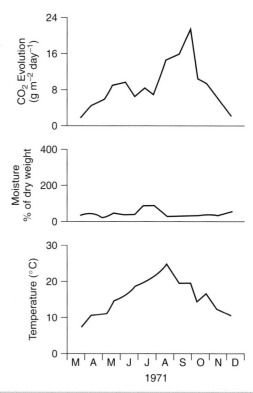

Figure 15.13 Seasonal patterns of carbon dioxide production

generate considerable physical stress within a mineral. Hydration is responsible for the conversion of the iron mineral haematite to limonite.

Isomorphous replacement

This involves the exchange of ions within minerals with others brought into contact with them through water flow across the mineral surfaces. Although in some cases this process may be reversible, it fundamentally changes the composition of a mineral and is a basic process of clay formation. In general terms, it involves the exchange of ions of similar size between a mineral and dissolved products within water. For example, silica has a diameter of 0.39 nanometres and may be replaced by aluminum, which is slightly larger, with a diameter of 0.57 nanometres, which in turn may be substituted by magnesium, at 0.78 nanometres. Not only does this process fundamentally change the chemical composition of a mineral, but the inclusion of larger ions in a silicate lattice induces physical stress within it. In this way, both mechanical and chemical weathering occur. Isomorphous replacement, along with solution and hydrolysis, gradually transforms the mineral, creating a secondary clay mineral. Clay minerals with humus, known as colloids, are particularly important in providing a reservoir of exchangeable ions that may be involved in further weathering processes. Colloidal material can be thought of as a particle with numerous negative charges, which attract free positively-charged ions (cations). These cations may be exchanged and commonly consist of hydrogen, potassium, calcium, magnesium and aluminum. The negative charges on the clay mineral surfaces are derived in a number of ways. For example, the replacement of Si^{4+} with Al^{3+} at the centre of a tetrahedron by isomorphous replacement will result in an excess of negative charge. Cations may be absorbed between layers of clay minerals, or through their outer surface, and may also be exchanged if water containing other cations passes across the clay surface. In the context of mineral weathering, the process is critical in determining the composition of percolating water, and therefore its weathering potential.

Oxidation and reduction

These involve the loss or gain of an electron. When an atom or ion loses an electron, oxidation occurs. When this process is reversed, it is referred to as reduction. By far the most common oxidizing agent is oxygen dissolved in water.

Oxidation is best known and illustrated by the rusting of iron. This involves the addition of water and air, which convert iron from its bivalent (Fe^{2+}) to trivalent (Fe^{3+}) form. Most naturally-occurring iron has a bivalent form and so oxidation disturbs the neutral charge of the crystal structure, which can only be regained by the addition of further cations, a process which may lead to the collapse of the mineral structure. Not only does oxidation have a potential impact on crystal structure, but a change in redox state may render a mineral more vulnerable to other weathering processes, such as solution. Particularly vulnerable elements to oxidation are iron, titanium, magnesium and sulphur. Bacteria have an important role in the oxidation of metals. These may be released from their containing structures and subsequently precipitated. They are chemolithotropic, which means that they obtain their energy from the enzymatic oxidation of inorganic compounds. This process is so effective that it has been used in the processing of inorganic compounds and metal ores.

The process of cation exchange (see Chapter 25) is an extension of isomorphous replacement in as far as it involves the exchange or replacement of one cation for another. However, it is usually restricted to a mineral surface. Clay minerals have a particular propensity for the exchange of cations between their lattices and water. This process is central to the formation and successive modification of clay minerals.

15.3.3 Products of chemical weathering

There are three basic products of chemical weathering: a residuum of unweathered minerals, clay minerals and solutes. The clay minerals are sometimes referred to as phyllosilicates because of their sheet-like structure. This structure results from combinations of layers of silica tetrahedra and aluminum or magnesium hydroxide, which are organized as octahedra. Kaolinite is the most common clay mineral and has a one-to-one layer structure, with alternating tetrahedral and octohedral layers. In contrast, illite, smectite (montmorillonite), vermiculite and chlorite have two silica sheets sandwiched between each alumina sheet. Weathering may lead to a specific clay mineral, or over time may result in successive minerals. The clay minerals that result from chemical weathering depend on such variables as: (a) the composition of the circulating pore water, especially in terms of its silica content and the range of cations present; (b) the mineralogy of the bedrock; (c) the prevailing pH and oxidation state of the soil; (d) the rate of water flow through the soil (rate of leaching) (see Chapter 25). Some of these environmental controls are illustrated in Table 15.5. The importance of water flow in determining the clay minerals that are produced can be illustrated via solution of albite (a sodium feldspar):

1. **Poor drainage with slow water flow:** Solvent saturation is quickly achieved and not all the sodium and magnesium ions are flushed away. They therefore remain to react with aluminum and silica to give montmorillonite.

$$\text{albite} + \text{magnesium} + \text{water} \rightarrow \text{montmorillonite} + \text{hydrogen sulphate}$$

2. **Better drainage with faster water flow:** Sodium and magnesium ions are dissolved and removed, while the more slowly dissolving aluminum and silica ions remain to form kaolinite.

$$\text{albite} + \text{water} \rightarrow \text{kaolinite} + \text{sodium} + \text{hydrogen sulphate}$$

3. **Rapid drainage, with very fast water flow:** Sodium, magnesium and silica are removed, while aluminum remains to form gibbsite.

$$\text{albite} + \text{water} \rightarrow \text{gibbsite} + \text{sodium} + \text{hydrogen sulphate}$$

Table 15.5 Mineralogy of weathering products in relation to environmental controls. (Source: after Loughnan, 1969)

Environment	pH	Eh	Behaviour of major elements	Mineralogy of weathering products
Non-leaching. Mean annual precipitation 0–300 mm. Hot	Alkaline	Oxidizing	Some loss of Na^+ and K^+. Iron present in ferric state	Partly decomposed parent minerals. Illite, chlorite, smectite and mixed-layered clay minerals. Hematite, carbonates, secondary silica and salts. Organic matter absent or sparse
Non-leaching below water table	Alkaline to neutral	Reducing	Some loss of Na^+ and K^+. Iron present in ferrous state	Partly decomposed parent materials. Illite, chlorite, smectite and mixed-layered clay minerals. Siderite (iron carbonate) and pyrite (iron sulphide). Organic matter present
Moderate leaching. Mean annual precipitation 600–1300 mm. Temperate	Acid	Oxidizing to reducing	Loss of Na^+, K^+, Ca^{2+} and Mg^{2+}. Some loss of SiO_2. Concentration of Al_2O_3, Fe_2O_3 and TiO_2	Kaolinite with or without degraded (K-deficient) illite. Some hematite present. Organic matter generally present
Intense leaching. Mean annual precipitation >1300 mm. Hot	Acid	Oxidizing	Loss of Na^+, K^+, Ca^{2+}, Mg^{2+} and SiO_2. Concentration of Al_2O_3, Fe_2O_3 and TiO_2	Hematite, goethite, gibbsite and boehmite with some kaolinite. Organic matter absent or spare
Intense leaching. Mean annual precipitation >1300 mm. Cool	Very acid	Reducing	Loss of Na^+, K^+, Ca^{2+}, Mg^{2+} and some iron and Al_2O_3. SiO_2 and TiO_2 retained	Kaolinite, possibly with some gibbsite or degraded illite. Organic matter abundant

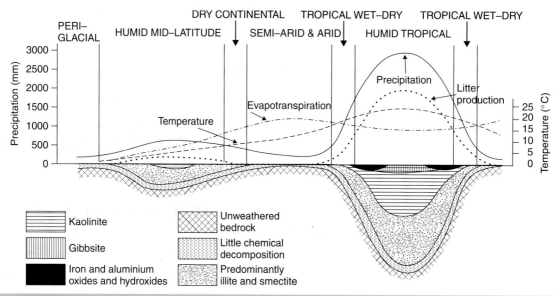

Figure 15.14 Relationship between temperature and rainfall seasonality and type of clay mineral. (Source: after Strakov, 1967)

The intensity of the leaching process, along with the length of time over which it is active, is therefore critical to the clay minerals produced. Temperature and the seasonality of rainfall are likewise critical in determining the assemblage of clay minerals produced, as illustrated in Figure 15.14. This is a point reinforced by the field data, which show that the weathering profiles of two soils developed on similar bedrock but with different climatic regimes are different (Figure 5.15(a),(b)). There is a good example in Figure 15.15(b) illustrating that weathered products can be of

great age. Here the basalt bedrock was intensely weathered in a hot, wet, tropical Tertiary climate when Britain was much closer to the equator, contrasting markedly with the weathering of basalt in today's temperate climatic regime. These data also show that the intensity of weathering is not uniform within a soil profile; for example, leaching varies with depth and therefore the composition of the clay minerals and the intensity of weathering may vary.

An example from late Tertiary volcanic ash from Wyoming, USA, illustrates the significance of organic action in the water

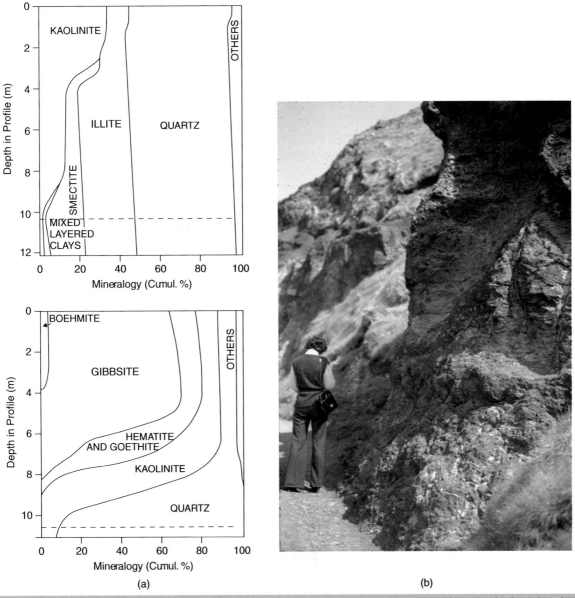

Figure 15.15 (a) Variation in mineralogy for weathering profiles on clay slates at Goulburn, New South Wales, Australia under a humid, warm temperate climate (top) and on a sandstone at Weipa, Queensland, Australia under a hot, monsoonal climate (bottom). The transition between bedrock and weathered material is indicated by a dotted line. (b) Laterite developed in a Tertiary hot, wet climate. The basalt bedrock has been completely weathered on the road above the Giant's Causeway, Antrim, Northern Ireland. This is in marked contrast with weathering on this rock type in today's climatic regime. (Source: (a) after Loughnan, 1969)

percolating through this rock. Soluble organic compounds derived from vegetation were controlled by the release and transport of solutes by complexing aluminum and iron and by causing low pH values (between 4.3 and 6.5). The dissolved organic concentrations, as humic acid, oxalate, formate and acetate, correlate with concentrations of major cations and follow a seasonal pattern.

A weathering profile produced by chemical weathering has a distinctive form, although its detailed form may reflect the properties of the rock mass being weathered. Chemical weathering attacks joints and fissures, and consequently it is along these discontinuities that weathering is concentrated. In some cases this may leave corestones, or unweathered rock, surrounded by weathered material. Good examples can be seen on granite in Hong Kong and Dartmoor. In some cases the original form and structure of the unweathered rock remains; this is referred to by the term saprolite. Here isovolumetric weathering takes place (weathering without any change in volume). The term regolith is used as a more general descriptor for a weathered debris mantle, and weathering front refers to the depth to which weathering reaches. The thickness of a weathering profile at any one site reflects a balance between the rate of weathering and the rate of erosion from the site by other processes. Even if weathering is extremely rapid, a weathered profile may be limited if all the weathered regolith is quickly eroded.

15.3.4 Controls on the intensity of chemical weathering

The different controls on the intensity of weathering can be divided into those that are intrinsic to the particular rock mass (for example the susceptibility of the minerals present to weathering) and those that are extrinsic (for example the influence of rainfall and temperature).

Intrinsic factors

There are two basic intrinsic variables: the presence of discontinuities and mineral susceptibility. Discontinuities within the rock mass are essential to effective weathering since they allow the penetration of air and water. They operate first at the scale of the rock outcrop, in terms of such discontinuities as bedding, joints, faults or foliation, and second at a mineral scale, where mineral alterations tend to follow microscopic cracks and fissures, or penetrate along cleavage planes. In general, rock bodies that are massive and free of fissures will weather more slowly than those that are highly fractured, irrespective of the mineral composition. In other words, fracture density is more important than mineralogical composition. Mechanical weathering has an important role to play in fracture propagation and facilitating access by water and air into the fabric of the rock. Rock mineralogy also has an important impact on the intensity of weathering, as some minerals are more susceptible than others. Mineral susceptibility has been approached in a range of different ways, with the aim of establishing a weathering series, or ranking of mineral stability. For example, a weathering series based on field observations in Minnesota identified quartz as the most stable and olivine, biotite and plagioclase as the least stable minerals. This is very similar to the order of crystallization found from a silicate melt (Figure 15.16). Quartz, as the most stable mineral, crystallizes at the lowest temperature. It is interesting to note that the most stable mineral is the one with the least difference between the temperature of crystallization and typical surface temperatures: it is the least out of equilibrium with surface conditions. However, more recent investigations have suggested that mineral stability is heavily influenced by environmental conditions, in particular the presence of organic matter. Where ample organic acids are present – derived from bacteria, algae, fungi and lichens – the weathering sequence of Goldrich (1938) appears to apply, but in near-sterile environments it does not, and in fact may even be reversed.

Another way of approaching mineral susceptibility is to examine the relative stability of different cations. In one study the concentration of cations in solution in ground water was compared with their abundance in unweathered rock within a granite landscape in California. Sodium, calcium and magnesium were the most abundant in solution and therefore the most mobile, while aluminum and iron were the least mobile. This study and similar examples allow general mobility sequences to be established, such as calcium $\geq$ sodium $\geq$ magnesium $\geq$ silica $\geq$ aluminum $\geq$ potassium $\geq$ iron $\geq$ titanium. The concentration of these cations in a mineral may help determine its stability to weathering.

The importance of parent material composition is illustrated in relation to the solution or carbonation of limestone. For example, calcite increases solubility with traces of magnesium; other carbonate minerals, such as aragonite, are much more soluble; and dolomite is less soluble. This is complicated in practice by other variables; for example, sparite is a highly soluble limestone but tends to form more resistant elements in the landscape because of its greater mechanical strength. Rock texture overrides other factors. For example, a coherent coral structure, composed of more soluble aragonite, may form more upstanding elements than a less soluble but loosely cemented calcite calcarenite. Solubility and erodibility are not necessarily linked. Mineral susceptibility is therefore often misleading and characteristics of the rock mass are more important.

Extrinsic factors

Environmental conditions, such as climate, vegetation and topography, impact on the weathering intensity, as we have seen in Figure 15.14. This is shown schematically in Figure 15.17. These factors essentially control such variables as temperature, water dynamics and water chemistry, the impacts of which are discussed below.

Field data show the potential impact of temperature on the weathering rate, because chemical reactions occur faster at higher temperatures. In general, for every 10 $^{\circ}$C rise, the rate of chemical

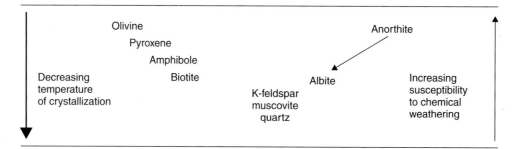

Figure 15.16 Bowen's reaction series and mineral susceptibility to chemical weathering

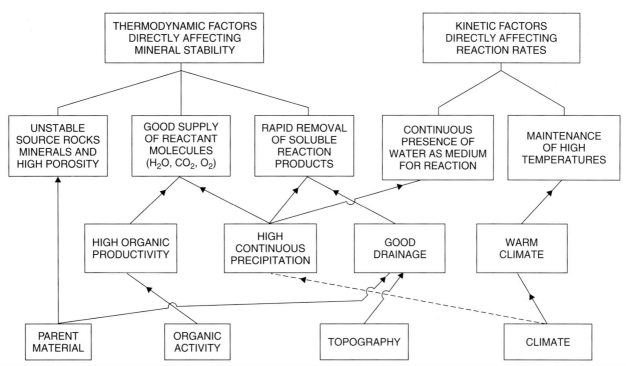

Figure 15.17 Schematic illustration of the way in which environmental factors can contribute to high rates of chemical weathering. (Source: after Curtis, 1976)

reactions increases by a factor of two. However, this simple picture is complicated when dealing, for example, with the solution of calcium carbonate by carbonic acid, formed by the reaction of carbon dioxide with percolating soil water. This is because the solubility of calcium carbonate increases as temperature falls. Consequently, although the rate of chemical reactions may be slower in cold regions, the solubility of carbon dioxide is enhanced. In practice however this does not influence the weathering rate since the biological productivity of cold regions is limited and therefore so is the carbon dioxide content of the soil. It has also been noted that solute concentrations on limestone fall with altitude as it becomes colder (Figure 15.18).

The chemistry of the weathering solution is controlled by pH, temperature and the nature of the soil through which the water must pass to reach the bedrock. Leaching of a soil may influence its pH, so for example a calcareous soil will help neutralize, or at least reduce, the acidity of percolating soil water. Idealized models of the effect of soil leaching on rockhead weathering for a peat soil overlying limestone show that initially a small rainfall event causes dilution of soil solutions. This may cause a rise in pH, but as the soil becomes saturated, carbon dioxide will begin to accumulate and increase water acidity. The downward movement of this soil water will leach the calcium from the soil, causing its pH to fall. In this way the soil may be acidified to the rockhead and dissolution of the underlying rock may then occur. As the soil dries out, capillary rise will recalcify the soil. In contrast, peat soil overlying siliceous sandstone will show little variation in soil pH during wetting and drying. The essential point is that soil chemistry has a profound effect on the chemistry and potential for weathering of percolating water.

The dynamics of water flow through a soil or rock are also an important variable in determining weathering intensity. Water flow through a soil is a function of rainfall minus losses due to evaporation or evapotranspiration, and may also be augmented by baseflow from ground water. As Figure 15.19 illustrates, there is a strong relationship between the solution of calcium carbonate and runoff, and as we saw earlier, the importance of water flow can determine whether a process is reaction- or transport-limited. As illustrated by the graphs in Figure 15.20, there is a strong correlation between mean annual precipitation and the clay minerals or residual metal hydroxides which result. The conclusion here is that the clay minerals produced from a given primary mineral are a function of climate, primarily rainfall, and the associated intensity of leaching. Nevertheless, Figure 15.21 reinforces the fact that the rock type has an influence too. Patterns of water flow are also influenced by topography and microclimate. For example, on flat slopes vertical water percolation and leaching are enhanced, whilst on steeper slopes a large component of subsurface flow is directed downslope. Field investigations suggest that in limestone areas weathering increases upslope, since soil leaching and therefore acidification occur more effectively, while the downslope movement of calcareous bedrock and soil means that soils offer a greater buffer to underlying limestone. The microtopography of various solutional forms, such as dolines (see Chapter 18), may influence moisture levels, carbon dioxide levels within a soil and ultimately soil temperature. The collective impact of these variables is manifest in the variation

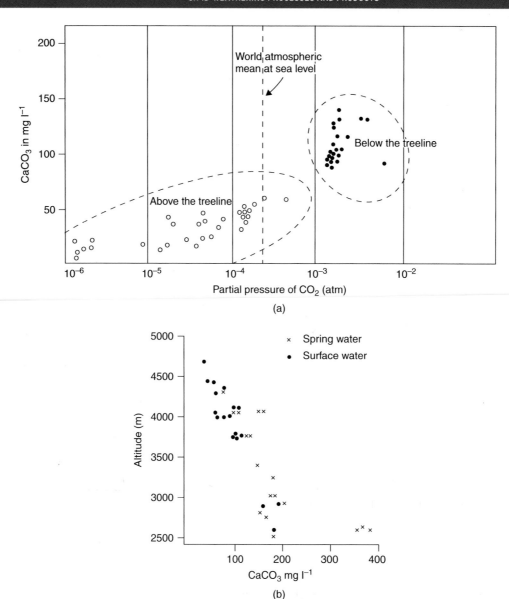

Figure 15.18 (a) Changes in solute concentrations above and below the treeline in the Canadian Rockies. CO_2 and solute levels are lower at higher, colder altitudes above the treeline. (b) Decreasing solute concentrations with altitude in the Himalayas

in the depth of the weathering mantle when charted against latitude, as the greatest depth of weathering is achieved in the humid tropics, where abundant precipitation, organic matter production and high temperatures combine.

15.4 Measuring weathering rates

Numerous studies have attempted to quantify the rate of weathering and its spatial variation. Such studies are not only of academic interest but are relevant in deciding the choice of geomaterial and the use made of different rocks. As long ago as 1880, Geikie made a survey of Edinburgh churchyards, comparing the weathering rate of different lithologies used in headstones and monuments.

He found that sandstone weathered very little in 200 years, maintaining a clear inscription, as did slate over 90 years, whilst marble showed rapid deterioration over the same time scales.

There have been many similar studies using a diverse range of strategies. Some of the methods used to determine the rate of weathering include: dated buildings, inscriptions and monuments; mass loss tablets; weathering rinds; surface roughness; and strength and solute levels.

15.5 Weathering landforms

Weathering is fundamental to the changing form of the Earth's surface and plays a central role in preparing rocks and sediments

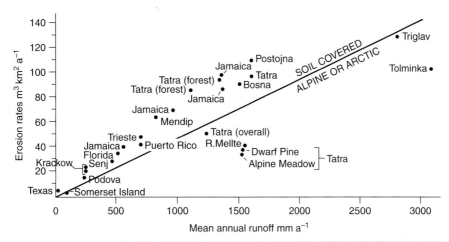

Figure 15.19 Relationship between mean annual runoff and erosion rates in limestone. (Source: after Trudgill, 1976)

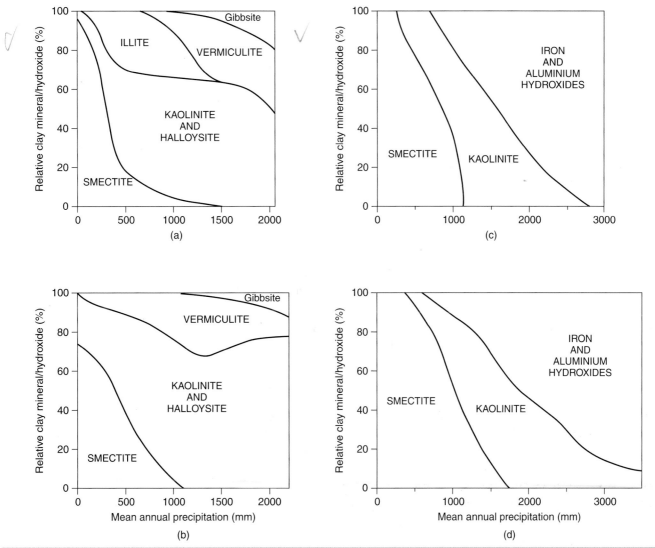

Figure 15.20 Variations in the clay mineral and residual hydroxide composition of soils in relation to mean annual precipitation: (a) on feldspar and quartz-rich igneous rocks; (b) on olivine, amphibole and pyroxene-rich igneous rocks; (c) in soils developed on basalt under an alternating wet and dry climate; (d) in a continuously humid climate. (a) and (b) are from California and (c) and (d) are from Hawaii. (Source: after Sherman, 1962)

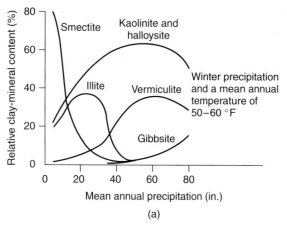

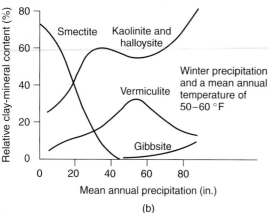

Figure 15.21 Clay minerals formed by weathering from (a) acid igneous rocks and (b) basic igneous rocks in California as a function of precipitation. (Source: after Drever, 1997)

before the processes of erosion. Hence it could be argued that weathering is manifest in some form in almost all landforms. It is possible though to identify a series of landforms that can be linked directly to weathering processes, and these are reviewed below. Most are relatively small-scale phenomena on rock surfaces, including weathering pits and honeycombed rock surfaces. At a larger scale, tor landforms (reviewed in Chapter 21) are directly linked to weathering, but with erosional processes playing a major role in removing the weathered debris. Perhaps the most widespread expression of weathering is the formation of karst landscapes (named after the type locality in the former Yugoslavia), which develop on carbonate rocks as a result of their rapid solution, although a complex set of other processes are in operation here too. These karst landscapes are reviewed in Chapter 18.

15.5.1 Weathering rinds

These may develop on the surfaces of boulders or exposed rocks, taking the form of a weathered surface horizon, in which the

chemistry of the rock has been altered. The thickness of the rind is proportional to the length of exposure to weathering and the rate at which it has proceeded, and the rinds provide an opportunity to study changes in rock or mineral chemistry as weathering develops. In many cases the rind is weakened as a result of this chemical alteration, but occasionally chemical weathering may lead to surface or case hardening of the rock. For example, the concentration of iron and manganese hydroxides within weathering rinds may lead to case hardening. The minerals are either dissolved from the rock's interior and drawn to the surface by capillary action, where they become precipitated, or else derived externally from precipitation or airborne aerosols. The concentration of water flow along joints or fissures may lead to preferential case hardening of a rock-joint network. On subsequent erosion or weathering, the inter-joint areas are preferentially removed and the joint pattern may be seen as a raised boxwork.

15.5.2 Weathering pits

Rock surfaces may also contain a detailed relief of small-scale depressions, which often provide an intricate surface ornament. On relatively flat surfaces, weathering pits and enclosed hollows may form, varying from a few centimetres to several metres in diameter and depth. On steeper rock surfaces multiple depressions or hollows, in close juxtaposition, may form distinct networks. These are given the general name 'cavernous weathering', but there are a plethora of poorly-defined terms used to describe various subtypes, including 'honeycomb' or 'alveolar' weathering, where there are numerous closely-spaced pits or depressions on the rock faces, and 'tafoni' (Figure 15.22), which are hollows cut into steeply sloping rock faces. Chemical weathering of susceptible

Figure 15.22 Tafoni caused by cavernous weathering from San Francisco, California. (Photo: Andrew Alden, 2002, geology.about.com)

minerals, the action of lichens, algae and fungi, and salt weathering have all been implicated in the formation of these forms. The simplest of depressions may form as a result of spalling of small patches on a rock's surface. However, the processes by which such depressions are maintained and enhanced are not clear. In some cases there is evidence to suggest that case hardening of a rock exterior, in the form of a weathering rind, may play a role. If this outer layer is breached, weathering may exploit the weaker rock beneath. An example model for tafoni development is illustrated in Figure 15.23, where the cavernous form may result from the breaching of a case-hardened exterior, and there is usually strong evidence for salts present on the backwalls, coming from within the rock itself, or if in a coastal location, from spray.

15.5.3 Cavernous weathering

This tends to be associated with semi-arid or coastal environments, which are characterized by both a supply of salt and by wetting and drying conditions. Consequently, salt weathering has been advanced as the primary cause. For example, the seawalls of Weston-super-Mare, UK, were completed in 1888 and capped by local sandstone, which has since been weathered at a rate of approximately 1 mm per year and shows alveolar weathering. Weathering intensity was measured along the seawalls using a qualitative scale (weathering at each site was ranked on a scale of 1–8) and then correlated with such variables as the distance from high water mark, windward/leeward exposure and azimuth. The results show a close relationship between the intensity of the weathering and the location of the spray zone on the seawall, characterized by multiple wetting and drying, and abundant salt. For example, weathering intensity decreases with distance from the high water mark, and weathering grade increases with exposure to spray and salt-laden winds (see Table 15.6). Although studies like this indicate a strong correlation between conditions which favour salt weathering and alveolar forms, they do not explain the nature of the positive feedback loop which must operate for

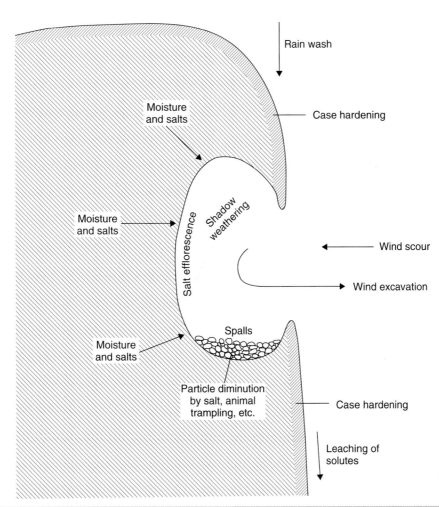

Figure 15.23 Model of tafoni development. (Source: Goudie and Viles, 1997)

Table 15.6 Weathering and chloride content of four sections of a sea wall at Weston-super-Mare, UK. (Source: after Mottershead, 1994)

	Sites			
	Birkett (1)	**Knightstone (3)**	**Anchor (2)**	**Royal Parade (4)**
Elevation (m above MHWST)	8.5–9.5	2.5	2.8–3.5	2.5
Exposure	Exposed	Sheltered (W); exposed (E)	Exposed	Partly exposed
Foreshore at HWM	Rocky	None	Rocky	Sandy beach
Distance to HWM (m)	0–5	0	0–5	10–50
Orientation	W	SW	S–W	W
Weathering grade (mean values)	1.95 ($n = 25$)	Exposed (E) 6.12 ($n = 31$); sheltered (W) 1.45 ($n = 19$)	Exposed 6.45 ($n = 9$); sheltered 2.55 ($n = 8$)	4.5 ($n = 55$)
Mean chloride content (mg l^{-1})	16.9 ($n = 9$)	13.6 (W); 20.2 (E); 16.9 (overall) ($n = 12$)	29.7 ($n = 8$)	16.9 ($n = 12$)

Figure 15.24 Honeycomb weathering caused by salt crystallization from Yehlin, Taiwan. (Photo: Stephen Codrington, Wikimedia Commons)

Figure 15.25 Honeycomb weathering from coastal sandy limestone, Gozo, Malta

(a)

(b)

Figure 15.26 (a) Spheroidal weathering in the Alabama Hills, near Lone Pine, USA. (b) Spheroidal weathering of basalt above the Giant's Causeway, Antrim, Northern Ireland. (Source: (a) http://www.alancolville.com/weathering.html)

Figure 15.27 Chemical weathering attacks granite along joints. (Source: http://www.alancolville.com/weathering.html)

the individual hollows to be formed, maintained and enlarged. Examples of this type of honeycomb or alveolar weathering are illustrated in Figures 15.24 and 15.25.

Other examples of weathering landforms include the spheroidal landforms caused by weathering of individual boulders, where the surface skin peels away, often to reveal the central, relatively unweathered, core (Figure 15.26). Tors (covered in more detail in Chapter 21) rely on weathering along joints in the granite (Figure 15.27), with the weathered products transported away. There are many other forms of small-scale solutional weathering landforms on limestone bedrock, which are discussed in Chapter 18.

Exercises

1. Define weathering and describe the general characteristics of the weathering of rocks and sediments.

2. How can volumetric changes in a rock mass occur and what do they lead to?

3. Explain the physical processes associated with frost weathering.

4. What factors are thought to control salt weathering?

5. Explain why solution is very important as a chemical weathering process.

6. What are the controls on clay mineral formation?

7. Explain why rock mineralogy has an important impact on the intensity of rock weathering.

8. Explain how tafoni and cavernous weathering are thought to form.

9. How can rates of weathering be measured?

10. Explain how environmental conditions such as climate, vegetation and topography can impact on weathering intensity.

References

Bland, W. and Rolls, D. 1998. Weathering: An Introduction to the Scientific Principles. London, Arnold, London.

Curtis, C.D. 1976. Chemistry of rock weathering: Fundamental reactions and controls. In: Derbyshire, E. (Ed.). Geomorphology and Climate. London, John Wiley & Sons, Ltd. pp. 25–57.

Drever, J.J. 1997. The Geochemistry of Natural Waters. 3rd edition. Eaglewood Cliffs, NJ, Prentice-Hall.

Goldrich, S.S. 1938. A study on rock weathering. *Journal of Geology.* **46**, 17–58.

Goudie, A. 1974. Further experimental investigation of rock weathering by salt and other mechanical processes. *Zeitschrift für Geomorphologie Supplementband.* **21**, 1–12.

Goudie, A. and Viles, H. 1997. Salt Weathering Hazards. Chichester, John Wiley & Sons, Ltd.

Hall, K. and Hall, A. 1991. Thermal gradients and rock weathering at low temperatures: Some simulation data. *Permafrost and Periglacial Processes.* **2**, 103–112.

Loughnan, F.C. 1969. Chemical Weathering of the Silicate Minerals. Amsterdam, Elsevier.

McGreevey, J.P. 1985. Thermal properties as controls on rock surface temperature maxima, and possible implications for rock weathering. *Earth Surface Processes and Landforms.* **10**, 125–136.

Mottershead, D.N. 1994. Spatial variations in alveolar weathering of a dated sandstone structure in a coastal environment, Weston-super-Mare, UK. In: Robinson, D.A. and Williams, R.B.G. (Eds). Rock Weathering and Landform Evolution. Chichester, John Wiley & Sons, Ltd. pp. 151–174.

Sherman, G.D. 1962. Problems of Clay and Laterite Genesis. American Institute of Mining and Metallurgical Engineering.

Strakov, N.M. 1967. Principles of Lithogenesis. Edinburgh, Oliver and Boyd.

Trudgill, S.T. 1976. The erosion of limestones under soil and the long term stability of soil-vegetation systems on limestone. *Earth Surface Processes.* **1**, 31–41.

Further reading

Bridge, J.S. and Demicco, R.V. 2008. Earth Surface Processes, Landforms and Sediment Deposits. Cambridge, Cambridge University Press. Chapter 3.

Selby, M.J. 1993. Hillslope Materials and Processes. Oxford, Oxford University Press. Chapters 8 and 9.

16 Slope Processes and Morphology

Debris flow scars in 1968–1969 in greater Los Angeles, only a few months after the flow events occurred. Photo: USGS, http://geology.wr.usgs.gov/wgmt/elnino/scampen/valverde.html

Earth Environments   David Huddart and Tim Stott
© 2010 John Wiley & Sons, Ltd

Learning Outcomes

After reading this chapter and completing the exercises you will be able to:

➤ Understand why slopes occur in the landscape.

➤ Understand how mass movements form, the types and classification of mass movements and the factors that allow them to take place, including processes such as soil creep, granular disintegration, rockfall, toppling and slab failure, rock slides and rock avalanches, debris and mudflows.

➤ Understand the Factor of Safety and what controls slope strength and slope stress.

➤ Understand the role of water in the creation of slopes, including rainsplash, overland flow and ground water flow.

➤ Be able to describe slope morphology and relate process to form.

➤ Understand the controls on slope evolution such as lithology, climate and antecedence.

16.1 Introduction

Slopes are some of the most basic elements of the landscape and, along with plains, terraces and plateaux, constitute the Earth's land surface. Consequently, understanding the processes that operate upon them, and the resulting landforms, is one of the most fundamental aspects of geomorphology. Slopes are shaped and modified by two broad types of process, one being mass movement, which are the downslope movement of soil and/or rock under the influence of gravity, and the other being the action of water flowing down a slope, towards a channel. Water flows down a hillslope because gravity acts on its mass; this effect only disappears when the water reaches a flat surface. The same pull acts on the solid hillslope surface material, but there is no continuous flow of rock and sediment towards areas of level surface. Why is this?

It is because hillslope materials have internal resistance, or shear strength, which acts against the gravity force. The result is that sloping surfaces can occur in solid masses without any movement of the material. Naturally there is a limit to the steepness and length of these sloping surfaces, which depends on the shear strength. Some clays, for example, contain little more strength than a mass of water and therefore will only be stable at very low slope angles. Once a critical angle is reached, mass movements of the slope occur, but some solid rocks possess so much strength that high, vertical slopes are stable over the short term. However, over geological time the stability of a slope is only a temporary feature and as a result there is not a continuous flow of material downslope but an intermittent succession of mass movements. Over a long time period the effect of these intermittent mass movements is to produce a slope of gradually lower angle.

Hillslopes are primary irregularities in the land surface exposed to weathering, and no newly-exposed surface is completely flat. There is always a gravitational gradient or slope component in the landscape. Runoff will occur in this direction and the gravitational force will always be acting in a downslope direction. So there has to be a primary slope because if this were not the case then runoff would not be concentrated and would not flow. Rivers would not transport, no valleys would be eroded and no slopes would develop.

In this chapter we will first review the processes of mass movement, before exploring the impact of water on slopes and then dealing more generally with the morphology of slopes and their evolution as a consequence of the processes that act upon them.

16.2 Slopes: mass movement

In this section we will explore the processes of mass movement, which are not only of geomorphological importance but also major natural hazards. Mass movement occurs in every climatic environment and is considered the dominant erosional process in some environments, simply because it is so obvious. For example, in one Japanese drainage basin it has been estimated that 83% of all the sediment produced is derived from mass movement. In this case, the high rates reflect the occurrence of steep slopes and the mean annual precipitation is in excess of 2.5 m. In Arctic regions, mass movement is also often the dominant geomorphological process. For example, at Kärkevagge in Sweden it has been estimated that 49% of erosion is due to rapid mass movement, while the remaining 51% is the product of slow failures. It is important to emphasize that mass movement is not simply a process of extreme environments but is also active in temperate regions, such as Britain. A recent inventory of all known mass movement in Britain identified over 8365 major failures. In the Upper Rhine Valley of Switzerland, rapid failures are responsible for about 85% of the mean annual displacement of material, which compares well with the 0.001% attributable to river action. Mass movement is clearly a very important geomorphological process and has the potential to threaten human life and infrastructure. For example, between 1947 and 1988 there were 1062 natural disasters involving over 100 fatalities, of which 29 were due to mass movement (Table 16.1). Similarly, between 1979 and 2000 the estimated cost of the damage from mass movements in the State of California was estimated to exceed US$10 billion. However, it is more appropriate to estimate hazard potential in terms of economic cost. In the late 1970s in the USA, mass movements were costing over US$1000 million per year. An example from Britain is documented in Box 16.1 (see in the companion website, The Mam Tor landslide, Derbyshire).

Table 16.1 Some of the most important mass movement disasters of the twentieth century and their impact in terms of deaths. (Source: after Jones and Lee, 1995)

Place	Date	Type	Impact
Java	1919	Debris flow	5100 killed, 140 villages destroyed
Kansu, China	1920	Loess flow	c. 200 000 killed
California, USA	1934	Debris flow	40 killed, 400 houses destroyed
Kure, Japan	1945		1154 killed
Tokyo, Japan	1958		1100 killed
Ranrachirca, Peru	1962	Ice & rock avalanche	3500+ killed
Valont, Italy	1963	Rockslide into reservoir	c. 2600 killed
Aberfan, UK	1966		144 killed
Rio de Janeiro, Brazil	1966		1000 killed
Rio de Janeiro, Brazil	1967		1700 killed
Virginia, USA	1969	Debris flow	150 killed
Yungay, Peru	1970	Earthquake-triggered debris avalanche	25 000 killed
Chungar	1971		259 killed
Hong Kong	1972		138 killed
Kamijima, Japan	1972		112 killed
S. Italy	1972–73		100 villages abandoned, affecting 200 000 people
Mayunmarca, Peru	1974	Debris flow	Town destroyed, 451 killed
Mantaro Valley, Peru	1974		450 killed
Mt Semeru	1981		500 killed
Yacitan, Peru	1983		233+ killed
W. Nepal	1983		186 killed
Dongziang, China	1983		4 villages destroyed, 227 killed
Armero, Colombia	1985	Lahar	c. 22 000 killed
Çatak, Turkey	1988		66 killed

16.2.1 What is mass movement?

Sediment transfer on hillslopes occurs through various processes, such as semiconcentrated or concentrated water movements, where sediment is moved as dispersed grains (which can influence the whole slope), or in a dispersed form in subsurface water movements through the soil, affecting solutes and very fine sand. Mass movements are the downslope movement of rock and/or soil under the influence of gravity, where the sediment moves as a mass, and grain-to-grain contacts are maintained. Rockfalls are included too. An individual rockfall event maintains its form until it hits the ground, where it may or may not break. Here a mass movement is a mobile mass of sediment, not of individual grains; these can be dramatic, but the slower processes of soil creep are mass movements too. Mass movement encompasses a range of processes frequently referred to as landsliding. However, this term is incorrect, since a landslide is simply one type of mass movement. Landslides take place when the soil or rock on a slope is weathered enough to lose some strength and slips or fails. The movement of the failed material is rapid in geomorphic terms but the processes leading up to failure (the weathering processes) control the rate of surface lowering. Hence we say that landslides are weathering-limited. There are exceptions as if the flow medium (a stream, a glacier or the sea) at the base of the slope is not powerful enough to remove the failed material then it will accumulate on the slope and stop the landsliding process. This is a positive feedback mechanism and the landsliding becomes transport-limited: it is limited by the rate at which material is removed from the slope base.

Mass movement may occur in a range of different ways, at a range of different rates, and involve anything from individual soil particles to large volumes of rock and soil. Types of mass movement include soil creep, landslides, rockfalls and topples and mud/debris flows, all of which occur at a range of different scales. Mass movement occurs in almost all environments but vary in frequency depending on a combination of geological and climatic factors. It is useful to recognize that on hillslopes various types of process occur, which may be:

- **Continuous and ubiquitous:** The process acts all the time, everywhere in the system. Here the rate of surface lowering on the slope is constant with time and constant spatially. Solutional loss is the only process in this category.

- **Discontinuous and ubiquitous:** The action is concentrated in particular events in time but acts uniformly everywhere. Soil creep processes are good examples.

- **Continuous but localized:** Continuous action is localized to a particular area, for example flow in a stream channel.

- **Discontinuous but localized:** The process acts discontinuously in time and space. Most landslides fall into this category.

16.2.2 Causes of mass movement: Why does failure occur?

The causes of mass movement can be understood by examining the concept known as the **Factor of Safety** (FS). This expresses the balance between the forces driving or causing failure and those resisting it; that is, the stresses acting on a slope versus the strength of the slope:

$$\text{Factor of safety} = \frac{\text{Slope strength}}{\text{Stress acting on the slope}}$$

If the FS is less than one, the slope is unstable; if it is equal to one, it is in critical stability; and if it is greater than one, it is stable. Most slopes are stable (that is, $FS \geq 1$) and become progressively unstable with time (that is, FS falls with time). It is possible to reduce the FS and thereby cause a mass movement either by reducing the strength of the slope or by increasing the stresses acting upon it. Each of these will be examined in turn.

Slope strength

The strength of a slope depends primarily on the materials from which it is constructed and a distinction is made here between engineering soils and rocks. Engineering soils are defined as any unlithified sediment and may include both soil, as studied by pedologists, and unlithified rocks, as studied by geologists. Factors which determine slope strength in rock slopes are primarily a function of lithology and geological structure and are normally estimated in terms of rock mass strength. Traditionally, the strength of rock slopes is determined by extracting intact rock samples and measuring the compressive strength in the laboratory. However, a rock is only as strong as the weakest component within it and these weak links are normally provided by joints and discontinuities. As a consequence, a range of engineering classifications for rock slopes has been developed to estimate their *in situ* strength, joints and all. This approach was modified by Selby (1980) to produce a rock mass strength classification for geomorphological use (Table 16.2). Selby suggested there were eight factors which determine a rock's strength. These are:

- **Intact rock strength:** The strength of a rock between any joints, fissures or other discontinuities. This can be determined by extracting samples of intact rock and testing them in the laboratory, or in the field using a Schmidt hammer. This is a simple handheld device which drops a known weight from a known height and measures the rebound: the greater the rebound, the stronger the rock. Schmidt hammer values correlate well with measurements of compressive strength made in the laboratory. Some hard rocks are extremely strong and possess great cohesion, and the stability analyses of these rocks suggest that they should be very stable and that vertical cliffs in excess of 1000 m should be possible. For example, granite has a compressive strength of between 1000 and 2500 kg cm^{-2} and theoretically vertical cliff heights should be stable at between 4000 and 10 000 m. Quartzite has compressive strengths between 1500 and 3000 kg cm^{-2} and theoretically vertical cliff heights between 6000 and 11 000 m should be possible. These cliffs are never seen in nature because the rocks are jointed, faulted and bedded, and these weaknesses are more important in allowing failures to occur than the actual rock strength.

- **Weathering:** The more weathered a rock, the weaker it is. This can be estimated in the field using a visual classification of the degree of weathering, such as that shown in Table 16.3.

- **The spacing of partings:** A parting is a joint, fault, bedding surface or discontinuity within a rock. The more closely spaced its partings, the weaker a rock will be.

- **The width of partings:** Partings that are simply hairline fractures are the strongest, since the rough areas on either side are closely interlocked, developing a large amount of friction. Wide partings have little contact between their two sides and develop little friction, and are consequently weaker.

- **Continuity of partings:** Partings that link up to form a continuous network give a much weaker rock mass than those that do not join up.

- **Orientation of partings:** Partings that dip out of a slope may act as failure surfaces over which blocks may slide out of a slope. In contrast, horizontal surfaces or those that dip into the slope do not provide potential failure surfaces.

- **Outflow of water:** Partings that have a significant flow of water along them may be lubricated by the water and water pressure

Table 16.2 Rock mass strength classification and ratings. (Source: after Selby, 1980)

Parameter	(1) Very strong	(2) Strong	(3) Moderate	(4) Weak	(5) Very weak
Intact rock strength (N-type Schmidt hammer 'R')	100-60	60-50	50-40	40-35	35-0
	r = 20	r = 18	r = 14	r = 8	r = 5
Weathering	Unweathered	Slightly weathered	Moderately weathered	Highly weathered	Completely weathered
	r = 10	r = 9	r = 7	r = 5	r = 3
Spacing of joints	>3 m	3-1 m	1-0.3 m	0.3-0.05 m	<0.05 m
	r = 30	r = 28	r = 21	r = 15	r = 8
Joint orientations	Very favourable. Steep dips into slope. Cross-joints interlock	Favourable. Moderate dips into slope	Fair. Horizontal dips, or nearly vertical (hard rocks only)	Unfavourable. Moderate dips out of slope	Very unfavourable. Steep dips out of slope
	r = 20	r = 18	r = 14	r = 9	r = 5
Width of joints	<0.1 mm	0.1-1 mm	1-5 mm	5-20 mm	>20 mm
	r = 7	r = 6	r = 5	r = 4	r = 2
Continuity of joints	None continuous	Few continuous	Continuous, no infill	Continuous, thin infill	Continuous, thick infill
	r = 7	r = 6	r = 5	r = 4	r = 1
Outflow of groundwater	None	Trace	Slight (<25 l/min/10 m²)	Moderate (25-125 l/min/10 m²)	Great (>125 l/min/10 m²)
	r = 6	r = 5	r = 4	r = 3	r = 1
TOTAL RATING	100-91	90-71	70-51	50-26	<26

Table 16.3 A scale of mass weathering grades. (Source: after Selby, 1980)

Grade	Class	Description
VI	Residual soil	A pedological soil containing characteristic horizons and no sign of original rock fabric
V	Completely weathered	Rock is discoloured and changed to a soil but some original rock fabric and texture is largely preserved. Some corestones or corestone ghosts may be present
IV	Highly weathered	Rock is discoloured throughout. Discontinuities may be open and have discoloured surfaces and the fabric of the rock near to the discontinuities may be altered so that up to one half of the rock mass is decomposed and disintegrated to a stage in which it can be excavated with a geological hammer. Corestones may be present but not generally interlocked
III	Moderately weathered	Rock is discoloured throughout most of its mass, but less than half of the rock mass is decomposed and disintegrated. Alteration has penetrated along discontinuities which may be zones of weakly-cemented alteration products or soil. Corestones are fitting
II	Slightly weathered	Rock may be slightly discoloured, particularly adjacent to discontinuities, which may open and will have slightly discoloured surfaces. Intact rock is not noticeably weaker than fresh rock
I	Unweathered fresh rock	Parent rock shoing no discolouration, loss of strength or any other weathering effects

developed along them may help reduce friction, facilitating failure.

- **Infill along partings:** Partings that have roots or other vegetation along them, or are lined with clay or soil, may develop less friction along their length than ones that are clean.

Selby (1980) ranked these factors in order of importance and devised a scoring scheme with which to estimate the rock mass strength of a rock slope (Table 16.2). The idea is that a section of rock is integrated using this classification and its score is summed to determine its rock mass strength. This scheme has been widely used in geomorphology to estimate rock strength and appears to work well (See Box 16.2 Rock mass Strength Applications in the companion web site).

Where the dominant slope material is engineering soil, such as Quaternary sediments, organic soils, weathered regolith or clays, strength is determined by the properties of the slope. The strength of a soil is determined by two factors: (1) the angle of internal friction and (2) soil cohesion. Cohesion is not present in all soils and as a consequence soils can be classified as cohesionless or cohesive.

The angle of internal friction is a function of the way in which individual soil grains are packed and interlock. It is the angle at which a unit of soil can be tilted before it fails, and it varies with sediment sorting, grain size, sediment shape and particle packing. In contrast, cohesion is a force which binds individual soil particles and is derived in a range of different ways, which include:

- **Capillary tension:** If the pore spaces within a soil are not completely filled with water then a suction force is exerted, which tends to draw the soil grains together. It is this force which enables sandcastles to be built out of damp sand but not out of dry sand.

- **Electrical tension:** Clay minerals have a sheetlike structure that forms platelets, the edges of which carry small electrical charges. Opposite charges attract and therefore clay platelets may attract one another. In addition, a range of other chemical charges and bonds may contribute to cohesion, not all of which are fully understood.

- **Cementation:** Sediments that have undergone partial cementation may have strong chemical bonds binding them together.

- **Overconsolidation:** Cohesion can develop in clays that have previously been loaded. These clays have a much higher level of cohesion than is found in similar materials that have not been loaded. Loading may occur due to burial and subsequent removal of the overburden to by erosion, for example in the London Clay. Alternatively, loading may occur beneath a glacier, and most glacial tills are consolidated to some degree. Loading packs sediment more efficiently, drives out pore water and therefore brings clay minerals and other materials into closer contact, strengthening the bonds and attractive forces between them. Figure 16.4 illustrates that loading is only one

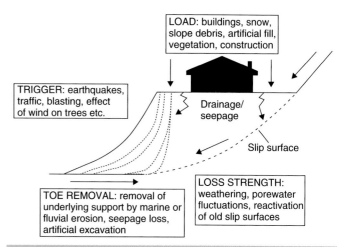

Figure 16.4 Diagrammatic representation of factors that can cause slope instability

factor of many that can cause failure, while Figure 16.5 illustrates the effects of logging forested slopes, which can also cause lower shear strengths in slope soils. The changes that can take place over time to turn a stable slope into an actively unstable one are shown schematically in Figure 16.6 and the factors that can cause mass movements are illustrated in Table 16.4.

It is important to note that the strength of some engineering soils, particularly clays, is affected by remoulding. A natural undeformed soil will have a peak shear strength and when the applied stress exceeds this level failure will occur. However, once failure has occurred the particles within the soil become remoulded and when the soil is loaded again it will fail more easily, at what is known as the residual shear strength. The importance of this is that mass movement deposits will fail more easily than the equivalent *in situ* material. This helps explain how shear planes or failure surfaces within an engineering soil may be reactivated by subsequent loading.

The strength of a slope can be reduced to two principal factors: (1) weathering and (2) variations in water content. Weathering reduces a soil's strength rapidly, since the surface area of sediment – and consequently the area susceptible to chemical attack – is much greater than that in lithified rock. Some of the major products of chemical weathering are secondary minerals, such as clay minerals, and consequently the amount of clay may change with time. Variation in water content is also critical, particularly in soils. In a dry unit of soil, the grains interlock to create intergranular friction. As moisture content increases, sediment cohesion may at first increase due to capillarity, but soil strength will ultimately begin to decrease as the soil becomes saturated. In a saturated soil, water in the soil's pores begins to push apart individual grains, reducing intergranular contact and therefore internal friction.

The changes in soil properties with increasing moisture content are measured via the Atterberg limits. The moisture content required to make a soil malleable is known as the plastic limit, above which it can be moulded in a plastic fashion. The moisture

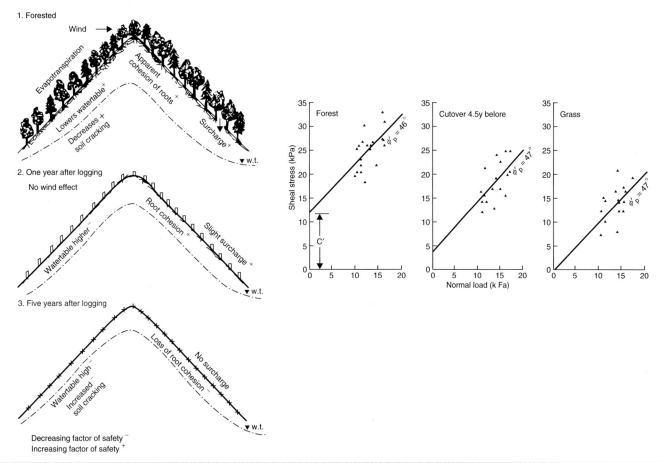

Figure 16.5 Effects of deforestation on the Factor of Safety with time. (Source: after Selby, 1993)

content required to make a soil flow as a liquid is known as the liquid limit. The difference between the plastic and liquid limits is the plasticity index, which is the range of moisture content over which a soil can behave as a plastic medium. Atterberg limits are frequently used to classify engineering soils. Cohesionless soils have low plastic and liquid limits, where only small additions of water are required to overcome the intergranular friction and liquefy the soil. In contrast, clay soils have much higher plastic limits and plasticity indices due to cohesion and require more moisture to make them behave in a plastic fashion, but remain plastic or 'active' over much wider moisture ranges. As shown in Figure 16.7, the Atterberg limits can be used to classify different types of soil movement.

In hardrock slopes, moisture variation may also help reduce strength by reducing friction along potential slip surfaces such as joints, faults and bedding surfaces. The importance of variations in water content within slopes explains the causal link between periods of heavy rainfall and mass failure. The link is particularly clear with shallow soil failures or slides. As an example, Figure 16.8 shows cumulative rainfall data for a selection of rainfall gauges near the Santa Monica and San Gabriel mountains in California during a period of heavy rainfall in January 1969. Also shown is the duration of shallow soil failures (debris flows and mudslides) during this period. Two things are apparent from these data:

1. There appears to be a threshold level of antecedent rainfall below which failure does not occur. This threshold, of approximately 10 inches, is the amount of water required before the soil slopes are sufficiently saturated to fail. It is evident from Figure 16.8 that these exceed the normal seasonal total.

2. Rainfall intensity in excess of 0.25 inches per hour is required to induce failure. Rainfall intensity is the amount of water in a given time. If it is low, the water pressures within the soil will not rise, since infiltration and throughflow losses will keep pace with the rate of input via the rainfall. However, if it exceeds the rate at which water can be lost by the soil, water pressures will rise, pushing apart the soil grains, decreasing intergranular friction and thereby leading to slope failure.

A similar point is illustrated in Figure 16.9, which shows data for mudflows/mudslides on the Antrim coast of Northern Ireland. Note the strong correlation between the movement of the mudslide and the rainfall record, which we refer to later in this chapter. This kind of relationship can be seen in Box 16.3 (see in the companion website), which shows the way landslides can be monitored and the relationship between landslide movement and infiltration of rainfall or rapid snowmelt increases.

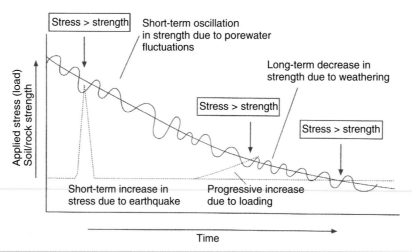

Figure 16.6 Diagrammatic view of changes that can turn a stable slope into an actively unstable slope

Table 16.4 Factors contributing to mass movement in soils

A. Factors contributing to high shear stress

Types	Major mechanisms
1. Removal of lateral support	(i) Stream, water or glacial erosion
	(ii) Subaerial weathering, wetting, drying and frost action
	(iii) Slope steepness increased by mass movement
	(iv) Quarries and pits, or removal of toe slopes by human activity
2. Overloading by:	(i) Weight of rain, snow, talus
	(ii) Fills, waste piles, structures
3. Transitory stresses	(i) Earthquakes – ground motions and tilt
	(ii) Vibrations from human activity – blasting, traffic, machinery
4. Removal of underlying support	(i) Undercutting by running water
	(ii) Subaerial weathering, wetting, drying and frost action
	(iii) Subterranean erosion (eluviations of fines or solution of salts), squeezing out of underlying plastic soils
	(iv) Mining activities, creation of lakes and reservoirs
5. Lateral pressure	(i) Water in interstices
	(ii) Freezing of water
	(iii) Swelling by hydration of clay
	(iv) Mobilization of residual stress

Table 16.4 (*continued*)

6. Increase of slope angle	(i)	Regional tectonic tilting
	(ii)	Volcanic processes

B. Factors contributing to low shear strength

Types	Major mechanisms	
1. Composition and texture	(i)	Weak materials such as volcanic tuff and sedimentary clays
	(ii)	Loosely-packed materials
	(iii)	Smooth grain shape
	(iv)	Uniform grain sizes
2. Physicochemical reactions	(i)	Cation (base) exchange
	(ii)	Hydration of clay
	(iii)	Drying of clays
	(iv)	Solution of cements
3. Effects of pore water	(i)	Buoyancy effects
	(ii)	Reduction of capillary tension
	(iii)	Viscous drag of moving water on soil grains, piping
4. Changes in structure	(i)	Spontaneous liquefaction
	(ii)	Progressive creep with reorientation of clays
	(iii)	Reactivation of earlier shear planes
5. Vegetation	(i)	Removal of trees
	(a)	Reducing normal loads
	(b)	Removing apparent cohesion of tree roots
	(c)	Raising of water tables
	(d)	Increased soil cracking
6. Relict structures	(i)	Joints and other planes of weakness
	(ii)	Beds of plastic and impermeable soils

In summary, therefore, weathering and variation in soil moisture content may reduce the strength of a slope.

Slope stress

There are a wide range of ways in which the stresses or forces driving slope failure can be increased. These can be divided into those which are transitory in nature and those which are more sustained. Transitory stress increases include ground vibrations, such as those caused by earthquakes. More sustained increases include the following:

- **Removal of lateral support:** This is achieved in most cases by removal of the slope bases, which may occur naturally via erosion or by artificial slope excavation (Figure 16.4). For example, erosion through the lateral migration of river meanders may erode a valleyside while marine erosion may steepen coastal slopes. This type of situation is well illustrated by coastal mass movements in southern England, which can be seen as an episodic process driven by coastal erosion, as illustrated in Figure 16.11. This is a particular issue in urban areas built on steep slopes, which are consequently dependent on these slopes remaining stable. There is often a marked contrast in the intensity of rainfall incidents in a concrete urban area compared with a vegetated rural area (Figure 16.12) and usually drastic stabilization measures have to be taken, as in Hong Kong (Figure 16.13).

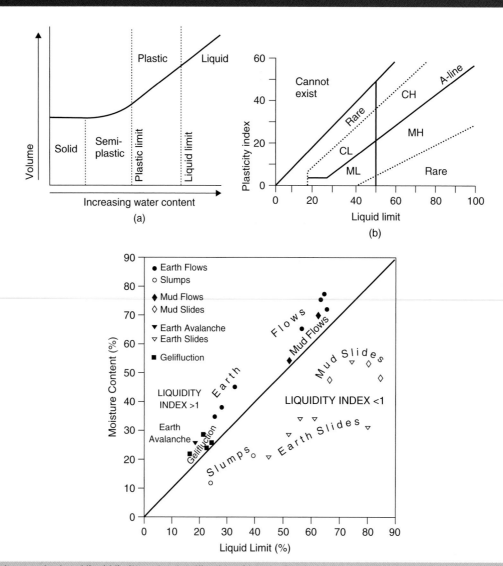

Figure 16.7 Moisture content and liquid limits and a classification of types of soil movement. A-line separates inorganic clays from silty and organic soils and is the Activity of a soil. C refers to clay, M:silt, H:high and L:low.

- **Removal of underlying support:** This involves the removal of underlying layers, leaving the ground surface unsupported. This type of process is particularly well illustrated by cantilever failures along some river banks (Figure 16.14(b)). The surface soil is bound together by grass roots and humus, while the subsoil is not. Water flow along the bank can erode the subsoil faster than the surface soil, which in time is left unsupported, leading to failure. Removal of underlying support is also common in certain ground water conditions. For example, solution of carbonate rocks beneath overlying soils or cap rocks may lead to subsidence and mass failure (Figure 16.14(a)). At a more subtle level, water discharge from springs or along junctions between permeable and impermeable strata may also lead to removal of underlying support or slow subsidence on sandy soils, or dropout failure on clay soils (Figure 16.14(c)), or where shakeholes or sinkholes are created in limestone areas by the solution of joints below an overlying drift cover.

- **Slope loading:** Loading of slopes, either naturally or unnaturally, is a major factor in driving mass failure. Increasing water content will increase load, as will snowfall. Similarly, the downslope movement and accumulation of soil may gradually vary with the distribution of load experienced on a slope. Vegetational changes also have an important role to play: for example, as trees grow, the load they impose on a slope increases. However, the most effective way in which a slope can be loaded is through construction, and this is a serious cause of mass movement in urban areas built on steep terrain.

- **Lateral pressure:** This involves outward pressure, forcing blocks or units of rock or soil outwards. An example would be the penetration of cracks by roots which then grow, wedging the cracks apart and imposing an outward force (Figure 16.15). A similar situation is provided by the process of freeze–thaw. As water-filled cracks freeze and the water turns to ice, it expands by 9% in volume. This type of process is linked to

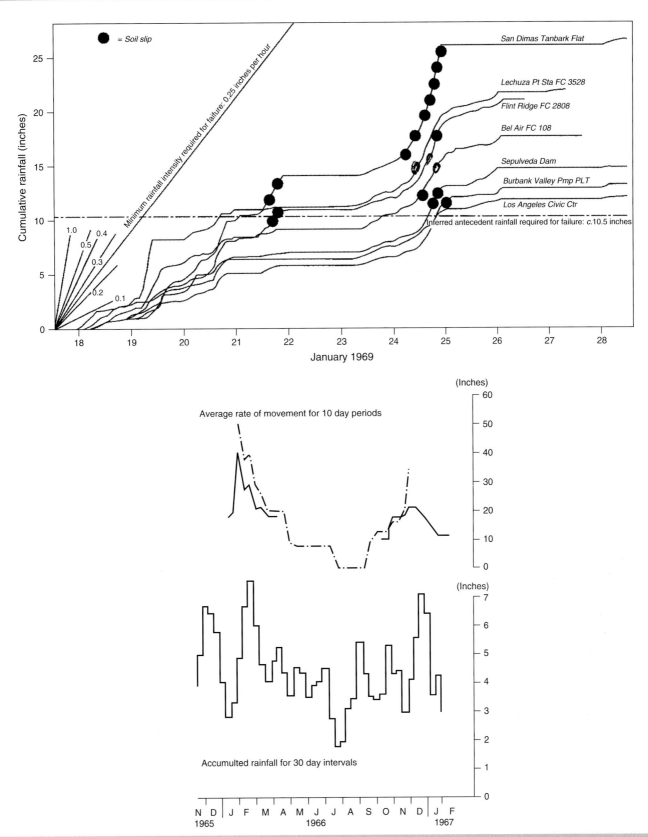

Figure 16.8 Cumulative rainfall at and near the Santa Monica and San Gabriel mountains (California), showing times of nearby slope failures, 18–29 January 1969. (Source: Campbell, R.H. 1975 Soil slips, debris flows and rainstroms in the Santa Monica Mountains and vicinity, southern California.United States Geological Survey Professional paper 851.)

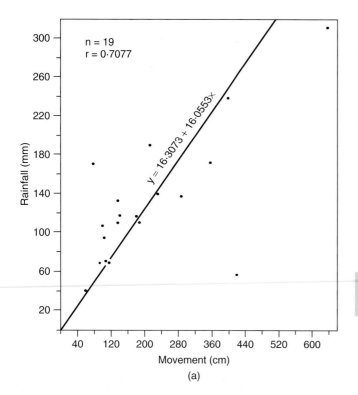

n = 19
r = 0·7077

y = 16·3073 + 16·0553x

(a)

D. B. PRIOR, N. STEPHENS, AND G. R. DOUGLAS

(b)

Figure 16.9 (a) Antrim mudflow movement and rainfall events. (b) Composite mudflows, Minnis North, County Antrim, Northern Ireland. (Source: after Prior et al., 1968)

Figure 16.11 Coastal mass movement east of Lyme Regis, near Black Ven, Dorset. (Source: Wikimedia Commons, GNU Free Documentation Licence)

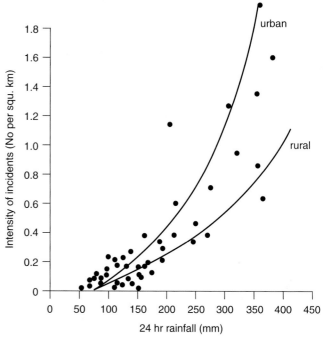

Figure 16.12 Relationship between intensity of rainfall incidents and rainfall totals in urban and rural areas

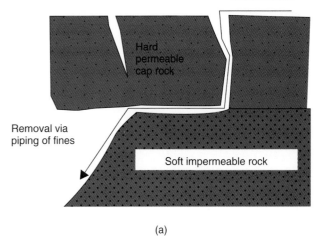

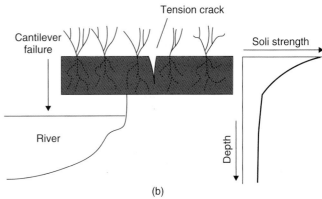

Slow subsidence on sandy soils/sediments

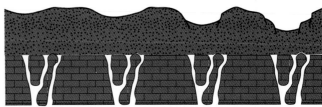

Dropout failure on clay soils/sediments

(c)

(a)

(b)

Figure 16.13 Slope stabilization techniques in Hong Kong urban areas: (a) warning sign; (b) concreting of slope and runoff channels

Figure 16.14 (a) Cantilever failure along river banks. (b) Solution below hard cap rock. (c) Shakehole/sinkhole development from solution of limestone along vertical joints

Figure 16.15 Wedging by tree roots, Chalcatzingo, Central Mexico

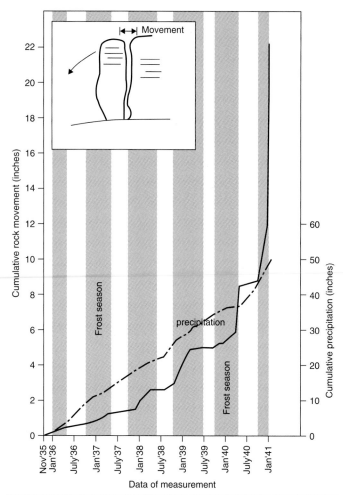

Figure 16.16 Threatening rock, showing time against cumulative movement

frost weathering and can be an important factor in rockfall. A particularly good example is provided by the famous failure of Threatening Rock in Chaco Canyon National Park in New Mexico. Figure 16.16 shows the cumulative movement of the rock tower prior to failure and notes the causality between periods of movement and the months in which freeze–thaw cycles occurred.

The principal causes of mass movement are illustrated in Figure 16.17. This is a subject best viewed in terms of those processes that reduce slope strength and those processes that increase the stresses acting upon them.

Having examined the principal causes of mass failure, the main types of failure will now be presented.

16.2.3 Types of mass movement

Mass movements can be classified in a range of different ways. In this section they are classified on the basis of the type of rock/soil involved and the process of movement. As Figure 16.17 shows, material may move as: (1) a series of step-like movements,

termed heave, in which individual particles move independently of one another; (2) sliding or free fall (in the case of sliding, all the movement occurs at one level, usually along a slip plane); (3) a flow, in which pore water between individual soil or rock particles is sufficient to move the sediment as a plastic or liquid mass. Rocks and soils which are dry tend to move either by heave, slide or fall, while moisture or water is essential to most types of flow. These elements can be used as a basis for classification. The three main types of movement define the three corners of the triangular classification matrix, while moisture content and the speed of the movement are used along two of the axes. The main types of mass movement are plotted on this classification, but it is important to note that a range of different terms are used within the literature to describe similar mass movement or subtle varieties thereof, and that most large mass movements will be composite in nature, involving more than one of these processes. The main types of mass movement defined here are:

- soil creep

- granular disintegration, rockfalls, toppling failure, slab failure and rock slides/avalanches

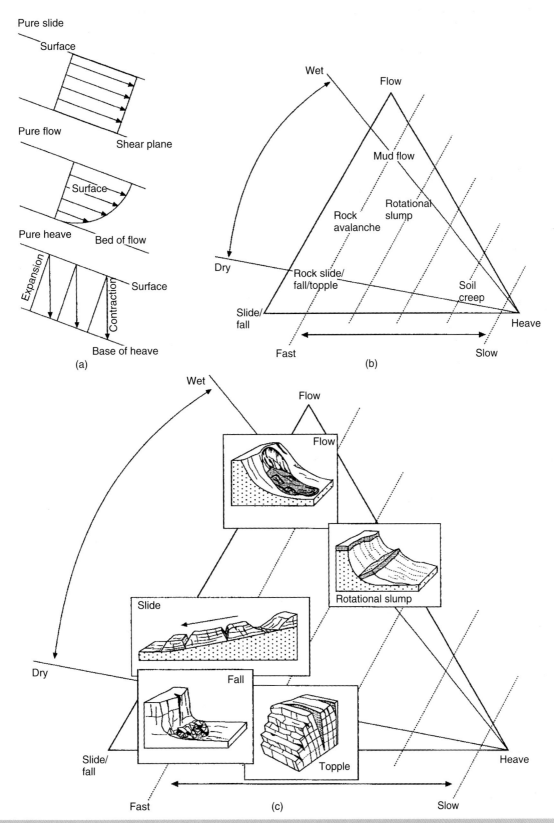

Figure 16.17 Classification of mass movements

- soil falls, topples and slides

- soil flows: debris, Earth and mud flows

- rotational slumps.

Soil creep

This is the slow downslope movement of individual soil particles under the influence of gravity. It is imperceptible except to long-term measurements and there is little mechanical disruption of the material, which can form small-scale soil creep terracettes (Figure 16.18). Creep in the rheological sense (rheology being the science of flow and deformation of material) implies a continuous deformation without failure under the influence of a steady stress. Soil creep may sometimes be this sort of movement, but this is not always the case. It can be caused by soil moisture and temperature variations, random soil movement by animals or microseisms, as well as by the steady application of a stress. Creep may be divided into two movement types, although they often occur together:

- **Continuous creep:** This commonly occurs before and after a major landslide or other sort of mass failure. It is often part of the process of progressive failure and the sort of processes responsible for this kind of creep are frequently also those responsible for progressive remoulding of overconsolidated clays. The process is more or less confined to clay soils or soils with an appreciable clay content, and is almost completely absent in sandy soils. Movement can take place on slopes as low as 9°. A comparison between sandy and clay-rich soils can be seen in Figure 16.19. With sands, the soil either fails with rapid deformation or does not fail at all, with nowhere in between, whereas with clays the rate of deformation is strongly dependent on the applied stress. Movement can take place between the lower yield strength and the upper yield strength without any failure taking place.

- **Seasonal creep:** This is where movements are triggered by seasonal or climatic variations in the soil. They are usually related to temperature changes, which produce expansions and contractions such as freeze–thaw and soil moisture changes. Solifluction could be included here, referring to water-lubricated sediment movement. Periglacial gelifluction (the periglacial equivalent) is discussed in Chapter 21.

- **Nonseasonal or random creep:** This can occur by the random displacement of particles by roots or animals. Root holes, animal holes, worm casts and molehills all involve a preferential downslope movement (Figure 16.20), which is reinforced when they collapse. Although this is often considered to be a fairly insignificant process, an example from some Malaysian slopes shows that a species of palm tree completely reworks the whole hillslope area to a depth of over 1 m every 5000 years, through processes of root growth and tree collapse. This involves a very rapid rate of creep downslope. However, 'creep'

Figure 16.18 Soil creep terracettes, Northern Pennines, UK

is not a good term to use in this case because in the physical sense it has nothing to do with rheological creep.

Creep has often been described as a heave process, in which slope expansion occurs normal to a slope, while slope contraction under the influence of gravity tends to occur more vertically (Figure 16.21(a)). Slope expansion is caused by wetting and drying, or by the growth of frost crystals. In practice, however, soil creep probably occurs via a range of processes, including small soil slips and flows, as well as displacement by burrowing organisms. One of the problems for understanding the detailed mechanics of soil creep is the lack of long-term observations on the rate of soil creep from a range of different environments. Figure 16.21 shows some experimental data. In theory, soil creep should increase with the slope angle, although this relationship has not so far been supported by experimental observations. Data from volcanic soils suggest that the type of vegetation cover may also be important. In the case of the shrub vegetation, the maximum root density occurs at a depth of about 25 cm, explaining the retardation of creep at this point. The higher rates under grassland may also reflect the lower bulk density of this soil, which is not loaded to the same degree as that with shrubs on it.

The classic qualitative indices for the presence of soil creep on a slope are summarized in Figure 16.22. However, there are objections to all of these indices, to the extent that none prove conclusively that overall soil creep is taking place, and furthermore creep may occur without these indices being present at all. The result has been attempts at quantitative assessments of soil creep, as theoretically rates of creep ought to be measured throughout the whole soil profile. However, the techniques involve placing markers in the soil through digging pits and an assumption has to be made that whatever is inserted moves with the soil (like marker columns of wood or clay). If there is a tendency for movement to occur round the marker then the rate of creep can be underestimated. Young pits, named after the originator, where nails or pins are inserted in the pit side, can theoretically be excavated any number of times since no disturbance of the

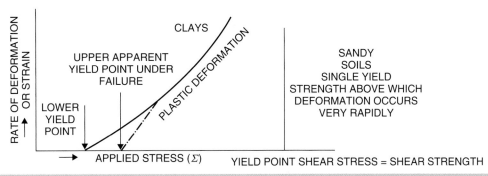

Figure 16.19 Applied stress and rate of deformation for clay-rich and sandy soils. Failure in sands/clays

Figure 16.20 Breakup of molehills downhill on a slope, close to the path to Long Churn cave, Yorkshire Dales, UK

soil in which the pins are inserted occurs. In practice though considerable soil water flow disturbance can be imagined and it is also possible that the pins could allow the soil to flow past them. No short-term measurements are possible here and it has been estimated that the errors in making the measurements may be of the same order as the soil movements. Much better are soil creep gauges such as strain gauges, linear motion transducers and the like, but again the greatest problem in measuring such small changes in movement is the potential disturbance from the instrumentation.

The importance of soil creep is that it is a low-magnitude process, but one which occurs very frequently, and as a consequence is geomorphologically very significant. Rates decline with depth because the soil is anchored at the base due to interlocking with the bedrock. Values of $1-6\,cm^3$ per cm per year seem to be normal, except where continuous creep extends to some depth. British upland soils seem to show movements of $1-3\,cm^3$ per cm per year. In deep Californian soils in the Berkeley Hills there have been recorded figures of $650\,cm^3$ per cm per year, but these are exceptional. Soil creep is very variable because the processes causing it are variable, and some very small movement rates have been suggested, such as wedging by grass roots and refilling at $0.003\,cm^3$ per cm per year, worms (burrowing and refilling) at $0.15\,cm^3$ per

cm per year and rabbits (burrowing and refilling) at $0.1\,cm^3$ per cm per year. How realistic these figures are is debateable. However, on landslide-stable slopes with good vegetation cover so that rainsplash and wash processes are absent, soil creep may be the only significant slope process. On other slopes, other processes may seem to dominate soil creep in terms of hillslope form, but the fact is that creep is important because it is ubiquitous.

Granular disintegration, rockfalls, toppling failure, slab failures and rockslides/avalanches

Granular disintegration occurs because some rocks do not possess a small-scale fracture pattern and so there is no rockfall. Such rocks are weak mechanically; see for example sandstones in the south-western United States, where any rock strength is due to the cement binding the grains together. When weathering occurs there is a very slow release of individual grains from the surface through the destruction of the cement bond. This usually occurs through solution of the cement and leaves the grains susceptible to gravitational force. This occurs particularly when the rock dries out after wetting and loses cohesion, which was caused by capillary water bonding. There can also be granular disintegration by thermal effects, when there is differential absorption of heat by different minerals, or by frost action and the impact of falling water (rainbeat).

Rockfalls involve the freefall of individual particles of rock under gravity. These can vary from small rock fragments to blocks several metres in diameter. The size of particle involved depends primarily on the spacing and orientation of joints within the rock mass. Rockfall is favoured by a densely-jointed rock mass of intersecting joints which dip out of the slope. The rockfalls occur from a very narrow, superficial part of the rockwall. Rockfall is particularly dominant in areas of active freeze–thaw and this is often the principal mechanism driving failure. This can be seen in Figure 16.23, showing a plot of chalk falls, frost and effective rainfall on the Kent coast, and a plot of rockfalls in Norway against the time of year in which they occurred, with peaks in April and October, when there are continual changes of temperature about the freezing point.

The importance of insolation weathering in creating rockfall is also emphasized in various parts of the world. In County Antrim, Northern Ireland, rockfall events could be arranged

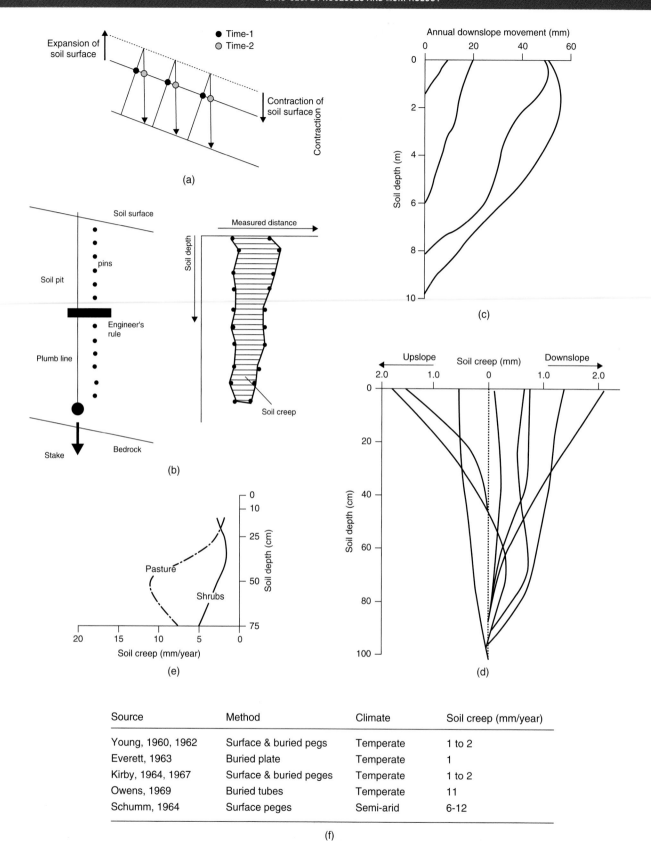

Figure 16.21 Movement in soil creep: (a) movement during soil creep; (b) Young pits before measurement and soil creep movement after two readings; (c) and (d) downslope movement and depth; (e) soil creep under pasture or shrubs; (f) observed rates of soil creep

The classic qualitative indices of the presence of soil creep can be briefly summarised in a diagram as follows:

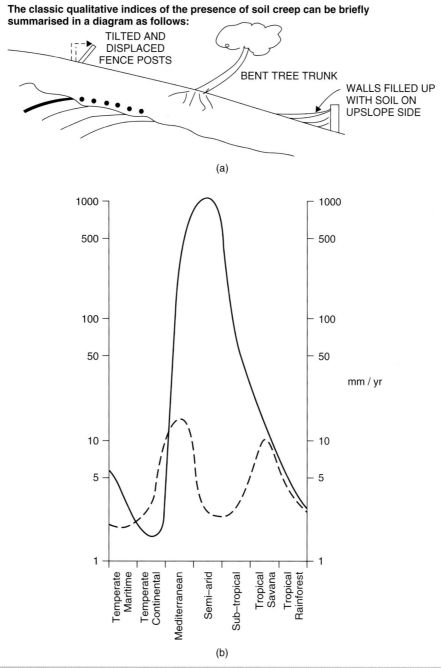

(a)

(b)

Figure 16.22 (a) Qualitative indices for soil creep. (b) Variation with climate of rates of soil creep and surface wash. Soil creep rates are indicated by the pecked line and surface creep by the continuous line

into two categories: those over 0.21 kg, which were seasonal, and those under 0.21 kg, which occurred the whole year round. The latter accounted for over 60% of all the events and therefore another process apart from freeze–thaw must be taking place. In Ellesmere Island in the Canadian Arctic, rockfall events were found to preferentially occur on south-facing slopes where the temperature sometimes reached as high as 39.7 °C, and therefore in both cases it is suggested that insolation weathering has a role to play in generating at least a significant number of rockfall events.

Usually a coarse, bouldery sediment is produced from rockfall, although the falls may land directly in a transit medium like a river, the sea or a glacier, and be transported away (Figure 16.24(a)). The end product though is often a scree slope. Screes are generally accumulation slopes (Figure 16.24(b),(c)), where the amount of sediment increases through time. However, if the lower boundary is a transit medium then an equilibrium form might be reached where the outputs equal the inputs. In this case the scree is a transit slope and the quantity of sediment is constant through

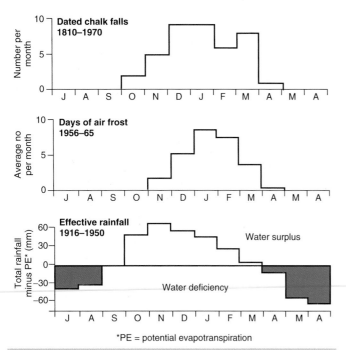

*PE = potential evapotranspiration

Figure 16.23 The incidence of rockfall, frost and effective rainfall on chalk, Kent coast. (Source: after Hutchinson, 1972)

angle of slope lower than the repose slope. The concavity at the base of the scree is caused by faster particles running out as a tail onto the accumulation surface. There is sorting due to the sievelike nature of the scree surface, and starting with a random surface of uniform size distribution, the small particles get trapped more easily by interlocking, whilst the larger particles slide over the scree surface and travel to the base of the slope (Figure 16.25).

There is a size sorting of the particles on screes and in general most screes show a downslope increase in grain size (Figure 16.25). This relationship has been examined both in the laboratory and in the field, to test the hypothesis that there is a cyclical development of sorting on screes depending on the headwall height and the frictional relative roughness to moving particles. Initially, sorting should be poor, the scree slope short and the headwall high.

time. There are therefore several possibilities for scree formation:

- **Rapid rockfall input and rapid output:** In this type the output causes avalanching, where all the particles move as a unit and the result is a straight scree at the angle of repose for the material falling. These screes are rare.

- **Low rockfall input and rapid output:** Again there is a straight scree slope formed, but this tends not to last very long as the low rockfall input cannot keep pace with the rapid output.

- **Low rockfall input and no output:** These screes are the most common type but they do not possess characteristics compatible with the angle of repose model. They show a basal concavity, a straight scree slope under the angle of repose and a coarse downslope sorting. The angle should be c. 39–42° but the screes are usually at a maximum slope of 32–35°; that is, 4–10° less than the angle of repose. This means that the screes must be very stable to mass avalanching. The straight section cannot be considered as a repose slope. There is also a basal concavity and the repose slope should be straight. The reasons for this difference in form are that these screes are formed by one-at-a-time rockfall inputs, not by mass avalanching. The boulders have some input energy (mass times the distance they have fallen) when they hit the scree surface. Some of this energy is transferred into downslope movement. However, the angle of repose is related to the coming-to-rest of a sliding mass of particles, which has a velocity much less than the downslope velocity of a falling boulder. Therefore the boulder moves itself and others by impact downslope and keeps the

(a)

(b)

Figure 16.24 (a) Chute slopes on Storglacieren and the rapid transport away of rockfall material. (b) Gordale screes, Yorkshire Dales, UK, where the screes are accumulation slopes. (c) Accumulation slope in Leirdalen, Central Norway. (d) Rockfall in Jurassic Limestone, St Alban's Head, Dorset, UK. Note the reactivation of fossil screes by the sea undercutting

(c)

Figure 16.25 Sorting on scree, Barf, Lake District, UK

(d)

Figure 16.24 (continued)

Even the finest rockfall particles have sufficient momentum to reach the base of the scree. There is then an intermediate stage where sorting reaches an optimum as there is a balance between the initial momentum and friction on the scree becomes finely adjusted. The third stage occurs as momentum diminishes as a result of the reduction in the headwall height and sorting is reduced. This model was tested in the field on Cader Idris, where nine screes were surveyed, with six sampling sites on each scree 20–80 m apart. The a, b and c axes were measured on fifty particles. The screes appeared to form four distinct groups, defined by the lithology, the height of the headwall, the length of the slope and the degree of concavity of the footslope. Particle size was compared with distance downslope, and for three of the four groups there was a positive correlation. The one that did not show a significant correlation had a high headwall and a short slope that was in the early stages of development, when the degree of sorting is limited.

Marked differences were noted when the Black Mountain scarp scree slopes were investigated. The steep and straight scree slopes that we have discussed were being replaced by a series of low-angled, concave debris flow cones from gullies. It was suggested

that these had probably been initiated by sheep-grazing-induced pressure in the Middle Ages. All the sediment movement down the gullies is by debris flows, which are initiated by heavy rainfall events. Note that not all such events cause debris flows, although they do act as a trigger. For debris flows to form there must be a minimum volume of sediment and therefore the slope development is controlled by the rate at which sediment is produced by gulley-side lowering.

Spitsbergen also shows excellent examples of scree talus accumulation (Figure 16.26). On Griegaksla Mountain the openly-exposed, west-facing slopes are dominated by high-frequency, low-magnitude rockfalls, which are caused by variations in direct solar radiation. On the east-facing slope, lying in the shadow during most of the day all year round, the slope processes are dominated by a lower frequency of rockfalls but a higher frequency of debris flows and slush avalanches, which are mainly connected to rainfall events. Elsewhere in central Spitzbergen, debris flows are common where there is a concentration on east- and north-facing slopes and/or in narrow valleys in which the slopes lie in shadow. A shallow active layer, up to a metre thick, with large amounts of frost and ice to melt above the frost table, causes plenty of melt water during the summer, lubricating the debris on the slopes. The trigger is high-intensity rainfall events. The morphologies of screes formed by these processes show major differences from those formed by rockfall. They have lower slope gradients, boulder lobes and tongues, channels and levees, hummocky debris flow deposits, an absence of upslope decrease in grain size, basal accumulations of fine sediment fans as low-energy flows wash the finer sediment from the sediments and a wide range of particle sizes at any given point, giving poor sorting (Figure 16.27).

Rock topples, or toppling failures, involve the rotation of the falling block around a pivot point and occur at a range of different scales (Figures 16.28 and 16.29). There is collapse along vertical fissures or joints, or along vertical bedding planes where rock has been tectonically deformed. Here frictional strength does not operate and the block is slowly wedged out by a range of processes, such as freeze–thaw. At a small scale, this process may involve the

Figure 16.26 Debris flow slopes, Longyearbyen valley, Spitsbergen. Note the relatively steep slope, debris flow tracks and lobes. These occur in Tertiary bedrock, which supplies the fine-grained sediment which weathers to mud as the matrix for the flows

slow deformation or creep of rock strata under the influence of gravity or the movement of much larger blocks. Threatening Rock in Chaco Canyon, New Mexico, provides an excellent example; its movement was measured away from the face from 1935 until the threat became too great and it fell in 1941 (Figure 16.16). Examples from Britain are caused by glacial undercutting, such as Eagle's Rock in Glenade, Sligo, and Nant Peris in Snowdonia. River undercutting has caused small examples along the Wye valley and in Swaledale (Figure 16.29). At a much larger scale, rock topples may involve large blocks of rock moving downslope from a cliff or valley side. This is well illustrated by Peak Scar in the Hambleton Hills. Peak Scar overlooks a river valley and consists of the Lower Calcareous Grit Formation (limestones and sandstones), which overlies the Oxford Clay. A large mass movement block is located in front of the cliff and appears to have rotated out of the slope as a large rock topple. This movement was probably facilitated by two factors: (1) a well-jointed rock mass in the form of the Lower Calcareous Grit Formation and (2)

a ductile substrate, the Oxford Clay, over which the block could move under the influence of gravity towards the valley floor.

Slab failure occurs where a mass of rock slips along a bedding plane, in which one bed slides over another. They can occur where the macrojoint pattern is more important than the microjointing. Friction is the major element of rock strength involved but water or clay can be washed in along the bedding plane to reduce this strength. Undercutting is needed by a river, a glacier or the sea, and the rock needs joints and bedding planes. Examples from Norway are shown in Figure 16.30. A good example in Snowdonia occurs in Cwm Graianog, where on the north wall the Festiniog Grit has failed, one slab failure being 200 m high over a width of 60 m.

A rock slide involves the movement of a large rock block over a shear plane, which usually follows a structural plane in the rock mass, such as a plane of foliation, a bedding plane or a master joint. If the failure surface is planar, it occurs rapidly and involves deformation of the rock mass then it is referred to as a 'planar slide'. One of the most famous examples of a planar slide was the Vaiont Dam Failure in Italy in 1963. In this case, a slide occurred parallel to the direction of bedding and within previous mass movement deposits. The debris entered a reservoir dammed by the Vaiont Dam, causing a displacement wave, which overtopped the dam by over 100 m, causing a major downstream flood, which in turn caused widespread destruction and the loss of 2600 lives. The Turtle Mountain failure in Alberta also provides an example of a planar slide, in this case occurring parallel to both bedding and joint surfaces (Figure 16.31(a)). A similar rock slide from Guerrero, Mexico, is shown in Figure 16.31(b).

When the rock mass is rafted on weak horizons such as soft clay beds, the failure is commonly referred to as a rock glide. In clay-rich rocks such as mudstones, planes of failure are commonly curved and the resulting rotational movement is referred to as a slump and is similar to rotational slumps in soil.

Some rocks are cut through by less systematic joint patterns that are almost random. They can undergo rock avalanching, where failure takes place along the variable weakness planes in the rock. Often this involves a thick wedge of rock at failure, and not just the surface layer as in rockfall. A wide zone of the rock mass can be affected, as in the Mount Huarascaran rock avalanche in Peru in 1970. This rock avalanche was earthquake-generated and involved rock, snow and ice, a mass of 2 million m^3, which fell 650 m vertically. It then moved 14.5 km on a 23° slope at 400 km per hour and then another 50 km on a 5° slope at 25 km per hour. The Mount Sherman avalanche in Alaska in 1964 was also earthquake-generated but was much bigger at 30 million m^3. It moved fast too, at between 80 and 300 km per hour. The mechanics of these fast runout avalanches are currently poorly understood, but one possibility is that they incorporate a basal air layer which fluidizes the base of the avalanche and flows in a way similar to a hovercraft. This would account for the lack of erosion noted in the areas over which they have moved.

The Kolka–Karmadon rock and ice avalanche occurred on the northern slope of the Kazbeck massif in North Ossetia on 20 September 2002, following a partial collapse of the Kolka Glacier. It started on the north-north-east wall of Dzhimarai-Khokh (4780 m above sea level) and seriously affected the valley of

Figure 16.27 (a) Debris flow lobes, the south side of Morsadalur, Iceland. (b) Debris flows, Southern Iceland. (c) Debris flows and rockfall run-out, Southern Iceland. At least two hillslope processes are operating on this slope, which is the case on most slopes. This means that process–form relationships are complex. (d) Debris flow track on screes, Gasgale Ghyll, Lake District, UK, on Skiddaw Slate. (e) Small debris flow lobes on scree, Grasmoor, Lake District, UK, on Skiddaw Slate

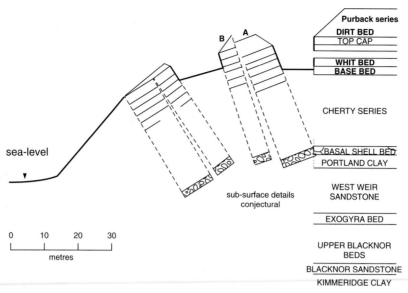

The Great Southwell Landslip, Isle of Portland: and example of toppling failure (after Bromhead. 1986).

(a)

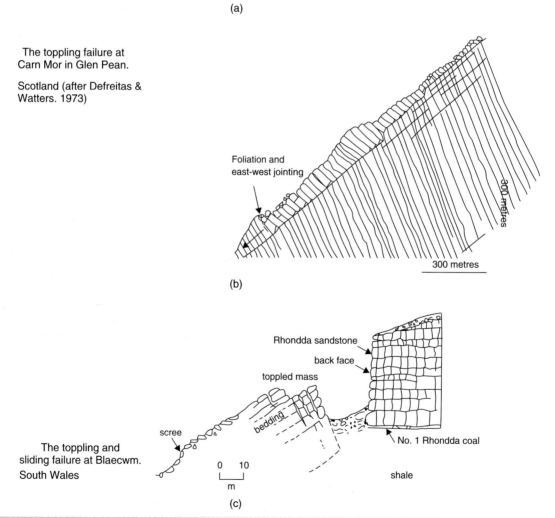

The toppling failure at Carn Mor in Glen Pean.

Scotland (after Defreitas & Watters. 1973)

(b)

The toppling and sliding failure at Blaecwm. South Wales

(c)

Figure 16.28 Toppling failure examples: (a) the Great Southwell Landslip, Isle of Portland; (b) Carn Mor in Glen Pean, Scotland; (c) toppling and sliding failure at Blaecwm, South Wales

(a)

(a)

(b)

Figure 16.29 Toppling failure examples: (a) Swaledale in limestone; (b) Symonds Yat, Wye valley, in limestone

Genaldon and Karmadon. The resulting avalanche and mudflow killed 125 people. Debris and ice filled the Genaldon Valley from the Kolka Glacier Cirque to the Gates of Karmadon, 18 km away. A 150 m thick chunk of the Kolka Glacier travelled 20 miles (32 km) down the Karmadon Gorge and Koban Valley at over 100 km per hour. The outflow of mud and debris measured 200 m wide and 10–100 m thick. Two villages along the gorge were under surveillance as flood waters backed up along the choked rivers. It finally came to rest in the village of Nijni Karmadon, burying most of the village in ice, snow and debris. A more detailed discussion of similar earthquake-triggered rock avalanches triggered by the November 3rd 2002 Denali Fault earthquake, Alaska, USA is given in Box 16.4 (see in the companion website). An example of a rotational rock slump is provided by the Folkestone mass movements on the south coast of England. Marine erosion at the base of the coastal slope has driven a series of rotational slumps developed in the Gault Clay, which has been weakened by the seepage of ground water through the overlying glauconitic sandstones.

The distinction between rock slides and topples is often very subtle and in practice large mass movements may involve both components. At a smaller scale it is possible to distinguish the

(b)

Figure 16.30 Slab Failure, Norway: (a) Tunsbergdalen, Central Norway, in mica schist with the sliding taking place because of the well-developed planar surface dipping out of the slope and the vertical jointing; (b) Naeroyfjord, location of large-scale slab failure, which is now in the fjord. The obvious failure plane is notable

susceptibility to sliding versus toppling on the basis of the relationship between a block's dimensions and the inclination of the failure surface. Wide, low blocks on shallow surfaces will tend to slide, while tall, narrow blocks on steeper surfaces will tend to topple. The conditions for stability, sliding and toppling failure of a block are shown in Figure 16.33.

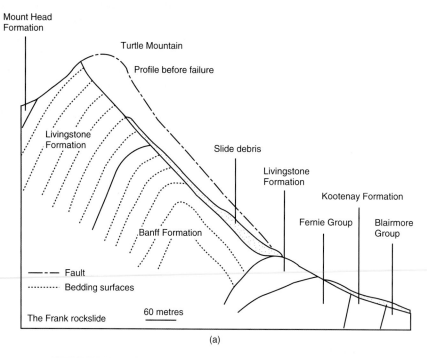

(a)

(b)

Figure 16.31 Rockslides: (a) Frank rockslide, Turtle Mountain, Alberta; (b) Guerrero, Mexico, August 1989. (Source: Wikimedia Commons)

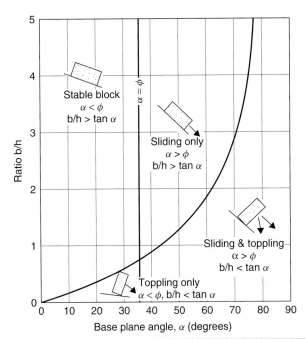

Figure 16.33 Conditions for stability, sliding and toppling failure of a block. (Source: after Hoek and Bray, 1977)

Soil falls, topples and slides

Falls, topples and slides may also occur in engineering soils. Examples of falls and topples are common along eroding river banks and on low coastal cliffs. Frost action and basal scour are often the critical factors in driving this type of soil failure.

Translational slides are by far the most common form of failure in soil slopes (Figure 16.34). They involve the movement of a shallow layer of soil and invariably have a planar slip plane located at the rock, or along the boundary between soil layers that have different densities and permeabilities. This type of failure is commonly associated with heavy rainfall and is common in tropical urban regions, such as Hong Kong and around Rio de Janeiro. In the context of Hong Kong, translational slides are a major problem and drastic attempts to counteract the problem are often undertaken (Figure 16.35). Here the number of movements is a function of not only the rainfall intensity but also urbanization. Rainfall intensity determines soil saturation. If the input of water to a soil via rainfall exceeds that which can drain away via infiltration and throughflow, the soil will become saturated and fail. Equally, urbanization concentrates water along drainage lines and, more importantly, it provides an increased number of steep unstable slopes in the form of cut and fills on slopes.

Rotational slumps involve failure along curved slip planes and are common in clay soils around the coast in south-east England. Similar failures also occur in more competent rocks, as discussed earlier. They have a well-defined arcuate scar in which a series of back-tilted slump blocks are located. The toe of the slump may become disturbed, forming a sediment flow.

Drainage is often ponded by the slump blocks, leading to the development of secondary sediment flows, which may play an important role in the downslope movement and destruction of the slump blocks. This type of failure is a particular feature of coastal slopes in south-east England, where the process is driven by coastal erosion. Similar rotational slumps are common in shallow landslides in weathered sediment in upland Britain, where there are steep slopes. These can occur in weathered till, as in Glencullen, County Wicklow, Ireland, or in weathered till and/or periglacial slope deposits, as in the Lake District and the northern Pennines (Figure 16.36). The movement almost always takes place at the soil–bedrock junction, which is the weakest part of the profile, as soil water and clay can get into the soil and lubricate this usually impermeable junction.

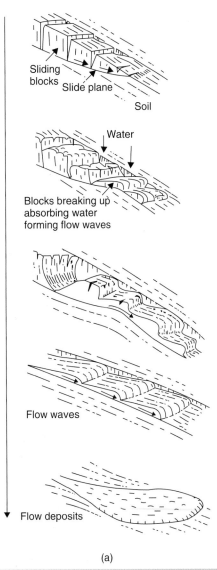

(a)

Figure 16.34 Translation slides: (a) transition from a translational slide to a soil flow; (b) styles of flow; (c) comparison of debris and river flows

Fluid flow

Plug flow

Sliding flow

Cross-section
Levée   Slow   Fast
Channel

Cross-section
Levée   Plug   Shear zone
Channel

Side elevation

No channel

Increasing water content

Increasing flow velocity

(b)

Flow	Sediment load by weight (%)	Bulk density (Mg/m3)	Shear strength (N/m2)
Stream flow	1 to 40	1.01 to 1.3	<10
Hyperconcentrated flow	40 to 70	1.3 to 1.8	10 to 20
Debris flow	70 to 90	1.8 to 2.6	>20

(c)

Figure 16.34 (*continued*)

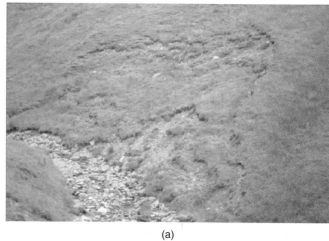

(a)

(b)

Figure 16.35 Attempts at slope stabilization, Hong Kong

(c)

Figure 16.36 Rotational slumps: (a) in the Pennines, caused by stream undercutting; (b) old and new slumps in the northern Pennines, again caused by stream undercutting; (c) large rotational slump, Skiddaw massif, north of Dash Beck, English Lake District

Soil flows: Debris, Earth and mud flows

Flows occur when coarse debris, fine-grained soil or clay is liquified, when the moisture content exceeds the liquid limit of a given soil. There are a plethora of confusing terms to cover this type of mass movement. 'Debris flow' is generally used to cover coarse-grained flows, 'Earth flow' the medium-sized fraction and 'mudflow' the finest fraction. In practice, however, all these terms have been used interchangeably. There are soil flows too.

A soil flow is a mix of sediment and water and can be seen as one end of a continuum in which river water forms the other extreme. The dynamic and sedimentary properties of a soil flow depend on the grain size of the material and the relative proportions of sediment to water. Fluid soil flows tend to have high flow velocities and be thin, and are often erosive. At the other extreme, soil flows may creep slowly forward, be relatively thick, bulldoze material in front of them and be nonerosive. The water content and sediment type also control the type of flow mechanism. Soil flows may behave as a Newtonian fluid, like water, or alternatively as more of a more plastic medium, such as a Bingham fluid.

Flows are derived in a number of ways. They commonly form as a consequence of translational soil slips, where the dislocated block becomes disaggregated to form a fluid flow. In other cases, debris-choked gullies, ravines or narrow valleys may become flushed out by storm flows to form debris flows. Morphologically, soil flows tend to have three distinct regions: (1) a source area; (2) a track; and (3) a depositional zone. The source area depends on the debris source for a flow but is often some form of shallow scar. The track varies in depth depending on the erosivity of the flow but is usually well-defined, with distinct levees. The levees are formed by material being pushed to the margins of the flow and stranded on the channel sides. They may also be formed by successive flow pulses dissecting earlier depositional lobes. Deposition occurs either when the slope angle is reduced or, more commonly, through the dewatering of the flow. Dewatering is particularly important where flows pass over porous substrates such as gravels. In the depositional zone the debris tends to spread out to form a distinct lobe. The shape of this lobe may be defined by the channel if the flow comes to a stop in the channel. Alternatively, the flow may spread out to form a much broader depositional lobe.

Debris flows are particularly destructive and are a major natural hazard. For example, in the San Francisco Bay area of the USA extreme rainfall events in which 200 mm of rain falls in as little as 15–100 hours are not unusual. After one particular storm, over 18 000 landslides transformed downslope into debris flows, killing 15 people and causing in excess of US$280 million in as little as 32 hours.

The mudflows on the Antrim coast of Northern Ireland are caused by the breakdown of Liassic shales to mud. Their movements have been shown to be seasonal, usually confined to the winter, with the maximum movement measured at 12.7 cm per day during January 1966. The importance of rainfall in controlling this movement was shown in Figure 16.9. The clays contain montmorillonite, which is known to have a high absorptive

(a)

(b)

Figure 16.37 Mudflows, Antrim coast road, Northern Ireland: (a) bowlslide or source area; (b) multiple flow lobes and the flow track

capacity for ions and water. When hydrated, and especially in the presence of sodium ions, these clays swell to several times their dry volume. This swelling is accompanied by increased plasticity, and therefore the association of high rainfall events, the high moisture content of the clays and the extremely plastic behaviour results in rapid movement. Records show that flow velocities may be constant over long periods, but the movement is made up of almost instantaneous jerks and therefore is not continuous. The movement is also probably a basal slip on a shear plane, and is more of a shallow slide than a flow. The well-developed morphology of a source area – the well-defined flow track, delimited by pronounced subparallel marginal shear planes and flow lobes – can be seen in Figure 16.37.

16.3 Hillslope hydrology and slope processes

Within a drainage basin, slopes intercept much of the precipitation that falls, whether it is rain, snow or hail. The transfer of this moisture to channels or other downslope water bodies may result in the geomorphological modification of the slope over which the water moves. In order to explore these processes it is necessary to identify the main routes of water transfer. For simplicity, rainfall will be the focus, but what follows is just as applicable to other types of precipitation.

In humid areas, much of the rainfall is intercepted by vegetation. Some of this water will be evaporated directly back into the atmosphere, both during and after a storm, while some of it may reach the ground via dripping leaves or stem flow. In areas of more partial or limited vegetation cover, precipitation may reach the ground directly. In areas with bedrock at or close to the surface, water will move directly, or after melting in the case of snow, downslope without significant loss, to collect in streams or depressions.

However, on most slopes the presence of soil or sediment may lead to infiltration. The rate of infiltration is a function of the sedimentary properties of the soil or sediment and the amount of organic matter present, especially on the soil surface. Where the intensity of rainfall (volume in a given time) exceeds the rate of infiltration, surface runoff or overland flow (Figure 16.38) will occur (Hortonian overland flow). Where the rate of infiltration exceeds the intensity of rainfall, all the moisture will enter the soil or sediment. Some of this water will be returned directly to the atmosphere by evapotranspiration as plants use the available water in the soil, while some will enter the soil to be stored or transferred. Water movement within a soil or sediment occurs by pore water flow and may be aided by the development of pipes or other zones of enhanced water flow within the soil profile. In general, water is transferred towards downslope water bodies, or it may infiltrate into underlying permeable strata and thereby recharge ground water. Close to a stream or water body where the soil is already saturated with moisture further infiltration will not occur and all the falling precipitation will flow overland as saturated overland flow.

Three basic methods can be identified by which the transfer of water on a hillslope may modify its geomorphology. These are: (1) the direct impact of falling precipitation on a soil or sediment surface, known as rainsplash; (2) during overland flow; (3) via subsurface flow within near-surface sediment or soil horizons.

16.3.1 Rainsplash

On bare or partially-vegetated slopes the direct impact of a raindrop on a soil or sediment surface may cause significant disturbance and/or movement of near-surface grains. Where the slope is subhorizontal, displacement tends to occur equally in all directions to produce a crater, but as the slope angle increases, gravity ensures that displaced grains tend to move predominantly downslope. For example, experimental studies show that on a 5° slope about 60% of the displaced material is moved downslope, compared to just 40% moved upslope, and the proportion moved downslope increases to 95% on a 25° slope. Consequently, rainsplash alone has the potential to modify slope form. In the south-western United States, rates of downslope transport via

Figure 16.38 Overland flow or surface runoff. (Source: http://ehp.niehs.nih.gov/docs/2001/109-12/forum.html, Wikimedia Commons)

rainsplash can exceed 2.6 cm^3 per cm per year. Rainsplash may also help dislodge surface particles or grains, thereby facilitating their removal by overland flow. Figure 16.39 shows an experiment in which episodes of surface water flow were alternated with episodes of rainfall. Initially, water moving over the surface causes significant erosion, but as the loose surface material is removed, the rate of erosion falls. However, periods of rainsplash may dislodge new material, thereby reinvigorating the rate of erosion by overland flow. Rainsplash may also affect the surface infiltration rate of a soil. The direct impact of a drop may create a small crater and, as aggregates on the soil break down, finer-grained particles may infill surface pores to create a crust, which impedes infiltration and may increase subsequent rates of runoff.

The potential of rainsplash is controlled by several factors, including the erodibility of the surface, the vegetation cover and the intensity of the rainfall event. The rate of rainsplash tends to fall with increasing vegetation cover, since vegetation not only intercepts the rainfall but may help bind the soil surface and soil organic matter can improve the stability of surface peds or aggregates in the face of the impact of raindrops. The intensity of the rainfall is also critical and is expressed as the unit volume of

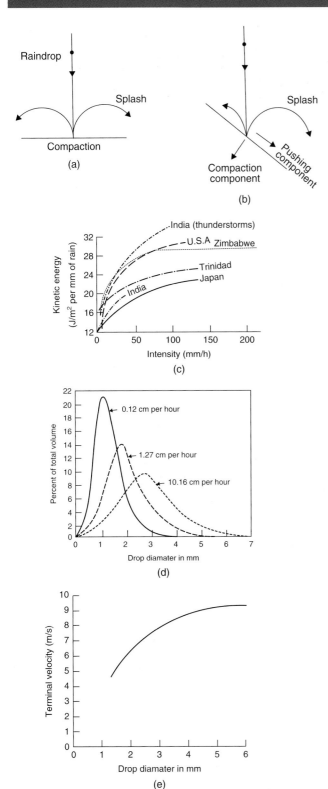

Figure 16.39 Rainsplash experiment: (a) forces involved in rainsplash transport; (b) relationships between kinetic energy and rainfall intensity. Each curve extends to the highest intensity recorded; (c) distribution of raindrop sizes in storms of given intensities; (d) the terminal velocities of raindrops in various sizes; (e) a generalized view of slopes undergoing slopewash

rain which falls in a given period. Rainfall intensity controls the size frequency distribution of raindrops that fall. High-intensity events tend to produce larger drops over a greater range of sizes, while low-intensity rainfall is associated with a small range and size of drops (Figure 16.39). Raindrops rarely exceed 5 mm in diameter, but in general larger drops have greater terminal velocities and consequently a greater geomorphological impact. Typical rainfall intensities vary from 75 mm per hour in temperate climates to 150 mm per hour in more tropical regions, although one recorded case in Africa exceeded 340 mm per hour. Each drop that falls has a finite amount of kinetic energy, proportional to the drop size and its terminal velocity. By adding up all the kinetic energy for each drop that falls, an estimate of the total energy impacted onto the ground during a storm can be made. The energy involved increases with rainfall intensity (Figure 16.39).

Earth pillars or pedestals on various scales can be formed by the undermining on the downslope side of a stone caused by the removal of the surrounding finer sediment until the stone overhangs and then topples downslope. Figure 16.40 illustrates such examples on very different scales.

(a)

(b)

Figure 16.40 Earth pillars/pedestals: (a) large scale in the Swiss Alps; (b) small scale in Aberdaron Bay, Lleyn Peninsula, North Wales

16.3.2 Overland flow

Runoff can be generated either under high-intensity conditions when the supply of water to the soil surface exceeds the rate of infiltration or, alternatively, when the ground is already saturated and the excess generates runoff. The importance of these two types of runoff will vary with the rainfall intensity, the antecedent rainfall conditions and perhaps with time during the course of a storm event. As illustrated in Figure 16.41, the proportion of a drainage basin in which saturated soil occurs gradually increases as a storm increases and consequently the importance of saturated overland flow will increase. The rate at which water is delivered to a stream and therefore the rise in its flood hydrograph will also be affected by the relative importance of the two main types of overland flow. Whatever the source of the surface runoff, the potential for erosion is the same.

In practice, water rarely moves as a sheet but becomes concentrated into distinct flows in the face of surface irregularities such as larger particles, plants or barriers, and check-dams composed of fallen vegetation. Particle entrainment and therefore erosion is a function of the water velocity and is determined by such variables as the slope angle, the length of the slope and the depth of the water relative to the surface roughness of the slope, but the rougher the surface, the slower the water flow velocity.

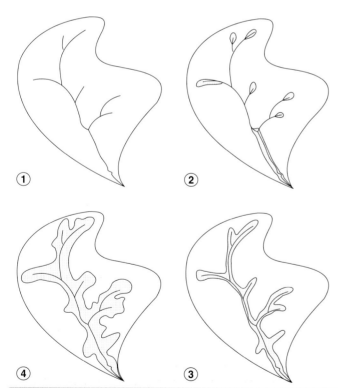

Figure 16.41 Overland flow and characteristics of the drainage basin: proportions of water in various parts of the drainage basin vary with time from the onset of rain and with the characteristics of the basin. Throughflow and saturated overland flow are included in soil moisture. The greater the Hortonian surface runoff, the greater the chance of soil erosion. (Source: after Selby, 1993)

Consequently, most erosion will occur on smooth, steep slopes with a substantial water layer. Slope angle is also an important variable and one of the intriguing aspects of these data is that they seem to imply that small increases of angle on relatively flat slopes have a much greater impact on the rate of erosion than does a similar increase on a steep slope. This is difficult to explain but is probably due to the fact that as slope angle increases, the length of slope exposed to vertically-falling precipitation decreases, so that as the slope angle increases, the vertically-falling water is dispersed over a larger surface area. Probably the most important variable in controlling the effectiveness of soil removal by overland flow, apart from slope angle, is the vegetation cover, since this affects not just the erodibility of the substrate but also the surface roughness. On cultivated slopes, the land management practice is critical to the vegetation cover and its variation through the year. The erodibility of the surface soil is also critical to the effectiveness of the process. Critical variables are the range of grain sizes present, the organic and clay content and the size of any peds or aggregates.

On most slopes overland flow quickly becomes concentrated into a series of small ephemeral channels known as rills. These channels are usually discontinuous, may have no connection to a stream system and are often destroyed between storm events. They may evolve rapidly through a single storm event. They are often best displayed on new road cuts or on mining spoil heaps. Rills may enlarge downslope and their frequency declines as a few large rills capture the available drainage. This occurs because water tends to concentrate in the lowest rill, and divides between rills are often overtopped or may collapse, allowing one rill to capture the water of another. On gentle slopes they tend to have a more dendritic form, becoming more parallel on steeper slopes. Drainage density also increases with slope angle. This can be seen in Figure 16.42(a), where a series of maps shows the drainage patterns that develop as slopes increase in inclination from about 2 to 7%. The rill patterns change from a dendritic type of pattern to a parallel pattern at the higher slopes. The drainage density of the channels on the experimental surface also changes with slope (Figure 16.42(b)). The amount of soil removed increases with the angle of the slope and where the three-dimensional geometry of a slope causes water convergence, for example at a valley head. Erosion is lowest on diverging slopes, such as those on spurs. Larger rills may evolve into more permanent gullies, although gullies may also form independently.

Gullies are defined arbitrarily as recently-extended drainage channels that transmit ephemeral flow and have steep sides, a steeply-sloping or vertical head scarp, a width of >0.3 m and a depth >0.6 m. They often link with the main channel or stream system within a catchment and tend to be a permanent feature. Gullies may form at breaks of slope or where changes in the vegetation cover occur on slopes composed of unconsolidated and mechanically-weak sediment, such as loess, volcanic ejecta or sand and gravel. They are also common on argillaceous rocks such as shales, mudstones and phyllites, where they may channel debris flows. Gullies also form as the result of the collapse of subsurface natural pipes. In general they are features of accelerated slope erosion and are usually regarded as the products

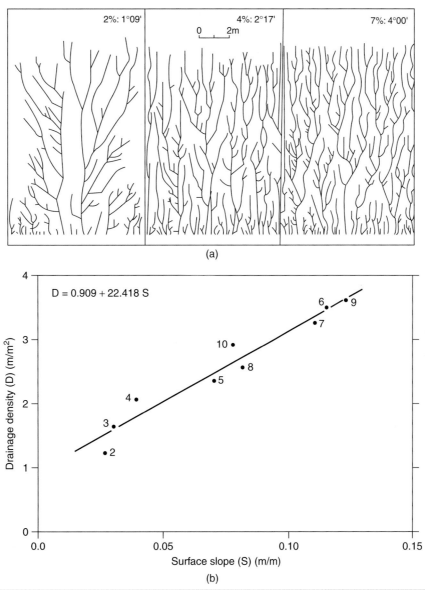

Figure 16.42 Rills: (a) experimental development of equilibrium rill systems under artificial precipitation on planar surfaces of differing original slope; (b) drainage density/slope relationships: mean drainage density as a function of surface slope for numbered surfaces 2–10

of rapid environmental change, such as faulting, burning of vegetation, overgrazing, climate change, extreme storm events or mass movements; in short, anything that disrupts the natural vegetation cover and exposes the soil beneath.

Morphologically, gullies may have a distinct head scarp (Figure 16.43), although this is not always present and may have either a continuous or discontinuous longitudinal profile. The presence or absence of a head scarp is partly a function of a gully's origin and the nature of its substrate. For example, a gully which forms from the enlargement of a rill on a noncohesive material, or within the scar of a mass movement, is unlikely to have a head scarp. In contrast, where a gully originates at a scarp or steep slope break and is cut into cohesive sediment, a head scarp is likely. Equally, if noncohesive material is capped

by a resistant layer such as root-bound soil then a scarp may also form. Retreat and maintenance of the scarp results from sapping as seepage of water loosens material from the face and it is removed from the gully floor by water flow. The head scarp also forms a small waterfall or step and the associated plunge pool may help maintain the scarp. A gully may have a continuous long profile, in which case it tends to increase in width and depth downslope as the runoff increases and tributary gullies contribute to the master. In the lower reach, as the hillslope decreases, zones of deposition extend as coalescing fans beyond the gully mouth. Discontinuous gullies are common in dry valleys with partial vegetation cover and a head scarp forms and retreats wherever the cover is broken. As these separate head scarps retreat, they

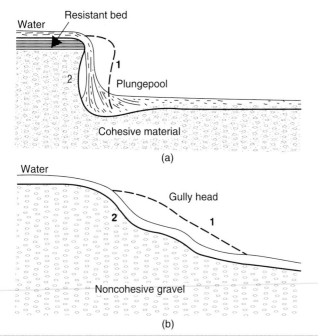

Figure 16.43 Types of gully head: (a) cohesive sediment with a resistant cap of root-bound soil supporting a head scarp; (b) non-cohesive sediment with no head scarp

may begin to coalesce over time to produce a more continuous form (Figure 16.44).

16.3.3 Subsurface flow and piping

In densely-vegetated regions, surface flow is often minimal and most of the precipitation that falls infiltrates the soil and moves within it. Water within a soil may move in all three dimensions. The retention and movement of water in a soil is governed by the energy available. This energy may be potential, kinetic or electrical and is generally referred to as the 'free energy'. Water will tend, like all substances, to move from a state of higher to lower free energy. This is manifest, for example, in the fact that water moves from wetter to dryer zones within a soil. Water may also rise vertically within a soil as a result of capillarity, which is the tendency for water to rise vertically within narrow fissures as a consequence of adhesion of water to the fissure walls. The narrower the fissure, the greater the rise, which is only limited as the weight of the water column balances the cohesive and adhesive forces within the water. In the case of soils, the fissures are provided by the pores and interstices between the particles or grains within the soil.

Liquid water may move within soil as either saturated or unsaturated flow. Saturated flow occurs when all the pores or voids within a soil are water-filled and water flow is driven by the presence of a hydraulic gradient. If one considers a horizontal soil saturated to different amounts along a given length, clearly the difference of water head – depth of saturation – between two points defines a gradient and water will flow from areas of high to low potential head at a speed proportional to the hydraulic

conductivity of the soil. Saturated flow is restricted to waterlogged ground, periods during and following storm events, and areas below the water table. However, most surface soils are rarely saturated all the time and water movement is more commonly achieved via unsaturated water flow, which is not driven by a hydraulic gradient. Water is usually absent from the large pores, which are commonly air-filled, and is restricted to finer fissures and the water films which surround the grains. Water is driven by variation in potential (high to low) within the soil, derived from such things as difference in saturation, vertical elevation and changes in grain size, and therefore capillarity. The mobility of water within a soil decreases with its saturation and is also influenced by grain size. Sandy soils have high conductivities near saturation because of their large pores. Near the air-dry state they have very low conductivities because they lack fine pores. In contrast, clays have very low conductivity when saturated and higher conductivity when unsaturated, due to the presence of numerous fine pores along which capillary movements may occur.

As water infiltrates into a soil – a process which may be enhanced by any surface fissure such as desiccation cracks – the rate of downward movement will vary as the soil properties change. For example, if the soil contains a horizon with a lower conductivity, water will begin to concentrate above it as the rate of vertical infiltration is checked, and this water may begin to move laterally as through flow or subsurface flow, particularly if the soil is on a slope. In areas with a relatively thin soil, this may occur at the rockhead, particularly if the rock is impermeable. This lateral movement may be enhanced by the development of pipes, which can occur naturally via positive feedback: as water moves through a soil, grains may be removed or dissolved along flow paths, and perhaps the random juxtaposition of several large pores allows a slight acceleration in water flow, which in turn enhances grain removal and expansion of the conduit or pipe. This process may be enhanced by the presence of animal burrows, roots or other zones of disturbed ground. Subsurface pipes are common within most catchments and have been documented in both semi-arid and humid regions, and in some cases they may contribute up to 50% of the total storm runoff. The collapse of pipes may lead to the formation of gullies. Natural pipes typically have diameters in the range of 0.02–>1 m and may have lengths from as little as a metre to in excess of one kilometre. The formation of pipes is favoured by such things as: (1) a seasonal or highly variable rainfall regime; (2) a soil subject to desiccation and cracking during dry periods; (3) a reduction of vegetation cover; (4) a soil with strong horizonation which includes an impermeable layer and which is sufficiently cohesive to support the roof of a developing pipe; (5) the existence of a lateral hydraulic gradient.

16.3.4 Ground water flow

As water infiltrates through the soil profile, or, for that matter, falls directly on exposed rock, it may infiltrate and enter the ground water system. Ground water conditions are determined by the porosity and permeability of the rocks present within an area. Porosity is a measure of space and volume within a rock

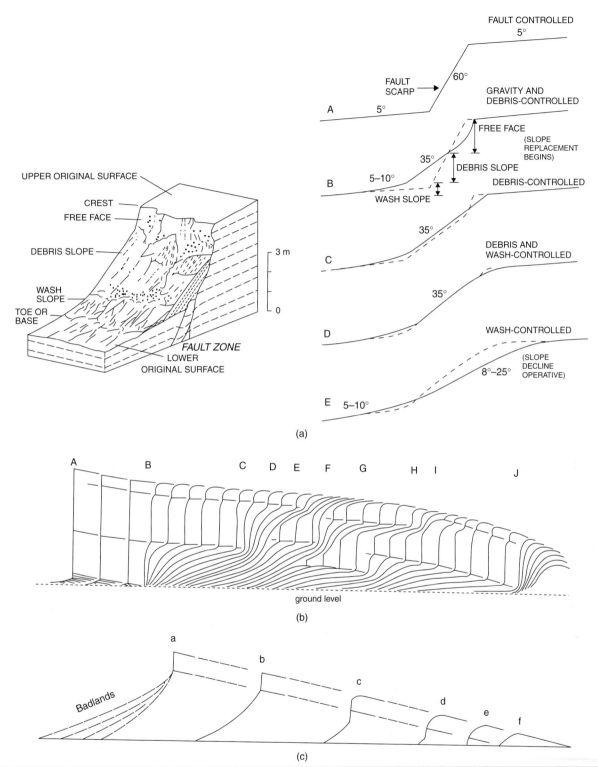

Figure 16.44 Head scarp retreat: (a) proposed sequence of changes in a fault scarp in Nevada. Incremental change is illustrated by the broken line, which represents the full line of the previous profile; (b) development of scarp forms in massive sandstone through time and space. For simplicity a constant ground level is assumed during scarp retreat, along with equal thicknesses of removal from major slab walls in each unit of time. At A. a cliff is present due to sapping above a thin-bedded substrate. At B. the thin-bedded substrate passes below ground level, scarp retreat slows and effective intraformational partings assume control of scarp form; (c) schematic illustration of the change in scarp form as the downdip retreat causes the caprock to become an increasingly important component of the scarp face. (Sources: (a) after Wallace, 1977; (b) after Oberlander, 1977; (c) after Schumm and Chorley, 1964)

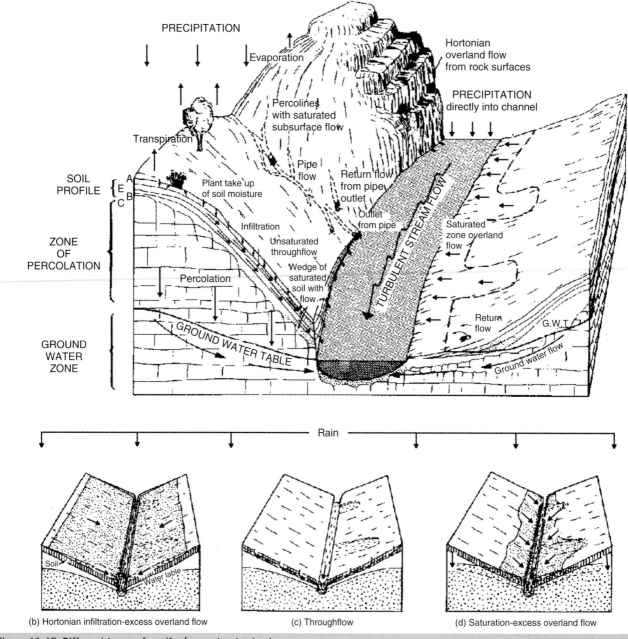

Figure 16.45 Different types of aquifer/ground water basin

body, and in terms of ground water can be considered as the amount of water a body of soil or rock can hold within itself. Two types of porosity can be identified: (1) primary porosity, a function of the pores or gaps between grains in sedimentary rocks, which is therefore usually absent in crystalline rocks and (2) secondary porosity, a function of fractures, joints or solution cavities within a rock, which is potentially present in all types of rock. In contrast, permeability expresses the ease with which water or fluid can pass through a rock and is a function of the connectivity of pores, joints, fractures and other water-bearing voids within a rock. It is important to emphasize that some rocks may have a high porosity but low permeability, such as clay.

Water entering the ground water system infiltrates under gravity first through an unsaturated (vadose) zone until it reaches a depth at which all the voids are water-filled (phreatic zone). The upper level of saturation is known as the ground water table. Here the water flows under the influence of a hydraulic head determined by the shape of the ground water table, or the piezometric surface (Figure 16.45). Ground water flow is always directed perpendicular to lines of equal hydraulic potential, known as equipotential surfaces. The surface of the ground water table, in uniformly-permeable rock, follows in a subdued fashion the surface topography. In such uniform conditions, the pattern of equipotential surfaces, and therefore of the ground water flow lines, is simple. In practice, however, the condition of uniform permeability is rarely met and the pattern of ground water flow may be much more complex and need not reflect surface water flow.

Ground water is in constant motion, flowing from areas of high hydraulic potential to areas of low potential, and without recharge the surface of the ground water table will become more subdued over time. The rate of water flow within a permeable horizon is determined by Darcy's law, which, at its simplest, states that the rate of flow is determined by the piezometric or hydraulic gradient and the intrinsic permeability of the sediment or rock involved. The greater the hydraulic gradient, or the greater the permeability of the rock/sediment concerned, the faster the rate of water flow.

Permeable and water-bearing rocks are known as aquifers, while impermeable horizons are referred to as aquicludes. Ground water movement is restricted to aquifers and ground water basins (hydrological units), and is confined and defined by aquicludes. There are a wide range of different types of hydrological unit or aquifer. Common ones include: (1) alluvial deposits, usually restricted to narrow belts along watercourses; (2) glacial deposits, such as moraines, eskers and kames; (3) fissured hard rock, particularly fracture zones; (4) confined or artesian ground water, in which an aquifer is sandwiched between two aquicludes; (5) karst (limestone); (6) sedimentary basins of permeable and porous rock; (7) thermal and mineral water rising along faults from either active or inactive tectonic or volcanic zones, where such water is usually of considerable economic value due to its perceived medicinal properties.

16.4 Slope morphology and its evolution

So far a range of slope-forming processes have been reviewed, from mass movements to those which act as part of a slope's hydrological system. The next stage is to consider whether these processes result in specific slope morphologies, and in particular to identify patterns of temporal slope evolution. Slope evolution has been the subject of intense debate and until the 1950s provided the central focus for geomorphological research. Opinion fluctuated between those who advocated universal models of slope decline (e.g. W.M. Davis) those who favoured the sequential replacement of slope facets with lower-angle ones (e.g. W. Penck) and those who favoured a process of parallel retreat (e.g. L.C. King). These three classic models are illustrated in Figure 16.46. Although the debate between these various theories did much to shape the early history of the geomorphological discipline, they were not based on detailed empirical observations and the study of process. In practice, slopes may evolve via all three models depending on the processes that operate in any given environment.

Given that each of the main slope-forming processes is constrained in some way by environmental conditions, it follows that different environments should be reflected in different slope characteristics. This might, for example, be manifest in the recognition

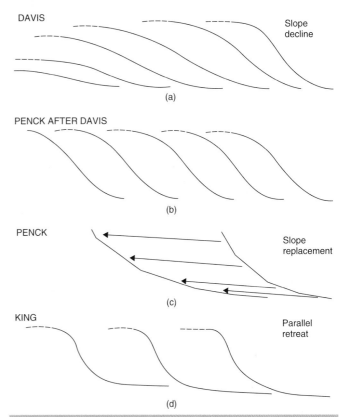

Figure 16.46 Classic models of slope development through time: (a) Davis' model of slope decline; (b) Davis' serious misinterpretation of Penck's model, indicating parallel retreat. In fact, rather than parallel retreat, Penck advocated slope replacement; (c) Penck's model of slope replacement, where rates of denudation are declining hillslopes evolve through a process of flattening from the base upwards. In effect, each part of the slope profile is replaced by a slope of lower gradient as it retreats, and through this process of slope replacement a concave profile is produced; (d) King's model of parallel retreat, where he rejected Davis' idea that slope gradients decline through time. He suggested that the free face retreats parallel to itself as sediment is weathered and then removed by rill erosion and a low-angle slope or pediment grows at its base

of morphoclimatic slope forms, or the association of particular slope morphologies with specific geological conditions. In order to explore this broader concept, it is first necessary to examine the link between process and slope form.

16.4.1 Measuring the slope morphology

The geometry of a slope may be described through maps, slope profiles or increasingly via digital elevation models. The simplest is the slope profile, which is a two-dimensional graph of elevation and distance, normally surveyed at an angle perpendicular to the maximum slope. Profiles may be divided into straight, concave, convex or vertical (free face or cliff) components. These can be quantified in terms of length, angle and degree of curvature, expressed as the rate of change in slope angle. Figure 16.47 provides an example of an annotated profile. The alternative is

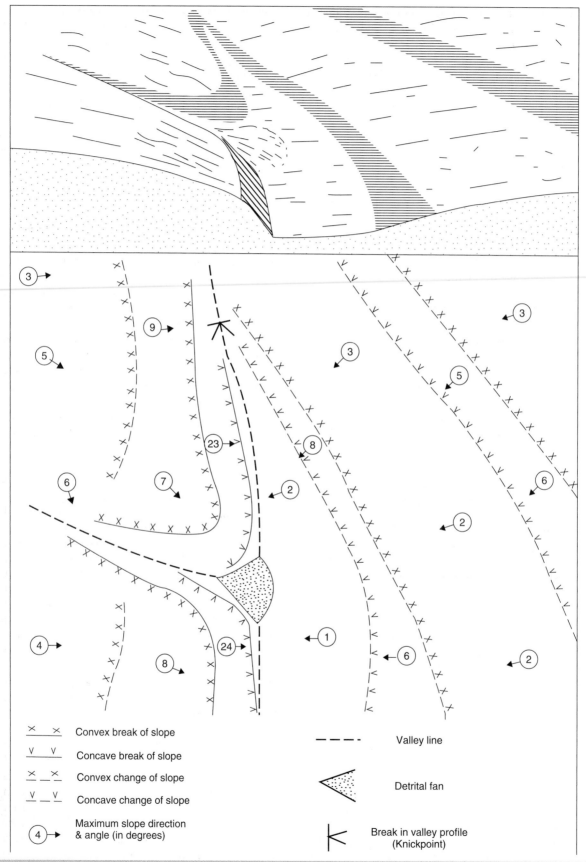

× ×	Convex break of slope
v v	Concave break of slope
× _ ×	Convex change of slope
v _ v	Concave change of slope
4 →	Maximum slope direction & angle (in degrees)
– – –	Valley line
(detrital fan symbol)	Detrital fan
⊢	Break in valley profile (Knickpoint)

Figure 16.47 Annotated slope profile

a geomorphological map, which delineates breaks of slope (a morphological map) and classifies a three-dimensional surface into a series of slope segments (Figure 16.48). Today it is possible to obtain digital elevation data from a range of sources at a range of scales, using satellite images, aerial photographs, electronic atlases and maps accessed via the Internet. Such models not only allow very sophisticated empirical data to be collected about slopes but, more importantly, facilitate slope visualization (Figure 16.49).

16.4.2 Process and form

It is useful to distinguish between two broad types of hillslope when considering the link between process and form. In rockslopes, the rate of erosion is limited by the availability of loose material derived from weathering, and such slopes can be classed as weathering-limited. The morphology of this type of slope is dependent on the nature of the rock mass and the spatial variation in its susceptibility to weathering. At the other extreme are slopes which contain an almost infinite supply of weathered material and whose morphology is a function of the capacity of the transport processes that act on them, effectively forming transport-limited slopes. On these slopes the transport processes and strength characteristics of the weathered regolith, rather than of the rock mass itself, are critical in determining the pattern of erosion and slope modification.

In practice these two extremes lie along some form of detachability continuum, as illustrated in Figure 16.50. Most studies of process–form causality are concerned with the transport-limited end of this spectrum. These slopes have been examined via a range of field studies and both physical and numerical simulations.

One of the classic field studies looking at slope evolution and the links between process and form was undertaken on two different rock formations (the Brule and Chadron formations) within the shale Badlands of South Dakota, where observations revealed two very different slope morphologies, the result of two different processes. Slopes on the Brule formation are rectilinear (Figure 16.51) and retreat was by parallel slope replacement; the slopes are dominated by erosion from overland flow. In contrast, slopes on the Chadron formation are convex and retreat was via a gradual decline in slope angle, as a result of erosion by both rainsplash and creep. These are two very different slope forms and evolutionary patterns within an area of similar climate and vegetation.

It has been shown that parallel slope retreat caused by wash erosion occurs because erosion is essentially uniform across the whole slope. If you imagine a slope with a layer of particles A–D overlying a layer W–Z (Figure 16.52), erosion by overland flow may act on any particle A–D and remove it from the slope during one or more storm event. Erosion is concentrated on the steepest part of the slope. Therefore, if, for example, particle A is removed, the base level for the particles above is lowered and they will quickly be eroded. Erosion by overland flow tends, therefore, to move a uniform thickness of material from a straight slope. In short, particles A–D will be removed before the underlying layer (particles W–Z) can be. In contrast, where creep or rainsplash is the dominant process, as is the case on the Chadron formation,

a convex profile evolving via a gradual decline in slope angle is observed. Consider the rectilinear slope in Figure 16.52, with a layer of particles A–D resting on a layer W–Z. In this case, a particle engaged in creep or dislodged by rainsplash, say particle D, will not be removed from the slope at the same time as or before particle A, but must be moved down the whole slope length behind the particle that lies downslope of it. The order of particle removal is therefore A, B, C, D, . . . Z. As particle D moves under the influence of rainsplash or creep it travels downslope to reveal the underlying layer (particle Z), while downslope particles C and D must move before the next layer of particles is revealed (particle Y). Before the lowest point on the slope can be reached, all particles have to pass through this point (A–D) in order to uncover the next layer, and as a result most of the available energy is consumed in moving material derived from upslope. Consequently, erosion is concentrated in the upslope area of the slope, while transport and deposition dominate on the mid and lower slopes, even though these may become the steeper slopes. The result is a convex slope profile. This study illustrates how different slopes in a similar region may evolve in very different ways, with very different profile morphologies, as a result of different dominant processes. In this case, the operating process appears to be a function of lithology.

A great number of field studies have also looked at the frequency distribution of slope gradients found on soil-mantled slopes. These studies appear to suggest that straight (rectilinear) slope units tend to cluster around a modal angle, which varies with climate and lithology from region to region, at a site between slopes that are being actively undercut and those that are not. These modal slopes appear to be linked to the threshold angle of stability for different kinds of slope material. A slope formed by river incision, or steepened by undercutting, will decline by mass movement processes until the slope angle becomes stable. This is approximately equal to the strength of the regolith, a function of its angle of internal friction and pore-water content. A good example of this link between slope angle and strength is provided by data from the Verdugo Hills in the Californian Coast Ranges. This gneissic terrain is characterized by steep, straight valley-side slopes covered by regolith up to 1 m thick. Slopes subject to basal undercutting by streams have a mean slope angle of 44.8°, while slopes that are not undercut have an angle of 38.2°. The undercut slopes undergo active failure; fines are removed and the debris mantle is relatively unweathered. In contrast, the protected slopes are finer-grained and show greater weathering maturity, and as a consequence they have lower strength thresholds. Comparison of strength with slope angle for these two sets of slopes shows a strong correlation.

The frequency distribution of straight slope segments on hillsides underlain by both Millstone Grit (Pennines, UK) and Exmoor Slate (Devon, UK) has been noted and it was found that distinct modes of slope angle were present, each linked to a particular weathering state and therefore to a particular strength of the regolith present. The data were used to argue that slopes undergoing mass movement declined by slope replacement rather than slope decline. If a process of hinged decline occurred then a greater range of slope angles should be present. These studies

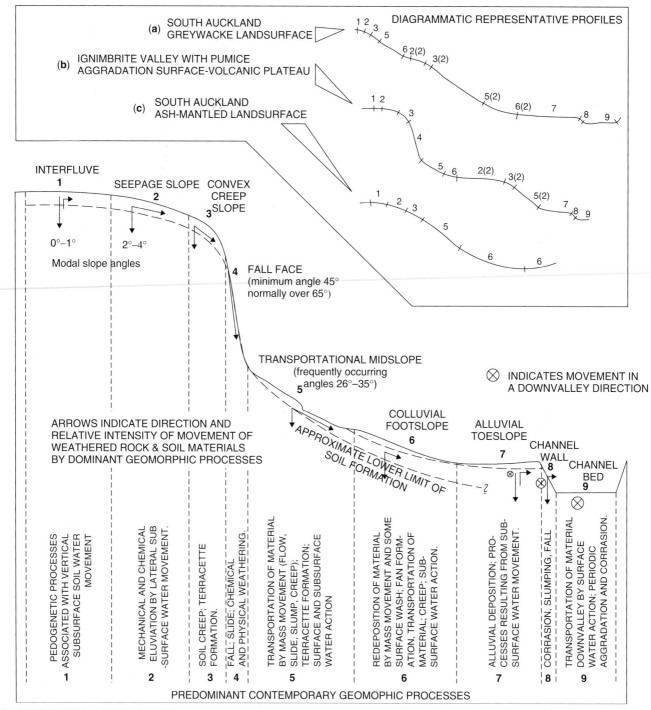

Figure 16.48 Descriptive geomorphological map: slope segments. (Source: after Small, 1991)

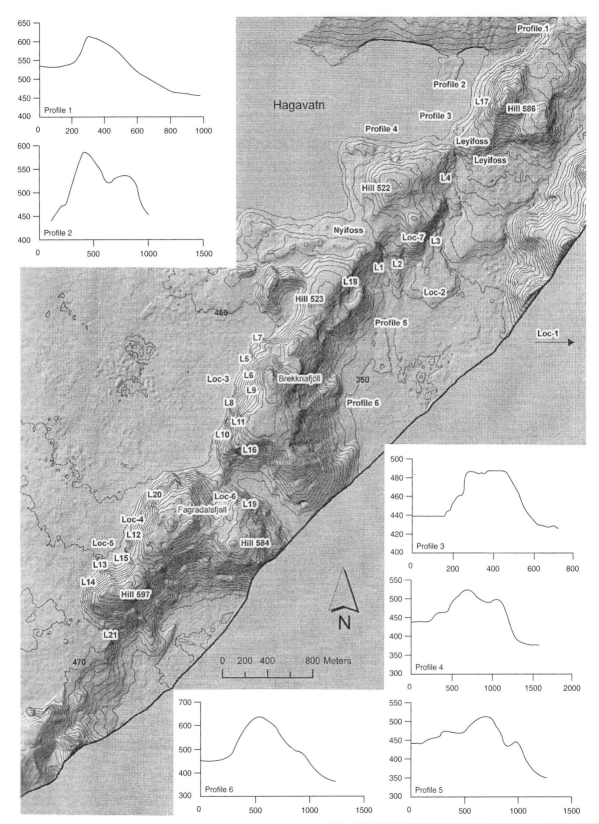

Figure 16.49 Digital elevation model, Jarlehettur, Central Iceland, with slope profiles. Ridge created by a subglacial, linear-fissure, volcanic eruption (after Bennett *et al* 2008)

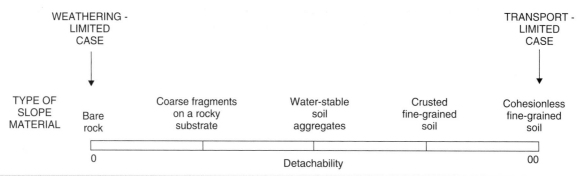

Figure 16.50 Continuum between weathering-limited and transport-limited slopes

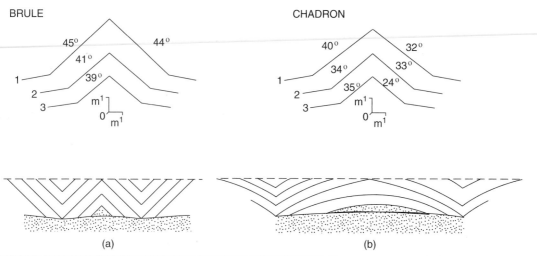

Figure 16.51 Slope processes. (Source: after Schumm, 1956)

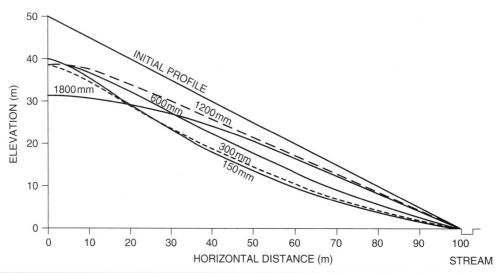

Figure 16.52 Model for the simulation of slope processes: comparison of simulated hillslope profiles after 100 000 years of different annual rainfalls. For all profiles rainfall/rainday = 15 mm and mean annual potential evapotranspiration = 200 mm. (Source: after Kirkby, 1976)

illustrated that soil-mantled slopes shaped by mass movement tend to have a characteristic slope angle and form, determined by the strength of the regolith.

Field observations, such as those reported above, have been supplemented by numerical simulations designed to explore the link between process and form. These models start with a known profile, usually a rectilinear slope, and apply some form of transport equation. The profile is divided into units and the transport into and from each unit is calculated. Where transport in exceeds transport out, deposition and profile modification occurs. Alternatively, if transport in is less than transport out, erosion will occur at that point on a profile. The results from one such model are shown in Figure 16.44 and illustrate how soil creep and rainsplash result in a convex profile, erosion by overland flow without gullying gives a linear slope, while erosion by overland flow with gullying gives a concave profile, since erosion by gullies increases with their drainage area and therefore with distance from the crest of the slope. Typical process–form inferences include: creep processes, which produce an expanding upper convexity on a slope; erosion by overland flow, especially by gullying, which produces an increasing lower concavity; uniform solution, which produces a parallel downwearing; and shallow landslides, which produce parallel slope retreat, associated with a lower concavity in which the slide debris accumulates. Selby (1993) provides a useful summary of the main conclusions derived from these numerical simulations:

- Processes involving downslope soil transportation (soil creep and rainsplash) tend to retreat by a decline in slope angle. These slopes are often transport-limited.

- Processes involving direct removal of material from the slope (overland flow and solution) tend to cause parallel retreat and may become weathering-limited.

- Downslope transportation at a rate that is only dependent on the slope angle (creep and rainsplash) produces a smooth slope convexity.

- Downslope transportation at a rate that is dependent on both slope and distance from the divide (erosion by gullying and pipes) tends to produce a basal concavity.

- Stream incision at a rate in excess of the rate of transportation on the slope produces a steep basal slope and may lead to mass movement.

These models of slope evolution have been examined using ergodic reasoning: the substitution of space for time in fault scarps, river bluffs and bluffs from ancient shorelines. Artificial slopes, such as spoil tips, have also been used in the same way. The basic idea is that if a range of slopes within a region is sampled and assembled in terms of absolute or relative age then it is possible to construct some form of temporal model. For example, the evolution of fault scarps in alluvium was studied in Nevada (Wallace, 1977). The original fault scarp was steep (50–90°) and mass movements dominated its initial modification. Rockfall led to the accumulation of a talus or debris slope. This debris built up and over the fault scarp to produce a debris-mantled slope, which initially lay at the angle of repose but has slowly been reduced by water erosion. The key conclusion is that slope process and slope form change through time as the slope system evolves (Figure 16.44). An example of Slope Evolution on Mississippi River Cliffs is documented in Box 16.5 (see companion web site).

A similar approach was used to look at wave-cut bluffs on modern Lake Michigan and compare them to forms abandoned along Glacial Lake Algonquin around 10 500 years BP and by Nipissing Lake 4000 years BP. Comparison of these profiles shows clearly a decrease over time in the inclination of the bluff mid-slope, with an increase in the extent of the convex crest and concave base of the bluff (Figure 16.54) These studies support the idea that mass failure occurs until a critical slope angle is obtained, determined by the material's strength. This is a process that occurs by parallel retreat and slope replacement. Thereafter, slopes decline in gradient and become more concave and convex as hydrological processes shape their forms.

16.4.3 Controls on process and slope form

We saw above how a specific process may give rise to a specific slope morphology. In practice, simple process–form relationships are often difficult to observe in the field, not least because several processes may operate on any given slope. Despite this, the next logical question is to ask what controls the effectiveness of a particular process. Is there a tool kit of slope-forming processes in which the choice of tool and therefore the form that results is determined by the prevailing environmental conditions? Figure 16.56 illustrates a situation where in six regions there are six different characteristic slopes, expressed here rather crudely as histograms of the slope angle. The slope system which results in a particular region or area will be a function of such variables as lithology, climate (vegetation and soil) and the geomorphological history, or antecedence.

Lithology

Lithology may influence process and form in two ways: (1) as a control on slope strength and erodibility; and (2) as a control on hillslope hydrology.

Perhaps the most obvious way to illustrate its influence on the strength and erodibility of a slope is to contrast the characteristics of mass movements that occur on a clay slope with those that occur on one underlain by more competent sandstone. Different lithologies have different susceptibilities to weathering and may produce very different types of weathering product, or slope-mantling regolith. This is illustrated by the evolution of debris-mantled slopes in semi-arid areas of the American Southwest. These hillslopes are characterized by gradients between 10 and 38° and are mantled by a thin layer of coarse debris. Typically, they are slightly concave in profile, with a more rectilinear upper slope. They may form below cliffs, but such features are often

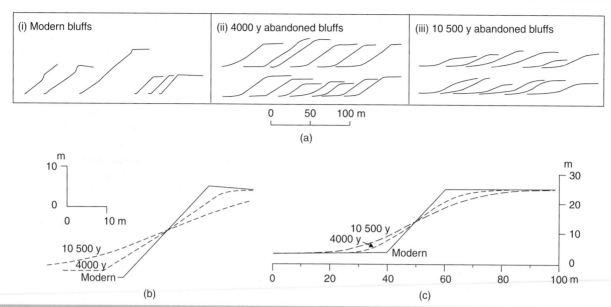

(a)

(b)   (c)

Figure 16.54 (a) Lake bluffs: measured profiles for modern shorelines, Lake Michigan, Nipissing Great Lake and Lake Algonquin. (b) Superimposed typical profiles. (Source: after Nash, 1980)

absent, having been removed by weathering and erosion. A strong relationship between the gradient of these slopes and the grain size of the debris mantle has been recorded in some areas, but not all. Slopes which form below cliffs will evolve via debris fall and should lie close to the angle of repose. Once the cliffs have declined, the debris slope should become adjusted to the hydraulic processes acting on it. These processes will act on the slope until they reach an angle at which they are no longer competent to move the coarsest material on the slope. This grain size will then form a protective mantle and will result in there being a correlation between the grain size of the mantling debris and the slope gradient. This relationship was found to be dependent on lithology, and varied from one rock type to another, primarily according to their different transportability. However, not all slopes show a positive correlation between grain size and gradient, suggesting that they are not all adjusted to the prevailing hydraulic processes. Further analysis of these slopes has shown that there is a continuum, from slopes that are underlain by rocks with widely-spaced joints, which give large blocks that cannot be moved by hydraulic processes and consequently have no slope adjustment, to slopes that are underlain by more closely-jointed rocks, which do show adjustment. In this lithology, joint spacing in particular is critical to the adjustment of these slopes. If the spacing is too large then the slopes will be effectively relict and the material will lie close to the angle of repose, whilst if the spacing produces a block size on weathering that can be moved then slope adjustment to the prevailing hydraulic processes will occur.

The outcrop pattern and stratigraphy of different lithologies may also have an influence on slope evolution. A line of cliffs formed by faulting or erosion will normally retreat and decline, to be replaced by a more gentle slope, unless the process of uplift or vertical incision is maintained. This process of cliff decline may be hindered by the presence of a surface cap rock or an indurated surface horizon. In humid climates, such slopes will quickly be

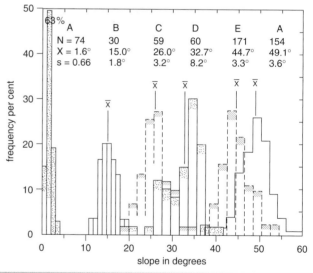

Figure 16.55 Histograms of frequency distributions of slope angle for six regions: (a) Steenvoorde, France; (b) Rose Well gravels, Arizona; (c) Bernalillo, New Mexico, Santa Fé formation; (d) Hunter-Shandaken area, Catskill Mountains, New York; (e) Kline Canyon area, Verdugo Hills, California; (f) dissected clay fill, Perth Amboy, New Jersey. (Source: after Strahler, 1950)

reduced via multi-mass movements, however in more arid regions slope escarpments may be maintained and can retreat over considerable distances. The Colorado Plateau provides one of the best examples where sub-horizontal sandstones overlie impermeable shales. The geometry of these escarpments is controlled by the characteristics of the cap rock – geometry, thickness and rock mass strength. Figure 16.55 shows the evolution of a typical scarp and the influence of cap rock geometry and competence.

Perhaps the ultimate illustration of the role of rock strength in controlling slope morphology is given in slopes that have evolved so that there is equilibrium between strength and slope. Such slopes are free from undercutting, talus and soil cover and have retreated sufficiently for any inherited form to have been removed. As a consequence the only control is the rock mass strength of the geological units involved. On some slopes the equilibrium between strength and slope may be influenced by cliff height and the overburden placed on the lowest units within a cliff. Some rock slopes have a characteristically concave profile in which the lowest units do not show any relationship between rock mass strength and slope. This is probably due to the fact that joints or fractures at the cliff base may fail more easily than those higher in the cliff, because the overburden exerts sufficient force to facilitate the crushing of asperities along a fracture that locks its two faces together.

The second influence of lithology on process and form is through the effect of rock type on controlling hydrology. For example, all the runoff from an impermeable lithology must occur through subsurface or overland flow, while on a permeable substrate water loss may occur through ground water flow. The drainage patterns on several different lithologies in which impermeable rocks occur show higher drainage density and consequently steeper valley side slopes as a result of the greater level of incision and relief. The classic study of the South Dakota Badlands by Schumm (1956) also reinforces this point. Two different

lithologies with different hydrological properties but similar climate and vegetation have different slope processes operating upon them and evolve differently through time (Figure 16.51). The margins of the Colorado Plateau in the USA are incised and fretted by numerous steep-walled canyons with amphitheatre headwalls and hanging tributaries. These canyons are primarily cut by fluvial processes and not by slope sapping. The porous and permeable cap rock of sandstone is undermined by ground water flow along the upper surface of the underlying shale. Water flow is concentrated on joint networks within the sandstone and sapping, where undermining of the competent sandstone by solution and erosion of the underlying shale, which leads to cliff failure, occurs in the direction of water flow and is therefore controlled by the geometry of the surface between the sandstone and shale (Figure 16.56).

Climate

Clearly there is a direct link between the erosivity of a particular process and climate. But there is also an indirect link, since climate controls vegetation and soil development, which in turn may influence a slope's erodibility. The question of this link has been addressed in a number of ways. Several authors have attempted to produce a synthesis of process rate data by climatic zone (Figure 16.57), which clearly shows how the effectiveness of different processes varies from one climatic region to another, despite the limitations of the data sets available. Kirkby

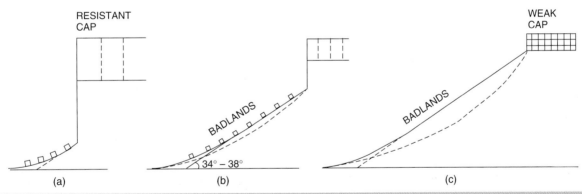

Figure 16.56 Colorado Plateau evolution of a typical arid scarp under the influence of three caprock types: (a) a resistant caprock overlying shale; (b) a less resistant caprock; (c) a thin, weak caprock. All three are from actual rock-type sequences in various locations. (Source: after Chorley *et al.*, 1984)

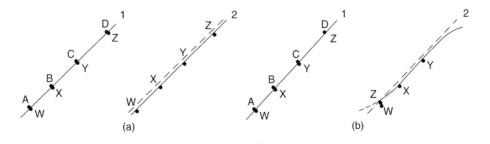

Figure 16.57 Erosion of particles on slopes undergoing rainwash and creep. (Source: after Schumm, 1956)

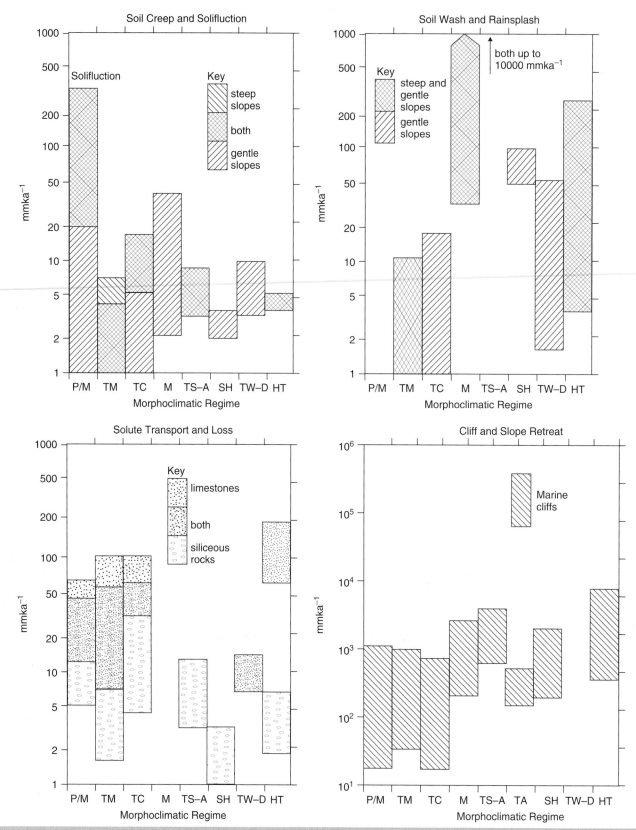

Figure 16.58 Climate zone : process rate data. Based on short-term field measurements. P/M, periglacial and montane; TM, temperate maritime; TC, temperate continental; M, Mediterranean; TS-A, tropical semi-arid; TA, tropical arid; SH, subtropical humid; TW-D, tropical wet/dry; HT, humid tropical

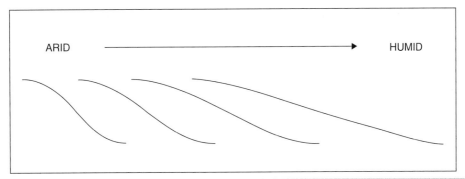

Figure 16.59 Conceptualized hillslope profiles from arid to humid areas across the USA. (Source: after Toy, 1977)

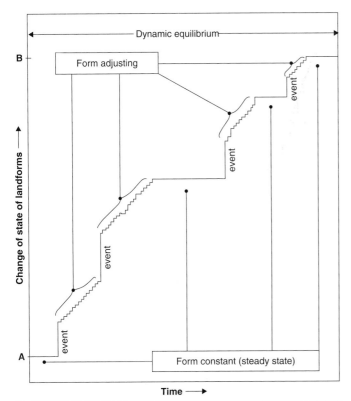

Figure 16.60 Slope evolution: dynamic equilibrium/extreme events. Within the term 'dynamic equilibrium' there are subsumed three states: (a) a landforming event; (b) the adjustment of form which follows after such an event; (c) a period of steady state in which there is virtually no adjustment. The curve representing the change of landforms with time may rise very steeply, gradually or hardly at all depending on the magnitude and frequency of the dominant process

(1976) used a numerical process–response model to examine the development of an initially straight slope under different rainfall conditions. As Figure 16.58 illustrates, the profiles within this model become increasingly convex as total rainfall increases. Other studies have noted that the basal concavity of a slope may be a function of climate, with the concavity being more abrupt and marked in semi-arid regions, where there is a strong relationship between the grain size of the debris mantle and the slope angle, such that steeper slopes are usually associated with coarser debris. As debris size declines at the slope base, so does slope gradient, which results in marked concavity. In more humid regions a greater range of grain sizes is present and the link between grain size and slope angle is less marked. Consequently, the basal slope concavity is less prominent.

Toy (1977) designed an experiment to evaluate the influence of climate on slope morphology. He selected a series of slope profiles along two transects across the USA, one along the 37th parallel, designed to pick out variation in rainfall, and one along the 105th meridian, designed to pick out variations in temperature. Other variables, such as lithology, were held constant by only selecting sites with similar properties, and all sites were chosen for their proximity to weather stations. At each site slope profiles were surveyed and summarized numerically, and compared statistically to the meteorological data. The results, summarized in Figure 16.59, suggest that hillslopes in arid regions are shorter, steeper and more convex, while those in humid areas are longer, shallower and more rectilinear. Toy (1977) explained the convex–concave slopes of arid regions in terms of the dominance of rainsplash on drainage divides and of overland flow on lower slopes, while the more rectilinear slopes of humid regions were attributed to the greater importance of mass movement. Although Toy's work has been criticized for not isolating all the other variables that might control slope morphology, it illustrates the potential for a link between climate and slope form.

Antecedence

Time, and the distribution of geomorphological events through time, is important in understanding slope form and its adjustment to process. A given event and its consequent geomorphological impact have a recurrence interval dependent on the event's magnitude. A recurrence interval is simply a probabilistic statement that an event of given magnitude is likely to occur during a given length of time. For example, a storm event with a recurrence interval of one in ten years has a probability of occurring in any given year of one in ten. Larger events tend to have greater geomorphological impact but also tend to occur with lower frequencies. The relationship between magnitude and frequency will

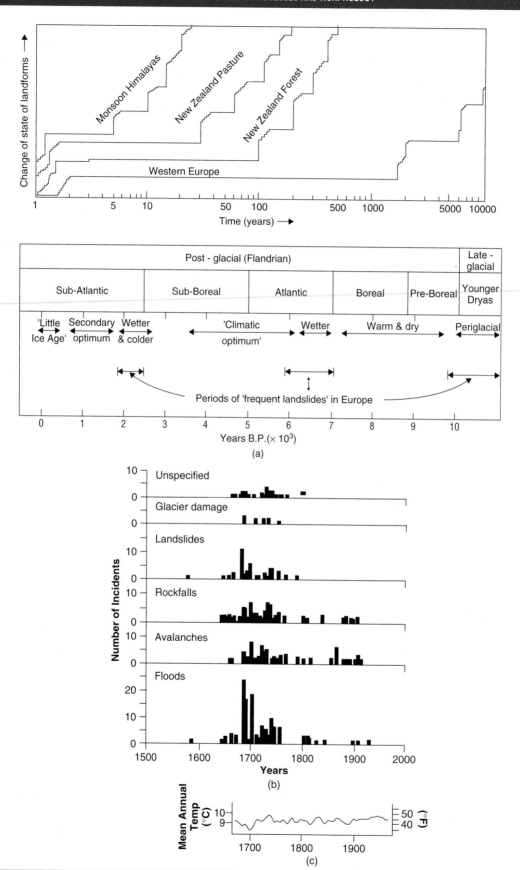

Figure 16.61 (a) Magnitude and frequency of landsliding in different environments. (b) Starkel's (1966) periods of frequent landslides in Europe. (c) Effects of the Little Ice Age in Norway. (Source: (c) after Grove, 1972)

vary between different processes and from one climatic region to the next.

The evolution of a slope can be seen as the result of two types of event: (1) high-magnitude or catastrophic events which occur with a periodicity dependent on the prevailing environmental conditions; and (2) low-magnitude but high-frequency events. The relative magnitude and frequency of these two types of event will determine the form of a slope. Slope evolution can be viewed in terms of a dynamic equilibrium punctuated by extreme events separated by periods of form adjustment (Figure 16.60). If the high-magnitude events occur with a frequency sufficient to prevent all evidence of their presence being removed from a slope then they will tend to dominate slope form. However, if these events occur with less frequency, or if erosion rates during periods of form adjustment are sufficiently high to remove all evidence of the events, then slope form will reflect the low-magnitude, high-frequency form. For example, a severe storm in a humid climate may cause shallow landslides that disrupt the vegetation and cause slope adjustment. However, these slopes are quickly eroded and vegetated again until little or no trace of them remains. In contrast, in a semi-arid area a severe episode of wash erosion may remain as a scar on the landscape for a long period, since the normal rates of denudation are so low.

The point can be viewed in terms of the ratio of denudation during an extreme event to that during normal conditions. In the humid tropics extreme events have little impact because the annual rate of denudation is so high. In contrast, within semi-arid deserts the geomorphological impact of extreme events far exceeds that achieved in a normal year. It is also important to note that the frequency of extreme events varies from one region to the next. As the plot in Figure 16.61(a) shows, the frequency of extreme events in the Himalayas under monsoon conditions far exceeds the frequency for Western Europe. Climate change will also impact on the magnitude/frequency relationship as a particular event may have had a far greater frequency in the past than it does at present (Figure 16.61(b),(c)) and consequently the observed slope form may be adjusted to the antecedent processes rather than those active today.

In summary, therefore, a slope may not always be adjusted to the processes active in the landscape today, or it may be adjusted to extreme events that will not necessarily be observed during a relatively short period of investigation.

It is clear from the above that there are many different types of slope process, operating at various speeds, and any individual slope may illustrate several processes operating in a complex manner involving feedback loops. However, the ranges of velocity are great, as can be seen in Figure 16.62(a), and discussions of the importance of these processes on individual slopes and in different climatic regimes are instructive as the general conclusion is that overall the fast, rare processes of hillslope modification are not as important as the hydrological processes, and that solution is the dominant process in hillslope change (Figure 16.62(b)).

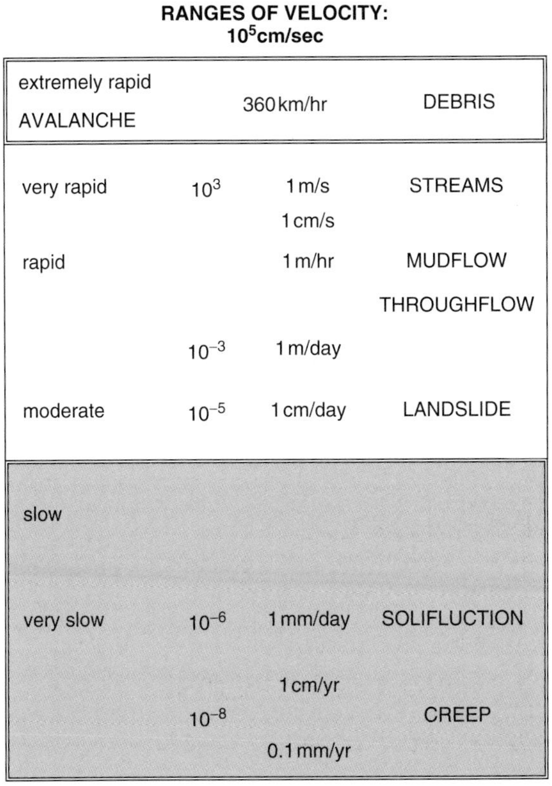

RANGES OF VELOCITY:
10^5 cm/sec

extremely rapid AVALANCHE		360 km/hr	DEBRIS
very rapid	10^3	1 m/s	STREAMS
		1 cm/s	
rapid		1 m/hr	MUDFLOW
			THROUGHFLOW
	10^{-3}	1 m/day	
moderate	10^{-5}	1 cm/day	LANDSLIDE
slow			
very slow	10^{-6}	1 mm/day	SOLIFLUCTION
		1 cm/yr	
	10^{-8}		CREEP
		0.1 mm/yr	

KARKEVAGGE (LAPLAND, N.SWEDEN) after Rapp (1960)

Process	Annual Volume	Annual Av. Movement	Annual Index of Denudation *
	cub.m	m.	
rockfall	50	225	19565
avalanche	88	150	21850
slides	580	0.5-600	76300
talus creep	300000	0.01	2700
solifluction	550000	0.02	5300
solution	150	700	136000

* tons of material x vertical distance of transport in metres
This study and others from many environments show that overall denudation of the landscape must have HYDROLOGICAL PROCESSES as a top priority

Figure 16.62 (a) Speeds of operation of processes on hillslopes. (b) Relative importance of processes on hillslopes

One of the key elements in slope evolution is the link between transport processes at a slope base, which determines the basal level, and the slope system. If we consider a small upland stream located in close proximity to the surrounding slopes, vertical incision or lateral erosion of the stream channel will either reduce or steepen the base level for adjacent slopes; either way they are likely to experience accelerated erosion, probably by some form of mass failure. In doing so, not only is the slope modified by the delivery of the debris to the stream, but this debris in turn will impact on the stream's transport processes and perhaps result in changes to its morphology. There is, therefore, a strong link between the slope system and the transport system which acts at the slope base, at least until these two systems become separated (by a broad floodplain, for example). This is also particularly true when one considers the morphology of coastal slopes, which often pass through a cycle of morphological change as the system switches between the transport and weathering end members.

Relief, as expressed by the difference in elevation between a hilltop and valley bottom, is important in determining the rate of slope modification. In a mountain region experiencing strong crustal uplift, incision by rivers and streams will constantly oversteepen the rising slopes. These slopes will be stripped of regolith and will be weathering-limited, and consequently their morphology will be primarily determined by the rock mass strength and its spatial variation within the hillside. In contrast, in areas of limited uplift or vertical stream incision hillslopes will tend to become divorced from the river system at their base, perhaps through a broad floodplain, and will tend to evolve a more stable morphology, largely determined by the transport processes which operate upon these slopes.

Exercises

1. Why does failure on a slope occur?

2. What are the factors that define a rock's strength?

3. How is cohesion in soils achieved?

4. Outline the factors that can increase slope stress.

5. What are the types of soil creep and how can they be recognized in the field?

6. What are the morphological characteristics of the various types of scree?

7. Outline the importance of lithology in slope development.

8. Summarize the main conclusions derived from numerical simulations of slope processes.

9. Why is the ground water system important in the context of slope processes?

10. Outline the importance of slopewash in modifying slope morphology.

References

Chorley, R.J., Schumm, S.A. and Sugden, D.E. 1984. Geomorphology. London, Methuen.

Everett, K.R. 1963. Slope movement, Neotoma Valley, Southern Ohio. Ohio State University Report 6, Institute of Polar Studies.

Hoek, E. and Bray, J. 1977. Rock Slope Engineering. 2nd edition. London, Institute of Mining and Metallurgy.

Grove, J.M. 1972. The incidence of landslides, avalanches and floods in western Norway during the Little Ice Age. *Arctic and Alpine Research*. **4**, 131–138.

Hutchinson, J.N. 1972. The preparation of maps and plans in terms of engineering geology. *Quarterly Journal of Engineering Geology*. **5**, 293–381.

Jones, D.K.C. and Lee, E.M. 1995. Landsliding in Great Britain. London, HMSO.

Kirkby, M.J. 1964. A study of rates of erosion and mass movements on slopes with special reference to Galloway. Unpublished PhD dissertation. University of Cambridge.

Kirkby, M.J. 1967. Measurement and theory of soil creep. *Journal of Geology*. **75**, 59–78.

Kirkby, M.J. 1976. Hydrological slope models: The influence of climate. In: Derbyshire, E. (Ed.). Geomorphology and Climate. Chichester, John Wiley & Sons, Inc. pp. 247–267.

Nash, D. 1980. Form of bluffs degraded for different lengths of time in Emmet County, Michigan, USA. *Earth Surface Processes and Landforms*. **5**, 331–345.

Oberlander, T.M. 1977. Origin of segmented cliffs in massive sandstones in southeastern Utah. In: Doehring, D.O. (Ed.). Geomorphology in Arid Regions. Boston, Allen and Unwin. pp. 79–114.

Owens, I.F. 1969. Causes and rates of soil creep in the Chilton Valley, Cass, New Zealand. *Arctic and Alpine Research*. **1**, 213–220.

Prior, D.B., Stephens, N. and Archer, D.R. 1968. Composite mudflows on the Antrim Coast of north-east Ireland. *Geografisker Annaler*. **50A**, 65–78.

Schumm, S.A. 1956. Evolution of drainage systems and slope sin badlands at Perth Amboy, New Jersey. *Bulletin of the Geological Society of America*. **67**, 597–646.

Schumm, S.A. 1964. Seasonal variations of erosion rates and processes on hillslopes in Western Colorado. *Zeitschrift für Geomorphologie Supplementband.* **5**, 251–258.

Schumm, S.A. and Chorley, R.J. 1964. The fall of Threatening Rock. *American Journal of Science.* **262**, 1041–1054.

Selby, M.J. 1980. A rock-mass strength classification for geomorphic purposes: With tests from Antarctica and New Zealand. *Zeitschrift für Geomorphologie.* **24**, 31–51.

Selby, M.J. 1993. Hillslope Materials and Processes. 2nd edition. Oxford, Oxford University Press.

Small, R.J. 1991. The Study of Landforms. 2nd edition. Cambridge, University of Cambridge Press.

Starkel, L. 1966. Post-glacial climate and the moulding of European relief. *Proceedings of the International Symposium on World Climate 8000 to 0 BC, Royal Meteorological Society, London.* 15–32.

Strahler, A.N. 1950. Equilibrium theory of erosional slopes approached by frequency distribution analysis, part 1. *American Journal of Science.* **248**, 673–696.

Toy, T.J. 1977. Hillslope form and climate. *Bulletin of the Geological Society of America.* **88**, 16–22.

Wallace, R.E. 1977. Profiles and ages of young fault scarps, north-central Nevada. *Bulletin of the Geological Society of America.* **88**, 1267–1281.

Young, A. 1960. Soil movement by denudational processes on slopes. *Nature.* **188**, 120–122.

Further reading

Bridge, J.S. and Demicco, R.V. 2008 Earth Surface Processes, Landforms and Sediment Deposits, Cambridge University Press, Cambridge, chapter 8 Movement of sediment by gravity, 255–277.

Carson, M.A. and Kirkby, M.J. 1972. Hillslope Form and Process. Cambridge, Cambridge University Press.

Carson, M.A. and Petley, D.J. 1970. The existence of threshold slopes in the denudation of the landscape. *Transactions of the Institute of British Geographers.* **49**, 71–92.

Evans, D. 2004. Slope Geomorphology. Critical Concepts in Geography Volume 2. Taylor and Francis.

Evans, S.G., Scarascia Mugnozza, G., Strong, A. and Hermanns, R.L. 2006. Landslides from Massive Rock Slope Failure. NATO Science Series: IV Earth and Environmental Sciences Volume 49. Berlin, Springer Verlag.

Strakhov, N.M. 1967. Principles of Lithogenesis. Edinburgh, Oliver and Boyd.

17 Fluvial Processes and Landform-Sediment Assemblages

The Skeidararsandur Braided Proglacial Outwash flowing from Vatnajökull, Southern Iceland

Earth Environments   David Huddart and Tim Stott
© 2010 John Wiley & Sons, Ltd

Learning Outcomes

After reading this chapter and completing the exercises you will be able to:

➤ Understand the physical controls on river flow in channels.

➤ Describe the transport and types of sediment load.

➤ Understand how rainfall is organized into drainage channels, systems and basins.

➤ Describe the river erosion processes and the controls on landform characteristics in rock channels.

➤ Understand why channel adjustments take place in cross-section and long profile.

➤ Understand why downstream size and sorting changes take place.

➤ Describe the fluvial landform-sediment assemblages in meandering, braided, alluvial fan and deltaic systems.

17.1 Introduction

Rivers are the routeways by which water and sediment are transported from the Earth's landmasses to the sea. They can be looked upon as the landscape's self-formed gutter system and are agents of both erosion and deposition. Gravity is the main controlling force by which excess water and moveable debris are carried from higher to lower elevations. The weathering products are carried down slopes into rivers, which are highly organized systems of physical and hydraulic variables and are complexly related to one another. At first sight there would appear to be an infinite variation within rivers, due to differences in local bedrock, topography, climate, land use and vegetation, but we must realize that although there is widespread variation in river morphology and fluvial landscape development, the physical principles controlling the fluvial processes are the same. The loose boundary hydraulics operate in the same way whether the river system is in the Arctic tundra, the high mountain Himalayas or the equatorial lowlands, and whether it is a contemporary river or was a Devonian river system.

In-depth treatments of these processes, the resultant geomorphic changes and the landforms and sediment assemblages produced can be found in specialist texts like Leopold *et al.* (1964), Best and Bristow (1993), Miall (1996) and Knighton (1998). The important applied aspects of fluvial geomorphology can be found in detail in texts such as Brookes (1988), Boon *et al.* (1992), and Thorne *et al.* (1997).

17.2 Loose boundary hydraulics

The basic principles of fluid hydraulics give a foundation for the study of the variables involved in fluvial geomorphology, the magnitude of these variables and the interaction between them. A river is a fluid and as such has no strength, as it can be deformed by shear force. It cannot resist stress: any stress on it causes movement, or strain. The forces that act on fluids are vectors, which can be resolved into components normal to and parallel with the fluid. Those components normal to the fluid surface are pressure and those parallel to the surface are shear stress (τ). The water flowing in a river channel is subject to two external forces: gravity and friction. Gravity is the basic force and is the tangential component of the weight of the water. It allows the downslope water movement. Friction on the other hand resists the downslope movement and occurs between the channel boundaries, which confine the flow. Shear forces oppose the water flow, caused by shear resistance along the banks and, to a much smaller extent, along the air–water interface.

The ability of rivers to cut a channel, erode the landscape and transport sediment depends on these forces. It can be said that climatic factors affect how much water the rivers in a particular area carry, but the resisting forces are controlled by the effects of bedrock variation, topography and vegetation type.

17.2.1 Types of flow in open river channels

Laminar flow

If water flows along a smooth, straight channel at very low velocities, it will probably move as laminar flow. Here parallel layers of water shear over one another, streamlines are parallel and separate from one another, flow moves parallel with the boundary and there is no mass transfer between layers. Any flow components other than those in the main flow direction are negligible and the velocity distribution is parabolic (Figure 17.1(a)). In laminar flow the layer of maximum velocity lies below the water surface. Velocity decreases with distance towards the bed, where it is zero. There is no mixing of fluid particles and it cannot support solid particles in suspension. However, it is probably not found in natural rivers, except perhaps near the bed and banks in the boundary layer.

Turbulent flow

When velocity exceeds a certain critical value the water flow becomes turbulent, where it is characterized by a variety of chaotic movements, with secondary eddies superimposed on the main flow (Figure 17.1(b)). Each fluid element follows a complex movement and mixing pattern. There is some water movement too, transverse to the main flow. Streamlines are complex and constantly changing through time so that, instead of well-defined water movement in sheets, there are complex eddies, giving an irregular, random motion to the general flow. The velocity gradient here is much greater nearer the boundary and much smaller away from it than in laminar flow. The variations in velocity are illustrated in Figure 17.1(b).

What controls whether fluid flow is laminar or turbulent?

The factors that affect the critical velocity at which laminar flow becomes turbulent are the viscosity and density of the fluid, the water depth and the roughness of the channel bed. The expression most commonly used to distinguish between laminar and turbulent flow is the Reynolds number, which is dimensionless. It is the ratio of the inertial force of a fluid to its viscous force and as such it represents the ratio between a driving force and a retarding force:

$$RN = \rho \, VR/\mu$$

where V is mean velocity, R is hydraulic radius (which is the cross-sectional area divided by the length of the wetted perimeter), ρ is the density and μ is the viscosity.

The ratio of this density divided by the viscosity is a fluid property called the kinematic viscosity, ν, and therefore $RN = VR/\nu$. For laminar flow, viscous forces are dominant (values under 2000) but in turbulent flow the dynamic viscosity is of less importance to the flow pattern (values are greater than 10 000). Most channel flow is turbulent and therefore has higher Reynolds numbers.

Types of turbulent flow

However, there are two types of turbulent flow: (i) streaming, the ordinary turbulence found in most rivers and (ii) shooting, which is found at higher velocities, for example in rapids, giving a great increase in the erosion rate. For any given boundary, liquid or temperature, there will be a range of values that define a zone of transitional flow between the two types. This transition zone also depends on the geometry of the flow and the surface roughness of the boundary. Whether turbulent flow is shooting or streaming is determined by the Froude number:

$$FN = V/\sqrt{gD}$$

where V is the river velocity, g is the gravity force and D is the water depth.

The Froude number is the ratio of the inertia force to the gravity force. If it is under 1 the river is in the streaming, subcritical or tranquil flow regime, but if it is over 1 the river is in rapid, supercritical or shooting flow. The boundary between the two is called critical flow, where FN = 1. Depth and velocity are the flow characteristics which determine the state of the turbulent regime. Two flows are dynamically similar if both the Reynolds

and Froude numbers are the same and this dynamical similarity is important for modelling as it allows the scaling down of large river systems so that they can be studied in laboratory flumes.

17.3 The energy of a river and its ability to do work

A river's energy can be measured as the product of a force (mass × acceleration) multiplied by the distance through which the force acts. Energy per unit volume, or the length of the river, is therefore equal to M × g. The river's turbulence and velocity are closely related to the amount of work it does, or is capable of doing, and this work is measured by the energy.

There are two types of energy: potential and kinetic. The former is converted to the latter by the action of the river flowing downstream. However, some of it is lost by friction, where the frictional heat losses depend on the characteristics of roughness, straightness, the amount of internal shearing by eddies and cross-sectional form. The energy which is not lost is available for work; that is, for erosion of the bed and banks and sediment transport. In the mountains, all energy is potential, where this is equal to the weight of the water × the head (the difference in elevation between the highest and lowest points). 95–97% of this energy is lost by heat generation and therefore only a small percentage is available for work. In between the source and the sea there is a conversion to kinetic energy, equal to half the mass of water × the square of the velocity at which that water is moving. So $Ep = Wz$ (weight × the head) which is converted by flow to $Ek = MV^2/2$.

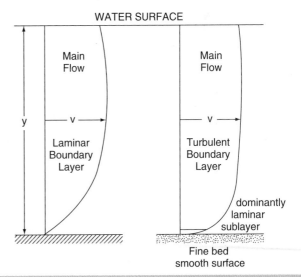

Figure 17.1 (a) Flow types in fluvial systems: (i) laminar; (ii) turbulent. (b) Variations in velocity (Source: after Knighton, 1998)

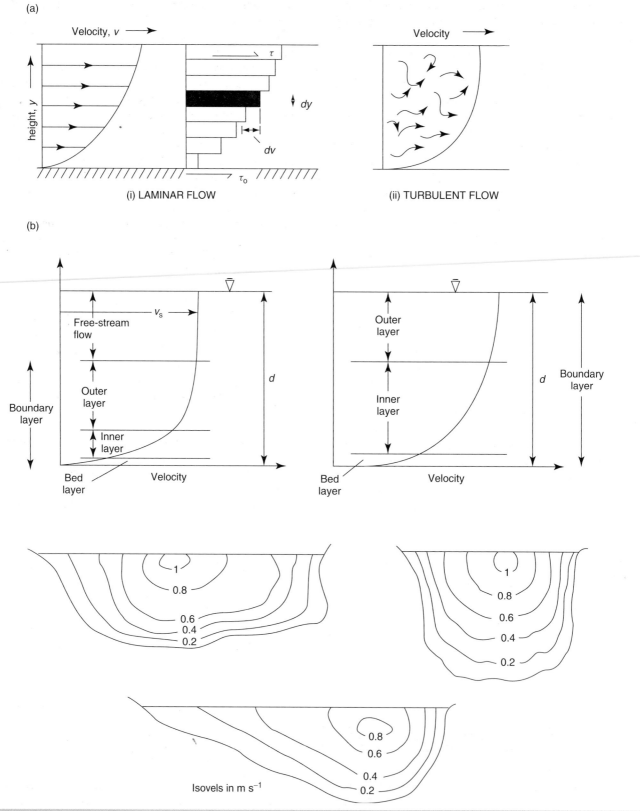

Figure 17.1 (*continued*)

The energy loss occurs through several processes. Skin resistance of the bed and banks depends for any given shape and size of the cross-sectional area on the square of the flow velocity and the boundary surface roughness. Internal distortion resistance is caused by discrete boundary features that set up secondary flow circulations and eddies, such as bars, bends, individual boulders and bedforms. It is proportional to the square of the velocity. Finally there is spill resistance, which occurs locally at particular places in open channels. Energy is lost by local waves and turbulence when a sudden velocity reduction is forcibly imposed on the flow. It is associated with local high velocities, such as when water banks up behind an obstruction, like a block of slumped bank sediment, and then spills over into lower-velocity flow.

The total energy is therefore influenced mainly by the velocity, which in turn is mainly a function of river gradient, volume of water, water viscosity and the characteristics of the channel cross-section and the bed. This relationship has been expressed in several equations, such as the Chèzy formula, which expresses the velocity as a function of the hydraulic radius and slope:

$$V = c\sqrt{RS}$$

where V is the mean velocity, R is the hydraulic radius, S is the slope and c is a constant which depends on gravity and other factors contributing to the friction force. In wide, shallow channels where R is almost equivalent to the mean depth, d, the mean velocity is proportional to the square root of dS. This formula is very difficult to use directly in judging river velocities that are very important in river erosion, so in 1889 the Manning formula was written in an attempt to refine the Chèzy formula in terms of its constant:

$$V = R^{2/3} S^{1/2}/n \text{ or } Q = A\, 1.49/n\, R^{2/3}\, S^{1/2}$$

where n is a roughness factor, which has to be determined empirically and varies not only between different streams but also within the same stream under different conditions and different times. It tries to take into account the frictional resistance due to the relative bed roughness (ratio of particle size to depth of the flow), bank resistance, channel sinuosity, bedforms, vegetation, eddy viscosity due to turbulence and the molecular viscosity of the water itself. The effects are not usually quantifiable and therefore wide estimators of roughness, as in the Manning formula, are used.

17.4 Transport of the sediment load

A river's ability to pick up (entrain) and transport sediment depends on its energy. It has been seen that most of the energy is converted to heat by the turbulent flow and bed and channel wall friction and hence is lost, but nevertheless sediment transport is an extremely important process. To move any sediment grain on the bed, a force must be exerted upon it, the normal upward component of which is equal to the downward normal component of the immersed weight. The force that is exerted on the grain is usually thought of as a drag stress, or critical shear stress (Figure 17.2(a)), proportional to the exposed area of the grain. As in rivers, turbulent eddies near the bed cause a local flow velocity fluctuation at any point at which there is a random chance that any given grain will be transported rather than its neighbours. Before we look at the drag force or the critical tractive force, we must point out another force which acts on grains normal to the main flow: the lift force (Figure 17.2(b)). Water flow past a grain on a level bed is faster over the top than around its sides, and at the contact between the grain and the boundary, the water is stagnant. It follows that the water pressure is greater on the underside of the grain than on the upper side, and as a result there is a net upward force exerted on the grain. This is capable of initiating transport of sediment. This idea of a lift force explains the movement of grains by saltation. If you temporarily neglect the drag force, it is easy to show that at high velocities grains would be expected to move up and down in a small water thickness close to the bed. The reason is that the lift force rapidly weakens with increasing height above the bed. At the bed, the lift force at high velocities may be greater than the submerged weight of the grain and, according to Newton's Second Law of Motion, the grains will accelerate upwards. As the grain moves away from the bed, the lift force decreases, and when this force is smaller than the submerged weight, deceleration and eventually acceleration back to the bed occur. If the drag force acting on the grain is superimposed onto this motion, the result is a bouncing motion in the water-flow direction. High-speed photography shows the initial movement direction is normal to the bed: the lift force is the actual initiating mechanism for grain movement.

Another sediment transport process is rolling. It is impossible to initiate sufficient lift force to create saltation in gravels, but sufficient tractive force is often available to produce rolling, especially in steep channels. Here the drag force is accompanied by the downslope weight component of the particle.

The third main transporting mechanism is suspension, which can be regarded as a simple extension of saltation. In this case the ratio of the lift force to the submerged weight is so great that the grains remain suspended above the bed for as long as the flow conditions prevail.

The critical tractive force (or the critical shear stress or drag force), as we have seen, is that force required to entrain a given particle. The velocity at which this force operates on a given slope is called the erosion or critical velocity. It is obvious that a greater tractive force is required for the entrainment of larger particles. This erosion velocity is sometimes known as the competent velocity and is the lowest velocity at which particles of a given size on the river bed will move. Much lower erosion velocities are needed to move sand than to move silt or gravel. For particles over 0.5 mm diameter the erosion velocity increases with grain size. For particles under 0.5 mm it decreases with grain size.

These ideas were expressed by Swedish geomorphologists following laboratory work and can be seen in the Hjülstrom and Sundborg diagrams (Figure 17.3). The erosion velocity curve is shown as a band because the value of the erosion velocity varies depending on the characteristics of the water and the grains to

(a)

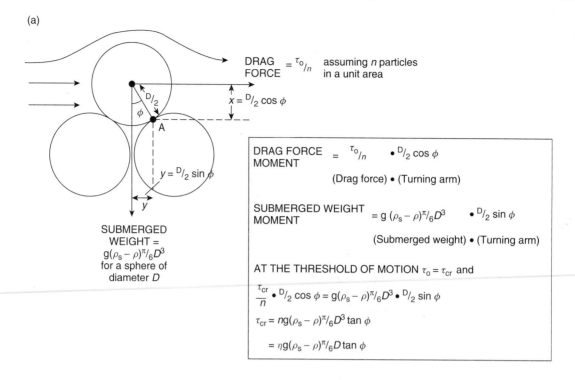

(b)

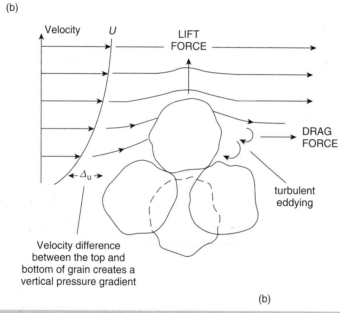

(b)

Figure 17.2 Critical shear stress, lift and drag forces

be moved. Grains of the same size but with different densities, lying on different beds and with different bed distributions, need different velocities in order to be moved. The velocity also varies with the water depth and density. The Hjülstrom diagram outlines broadly the realms of erosion, transport and deposition in terms of velocity and grain size, and two important conclusions can be drawn from it: sand is easily eroded, while silts, clays and gravels are more resistant. For silts and clays there are cohesive forces binding the grains together. Also, the bed tends to be smoother.

Gravels are harder to entrain because of their size and weight. The curves also show that once silts and clays have been entrained, they can be transported at much lower velocities. For example, grains 0.01 mm in diameter are entrained at a critical velocity of about 60 cm s^{-1}, but remain in motion until the velocity drops even below 0.1 cm s^{-1}.

The settling velocity is an important concept which explains why sediment is kept in suspension. Stokes' Law of Settling Velocity applies to small grain sizes and follows from an analysis

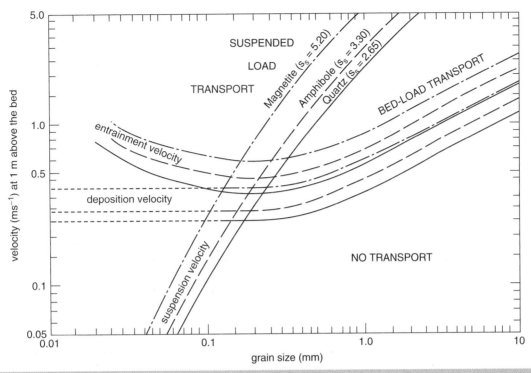

Figure 17.3 Hjülstrom and Sundborg diagrams

of the forces which oppose each other and affect a grain in suspension. The resistance a fluid offers a spherical grain falling through it depends on the surface area of the grain, $6\pi r$, the viscosity of the fluid, μ and the fall velocity, V. The buoyant force tending to hold the grain up in the fluid is equal to the volume of the sphere, $4/3\pi r^2$, multiplied by the fluid density, ∂_2, and the gravity force, g. The settling rate depends mainly on the grain size and the density, because the factors of gravity, viscosity and water density are constant for a given time at a given point in a river. For small grains we can say that settling velocity is proportional to the square of the grain diameter, whilst for larger particles the settling velocity is proportional to the square root of the diameter. It is clear that the tractive forces and settling velocities control whether particles will be transported or deposited. Deposition occurs when the river is no longer competent to carry the sediment load. This might be because of decreasing gradient, decreasing water volume, increasing load calibre or a damming of the channel. The gradient might change from one rock type to another, and it will suddenly decrease when a river flows from a mountainous area onto a plain, or when it flows into a lake or the sea. A decreasing volume can occur when an increase in vegetation cover promotes infiltration and the detention of rain water through storage and vegetative use. It might also occur after climatic change, where less rainfall is supplied to a river basin, and in arid regions the volume of water decreases downstream as the river loses water by evaporation or seepage. Deposition may also occur when the load supplied to a river exceeds its competency or capacity, for example the load supplied from a glacier or from volcanic ash on volcano

slopes. Often a more turbulent, steeper tributary carries a load which the main river is unable to transport. The change in load calibre may also cause deposition when the grain size is increased beyond the competency; for example, when the river incises into its channel it may erode sediment of a greater grain size. It might also be competent to entrain at one location but incompetent downstream, where deposition will occur.

17.5 Types of sediment load

There are various categories of sediment load in rivers. Some rivers carry more dissolved load (or wash load) than they do solids. Whether this is the case depends on the relative contribution of ground water and surface runoff to river discharge. If river discharge is stable, resulting mainly from ground water flow, then the concentration of dissolved material will generally be high. When ground water flows slowly through the ground it has more time to pick up chemical elements released by weathering and by the breakdown of organic matter than does water flowing over the surface. However, when surface runoff is the main contributor to flow the concentration is usually low. Due to their small size, the particles have such a low settling velocity that they can pass through the fluvial system relatively unaffected by the hydraulic conditions of any given river reach.

The solid sediment load is divided up into bedload and suspended load. Bedload moves by rolling and saltation and is that portion of the total sediment load whose immersed weight is carried by the solid bed. Suspended load on the other hand is that part of the total load whose immersed weight is carried by the fluid: that carried by the upward momentum in the turbulent flow eddies. It is therefore finer than the bedload. Once particles are entrained and part of the suspended load, little energy is required to transport them and they can be carried along by a flow that has a velocity less than the critical velocity needed for their entrainment. Moreover, the suspended load decreases the inner turbulence and reduces the frictional energy losses, making the river more efficient. Usually the concentration is highest nearest the bed and rapidly decreases vertically. There is a good correlation between suspended load and river discharge.

Two related concepts are the competence and the capacity of a river. The competence refers to the maximum size of sediment that the river is capable of transporting and the capacity is the total load actually transported. When the velocity is low, only small grain sizes can be moved, but as the velocity increases, larger grains can be entrained. Rivers in flood can shift considerable amounts of sediment because the maximum particle mass that can be moved increases with the sixth power of the velocity. Much sediment in upland rivers is not moved even by the highest flood, however, and the boulder gravels are likely to be residual in present-day conditions and will only be transported if they are worn down by collision impact and abrasion. The source is largely from glacial till, mantling the valley sides, or from glaciofluvial sediments, which were originally deposited under extremely high melt-water flow velocities, or even from subglacial water flow under high pressure.

17.6 River hydrology

Rivers are natural watercourses that form the land-based part of the global water cycle (where water circulates around the Earth, between the oceans, the atmosphere and the land). Without circulation of this water, driven by energy from the Sun, rivers would not flow. After rainfall the water droplets reaching the ground will pass into the soil by infiltration, if there are air spaces for them to pass through. If the air spaces become full of water, the soil will no longer be able to absorb any more and the water will have to flow over the surface as overland flow until it either eventually passes into the soil or reaches a river channel. The rate at which rainfall can infiltrate into the soil is known as the infiltration capacity. Depending on the ground surface characteristics and types of land use, this can vary markedly (Table 17.1). Water which gets into the soil moves down the slope as throughflow, either between the pores or through soil pipes. If the bedrock is porous or permeable, soil water can pass into the ground-water system by percolation. Occasionally in Britain snowmelt can generate significant rises in river levels, but in high mountains water storage in snow and glacier ice responds to seasonal temperature change and controls river levels.

17.6.1 Rainfall-runoff responses

If rainfall takes a fast route (Table 17.2), whereby it does not infiltrate into the soil, and provided that it is heavy and continuous,

Table 17.1 Infiltration capacities for various surfaces. (Source: after Stott, 2000)

Land use	Infiltration rate (mm per hour)
Old permanent pasture	57
Moderately-grazed pasture	19
Heavily-grazed pasture	13
Weeds or cereals	9
Bare ground (baked hard by Sun)	6

Table 17.2 Water pathways through a drainage basin. (Source: after Stott, 2000)

A Fast route	B Medium route	C Slow route
rain	rain	Rain
overland flow	infiltrates soil	infiltrates soil
stream channel	throughflow in soil	percolates rock
flows into sea	stream channel	flows from spring
	flows into sea	into stream channel
		flows into sea

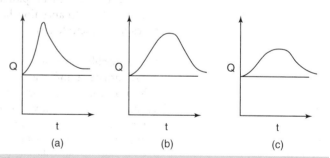

Figure 17.4 Storm hydrographs for the types of water pathway in a drainage basin: (a) fast; (b) medium: (c) slow. Q, discharge; t, time

rivers can rise rapidly and flooding can result. The rainfall–runoff response rate is rapid, as can be seen from the stream hydrograph in Figure 17.4(a), which rises and falls relatively quickly and has a flashy regime. Other types of hydrograph are shown in Figure 17.4(b) and (c). That portion of the stream flow below the horizontal line is known as baseflow and is maintained by soil throughflow and ground water when it does not rain.

The relationship between rainfall and discharge can be affected not just by the infiltration capacity but by many other factors. For example, the type of precipitation is crucial as heavy, prolonged rainfall can cause ground saturation, surface runoff and rapid river rises. Heavy snowfall means that precipitation is held in storage and river levels drop, but as the temperature rises there can be snow melt-water floods. If the basin is small, it is likely that precipitation will reach the main channel more quickly than it would in a larger drainage basin. This means that the lag time between the peak rainfall and the peak discharge is shorter in a small basin. A more circular basin will also have a shorter lag time because all points in the basin are roughly equidistant from the outlet. In mountainous catchments, where slopes are steep, water is likely to reach the river more quickly than in basins which have gentle gradient slopes.

Extremes of temperatures, both frozen ground in winter and baked, hard, dry ground in summer, can restrict the infiltration and hence increase runoff. Land use might be a factor as vegetation intercepts rainfall and stores moisture on leaves and branches before it evaporates back to the atmosphere. This means less water will reach the river channel. For example, dense forest can intercept up to 80% of rainfall, whereas arable farmland may only intercept 10%. Figure 17.5(a) illustrates the effects of afforestation on the stream hydrograph in the North Pennines. Here drainage ditching is carried out on the wet, upland catchments in order to drain the land before planting conifers. The lag times are reduced and increases occur in peak discharge. Figure 17.5(b) shows the effect of mature coniferous plantations on the storm hydrographs of the River Wye (moorland land use) and the River Severn (mature coniferous woodland); the geology and precipitation are the same in both basins. The peak flow is delayed and is lower in the Severn, and runoff is less rapid due to interception and evaporation by the trees. Harvesting of such coniferous woodlands means that interception will be considerably reduced and there will be a return to a more rapid response. This is further explored in Box 17.1 (see in the companion website). Urbanization has increased flood peaks too, as water cannot penetrate tarmac and concrete.

The geology can be important as permeable rocks allow water to percolate into them, causing less surface runoff and fewer streams, as in limestone terrain. This is in contrast to impermeable rock types like granite, which do not allow much water to pass through them and so produce more surface runoff and many streams. The soil type controls the infiltration speed, the storage and the throughflow rate. This is because sandy soils, which have large pore spaces, allow rapid infiltration and throughflow, whereas clay soils have small pore spaces and sometimes swell when wetted, reducing infiltration and throughflow and encouraging surface runoff. The drainage density is also a factor and refers to the number of streams in a given area. The higher the number of streams, the greater the chance of a flash flood.

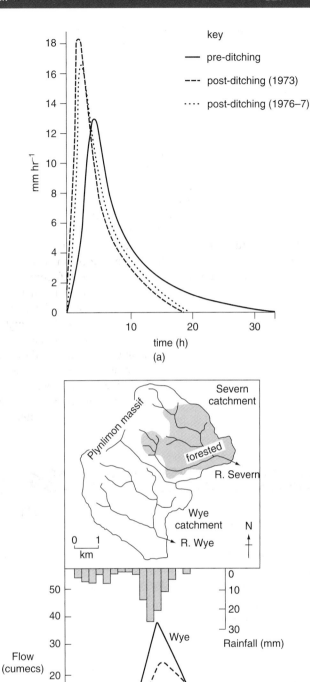

Figure 17.5 Impact of forestry activities on stream hydrographs. (a) Impact of ditch excavation on a forested catchment at Coalburn in the Northern Pennines. (b) The effect of mature coniferous plantations in upland Britain on the storm hydrographs of the River Wye (moorland land use) and the River Severn (mature coniferous forest). (Source: after Collard, 1988)

17.7 The drainage basin

River systems consist of channels: the troughs in which the actual water flows. The valley is a much larger landform and may not be related to the river itself; for example it could be a glacially-eroded trough. Streams join together to form networks, which are a pattern of streams draining a specific area through a common outlet. The valley area drained by a stream, or a network of streams, is defined as the drainage basin. Its boundary is marked by a ridge of high land beyond which any precipitation will drain into the next adjoining drainage basin. This boundary is the watershed, or divide. This drainage basin has inputs, transfers, storage points and outputs, as indicated in Figure 17.7. It is an open system and forms an important part of the water cycle.

The water balance shows the state of equilibrium in the drainage basin between the inputs and the outputs. It can be expressed as:

$$P = Q + E \pm \text{changes in storage}$$

where P is the precipitation (using rain gauges), Q is the discharge (measured by weirs or flumes in the river channel) and E is evapotranspiration, which is more difficult to measure but is usually quantified either by direct measurements using evaporation pans

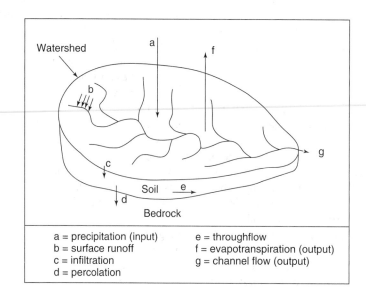

a = precipitation (input)
b = surface runoff
c = infiltration
d = percolation

e = throughflow
f = evapotranspiration (output)
g = channel flow (output)

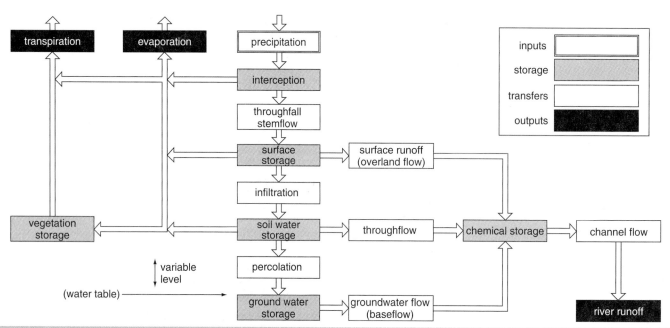

Figure 17.7 Drainage basin components. (Source: after Waugh, 1990)

and lysimeters, or by meteorological formulae or moisture budget methods (see Briggs and Smithson, 1985). Precipitation is hence the driving force for this system.

17.7.1 Origin of channels and drainage basins

There is a strong interrelationship between drainage basins and hillslope processes, as you have to assume that a slope will always be present as part of any land surface, which means water will flow downslope and there will be a concentration of this flow at some point. On a sparsely-vegetated, new land surface, erosion takes place through wash processes where a concentrated wash occurs on the primary slope to produce a system of rills. However, there will be a small area between the divide and the top of the rills in which there is little or no slopewash because the depth of overland flow is not great enough or the velocity is not fast enough to cause erosion (Figure 17.8(a)). Hence the initial stage is a pattern of more or less parallel rills of variable length which may be destroyed before the next storm or may stay in position (Figure 17.8(a)). In the second stage, if the rills are constant in position then they may develop further. If one rill is longer than another, it collects new drainage, and therefore erodes deeper than its neighbour. When there is overland flow, the rill divides are overtopped and there is a tendency for the flow to occur towards the deepest rill, breaking down the divide. Therefore master rills form, where rills join up and increase the flow in the next rill, and this tends to focus slopes towards the master rill.

This development of slopes across the general gradient is called cross-grading.

Further development is a lottery, a random process in which the rill gaining the most drainage captures more rills and eventually captures them all, thus winning the game. This is how it is often thought the typical drainage network develops (Figure 17.8(b)). However, there are problems explaining river networks in this manner. It seems that the processes outlined above are confined to rill systems that are very small-scale on largely nonvegetated surfaces, whereas rivers are comparatively large. Very rapid change can affect rill systems, whereas river changes are slow and gradual. Therefore, although rill cross-grading is responsible for the small-scale organization of drainage on newly-exposed ground, it is unlikely to be the cause of overall river channel development.

One of the big differences between rill systems and river systems is the constancy of the flow, as rill systems tend to dry up after a storm, and no water is stored. Rivers, on the other hand, flow almost all the time and therefore water storage – and the storage location – is likely to be of importance to their origins. One way in which larger-scale drainage elements may develop is through the action of subsurface seepage, where water moves slowly through the soil and is stored there, effectively from one rainstorm to the next. A slight concentration of flow, possibly due to a hollow in the bedrock surface below the soil, may cause flow concentration along one line down a slope (Figure 17.8(c)). This gives a definite zone, or percoline, in which water movement through the soil occurs, and this is flanked by other zones with rather less flow. There is increased chemical weathering of the bedrock surface here, which tends to increase the channel in

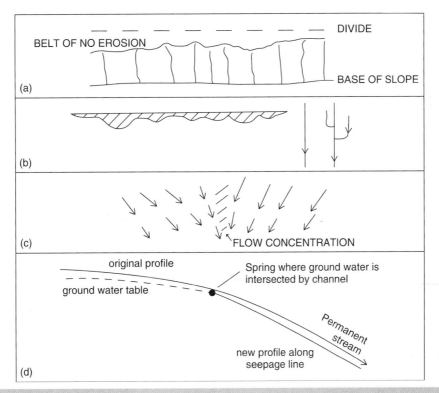

Figure 17.8 Origin of rill systems

the bedrock. The soil strength is therefore reduced and there is a tendency for faster creep and landslipping, and therefore a removal of soil cover. So it seems likely that channels develop along enhanced seepage zones in the soil.

Eventually the channel caused by the increased bedrock erosion and soil erosion along a seepage line will intersect the groundwater table and a spring will occur. Below this spring a permanent stream can develop (Figure 17.8(d)), and eventually a drainage basin, due to the diversion of flows towards the new channel. In this way a permanent stream can develop below the soil and plant cover and can occur on a scale comparable with river networks. This means that rills and rivers are different in origin, even though their patterns are the same. Figure 17.9 illustrates the development of a rill system on a coal waste tip.

The extension of the drainage network can be caused by: (a) surface runoff with extension of the rill system; (b) headward erosion of existing streams by spring sapping; or (c) the initiation of new channels by slipping, or by percoline activity. New channels can be created in catastrophic rare storms, when landslide scars are eroded and mudflow channels provide new drainage lines. However, these will only be maintained if sufficient flow is present, or if it is diverted to these areas, so that permanent water movement occurs in them. Otherwise they will be filled in again by slope processes.

The limits to drainage development seem to be reached fairly quickly in any given area as the form and drainage density of small basins are usually similar to those of large ones. Predetermined watersheds (divides) are a limit because these determine the area drained by a stream. If a watershed stays in roughly the same position through time, the slope will get lower as erosion proceeds, but the drainage area will remain approximately constant. Hence headward growth is limited by the position of the watershed. Assuming roughly constant climatic conditions, headward channel extension will occur very quickly and then stop

when an equilibrium has been reached, because concentration of subsurface flow will cause channel formation to occur below the point at which the concentration is effective enough to cause increased erosion. Above that point no channel formation will occur, so that when the permanent stream channel has extended to that point, no further extension will take place. Catastrophic events can cause gullying above the permanent stream position but flow will not be enough to maintain it and it will eventually fill in. Thus these events are likely to be relatively minor fluctuations of the stream network.

The limit to drainage development, other things being equal, seems to be the dynamic equilibrium between hydrology, weathering and hillslope processes. The drainage network is adjusted to carrying away sediment and particularly water which passes into it. There is however slow evolution over long time periods, as the hydrology, climate, topography and weathering processes may all change and these changes may require the growth or contraction of the river pattern. Hence, over geological time, rivers are not in equilibrium and most of the important elements in the system change.

17.8 Drainage patterns and their interpretation

The drainage pattern that evolves in an area depends on many variables, but the geology can give rise to specialized types of pattern in which there is a marked structural control. However, the dendritic drainage pattern (Figure 17.10(a) and (b)) is the simplest and most common form. It results from fluvial processes operating in homogeneous terrain where there are no strong geological controls. However, a parallel pattern can develop in which there is a steep regional dip that gives a marked drainage direction. A trellis pattern indicates both a regional dip and strong geological control, where folded sedimentary rock occurs. A rectangular pattern develops where there is strong right-angled jointing and faulting, whereas a radial pattern occurs around an eroded structural dome or volcano (Figure 17.10(c)). The annular pattern is also associated with an eroded dome, but here the fluvial system develops along weaker strata in bedded rocks. Where there are hummocky sediments, such as those sometimes left by glacial deposition, or where there is limestone solution in an area, multibasinal drainage occurs. Finally, contorted drainage occurs where the landscape is heavily influenced by neotectonics and metamorphic activity.

17.9 Fluvial channel geomorphology

A river system creates distinctive landscapes where there are different controls upon it, for example high mountain rivers where rivers move from mountains onto plains, or low gradient systems

Figure 17.9 Rill system development on a coal waste tip, Lowca, West Cumbria, UK

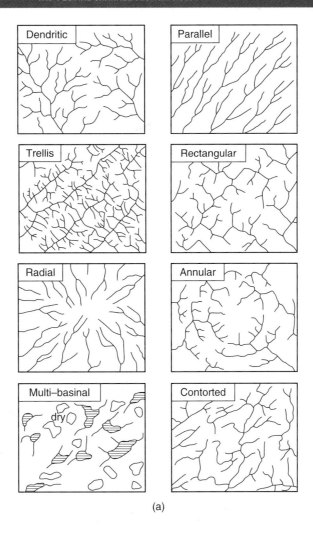

(a)

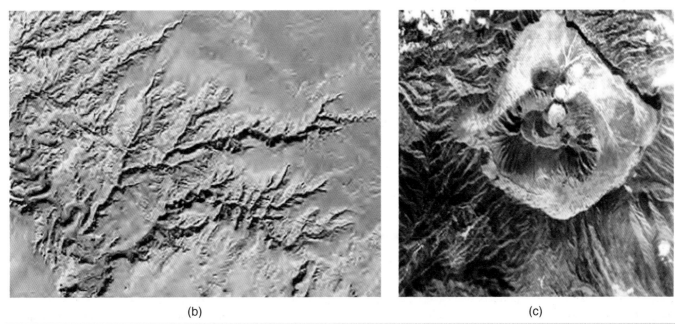

(b)   (c)

Figure 17.10 (a) Drainage patterns. (b) Dendritic drainage, Dirty Bend, Utah, USA. (c) Radial drainage, Bromo volcano, Java. (Source: http://rst.gsfc.nasa.gov)

closer to the sea. This is despite the fact that the fundamental loose boundary hydraulic processes are the same. So there are distinctive river landform–sediment assemblages, which we will consider in the next section.

Here we will consider river erosion processes, the morphological activities of rivers in terms of small-scale sediment transport and river channel long-profile and cross-section adjustments. The morphology of the river channel, including its cross-sectional shape, size, longitudinal profile and planform pattern, is the result of sediment erosion, transport and deposition processes taking place within the controls imposed by the geology and terrain of the drainage basin. Rivers are constantly evolving and adjusting as a response to the sequence of normal flow, flood peaks and droughts, which are controlled by the regional climate, local weather and catchment hydrology. In this respect the channel geomorphology can be best explained if distinctions are made between those factors which drive the fluvial system in producing the channel, those which characterize the physical boundaries of

the channel and those which respond to the driving and boundary conditions and define the channel form.

17.9.1 River erosion processes

There are four principal erosion processes:

1. Corrasion takes place when the river picks up sediment, which acts like sandpaper and wears away the rock by abrasion. It is effective during flood stages and is the major process by which rivers erode horizontally and vertically. It can produce bedrock hollows and potholes, where gravels can be trapped in pre-existing holes, and because the current is turbulent, the gravel can be swirled around in these holes which enlarge vertically.

2. Corrosion or solution is where the rock is dissolved in the water. It occurs ubiquitously and continuously and takes place independent of river discharge or velocity, although

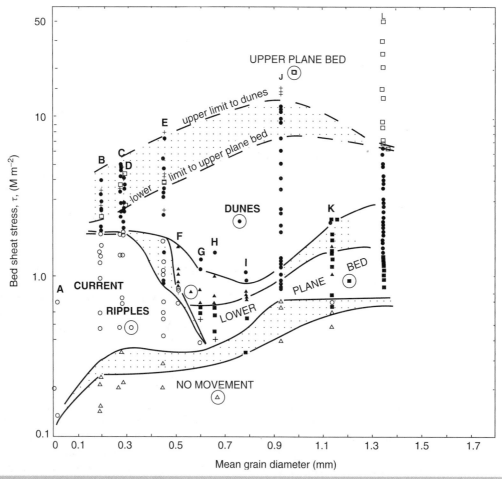

Figure 17.11 Bedform phase diagram to show the stability fields of the various bedforms developed by uniform steady flow over granular beds in straight laboratory channels. Note the degree of overlap (stippled) between fields. (Source: Recalculated data from Costello, 1974: F,G,H,I,K; Guy *et al*, 1966: B,C,D,E,J; Mantz, 1978: A and Williams, 1970:L. Adapted from Leeder, 1982)

it can be affected by the chemical composition of the water, particularly from the soil system. For example, where there is limestone bedrock the soil carbon dioxide and humic acids can produce high calcium concentrations in the water.

3. Attrition occurs where the sediment particles being transported, particularly as bedload, collide with one another, which may cause breakage into smaller fragments. This accounts for originally angular material becoming progressively more rounded with time and distance downriver.

4. Hydraulic action takes place where the sheer force of the turbulent water hitting river banks, as on the outside of meander bends, is forced into cracks. Here the air in the cracks may be compressed, increasing pressure and causing expansion of the cracks and eventually bank collapse. Cavitation is part of this hydraulic action, where air bubbles collapse, producing shock waves which weaken the channel perimeter as they hit. See Box 17.2 in the companion web site for recent work on Alpine proglacial suspended dynamics in warm and cool ablation seasons and the implications for global warming.

17.9.2 Morphological activities of rivers

Sand-bed rivers

A river bed offers resistance to flow, which is a function of both the bed sediment and the nature of the flow. The total resistance can be divided into two components: the form resistance of the individual particles on the bed and the shape resistance due to the bedforms into which the particles have been moulded by the flow. These bedforms reflect the deformable nature of a bed formed of noncohesive sediments and it has been found by observations in natural sand-bedded rivers, and particularly in laboratory flume experiments, that there is a variety of these bedforms under changing flow conditions. There is a characteristic bedform sequence with ascending flow power, shear stress and flow velocity, accompanied by sediment transport over the bed. These stability fields for different bedforms can be seen in Figure 17.11. The sequence from no movement on a plane bed to transitional forms occurs in tranquil flow, and the sequence from upper plane bed with sediment movement to antidune bed waves is confined to rapid flow.

The types of sand bedform are classified as follows (see also Table 17.3):

1. **Small-scale current ripples:** These are small, triangular bedforms with gentle upstream (stoss) and steeper lee slopes, which are under 4 cm in height, under 60 cm in chord and have a vertical form index in the range 5–20. They migrate by erosion on the stoss side and deposition on the lee side. Their height is independent of water depth. Six basic patterns are recognized, based on crestline shape (Figure 17.12). Examples are given in Figure 17.13(a)–(c).

Table 17.3 Sequence of bedforms with ascending flow power

	1. Plane bed with no sediment movement	
1–5 Tranquil flow	2. Small-scale current ripples	Grain size under 0.6 mm; stream power 100–1000 ergs per cm^2 per second; lower-phase plane bed where grain size over 0.6 mm
	3. Large-scale current ripples (dunes) with superimposed small-scale current ripples	Stream power 1000–2000 ergs per cm^2 per second
	4. Large-scale current ripples (dunes)	
	5. Transitional between dunes and plane beds	
6–7 Rapid flow	6. Plane beds with sediment movement	Over 2000 ergs per cm^2 per second
	7. Standing waves and antidunes	

2. **Large-scale current ripples or dunes:** These are large, triangular bedforms which occur when the boundary shear stress and stream power associated with ripples is increased. They are over 60 cm in chord, over 4 cm in height and have a vertical form index 10–100. The maximum amplitude they can develop is the approximate average depth. In plan they vary from linguoid, through straight, to lunate. Examples are given in Figure 17.13(d) and (e).

3. **Transitional forms of bed roughness:** Included here are washed-out dunes, flat sand bars and plane beds.

4. **Upper-phase plane beds:** A plane bed is a bed without elevations or depressions larger than the maximum size of the bed sediment, where there is a relatively low resistance to flow by the bed and a high sediment transport rate.

5. **Antidunes:** These bedforms occur in trains of inphase symmetrical sand and surface water waves. The height and length of these waves depends on the scale of the flow system and on the water characteristics and bed sediment. They do not exist as a continuous train of waves that never changes shape, but rather gradually form from a plane bed and a plane water surface (Figure 17.14). They can grow in height until they become unstable and break, or they may gradually subside. As the water and bed surfaces are in phase this is a positive indication that the flow is rapid.

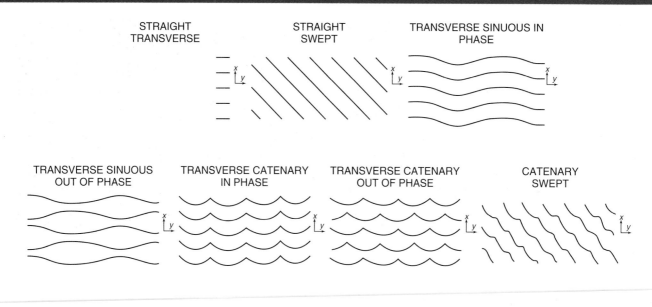

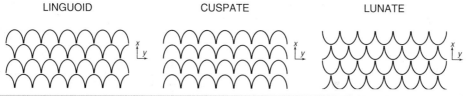

Figure 17.12 Various planforms recognized for ripples and dunes (after Allen, 1968)

However, the terminology of bedforms is controversial, with the major differences being due to the large variation in scale observed in the laboratory flume and in natural field conditions. For example, Coleman (1969) had his own bedform classification on the Brahmaputra, and similar scale bedforms have been noted in the Mississippi:

1. **Ripples:** Up to 30 cm in height, with a very variable form index.

2. **Megaripples:** 0.3–1.5 m in height. Height independent of water depth. Most common during low water and rising stage.

3. **Dunes:** 1.5–7.5 m in height, vertical form index 30–60. Most common bedform during peak flood.

4. **Sand waves:** 7.5–15 m in height. Develop most frequently at peak flood and falling stage. Smaller bedforms are superimposed on their backs, generally megaripples and dunes.

5. **Large-scale lineations:** Closely-spaced parallel ridges and grooves, 7.5–30 m apart. Individual ridges vary in width from 1 or 2 m to 7 m and in height from 1 to 2 m. Commonly they are over 1000 m in length, with no evidence to suggest whether the crests are built up or the troughs scoured.

Flow regime concept

All these bedforms produced by sediment transport in sand-bed rivers produce an internal stratification which can be related to the bedform that produced them. In interpreting fluvial depositional environments of whatever age, the relationships between stratification and the flow conditions are very important. This brings us to the flow regime concept. If the type of stratification is considered instead of the bedform then for a given grain size, any change in stream power can be recorded by simply recording variations in stratification. However, changes in slope, flow velocity and depth cannot be related simply to stratification changes. Simons *et al.* (1965) expressed a simplified equation relating the many variables to bed roughness:

$$\text{Form of bed roughness} = f\,(\text{S, D, d, w, z, p})$$

where S is the slope of the energy grade line, D is the depth, d is the median diameter of the bed sediment w is the width, z is a measure of the size distribution of the bed sediment and p is the density of the sediment–water mixture. Any single bedform change in a fluvial sedimentary sequence, shown by the internal stratification (examples shown in Figure 17.13(f),(g),(j)), could be the result of several combinations of change in the above variables. The flow

Figure 17.13 Bedforms and cross-stratification types: (a) low flow ripples; (b) ripples with a silt drape; (c) ripples in Old Red Sandstone (Devonian); (d) dunes, Wells-next-the-Sea, East Anglia; (e) dunes, river estuary near Morfa Dyffryn, western Wales; (f) tabular cross-stratification, Venus Bank glaciofluvial, Shropshire; (g) trough cross-stratification, Carboniferous Whitehaven Sandstone, Saltom shore, Whitehaven; (h) horizontal stratification in coarse sand and pebble gravel, Harrington proglacial sandur, West Cumbria; (i) small antidunes in runnel on beach at Keem bay, Achill, western Ireland; (j) parallel lamination and small-scale ripples, Tunsbergdalen, central Norway

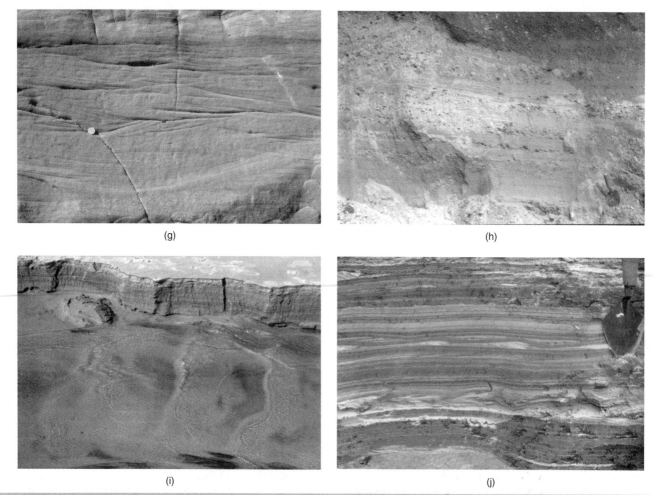

(g)

(h)

(i)

(j)

Figure 17.13 (*continued*)

regime concept, which is based on the integrated resultant of all these variables, tries to circumvent this problem. Fluvial channels have been classified into lower and upper flow regimes, with a transition in between. The classification is based on the form of the bed configuration, the mode of sediment transport, the process of energy dissipation and the phase relationship between the bed and the water surface.

In the lower flow regime the resistance to flow is large and sediment transport is small. The bedforms are either ripples or dunes, or a plane bed with no sediment movement where the grain size is over 0.6 mm. The water surface undulations are out of phase with the bed surface and the most common type of sediment transport is of individual grains up the back of a ripple or dune and avalanching down its lee face. Velocity is low and the water surface placid.

In the upper flow regime resistance to flow is small and sediment transport is large. The usual bedforms are plane beds or antidunes and the mode of transport is the individual grains rolling almost continuously downstream in sheets a few diameters thick, or in thin bed waves. Velocity is high.

There is a transition between the two flow regimes, where the bed roughness may vary from that typical of the lower flow regime

to that typical of the upper flow regime, depending mainly on the preceding conditions.

Harms and Fahnestock (1965) related the hydrological regime of the Rio Grande to its bedforms and showed that the basic sequence of changes in bedforms with increasing flow power in flumes was correct. However, in the Brahmaputra, Coleman (1969) showed that during the period of increasing discharge and velocity the bed underwent a definite sequence of changes. At first smaller bedforms (megaripples) were present and the water surface was relatively smooth. Dunes appeared as discharge increased. This was initially poorly developed, with the bed very irregular in both transverse and longitudinal section, and the water surface was randomly broken with small surface boils. After a period of time a well-developed dune field existed and surface turbulence assumed a regular pattern controlled by the position of the dune crests. Where the velocity reached a maximum, sand waves developed in some areas, and in others a plane bed existed. Water turbulence reached a maximum intensity, forming a regular pattern orientated parallel to bedform crests when sand waves were present. The sequence is similar in many respects to that indicated from flume work, but the scale is very different, although the repetitive sequence is apparent.

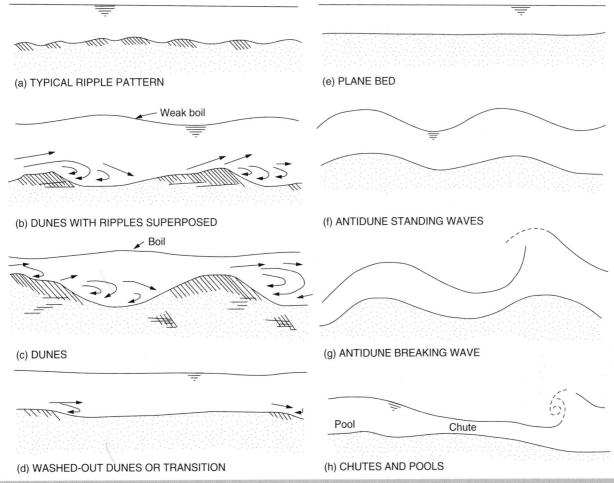

(a) TYPICAL RIPPLE PATTERN

(b) DUNES WITH RIPPLES SUPERPOSED

(c) DUNES

(d) WASHED-OUT DUNES OR TRANSITION

(e) PLANE BED

(f) ANTIDUNE STANDING WAVES

(g) ANTIDUNE BREAKING WAVE

(h) CHUTES AND POOLS

Figure 17.14 River bedforms and surface water morphology in an alluvial channel

17.9.3 River channel adjustments in cross-section

It has been shown that river width, depth and velocity are related to discharge. It was suggested that the term 'hydraulic geometry' be used to describe graphs showing how these parameters change (Figure 17.15). In the downstream direction, hydraulic geometry graphs have the following relationships:

$$\text{width} \propto Q_w\, 0.5$$
$$\text{depth} \propto Q_w\, 0.4$$
$$\text{velocity} \propto Q_w\, 0.1$$

These shape characteristics appear to be the ones that distribute available energy within the system in such a way so as to bring about a balance between input and output of water and sediment and a stability of form. The hydraulic geometry equations are maintained by negative feedback in the system. If any of the above three characteristics change locally in a river, a set of reactions occurs among the variables in order to absorb the effect of the initial change and restore the original condition. River velocity tends to increase downstream for flows of a particular

frequency – maybe once a year. It does not decrease downstream because slope does. Velocity is also controlled by depth and frictional effects related to the channel roughness, so it is proportional to depth × slope/roughness. In the downstream direction channel depth increases, while both slope and roughness tend to decrease. Only slope changes would mean that velocity decreases downstream. The other factors increase velocity downstream and overcompensate for any slope reduction.

17.9.4 River channel adjustments: The long profile

As more water joins the main river, the discharge, the channel cross-sectional area and the hydraulic radius will all increase. The result is that the river flows over a gradually decreasing gradient, which in profile shows a characteristic concave long profile (Figure 17.16(a)), although there are often irregularities in the long profile, especially closer to the source (Figure 17.16(b)). The lowest point to which river erosion can take place is the base level, the current sea level. However, there can be local base levels on the long profile too, such as where the river flows into a

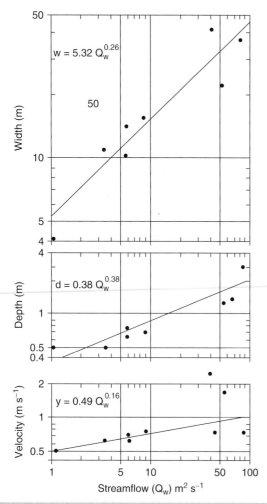

Figure 17.15 Graph of downstream hydraulic geometry for the Palouse River, Washington State, USA, derived from analyses of data at eight stations along the river. (after Leopold and Maddock 1953)

lake, where there is a resistant rock unit across the river valley, or where a tributary joins the main river. A graded river has a gently sloping long profile, with the slope decreasing towards the river mouth (Figure 17.17(a)), and this is where a river exists in a state of balance, or dynamic equilibrium, with the rate of erosion being equal to the rate of deposition. The balance is always transitory as changes in velocity, water volume and load increase either the rate of erosion or the rate of deposition until a state of equilibrium has again been reached. If, for example, the long profile of a river has a waterfall and a lake, erosion is likely to be greatest at the waterfall, whilst deposition will slowly fill in the lake (Figure 17.17(b)). In time both landforms will be eliminated and a graded profile will result. Grade not only affects the long profile but is a balance in the river's cross-section and channel roughness too. All aspects of the river's channel are adjusted to the discharge and the load of the river at a given point in time.

17.9.5 Changes in base level

Two groups of factors can influence changes in base level: climatic, with the effects of glaciation and changes in rainfall (either an increase or a drought), and tectonic, where there is crustal uplift. Base levels can be either positive or negative. Where there is a sea-level rise in relation to the land, the result is a decrease in the river slope, with a consequent increase in deposition and flooding in coastal zones. Negative base-level change occurs when sea level falls in relation to the land. This causes land to emerge from the sea, an increase of river gradient and therefore an increase in river erosion. This process is called rejuvenation. During the Pleistocene period, sea level has risen and fallen due to deglaciation/ice melting and depression of the crust by the build-up and weight of the ice. This has resulted in rejuvenation caused by isostatic uplift on several occasions, and many rivers show partly graded profiles (Figure 17.17). If the land rise is rapid, the river does not have time to erode vertically to the new sea level and so it may descend over waterfalls. In time the river cuts backwards and the waterfall, or knickpoint, retreats upstream and marks the maximum extent of the newly-graded profile. Excellent examples of rejuvinated rivers are those associated with the Craven faults in north-west Yorkshire, especially in the Ingleton Glens, where there are waterfall knickpoints (Figure 17.18(b)), incised gorges and valley-in-valley landforms. If the uplift of the land, or the sea-level fall, continues for a long period then the river may cut down through bedrock to form incised meanders, like the River Wear at Durham. These are entrenched, incised meanders with a symmetrical cross-section which results from rapid incision, or from valley sides resistant to erosion. Ingrown meanders occur where the uplift of land or incision by the river is less rapid, allowing the river more time to erode laterally and produce an asymmetrical valley shape (for example, the River Wye at Chepstow).

River terrace sequences show that former floodplains of the river are left higher than the contemporary river after vertical erosion. Excellent examples are seen in the Thames Valley (Bridgland, 1994); two parts of the valley are shown in Figure 17.19. There has been much debate about the possible correlation between the formation of these river terraces and Pleistocene climatic fluctuations. The terrace aggradations in the upper parts of river valleys were considered to be the product of cold-climate environments, while those in the lower reaches have frequently been attributed to aggradation in response to relative rises of sea level during interglacials. Direct correlation between these two types of terrace should therefore not be possible. However, it has become apparent that in the Thames the cycles of river aggradation and rejuvenation were superimposed upon a general decline in sea level since the Pliocene, with immense thicknesses of sediment eroded in Britain during this period. This resulted in gradual tectonic readjustment, which explains the progressive lowering of base level that is needed for the formation of a terrace sequence. For the Thames Valley it has been possible to suggest a climatic

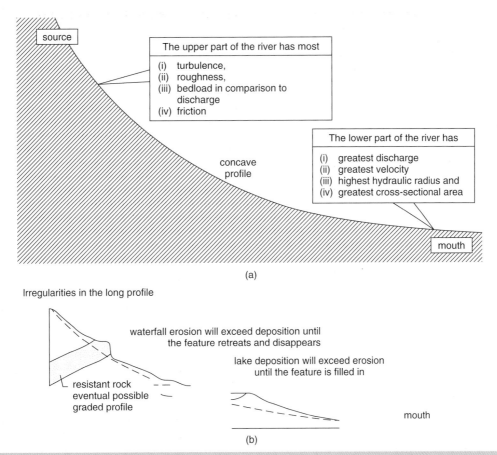

Figure 17.16 (a) Typical river long profile. (b) Irregularities in the long profile (Source: after Waugh, 2000)

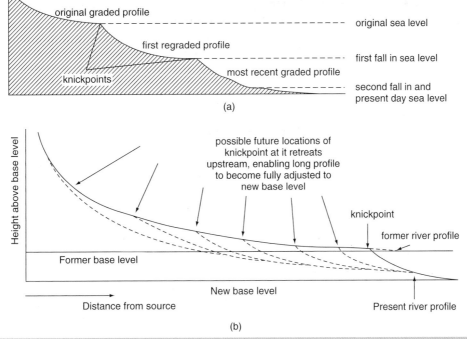

Figure 17.17 (a) The effect of rejuvenation on the long profiles of a river. (b) The adaption of a river's long profile to an increase in energy. (Sources: (a) after Waugh, 1990; (b) after Collard, 1988)

(a)   (b)

Figure 17.18 (a) High Force: where the River Tees crosses the Whin Sill, the rock system followed by Hadrian's Wall. The waterfall itself consists of an upper band composed of Dolerite (the Whin Sill), a hard rock which the river takes a lot of time to erode. The lower section is made up of Carboniferous Limestone, a softer rock which is more easily eroded by the river. The wearing away of rock means that the waterfall is slowly moving upstream, leaving a narrow, deep gorge in front of it. The length of the gorge is currently about 700 m.
(b) Thornton Force, Ingleton Glens: here the lower, older rocks are divided from the upper Lower Carboniferous Limestone by a major unconformity. There is a plunge pool below the force

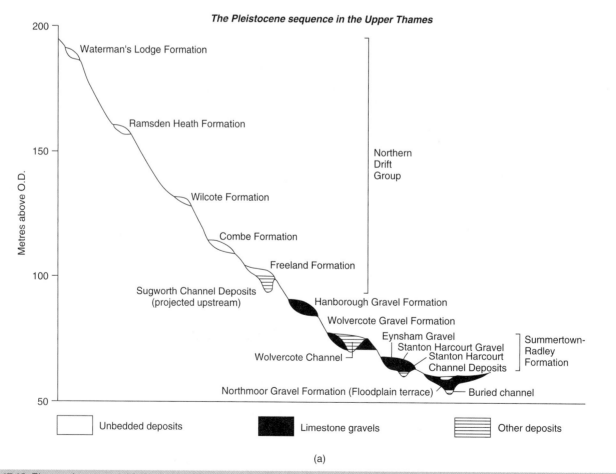

(a)

Figure 17.19 Thames terraces: (a) in the upper Thames (Evenlode); (b) in the classic Middle Thames (after Bridgland, 1994)

The Middle Thames

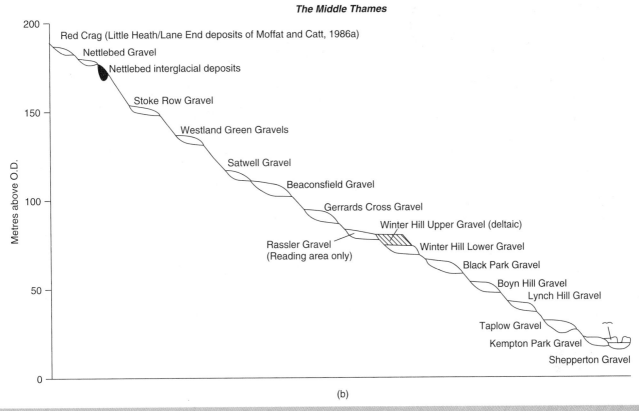

(b)

Figure 17.19 (continued)

model for terrace formation in which the aggradation of sands and gravels occurs both at the beginning (phase 4) and end (phase 2) of cold climatic episodes. The latter is probably the main aggradational phase, represented by most of the classic terrace gravel accumulations within the Thames Valley.

- **Phase 1:** Downcutting by rivers during a time of high discharge, under cold climate conditions. The limits of this rejuvenation would be controlled by base level.

- **Phase 2:** Aggradation of sand and gravel and the formation of floodplains at the new level. Sedimentation now exceeds erosion, leading to a vertical accumulation of sediment.

- **Phase 3:** Limited deposition by less powerful rivers under temperate interglacial conditions. Single-thread channels usually, but overbank sediments may be more extensive. Estuarine sediments accumulate above phase 2 deposits in the lower valley reaches.

- **Phase 4:** Climatic deterioration results in increases in discharge coupled with enhanced sediment supplies, brought about by the decline in interglacial vegetation and increases in erosion

and mechanical weathering. This causes the removal and/or reworking of existing floodplain deposits and the renewed aggradation of sand and gravel.

- **Then:** Discharge exceeds sediment supply, causing renewed downcutting (repeat of phase 1).

Although it is possible to have paired terraces of equal height created by rapid erosion by the river, it is more common that the river cuts down relatively slowly, enabling it to swing from side to side of the valley at the same time. This results in the terrace to one side of the river being eroded as the river migrates. Another good example of river terrace development is taken from the river Hodder at Burholme Bridge (Figure 17.20), close to the Forest of Bowland (Lancashire Pennines).

17.9.6 Channels in planform

There are four channel planform patterns: straight (Figure 17.21), braided (Figure 17.22), anastomosing and meandering (Figure 17.23(b)), but there is a continuous gradation between these types and in a single river it is possible to find more than one. Despite this there have been a number of attempts to

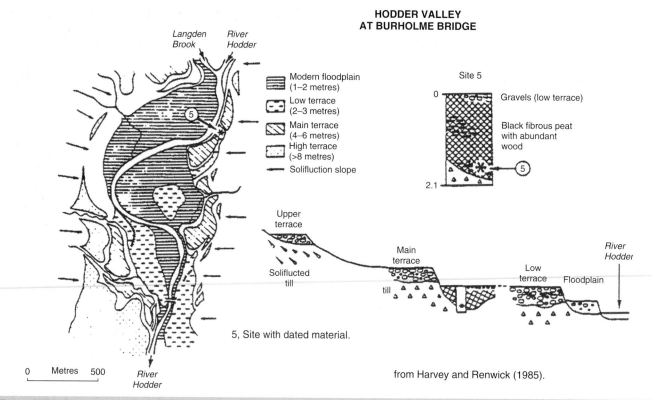

Figure 17.20 River terraces in the Hodder valley at Burholme Bridge, near the Trough of Bowland, Lancashire Pennines

discriminate between single-thread (meandering and straight) and multi-thread (braided) on the basis of parameters such as discharge and channel slope, as shown in Figure 17.24. For any one channel capacity and water discharge, they predict braids under conditions of high slope and large sediment load, and single-thread rivers where the channel has lower slope and smaller sediment load. However, long, straight channels are rare and river reaches are nearly always sinuous to some extent (Figure 17.25). In flumes, straight channels are abnormal and unstable. Where straight channels occur they have a negligible sinuosity over many times the channel width, but nevertheless the thalweg, or line of maximum depth, almost always wanders from one bank to the other (Figure 17.25). As a result, lateral bars are formed alternately along the banks. The longitudinal profile therefore shows an alternation between deep pools, opposite the bars, and shallow riffles, spaced typically at 5–7 channel widths, between the bars. Flow in straight channels tends to wander from bank to bank and there is a tendency for two rotating circulation cells in the water, but it is difficult to explain this pattern. This means that the surface currents converge in the centre of the river and plunge below, and there is a tendency for the bed to be slightly less deep in the channel centre due to this circulation pattern (Figure 17.25). Other characteristics of straight channels are steep river gradients, large variations in stage, aperiodic floods, a small bedload and a tendency to occur for only short lengths of the river (10× the channel width).

Meandering channels are very common, and all channels show meandering to some extent, on all scales from rills to the Mississippi. A meandering channel is defined as a channel where the sinuosity is >1.5, and is the channel length (A) divided by the valley length (B) (Figure 17.23). Straight channels have sinuosity are <1.5. The geometry of meanders has been studied and it has been found that the wavelength is proportional to both the mean discharge and the channel width. There is also a more or less linear relationship between the meander amplitude and channel width.

For any given discharge, braided rivers, where the planform shows multiple alluvial islands at low water, occur on steep slopes, whilst meandering channels are found on gentler slopes. It can also be seen in flumes that sinuosity can be reduced by replacing homogenous bank sediment with heterogeneous sediment, keeping the other factors constant. It remains far from clear why some rivers meander and others do not, but it appears that meandering rivers have their origins in the relatively straight sections where the pools and riffles develop. As the pool is an area of greater erosion, where the available energy builds up because of reduced friction, velocity and erosive capacity increase. Across the riffle a higher total percentage of the energy is used in overcoming friction, velocity and erosive capacity and further deposition may take place. In order to avoid the riffles the main current swings from side to side in a sinuous course and so the maximum discharge and velocity are directed towards one channel side.

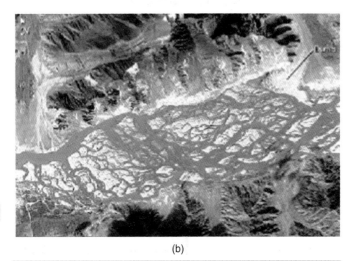

Figure 17.21 River Wye at Symonds Yat

Figure 17.22 (a) Proglacial braided outwash, Kangerdluarssuk, SW Greenland. (b) Braided Brahmaputra River in its upper section, India. (Source: http://www.rst.gsfc.nasa.gov)

This will be eroded, whilst on the opposite bank, where volume and velocity are at a minimum, deposition will occur. In time the meandering pattern develops.

Braided streams have unstable, shallow channels, separated by actively migrating and accretionary braid bars (Figure 17.22). They are favoured where valley gradients are high, where discharge is large and highly variable, where banks are noncohesive and lack the stabilizing influence of vegetation, and where bedload discharge is high. In the wide, shallow channels, the secondary circulation cells that are a quasi-stable feature of flood flow in narrow deeper channels become unstable and break into a number of smaller cells. This leads to accretion, where flows converge and rise from the bed, and the development of braid bars (Figure 17.26). Diversion of flow around the emerging bars causes bank erosion and increases the channel width, which in turn leads to an increase in the number of secondary circulation cells, the development of further bars and so on. Bars are emergent only under low flow but larger grain sizes do tend to become lodged at the proximal end of each bar (Figure 17.27) and a strong grain-size gradient towards finer sizes develops at the distal bar ends (Ashworth and Ferguson, 1986).

The relationships between sediment load and channel form were noted by Schumm (1981, 1985) in a widely-used classification (Figure 17.28). This classification summarizes many of the most important trends, for example that rivers transporting a coarse sediment load are more likely to be of the low-sinuosity, multi-thread channel type and that they are more unstable and prone to avulsion (where there is large-scale movement of the channels of the floodplain and abandonment of others) than are rivers with a fine-grained suspension load.

Straight channels can occur where the sediment load is dominantly of bed-load, mixed-load or suspended-load type. Braided channels are of bed-load or mixed-load types. Rust (1978) showed that two simple parameters could be used to define channel style for most rivers in a fourfold classification (Figure 17.29). These parameters are channel sinuosity and a braiding parameter. The

Sinuosity of Natural Channels

Plan Patterns

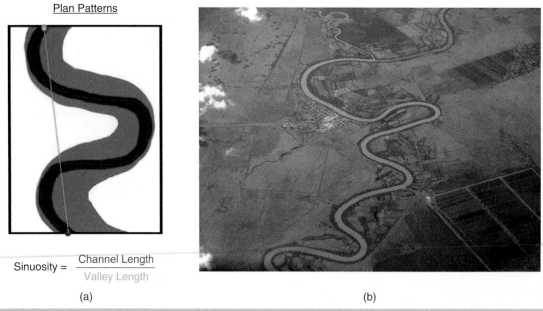

$$\text{Sinuosity} = \frac{\text{Channel Length}}{\text{Valley Length}}$$

(a)   (b)

Figure 17.23 (a) Sinuosity in river channels. A river that does not meander has a sinuosity of 1; the more it meanders, the closer the sinuosity gets to 0. (b) Strongly meandering and highly sinuous river in the Rio Cauto at Gaumo Embarcadero in Cuba. (Source: Wikimedia Commons)

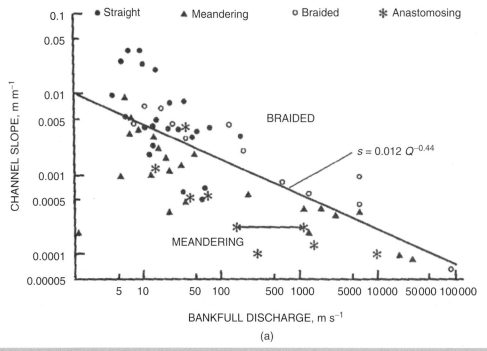

(a)

Figure 17.24 Bivariate plots used to discriminate between single-thread and braided channels on the basis of: (a) channel slope and bankfull discharge (after Leopold and Wolman, 1957); (b) slope of stream bed/Froude Number (Sb/Fr) and water depth/width (y/w). (Source: (b) after Parker, 1976)

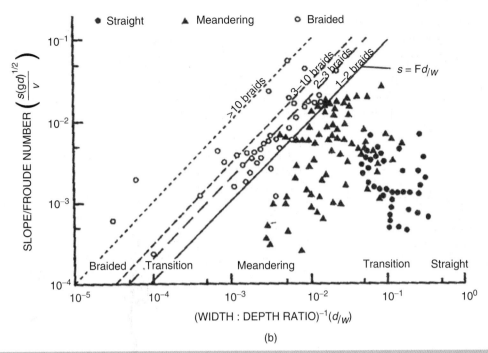

Figure 17.24 (*continued*)

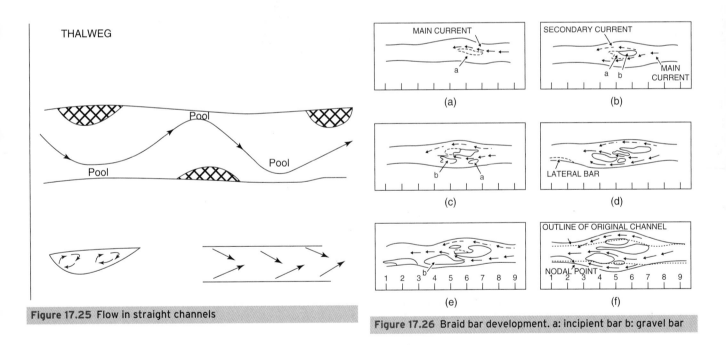

Figure 17.25 Flow in straight channels

Figure 17.26 Braid bar development. a: incipient bar b: gravel bar

Figure 17.27 Rhomboid bar, Kaldidalur, Thorisjökull. Coarsest gravel is at the upstream end of the bar, in the foreground

all had a braiding index of >8.0. Miall (1994) suggests that a range of 1.0–3.0 be used to define the wandering category.

17.9.7 Downstream size and sorting changes

As a general rule there is a downstream size reduction with progressive distance from the headwaters of a river, along with an increase in clast roundness and better size sorting (Figure 17.30). The cause of this fining is either the gradual attrition of the clasts or their selective entrainment by variably competent flows (cf. Dawson, 1988; Huddart, 1994). In laboratory flume experiments abrasion was shown to produce progressive rounding and comminution, and there are differences caused by rock types too (Kuenen, 1956). These changes can be seen in Figure 17.30, from both laboratory experiments and actual rivers.

It is generally thought that whilst both processes go on in rivers, selective entrainment is the more important. For example, McPherson (1971) found that in the Canadian Rockies b-axis measurements decreased by 9.8 mm per kilometre, while roundness did not change downstream. Bradley (1972) believed that an 87% mean size reduction in the Knik River gravel over 25.6 km was caused by sorting. His abrasion tank studies suggested that sorting was responsible for 90–95% of this, with only 5–10%

latter expresses the number of bars per meander wavelength of the river. However, it has become apparent that this classification needs revision as, for example, there is a type of gravel-bed river, termed wandering, which is intermediate between braided and meandering. Rivers like the Bella Coola and the Squamish Rivers, both in British Columbia, have braiding parameters of between 2.3 and 2.5, whereas the braided rivers discussed by Rust (1978)

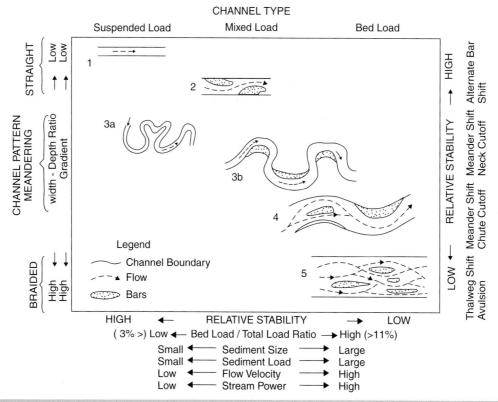

Figure 17.28 Classification of alluvial channels based on sediment load (after Schumm, 1981, 1985)

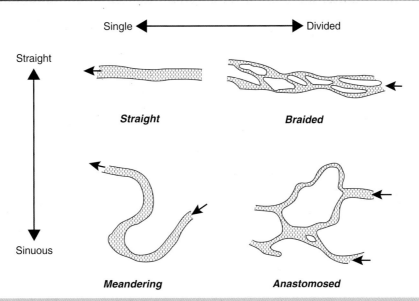

Figure 17.29 Classification of channel pattern based on sinuosity and degree of channel division. (Source: after Rust, 1978)

due to abrasion. Further evidence in support of sorting is the downvalley convergence in grain size.

However, whilst selective sorting seems to be the dominant cause of downstream clast changes in most rivers, this might not be the case in high-energy rivers, or those transporting less-resistant rock types. This view gains some support from Pearce (1971), who found evidence of short-distance fluvial rounding of volcanic gravels in less than 3.2 km of transport. Wet sand blasting by smaller particles was suggested to be important in this abrasion. Bridgland (1986) suggested that a better rounding might reflect a higher proportion of weathered material, rather than a longer transport distance, and both Mills (1979) and Huddart (1994) have evaluated the role of rock type as the most important parameter in controlling downstream distance changes.

Not all studies of sediment grain size have reported downstream fining, and some even show downstream coarsening. One explanation is the addition of sediment by tributaries, which may have steeper slopes and hence greater competence than the main stream at the point of confluence (Knighton, 1980; Ichim and Radoane, 1990).

17.9.8 Fluvial landform-sediment assemblages

Fluvial sediments can be composed of two major types of sediment:

1. Channel or substratum deposits formed in the lower part of a floodplain sequence, which include point bar and channel bar deposits and channel lag deposits left after riverbed winnowing. Bedload sediment dominates.

2. Overbank or topstratum deposits, where suspended load dominates. These include swale fill, levee, crevasse splay and floodbasin deposits. They overlie channel deposits and form the upper part of a typical floodplain. There are transitional deposits, including channel fill deposits, which generally include both bedload and suspended-load sediments.

Meandering river landform-sediment assemblages

A meandering river typically develops where gradients and discharge are relatively low, and today they are characteristic of humid, vegetated areas of the world, where seasonal discharge rates are fairly steady and sediment availability is relatively low. This is due to subdued topography and the impeding effect of vegetation, both on soil erosion and lateral erosion of channel banks. The classic system is the Mississippi (cf. Fisk, 1947; Aslan and Autin, 1999; Figure 17.31(a) and (b)), but another is in the Manchurian plain, Northern China (Figure 17.31(c)). A much smaller example is shown in Figure 17.32(a).

In this section the morphological components of some typical landforms will be described, followed in each case by their sedimentary characteristics.

Fine-grained point bars

Each meander loop encloses a point bar formed of scroll-shaped ridges (scroll bars) and swales, which are hollows between the scroll bars, roughly conformable with the curve of the channel. The scroll bars vary in transverse spacing from a few metres to a few hundred metres, and in height above the adjacent swales from a few hundred centimetres to three metres. Each ridge represents an aggradation of bedload sediment against the convex channel

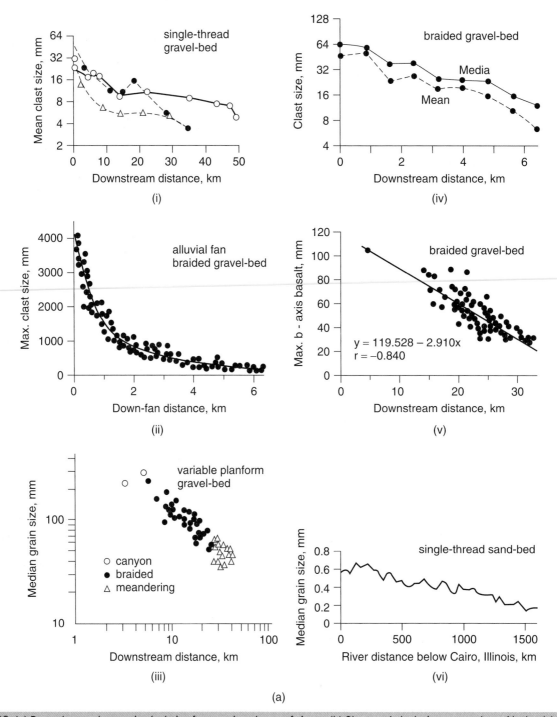

Figure 17.30 (a) Downstream changes in clast size from various types of rivers. (b) Slope and clast-size parameters, Alaska. (c)-(e) Downstream rounding of clasts and grain-size and sorting change with distance. (Source: Huddart, 1994)

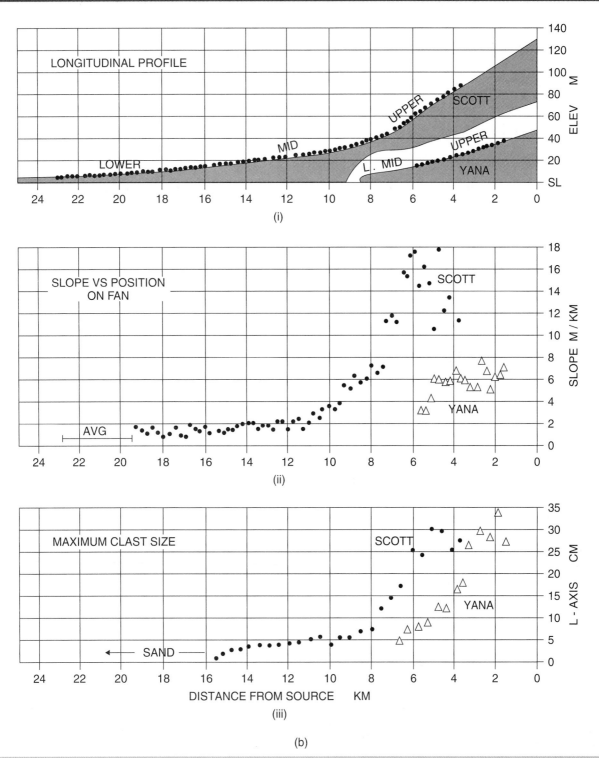

Figure 17.30 (continued)

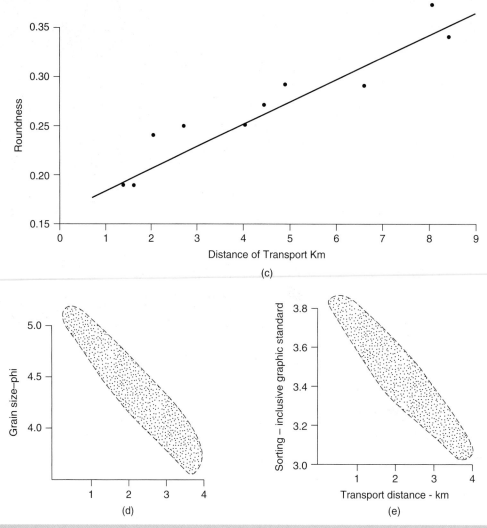

Figure 17.30 (continued)

bank during a flood, or a series of floods, when the concave bank was eroded back a similar amount.

A scroll bar begins to form at the upstream end of the meander loop and in the course of time the bar is built upwards and extended further downstream on a line parallel to the convex bank but separated from it by a swale. The cross-channel bedflow and the decline in flow intensity towards the convex banks are the important mechanisms for point bar accumulation. Due to lateral flow, coarse sediment is carried from the channel deeps to a relatively high position on the point bar, and due to the same process finer sediment is carried higher up the bar. Variations in discharge lead to interlayering of coarse and fine sediment, although there is a general fining upwards from the channel deep. These deposits extend up from lag deposits formed in scour pools and the deposits of the river bed. Their shape depends on the extent of the channel wandering and varies in ground plan from broad-based and triangular to lobe or pear-shaped. The grain-size variation is great but the sediments do represent the coarser debris

carried by the river. The fining upward sequence in point bars is attributed to a declining bed shear stress from the point bar toe to the point bar top and these deposits are preserved mainly in sand.

Sometimes the traces of former point bar depositional surfaces are preserved and can be recognized within the lower, coarser unit. These lateral accretion surfaces have been termed 'epsilon cross stratification' and are important in palaeohydraulic studies for estimating bankfull width and depth. Small-scale cross stratification and thin beds of silt are confined to the upper layers and are often sun-cracked and show rainprints and footprints, with thin rootlet and peat beds. Horizontally-stratified sands formed on plane beds can occur in any part of the point bar, although they are most common in thin flows in the upper levels. Swale fills are formed where there is aggradation within arcuate hollows on the point bar surfaces. Fill occurs at high stream stages when floodwaters submerge the bar, and at intermediate stages when the deeper swales open downstream into the main channels to form local

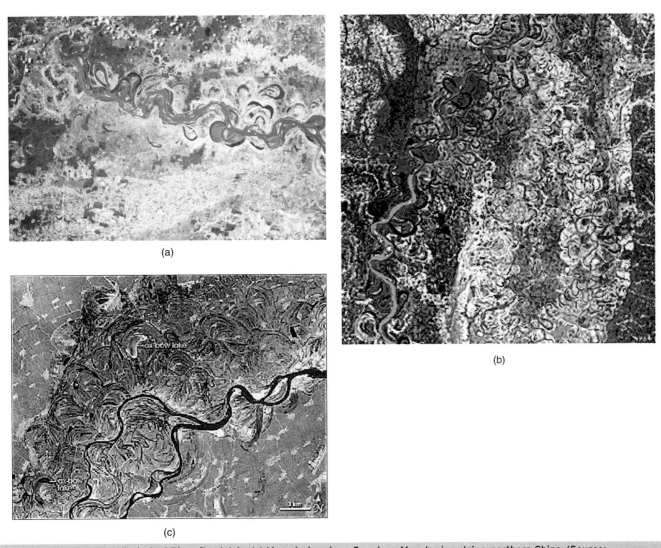

(a)

(b)

(c)

Figure 17.31 (a) and (b) Mississippi River floodplain. (c) Meandering river, Songhua, Manchurian plains, northern China. (Source: http://rst.gsfc.nasa.gov)

backwaters. They consist of narrow, arcuate bodies of sediment of prismatic cross-section, thickening downstream, with convex-downward bottom surfaces. Deposition is mainly from suspension and in the Mississippi consists of well-stratified sandy silts, clayey silts and silty clays.

Cutoff

This occurs whenever a river can shorten its course and thus locally increase its slope (Figure 17.32(b)). The frequency increases with river sinuosity. There are two types: chute cutoff occurs where the river shortens its course by eroding along a swale on the bar enclosed by the meander loop. Enlargement of the new channel and plugging of the old proceed gradually. The plugging is usually by bedload sediment and filling is finally by suspended sediment from overbank floods. Neck cutoff is mainly responsible for the abandonment of meander loops. Such a cutoff occurs late in the development of the loops by the gouging of a new channel across

the narrow neck of land between the two loops. Bedload sediment rapidly plugs the end of the abandoned channels to give an oxbow lake and filling is completed by sediment from overbank flows. The sediments resulting from chute cutoff are shorter, less strongly curved and in cross-section are prismatic and U-shaped to triangular. Bedload sediments are important because the river continues to flow through the old channel. After closure there is further filling by fines only. In neck cutoff the sediment units are longer and more strongly curved and bedload sediments are only important as plugs.

Natural levees

These are wedge-shaped ridges of sediment bordering river channels. They vary in width between a half to four times the channel width, and in elevation range from a few hundred centimetres to as much as eight metres, depending on the river size and the load calibre. Their height is greatest at or close to

(a)

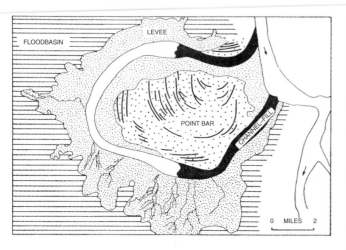
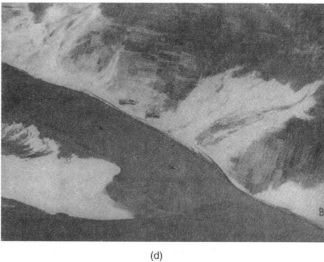

(b)

(d)

(c)

Figure 17.32 Meandering river environments: (a) River Cuckmere near Cuckmere Haven, Sussex; (b) abandoned meander; (c) levee sequence; (d) and (e) crevasse splays

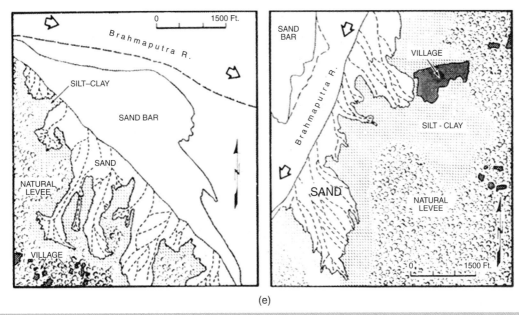

(e)

Figure 17.32 (*continued*)

the channel edge and they are best developed on concave banks. Overbank deposition occurs when the velocity is checked, so that not all of the load can be carried. The coarsest sediment is closest to the channel and the finest sediment is further down the levee, giving the landform its gradient.

Levees occur as sinuous, ribbon-like, prismatic bodies of triangular cross section and are among the coarser sediments of the topstratum deposits, which are a reflection of their deposition in a position proximal to the active river. In the Mississippi they are formed of sandy silts and silty clays and they fine away from the channel locally and regionally downstream. They generally consist of a vertical alternation of fine sands with silts and clays, and beds of plant remains are common. Any sands generally have erosional bases which may have Sun cracks. Usually sand shows small-scale cross-stratification, although there can be some horizontal and large-scale cross-stratification too. There is a vertical grading from coarse to fine. A sharp base and a gradational top are characteristics of sand layers in levees. Silts and clays vary from massive to laminated and any soil beds come in the form of mottled and homogenized deposits, with roots and plant litter. As porosity and permeability are usually good, there is local oxidation and brown colours in these sediments (see Figure 17.32(c)).

Crevasse splays

These form where deposition beyond the main channel occurs because of crevassing at a high stage, where the water can leave the channel through isolated low sections or breaks in the natural levees. Once a crevasse is initiated, the flood waters deepen the new course and develop a system of distributary channels on the upper slopes of the levees. Deposition takes over from erosion on the lower slopes and a tongue-shaped sediment unit, or crevasse splay, is deposited from the repeatedly dividing streams. Since the master channels of the crevasse systems often tap sediment at

relatively low levels in the major river channel, the splay deposits are often coarser than the associated levee sediments.

Splays can be of two types. In the first, the water leaves the river system via a single channel, crosses the levee and terminates in a delta-like fan consisting of multiple channels. Here little coarse sediment extends into the adjacent flood basin. In the second type, the water leaves the river through a multi-channelled, anastomosing system. In this case a broad zone of coarser sediment gets into the flood basin (Figure 17.32(d) and (e)). The deposits are narrow to broad localized sediment tongues that are sinuous to lobate in plan and up to 3 m thick. They are usually a little coarser than the levee sediments and are formed of sands which have a sharp, channelled base and an upward grading from coarse to fine.

Flood basin

This is the lowest part of the floodplain. It is a poorly drained, flat, relatively featureless area, with a length much greater than its width. It acts as a stilling basin in which the suspended fines can settle from overbank flows after the coarser sediment has been deposited on the levees and crevasse splays. It possesses an internal drainage system, composed of small channels which conduct water from the meander belt during flood stage and back into the active channels, with the flow ponded in the basins, during falling stage. There is a downstream tendency for the basins to increase in area relative to the levee and channel, which is probably a reflection of the downstream fining of the river sediment. The sediments form elongated, tabular, prismatic sediment bodies of rectangular cross-section. The thick sequences are predominantly of laminated, or occasionally rippled, silt and clay, with rare sand. Often they are rich in peat layers, and dark colours are characteristic. In dry climates there is a lack of organic material, brown or red sediments, which form as carbonate caliches, or ferruginous laterites, while suncracks, rain prints and precipitated salts are common.

Floodplain construction

Lateral accretion of the river bedload on the sideways migration of the channel results in point bars, channel bars and alluvial islands. Vertical accretion of suspended load after overbank flows leads to construction of levees, crevasse splays and floodbasins on top of the lateral accretion deposits. Whether significant vertical accretion takes place depends on the size of the suspended load, the general calibre of the total stream load, the rate of channel migration and the speed of overbank flows. It is also affected by external factors such as changes in river base level or in land level due to subsidence or uplift. However, suspended sediment in overbank flows is necessary for the formation of vertical accretion deposits and Fisk (1947) refers to the poor development of levees along certain Mississippi tributaries as being caused by the small suspended load at stages well below bankfull stage, and not at flood stage. The external factors have been suggested to be important, as in the recent past a combination of subsidence and rising base level has led to deep alluviation in the lower Mississippi valley and in streams flowing into the Gulf of Mexico. The stream gradients have been lowered, the load calibre decreased, and there has been a trend towards the creation of floodplain relief. The sedimentary regime of river floodplains has as its most basic feature the repeated submergence and emergence of the sedimentary surface. There seems to be a maximum of four stages recognized in this cycle:

1. Spilling of flood water into the empty floodbasin from the main channel. The floodwaters leave by way of unsealed crevasse channels and then alter over uncrevassed stretches of the levee. Flow down the levees is fast and destructive, and results in sheet erosion, channel widening and the widespread destruction of any plant cover.

2. Filling up of the floodbasins. At first this is marked by the filling up of the subsidiary floodbasins and their interconnection through the submergence of sills, until the floodwaters obtain free passage down the full length of the floodplain.

3. Decay of the flood wave. As the flood subsides, velocity of flow over the floodplain declines until sills reappear and ponding takes place. Here the finest sediment settles out.

4. Drying out of the floodbasin and the modification of the newly deposited sediment by raindrops and animals' and birds' feet, although the sediment rapidly hardens under the Sun and wind and desiccation cracks form.

Through time, build-up of the river valley level and channel level can take place, causing aggradation. This can be due to a change in the sediment yield. If so, overbank deposition will occur until river level and valley level are adjusted to transmit the sediment received. This is a feedback mechanism to adjust factors of slope at any point in a channel to an increase in sediment yield. It is a self-destructive feature and when input equals output, the aggradation stops. On the other hand, there can be a scouring out of the whole valley, or degradation. Again there is a feedback mechanism in any particular channel reach. An increase in river discharge can cause this. Here more sediment can be carried by the flow until the slope lessens, or until the width and depth of the river adjust to the new water discharge. This is how river terrace sequences form through successive cycles of aggradation and degradation.

Periodically in meandering rivers there is a sudden abandonment of part or the whole of a meander belt for a new course at a lower level on the floodplain. This process is known as avulsion. Traditionally the fine-grained floodplain deposits of meandering rivers are thought to represent repeated episodes of widespread overbank deposition and slow rates of sediment accumulation, but recent investigations suggest that large volumes of clay-grade sediments and lesser amounts of sand accumulate repeatedly in floodplain depressions during crevassing and avulsion, and in periods of rapid floodplain aggradation. It has been found that during the Holocene, floodplain processes and fluvial styles in the Lower Mississippi changed in response to decreasing rates of floodplain sediment accumulation and decelerating sea-level rise, for example. Avulsion has played a major role in floodplain construction (Aslan and Autin, 1999).

The floodplain development near Ferriday, Louisiana is shown in Figure 17.33 and the two models of deposition are shown in Figure 17.34. In Figure 17.34(b), because the processes deposit sediments on the floodplain at virtually all stages of main-channel flow, the sediments accumulate continuously rather than during infrequent floods. This leads to rapid sedimentation and filling of regional depressions with lake and backswamp muds, like in the Atchafalaya basin, and once a depression is filled, subsequent avulsion initiates rapid sedimentation and aggradation elsewhere on the floodplain. Repetition of this sequence produces interfingering sheets, or wedges of floodplain clays and silts and isolated sand bodies, with little evidence of soil formation. These fine-grained channel sands, between 5 and 8 m thick, represent small, sinuous streams that only migrated laterally over distances of hundreds of metres, as indicated by the width of the sands, from 0.5 to 1 km (compare the lateral migration of the present-day Mississippi, which produces sand bodies that are 5–15 km wide and 20–30 m thick). In this view, large overbank floods are not as important as originally thought for the total volume of Holocene fine sediments in the southern Lower Mississippi valley. An important point to be made here is that the earlier model for Mississippi fine-grained fluvial sedimentation was widely used for the interpretation of modern and ancient fluvial deposits. Following the work discussed in Aslan and Autin (1999) it is now realized that there is another model for fine-grained floodplain construction formed during episodes of base-level rise.

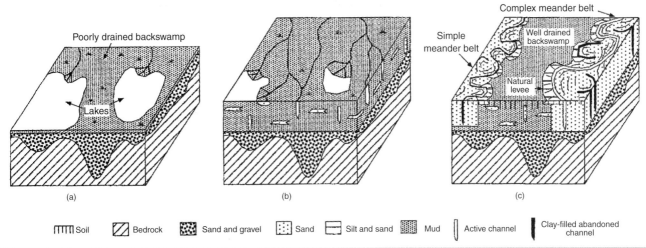

Figure 17.33 Summary of Holocene floodplain development, near Ferriday, Louisiana, USA. (Source: after Aslan and Autin, 1999)

Braided river landform-sediment assemblage

Bar development

Although a river that divides into two channels around a single braidbar (an alluvial island) can be called braided, it is more common for a braided river to show many bars across its line of flow. The braided channels are composed of several successive divisions and rejoinings of flow around bars. The latter are submerged at high flow. The best examples of braided river systems are the proglacial outwash plains (Figure 17.35(b) and (c)), but upland rivers in general and rivers flowing from a mountain front to a lowland also show this pattern (Figure 17.35(a)).

The building of islands, or bars, is the main characteristic of these river systems. Building can occur in two different ways: through the aggradation of longitudinal bars and through the dissection of transverse bars. The longitudinal bars are lozenge-shaped and have a marked elongation parallel with flow. There is a more or less regular transverse spacing which, together with the scale of the bars, depends on discharge and the calibre of the bedload. Their formation involves the build-up of channel ridges in poorly-sorted sediment. When the river is unable to move its coarsest load, the finer clasts are trapped. By a succession of entrapments a ridge builds up laterally and vertically. In the later stages of growth, avalanche faces sometimes develop on the downstream ends of bars. The bars are characteristically long and narrow, slightly inclined on their top surface, usually with coarse upstream ends, which grade to finer-grained gravels or sand downstream (Figure 17.35(d)). These bars commonly override and merge laterally with each other, resulting in complex bars with a variety of shapes and surface grain-size distributions. Internally the bars show horizontal stratification, gentle dips downstream of $1-2°$, imbrication of gravels, with a dip upstream, and relatively poor sediment sorting. In braided river channels therefore many

of the bars are erosional remnants with a long and complex history and they show terracing of bars during lower flows, when they are dissected.

Transverse bars are tabular sediment units or bars which grow through the migration of foresets, more or less perpendicular to current direction and generally in a distal location (Figure 17.35(e)). They develop in shallow depressions on the river bed and form by aggradation to a profile of equilibrium. Sand moves along the bed until it encounters a depression, where because of the increased depth the tractive shear stress and velocity are less than critical, with the result that the sand is deposited as a delta, upwards from the depression floor. This build-up constricts the flow and increases velocity until it re-attains critical values. The top of the bar then becomes the channel floor and sand is transported again. When the water level decreases enough, the bars become exposed and begin to be eroded by small channels and braiding starts. These bars are usually restricted to more distal braided river systems. The sediments are sands, better-sorted and finer-grained than in the longitudinal bars.

However, in the Kicking Horse River in British Columbia, Hein (1974) recognized four types of unmodifed bar, with the addition of diagonal and point bars (Figure 17.36). Observations here suggested that the bars start off from diffuse gravel sheets consisting of the coarsest bedload and that these move at the highest discharges. At slightly lower flows the diffuse gravel sheets stop and form a lag pavement, which is the nucleus for bar growth. If the water and sediment discharge remain high after the lag has been deposited, the bar will tend to grow downstream faster than it aggrades vertically. Under these conditions no foreset slope will develop and stratification will tend to be horizontal. If water and sediment discharge decline rapidly after the lag is deposited, the bar may aggrade faster than it grows downstream. In this case an avalanche face stands more chance of developing; cross-stratified gravels or sands may also develop. Within each bar an upward

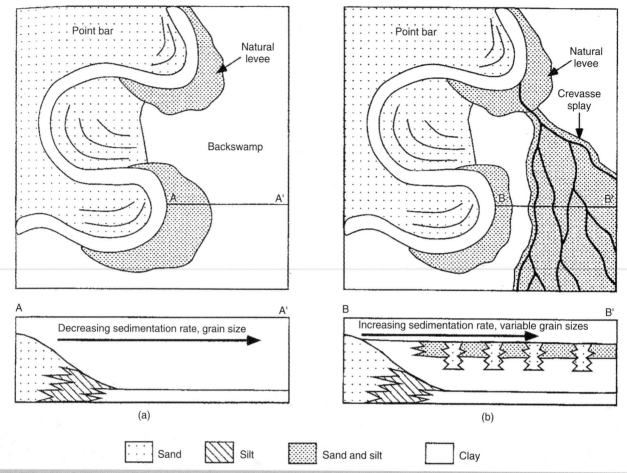

Figure 17.34 Floodplain maps and cross-sections showing lateral changes in sediment grain sizes and sedimentation rates associated with two models of fine-grained floodplain deposition. (a) Floodplain deposits formed by overbank flooding and sedimentation. (b) Floodplain deposits related to crevassing and avulsion. (Source: after Aslan and Autin, 1999)

fining has been noted. In Figure 17.35(e) various mechanisms are put forward for the formation of braid bars.

Causes of braiding

There has been much discussion as to the underlying cause of braiding. The suggested explanations include erodible banks, rapid and extreme fluctuations of discharge, high regional slopes, abundant sediment load and the local incompetence of the river. It was found for example in the Mississippi valley that there was a tendency to braid where bank caving was active and where sands were present as easily-erodible sediments in the banks. It was also found in Idaho that the Wood River showed a braided channel pattern where it flowed through prairie but a meandering pattern where it flowed through forest. The difference was due to bank resistance, which was affected by the presence or absence of bank vegetation. It is evident that if the banks are unerodible and the channel width confined, the capacity of the reach for transport will be increased, reducing the likelihood of deposition. In addition, any bars that form will be eroded as the flow increases, since bank erosion cannot take place. Thus for bars to become stable and divert the flow, the banks must be sufficiently erodible so

that they, rather than the incipient bar, are eroded as the flow is diverted around the bar that is being deposited.

However, on the White River in Alaska it was found that both meandering and braided reaches were formed frequently at the same time and at the same discharge. At times the number of channels was increasing in one area and decreasing in another. Not all such changes are explained in terms of changing erodibility of the banks; they must be related to factors that are more easily altered.

It was suggested that where there were large and sudden variations in discharge there were braided rivers and where the flow was more regular there were meandering rivers. Rapidly-fluctuating discharges contribute to the instability of the transport regime and to erosion of the banks, but perhaps this a contributory factor and not an essential element of the braided river environment. Against its being an important factor is the frequency with which braiding and meandering reaches are interspersed and the fact that laboratory studies have shown that braiding can be produced with no variation in discharge. This seems to indicate that rapid discharge fluctuations can be ignored as a cause of braiding in most rivers.

High regional slopes, or changes in slopes, have been suggested as a factor in braiding, but they do not explain, for example, the fact that on the White River braiding developed on slopes ranging from 0.01 to 0.2 but only coincident with bedload movement of the coarse material. The river frequently braids in one part of the valley on slopes both higher and lower than those of other parts, where the river has only one channel. In some cases slope may aid in setting in motion enough material to form a braided pattern, but in others it serves only to maintain velocities so high that the deposition of bars does not take place.

The sediment abundance appears to be the factor emphasized by most workers, coupled with the heterogeneity of the bedload. The latter creates irregularities in the movement of sediment and might be a contributory factor in braiding, but it has not been suggested that a braided channel pattern can be developed without an appreciable bedload. In all the explanations for braiding the common element seems to be that movement of bedload exceeds local competence. The result is deposition in the channel, which causes diversion of flow. The rate of change of the pattern in a river that is not restricted by resistant banks is controlled by the amount of bedload. Erodible banks and discharge fluctuations will help in this process.

It has been suggested therefore that braiding is an equilibrium channel form occurring in areas of erodible bank material and highly-fluctuating discharges, and in rivers with large sediment loads, especially bimodal loads. It is a channel pattern designed to increase sediment transport for a given river discharge.

A model for braid development is illustrated from laboratory experiments in Figure 17.26.

The proglacial outwash plain, or sandur, as an example of a braided river system

There are two types of sandur, depending on whether the proglacial deposition from a valley glacier is confined by steep valley walls to give a valley train, as in Morsardalur, Iceland, or whether there is deposition from a wide ice margin terminating on a broad lowland where the melt-water streams can spread extensively laterally and build up an extensive surface, as on Skeidararsandur, Iceland (frontispiece to this chapter). In this type of environment the debris is usually very poorly sorted, with a wide

(a)

(b)

(c)

(d)

Figure 17.35 Braided rivers systems: (a) Irthing valley, north of Brampton, Cumbria, UK; (b) Morsardalur braid bars, Southern Iceland; (c) Kaldidalur, Thorisjökull, longitudinal gravel bar; (d) Kaldidalur, Thorisjökull, distal transverse bar in sand. (e) Mechanisms of braid development. (after Leopold, Wolman and Miller 1964)

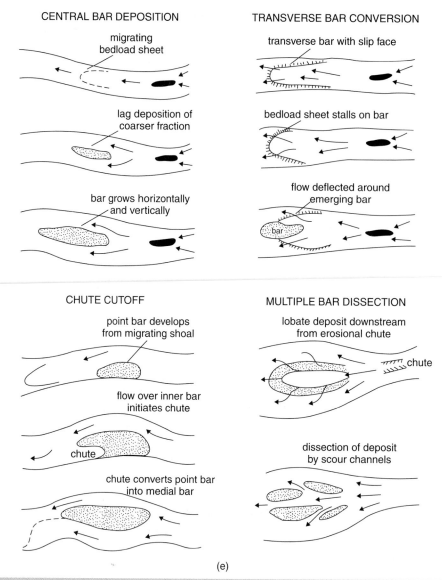

(e)

Figure 17.35 (*continued*)

range of grain sizes supplied by glacial ice. The coarsest debris is nearest the ice margin and the finer grain sizes are carried further downstream. The river discharge rapidly fluctuates on a daily and seasonal basis and is hence extremely variable. The rate at which aggradation takes place depends mainly on discharge and the load of the rivers, and the level of the sandur is often linked to the position of a moraine to which the outwash is graded. It appears therefore that outwash aggradation can take place most effectively when the ice front advances and then remains stationary for long enough for a terminal moraine to form. Under these conditions the melt-water must flow upwards from beneath the ice to reach the level of the developing sandur. In the proximal zone close to the ice the channels are deeper, often incised, and well-defined, often with a distinct main channel. There are sometimes kettled sandur surfaces where river sediment has been deposited on top of ice to give stagnant ice, which kettles the

surface when it melts. The zone further downstream shows shallow, wide channels where the banks are less well defined. In the most distal zone there are intertwined streams in a shallow bay, where there is very shallow water. As the discharge fluctuates the boundaries of these zones fluctuate over great distances. There is never a braided river pattern over the whole sandur all the time, however, as large areas are dry, vegetated and abandoned.

There may be slightly higher terraces of differing ages and vegetation cover across a sandur. It is common to recognize two major categories in all braided rivers: proximal and distal. In the proximal channels, channels and longitudinal bars dominate, the sediment is coarser and more poorly sorted, and the slopes are higher. In the more distal environment there are transverse bars and the sediment is finer and better sorted, with lower river gradients. Thin mud lenses and/or organic sediments are common to both categories and represent periods of quiet deposition in

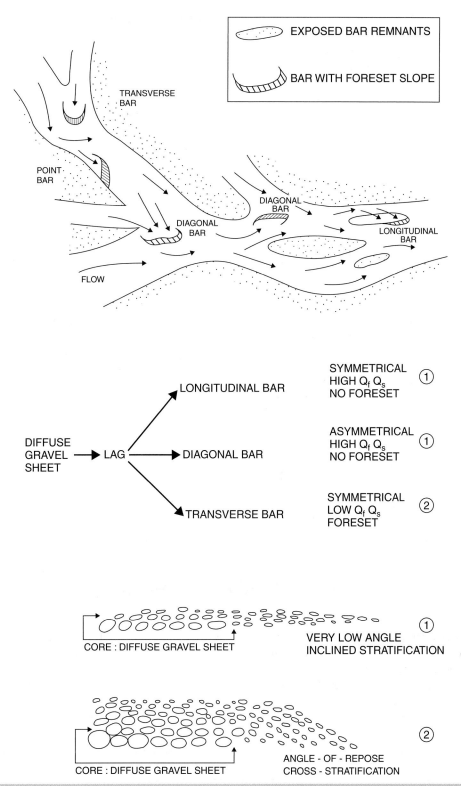

Figure 17.36 Gravel bar types, the relationship between bar types and stratification, and the development of braid bars. (Source: after Hein, 1974)

abandoned channels during low water stages. The highly-variable discharge tends to rework most of this finer sediment with time. In fact, many braided river systems can be affected by subglacial volcanic floods, as seen in the frontispiece to this chapter, and glacier surge floods (Chapter 20) clearly affect the variable discharge and the development of braiding.

Types in between meandering and braided

Instead of the standard fining-upward point bar sequence found in meandering and braided rivers, some rivers show coarse-grained point bars where there is no grain size trend. Examples of this river type are the Colorado (Texas) and the Amite (Louisiana), which are characterized by high bedload, a relatively low sinuosity, high gradient, an excessively large discharge of short duration and banks that are stabilized by vegetation and resistant to erosion (see Table 17.4 for a comparison of fine- and coarse-grained point bars). These are bedload rivers with weak banks that are stabilized. If the vegetation were absent, or sparse, then these systems would be braided. They occupy a position somewhere between meandering and braided in the fluvial system, and during extreme flood a variety of morphological features not

found in the channel patterns described so far will form. The point bars consist of the following morphological zones (see Figure 17.37):

1. A concave bank which is heavily vegetated. Sediment accumulates only during extreme flood, when topstratum deposits of clayey silt to silty fine sand are deposited. These sediments are mixed by plant roots, or show alternating wavy lamination in clayey silt/sand and plant debris lines. Any lateral erosion is retarded by the dense vegetation.

2. A scour pool: the deeper part of the channel adjacent to the concave bank.

3. Scour troughs on the channel floor, which commonly develop downstream from large trees that have slumped into the scour pool from the adjacent concave bank. These are filled by trough cross-stratification.

4. A lower point bar, which is a broad flat area between low water level and the toe of the lowest chute bar. It slopes $1-2°$ towards the scour pool, but near the junction of the lower point bar and the low water level the slope increases to around $8°$. Three stratification types are present and all denote tranquil flow: trough and tabular cross-stratification, formed under similar flow conditions, and parallel lamination, at a later stage when the flood has decreased. However, most of the bar surface is featureless, with no preserved bedforms.

5. Chutes develop on the convex side of the river under extreme flood, when the thread of the maximum surface velocity shifts from the concave towards the convex bank. Many seem to be initiated as scours downstream from uprooted trees. They are characterized by relatively steep sides, flat bottoms and slightly sinuous profiles. They are deepest at the junction with the main channel and get shallower downcurrent. Here erosion gives way to deposition. They are gravel-floored and during extreme flood most of the coarse-grained traction sediment is transported through the chute, but the fine-grained sediment, which is characteristic of the chute fill, accumulates during falling flood stage. This sediment is carried in suspension because the stratification parallels the channel cross-section.

6. Chute bars are lobate in plan (Figure 17.38(a)) and form downcurrent from chutes, but only under extreme floods (Figure 17.38(b)). They are located at various levels along the convex bank, determined by the flood height, and those occurring during the maximum flood achieve the maximum height and areal distribution. They are composed of simple accretionary sequences consisting of thin, parallel laminated-to-wavy bedding as topstrata, which grade laterally and downcurrent into large-scale foresets, dipping at $15-20°$, which in turn grade into parallel laminated bottomsets. The scour of the chute and the initial bar construction are contemporaneous. It is thought that the gravel and coarse sand are transported as a heavy fluid layer where there is a

Table 17.4 Comparison of fine- and coarse-grained point bars. (Source: after McGowan and Garner, 1970)

	Fine-grained, e.g. Brazos River (Texas)	Coarse-grained, e.g. Colorado River (Texas)
Channel stability	High	Slight
Channel cross-section	Relatively narrow; asymmetrical	Symmetrical or asymmetrical
Sinuosity	High	Low
Gradient	Low	Moderate
Sand facies geometry	Multistorey	Multilateral
Vertical sedimentary structure sequence	1. Parallel lamination and thin tabular cross-sets	5. Climbing ripple lamination
	2. Thick tabular cross-sets	4. Small-scale trough sets
	3. Parallel lamination	3. Large-scale troughs or homogenous sediment
		2. Parallel lamination, trough cross-sets and tabular cross-sets
		1. Large scale trough cross-sets
Grain size	Upward fining	No trend

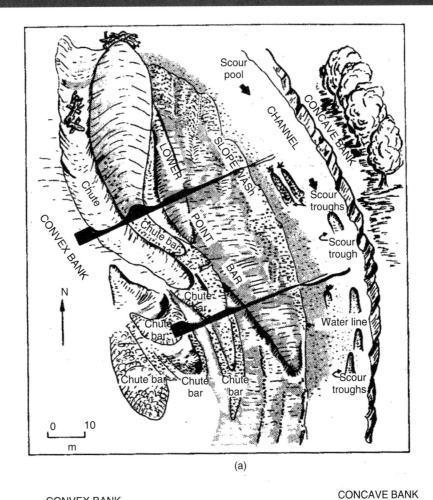

(a)

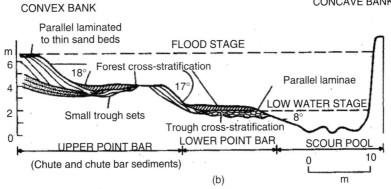

(b)

Figure 17.37 Coarse-grained point bar on the Amite River, near Magnolia, Louisiana, USA, showing the morphology, profile and cross-stratification types. (Source: after McGowan and Garner, 1970)

concentration of both bedload and sediment suspension near the river bed, when the flow is confined to the chute. Sand is carried in suspension beyond the foreset face into a backflow zone where it accumulates as parallel laminated and rippled sand. Flow spreads radially away from the distal chute and the sediment disperses to form a lobate sand body. Older chute bars commonly show more than one accretionary phase.

7. Floodplain sediments are deposited contemporaneously with the chute bars and merge laterally with them. They consist of foreset cross-stratification in granule to pebble gravel that grades into alternating parallel laminated sand and gravel, fining upwards into parallel laminated sand. This sequence is formed in extreme flood in both the rising and the waning stage. Scour fills of gravel and gravelly sand succeed these sequences and form as the flood stage is lowered, and indicate deposition in gullies that trend toward the river. Most of the sediment is brought in as traction load because during the extreme flood the floodplain is part of a wide, shallow river.

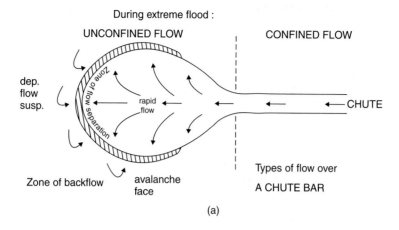

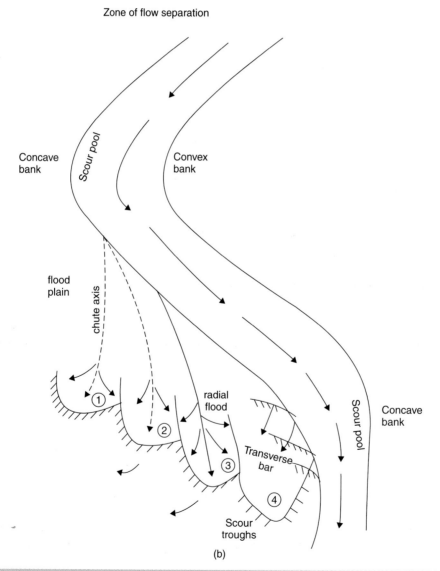

Figure 17.38 (a) Morphology of chute bars. (b) Flow through chutes and across chute bars. (Source: after McGowan and Garner, 1970)

Alluvial fan sediment-landform assemblage as part of the fluvial piedmont

This is such a distinctive geomorphological and sedimentological system that it will be considered separately. An alluvial fan is a sediment unit built up by a mountain stream that flows from a steeply-dissected mountain front on to an area of lower gradient. Deposition is favoured here because at the canyon mouth there is a decrease in stream gradient and all streams undergo a velocity reduction and loss of transporting power. Alluvial fans are the result of this deposition. They occur in all climatic zones but seem to be best-developed in areas of low and intermittent rainfall with poor vegetation cover, like the North American south-west, the Near and Middle East and central Australia. They also occur in humid areas such as Japan, the Himalayas and the Alps, where steep mountain torrents join broader, gentler valleys, and Arctic regions, such as Northern Scandinavia, Alaska and Arctic Canada. The smaller ones are sometimes called alluvial cones and can be found in many upland areas of the British Isles, where valleyside tributaries deposited them on formerly glaciated, relatively flat valley floors, like the Little Hareden fan in the Langden Brook valley in the Forest of Bowland. They commonly form coarse, detrital aprons to mountain masses that are fault-bounded, but they are not totally restricted to this geological environment.

They are perhaps best developed in deserts, where there is a characteristic profile, including the mountain, a bedrock pediment and alluvial plains. The latter include base-level plains, such as playas and bajadas (alluvial plains of considerable complexity, both stratigraphically and topographically), and alluvial fans, which can coalesce to form the bajada. This profile is known as a piedmont slope. The following conditions in deserts explain the marked occurrence of fans: (a) the lack of vegetation, which means that the position of the streams is relatively unfixed; (b) long periods of debris accumulation in mountains and occasional heavy thunderstorms, which lead to rapid and massive debris evacuation from the mountain front; (c) a small ratio of depositional area to mountain area; that is, large highlands bordering small lowlands.

However, Stanistreet and McCarthy (1993) have described a new class of large, low-energy fan based on the modern Okavango fan in Botswana and have erected a simple alluvial fan classification (Figure 17.39). They recognize that various fluvial processes may occur in different proportions, depending on controls like climate and the nature of source sediment, and consequently employ a triangular classification, showing the importance of the three main processes: sediment gravity flows, braiding and meandering. Their first class of fan is the debris-flow-dominated fan, which is the only true fan accepted by Blair and McPherson (1994).

Morphology

The plan view is fan-shaped (Figure 17.40(b)). Each fan geometrically resembles the segment of a cone, with the apex where the stream issues from the mountain front, through a ravine or canyon. Usually the erosional products are transported in a single stream through this fanhead or fanbay area. The streams are commonly entrenched in fanhead trenches. From this apex the fan dips radially away towards the toe, which is the outer or lowest part of the fan, with the fan slope gradually declining. This base is linked to the middle of the fan across the mid-fan by an area of moderate slope. Attempts have been made to describe the shape precisely using simple equations, for example the surface of the fan has been compared to that of the segment of a cone, characterized by concavity of the radial profile (the longitudinal profile) and convexity of the transverse profile. The scale varies from those over 60 km across to the much smaller alluvial cones, which are maybe several hundred metres at the most. The slope angle varies, but few fans have gradients over 10°. Downstream from the fanhead trench the stream channel divides into a shallow, braided pattern. New sediment is added to the fan along narrow radial bands because the area of deposition shifts about continually on the fan through time. Individual channels are blocked by sediment or are back-filled, or large individual boulders block the channels, so that in flood periods there is a continual shifting around of the depositional zones. Beyond the fan base area the braided channels die out either into the silt/evaporite playa lake environment, as in Death Valley, California, or into a normal floodplain with a perennial river flowing to the sea, like the Mackenzie River delta area in Arctic Canada.

Fan area

Alluvial fans are found to be approximately 1–3 times larger in area than the source drainage basins which feed them. The precise relationship between the two depends on the bedrock and the climate. Several workers have recognized that with increasing drainage basin area, the area of alluvial fan increases and the fan slope decreases. This can be expressed in terms of log plots of fan area versus drainage basin area: Af = c A dn, where Af is the fan area, A is the drainage basin area of 1 km^2, c is a constant varying from one drainage basin to another, depending on source area lithology and the tectonic environment, and n is the slope of the regression line.

To illustrate further the effects of bedrock type, Hooke (1967) found that in the Deep Springs Valley, fans with a predominance of quartzite were only about one third of the size of fans with source areas underlain by dolomite and quartzite. Bull (1964) found that fans derived from mudstone/shale sources were about twice as large as fans derived from mainly sandstone sources of comparable size. It has been argued that the general equation results from a tendency towards a steady-state area among coalescing fans in similar environments, that it is part of a space-sharing system. The argument rests on both laboratory and field observations suggesting that relatively uniform deposition may occur over a fan surface. If a fan is too small for the debris volume supplied to it, it will increase in thickness and area faster than adjacent fans, until a steady state is established. As the debris volume being supplied increases with drainage area, larger drainage areas

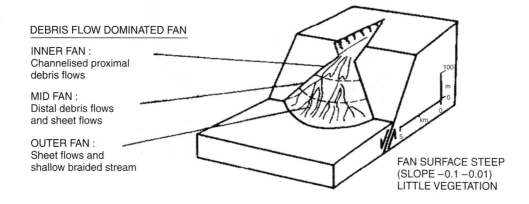

DEBRIS FLOW DOMINATED FAN

INNER FAN :
Channelised proximal
debris flows

MID FAN :
Distal debris flows
and sheet flows

OUTER FAN :
Sheet flows and
shallow braided stream

FAN SURFACE STEEP
(SLOPE −0.1 −0.01)
LITTLE VEGETATION

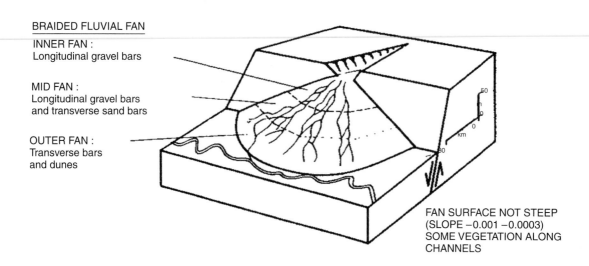

BRAIDED FLUVIAL FAN

INNER FAN :
Longitudinal gravel bars

MID FAN :
Longitudinal gravel bars
and transverse sand bars

OUTER FAN :
Transverse bars
and dunes

FAN SURFACE NOT STEEP
(SLOPE −0.001 −0.0003)
SOME VEGETATION ALONG
CHANNELS

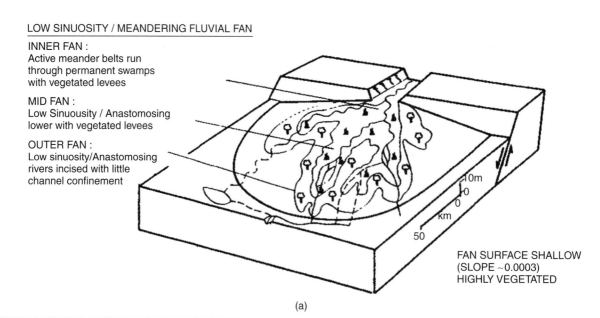

LOW SINUOSITY / MEANDERING FLUVIAL FAN

INNER FAN :
Active meander belts run
through permanent swamps
with vegetated levees

MID FAN :
Low Sinuousity / Anastomosing
lower with vegetated levees

OUTER FAN :
Low sinuosity/Anastomosing
rivers incised with little
channel confinement

FAN SURFACE SHALLOW
(SLOPE ~0.0003)
HIGHLY VEGETATED

(a)

Figure 17.39 (a) A triangular classification of alluvial fans, showing the three main depositional styles and the models developed for each. (b) A comparison of the sizes of modern alluvial fan systems. (Source: after Stanistreet and McCarthy, 1993)

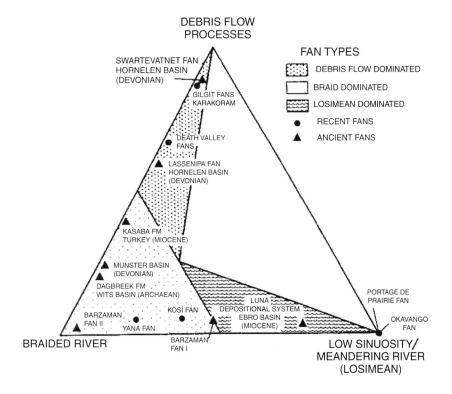

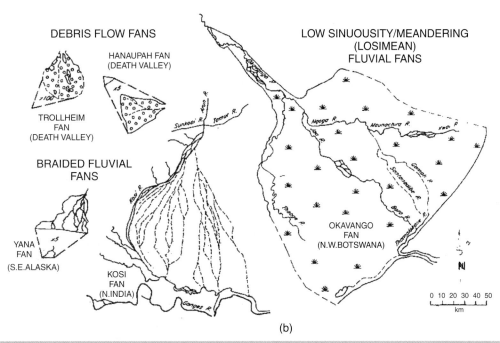

(b)

Figure 17.39 *(continued)*

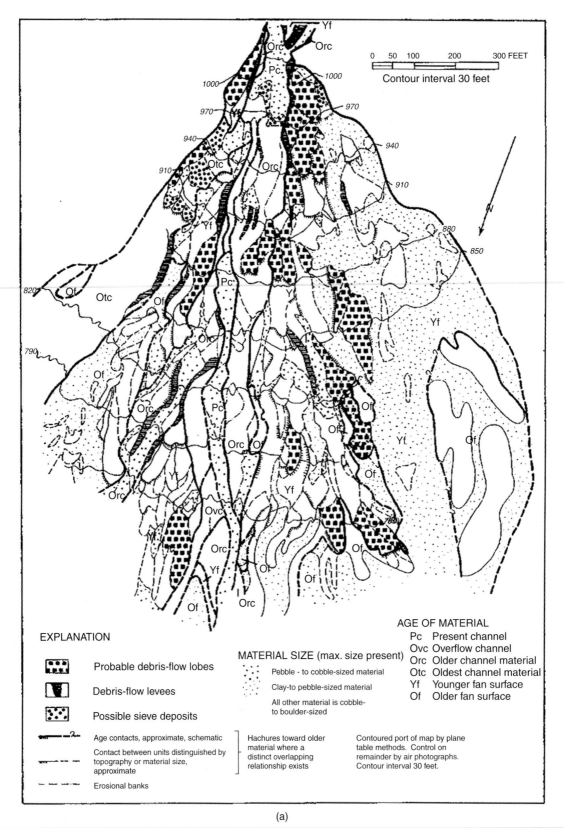

(a)

Figure 17.40 Alluvial fans: (a) an example of a fan dominated by sediment-gravity flows: the Trollheim fan, Death Valley, California. In other fans where these processes operate, traction current deposition is also important; (b) southern Taklimakan desert, Sinkiang province, China. (Source: http://rst.gsfc.nasa.gov)

(b)

Figure 17.40 *(continued)*

are generally associated with larger fans. This idea requires that the area of each fan is proportional to the volume of material supplied to it per unit of time. Five important factors other than area affect c:

1. The ratio of depositional area to erosional area, with which c is directly and positively correlated.

2. Sediment yield, as determined by rock types and relief in the drainage basin. That is, fans derived from easily-eroded rock types are larger than those derived from resistant rocks.

3. Tectonic tilting, which alters basin area and relief. In Death Valley, for example, fans associated with the dip-slope of upfaulted blocks tend to be larger than those adjacent to fault-scarp mountain fronts.

4. Climate on a regional scale.

5. The amount of space available for fan deposition on a local scale.

The value of c is usually under 1, implying that large drainage basins yield proportionally less sediment on to fans than smaller ones. This seems to be because large basins are less frequently covered by a single storm; more sediment may be stored on valleyside slopes or in the channels of large basins, and the valleyside slopes tend to be lower in larger basins and therefore to yield less sediment.

Fan slope

This rarely exceeds $10°$ and is determined by several variables, such as the grain size, water discharge and the types of depositional process. The longitudinal profile of an unsegmented fan is generally a smooth exponential curve. It seems likely that this longitudinal profile will be steeper for fans where source areas have high rates of sediment production than for those in areas with low rates of sediment production. Fan slope is generally inversely proportional to fan area, drainage-basin area and discharge. The latter is especially important because large discharges can transport debris on a lower slope than small discharges as they have higher flow velocities and bed shear stresses. It also follows that for smaller discharges, the deposited sediment will be primarily near the fanhead, whilst larger discharges will deposit sediment nearer the toe. So fan slopes vary directly with grain size, the steeper slopes occur where depositional processes are associated with the production of coarse sediment, like debris flows and sieve deposits, and gentler slopes form where fluvial processes alone dominate. Fans with flows of higher sediment concentration generally have steeper slopes.

Processes of formation

There are three processes responsible for fan deposition: stream flow, debris flow and mudflow, and the balance between them is chiefly determined by climate. Most fans consist of interbedded water-transported and debris-flow sediments, but as the rainfall becomes more intense and the climate more semi-arid, the percentage of debris flows and mudflows becomes greater.

Stream flows   These predominate in areas of moderate year-round rainfall. Transport and deposition occur in braided, perennial streams. The water-laid sediments consist of sheetflood sediments, stream channel sediments and sieve deposits. Most are formed of sheets of sand, silt and gravel deposited by the braided network of distributary channels. Deposition is caused by a widening of the flow into shallow sheets with lower depths and flow velocities. Water depths are generally under 30 cm and these shallow distributary channels rapidly fill with sediment and shift location, and the resultant deposition consists of sheets of sand and gravel traversed by shallow channels, which repeatedly divide and rejoin. In general the sediments are well-sorted, well-bedded, sandy gravels and sands, and the bedforms are mainly low bars.

Stream-channel sediments consist of fillings in stream channels that were temporarily entrenched into the fan. They are generally coarser-grained and more poorly sorted, and the bedding may not be so well defined.

Finally, sieve deposits are formed when the surface-fan sediment is so coarse and permeable that even large flood discharges infiltrate into the fan completely before reaching the fan toe. They occur where the fan source area supplies little sand, silt and clay to the fan. This infiltration promotes the deposition of gravel as lobate deposits, which further decreases the slope and promotes additional deposition. As the water passes through, rather than over, such deposits, they act as sieves, permitting the water to pass but holding back the coarser sediment in transport.

Debris flows

The early workers on alluvial fans assumed that they were formed by running water alone, and the importance of debris flows in their formation was not realized until after Blackwelder's (1928) discussion of the process.

Water flows can selectively deposit part of their sediment load as a result of a decrease in velocity or flow depth. However, when a flow incorporates sufficient sediment, sediment entrainment becomes irreversible and the flow behaves more like a plastic mass rather than a Newtonian fluid. Debris flows have a high density (2.0–2.4) and viscosity (over 1000 poise) compared with stream flows because of the debris content in suspension. Streams have a viscosity of about 0.01 poise and the density of a stream carrying suspended sediment is only slightly over 1. Owing to the high density of mud, forces on rocks in debris flows are very different from those on comparable rocks in stream flows. The submerged weight of a rock in a debris flow is reduced by perhaps as much as 60% relative to its submerged weight in water. Also, the combination of the low-density contrast between rocks and mud and the high density of mud reduces settling velocity. Thus gravity forces tending to prevent motion are reduced. With the reduced settling velocity in debris flows, coarser material is retained in suspension. As mud has a finite yield strength, debris flows stop when the shear stress on the bed no longer exceeds the yield strength of the mud. This may result from the loss of water to the underlying dry fan bed, increasing the yield strength, or from a decrease in either the fan slope or flow depth as the flow moves downfan and spreads out.

So a debris flow is a mass flow in the viscous region, with an appreciable content of suspended clasts. This viscous flow is characterized by a dispersive pressure, which results from the distortion of the fluid between grains. A lot of the experimental work on debris flows in the past was carried out using clay slurries, with no coarse grains, but some was done on flows of sewage sludge, which clearly had coarser clasts. Enos (1977) suggested that debris flows move in laminar flow rather than through turbulent flow, although large-scale circulation does occur near the front of a flow or because of channel irregularities. Hence some slow mixing is possible. However, laminar flow is indicated by the preservation of fragile clasts, like shales; by the rafting of blocks projecting above the flow zone, because these are rafted gently rather than being tossed about; and by the lack of scour marks, such as flutes, which could be attributed to turbulent eddies impinging on the substrata. Debris flows appear to have similar regular flow cycles, with a stream of muddy water that gradually becomes deeper, with more and more sediment. Then a debris wave occurs, which consists of a front largely made up of the coarser material in transport, with the more fine-grained sediment behind it – where the boulders

are more widely separated – and then the original muddy water state behind that.

Debris flows are most common in relatively dry areas, near the fan apex, where there is a scattered but intense rainfall. The factors that promote them are:

1. Abundant water as intense rainfall over a short time period, at irregular intervals.

2. Steep slopes which have sufficient vegetation cover to prevent rapid erosion.

3. A source rock to provide the mud matrix.

In morphology, debris flows have flat tops, well-defined steep sides and a lobate form, extending from sheet-like masses. Coarse sediment frequently accumulates at the front of a flow and is shoved aside by the advancing snout, forming levees that confine the rest of the flow. These also accumulate at peak discharges when the debris flow overtops the channel banks. Usually there is a distinct sorting of clasts on the levees. However, rain, rill erosion, soil creep and weathering processes all gradually modify their distinctive morphology and internal structure. Eventually the only remaining evidence may be cobble and boulder accumulations, with low relief.

In internal composition, debris flows are composed of cobbles and boulders in a fine matrix. The bedding is not well defined but sometimes bedding planes and finer-grained sediment washing lines can be picked out between flows. A fluid debris flow will have graded bedding and an imbrication in the tabular gravel clasts, but more viscous flows have their larger clasts more uniformly distributed. Debris flows are characteristically poorly sorted. For example, in the Diablo Range, California, braided stream deposits have sorting values between 0.48 and 2.4, with a mean of 1.0; stream channel deposits have values between 0.82 and 3.4, with a mean of 2.0; whilst debris flows have values between 4.1 and 6.2, with a mean of 4.7. One example is the Trollheim fan in Death Valley, California (Figure 17.40(a)).

Mudflows

Mudflows are finer types of debris flows composed of highly-concentrated mixtures of clay, silt, sand and occasional gravel. Their production on fans is favoured by clay-bearing rocks in the source area. Depending on the slope angle and viscosity of the flow, they either travel rapidly as surges down channels or glide along at very slow speeds. They produce poorly-sorted, ungraded sediment and usually have polygonal desiccation cracks on the surface of each flow.

Summary of the characteristics of alluvial fan sediments

Alluvial fans can have greatly differing grain sizes and surface characteristics. Some consist of organic silt, like at Aklavik (North West Territories, Canada), which exists in a permafrost environment, with polygonal ground and frost mounds, where the source rocks are fine- and coarser-grained sandstones, shales and siltstones that have been broken down. Others consist of pebble- and boulder-sized sediment, with virtually no fines. There is usually more than one depositional process occurring on most fans and the proportions can vary both vertically and in the downslope direction from the fan apex. The beds are characteristically sheet-like in form, with dimensions between 10 and over 100 times the width of the channel that transported the sediment to the fan. Fans show a high consistency of current flow directions, and the distribution of maximum and mean particle sizes shows that particle size decreases downslope with increasing distance from the fan apex. Their stratigraphic geometry reflects both the accumulation of vast numbers of beds of differing extents and thicknesses and the changes in the loci of deposition, caused by the entrenchment and backfilling of the trunk mainstream.

The fans abutting upland areas are composed of successive narrow sheets, which represent coalescent alluvial fans. In vertical section the individual sheets are wedge-shaped, the deposits rapidly thinning and fining away from the uplands. They often pass into finer-grained, laminated, playa lake deposits. Often alluvial fan deposits are orogenic deposits found in areas where recent uplift has occurred along a fault line, which causes rapid accumulation of fan sediments adjacent to a mountain front. The optimum conditions for accumulation of thick sequences of fan deposits occur where the rate of uplift exceeds the rate of downcutting of the trunk stream channel at the mountain front. Fans can be important economically too, as the deposits form ground-water reservoirs in many areas and the recharge of many ground-water basins is through alluvial fans fringing them.

Fluvial landform characteristics in rock channels

In upland areas and high mountains, rivers usually show an increased gradient, increased velocity of water and hence increased erosion power, but to counteract this there is increased frictional resistance, too. These river characteristics, plus the fact that there are often varied rock types and geological structures (like joints and faults) in which river valleys have been eroded, result in rapids, cascades (where water descends a series of rock steps), waterfalls and gorges.

Waterfalls are created because the geology is not uniform in a particular area, as a river flowing for the whole of its course over rocks with similar lithological and physical characteristics would have a perfectly smooth sloping bed and its gradient would be represented by a gradually flattening curve. If, however, the river were to flow over rocks that differed from each other in their capacity to resist erosion, the gradient would be irregular or uneven because the water would be able to lower its bed more rapidly in soft rock types than in those that were hard and more resistant to erosion. A river will also be able to erode more rapidly in rocks that, although resistant, have a marked joint system that

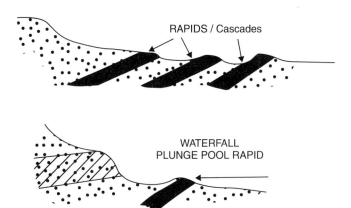

Figure 17.41 Development of rapids

it can exploit than in rock types that are massive and relatively free from geological weaknesses. So when a river flows from relatively hard to softer, less-resistant rocks it will lower its bed more rapidly, and where there are many alternating hard and soft rock types in an area, a set of rapids will develop (Figure 17.41). Erosion tends to be greatest where the gradient is steepest and lower where the gradient is more gentle; therefore, erosion is greatest where rapids occur. Over time the riverbed gradient will lose its irregularities and flatten out. This means that rapids and waterfalls are temporary features of the landscape.

A waterfall's location is determined by the presence of hard or massive rocks overlying softer rocks, but its actual form is controlled by a variety of factors, which are primarily structural/tectonic geological controls, like faults, joint systems, the inclination and dip of the beds, and whether there are bedding planes in the rocks. Examples of typical waterfalls from Iceland, showing some of these controls, are given in Figure 17.42 (see also Figure 17.18 for examples from the Pennines).

Where water passes over alternating shale and sandstone and drains along bedding planes, joints and any fault lines, erosion of the shale occurs, leaving the harder sandstone as an overhanging ledge. This is gradually eroded and the waterfall retreats upstream, giving a steep-sided gorge. All waterfalls show this tendency to upstream retreat and gorge formation.

Often, the rock stratum just below the more resistant shelf will be of a softer type, meaning that there will be undercutting due to splashback. This forms a shallow, cave-like formation or underhang and a plunge pool beneath and behind the waterfall. Eventually, the outcropping, more-resistant cap rock will collapse under pressure, adding blocks of rock to the base of the waterfall. These blocks are then broken down into smaller boulders by attrition as they collide with one another, and erode the base of the waterfall by abrasion, creating a deep plunge pool.

Gorges form for other reasons too. If there is movement on a fault and uplift, the result can be rejuvenation of the rivers and downcutting, and gorges can form in rock. One example is the Ingleton Glens in north-west Yorkshire, which are associated with movement on the Craven Faults. In areas with limestone bedrock there can be cave collapse or subglacial erosion under an ice sheet, as for example in the Goredale and Malham areas (north-west Yorkshire), or perhaps rapid erosion associated with snowmelt, as has been suggested for Cheddar Gorge in the Mendips. Finally, and usually on a much bigger scale, gorge formation can be caused by a process known as antecedent drainage development, where in a high mountain area that is being uplifted over a long period of geological time, like the Himalayas, rivers continue to cut vertically as the mountains rise upwards, resulting in a gorge that cuts across the growing mountain range.

Deltaic processes and environments

Superficially, deltas are not unlike alluvial fans in appearance, as both have a radial channel network, look like the Greek letter delta and owe their origin to the fact that a single stream deposits its transported load suddenly where the gradient is lowered. Deltas

vary greatly in size, but about 30 of the world's major deltas are over 5000 km², and they occur in all climatic regimes, from cold and partly-frozen seas (the Lena and Mackenzie deltas) right through to the tropics (the Niger and Orinoco deltas). Some are in virtually tideless, landlocked seas, like the Danube and the Rhone deltas, and some are in lakes, like the Volga (Caspian Sea) and the Amu Darya (Aral Sea). Their wide geographical distribution inevitably means differences in the characteristics of the deltaic processes, landforms and sediments. For further detailed description of this variability, and a discussion of the processes, environments and sediment facies, see Morgan (1970), Broussard (1975), Coleman and Prior (1982), Elliot (1986), Whatley and Pickering (1989) and Colella and Prior (1990). Four main factors are important in changing deltaic sedimentation:

1. River regime (where variations influence the sediment load and the transport capacity).

(a)

(b)

(c)

(d)

Figure 17.42 Waterfalls in Iceland: (a) Dettifoss, a waterfall in Jökulsárgljúfur National Park in north-east Iceland, not far from Mývatn. It is on the Jökulsá á Fjöllum river, which flows from the Vatnajökull glacier and collects water from a large area in north-east Iceland. The falls are 100 m wide and have a drop of 44 m down to the Jökulsárgljúfur canyon. Dettifoss is reputed to be the largest waterfall in Europe in terms of discharge, having an average water flow of 200 m³ s⁻¹; (b) gorge below Dettifoss; (c) Gullfoss, east of Langjökull; d) Godafoss, on the south coast

2. Coastal processes (wave energy, tidal range and current strength).

3. Structure of the region (stable, subsiding or elevating).

4. Climate (which affects factors such as vegetation).

A commonly-used classification of deltas is based on the relative importance of fluvial processes versus basinal processes (waves and tides); this is shown in Figure 17.43(a). Representative modern examples are illustrated in Figure 17.43(b) and a Landsat image of the Lena delta is given in Figure 17.43(c).

Simple delta types

Gilbert delta   The basic concepts of deltaic deposition were presented by G.K. Gilbert in the late nineteenth century from his study of the fluvial/lacustrine sequences around Pleistocene Lake Bonneville in the American south-west. He recognized that the main characteristics of deltas were the prograding nature of the river system and the sediment accumulation into the standing water body. This led to the succession: bottomset, foreset, topset beds, which were transitional with one another, although all could be recognized in cross-section. Pulsatory density underflows can be caused by the high sediment concentrations flowing along the lower delta slopes, which can deposit various types of ripple drift sequence in fine sand and silt. Some underflows may be triggered by slumps on the foreset slope, but the high sediment concentration is probably far more important in flow initiation. Such flows were very common in Pleistocene glacial lakes and still are in glacial lakes today, particularly in summer. However, in the delta of a large river the sediment accumulation is much more massive, which can lead to regional downwarping, as well as local differential compaction; spatial relationships are much more complex between different environments and grain-size relationships are much more complex in large modern deltas. In general therefore the Gilbert terminology serves no useful purpose for the study of modern deltas but has been very useful for describing Pleistocene and modern glacial-lake deltas (see Section 20.5.5).

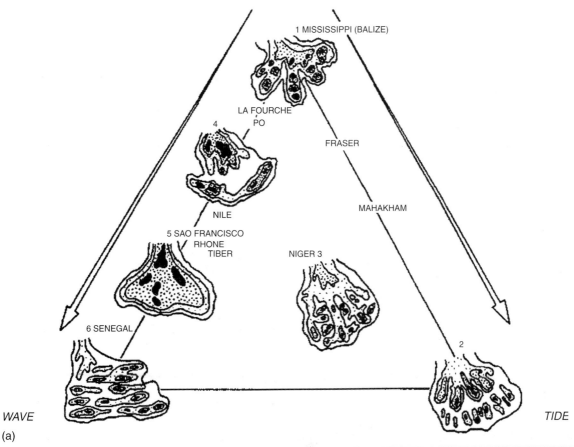

(a)

Figure 17.43 (a) Sand body geometries of the six delta types of Coleman and Wright (1975), plotted on the classification of Galloway (1975). (b) Modern examples of river-dominated, wave-dominated and tide-influenced deltas. (c) Lena river delta, Siberia; false colour image made using shortwave infrared, infrared and red wavelengths. (Sources: (a) and (b) after Walker and James, (1992); (c) from http://visibleearth.nasa.gov/view_rec.php?id=3451

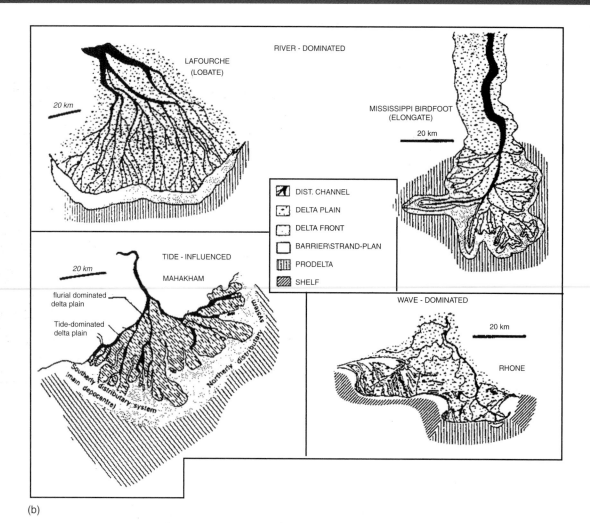

(c)

Figure 17.43 (*continued*)

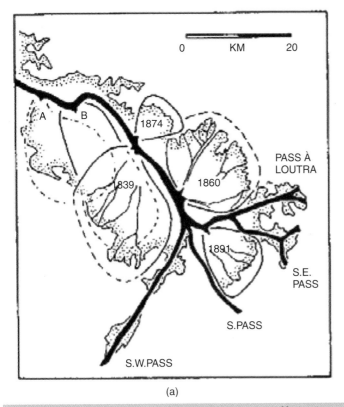

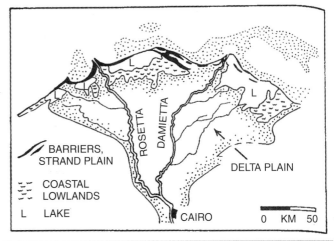

Figure 17.44 Bird's foot delta, Mississippi. (Source: http://rst.gsfc.nasa.gov)

Bird's foot delta   The classic example of this form is in the modern Mississippi delta, where the energy level is very low and the sediment is not transported away from the input area very quickly. Hence narrow strips of land are built out, bounded by levees for many kilometres into the water body, with shallow bays between them (Figure 17.44). In the Mississippi and the Atachafalaya (200 km to the west) the sediment input is high, and together they transport over one million tons of sediment into the Gulf of Mexico each day, but over 75% of this is silt and clay. The Mississippi delta is discussed in more detail in Box 17.3 (see in the companion website).

Classical arcuate delta   The Nile and Niger deltas are good examples of this form (see Figure 17.45). The latter faces the Atlantic Ocean and is affected by the trade winds and by ocean swell waves with heights of 2 m and periods of 10–12 seconds, which are very powerful and act throughout the year. These give the delta a smoothly-curved shoreline, but along the coast there are a series of sandy barrier islands, broken intermittently by distributary channels. The sediment input is relatively small, with sand more important than in the Mississippi, but the water body has a high energy.

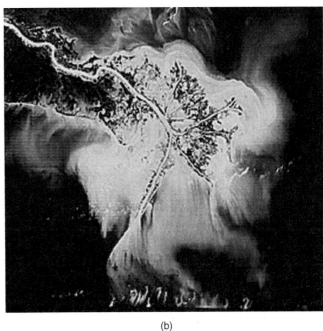

Figure 17.45 Nile delta environments. (Source: after Walker and James, 1992)

Estuarine delta   Here the shoreline is characterized by several distributaries emptying into the sea through funnel-shaped estuaries. Good examples are the Ganges/Brahmaputra and the Orinoco. For Geological Case Studies of Fluvial Systems see Box 17.4. in the companion web site.

Exercises

1. How does energy loss occur in a river?

2. How is sediment transported in rivers?

3. What are the main erosion processes in rivers?

4. After water falls on a hillslope, what are the stages in the development of a river system on that slope?

5. Describe the main types of bedform that form in sand-bed rivers.

6. How do river terraces form?

7. Describe how the components of the meandering river landform–sediment assemblage form and give the characteristics of each morphological part.

8. Why do bars develop in braided rivers? Describe their characteristics.

9. What are the morphological and sedimentological differences between the Mississippi shoal-water deltas and the modern bird's foot delta?

10. Describe the morphological and sedimentological characteristics of alluvial fans in different climatic regimes.

References

Allen, J.R.L. 1968. Current Ripples. Amsterdam, North-Holland.

Allen, J.R.L. 1983. River bedforms: Progress and problems. In: Collinson, J.D. and Lewin, J. (Eds). Modern and Ancient Fluvial Systems. International Association of Sedimentologists, Special Publication 6. Oxford, Blackwell Scientific Publications. pp. 19–33.

Ashworth, P.J. and Ferguson, R.I. 1986. Interrelationships of channel processes, changes and sediments in a proglacial river. Geografiska Annaler. 68, 361–371.

Aslan, A. and Autin, W. 1999. Evolution of the Holocene Mississippi river floodplain, Ferriday, Louisiana: Insights on the origin of fine-grained floodplains. Journal of Sedimentary Research. 69, 800–815.

Best, J.L. and Bristow, C.S. 1993. Braided Rivers. Geological Society of London, Special Publication 75.

Blackwelder, E. 1928. Mudflow as a geologic agent in semi-arid mountains. Geological Society of America Bulletin. 39, 465–484.

Blair, T.C. and McPherson, J.G. 1994. Alluvial fans and their natural distinction from rivers based on morphology, hydraulic processes, sedimentary processes and facies assemblages. Journal of Sedimentary Research. 64A, 450–489.

Boon, P, Petts, G. and Calow, P. 1992. River Conservation and Management. Chichester, John Wiley & Sons, Inc.

Bradley, W.C. 1972. Coarse sediment transport by flood flows on Knick River, Alaska. Geological Society of America Bulletin. 83, 1261–1284.

Bridgland, D.R. 1986. Clast Lithological Analysis. Quaternary Research Association Technical Guide 3. Cambridge.

Bridgland, D.R. 1994. Quaternary of the Thames. Geological Conservation Review Series. London, Chapman and Hall.

Briggs, D.J. and Smithson, P. 1985. Fundamentals of Physical Geography. London, Hutchinson.

Brookes, A. 1988. Channelized Rivers: Perspectives for Environmental Management. Chichester, John Wiley & Sons, Inc.

Broussard, M.L. 1975. Deltas, Models for Exploration. Houston, Houston Geological Society.

Bull, W.B. 1964. Geomorphology of segmented alluvial fans in western Fresno County, California. United States Geological Survey. 352E, 89–129.

Colella, A. and Prior, D.B. 1990. Coarse-grained deltas. International Association of Sedimentologists, Special Publication 10.

Coleman, J.M. 1969. Brahmaputra River channel processes and sedimentation. Sedimentary Geology. 3, 129–239.

Coleman, J.M. and Prior, D.B. 1982. Deltaic environments. In: Scholle, P.H. and Spearing, D.R. (Eds). Sandstone Depositional Environments. American Association of Petroleum Geologists, Memoir 31. pp. 139–178.

Coleman, J.M. and Wright, L.D. 1975. Modern river deltas: Variability of processes and sand bodies. In: Broussard, M.L. (Ed.). Deltas, Models for Exploration. Houston, Houston Geological Society. pp. 99–149.

Collard, R. 1988. The Physical Geography of Landscape. London, Collins Educational.

Costello, W.R. 1974. Development of bed configurations in coarse sands. Cambridge, Massachusetts Earth and Planetary Science department, MIT, Report 74-1.

Dawson, M. 1988. Sediment size variation in a braided reach of the Sunwapta River, Alberta, Canada. Earth Surface Processes and Landforms. 13, 599–618.

Elliot, T. 1986. Deltas. In: Reading, H.G. (Ed.). Sedimentary Environments and Facies. Oxford, Blackwell Scientific Publications. pp. 113–154.

Enos, P. 1977. Flow regimes in debris flows. *Sedimentology*. 24, 133–142.

Fisk, H.N. 1947. Fine-grained Alluvial Deposits and their Effect on the Mississippi River Activity. Mississippi, Vicksburg.

Galloway, W.E. 1975. Process framework for describing the morphologic and stratigraphic evolution of deltaic depositional systems. In: Broussard, M.L. (Ed.). Deltas, Models for Exploration. Houston, Houston Geological Society. pp. 87–98.

Guy, H.P., Simons, D.B. and Richardson, E.V. 1966. Summary of alluvial channel data from flume experiments 1956-1961, United States Geological Survey Professional paper 462–1.

Harms, J.C. and Fahnestock, R.K. 1965. Stratification, bedforms and flow phenomena (with an example from the Rio Grande). In: Middleton, G.V. (Ed.). Primary Sedimentary Structures and their Hydrodynamic Interpretation. Society of Economic Palaeontologists and Mineralogists Special Publication 12. pp. 84–115.

Harms, J.C., Southard, J.B., Spearing, D.R. and Walker, R.G. 1975. Depositional Environments as Interpreted from Primary Sedimentary Structures and Stratification Sequences. Society of Economic Palaeontologists and Mineralogists, Short Course Notes 2. Tulsa, Oklahoma.

Hein, F.J. 1974. Gravel transport and stratification origins, Kicking Horse River, British Columbia. Unpublished MSc Thesis. Geology Department, University of Hamilton.

Hooke, R. LeB. 1967. Processes on arid-region alluvial fans. *Journal of Geology*. 75, 438–360.

Huddart, D. 1994. Rock type controls on downstream changes in clast parameters in sandur systems in southeast Iceland. *Journal of Sedimentary Research*. 64A, 215–225.

Ichim, I. and Radoane, M. 1990. Channel sediment variability along a river: A case study of the Siret River (Romania). *Earth Surface Processes and Landforms*. 15, 211–225.

Knighton, A.D. 1980. Longitudinal changes in size and sorting of stream-bed material in four English rivers. *Geological Society of America Bulletin*. 91, 55–62.

Knighton, D. 1998. Fluvial Forms and Processes. London, Arnold.

Kuenen, P.H. 1956. Experimental abrasion of pebbles 2: Rolling by currents. *Journal of Geology*. 64, 336–368.

Leeder, M. 1982. Sedimentology: Process and product, Harper Collins Academic, London.

Leeder, M.R. 1983. On the interactions between turbulent flow, sediment transport and bedform mechanics in channelized flow. In: Collinson, J.D. and Lewin, J. (Eds). Modern and Ancient Fluvial Systems. International Association of Sedimentologists, Special Publication 6. pp. 5–18.

Leopold, L.B. and Wolman, M.G. 1957. River channel patterns: Braided, meandering and straight. United States Geological Survey Professional Paper 282-B.

Leopold, L.B., Wolman, M.G. and Miller, J.P. 1964. Fluvial Processes and Geomorphology. San Francisco, Freeman.

Mantz, P.A. 1978. Bedforms produced by fine cohesionless, granular and flakey sediments under subcritical water flows. Sedimentology 25, 83–104.

McGowan, J.H. and Garner, L.E. 1970. Physiographic features and stratification types of coarse-grained point bars: modern and ancient examples. *Sedimentology*. 14, 77–112.

McPherson, H.J. 1971. Downstream changes in sediment characteristics in a high energy mountain stream channel. *Arctic and Alpine Research*. 3, 65–79.

Miall, A.D. 1992. Alluvial models. In: Walker, R.G. and James, N.P. (Eds) Facies Models: Response to Sea Level Change. Geological Association of Canada, GeoTexts 1. pp. 119–142.

Miall, A.D. 1994. Reconstructing fluvial macroform architecture from two-dimensional outcrops, examples from the Castlegate Sandstone, Book Cliffs, Utah. *Journal of Sedimentary Research*. B64, 146–158.

Miall, A.D. 1996. The Geology of Fluvial Deposits. Berlin-Heidelburg, Springer-Verlag.

Mills, H.H. 1979. Downstream rounding of pebbles: A quantitative review. *Journal of Sedimentary Petrology*. 49, 295–302.

Morgan, J.P. 1970. Deltaic sedimentation modern and ancient. Society of Economic Palaeontologists and Mineralogists, Special Publication 15.

Pearce, T.H. 1971. Short distance fluvial rounding of volcanic debris. *Journal of Sedimentary Petrology*. 41, 1069–1072.

Rust, B.R. 1978. A classification of alluvial channel systems. In: Miall, A.D. (Ed.). Fluvial Sedimentology. Canadian Society of Petroleum Geologists, Memoir 5. pp. 187–198.

Schumm, S.A. 1981. Evolution and response of the fluvial system, sedimentological implications. In: Etheridge, F.G. and Flores, R.M. (Eds). Recent and Ancient Non-marine Depositional Environments: Models for Exploration. Society for Economic Palaeontologists and Mineralogists, Special Paper 31. pp. 19–29.

Schumm, S.A. 1985. Patterns of alluvial rivers. *Annual Review of Earth and Planetary Science*. 13, 5–27.

Simons, D.B., Richardson, E.V. and Nordin, C.F. 1965. Sedimentary structures generated by flow in alluvial channels. In: Middleton, G.V. (Ed.). Primary Sedimentary Structures and their Hydrodynamic Interpretation. Society of Economic Palaeontologists and Minerologists Special Publication 12. pp. 34–52.

Stanistreet, I.G. and McCarthy, T.S. 1993. The Okavango fan and the classification of subaerial fan systems. *Sedimentary Geology*. **85**, 115–133.

Stott, T.A. 2000. The River and Waterway Environment for Small Boat Users. Nottingham, British Canoe Union.

Thorne, C.R., Hey, R.D. and Newson, M.D. 1997. Applied Fluvial Geomorphology for River Engineering and Management. Chichester, John Wiley & Sons, Inc.

Walker, R.G. and James, N.P. 1992. Facies Models: Response to Sea Level Change. Geotext 1. St John's, Newfoundland, Geological Association of Canada.

Waugh, D. 1990. Geography: An Integrated Approach. Walton-on-Thames, Nelson Thornes.

Whatley, M.K.G. and Pickering, K.T. 1989. Deltas: Sites and Traps for Fossil Fuels. Geological Society of London, Special Publication 41. Oxford, Blackwell Scientific Publications.

Williams, G.P. 1970. Flume width and water depth effects in sediment transport experiments. United States Geological Survey professional paper, 562-H.

Further reading

Bridge, J.S. and Demicco, R.V. 2008. Earth Surface Processes, Landforms and Sediment Deposits. Cambridge, Cambridge University Press.

Harvey, A.M., Mather, A.E. and Stokes, M. 2005. Alluvial Fans: Geomorphology, Sedimentology, Dynamic. Geological Society, Special Publication 251.

Robert, A. 2003. River Processes: An Introduction to Fluvial Dynamics. London, Arnold.

18 Carbonate Sedimentary Environments and Karst Processes and Landforms

Island of Bora Bora (French Polynesia) in the south-west Pacific, showing the development of a reef-rimmed, tropical platform.
Source: www.zvs.net/.../11/Bora-Bora-island_Papeete.jpg

Earth Environments   David Huddart and Tim Stott
© 2010 John Wiley & Sons, Ltd

Learning Outcomes

After reading this chapter and completing the exercises you will be able to:

➤ Understand the controls on carbonate sedimentary environments, including several types of carbonate factory.

➤ Describe and classify the various types of carbonate rocks and carbonate facies.

➤ Understand the controls on the solution of carbonates, including the variations in carbon dioxide pressure, temperature, the pH of rain water and the geological characteristics of the rock.

➤ Understand how karst processes modify such carbonate rocks to give a characteristic landform assemblage at the surface, including karren, solutional depressions and limestone pavements.

➤ Understand how underground caves form in limestone areas and the landforms associated with their development, including phreatic and vadose systems and cave speleothem development.

➤ Understand how caves can develop through geological time and the influence of glacial and tropical processes in the development of karst landscapes.

18.1 Introduction

Carbonates are soluble, with limestone the most common of the soluble rocks, covering a twelfth of the Earth's land surface. As such, carbonates form a distinctive and characteristic landscape, to which we give the name 'karst', a general term describing surface and underground scenery associated with this type of rock. 'Karst' is derived from the Slovenian word *kras*, which indicates dry, bare, stony ground; some of the typical components of the limestone scenery were studied and named in this part of Europe. Distinctive karst areas can also be found in the Arctic (e.g. Canada and northern Norway), in the equatorial tropics (Borneo and, spectacularly, China (Figure 18.1)), in the Mediterranean (southern France and Majorca) and in parts of Britain and Ireland (where we find a distinctive glacio-karst scenery). In fact, solution of limestone produces distinctive scenery, in some cases showing subtle differences between different areas, and in others major differences, depending on a combination of factors such as different types of limestone, the past and present climates and recent glacial history.

Before we look at some of the karst processes in more detail we need to understand the formation of carbonate rocks and how certain characteristics of these rocks can influence karst scenery and processes. We will therefore begin this section by reviewing carbonate sedimentary environments and the characteristics of carbonate rocks.

18.2 Carbonate sedimentary environments and the creation of carbonate rock characteristics

Three basic facts need to be discussed in explaining the nature of carbonate depositional systems. These are that carbonate sediments are largely of organic origin; that the systems can build wave-resistant structures; and that they can undergo extensive diagenetic alteration as the original minerals are metastable. The implications of these basic facts are pervasive throughout any discussion of carbonate rock characteristics and how they influence karst processes. The chemical and physical carbonate rock characteristics act to influence how effectively weathering and erosion

(a)

(b)

Figure 18.1 Tower karst landscapes, Guilin, China. (Source: Microsoft Clip gallery)

(c)

Figure 18.1 (*continued*)

act in the landscape. We will encounter these in other chapters of this book and in later parts of this chapter, but initially we will briefly review the production of carbonate material, the types of carbonate factory, the production of sediments associated with coral reef growth and the basic patterns of carbonate sediment accumulation in other depositional environments.

18.2.1 Production of carbonate material

The material for carbonate sediments is taken from the dissolved load of the sea. In this respect, carbonates resemble evaporites and differ from siliciclastics, which depend on sediment supply from an eroding hinterland. Throughout the Phanerozoic, the sea was the prime locus of carbonate precipitation and organisms were the dominant drivers; inorganic precipitation was (and still is) important, but clearly took second place behind fixation of calcium carbonate as elements of organic skeletons (over 90% of carbonates formed in modern environments are marine and form biologically). Thus 'carbonates are born, not made' and carbonate growth and production are intimately tied to the ocean environment, with light, temperature and nutrients being the most dominant controlling factors. In different ways, abiotic precipitation, too, is controlled by the ocean environment. Shoal-water carbonate sediments are oceanic sediments, much like the pelagic sediments of the deep-sea floor.

What we need to bear in mind is that carbonate precipitation in the marine environment occurs via abiotic and biotic pathways and that three modes can be distinguished according to the influence of organisms on the precipitation process:

1. The abiotic mode, where the influence of organisms is negligible. Precipitation is controlled by the thermodynamic saturation state and the reaction kinetics of the aqueous solution. Formation of fibrous aragonite cements and formation of ooids would be good examples.

2. The biotically-induced mode, where organisms induce precipitation. The minerals created differ little from the abiotic precipitates. Micritic carbonate precipitated by bacteria would be a good example.

3. The biotically-controlled mode, where the organisms determine the composition and form of the crystals, as well as the onset and termination of precipitation. Almost all carbonate skeletons, such as bivalve shells, coral skeletons and foraminifera, produce distinctive carbonates.

In the next section, these modes, their most important environmental controls and their combination into geologically-significant carbonate factories will be discussed.

18.2.2 Types of carbonate factory

Figure 18.2 shows the principal carbonate precipitation modes and carbonate factories. Marine carbonate precipitation may be free of organic influence (abiotic), induced by organisms but not further influenced by them (biotically-induced) or totally regulated by organisms (biotically-controlled). On large scales, the precipitation modes combine to give three different types of carbonate production system or 'factory', along with variations in the erosional regime. These are: the tropical factory, dominated by biotically-controlled precipitation by photosynthetic organisms; the cool-water factory, dominated by biotically-controlled precipitation by largely nonphotosynthetic organisms; and the mud-mound factory, dominated by biotically-induced precipitation, mainly by bacteria.

Tropical factory

The tropical factory lies in the photic environment of the low latitudes, about 30–35° above and below the equator, where production is dominated by skeletal precipitation, with a significant abiotic contribution from ooids. Large amounts of marine cements are added during early diagenesis within the depositional setting. The microbial mode may contribute via oncoid grains and stromatolites. This tropical factory shows the highest growth potential of all carbonate factories. In addition, it has the strongest tendency to build rims at the seaward limits of the shoal-water domain. These rims are either barrier reefs or sand shoals with extensive syndepositional lithification and both can resist high water energy and build into the supratidal zone.

Light is the most important property for carbonate precipitation in this tropical factory, as most skeletal carbonate is secreted by organisms that perform photosynthesis themselves or live in symbiosis with photosynthetic algae. Photosynthesis is a complex process but the basic reaction may be simplified as:

$$CO_2 + H_2O + \text{solar energy} = HCHO + O_2$$

where HCHO represents a simple summary formula of the organic matter produced in a photosynthesizing organism. The formula clearly illustrates the link between photosynthesis and carbonate chemistry, as photosynthesis extracts CO_2 from the sea water, so increasing its carbonate saturation and facilitating precipitation of carbonate minerals. Precipitation in the form of $CaCO_3$ has the added advantage for organisms that they can remove potentially deleterious Ca^{2+} ions from the system and build a protective skeleton.

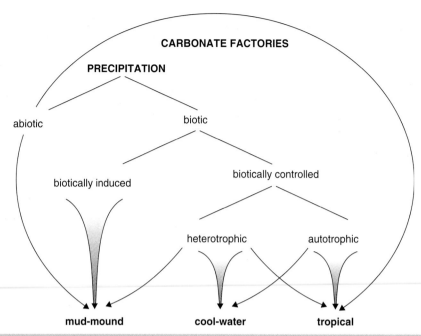

Figure 18.2 Carbonate factories

The link between skeletal carbonate fixation and photosynthesis (and thus light) explains the decrease in skeletal carbonate production with water depth, particularly in tropical environments (Figure 18.3). The typical general pattern is shown in Figure 18.4. The growth–depth curve shows a shallow zone of light saturation, followed by a zone of rapid decrease and then a third, deep zone where growth asymptotically approaches zero. This growth curve can be derived from the well-established exponential decrease of light intensity with depth via a hyperbolic function. The growth–depth patterns of most other carbonate-producing organisms are less well known but seem to follow similar trends. Light saturation means that light is not a growth-limiting factor. Other controls on growth forms and patterns by water energy and sedimentation rates are illustrated in Figure 18.5.

Temperature rivals light in its effect on skeletal carbonate production. Generally, warmer is better, but there exist upper temperature limits for the various carbonate-secreting organisms. Thus, the calcifying benthos functions in a temperature window that is different for different organisms. Most hermatypic (i.e. symbiotic) corals function in the range 20–30 °C. The upper temperature boundary sets important limits to carbonate production, particularly in restricted lagoons. The most important effect of temperature, however, is the global zonation of carbonate deposits by latitude (Figure 18.6). In spite of what has just been said about the role of light, the boundary between the northern and southern limits of coral reefs in the modern oceans, and thus the boundary between tropical and cool-water carbonates, is controlled by winter temperature rather than radiation. This indicates that the temperature limit for hermatypic coral growth is currently reached before the light limit. However, in the geologic record, this may not always have been the case. A generalized comparison between modern warm-water shallow

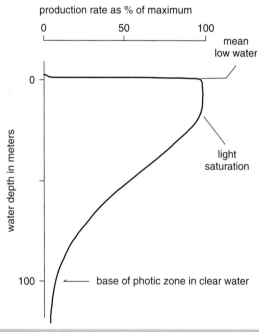

Figure 18.3 Production of the tropical factory from terrestrial elevations to subphotic depth. In most terrestrial environments, carbonate rocks are being eroded, i.e. production is negative, because rain water is undersaturated with respect to carbonate. Maximum production is in the upper part of the photic zone, whence it decreases approximately exponentially with depth

marine reef and cold/cool-water deeper marine coral reefs is given in Table 18.1.

Salinity varies relatively little in the open marine environment but the effects of these subtle variations on carbonate production

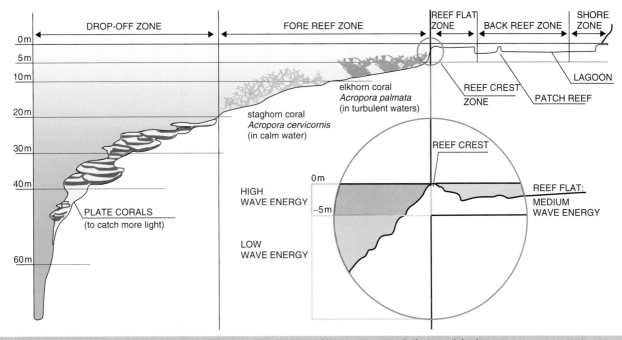

Figure 18.4 General pattern of corals in different zones. (Source: http://www.encora.eu/w/images/f/f7/Zonation_coral_reefs.jpg)

are not well known. Where access to the open ocean is restricted, salinity varies greatly and significantly affects the diversity of the biota, and the combined effects of salinity and temperature variations allow a subdivision of carbonate environments. The effect of salinity on modern plants and animals is shown in Figure 18.6, where it can be seen that a decrease or increase in salinity severely limits the diversity and hence will have an effect on carbonate environments.

Contrary to common expectations, high-nutrient environments are unfavourable for many carbonate systems. Nutrients, nevertheless, are essential for all organic growth, including that of carbonate-secreting benthos. However, the carbonate communities, particularly reefs, are adapted to life in submarine deserts. They produce their organic tissue with the aid of sunlight from the dissolved nitrate and phosphate in sea water and are very efficient at recycling nutrients within the system. In high-nutrient settings, the carbonate producers are outpaced by soft-bodied competitors such as fleshy algae, soft corals or sponges. Furthermore, the destruction of the reef framework through bioerosion increases with an increasing nutrient supply.

Cool-water carbonate factory

The cool-water factory extends poleward of the tropics, in some places to beyond the polar circle. It is dominated by skeletal material of nonphotosynthetic biota (that is, biotically-controlled precipitation). Mud is a minor component derived largely from abrasion of skeletal parts. Organic framebuilders and early cementation are scarce or even absent. Consequently, there is no rim-building capability and the depositional geometry is that of a ramp (Figure 18.7). Accumulation rates are significantly lower than in the tropical factory.

Mud-mound factory

The mound factory is dominated by microbially-induced precipitation. A large number of the micritic carbonates are stiff or hard upon precipitation and support large vugs (holes) that are filled by marine cement during very early diagenesis. Thus, abiotic precipitates contribute significantly to the mud-mound factory. In addition, skeletal material of the cool-water type often contributes to the mud-mound deposits. The mound factory is capable of building elevated structures – the mud mounds – but does not seem to be able to raise them into the high-energy zone of permanent wave action. Here sedimentation rates are comparable to those of the tropical factory. However, the mounds export less sediment than tropical reefs and platforms and so the overall production of the mound factory is probably less than that of the tropical factory.

Erosion in the carbonate factories

Discussion of carbonate production would be incomplete without considering erosion as sedimentary rock is removed from its original place of deposition. Mechanical erosion is a key process in altering the original production output, but in carbonates two other processes are also important: chemical dissolution and bioerosion by boring, etching or rasping organisms.

1. **Mechanical erosion:** This is distributed very unevenly over carbonate depositional environments, although it is most intensive at platform margins where the construction of reefs and lithified sand shoals drastically alters the equilibrium profile between the sediment pile and the wave energy present in the water column. Siliciclastic systems that easily adjust to the

Growth pattern

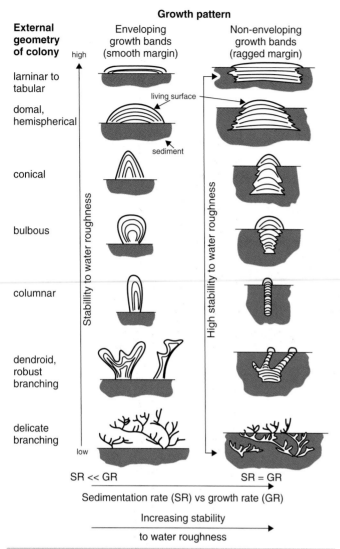

Figure 18.5 Controls on growth forms and growth patterns of colonial reef builders by water energy and sedimentation rates. The relationship between the external shape and the external growth-banding geometry of the colonies can be used to infer water roughness and the relative rates of sedimentation. Ragged margins point to high-energy conditions, smooth margins to low-energy conditions. (Source: after James and Bourque, 1997)

2. **Chemical erosion:** Today the surface waters of the ocean are nearly everywhere saturated with respect to calcite. Dissolution in shoal-water carbonate settings is restricted to the most soluble mineral phases, magnesian calcites with very high magnesium contents and aragonite grains of very small size (1 micron or less). However, sea-floor dissolution is intensive in the deep ocean.

3. **Bioerosion:** Erosion by organisms is much more intensive in limestones and dolomites than in silicate rocks. There are several reasons for this, but carbonate minerals generally are more soluble than silicates and can easily be etched by organic acids. They are also softer than silicates and thus it is easier to abrade their organic hard parts. Finally, skeletal carbonate contains abundant food in the form of organic matter between the carbonate minerals. The intensity of bioerosion varies greatly but is most intensive in the intertidal zone of the tropics, where it acts to cut horizontal notches in the limestones. Bioerosion also increases with the nutrient content of the sea water and this causes much damage to reefs that are close to a hinterland with agriculture because runoff of dissolved agricultural fertilizers increases the nutrient content of the sea, greatly intensifying bioerosion.

18.2.3 Growth potential of shoal-water carbonate systems

Siliciclastic depositional systems depend on outside sediment supply for their accumulation, but in carbonate settings the ability to grow upward and produce sediment is an intrinsic quality of the system, known as the growth potential. Conceptually, one should distinguish between the ability to build up vertically and track sea level – the aggradation potential – and the ability to produce and export sediment – the production potential. The aggradation potential is particularly critical for survival or drowning of carbonate platforms and the production potential is a crucial factor for the progradation and retreat of carbonate platforms and the infilling or starvation of basins.

The growth potential is different for the different factories, with the tropical factory having the highest and the cool-water factory the lowest. Carbonate sedimentation rates decrease with the increasing length of the time span in which they have been measured, and this is a universal pattern of the sediment record. This pattern reflects the fact that sedimentation and erosion are episodic or pulsating processes and the record is riddled with hiatuses of highly variable duration. It is assumed that the upper limit of observed sedimentation rates is an approximation of the growth potential of carbonate factories.

The growth potential also differs for different facies belts of a factory; the tropical factory shows many examples of this kind. Most important is the difference between the growth potentials of the rim and the interior of tropical carbonate platforms, with that of the rim significantly higher than that of the platform interior. When confronted with a relative sea-level rise that exceeds the

ocean's energy regime develop shelf breaks at around 100 m on average, and reefs and carbonate sand shoals may build to sea level at the same position. In fact, modern coral reefs are able to withstand all but the most intensive oceanic wave regime in the tropics. It is obvious that mechanical erosion is intensive in these settings and constant frame-building is needed to repair the erosional damage. In considering the effects of mechanical erosion it should be noted that sea-floor lithification is common and geologically coeval with deposition. This greatly reduces the rate of mechanical erosion and makes piles of carbonate sediment more resistant to lateral displacement.

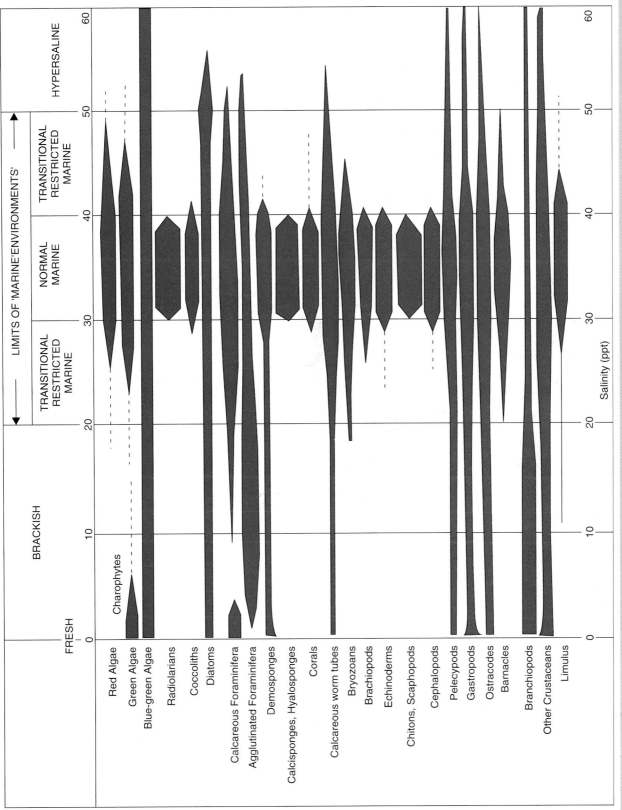

Figure 18.6 Effect of salinity on modern animals and plants in the marine environment. (Source: after Heckel, 1972)

Table 18.1 Comparison of modern warm-water shallow marine reefs and cold/cool-water deeper marine coral reefs. (Source: after Flügel, 2004)

Attribute	Warm shallow-water reefs	Cold deeper-water reefs
Sea water temperature range	20–28 °C	4–12 °C
Bathymetrical range	–50 m	250 to >1000 m
Light	Photic zone	Aphotic to dysphotic zones
Zonation patterns	Windward–leeward	Current zonation
	Light zonation	Taphonomic zonation
Nutritional regime	Oligotrophic	Eutrophic
Trophic modes	Photo- and heterotrophy	Heterotrophy
Diversity of coral framework builders	High	Low
Diversity of associated fauna	High	High
Diversity of associated flora	High	Non-existent
Biological framework stabilization	Corralline algae	Encrusting sponges
Bioerosional impact	High	High
Diagenetic impact	High	Low
Interskeletal cementation	Very common	Rare
Intraskeletal cementation	Very common	Low to common
Internal sediment	Very common	Very common
Automicrite	Common	Very common

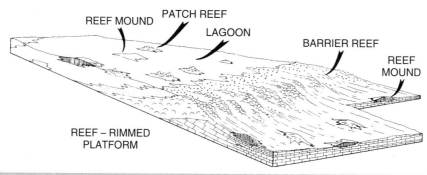

Figure 18.7 Ramp environment

growth potential of the interior, a platform will raise its rim and leave the lagoon empty.

18.2.4 Geometry of shoal-water carbonate accumulations

Interpreting depositional geometry is an important step towards predicting the anatomy of sedimentary rocks in the subsurface. One advantage of sediment analysis by depositional geometry is that it can be performed not only on easily-accessible outcrops but also on seismic or georadar profiles and photos of inaccessible outcrops. This geometry contains significant information on the internal structure and the depositional history of a formation. Below there is a discussion of the basic controls on the geometry of carbonate accumulations and a suitable terminology for their description, followed by a description of the characteristic patterns associated with the three carbonate factories or specific depositional environments.

The geometry of carbonate deposits results from the spatial patterns of production and the superimposed effects of sediment redistribution by waves and currents. Four commonly-occurring patterns in carbonate geometry are directly related to principles of carbonate production and the hydrodynamics of the water column:

1. Carbonate factories build up and tend to rise above the adjacent sea floor. The reason for this is the benthic origin of most carbonate sediment; benthic organisms do not like being buried alive. In fact they thrive in areas where lateral influx of sediment is modest, such that *in situ* carbonate production can outpace it and raise the sea floor in the zones of intensive production.

2. Carbonate production is highest in the uppermost part of the water column, but the terrestrial environment immediately above is detrimental to carbonates. Consequently, carbonate accumulations tend to build flat-topped platforms close to sea level and small differences in production are levelled out through sediment redistribution by waves and tides.

3. Rims are present at the platform margins. The boundary between the undaform domain, which is shaped by waves, and the clinoform domain, which is shaped by gravity transport, is significant in all depositional systems. In tropical carbonates it is the preferred location of frame-builders and thus of barrier reefs that form a rim. However, the platform rim can also consist of sand shoals that form wave-resistant barriers at the platform margin. Both reefs and sand shoals may be strengthened by abiotic cementation. The production of the platform rim is higher than that of the platform interior, such that the rim rises above the lagoon and sheds its excess sediment both downslope and toward the lagoon. Cool-water carbonates have no real rims and develop seaward-sloping profiles in equilibrium with wave action. The accumulations of the mound factory mostly form below intensive wave action and thus are convex rather than flat.

4. Steep slopes. Carbonate slopes steepen with height and are generally steeper than those of siliciclastic accumulations. Several effects contribute to this trend: shoal-water carbonate production includes much sand and rubble and these sediments cannot be transported far and have high angles of repose; slope lithification and cementation also retard slumping and stabilize steep angles once they are formed.

18.2.5 Chemistry and minerology of carbonate rocks

Calcite and aragonite are the only volumetrically important minerals in marine carbonate precipitates, whether organically or inorganically controlled. Calcite has highly variable admixtures of magnesium, but the properties of low-magnesian calcite and high-magnesian calcite are so different that the two varieties are

often treated as separate minerals. Dolomite on the other hand is to all practical intents and purposes a product of diagenetic alteration. The chemical, biological and mineralogical aspects of the precipitation of these carbonate minerals in the marine environment are complex and we will not take them any further here. In most recent and subrecent carbonate sediments, two calcium carbonate minerals predominate: aragonite and calcite. The calcite can be divided into two types – high and low magnesium content – but the mineralogy depends on the carbonate skeletons of the forming organisms and the nonskeletal grains. Over time, as aragonite is metastable, high-Mg calcite loses its Mg and all carbonate sediments are changed to low-Mg calcite and dolomite Ca Mg $(CO_3)_2$ can develop within both sediments and rocks. There can be a noncarbonate component too, such as quartz, pyrite and other minerals, and evaporite minerals, especially gypsum-anhydrite, can be found in association with limestones. The major diagenetic environments and the major processes are summarized in Table 18.2; greater detail can be found in Flügel (2004).

18.2.6 Textures of carbonate rocks

The components of limestones are extremely varied; they are discussed in detail in Tucker (1981) and include ooids, peloids, skeletal material from mollusca, echinoids, bryozoa, foraminifera, sponges, coccoliths and algae, which can form algal mats (stromatolites: see Figure 18.8). These can affect the textures of the rocks produced and the grain size variation in terms of overall size, shape and sorting. Many limestones are composed of fine-grained carbonate. This rock type is called micrite (microcrystalline calcite, with a grain size under 4 µm) and is formed largely from the break-up of calcareous green algae. It can form in deep basins, tidal flats and shallow lagoons.

On the other hand, in many limestones a cement occupies most of the original pore spaces. This is a clear, equant calcite called sparite. Its characteristics indicate its cement origin: the location between grains and skeletons and within original cavities; its generally clear nature, with few inclusions; its increase in crystal size away from the substrate or cavity wall; and crystals with a preferred orientation of optic axes normal to the substrate. The location of sparite formation is under debate, but some of the main environments of cementation of carbonate sediments can be seen in Figure 18.9.

18.2.7 Classification of carbonate rocks

Three systems of classification are commonly used, each with a slightly different emphasis. There is an excellent discussion of the benefits and disadvantages of the various schemes in Flügel (2004).

1. A simple scheme based on grain size which divides limestones into calcirudite, where most grains are over 2 mm,

Table 18.2 Major diagenetic environments and major processes occurring in different diagentic environments. (Source: after Flügel, 2004)

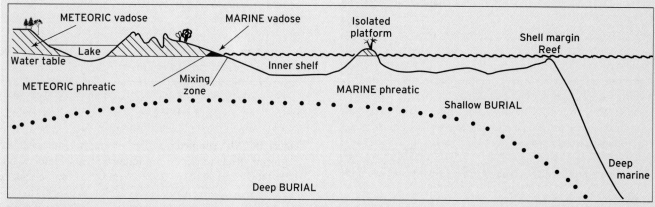

(a)

Diagenetic environment	Location	Pore Filling	Processes	~ Time needed
Meteoric vadose environment	Above water table, between land surface and meteoric phreatic zone	Pores filled with freshwater and/or air	*Solution zone* (soil): Extensive solution; removal of aragonite; formation of vugs. *Precipitation zone* (near surface): Minor cementation	$10^3 - 10^5$ years
Meteoric phreatic environment	Below water table, may tend downwards 100s of meters	Pores filled with freshwater	*Solution zone* (e.g. sinkholes, caves): Solution; formation of molds and/or vugs. *Active zone* (upper part of meteoric phreatic environment): Dissolution of aragonite and Mg-Calcite; rapid and diverse cementation; precipitation of calcite; creation of molds and vugs. *Stagnant zone* (deeper part and in arid climates): Little cementation; stabilization of aragonite and Mg-calcite	$10^3 - 10^5$ up to $10^3 - 10^7$ years
Marine phreatic environment	On the shallow or deep sea floor or just below	Pores filled with marine water	*Shallow-marine environment*: Waters oversaturated with respect to $CaCO_3$: rapid cementation by aragonite and Mg-calcite; diverse cement types. *Deep-marine and cold-water environments*: Waters undersaturated with respect to $CaCO_3$; strong dissolution of aragonite and calcite at two dissolution levels	$10^1 - 10^4$ years
Burial environment	Subsurface beneath reach of surface-related processes, down to realm of low-grade metamorphism. May tend downwards 1000s of meters	Pores filled with brines of varying salinity, from brackish to highly saline	*Shallow burial* (first few meters of tens of meters) and *deeper burial* (sediment overburden of hundreds to thousands of meters): Physical compaction; chemical compaction (pressure solution); cementation; porosity reduction	$10^6 - 10^8$ years

(b)

calcarenite, where most grains are between 62 μm and 2 mm, and calcilutite, where most grains are under 62 μm.

2. Folk's classification, based mainly on composition (Figure 18.10), which recognizes three components: (a) the allochems (grains or particles), (b) the matrix, chiefly micrite, and (c) cement, mainly drusy sparite. An abbreviation for the allochems is used as a prefix to micrite or sparite, whichever is dominant: bio- for skeletal grains; oo- for ooids; pel- for peloids; intra- for intraclasts.

3. Dunham's (1962) classification (Figure 18.11(a)), categorizes limestones on the basis of texture into grainstone (grains without a matrix), packstone (grains in contact, with matrix), wackestone (with coarse grains floating in a matrix) and mudstone (micrite with a few grains). Later, additional terms were added to give an indication of coarse grain size (floatstone and rudstone) and of the type of organic binding in the boundstones during deposition (bafflestone, bindstone and framestone). The terms can be qualified to give information on composition, for example oolitic grainstone or crinoidal

Figure 18.8 Stromatolites: marginal fresh-water forms from Coahuila, Northern Mexico, found in pozas (pools) fed from springs

rudstone. An interpretive sketch of the different types of reef and mound limestone recognized by Embry and Klovan (1971) is given in Figure 18.11(b).

18.2.8 Carbonate facies models and sedimentary environments

Carbonate sedimentary environments

Shallow platforms

Figure 18.12 shows the shore-to-slope profiles of siliciclastic, cool-water carbonates and rimmed carbonate platforms, and a modern case study from the Bahama Banks is discussed in Box 18.1 (see in the companion website).

1. **Siliciclastics:** Have abundant sediment supply, where the result is a seaward dipping surface in equilibrium with deepening wave base.

2. **Rimmed carbonate platforms:** On tropical platforms, the wave-equilibrium profile is grossly distorted by the construction of wave-resistant reefs and quickly-lithifying sand shoals, both occurring mainly at the platform margin. Platforms are basically dish-shaped and equilibrium profiles develop only locally in parts of the lagoon. As can be seen in Figure 18.12(b), rimmed margins are common on carbonate platforms. Reefs (facies 5) are important rim-builders, but early-lithified, stacked sand shoals (facies 6) can be equally effective in defending platforms. A rim's effectiveness in protecting the interior depends largely on its continuity. Reef and sand shoals may also occur together on a platform margin. The growth potential of the platform where there is a seaward sloping ramp may precede the growth of a rimmed profile.

3. **Carbonate ramps:** Accumulations without rims that resemble the siliciclastic equilibrium profile. Cool-water carbonates follow this pattern. Tropical platforms commonly show ramps as a transient stage during rapid transgressions before a rim can develop. So ramps are shoal-water carbonate systems that lack the steep slope seaward of the platform margin on rimmed platforms. They show a seaward-sloping surface with dips of $0.1-1.5°$ instead. Ramps may or may not have an offshore rim and a lagoon. This complicates the definition of a ramp. Originally, ramps were depicted as systems devoid of an offshore rim. Under these conditions, the distinction of ramps and rimmed platforms is easy: rimmed platforms have a high-energy facies belt offshore. Ramps lack this belt and their only belt of high-energy sediments is in the littoral zone, close to shore; farther offshore is the zone of storm wave action, and finally the zone of deeper-water mud facies below storm wave base.

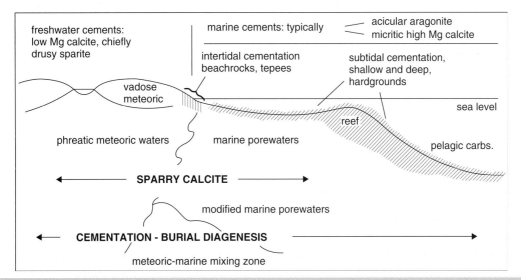

Figure 18.9 Main environments of cementation of carbonate sediments. (Source: after Tucker, 1981)

Principle allochems in limestone	Limestone types			
	cemented by sparite		with a micritic matrix	
skeletal grains (bioclasts)	biosparite		biomicrite	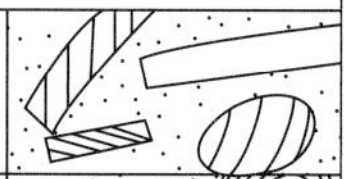
ooids	oosparite		oomicrite	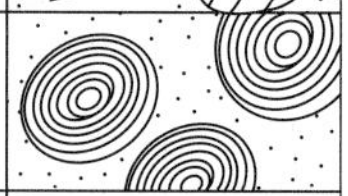
peloids	pelsparite		pelmicrite	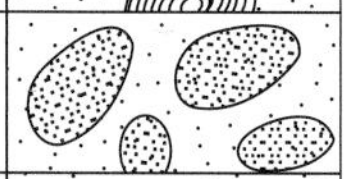
intraclasts	intrasparite		intramicrite	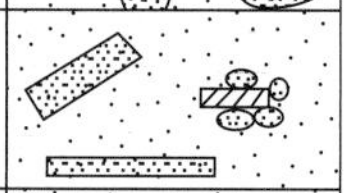
limestone formed in situ	biolithite	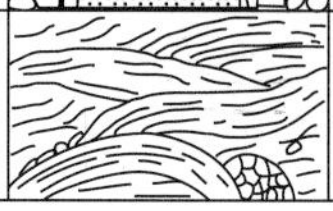	fenestral limestone -dismicrite	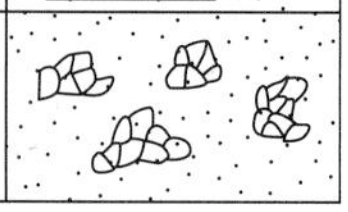

Figure 18.10 Folk's classification of carbonates based on composition

Original components not organically bound during deposition						Original components organically bound during deposition		
of the allochems, less than 10% > 2 mm diameter				of the allochems more than 10% > 2 mm		boundstone		
contains carbonate mud (particles less than 0.03 mm diameter)			mud absent	matrix supported	grain supported	organisms acted as baffles	organisms encrusting and binding	organisms building a rigid framework
mud-supported		grain supported						
less than 10% grains	more than 10% grains							
mudstone	wackestone	packstone	grainstone	floatstone	rudstone	bafflestone	bindstone	framestone

(a)

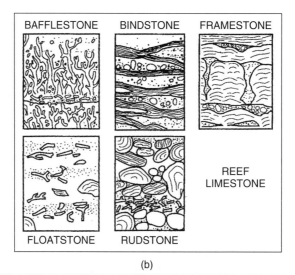

(b)

Figure 18.11 (a) Classification of limestones based on depositional texture (after Dunham, 1962, with modifications of Embry and Klovan, 1971). (b) Interpretive sketch of the different types of reef and mound limestone recognized by Embry and Klovan (1971)

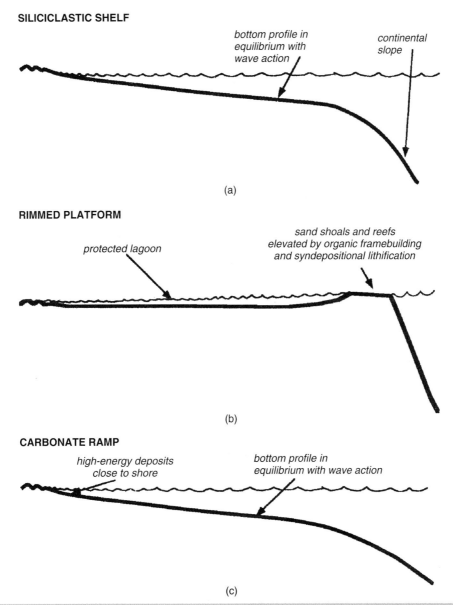

Figure 18.12 Shore-to-slope profiles of siliciclastic, cool-water carbonates and rimmed carbonate platforms

Tidal flat and lagoonal environments

Tidal flat deposition is dominated by micritic limestone deposition, which often contains stromatolites, desiccation cracks and evaporitic mineral sequences. Similar lagoonal successions tend to be dominated by micrites, biomicrites and pelmicrites, but there is occasional evidence of tidal processes and storms.

Reefs and mounds

Reefs In carbonate sedimentology, the term 'reef' denotes a solid, organic, aragonitic, wave-resistant build-up formed by the interplay of organic frame-building, erosion, sedimentation and cementation. Ecologically, reefs are defined by texture and composition. The geometry of reefs is only a secondary attribute, but an important one.

In a modern reef there is an organism sediment mosaic composed of framework organisms, including encrusting, attached, massive and branching metazoans; internal sediment, which infills primary growth; bioeroders, which break down the reef elements by boring, rasping, or grazing and therefore produce sediment for the area surrounding the reef; and the internal reef deposits and cement, which actively lithify the reef. Today corals and red algae construct the framework of the reef, but reef organisms have undergone progressive evolution through geological time and hence the nature of reef development has changed.

In most reefs today the predominant organisms are stony corals, colonial cnidarians that secrete an exoskeleton of calcium carbonate. As the coral head grows it deposits a skeletal structure that encases each new polyp. This coral skeleton erodes and the particles settle to fill spaces in the reef. Waves, grazing fish such as parrotfish, sea urchins and sponges are some of the breakers-down of the coral. Corals can be found in temperate waters, such as off western Ireland and in various parts of the Arctic, but reefs only form in a zone from 30° north to 30° south of the equator, where there is usually warm water (over 20 °C). Also, corals do not grow at water depths over 30 m. The coralline algae are important because they add stability to the reefs by depositing sheets of calcium carbonate over their surfaces.

Coral reefs can take a variety of forms:

- **Fringing reef:** A reef that is either directly attached to a shore or borders it with an intervening shallow channel or lagoon (Figures 18.14 and 18.15).

- **Barrier reef:** A reef separated from a mainland or island shore by a deep lagoon, such as the Great Barrier Reef in Australia (Figures 18.16 and 18.17). Associated with the barrier reef are other environments like beaches and lagoons (Figure 18.18).

- **Patch reef:** An isolated, often circular reef, usually within a lagoon or embayment.

- **Apron reef:** A short reef resembling a fringing reef, but more sloped, extending out and downward from a point or peninsular shore.

Figure 18.15 Fringing reef off the coast of Eilat, Israel. (Photo: Mark A. Wilson, Wikimedia Commons)

- **Bank reef:** Linear or semicircular in outline. Larger than a patch reef.

- **Ribbon reef:** A long, narrow, somewhat winding reef, usually associated with an atoll lagoon.

- **Atoll reef:** A more-or-less circular or continuous barrier reef extending all the way around a lagoon without a central island.

- **Table reef:** An isolated reef, approaching an atoll type, but without a lagoon.

Coral reefs support an extraordinary biodiversity, even though they are located in nutrient-poor tropical waters. The process of nutrient cycling between corals, zooxanthellae and other reef organisms provides an explanation for the fact that coral reefs flourish in these waters. Recycling ensures that fewer nutrients are needed overall to support the community (the zooxanthellae are single-celled algae which have a symbiotic relationship with the coral polyps, carry out photosynthesis and produce excess organic nutrients, which are used by the coral polyps).

Cyanobacteria also provide soluble nitrates for the coral reef through the process of nitrogen fixation. Corals absorb nutrients, including inorganic nitrogen and phosphorus, directly from the water, and they feed upon zooplankton that are carried past the polyps by water motion. Thus, primary productivity on a coral reef is very high, resulting in the highest values per square metre, at $5-10\,\mathrm{g\,C\,m^{-2}\,day^{-1}}$. Producers in coral reef communities include the symbiotic zooxanthellae, coralline algae and various seaweeds, especially small types called turf algae.

Coral reefs are home to over 4000 species of fish, such as colourful parrotfish, angelfish, damselfish and butterflyfish. It has been suggested that the fish species that inhabit coral reefs are able to coexist in such high numbers because any free living space is

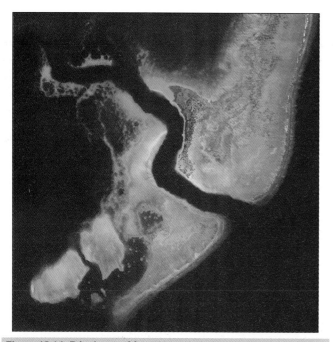

Figure 18.14 Fringing reef from Mayotte between Madagascar and the African mainland. (Source: http:/www//veimages.gsfc.nasa.gov/1648/ikpnos_mayotte.jpg, image courtesy of Serge Anrdreformet)

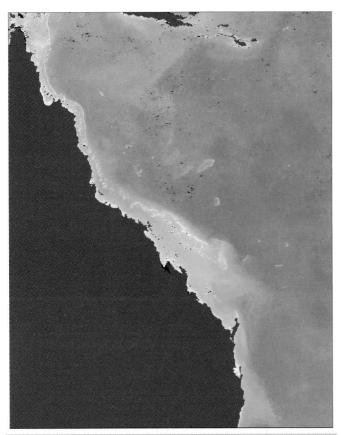

Figure 18.16 The Great Barrier Reef, Australia.
(Source: http://visibleearth.nasa.gov/view_rec.php?id=15689)

Figure 18.17 Satellite image of part of the Great Barrier Reef.
(Source: NASA)

rapidly inhabited by planktonic fish larvae, which then stay there for the rest of their lives. The species that inhabits a free space is random and this process has therefore been termed a 'lottery for living space'.

Reefs are also home to a large variety of other organisms, including sponges, cnidarians (including some types of coral and jellyfish), worms, crustaceans (including shrimp, spiny lobsters and crabs), molluscs (including cephalopods), echinoderms (including starfish, sea urchins and sea cucumbers), sea squirts, sea turtles and sea snakes. Aside from humans, mammals are rare on coral reefs, with visiting cetaceans such as dolphins being the main group. A few of these varied species feed directly on corals, while others graze on algae on the reef and participate in complex food webs.

A number of invertebrates, collectively called cryptofauna, inhabit the coral skeletal substrate itself, either boring into the

Figure 18.18 (a) Hill Inlet and Whitehaven Beach, Great Barrier Reef. (b) Whitehaven Beach, Whit Sunday islands. (Sources: (a) http://www.quenslandholidays.com.au; (b) http://travel.ciao.co.uk). Courtesy U.S. National Oceanic and Atmospheric Administration (NOAA)

skeletons (through the process of bioerosion) or living in pre-existing voids and crevices. Those animals boring into the rock include sponges, bivalve molluscs and sipunculans. Those settling on the reef include many other species, particularly crustaceans and polychaete worms.

Due to their vast biodiversity, many governments worldwide take measures to protect their coral reefs. In Australia, the Great Barrier Reef is protected by the Great Barrier Reef Marine Park Authority, and is the subject of much legislation, including a Biodiversity Action Plan.

Throughout the Earth's history, from a few million years after hard skeletons were developed by marine organisms, there have almost always been reefs, formed by reef-building organisms in the seas. The times of maximum development were in the Middle Cambrian (513–501 Ma), Devonian (416–359 Ma) and Carboniferous (359–299 Ma), due to the Order Rugosa (extinct corals), and the Late Cretaceous (100–65 Ma) and all the Neogene (23 Ma–present), due to the Order Scleractinia corals.

However, not all reefs in the past were formed by corals, as in the Early Cambrian (542–513 Ma) they resulted from calcareous algae and archaeocyathids (small animals with a conical shape, probably related to sponges), and in the Late Cretaceous (100–65 Ma) there also existed reefs formed by a group of bivalves called rudists. The hypothetical zoned, marginal coral (mainly Cenozoic), stromatoporid (mainly Siluro-Devonian) and stromatolite (mainly early Proterozoic) reefs are illustrated in Figure 18.19(a), taken from James and Bourque (1997).

Mounds

James and Bourque (1997) divide organic carbonate build-ups into reefs and mounds (Figure 18.19(b) and 18.18(c)). In this classification, 'mound' denotes a rounded, hill-like, submarine structure composed of skeletal material or micrite of predominantly microbial origin. Mounded accumulations of skeletal material can be interpreted as hydrodynamic structures, just like any other detrital accumulations. Siliciclastic sediment drifts in the deep sea indicate that muddy sediments may still exhibit bedding structures related to bedload transport.

Carbonate slopes

Slopes and debris aprons around platforms are important elements of the edifice and largely determine the extent and shape of the top. These areas also act as sinks for much of the excess sediment produced by the platform top. In sequence stratigraphy, platform margins and slopes play a crucial role, as they hold much of the information on lowstands of sea level.

Geometry and facies of slopes and rises are governed by several rules:

1. The volume of sediment required to maintain a slope as the platform grows upward increases as a function of platform height. The increase is proportional to the square of the height for conical slopes of isolated platforms, such as atolls, and is proportional to the first power of the height for linear platform slopes.

2. The upper parts of platform slopes steepen with the height of the slope, a trend that siliciclastics abandon at early stages of growth because they reach the angle of repose of mud. As a consequence, the slopes of most platforms, notably the high-rising ones, are steeper than siliciclastic slopes. Changes in slope angle during platform growth change the sediment regime on the slope, shifting the balance between erosion and deposition of turbidity currents. This in turn changes sediment geometry on the slope and the rise.

3. The angle of repose of loose sediment is a function of grain size. Engineers have quantified this relationship for manmade dumps, and the same numerical relationships have recently been shown to apply to large-scale geological features. The most important control on the angle of repose is the degree of cohesion of the sediment.

The change in depositional regime with changing slope angle is largely caused by the degree of sediment bypassing on the slope. The sediment source of the vast majority of carbonate slopes is the platform. Thus, sediment enters the slope on the upper end and is distributed downslope by several sediment gravity transport processes. At gentle slope angles, the competence and capacity of the transporting agents decreases steadily away from the sediment source at the platform margin. Consequently, sedimentation rates decrease and time lines converge basinward. As slope angle increases, the power of sediment gravity flows increases such that significant volumes of sediment bypass the steep slope and come to rest on the adjacent basin floor. Basinward of the bypass slope a sediment apron develops as a series of laterally-coalescing turbidite fans. These aprons are comparable to the continental rises of deep ocean basins and are dominated by turbidites.

Steepening of slopes beyond the realm of bypassing produces erosional slopes, where the sediment budget is entirely negative and more material is exported by erosional turbidity currents and by slumping and basal slope failure than is received from the platform.

18.3 Evaporites

These are the chemical sediments that have precipitated from water following the evaporative concentration of dissolved salts. The most common are gypsum ($CaSO_4 \bullet 2H_2O$), anhydrite ($CaSO_4$) and halite ($NaCl$). These types of mineral are formed in low latitudes, where the temperatures are very high, relative humidity is low and evaporation far exceeds precipitation. The main depositional environments (Figure 18.20) are: subaqueous precipitation from a shallow-to-deep standing water body on to the floor of a lake, sea arm or marine-barred basin and subaerial precipitation of evaporitic minerals within sediments (e.g. sabkhas along the Trucial coast of the Arabian Gulf) or in very shallow brine pools or salinas (as in the American south-west). In the sabkha environment the typical facies, illustrated in Figure 18.21(a), are nodular (chicken-wire) and enterolithic

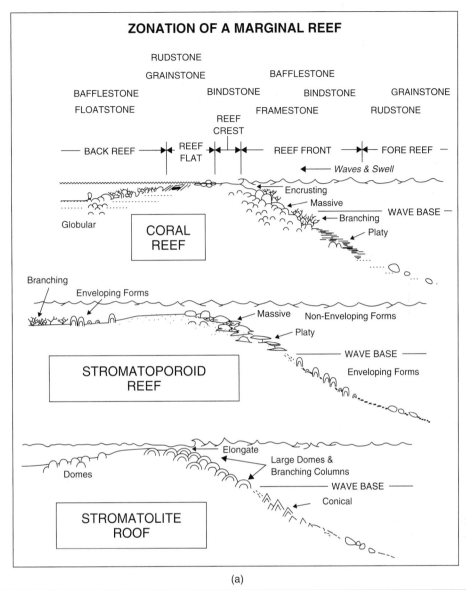

(a)

Figure 18.19 (a) Cross-sections through hypothetical zoned, marginal coral, stromatoporid and stromatolite reefs. (b) A conceptual classification of reefs and mounds. (c) Different facies that make up a reef or mound and the organism–sediment mosaic which typifies a growing reef, whose composition is the sum of the four processes illustrated in the inset. (Source: after James and Bourque, 1997)

anhydrite, which usually form as part of a sabkha cycle. The shallow water and intertidal sedimentary structures show the environment, and as a result of net deposition over the tidal flat surface the sabkha gradually progrades seawards over the intertidal sediments to give a cycle showing supratidal evaporitic sediments overlying intertidal and subtidal carbonates, which may be repeated many times as in the St Bees evaporites in the Permian sediments of west Cumbria.

In order to get the evaporitic precipitation within deep-water basins it is necessary to have an extremely arid climate, periodic replenishment of sea water and a barrier giving near-complete isolation from the main mass of sea water. The classic Mediterranean Messinian basin is an example of this, and the model allows for the precipitation of laminated/varved evaporites, graded anhydrite beds (turbidites), contorted and brecciated evaporites (slumps) and the development of a sequence, as in Figure 18.21(b).

18.4 Carbonate facies models

At first sight carbonate rocks can appear very complex, as there is a tremendous variety of textures, structures and sediment grain types. Irregular diagenesis adds to the impression of almost chaotic diversity and lack of pattern. However, if carbonate

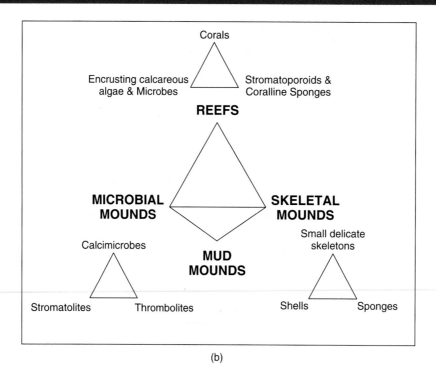

(b)

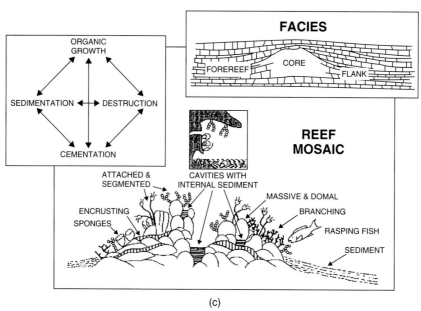

(c)

Figure 18.19 (continued)

sediments are characterized by sedimentary structure, texture and grain size, there is a recurring succession of facies belts that can be recognized in shore-to-basin transects. These facies appear throughout the Phanerozoic and, with only slight modification, in the late Precambrian. This surprising persistence indicates that the evolution of organisms in this time interval had only a minor modifying effect on the basic carbonate facies. The standard carbonate facies seem to capture trends dictated by other parameters, such as the distribution of growth rates as a function of depth and distance from shore, the degree of

protection from the action of waves and tidal currents, and the degree of restriction in the water exchange with the open sea. On the slopes, the angle and the balance between sedimentation and erosion are crucial controls.

We saw above that the tropical carbonate factory is by far the most productive carbonate system, producing nearly all of its sediment in a narrow depth range that extends only a few tens of metres down from sea level. The seaward perimeter of this highly-productive zone tends to be protected by an elevated, wave-resistant rim. This carbonate growth function generates

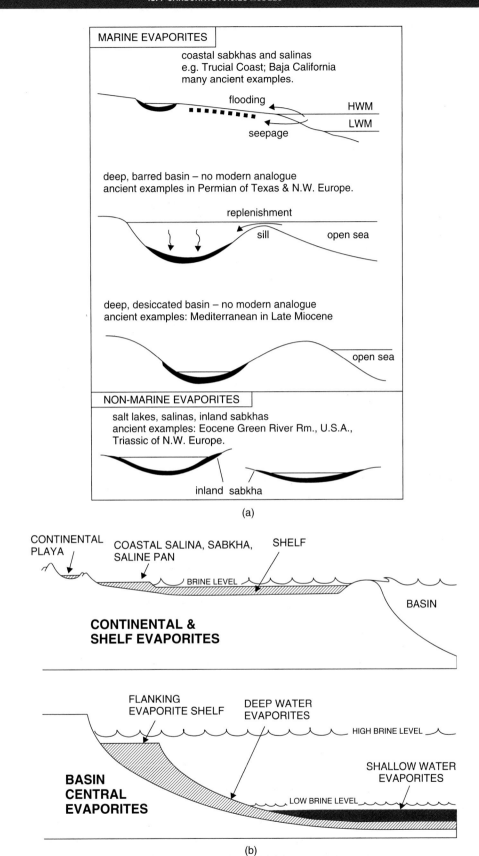

Figure 18.20 (a) The main types of depositional environment of evaporites. (b) Locations of the evaporite factory. (Sources: (a) after Tucker, 1981; (b) after James and Kendall, 1997)

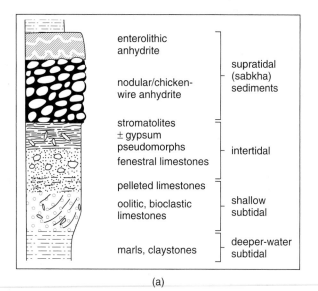

(a)

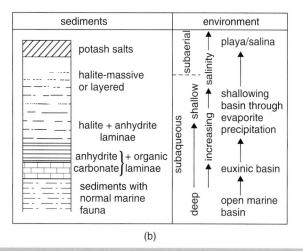

(b)

Figure 18.21 Evaporite sequences: (a) a sabkha cycle; (b) a cycle formed in an initially deep, marine-barred basin, with periodic replenishment with fresh water. (Source: after Tucker, 1981)

a platform geometry with two basically different depositional settings: the very flat top swept by waves and tidal currents, and the steep slopes shaped by gravity-driven sediment transport. The terms 'undaform' and 'clinoform' aptly describe these settings. If we add the 'fondoform' environment of the flat basin floors, we obtain a basic subdivision of carbonate depositional facies.

Within the undaform platform top, two parameters control the further subdivision of depositional environment and facies: the degree of protection from waves and currents, and the degree of restriction in the exchange of water with the open sea. In the slope (or clinoform) environment, the most important control on both geometry and facies is the balance between sediment input from above and sediment output on to the basin. Accretionary slopes receive more sediment than they give up to the basin, and on erosional slopes the sediment budget is negative. A transitional

stage, the 'bypass slope', can be recognized, where coarse material bypasses the slope in turbidity currents but mud accumulates.

18.4.1 Facies patterns: From ramp to rimmed platform

The shore-to-slope profile of shoal-water carbonate accumulations is a spectrum between two end members: the seaward-dipping profile in equilibrium with wave action and the rimmed platform with horizontal top, wave-resistant rim and steep slope. It has been suggested that the seaward-sloping equilibrium profile is analogous to that of siliciclastic systems, whereas the morphology of rimmed platforms is peculiar to tropical carbonates.

The same situation exists with regard to depositional facies: the succession of textures and sedimentary structures on carbonate ramps can be directly matched with siliciclastic shelf models. More specifically, one can correlate the facies pattern of attached ramps with that of the siliciclastic beach–open shelf model and the facies succession of detached ramps with the siliciclastic shelf–lagoon model. However, there is no siliciclastic analogue for rimmed platforms, but the facies pattern of rimmed platforms can be derived from the siliciclastic beach–shelf model or from the carbonate ramp model by inserting a wave-resistant rim (composed of reefs or partly lithified sand shoals) and adjusting the facies for the effects of this rim.

In the climax stage of platform evolution, the wave-resistant rim sits at the platform edge, directly atop the slope. This position is rather stable because of the dynamics of platform growth: a rim that originally forms in a more bankward position will rapidly prograde if it produces more sediment than is needed to match relative sea-level rise. Such excess production is likely considering the high productivity of rims. At the platform edge, the rate of progradation slows down because the high slope requires much larger sediment volumes for the same amount of progradation. Thus, further progradation will be slower but the slope will tend to steepen to the angle of repose due to the high sediment supply from the productive rim.

On the platform side, the rim sheds excess sediment ranging in size from clay to boulder. Beyond the reach of rim debris, an open lagoon may develop, with a bottom profile in equilibrium with wave action. However, at the climax stage of platform evolution this lagoon is largely filled and replaced by tidal flats that expand seaward almost to the platform rim.

The succession of facies on rimmed tropical platforms has been placed into a standard facies model by Wilson (1975), based on many case studies by numerous researchers. This model of standard facies has passed the test of time and is reproduced in Figure 18.22. Despite its remarkable success, certain modifications can be made. For example, the standard model is overcomplete in that it contains more facies belts than one normally finds on any one platform. Platforms with a reduced number of facies belts may be perfectly normal and healthy. In the clinoform and fondoform settings, facies 2 (deep shelf) will only be present in

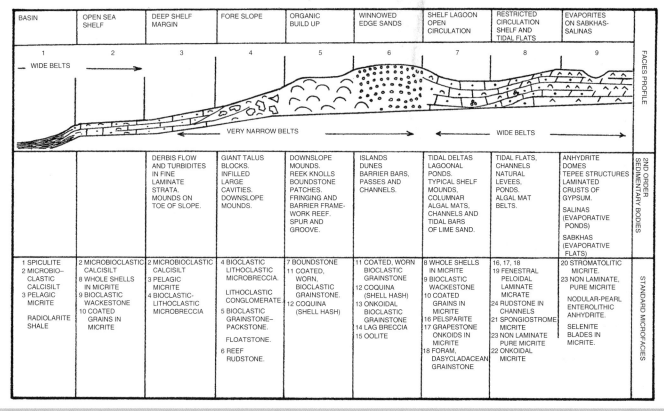

BASIN	OPEN SEA SHELF	DEEP SHELF MARGIN	FORE SLOPE	ORGANIC BUILD UP	WINNOWED EDGE SANDS	SHELF LAGOON OPEN CIRCULATION	RESTRICTED CIRCULATION SHELF AND TIDAL FLATS	EVAPORITES ON SABKHAS-SALINAS	
1	2	3	4	5	6	7	8	9	FACIES PROFILE
		DERBIS FLOW AND TURBIDITES IN FINE LAMINATE STRATA. MOUNDS ON TOE OF SLOPE.	GIANT TALUS BLOCKS. INFILLED LARGE CAVITIES. DOWNSLOPE MOUNDS.	DOWNSLOPE MOUNDS. REEK KNOLLS BOUNDSTONE PATCHES. FRINGING AND BARRIER FRAME-WORK REEF. SPUR AND GROOVE.	ISLANDS DUNES BARRIER BARS, PASSES AND CHANNELS.	TIDAL DELTAS LAGOONAL PONDS. TYPICAL SHELF MOUNDS, COLUMNAR ALGAL MATS, CHANNELS AND TIDAL BARS OF LIME SAND.	TIDAL FLATS, CHANNELS NATURAL LEVEES, PONDS. ALGAL MAT BELTS.	ANHYDRITE DOMES TEPEE STRUCTURES LAMINATED CRUSTS OF GYPSUM. SALINAS (EVAPORATIVE PONDS) SABKHAS (EVAPORATIVE FLATS)	2ND ORDER SEDIMENTARY BODIES
1 SPICULITE 2 MICROBIO-CLASTIC CALCISILT 3 PELAGIC MICRITE RADIOLARITE SHALE	2 MICROBIOCLASTIC CALCISILT 8 WHOLE SHELLS IN MICRITE 9 BIOCLASTIC WACKESTONE 10 COATED GRAINS IN MICRITE	2 MICROBIOCLASTIC CALCISILT 3 PELAGIC MICRITE 4 BIOCLASTIC-LITHOCLASTIC MICROBRECCIA	4 BIOCLASTIC LITHOCLASTIC MICROBRECCIA. LITHOCLASTIC CONGLOMERATE. 5 BIOCLASTIC GRAINSTONE–PACKSTONE. FLOATSTONE. 6 REEF RUDSTONE.	7 BOUNDSTONE 11 COATED, WORN, BIOCLASTIC GRAINSTONE. 12 COQUINA (SHELL HASH)	11 COATED, WORN BIOCLASTIC GRAINSTONE 12 COQUINA (SHELL HASH) 13 ONKOIDAL BIOCLASTIC GRAINSTONE 14 LAG BRECCIA 15 OOLITE	8 WHOLE SHELLS IN MICRITE 9 BIOCLASTIC WACKESTONE 10 COATED GRAINS IN MICRITE 16 PELSPARITE 17 GRAPESTONE ONKOIDS IN MICRITE 18 FORAM, DASYCLADACEAN GRAINSTONE	16, 17, 18 19 FENESTRAL PELOIDAL LAMINATE MICRATE 24 RUDSTONE IN CHANNELS 21 SPONGIOSTROME MICRITE 23 NON LAMINATE PURE MICRITE 22 ONKOIDAL MICRITE	20 STROMATOLITIC MICRITE. 23 NON LAMINATE, PURE MICRITE NODULAR-PEARL ENTEROLITHIC ANHYDRITE. SELENITE BLADES IN MICRITE.	STANDARD MICROFACIES

Figure 18.22 Standard carbonate facies. (Source: after Wilson, 1975)

addition to facies 1 if the platform has recently backstepped or has been structurally deformed. If a deep shelf exists, the structure of the lithosphere virtually dictates that it be connected with the deep basin floor of facies 1 by a slope. In epeiric seas, the facies succession may terminate with facies 2. In the undaform setting, the rim need not consist of facies 5 plus 6. Many perfectly healthy platforms have either facies 5 (reefs) or facies 6 (sand shoals) as a rim.

In at least one instance the Wilson model lacks a facies belt. Facies 9 of Wilson (1975) was defined for arid climates only. However, equivalent deposits in humid climates have been observed and are added here as facies 9(b). It should be noted that facies 9(a) and 9(b) are alternatives that cannot occur side by side in one shore-to-basin succession.

Another characteristic of the Wilson model is that it uses a discrete horizontal scale with sharp boundaries between facies. In nature, these facies boundaries may be gradational and irregular. For instance, the subdivision of slopes in facies 3 and 4 is often impossible, and a combined belt 3/4 may be more appropriate. The boundary between facies 7 and 8 is often very gradual. In these instances it is preferable to designate a combined facies belt 7/8 and express increasing restriction by biotic indices, or a larger number of subfacies.

The standard model says nothing about windward–leeward differentiation. Most platforms develop asymmetries in response to dominant wind directions. In fact seismic surveys reveal these asymmetries better than most other techniques.

18.4.2 The standard facies belts

1(a) Deep sea

Setting: Below wave base and below the euphotic zone; part of deep sea, i.e. reaching through the thermocline into the realm of oceanic deep water.

Sediments: Entire suite of deep-sea sediments such as pelagic clay, siliceous and carbonate ooze, and hemipelagic muds including turbidites. Adjacent to platforms we find mixtures of pelagic and platform-derived materials in the form of peri-platform oozes and muds.

Biota: Predominantly plankton; typical oceanic associations. In peri-platform sediments up to 75% shallow-water benthos.

1(b) Cratonic deep-water basins

Setting: Below wave base and euphotic zone, but normally not connected with the oceanic deep-water body.

Sediments: Similar to 1(a) but in Mesozoic-Cenozoic rarely ever pelagic clay. Hemipelagic muds very common. Occasionally anhydritic. Some chert. Anoxic conditions fairly common (lack of bioturbation, high organic content).

Biota: Predominantly nekton and plankton. Coquinas of thin-shelled bivalves. Occasionally sponge spicules.

2 Deep shelf

Setting: Below fair-weather wave base but within reach of storm waves, within or just below euphotic zone, forming plateaus between active platform and deeper basin (these plateaus are commonly established on top of drowned platforms).

Sediments: Mostly carbonate (skeletal wackestone, some grainstone) and marl, some silica. Well bioturbated, well bedded.

Biota: Diverse shelly fauna, indicating normal marine conditions. Minor plankton.

3 Toe-of-slope apron

Setting: Moderately inclined sea floors (over $1.5°$) basinward of a steeper slope.

Sediments: Mostly pure carbonates, rare intercalations of terrigenous mud. Grain size highly variable; typical are well-defined graded beds or breccia layers (turbidites or debris-flow deposits) intercalated in muddy background sediment.

Biota: Mostly redeposited shallow-water benthos. Some deep-water benthos and plankton.

4 Slope

Setting: Distinctly inclined sea floors (commonly $5°$ to near-vertical) seaward of the platform margin.

Sediments: Predominantly reworked platform material with pelagic admixtures. Highly variable grain size. End members are gentle muddy slope with much slumping and sandy or rubbly slope with steep, planar foresets.

Biota: Mostly redeposited shallow-water benthos. Some deep-water benthos and plankton.

5 Reefs of platform margin

Setting: (a) Organically stabilized mud mounds on upper slope, (b) ramps of knoll reefs and skeletal sands or (c) wave-resistant barrier reefs rimming the platform.

Sediments: Almost pure carbonate of very variable grain size. Most diagnostic are masses or patches of boundstone or framestone, internal cavities with fillings of cement or sediment, multiple generations of construction, encrustation and boring and destruction.

Biota: Almost exclusively benthos. Colonies of framebuilders, encrusters, borers, along with large volumes of loose skeletal rubble and sand.

6 Sand shoals of platform margins

Setting: Elongated shoals and tidal bars, sometimes with aeolianite islands. Above fair-weather wave base and within euphotic zone, strongly influenced by tidal currents.

Sediments: Clean lime sands, occasionally with quartz: partly with well-preserved cross bedding, partly bioturbated.

Biota: Worn and abraded biota from reefs and associated environments. Low diversity in fauna adjusted to very mobile substrate.

7 Platform interior: Normal marine

Setting: Flat platform top within euphotic zone and normally above fair-weather wave base. Called a lagoon when protected by sand shoals, islands and reefs of platform margin. Sufficiently connected with open sea to maintain salinities and temperatures close to those of adjacent ocean.

Sediments: Lime mud, muddy sand or sand, depending on grain size of local sediment production and the efficiency of winnowing by waves and tidal currents. Patches of bioherms and biostromes. Terrigenous sand and mud may be common in platforms attached to land, and absent in detached platforms such as oceanic atolls.

Biota: Shallow-water benthos with bivalves, gastropods, sponges, arthropods, foraminifera and algae particularly common.

8 Platform interior: Restricted

Setting: As for facies 7, but less well connected with open ocean, so that large variations of temperature and salinity are common.

Sediments: Mostly lime mud and muddy sand, some clean sand. Early diagenetic cementation common. Terrigenous influx common.

Biota: Shallow-water biota of reduced diversity, but commonly with very large numbers of individuals. Typical are cerithid gastropods, miliolid foraminifers.

9(a) Platform interior: Evaporitic

Setting: as in facies 7 and 8, yet with only episodic influx of normal marine waters and an arid climate so that gypsum, anhydrite or halite may be deposited, along with carbonates. Sabkhas, salt marshes and salt ponds are typical features.

Sediments: Calcareous or dolomitic mud or sand, with nodular or coarse-crystalline gypsum or anhydrite. Intercalations of red beds and terrigenous aeolianites in land-attached platforms.

Biota: Little indigenous biota except blue-green algae, occasional brine shrimp, abnormal- salinity ostracodes, molluscs.

9(b) Platform interior: Brackish

Setting: Poor connection with the open sea, just like 9(a), but with a humid climate such that fresh water runoff dilutes the small bodies of ponded sea water and marsh vegetation spreads in the supratidal flats.

Sediments: Calcareous marine mud or sand with occasional fresh-water lime mud and peat layers.

Biota: Shoal-water marine organisms washed in with storms, plus abnormal-salinity ostracods, fresh-water snails and charophytic algae.

Wilson's (1975) standard facies are also capable of representing the situation on carbonate ramps. On attached ramps, the succession includes facies 7 passing into facies 2. On detached ramps, the succession starts on the landward side with facies 8 or 7, and, depending on the degree of restriction, includes the barrier as facies 6, followed on the seaward side by facies 2. The critical difference to rimmed platforms remains, even if the same facies categories are used to describe the situation. The diagnostic criteria are the slope facies 3 and 4. On rimmed platforms, facies 2 is separated from rim facies 6 by a slope with facies 3 or 4. Homoclinal ramps lack slope facies altogether, whereas distally-steepened ramps show slope facies 3 or 4 seaward of facies 2. Using Wilson's (1975) standard facies for description of ramps is advantageous because ramps often grade upward into rimmed platforms. Rimmed profiles and ramp profiles may also alternate in space where the rim index is low.

18.5 Karst processes

The processes that take place in terrain that is composed of carbonate rocks are complex, first of all because of the variation in the types of carbonate that can form in a distinctive set of depositional environments, as we have just outlined: for example from fine-grained micrites to sparites and coarse-grained reef talus deposits. These carbonates too can be changed into dolomites

as part of diagenesis. Tectonic processes can give several joint patterns in the carbonates, which can control the direction in which the ground water in karst areas can flow; the underlying bedrock influences where this ground water flows again to the surface, or resurgences. There is a bedrock control too, depending on the type of rock, or Quaternary sediment, which overlies the carbonate rock. It is clear that because there is underground water flow through the carbonates, there are karst processes which are subsurface processes as well as those which take place on the surface. In some parts of the world the influence of glacial processes is superimposed on the karst processes and there is an interaction between the two, providing a distinctive glacio-karst landscape, as in the Yorkshire Dales in northern England, or the Burren in western Ireland.

Base-level changes, whether caused by fluvial erosion, or more likely by glacial erosion, can affect the landscape in karst areas, and often there are water-table drops, which can fossilize parts of the system that formed in the phreatic zone and create new phreatic zones lower in the topography. This means that underground processes can often take place over many hundreds of thousands of years, and there is the potential for complex underground morphology, flowstone development and cave sediments to form, which allow, if they are fully understood, a detailed understanding of the landscape development over a considerable time period. This can be achieved with the help of a range of dating techniques, such as uranium-series dating of flowstone sequences, and by understanding the morphological development of the cave and the stratigraphy in cave sediments. Some of the complex controls on karst processes are summarized in Figure 18.23.

18.5.1 Solution of carbonates

In Section 15.3.2 we outlined the importance of solution in the chemical weathering of limestone, because most limestones are over 90% calcium carbonate and therefore effectively monomineralic. In this section we treat the solution of limestones in more detail, as it is fundamental to the development of karst landscapes throughout the world. One noted phenomenon in limestone regions is that the average loss of $CaCO_3$ in rivers is between 150 and 300 ppm, whereas when pure water is saturated with this mineral the figure is only $12-14$ ppm $CaCO_3$. When compared with the loss of silica from terrain composed of siliceous rocks, which is around $15-25$ ppm, the rate of solution on limestone is $8-10$ times higher. The landscape that is produced has distinguishing characteristics which reflect this accelerated solution, such as underground river and cave systems, sink holes and other types of closed depressions and minor-scale solution landforms, such as karren.

This increased erosion has been explained in several ways in the past. Humic acids were thought be responsible for the increased solution; to date this has not been completely disproved. On the other hand, colloidal suspension of $CaCO_3$ can occur. But the generally accepted major reason is the presence of atmospheric and soil carbon dioxide. The latter is thought to be the most important because there is only 0.03% carbon dioxide in the

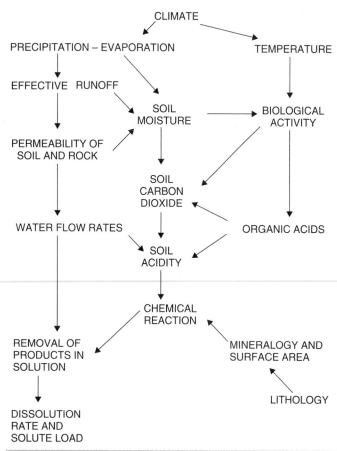

Figure 18.23 Complex controls on karst processes. (Source: after Trudgill, 1985)

atmosphere, whereas in the soil there is up to 1%. It seems likely therefore that most limestone solution takes place at the soil–limestone contact.

Carbon dioxide is soluble in pure water and the variation of solubility with pressure follows Henry's Law. This states that the solubility of carbon dioxide increases with a decrease of temperature and an increase of pressure. So all other things being equal, carbon dioxide is more soluble at lower temperatures than at higher ones. We saw earlier that the solubility of calcium carbonate is given by the general reversible equation:

$$CaCO_3 + CO_2 + H_2O = Ca(HCO_3)$$

In fact there is a family of equations which register the formation of calcium bicarbonate, and all are reversible. These are the equilibrium of hydration ($CO_2 + H_2O$ giving carbonic acid H_2CO_3); the dissociation of carbonic acid (H_2CO_3 dissociates to $H^+ + CO_3$); the combination of hydrogen and carbonic ions to give carbonic acid ($H^+ + CO_3^{--}$); the dissociation of the carbonate mineral $CaCO_3$ ($Ca^{++} + CO_3$); and the ionic dissociation of water ($H^+ + OH^-$). We can see here that the solubility of calcium carbonate forms a series of reactions and any variation in the rate of one affects the rate of the others.

The carbon dioxide dissolved in water, which regulates the equilibrium of the calcium carbonate, is called the carbon dioxide of equilibrium. Where the pressure of the carbon dioxide dissolved in water is lower than the pressure of the free gaseous carbon dioxide, the solution is not in equilibrium and there is precipitation of calcium carbonate. On the other hand, if the pressure of the dissolved carbon dioxide is greater than that of the free carbon dioxide, the water will attack and dissolve limestone. Here we have surplus carbon dioxide, called aggressive carbon dioxide. In fact the precipitation of calcium carbonate as flowstone can occur through various factors. In temperate regions the diminution of the carbon dioxide that is present is the most important factor but in the tropics the evaporation of water is most important. Locally the effects of bacterial and organic processes can be important, for example in the formation of cauliflower stalactites, or where tufa is precipitated by spring action on the surface.

Several factors can affect the solubility of carbonates but the main one in the solution is the variations of carbon dioxide pressure. Carbon dioxide is more soluble in water of lower temperature than at higher temperatures and so more calcium carbonate should be dissolved in colder climates. However, we see that the solution load in high Arctic limestone areas is only about 90 ppm for saturation in streams, yet in these areas there is little or no soil, so perhaps there is more of an emphasis on the importance of soil air carbon dioxide in the solution. It is possible that the variation of rain-water acidity which falls on carbonates could be a factor as the pH of rain water varies around the world from 4.0 to 7.1, although it usually averages 5.0–5.5. Locally it can vary markedly, and in West Yorkshire for example the rain is more acidic when the wind is from the south-east and east (pH 4.0) than when it is from the west or north-west (pH 6.6). This may have been important historically mainly because of industrial pollution from the south-east, but in general the role of carbon dioxide in rain water has likely been overrated as an important factor.

We have emphasized that the most important environment for an increase of carbon dioxide pressure is in the soil, where there can be a pressure many times greater than that in the air. This is related to vegetation and microbial activity in the soil and can show a seasonal activity: it is greatest in the spring and summer. This has been seen in South Wales, Slovenia and Virginia, where there are distinctly greater values in summer than in winter. Yet evidence from the Mendips suggests that there is no seasonal variation there, with spring waters showing 235 ppm calcium carbonate regardless of the season. Organic acids may also have a role to play, but the extent to which they are efficient limestone dissolvers is not proven. There is other biotic activity too by lichens, algae and marine organisms along carbonate coasts and in fact the amount of organic solution is thought to be considerable in some locations.

Carbon dioxide pressure is increased under snowbanks and the gas is more soluble in cold water than warmer water. Daily thawing and freezing of the snow surface leads to a greater concentration of carbon dioxide in the air content within the snowbanks, and as this gas is heavier than air it tends to be concentrated towards the snowbank base. The result is that rain and snow melt water passing through the snowbank can become highly charged with carbon dioxide. However, a rise of temperature accelerates the solutional rate and hence negates the effect of greater solubility of carbon dioxide in low-temperature waters. This can be illustrated by the fact that the tropical karst areas have 400% more solution than Arctic or Alpine areas.

A further factor is the water volume passing through a given limestone area. It is well known that the calcium carbonate content varies inversely with the stream discharge because at low water discharges there is greater contact with the rock. For example, in north-west Yorkshire at low discharge it is 180 ppm, whilst high discharge shows only 60–80 ppm. Nevertheless there is more limestone in total quantity dissolved under high discharges, despite the lower calcium carbonate concentration. We can see here that the more water is available, the more limestone will be dissolved and taken away in solution.

In dry climates such as polar and tropical deserts limestone corrosion will be at a minimum, whereas the temperate and tropical wet regions will have the greatest corrosion of this rock type. How far the carbon dioxide-charged water has percolated through the bedrock can be significant too and can affect the solution amount. For example, in GB cave in the Mendips percolation water dripping from the roof at a depth of 38 m from the surface has values of 120 ppm, whereas at 130 m the figure is 240 ppm. So the water percolating through the greater thickness of rock has the greater value because it has been in close contact with that rock for longer. It is also recognized that the limestone solution is affected by small quantities of chemical impurities in the rock, such as traces of some heavy metals, including lead, scandium, copper and manganese, which can have an inhibiting effect. On the other hand, lead in lead sulphide or iron sulphide, which commonly occur in shale units between limestone beds, will increase the carbonate solution as the sulphides oxidize to sulphuric acid. It is also believed that sodium and potassium will increase the solubility of calcite and that salts formed by the association of alkali cations with anions like Cl^- and SiO_4 will increase the solubility.

The geology of the limestone beds, including the texture and structure of the beds, affects the solution of the rock. For example, more-porous limestones are more soluble because more water containing carbon dioxide can permeate them. How easily the water can get through is important too, and here tectonic joints are crucial, allowing easy penetration by water. Where there is a high percentage of sparry calcite (sparite) in the limestone, the rock tends to be more resistant to solution. Here the interlocking crystals make the rock very impermeable, and the same applies to micrite because of the fine-grained crystals.

18.5.2 Other geomorphic processes affecting carbonates

Many other geomorphic processes can affect carbonate rocks, such as the frost shattering of thinly-bedded limestones on limestone pavements and the deposition of pockets of windblown loess in the solutional depressions (e.g. in the Morecambe Bay area of north-west England), or the opening up of joint fractures by biotic wedging. However, most of these processes are subsidiary to the main processes of carbonate solution and, in some places, glacial and glaciofluvial erosion.

18.5.3 Surface solutional landforms

Solutional landforms in limestone regions can be very variable in scale, from the small-scale solutional karren landforms to the large-scale poljes in the Dinaric karst and the so-called tropical cockpit and tower karst in Jamaica, as well as other examples in China and Vietnam (See Figures 18.1, 18.24 and 18.25).

Karren

These are small-scale solutional landforms that are often described using a series of German names, although the classification of solutional microforms developed on limestones and derived from Jennings (1985) is given in Table 18.3. There seem to be four major types, although many others have been suggested (see Box 18.3, see in the companion website):

1. **Rundkarren:** Rounded, smooth-bottomed grooves usually up to 25 cm in depth and width and up to 2 m in length (Figure 18.29). They are initiated by solution from peaty soil water under glacial till, as on Scar Close, just above Great Douk Cave in north-west Yorkshire.

2. **Rillenkarren:** Small runnels and grooves with rounded troughs and sharp, crinkled fine ridges, 1–12 cm deep and 1–2 cm wide, and usually under 50 cm long. They occur by solution from direct rainfall and surface runoff on slopes, particularly in the Mediterranean regions and in desert regions, like the Coahuila basin in Northern Mexico (Figure 18.30). See Box 18.4 (see in the companion website) for an experimental approach to their formation.

3. **Kluftkarren:** These occur on limestone pavements where there is solution along the edges of the clints down the joints, as can be seen from Figure 18.29(b).

4. **Trittkarren:** These occur on flat limestone surfaces and form steps with a flat tread and steep back slope.

Kamentiza (a Serbo-Croat term) or solutional basins form on the horizontal limestone clints where water collects in small basins and solution progressively enlarges them (Figure 18.31). They can be up to 50 cm deep, have a round or oval shape and steep sides with flat bottoms, and occasionally overhanging sides. Their formation is probably aided by humic acids from the peat above the limestone. Mosses, lichens and other plants growing on the limestone can cause the formation of fine pits and holes which may produce entry points for raindrops and gradual further solution.

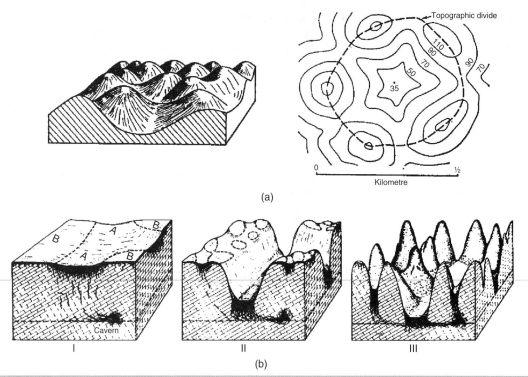

(a)

(b)

Figure 18.24 Tropical karst landforms: (a) cone karst with an associated map of cockpit karst; (b) three stages in the development of tropical tower karst. (Source: after Williams, 1969)

Figure 18.25 Tropical karst and the doline lake in the middle of Co Do Island, Vietnam. (Source: Waltham, 2005)

Marine karren are found on carbonate coasts. They are produced by solution from sea water, sea-water spray and marine organisms. They form fretted and rough pits and hollows, separated by sharp pinnacles, and there is often a sequence of morphological forms from low tide to above the high-tide line. At low tide the rock surface is cockled and fretted with small pits and pinnacles and trough-like, elongated hollows along the joints. The rough surface of the rock becomes smoother towards the high-water mark, and rounded hollows and widened joints become more common.

Solutional depressions

There are many types of surface depression in karst regions, ranging from small dolines (a Slovenian word meaning 'closed valley') of various origins to large-scale poljes up to 60 km long, found in the Dinaric karst and in southern China. The small dolines give a pitted relief to the surface topography. They are crater-like and many are given regional-local names depending on their origins. The forms are as follows:

1. **Solutional dolines:** Where water infiltrates into joints and fissure in limestone, solution takes place and then there is settling and lowering to give closed depressions. In Alpine areas snow can be an important control on their formation, as any hollow or joint intersection in the limestone in which snow is able to lie becomes deepened during the winter months. The forms tend to be steep-sided and rocky (Figure 18.32). Some karst areas, like the Burren and north-west Yorkshire, have few solutional dolines due to glacial erosion in the last glaciation, whilst other areas, such as the Mendips, have many of these landforms as they were not glaciated.

2. **Uvulas:** When the depressions coalesce they are then known as uvulas and show a characteristic form that is irregular in shape, with hollows with undulating floors made up of more than one doline. An example is Douky Bottom in Littondale (north-west Yorkshire). These landforms can also occur on beds overlying limestones such as shales or sandstones, where there can be subsidence of the overlying beds into fissures such

Table 18.3 Classification of solutional microforms. (Source: after Jennings, 1985)

		Form	Typical dimensions	Comments
Forms developed on bare limestone	Developed through areal wetting	Rainpit	<30 mm across, <20 mm deep	Produced by rain falling on bare rock. Occurs in fields on gentle rather than steep slopes. Can coalesce to give irregular, carious appearance
		Solution ripples	20-30 mm high; may extend horizontally for >100 mm	Wave-like form transverse to downward water movement under gravity. Rhythmic form implies that periodic flows or chemical reactions are important in their development
		Solution flutes (rillenkaren)	20-40 mm across, 10-20 mm deep	Develop due to channelled flow down steep slopes. Cross-sectional form ranges from semi-circular to V-shaped but is constant along flute
		Solution bevels	0.2-1 m long, 30-50 mm high	Flat, smooth elements usually found below flutes. Flow over them occurs as a thin sheet
		Solution runnels (rinnenkarren)	400-500 mm across, 300-400 mm deep, 10-20 m long	Down-runnel increase in water flow leads to increase in cross-sectional area. May have meandering form. Ribs between runnels may be covered with solution flutes
	Developed through concentration of runoff	Grikes (kluftkarren)	500 mm across, up to several metres deep	Formed through the solutional widening of joints or, if bedding is nearly vertical, of bedding planes
		Clints (flackkarren)	Up to several metres across	Tabular blocks detached through the concentration of solution along near-surface bedding planes in horizontally-bedded limestone
		Solution spikes (spitzkarren)	Up to several metres	Sharply-pointed projections between grikes
Forms developed on partly-covered limestone		Solution pans	10-500 mm deep, 0.03-3 m wide	Dish-shaped depressions usually floored by a thin layer of soil, vegetation or algal remains. CO_2 contributed to water from organic decay enhances dissolution
		Undercut solution runnels (hohlkarren)	400-500 mm across, 300-400 mm deep, 10-20 m long	Like runnels, but become larger with depth. Recession at depth probably associated with accumulation of humus or soil, which keeps sides at base constantly wet
		Solution notches (korrosionkehlen)	1 m high and wide, 10 m long	Produced by active solution where soil abuts against projecting rock, giving rise to curved incuts
Forms developed on covered limestone		Rounded solution tunnels (rundkarren)	400-500 mm across, 300-400 mm deep, 10-20 m long	Runnels developed beneath a soil cover which become smoothed by the more active corrosion associated with acid soil waters
		Solution pipes	1 m across, 2-5 m deep	Usually become narrower with depth. Found on soft limestone such as chalk, as well as mechanically-stronger and less-permeable varieties

as on the Pennant Grit overlying Carboniferous limestone in South Wales, or the spectacular lines of shakeholes on the shales overlying Yoredale limestones in the Yorkshire Dales (Figure 18.33).

3. **Collapse dolines:** These can occur where cave chamber roofs collapse, such as in Great Douk, where there are abrupt, steep-cliffed walls, many large collapsed limestone blocks where the cave stream disappears and there is evidence for slickensides on the cave entrance-passage vertical cliff. This collapse fits in with evidence from the lower part of the cave, where there are major collapse rifts in the cave roof just below Little Douk Pot. These pots may also have been formed by stream sinkholes prior to the last glacial phase and subsequent erosion of the shale by glacial erosion back to its current position.

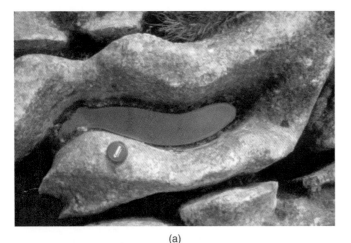

(a)

(b)

Figure 18.29 Rundkarren: (a) on Scar Close National Nature Reserve, Western Ingleborough; (b) on Gait Barrow National Nature Reserve, Cumbria

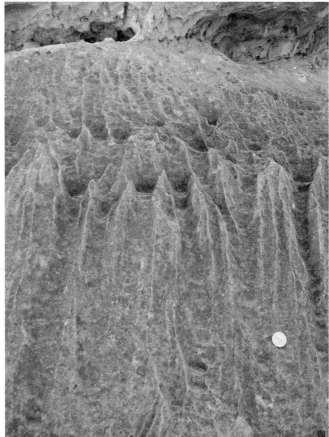

Figure 18.30 Rillenkarren from limestone margins to the Cuatrocienegas basin, Coahuila, Northern Mexico. There is evidence that the climate was once wetter in this area from the cave systems and deep gorges and valley systems in the mountain margins

Figure 18.31 Kamenitza basins on Scar Close, Western Ingleborough, north-west England

The classification of dolines adapted from Jennings (1975) is illustrated in Figure 18.34, and the point that most dolines and sinkholes are formed by a combination of solution and collapse is emphasized in Figure 18.35(a). Depressions can be isolated in the landscape, but more often, where exploitable joints are found with greater frequency, development occurs over time. The end product in the landscape is an 'egg box' topography or polygonal karst (Figure 18.35(b)). In tropical areas with high rainfall these types of landscape are common, for example the 'cockpit' country of Jamaica, or New Guinea.

Poljes, on the other hand, are large oval depressions with flat floors. Many appear to be fault-guided and flood during the winter but are dry in summer. This is partly due to rainfall where they are river-fed and to the degree of openness of the limestone fissures where they are spring-fed. They are usually river-drained, the floor covered in alluvium, and they usually drain through depressions or swallow holes in the rock. It appears that they form over a long time period and can have a complicated origin.

In Ireland they are known as turloughs (dry lakes). The Carran depression in County Clare is a good example; it is approximately 4.5 km long and 2.2 km wide (Figure 18.36). Close to the southern shore of Lower Loch Erne are two other examples, the Fardrum and Roosky turloughs (Figure 18.37).

Figure 18.32 Solutional doline (sinkhole) from the Wood Buffalo Park, Franklin Mountains, Canada. Note the likely importance of snow in this landform's formation. (Photo: Raymond Griguere, from the Canadian Encyclopedia)

Turloughs are seasonal lakes occurring in lowland limestone which owe their changing level to fluctuations in the local water table as ground water falls and rises, reflecting the rainfall input locally. They are usually grassy hollows, sometimes extending over many acres, which during wet weather fill with water through subterranean passages in the rock, and empty by the same means. The rate of the rise and fall and the duration and frequency of the flooding vary in different turloughs. They are dependent on several factors connected with underground drainage. These changes are in many cases quite rapid, where a large amount of water can collect in one hour, and disappear in an equally short time. Some turloughs hold permanent water on their floors, while others show no water at all when flooding is absent. Often swallow holes set among the muddy, moss-covered rocks indicate the point of inflow and outflow.

Swallow holes, sinkholes or stream sinks (all the same landform) can be formed by the enlargement of bedding planes by river processes into caves such as the entrances to Borrins Moor and Long Churn caves in the Yorkshire Dales. Potholes can form where vertical joints, and occasionally faults or beds that are steeply dipping, develop in strong, hard, sparitic limestone. A good example here would be Alum Pot or Gaping Gill in the Yorkshire Dales. However, some swallow holes have little or no surface topographic expression, for example where a river disappears slowly into its bed and gradually becomes dry downstream and less incised. There appear to be no visible depressions, but the water sinks into bedding planes and joints, as in the Manifold valley in the Peak District (Derbyshire).

Stream resurgences and surface valleys

Stream resurgences or risings are often distinctive landforms in karst areas and normally form where water flowing through limestone beds hits an impermeable series of rocks. This occurs in Greta Dale in north-west Yorkshire, where Eller Keld and God's

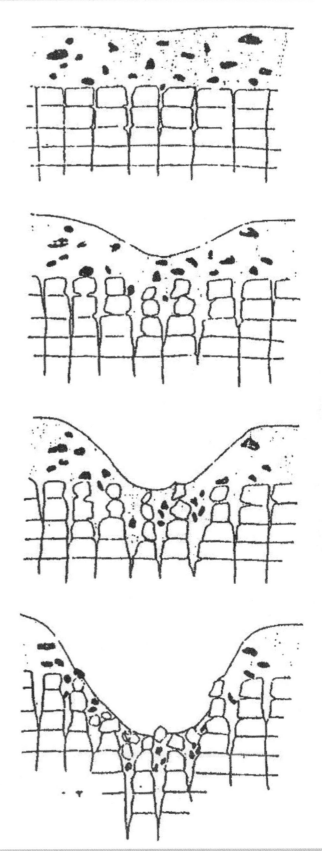

Figure 18.33 Shakehole development. (Source: after Trudgill, 1985)

DOLINES

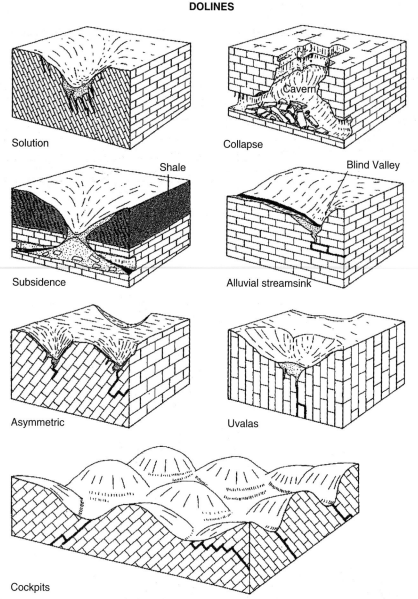

Figure 18.34 Classification of dolines. (Source: after Jennings, 1975)

Bridge are good examples of normal resurgences (Figure 18.38). As cave streams converge underground there are more sinks than resurgences. However, sometimes Vauclusian springs form, named after the type locality in France, where the water flows from the limestone under pressure and often uphill.

Surface valleys can have several origins in limestone areas and strictly speaking are not karst in origin because they are formed by fluvial processes, although they do form a distinctive set of landforms in many karst regions. Valleys which flow through limestone areas tend to be narrow, cliffed, gorge-like, sometimes with natural bridges, and steep-sided. The water in the rivers tends to be clear because of the solutional load. Potholes are common on the massive, sparite limestones and there is differential solution along joints and bedding planes. Peak discharge water can sink underground and the valleys are dry for part or most of the year, for example Kingsdale in north-west Yorkshire. So-called pocket valleys occur in association with large springs that resurge at the base of a limestone massif. They are flat-bottomed, steep-walled and often U-shaped, with a steep abrupt cliff at their head; examples in Britain would be Malham Cove in the Yorkshire Dales and Wookey Hole in the Mendips.

There are many dry valleys in karst areas that are steep-sided and gorge-like, with often steep scree slopes on their sides. Examples are the dry valley network converging on Cheddar gorge in the Mendips and the Clapham Bottoms in Ingleborough. Some are associated with faults in north-west Yorkshire, particularly

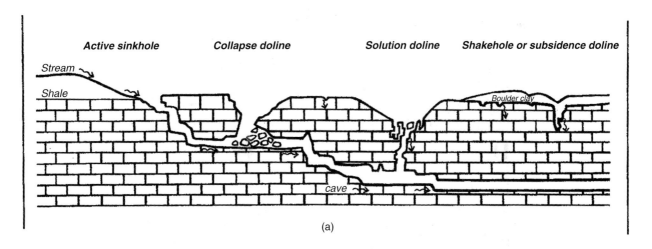

Active sinkhole   **Collapse doline**   **Solution doline**   **Shakehole or subsidence doline**

Stream

Shale

Boulder clay

cave

(a)

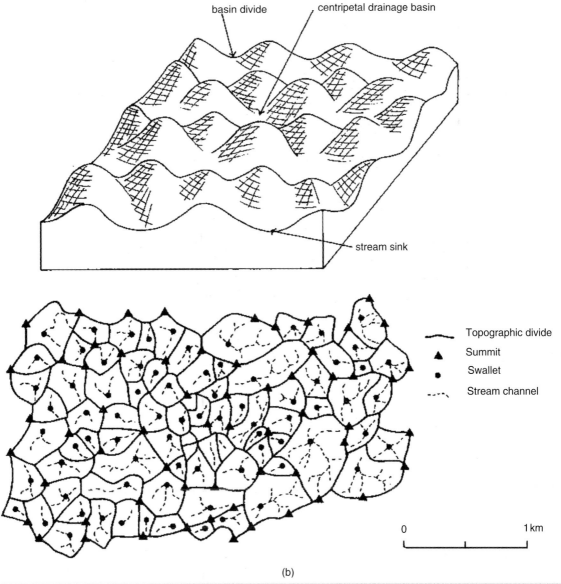

basin divide   centripetal drainage basin

stream sink

Topographic divide

▲ Summit

● Swallet

- - - Stream channel

0   1 km

(b)

Figure 18.35 (a) Active sinkhole, dolines and shakehole. (b) A polygonal karst showing a three-dimensional representation (top image) and a polygonal karst in New Guinea in plan (bottom image). (Source: after Trudgill, 1985)

Figure 18.36 Turlough or polje, Carran depression, western Ireland. (Source: Plantsurfer, Wikimedia Commons)

Figure 18.37 Fardum and Roosky turloughs, near Enniskillen, Co. Fermanagh, Northern Ireland. (Source: Wikimedia Commons)

along the Craven fault lines, where snowmelt and/or frost action picked out weaknesses in the limestone. Permafrost conditions prior to the last glacial may have helped their formation, as water could not flow underground, but the major cause of these dry valleys in glacio-karst regions is subglacial melt-water erosion during deglaciation, forming subglacial melt-water channels. Classic examples here would be Gordale and the valleys above Malham Cove in the Yorkshire Dales (Figures 18.39 and 18.40), and Trow Gill, south-east of Ingleborough.

Limestone pavements

These are distinctive bare landforms in areas of glacio-karst and are initially caused by the effects of glacial erosion on a series of limestone beds with different strengths. The stripped horizontal limestone is covered by glacial till during the glacial and subsequently solution of the limestone occurs beneath the drift cover, often supplemented by soil water from acidic peat development on top of the till. Gradually through time the drift cover is partially eroded, leaving indications of the former cover in glacial erratics on the pavements and in the joints and limestone pedestals on the pavements. The bare rock outcrops that are exposed on horizontal beds are the limestone pavements and

are composed of flat-topped limestone outcrops known as clints, separated by widened joints, or grikes (Figure 18.41). They tend to be best developed on the massive, sparry limestones and their size depends on the joint density. This density increases the closer the area is to major faults; this can be seen down Greta Dale, where the clints are smaller closer to the Craven faults.

The clint surfaces are often fretted by karren landforms and in a glacio-karst area these landforms are dominated by rundkarren, kluftkarren and kamentiza basins. Once exposed on the surface, pavements are usually modified by subaerial processes and often frost shattering of the limestone is evident. Depending on how long the pavements have been exposed to surface processes, which can be affected by past deforestation, sheep grazing and soil erosion, two pavements in the same area can be very different. This is clear from the Scar Close pavement just above Great Douk, north-west Yorkshire, and the pavement just below that cave, although here the likelihood is that because the pavements are in different units of limestone there is a geological factor involved too (Figure 18.42(a),18.42(c)). In Britain, limestone pavements can typically be found in north-west Yorkshire on either side of Greta Dale, around Morecambe Bay and on the Burren in western Ireland (Figure 18.42(b)).

18.5.4 Cave formation processes and landforms

A cave is formed below ground by solution of the limestone by running water. It can have both horizontal and vertical passages, which morphologically can be extremely complicated (Figure 18.43(a)). As soon as a limestone bed is exposed to rainfall its network of fissures fills with water, but for significant cave development there must first be a complete through-system of micropassages that contain an efficient drainage system. When this water flow is created through the system, solution takes place and caves form. The underground drainage routing is controlled by a number of factors, but to a great extent its position is dependent on a major amount of water inflow and therefore cave entrances are adjacent to the outcrop boundary with the overlying rock type, as we can see from Western Ingleborough in the Yorkshire Dales (Figure 18.38).

A difficult question to answer is why do some fissures and joints become enlarged into caves when others do not? It is not just solution as water becomes rapidly saturated with calcium carbonate, especially as during the initial stages small water discharges are necessary. We have seen that carbon dioxide is required for solution to take place; this could come from inflowing water, cave air or organic substances, but these sources cannot penetrate down into the phreatic ground-water zone where caves initially develop. In some cases it has been suggested that the extra solution necessary comes from bacterially-assisted oxidation of organic matter in the ground water, or from sulphide minerals in the rock. One important suggestion is Bögli's Mixture Corrosion hypothesis, where two water streams which differ in

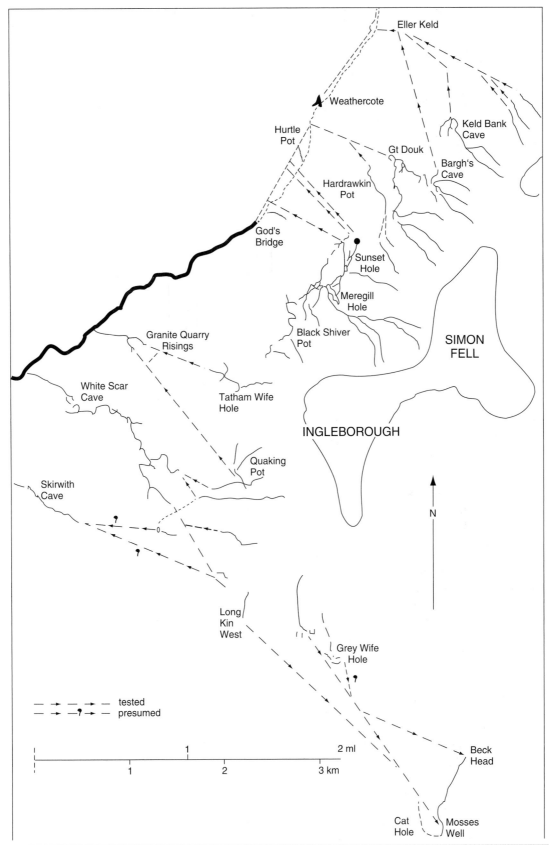

Figure 18.38 Resurgences on the eastern side of Chapel-le-Dale and Southern Ingleborough, north-west Yorkshire. (Source: after Waltham, 1974)

(a)   (b)

Figure 18.39 (a) Gordale dry valley and waterfall site. (b) Gordale dry valley, fossil and active screes, north-west Yorkshire

Figure 18.40 Watlowes dry valley (subglacial melt-water channel), above Malham Cove, north-west Yorkshire

the amounts of limestone that they are potentially able to dissolve meet, resulting in an increase in the total amount of limestone than can be dissolved. This, coupled with Renault's (1967–1969) hypothesis, seems likely to be the answer.

Renault suggested that in temperate regions the limestone surface region is cold in winter but at depth is relatively warm. As rain hits the cold surface rock, it dissolves the limestone and quickly becomes saturated. As this water passes into the warmer zone there is a loss of carbon dioxide and redeposition of calcium carbonate. As it flows further into the rock it therefore becomes more aggressive and dissolves more rock. There is some supporting evidence in that in the equatorial regions and Arctic periglacial areas there is a lack of deep caves.

The first stage of cave development is a widespread mesh of very small, open, randomly-distributed solutional features called phreas. Because only a small percentage of the original fractures in limestone develop into caves, it must be these that have the greatest hydraulic gradient. If the water discharge is sufficient, turbulent flow instead of thin laminar flow is created and the solution rate is increased, and hence the phreas are enlarged most. It seems that when the diameter of these phreas becomes over 5.0 mm, turbulent flow starts. When one passage offers easier flow conditions it must capture the drainage from others and must increase in size. The others are then abandoned and left dry, but in these early stages of development water speed is slow, as is the rate of development.

Originally it was thought that caves developed by solutional and erosional processes in underground rivers that behaved like surface rivers: free-flowing and under no hydrostatic pressure. These were called vadose streams and caves. However, many cave characteristics cannot be explained unless the passages are completely filled with water under hydrostatic pressure (phreatic flow). This means that vadose caves develop above the water table and phreatic caves below the water table, and that usually any cave has these two components at any moment in time. Due to base-level changes, one part of the system can be super-imposed on the other, and extremely complicated development results.

(a)

(b)

Figure 18.41 (a) Clints and grykes above Malham Cove.
(b) Pavement: Moughton Scars

Phreatic cave passages

These types of passage are characterized by the following morphology:

1. They have a circular or elliptical form, which offers the least resistance to the water. This is guided by fissures (Figure 18.43, 18.44(b)).

2. The top and base of the passages may be the same. Some may have an uphill gradient (directions shown from scallops) and there may be siphons and Vauclusian springs where there is evidence for water flow uphill under pressure, for example in Sleets Gill cave in Littondale, north-west Yorkshire).

3. Spongework or honeycombing may be present. These are complicated patterns of minor cavities on the cave roof and/or walls indicating solution by slow-moving water.

(a)

(b)

(c)

Figure 18.42 Limestone pavements, a glacio-karst landform: (a) below Scar Close, Western Ingleborough; (b) Burren. Far crest is Cathair Dhuin Irghuis: large, flat clints with widely-spaced grykes; (c) below Scar Close, Western Ingleborough. (Sources: (a) http://www.dales-photos.co.uk/linton-bridge.htm; (b) Dr Charles Nelson, reproduced under Creative Commons Licence)

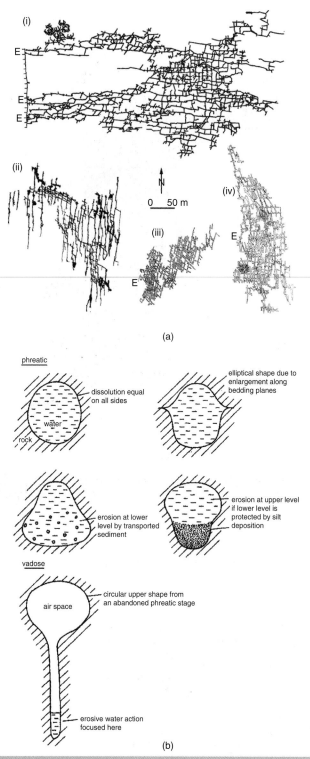

(a)

Figure 18.43 (a) Complicated cave systems: (i) Botorskaya Cave, Siberia, 57 km; (ii) Fuchslabyrinth Cave, Germany, 6.4 km; (iii) Moestrof Cave, Luxembourg, 4 km; (iv) Knock Fell Cavern, Northern Pennines, UK, 4 km. (b) Cave morphology under vadose and phreatic stages. (Sources: (a) http://www.speleogenesis.info/archive/index.php; (b) after Trudgill, 1985).

4. There are anastomoses along bedding planes, which are minor curvilinear, tube-like solution cavities, often of an intricate pattern.

5. A roof cavity is sometimes present, which is a deep and narrow slot dissolved along joints.

6. Rock pendants can be present, caused by differential solution.

7. Scallops can form on the floor and walls, and many show flow against the gradient under pressure. These are spoon-shaped hollows that have been eroded by turbulent eddies in the flow. In phreatic passages, because the water movement is slow the scallops are large, whereas in vadose flow the water has a greater velocity, eroding smaller scallops (Figure 18.45). The direction of flow can be worked out from the scallop morphology, as the upstream ends of the scallops have steeper, sharper lips.

In north-west Yorkshire there are such caves in Swaledale (Figure 18.44(b)), much of the Gaping Gill system (Figure 18.46), the Kingsdale Master cave (Figure 18.47) and the entrance passage to Great Douk cave. These caves are out of phase with the present-day topography as they are far above any present-day phreatic ground-water systems. This is reinforced by the presence of glacial sediments in some of them, which suggest at least a period of initial formation before the last glaciations, and there is evidence for much older developmental stages.

Vadose cave passages

When the water level in a cave drops so that air enters the passage and the water flow is free and no longer confined, the passage shape changes. Rivers take the easiest downhill course within the confines of the overall geological structural control. Erosion deepens the passage to give a rectangular shape, and often a canyon or gorge is cut into the earlier phreatic phase (Figure 18.48). Other characteristics of vadose cave systems include:

1. Vertical shafts eroded down joints. These can be up to 100 m deep and between 1 and 10 m across.

2. Meanders, cutoffs, potholes and waterfalls. See for example the upper section of Great Douk Cave, north-west Yorkshire.

3. Flat uneroded roofs, usually along a bedding plane, marking the limestone bed above.

4. A continuous downhill gradient, usually gently inclined.

5. Dimensions generally smaller than the main phreatic caves in a karst area, and rarely over a metre wide.

6. An obvious line marking the flood level of peak discharge, as flowstone or sediments cannot build up below that level. In

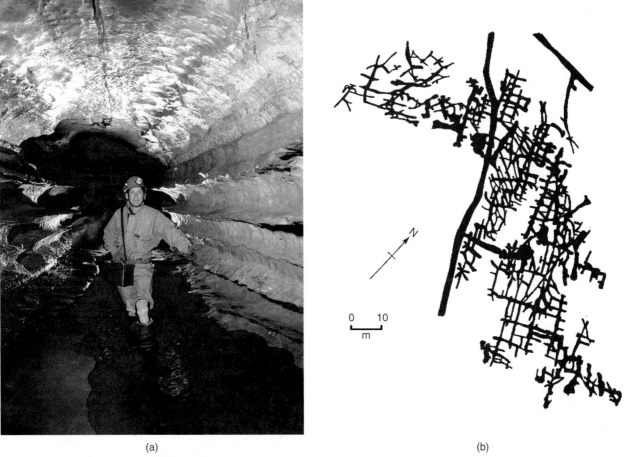

(a)   (b)

Figure 18.44 (a) Phreatic cave in Ogof Clogwyn, Clydach Gorge, South Wales. (b) Devis Hole Mine Cave, Swaledale, north Yorkshire. (Source: (a) http://www.dudleycavingclub.org.uk)

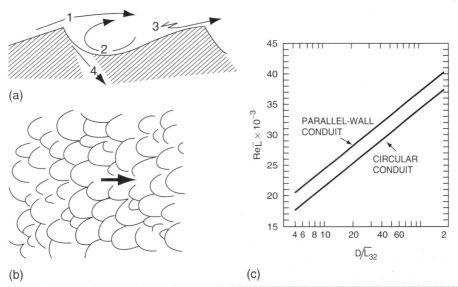

(a)

(b)   (c)

Figure 18.45 Scallops. (a) Section through a scallop showing: 1. the saturated boundary layer detaching; 2. turbulent eddy with the maximum locus of dissolution; 3. diffusion, mixing and reattachment; 4. the course of the steepest cusp following further dissolution (the scallop migrates downstream). (b) The characteristic appearance of a scallop pattern, with individual scallops overlapping. (c) The predicted relation between Reynolds Number and the ratio, conduit width diameter (D) to Sauter mean scallop length, L_{32}. (Source: Curl, 1974)

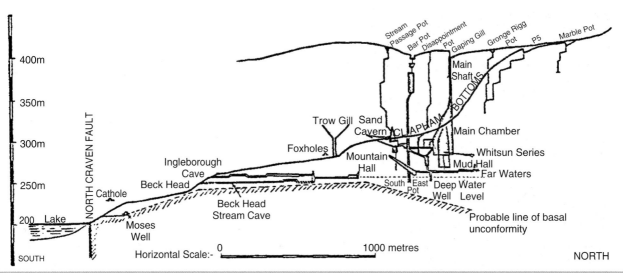

Figure 18.46 The Gaping Gill cave system, the Clapdale valley, north-west Yorkshire, showing the impermeable basement, vertical sections and horizontal sections of this system

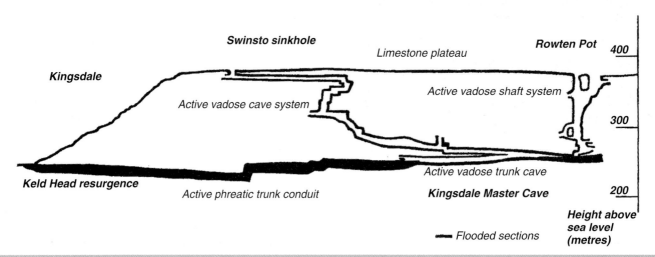

Figure 18.47 Kingsdale Master cave: the system from Swinsto Hole to the Keld Head resurgence, beneath the western flank of Kingsdale, north-west Yorkshire. The active vadose system and the active trunk phreatic conduit exist at the resurgence. (Source: after Waltham, 1974)

fact, sediments can only start to build up in a cave when the water abandons a certain section for lower levels.

A typical Yorkshire Dales cave system is shown in Figure 18.47, where the system from Swinsto Hole to the Keld Head resurgence in Kingsdale includes vadose canyons and shafts leading to a trunk passage: the Kingsdale Master cave, which is a phreatic tube for most of its length to the resurgence.

Factors influencing cave development

Many factors influence cave development in a karst area. Some are geological characteristics of the limestones, others are related to the regional geomorphological development, still others are climatic factors and some may be influenced by man's activities. The character of the limestone is important because there can be major differences in the structure, texture and porosity and the more soluble beds are preferentially eroded. Caves often originate at the contact of relatively impure beds with purer units, and shale beds are thought to be particularly important in north-west Yorkshire. This is because they tend to retain water and may contain iron pyrites, which weathers to give sulphuric acid, which can dissolve the limestone in contact with it more rapidly. Mechanically too, shale is more easily eroded by water than limestone and so bedding-plane passages are common. The influence of the shale bands in Swinsto Hole, (north-west

(a)

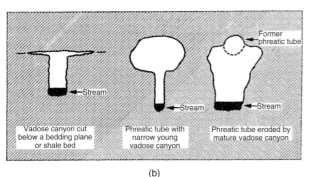

(b)

Figure 18.48 (a) Vadose cave passage, Mammoth Cave, Kentucky. Note the canyon shape and the differential erosion laterally, caused by the differing resistances of the beds of limestone. (b) Typical vadose cave systems. (Sources: (a) Dave Bunnell, http://en.wikipedia.org/wiki/File:Mammoth_cave_canyon.jpg; (b) after Hindle, 1980)

narrow, winding, vertical slits and are often called corridor caves, as in the middle section of Great Douk cave and in parts of Long Churn cave in north-west Yorkshire. Sometimes they are triangular in shape because solution works laterally along another bedding plane at the base of the joint. Some of the vertical sections in caves can be explained by faults, as at Alum Pot, north-west Yorkshire. Both fault and joint control can be seen in the Gaping Gill system in north-west Yorkshire in Figure 18.52. The joint directions on the limestone pavement above the cave at Great Douk control the passage orientation, including right-angle bends.

Where there is a relatively thin limestone series, caves tend to occur towards the base or at the base of the outcrop along the contact with the impermeable beds. The local geomorphology and regional topography can affect cave development, as in north-west Yorkshire where the landscape is deeply dissected by glacial erosion, with the creation of U-shaped valleys and eroded benches. Here the active vadose caves were superimposed on an extensive network of abandoned phreatic passages when the limestone was not so eroded. The area then contained a more restricted water circulation and underground drainage system, but it was progressively eroded and there are probably preglacial, interglacial and post-glacial caves, or cave sections formed at different periods, superimposed on earlier phases to give a complex system (Figures 18.53 and 18.54).

In tropical areas where there has been no glaciation, uplift and deep incision have been postulated as important controls, as for example the Finim Tel caves in the Hindenburg Ranges of New Guinea (Brook, 1977). In this area, uplift and tilting have been responsible for the cave depth and extent (Figure 18.55). The caves increased in depth as the height above sea level increased in relation to uplift.

Climatic factors are also important in cave development. They usually form in areas with high rainfall and therefore high discharges; in arid and semi-arid regions cave development is restricted, but caves did develop in these areas during wetter phases in the Quaternary, as in the Cuatrocienegas area of Coahuila (Northern Mexico). The climatic changes associated with the Quaternary are thus important. Passage shape can change with discharge, the temperature variations can affect the solubility of the limestone and, as we have seen, there can be major phases of glacial downcutting in some parts of the world, causing water-table drops. All of these complex changes can be interpreted from the cave morphology and from cave sediments and flowstone development.

Man's activities can influence cave development too. For example, where deforestation occurs there is much less organic carbon dioxide and limestone solution is decreased. This has occurred throughout the upland limestone areas of Britain since the Mesolithic but is perhaps counterbalanced by more recent drainage of the peat areas above the limestone beds by farmers, accelerating water discharge and speed of flow in the caves below the drainage ditches.

A hypothetical cross-section of a cave passage, showing its stages of development over the last 750 000 years, is shown in

Yorkshire) can be seen in Figure 18.49. In County Clare, western Ireland, these bedding-plane passages are the main entry point for water. They are thin shelves up to 30 m wide but usually only up to 1 m high. The caves here are low and wide because the limestone is not steeply dipping but fairly horizontal.

The joint frequency in the beds can be an important control as the sparites are strong, with few joints, whereas the micrites and biomicrites break down more easily, having a greater number of joints. This structural control can be seen in Ogof Ffynnon Ddu and Dan-yr-Ogof in South Wales (Figure 18.50) and there can be preferential solution along bedding planes and where there are different limestone types (Figure 18.51). Joint plane caves are high,

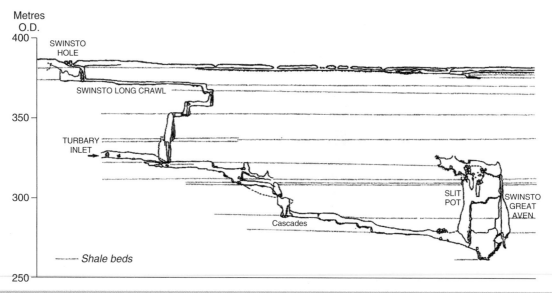

Figure 18.49 Influence of shale bands in Swinsto Hole, north-west Yorkshire. (Source: after Waltham, 1974)

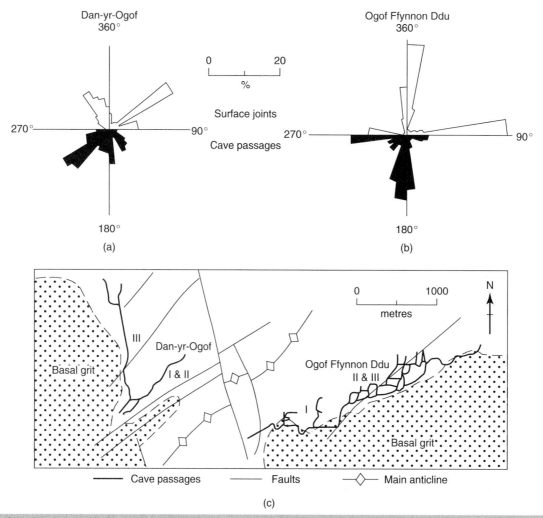

Figure 18.50 Structural control in South Wales caves: (a) Dan-yr-Ogof cave; (b) joint orientations on the surface (unshaded) and in the Ogof Ffynnon Ddu cave; (c) regional tectonics and the cave systems. (Source: after Weaver, 1973)

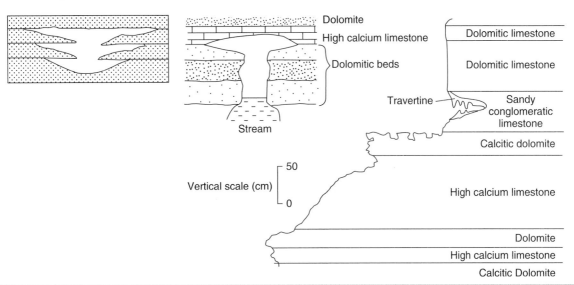

Figure 18.51 The influence of preferential solution along bedding planes and of rock-type differences on cave shape. (Source: after Tratman, 1969; Charity and Christopher, 1977)

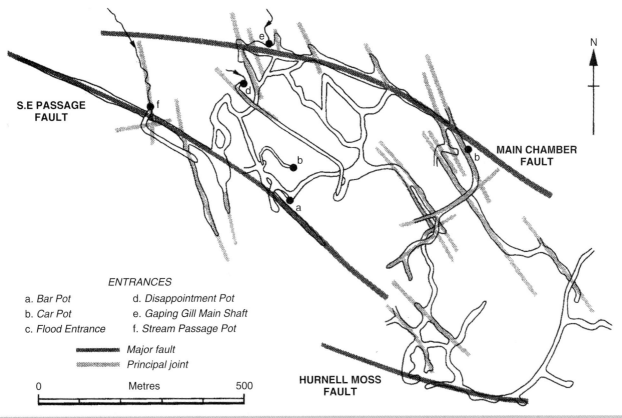

Figure 18.52 Fault and joint control in the Gaping Gill system. (Source: after Glover, 1974)

Figure 18.53, and a simple example from the Yorkshire Dales can be seen in Figure 18.54, taken from Hindle (1980). This is based on the Gragareth caves. In reality the development is much more complex, but we can now put real dates on certain phases with much more accuracy.

It is generally recognized today that there is not one single general mechanism for cave development, as older theories wanted. We now distinguish three common cases: the vadose cave, the deep phreatic cave and the water-table cave. The types that form in a given area are governed by the frequency of fissures significantly

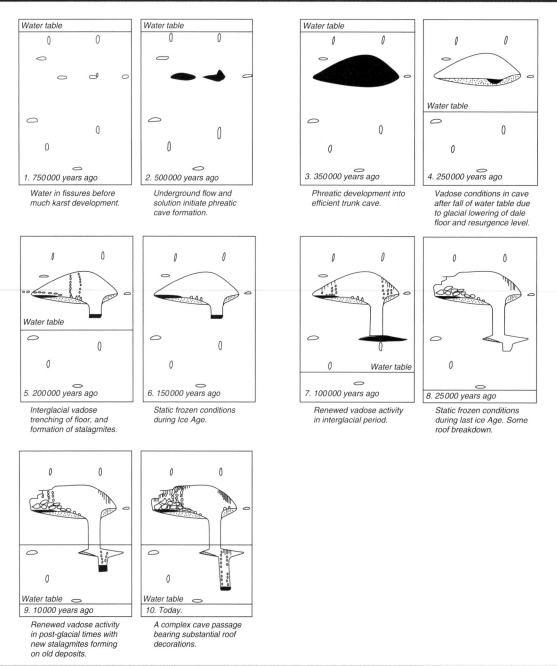

Figure 18.53 Cave complexity and time in the Yorkshire Dales over the last 750 000 years. (Source: after Hindle, 1980)

penetrated by ground water and by the joint-to-bedding plane ratio. Together these two characteristics combine to form the concept of hydraulic conductivity, which is the coefficient of proportionality describing the rate at which water can move through a permeable medium like limestone.

Water-table caves are common in flat-lying rock, where the perching of ground water occurs because of the presence of more-resistant rock layers. Deep penetration is inhibited by the presence of shallow, open bedding planes which are continuous to springs. We have seen that vadose caves develop where sufficient numbers of streams collect above sink points and move

water to the water table or a spring. Deep phreatic caves reach their maximum development in steeply-dipping rocks because continuous bedding planes move water to greater depths.

The larger passages of many cave systems show a succession of levels, with the youngest and still-active caves lying at the lowest elevations, so that the level in which enlargement occurs seems to be concentrated at or near the contemporary river levels, as in Kentucky, the Yorkshire Dales and the European Alps (see Box 4.5, see in the companion website). The generalized profiles of cave levels in Mammoth Cave in Kentucky are shown in Figure 18.56. Cave development seems possible above, at or below the water

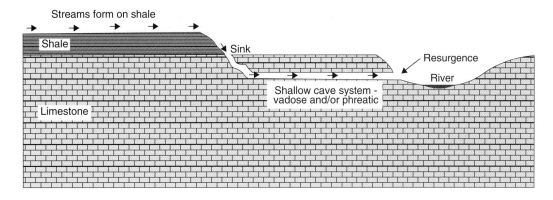

Streams form on shale

Shale

Limestone

Sink

Resurgence

River

Shallow cave system - vadose and/or phreatic

(a) PRE – GLACIAL

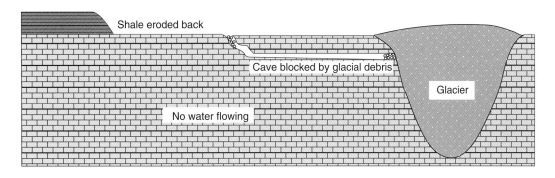

Shale eroded back

Cave blocked by glacial debris

No water flowing

Glacier

(b) GLACIAL

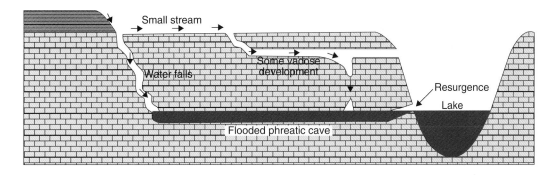

Small stream

Water falls

Some vadose development

Resurgence

Lake

Flooded phreatic cave

(c) INTERGLACIAL

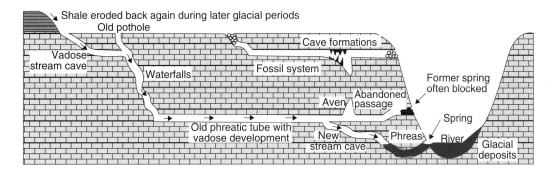

Shale eroded back again during later glacial periods

Old pothole

Vadose stream cave

Waterfalls

Cave formations

Fossil system

Former spring often blocked

Aven

Abandoned passage

Spring

Old phreatic tube with vadose development

New stream cave

Phreas

River

Glacial deposits

(d) PRESENT

Figure 18.54 Stages of cave development in the Yorkshire Dales over the last 750 000 years. (Source: after Hindle, 1980)

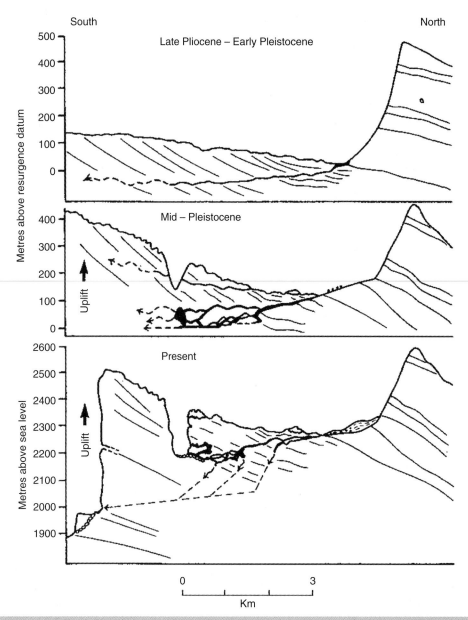

Figure 18.55 Inferred cave evolution in the Hindenburg Ranges caves, New Guinea. (Source: after Brook, 1977)

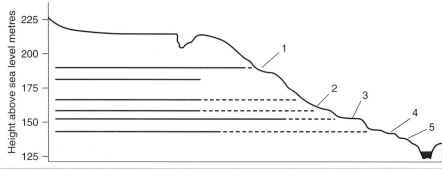

Figure 18.56 Profiles of cave levels, Mammoth Cave, Kentucky, USA: 1. upper terrace (preglacial?); 2.-3. pre-last glacial terraces (?); 4. upper last glacial; 5. lower last glacial. (Source: from Trudgill, 1985)

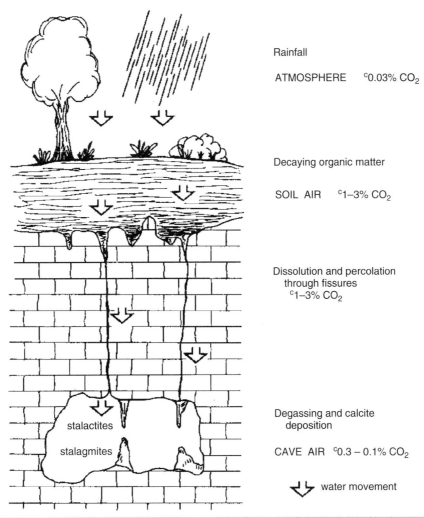

Rainfall

ATMOSPHERE ^{c}0.03% CO_2

Decaying organic matter

SOIL AIR ^{c}1–3% CO_2

Dissolution and percolation
through fissures
^{c}1–3% CO_2

Degassing and calcite
deposition

CAVE AIR ^{c}0.3 – 0.1% CO_2

water movement

stalactites

stalagmites

Figure 18.57 Model for the formation of calcite deposits in cave systems by degassing. (Source: after White, 1976)

table, and generally the location of cave formation depends on local geology and hydrology. It is possible for one cave to have passages formed above, at and below the water table. However, it is likely that much cave solution takes place at the water table in any one time period.

Cave speleothems

Water flow, which creates caves, also becomes an interior decorator of caves. Particularly at later stages of any cave's developmental history, calcite can precipitate from the carbonate-laden water as it emerges from fissures or pores in the limestone on to the cave roof or down the cave walls. There are many types of speleothem (derived from the Greek for 'cave', *spelaion*, and 'deposit', *thema*) or flowstone; they can be a sparkling white if pure calcite, or stained brown from iron, black from manganese or green from copper in the cave waters. Speleothems form by degassing, where the CO_2 is derived mainly from the soil air (Figure 18.57). The simplest form is the straw stalactite (Figures 18.58 and 18.59), which is a delicate tube growing from a pore in the cave roof.

Single drips of water deposit a ring of calcite 5 mm in diameter, which is the water droplet size. Water continues to flow down the lengthening hollow tube, adding calcite crystals to the end before splashing to the cave floor. As this happens, even though the water already gave up some of its carbon dioxide on bursting as it hit the floor, more carbon dioxide is lost to the cave air and more calcite is developed, and so a stalagmite forms with time. Columns can occur when the stalactite growing down meets the stalagmite growing up. The process of development is a slow one but must depend on the flow rate of the feeding water, the amount of calcium bicarbonate in the water and the air conditions. Thus speleothems can be dated by Uranium Series or carbon dating.

Helictites, on the other hand, defy gravity and come in a great number of shapes and sizes. The hypotheses as to how they form are almost as varied as the forms themselves (Figure 18.60). These range from the influence of winds or draughts blowing along cave passages, to pressure within cracks, to the effects of impurities or electrical charges. Curtains can form when rivulets of carbonate-saturated water run down the cave walls and give off their calcium carbonate (Figure 18.61(a)). Rimstone pools can

(a)

(b)

Figure 18.58 Stalagmites: (a) Witch's Finger, Carlsbad Caverns, New Mexico; (b) Hall of the Giants, the Big Room, Carlsbad Caverns, New Mexico. (Sources: (a) Peter Jones, Wikimedia Commons; (b) Wikimedia Commons)

(a)

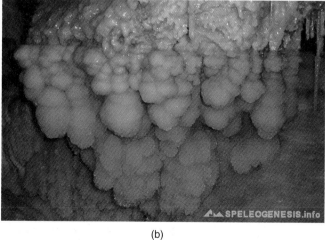

(b)

Figure 18.59 (a) Flagged Soda Straws, Castleguard Cave, Banff National Park, Canada. (b) Straw Stalagmites in Gardners Gut Cave, New Zealand. (c) 'Bottlebrushes', where the stalactites have become immersed in cave pool for a long time. Gruta de Salir, West Tiphonc valley, Portugal. (Sources: (a) http://en.wikipedia.org/wki/Catlegiard_Cave; (b) Wikimedia Commons; (c) Sofia Reboleira, Atlas of Cave Morphs)

Figure 18.60 Helictites: (a) from Jenolan caves, Australia; (b) 'Ice Cream Wall' flowstone curtains, Lost World Caverns, West Virginia (c) Helictites in Diamond Cave. (Sources: (a) Jason Ruck, Creative Commons; (b) Wikimedia Commons; (c) Wikimedia Commons)

Figure 18.61 (a) Curtains in Grotte de Soreq, Israel. (b) Curtains, the "Ice Cream Wall" Lost World Caverns, West Virginia. (c) Flowstone Gunns Plains Cave, Tasmania. (d) Flowstone at Hierve El Agua, Oaxaca, Mexico. (Sources: (a) Wikimedia Commons, (b) Wikimedia Commons, GNU Free Documentation License, (c) Wikimedia Commons)

(c)

(d)

Figure 18.61 (*continued*)

(a)

(b)

Figure 18.62 Rimstone pools: (a) Grotte de St. Marcel Ď Ardeche, France (b) Hierve El Agua, Oaxaca, Mexico.

form when water fills any hollows then overflows (Figure 18.62); the thin water film flowing over the edge gives off more carbon dioxide and so more calcite is deposited. Over time these rims grow into dams known as gours and a sequence of terraces of rimstone pools may develop, with calcite precipitation out on the surface of the water to be collected along the rims or dropping to the base of the pool. These landforms can form on a variety of scales and both on the surface from springs and underground along old meander passages, as in Great Douk cave in the Yorkshire Dales. Of course, calcium carbonate can be precipitated at the surface too as tufa, as in Janet's Foss in Gordale, or as spectacular flowstone waterfalls, as at Hierve El Agua in Oaxaca in south-western Mexico (Figure 18.62(d)).

Fluvial erosion landscapes in karst areas

Although we briefly discussed the origin of gorges in glacio-karst landscapes eroded by subglacial melt water earlier, there are also specialized morphologies caused by fluvial erosion. Initially deep dolines open up along joints and fissures, as in the top panel of Figure 18.63. Here the vertical incision is greater than lateral erosion. By enlargement of these landforms, deep box canyons or karst streets form. As these streets widen and extend, the rock ridges in between are dissected and ultimately destroyed, leaving closed depressions with vertical walls called platea, which can become covered with alluvium. As they expand, they become poljes, and rock towers can be left rising above flat plains, as in the

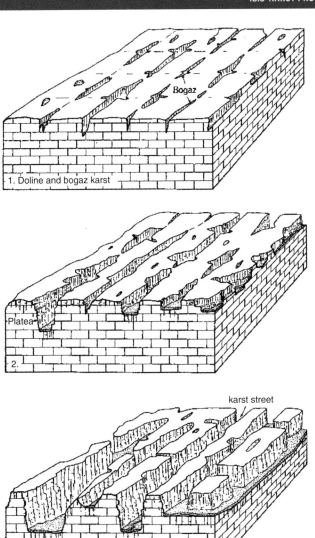

Figure 18.63 Development of karst streets and tower karst of the Mackenzie Mountains type, Canada

(modified from Williams, 1978b)

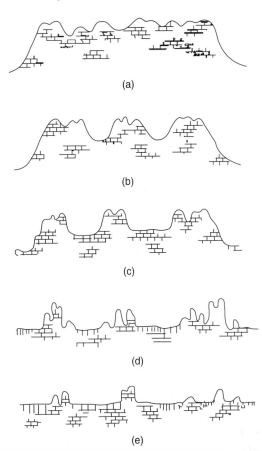

Figure 18.64 Development sequence in Chinese karst: (a) high erosion surface, polygonal karst; (b) 50% incision; (c) wider valleys present; (d) clusters of towers with wide valleys and plains, alluvial soils; (e) plain with isolated towers. (Source: after Williams, 1978)

Australia and Tanzania. In fact, tropical karst landscapes are distinctive, with cone karst and tower karst in various stages of development, and large alluvium-covered poljes with residuals (Figure 18.24 and 18.25). However, some caves appear to be preserved for long periods of time, rather gradually eroding as would be expected. Box 18.6 (see in the companion website) discusses some such preserved caves in eastern Australia.

Exercises

1. Outline the types of carbonate factory and why there are differences between the sediments that can be produced in them.

2. Why does erosion take place in carbonate factories?

3. What are the major processes of diagenesis in carbonate successions and why do they occur?

bottom panel of Figure 18.63. These types of landform have been described for Canadian karst by Brook and Ford (1977) and for Chinese karst by Williams (1978), as can be seen in Figure 18.64. It is likely that because of the high rainfall these types of landform will develop best in tropical areas, where they have been described in the Celebes, New Guinea, Sarawak, Arnhem Land in Northern

4. What are the different types of coral reef? Outline the characteristics of the sediment successions that are produced in such environments.

5. Describe the factors that can affect the solubility of carbonates.

6. What is the process of condensation corrosion? Explain the role it plays in the formation of röhrenkarren.

7. Explain the various origins of dolines in karst areas.

8. How do caves form in the phreatic zone and what are the characteristics of such caves?

9. Explain how geological controls influence cave development.

10. Why are there so many different types of cave speleothem?

References

Brook, D. 1977. Caves and karst of the Hindenburg Ranges. *Geographical Journal*. **143**, 27–41.

Brook, G.A. and Ford, D.C. 1977. The sequential development of the karst landforms in the Nahanni region of northern Canada and a remarkable size hierarchy. *Proceedings of the 7th International Speleological Congress, Sheffield 1977*. British Cave Research Association, ISU. 77–81.

Curl, R.L. 1974. Deducing flow velocity in cave conduits from scallops. *Bulletin of the National Speleological Society of America*. **36**, 1–5.

Charity, R.A.P. and Christopher, N.S.J. 1977. The stratigraphy and structure of the Ogof Ffynnon Ddu area. *Transactions of the British Cave Research Association*. **4**, 403–416.

Dunham, R.J. 1962. Classification of carbonate rocks according to depositional texture. In: Ham, W.E. (Ed.). Classification of Carbonate Rocks: A Symposium. American Association of Petroleum Geologists Memoir 1. pp. 108–171.

Embry, A.F. and Klovan, J.E. 1971. A late Devonian reef tract on northeastern Banks Island, NWT. *Bulletin of the Canadian Petroleum Geologists*. **19**, 730–781.

Flügel, E. 2004. Microfacies of Carbonate Rocks: Analysis, Interpretation and Application. Berlin, Springer-Verlag.

Glover, R.R. 1974. Cave development in the Gaping Gill system. In Waltham, A.C. (Ed.). Limestones and Caves of North-West England. Newton Abbot, David and Charles. pp. 343–384.

Heckel, P.H. 1972. Recognition of ancient shallow marine environments. In: Rigby, J.K. and Hamblin, W.K. (Eds). Recognition of ancient sedimentary environments. Society of Economic Palaeontologists and Mineralogists, Special Publication 16. pp. 226–286.

Hindle, B.P. 1980. Cave formation in northern England. Dent, Lyon Ladders.

Jennings, J.N. 2003. Karst Geomorphology: An Introduction to Systematic Geomorphology. Volume 7. Massachusetts, MIT Press.

James, N.P. and Bourque, P.-A. 1997. Reefs and mounds. In: Walker, R.G. and James, N.P. (Eds). Facies Models: Responses to Sea level Change. Geotext 1. 3rd edition, Sherbrooke, Quebec, Geological Association of Canada. pp. 323–348.

James, N.P. and Kendall, A.C. 1997. Introduction to carbonate and evaporite fades models. In: Walker, R.G. and James, N.P. (Eds). Facies Models: Responses to Sea level Change. Geotext 1. 3rd edition, Sherbrooke, Quebec, Geological Association of Canada. pp. 265–276.

Jennings, J.N. 1985. Karst Geomorphology. Oxford, Blackwell.

Renault, P. 1967–1969. Contribution a l'etude des actions mécaniques et sédimentologiques dans la spéléogenese. *Annales de Spéléogie*. **22**, 5–17; **23**, 259–307, 529–596; **24**, 317–337.

Tratman, E.K. 1969. The Caves of NW Co. Clare. Newton Abbott, David and Charles.

Trudgill, S. 1985. Limestone Geomorphology. London, Longman.

Tucker, M.E. 1981. Sedimentary Petrology: An Introduction. Oxford, Blackwell Scientific Publications.

Waltham, A.C. 1974. The Limestones and Caves of North West England. Newton Abbott, David and Charles.

Waltham, T. 2005. Karst and caves of Ha Long Bay, speleogenesis and evolution of karst aquifers. *The Virtual Scientific Journal*. **3**.

Weaver, J.D. 1973. The relationship between jointing and cave passage frequency at the Head of the Tawe Valley, South Wales. *Transactions of the Cave Research Group of Great Britain*. **15**, 169–173.

White, W.B. 1976. Cave minerals and speleothems. In: Ford, T.D. and Cullingford, C.H. (Eds). The Science of Speleology. Academic Press. pp. 267–327.

Williams, P.W. 1969. The geomorphic effects of ground water. In: Chorley, R.J. (Ed.). Introduction to Fluvial Processes. Methuen. pp. 108–123.

Williams, P.W. 1978. Karst research in China. *Transaction of the British Cave Research Association*. **5**, 29–46.

Wilson, J.L. 1975. Carbonate Facies in Geological History. Berlin, Springer-Verlag.

Further reading

Atkinson, T.C. and Smith, D.I. 1976. The erosion of limestones. In: Ford, T.D. and Cullingford, C.H. (Eds). The Science of Speleology. Academic Press. Chapter 5.

Fogg, P. and Fogg, T. 2001. Beneath Our Feet: The Caves and Limestone Scenery of the North of Ireland. Belfast, Natural Heritage.

Ford, D.C. and Williams, P.W. 2007. Karst Geomorphology and Hydrology. Chichester, John Wiley & Sons, Inc.

Goldscheider, N. and Drew, D. 2007. Methods in Karst Hydrogeology. IAH International Contributions to Hydrogeology. Taylor and Francis.

Gunn, J. 2004. Encyclopedia of Caves and Karst Science. Fitzroy Dearborn.

Harmon, R.S., Wicks, C.M., Ford, D. and White, W.B. 2006. Perspectives on Karst Geomorphology, Hydrology and Geochemistry: A Tribute Volume to Derek C. Ford. Geological Society of America.

Tegethoff, F.W, Rohleder, J. and Kroher, E. 2001. Calcium Carbonate: From the Cretaceous Period into the 21st Century. Birkhauser Verlag AG.

Waltham, T. 1987. Karst and Caves. Yorkshire Dales National Park Committee in conjunction with the British Cave Research Association.

Waltham, T., Simms, M.G., Farrant, A. and Goldie, M.S. 1996. Karst and Caves of Great Britain. Geological Conservation Review Series. Kluwer Academic Publisher.

19 Coastal Processes, Landforms and Sediments

Sandwood Bay, north-west Scotland, showing the sandy foreshore and runoff channels with ripples at low water

Earth Environments David Huddart and Tim Stott
© 2010 John Wiley & Sons, Ltd

Learning Outcomes

After reading this chapter and completing the exercises you will be able to:

➤ Understand the physical controls on waves, tides and tsunami and their importance in developing coastal landforms and systems.

➤ Understand the formation of coastal landform assemblages such as rocky coasts and platforms, beaches, coastal dunes, estuaries, barrier coasts and carbonate coastlines.

➤ Understand the controls on the distribution of coastal landsystems.

➤ Understand the importance of sea-level change in creating coastal landsystems.

➤ Understand the importance of future climate change in influencing coastal landsystems.

19.1 Introduction to the coastal zone

The coastal or littoral zone is one of the Earth's most dynamic exogenic environments. It has existed on the Earth since the first oceans formed and is well recorded within the sediments of the geological record. Interpretation of the sedimentary record of coastal environments allows a reconstruction of the Earth's changing palaeogeography through time. An understanding of coastal processes and environments is therefore important for an understanding of the Earth's changing environments. More importantly, the coastal zone provides environmental geoscientists with one of their greatest challenges today, since the dynamic coastal zone is also a focus for urbanization, commerce, industrialization and recreation, and shows the impact of sea-level change.

The littoral zone is narrow when compared to the size of the oceans, which cover 70% of the Earth's surface, and is frequently less than a few hundred metres wide. It occurs landward of the point at which waves first stir bottom sediments and may extend inland to a cliff, coastal dune system or estuary head. It is subdivided by an array of terms, many of which are difficult to apply and frequently mean different things to different people (Figure 19.1). The major primary components of this coastal system are illustrated in Figure 19.2, where the feedback loop between morphology and processes is responsible for the complexity in coastal evolution. It is also important to note that over geological time scales the littoral zone may move either seaward or landward with changes in relative sea level, and that a variety of time scales control the processes that operate in the coastal system and shape the coastal landscape, from geological to instantaneous.

Within the coastal zone, energy is derived from waves and tidal currents, which move across the oceans as a function of both the wind and the imbalance in solar energy which generates it and also of the gravitational interplay between the Sun and Moon and the Earth's hydrosphere. The coastal zone is an energy sink or buffer between the mobile and energy-rich oceans and the more static islands of land within them. The distribution of different coastal environments, such as deltas, beaches, eroding cliffs, mudflats, dunes and salt marshes, reflects the uneven distribution of energy along a shoreline and the movement of sediment in response to these inequalities in energy distribution. In a stable world and over time a coastal zone will evolve so that the distribution of energy is more uniform. In practice, little remains stable in this dynamic environment, but as a concept it serves to help understand the distribution and evolution of different coastal environments.

A section of coast can be subdivided laterally into a series of coastal cells based on the longshore movement of sediment. Coastal boundaries are placed at points where longshore sediment transport is zero, perhaps at a headland or estuary mouth, and a hierarchy of lesser boundaries – permanent, seasonal or more ephemeral – can be defined between these more permanent cell boundaries to provide an order of subcells. These sediment cells are of importance in understanding and managing the coastal zone, since they provide a unit with which to assess sediment budgets. Each cell has a distinct sediment budget, with inputs of sediments derived from coastal erosion of offshore sand banks or rivers, and outputs associated with offshore movement of sediment or its landward movement under the influence of the wind. This budget can also be influenced by human activity; for example, the removal of beach sand or gravel for construction may cause a net loss of sediment from a cell. Within a cell, sediment may move alongshore or offshore, and certain parts may become depleted in sediment, perhaps leading to coastal erosion, while other areas may have a surplus, causing the progradation, or build-up, of a beach or spit.

In many cells, even though the local distribution of gains and losses may be uneven, overall there is a balance between gains and losses of sediment. Alternatively, there may be a natural drain on sediment which means that a cell always has a negative balance and that more areas of the coast will show evidence of sediment loss than of coastal accretion. These trends may be either amplified or dampened by human activity such as coastal management. To gain an understanding of the coastal zone one must have knowledge of the spatial and temporal distribution of energy along a coast, controlled by wave climate and tidal regimes, and the distribution of sediment sources, sinks and losses within a coastal cell, which controls its overall sediment budget. Before we can review the morphology and sedimentology of different coastal environments and the energy regimes in which they exist, we need to understand something of the mechanics of waves and tides, because it is this energy that generates near-shore currents, sediment transport and morphological change.

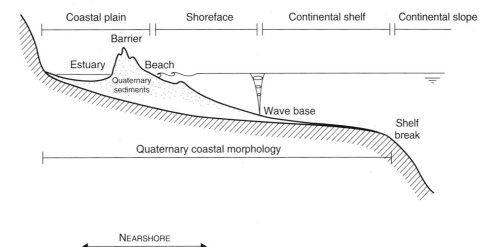

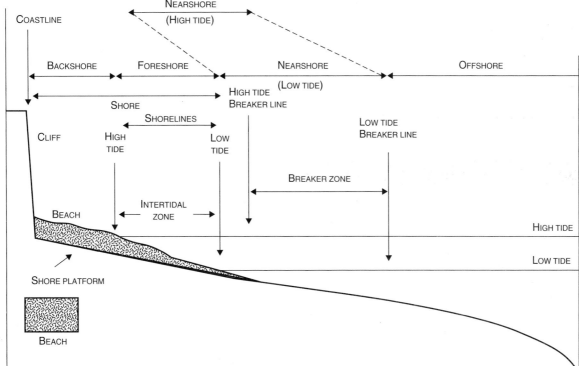

Figure 19.1 Coastal zone terminology and boundaries. (Source: after Bird, 2000 and Masselink and Hughes, 2003)

19.2 Sea waves, tides and tsunamis

19.2.1 Sea waves and shore-normal and longshore currents

In deep water, waves are generated by the movement of wind across a water surface transferring energy from one fluid body to another. The amount of energy transferred shows itself in the height of the wave and is a function of wind velocity, duration and the distance over which the wind blows in contact with the water body. This is known as the fetch. Large sea waves are therefore generated by strong winds of long duration that blow in the same direction over long sea surface distances. Consequently, those coasts with an ocean prospect, downwind of global westerlies or easterlies, will receive the highest waves, while those with a protected coastline, with a limited fetch and only local waves, will receive much lower waves.

A wave moves energy, not water, because as a wave form passes a given point, an individual package of water simply rotates around an orbit equivalent to the height of the wave (Figure 19.3). As with any wave, a sea wave can be characterized by its height (the vertical distance between a crest and a trough, which is a function of wave energy), wavelength (the distance

(a)

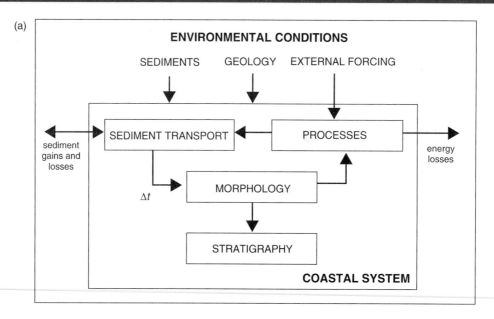

(b)

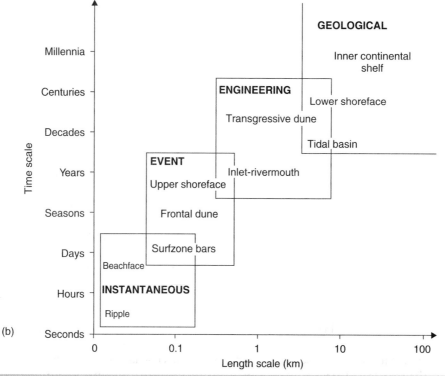

Figure 19.2 Major components involved in coastal morphodynamics and the spatial and time scales involved in coastal evolution. (a) Primary components. The feedback loop between morphology and processes is responsible for the complexity in coastal evolution. (b) Spatial and time scales in coastal evolution, where large-scale coastal landforms evolve over a long time period and small-scale coastal features respond over a short time

between two troughs or crests), period (the time interval between two consecutive wave crests or troughs passing a fixed point), velocity or celerity and steepness (the ratio of wave height to length) (Figure 19.3).

Longer-period waves, with longer wavelengths, have greater celerity. This is important as it explains how waves tend to become sorted by size as they travel. As the wind blows over a stretch of

water, perhaps during a storm, it generates waves with different heights, wavelengths and periods, due to fluctuations in the wind velocity and direction. Since longer-period waves tend to move faster, they will begin to outdistance the shorter-period waves as the wave group generated by the storm travels over the ocean surface, a process known as wave dispersion. With time, the wave group will become sorted, with longer-period waves at the

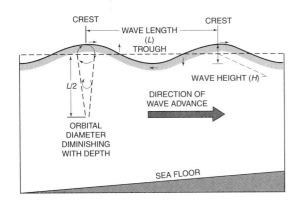

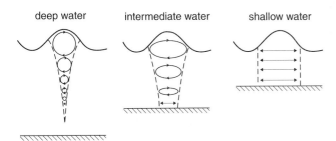

Figure 19.3 Wave motion from deep to shallow water: (a) pattern of currents as wave crests and troughs move shoreward; b) motion of water particles under waves according to linear wave theory. In deep water, water particles follow a circular motion, with the radius of the orbits decreasing with depth beneath the surface. In intermediate water the orbits are elliptical and become flatter as the sea bed is approached. In shallow water all water motion consists of horizontal movements to and fro that are uniform with depth

front and shorter-period waves at the back. Wave attenuation (energy loss) is small and occurs close to the source area but is most marked for short-period waves. Consequently, over time the spectrum of waves generated by the storm will become less as wave dispersion and attenuation occur. This helps explain how a uniform swell is generated on coasts with an ocean prospect, while in more enclosed seas, such as the Mediterranean or North Sea, with limited travel distances, a more chaotic wave regime exists.

As a wave moves onshore it shoals and the oscillating water packages which make up the wave form begin to interact with the sea floor. At this point a wave undergoes a series of transformations, which ultimately lead to its breaking and the creation of shore-normal and longshore currents, in which the energy within a wave is used to drive water flow. In deep water a wave has a nearly sinusoidal form, with low rounded crests, but as the wave enters shallower water the wave velocity begins to fall due to frictional retardation by the sea floor. Given that the wave in front is always in shallower water and therefore travels slower than the wave behind, the distance between two waves – the wave length – must fall, just as the distance between a series of cars becomes less as they draw up at a set of red traffic lights. Wave height increases as waves shoal, and consequently waves become

steeper. The only parameter which remains unchanged is the wave period. This results in a series of steeper waves, with peaked wave crests, separated by relatively flat troughs. In other words, wave shoaling causes a wave to steepen until it is no longer stable and it breaks. Wave orbits become compressed by shoaling, such that a wave is essentially composed of two shore-normal currents, one onshore and one offshore. These currents become increasingly asymmetrical in velocity and duration as shoaling proceeds; the onshore component increases in velocity but decreases in duration, while the offshore component increases in duration but is slower. Onshore and offshore water continuity is maintained. A wave that breaks well offshore will have almost symmetrical onshore and offshore currents, while one that breaks further onshore will have more asymmetrical currents.

The rate of shoaling is influenced by offshore topography. Offshore topography is not always uniform and parallel to the incident wave crest, and consequently a wave moving onshore over an irregular submarine topography, or simply at an oblique angle to the submarine slope, will be influenced by the bottom topography and will slow along different sections of its crest at different times. This will cause the orientation of the wave to change, a process known as wave refraction. The importance of wave refraction is in its impact on the distribution of wave energy along a section of coast (Figure 19.4). Where wave convergence occurs, the amount of energy per unit length of wave increases and manifests in higher waves, but conversely where wave divergence occurs the amount of energy in a wave is spread out over a longer section of coast and wave height falls. Offshore topography therefore has an important role in determining the longitudinal energy gradients present on a section of coast. This may change with time, as offshore bars migrate along a coast, or with changes in sea level.

As a wave shoals, it steepens until it finally becomes so unstable that it breaks. The way in which a wave breaks and the point at which it does so on a beach is determined by the steepness of the incident wave, the water depth and the beach gradient. Spilling breakers (Figure 19.5) tend to occur on beaches with a shallow gradient, irrespective of wave steepness, and are characterized by turbulent water, which dissipates the wave energy. Plunging breakers occur on steeply-inclined shores, with steep waves that curl over to form a classic surfer's wave, plunging forward and downward as an intact mass of water. Surging waves are low waves which tend to occur on steep beaches and collapse, rather than break, to give a surging up-rush of water characterized by little foam or water turbulence.

The type of breaker present on a given beach will change with the wave regime and with changes in the beach profile. Spilling breakers tend to break far offshore and are therefore associated with symmetrical onshore and offshore currents, while surging waves break further onshore and are associated with more asymmetrical shore-normal currents. It is important to note that waves do not necessarily break at one point on a beach. If the beach slope is fairly continuous, there is likely to be a uniform dissipation of the waves as they cross the width of the surf zone, with waves breaking at all depths. However, if the profile is characterized by a series of bars and troughs, breaking will tend

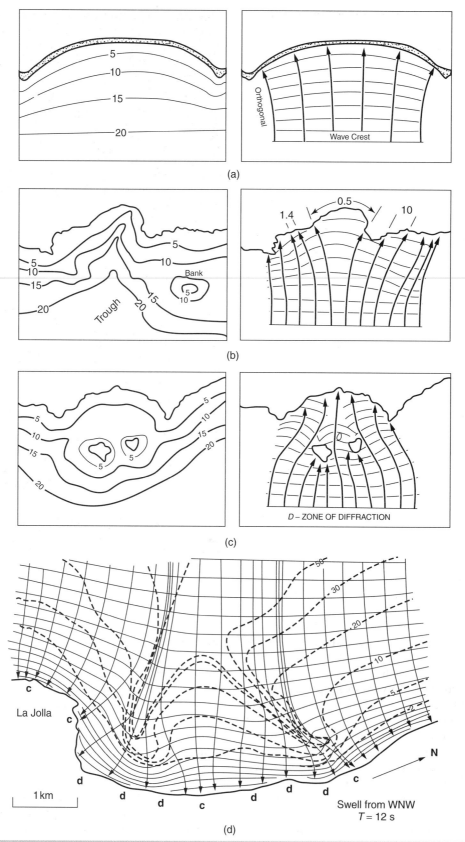

Figure 19.4 Impact of wave refraction on energy. (a) Refraction of waves moving into a bay. (b) Refraction of waves over a trough (refraction coefficients indicated). (c) Refraction around and diffraction between offshore islands. (d) Wave refraction over submarine canyons and along the headland of La Jolla, California. 'd' and 'c' refer to wave divergence and convergence, respectively

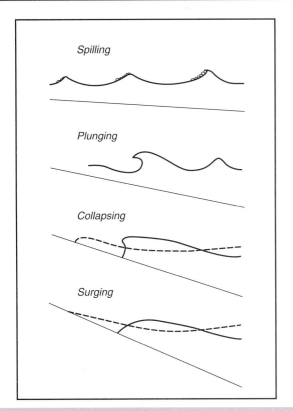

Spilling

Plunging

Collapsing

Surging

Figure 19.5 Four types of breaking wave

(a)

(b)

Figure 19.6 Examples of waves: (a) destructive waves breaking on a sandy beach forming a smooth foreshore; (b) spilling breakers and backwash, Formby Point, UK

to be concentrated in the shallower water above the bars. Waves may reform having broken on an outer bar, before breaking again on subsequent bars or on the upper beach face. Examples of destructive, erosional waves on a sandy foreshore are illustrated in Figure 19.6, and an overview of wave processes is given in Table 19.1.

When waves break on a beach they are associated with wave or water setup, a rise in the mean water level above the still-water elevation of the sea (Figure 19.7). This is a piling up of water against the shoreline due to the waves and is caused by the breaking waves being driven shorewards. The degree to which wave setup extends offshore is a function of wave height, with higher waves giving greater setup. The water dynamics within this zone, which is shoreward of the point at which most waves break, are extremely complex. Early theories suggested that once a wave had broken, a bore of water moved shoreward, the velocity of which was a function of a wave's steepness or energy and the beach slope. However, orbital wave motions are now believed to contribute to water dynamics within the surf zone and it has been suggested that the onshore–offshore current asymmetry seaward of the break point continues within the surf zone and increases onshore as the overall magnitude of the current falls with water depth.

The surf zone is also associated with waves that have periodicities ranging from 20 seconds to several minutes – in excess of those produced by wind – known as infragravity waves. The energy produced is trapped between the beach face and the

breaker zone and is manifest on many beaches as certain types of infragravity wave known as edge waves. These are thought to be responsible for many of the nearshore features, such as rip channels and multiple bars. The energy within a wave can either be dispersed by water turbulence or reflected seaward by the beach face.

Reflected waves move seaward and may either be lost offshore (leaky edge waves) or refracted by the submarine slope into a shore parallel direction (trapped edge waves). The amplitude of trapped edge waves decreases rapidly offshore and these waves may either move alongshore as progressive waves or more likely become trapped by the interaction of a wave with its reflection to form a standing wave. On any given beach a range of edge waves, with a range of different periodicities, may exist and can interact with incoming incident waves from offshore to give regular variation in wave setup along a beach.

If standing edge waves have similar periodicities to the incident waves then at points at which the two wave crests are

Table 19.1 Overview of wave processes (where h = water depth, L_o = deep-water wavelength, H_o = deep-water wave height)

Process	Description of process	Types of waves
Deep water: $h/L_o > 0.5$ (waves are unaffected by sea bed):		
Wave generation	Waves are generated by wind, Wave height and period increase with increase in wind speed and duration	Sea
Wave dispersion	In deep water long-period waves travel faster than short-period waves. This results in a narrowing of the wave spectrum	Swell
Intermediate and shallow water: $h/L_o < 0.5$ (waves are affected by sea bed):		
Wave shoaling	The wave length shortens with decreasing water depth. This results in a concentration of the wave energy over a shorter distance and an increase in the wave height	Shoaling waves
Wave asymmetry	Shoaling waves become increasingly asymmetrical and develop peaked crests and flat troughs	Asymmetric waves
Wave refraction	Wave refraction causes wave crests to become more aligned with the coast, resulting in a decrease in the wave angle	Refracted waves
Wave diffraction	Leakage of wave energy along the wave crests into shadow areas	Diffracted waves
Surf zone: $h < 1H_o$ (regular waves) or $h < 2H_o$ (irregular waves):		
Wave breaking	When the horizontal velocities of the water particles in the wave crest exceed the wave velocity, the water particles leave the wave form, and the wave breaks	Spilling plunging, surging breakers
Energy dissipation	As breaking/broken waves propagate through the surf zone, they progressively decrease in height as a result of wave energy dissipation	Surf zone bores
Wave reflection	On steep beaches a significant part of he incoming wave energy is not dissipated in the surf zone, but is reflected back to sea	Standing waves
Infragravity wave energy	A significant part of the wave energy in the surf zone is at very low (infragravity) frequencies. Infragravity waves are particularly energetic during storms	Infragravity waves, long waves, edge waves
Swash zone: $h = 0$:		
Runup	The maximum water level attained on a beach is higher than the still water level. This vertical displacement of the water level is known as runup and consists of a steady component (wave set-up) and a fluctuating component (swash)	Wave runup, swash, wave set-up

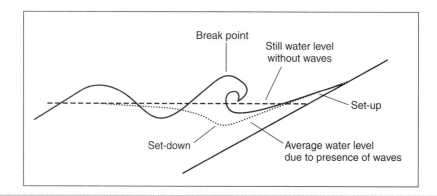

Figure 19.7 Wave setup

superimposed, the wave height of the resultant incident wave and the water setup beyond it will be increased. Equally, where a trough and a crest are superimposed the resultant incident wave has a reduced height and the water setup associated with it at that point is reduced. In this way, the height of an incident wave may vary systematically along its length in a shore-parallel direction, and the degree of wave setup associated with this wave will vary along the shore as well. The spacing between points of enhanced wave setup will be regular and a function of the steepness of the beach. Steep beaches reflect lots of wave energy and give high-frequency edge waves with short wavelengths, similar to the incident waves. In contrast, flatter beaches with a broader surf zone tend to result in lower-frequency edge waves with short wavelengths.

The height of the wave setup is one factor which determines the length of wave run-up on the beach face above still-water level. Wave run-up consists of two components, an onshore current known as the swash (uprush), which starts with a high velocity that declines onshore, and a return gravity-driven flow known as the backwash (downrush), which accelerates offshore (Figure 19.6(b)). Generally swash motion is asymmetrical, with greater uprush than downrush speeds, but swashes are important as their motion maintains the beach gradient. The magnitude of these currents is very difficult to predict but is a function of such variables as wave height, energy loss within the surf zone due to turbulence and friction, variation of water depth across the surf zone, the degree of orbital current asymmetry present beyond the break point and the beach slope.

The interaction of edge waves with incident waves of similar periodicity may cause short wavelength variation (1–10 m) in wave setup along a beach, resulting in regular variations in the extent of wave run-up. This may appear as a scalped edge to the swash limit. The spacing between waves and therefore between each swash and backwash flow is important to their geomorphological potential on a beach face. Equally, the permeability of a beach may influence the magnitude of each flow. For example, on gravel beaches most of the water infiltrates into the beach, making the backwash negligible. On sandy beaches the water table within the beach influences current strength. As water levels increase with rising tide, the water table within the beach also rises, but often lags behind the mean water level. Consequently infiltration may be high, reducing the backwash. As the tide turns, the water table within a beach will begin to fall, but at a slower rate than the tide. In this case the outflow of water from a beach may actually increase the magnitude of the backwash.

So far we have concentrated on shore-normal currents, but two types of longshore current can be identified: near-shore circulation cells and longshore currents due to oblique waves (Figure 19.8). Near-shore circulation cells are believed to be the result of the interaction of low-frequency edge waves with incident waves, giving regular longshore variation in wave setup, with wavelengths of tens to hundreds of metres. The spacing of rip-currents is controlled by locations at which the wave setup is small relative to the intervening areas of higher wave setup. Given that, water flows from high to low water and is driven alongshore and fed into rip-currents to form a distinctive cell. The spacing of rip-currents is not always determined by the frequency of edge waves and offshore topography may play a role in determining their location and spacing. Oblique waves may also generate longshore currents. The energy component within an oblique wave can be resolved into two vectors: a shore-normal vector and an alongshore vector. The degree to which energy is partitioned between these two vectors is a function of the angle of incidence between the incoming wave and the beach. The greater the angle, the greater the longshore current component. Longshore currents due to oblique waves may interact with near-shore circulation cells to give a modified pattern of circulation, as shown in Figures 19.8 and 19.9.

19.3 Tides

Tides are regular fluctuations in water level due to the gravitational attraction of the Sun and the Moon on the Earth's hydrosphere, but their effects are hardly noticed in deep water. This is not the case in the near-shore areas and tidal currents have an important role to play in sediment movement. By way of introducing their importance we will outline the fundamentals of their generation.

Tides are in fact waves with extremely long wavelength, but unlike sea waves are unable to progress without astronomical forcing due to the shallowness of the oceans. Since gravitational force is a function of a body's mass and declines with distance, the Moon is of greater importance than the Sun to the Earth's tides. Although the Sun is much bigger than the Moon, its pull is less than half that of the Moon because it is so much further away.

The basic tidal pattern is a semi-diurnal one, with two high tides in 24 hours. However, the relative magnitude of the two high tides may vary from location to location, giving a more complex tidal signature. In fact, in some cases, the second high tide is so small that the tidal pattern is effectively a diurnal one. In other areas, the tidal signature changes from a diurnal to a semi-diurnal pattern during a lunar month, which is the time taken for the Moon to rotate around the Earth's axis (29 days) (Figure 19.10). The timing of a high tide progresses by 50 minutes every 24 hours, so high tides do not occur at a constant time each day. The magnitude of high tides varies over a lunar month and over a year. High tides are often much greater close to the shore, in shallow water, than offshore in the ocean.

Tidal range in some estuaries or bays is extremely high relative to that in adjacent sea or ocean basins. This adds to the scale of regional and local variations in tidal range.

Tides form when ocean water on the surface of the Earth passing under the Moon or the Sun is pulled towards it, forming a bulge. A bulge also forms on the opposite side of the Earth as a result of centrifugal force. This can be explained as follows: the Earth and the Moon rotate around a common axis and do not collide as a result of their mutual gravitational attraction due to the centrifugal forces generated by their rotation; the gravitational forces are equal and opposite to the centrifugal force, which gives

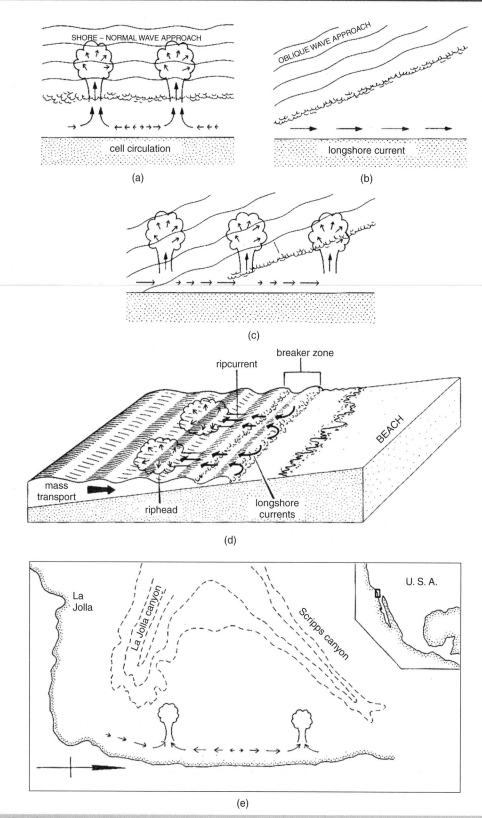

Figure 19.8 Interaction of near-shore and longshore circulation. Shore-normal wave approach (a) forms a series of rip cell currents. The oblique wave approach shown in (b) results is a steady unidirectional longshore velocity. Addition of these two current types results in the circulation diagram shown in (c). Here rip-currents are fed by unidirectional longshore currents whose velocity increases towards the rip base. (d) Cell circulation in the near-shore zone, where the slow onshore mass transport is transferred into shore-parallel currents inshore of the breaker zone. These in turn feed the rip-currents. (e) Effects of irregular bottom topography. This causes a sequence of wave convergence and divergence at the shore, resulting in a cell circulation with rip-currents

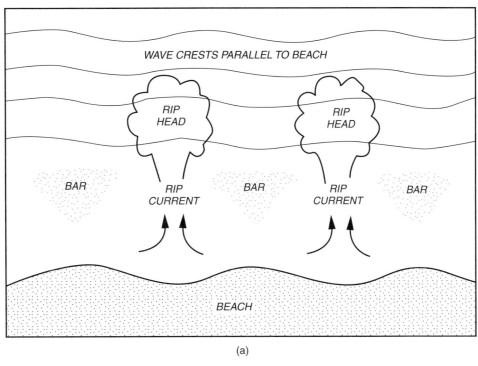

(a)

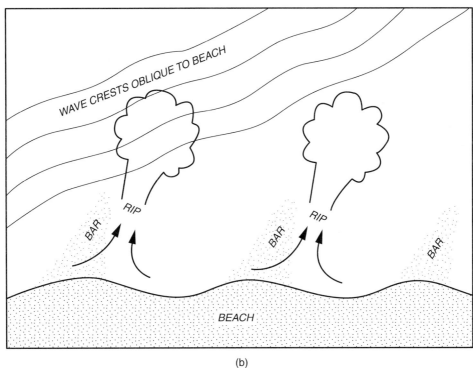

(b)

Figure 19.9 Rip cell circulation and bar formation causes a discontinuous longshore bar in (a), whilst the cell circulation caused by an oblique wave approach, as in (b), forms a series of transverse bars attached to the shore

stability. This balance is only a planetary average, however. All objects on the Earth experience the same centrifugal force, but those closest to the Moon at a particular point in time experience a slight excess of gravitational attraction, leading to the first water bulge on the surface of the ocean. At the opposite side of the globe, where the gravitational attraction of the Moon is at its least, there is a slight excess in centrifugal force over gravitational attraction, which gives the second and complementary bulge in the water surface. The consequence of this double bulge is to give a semi-diurnal periodicity to the tide.

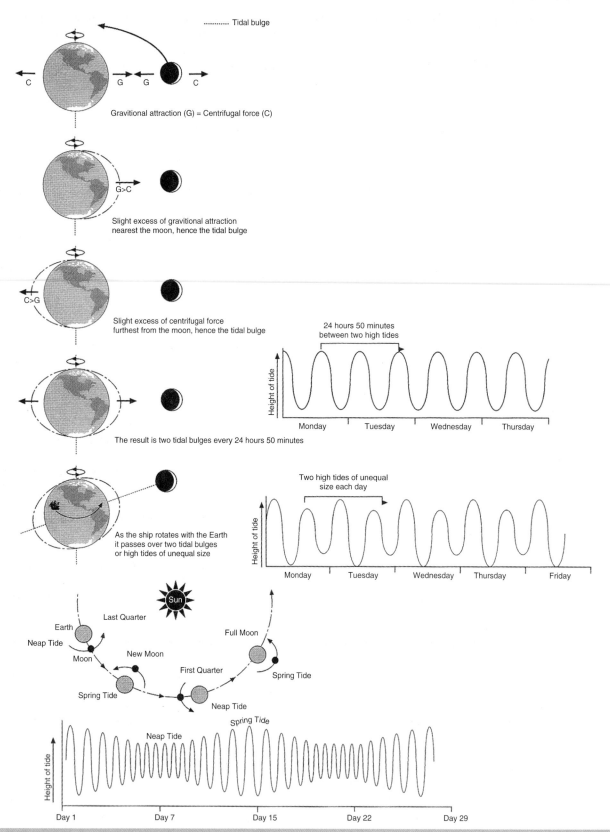

Figure 19.10 Formation of tide patterns

The bulge is stationary relative to the Moon, yet the Earth is rotating. Consequently a ship on the Earth's ocean surface will experience a water bulge twice in every 24 hours, effectively experiencing two high tides. The magnitude of the two high tides is not equal at all locations on the Earth, since the angle between the Earth and the Moon, the declination, varies over a month from 28° S to 28° N of the equator. Consequently, the two tidal bulges are tilted at an angle relative to the Earth's axis of rotation. This means that a ship rotating with the Earth will pass over two different thicknesses of tidal bulge, with one larger than the other, according to its geographical location (Figure 19.10): in some locations the two high tides are of similar magnitude, while at others the second high tide may effectively disappear, giving a diurnal tidal signature. Because the declination varies over a lunar month, so does the relative magnitude of the two high tides, which explains why in some locations tidal signature seems to be both diurnal and semi-diurnal.

High tide does not occur at the same time every day but progresses by 50 minutes in every 24 hours. This is because the Moon is also rotating around the Earth's axis, and consequently so are the two tidal bulges. Our ship completes a rotation with the Earth every 24 hours but during that time the Moon has moved slightly forward in its rotation of the Earth and the ship must rotate for an extra 50 minutes to catch up with the Moon and its tidal bulges. Consequently there are two high tides in every 24 hours 50 minutes (Figure 19.10).

The magnitude of a high tide varies over a lunar month as the Sun and the Moon line up to either amplify or dampen the gravitational attraction. When the Moon and Sun are in line, their combined gravitational attraction gives a higher tide, known as a spring tide, but when they are at right angles, the bulge under the Moon is reduced by a slight rise under the Sun (Figure 19.10). The alignment between the Sun and the Moon and the distance between the Sun and the Earth varies over a year as the Earth rotates around the Sun and consequently the relative magnitude of spring tides also varies. At the equinoxes (21 September and 21 March), when the Sun is overhead at the equator, the Sun–Moon–Earth alignment is almost straight, giving a maximum tidal rise. At the solstices (21 June and 21 December), this alignment is less perfect and the tidal rise is less. In addition, the Earth is closer to the Sun during the northern-hemisphere summer, such that winter tides in the northern hemisphere tend to be higher. The autumnal equinox is, in fact, associated with the highest tides of the year (Figure 19.10).

A tidal bulge is a wave and as such has the properties of a wave. Although its wavelength is such that it is unable to progress without astronomical forcing, a tidal bulge, like sea waves, will shoal in shallow water. Consequently, wave height increases as the wave slows in shallow water, explaining why tidal ranges increase onshore and are often higher than those in adjacent ocean areas. For example, ocean tides are rarely greater than 50 cm or so, yet in shallow water they may increase to 2.5 m, and in more enclosed regions may be amplified to over 15 m.

In some oceans and seas, high tides appear to rotate, clockwise in the northern hemisphere and anticlockwise in the southern hemisphere, around points of no tidal range (amphidromic points). This results from the interaction of a tidal wave with the margins of ocean or sea basins. This reflected wave may interact with the next tidal wave. At points at which the wave crest is superimposed on a trough the two waves will cancel each other out, producing a line along which no change in water elevation occurs. Where wave crest meets wave crest, the two waves will be amplified and the water surface will be elevated even further. The interaction of the two waves creates a standing tidal wave, in which the water surface is seen to simply rise and fall either side of the nodal line.

In practice, due to the Earth's rotation and the Coriolis force created thereby, this nodal line is reduced to a point, the amphidromic point (Figures 19.11 and 19.12). Effectively, the standing wave crest is swirled to the right in the northern hemisphere and to the left in the southern hemisphere by the Earth's rotation. This results in the wave crest being rotated around the shores of the ocean basin – anticlockwise in the northern hemisphere – while the centre of the basin remains at a constant level. The tidal range in such a system increases from the amphidromic point, where it is zero, outwards. Four examples of tidal variability from monthly tidal records are illustrated in Figure 19.13.

Finally, we must explain how the tidal range in some estuaries or bays is amplified. Ocean tides, even in shallow water, rarely exceed 2.5 m, yet in the Bristol Channel the tidal range exceeds 12 m, while in the Bay of Fundy it is over 15 m. This amplification is a result of tidal resonance: the rhythmic beat of ocean tides at the mouth of a bay, estuary or enclosed sea may be amplified by resonance. For example, if an ocean tide at the bay mouth generates a tidal wave in the bay which progresses across the bay before being reflected seaward, and this reflected wave happens to combine with a new tidal wave at the bay, more amplification occurs and the tidal range within the bay may increase. Consequently, the interaction of ocean tide waves with coastal topography may lead to either the amplification or the dampening of a coastal tidal system.

In summary, tides vary both geographically and temporally, as a result of changes in the gravitational forcing over a lunar month or solar year and due to the interaction of tidal waves with coastal topography. One of the key influences of tides is on the vertical distribution of wave action, as the greater the tidal range the less time the wave acts on any one level of the coast. In areas of large tidal range, currents dominate the geomorphological system. Wind waves still occur in these areas but they only have a minor effect on the coastal landforms since they are continually being shifted up and down by the tide. The concept can be embodied in that of tidal duration, which is the length of time that any one point is at still-water level, where wind waves can be effective. A more detailed account of tides can be found in Chapter 3 of Masselink and Hughes (2003).

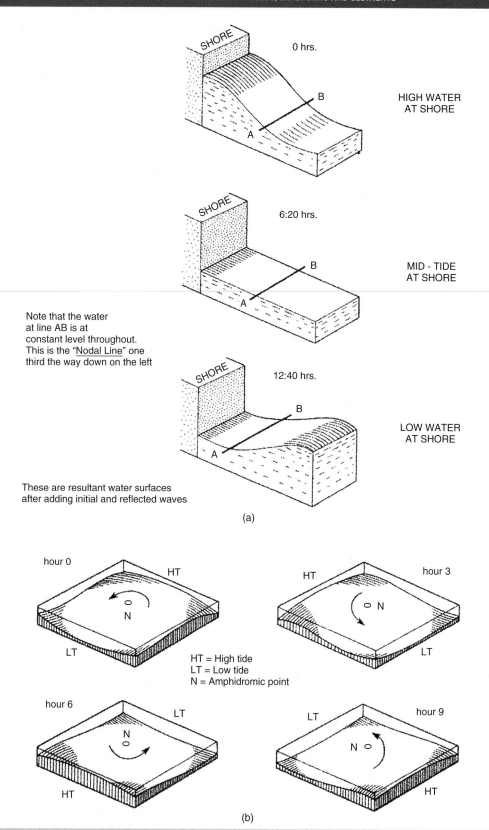

(a)

Note that the water at line AB is at constant level throughout. This is the "Nodal Line" one third the way down on the left

These are resultant water surfaces after adding initial and reflected waves

HT = High tide
LT = Low tide
N = Amphidromic point

(b)

Figure 19.11 Development of a rotating tidal system. (a) High and low water at shore and the concept of the nodal line. Nodal line AB experiences no change in water level as the water surface goes up and down at high and low tides at the shore (known as a standing wave). (b) High and low tides and the amphidromic point. When the Earth rotates the Coriolis force swirls the standing wave crest to the right in the northern hemisphere. This results in the wave crest rotating around the shores of the sea while the centre of the sea remains at a constant level. The central point is the node and is known as the amphidromic point, while the tidal system produced is known as a rotating tide

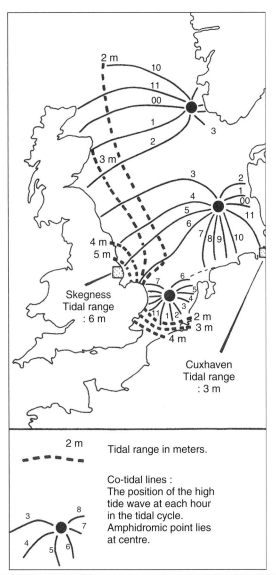

Figure 19.12 Tidal systems in the North Sea. Three amphidromic systems are shown, the tidal range at the coast being partly due to the distance from the central amphidromic point

Labels within figure:
- 2 m
- Skegness Tidal range : 6 m
- Cuxhaven Tidal range : 3 m
- 2 m Tidal range in meters.
- Co-tidal lines : The position of the high tide wave at each hour in the tidal cycle. Amphidromic point lies at centre.

19.4 Tsunamis

Tsunamis are long-period (typically with wavelengths of several hundred kilometres) high-velocity (up to 800 km per hour) waves that are less than 2 m in height, produced by a disturbance of the seafloor by an underwater landslide, volcanic eruption, earthquake or meteor impact. After they have travelled quickly over the oceans they may shoal and break on the surrounding continental shelves and coasts, with the result that they slow down, become shorter in wavelength and achieve heights of over 30 m. They are extremely destructive, life-threatening and can catastrophically reshape coastlines by eroding beaches and flooding coasts. Their after effects can include coastal erosion or deposition that would not otherwise have occurred. They are most

common in the Pacific Ocean, where they are associated with the zones of crustal instability. A well-documented example from coastal Sumatra is described in Box 19.1 (see in the companion website), along with its effects throughout the western Pacific and Indian Oceans.

It has been argued that tsunamis can throw large boulders up onto cliffs, as seen on the coast of south-east Australia (Bryant *et al.*, 1996), and it is likely that an extremely large tsunami occurred in the south-west Pacific around 105 000 years ago as a result of a massive submarine landslide in Hawaii, producing erosional and depositional effects on bordering coasts. Whelan and Kelletat (2003) reviewed mega-tsunamis generated by submarine slumps from volcanic islands throughout the Quaternary and a special issue of *Marine Geology* (volume 203) was dedicated to submarine slump-generated tsunami (cf. McAdoo and Watts, 2004; McMurtry *et al.*, 2004). There has also been discussion as to the correct identification of tsunami deposits (as opposed to storm-surge deposits) in coastal sequences (cf. Williams *et al.*, 2005; Switzer *et al.*, 2005).

19.5 Coastal landsystems

The coastal zone is an energy buffer between the energy-rich ocean and the basin margins. Coastal morphology evolves in response to applied energy and each landform, whether it be an eroding cliff, beach, dune or salt marsh, develops within a specific energy niche located along an energy gradient, which may run either parallel with or normal to the shore. These energy gradients may be the result of changes in variables such as water depth, which concentrates wave energy at certain locations, while creating lower coastal strips elsewhere. Energy gradients change through time as a result of coastal erosion, sedimentation and changes in wave climate or sea level. For example, the sediment availability within a coastal cell may change the energy gradient. A cell with a negative sediment budget may, for example, be associated with a loss of beach volume and thus an increase in wave energy, which will lead to accelerated coastal erosion. Equally, as sea level rises, energy regimes may increase, leading to a shoreward or lateral movement of a particular environment or landform component, given sufficient space and time.

On geological time scales, changes in sea level and the energy gradients within them are recorded by transgressive and regressive sedimentary facies patterns, which show onshore or offshore migration of beaches, lagoons, marshes or deeper water, in which each environment represents a specific energy niche. The adjustment of coastal morphology to changes in energy regime can also be seen at smaller scales in the adjustment of individual landforms. For example, it will be seen below how beach morphology is closely related to the incident energy regime, with flatter beaches occurring under higher energy regimes.

It is important to emphasize that, although coastal environments like salt marshes or mud flats form at the lower end of the energy continuum, they are not necessarily just quiet backwaters and can often be associated with energetic tidal currents and rapid changes in water volume. Essentially, these environments form

IMMINGHAM : semidiurnal form

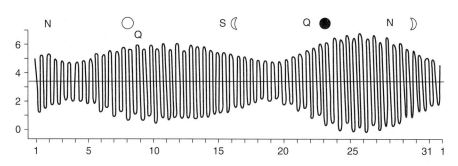

SAN FRANCISCO : mixed, predominantly
semidiurnal form

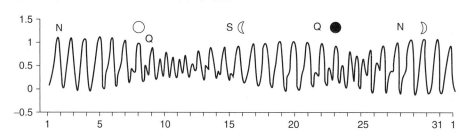

metres

MANILA : mixed, predominantly
diurnal form

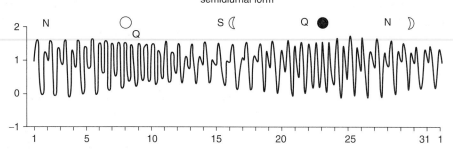

DO – SON : diurnal form

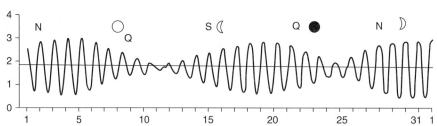

Figure 19.13 Examples of tidal variability from monthly tidal records: (a) Immingham, eastern England, is a semi-diurnal tide, F = 0.00–0.25; b) San Francisco, California, is a mixed, predominantly semi-diurnal tide, F = 0.25–1.50; (c) Manila, Philippines, is a mixed, predominantly diurnal tide, F = 1.5–3.0; (d) Do-San, Vietnam, is a diurnal tide, F = >3.0; where F is the ratio of the sum of the amplitudes of the two main diurnal components (k and O_1) to the sum of the amplitudes of the two main semidiurnal components (M_2 and S_2). A high value of F (e.g. over 3.0) implies a diurnal tidal cycle. Fluctuations in the tidal range are largely due to the Moon's declination. Low values of F (e.g. under 0.25) imply a semidiurnal tide and the main fluctuations in tidal range are due to the relative positions of the Sun and the Moon, giving a spring and neap variation

in low external energy niches but are often associated with a high internal energy regime. Understanding medium- and long-term coastal evolution is about understanding shifts in the energy gradients present and the rate at which coastal environments can migrate and adapt to these shifts.

In order to structure a discussion of coastal morphology, four landsystems are recognized. These are not necessarily mutually exclusive and each contains a subset of landforms and environments which are often located in response to internal and external energy gradients. The four landsystems are: (1) wave-dominated assemblages, (2) tide-dominated assemblages, (3) biogenic assemblages and (4) river-dominated assemblages.

19.5.1 Wave-dominated coastal landform assemblages

Coastal slopes, cliffs and shore platforms

Steep coastal slopes, cliffs and associated shore platforms dominate the erosional, high-energy environments within the wave-dominated landform assemblage. The morphology of a coastal slope is a function of the rate at which subaerial processes, usually mass movements, deliver sediment to the foot of the slope versus the rate at which this sediment is removed by marine transport (Figure 19.17). If the supply of sediment exceeds the rate at which it can be removed by the sea then the angle of the slope will decline and may ultimately become stable. Conversely, if the rate

of marine transport exceeds the rate of delivery, such that the sea can actually erode the slope foot, accelerating collapse, a steep cliff will remain. The emphasis here is on the interplay between subaerial slope processes and marine transport and erosion at the foot of the slope, and it is often difficult to tease apart the respective roles of subaerial processes from the action of the sea.

The relative importance of marine versus subaerial processes may help explain the global distribution of cliffed coasts, which, despite numerous exceptions, tend to be more common within temperate latitudes. Pethick (2001) suggests that the form of a coastal slope is linked to latitude. The low wave energy of humid tropical coasts combines with low sediment yields from subaerial processes, despite intense chemical weathering, since vegetation extends down to the high tide level and coastal slopes are characterized by relatively low slopes. High-latitude coasts are dominated by subaerial processes which supply large amounts of rock debris and talus to the foot of the coastal slope but, often due to the presence of sea ice, marine removal is limited. However, in temperate regions, steep coastal cliffs are common, since here rapid debris removal, promoted by the high wave energy, prevents the build-up of slope foot debris, while, at the same time, active cliff erosion promotes cliff failure, producing a steep coastal slope.

There are several morphological elements present on steep coasts, in which marine erosion and transport exceed the rate of subaerial debris supply. These are: (1) cliffs, (2) shore platforms and (3) caves, arches and stacks.

Cliffs come in all shapes and sizes, from a few metres to in excess of 500 metres. Some are vertical, while others are not. In

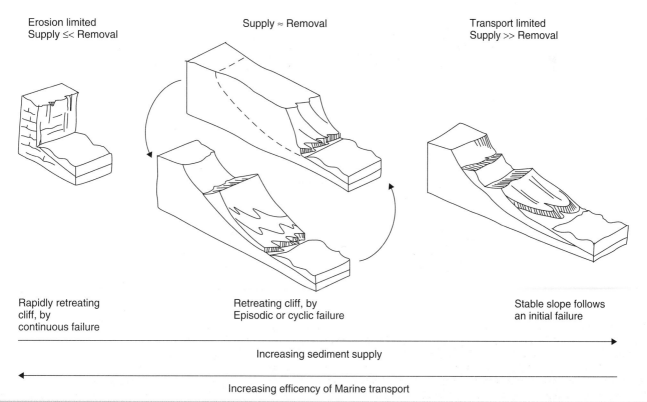

Erosion limited
Supply ≤< Removal

Supply ≈ Removal

Transport limited
Supply >> Removal

Rapidly retreating cliff, by continuous failure

Retreating cliff, by Episodic or cyclic failure

Stable slope follows an initial failure

Increasing sediment supply

Increasing efficency of Marine transport

Figure 19.17 Morphology of coastal slope, sediment supply and efficiency of marine transport

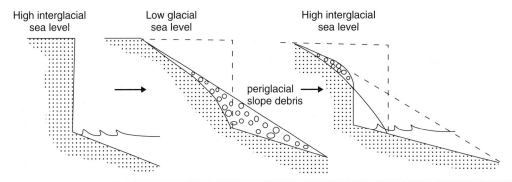

Figure 19.18 Development of a bevelled cliff. (Source: after Trenhaile, 1987)

fact, a precise definition is difficult, but any steep coastal slope is usually referred to as a cliff. They occur along approximately 80% of the world's coastlines. The detailed morphology of a cliff (height, angle and the number and frequency of slabs, buttresses, arêtes and gullies) is largely a function of the characteristics of the rock mass, such as the orientation, spacing and frequency of rock discontinuities. The form may also be modified by the nature of the processes and on its antecedence. For example, common around the south-west coast of Britain are composite or bevelled cliffs, in which a convex slope of soil and weathered regolith sits above a steep rock-cut vertical or free face.

Their formation involved three stages (Figure 19.18). A vertical cliff formed due to marine processes during the last high interglacial sea level. The convex slope was part of a coastal slope formed under periglacial conditions during the last glacial, which graded to a much lower sea level than is present today, abandoning the cliff. Periglacial processes progressively degraded this abandoned cliff and covered it with geliflucted debris. As sea levels rose during the current interglacial, due to the melting ice, erosion by wave attack trimmed this slope, removing the debris to produce the lower cliff, leaving the regolith-covered remnant above.

More complicated explanations are necessary for multi-storeyed cliffs, which require several sea-level changes for their formation, but they can also be explained by alternating lithologies of different erosional resistance. Consequently, the morphology and cliff profile is an integration of different coastal processes, active at different times in different climatic conditions, lithologies and structures, and sea-level histories.

Cliff erosion and retreat leaves behind the stump of the old cliff or slope, marking the lowest level to which the erosion reached, and in this way a shore platform is formed. Shore platforms have a diverse range of morphological forms and vary from region to region. Their morphology is heavily influenced by the orientation of bedding, or by similar discontinuities within the rock mass (Figures 19.19 and 19.20). The formation of these platforms has long been debated and the relative importance of marine versus subaerial processes has been hotly discussed.

Caves, arches and stacks are common on some steep rocky cliffs and form spectacular geomorphological forms (Figure 19.19). Caves form as a result of marine erosion exploiting a line of weakness within an otherwise competent rock mass. If the rock

mass were not competent, the cave would quickly collapse to form a steep-sided inlet. The evolution of stacks and arches has often been examined using old photographs and historical records, for example Shepard and Kuhn (1983) used a collection of postcards and photographs to examine the coastal arches around La Jolla in California. However, although spectacular, these features probably owe more to the characteristics of the rock mass than they do to marine processes.

In reviewing processes of coastal erosion, there are two important variables: one is the erodibility of the cliff, in terms of both its propensity to mass failure and its susceptibility to marine erosion, and the other is the transport and erosional potential of the sea itself. The controlling factors for coastal erosion are summarized in Figure 19.21. The erodibility of a cliff can be expressed in terms of its rock mass strength. Cliffs with a low rock mass strength, for example, being composed of weathered, well-jointed, out-dipping strata, are much more likely to fail than those with a higher rock mass strength, which are composed of more competent beds that dip into the cliff.

Many studies have shown how coastal geometry is often controlled by the location of master joints, faults or other discontinuities, which act as lines of weakness along which marine processes can penetrate. This is particularly true for coastal cliffs with caves, arches or stacks (Figure 19.19). Here you need rock that has a high rock mass strength, and can therefore maintain a steep face or roof, but is cut by zones of weaker rock, perhaps along faults or due to the intersection and concentration of joints. In some extreme cases, the coastal cliffs may form along fault scarps and owe more to fault movement than coastal erosion. For example, many of the high, steep cliffs on the arid coast of Pakistan are attributable to fault scarps, whilst on the Isle of Skye, faulted coasts in fairly sheltered water have only been very slightly modified by waves. There are several similar examples from North America, such as the north-eastern coast of San Clemente Island, California.

Lithology and rock mass character are equally important to the type of mass movement that operates on a coastal slope. For example, large rotational slumps occurring predominantly on fine-grained clays such as the London Clay which outcrops on the Essex and Kent coasts of the UK, while rockfall dominates the 600 m-high black lava cliffs of Tenerife. Lithology and the pattern of outcrop may also influence the erodibility of a cliff base

(a) Sloping shore platform

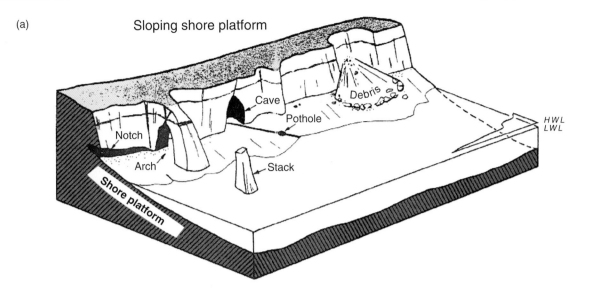

(b) Sub-horizontal shore platform

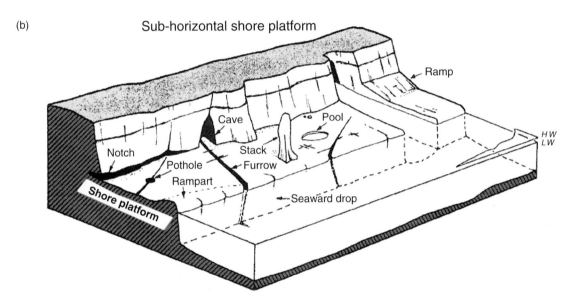

(c) Plunging cliff

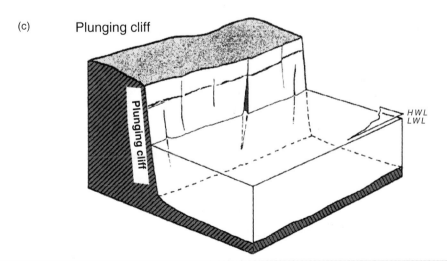

Figure 19.19 Major morphologies of rocky coasts, with the characteristic erosional landforms. (Source: after Sunamara, 1992)

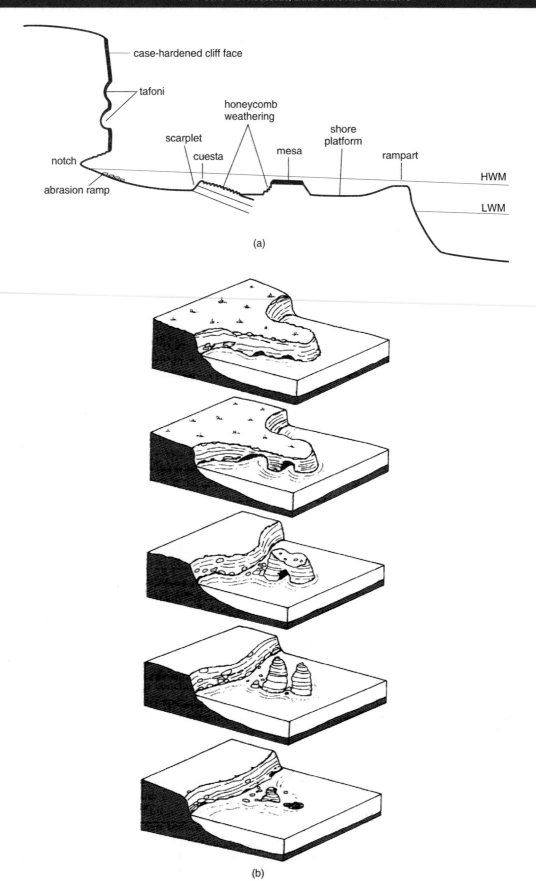

Figure 19.20 Coastal erosion landforms: (a) cliff and shore platform landforms; (b) erosion of a rocky headland to form caves and natural arches, which collapse to form stacks, which in turn are reduced by erosion to form stumps

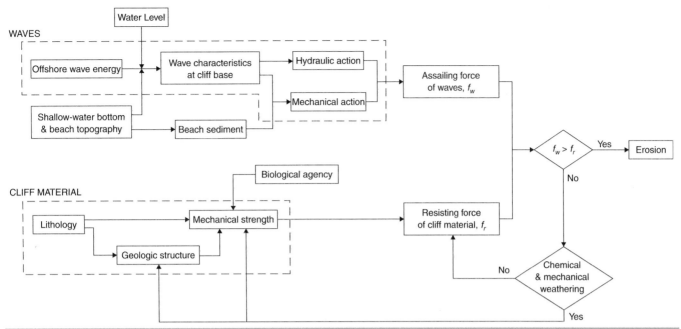

Figure 19.21 Factors affecting wave-induced cliff erosion. The controlling factors are the assailing force of waves and the resisting force of cliff material. Erosion is determined by the relative intensity of the two. (Source: after Sunamara, 1992)

by marine processes. For example, limestone outcrops favour solution and facilitate erosion by intertidal organisms. Similarly, highly-fractured rock contains numerous discontinuities that can be exploited by the sea and broken down into smaller blocks more susceptible to transport.

The other element here is the transport and erosional potential of the sea. On a global scale this is controlled by the distribution of large waves. The highest waves tend to be concentrated on temperate coasts, since high-latitude coasts are sheltered by sea ice and the wind regime of the tropics does not necessarily favour large waves. On a more regional scale, variations in coastal exposure, due to changes in fetch, for example, or the sheltering effects of offshore islands, may cause longshore energy gradients, while offshore topography controls wave refraction during shoaling and therefore the concentration or dispersal of wave energy along a coast. On a local scale, the importance of the presence of surge channels in the formation of caves has been emphasized. These surge channels are topographic depressions which funnel and concentrate waves on specific sections of headland.

Tides, and in particular tidal range or duration, have a significant role to play. The tidal range determines the vertical range over which wave action can operate. In a microtidal regime, with little or no tidal variation, wave action is concentrated at one zone on the coastal slope; in such regimes the duration of wave attack at any one level on a coastal slope is limited. This concept has been used to explain some of the morphological diversity present within shore platforms. A strong correlation has been found between shore platforms with a low shore-normal gradient and areas with low tidal ranges, whilst a high tidal range is associated with steep platform gradients. Although strong, this gradient is complicated by geological factors, which also strongly influence

platform morphology (Figure 19.23). As the erosion is generally imperceptible in human lifetimes it is difficult to evaluate the relative and absolute contributions to it and determine whether the development of shore platforms is mainly the result of marine or weathering processes. This is reviewed in Box 19.2 (see in the companion website).

The relative importance of these two variables, erodibility versus wave energy, is difficult to assess and has been the subject of extensive debate. It no doubt varies from one location to the next. However, the work of Benumof et al. (2000) on coastal sea cliff erosion rates in San Diego County, California, is instructive (Figure 19.22 and Table 19.2). Serial air photographs were used to determine the rate of coastal erosion between 1932 and 1994 along a section of the San Diego coast. The rock mass strength of the cliffs at a range of sites was also determined from field observation, while the average wave energy present was calculated from offshore wave records. The researchers found a strong negative correlation between the rate of erosion and rock mass strength: weak cliffs erode more quickly. However, they also found a negative correlation between wave energy and erosion, which counter-intuitively implies that erosion rates increase with a fall in the wave energy regime (Figure 19.22). Benumof et al. (2000) conclude that the primary control on coastal erosion is rock mass strength and that in this case the distribution of wave energy is not critical. These conclusions are broadly supported by the distribution of accelerated coastal erosion within England, which follows the outcrop of soft engineering soils, clays and sands.

Irrespective of the relative importance of erodibility versus wave action, it has to be remembered that if the sea did not remove the eroded debris, sea cliffs would disappear. Equally, without

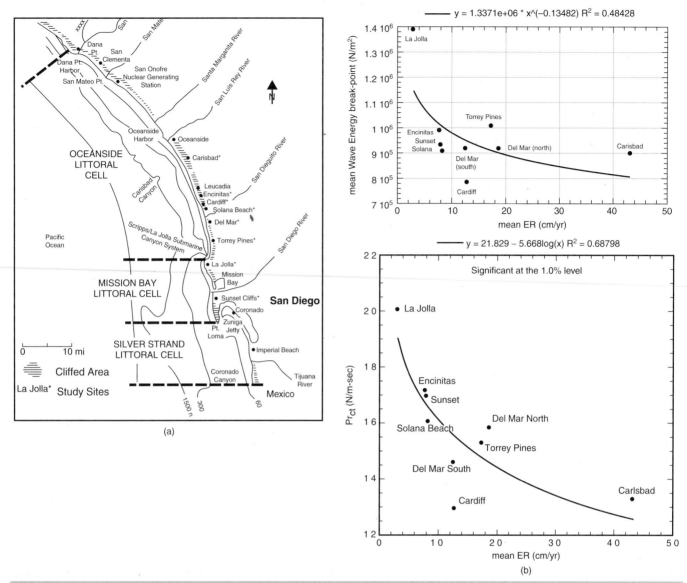

Figure 19.22 Sea cliff erosion rates: (a) locations of study sites and major littoral cells; (b) relationship between sea cliff erosion rates and mean wave energy at the break point (top graph) and relative wave power at the cliff toe (bottom graph). Benumof *et al.* (2000) illustrated similar relationships between sea cliff erosion rates and mean wave power and mean wave energy in 10 m of water

coastal erosion at the base of the cliff, failure would not occur. The processes of coastal erosion are not well-understood however, in part due to the difficulty of making reliable observations in such a hostile environment. Most authors list the following processes of coastal erosion; a summary is given in Table 19.3.

Quarrying

This involves the physical removal of a block or fragment of rock by the sea. This process is dependent on two parameters: first, the presence, spacing and continuity of joints, fractures or other discontinuities within the rock which allow a block to be defined; second, the water shear necessary for block extraction and removal. Fracture propagation may occur by any combination of the following: repetitive injection of water into fractures;

the repetitive compression of air within a fracture during wave breaking; hydrostatic water pressure; percussion caused by erosional unloading, weathering and the impact and vibration caused by repetitive wave breaking. In practice, the key to the process is the presence, spacing and continuity of preexisting fractures, since fracture propagation is likely to be slow and to require an initial discontinuity to exploit. This illustrates the importance of rock lithology to wave erosion. The ease with which a block can be removed will be a function of the fluid shear either within the wave or within the wave run-up post-breaking. Offshore topography is critical here in determining the degree of wave transformation due to shoaling and the position of the break point in respect to the cliff foot, all of which may change over a tidal cycle as water depth varies in front of the cliff.

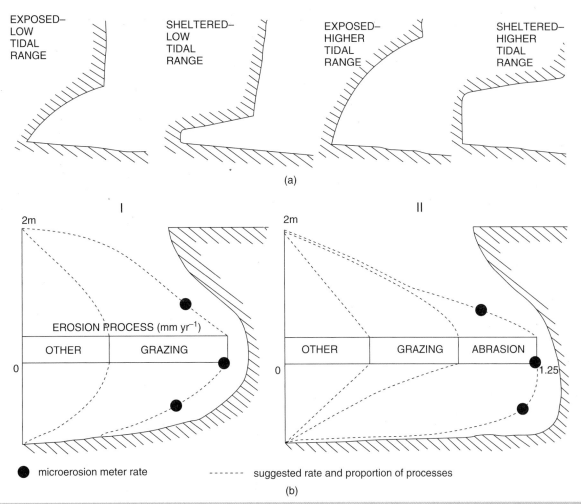

Figure 19.23 Controls on erosional notches: (a) the effect of tidal range and exposure on the form of the notch in Indonesia; (b) erosion rates and processes in notches on Aldabra Atoll, where (i) sand is absent, (ii) sand is present. (Source: after Trudgill, 1976)

Sunamara (1975–1981) made observations pertinent to this discussion within a wave tank with an artificial cement cliff at one end. He found that erosion at the cliff base, in the form of a notch, was greatest with waves that broke against the cliff rather than waves that broke before reaching it. The critical point here is that shoal-wave transformations and therefore cliff-foot topography and associated water depth are key factors in determining the rate of erosion. A cliff that plunges into deep water will simply reflect waves seaward and little erosion will occur. However, a cliff with a narrow beach or rock platform at its base which facilitates waves breaking at the cliff foot will be associated with significant cliff erosion. As a beach or platform grows, or as water depth falls during a tidal cycle, wave breaking will occur further offshore and only broken waves will impact on the cliff foot. Consequently the rate of erosion will fall.

Abrasion

The scouring effect of water charged with sand or shingle will abrade a cliff as it impacts and shears past it. Sunamara (1977) illustrated this point with one of his physical wave-tank models, where he showed the rate of cliff erosion within the wave tank

with time against the size of the beach formed by the eroded debris. Initially erosion is negligible since there is little abrasive sediment within the water. As a small beach builds up as erosion proceeds, the rate of erosion accelerates, since the beach provides a ready source of abrasive sediment for the breaking waves. In time, the beach builds up sufficiently to buffer the cliff from the waves and the rate of erosion declines.

Corrosion

Solution of calcareous rock, such as limestone or chalk, may be a significant erosional process on some cliffs, particularly within warmer climates. It has been proposed that the form of the littoral zone in limestone regions could be classified according to temperature and tidal range, describing four main types of coastal limestone landform (Figure 19.24). In cool temperate regions, such as on the Carboniferous limestone of South Wales, or in southern Ireland, the spray zone is pitted by small corrosion hollows a few millimetres in depth and <5 mm in width.

Lower down the platform, as the frequency of inundation increases, there are solution pools perhaps 10–30 cm deep, while towards the low water mark sharp pinnacles or lapies (marine

Table 19.2 Erosion at San Diego sea cliff sections: (a) generalized litholgic, strength and structural characteristics of each of the nine San Diego County sea cliff sites investigated; (b) locations of wave grid cells and corresponding sea cliff erosion and wave exposure data. CRL, Carlsbad, unlithified sand; ENC, Encinitas, sandstone; CRDF, Cardiff, sandy claystone; SB, Solana Beach, sandstone; DMN, Del Mar Noth, unlithified sand; DMS, Del Mar South, sandy claystone; TP, Torrey Pines, shale; LJ, La Jolla, sandstone and shale; SSC, Sunset Cliffs, sandstone and shale. (Source: after Benumof et al., 2000)

(a)

Parameter	CRL	ENC	CRDF	SB	DMN	DMS	TP	LJ	SSC
Intact rock strength	Very weak	Strong–very strong	Moderate	Strong	Very weak	Weak	Weak–moderate	Very strong	Very strong
Weathering	High	Moderate–slight	Moderate	Moderate–slight	High	Moderate	High	Moderate–slight	Moderate–slight
Spacing of joints (m)	'Infinite'	0.3–3.0	0.05–0.3	0.3–3.0	'Infinite'	0.05–0.3	0.05–0.3	0.3–3.0	0.05–0.3
Joint orientation	Extremely unfavourable, unconsolidated	Steep dips out of slope	Steep dips out of slope	Steep dips out of slope	Extremely unfavourable, unconsolidated	Steep dips out of slope	Steep dips out of slope	Steep dips out of slope	Steep dips out of slope
Width of joints (mm)	Unconsolidated	1.0–5.0	1.0–5.0	1.0–5.0	Uncosolidated	1.0–5.0	5.0–20.0	1.0–5.0	1.0–50
Continuity of joints	Continuous, unconsolidated	Continuous with thin infill	Continuous with thin infill	Continuous with thin infill	Continuous, unconsolidated	Continuous with thin infill	Continuous with thin infill	Few continuous/partly cemented	Continuous with thin–zero infill
Groundwater outflow	Slight	Slight	Moderate	Slight	Slight	Moderate–slight	Moderate	Slight–trace	Slight

(b)

Site	Location[1]	Erosion rate[2] (cm/year)	Stdev erosion rate[3] (cm/year)	Exposure[4] (degrees)
Carlsbad	117°19'23.7818"W 33°06'07.3828"N	43.02	8.23	249
Encinitas	117°18'13.8536"W 33°03'05.9573"N	7.70	2.31	252
Cardiff	117°17'27.2443"W 33°01'22.2876"N	12.69	3.00	247
Del Mar	117°16'25.0891"W 32°57'29.0342"N	18.73 (North) 12.54 (South)	4.84	255
Torrey Pines	117°15'22.9340"W 32°53'29.2850"N	17.36	4.55	265
La Jolla	117°16'40.6347"W 32°51'06.7373"N	3.06	1.50	316
Sunset	117°15'46.2250"W 32°43'13.7348"N	7.88	3.06	260

[1] Location of wave model cell grids in 10 m water;
[2] Mean sea cliff erosion rate;
[3] Standard deviation of sea cliff erosion rates;
[4] Shore-normal coastline exposure to waves

Table 19.3 Processes of erosion on rocky coasts

Process	Description	Conditions conducive to process
Mechanical wave erosion:		
Erosion	Removal of loose material by waves	Energetic wave conditions and microtidal tide range
Abrasion	Scouring of rocks surfaces by wave-induced flow with mixture of water and sediment	'Soft' rocks, energetic wave conditions, a thin layer of sediment and microtidal tide range
Hydraulic action	Wave-induced pressure variations within the rock causes and widens rock capillaries and crack	'Weak' rocks, energetic wave conditions and microtidal tide range
Weathering:		
Physical weathering	Frost action and cycles of wetting/drying causes and widens rock capillaries and cracks	Sedimentary rocks in cool regions
Salt weathering	Volumetric growth of salt crystals in rock capillaries and cracks widens them	Sedimentary rocks in hot and dry regions
Chemical weathering	A number of chemical processes remove rock material. These processes include hydrolysis, oxidation, hydration and solution	Sedimentary rocks in hot and wet regions
Water-layer levelling	Physical, salt and chemical weathering working together along the edges of rock pools	Sedimentary rocks in areas with high evaporation
Bioerosion:		
Biochemical	Chemical weathering by products of metabolism	Limestone in tropical regions
Biophysical	Physical removal of rock by grazing and boring organisms	Limestone in tropical regions
Mass movements:		
Rock falls and toppling	Rocks falling straight down the face of the cliff	Well-jointed rocks, undercutting of cliffs by waves
Slides	Deep-seated failures	Deeply-weathered rocks, undercutting of cliffs by waves
Flows	Flowing of loose material down a slope	Unconsolidated material, undercutting of cliffs by waves

karren) coexist with pools. These lapies have formed by solution of rock by running or standing water. In warm tidal seas, jagged lapies are found further up the platform in the spray zone, while vasques, which are shallow, flat-bottomed pools forming networks organized into a series of terrace-like steps, occur in the upper portions of the intertidal zone and become overdeepened with overhanging sides seawards. The pools are separated from each other by sinuous, narrow-lobed ridges. In warm, tide-free seas, lapies and pools occur in the spray zone, while a notch forms at the high tide level, with an upper lip which may extend seaward by up to 2 m. Corniches are organic protrusions, generally formed of calcareous algae, that grow out from steep rock surfaces in the intertidal zone. Finally, in very warm seas such as in Hawaii, the Red Sea or the Caribbean, the zonation consists of: lapies in the spray or splash zone; a lip and associated notch with a depth of 1–2 m at about the level of high tide; a platform with vasques in the intertidal zone; a low tidal cliff. These types of littoral limestone zonation are illustrated in Figure 19.24. There are many local variations on this theme but the coastal morphology

at any one site is often heavily influenced by bioerosion, which is particularly important on calcareous lithologies. The solution notch is the defining component and its form can be related to tidal range and coastal exposure, as shown in Figure 19.23.

Rock weathering

The importance of weathering within coastal environments has been frequently debated, not least because of the many suitable influences that will help erosion of coasts. Spray, splash and tides subject coastal cliffs and intertidal rock platforms to alternate cycles of wetting and drying and exposure to salt. Frost action and thermal expansion may operate in some climatic regions, but salt weathering is probably the most important process of physical weathering in the coastal zone. Chemical weathering, in particular solution and alternating hydration–desiccation, is also of relevance. Collectively, weathering processes on the coast have been referred to as 'water-layer levelling' and tend to occur in the spray or splash zone, where pools of standing water may accumulate.

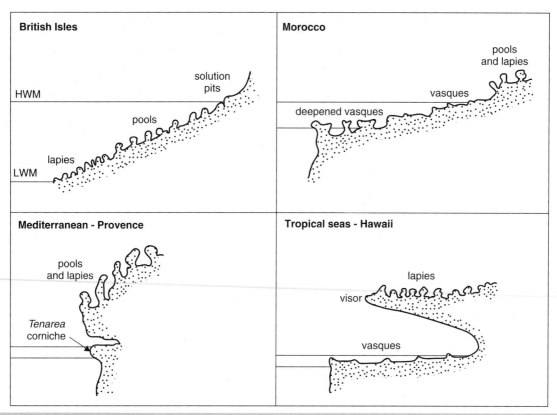

Figure 19.24 Littoral limestone zonation, distinguished on the basis of temperature and tidal regime. (Source: after Guilcher, 1953)

Bioerosion

This involves the abrasion of rock by browsing invertebrates and fish, and the chemical action of secreted fluids from a range of organisms. It is probably of greatest importance in the tropics, where an enormously varied marine biota lives on calcareous substrates. Outside the tropics, the process is less effective due to less-suitable lithology, more vigorous wave action, which hinders biodiversity, and cooler climates, which again tend to limit the biodiversity. Microfauna are particularly important since they provide the habitat and food for high orders. These include algae, fungi, lichens and bacteria, all of which may secrete fluid and thereby erode the substrate. Microflora is grazed by a range of vertebrate and invertebrate fauna. The process of grazing can, in some cases, cause erosion as animals mechanically rasp the rock surface, often already chemically weakened by fluids secreted from the microfauna while grazing the flora present. On Aldabra Atoll, for example, it was calculated that grazing organisms account for about 60% of the surface erosion within the intertidal zone when sand is absent (that is, where there is limited abrasion). In addition, many of these marine invertebrates live within borings, the excavation of which by both chemical and mechanical means causes substantial amounts of erosion.

The main factor that controls cliff erosion is the resistance to marine erosion processes, and this is largely controlled by lithology. Sunamura (1992) correlated cliff erosion rates with lithology as follows: <0.001 m per year for granitic rocks; 0.001–0.01 m per year for limestone; 0.01–0.1 m per year for shale and flysch;

0.1–10 m per year for Quaternary sediments; >10 m per year for volcanic pyroclastics. However, despite these generalizations, time-averaged and space-averaged rates of erosion are extremely variable and can be measured by a range of techniques, such as comparing maps, old photos and aerial photos, direct measurements of the exposure of steel pegs or nails driven into the cliff, and micro-erosion meters.

Beaches

To many people, the word 'beach' conjures up pictures of the Sun, tanned bodies and crashing waves, but beaches occur in most climates and are not always blessed by sunshine. A more precise definition is: a wave-deposited, three-dimensional body of cohesionless and unconsolidated sediment (sand or gravel) along a coast extending above the lowest water mark (spring low tide), or a line of breakers at low tide to cliff or dune. It is the upper part of the shoreface and has a concave-upwards profile. The key to this definition is the word 'cohesionless', which implies that the sediment is free to move and be reshaped by the incident waves if the current velocities generated by them exceed the transport thresholds of the grain sizes present on the beach.

Beaches can be composed of a range of particles sizes, from fine sand to cobble, or even boulder gravel, and all beaches have one thing in common: an ability to dissipate wave energy. This can be viewed conceptually in terms of the availability of excess energy for sediment transport. Sediment is moved on a beach by waves and the currents they generate in such a way as to reduce the excess energy present. If excess energy is high, perhaps during a storm event, sediment is transported offshore and the beach profile flattened, which increases the effective energy dissipation on the beach and therefore the excess energy present. In this way, a beach's morphology is brought in line with the prevailing energy conditions. Conversely, if energy is dissipated too effectively then there is little excess energy for sediment transport and deposition occurs, which tends to result in a net onshore movement of sediment and a steeper beach profile: energy dissipation is reduced. The morphology of a beach is therefore a response to the prevailing energy conditions, which change from day to day as storms pass. In fact, beaches are one of the most effective natural buffers for wave energy.

In this section beach form will be examined as a response to changes in energy input in two dimensions, shore-normal and longshore, before these elements are combined to form a three-dimensional model of beach form.

One of the most fundamental questions about shore-normal beach profiles is: Why are some beaches characterized by steep shore-normal profiles when others have flat profiles, and why, over time, can a beach vary between these two extremes? Some authors have referred to these two extreme profile forms as summer or swell and winter or storm profiles, with steeper profiles generally being associated with quieter wave conditions (Figure 19.25(a) and (b)). In other words, these profiles are seasonal. In practice these terms are misleading, since storms occur both in summer and winter, the beach response is cyclic as opposed to seasonal and several sequences of barred and nonbarred profiles can occur in any year depending on the wave climate storminess. The problem is further complicated by the fact that beach gradient tends to increase with particle grain size and sorting of beach material, so that beaches composed of well-sorted shingle or gravel tend to have steeper profiles than ones composed of sand (Figure 19.25(c)). However, even a steep shingle beach will tend to oscillate between a flat and a steep profile over time.

Steep beach profiles usually possess a marked landward ridge or bar, known as the berm (Figures 19.25(a) and 19.26), which forms at the swash limit and historically has sometimes been referred to as a swash-bar. Steep beaches may also possess a pronounced break of slope at the position of the breaking waves, sometimes referred to as the beach step. In contrast, flatter beaches have a longshore bar, just below low tide level, shoreward of which there is often a wide, flat low-tide terrace, which may contain a series of shore-parallel ridges and runnels (Figure 19.26), as on the Sefton coast (south-west Lancashire).

The transition between a steep and a flat beach is associated with the erosion of the berm and its seaward transport to be redeposited as a longshore bar. Longshore bars have a trough on the shoreward side and may extend alongshore for considerable distances, in some cases up to 50 km. In plan form, these bars may take on a variety of crescentic and rhythmic forms, and multiple bars may form in some locations (Figures 19.27 and 19.28). Longshore bars appear to form in many different ways. In some cases they are a function of breaking waves and consequently form at the break point, and multiple bars may result from multiple break points, associated with a range of incident wave heights, or due to tidal variations. However, some longshore bars form seaward of the break point even at low tide, and an alternative mechanism is required to explain this. One consequence of the formation of a longshore bar is to increase wave reflection and it has been suggested that the interaction of reflected and incident waves may set up standing waves in a shore-normal direction. This, in turn, tends to result in stationary sediment transport cells, which could lead to the formation of multiple bars.

Consequently, once one bar forms in-shore, by whatever mechanism, wave reflection may create further bars. Longshore bars are similar to ridges and runnels, which have been described from a range of beaches with a high tidal range (e.g. Blackpool or Dieppe). They tend to be smaller and lower in profile and are cut by frequent channels, which drain water from the runnels. There is some evidence that inner bars migrate onshore, thereby facilitating a steepening of the beach face, and onshore bar migration may be one process by which beach recovery occurs. In other cases, the bars are simply eroded and the sediment moved onshore.

Cycles of variation

There have been numerous studies of serial beach profiles, with repetitive surveys of the same profile at successive intervals taken in order to examine profile changes. This type of analysis has been facilitated by the application of eigenfunctions, which allow the principal components of change in the beach profiles to be identified. An example can be seen in Figure 19.29, with a two-year series of beach profiles collected from Torrey Pines Beach in California, where it was found that the profile variation could be accounted for by three eigenfunctions. The largest effectively represents the mean profile. The second eigenfunction records oscillation between a berm and a longshore bar. It has a maximum at the location of the summer berm and a minimum at the location of the water bar. This function shows a distinct seasonality and records the oscillation in the profile between a winter storm profile and a summer swell profile. The third eigenfunction corresponds to profile variations caused by the development and destruction of a low tide terrace between the beach face and longshore bar. This function does not show any distinct seasonality and its presence is therefore due to other factors.

This type of analysis clearly records how profile oscillation occurs between steep and flat beaches associated with the shift in the location of the berm bar. It picks out not only seasonal changes but also longer-term cycles of change such as those associated with the longshore movement of sediment by waves, which might change beach volumes for a while. There are also the longer-term cycles of erosion and recovery, whereby, for example, a hurricane may cause severe beach erosion and a slow recovery over several years, superimposed on which may be annual oscillation. On much longer time scales, sea-level change or climate change may be recorded in profile changes, but in practice beach profile records do not exceed about 30 years, which is too short to pick out some of these trends. From Figure 19.25(e) it can be seen that the cycles of beach change can vary from tidal to decadal cycles.

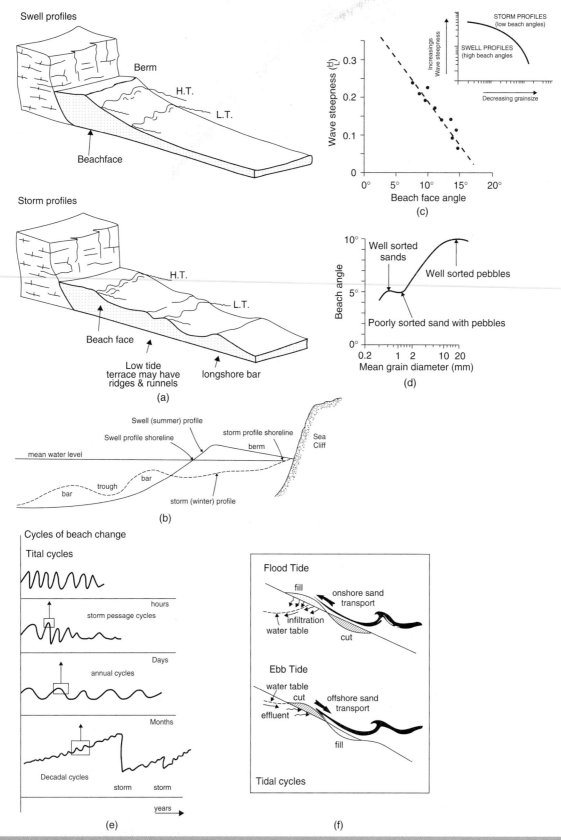

Figure 19.25 Beach changes in swell and storm conditions: (a) swell and storm profiles; (b) swell and storm profiles: cross-section; (c) relationship between wave steepness, grain size and beach-face angle; (d) beach-face angle depends on sediment sorting as well as mean grain size. The poorly sorted sediment in the centre of the graph is associated with low beach angles, probably due to reduced percolation rates; (e) cycles of beach change from tidal to decadal; (f) tidal cycles of change

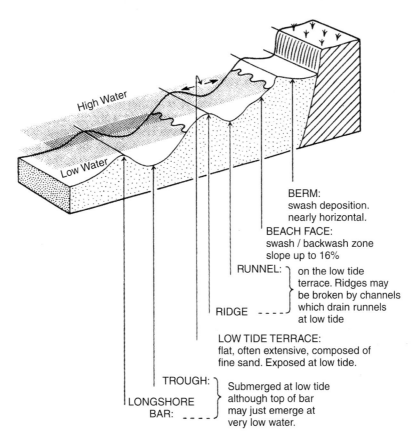

BERM:
swash deposition.
nearly horizontal.

BEACH FACE:
swash / backwash zone
slope up to 16%

RUNNEL: ⎤ on the low tide
            │ terrace. Ridges may
            │ be broken by channels
            │ which drain runnels
RIDGE ---- ⎦ at low tide

LOW TIDE TERRACE:
flat, often extensive, composed of
fine sand. Exposed at low tide.

TROUGH: ⎤ Submerged at low tide
          │ although top of bar
LONGSHORE │ may just emerge at
BAR: ---- ⎦ very low water.

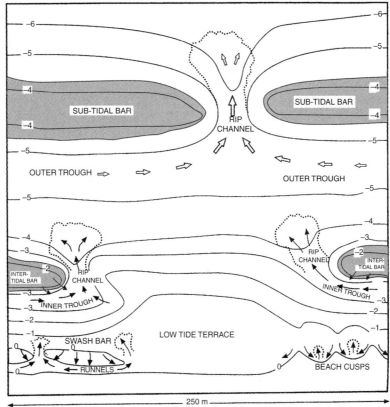

Figure 19.26 Dominant morphology and terms on a beach: (a) cross-sectional beach profile; (b) contour plot of beach morphology

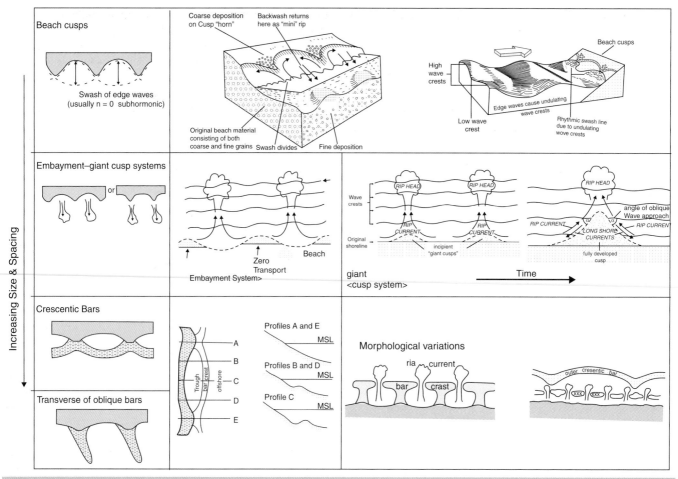

Figure 19.27 Hierarchy of rhythmic beach landforms

Explaining this variation in beach profile is not straightforward. Why does net offshore transport occur under high-energy conditions while net onshore transport occurs under calmer conditions? Early wave-tank studies found strong relationships between the angle of the beach face and wave steepness (Figure 19.25(c)), although the threshold wave steepness varies with the mean grain size of the beach. But why should steep, high-energy waves lead to offshore transport?

Conceptually, this is probably a function of wave asymmetry. As we saw earlier, the asymmetry of offshore and onshore currents increases as wave shoaling occurs. Waves that break far onshore have strongly asymmetrical currents. Under low wave conditions, wave breaking tends to occur via surging or collapsing waves far onshore and current asymmetry is well developed, but the threshold for sediment transport is only exceeded by the high velocity, though short-duration, onshore currents. Consequently, the net direction of sediment movement is onshore. This steepens the beach, allowing waves to break even further onshore, thereby exacerbating the current asymmetry. If wave height and energy now increase the velocity of the shore-normal current will also increase, such that both the onshore and the offshore components may exceed the transport threshold for the grain sizes present on the beach. In this case, the longer duration of the offshore

component relative to the onshore current will lead to net offshore transport. This will flatten the beach profile so that breaking takes place further offshore and current asymmetry declines, which ultimately limits the degree of offshore movement. There are problems with this model but it does provide a conceptual solution which emphasizes the interplay of wave asymmetry, beach slope and the position of the break point.

Beaches, ridges, spits, cuspate forelands, barriers and cheniers

If we now turn our attention to the longshore morphology of a beach, we can recognize several elements: (1) rhythmic topography, (2) shoreline beach and (3) detached beach. Rhythmic beach forms are landforms that occur repetitively along a beach and have been described over a range of scales, from a few metres to several kilometres. A hierarchy of forms can be recognized (Figure 19.27), which can be listed with increasing size as: (1) beach cusps, (2) rip-current embayments or cuspate shorelines, (3) crescentic bars and (4) transverse and oblique bars. At the smallest scale are beach cusps, which are horn-shaped cusps of coarser sediment separated by small bays of finer sediment between 1 and 60 m wide (Figure 19.30). Wave run-up divides both sides of the horn to form two currents, one either side, which

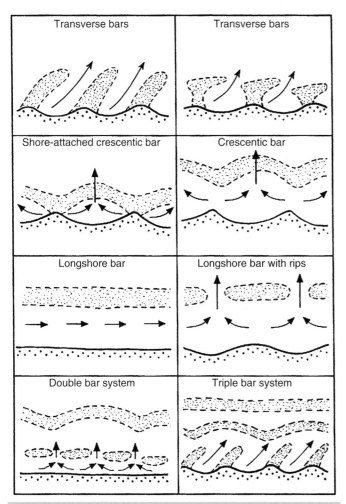

Figure 19.28 Different types of near-shore bar morphology

to slow and repel swash currents. Over time, feedbacks produce, at least in theory, a regular set of beach cusps.

Theoretical analysis of this self-organizing swash system suggests that cusp spacing is determined by swash length, and Coco *et al.* (2000) found a strong correlation between spacing and swash length. They used computer modelling to simulate a situation where an initial planar beach with small random perturbations is subject to swash action. Some of these perturbations became enhanced while others were suppressed. Ultimately the feedback processes between morphology and swash hydrodynamics resulted in the formation of beach cusp morphology, which is a self-organizing feature.

In summary, both processes are able to explain cusp spacing and could therefore lead to cusp formation. Coco *et al.* (2004) examined the role of tides in beach cusp development and found that cusp dimensions were tidally modulated. It is possible that cusps form by more than one process, providing another example of the concept of equifinality.

The next level in our hierarchy of rhythmic forms is embayments generated by rip-currents. These features are large cuspate forms associated with rip-currents and essentially they represent rhythmic variations in beach width. The regular beach forms occur in one of two locations, either between rip-currents or shoreward of them. It has been suggested that seaward-flowing rip-currents would tend to erode a channel and that points of zero transport should occur between each rip-current. Over time this should lead to a crenulated or cuspate beach form, and indeed this scenario is widely observed. However, in wave-tank experiments it was found that giant cusps formed behind rip-currents and that embayments tended to form between each rip, which was the exact opposite of the observations of earlier workers. The tank observations suggest that the forms may develop until they are sufficiently large to induce a reverse longshore current due to the oblique angle of incidence. Field examples of both types have been observed and their impact on cliff erosion noted.

Longshore bars may have a crescentic or a lunate form. Crescentic bars vary in length from 100 to 2000 m but are typically only 200–300 m long. They occur singly or as multiples and are commonly dissected by rip-currents. The key aspect of these bars is their regularity and even spacing, and consequently standing-edge waves have been proposed as their most likely cause. It has been suggested that edge waves cause a net sediment movement alongshore and that this transport is minimal at wave nodes. The bars are attached to the beach at this point.

Transverse sand bars and oblique bars also occur with regular spacing along beaches. A variety of forms exist and a range of explanations have been proposed. One of the more plausible is that under oblique waves, near-shore circulation cells may become realigned and associated crescentic bars may become elongated between the rip-currents.

Although a beach may contain a range of rhythmic forms, its overall planform may be quite independent. Beaches tend not to reflect the individual irregularities of cliffs or of the coastal slope, but instead form a smooth seaward curve. A beach between two headlands, known as a pocket beach, may become aligned to the prevailing wave crests. This is due to the fact

sweep the bay and return seaward at its centre, effectively forming a very small near-shore circulation cell.

Explaining the formation of cusps is a question of explaining the regularity in their spacing. Two alternative models have been proposed: cusps are the product of edge waves, or they are the product of self organization in the swash zone. The presence of edge waves can cause regular variation in wave set-up along a beach. Higher waves would give greater wave run-up and tend to set up a pattern for water circulation similar to that observed on cusps. If this theory is correct then the spacing of the cusps should correlate with the frequency of standing-edge waves predicted for any given beach. Coco *et al.* (2000) compiled all the available data on cusp spacing and edge-wave wavelength and reported a strong correlation.

However, cusps appear to occur on beaches that should, according to the theory, not be associated with regular standing-edge waves with short wavelengths. This has given support to the alternative model of cusp formation, which emphasizes the positive feedback on swash velocity of preexisting topographic depressions on the beach slope. Here beach depressions tend to confine and accelerate swash flows, while topographic highs tend

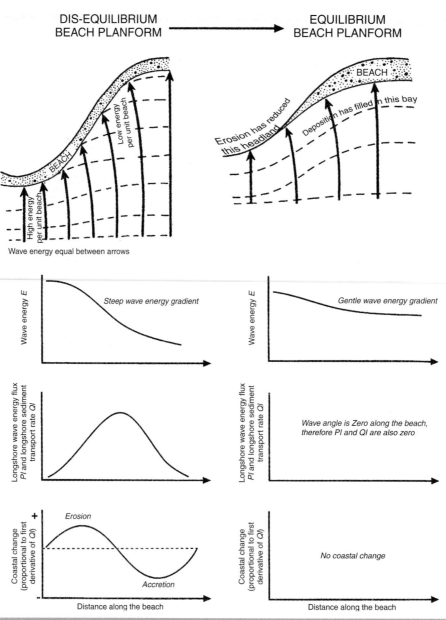

Figure 19.29 Beach profile variation

that longshore transport due to oblique waves is a function of the angle of incidence. If the angle is high, longshore transport is high. Consequently, sediment will tend to move from areas with a high angle of incidence to those with lower or zero angles, and in this way sediment is moved to minimize the angle of incidence. Given time and a consistent wave regime, the planform of the beach will parallel the incident wave crests and longshore transport will effectively cease. Beaches that are more open, and therefore receive a longshore input and longshore output of sediment, will become aligned such that the angle of incidence between them and the incident waves is just sufficient to move the sediment

in transit along the coast. In practice, wave regimes may not be sufficiently stable for long enough for a beach to become aligned in this way.

Detached beaches such as spits, cuspate forelands and barrier beaches are spectacular coastal features (Figure 19.31). Spits form when a longshore current suffers a rapid shift in direction or strength, usually as a result of an abrupt turn in the coastline landwards, for example at the mouth of a bay or estuary (Figures 19.32 and 19.33). The currents supply sediment faster than it can be removed along the new coastal orientation and thereby extends the beach into deeper water. As a spit becomes established, the

(a)   (b)

Figure 19.30 Beach cusps: (a) Clogga Strand, Wexford, southern Ireland; (b) storm beach profile and beach cusps, Sidmouth, Devon, UK

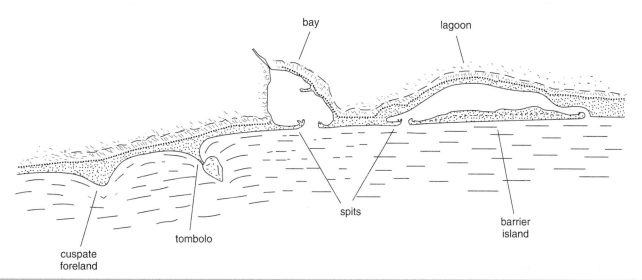

Figure 19.31 Large-scale detached beach morphology

continuity of lateral sediment transport becomes broken and the spit extends rapidly. The development of the spit will be limited by the rate of longshore sediment supply to it and by the depth of the water into which it advances. Wave refraction around the tip of the spit may cause it to curve inwards and become a recurved spit (Figures 19.32 and 19.33). The proposed evolution of Spurn Head in eastern England is shown in Figure 19.34.

'Cuspate foreland' and 'tombolo' are terms which apply to a range of different forms. At a small scale (0.5–1 km), cuspate forelands can form as a result of rhythmic forms, as discussed

above, but generally the term is reserved for much larger triangular features, such as the Dungeness Foreland in southern England (Figure 19.35) which has a total length of around 30 km (see Box 19.3 in the companion website for details of recent work on this foreland). The capes of the Carolina coast or Cape Canaveral in Florida, on the eastern seaboard of North America, may also be classed as large cuspate forelands.

A range of explanations have been advanced, and most are site-specific. Wave refraction around an offshore island or shoal may create a wave shadow landward of the obstruction, leading to

Figure 19.32 El Mogote recurved spit, La Paz, Baja California Sur, Mexico

(a)

(b)

Figure 19.33 Spit morphology: (a) Spurn Head Spit, eastern England from the air (1979) from Creative Commons, image owned by Stanley Howe (b) Dungeness Spit, Washington from Flikr.com under Creative Commons License

the development of a spit of sand between the shoal or island and the beach, known as a tombolo. An example is the 28 km shingle Chesil Beach in Dorset, which runs from Abbotsbury to the Isle of Portland (Figures 19.36 and 19.37). This is a swash-dominated beach subject to longshore drift when waves arrive obliquely at the shore. It grades from granule gravel at the western end to pebbles and cobbles near Portland Bill. Grading is correlated with a lateral decrease in wave energy, with larger waves coming in through deeper water to deposit the higher and coarser beach at the south-eastern end. It must have formed and been carried shorewards during the Late Quaternary transgression as it protects the submerging land margin from the open sea waves. The shingle barrier is permeable, so during storms and high tides there are sea-water seepages through to the Fleet, washing out fans of gravel. It is still being transported shorewards by overwash during strong storms, and peaty sediments which formed in the lagoon now crop out on the shore. In practice, such examples are relatively rare and the presence of offshore shoals cannot explain the formation of large cuspate forelands like that at Dungeness or along the Carolina coast. In these cases, it appears that the interaction of rivers, sea-level rise and spits is the key to their formation.

Barrier beaches occur extensively on the eastern coast of North America and along the northern coast of the Netherlands (Figure 19.38) and it has been estimated that they may occur along 13% of the world's coastline. They are also well recorded in the geological record. Essentially, they consist of detached beaches forming islands elongated parallel to the shore. They are usually no more than 1 km wide and may extend laterally along the shore for several hundred kilometres in extreme cases. Where sand dunes are present they may be 100 m or so high, although they are generally very low. They are broken alongshore by tidal inlets and often look in plan like a pecked line parallel to, but detached from, the shore when viewed from above. They are backed by a lagoon and are generally composed of cross-bedded sand and pebble gravel capped by windblown sand. Seaward, the sand grades into silts and fine sands and, ultimately, into muds of the continental shelf. Landwards, the lagoon may contain a range of facies from silts, clays or peats.

There is no clear consensus about the formation of barrier beaches and many of the proposed explanations are site-specific. However, there are three main models (although several others have been proposed): dissection of a longshore spit by storm waves to form a series of discrete islands; shoreward movement of a beach ridge with a sea-level transgression and drowning of sand dunes; beach berms forming offshore islands that become the focus for subsequent wave action and sedimentation. In practice, examples of all three can be found and the evolution of a barrier island, however it is initiated, is a function of the interplay between the following variables, which help explain the

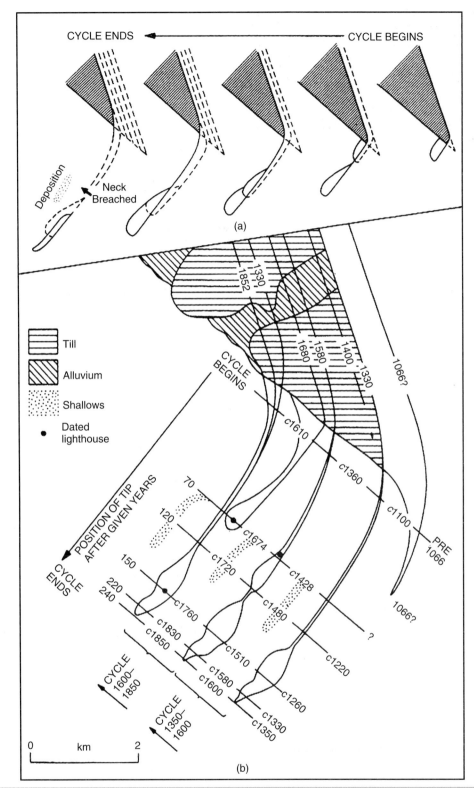

Figure 19.34 Proposed evolution of Spurn Head Spit: (a) schematic cycle; (b) actual development through at least two cycles

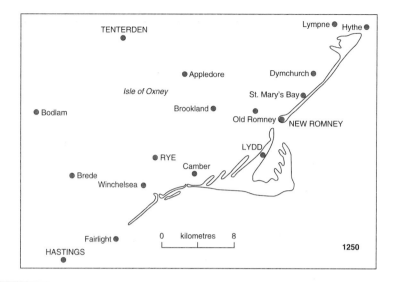

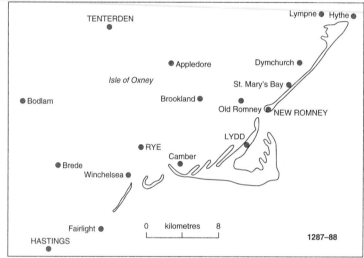

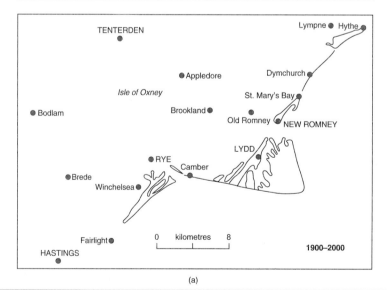

(a)

Figure 19.35 Dungeness cuspate foreland: (a) showing stages of growth in front of a former cliffed embayment; (b) locations on the foreland; (c) historical shorelines of Rye Bay and Dungeness. (Sources: (b) after Long *et al.*, 2004)

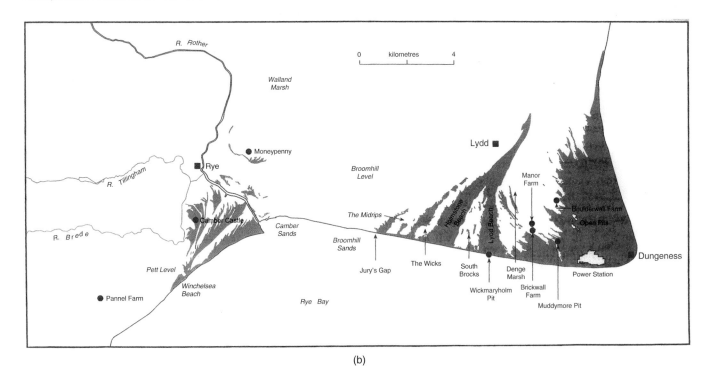

(b)

Figure 19.35 (continued)

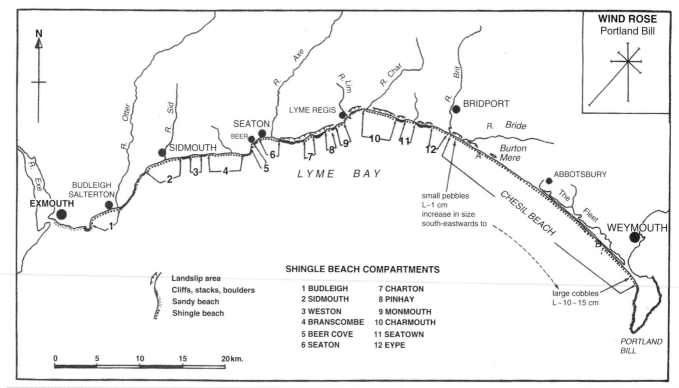

Figure 19.36 Shingle beach compartments in Lyme Bay and the location of Chesil Beach, a laterally-graded tombolo enclosing a lagoon (the Fleet)

Figure 19.37 Chesil Beach and the Fleet, southern England

complex history of many barriers: the history of sea-level rise; the supply and distribution of sediment sources and sinks; the water exchange through the tidal inlets that flank each barrier; and the near-shore wave regime.

Post-glacial eustatic sea-level rise during the Holocene is seen as one of the key variables in the formation of barrier islands. As a result, many barrier islands have migrated landwards across the coastal plain and their elevation rise prevents submergence. This process is episodic and is driven by storm waves, which erode sediment from the front face of the barrier and cross the island to form an overwash fan along the landward side (Figure 19.39(a)). This moves the barrier both landward and upward, as reflected in the barrier's internal architecture, which tends to consist on the landward side of interdigitated sheets of sand (overwash) and muds/organic muds (lagoonal).

This landward migration as a result of sea-level rise does not always occur and there can be stable sea levels at times (Figure 19.39(b)). In exceptional cases barriers may in fact move seaward if sediment supply is high.

Sometimes a transgressive migration may have occurred early in the Holocene, when sea level rose rapidly before being replaced by a regressive migration due to high sedimentation and a slower rate of sea-level rise (e.g. the coast of Nayarit, Mexico; Figure 19.39(c)). Transgressive and regressive barrier islands may be found in close proximity, and even on different parts of the same barrier, due to longshore variations in sediment supply. On the Gulf Coast of the United States, barriers can be classified as either high- or low-profile barrier islands. Low-profile barriers occur where sediment supply is limited and the island is of low relief, and is therefore frequently overtopped and consequently undergoing landward migration. High-profile barrier islands are close to the sediment source and consequently have higher relief,

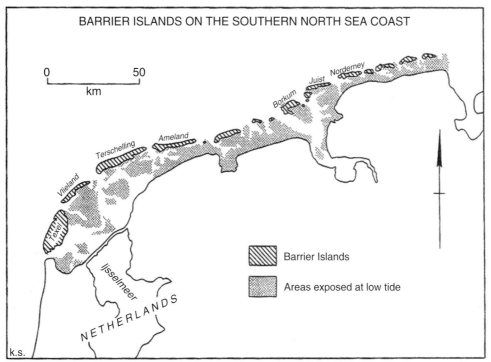

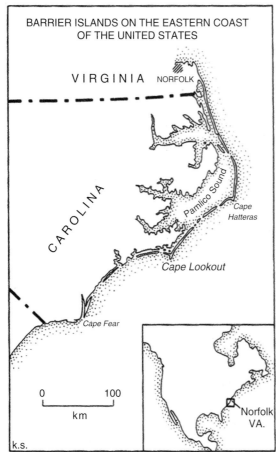

Figure 19.38 Barrier islands on the Netherlands coast and the eastern coast of the United States of America

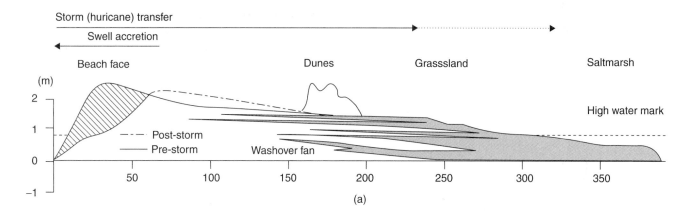

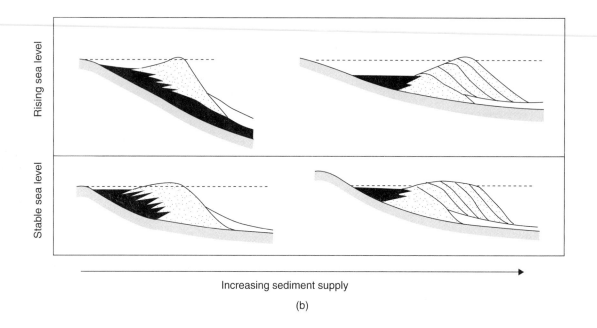

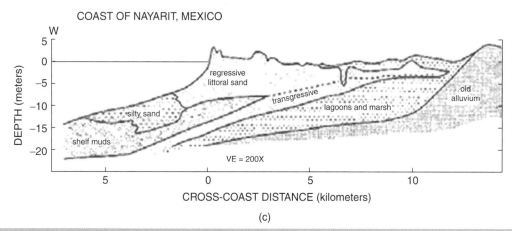

Figure 19.39 Stratigraphic models and development of barrier islands. a) transgressive barrier, sequence of underlain by washover and lagoonal facies b) regressive barriers commonly contain prograding beach ridges c) Sediments on coast of Nayarit. Mexico (after Curray et al, 1969) formed first by transgression of the sea when the Holocene sea level rise was rapid, followed by a regression when the rate of rise was slower so that the supply of sand from nearby rivers could build out the shore.

Figure 19.40 Chenier ridges on the Dengie Peninsula, Essex

thereby minimizing overwash and landward migration. Understanding the interplay of sedimentation and sea-level change is very important to understanding geological examples in which inferences about palaeogeography, climate and sea level are made.

Cheniers are coastal ridges formed of shelly sand and gravel overlying muddy or marsh deposits and surrounded by mudflats and marshes. They were originally described from the delta coastal plains of Louisiana and named after the fact that oak trees (chênes) grow on them. There has been much debate over the conditions necessary for the episodic deposition of fine- and coarse-grained sediments on chenier plains. It seems that mudflat sedimentation occurs during times of abundant sediment supply, whereas chenier ridges form during periods of diminished sediment supply, when wave winnowing and concentration of the shelly sand/gravel fraction occurs from the muds, or from sand moved by longshore drift from a nearby sediment source. The chenier plain develops through an alternation of these two processes, for example by switching of delta lobes, as along the Louisiana coastal plain, or by variations in storminess as the low-lying terrain progrades. The examples illustrated in Figure 19.40 have been driven onto salt marshes in East Anglia from swash action in successive storm surges, where they were described from the Dengie Peninsula in Essex. They have also been described from the tropical and subtropical coasts of Guyana, Surinam, north Australia and north-east China.

Beach classification

A beach classification can prove to be a useful framework for the study of beach morphodynamics and change. The gradient is important as it influences the amount of wave energy that reaches the beach and the beach's shape: some beaches are steep and others are flat. An extensive analysis of these characteristics in Australia has led to a classification that has been applied throughout the world. There are two end members to this classification: dissipative and reflective beaches.

A dissipative beach has a gentle gradient in the shallow subtidal and intertidal areas, where wave energy is gradually dissipated

(Figure 19.41). Reflective beaches on the other hand have relatively steep gradients, where a significant percentage of wave energy is reflected back towards open water, with beach cusps and/or beach step morphology. Dissipative beaches tend to have multiple sand bars that cause waves to break and dissipate wave energy as the waves move shoreward; they are accreting. Reflective beaches do not have bars and have such steep gradients that little wave energy is lost as waves move across the near-shore zone; they are erosional. There is an intermediate category of beach between the two types, which may also have near-shore bars, eroded through by rip-current channels.

These models (Figure 19.41) do not apply to gravel beaches, and although these are reflective they have several morphodynamic characteristics that set them apart from sandy reflective beaches. For example, they never develop bar morphology, even in extreme storms.

Sandy beaches also show a sequence of bedforms on a small scale, which indicate the immediate antecedent conditions at their location. These bedforms vary according to the mean flow or orbital velocities and the grain size of the sediment (Figure 19.42(a)). The ripples at low velocities show a variety of crest patterns, from straight-crested to asymmetric and ladderback patterns (Figures 19.42(a) and 19.43(a)), and at high velocities low-amplitude antidunes are formed on the beach (Figure 19.42(b)) and plane beds with rapid sediment movement and horizontally-stratified sediment are deposited. When the sandy beach is exposed a whole series of landforms appear as wind dries out and erodes the sand (Figure 19.43(b)), and depositional aeolian ripples and small aeolian dunes can form (Figure 19.43(c)).

Coastal dune systems

Coastal dune systems occur where strong onshore winds coincide with abundant sediment supply and the availability of colonizing vegetation. There is usually a sequential development of these dunes (Figures 19.44 and 19.45(c)). Low near-shore slopes combined with a large tidal range provide wide expanses of sand, which dry at low tide and form an abundant sediment resource, such as that around Liverpool Bay in the eastern Irish Sea basin.

Embryonic and foredunes occur in the initial stages (Figure 19.45), with the sand trapped behind flotsam and jetsam at the back of the beach at the high-tide limit, promoting the formation of shadow dunes. These are stabilized by pioneer annual species of plants, like sea rocket, and then adapted grasses, like marram or lyme grass. The sand trapping allows the shadow dunes to grow upwards as embryo dunes, and these coalesce to form a foredune ridge. The generally onshore movement at right angles to the prevailing wind direction is reversed by erosion of the seaward face of dunes during storms. As a result, dunes can be seen as temporary stores of sediment, with frequent recycling occurring.

The basic physics of windblown sand are similar whether the dune system be in a desert or on a coastal strip; these processes

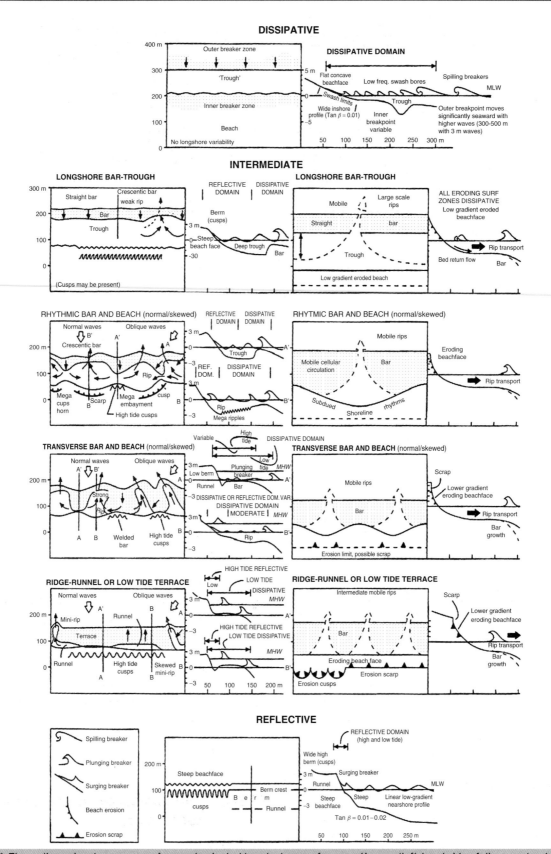

Figure 19.41 Three-dimensional sequences of wave-dominated beach changes for accretionary (left-hand side of diagram, top to bottom) and erosional (right-hand side of diagram, bottom to top) conditions

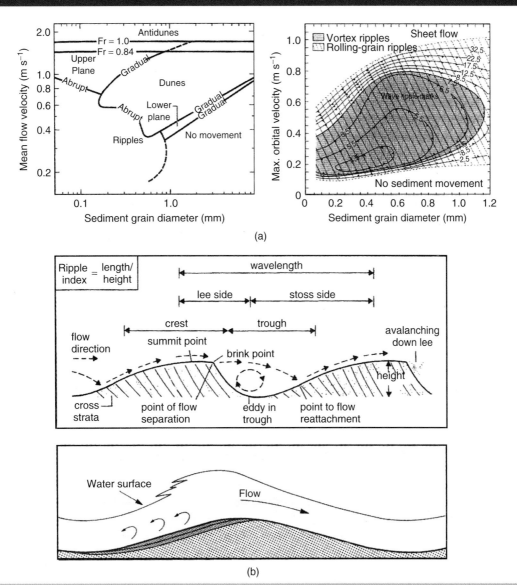

Figure 19.42 Sediment transport nomenclature and types on sandy beaches. (a) Bedform stability diagram for unidirectional flow (left) and oscillatory flow (right) beneath wind waves and swell. (b) Nomenclature, flow and sediment transport bedforms in unidirectional flow: ripple or dune (top) and antidune (bottom)

are explored further in Chapter 22. However, the morphology of coastal dunes is very different, primarily due to the presence of abundant vegetation. Coastal dunes usually exist in a wide zone bordering the high-tide mark and extending inland for anything up to 10 km. This zone of sand deposition can have a relatively straightforward morphology of sand ridges running parallel to the shoreline and separated by troughs. However, in other cases the system is much more complex, with ridges running obliquely to the coast, or forming isolated U-shaped masses of sand. Generally coastal dune heights are between 1 and 30 m. Wind erosion can occur down to the water table and dune slacks form in these blowouts, which again are ephemeral landforms in the ever-changing dune landscape (Figure 19.46(a) and 19.46(b)). Onshore winds are channelled into the blowout, resulting in flow acceleration and scouring at the base down to

the wet sand at the water table and along the depression sides. Often the dunes are stabilized by vegetation at the margins of eroded areas, and parabolic or transverse dunes form and migrate inland (Figure 19.46). Flat sand sheets can be a later stage of development as the dunes are flattened; these are often colonized by acidic vegetation to form dune heathland. A classification of coastal dunes is given in Table 19.4.

19.5.2 Tide-dominated coastal landform assemblages

These tend to occur in more sheltered locations within coastal embayments or estuaries and contrast with wave-dominated assemblages mainly in the nature of their grain size.

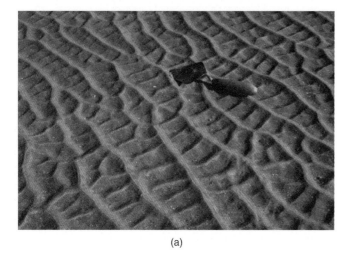

(a)

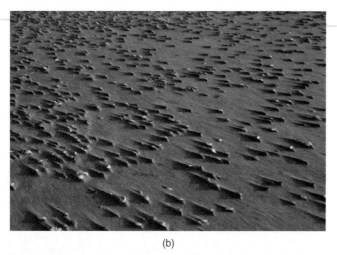

(b)

(c)

Figure 19.43 Sediment transport and erosion on sandy foreshores: (a) ladder-back ripples, Formby Point, south-west Lancashire; (b) wind erosion at the back of a sandy foreshore, with accumulation tails around shell debris, Formby Point, south-west Lancashire; (c) small aeolian dunes with wind ripples on their backs migrating across an eroded sandy foreshore, Raven Meols shore, Formby Point, south-west Lancashire

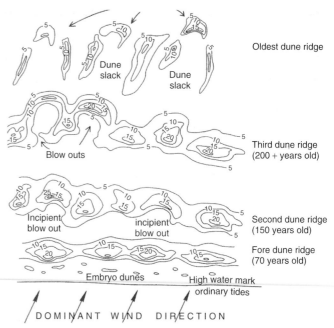

Figure 19.44 Sequential development of a dune system. The blowouts of the second and third dune ridges eventually form into parabolic dunes in the oldest ridges

The landforms of the tide-dominated assemblage are composed primarily of silt and clay. They tend to occur in areas of medium-to-large tidal range (>3 m) that are sheltered from wind waves, although if sediment supply is very high they can occur on more exposed coasts. Tidal flats may occur in the shelter of spits, barrier beaches or in estuaries.

We will start by examining the processes and morphology of estuaries, before examining the tidal sediments and landforms that tend to infill them.

Controls on estuaries

To most of us, an estuary is simply the tidal mouth of a river, but a more precise definition is a semi-enclosed coastal water body which has a free connection with the open sea and in which both tidal and fluvial processes operate. The shapes of estuaries and the sediment fills they contain are therefore the result of a complex interplay of tidal and fluvial processes.

The sediment fill is dominated by mudflats, salt marshes or mangroves. The detailed processes and morphology of landforms are discussed in the next section, but here the focus is on the large-scale morphology and currents which define the patterns of sedimentation. It is important to emphasize that without a preexisting river mouth or embayment, tidal flows would not exist, and they are therefore channel-dependent.

Over time the embayment or channel will undergo adjustment by estuarine processes so that co-adjustment between process and form occurs, giving an estuary whose morphology is in equilibrium with the processes operating within it. There are therefore

Figure 19.45 Embryonic and foredunes: (a) embryonic dune around a Christmas tree used for sand trapping along the eroded dune front, Formby coast, south-west Lancashire; (b) embryonic dunes, Malahide, Dublin Bay, southern Ireland; (c) sequence of dunes, Morfa Harlech, north-west Wales

Figure 19.46 Blowouts in dunes and transverse dunes: (a) dune slack in blowout, Morfa Dyffryn, north-west Wales; (b) Devil's Hole blowout, Raven Meols, south of Formby Point, south-west Lancashire; (c) transverse dune, east of the Devil's Hole blowout, Formby Point

Table 19.4 Classification of coastal dunes. (Source: after Davies, 1980)

Type	Forms	Orientation
Primary dunes (sand derived from beach)		
Free dunes (vegetation unimportant)	Transverse ridge, barchans, oblique ridges	Wind-orientated and generally perpendicular to direction of constructional winds
Impeded dunes (vegetation important)	Frontal dunes, sand beach ridges, dune platforms	Nucleus-orientated and generally parallel to rear of source beach
Secondary dunes (sand derived from erosion of impeded dunes)		
Transgressive dunes	Blowout dunes, parabolic dunes, longitudinal dunes, transgressive sheets	Wind-orientated and generally parallel to directon of constructional winds
Remnant dunes (eroded remnants of vegetated primary dunes)		

two important elements here: the antecedent morphology and the adjustment and modification by the currents within it.

Three classes of antecedent morphology are defined: (1) Bar-built estuaries, where the sea does not penetrate far inland, due to steep river gradients or small tidal ranges. The mouth of the river tends to become dominated by spits and bars, forming a small estuary or lagoon in the mouth of the river. (2) Drowned river valleys, where post-glacial sea-level rise during the Holocene has caused a widespread sea-level transgression and the associated drowning of river valleys. In extremely dissected terrain, these drowned valleys may penetrate far inland to form rias. (3) Fjords, which are drowned glacial troughs that may form extremely deep estuaries, often with very steep sides.

These antecedent morphologies effectively form basins, which are infilled by sediment and provide the detailed morphology of any estuary. The pattern of sedimentation is controlled by two sets of water currents: tidal currents and residual currents set up by the mixing of fresh and saline water. Tidal currents result from the movement of a tide wave into and up an estuary (Figure 19.47). Lateral water displacement is minimal, since it is effectively the wave form which is moving up the estuary. As the wave moves up-estuary, it causes water level to rise (flood tide) and fall (ebb tide). The wave declines in height and tidal range with the distance travelled up-estuary due to the effects of frictional drag from the estuary sides, until it is completely attenuated. The inland penetration of a tide wave is therefore dependent on its original magnitude and the geometry of the estuary.

In very long rivers with a large tidal influence several tide waves may move upriver at one time, giving a complex pattern of high and low tides. As the tide wave is attenuated upstream it also becomes increasingly asymmetrical, with the flood tide increasing in velocity and decreasing in duration relative to the ebb tide. This means that the upper reaches of an estuary become net sediment sinks, since the dominant tidal current is associated with the flood or onshore tide. For example, in the upper reaches of the River Mersey in the UK, in excess of 68 million tons of sediment has been deposited in the last 20 years alone.

The timing of peak current velocity within a tide wave with respect to high and low tide also changes as the wave progresses up an estuary. In the open sea, current velocity within the wave is at a maximum at either low or high water, at the crest or trough of the wave. However, as the tide wave moves inland, the maximum current velocity tends to shift to a few hours either side of low and high tide, and may in fact occur close to mid-tide in the upper reaches of an estuary. The reasons for this are complex but involve tide energy, reflected by the estuary sides, interacting with a following tide wave.

This lag between high- and low-tide water level and maximum current flow is of importance to the cross-sectional morphology of an estuary fill. At high tide, when the maximum area of the estuary is flooded, current velocity is minimal, leading to rapid sedimentation. Close to mid-tide, current velocities are at a maximum and maintain a distinctive channel, whose cross-sectional morphology is adjusted to these high currents. Residual currents result from the mixing of saline and fresh water within an estuary. Fresh water is less dense than saline water and tends to rise up above it to float near the surface, while the salt water may form a wedge-shaped water body on the estuary floor. Dilution of salt water by mixing with fresh water causes a net loss of saline water, which is replaced by an onshore flow known as a residual current (Figure 19.48). The degree and pattern of mixing determines the strength of these residual currents and is dependent on the relative importance of fluvial and tidal currents.

Types of circulation and estuary fill

Dyer (1973) recognizes three main types of circulation, which are associated with three distinct types of estuary fill: (1) salt-wedge or microtidal estuaries; (2) partially-mixed or mesotidal estuaries; (3) fully-mixed or macrotidal estuaries.

Salt-wedge estuaries (Figure 19.48(a) and 19.48(b)) occur as a result of microtidal conditions, in which the tidal range is usually less than 2 m. As a result, the estuary is dominated by fresh water flow rather than tidal currents. Vertical stratification of the water

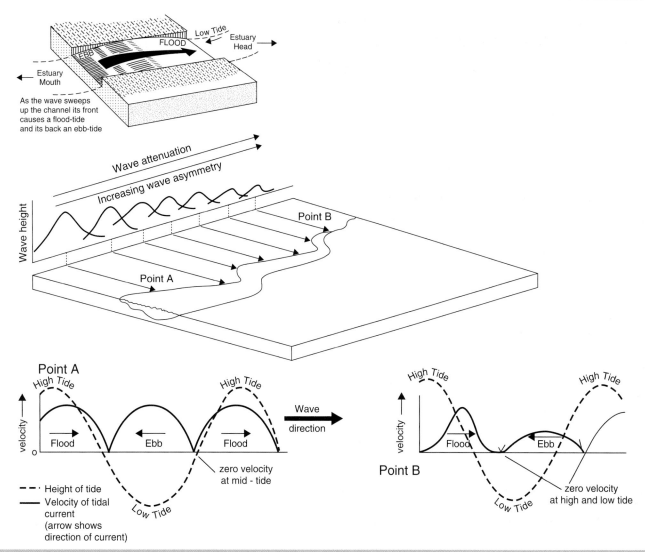

As the wave sweeps
up the channel its front
causes a flood-tide
and its back an ebb-tide

Figure 19.47 Tidal wave heights and velocities as a tide penetrates an estuary

column is strong, with salt water forming a wedge-shaped water body at the bottom of the estuary channel and the fresh river water flowing above. The tip of this salt wedge moves in and out of the estuary with each tidal cycle. Tidal currents are minimal and consequently water mixing is limited. Loss of saline water from the upper boundary of the salt wedge, via dilution and mixing with the freshwater, results in a small upstream residual current. This is relatively unimportant compared to the fluvial discharge, however. As the river water enters the estuary it may transport large amounts of sediment in traction. However, as the fresh water rides up and over the salt wedge, contact with the bed is lost and rapid sedimentation occurs. These are ideal circumstances for the construction of estuary bars or deltas. The Mississippi provides a good example of this type of situation, as do the Rhone, Niger and Orinoco rivers. The pattern of sedimentation within this type of estuary is very distinct and is characterized by the predominance of sediment derived from inland sources and an upstream increase, unlike almost all other tidal sediments. The salt-water tip in the estuary mouth is a focus for sedimentation in the form of a bar or delta.

Partially-mixed estuaries occur in areas with a larger tidal range (mesotidal range 2–4 m) and have a smaller fluvial input. Vertical stratification of the water body still occurs, with fresh water sitting above the salt water, but the level of mixing is much more pronounced. Consequently, the onshore residual current within the salt wedge is much stronger. This current tends to decline towards the tip of the salt wedge and as a result grain size declines up-estuary to this point. The combination of a strong up-estuary residual current and increased tidal flows means that a large quantity of marine sediment is moved up-estuary. Tidal penetration is still limited by the relatively modest tidal range and consequently this estuary tends to be relatively short in length. It is characterized by two distinct morphological features: ebb and flood deltas and a meandering tidal channel. Ebb and flood deltas are essentially large, submerged sand or silt bodies which form either side of the estuary mouth; ebb deltas offshore of the estuary entrance and flood deltas onshore of the entrance. These submerged sand bodies result from the interaction of ebb and flood currents. The cause of the meandering planform of the tidal channel in the

WATER MOVEMENT IN A SALT WEDGE ESTUARY

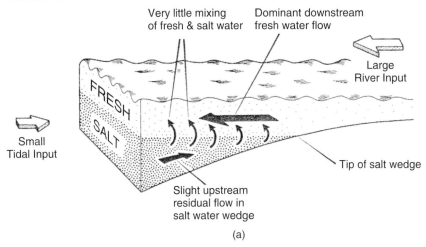

(a)

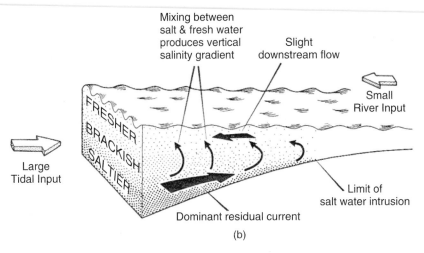

(b)

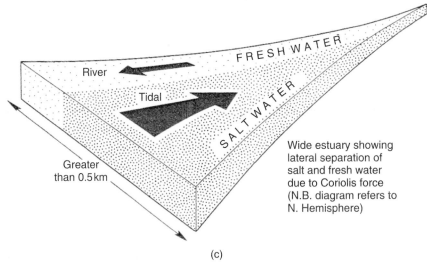

(c)

Figure 19.48 Water movement in a salt-wedge estuary with: (a) small tidal inputs; (b) large tidal inputs; (c) separation of salt and fresh water in wide estuaries

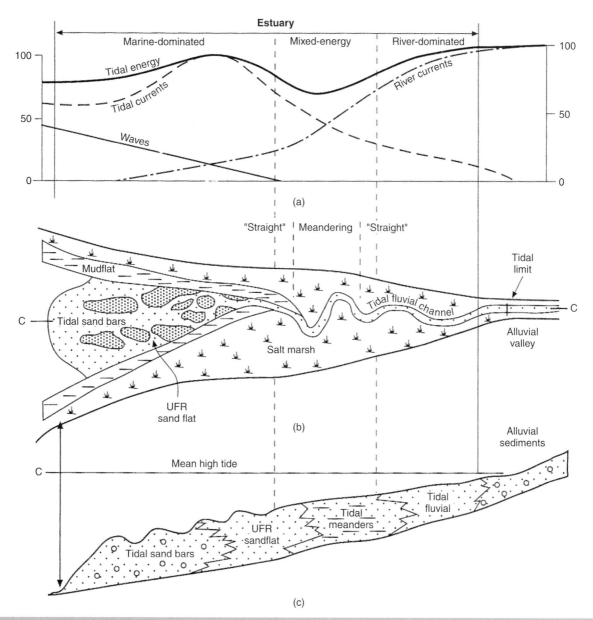

Figure 19.49 Characteristics of well-mixed, tide-dominated estuaries. (a) Chart illustrating the change in energy regime along the estuary axis. (b) Plan view of an estuary, showing the main morphology. (c) Cross-sectional view along the estuary axis, showing the stratigraphy. (Source: after Dalrymple *et al.*, 1992)

upper reaches of these estuaries is not clear, and a range of ideas have been proposed. Examples of this type of estuary include the James River in the USA and the River Mersey in the UK.

Fully-mixed estuaries occur where the estuary is particularly wide and the tidal regime large (macrotidal estuaries >4 m). Vertical stratification of the water column is absent; instead the Coriolis force tends to swing the salt-water flow and the fresh-water flow to their right in the northern hemisphere and their left in the southern hemisphere (Figure 19.47(c)). Some mixing may occur, enhancing the powerful tidal currents with a residual flow. One bank of the estuary will be dominated by fluvial sediment derived from inland, while the other will contain abundant marine

sediment. Ebb and flood deltas are absent and instead the central channel near the estuary mouth is occupied by a linear sand bar, which parallels the tidal flow. In planform, these estuaries are usually characterized by a distinct 'trumpet-like' shape. Examples include the Firth of Forth and the Severn and Humber estuaries in the UK, the Delaware in the USA and the Rio de la Plata in Argentina. The characteristics of well-mixed, tide-dominated estuaries are shown in Figure 19.49.

Mudflats and saltmarshes

Deposition within lagoons and estuaries is dominated by the tidal currents and involves the sedimentation of fine-grained sands, silts and muds. Three basic morphological elements can be recognized (Figure 19.50): (1) high-tide mud flat; (2) an intertidal slope, composed of sandy mud; (3) a subtidal surface or channel, which may be submerged at low tide. In cross-section, these three elements define a lenticular body of sediment whose sediment fines vertically and onshore is inundated with less frequency vertically. The upper onshore surface may become colonized to form a saltmarsh or mangrove swamp as the level of inundation falls. In planform, the high-tide flats are dissected to varying levels by a dendritic network of channels or tidal creeks.

The average grain size of sediments on these mudflats ranges from 0.001 to 0.020 mm and the deposition of such fine-grained sediments poses something of a hydrodynamic paradox. The settling velocity for such particles is extremely slow as it would take 57.8 hours for a clay-sized particle (0.002 mm) to settle through 0.5 m of perfectly still water. The water column is never still for sufficient time for this to occur, so how is sedimentation achieved?

Effectively, grain size is increased by flocculation, the process by which small particles form a larger agglomeration, or floc, with a consequent increase in settling velocity. Clay particles possess electric charges, which in fresh water repel each other, thereby preventing flocculation. However, in the presence of salt, these charges may be neutralized, facilitating flocculation. Flocculation can also be achieved by the presence of vertebrates, which ingest clay and silt-sized particles while feeding and secrete faecal pellets. As a consequence, sedimentation rates may be high; for example, it was calculated that the rate of sedimentation on mudflats in the Wash, UK, was of the order of 3.6 cm per year. The presence of vegetation on high mudflats also aids sedimentation and may reduce wave energy by up to 90%. In addition, fine-grained silts and clays are cohesive and consequently, once deposited, are extremely resistant to erosion. Cohesion is a result of the attractive forces between clay particles, the water content of the sediment and the presence of biological mucus or plants with a mat-forming habit.

The morphology and distribution of sediment on mudflats reflect the relationship between current velocity and tidal stage discussed in the previous section. We saw how the maximum current velocity within an estuary occurs close to mid-tide, rather than at high or low tide as it would in the open sea. We will now

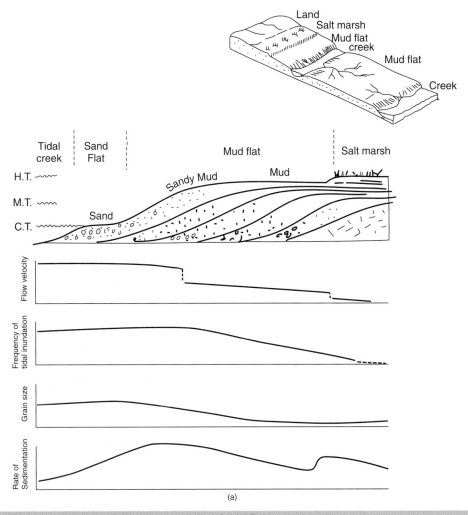

Figure 19.50 Morphological and sedimentological characteristics of tidal flats

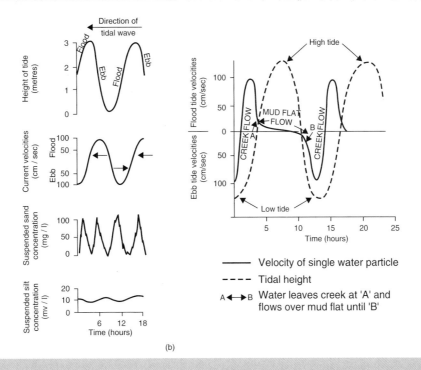

(b)

Figure 19.50 (*continued*)

consider how current velocity will vary with tidal stage across a typical section of mudflat, such as that in Figure 19.49.

At low tide, water velocity is zero and fine-grained sediments will settle out in the floor of the tidal creek. As tidal levels rise towards the mid-tide level, current velocity will rise, sweeping the creek clear of all but the coarsest sediment. Above the mid-tide level, current velocities will begin to fall, and in addition the water will begin to spread out over the upper surface of the mudflat, thereby increasing the channel cross-section and accelerating the fall in current velocity. Only the finest suspended sediment (silts and clays) will remain in transport. Current velocity will finally fall to zero at high tide and fine sediment will begin to settle. As the tide ebbs, the current will slowly increase towards mid-tide, a process that will be facilitated by the decline in the wetted cross-section. Fine sediment deposited at high tide in the creek and on the adjacent banks will be swept clear.

This helps explain the distribution of grain size over a mudflat. Sedimentation occurs preferentially on the upper mudflat surface and, given that vertical accretion is limited by the high-tide level, deposition will tend to occur as lens-like layers which prograde towards the centre of the tidal creeks. As the mudflat approaches the high-tide level, the frequency of inundation will fall and colonization by vegetation may occur.

Colonization of mudflats leads to the development of salt marshes, or in tropical environments, to mangrove swamps. Colonization is controlled by such variables as the availability of colonizing species, the water salinity, the velocity of tidal currents, which might prevent seedlings from germinating, and the availability of light, which is controlled by the frequency of tidal inundation and associated water turbidity. Vegetation enhances the vertical accretion of a mudflat surface. Stems and leaves act as current baffles, which may reduce current velocity, while other plants extrude salt, which facilitates flocculation. Algal mats not only trap sediment but increase the surface cohesion, limiting sediment loss and ultimately organic matter, which adds to the sediment pile. As the surface accretes, the original colonizers are successively replaced by competition from less tolerant plants. With time, the rate of vertical accretion on a salt marsh will fall as the frequency of tidal inundations decreases. For example, on young marshes (<100 years) rates of sedimentation may exceed 10 cm per year, while on older marshes (>200 years) rates of sedimentation may be only 0.001 cm per year.

Mangrove swamps tend to develop instead of salt marshes in tropical and subtropical regions, typically 30° N and S of the equator. They are tidally-submerged tropical coastal woodlands with low trees and shrubs, characterized by an entanglement of arching prop roots, which entrap sediment. They can form over extensive areas; for example in the Philippines and south-west Florida they can form coastal strips over 20 km wide. The principal difference between these swamps and the salt marshes of more temperate latitudes is the increased amount of organic matter involved.

19.5.3 River-dominated coastal landforms

In areas of limited tidal range with restricted wind and wave regimes, or where sediment supply is very high, a coast may become dominated by a strong fluvial influence. In most cases, this is shown by the development of a delta. A delta is a mass of fluvial sediment extending out from a shoreline into a water body; as sediment is delivered to the river mouth it is deposited and builds seawards – if it is coarse enough – to form a giant cross-bed or foreset. By the addition of foresets, a delta progrades into deeper water. Within classic deltas more usually recognized in glacial lakes, three elements can be recognized: (1) topsets, which are sub-horizontal beds formed by the extension of the river bed or coastal plain seawards as a delta progrades; (2) inclined foresets, formed as a delta progrades by the addition of material down the delta slope; (3) gently-angled bottom sets, which form as the distal continuation of foresets. In reality deltas are more complex and may contain a bewildering range of sedimentary environments, which collectively form a prograding package of sediment.

Deltas form whenever the current velocity of a river is checked by flow into a standing water body and can occur on a microscale in puddles, lakes, enclosed seas, estuaries, coastal lagoons and more exposed coasts, wherever the rate of sediment supply exceeds the rate of removal by the action of wave or tidal currents.

Within a river, sediment usually moves in traction along the bed (bedload) and as suspended sediment. As a river discharges into the sea, the velocity is reduced rapidly by the standing water, leading to deposition of the bedload. The rate at which this process occurs is a function of the inertia or ability of the river water to continue to flow seaward before it slows (dependent primarily on discharge and river gradient or stream power) and of frictional retardation of river water by the channel and sea floor (determined by water depth and both coastal and fluvial configurations). The behaviour of the suspended sediment is more complex and is dependent on the density (determined by sediment concentration, salinity and temperature) of the river water relative to that of the sea. If the river water has a similar density to the water body into which it flows, rapid vertical and horizontal mixing will occur, leading to rapid sedimentation close to the injection point (homopycnal flow). This is often the case when a river discharges into a lake, for example. Alternatively, if the river water is denser than the water into which it flows, perhaps if it is very cold or has a particularly high sediment concentration, it forms a distinct bottom current, which may move sediment basinward and suppress development of a delta (hypercynal flow). In contrast, if the river water is less dense than the water body into which it flows, which is usually the case when fresh water flows into the sea, the river water rises to the surface as a buoyant plume, which may move seaward, spreading the scatter of suspended sediment (hypopycnal flow).

Combining the effects of river inertia, friction and the buoyancy of the injected water, four broad delta environments can be recognized (Figure 19.51):

1. **Inertia-dominated river mouths:** Rivers with high stream power, due to steep coastal slopes or high discharge, establish a strong seaward jet, which dominates deposition. Typically, flows are homopycnal and rapid turbulent mixing leads to the development of a uniform delta lobe.

2. **Friction-dominated river mouths:** Where the sea is shallow, friction between the bed and the water jet limits seaward penetration and the sediment deposit tends to spread laterally along a coast. Rapid deposition with channel mouths gives distributary or mid-ground bars, which may lead to channel

bifurcation and are an important building blocks within some deltas.

3. **Buoyancy-dominated river mouths:** Where fluvial inertia and tidal regimes are low, hypopycnal flows dominate and seaward penetration of the river-derived sediments is greater. Rapid deposition of bedload occurs close to the river mouth, while settling of suspended sediment in the freshwater surface plume often leads to a seaward fining succession of bottom sediments.

4. **Hyperpycnal-dominated river mouths:** These occur if the injected water is very dense and may form a concentrated bottom current in which both bedload and suspended sediment are transported basinward. This scenario tends to only occur in freshwater lakes, although the Huangho delta provides spectacular marine examples, where the river is so charged with sediment, derived from the Chinese loess, that the water jet simply sinks without trace on reaching the sea.

All of these forms can occur within a single large delta, depending on local conditions, and a delta may evolve from one characteristic form to another with changes in sea level or climate. Delta morphology is also influenced by both tide and wave processes. A summary of distributary mouth processes is given in Table 19.5.

The interaction of factors responsible for the global variation in morphology is illustrated in Figure 19.52. At one extreme are deltas dominated by fluvial processes (Figure 19.53), such as the Mississippi Delta (Figure 19.54) in the USA, where sedimentation occurs around distributary channels and is not reworked significantly by marine processes. Where wave processes are more effective, the delta front has a smooth arcuate form, which is the result of littoral redistribution of fluvial sediment (e.g. Senegal Delta, Senegal; São Francisco Delta, Brazil). Tidal-dominated deltas (Figure 19.53) have channels in which the fluvial sediment is reworked into ebb and flood deltas and into a range of tidal channel fills (e.g. Fly Delta, Papua New Guinea; Ganges-Brahmaputra Delta).

So far we have concentrated on deltas formed by large rivers transporting fine-grained sediment. In areas of high relief, perhaps with active tectonics, or with river systems dominated by coarse-grained sediment such as that found in glacial or desert environments, alluvial fans or coarse-grained braided river systems may deliver sediment to the sea or a lake to form a fan delta (Figure 19.55(a)).

The morphology of these deltas is controlled by the nature of the feeder system, whether point-specific or distributed, and by water depth. In shallow water, the delta forms a continuation of the subaerial fan or delta plain. In contrast, in deep water, a steep fan dominated by the downslope resedimentation of fluvial sediment via slumping and subaqueous sediment gravity flows is found below water. A summary of sediment transport processes for coarse- and fine-grained deltas is given in Table 19.6. Between the two extremes are so-called Gilbert deltas (named after G.K. Gilbert, an early American geologist who recognized these forms

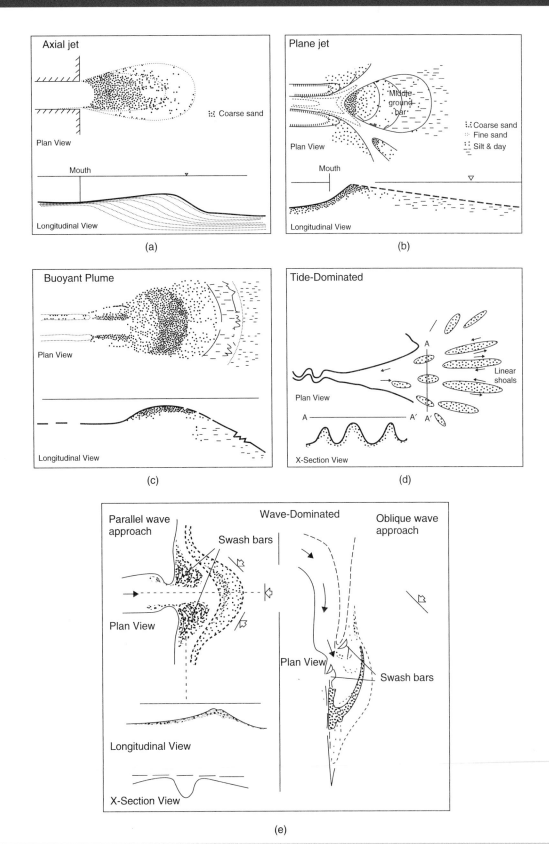

Figure 19.51 Morphology of distributary mouths, dominated by: (a) axial jets; (b) plane jets; (c) buoyant plumes; (d) strong tides; (e) strong waves

Table 19.5 Summary of distributary mouth processes

	Jet		Plume		Tides	Waves
	Axial	Planar	Surface	Near-bed		
River-current velocity	Large	Large	Moderate	Moderate	Small	Small
Offshore slope	Steep	Gentle	Relatively steep	Relatively steep	Relatively gentle	Relatively steep
Density conditions	Homopycnal	Homopycnal	Hypopycnal	Hyperpycnal	Homopycnal	Homopycnal
Morphology	Narrow, lunate bar	Broad, radial bar or middle-ground bar and bifurcating channel	Subaqueous levees and distal bar	Gullies and channels from mudslides and turbidity currents	Linear channels and shoals aligned perpendicular to the coast	Swash bars and beach ridges aligned parallel to the coast

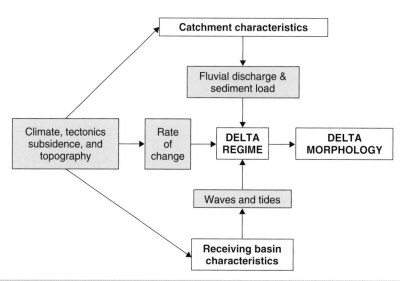

Figure 19.52 Interaction of factors responsible for the global variation in delta morphology

in Lake Bonneville, Utah), which possess classic delta morphology in the form of topsets, foresets and bottom sets. In practice, delta morphology may change with variations in sea level, wave energy or sedimentation.

19.5.4 Carbonate coastal landforms and sediments (see Chapter 18 for a more detailed discussion)

Most limestones, chalks and other carbonate-rich rocks are derived from marine carbonates. Therefore knowledge of coastal carbonate environments is important tool for understanding the geological record and life on Earth. However, as with other coastal environments, the key to the past is not always the present

when dealing with carbonates, since the building blocks – the organisms – have changed through time and modern carbonate reefs look very different from those of, for example, the Early Palaeozoic.

On carbonate coasts, organisms have several roles, which include: (1) precipitation of mineral matter; (2) the binding of sedimentary particles; (3) the provision of an architectural core, like the steel girders in a building. Many animals and plants living in the sea produce aragonite and calcite as skeletal or other strengthening structures. After death, this material is broken up or comminuted to varying degrees to form carbonate grains. In some cases the abrasion and comminution is such that little remains to identify the primary origin of the grain. In other cases, the carbonate may remain largely *in situ*, forming a series of building blocks around which other sediments may accumulate,

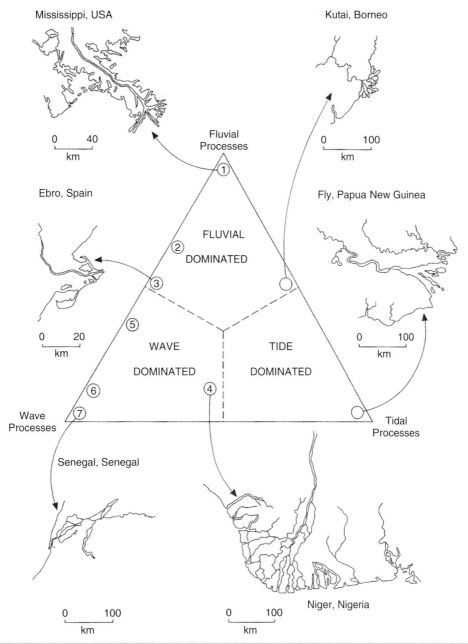

Mississippi, USA

0 40 km

Ebro, Spain

0 20 km

Kutai, Borneo

0 100 km

Fluvial Processes

FLUVIAL DOMINATED

WAVE DOMINATED

TIDE DOMINATED

Fly, Papua New Guinea

0 100 km

Tidal Processes

Wave Processes

Senegal, Senegal

0 100 km

Niger, Nigeria

0 100 km

Figure 19.53 Ternary diagram of delta types related to the importance of fluvial, wave and tidal controls

or, in some cases, be trapped. Carbonate grains contribute to such landforms as beaches and mudflats but may also result in distinctive accumulations and landforms, perhaps the most obvious of which is the coral reef.

The key to a carbonate coast is biological productivity, which is favoured by warm, shallow and clear water and is therefore most pronounced in tropical and subtropical environments, although carbonate coasts may occur in more temperate latitudes. For example, cool-water carbonate landforms occur widely on the New Zealand and southern Australian shelves. The sensitivity of biological production to limiting environmental conditions

is illustrated by hexacorals, the main reef-builders of today, which require very specific environmental conditions. This helps explain the present distinctive global distribution of coral reefs (Figure 19.56).

The morphology of carbonate accumulations is the result of the interplay between waves, tides and biological productivity. Carbonate accumulations such as coral reefs are not necessarily quiet water features and their morphology owes much to wave action and the redistribution of carbonate sediments. One can recognize two broad morphological forms, platforms and ramps, although in practice the distinction is somewhat artificial.

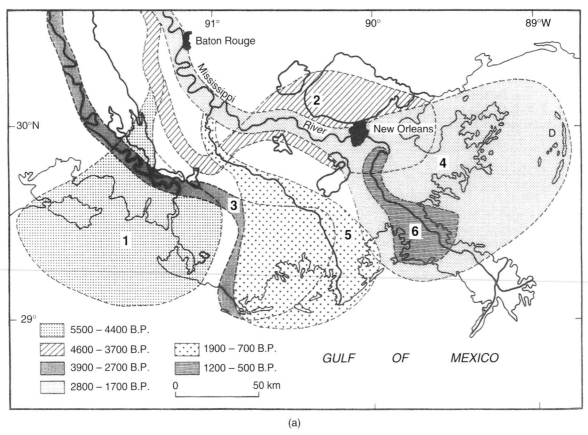

(a)

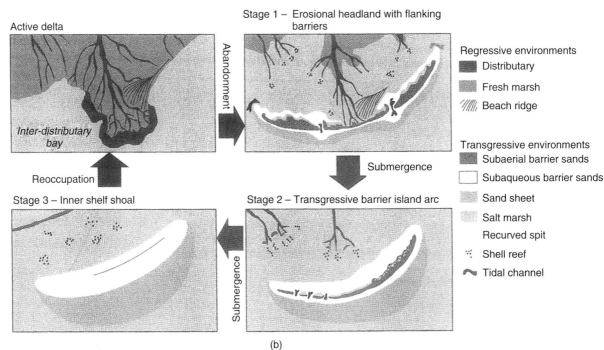

(b)

Figure 19.54 Mississippi Delta development: (a) subdelta lobes, now partly submerged and dissected because of subsidence; (b) barrier evolution on the delta. (Source: after Penland *et al.*, 1988)

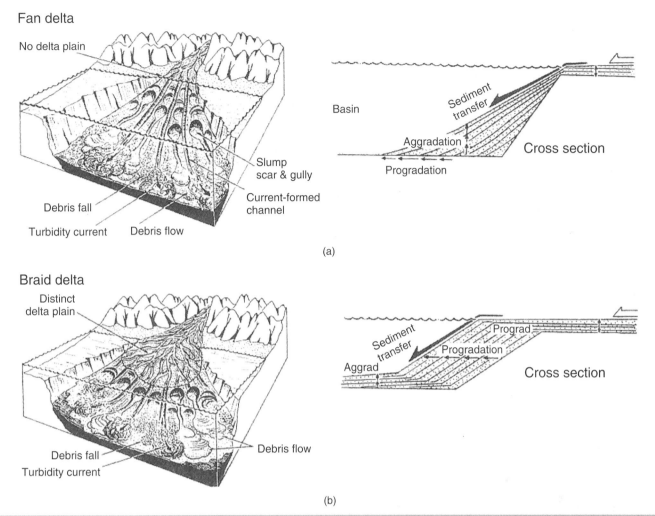

Figure 19.55 Two styles of coarse-grained delta. (a) Fan deltas are a subaqueous extension of subaerial alluvial fans, in which there is no delta plain and no subaerial progradation. (b) Braid deltas show a distinct delta plain and undergo subaerial progradation

Table 19.6 Summary of sediment transport mechanisms for coarse- and fine-grained deltas

Morphological unit	Fan deltas	Braid deltas	Fine-grained deltas
Delta plain	Land-derived slope failure and avalanches; bedload transport by unconfined river flow	Land-derived slope failure and avalanches; dedload tranport by poor-to-well-confined river flow	Bedload and suspended-load transport in well-confined channels
Delta front	Land- and delta front-derived, unconfined avalanches and debris flows	Bedload transport by friction-dominated jets; delta front-derived, poor-to-well-confined turbidity currents; land-derived avalanches	Bedload and suspended-load transport by waning jets; suspended-load tranport by plumes; delta front-derived slumps and channelized turbidity currents
Prodelta	Land- and delta front-derived, unconfined avalanches and debris flows	Delta front-derived debris flows and poorly-channelized turbidity currents	Suspended-load transport by plumes; delta front-derived turbidity currents

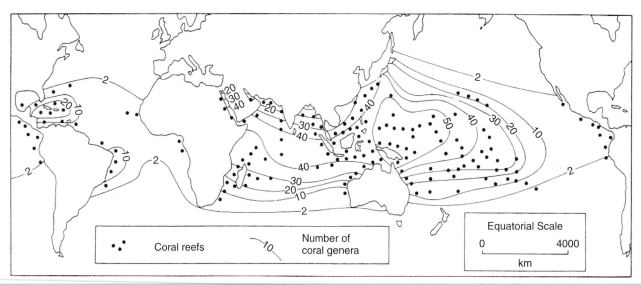

Figure 19.56 Global distribution of coral reefs today and the abundance of reef-building coral in general. (Source: after Davies, 1980 and Stoddart, 1969)

Carbonate platforms

These consist of a shallow platform with a steep offshore slope and a fringing reef, often of coral. These platforms may be attached to land, surround islands or enclose landless lagoons (atolls). Platforms often rest on faulted basement rocks and develop as a result of the increased biological productivity with falling water depths. The fringing reefs develop because coral productivity tends to peak at the shelf edge, since there is little terrestrial sediment and the water is well-oxygenated due to the waves breaking on the platform edge. The Bahamas banks provide an excellent modern example of a carbonate platform. The development of such platforms was substantially enhanced by the evolution of the reef-building hexacorals in the Mesozoic.

Platform morphology is determined by sea conditions and the types of organism present, which have changed through geological time. In quiet seas, particularly within the Palaeozoic, the platform edge and offshore ramp is often associated below the wave base with the build-up of carbonate mounds derived from biologically-precipitated muds, organic detritus and *in situ* remains. Above the wave base, this carbonate sediment is reworked into a barrier beach enclosing a lagoon of fine-grained carbonate mud. As the coastal energy regime increases, reef knolls may occur on the platform edge, and in high-energy conditions robust hexacoral may form distinct offshore barrier reefs, landward of which there may be patches of reef.

Landward of the platform edge and the associated reefs, lagoon sediments form, comprising pelleted, lime muds excreted from a range of organisms, or derived from microfossils such as coccoliths. Fine-grained detrital carbonate also accumulates in these quiet water conditions.

Coral cays may accumulate on some platforms. These are accumulations of carbonate sediment located where wave-induced currents converge on a platform as a result of wave refraction. The cays may start as shoals or sand banks but can, in some cases, emerge above mean water level. They are extremely responsive to change in the wave regime on a platform.

In more-sheltered, shallower water locations lagoonal sediments may accumulate. These sediments range from fine-grained, detrital carbonates to organic accumulations which can trap sediment. The nature of lagoonal sedimentation is influenced by the water balance within a lagoon. In arid areas, for example, with high levels of evaporation, extremely saline tidal flats (sabkhas) are formed in which sedimentation is dominated by evaporites and the growth of stromatolitic algal mats. Stromatolites are algal mats which trap fine-grained sediment and form dome-shaped mounds in the intertidal zone.

In more humid areas, high levels of rainfall may produce lagoons, which periodically have very low salinities following rainfall events, causing biological stress. In these environments, for example in the Bahamas, extensive algal marshes may form just above high water.

Carbonate ramps

These consist of a uniform slope, usually less than about 5°, which grades from the shore face offshore. Ramps often occur where excess nutrient production, due to ocean upwelling or other adverse conditions, tends to inhibit the development of hexacorals. For example, the Arabian Gulf is characterized by ocean upwelling and a carbonate ramp.

Responses to sea-level change

One of the key features of carbonate platforms and their associated reefs is the ability of the landforms to respond to changes in sea level. If sea-level rise is very rapid, such as during the early Holocene following the last postglacial, even fast-growing corals may be overwhelmed, leading to the abandonment of reefs. However, if changes in sea level are more gradual then coral reefs

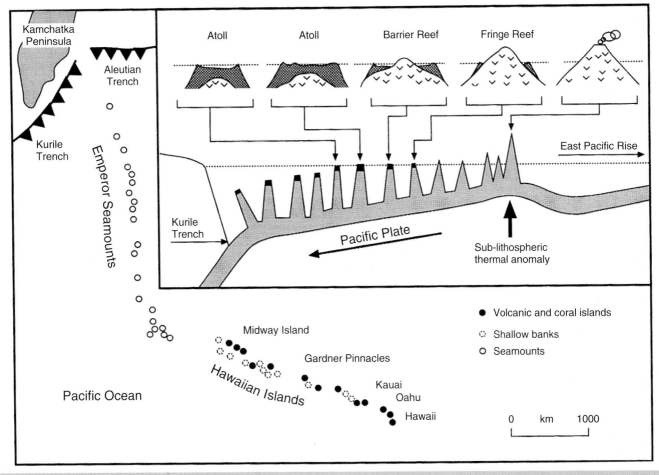

Figure 19.57 Model of reef evolution on the Pacific Plate

may be able to keep pace with the rate of change, as shown in Figure 19.57. Typical upward growth rates are 1.8 m in 1000 years.

Coral atolls illustrate this relationship. A coral atoll consists of a ring of reefs around a platform that does not contain a central land island. This was first described by Darwin in 1842 from the Pacific, and was explained as a consequence of volcanic island subsidence: as the island subsides, the coral reefs build upwards (Figures 19.57 and 19.58). Although these ideas, and the relative importance of subsidence versus Quaternary sea-level fluctuations, have been debated over the years, they are now generally accepted. As the sea floor in the Pacific moves westwards and cools from the axis of the East Pacific Rise (divergent plate margin) at a rate of about 100 m in 1000 years, volcanic islands and seamounts subside at about 0.02–0.03 m per 1000 years, with which the upward growth of coral reefs is able to keep pace (Figure 19.57).

Many island chains, such as the Hawaiian and Society Islands, demonstrate the Darwinian sequence of coral reef formation along their length. Coral reefs form on these slowly migrating and subsiding islands and fringing reefs are transformed into barrier reefs, followed by coral atolls. As the coral atolls continue to migrate towards the west-north-west, the temperature reduction in the water results in a decline in coral growth rates. At some stage vertical reef growth cannot keep up with the subsiding substrate and the coral atoll drowns, forming a guyot.

19.6 Distribution of coastal landsystems

Four coastal landsystems have been defined so far in this chapter. Their distribution in space and time is a function of a hierarchy of factors. Perhaps the most important of these is plate tectonics, which determines the gross coastal topography and the nature of the coastal hinterland. Inman and Nordstrom (1971) proposed a tectonic classification of coasts in which the plate tectonic setting for coastlines was emphasized. They recognized three basic coastal types: collision coasts, trailing-edge coasts and marginal sea coasts.

• **Collision coasts or convergent-margin coasts:** These occur at convergent plate boundaries, where two basic types can be recognized: (1) continental collision coasts, where a continental plate moves over an ocean plate (e.g. the west coast of North and South America); (2) island arc coasts, where two ocean plates collide (e.g. the Aleutian Islands).

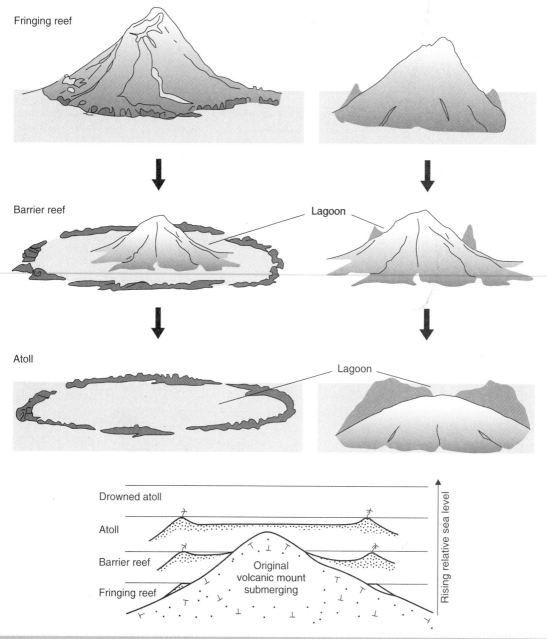

Figure 19.58 Darwin's theory of atoll formation in three stages. The developmental sequence is a continuum taking place as volcanic islands subside over the ocean crust

- **Trailing-edge coasts or passive-margin coasts:** These are divergent plate boundaries. Three basic types can be recognized: (1) Neo trailing-edge coasts or new trailing-edge coasts, formed along a recent rift or spreading centre (e.g. the Red Sea or the Gulf of California); (2) Amero trailing-edge coasts, the trailing edge (passive margin) of a continent with a collision coast on its opposite side (e.g. the east coast of America); (3) Afro trailing-edge coasts, where trailing-edge coasts occur on both sides of the continent (e.g. the east and west coasts of Africa).

- **Marginal sea coasts:** These occur where coasts fronting onto marginal seas are protected from the open ocean by island arcs (e.g. in Korea).

Collision coasts are all relatively straight and mountainous and are generally characterized by sea cliffs, raised terraces and narrow continental shelves (see Table 19.7(a)). Fluvial influence is limited because of the small mountainous island catchments, and wave action tends to dominate (Table 19.7(b)). The narrow continental shelf also precludes the accumulation of large

Table 19.7 Relationship of (a) coastal features and (b) continental shelf width to tectonic setting

(a)

| Shelf width | Tectonic setting | | | | | |
| | Convergent-margin coasts | Passive-margin coasts | | | | % of world coastline |
		Nascent	Mature	Mature distal orogen	Marginal sea	
Narrow (<50 km)	23.9	6.5	11.2	12.6	3.3	57.5
Wide (>50 km)	1.4	0.1	3.6	29.1	8.3	42.5
Total	23.5	6.6	14.8	41.7	11.6	100.0

(b)

| Coastal features | Tectonic setting | | | | | | |
| | Convergent-margin coasts | Passive-margin coasts | | | | Total | % of world coastline |
		Nascent	Mature	Mature distal orogen	Marginal sea		
Wave erosion	47.9	5.6	3.8	4.9	37.8	100	44.7
Wave deposition	15.5	12.4	21.4	33.3	17.4	100	11.6
Fluvial deposition	5.7	6.4	9.9	62.4	15.6	100	3.2
Aeolian deposition	1.9	18.8	79.3	-	-	100	1.2
Biogenic activity	36.1	21.0	11.3	15.8	15.8	100	3.0
Glaciated coasts	6.4	-	7.2	86.4	-	100	30.7
% of world coastline	39.1	4.3	6.8	35.4	8.8		94.4*

*Excluding the Antarctic coastline
Note: all figures are percentages

coastal sediment bodies. In contrast, Amero trailing-edge coasts have broad continental shelves on which big rivers supply large amounts of sediment and consequently they tend to be associated with barrier islands and low-lying coastal estuaries. Afro trailing-edge coasts are also associated with broad continental shelves, but fluvial sediment supply is lower due to the absence of inland mountains. Neo trailing-edge coasts occur along recently-rifted shorelines and consequently the coastal slope is steep and wave action is often limited by the small fetch. Marginal sea coasts have a diverse morphology but tend to be dominated by local processes, having a limited fetch. Inman and Nordstrom (1971) tested their classification by looking at the distribution of key coastal landforms; for example, the 28 largest rivers all discharge across trailing-edge coasts, the 17 largest deltas all occur on marginal sea coasts and 49% of all barrier islands occur on trailing-edge coasts.

A summary of the relationship of coastal features and continental shelf width to tectonic setting is given in Table 19.7.

Plate tectonics may therefore help shape the gross morphology of a coast and the likelihood that any given landform is present. However, the lower-order variables control the distribution. These include wave conditions, climate and tidal range.

We have already seen that certain landforms are favoured by large tidal range, for example salt marshes and tidal mudflats are favoured by large, macrotidal ranges, while spits and beaches tend to develop in areas with a more limited microtidal regime. Wave conditions are also critical as the occurrence of large oceanic fetches with easterlies and westerlies tends to favour large waves. Wave heights are more limited in the tropics and in polar regions, where sea ice may dampen ocean swell and protect coastlines. This explains why steep-cliffed coasts tend to be a feature of temperate latitudes. Climate is important in controlling the distribution of coral reefs and of carbonate coasts, which tend to favour more tropical waters. It may also influence the presence of sea ice or the rates of subaerial weathering and coastal debris supply. At a local level it is the energy regimes along a given coast that are critical to the local distributions of sediment and landforms.

19.7 The impact of climatic change on coastal landsystems: What lies in the future?

19.7.1 Introduction

Global warming and its associated climate changes (see Chapter 27) will cause worldwide sea level to rise in this century and beyond, and although dramatic sea-level change has been common throughout the Quaternary, and even in the Holocene, it will impact in a major way on coastal landsystems and the populations that inhabit the coast. The aim of this section is to provide a summary of the likely effects of sea-level rise on the various types of coast and then undertake a detailed case study to examine the effects in more detail.

Currently there are several calculations of the extent of future sea-level rise, but most predict a rise of 10–15 cm by 2030, accelerating to 30–80 cm by 2100. Such rises will cause major coastal changes around the world. Eventually all of the world's coastlines will be submerging coastlines, even in areas that are still emerging after the effects of the last glacial, like the Gulf of Bothnia. There will be extensive submergence of low-lying coastal areas and current intertidal areas will be no more. On rocky coastlines sea-level rise will just raise the high- and low-tide levels, although the amount of marine erosion is likely to increase. As sea-level rises, coasts that are now stable will be increasingly eroded, and this erosion will accelerate on coastlines that are already receding. This will be exacerbated by more-frequent and higher-energy storms than are now apparent, which will cause surges further inland than they have appeared in the past. In particular, coasts already suffering storm surges due to hurricanes, like the Gulf and Atlantic coasts of North America, will have more frequent, persistent and extensive marine flooding.

However, although there will generally be a deepening of near-shore environments, some coasts may receive an increased sediment supply from longshore or river sources, which may maintain or even diminish the near-shore profile as sea level rises. Wave energy here will not intensify and there might even be progradation. So predictions are usually speculative when it comes to precise impacts, but nevertheless some generalizations are possible and useful.

The impacts of global warming will also cause a migration of climate zones. Tropical cyclones will become more frequent and severe and will extend to higher latitudes, for example. There will be changes in coastal currents like the El Niño Southern Oscillation (see Chapter 8). Some areas will receive more rainfall, increasing river flooding, which will result in a water-table

rise and add to that caused by sea-level rise. These climatic changes will have a major impact on coastal ecosystems and their distribution.

19.7.2 Impacts of a rising sea level on coastal ecosystems

Rocky, cliffed coasts (Figure 19.59(a)) are mainly a result of the relatively stable sea levels for the last 6000 years, so with sea-level rises there will be a submergence of shore platforms and increased erosion on some cliffs, except in highly-resistant rock types. The frequency of coastal landslides will increase in some areas, depending on the increase or decrease of rainfall. In areas of increased landsliding there will be an additional sediment supply to local and downdrift beaches, offsetting at least temporarily the effects of the rise.

On beaches, spits and barrier coasts the beach erosion will intensify as larger waves break onshore (Figure 19.59(b)). On beaches that have been prograding the seaward movement will be halted, and on beach-ridge plains and barriers accelerated erosion will occur along their seaward margins. Barriers that are already transgressive will continue to migrate with sea-level rise and some that were stable will become transgressive, with washovers and landward movement of dunes. Some beaches may be maintained, depending on increased sediment supply along the coast despite the sea-level rise, but by later in this century most beach-fringed coasts will have retreated substantially and will be eroding more rapidly than today.

On dune coasts more blowouts will occur and transgressive dunes will develop, although this may depend on associated climatic changes. If there are drier and windier regimes then dunes that are currently stable may become unstable as the vegetation cover is weakened and the sand becomes more mobile, but if the climate becomes wetter or less windy then dune stabilization could result, even where the coastal fringe is eroded.

On salt marshes, mangroves and intertidal coasts (Figure 19.59(c)) a sea-level rise will submerge the intertidal areas, there will be accelerated erosion on microcliffs along the seaward margins of salt marshes and mangrove terraces and tidal creeks will widen, become deeper and extend headward. Gradually new intertidal areas will form to landward and vegetation zones will move laterally, depending on the inland topography. Precise changes to the extent of the subenvironments will depend on the extent to which submergence is offset by continuing accretion of sediment, possibly supplied by nearby rivers (especially if there is an increased rainfall) or shallow seas.

Estuaries and lagoons will generally widen and deepen and may transgress inland as sea level rises (Figure 19.59(d)). Tides will get further upstream, tide ranges may increase and there will be changes in shoal deposition patterns. Flooding will become more extensive and persistent and a higher percentage of sediment will be kept in the estuary. Coastal lagoons will generally enlarge and deepen as sea level rises, with submergence and erosion of their shores and fringing swamp areas, and widening and deepening of tidal entrances, increasing the inflow of sea water. Eventually the

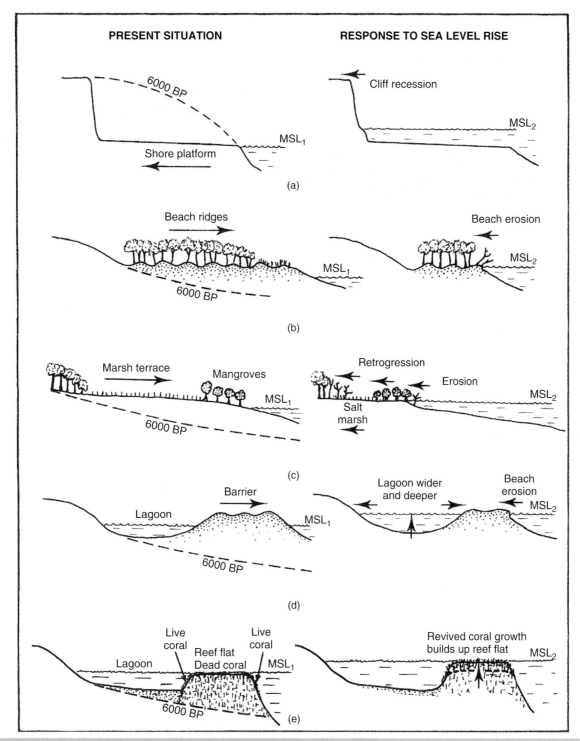

Figure 19.59 Changes in response to a sea-level rise on: (a) a cliffed coast with a shore platform; (b) beach ridges; (c) a marsh terrace; (d) a barrier-fringed lagoon; (e) a coral reef. (Source: after Bird, 2000)

lagoons will be reopened as marine inlets and embayments. New lagoons will also be created by sea-water incursion into low-lying areas behind dune fringes on coastal plains.

A rising sea level will cause submergence and erosion of low-lying deltas and coastal plains, and delta progradation will stop, especially where there is little or no compensating sediment accretion. However, increased sediment yields from greater runoff may build some deltas.

Corals and algae on the surface of intertidal reef platforms will be stimulated by a rising sea level, which will lead to an upward growth of reefs (Figure 19.59(e)), initiated by the expansion and dispersal of presently sparse and scattered living corals on reef platforms. However, increasing sea temperatures are likely to impede coral growth and reef aggradation, and higher rainfall and greater freshwater discharge will also impede coral growth. A slow sea-level rise should stimulate the coral growth on reef platforms, but an accelerating rise may lead to the drowning and death of some corals and the eventual submergence of inert reefs. It has been suggested that reefs are likely to keep up with a sea-level rise of less than about 1 cm per year and to be drowned if it reaches over 2 cm per year, but there are many other environmental controls too, such as changes in sea temperature, salinity and the effects of tropical cyclones. With the suggested rates of sea level rise per year by the mid part of this century, coral reefs are likely to be modified in a major way.

19.7.3 Case study: The likely effects of climatic change on the coastal geomorphology of the Sefton Coast, south-west Lancashire

The Sefton coast (Figure 19.60), extending over 34 km, has always been dynamic. It was subject to changes caused by wind, waves and tidal currents during the Holocene sea-level rise and stabilization. The sandy foreshore and dune system, with no rock exposures, extends from the Ribble to the Mersey, and although there has been increasing urbanization, the coast is one of the most important nature conservation areas in Europe and an important tourist destination. Coastal change since 1848 has been documented by Lymbery *et al.* (2007).

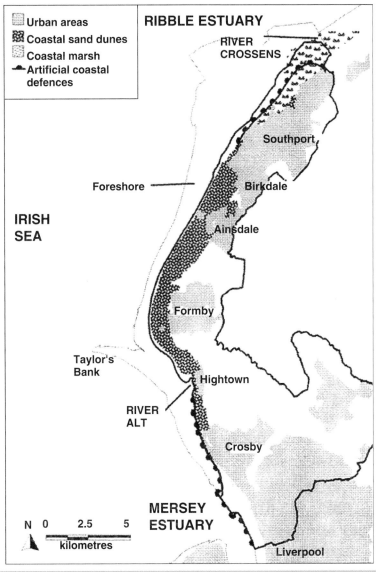

Figure 19.60 Sefton coast, south-west Lancashire, showing generalized landscape types. (Source: after Lymbery *et al.*, 2007)

Table 19.8 Recent and predicted climate changes in north-west England

Summary of recent changes in North West (or UK) Climate

- Average mean temperatures have risen by 0.4 °C at Manchester Airport between 1988 and 1997 when compared to the 1961–1990 average. This equates to 2.65 °C over a century (SNW, 1998).

- Globally the 1990's were the warmest decade in the last century, with 1998 being the hottest year on record and 2001 the third hottest (UKCIP, 2002).

- The thermal growing season for plants in central England has lengthened by about one month since 1900 (UKCIP, 2002).

- Seasonal rainfall has varied by as much as 15% from the average in the last 30 years (SNW, 1998).

- A decrease in summer rainfall during the last century of up to 20% (SNW, 1998).

- Increases in high intensity winter rainfall have been experienced since the 1960's (SNW, 1998).

- An increase in flooding of some major rivers in the region in the last few decades (SNW, 1998).

- Sea level around Liverpool has risen by about 6 cm in the last 50 years (SNW, 1998), over the last 100 years the UK average has risen by about 10 cm (UKCIP, 2002).

Summary of predicted changes to the North West (or UK) Climate.

- The warming in the North West varies between 0.7 °C and 2.1 °C by the 2050's (Holman et al, 2002) when compared to the 1961–1990 average. Average annual temperatures across the UK may rise by between 2 °C and 3.5 °C by the 2080s, depending on the scenario (UKCIP, 2002)

- There may be more warming in summer and autumn than in winter and spring (UKCIP, 2002).

- A small increase in annual rainfall of 3–4% is predicted but winters will become wetter than at present by between 10 and 35% whilst the summers may become drier (UKCIP, 2002).

- By the 2080s across the UK snowfall amounts will decrease by between 50 and 90% (UKCIP, 2002).

- More depressions may cross the UK in winter which could lead to stronger winds especially across the central and southern part of the country. However, predictions for changes in wind are very uncertain and must be treated with caution (UKCIP, 2002).

- The contrast between winter and summer climate will increase for all scenarios. Winters will become wetter and summers may become drier (UKCIP, 2002).

- Sea level rise predictions give a change of 7 and 36 cm by the 2050s, and by between 9 and 69 cm by the 2080s (UKCIP, 2002).

- Extreme sea levels, occurring through combinations of high tides, sea-level rise and changes in winds, may be experienced more frequently in many coastal locations (UKCIP, 2002).

Many recent changes in the north-west England climate may add to the effects of the sea-level rise, and a series of predicted future changes are summarized in Table 19.8, with predicted sea-level rises summarized in Table 19.9. It is suggested that the potential impacts on the coast can be summarized as follows:

1. There will be a substantial loss of area by 2105, although the threat to residential areas is relatively low.

2. The beaches and mudflats act as buffers to wave energy, absorbing it before it can impact on the coast, but with rising sea levels and increases in winter depressions these features will be reshaped and eroded, which will cause more storms (see Table 19.10).

3. The dunes stretch over 17 km and cover 2100 ha, are of major conservation importance and have a vital sea-defence

Table 19.9 Prediction of sea-level rise in north-west England. (Source: from UKCIP, 2002 and Defra, 2006)

	Low Emissions			High Emissions		
Year of Prediction	2020s	2050s	2080s	2020s	2050s	2080s
Net sea level change (cm)	4	7	9	14	36	69

Predicted sea level rise of the North West (UKCIP, 2002)

Administrative or Devolved Region	Assumed Vertical Land Movement (mm/yr)	Net sea level rise (mm/yr)				Previous allowances
		1990-2025	2025-2055	2055-2085	2085-2115	
NW England, NE England, Scotland	+0.8	2.5	7.0	10.0	13.0	4 mm/yr* constant

*Updated figures now reflect an exponential curve, and replaces the previous straight line graph representations.
Defra supplementary guidance on sea level rise.

Table 19.10 Potential impacts of sea-level rise on beaches and mudflats. (Source: after Wisse and Lymbery, 2007)

Changes in species composition of the intertidal flora and fauna may occur with species migration in a northerly direction. This could see species with a more northerly distribution losing out whilst species with a southerly distribution becoming more common. This could have a common effect or other species that feed on the foreshore.

With predicted increase in heavy winter rain events there could be more localised flooding and interstitial water from run off.

Linked to both rainfall, temperature and sea level rise there is the potential for the amount of interstitial water (the water that occupies the space between grains of sediment) to change are the salinity of this water to be altered, which could again affect the species composition within the sand and mud.

Rising sea levels could increase the amount of energy coming onto the shore. This could alter the steepness of the foreshore and sediment composition. A higher energy environment would see less deposition of finer sediments, this could lead to mud flats becoming sandier and the sand on the beaches becoming coarser. This in turn could affect wader species with those favouring mudflats such as redshank and dunlin reducing whilst other such as oystercatcher increasing.

Potentially alter sediment supply by changing coastal and estuary dynamics.

Could deprive sand dunes of wind blown sand as the rising sea water floods more of the beaches and reduces the amount of time that the beaches are exposed to the air.

An increase frequency of storm events could cause more redistribution of sediment across the shore and more significant changes in the foreshore topography.

function. However, they will be eroded and there will be a landward realignment, which will intensify because of increased wave energy and frequency due to increased storms. A potentially lower water table might lead to larger blowouts and increased sand mobilization. The potential impacts are given in Table 19.11.

4. The salt marsh will be further eroded and tides will rise higher and impact more frequently on the frontal marshes, flooding them for longer. Hence the vegetation will change and the marsh will realign landward, with erosion of the front edge.

The potential impacts to salt marshes are given in more detail in Table 19.12.

5. There will be habitat loss on the coast, and the sea walls at Crosby and Southport will suffer increased sea-overtopping events.

It is clear that there are a large number of potential consequences of climatic change and sea-level rise on this coast. Continued monitoring of coastal processes and their geomorphic effects will be important in predicting the scale and rate of change and the coast's future management and evolution.

Table 19.11 Potential impacts of sea-level rise on sand dunes. (Source: after Wisse and Lymbery, 2007)

Potential impacts to Sand dunes

- Changes to evaporation rates could affect the amount of dry sand on the foreshore and consequently the amount available to be blown into the dunes.

- Changes in species composition may occur due to increases in temperature with species migrating in a northerly direction. This could see species with a more northerly distribution losing out whilst species with a southerly distribution becoming more common.

- Earlier and longer growing seasons are likely to occur (UKCIP, 2002) but the increased likelihood of extreme summer temperatures (UKCIP, 2002) may increase stress on vegetation during the summer.

- With increased summer temperatures and longer sunshine hours there are likely to be more people visits to the coast. The increased visits could, however, negatively impact on the fragile frontal dunes and increase the erosion and amount of mobile sand if not correctly managed.

- Changes in the rainfall patterns could significantly affect the groundwater table and thus dune slacks and the species reliant upon them. It is likely that the water table will have an increased variability both within and between seasons and possibly for long periods it could be 1m lower than at present. However, there could also be occasions with a very high water table but becoming less frequent. The groundwater table in the dunes is vital for many plant species and wet slack habitats.

- With predicted increases in heavy winter rain events there could be more localised flooding.

- With lower rainfall in summer the foreshore and dunes will become drier making more sand available to be blown into the dune system.

- Similarly to changes in temperature, changes in rainfall could alter species composition in the dunes, with more drought tolerant species becoming dominant.

- Linked to both rainfall and temperature the reduced availability of soil moisture in the dunes could create a more stressful environment for vegetation.

- Rising sea levels could cause a landward realignment of the sand dunes (Cook and Harrison, 2001). Under all climate change scenarios coastal erosion will increase (Office of Science and Technology, 2004). Coastal squeeze could be a problem for the sand dunes in certain areas.

- Rising sea levels could potentially alter sediment supplies and process patterns by changing estuary dynamics.

- An increase frequency of storm events could have significant impacts on sand dunes by preventing the sand dunes recovering sufficiently between the events and increasing rates of erosion.

Table 19.12 Potential impacts of sea-level rise on salt marshes. (Source: after Wisse and Lymbery, 2007)

Potential impacts to Salt marshes

- Charges to evaporation rates that could affect the amount of moisture maintained and dry it more often creating a more stressful environment for vegetation.

- Charges in species composition may occur with species migrating in a northerly direction. This could see species with a more northerly distribution losing out whilst species with a southerly distribution becoming more common. It is predicted that salt marsh grass will begin to lose climatic space, this loss is more evident on the East coast with the areas of loss extending from the Thames to the Humber. However, it is expected that sea purslane will increase its range.

- Earlier and longer growing seasons are likely to occur (UKCIP, 2002) but the increased likelihood of extreme summer temperatures (UKCIP, 2002) may increase stress on vegetation during the summer.

- With predicted increase heavy winter rain events there could be more localised flooding.

- Similarly to changes in temperature, changes in rainfall could alter species composition vegetation during the summer, with more drought tolerant species becoming more established.

- Rising sea levels could cause a landward realignment of the salt marsh. Under all climate change scenarios coastal erosion will increase. Coastal squeeze could be a problem for the salt marsh in certain areas.

- The more frequent flooding of the marsh could lead to changes in species composition with more salt tolerant species becoming dominant.

- Potentially alter sediment supplies by changing estuary dynamics.

- An increase frequency of storm events could have significant negative impacts on salt marsh by preventing the salt marsh recovering sufficiently between the events and increasing rates of erosion.

Exercises

1. Describe the type of wave that can occur in a coastal system and the changes that can occur when it breaks onshore.

2. Explain why tides occur and why they vary both geographically and temporally.

3. What are tsunamis and how do they form?

4. What are the controlling factors for coastal erosion?

5. Explain why the beach profile varies through time.

6. How do beach cusps form?

7. Describe the morphology and sedimentary characteristics of beach barriers.

8. What are the types of water circulation associated with estuaries? Explain how they give rise to distinct types of estuary fill.

9. What are the controlling factors in the development of a coastal dune system?

10. What is the likely impact of climatic change on coastal landsystems in the next 50 years?

References

Benumof, B.T., Storlazzi, C.D., Seymour, R.J. and Griggs, G.B. 2000. The relationship between incident wave energy and seacliff erosion rates: San Diego County, California. *Journal of Coastal Research*. **16**, 1162–1178.

Bird, E. 2000. Coastal Geomorphology: An Introduction. Chichester, John Wiley & Sons, Inc.

Bryant, E.A., Young, R.S. and Price, D.M. 1996. Tsunami as a major control on coastal evolution, southeastern Australia. *Journal of Coastal Research*. **12**, 831–840.

Coco, G., O'Hare, T.J. and Huntley, D.A. 2000. Investigation of a self-organisation model for beach cusp formation and development. *Journal of Geophysical Research*. **105**, 21991–22002.

Coco, G., Burnet, T.K. and Werner, B.T. 2004. The role of tides in beach cusp development. *Journal of Geophysical Research*. **109**, C04011.

Dalrymple, R.W., Zaitlin, B.A. and Boyd, R. 1992. Estuarine facies models: Conceptual basis and stratigraphic implications. *Journal of Sedimentary Petrology*. **62**, 1130–1146.

Davies, J.L. 1980. Geographical Variation in Coastal Development. New York, Longman.

Defra. 2006. Flood and Coastal Defence Appraisal Guidance. FCDPAG3 Economic Appraisal. Supplementary Note to Operating Authorities: Climate Change Impacts, October 2006. http://www.defra.gov.uk/environ/fcd/pubs/pagn/Climatechangeupdate.pdf.

Dyer, K.R. 1973. Estuaries: A Physical Introduction. New York, John Wiley & Sons, Inc.

Guilcher, A. 1953. Essai sur la zonation et la distribution des formes littorals de dissolution du calcaire. *Annales Géographique*. **62**, 161–179.

Inman, D.L. and Nordstrom, C.E. 1971. On the tectonic and morphological classification of coasts. *Journal of Geology*. **79**, 1–21.

Lymbery, G., Wisse, P. and Newton, M. 2007. Report on Coastal Erosion Predictions for Formby Point, Formby, Merseyside. Bootle, UK, Sefton Council.

Long, A.J., Plater, A.J., Waller,, M.P., Roberts, H., Laidler, P.D., Stupples, P. and Schofield, E. 2004. The Depositional and Landscape Histories of Dungeness Foreland and the Port of Rye: Understanding Past Environments and Coastal Change. Durham, University of Durham. http://www.geography.dur.ac.uk/information/staff/personal/long/Dungeness.pdf.

Masselink, G. and Hughes, M.G. 2003. Introduction to Coastal Processes and Geomorphology. London, Hodder Arnold.

McAdoo, B.G. and Watts, P. 2004. Tsunami hazard from submarine landslides on the Oregon continental slope. *Marine Geology*. **203**, 235–245.

McMurtry, G.M., Watts, P., Fryer, G.J., Smith, J.R. and Imamura, F. 2004. Giant landslides, mega-tsunamis, and paleo-sea level in the Hawaiian Islands. *Marine Geology*. **203**, 219–233.

Penland, S., Boyd, R. and Suter, J.R. 1988. Transgressive depositional systems of the Mississippi Delta plain: Model for shoreline and shelf sand development. *Journal of Sediment Petrology*. **58**, 932–949.

Pethick, J. 2001. An Introduction to Coastal Geomorphology. 2nd Edition. London, Arnold.

Shepard, F.P. and Kuhn, G.G. 1983. History of sea arches and remnant stacks of La Jolla, and their bearing on similar features elsewhere. *Marine Geology*. **51**, 139–161.

Stoddart, D.R. 1969. Ecology and morphology of recent coral reefs. *Biological Reviews*. **44**, 433–498.

Sunamara, T. 1975. A laboratory study of wave cut platform formation. *Journal of Geology*. **83**, 389–397.

Sunamara, T. 1977. A relationship between wave induced cliff erosion and erosive force of waves. *Journal of Geology*. **85**, 613–618.

Sunamara, T. 1981. A predictive model for wave-induced cliff erosion, with application to Pacific coasts of Japan. *Journal of Geology*. **90**, 167–178.

Sunamara, T. 1992. Geomorphology of Rocky Coasts. Chichester, John Wiley & Sons, Inc.

Switzer, A.D., Pucillo, K., Haredy, R.A., Jones, B.G. and Bryant, E.A. 2005. Sea level, storm or tsunami: Enigmatic sand sheet deposits in a sheltered coastal embayment from southeastern New South Wales. *Journal of Coastal Research*. **21**, 655–663.

Trenhaile, A.S. 1987. The Geomorphology of Rock Coasts. Oxford, Oxford University Press.

Trudgill, S.T. 1976. The marine erosion of limestones on Aldabra Atoll, Indian Ocean. *Zeitchrift für Geomorphologie. N.F. Supplement Bund*. **26**, 164–200.

UKCIP (UK Climate Impact Programme). 2002. Climate Change Scenarios for the United Kingdom. Oxford.

Whelan, F. and Kelletat, D. 2003. Submarine slides on volcanic islands: A source for mega-tsunamis in the Quaternary. *Progress in Physical Geography*. **27**, 198–216.

Williams, H.F.L., Hutchinson, I. and Nelson, A.R. 2005. Multiple sources for late-Holocene tsunamis at Discovery Bay, Washington State, USA. *The Holocene*. **15**, 60–73.

Wisse, P. and Lymbery, G. 2007. Climate change and the Sefton Coast: Implications for coastal geomorphology. Merseyside, UK, Sefton Council.

Further reading

Davis, R.A. Jr. and Fitzgerald, D.M. 2004. Beaches and Coasts. Oxford, Blackwell Science.

Schwartz, M.L. 2005. Encyclopaedia of Coastal Science. Berlin, Springer.

20 Glacial Processes and Landsystems

Tidewater and surge-type glaciers Kongsbreen and Kronebreen in Kongsfjorden, north-west Spitsbergen. Source: with permission of Norsk Polarinstitutt

Earth Environments   David Huddart and Tim Stott
© 2010 John Wiley & Sons, Ltd

Learning Outcomes

After reading this chapter and completing the exercises you will be able to:

➤ Understand the concept of glacier mass balance and how snow is transformed into glacier ice in various glacier types.

➤ Know the mechanisms for glacier flow, including hypotheses for surges.

➤ Understand how glacier ice can erode the landscape and the types of sediment and landforms that can be produced.

➤ Understand how glaciers and ice sheets can deposit sediments in a range of glacial landforms and landsystems.

20.1 Introduction

Today ice, in the form of ice sheets and glaciers, covers approximately 11% of the Earth's surface, a global total of over 15 million km², with 99% residing in the ice sheets of Antarctica (91%) and Greenland (8%). Just 18 000 years ago, during the last glacial episode of the Cenozoic Ice Age, this figure was as high as 30%, although the current glacial era began in Antarctica as early as 40 million years ago. The bulk of this increase in ice cover occurred within the northern midlatitudes, where large ice sheets expanded over North America (Laurentide ice sheet), Fenno-Scandinavia, the Barents Sea and the British Isles. These ice sheets were the scribes of the Ice Age and have left a legacy of landforms and sediments that dominate the landscape of these regions today.

Large ice sheets have also been important in earlier geological intervals (Figure 20.1).

However, although ice sheets have been, and continue to be, the most important shapers of the landscape, there are many other scales and types of ice mass that also influence the appearance of the landscape. Ice masses can be classified morphologically based on the size and characteristics of their environment:

1. **Niche, cliff or wall-sided glaciers:** These are triangular wedges of ice lying in a shallow funnel-shaped hollow in the upper part of a hillslope (Figure 20.2(a)). They have steep slopes and are often associated with rock benches. They originate as snow-patch nivation hollows and are an early stage of corrie or cirque glaciers.

2. **Corrie or cirque glaciers:** These occupy an armchair-shaped hollow, with a steep back wall, and are surrounded by arête ridges (Figure 20.2(b)).

3. **Valley glacier of the Alpine type:** These are formed when ice flows out of a corrie basin downvalley, although sometimes ice moves from several corries, as in the Grosser Aletsch glacier in Switzerland. These glaciers are confined within valley walls throughout their length and terminate in a narrow tongue.

4. **Valley glaciers of the outlet type:** These are fed from ice caps, not corrie glaciers. Examples are found in Iceland and Norway, although the biggest is the Lambert glacier in Antarctica, which is 700 km long. The altitudinal range of these valley glaciers is high in relation to their size, so the bedrock slopes are probably steep and the ice turnover is vigorous. The ice moves over an ice fall and flows in a rock-walled trough.

5. **Transection glaciers:** These occupy much of a mountain area and glaciers flow from them in different directions in radiating valleys. The accumulation area at a high elevation is

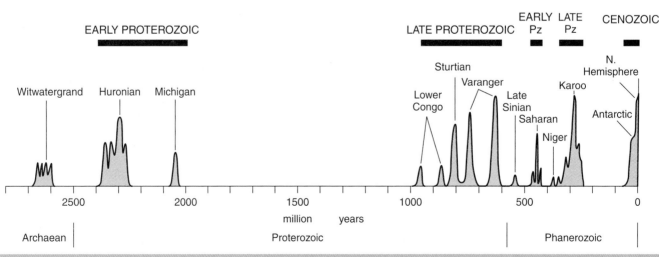

Figure 20.1 Periods of glaciation in the geological record. (Source: after Hambrey, 1992)

Figure 20.2 (a) Niche or wall-sided glacier, Kaskasatjokkglacieren, Kebnekaise, northern Sweden. (b) Cirque glaciers, western Spitsbergen. (c) Malaspina piedmont glacier, Alaska, 1994. (d) Lambert glacier, Antarctica ice streams. (Sources: (c) NASA; (d) USGS)

not large enough to be called a mountain ice cap. An example is the Lowenskjold glacier in Spitsbergen.

6. **Piedmont glaciers:** These are formed when valley glaciers move out from the mountain front onto a lowland beyond. The classic examples are the Malaspina and Bering glaciers in Alaska (Figure 20.2(c)).

7. **Floating ice tongues and shelves:** These occur where the ice flows into the sea beyond. Today they mainly occur in Antarctica and Greenland. They are often fed by long valley glaciers and ice streams (Figure 20.2(d)).

8. **Mountain ice caps:** These occur at high altitudes, where ice flows from the accumulation area to lower levels. An example is the Svartisen ice cap in northern Norway.

9. **Glacier cap or lowland ice cap:** These occur where relatively small ice caps develop at fairly low altitudes on flattish land in the high Arctic. An example is the Barnes ice cap in Canada.

10. **Ice sheets:** Here the ice builds up over a continental scale. Examples include today's Antarctic and Greenland ice sheets and many others in the geological record.

However, a much simpler morphological classification recognizes three fundamental types, based on the way in which their morphological expression reflects an interaction between the glacier ice and the topography. There is an ice sheet or ice cap, which is superimposed on the underlying topography, which it largely submerges. In this case the direction and flow of ice reflect the ice's size and shape rather than the shape of the ground. Then there are ice shelves, which are floating ice sheets, only loosely constrained by the shape of the coastline. Finally there are glaciers, constrained by the topography, which are strongly influenced in both form and flow direction by the ground shape.

In this chapter we examine the mechanics of the glaciers that carved these landscapes and review the landform and sediment assemblages that they left behind. Before considering these landform–sediment assemblages, the controls on glacier growth and decay and the mechanics of glacier flow will be examined.

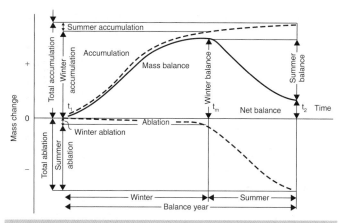

Figure 20.3 Total accumulation and total ablation curves for a glacier

20.2 Mass balance and glacier formation

20.2.1 How are glaciers formed?

Glaciers are formed of ice crystals, snow, air, water and rock debris and will form whenever a mass of snow accumulates over successive years, compacts, and turns to ice. This can occur in any climatic zone in which the input of snow exceeds the rate at which it melts, but generally glaciers are located in high mountains, high latitudes and on the western sides of continents. Glacier ice can be defined as ice made by the recrystallization of fallen snow and refreezing of melt water that has undergone deformation.

The inputs to the glacier system are snowfall (which is transformed to glacier ice), the potential energy of elevation and rock debris. These inputs are cascaded or transferred from one climatic environment to another when there is a progressive loss of mass by evaporation, melting and deposition, and also dissipation of energy from the system as heat. The output is expressed in terms of the modification of the drainage basin by the ice and its associated melt waters, by either erosion or deposition. Once established, a glacier's survival will depend on the balance between its accumulation and mass loss (ablation). This is known as the glacier's mass balance and is controlled primarily by climate, through a chain of processes. There is a local exchange of mass and energy at the glacier surface, the dynamic response of the glacier, which affects its net budget and results in some enduring evidence of the position of the glacier terminus.

Figure 20.3 shows how the total amount of accumulation and the total amount of ablation each year define the annual mass balance of a glacier. Ablation, by melting, will tend to dominate in the warm summer months, and accumulation in the winter months. If the amount of ablation equals the amount of accumulation over a year then the net balance of the glacier will be zero and its size will remain constant. On the other hand, if there is more accumulation than ablation then the net

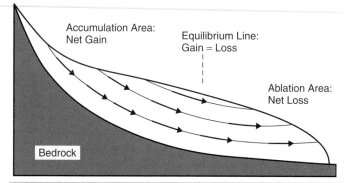

Figure 20.4 Accumulation and ablation subsystems of a glacial system

balance will be positive and the glacier will grow and expand. If it has a negative mass balance then the glacier will shrink rapidly and may ultimately disappear, as is the case at present under the influence of global warming. Hence the glacial system has two recognizable subsystems, the accumulation and ablation subsystems (Figure 20.4), which are separated by a conceptual line (the equilibrium line). Glaciers experience changes of input and respond to these, but there is always a lag or response time. So we can say that glaciers are usually out of equilibrium in the short term.

The study of glacier mass balance is therefore the study of inputs to the glacial system. Inputs to a glacier's mass balance include snow, hail, frost, avalanched snow and rainfall where it freezes within the glacier. If these inputs survive summer ablation they will begin a process of transformation into glacier ice. Snow is the building block of glacier ice and each ice crystal is formed of hundreds to thousands of snowflakes welded together into a single homogeneous structure. The transformation from snow to glacier ice takes place as a continuous process, but several well-defined stages can be recognized:

1. **Newly-fallen snow crystals to granular or old snow:** The irregular shape of the snowflakes gives snow an enormous

surface area, which gradually decreases through time. In dry, fresh snow, air occupies up to 97% by volume. Initially the parent material is a light, fluffy, loose aggregate of snowflakes, which accumulate in successive layers, each of which represents a single storm or different phases within a storm. These layers show stratification due to differences in density, grain size, porosity and compactness, which depend on the conditions of precipitation and deposition of the snow. Almost as soon as it is deposited the snow begins to change its physical characteristics through the processes of sublimation (which is the conversion of a solid directly into a vapour and subsequent condensation without melting), local melting, and crushing and compaction. Initially the snow density is 0.05–0.08 but this increases rapidly with time. Within a few days to a few weeks it becomes a loose, still highly-pervious, granular aggregate known as old snow, in which the density increases to >0.3. This density increase is caused by compaction and settling due to melting of the snow crystal points. This leads to an increase in the average grain size as smaller grains tend to melt before larger ones, and grains may join together by regelation, which is refreezing after pressure melting. This is particularly the case in the surface layers, which undergo a daily cycle of freezing and thawing. Meltwater facilitates the grain packing by lubricating the grains, since the surface tension of a water film tends to pull the grains together. Gradually the air spaces between the crystals are eliminated. There are important temperature effects too, as crystals are smaller where temperatures are lower, and as temperatures increase many crystals stick together and fall as large flakes. Where temperatures are close to $0\,^{\circ}$C newly fallen snow becomes coarse and granular within a few days, but in very cold, polar climates the process can be delayed for years. In inter-tropical areas in high mountains the snow often falls as soft hail, which partially melts during the day and refreezes at night, and is rapidly transformed to old snow.

2. **Old snow to firn or névé:** That is, snow that has survived a summer melt. The change occurs where the density reaches 0.4–0.55; the latter figure is the greatest that can be achieved in old snow by simply shifting the grains so that they fit snugly together. Any change beyond that involves deformation, melting, refreezing and recrystallization, which allows the grains to fit more closely, reduces the pore space and permeability, and increases the density and compactness. The grain size increases to >0.7 mm, up to several millimetres.

3. **Firn to glacier ice:** This occurs when all the interconnecting air passages between the grains are sealed; that is, when the permeability to air is 0, and therefore when the permeability to water is also 0. When this happens the density is between 0.8 and 0.85. This is normally achieved after burial by accumulation to depths of >30 m, but the time taken is very variable; for example in the Claridenfirner in Switzerland it can take anywhere between 25 and 40 years, and in Greenland the period is about 150–200 years. Gradually all the air

bubbles slowly transform and reduce in size, the crystals grow in size to over 1 cm and the material becomes hard, blue ice.

4. **Slow plastic deformation of the ice under its own weight:** Reductions in pore space take place and density increases up to values of 0.89–0.9. The ultimate density of pure ice is 0.917, but this is rarely reached, except in individual ice crystals.

We have seen that the transformation involves compaction, the expulsion of air and the growth of interlocking ice crystals, and the rate at which it takes place is dependent on climate. For example, in the interior of the eastern Antarctic ice sheet, where there is very little accumulation and even less melt due to the continentality of the climate, the transformation may take up to 3500 years, at a depth of 160 m. By contrast, on the Upper Seward Glacier in the Yukon the transformation is achieved in as little as 3–5 years, at only 13 m depth, since the maritime climate gives high accumulation and melt rates.

Mass balance studies have to do with changes in glacier mass and how these occur in both time and space. More particularly, they measure the mass change in any given year. The concepts are important because this is the link in the chain between climate and glacial advances or retreats. It has important practical importance too, as melt streams are used extensively by hydro-electric power stations. The results of mass balance measurements become particularly important where there is a long sequence of results through time, as at Störglacieren in Kebnekaise (northern Sweden) or the South Cascade glacier in the USA, where trends can become apparent (Figure 20.5). The mass balance of a glacier can be likened to your own personal income or a country's income; it reflects the health of a glacier (Figure 20.6).

20.2.2 How is mass balance measured?

Depending on the difficulty of access of the glacier studied, the technique might be direct measurement of snow and ice gain or loss, photogrammetric methods or hydrological methods.

Direct methods

The usual method is to measure the net balance at representative positions on the ice. The data can be plotted on a map and contours of equal balance can be drawn. The accuracy is difficult to establish because there may be sampling problems, for example crevassed areas that are hard to sample, and it is difficult to establish how representative the points that have been used are. The mass of snow and ice accumulated during the current balance year that remains at the end of the year is measured in the accumulation area. Pits have to be dug or cores taken to measure this net balance and previous summer surfaces (the time of minimum thickness) must be identified, which may be indicated by a layer of dirt in the firn, or a sudden density, hardness or grain-size change (Figure 20.7). So in a pit, the thickness of the snow accumulation and the density of the layers

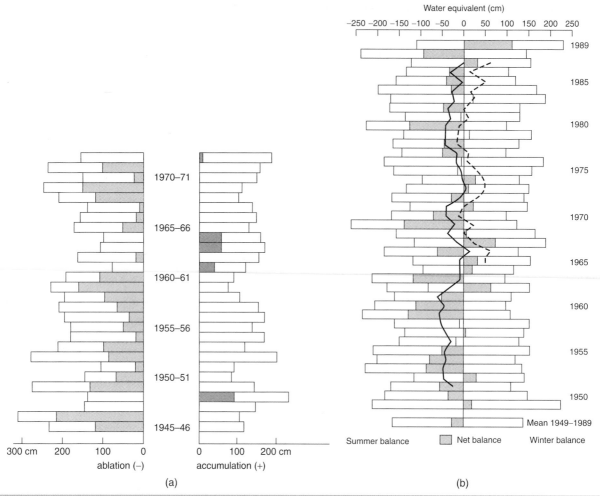

Figure 20.5 Long-term mass balance records for: (a) Störglacieren, Kebnekaise, northern Sweden; (b) South Cascade glacier, USA

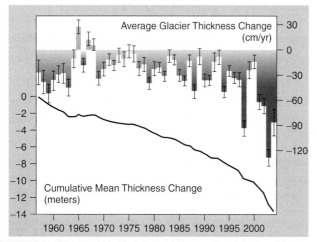

Figure 20.6 Mass balance of world glaciers, showing the cumulative mean thickness change. (Source: from data prepared by Robert R. Rohde of the National Snow and Ice Center. Image from Global Warming Art Project, http://www.globalwarmingart.com/wiki/image:Glacier_Mass_Balance.png, GNU Free Documentation Licence)

Figure 20.7 Mass balance measurement pit in the accumulation area of Störglacieren, Kebnekaise. Note the previous summer layer

must be measured by weighing a known volume and assuming an ice density of 0.89. The measurements at points are usually expressed as equivalent volumes of water per unit area.

The mass of ice lost from the ablation area has to be measured by stakes drilled into the ice as references. The distance between the top of a stake and the ice surface is measured at the beginning and the end of the balance year. The difference between the two figures is multiplied by 0.9 to convert to water equivalent balance. There are difficulties of sampling here too; in particular, the stakes can melt out over an ablation season.

Photogrammetric methods

Here accurate glacier maps are made at intervals of one or a few years and comparisons of two maps are used to determine volume change in the intervening period. This can be converted into a change in mass by using average densities for firn and ice. As the annual change – except near the glacier front – may be only a few tens of centimetres, a high standard of contouring accuracy is needed, and the timing of the photo runs is crucial. The map must show the glacier at the end of the balance year.

Hydrological methods

These only determine the net balance for the whole glacier. Measurements must be made over the whole of the drainage basin in which the glacier lies. Stream gauges are used to measure the total runoff from the basin (R). Precipitation (P) has to be measured, which is difficult. The amount of water, ice and snow lost by evaporation (E) must also be measured or estimated. If each of these quantities is totalled over the whole balance year and expressed in volumes of water then the net balance (B_n) can be obtained from the equation $B_n = P - R - E$.

The mass balance changes result in thickening or thinning and advances or retreats over time. Numerous studies have examined the relationship between glacier mass and climatic inputs, and correlations between mass changes for the Rhonegletcher in Switzerland with climatic records for the period 1882–1987 show that most of the mass loss is attributable to temperature increases, particularly since the 1940s. As a general rule, summer temperatures are more important than precipitation in controlling mass changes of mountain glaciers in humid maritime areas. Some workers have suggested that precipitation is more important, however. Important links between large-scale atmospheric and oceanic circulation and glacier mass balance have been identified; for example, years of low mass balance on Peruvian and Bolivian glaciers coincide with the occurrence of El Niño.

20.2.3 Glacial melt water

Outputs from the mass balance system are collectively known as ablation. Ablation may occur in one of three main ways: by ice melt, by iceberg calving or by sublimation. Each will be examined in turn.

Ice melt

Glacial melt water is derived from direct melting of ice on the surface of or within a glacier. On the surface this is a function of solar radiation received, while within and at the base of the glacier heat is supplied by friction due to ice flow and heat derived from the Earth's crust beneath the glacier (geothermal heat). Melting can therefore occur both on the surface of the glacier and within the glacier itself. Surface melting is primarily a result of warm air temperatures and is therefore highly seasonal, whereas melting within the glacier is not. It is important to emphasize that melting is not confined to the ice margin but may occur across the whole of the glacier.

As melt water is produced it drains from the glacier via a number of different routes, ranging from surface channels to subglacial conduits or tunnels. The flow routing adopted by any given glacier is dependent on the thermal characteristics of the glacier ice involved. If the glacier ice is either actively melting or at the pressure melting point then it is said to be warm and wet. If these conditions prevail at the glacier bed then the glacier is a warm-based glacier and its bed contact is likely to be well-lubricated by water. The term 'pressure melting point' reflects the fact that the melting point of water varies with pressure. On the surface of a glacier close to sea level ice melts at $0\,°C$, but at depth within the glacier the weight of overlying ice depresses the melting point slightly. In contrast, where the temperature of the ice is far below the pressure melting point and the ice is frozen to its bed the glacier is said to be a cold-based glacier. In practice, glaciers may possess mixed thermal regimes, in which certain sections are warm-based and others are cold.

If we focus simply on the two extremes, each has a very different style of melt water drainage. In a warm-based glacier, surface melt water quickly finds its way to the glacier bed. Surface melt water becomes concentrated into surface or supraglacial channels (Figure 20.8(a)), which are quickly intersected by crevasses that route the water into the body of the glacier. Crevasses that capture this supraglacial water may evolve into vertical shafts known as moulins (Figure 20.8(b)), which direct the water into englacial conduits that ultimately drain towards the glacier bed. Melt water generated at the glacier bed or routed to it from the surface may flow either in a distributed network of channels and water-filled cavities or within a series of arterial conduits or tunnels at the glacier bed. Subglacial channels (Figure 20.8(c)) may be cut either into the underlying substrate, in which case they are known as Nye-channels, or upwards into the base of the glacier, where they are known as Röthlisberger-channels (both named after glaciologists). Arterial drainage systems tend to have a dendritic pattern, where the number of channels decreases progressively down-glacier as they converge and merge. The direction of englacial or subglacial conduits within a glacier is determined by the pressure gradient imposed by the weight of the glacier. Consequently, the direction of flow is controlled by variations in ice thickness, which in turn are determined by the surface topography of the glacier and the gradient of its bed. Englacial conduits cut into the ice, or subglacial channels, are maintained by the balance between the melting of the tunnel

Figure 20.8 Glacier drainage systems: (a) supraglacial pattern, the Bersaerkabrae, Stauning Alps, east Greenland; (b) supraglacial stream, Fabergstolsbreen, central Norway; (c) moulin and crevasse in Antarctica. (Source: Wikimedia Commons)

walls caused by the frictional heat generated by the water flow and their closure by ice flow.

Discharge from warm-based glaciers often shows a distinctive annual pattern, associated with the evolution of the internal plumbing of a glacier each year (Figure 20.9). During the winter, melt water flow is low and therefore englacial and subglacial conduits and tunnels tend to close. Supraglacial streams are also hidden beneath fresh snow. As the spring melt begins, water starts to accumulate within the glacier, since the drainage system has not yet evolved. Gradually during the spring and early summer a system of supraglacial channels and englacial and subglacial conduits evolves and the rate at which glacial melt water is discharged from the glacier increases. This often results in a peak of water flow in early summer as the water stored within the glacier is drained. Thereafter discharge levels are closely linked to the rate of melting within the glacier.

In contrast, cold-based glaciers have a drainage pattern dominated by water flow in supraglacial channels, since water penetrating the glacier surface tends to become frozen. Englacial and subglacial conduits can exist, particularly close to the ice margin,

if the discharge of surface water and therefore the heat generated by it is sufficient to combat the sub-freezing temperatures of the ice.

Iceberg calving

Where a glacier terminates in water (Figure 20.10(a)), either in the sea or in a lake, blocks of ice will break from its front (snout or terminus) as icebergs. This process is known as iceberg calving and can be a particularly rapid way for a glacier to lose mass. The rate of glacier calving is a function of several factors. For example, the presence or absence of a tidal range can be important, since the rise and fall of water may cause the ice margin to flex, encouraging it to calve. However, the principal factor is the geometry of the water body into which the glacier terminates. In particular, water depth is critical since the snout is more likely to either be floating or close to floating in deep water, and therefore the glacier is likely to calve more rapidly. The water-based glacier margins tend to be located at pinning points, which are points at which the

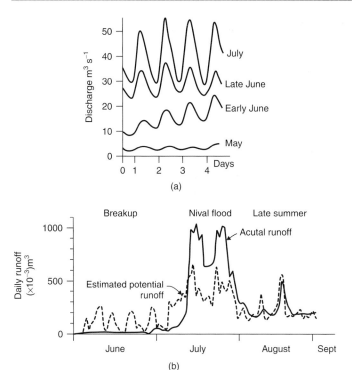

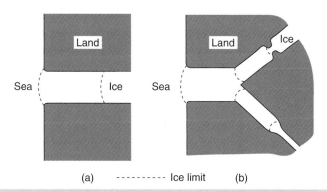

Figure 20.9 Annual pattern of discharge from glaciers, showing the diurnal and seasonal fluctuations: (a) development of stream hydrographs from May to July, showing the development of the diurnal cycle; (b) subdivisions of daily runoff from break-up through the nival flood to late summer

Figure 20.10 Control of fjord geometry on the location of ice margins. (a) In a fjord of constant width and depth a glacier experiencing a climatic deterioration will advance to the fjord mouth, since only there can the ice margin increase in size, thereby increasing the rate of calving and hence ablation. (b) Examples of pinning points

water depth or ice front width is lower, providing more stable conditions.

Within Alaskan tide-water glaciers, cycles of behaviour have been detected that are out of synchroneity with changes in mass balance and are the result of instabilities caused by calving margins. As a tide-water glacier advances into deeper water it deposits sediment, which reduces water depth, in turn reducing the rate of calving. If the glacier loses less mass it will continue to have a positive balance and will therefore continue to advance until the rate of calving matches the rate of accumulation. This may occur in deep water. If the glacier now becomes decoupled from its moraine bank, by a large calving event or some other random event, the margin will be in deep water and will calve rapidly. Retreat may occur until the glacier reaches the head of the fjord, at which point its margin will not be in phase with its mass balance, causing it to advance again. In this way tide-water glaciers may show cyclic episodes of advance and decay that are not in phase with climate.

This capacity for a nonlinear response to changes in mass balance was noted in a simple conceptual model, shown in Figure 20.10. Here the rate of glacier calving is proportional to the width of the calving front. Consider two glacier lobes, where one terminates in a narrow fjord and the other on land. If both glacier lobes experience a deterioration in climate and a shift towards a more positive mass balance, both their ice margins will begin to advance. The land-based ice margin will advance until its snout is at a lower elevation, or the surface area of its ablation zone is increased sufficiently to balance the mass gain due to the deterioration in climate and the increased accumulation. The advance of the ice margin is likely to be modest and in proportion to the scale of the climatic deterioration. In contrast, the tide-water margin will already be at at sea level and the rate of ablation will largely be controlled by the rate at which calving can occur from the snout. The rate of calving is restricted by the width of the calving margin from which icebergs can break off. In this case the glacier must advance to a point in the fjord at which the width of the calving front can increase, assuming a uniform water depth. If this point is some distance down the fjord then the advance is likely to be considerable and out of proportion with the deterioration in climate. This simple model illustrates that the water-based margins are not necessarily stable and that small changes in mass balance can cause dramatic changes in the glacial geometry.

Sublimation

In very cold and dry environments, such as those in the interior of Antarctica, mass may also be lost through sublimation. This process, although rare, is important, since the sediments produced tend to preserve the sedimentary structures within basal sediment layers.

Effects on mass balance

The growth and decay of large ice sheets is a complex problem. Traditionally, ice sheets are considered to grow – in response to a positive mass balance – through a sequence of larger and larger ice bodies, from snow patches, through cirque glaciers, and on to valley glaciers, ice fields, ice caps and ice sheets. They develop first in upland areas and then expand into lowland regions as the ice masses merge and grow. More recently, computer models have suggested that the sequence of growth may follow a slightly different pattern, from snow patches, through cirque glaciers and valley glaciers, and into piedmont lobes and then small ice sheets.

In this scenario valley glaciers first develop in mountainous areas and then flow out into low-lying areas, where the ice spreads out as large lobes, known as piedmont lobes. These lobes merge and thicken rapidly in an unstable fashion, giving small ice caps which merge to form larger ice sheets. The piedmont lobes thicken and grow dramatically, since they have low gradients and consequently low ice velocities, and cannot therefore discharge the ice pouring into them from the fast-flowing valley glaciers that drain the steep mountainous areas behind. This situation leads to instability and the rapid growth of the initial ice cap. The presence of instability within a system – in this case rapid increase in ice volume out of proportion with the change in mass balance – makes it both unpredictable and extremely dynamic.

A similar situation can occur as a consequence of topography. A computer model of the Younger Dryas ice cap in the Scottish Highlands illustrates this point well. It shows the sensitivity of this ice cap to the mountain topography of the Highlands, where certain types of topography accelerate its growth. For a given deterioration in climate, parts of the ice cap which advance into large basins grow more dramatically than those centred on topographic ridges. Consequently the location of large topographic basins has a dramatic effect on the rate at which ice sheets grow, and their centres are not necessarily located on areas of maximum topographic elevation. In these locations small deteriorations in climate can have a dramatic effect on the size of the ice mass.

Deglaciation, or the decay of a glacier or ice sheet, occurs as a result of a sustained shift to a negative mass balance, something that can be achieved in one of three ways:

1. **Varying precipitation:** Increasing or decreasing precipitation can affect mass balance. For example, the glacier may begin to decay (a negative mass balance) due to a decrease in precipitation. It has been suggested that the last ice sheet in Britain during the last glacial maximum began to decay as a consequence of precipitation starvation, due to the polar front being located well to the south of Britain, which limited the number of snow-bearing depressions.

2. **Varying air temperature:** Air temperature is linked to the amount of ablation, where an increase in air temperature may lead to a negative mass balance through enhanced ablation.

3. **The proportion of maritime ice margin:** That is, the length of ice margin at which iceberg calving can occur. This is controlled by topography and relative sea level. For an ice sheet close to sea level, this may increase as a consequence of eustatic sea-level rise or isostatic depression. For example, rapid deglaciation of the Irish Sea during the last glacial maximum has been suggested to be a consequence of eustatic sea-level rise while the British Ice Sheet was at its maximum (and the maximum isostatic depression). This implies that ice sheets in different parts of the world do not necessarily respond in a synchronous fashion.

Variation in any or all of these three variables may result in a change in the mass balance of a glacier. This cautions against assuming a simple causal link between temperature rise and ice-sheet decay. It is also important to emphasize that when terminating in water, glaciers may show marginal fluctuation as a result of mass balance changes that are not driven solely by climate.

20.3 Mass balance and glacier flow

So far we have considered the mass balance of a glacier as a whole, but in practice the distribution of mass gains and losses on a glacier is uneven. In a simple valley glacier, accumulation occurs in its upper reaches, since temperatures decline with elevation and precipitation may be initiated by topography. In contrast, ablation will tend to dominate over accumulation at the snout of a glacier, as it is at a lower elevation and may terminate in water. Consequently, mass is added in the upper reaches and lost at the snout. It is this imbalance which drives glacier flow. It is possible to define zones on a glacier or ice cap, as we have seen: one in which accumulation dominates and another dominated by ablation. The line between these two zones defines the equilibrium line for the glacier, the line along which mass gain and loss are balanced. The elevation of this line will reflect the prevailing climate and geometry of the glacier basin.

The addition of mass in the upper reaches of a glacier and its loss at the snout will steepen the glacier's longitudinal gradient. Consider a valley glacier: during a balance year a wedge of snow and ice is added above the equilibrium line and one is removed from below it, causing the gradient to steepen. It is this steepening of the ice-surface gradient that provides the driving force for glacier flow. The shear stress at the base of the glacier is proportional to its thickness and gradient, such that an increase in the ice-surface gradient will increase the gravitational shear stress imposed on the ice. Ice must be transferred from the accumulation zone to the ablation zone to counter this.

Glaciers may flow in response to any or all of the following processes:

1. **Internal deformation or ice creep:** Glacier ice is polycrystalline, as it is composed of numerous crystals of different sizes, shapes and often orientations, and may deform or strain under stress. This occurs due to the deformation of individual crystals via the dislocation of atomic layers within an ice crystal. A simple explanation suggests that ice crystals consist of a series of hexagonal atomic rings which glide over one another in response to an applied stress, like a stack of playing cards. When a stress is applied to a single ice crystal, the crystal immediately deforms by a small amount. If this is below the yield strength, or elastic limit, of ice, the crystal will return to its normal form when the stress is removed. However, if it is above the elastic limit then permanent deformation will occur. The rate of ice creep is determined by Glen's Flow Law, in which the strain rate is proportional to an ice hardness

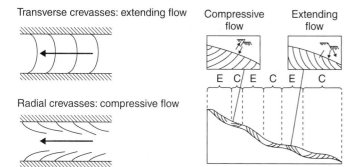

Figure 20.11 Extending and compressive flow within glaciers. The surface crevasse pattern is different and compressive stress is associated with a decrease in the subglacial slope, whilst extensional stresses occur where the glacier bed steepens

factor and to the applied stress to the power of three. This means that a small increase in stress causes a large increase in the rate of strain. The ice hardness factor is dependent on a range of factors, including ice temperature, water content, the impurity content, crystal size and dirt inclusions. Of these, temperature is perhaps the most important. At melting point a stress of 1 bar will lead to a strain of 30% per year, but at $-15\,^{\circ}$C to a strain of only 3% per year. Deformation may also occur as a result of folding and faulting of ice. Ice deformation can be expressed in terms of extending and compressive flow, and crevasse patterns on the glacier reflect these different modes (Figure 20.11).

2. **Basal sliding:** Where glaciers are unfrozen at their beds, sliding may occur along this basal boundary. This process is controlled by a range of different factors, but most strongly by the degree of coupling between the glacier sole and the bed. This is a function of the bed's roughness, of subglacial water pressure and of the organization of the subglacial water system. Basal sliding has been modelled in terms of two processes by which ice can overcome basal obstacles or roughness: (a) regelation slip, which occurs where ice close to the melting point encounters an obstruction to flow; (b) enhanced creep due to the concentration of shear stress against the upstream face of obstacles. Upstream of an obstacle at the glacier bed, basal stress is increased, while downstream it shows a slight decrease. This gives rise to slight differences in the pressure melting point of the ice. The melting point will be depressed upstream of the obstacle, due to the enhanced pressure. Consequently, ice close to pressure melting will melt upstream of the obstacle and the water will pass around the obstacle to refreeze downstream. This process is known as regelation slip and works best for small obstacles, since the heat liberated by the refreezing of the water (release of latent heat) can pass through the obstacle to compensate for the depression of ice temperature due to melting (consumption of latent heat). In contrast, larger obstacles concentrate more stress on their upstream faces, which causes a local acceleration in the rate of creep, which as we saw above is very sensitive to the applied

stress. In this way, ice may move around both small and large obstacles, although it is worth noting that bed roughness of an intermediate size may pose the maximum resistance to sliding. The rate of basal sliding is also influenced by the contact area between the bed and the glacier sole, which is determined by basal water pressure. Basal friction is a function of the weight or thickness of the overlying ice, as thin ice imposes less basal friction than thick ice. However, one must also consider basal water pressure, since high levels of basal water pressure can effectively 'jack-up' or help support the weight of the ice. The greater the water pressure, the less the basal friction and consequently the easier it is for basal sliding to occur. This is shown by strong relationships on many valley glaciers between summer melt or rainfall events and episodes of enhanced basal sliding. The organization of basal drainage is also important. As discussed above, a glacier may adopt one of two types of basal drainage system: a system of linked cavities or an arterial drainage system. In the latter case the contact area between the bed and the glacier sole is reduced, allowing basal sliding to occur. A small proportion of glaciers show unstable modes of glacier flow, known as surges, in which they experience episodes of accelerated flow, punctuated by periods of quiescence. One of the explanations proposed to explain such behaviour is the switching between arterial and distributed drainage systems, the latter allowing rapid glacier surges.

3. **Subglacial deformation:** Where glaciers rest on sediment rather than bedrock, movement may be accommodated below their base by deformation within the sediment. Movement of markers in a borehole drilled into subglacial till from an englacial conduit at the margin of Breidamerkurjökull in Iceland was reported. Over a period of 244 hours, 88% of the glacier's forward movement occurred within the subglacial sediment layer. These observations have had a profound effect on our understanding of glacier motion, and of glaciology in general. The idea is that saturated basal sediments may deform at a lower yield stress than ice, and thereby accommodate much of the strain within the system. The more saturated the sediment, the weaker it is likely to be, and the faster the glacier will flow. Deformation is therefore favoured by high water supply, or poor drainage, and by fine-grained sediments.

A glacier may flow by any or all of these processes, and their relative importance is often fiercely debated. In general, the basal thermal regime of a glacier is critical to the nature of the flow processes that will operate. Traditionally, glaciers that are cold-based – that is, frozen to their beds – are considered to flow only by internal creep, while warm-based, wet-melting glaciers may flow by basal sliding and bed deformation, given a suitable substrate. This suggests that cold-based glaciers are slow-moving and do little to modify their beds. However, recent observations have begun to challenge these assumptions. Cold-based glaciers are now known to experience basal sliding, and where they rest on thick layers of permafrost or frozen ground, they may cause these to deform.

20.4 Surging or galloping glaciers

Some glaciers, after showing few signs of activity for many years, suddenly start to move rapidly. Typically the ice in the lower reaches moves several kilometres in a few months, or two years at most. This is a surge. Before the surge the glacier consists of stagnant ice, covered with thick supraglacial debris, for several kilometres above its terminus. The ice becomes reactivated and thickens greatly during the surge and the boundary between active and stagnant ice advances as the surge proceeds.

This is an important topic in glaciology because the mechanism for rapid flow was unresolved until recently, and surges produce a distinct landsystem. It is now thought that Pleistocene ice sheets are likely to have surged.

Glaciers that have surged include the Muldrow glacier in Alaska. During the winter of 1956–57 this glacier moved 6.5 km in around nine months, which was 100 times the normal velocity. In the upper part of the glacier the surface was lowered by 50–60 m, fringes of ice were left hanging on valley walls and tributary glaciers were sheared off and left as vertical ice cliffs. In the lower reaches the glacier thickened by 50–60 m.

The Steele glacier in the Yukon surged in the summer of 1966, when an ice wall advanced 12 km, and then a further 3 km in 1967. For at least 30 years before that the lower part of the glacier was relatively inactive, virtually stagnant, and had downwasted for decades.

The Otto glacier on Ellesmere Island in the Canadian Arctic is a rare case of a high Arctic glacier that has surged. It advanced 3 km as a floating tongue, with the calving of many bergs, and at one stage in July and August 1959 it was flowing at 7.7 m per day, whereas a tributary glacier was moving at only 0.83 m per day in the same period.

On the Medvezhii glacier in the Pamirs in 1963 the boundary between its affected and unaffected parts was marked by a large transverse fracture, where the ice on the downstream side was displaced by around 80 m vertically relative to the upstream section, and in addition a continuous fracture 8 km long and 10 m wide, which may have reached the bed, extended along the glacier near each side. Four shallow waves about 2 km in length and 50–70 m high were also observed on the ice surface as the ice increased in the lower section and decreased in the upper part.

Bråvellbreen in Svalbard is a modern ice cap that surged 21 km in 1935–1938.

A surging glacier can be defined as one which periodically (15–100 or more years) discharges an ice reservoir by means of a sudden, brief, large-scale ice displacement, which moves between 10 and 100 times faster than the glacier's normal flow rate between surges.

20.4.1 How can surging glaciers or glaciers that have experienced surges be recognized?

On aerial photos, two phases can be recognized: an active and an inactive phase.

- **Active phase:** Chaotically crevassed surface (Figure 20.12) (broken blocks, spires and columns). Large-scale features not obliterated and moving downglacier (e.g. moraines on the surface). Rapidly opening crevasses. Sheared margins and sheared-off tributaries. Folds (Figure 20.13). Large vertical and horizontal displacements. Bulging, overridden advancing fronts (Figure 20.14).

- **Inactive phase:** Distinctive surface features as a result of the periodic flow velocity changes: repeated moraine loops, folds in the medial moraines, bulbous-like moraine loops. These are formed by ice flow from a tributary glacier whilst the main glacier is inactive; the surge of the main glacier carries the loop several kilometres down-glacier. The tributary glacier then forms a new loop after the surge. Contorted ice foliation. Stagnation of large parts of the lower glacier (Figure 20.15). Longitudinal profiles locally steeper than the profiles of adjacent lateral moraines, or trimlines, which may

Figure 20.12 Chaotically crevassed surface of a surging glacier, Fridjovbreen, western Spitsbergen

Figure 20.13 Large-scale folding in a surging glacier, Arebreen, Reindalen, Spitsbergen

Figure 20.14 Bulging, overridden advancing front of a surging glacier, Arebreen, Reindalen, Spitsbergen

Figure 20.15 Stagnant lower part of the Cantwell glacier, Alaska

have been present before the surge. Reduction in the overall glacier gradient.

The general characteristics of surging glaciers are important therefore, and have to be explained in any hypothesis of their formation. All surging glaciers have surged repeatedly, based on the old moraine loops. Most surges are uniformly periodic and the cycle length appears to be fairly constant for a given glacier. However, there are large variations in cycle length between glaciers and between glaciated regions. For example, in Svalbard the active phase typically lasts 4–10 years, compared with only 1–3 years for Iceland, north-western North America and the Pamirs. All glacier surges take place in a relatively short time period, after which the glacier lapses into an inactive phase, which lasts a much longer time (anywhere between 15 and over 100 years, but typically 20–40 years).

There seems to be no simple relationship between the time for a complete cycle or for the active phase and the length, area or speed of the glacier. Ice-flow speeds during the active phases are much greater than normal (as much as 5 m per hour) for

short periods of time and can average over 6 km per year at fixed localities for periods of a year or more. However, in Svalbard the active phase is longer but the ice velocities are comparatively slow, ranging from 1.3 to 16 m per day, compared with say the Variegated glacier in Alaska, where there were velocities of 50 m per day.

It has been recognized that nearly all types and sizes of glacier surge, even small ice caps. In the western United States the Bering glacier is the largest surging glacier (over 200 km long, with an area of 5800 km^2), whilst the smallest is only 1.7 km long, with an area of 4 km^2. Surging glaciers occur in all climatic environments in the western United States, including the maritime and more continental conditions, and both sub-polar and temperate glaciers surge, yet the surges appear to be restricted to certain areas. For example, in the western United States they are found in the St Elias Mountains, the Eastern Wrangell Mountains and the Alaska Range, but not in other glacierized regions. Although this suggests special conditions, there are no particular rock types, permeabilities or surface roughnesses involved. Some are found along the Denali Fault, but other fault valleys do not show surging glaciers. In Svalbard it was suggested that there was a tendency for surges to occur on sedimentary rocks.

Surges always have a negative effect on the mass budget as they transfer large amounts of ice from a higher to a lower elevation, where the ablation is more rapid. So a surge will be followed by a retreat of the glacier margin. The melt-water regime is affected, as large volumes of turbid melt water are released, which coincide with the end of the surge.

The quiescent-phase structures in the surge glaciers are mainly ductile – normally foliation – whereas the surge-phase structures are typically brittle. These include thrusts as the surge front moves forward and compresses the ice; wholesale crevassing as the ice behind the surge front comes under the influence of tensile stress; and intense shearing of the margins, which can form longitudinal foliation in the ice. Compression of the surge front causes folding, thrusting, crevassing and thickening of the basal debris sequences, and a distinct crevasse pattern on the Variegated glacier was found. In the upper regions, unaffected by the surging, transverse crevasses record primarily extending flow. In the middle zone a complex crevasse pattern is seen, including longitudinal crevasses formed by compressive flow in advance of the surge velocity peak and transverse crevasses formed by extending flow behind the velocity peak. The lower zone is characterized by longitudinal crevasses as this zone is situated down-glacier of the final position of the velocity peak and experiences only compressive flow. Ice in the lower area ablates largely by downwasting and a drainage network with distinctive potholes commonly occurs.

The periodic activity illustrates that the longitudinal profile of the surging glaciers is not stable. During the active phase the ice reservoir thickens; when the glacier becomes sufficiently thick and steep at the lower part of the reservoir the component of basal shear stress apparently reaches a critical value and the surge begins. Once this happens it propagates rapidly, but this active phase ends when the total bed shear stress reaches a critical low value. However, what really causes these changes in bed shear stress? There must be a change in the subglacial boundary conditions to cause the much-greater-than-normal basal sliding velocities, and then another change to stop the surge.

20.4.2 Theories to explain surging glaciers

In 1899, from a study of Alaskan surges, it was suggested that an earthquake theory could explain the surges as a delayed response to avalanching on to glaciers of large volumes of snow and ice during a major earthquake. However, there has been no real evidence for this since 1900 and no correlation in time or space between surges and earthquakes. Later it was suggested that temperature changes at the glacier bed might explain surges. If in a glacier frozen to its bed the temperature of the basal ice were raised, the velocity would be increased because the flow law of ice depends sensitively on temperature. The increased flow would increase the amount of heat produced and the temperature would continue to rise until the melting point was reached, which would result in bed slip. The causes of the initial temperature increase might be an increase in velocity due to an increase in mass balance, or an increase in the amount of heat conducted to the glacier bed after a series of warm summers. There are objections to these ideas as it seems likely that most surging glaciers are at pressure melting temperature, at least at their beds, whether a surge is in progress or not, and this flow should occur at velocities comparable with sliding velocities in other nonsurging glaciers, not 100 times as fast.

On the other hand, a theory of glacier sliding in which the sliding velocity is largely controlled by irregularities of certain size on the glacier bed suggests that an increase in the amount of water at the glacier bed causes surges. Any water layer at the bed is normally thinner than the controlling obstacle size and does not affect the sliding velocity very much. Suppose that the thickness of the water layer were to increase until it exceeded the controlling obstacle size. Then the sliding velocity would be controlled by obstacles larger than the critical size and much higher velocities would be possible. A glacier with a bed that is much smoother than normal is likely to surge because the controlling obstacle size is much smaller for such a bed. There also needs to be a watertight bed. These conditions of a thick water layer will cause the glacier to slide faster, due to a water cushion that makes the bed appear smoother than it actually is.

A mathematical analysis of the problem concluded that this mechanism could account for a surge under specialized conditions: the glaciers should be between 10 and 30 km long; the base should be at pressure melting point; water at the bed should flow as a sheet and not in channels; basal shear should be about twice the normal value; and large bumps on the glacier bed should be few and far between. Whilst it is granted that a surge must be possible only under restricted conditions – otherwise more glaciers would surge – it is considered that many of the above conditions are too restrictive, as surges have occurred in both smaller and larger glaciers; water may flow in discrete channels, not sheet flow; and this whole theory is based on the fact that all water at the glacier bed originates from the melting of ice there, with water from the surface ignored. Hence the whole theory is untenable.

The question of the controlling obstacle size has also been in dispute. For example, other researchers suggest a controlling size of 50 cm rather than the original 1 cm, so the original theoretical explanation is not entirely satisfactory.

In 1985 the rigid bed hypothesis was produced, based on observations of the 1982–83 surge of the Variegated glacier in Alaska, where the movement was strongly controlled by the storage, distribution and pressure of the water at the bed and 95% of the ice motion was by sliding. During quiescence there was a well-developed tunnel network under the glacier. The prelude to the surge was the closure of this tunnel system, trapping meltwater and increasing the basal water pressure. Hence friction at the bed was decreased and the rate of sliding increased. The glacier was separated from its rigid bed and there developed a system of linked cavities, which were controlled by bedrock irregularities as the surge velocities were attained (Figure 20.16). The linked cavities developed into a series of interconnected tunnels, which allowed the melt water to discharge more efficiently as a flood. The initiation of this surge during the winter led to arguments that rapid sliding was triggered by rising water pressures at the bed, resulting from closure of the conduit system that drained the glacier during the summer. Conduit closure occurred through ice creep at a threshold value of ice overburden pressure below the thickening reservoir area. This is thought to have trapped the subglacial water in a distributed system, leading to extensive cavity formation and accelerated sliding. This pressurized cavity system lasted until the following summer, when increased water discharge caused the reestablishment of efficient drainage, which reduced subglacial water pressures and halted rapid sliding. Geometric changes in the glacier reservoir area prevented renewed surge activity and the glacier then entered a quiescent phase. There was some support for this as the 1986/7 surge of Peters glacier started in winter and the 1987/8 surge of West Fork glacier in Alaska began in late autumn, after the main melt season.

However, although this is a persuasive model, it cannot be completely applicable as subpolar glaciers partially frozen to their beds also surge. For example, the Trapridge glacier in the Yukon has a frozen margin but up-glacier basal ice is at pressure melting point and rests on a bed of unfrozen, deformable sediment. Hence the deformable bed hypothesis was suggested. Since 1969, when the Trapridge glacier was first surveyed, an impressive, wave-like

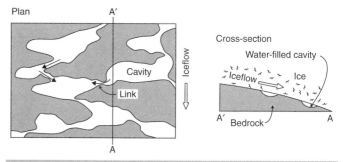

Figure 20.16 Network of linked basal cavities in plan and cross-section. Each channel is linked by subglacial channels. (Source: after Hooke, 1989)

bulge had formed on the glacier surface, forming an advancing lobe, with a steepening frontal edge. The bulge first developed above the boundary between the cold-based ice at the margin and the wet-based ice up-glacier, but it overtook the thermal boundary. Much of the forward bulge movement was due to subsole deformation, and as the bulge moved downstream the area of deforming till at the base expanded. It was suggested that a surge begins with the destruction of the subglacial drainage system and may be initiated by progressive changes due to water flow within permeable subglacial sediment, which as the progressive build-up of ice takes place leads to an increase in basal shear stress. This reduces the permeability of the till, and hence water pressure builds up and sediment becomes a deformable slurry, which allows the glacier to flow at an enhanced speed. In this model the surge ceases because redistribution of the ice leads to a lowering of the basal ice stress to a level that allows the subglacial sediment to return to a permeable state. One control here is the supply of erodible bedrock that is needed to create a thick enough deformable till layer.

Changes in subglacial water regime thus seem a prerequisite for surges to occur, but there appear to be at least two hypotheses in different types of glaciers that can explain them. The role of basal temperatures is another variable that might contribute to surges, but the evidence for any changes is not currently directly available. Observations comparing the Black Rapids glacier (Alaska Range) with those of the Variegated glacier (St Elias Mountains) and the Trapridge glacier (Yukon) indicate that there may be a variety of mechanisms to account for surge-type behaviour of glaciers. A complete theory currently remains elusive and awaits an expansion of the observational database.

20.5 Processes of glacial erosion and deposition

20.5.1 Glacial erosion

This can be achieved in one of two ways: either by abrasion or by quarrying. Glacial abrasion involves the wearing away of the glacier bed by debris, in the same way that a piece of sandpaper may abrade a block of wood. Using this analogy, it is possible to draw up a list of variables that are likely to be of importance in controlling the efficiency of glacial abrasion, which include:

1. **The contact force:** The force with which the abrading clast is pushed against the bed will control the friction between it and the bed and therefore the velocity of the clast relative to the ice. In addition, using the sandpaper analogy, the harder one presses, the greater the results.

2. **Clast velocity:** The velocity of the clast moving over the bed is important: the faster it moves, the more abrasion will result. The velocity of the clast relative to that of the basal ice will depend on the level of friction between the clast and the bed and is therefore partly dependent on the contact force.

It has been noted that there is more abrasion in glaciers with fast basal ice velocity than there is in slower glaciers.

3. **Debris concentration within the basal ice layer:** Debris-free ice is unlikely to achieve much abrasion, just as glass-free sandpaper is unlikely to abrade our block of wood. The greater the concentration of debris, the greater the number of tools with which to abrade the glacier bed, although at very high debris concentrations the debris may begin to act as a cushion between the bed and the glacier sole.

4. **Debris renewal:** As abrasion occurs, the clasts at the glacier bed will suffer comminution (crushing and abrasion) and will therefore need to be renewed by a sustained flux of debris towards the glacier bed. This will in part depend on the relative hardness of the clasts and the bed.

5. **Hardness of the clasts relative to the bed:** If the clast in the base of the ice is softer than the bedrock beneath the ice then it will be the clast rather than the bedrock that is eroded. The most favoured area for abrasion is where the glacier sole leaves a hard rock area and flows over a softer rock. If there is a layer of water at the bed then this could buoy up the ice and reduce abrasion, but of course it could also reduce friction and the sliding velocities might increase. It will all depend on the thickness of the water layer.

6. **Other controls:** It might be that highly permeable lithologies at the bed will reduce water flow at the base and limit the removal of rock flour. This will encourage the accumulation of clay-rich debris and hinder abrasion. It has also been considered that particle characteristics might influence abrasion, where angular clasts would produce more abrasion than round ones, and bigger clasts would produce more abrasion than small ones as they partly control the pressure on the bed.

There has been considerable debate about the relative importance of each of these variables and over the glaciological variables that control them. As a consequence, a number of different abrasion models have been proposed. One suggested that the contact force between the abrading clast and the glacier bed is determined by the normal effective pressure or load at the glacier bed. This was determined by the balance between the weight of the glacier above the clast, determined by the ice thickness, minus the counter force imposed by basal water pressure. This model envisages a continuum between abrasion and clast lodgement. As the normal effective pressure increases for any given glacier velocity, abrasion increases due to the increased contact force. However, as the normal effective pressure continues to rise, the rate of abrasion begins to slow and ultimately declines. This is because the increased pressure causes an increase in the friction between the clast and the bed and the clast slows. Ultimately the clast will stop moving and lodge as the normal effective pressure continues to rise (Figure 20.17). In this model, therefore, the rate of abrasion is a function of such variables as glacier

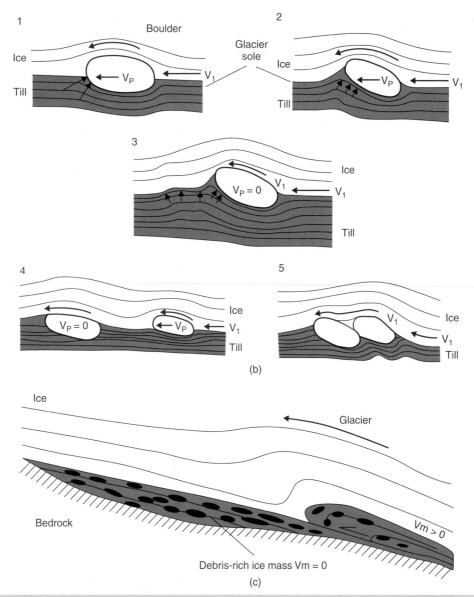

Figure 20.17 Particle lodgement beneath a glacier. (Source: after Bennett and Glasser, 1996)

thickness, basal water pressure and ice velocity. It is important to emphasize, however, that this is just one abrasion model and that other workers do not necessarily believe that clast contact pressure is determined by normal effective pressure. Instead, these models tend to emphasize the rate of basal melting that brings clasts into basal contact.

The evidence for abrasion is the ubiquitous rock flour in melt-water streams. For example, the erosion rate in small glaciers in central Norway was 1.18 mm per year in the period 1967–71, calculated from measurements of the rock flour in the melt-water streams. Here the floors of the glacial basins were being lowered by around 1 m every 1000 years. There have also been laboratory experiments to indicate that abrasion takes place, and there is evidence in the field of striations of varying scales that are cut into bedrock (Figure 20.18).

Glacial quarrying or plucking is the removal of blocks of rock by a glacier and involves two distinct processes: fracture propagation, or block loosening, and block entrainment, or removal. This follows logically from the fact that before one can remove a block it must be isolated by fracture surfaces. As a consequence, the properties of the rock mass are critical to the efficiency of the process. If a rock has few, widely-spaced and unconnected discontinuities such as joints, beds or other lines of weakness then it is unlikely to experience much in the way of glacial quarrying. In contrast, if a rock contains numerous, closely-spaced, wide and well-connected fractures then it is likely to be eroded by ice. It is also worth emphasizing that the size and geometry of the landforms produced by glacial quarrying will in part reflect the spacing and geometry of the discontinuities present within the rock mass.

Figure 20.18 Striations, coast north of Kap Petersen, eastern Greenland

In practice, for quarrying to occur some form of fracture propagation is required, in effect joining-up the discontinuities or fractures already present to produce an isolated block ready for entrainment. There are several methods by which this fracture propagation can occur. First, rapid fluctuations in basal water pressure can, if carried out repeatedly, wedge open fractures. Second, the ice itself can impose stress within the rock mass, leading to its fracture. If we consider a simple two-dimensional bump, the normal effective pressure will be elevated on its upstream side as the ice flows against it and will be depressed in its lee. This differential stress regime may encourage rock to fracture. Significantly, this stress regime is increased dramatically if the basal cavity between the bump and the glacier sole forms in the lee of the bump or obstacle. This will tend to occur where the ice is both thin and fast-flowing. Fracture propagation may also occur as a result of glacial erosion through stress release. As erosion takes place the rock overburden is removed and replaced by ice. The stress release involved in this process can cause the development of sheet joints that parallel the land surface.

The second component of glacial quarrying, after block isolation, is the entrainment of these blocks. Loose blocks and debris may be entrained within the regelation ice formed in the lee of obstacles as water melts on their upstream faces and then refreezes. In addition, blocks that fall into basal cavities may be entrained if the cavity closes as the ice surrounds the block and moves forward. Finally, and perhaps most importantly, blocks may be entrained via a simple 'heat-pump', which causes the glacier to locally freeze to its bed. The key to this process is the latent heat released or absorbed during the phase change between water and ice. We have already seen that pressure melting may occur on the upstream side of an obstacle at times of high pressure, for example during a pulse of ice flow. Provided that no water is lost from the system, the ambient heat loss due to consumption of latent heat during melting will equal the ambient heat gain during refreezing. However, if the water is transferred from one side of the obstacle to the other, or from one location to the next, net cooling may occur, which will cause the ice to locally freeze to its bed. Any

forward motion while the glacier is frozen to its bed may lead to particle entrainment. The evidence for quarrying is seen in boulders with freshly-sheared faces and chattermarks (Figure 20.19). The process is helped not only by the development of sheet joints but by periglacial conditions prior to glaciations, which produce plentiful frost-shattered debris, or preglacial or interglacial deep-weathering produced by chemical processes, although these are likely to produce finer-grained, clay-rich sediments.

20.5.2 Sediment entrained basally

Sediment entrained and transported at the base of a glacier has distinct characteristics. First, the grain-size distribution tends to contain two dominant modes: the larger grain-size mode (pebbles to boulders) is composed of rock fragments, while the finer grain-size mode is composed of a rock flour of mineral grains produced by the crushing of the larger rock fragments. Second, the larger rock fragments show extensive evidence of subglacial wear in the form of striated and facetted faces with subangular or even rounded edges.

In addition to the direct processes of glacial abrasion and quarrying referred to above, it is important to note that glaciers that are topographically constrained may act as large conveyors of debris derived from valley sides and mountains (nunataks) which extend above the ice surface. In the ablation zone this debris may remain on the ice surface to give a continuous debris cover, while in the accumulation zone it is buried by fresh snowfall and some of it may find its way to the glacier bed. Debris transported on the surface of a glacier, or within it, but not at its bed, has very different sedimentary characteristics to sediment transported at the glacier bed. These characteristics reflect the sediment's origin as rockfall and talus debris, and consequently it tends to be dominated by large angular clasts that show little evidence of glacial wear (Figure 20.20).

20.5.3 Glacial sedimentary environments

Glaciers deposit sediment in a wide variety of ways. Five distinct sedimentary environments or systems can be recognized.

Subglacial sediments

Sedimentation may occur subglacially beneath a glacier in several ways. First, sediment may lodge beneath actively-moving ice whenever the friction between the bed and the clast or particle exceeds the drag of the ice. This may result from high rates of basal melting, from increases in the normal effective pressure and from decreases in ice velocity. Normal effective pressure is not only controlled by ice thickness, but also by fluctuations in subglacial water pressure. In addition, clasts ploughing through basal sediment at the ice–bed interface may become lodged against one another, as shown in Figure 20.17. Sediment that accumulates in this way is referred to as lodgement till and is characterized by consolidated subglacially-transported sediment that has been subject to subglacial shear during its deposition.

(a)

(b)

(c)

Figure 20.19 (a) Chattermarks and striations, Tunsbergdalen rockbar, central Norway. (b) Bullet boulder, Iceland glacier forefield. (c) Bullet boulder, Hargreaves glacier, British Columbia. (http:tvl1.geo.uc.edu/ice/image/eropro/eropro/411-9.html)

Figure 20.20 Supraglacial debris, Bersaerkerbrae, Stauning Alps, eastern Greenland

The clasts within the sediment may show a preferred long-axis orientation (fabric) parallel to the direction of shear and therefore to the direction of local ice flow.

The second way in which sediment may accumulate subglacially is through the ablation melting, or sublimation, of debris-rich basal ice. Where the ice is not moving, this will give a subglacial meltout till. This sediment resembles lodgement till but does not contain evidence of post-depositional shear beneath moving ice. Basal melting is achieved through geothermal heat flux.

Finally, subglacial sediment may also be deposited as a consequence of subglacial deformation. This may involve already-deposited glacial till or nonglacial sediments over which the ice advances. In practice, the deformation till that results may be characterized by admixtures of glacial and nonglacial sediment, giving it a diverse range of properties. The degree of mixing or deformation will control the nature of the deformation till. Where the degree of deformation is small, the overridden sediment may show little sign of deformation, perhaps just a few

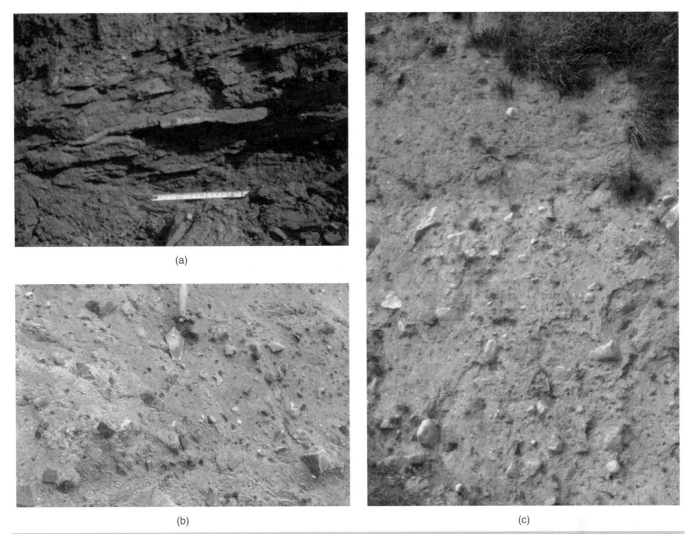

(a)

(b)

(c)

Figure 20.21 Homogeneous till: (a) in front of Thorisjökull, central Iceland, showing tectonic shear lenses; (b) Ballylurkin till, Wexford coast, Ireland; (c) derived from Mourne Granite, Northern Ireland

minor folds or faults. However, as the degree of deformation increases, the level of mixing will increase until a homogenous till results (Figure 20.21).

In practice subglacial sediment may form by all the three processes described above: lodgement, meltout and deformation, and the distinction between these processes and the resultant sediments can be extremely subtle. A comprehensive review of subglacial till, its formation, sedimentary characteristics and classification is given in Bennett and Glasser, (1996) and Evans *et al.* (2006).

Supraglacial sediments

Glacial sediment may also be deposited within supraglacial sedimentary systems. Within the ablation zone of a glacier, debris becomes concentrated on the glacier surface through ablation. Additional debris falling on to the glacier surface from the surrounding valley sides in the accumulation zone will also remain at the glacier surface.

Basal debris may also be elevated to the glacier surface within the accumulation zone via a variety of processes. The simplest of these involves the deceleration of glacier flow towards the glacier snout, which causes compression and the deflection of debris bands towards the glacier's surface. If this longitudinal compression is more pronounced, debris may be elevated along thrust faults or within folds. Enhanced longitudinal compression due to flow deceleration is particularly pronounced at a glacier with a warm-based interior and cold-based margin. Such glaciers are said to have a mixed basal thermal regime, or to be polythermal, and are common in sub-Arctic regions, such as the Svalbard archipelago.

Longitudinal compression may also be caused by a reverse bedrock slope beneath a glacier margin. The accumulation of debris on the surface of a glacier (supraglacial debris) leads to the creation of surface topography. As debris first melts out on a glacier surface it darkens it, which allows more solar radiation to be absorbed and so accelerates ablation. However, as the debris layer builds it begins to insulate the underlying ice, slowing its

(a)   (b)

Figure 20.22 (a) Medial moraines, Midre Lovenbreen, Kongsfjorden, Spitsbergen. (b) Medial and lateral moraines, Gornergratgletscher, European Alps. These are formed mainly from valley-side rockfall events, which create a lateral moraine; where two lateral moraines coalesce, as in the middle part of the photo, a medial moraine forms and is carried down-glacier by the ice flow

ablation. Consequently areas of thick debris cover, for example along medial moraines (Figure 20.22), the site of rockfall debris on the glacier surface, or along the outcrop of a debris-rich structure such as a thrust, will tend to be raised on the glacier surface relative to intervening areas where ablation is not hindered by a thick debris cover. This local topography may be spatially well-ordered due to the control of a dominant pattern of glacial structures, or it may be random. Once a surface topography is established it will change over time as the ice surface downwastes. Debris will tend to move on surface slopes from locations of high to low elevation, and topography may thereby be inverted several times during downwasting of the glacier.

These slope movements will often be achieved through a variety of different types of debris flow and will give rise to a deposit that is described as a flow till (Figure 20.23). The properties of flow till reflect the fluid content of the debris and therefore its mobility. The debris is likely to be unconsolidated and may have been transported both subglacially and supraglacially. Flow till may also contain a large amount of fine-grained material deposited in thaw lakes and supraglacial streams within the supraglacial topography of a glacier. This material may be interbedded with flow till, or reworked within it to give finer-grained flow tills. In general the environment is a complex one, with many different types of sediment.

20.5.4 Glaciofluvial deposition

This involves the reworking of glacial sediment by melt water and its deposition within, on and beyond a glacier in rivers and lakes. The processes are similar to those of any fluvial or lake system. However, in the proximity of a glacier the sediment loadings are extremely high and consequently give glaciofluvial systems distinct characteristics. There are fluctuations in discharge both on a daily cycle and during the year, and because the sediment is

Figure 20.23 Flow till within outwash, Newtownmountkennedy, Northern Ireland

easily erodible there are rapid grain-size variations and variations in sedimentary structures. It is also important to emphasize that glaciofluvial sediments are not just deposited beyond the ice margin but may form a significant component of supraglacial sedimentary environments, and may also be deposited within englacial or subglacial conduits.

The glaciofluvial sediments can be classified based either on morphology or on the environment of deposition, or a combination of both (see Figure 20.24). One simple classification is based on form – spreads, mounds or ridges – but it does not separate sediments formed in totally different environments, for example proglacial sandurs and lake plains. The environment of deposition can usually be divided into proglacial and ice-contact deposits, and if the relationship between the deposit and the ice mass can be determined then adjectives such as 'subglacial' and 'marginal' can be applied.

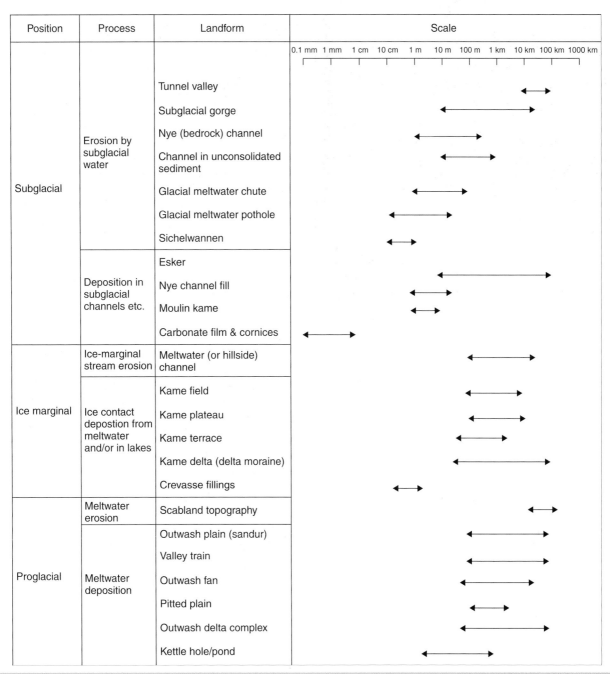

Figure 20.24 Classification of glaciofluvial sediments according to position, process and size. (Source: after Hambrey, 1994)

20.5.5 Glacial lakes

These can form in a variety of different situations. Lakes may be marginal, dammed in ice-free valleys by glaciers, proglacial in front of a glacier margin (particularly where there is a reverse bedrock slope), or may occur within proglacial landforms, particularly where buried ice has melted to give a thaw depression, or kettlehole (Figure 20.25(a)–(d)). Finally, subglacial lakes, often on a large scale (for example Lake Vostok, ~240 km long and 50 km wide), have been recognized beneath Antarctica. In fact,

over 100 such subglacial lakes beneath the ice sheet on that continent have been identified, and it is likely that all continental ice sheets had such lakes beneath them in the geological past (see Box 20.1, see in the companion website). The widespread occurrence of tunnel valleys within the limits of the Laurentide and Fennoscandian ice sheets suggests that large, transient discharge events periodically drained subglacial reservoirs.

Of particular interest here are those lakes that are influenced directly by glacier ice having an ice margin within them. Within these lakes the processes of glaciolacustrine sedimentation

(a)   (b)

(c)   (d)

Figure 20.25 Types of glacial lake: (a) ice marginal lake dammed by the very steep topography, Eastern Kangerdlluarsuk glacier, south-western Greenland ice sheet. This lake periodically drained beneath the ice margin and gave rise to marginal lake terraces; (b) ice marginal lake in dammed tributary valley. The lake in Vatnsdalur drains under the glacier Heinabergsjökull (southern Iceland) by periodic jökulhlaups; (c) proglacial lake in front of Heinabergsjökull, southern Iceland, held up by a terminal moraine; (d) kettlehole lakes, Vegbreen, Spitsbergen

dominate. The landforms produced can be subdivided into deltas, flat lake plains, beaches and shorelines, and cross-valley or de Geer moraines. It is important to emphasize that any sediment added to the lake will be disaggregated and that the characteristics of the resultant deposit will be influenced by the properties of the water body.

Several different depositional processes are recognized within lake basins, each of which gives a different type of sedimentary facies. The main processes are:

1. **Deposition from melt-water flows:** The way in which melt water enters a lake depends upon the density of the incoming water relative to that of the lake, controlled by temperature and turbidity. The density of lake water may change during a year. During winter, lake waters are often fairly homogenous in character, but in summer a distinct thermal stratification can develop, in which surface layers warm rapidly in contrast to deeper layers (Figure 20.26). If there is a significant difference in density between the incoming glacial melt water

and that in the lake, the sediment-laden water will retain its integrity as a plume, which can enter the lake in one of three ways: (a) as an underflow, if the plume is denser than the lake water; (b) as an overflow, if the plume is less dense than the lake water; (c) as an interflow, where the incoming water is denser than the surface water but not as dense as the bottom water within the lake. If the water is introduced as an overflow or interflow then the current will quickly be separated from the bed and therefore from any sediment in traction, causing rapid deposition of sediment in a fan, which if it builds up may form a delta. Alternatively, if the water is introduced as an underflow then the sediment-laden water may be injected further into the lake as a series of sediment gravity flows, such as turbidity currents.

2. **Direct deposition of supraglacial debris from the ice front:** Material may be transported to the snout of the glacier to fall over its calving front as ablation proceeds.

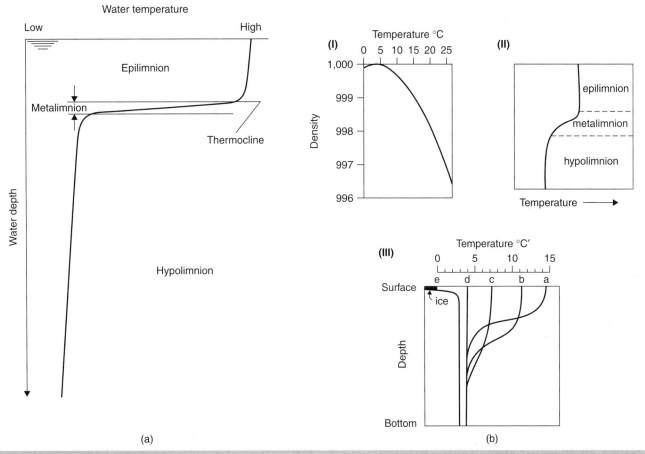

Figure 20.26 Thermal stratification in glacial lakes. (a) Temperature with depth in a lake, showing the position of the thermocline dividing the warmer epilimnion from the colder hypolimnion. (b) (i) The relationship between the temperature and density for pure water; (ii) the temperature profile as a result of surface heating and wind-mixing; (iii) hypothetical evolution of the temperature structure of a deep lake between midsummer (a) and winter (e): curve (d) illustrates isothermal conditions resulting from lake overturn in autumn and spring

3. **Iceberg rainout:** Calving of icebergs from an ice front may raft debris into the lake, which is then deposited as the iceberg melts. Differential melting of a berg may cause it to capsize, delivering pulses of supraglacial debris into the lake. This may generate dropstones, which are large clasts set within fine-grained sediment deposited by suspension settling or turbidity currents within the lake (Figure 20.27).

4. **Suspension:** This may occur throughout the lake, as suspension settling, although it will tend to concentrate close to the injection sites. This process will be most pronounced where melt water is introduced as an overflow or interflow.

5. **Resedimentation by sediment gravity flows:** Rapid deposition on the margins of a lake basin may quickly lead to unstable slopes that fail, moving sediment basinward in a variety of subaqueous mass movements, including slumps, debris flows and fine-grained turbidity currents.

In addition to the above processes, it is important to emphasize that lakes may be infilled by slope deposits and nonglacial or

Figure 20.27 Dropstones, Newbigging, southern Scotland

glacier-fed streams. A wide range of sediments are therefore characteristic of this environment, although many of them show some degree of size sorting associated with the water body in which they are deposited. It is important to emphasize that beneath the grounded ice margin the normal processes of lodgement, meltout and deformation will occur, and if a glacier is in retreat then these deposits may underlie the lake sediments.

Deltas

When a proglacial river flows into a glacial lake the velocity is checked and the sediment is deposited as a delta, usually in the form of a tripartite topset, foreset and bottomset sediment sequence. The relatively thin, topset facies is formed by this braided river system flowing over the delta surface and is composed of pebble gravel and pebbly sands in the form of erosional channels, imbricated horizontal gravel bars and the occasional collapse structure formed by ice-block melting. Lateral and vertical relationships are both gradational and sharp, and there is a downstream fining and vertical coarsening as the river builds up the delta.

Below and laterally is the foreset facies, which is composed of large-scale, cross-stratified sets, which dip between 30 and 10° and are formed in a proximal position to river flow. There is a continuum from these foresets through toesets to the bottomsets that formed in a distal position, which dip at between 5 and 2°. The characteristics of the foresets are as follows: the thickness of the individual beds decrease downdip; the maximum dip occurs nearest the sediment supply and decreases in the transport direction; the grain size decreases in the transport direction from proximal, pebbly coarse sand to distal, silty sands and silts; and their approximate thickness gives an indication of the lake depth (Figures 20.28 and 20.29). The horizontally-stratified coarse sand and gravels are caused by downslope, gravitational sliding of sediment into the lake due to shearing in the mass of sediment, irrespective of the direction of melt-water inflow.

The bottomsets, on the other hand, are dominated by parallel laminated sands and silts and various types of ripple-drift, small-scale cross-lamination (Figure 20.30). The ripple-drift forms demonstrate the importance of a high deposition rate from

Figure 20.29 Foreset facies, Holme St Cuthbert proglacial lake, north-west Cumbria

Figure 20.30 Bottomset facies, Blessington proglacial lake delta, County Wicklow, southern Ireland

Figure 20.28 Thin topsets and foresets, Galtrim moraine, County Meath, southern Ireland

suspension and the sinusoidal types indicate a relatively high rate of fines deposition and a consequent stabilization of the lake floor. With bed-load transport type A, ripple drift is formed, but parallel lamination indicates flow velocities under 11 cm per second and together with clay bands indicate deposition solely from suspension. The high sediment concentrations and the slumping on the foreset slope allow pulsatory, density underflow currents, occuring as turbidity currents, which deposit all these forms of ripple drift and occasional large-scale, tabular cross-sets. Occasionally slurries and flow tills from the ice interdigitate in the proximal foresets. There can be dropstones and dropstone pods from iceberg melting, and faulting can occur in the steeper foreset beds as part of the slumping.

Lake plain deep-water silts and clays

These sediments merge laterally and gradually from the bottom-sets to produce an almost horizontal lake plain. The sediments are laminated and are often called rhythmites because of the grain-size changes between successive thin layers, accompanied by a colour change, with the finest layers being the darkest. Individual fine and coarse laminae joined together are known as a couplet. These are usually referred to as a varved sediment, indicating annual deposition from coarser summer sedimentation and finer winter suspension sedimentation (Figure 20.31). Some varves contain silt balls and small stones in a coarse gravel layer (Figure 20.32).

There is some evidence that these deposits are annual in nature, as pollen and spores are present in the summer layers of Lake Barlow-Ojibway in North America and are absent or rare in the winter layers. There is agreement with de Geer's varve chronology for the Pleistocene chronology in Scandinavia, derived from C[14] dating of the same sequences. The approximate thickness of the annual sedimentary layers from modern Lake Louise in Alberta is known from the discharge and the sediment load of the melt-water streams flowing into the lake. The figure is 4 mm, and a core from

the deepest part of the lake shows faint banding, with a spacing of about 4−5 mm, which strongly suggests a seasonal rhythm.

These varved clays can be compared with deep-sea turbidites, except that they are much finer in grain size. However, individual units cover large areas without significant changes in thickness and texture. They do show silt balls (Figure 20.32), erosional markings at the base of some beds, sharply-defined lower boundaries, a decreasing grain size from unit bottom to top and graded bedding, and the most common sedimentary structure is horizontal lamination. Turbidity currents give rise to the characteristic A−E sediment sequence as the current decelerates (Figure 20.33). The graded interval decreases in thickness and at a certain point the current velocity is so low that horizontal lamination is formed at the bottom and a sequence B−E forms. From the starting point of deposition, types A−E will successively be formed.

This typical turbidite sequence can be applied to varve sequences. For example, in the Irthing valley in Cumbria, deep-water clays/silts laid down in proglacial Lake Carlisle during the Main Devensian ice retreat show proximal thicknesses up to 40−50 cm for individual couplets and B−E and C−E sequences in the lower varves, with D−E couplets in the upper varves. It is now recognized that varve couplets have three genetic parts: a sand/silt layer deposited from a density underflow; the suspension load from these currents; and the slow, continual suspension settling in the lake (indistinguishable from the suspension load).

The model for varve clay deposition suggests that summer melt water charged with sediment emerges from subglacial tunnels at the ice front. The coarsest sediment is deposited as a fan/delta near the tunnel mouths, but the finer sediment still in suspension, is carried along the foresets and delta bottomsets as a turbidity current and moved far out into the lake. The summer part of the varve is deposited by such annually-produced currents flowing along the lake bottom. The current spreads a great distance, is divided into lobes by bottom irregularities and halts because of either ponding before a counterslope or loss of density contrast through deposition.

Associated with glacial lakes are shoreline sediments in the form of beaches and erosional landforms such as arches and shoreline benches in bedrock (Figure 20.34). Occasionally these beach sediments may be deformed by later glacier advances, as at proglacial Hagavatn (Langjökull, Iceland). There has been much discussion as to the origin of glacial melt-water channels, and originally they were thought to be formed by lake overflows. Subsequently most are now thought to be subglacial in origin, but there are undoubtedly lake overflow channels that have been eroded by catastrophic overflows, such as the Hollywood Glens in County Wicklow (Republic of Ireland), from proglacial Lake Blessington, the overflows associated with Hagavatn, Iceland, or on a much bigger scale the channelled scablands of Oregon, from the overflows of glacial Lake Missoula (Figure 20.35).

Formation of de Geer moraines

These moraines generally occur as a series of closely-spaced ridges in association with subaqueous sediments (lakes or marine) and

Figure 20.31 Varved couplets, Uppsala, southern Sweden

Figure 20.32 Silt balls and stones in Varves, Carrow Hill, Petteril valley, Cumbria

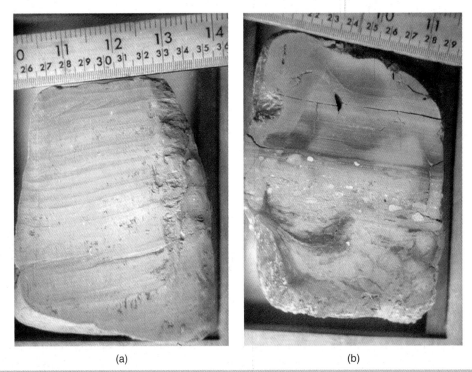

(a)   (b)

Figure 20.33 A–E sediment sequences in varves from turbidity currents at Great Easby, Irthing valley, Cumbria: (a) ripples, fine sands and parallel laminated silts; (b) coarse-grained units and convolute lamination

mark the intermittent retreat of glaciers that terminated in water. They are named after a famous Swedish geologist. Many models for their formation have been suggested. They could mark former ice-marginal positions at which sediment was either deposited or pushed up during a minor readvance, perhaps on an annual basis. Alternatively, they may be linked to squeezing of saturated sediment from beneath the ice margin (Figure 20.36). It has also been argued that they are formed in basal crevasses, where it is suggested transverse crevasses form in areas of extending flow and then widen during partial ice-margin flotation. Moraines are

Figure 20.34 Glacial lake shorelines and other geomorphic landforms: (a) Hagavatn, proglacial shorelines, Langjökull; (b) natural arch eroded by lake level and lake shorelines; (c) Heinabergsjökull, former marginal lake shorelines; (d) giant current ripples in Washtucna Coulee from the Lake Missoula floods; (e) Billy Clapp pothole and channelled scablands. (Sources: (d) and (e) http://iceagefloods.blogspot.com/2008_11_01_archive.html)

Figure 20.35 Glaciofluvial channels: (a) subglacial channel, western flanks of Corney Fell, south-west Cumbria; (b) submarginal, subglacial channel cut by a river under Tunsbergdalsbreen, central Norway; (c) subglacial chute, Osterdalsisen, northern Norway, eroded by water flowing directly down the valley side under the marginal ice; (d) subglacial chute, west Cumbria; (e) Newtondale, north Yorkshire Moors; (f) overflow from Hagavatn, Langjökull, Iceland; (g) channelled scablands, Oregon: Billy Clapp lake area, cut by floods estimated to be 250–300 m deep and flowing at approximately 15–20 m per second. Overflow from glacial Lake Missoula. (Source: http://iceagefloods. blogspot.com/2008_11_01_archive.html)

(g)

Figure 20.35 *(continued)*

thought to form when water level drops and the glacier is lowered to the waterlogged substrate, which is squeezed up into the basal crevasses (Figure 20.36).

20.5.6 Glaciomarine environments

A similar suite of processes and products occurs when glaciers terminate in the sea. These differ in several key respects, however. First, the incoming glacial melt water is fresh and is therefore always less dense than sea water, and consequently tends to be injected as an overflow, which forms a surface plume of fresh, sediment-laden water. Second, the rate of glacier calving is often much greater, due to increased water depth and the ice-marginal flexing caused by tidal variations. Third, settling from suspended sediment may be enhanced by the salt water. The saline water helps neutralize the electrical charges on clay particles, which allows them to flocculate, forming larger grains that settle faster. Finally, glaciomarine environments are less sterile, and a biological component may be present, especially with increasing distance from the ice margin.

How can glaciomarine sediments be recognized?

Modern examples have been described from the north-eastern Pacific Ocean and Baffin Island, and well-authenticated Pleistocene examples have been described from the lower Frazer valley in British Columbia, the Gastineau channel formation near Juneau in Alaska, Middleton Island in the Gulf of Alaska and the Puget Lowlands in Washington State. Their characteristics are summarized in Table 20.1.

Processes controlling glaciomarine sedimentation

Deposition from the glacier

The rate of sedimentation depends on the volume of ice being melted, the debris content of that ice and the movement of the ice front. Sedimentation is influenced by the discharge from glacial rivers into more dense marine waters. The submarine discharge is from sediment-laden rivers and takes the form of a jet, the behaviour of which depends on discharge (Figure 20.37). Inevitably the sediment plume reaches the surface, but suspension sedimentation takes place as it does so. Supraglacial drainage carries little sediment and, along with proglacial streams, enters the fjord as a buoyant overflow. Iceberg calving is important,

through the release of supraglacial sediment and sometimes basal debris, generating waves and controlling the position of the ice front in combination with the forward movement of the glacier. However, on Baffin Island it was found that 86% of the sediment was from a glaciofluvial source, 9% was supraglacial, 3.7% was subglacial and only 0.8% was from ice-rafting.

Many tide-water glaciers are of the surge type in the Arctic and Alaska. The style of sedimentation can change radically because of increased melt-water discharge, where there is a radical change in the subglacial melt system to a sheet flood, and due to erosion and reworking of glaciomarine sediments during a surge.

Floating ice

Dropstones are important in recognizing glaciomarine sediments, especially in the pre-Pleistocene geological record, but they can also form in lakes. As we have seen already, compared with deposition at the grounding line, deposition from floating ice is relatively minor. It is an important process off the continental shelves, however, and the main controls on this type of sedimentation are the concentration and distribution of debris in the source glacier, the residence time of the iceberg in the fjord, the volume of ice calved, the rate of iceberg drift and iceberg melting, and the amount of wave action, which influences the number of overturning events.

Patterns and rates of sedimentation in fjords

These are highly variable, but gravels and sands tend to accumulate near to the glacier, though they can occur patchily elsewhere, where diamictons are winnowed and there are dropstone pods (Figure 20.38). Subglacial streams deposit gravel, sand and mud and tend to be proximal. The ice-proximal mud tends to be laminated as a result of variations in discharge and tidal currents, and often exists as graded couplets. Mud extends from the glacier cliff to the distal parts of the fjord basin and beyond if there is no sill, but ceases to be laminated because of low sedimentation rates, flocculation and bioturbation.

Many fjords have more than one basin, which results in complex sedimentation patterns. In Svalbard in Kongsfjorden it was reported that sediments were thickest in the inner basin, where there was over 100 m of diamicton and compacted glacial sediments. Outside the inner basin there was between 20 and 60 m of sediment, consisting of diamicton, ice front and surge deposits. In southern Alaska there are much greater sediment thicknesses, from several hundred metres to over a thousand metres, due to rapid crustal uplift, which results in extremely active erosion. Both of these examples are young, but in Ferrar Fjord in Antarctica a

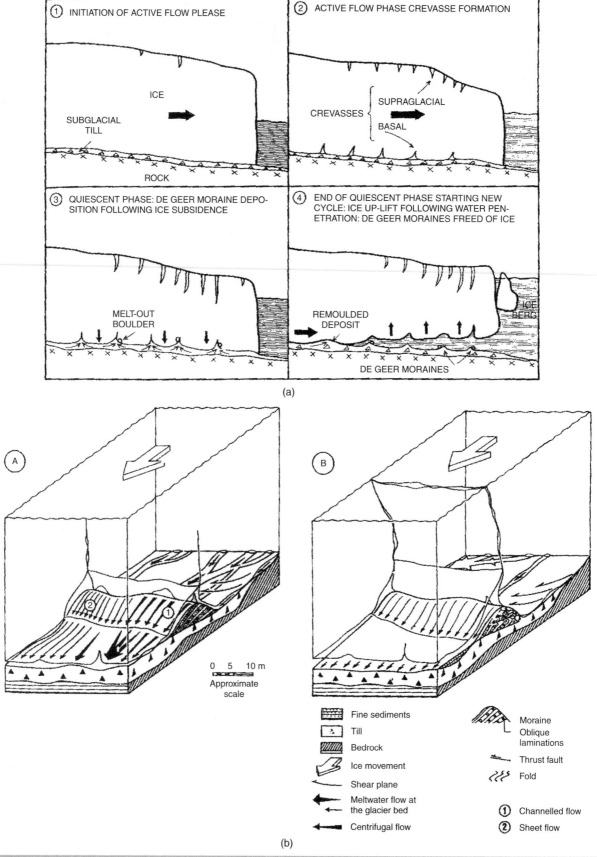

Figure 20.36 Models for the origin of de Geer moraines. (Source: after Benn and Evans, 2002)

Table 20.1 Characteristics of glaciomarine sediments

Delivery of sediment to the glaciomarine environment

Ice dynamics, diamict aprons, glacial meltwater; sea ice and river ice; biogenic material

Ice-ocean interactions

Reflect wide variety of ice margin types; tidal pumping especially at grounding line

Rates of ice cliff melting

Melting and freezing at the base of ice shelves

Iceberg calving

Iceberg melting rates, fragmentation and overturning

Iceberg drift paths

Oceanographic processes

Meltwater inflow and stratification

Sediment suspended in meltwater

Tides

Effects of sea ice on circulation (brine enrichments → unstable dense layer)

Depth of water

Bottom currents

Processes of reworking

Current activity

Subaqueous mass movement

Scour by icebergs and sea ice

Bioturbation

Compaction and recycling of sediment by ice readvances

Carbonate sedimentation

Diagenesis

Carbonate in ancient diamictite sequences used as evidence against glaciation

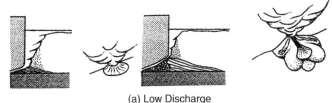

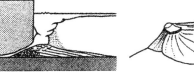

(a) Low Discharge

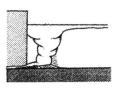

(b) Moderate water discharge, high sediment discharge

(c) High Discharge

Figure 20.37 Variations of the sediment plume according to discharge differences and the types of sediment fan associated with these differences. (Source: after Powell, 1990). Reproduced by kind permission of Tom Foster

Figure 20.38 Dropstone pods in glaciomarine sediment, Ossian Sarsfjellet, Kongsfjorden, north-west Spitsbergen

sedimentary record goes back to the Pliocene, at least to 4 million years, yet there is still only 166 m of sediment.

The rates of sedimentation are extremely variable: in Glacier Bay, Alaska, there is 9 m per year close to the ice front, but 0.5 m per year is perhaps more typical for temperate and cold glacial regimes. In the distal parts of the system there is an order of magnitude less, and sediment rates decline with distance from the ice front.

Depositional landforms in fjords

There are a variety of landforms and processes associated with the fjord glaciomarine environment (Figure 20.39). First, moraine banks are formed by a combination of lodgement, meltout, dumping, pushing and squeeze processes when a glacier terminus, grounded in water, is in a stable state. These banks are often associated with submarine outwash. They are composed of a variety of sediments, but pockets of diamicton enclosed in poorly-sorted sandy gravel (often with clasts up to boulder size) and gravelly sands are common. The foreslope is affected by processes such as sliding, slumping and gravity flowage. The upglacier slope may also show collapse once the glacier has receded.

Second, there are grounding-line fans where subaqueous outwash extends from the glacier tunnel that discharges subglacial melt water. Most of the sediment is deposited rapidly from melt water, which issues as a horizontal jet at or near the sea floor during full pipe flow. There is a range in particle sizes, most of

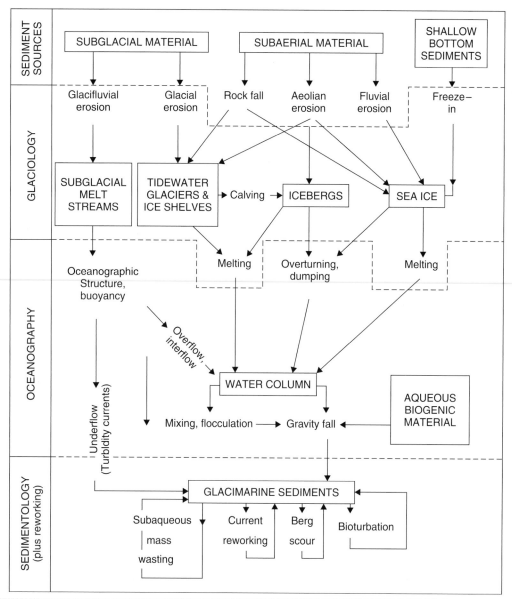

Figure 20.39 Complex process systems in the fjord glaciomarine environment. (Source: after Dowdeswell and Scourse, 1990)

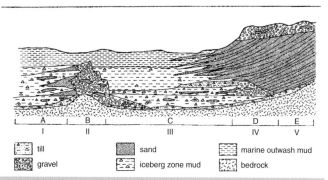

Figure 20.40 Hypothetical complex stratigraphy in a glaciomarine fjord complex, based on work in Alaskan tide-water glaciers. (Source: after Powell, 1981)

which are carried to more distal positions in the sediment plume. Grounding-line fans that develop during sustained advances are not well preserved, because of reworking. Nor are fans that develop during rapid recession, since there is insufficient time for them to grow. Hence they are best developed if the ice margin is at least quasi-stable.

Third, ice-contact deltas can develop where there is a stable ice margin for at least tens of years and sedimentation builds up to sea level.

Beyond these three landforms there are relatively flat proglacial laminates, produced by deposition of suspension sediments. These are rhythmically laminated due to the variations in stream discharge into the fjord, combined with the effects of tidal currents. Dropstones are common in this environment. Further from

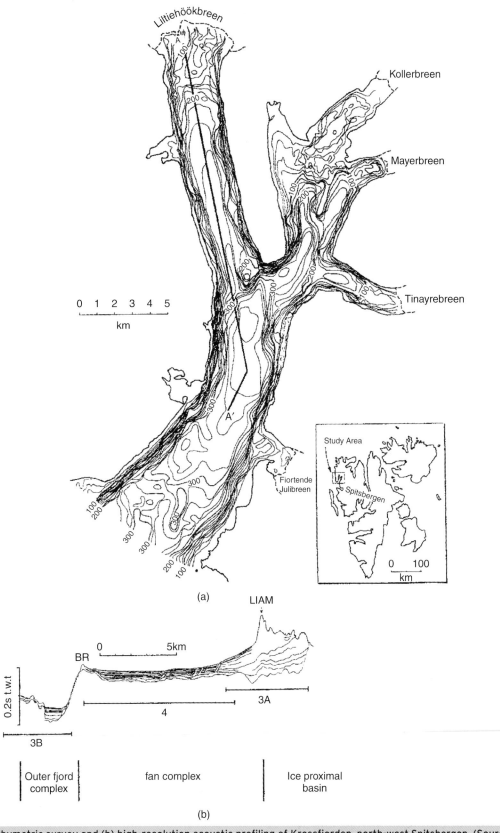

(a)

(b)

Figure 20.41 (a) Bathymetric survey and (b) high-resolution acoustic profiling of Krossfjorden, north-west Spitsbergen. (Source: after Sexton, 1992)

Table 20.2 Principal lithofacies in Alaskan-type fjords and their interpretation

Lithofacies	Interpretation		
Diamicton	(a)	Lodgement till	
	(b)	Subglacial and meltout till	
	(c)	Subaqueous sediment gravity flows	
	(d)	Silt and clay from meltwater streams + iceberg debris	
Gravel (poorly sorted)	(a)	Subaqueous outwash:	grounding-line fan
			ice-contact delta
			fluviodeltaic complex
	(b)	Gravity flow	
	(c)	Lag deposit from winnowed diamicton	
Gravel (well sorted)	(a)	Beach	
	(b)	River mouth bar	
Sand (poorly sorted)	(a)	Subaqueous outwash	
	(b)	Gravity flow	
Sand (well sorted)	(a)	Beach	
	(b)	River mouth bar	
Mud with dispersed clasts		Meltstream derived silt + clay, with minor ice-rafted debris ("iceberg zone mud")	
Mud	(a)	Meltstream derived silt + clay	
	(b)	Tidal flat sediment	
Rhythmites (laminated sand and mud)		Deposition from underflows generated from meltwater streams, and influenced mainly by tides (cyclopels and cyclopsams)	

the ice source are fjord-bottom complexes, where there are inter-stratified homogeneous muds derived from suspended glacial rock flour interbedded with the laminites. These may grade into diamicton or bergstone mud. Around the fjord there is likely to be a complex of nearshore fluvial, beach and tidal flat landforms.

These landforms are represented in a complex stratigraphic architecture for fjords, as can be seen in Figure 20.40, and geophysical techniques such as high-resolution acoustic profiling have been used, for example in Krossfjorden, Svalbard. Three main complexes are recognized: (a) ice-proximal, subaquatic hummocky terminal moraines or morainal banks; (b) a fan complex with structureless till or glaciomarine sediment at the base, overlain by well-laminated sand/silt produced by suspension and turbidity currents; (c) an outer fjord complex with a blanket of structureless or laminated sediment (Figure 20.41). Box 20.2 for a description of anomalous glaciomarine sediments in Eastern Ireland: the Mell Formation (See the companion web site).

Depositional glaciomarine sequences on continental shelves

On the continental shelves and beyond there is an important glaciomarine set of sedimentary environments, particularly around Antarctica and Greenland. The processes controlling sedimentation on glacier-influenced continental shelves are given in Table 20.2 and discussed fully in Chapter 8 of Hambrey (1994). Figure 20.42 illustrates examples of ice dynamics, sediment sources and sedimentary processes and products at the margin of the Antarctic ice sheet.

The sediment distribution patterns are variable, depending on the glaciological regime and the lithofacies preserved. For example, there are major differences seawards of the ice shelves in Antarctica from those areas bordering tectonically-active mountain areas with tide-water glaciers, like in Alaska, and those continental shelves influenced by cold glaciers, like the Barents

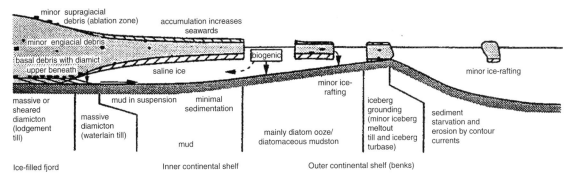

(a) Ice shelf in recessed state

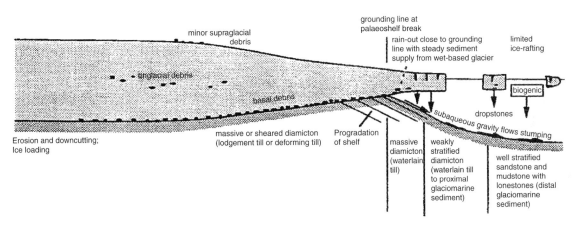

(b) Ice shelf advanced to edge of continental shelf

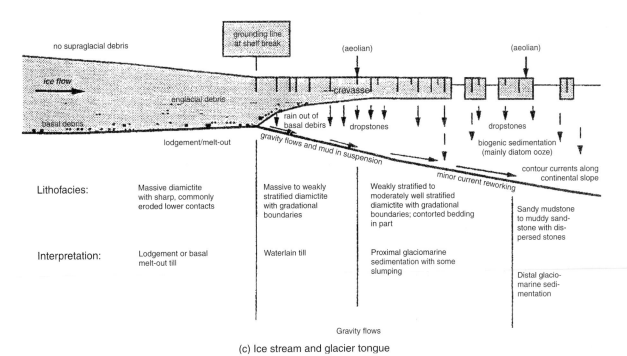

(c) Ice stream and glacier tongue

Figure 20.42 Ice dynamics, sediment sources and sedimentary processes and products at the Antarctic ice-sheet margin, and their interpretation. (Source: after Hambrey, 1994)

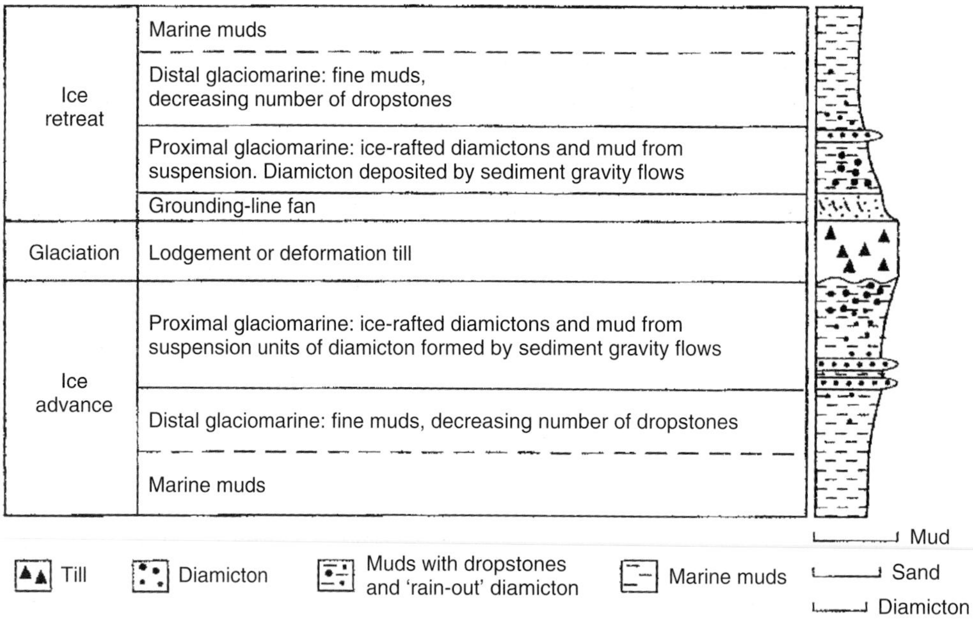

	Marine muds
Ice retreat	Distal glaciomarine: fine muds, decreasing number of dropstones
	Proximal glaciomarine: ice-rafted diamictons and mud from suspension. Diamicton deposited by sediment gravity flows
	Grounding-line fan
Glaciation	Lodgement or deformation till
Ice advance	Proximal glaciomarine: ice-rafted diamictons and mud from suspension units of diamicton formed by sediment gravity flows
	Distal glaciomarine: fine muds, decreasing number of dropstones
	Marine muds

▲▲ Till   Diamicton   Muds with dropstones and 'rain-out' diamicton   Marine muds   Mud Sand Diamicton

Figure 20.43 Hypothetical log deposited on the continental shelf during a single glacial cycle in which a grounded glacier advances and decays across a continental shelf. (Source: after Bennett and Glasser, 1996)

Shelf. This can be seen from the sedimentation rates; in the Weddell Sea and the Barents Shelf the rates are between 0.02 and 0.05 mm per year, whereas in the Gulf of Alaska the rate is 1.0 mm per year. Again, further details are provided in Chapter 8 of Hambrey (1994). An example of a hypothetical lithofacies association formed as a result of the advance and recession of a temperate glacier across a continental shelf is illustrated in Figure 20.43.

20.6 Glacial landsystems

Having reviewed some of the basic controls on the formation of glaciers and their modes of flow, we need now to turn our attention to the landforms and sediment assemblages produced by glaciers and ice sheets. Glaciers produce a diverse range of landforms and sediments, which can be combined in a variety of ways to give different glacial landsystems depending on the glacial dynamics and topographic location. Evans (2003) provides further detailed landsystem templates for glacial depositional environments and glaciated terrain. In this section, we focus on six landsystems, before considering the variables that control their distribution within a range of different ice sheets. The landsystems are:

- preservational, cold-based continental ice sheets
- warm-based continental ice sheets
- glaciomarine systems
- valley systems

- surging glacier landsystems
- glaciovolcanic landsystems.

20.6.1 Preservational, cold-based continental ice sheets

Glaciers modify the landscapes over which they move and tend to preserve that date from either earlier periods of glaciation or interglacial or pre-glacial times. There are several ways in which this preservation can occur. First, at the centre of large continental ice sheets ice velocities are either very low or negligible. This is due to the fact that ice flows away from ice divides or centres of mass, towards the margin, and because ice velocity accelerates from the divide to a maximum below the equilibrium line. Consequently, bed modification may be minor beneath the ice divide. Second, continental-scale ice sheets tend to have interiors that are frozen to their beds. This in part reflects the fact that ice flow is minimal, thereby generating little in the way of frictional heat.

Preserved landscapes may consist of earlier glacial landforms deposited during periods when the ice divides were located in different places, or of pre-glacial landforms. In upland areas, characterized by frozen bed conditions, landscapes of selective linear erosion may develop. Over the upland regions thin, cold-based ice may preserve the pre-glacial landscape. However, over valleys the deeper, and therefore warmer, ice may lead to localized erosion. A good example of this type of landscape is the Cairngorm Mountains in Scotland, where the plateaux tops preserve pre-glacial tors and weathered granite, while the adjacent valleys provide classic examples of glacial troughs.

Figure 20.44 Landscape of areal scour, north-west Highlands of Scotland

(a)

Figure 20.45 Knob and lochan scenery, north-west Highlands of Scotland

20.6.2 Warm-based continental ice sheets

Landscapes of areal scouring

Areas dominated by hard, nondeformable substrates are often subject to extensive subglacial erosion, giving landscapes of 'areal' scour (Figure 20.44). These landscapes consist of a succession of streamlined rock forms such as whale-backs, roche moutonnées and rock basins. In the north-west Highlands of Scotland this areally-scoured landscape is known as knob and lochan (Figure 20.45).

Roche moutonnées (Figure 20.46) occur at a range of scales, from a few metres to several kilometres, and consist of upstanding rock bodies with a smooth, streamlined, upstream face created by abrasional processes and a downstream face that shows evidence of glacial plucking. They tend to be indicative of fast, thin ice, since these conditions favour the formation of basal cavities essential for glacial quarrying. It is important to note that the detailed morphology of roche moutonnées is very variable, due

(b)

Figure 20.46 (a) Roche moutonnée, northern Iceland. Ice movement is from right to left. (b) Plucked rock surface, Maine

to the importance of rock-mass discontinuities in determining the patterns of quarrying.

Whalebacks are streamlined on all sides and are the product of abrasion; no cavities are required so they may form in thicker ice. These eroded surfaces may be mantled to varying degrees by glacial sediments and landforms formed as the ice margin retreats.

Subglacial bedforms

In areas of soft deformable sediment, subglacial bedforms may occur. There are three main subglacial bedforms: drumlins, mega-flutes and røgen moraines.

Drumlins are streamlined hills of varying size, usually in the range 100–1000 m long and 50–500 m wide. They tend to occur in drumlin fields, for example in Northern Ireland, Ribblesdale, the Eden valley and north and west Cumbria, and form a 'basket of eggs' topography (Figure 20.47). They have a blunt upstream size and tend to taper downflow, so they are ellipsoidal, streamlined bedforms. They have three main internal compositions: rock-cored drumlins, with a core of rock around which subglacial till has been moulded; till-cored drumlins, which have no rock core and may be composed of deformed sediment throughout; and drumlins cored by glaciofluvial sediment around which till is deposited.

The origin of drumlins has been the source of much controversy and has given rise to numerous hypotheses. However, the most widely accepted model of drumlin formation is based on the principles of subglacial deformation and the ability of such processes to erode and deposit sediment. According to this model, regions within a deforming layer that are stronger or stiffer than average will remain static or deform more slowly. In contrast, weaker intervening areas will undergo higher strain rates. In a glacier forefield with a strong grain-size contrast between the coarse gravels of the melt-water channels and the intervening fine-grained areas there are permeability differences in the sediment. The gravels are permeable and therefore have low pore-water pressures when deformed, leading to sediment strength, in contrast to the fine-grained intervening sediments, which are less permeable and therefore deform more easily. As deformation proceeds the coarse-grained sediments act as slowly-deforming obstacles around which the more mobile fine-grained sediment deforms. This type of model can be used to explain a range of different internal compositions for drumlins. A deforming layer

(a)

(b)

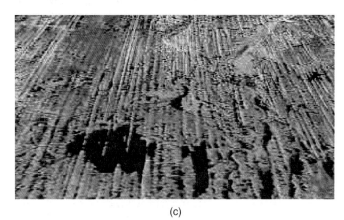

(c)

Figure 20.47 (a) Drumlins, County Leitrim, Republic of Ireland. (b) Drumlin field, north-eastern Ireland. (c) Megafluting, Nunavat Territory, Arctic Canada. (Source: USGS)

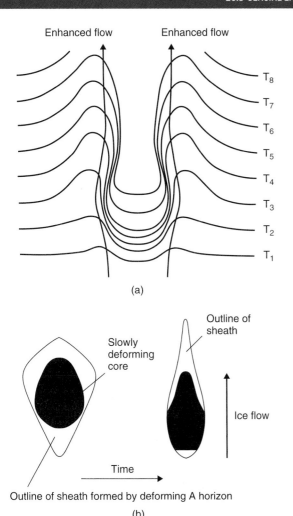

Figure 20.48 Drumlin formation by subglacial deformation. (Source: after Boulton, 1987)

Røgen moraines consist of streamlined subglacial ridges that form perpendicular to the direction of ice flow; they are 10–30 m high and 1–2 km long. They are intimately associated with other subglacial bedforms, such as flutings or drumlins, and often grade into them. A variety of theories have been proposed to explain their origin. They have been seen as the product of a change in ice-flow direction causing the modification of an existing subglacial fluting or drumlin. Alternatively, they have been attributed to the dislocation of blocks of subglacial sediment at the boundary between frozen and unfrozen beds within large mid-latitude ice sheets. It has been pointed out that most røgen moraines occur within zones believed to have been characterized by thermal transition and consequently it is believed that røgen moraines may be indicative of this type of transition.

Formation of eskers

Eskers are a feature of the wet-based portion of continental ice sheets and valley glaciers. They are linear, sinuous ridges of varying length and height, composed of stratified, glaciofluvial sediments. They form in a variety of different ways, but extensive eskers occur beneath continental ice sheets. These ridges can in some cases extend for hundreds of kilometres, as in Sweden and Canada, but they are never wider than 700 m, nor more than around 100 m high.

However, continental-scale subglacial tunnel fills may form sequentially during decay and it has been suggested that most eskers are probably built in short segments by streams extending only up to a few kilometres from the ice margin. As the ice margin retreats, the stream segment building the esker retreats, maintaining a more or less constant length by extending itself headward. The spacing of individual segments of the esker can be related to the rate of annual retreat and they often follow the latest ice-movement direction.

Morphologically they can form as single continuous ridges of uniform dimensions, they can be a series of mounds linked by low ridges or they can occur as complex braided systems with several tributaries. The sediment type ranges from boulder gravels to silts and clays in places. Their widespread distribution has often been used to argue against subglacial deformation, as how could such ridges survive if a bed were being actively deformed? Work on the distribution of eskers however suggests that they are predominantly restricted to those areas with hard glacier beds; in the case of the former mid-latitude ice sheets, that is the central shield areas.

There are several mechanisms for the formation of eskers, and they are polygenetic landforms. They can occur in ponded water where subglacial or englacial melt water emerges into a proglacial lake to form an esker delta; as the ice retreats during deglaciation a beaded esker is formed. In the bead there is deposition in an environment characterized by rapid flow deceleration. The proximal end of the bead is in contact with ice and will give rise to a steep, linear, ice-contact face on retreat; the distal end is the lake bottom (Figure 20.49). There is a relatively large current variability as there is an expansion of flow as the lateral constraints of the tunnel are removed. So the bead is built up as a delta during the

can deform against a rock-cored obstacle or against a slowly-deforming core formed by a fold within subglacial sediment. In all cases it is the deformation of a rapidly-deforming layer around a more rigid core that gives the drumlin form. Figure 20.48 illustrates the morphology formed by the deformation of a deforming layer around a cylindrical obstacle. This type of hypothesis has been modified and presented in several different forms but provides a strong candidate. It is important to note however that not all glacial geomorphologists agree about this process, and some question the spatial extent of subglacial deformation which the widespread distribution of drumlins implies, while others have attributed drumlin formation to the passage of large subglacial floods causing erosion of the ice bed.

Mega-scale glacial lineations are also probably formed by subglacial deformation and are similar in some respects to the spatial extent of individual spreads of subglacial sediment, effectively moving within the deforming layer as moving thickenings. More recently, drumlins have been modelled as the product of flow instabilities within the deforming layer; the significance here is that no obstacles are required.

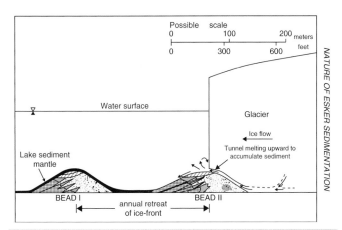

Figure 20.49 Esker formation in a proglacial lake

summer melt season, but during the winter there is little discharge, resulting in a gap in the esker ridge. As the ice front retreats there is therefore an annual periodicity. The most common facies proximally are deltaic or gravel bar fronts overlain by cross-beds, whilst there is a downstream change to distal rippled fine sand and graded sands and finally to rhythmically-laminated silts and clays. In this deltaic esker model there is a rapid downstream facies change, with interfingering from proximal gravel to silt and clay rhythmites, deposition from turbidity currents, the lateral persistence of any one unit, an annual cyclicity of units ending in clay or parallel laminated fine sand, and overall there is little evidence of erosion. Typical sediment types are illustrated in Figure 20.50.

Other mechanisms for esker formation take place when deposition occurs within a conduit, which can be either an open channel or a tunnel. It is often difficult to confidently distinguish between these two types because high-magnitude discharges in open channel flow can be similar to closed conduit flow. However, where eskers run uphill or are associated with glacial drainage channels that run upslope, both deposition and erosion are subglacial in origin, where the water flows under hydrostatic pressure. The characteristics of closed conduit flow eskers are as follows:

- Evidence of erosional activity of the subglacial streams, with incorporation of basal till and local bedrock into the lower sediments (Figure 20.51).

- Dropstones from the base of the ice (Figure 20.52) into the tunnel; often these are excessively big for the postulated flow discharge in the tunnel.

- Sometimes a threefold sedimentological sequence. There is a central channel gravel facies, composed of large-scale channels with fills composed of pebble gravel fining upwards through horizontally-stratified into cross-stratified and rippled sands. This type of channel fill indicates a change from the upper flow regime to the lower part of the lower flow regime, a waning current, and is probably caused by deposition from one flood cycle. There is, second, a marginal sand facies laterally on the ridge flanks, composed of cross-stratified and rippled sand sequences indicative of the lower flow regime and then, third, a back-water silt and clay facies composed of parallel laminated silt and clay and occasional rippled fine sands which infill abandoned channels. An example is shown in Figure 20.53.

- Occasionally, in small eskers, graded gravel units in the ridge centre, which grade upwards into coarse, horizontally-stratified sand. These units could represent cyclic deposition, with each cycle deposited during a summer's flood discharge, or shorter flood cycles where the ridge is built up quickly.

- Often, marginal faults indicating withdrawal of ice support and formation during ice melting.

- Possibly a thin cover of ablation till, although this may not always occur as many ice tunnels melt upwards into clean ice.

- Possibly a pseudoanticlinal structure of the bedding, which suggests that during deposition much or all of the surface of the esker must have been in contact with water rather than with ice walls. This requires secondary currents flowing

Figure 20.50 Fine-grained facies in esker bead, the Torkin, north-west Cumbria: (a) esker bead; (b) ripple drift and parallel laminated sequences

(a)

(a)

(b)

Figure 20.51 Erosional activity of subglacial streams: (a) sub-ice stream, Nigardsbreen, central Norway; (b) turbulent proximal proglacial stream, Nigardsbreen, central Norway

(b)

Figure 20.52 (a) Dropstones from the base of subglacial ice incorporated into esker gravels, Polremon, western Ireland. (b) Extremely coarse gravels from subglacial ice, Polremon, western Ireland

upwards along the flanks of the growing esker, since otherwise the sediment would move down the slopes and be deposited as flat beds. Such secondary flows are well known from turbulent flow in noncircular pipes, where they generally flow towards sharp corners from the central region of the pipe then outward along the walls. In subglacial passages with an arched cross-section this would mean flow upward along the side walls and inward across the floor of the passage and up the sides of the esker.

• Often debris-laden basal ice, providing sediment for the sub-glacial streams in response to the melting of the passage walls, especially when some esker sediments contain clasts from bedrock outcrops near the esker but not actually crossed by it.

• Sliding bed facies, indicating subglacial flow. Here there are thicknesses of poorly-sorted, sandy gravel, which are matrix-supported, where the fine-grained matrix was deposited contemporaneously with the coarser gravel clasts and not after the deposition of the gravels (Figure 20.54).

It has been noted from pipe-flow laboratory experiments of slurries that there is a steady flow of fluid–sediment mixtures as a sliding bed, where the whole sediment bed is in motion en bloc along the bottom of the channel. It has been suggested that the sliding bed may be the full-pipe equivalent of fluvial antidunes at high shear stresses. The driving force for the sliding is the pressure gradient on the flow through the bed shear stress on the surface of the bed, whilst there are resisting forces from the friction of the sediment against the conduit walls. It has been noted that mixed grain sizes are often more easily transported than one grain size alone, because the finer sizes have a buoyancy effect on the coarse grains. Hence the mechanism in eskers is the excess of hydrostatic pressure from the proximal to the distal ends of the subglacial tunnel. Gaps in the esker ridge can be explained as being formed at relatively low shear intensities but contemporaneously with poorly-sorted sands and gravels of the sliding bed facies.

Figure 20.53 Low Plains esker channels, Eden valley, Cumbria

Figure 20.54 Sliding bed facies in subglacial esker, Baronsland, County Mayo, western Ireland

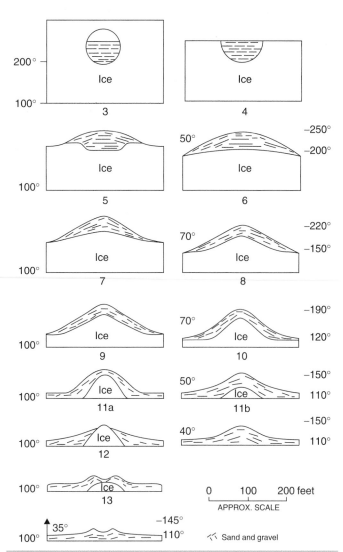

Figure 20.55 Model of deposition for subaerial esker deposition. (Source: after Price, 1963)

- A low variability of palaeocurrents, as the flow is constrained, and a general lack of fine sediments, as these are quickly flushed through the system by the high discharges flowing under high pressure.

- Aggradation of the esker by upwards melting of the tunnel roof and by lateral melting of the tunnels walls. The ability of flowing water to maintain a conduit far exceeds that of the ice to close the conduit by creep so it is no problem for the tunnel to survive, especially during the later stages of deglaciation, although there is nothing in the formation mechanics that requires eskers to be associated with stagnant ice. Copious supplies of melt water are necessary for esker formation and often very high discharges are shown by the grain sizes in fluvial transport.

Eskers can also form englacially, or from an original supraglacial river. This mechanism was originally criticized by many workers, who suggested that narrow strips of fluvial

Figure 20.56 Breidamerkajökull subaerial eskers, southern Iceland

sediment laid down on the floors of supraglacial rivers would not produce ridges after the ice around and beneath them had melted. For example, it was suggested that the supraglacial sediment on the Wolf Creek glacier in Alaska would be constantly redistributed by differential melting of the ice beneath the sediment, so that no specific landforms could develop.

However, complex esker systems associated with some modern glaciers in Iceland, Alaska and Norway are thought to have been formed from supraglacial or englacial positions in the glaciers. These eskers have markedly irregular crestlines, ice cores and disturbed internal stratification. A theory for their formation has been developed. Initially, if the supraglacial or englacial stream deposits survive the early stages of ice wastage they will prevent the ice beneath them from melting at the same rate as the clean ice on either side and this will result in the formation of an ice-cored gravel ridge. Little of the original bedding will be left as slumping will occur down the sides of the ice core. If this process continues for long enough there will be a series of very small ridges, but if the deposits arrive at the subglacial surface at any earlier stage then a distinct ridge form will be preserved. The irregular crestlines are the result of a varying thickness of ice forming the ice cores. If the ice core remains long enough for sufficient slumping to take place down each flank then two esker ridges will be produced. Alternatively, if the ice core wastes away rapidly then only one ridge will form.

Many of the eskers described have tributaries and it is apparent that the melt-water streams responsible for these landforms must have had access to a great amount of debris from the surface or in the ice. It is significant that most esker systems of this category are found on the lee side of rock ridges, where the ice was probably stagnant for a long period of time, allowing the development of a complex drainage system that was not subsequently destroyed by ice movement.

This model of deposition is illustrated in Figure 20.55 and examples of the morphology can be seen in Figure 20.56.

Two case studies from the modern high-Arctic glacier Vegbreen in Svalbard and from Carstairs in Scotland are presented in Boxes 20.3 and 20.4 (see in the companion website). These illustrate that eskers are polygenetic landforms that can be complex and difficult to interpret based on morphology alone. Good sectional evidence is necessary in order for sediments to be interpreted correctly.

Formation of kames

Similar interpretational problems apply to the glaciofluvial landforms referred to as kames. Many types of ice-contact ridge can be covered by this encompassing term. It means a steep-sided mound and is used to describe a large number of sand and gravel landforms where the genesis is not known. As such it serves a useful purpose when there is no sectional evidence. There are four morphological types recognized, each with a different origin:

- **Kame hillocks, conical kames or moulin kames:** These refer to the same landforms in stagnant ice, where there are numerous cavities, depressions and moulins in which sediments can accumulate. The result of ice wastage is often a kame belt or kame and kettle topography, in which there are numerous mounds and depressions. The conical or moulin kame can occur where supraglacial melt water flows down a moulin or a large hollow in debris-covered ice to reach the basal till or bedrock (Figure 20.61). This debris load is deposited in a chaotic manner and when the ice finally melts there is a mixed assemblage of grain sizes, till at the base and marginal faults. The so-called perforation hypothesis tries to explain kame hillocks, where it is thought that debris accumulates in surface pools on stagnant ice. As this pool warms it gradually melts its way down to the ground beneath, and as the surrounding ice melts the end product is an isolated kame mound. Later it was suggested that isolated kames develop where melt water flows into small ponds on or within stagnant ice. At the point of entry the stream deposits a small delta which would form isolated mounds when the ice has melted away.

- **Kame plateaux or flat-topped hills:** These might be formed by melt water flowing into a small lake surrounded by stagnant ice and can be recognized by foreset beds. They might also represent river deposition in a wide crevasse or ice-walled trench and can be recognized by their sediments and structure, and the fact that they are elongated. Ice-walled lake plains, as described in Dakota or the Eden valley in Cumbria, also have this morphology, and there would need to be sectional evidence to interpret them correctly. Kame ridges are linear landforms composed mainly of fluvial sediment and tend to be straight and fairly short. They might be formed as crevasse fillings.

- **Kame terraces:** These are flat-topped, elongated landforms found along valley sides. They often have a kettle-holed surface and are bounded by a steep, irregular ice-contact slope, which marks the position of the ice-contact margin at the time a particular terrace ceased to be formed (Figure 20.62). They can be caused by ice-marginal deposition in either a river

Figure 20.61 Conical or moulin kame, near Castle Carrock, Brampton Kame belt, north-east Cumbria

Figure 20.62 Kame terrace, south side of lower Wasdale, south-west Cumbria

or a lake, but the type of sediment will clearly differentiate this. Faults will form and mark the ice-contact slope. Some valleys, like the Eddleston valley in southern Scotland, show a sequence of kame terraces, which form along the valley side as the ice downwastes *in situ*.

• **Kettled sandar:** These are difficult to fit into any classification as they are both proglacial and ice-contact deposits. They are proglacial because they are part of a proximal sandur system and they are ice-contact because the sediments are laid down on top of glacier ice and contain large melted ice blocks which melt to form kettleholes. Usually remnants of the initial sandur surface survive the development of kettleholes and it is easy to distinguish between a kettled sandur and a kame complex.

Models for the development of kames are illustrated in Figure 20.63 and Table 20.3 and are discussed further in Box 20.5 (see in the companion website).

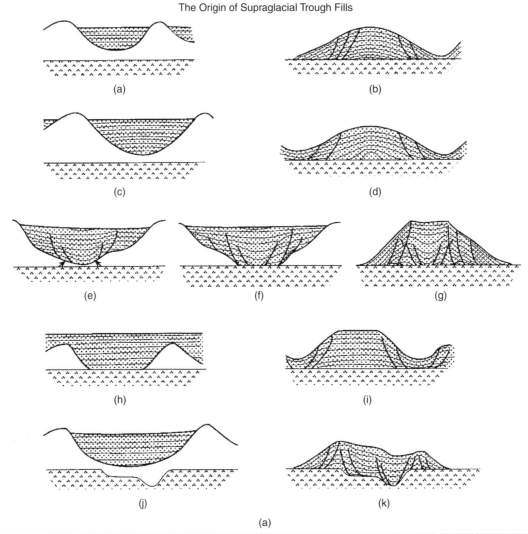

The Origin of Supraglacial Trough Fills

(a) (b) (c) (d) (e) (f) (g) (h) (i) (j) (k)

(a)

Figure 20.63 Models for kame formation: (a) sequences showing the potential variety of kame formation. Left-hand side shows ice and sediment, right-hand side shows the situation after the ice has melted and the typical sedimentary structures; (b) schematic models for the production of kames, kame terraces and kame plateaux. (Sources: (a) after Boulton, 1972; (b) after Brodzikowski and van Loon, 1991)

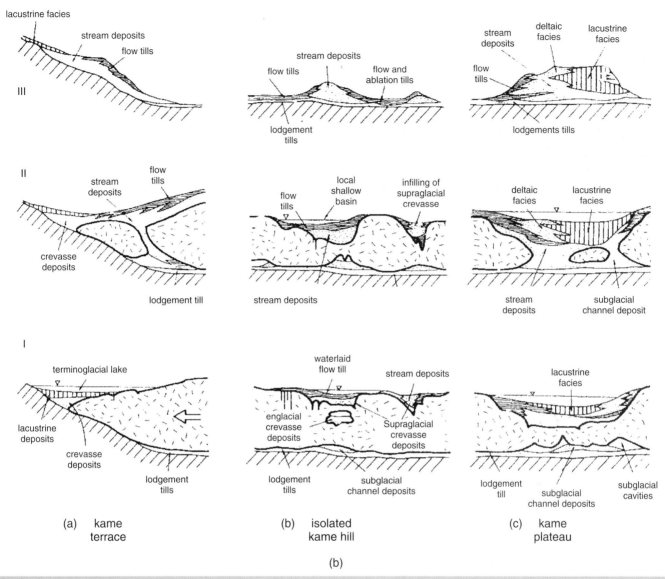

Figure 20.63 (continued)

Moraine formation

Around the ice sheet are large, lobate ice margins whose edges are defined by moraines. In some cases these moraines are extremely extensive and involve the formation of multiple-crested ridges. Particularly good examples are those of the Dirt Hills in Saskatchewan, Canada, which cover an area of 1000 km^2 and rise up to 150 m above the surrounding areas. They are associated with subglacial hydrogeology, which favours subglacial deformation. In other areas the moraines are more modest and pass into extensive belts of hummocky moraine, which in turn tend to grade into drumlins up-glacier. These extensive belts of hummocky moraine were once thought to be the result of rapid glacier stagnation. However, in practice it is difficult to see how such vast quantities of debris could be elevated to the ice surface.

A recent suggestion is that many of these hummocky moraines form by subglacial pressing. Rapid ice flow over a deforming bed leads to the formation of drumlins and also very low-gradient ice lobes, prone to stagnation as climate ameliorates. As the ice thins and stops flowing rapidly it settles into the underlying sediment, which is pressed into basal crevasses and fractures and may make its way to the surface. In this way large areas of stagnant terrain are produced with flat-topped mounds and a range of irregular forms. Other areas may form as consequence of sediment build-up on the ice margin, especially in zones of thermal transition, or where ice pushes against reverse bedrock slopes, as in Dakota. In fact, many of the lobes of the southern margin of the Laurentide ice sheet form in embayments within escarpments.

Much of the northern mid-latitudes was covered by large continental ice sheets during the height of the last glacial cycle. Today the ice sheets of Antarctica and Greenland represent

Table 20.3 Lithofacies observed and interpreted from drill sites in McMurdo Sound and Prydz Bay, Antarctica. (Source: after Hambrey, 1994)

Facies	Description	Interpretation
Massive diamictite	Non-stratified muddy sandstone or sandy mudstone with 1–20% clasts; occasional shells and diatoms.	Lodgement till (with preferred orientation of clast fabric) or waterlain till (random fabric).
Weakly stratified diamictite	As massive diamictite, but with wispy stratification; bioturbated and slumped; partly shelly and diatomaceous.	Waterlain till to proximal glaciomarine sediment.
Well-stratified diamictite	As massive diamictite with discontinuous and contorted stratification; occasional dropstone structures; abundant diatoms and shells.	Proximal glaciomarine/ glaciolacustrine sediment.
Massive sandstone	Non-stratified, moderately well-sorted to poorly sorted sandstone, with minor mud and gravel component; loaded bedding contacts.	Nearshore to shoreface with minor ice-rafting in distal glaciomarine setting; better sorted sands with loaded contacts are gravity flows; associated with slumping.
Weakly stratified sandstone	As massive sandstone, but with weak, contorted, irregular, discontinuous, wispy, lenticular stratification; brecciation, loaded contacts and bioturbation.	Nearshore with minor ice-rafting in distal glaciomarine setting; better sorted sands are gravity flows; associated with slumping.
Well-stratified sandstone	As massive sandstone, but with clear, but often contorted stratification.	Nearshore with minor ice-rafting in distal glaciomarine setting; some slumping.
Massive mudstone	Non-stratified, poorly sorted sandy mudstone with dispersed gravel clasts; intraformational brecciation and bioturbation; dispersed shells and shell fragments.	Offshore with minor ice-rafting in distal glaciomarine setting; some slumping or short-distance debris flowage.
Weakly stratified mudstone	As massive mudstone but with weak, discontinuous, sometimes contorted stratification defined by sandier layers; bioturbated.	Offshore to deeper nearshore with minor ice-rafting in distal glaciomarine setting: slumping common.
Well-stratified mudstone	As massive mudstone, but with discontinuous, well-defined stratification, with sandy laminae; syn-sedimentary deformation and minor bioturbation.	Deeper nearshore with minor ice-rafting in distal glaciomarine setting; some slumping.
Diatomaceous ooze/diatomite	Weakly or non-stratified siliceous ooze with >60% diatoms; minor components include terrigenouss mud, sand and gravel.	Offshore, with minor ice-rafting in distal glaciomarine setting.
Diatomaceous mudstone	Massive mud or mudstone with >20% diatoms and minor sand.	Offshore with sedimentation predominantly influenced by ice-rafting and underflows in distal glaciomarine setting.
Bioturbated mudstone	As massive mudstone, but stratification highly contorted or almost totally destroyed by bioturbation.	Offshore to deeper nearshore with minor ice-rafting: extensively burrowed.
Mudstone breccia	Non-stratified to weakly stratified, very poorly sorted, sandy mudstone intraformational breccia with up to 70% clasts; syn-sedimentary deformation and minor bioturbation.	Offshore to deeper nearshore slope-deposits with minor ice-rafted component, totally disrupted by debris flowage.
Rhythmite	Graded alternations of poorly sorted muddy sand and sandy mud; stratification regular on a mm-scale; dispersed dropstones.	Turbidity underflows derived from subglacial source, with ice-rafting in a proximal glaciomarine setting.
Congiomerate	Non-stratified to weakly stratified, poorly-sorted, clast- to matrix-supported sandy conglomerate; normal and reverse grading evident; clasts up to boulder size; intraclasts of mudstone frequently incorporated; loading and other soft-sediment features present.	Slope debris-flows derived directly from proglacial glaciofluvial material, or from subaqueous discharge from glacier; well-defined beds may be fluvial.

contemporary examples, although in very different latitudes. These large mid-latitude ice sheets have left a palimpsest of glacial landforms of different ages that form a range of glacial landscapes. Let us consider a north–south transect through one of these large mid-latitude ice sheets across an area of continental shield (Figure 20.65). The first thing to note is that the base of the ice sheet is divided into a series of thermal zones. The centre of the sheet is cold-based and frozen to its bed. Towards the southern maritime margin the ice sheet first gains thermal equilibrium, before becoming warm-based towards the margin. Towards the north, or continental margin, the ice sheet may warm to a state of thermal equilibrium, but it is cold-based and underlain by permafrost at the margin. Each thermal zone will give a distinctive landsystem, although one should note that as the ice sheet grows and decays the pattern of thermal zones may change. Equally, the centres of mass within the ice sheet may change, which can cause a shift in the distribution of processes. We can recognize several landsystems beneath this ice sheet: a cold-based continental interior; a zone of maximum erosion

and areal scour; a southern zone of deposition; and a zone of cold-based ice margins in the north.

In the interior of the ice sheet little geomorphological modification occurs, first because the rate of ice flow is minimal, increasing to a maximum beneath the equilibrium line, and second because the glacier is frozen to its bed. As a consequence the landscape may preserve evidence of earlier landforms and sediment systems, or elements of the pre-glacial landscape. Where the centre of mass of an ice sheet has changed over time, or where there have been radical changes in its thermal character, it is likely that elements of earlier glacial landsystems may be preserved. Further discussion is elaborated in Bennett and Glasser (1996, pp. 347–352).

In the case of the Laurentide ice margin, as it retreated large proglacial lakes developed, in part exploiting natural basins, but also as a result of isostatic depression, creating a natural basin between the ice margin and the fore-bulge. As a consequence, extensive glaciolacustrine basins are typical of these margins. These result in a diverse collection of sediments produced by ice-marginal processes. In addition, the drainage of these lakes

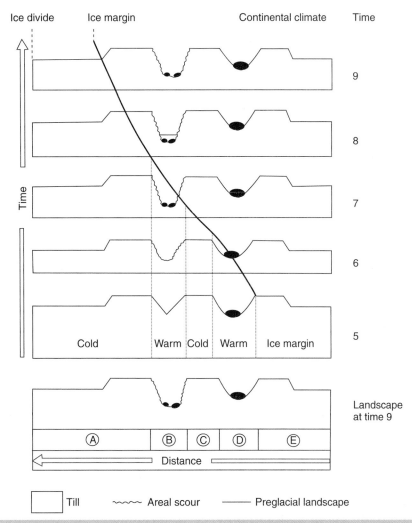

Figure 20.65 Conceptual model of the landscape evolution beneath a mid-latitude ice sheet. (Source: after Bennett and Glasser, 1996)

Table 20.4 Features and interpretation of lithofacies in the cored CIROS-2 fjord sequence, Ferrar Fjord, Victoria Land, Antarctica. (Source: from Barrett and Hambrey, 1992)

Facies	Sedimentary features	Interpretation
Massive diamictite	Non-stratified; clasts uniformly distributed; horizontal shear surfaces; occasional striated clasts	Lodgement till; less commonly basal meltout till or waterlain till
Weakly stratified diamictite	Weak, wispy stratification in matrix, and dispersed clasts, including dropstones	Proximal glaciomarine/glaciolacustrine sediment derived from basal debris melted out near the grounding-line; some winnowing
Well stratified diamictite	Well developed stratification in matrix and dispersed clasts, including dropstones	Proximal to distal glaciomarine/glaciolacustrine sediment; strong bottom reworking
Massive sandstone	Fine to very fine, uniform, unstratified; a few beds are graded	Aeolian or supraglacial sand deposited directly or washed off floating ice; graded beds may be turbidities
Stratified sandstone	Fine to very fine; occasional fine horizontal mm-cm laminae	As above, but with intermittent deposition
Mudstone	Moderately and weakly stratified to unstratified; little sand and few gravel sized clasts	Sedimentation from suspension well beyond ice margin, originating from basal debris close to the grounding-line; little from sea or berg ice
Rhythmite	Alternation of well sorted, very fine sand and clayey silt layers from a few mm to a few cm thick; no dispersed gravel clasts	Sedimentation from intermittent underflows of fines winnowed from basal glacial debris with background sedimentation of aeolian sand; some may be varves, no ice-rafting

produced catastrophic floods that created extensive areas of scabland in which sediment was stripped and the surface was scoured by flood waters (See for example the drainage of Glacial Lake Missoula).

There has been much controversy related to how much subglacial lake drainage can contribute to certain types of landform development, such as drumlins and large-scale bedrock grooving (see examples in Box 20.6, see in the companion website).

The moraine systems on the Prairies and Great Lakes region record highly-dynamic ice margins that deposited and reworked large quantities of sediment during multiple advance-and-retreat cycles. When the ice sheet retreated northwards on to the hard substrate of the Canadian Shield it continued to retreat actively and deposited overlapping concentric moraines and short discontinuous eskers. In many places these moraines are associated with large proglacial lakes ponded against the ice margin, such as Glacial Lake Agassiz. In Fennoscandinavia the large lakes consisted of much of the Baltic basin, which was fresh water for a time, and in this lake extensive esker-fed fans formed, some of which emerged to the water surface to form deltas. These fans merged along the ice margin to form moraine-like sediment masses, which mark the ice edge. The most famous are the Salpausselka moraines of southern Finland, just north of Helsinki.

20.6.3 Glaciomarine systems

The glaciomarine environment stretches from fjords close to where the ice reaches the sea to those areas affected by iceberg drifting only, which can be far distant from the ice source. In locations where the ice sheet is able to advance into the sea to the margins of the continental shelf, one landform of particular note along such margins is the trough-mouth fan. Where ice streams or corridors of enhanced ice flow punctuate the ice margin they deliver a disproportionate amount of the ice and sediment to the margin, which therefore form focal points for rapid sedimentation, which can effectively cause the edge of the continental shelf to advance via a series of steeply-dipping foresets.

It is in the glaciomarine system that sediments have the best chance of preservation, and it therefore preserves the most complete glacial record. It is a complex system, as we have already seen, but we can divide it up into two major components: fjord sedimentation and sedimentation on continental shelves.

Close to the ice in fjords there are a set of processes whereby debris is released directly from the ice, from subglacial streams, from suspended sediment and from icebergs. The facies and interpretations for two contrasting lithofacies associations form Alaskan-type fjords. Measurements from arid-type fjords in Antarctica, taken from Barrett and Hambrey (1992), are illustrated in Table 20.4. There are many depositional landforms, as was seen earlier, and the suggested stratigraphic architecture and lithofacies, combining bottom-sampling, drill-hole data and seismic surveys, is found in Figures 20.46 and 20.66 and Table 20.3.

Similarly, data from continental shelves are covered in Section 20.5.6, and more fully in Hambrey (1994).

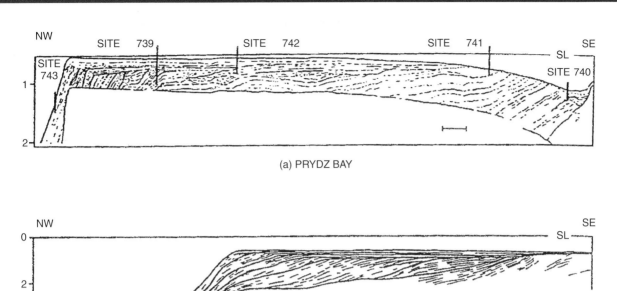

(a) PRYDZ BAY

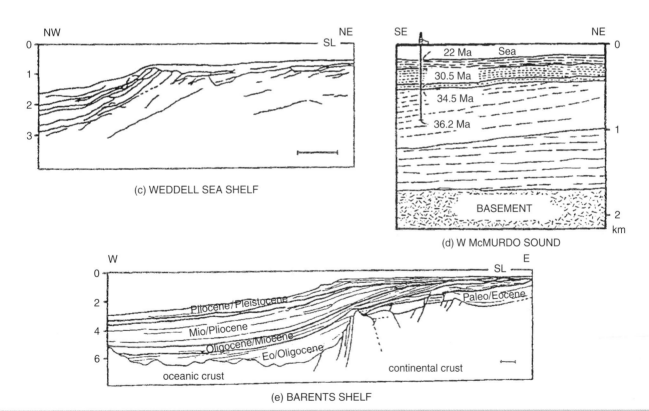

(b) OFF NW ANTARCTIC PENINSULA

(c) WEDDELL SEA SHELF

(d) W McMURDO SOUND

(e) BARENTS SHELF

Figure 20.66 Stratigraphic architecture of polar continental shelves. (Source: from Hambrey *et al.*, 1992)

20.6.4 Valley systems (glaciers constrained by topography)

Let us consider two end members of strongly-channelized ice flow, one in a wet, temperate location such as Iceland, and a second in a dry, Arctic location such as Svalbard.

The Icelandic landsystem will consist of a series of glacier tongues or valley glaciers constrained laterally by steep topography. This could equally apply to the European Alps or parts of Norway. Sometimes a distinction is made between Alpine valley systems, where the area of ice accumulation is overlooked by higher ground, and the Icelandic valley system, which is closed at one end by a trough head or end and is associated with erosion beneath ice caps or ice sheets. During the twentieth century these glaciers have been subject to relatively steady decay, revealing a glacier forefield rich in sediments and landforms. The annual position of each ice margin is often marked by a small seasonal moraine formed either by ice-marginal pushing or by the dumping of supraglacial material from the ice margin. The seasonal push moraines (Figure 20.67) form because of the imbalance between glacier flow, which occurs throughout the year (although it may accelerate in the wet summer months due to enhanced basal lubrication), and ice-marginal ablation, which tends to only occur in summer. The position of the ice margin is determined by the interplay of ice velocity, which moves new ice to the margin, and marginal melting, which removes it. In the winter, when melting is reduced, the glacier still flows and consequently the margin advances, only to retreat in the summer as melting exceeds the rate of ice flux due to ice flow. This seasonal ice-marginal fluctuation is typical of temperate, maritime ice margins.

Perpendicular to these seasonal push moraines are subglacial flutes (Figure 20.68). Flutes are long, linear, streamlined ridges of sediment usually only a few metres in width and height but extending over several hundred metres. They are formed at the glacier bed as it moves over large boulders or bedrock obstacles.

Subglacial sediment is moved under pressure into the lee cavity formed as the ice moves over the obstacle to form a linear flute. This simple system of landforms is dissected by large melt-water fans. These fans have their apex at a melt-water portal and fan out at a low gradient from this point. Where valley topography or the landform topography of the forefield is prominent, these fans may become highly constrained or incised. They may also build up and on to the ice margin, incorporating significant amounts of ice, which then melt on ice retreat to give a topography of irregular mounds and kettleholes at the fan apex (a kettled sandur). This type of topography is sometimes referred to as kame and kettle topography.

This landsystem is complicated further in areas of the ice margin with a high concentration of supraglacial debris. In these areas a topography of mounds and depressions, often referred to as hummocky moraine, may develop. The high levels of debris concentration on the ice lead to the creation of ice-cored topography, which through successive phases of debris movement, melting, and consequently repeated topographic inversion, leads to the deposition of an irregular topography of hummocks. In locations in which the surface concentration of debris on the ice margin is particularly high, hummocky moraine may dominate the landform assemblage.

In drier continental locations such as Svalbard this landsystem is somewhat different. Seasonal ice-marginal oscillations tend to be absent and consequently annual dump or push moraines are not present. Instead the ice margin is often dominated by large amounts of supraglacial debris. This is derived in part from adjacent valley sides, but mostly through the elevation of basal debris. This type of ice margin tends to be associated with mixed thermal regimes, in which the outer ice margin is frozen to the bed while the interior of the glacier remains at pressure melting. The consequence is that the interior of the glacier is fast-flowing while the margin is not, which leads to longitudinal compression. This compression may lead to folding or thrusting within the body of the ice, and both may lead to the elevation of debris-rich

(a)

(b)

Figure 20.67 Seasonal push moraines: (a) Fjallsjökull, southern Iceland; (b) Fingerbreen, Svartisen, northern Norway. (c) Model for the formation of seasonal push moraines. (Source: after Bennett and Glasser, 1996)

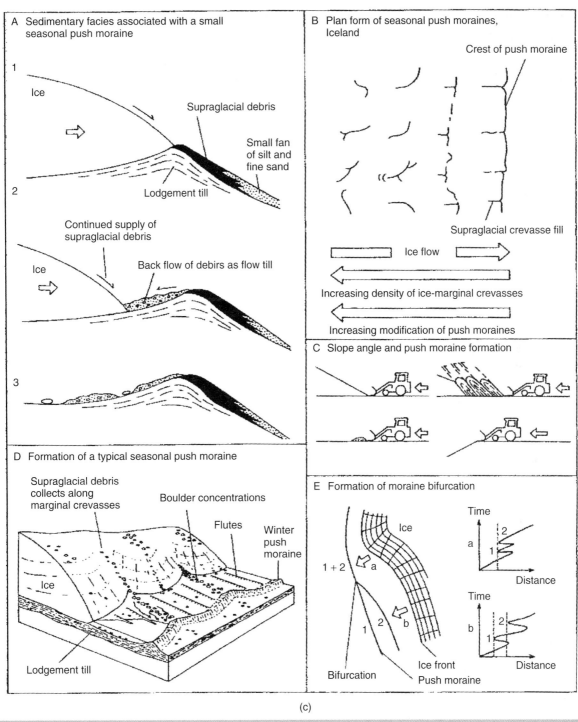

A Sedimentary facies associated with a small seasonal push moraine

1

Ice

Supraglacial debris

Small fan of silt and fine sand

Lodgement till

2

Continued supply of supraglacial debris

Ice

Back flow of debirs as flow till

3

D Formation of a typical seasonal push moraine

Supraglacial debris collects along marginal crevasses

Boulder concentrations

Flutes

Winter push moraine

Ice

Lodgement till

B Plan form of seasonal push moraines, Iceland

Crest of push moraine

Supraglacial crevasse fill

Ice flow

Increasing density of ice-marginal crevasses

Increasing modification of push moraines

C Slope angle and push moraine formation

E Formation of moraine bifurcation

Time

Ice

1 + 2   a

a   1   2

Distance

Time

1   2   b

b   1   2

Distance

Ice front

Bifurcation

Push moraine

(c)

Figure 20.67 *(continued)*

ice at the glacier bed. In fact, in some cases it has been argued that significant slabs of subglacial sediment may be elevated within the ice during the process of thrusting.

The build-up of large amounts of supraglacial debris causes the differential ablation of the ice margin. At ice margins where this build-up is concentrated close to the ice margin, large ice-cored moraines, sometimes referred to as ablation moraines, in which

masses of ice-cored sediment become detached from the glacier, can result. These ice-cored masses may then melt out through a process of continual topographic inversion to give an arcuate belt of moraine hummocks, often out of all proportion to the size of the original ice-cored moraine. Where the debris build-up is more continuous, a larger zone of ice-cored topography results. This topography is often controlled by the pattern of debris

Figure 20.68 Fluted ground moraine: (a) in front of Okstindbreen, northern Norway; (b) Bear Glacier, Alaska. Note the sinuous esker. (Source: http://tvl1.geo.uc.edu/ice/image/subland/Bearflute.html)

Figure 20.69 Ice-cored moraines, Svalbard: (a) Midre Lovenbreen, ice movement from left to right; (b) Ossian Sarsfjellet, north side of Kongsfjorden

outcrop on the glacier surface, which is a function of its tectonic structure: for example, the location of thrusts or folds. In this case a common feature is the formation of a series of ice-cored moraines that parallel the approximate trend of the ice margin (Figure 20.69).

Sedimentation within such zones is particularly complex and its character is largely determined by the amount of water flow present. If the water flux is high then the debris present will tend to be reworked as it slides or flows from topographic highs. Fluvial sedimentation may quickly infill the topographic lows, and these may be enhanced by the frictional heating of the water present. When the ice is finally melted the topography will tend to be inverted and what was once a channel or basin might become a high point. A complex topography of mounds, ridges and hollows will result, which may record some structure, either of the main water flow routes or of the constraining ice-cored ridges, and therefore the structural form of the glacier. Where

the water flux is much less, a complex pattern of sedimentation may result, in which the topographic lows are infilled by a complex facies of flow tills, lacustrine sediments formed in local ponds and basins, and fluvial sediments. The facies geometry of some of these inverted basin fills and stratigraphic sections is a tripartite sequence in which the basal layer consists of subglacial sediment, such as a lodgement till, above which there are fluvial sands and gravels capped by a flow till (Figure 20.70). It has also been suggested that in locations at which significant slabs of subglacial sediment are elevated within the glacier, moraine mounds or even discrete ridges may result from the preservation of the basal part of the thrust sediment slab (see Box 20.7, see in the companion website for a further discussion of the thrusting process).

A common feature of this type of terrain is the occurrence of ice-dammed lakes, particularly as the ice retreats to reveal ice-free valleys, which may have their drainage impeded by ice.

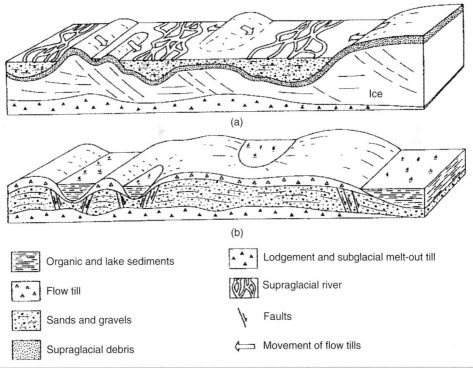

(a)

(b)

	Organic and lake sediments		Lodgement and subglacial melt-out till
	Flow till		Supraglacial river
	Sands and gravels		Faults
	Supraglacial debris		Movement of flow tills

Figure 20.70 Tripartite glacial sediment development. (Source: after Bennett and Glasser, 1996)

A simplified picture of one such basin is that the lake levels are determined by the height of the ice dam and the presence of low points, or cols, within the surrounding valley sides. Lake shorelines are often well developed despite the limited fetch within such lakes. This reflects a number of factors: first, the saturated lake margin is an ideal environment for processes of freeze–thaw, and second, the calving of icebergs produces large waves, which in the confined location of a lake have considerable erosional potential.

The location of the ice margin may be marked in a variety of different ways. Cross-valley moraines or de Geer moraines may form, which can be very similar to small push moraines, and may result from a variety of different processes, from the product of subglacial squeezing-out of saturated sediment to calving. Small lacustrine fans may also develop at melt-water portals associated with the rapid deposition of sediment within the water body. These fans are composed of sediment deposited both as tractional load close to the margin and as the product of sediment gravity flows into the lake. Ice-marginal moraine ramps may also form as a consequence of the direct dumping of sediment from the ice margin, or from the melt-out of debris from icebergs concentrated at the margin. The supply of subglacial sediment to the ice margin by subglacial deformation may also contribute to the formation of these marginal ramps. Glaciolacustrine fans may emerge from the lake bed and sedimentation may become subaerial, leading to the formation of a delta.

There are other types of ice-cored moraine associated with snowbanks, as illustrated in Figure 20.71.

Figure 20.71 Ice-cored moraine associated with a snowbank in front of a glacier, which is then incorporated into a proglacial moraine, Tarfalaglacieren, Kebnekaise

20.6.5 Landforms of glacial erosion associated with such systems

Major landforms of glacial erosion can be seen as modifications carried out to the glacial bed in such a way that the hydraulic geometry of the bed becomes more efficient for the evacuation of ice. A classification is given in Table 20.5.

Table 20.5 Classification of glacial erosion landforms according to process of formation, relief form and scale. (Source: after Hambrey, 1994 and Benn and Evans, 2002)

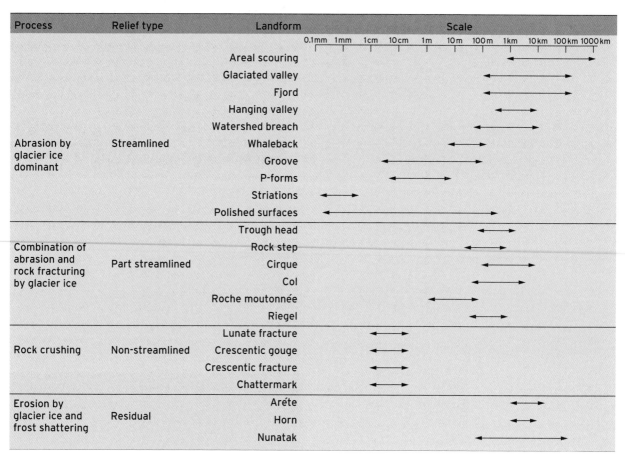

Corries or cirques are morphologically a hollow, open downstream but bounded upstream by a steep headwall, arcuate in plan, around a more gently-sloping floor, which often contains a lake in an overdeepened rock basin (Figure 20.72). There can be major size variations and altitude variations, but generally there is a concentration in upland areas. Often the size difference is due to different ages of occupancy of ice, but ice volumes and the bedrock type are important too. They have a preferential location in north-east to eastern quadrants in the northern hemisphere, simply because snow drifts to the lee of the mountains from the predominant westerly winds and the aspect protects this snow from melting. Hence they develop in pre-existing topographic hollows initially through the processes associated with nivation. The morphology of the rock surface is accepted as reflecting the contrast between glacial erosion of the floor and the base of the headwall and subaerial frost-shattering on the upper headwall and the surrounding arêtes or knife-edge ridges. The erosional features in the corrie form from rotational slipping of ice in the hollow, which erodes the rock basin, and depositional landforms, such as moraines or pro-talus ramparts, demarcate the ice–snow boundary at their downvalley end.

When the processes of glacial erosion and frost-shattering combine to erode back into the mountain massif, pyramidal peaks or horns are the end product (Figure 20.73).

U-shaped valleys are formed where the ice, or the major part of it, is discharged in a rock channel (Figure 20.74(a)). Occasionally the valleys are asymmetrical when an aspect control forms more debris on one side of a valley, but often they are symmetrical, with their shape more of a parabola. They are often overdeepened, with hanging tributary valleys, and have rock bars, basins or steps superimposed on their long profile. These undulations are caused by variations in rock type or frequency of jointing, a change in the volume of the ice or confluence/diffluence of ice streams. There are often marked trimlines and truncated spurs (Figure 20.75) associated with such valleys. A U-shaped valley is an equilibrium form related to the amount of ice discharged, and the most efficient shape for ice evacuation has been shown to be a semicircle, as this means the minimum of frictional resistance in relation to a given ice volume. The variations in long profile occur because there has not been sufficient time for the ice to erode away bedrock irregularities.

(a)

(b)

(c)

Figure 20.72 Corrie morphology: (a) Angle Tarn, Lake District, UK, showing overdeepened basin and ice-plucked back wall; (b) arête ridges and corrie backwalls, Arran, Scotland; (c) corries and arêtes, Chugach Mountains, north of Tonsina glacier (north-central Chugach), Alaska. (Source: Bruce F. Molina, http://www.usgs.gov/features/glaciers2.html)

Fjord systems are actively forming today in Antarctica, Greenland, the Canadian Arctic islands, Svalbard, Southern Alaska, Chile and South Georgia and have formed in the past in many glacially-modified parts of the world, such as Iceland (Figure 20.74(b)). The longest is probably the Lambert graben in Antarctica, which is 800 km long, 50 km wide and 3 km from crest to floor. Usually the fjords have shallow sills, multiple basins and form from progressive downcutting through several glacial cycles. They often follow weakness in the rock, such as faults, and hence are straight in long profile and in certain easily-eroded bedrock types.

20.6.6 Surging glacier landsystems

It has been noted that surging glaciers in Svalbard and Iceland contain much more debris than nonsurging glaciers in the same areas. Excessive debris entrainment probably occurs during surge activity and relates to the basal ice conditions, the glacial tectonics and the nature of the underlying surface, as discussed earlier. However, a general sediment and landform assemblage can now be recognized in areas that formerly had surging glaciers. Five landform types appear to be concentrated at surging glaciers: two developed during the surge, two formed after the surge and one produced both during and after the surge.

Flutes composed of till and/or glaciofluvial sediment are commonly present where glaciers have receded from their surge limits. The development of fluted topography is probably encouraged by the presence and abundance of saturated sediment at the glacier base and by the enhanced growth of basal cavities during the surge. Push moraines are common in surging Svalbard glaciers and usually form from proglacial, fine-grained sediment that has been tectonically deformed. The impact of a fast-moving glacier upon sedimentary layers overlying a pronounced décollement

(a)   (b)

Figure 20.73 Pyramidal peaks or horns: (a) Matterhorn, European Alps; (b) Stauning Alps, east Greenland

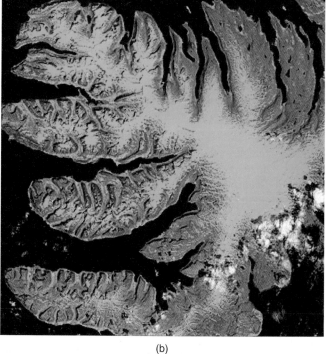

(a)   (b)

Figure 20.74 U-shaped valleys and fjords: (a) Nant Francon, Snowdonia, north Wales; (b) fjord systems in western Iceland. (Source: (b) UGGS Landsat 7)

Figure 20.75 Truncated spurs, Blencathra, English Lake District

plane (the top of the permafrost) compresses the sediments into arcs of Jura-like ridges.

After the surge narrow till ridges develop on the glacier surface, probably formed by squeezing of water-lubricated sediment into ice structures such as crevasses, shear planes and foliation lines. These are known as crevasse-fill ridges (Figure 20.76). A second type of landform produced after the surge is amorphously-shaped areas of hummocky till, which are commonly ice-cored and are associated with concentrations of surface angular debris from deformed medial moraines. Similar topography can develop through the slow ablation of ice beneath a cover of flow and melt-out till. This is one mechanism for the formation of 'hummocky moraine', commonly associated with ice-stagnation landforms such as kettleholes. The role of thrusting and landform development in the formation of moraine mounds in Svalbard associated with surging glaciers has been described in a series of papers from Hambrey and Huddart (1995), Huddart and Hambrey (1996), Bennett *et al.* (1998) and Glasser and Hambrey (2001) and is illustrated in Figure 20.77. Here slabs of basal sediment are thrust up within the margins of polythermal glaciers. This model has also

been applied to explaining the formation of hummocky moraine, the distinctive moraine-mound assemblages that occupy many valley floors in upland Britain, such as in Coire A'Cheud-cnoic (Torridon, north-western Scotland) and Ennerdale (the English Lake District). However, these ideas have proved controversial and stimulated much debate, with both supporters and detractors. (Box 20.7 in the companion web site discusses further the role of thrusting in the formation of moraines).

Ice-contact glaciofluvial sediments commonly survive as eskers, kames and kame terraces because there are extensive lower zones of surged glaciers that are inactive, with both melt water and debris abundant. A landform apparently found only with surging glaciers is the 'concertina' esker described from the surge-type Bruarjökull in Iceland. The indications are that during quiescent phases water drainage takes place in ice-walled and ice-roofed channels, and as the ice retreats, these form eskers. These eskers become deformed during a surge, which is reflected in the strong longitudinal compression of the ice in the terminal zone. The wavelength of one of the compressed eskers indicates that ice in a 4 km-wide margin was compressed by 50% of its original length. At Bruarjökull these eskers are found from the 1963–64 and 1890 surges but not from the 1810 one. It is clear that these eskers may not survive if overrun by later surges.

During and after the surge, extensive outwash fans begin to form when floods of basal melt water are released during the surge activity and continue to develop throughout the long ablation period. An example is show in Figure 20.78, from the 1991 surge of Skeidararjökull in Iceland. From the same surge there has been evidence for extensive englacial drainage systems preserved in ice that has surged, although the eventual preservation potential in this situation is zero.

A model for surge-moraine formation (Figures 20.78 and 20.79) at Fridjovbreen, Svalbard has been described and these moraines appear to be a distinctive type of landform, although it will be difficult to recognize them after deglaciation. When the ice margin advances against earlier lateral moraines, flow compression initiates thrusting. This entrains sediment into the glacier, and this tectonic wedge of ice and debris appears to act as a buffer between the glacier and the lateral moraine. Ice displacement occurs along the décollement or thrust surface up-glacier of the debris-rich ice wedge. This displacement may be facilitated by a rheological contrast between debris-rich ice and the cleaner ice within the main glacier. Deformation continues within the debris-rich wedge, the thrusts within it are gently folded and a complex mélange of ice blocks and folded debris rafts is produced.

However, few of the above landforms are exclusive to surging glaciers, though the spatial patterns and degree of development of the landforms probably are. In fact it may be that the inactive, ablating zones of surged glaciers provide the best real-world models for the study of landforms created during the dissipation of Pleistocene ice sheets in temperate regions like the British Isles.

The suggestion was made that ice streams in western Antarctica have surged, and since then many suggestions of surging mid-latitude ice sheets have been put forward. Attention has been focussed on the possible surging activity of the Hudson Strait

Figure 20.76 Crevasse-fill ridges, Hagafellsjökull Eystri, Langjökull, central Iceland

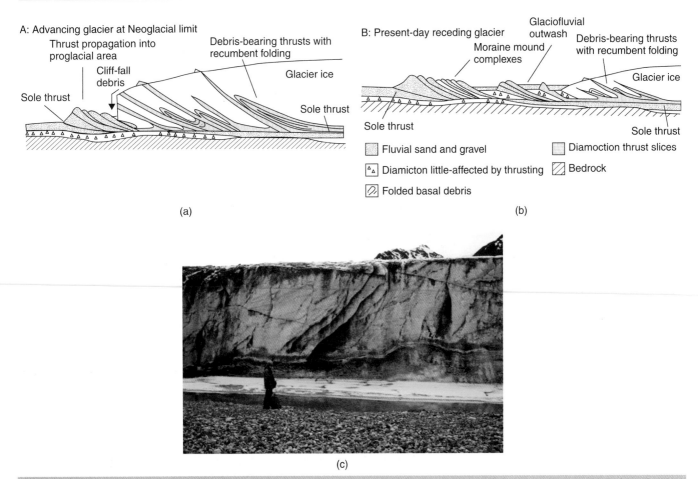

A: Advancing glacier at Neoglacial limit

Thrust propagation into proglacial area

Cliff-fall debris

Sole thrust

Debris-bearing thrusts with recumbent folding

Glacier ice

Sole thrust

(a)

B: Present-day receding glacier

Glaciofluvial outwash

Moraine mound complexes

Debris-bearing thrusts with recumbent folding

Glacier ice

Sole thrust

Sole thrust

☐ Fluvial sand and gravel

△△ Diamicton little-affected by thrusting

⌀ Folded basal debris

☐ Diamoction thrust slices

⧄ Bedrock

(b)

(c)

Figure 20.77 The role of thrusting and landform development in the formation of moraine mounds: (a) during advance; (b) during retreat. (c) Thrusts in Kongsvegen ice front, Kongsfjord, Spitsbergen. (Photo courtesy of Mike Hambrey)

Figure 20.78 Outwash fan from late surge floods, Skeidararjökull, southern Iceland

ice stream of the last Laurentide ice sheet as there is evidence from ocean cores for periodic advances of this ice stream and associated high discharges of icebergs and ice-rafted debris into the Labrador Sea and North Atlantic. These have been called Heinrich Events. Instability of ice streams has also been attributed to rapid motion due to subsole deformation and to periodic switching between periods of build-up and surge, called binge–purge cycles. Modelling studies have suggested that the ice stream could have alternated between cold-based, during ice build-up, and wet-based, during purges, with the transition occurring at some critical ice thickness and gradient. Whilst this modelling is compelling, there is evidence to suggest the Heinrich Events may be driven by climatic cycles unrelated to the internal dynamics of the Laurentide ice sheet. These are the so-called Bond cycles. There appears to be evidence of surging in the North Sea, producing the eastern Yorkshire till sequence, and in the northern Irish Sea basin.

20.6.7 Glaciovolcanic landsystems

Glaciovolcanic sequences have been described in many locations and there is a growing literature as research has attempted to interpret the palaeoglaciological significance of such evidence (Smellie, 2000, 2002). Although such evidence is geographically confined to volcanic regions that have been or are currently glaciated, it is often a significant component of the glacial landsystems of

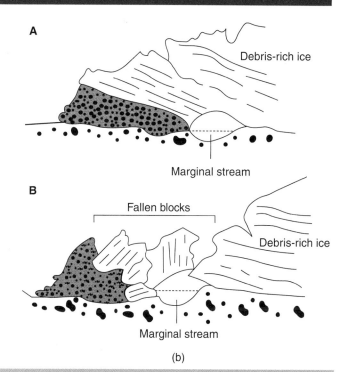

(a)

(b)

Figure 20.79 Surge-moraine formation at Fridjovbreen, western Spitsbergen: (a) surge moraine and glacier surge front; (b) model for the formation

these areas. For example, glaciovolcanic sequences are widespread as part of the glacial landsystem in Iceland and have also been described from the Antarctica Peninsula and extensively in parts of British Columbia. Most of this research has focused on basaltic magmas, but subglacial eruptions associated with rhyolite and obsidian sequences have also been described.

The classic model for subglacial basaltic volcanism (Jones, 1966, 1969, 1970) begins with pillow lava effusion in a subglacial, water-filled cavity below thick ice. As the volcanic pile builds and the ice overburden thins there is a transition to a more violent style of eruption in which the dominant geological products are pyroclastic deposits, including hyaloclastite, and volcanic breccias deposited in englacial lakes, which become open at the ice surface as the lake transforms from subglacial to subaerial but remains englacial. If the volcanic pile becomes emergent within an englacial lake then the sequence will become capped by subaerial lava resting on a passage zone of flow-foot breccia (Jones 1969, 1970; Smellie, 2002; Skilling, 2002). A morphological distinction is made between linear ridges (tindars) (Figure 20.80) and flat-topped hills (stapi or tuyas), reflecting both the nature of the vent and whether or not the volcano became emergent and capped by subaerial lava, as is the case with stapi or tuyas. Other morphological elements have been described from Iceland and British Columbia, formed by direct lava emplacement in ice-contact domes (see Chapter 13 for an example).

This classic model (Figure 20.81) has been enhanced by an increasing number of detailed geological case studies, which have been used not only to explore the impact of ice loading on volcanism but also increasingly to make palaeoglaciological

Figure 20.80 Linear subglacial volcanic ridges (tindar): part of the Jarlhettur system, Iceland

inferences from this geographically-restricted but regionally-important glacial landsystem. Despite this apparent wealth of geological description there is still widespread disagreement over the scale of the subglacial or englacial lakes involved (Werner and Schminke, 1999), their hydrology and connectivity to the rest of the subglacial hydrological system, whether these eruptions are the result of monogenetic or multiple eruptions (Moore and Calk, 1991; Werner and Schmincke, 1999) and the degree to which distinctive sequences can be linked to the style of eruption

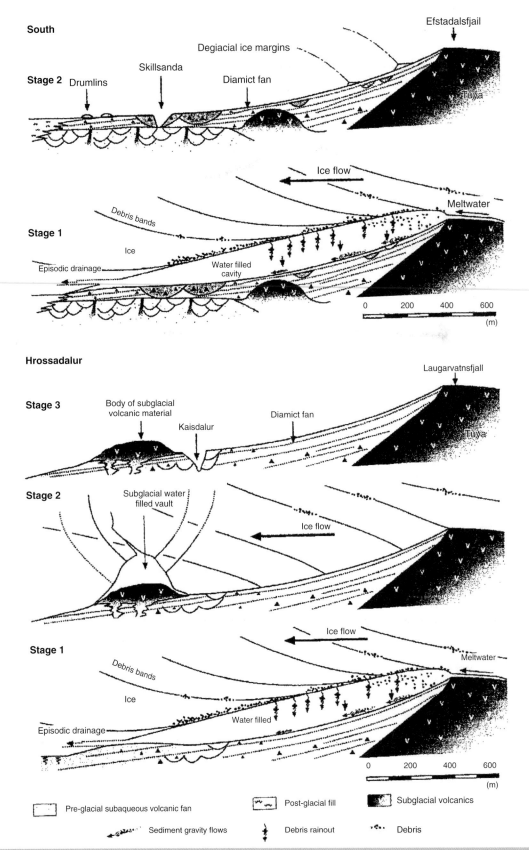

Figure 20.81 Schematic cross-sections and facies models of the diamict fans associated with Efstadalsfjall and Laugarvatnsfjall. (Source: after Bennett et al., 2006)

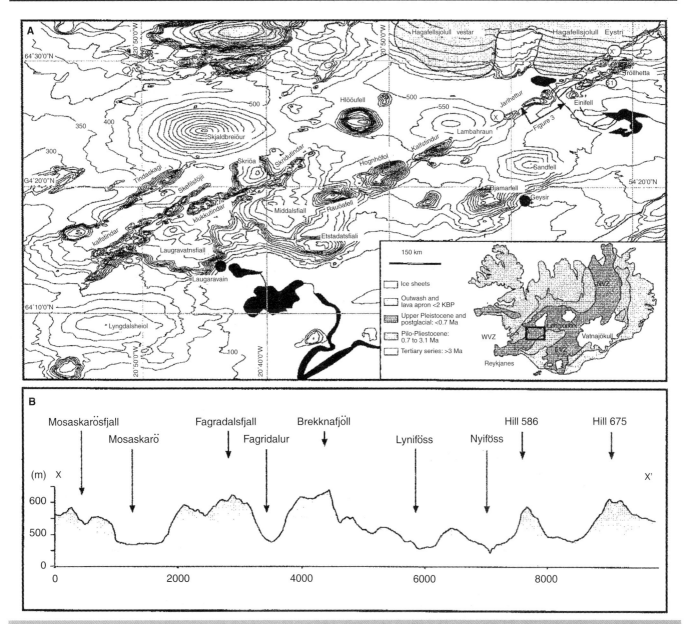

Figure 20.82 Topography of the Brekknafjöll-Jarlhettur linear ridge system, Langjökull, Iceland. (Source: after Bennett *et al.*, 2008)

(Smellie, 2000). There is also a focus on the volcanic rather than the glacial products associated with this landsystem, and glacial sediments are rarely described in detail as part of such systems.

Thin diamict drapes (<5 m) have been noted in association with tuyas and tindars at several locations in Iceland and glacial clasts have been recorded as a component of the volcanic sequences at Bláhnúkur (Tuffen *et al.*, 2001). However, Bennett *et al.* (2006) have addressed this issue by describing large glaciolacustrine fans composed of diamict and gravel in the down-ice lee of tuyas in the Laugarvatn area of Iceland (Figure 20.82). This work

hints at the potential for glaciovolcanic environments to be associated with more extensive glaciolacustrine sediment assemblages. The importance of glacial processes has been emphasized by Bennett *et al.* (2008) in an exploration of the glaciovolcanic sequences of the Brekknafjöll-Jarlhettur ridge in central Iceland (Box 20.8, see in the companion website; see also Figure 20.83), adjacent to the eastern margin of the Langjøkull ice cap, which not only contains a complex assemblage of glaciovolcanic products but is associated with deposits of glaciolacustrine diamict that are in places deformed to create large-scale glaciotectonic structures.

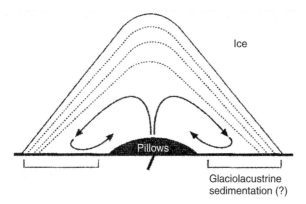

Ice

Pillows

Glaciolacustrine
sedimentation (?)

Development of a trapezoidal subglacial vault through the
convective circulation of fluids. beneath thick ice

(a)

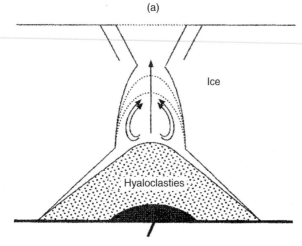

Ice

Hyaloclasties

Rapid vertical growth as a result of increased explosivity
and infilling of accommodation space with hyaloclastites
leading to the formation of an ice-walled chimney

(b)

Figure 20.83 Development of an ice-constrained englacial
chimney. (Source: after Bennett *et al.,* 2008)

Exercises

1. Although the process is continuous, what stages can
   be recognized in the transformation of snow to glacier
   ice?

2. How can the mass balance of a glacier be measured?

3. What are the morphological characteristics of surging
   glaciers?

4. Evaluate the view that changes in the subglacial hydrological regime seem to be a prerequisite for surges to
   occur.

5. What are the variables that control the efficiency of
   glacial erosion?

6. Compare and contrast the characteristics of
   supraglacially- and subglacially-transported debris.

7. How are glaciolacustrine deep-water silts and clays
   formed?

8. Describe the characteristics of esker landforms and
   sediments that might indicate a subglacial origin.

9. Evaluate the role of glacial thrusting in the formation
   of moraines.

10. Describe the characteristic glaciovolcanic facies that
    can occur and explain the differences between the
    formations of tindar and tuyas.

References

Barrett, P.H. and Hambrey, M.J. 1992. Plio–Pleistocene sedimentation
in Ferrar Fiord, Antarctica. *Sedimentology.* **39**, 109–123.

Benn, D.I. and Evans, D.J.A. 2002. Glaciers and Glaciation. London,
Arnold.

Bennett, M.R. and Glasser, N.F. 1996. Glacial Geology: Ice Sheets and
Landforms. Chichester, John Wiley and Sons, Inc.

Bennett, M.R., Hambrey, M.J., Huddart, D. and Glasser, N.F. 1998.
Glacial thrusting and moraine-mound formation in Svalbard and
Britain: The example of Coire a'Cheudchnoic (Valley of Hundred
Hills), Torridon, Scotland. *Quaternary Proceedings.* **6**, 17–34. Moun-
tain Glaciation edition, L.A. Owen. *Journal of Quaternary Science.*
13(Suppl1).

Boulton, G.S. 1972. Modern Arctic glaciers as depositional models
for former ice sheets. *Journal of the Geological Society of London.* **128**,
361–393.

Boulton, G.S. 1987. A theory of drumlin formation by subglacial
sediment deformation. In: Menzies, J. and Rose, J. (Eds). Balkema,
Rotterdam, Drumlin Symposium. pp. 25–80.

Brodzikowski. K. and Van Loon, A.J. 1991. Glacigenic Sediments.
Amsterdam, Elsevier.

Dowdeswell, J.A. and Scourse, J.N. 1990. Glacimarine Environments:
Processes and Sediments. Geological Society of London Number 53.

Evans, D.J.A., Phillips, E.R., Hiemstra, J.F. and Auton, C.A. 2006. Subglacial till: Formation, sedimentary characteristics and classification.
Earth Science Reviews. **78**, 115–176.

Glasser, N.F. and Hambrey, M.J. 2001. Styles of sedimentation beneath Svalbard valley glaciers under changing dynamic and thermal regime. *Journal of the Geological Society of London.* **158**, 697–707.

Hambrey, M.J. 1992. Secrets of a tropical ice age. *New Scientist.* 1 February, 42–49

Hambrey, M.J. 1994. Glacial Environments. London, UCL Press.

Hambrey, M.J., Barrett, P.J., Ehrmann, W.U. and Larsen, B. 1992. Cenozoic sedimentary processes on the Antarctic continental margin and the record from deep drilling. *Zeitchrift für Geomorphologie Supplementband.* **86**, 77–103.

Hambrey, M, Christoffersen, P., Glasser, N. and Hubbard, B. 2007. Glacial Sedimentary Processes and Products. International Association of Sedimentologists Special Publication 39.

Hooke, R. le B. 1989. Englacial and subglacial hydrology: a qualitative review. *Arctic and Alpine Research.* **21**, 221–233.

Jones, J.G. 1966. Intraglacial volcanoes of south-west Iceland and their significance in the interpretation of the form of the marine basaltic volcanoes. *Nature.* **212**, 586–588.

Jones, J.J. 1969. Intraglacial volcanoes of the Laugarvatn region, south west Iceland I. *Quarterly Journal of the Geological Society of London.* **124**, 197–211.

Jones, J.J. 1970. Intraglacial volcanoes of the Laugarvatn region, south-west Iceland II. *Journal of Geology.* **78**, 127–140.

Moore, J.G. and Calk, L.C. 1991. Degassing and differentiation in subglacial volcanoes, Iceland. *Journal of Volcanological and Geothermal Research.* **46**, 157–180.

Powell, R.D. 1981. A model for sedimentation by tidewater glaciers. *Annals of Glaciology.* **2**, 129–134.

Powell, R.D. 1990. Glacimarine processes at grounding-line fans and their growth to ice-contact deltas. In: Dowdeswell, J.A. and Scourse, J.N. (Eds). Glacimarine Environments: Processes and Sediments. Geological Society of London Special Publication 53. pp. 53–73.

Price, R.J. 1973. Glacial and Fluvioglacial Landforms. Edinburgh, Oliver and Boyd.

Sexton, D.J., Dowdeswell, J.A., Solheim, A. and Elverhøi, A. 1992. Seismic architecture and sedimentation in north-west Spitsbergen fjords. *Marine Geology.* **103**, 53–68.

Skilling, I.P. 2002. Basaltic pahoehoe lava-fed deltas: Large-scale characteristics, clast generation, emplacement processes and environmental discrimination. In: Smellie, J.L. and Chapman, M.G. (Eds). Volcano–Ice Interaction on Earth and Mars. Geological Society Special Publication 202. pp. 91–113.

Smellie, J.L. 2000. Subglacial eruptions. In: Sigurdsson, H. (Ed.). Encyclopedia of Volcanoes. San Diego, Academic Press. pp. 403–418.

Smellie, J.L. 2002. The 1969 subglacial eruption on Deception Island (Antarctica): Events and processes during an eruption beneath a thin glacier and implications for volcanic hazards. In: Smellie, J.L. and Chapman, M.G. (Eds). Volcano–Ice Interaction on Earth and Mars. Geological Society Special Publication 202. pp. 59–80.

Tuffen, H., Gilbert, J.S. and McGarvie, D.W. 2001. Products of an effusive subglacial rhyolite eruption: Bláhnúkur, Torfajökull, Iceland. *Bulletin of Volcanology.* **63**, 179–190.

Werner, R. and Schmincke, H.-U. 1999. Englacial vs lacustrine origin of volcanic table mountains: Evidence from Iceland. *Bulletin of Volcanology.* **60**, 335–354.

Further reading

Benn, D.I. and Evans, D.J.A. 1998. Glaciers and Glaciation. London, Arnold.

Evans, D.J.A. 2003. Glacial Landsystems. London, Arnold.

Evans, D.J.A. 2007. Glacial landsystems. In: Elias, S.A. (Ed.). Encyclopedia of Quaternary Science. Oxford, Elsevier. pp. 808–818.

Evans, D.J.A. and Russell, A.J. 2002. Modern and ancient ice-marginal landsystems. *Sedimentary Geology.* **149**. Special issue.

Knight, P.G. 2006. Glacier Science and Environmental Change. London, Blackwell.

Hubbard, B. and Glasser, N.F. 2005. Field techniques in glaciology and glacial geomorphology. Chichester, John Wiley and Sons, Inc.

Siegert, M.J. 2001. Ice Sheets and Late Quaternary Environmental Change. Chichester, John Wiley and Sons, Inc.

21 Periglacial Processes and Landform-Sediment Assemblages

Pingo degradation and melting ice wedge polygons near Tuktoyaktuk, North West Territories, Canada.
Photo: Emma Pike, Wikipedia Commons

Earth Environments   David Huddart and Tim Stott
© 2010 John Wiley & Sons, Ltd

Learning Outcomes

After reading this chapter and completing the exercises you will be able to:

➤ Understand the terms 'periglacial' and 'permafrost'.

➤ Understand how periglacial processes form landforms and (associated) sediments, such as frost wedging, frost heaving and thrusting, frost cracking, the growth of ground ice domes, nivation and cryoplanation, gelifluction, cambering, wind and fluvial action.

➤ Give an overview of the importance of periglacial processes in shaping both the upland and lowland landscapes of Britain.

21.1 Introduction to the term 'periglacial'

'Periglacial' is a term which has had many connotations since it was first used by Lozinski (1909, 1912). A periglacial zone was originally defined as one which was peripheral to Pleistocene glaciers and was cold, mountainous, proglacial or ice-marginal, sparsely vegetated and mid-latitude in nature. Such a periglacial zone is a specific and limiting type of periglacial environment, which is difficult to recognize in modern analogues today and is not typical of the vast majority of present-day periglacial environments.

Nor was it necessarily typical of Pleistocene periglacial environment to the south of the large continental ice-sheets either. The landscape was dominated by frost processes, but frost action also influences landscapes that were not at the margins of Pleistocene glaciers. For this reason a broader definition of 'periglacial' is needed.

French (2000) suggests that the term 'periglacial' should be regarded as synonymous with a cold and nonglacial environment and so should be applied to environments in which frost-related processes and/or permafrost are either dominant or characteristic. Permafrost, as discussed in the next section, does not have to be present as processes, and landform developments definitely regarded as periglacial, such as gelifluction, frost wedging, frost sorting and patterned ground development, can form in areas that lack permafrost. The nonglacial, periglacial environment has the following general characteristics:

1. Usually permafrost (perennially frozen ground).

2. An active layer with seasonally-thawed ground.

3. An incomplete vegetation cover of low shrubs, dwarf trees and herbaceous plants; that is, the tundra (see Chapter 25).

4. Ground that is free of snow for part of the year.

5. Often frequent fluctuations of the air temperatures across the $0\,°C$ isotherm, but in some areas the summer can have temperatures as high as $20\,°C$, though with extremely cold winters, down to $-50\,°C$ or more, especially in the continental interiors away from the coast.

This means that in a periglacial environment the effects of freezing and thawing of the ground surface modify that surface by displacement of soils, by the migration of ground water and by the formation of several landform–sediment assemblages characteristic of this type of environment. Over a third of the planet's terrestrial surface can be included in this kind of definition and so a detailed knowledge of such processes and environments is of major significance to exploitation of mineral resources, settlement, and pipeline and road building. This is particularly relevant in the light of global warming. A periglacial environment includes both low-latitude plateaux and mountain summits, as well as the cold, dry interiors of Asia and North America and the maritime, boreal coasts of Canada, Iceland, Northern Scandinavia, Svalbard and Alaska.

21.2 Permafrost

Often frozen ground is divided into two categories: that which annually thaws, referred to as seasonal frost; and permafrost. The penetration depth of seasonal frost is partly a function of the below-zero air temperatures and their duration, but also partly of antecedent above-zero air temperatures and their duration, as the latter warm the ground and this stored heat must be conducted away before freezing can start. Thus the depth of frost penetration in continental climate areas may be limited, despite low winter air temperatures, because of the relatively large amount of stored summer heat in the ground. Penetration of frost into the ground is also influenced by the amount of water that must be frozen since the latent heat of fusion of water is $333\,kJ\,kg^{-1}$, compared to the specific heat capacity of only $4.19\,kJ\,kg^{-1}$. This means that in wetter areas seasonal frost penetration will be much lower than in drier areas and both aggradation and degradation of permafrost will be slower.

Permafrost is found where the ground below the surface (rock, sediment and soil) remains frozen for a period of time greater than one year. With a mean annual temperature less than $0\,°C$, the depth of frost penetration exceeds the depth of thaw and if this type of climatic regime persists an increment of permanently frozen ground will be created each year. Thus permafrost has aggraded in the past by centimetres every year to depths of $1500\,m$ in Siberia, and common depths are several hundred metres throughout the Arctic (e.g. $305\,m$ in Svalbard and east Greenland; Thule $520\,m$; Resolute $455\,m$). The permafrost thickness represents an equilibrium between heat loss to the atmosphere and the increase in geothermal heat with depth. This means that the distribution of permafrost depends on the mean annual air temperature and the thermal properties of the Earth's materials.

Five different permafrost types have been recognized:

1. **Continuous:** Permafrost is found everywhere, except under deep lakes.

2. **Discontinuous:** Many scattered small thawed areas are found within a zone of permafrost, so for example permafrost is absent under water bodies and on warmer sites, for example south-facing slopes in Canada north of around 55° N. It usually exists as an extensive marginal zone at the edge of continuous permafrost.

3. **Sporadic permafrost:** Preserved in scattered locations, for example north-facing slopes or peat bogs, where the peat insulates and stops the summer thawing in generally unfrozen terrain.

4. **Subsea permafrost:** Frozen sediment on submerged sea bottoms, as in the Beaufort Sea and along the northern coastal edge of Russia, Alaska and parts of northern Canada.

5. **Alpine permafrost:** Found at high altitudes in areas where continuous and discontinuous permafrost are not common.

Originally in Canada a simple subdivision from south to north was given by Brown (1970), as shown in Figure 21.1. In the same country Smith and Riseborough (2002) defined the climatic and environmental conditions that determine the limits and continuity of permafrost. Whilst ultimately permafrost is a climatic phenomenon, the ground thermal conductivity ratio, depending on the subsurface material, is shown to be the critical factor in determining the southernmost extent of discontinuous permafrost. In contrast, snow cover is critical in determining the northern limit of discontinuous permafrost. Calculated-temperature-of-permafrost values increase gradually southwards towards the limit of permafrost occurrence as the effect of a rising mean annual air temperature is counteracted by an increasing thermal offset. This results in a diffuse geographical transition in the disappearance of permafrost. In contrast, there is a more abrupt transition to continuous permafrost at the northern limit of the discontinuous permafrost zone, associated with geographical changes in snow cover. The transition from discontinuous to continuous permafrost occurs between a mean annual air temperature of −6 and −8 °C, which agrees well with the air temperature limit often cited by other authors.

Ice is usually associated with permafrost because water is usually present when ground freezes. However, permafrost does not require the presence of ice as there is dry permafrost in Peary Land in north-east Greenland and in parts of Antarctica. If ice is present it is as pore or interstitial ice in the pore spaces and fractures, and hence it cements the soil–sediment matrix and forms as the freezing plane descends into the ground surface without displacing the soil. Although hard and rock-like, such ice-cemented permafrost is not totally impermeable as unfrozen

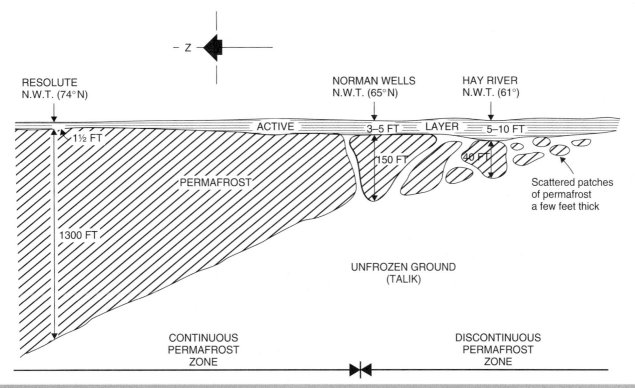

Figure 21.1 N–S vertical profile of permafrost in Canada, showing the decrease in thickness southwards and the relationship to continuous and discontinuous zones. (Source: after Brown, 1970)

water films remain, separating the ice from the mineral grains, along which some water migration is possible.

There is also segregated ground ice, which forms lenses, veins and wedges of pure ice as liquid water and vapour are attracted to the lower vapour pressure at the freezing plane. Soils with a high silt content – that is, intermediate porosity and permeability – allow the optimal storage and diffusion of water and thus the formation of segregated ground ice.

In heavy clays, while the soil may be granulometrically frost-susceptible, the low permeability can restrict water migration and promoting ice-lens growth. Much of the pore water occurs as thin films within which capillary and adsorption effects lower the freezing point by several degrees Celsius. Progressive freezing of such water results in a fall in Gibbs free energy and sets up a cryosuction, causing water to migrate towards the freezing front from unfrozen areas below. The result is that as a freezing front slowly advances downward from the ground surface, a phase change takes place in a partially-frozen zone, within which water progressively freezes to form lenses of clear ice. These lenses continue to grow, fed by cryogenic water migration. The movement of water through frozen soil has been demonstrated by the fact that ice lenses continue to grow behind the freezing front in soil that is several degrees Celsius below zero. Water transmission takes place through frozen soils but the permeability is lowered by several orders of magnitude as the soil temperatures fall a few tenths of a degree below freezing. There appears to be significant water movement at temperatures as low as $-2\,°C$, due to the difference in thermal potential rather than a hydraulic gradient. So water can migrate along a thermal gradient through frozen soil to allow thickening of ice lenses and continued frost heaving even after the ground is frozen.

Massive ice layers up to several metres in thickness can occur as a result of pressurized water flow towards the freezing front. This pressurized water responsible for the intrusive ice may be derived from ground water flow beneath permafrost driven by a hydraulic head (open system), or from pore-water expulsion ahead of a penetrating freezing front in saturated coarse sands and gravels (closed system). However, not all massive ice bodies within continuous permafrost originated as intrusive ice and there is some evidence that at least some represent buried, relict glacier ice in both Siberia and the western Canadian Arctic.

Permafrost covers about 25% of the world's nonglaciated land surface, covering 82% of Alaska, 50% of Canada and the former Soviet Union and 20% of China. Much of the mid-latitude permafrost is relict and has been inherited from the Pleistocene cold phases, when permafrost covered up to 40% of the planet's land surface. Most areas of permafrost have an upper active layer, between 1 and 3 m thick, which is subject to a cyclic thawing during the short summer melt season and refreezing again in the late autumn.

In some locations unfrozen layers or taliks are located underneath or within permafrost (Figure 21.2). In the continuous permafrost zone, taliks are found beneath lakes because of the water's ability to store and transfer heat energy vertically. Their extent beneath a lake is related to the depth and volume of the water body and it is clear that the larger the water body, the greater their ability to store and transfer more heat energy downward.

A closed talik is where unfrozen ground is located within the permafrost; these can develop where lakes fill in with sediment and become bogs. As the open water is removed, summer solar radiation begins to be received by a surface with a lower specific heat capacity and poor vertical heat transfer. As a result, soil near the surface begins to freeze solid, encasing a zone of unfrozen soil in permafrost. An open talik is an area of unfrozen ground that is open to the ground surface but is otherwise surrounded by permafrost. A through talik is unfrozen ground that is exposed to the ground surface and to a larger mass of unfrozen ground beneath it. This type clearly forms in the discontinuous permafrost.

The geomorphological significance of permafrost is that it confines water and frost to the active layer between the permafrost

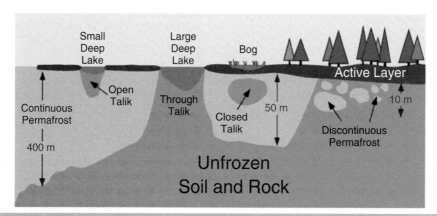

Figure 21.2 Vertical cross-section of the transition zone between continuous and discontinuous permafrost, illustrating the various types of talik (open, closed and through)

table and the ground surface. The descent of the freezing plane from the surface pressurizes the soil water, reduces the freezing temperature and maintains the thawed, active layer during the autumn freeze-up. The growth and decay of segregated ground ice causes heave and subsidence.

21.3 Periglacial processes and landforms

The family of periglacial processes and their resultant landforms is often complex; we will discuss the main groups in this section. Two are of prime importance:

1. The development of perennially frozen ground with ice segregation, combined with the thermal contraction of the ground under intense cold, the creep of ice-rich permafrost and the thaw of such ground.

2. The frost weathering of soils and bedrock, including the break-up of surface rock by frost wedging and/or a complex of physicochemical weathering processes.

21.3.1 Frost wedging (frost shattering or splitting)

In this process there is a prying apart of material by the force of water expansion on freezing. It is not necessarily confined to the force exerted by the 9% volume expansion accompanying the water freezing, but in porous sediment may be due to the directional growth of the ice crystals. The factor controlling this process is that water is required but the amount available

(a)

(b)

(c)

Figure 21.3 Breakdown of shales and mudstones by frost wedging

(d)

(e)

Figure 21.3 (*continued*)

can be critical as the bulk-freezing strain of a rock increases with the increased water content. The nature of the rock is important too and sedimentary rocks, such as siltstones and shale, which contain mica, or illitic clays, have fissility planes through which the water can preferentially migrate. Hence shales break down more readily than igneous rocks (Figure 21.3). Other important planes of weakness in rock are joints and bedding planes, where water can migrate, freeze and expand.

Other factors of importance to this process are temperature and time. Depending on the freezing rate, two different rocks can reverse their susceptibility to breakdown. Mellor (1970) found that the bulk-freezing strain of a rock increases not only with increased water content but also, for any given water content, with an increased freezing rate. He suggested that at water contents over 50% saturation, the freezing strain may be enough to cause internal cracking. Rapid freezing should seal the surface rock area and create a closed system, increasing the pressure effects. However, in fine-grained, uncracked rocks, slow freezing may promote frost wedging by allowing the water flow to the freezing front to build up disruptive ice crystals, whereas rapid freezing would inhibit water flow.

The critical temperature for the development of frost wedging is between -4 and $-15\,°C$, and it has been found that relatively slow cooling rates produce the greatest quantity of shattering stress. The number of freeze–thaw cycles is important, although it is not air temperature that is crucial but temperatures in the rock itself. The intensity and length are critical too. It has been shown that for marble and granite a cooling rate of between -0.1 and $-0.5\,°C$ per hour is the optimum. The thawing of snow and/or ice adjacent to dark rocks warmed by insolation is common at subfreezing air temperatures and is a potent factor in frost wedging when melt water seeps into joints and refreezes. The maximum effect of frost wedging is in the spring gauged by the release of rock fragments.

However, a more realistic model, particularly for soils, has been suggested: the segregation ice model of frost weathering (Walder and Hallet, 1985, 1986). The model treats freezing in water-saturated rock as closely analogous to slow freezing in fine-grained soils. Expansion is mainly the result of water migration to growing ice lenses and only secondarily of the 9% volume increase (Hallet *et al.*, 1991). Frost weathering is seen as the progressive growth of microcracks and relatively large pores wedged open by ice growth. Largely theoretical and untested in nature, this model still has a lot of evidence to support it, including direct evidence of water migration towards the freezing front in freezing rocks, and laboratory data from freeze–thaw experiments in which frost-induced spalling and/or expansion occured.

21.3.2 Landform-sediment assemblages associated with frost wedging

The products of frost wedging are usually angular, coarse, gravel-sized fragments, although there are particles down to silt and clay-grade sediment. However, Potts (1970) suggested that processes other than frost shattering produced the silt and clay-size particles that are found in gelifluction deposits. In other words there were chemical weathering processes going on as well to produce these fine sediments.

The importance of this process is that blockfields are produced on flat, plateau areas (Figure 21.4(a)) and blocky slopes (Figure 21.4(a)). It is also one of the dominant processes in rockfall and as a result helps produces characteristic upland landforms in Britain and Europe, such as scree slopes (Figure 21.4(b)). Extensive rocky outcrops are common in the periglacial regions of the world and the cold temperatures do not favour an extensive development of regolith and soils over that bedrock. The result is that the bedrock is exposed on plateau and slope areas to the extreme climate. It is known that extreme variations in temperatures can occur, so for example on Ellesmere Island in the Canadian Arctic temperatures as high as 39.7 °C were measured by Eichler (1981). In other areas, like the Scandinavian high mountains, oscillations across 0 °C are common, and it is likely that it is a combination of these temperature variations, both highs and lows, that produces frost wedging, though it is likely that insolation weathering contributes to rockfall as well. For example, Douglas (1980) found that over 60% of all rockfall events were independent of freezing conditions on the Antrim coast of Northern Ireland. Events over 0.21 kg in size occurred seasonally, as a result of freeze–thaw, but events under 0.21 kg were observed throughout the year.

These landform–sediment associations of coarse, angular debris can become relict and fossilized with a change in climate, and this can be seen extensively at lower elevation in British uplands such as the Lake District and the northern Pennines. Here there are fossil blocky slopes and screes formed at least as long ago as the Late Glacial Pollen Zone 111 climatic phase, and many of the summit blockfields in these areas may be mainly relict from the Late Pleistocene glacial phase, when some may have been nunatak zones, sticking up above the ice sheets and plateau ice caps. There are also lowland fossil scree slopes, like those around Arthur's Seat, below Salisbury Crags in Edinburgh, and in Cheddar Gorge in the Mendips, which probably date to this cold phase.

21.4 Frost heaving and frost thrusting

Taber (1929) showed that the pressure generated by the growth of ice crystals in soils is at right angles to the freezing plane as it penetrates into the soil. Hence pressures are predominantly upwards. This is frost heaving. However, there are pressures in various directions due to the heterogeneous sediments which form these soils and so the varying heat conductivities of different minerals influence the orientation of cooling surfaces and introduce lateral movement in places, as do topographic situations such as river banks. This lateral movement of the mineral soil during soil freezing is known as frost thrusting.

The evidence for frost heaving can be seen in the heaving of joint blocks on mountain summits. It is also common to see the upfreezing of objects in frozen soils and the common edgewise projection or vertical orientation of stones in periglacial soils. Experimentally, Corte (1966) demonstrated that stones can move

(a)

(b)

(c)

Figure 21.4 Landforms created by frost wedging: (a) Block slope, Great Dun Fell, northern Pennines; (b) Scree, Kangerdluarssuk, South West Greenland; (c) Blockfield on Bow fell, Lake District, UK.

vertically upwards through soils that are subject to freezing and thawing cycles. There is too the appearance of stones in summer- or autumn-cleared farmer's fields in cold areas like Wisconsin after winter and the vertical heaving of fence posts. Washburn

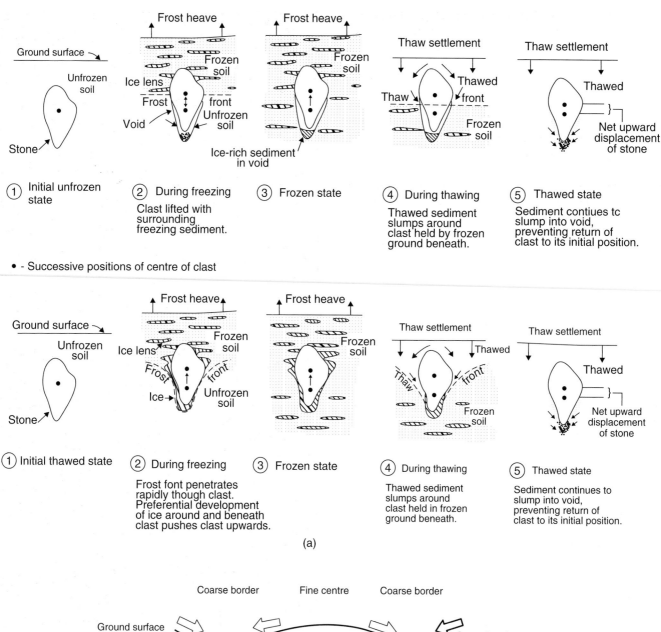

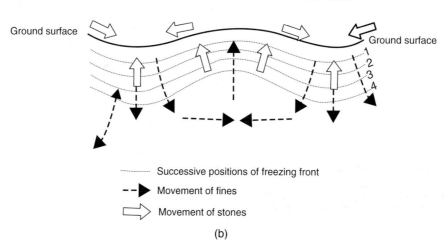

Figure 21.5 (a) Frost sorting of soils according to the frost pull (top) and frost push (bottom) mechanisms. (b) Movement of coarse and fine soil due to freezing and thawing. (Source: after Ballantyne and Harris, 1994)

(1969) also showed by placing targets in periglacial soils in the Mesters Vig area of east Greenland that the greatest heaves were in the wettest places and that where moisture conditions were similar, vegetation was the controlling factor. There were greater heaves in thin tundra vegetation than in more richly-vegetated tundra.

21.4.1 What are the mechanisms of this heaving?

Frost heaving results in the size sorting of the heterogeneous sediments because the larger clasts migrate towards the surface at a rate of up to 5 cm per year. The mechanics of this heaving are not that well understood but various ideas have been suggested, with two diametrically opposed schools developing: the frost pull mechanism and the frost push process (Figure 21.5(a)).

In frost pull the soil thickness expands, or is pulled up, as the freezing plane descends from the surface. However, with thawing from the surface, the cohesive matrix of fine sediment retracts and the thawed fines collapse around the clasts whilst the clast bases are still frozen, filling the spaces beneath clasts (Figure 21.5(b)). It has also been suggested that clasts, in addition to being hindered from dropping back by slumping-in of soil during thawing, would also be hindered because the cavities left by stones as they were heaved would tend to be narrowed by frost thrusting during the freezing process. This leaves them in a slightly elevated position relative to the preceding thaw season.

In frost push upfreezing is explained by the greater heat conductivity of the clast compared to the fines. Hence ice would first form around the stones, or at their bases, and force them vertically up. The seeping-in of fines during thawing would prevent the stones from returning to their original positions. In line with this idea is the better diffusivity of stones compared to fines. This would lead to earlier freezing of fines in contact with stones than of fines at the same level in the soil further away. Hence water would be drawn preferentially to stones during freezing, particularly to their bases from below, where there would be an uninterrupted water source. Thawing as well as freezing would first occur adjacent to the stones. These should drop back, but fines thawed around the upper portions of the clast, whilst their bases are still frozen, do not allow this. These fines would eventually fill in around the stones and prevent their return to their original position. Washburn (1970) thinks that both sets of processes are responsible for the upfreezing of stones in periglacial soils, and this seems likely.

21.5 Landforms associated with frost sorting

These processes result in frost sorting, which produces various types of patterned ground and movement of coarse and fine soils, as in Figure 21.6. For example, on flat areas there are sorted circles with fines in the middle and a coarser rim of stones at

(a)

(b)

Figure 21.6 Patterned ground landforms: (a) sorted circles, Kebnekaise, Arctic Sweden; (b) sorted circles, Baffin Island, Arctic Canada; (c) stone stripes, Mesters Vig, east Greenland; (d) stone stripes, Blue Screes, Blencathra, Lake District, UK

the edge (Figure 21.6(a),(b)), whilst where there is a gradient the sorting gives rise to sorted stone stripes, with alternations of relatively coarse and fine clasts (Figure 21.6(c),(d)). The mechanisms for this process are difficult to explain but are discussed in a case study from western Spitsbergen in Box 21.1 and extensively in Washburn (1969, 1973). In the latter's publications it

(c)

(d)

Figure 21.6 *(continued)*

is clear that patterned ground is polygenetic in origin; he recognizes the following forms: circles, polygons, nets, steps and stripes. In all categories there are both sorted and nonsorted varieties. Table 21.1 illustrates Washburn's (1973) genetic patterned ground classification.

21.6 Needle ice development

Needle ice in frozen soils where there is no vegetation cover is a minor form of ground ice that helps to loosen soil for other erosion processes and tends to move small particles to the soil surface. On slopes, needle ice can also enhance the creep of moving soil particles at right angles to the gradient. Needle ice, or pipkrakes, consists of groups of narrow ice slivers usually up to several centimetres long, but sometimes up to 40 cm, which can form in moist soils when temperatures drop rapidly, usually overnight. The needle ice extrudes from the soil pores as the soil water freezes and pushes soil particles with it. As the ice needles grow the soil becomes desiccated and disturbed and so is more susceptible to wind and water erosion after the ice has melted, and to erosion through human or sheep trampling pressure.

21.7 Frost cracking and the development of ice wedges

This process takes place where there is the winter thermal contraction of sediments and ice at very low soil temperatures that occurs with low air temperatures and where there is a lack of snow and vegetation cover. This forms a seasonal crack in winter and results in a polygonal crack system in the permafrost (Figure 21.9). Snow and hoar frost enters the open cracks, preventing their closure in summer when the ground temperature increases again, generating veins up to several metres into the permafrost.

The best correlation between air-temperature drops and ice-wedge cracking occurs in areas with thin snow cover. Water can seep from the active layer into the vertical cracks, which are usually up to three metres deep but can be much deeper. This water then freezes and cracks, since ice has less tensile strength than the frozen ground. The repeated cracking and incremental accretion of the ice in the cracks creates ice wedges with time (Figure 21.10), and these increase in size as the years pass and the same processes operate. Each incremental addition of ice can be recognized as narrow foliations.

As the active layer melts in the summer, sediment can also be washed into the ice wedges as a thin sediment line, which then subsequently gets frozen into the ice wedge. These ice wedges grow in both width and depth and occupy the polygonal network of the thermal contraction cracks. In northern Siberia and the lowlands of the western Canadian Arctic ice wedges can grow very big, often 3–4 m wide near the surface and extending downwards between 5 and 10 m. However, it has been found that under half of the wedges in any area crack annually and the ability of melt water to penetrate the crack before it closes in the early spring governs whether an ice vein will form. Local conditions, such as extra snowfall in any one year, can prevent cracking.

Mackay (1992) points out that the various theoretical discussions of thermal contraction cracking, such as Lachenbruch (1962), deal with simple uniform conditions. The real situation is that as ice wedges grow, complexity develops due to factors like relief changes, polygon development, vegetation and snow cover. It may be that cracking is a random process. Ice wedges join predominantly at right angles to form polygonal nets of patterned ground that cover extensive areas. The average dimensions of these polygons commonly range from 15 to 40 m. Lachenbruch (1962) concludes that the angular intersection of a polygonal network of frost cracks will exhibit a preferred tendency towards a right-angle pattern, but this contrasts with the many descriptions in which hexagonal or angular junctions dominate.

The visco-elastic model of the behaviour of frozen ground predicts that cracking is most likely during periods of rapid cooling at temperatures below $-20\,°C$, when the frozen ground is relatively brittle, and field measurements have confirmed this. In fact, slightly lower temperatures may be necessary in sands and gravels than in silts and clays.

Monitoring of thermal contraction cracking and growth of ice wedges following drainage of a thaw lake on Richard Island in the Mackenzie Delta showed great variability in wedge growth, with

Table 21.1 Washburn's (1973) genetic classification of patterned ground

Processes — **CRACKING ESSENTIAL** (Thermal Cracking; Frost Cracking) and **CRACKING NON-ESSENTIAL**.

Pattern	Sort	Desiccation Cracking	Dilation Cracking	Salt Cracking	Seasonal Frost Cracking	Permafrost Cracking	Frost Action Along Bedrock Joints	Primary Frost Sorting	Mass Displacement	Differential Frost Heaving	Salt Heaving	Differential Thawing and Eluviation	Differential Mass Wasting	Rillwork
CIRCLES	NONSORTED								Mass-displacement N circles	Frost-heave N circles	Salt-heave N circles			
CIRCLES	SORTED						Joint-crack S circles (at crack intersections)	Primary frost-sorted circles, incl.? Debris islands	Mass-displacement S circles, incl. Debris Islands	Frost-heave S circles	Salt-heave S circles			
POLYGONS	NONSORTED	Desiccation N polygons	Dilation N polygons	Salt-crack N polygons	Seasonal frost-crack N polygons	Permafrost-crack N polygons, incl. Ice-wedge polygons. Sand-wedge polygons	Joint-crack N polygons?		Mass-displacement N polygons?	Frost-heave N polygons?	Salt-heave N polygons?			
POLYGONS	SORTED	Desiccation S polygons	Dilation S polygons	Salt-crack S polygons	Seasonal frost-crack S polygons	Permafrost-crack S polygons	Joint-crack S polygons	Primary frost-sorted polygons?	Mass-displacement S polygons?	Frost-heave S polygons?	Salt-heave S polygons?	Thaw S polygons?		
NETS	NONSORTED	Desiccation N nets incl? Earth hummocks	Dilation N nets		Seasonal frost-crack N nets, incl.? Earth hummocks	Permafrost-crack N nets, incl. Ice-wedge nets and? Sand-wedge nets			Mass-displacement N nets, incl.? Earth hummocks	Frost-heave N nets, incl. Earth hummocks	Salt-heave N nets			
NETS	SORTED	Dessication S nets	Dilation S nets		Seasonal frost-crack S nets	Permafrost-crack S nets?		Primary frost-sorted nets	Mass-displacement S nets	Frost-heave S nets	Salt-heave S nets	Thaw S nets		
STEPS	NONSORTED								Mass-displacement N steps	Frost-heave N steps?	Salt-heave N steps?		Mass-wasting N steps	
STEPS	SORTED							Primary frost-sorted steps?	Mass-displacement S steps	Frost-heave S steps	Salt-heave S steps	Thaw S steps?	Mass-wasting S steps	
STRIPES	NONSORTED	Desiccation N stripes	Dilation N stripes?		Seasonal frost-crack N stripes?	Permafrost-crack N stripes?	Joint-crack N stripes?		Mass-displacement N stripes	Frost-heave N stripes	Salt-heave N stripes		Mass-wasting N stripes?	Rillwork, N stripes?
STRIPES	SORTED	Desiccation S stripes	Dilation S stripes		Seasonal frost-crack S stripes?	Permafrost-crack S stripes?	Joint-crack S stripes	Primary frost-sorted stripes?	Mass-displacement S stripes	Frost-heave S stripes	Salt-heave S stripes	Thaw S stripes?	Mass-wasting S stripes	Rillwork, N stripes?

N and S refer to Non-sorted and sorted.

a tendency for the rate to decrease through time. The ice-wedge polygons are preserved until climatic change occurs, when the ice in the wedges melts and they are replaced by ice-wedge casts as fine sediments are finally washed or blown into the cracks. Rates of wedge growth are low and it would take thousands of years for a 1 m-wide ice wedge to form.

Depending on the type of sediment in which they form, the ice-wedge casts can be composed of different sediment types, sands, gravels or tills (Figure 21.11(a)–(d)). In arid, cold, periglacial environments, some ice wedges can accumulate wind-blown fine sand within their winter cracks; these are called sand wedges, and sand-wedge casts when relict. All types of relict ice or sand-wedge cast are important in that they tell us a lot about the type of periglacial environment that operated in the past. They can only form when the air temperatures drop well below 0 °C and there is permafrost present. Based on the distribution of active wedges

(a)   (b)

Figure 21.9 Polygonal ground: (a) Mackenzie Delta, Canada; (b) Labrador, Canada

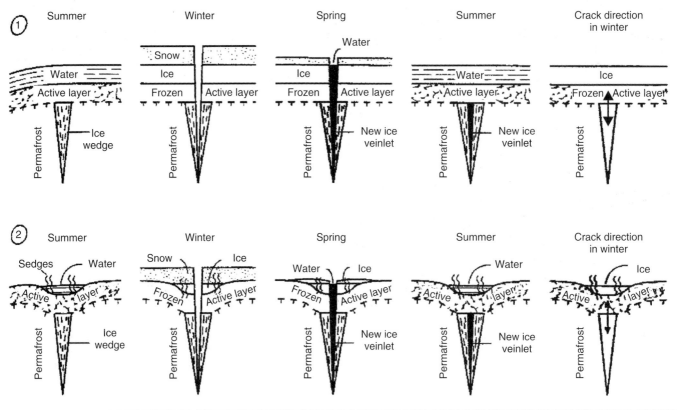

Figure 21.10 Growth of ice wedges. (Source: after Mackay, 1989)

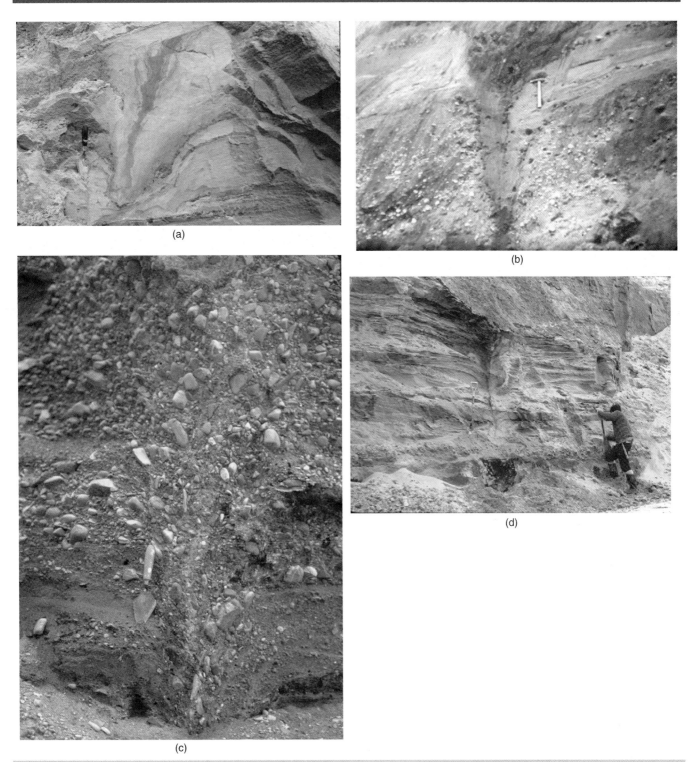

(a)

(b)

(c)

(d)

Figure 21.11 Ice-wedge casts preserved in various types of sediment

today it has been suggested that they denote temperatures at least as low as an average of $-6\,^{\circ}$C, with no vegetation cover (Figure 21.12).

In Alaska the mean annual air temperature ranges from approximately $-6\,^{\circ}$C in the south to $-12\,^{\circ}$C in the north, with a mean annual degree-days ($^{\circ}$C) freezing range of 2800–5400.

Minimum temperatures at the top of the permafrost here range from -11 to $-30\,^{\circ}$C. In the discontinuous permafrost zone where there are inactive wedges, the mean annual air temperature ranges from about $-2\,^{\circ}$C in the south to $-6\,^{\circ}$C in the north and the mean annual degree-days freezing range is 1700–4000.

However, Harry and Gozdik (1988), Burn (1990) and Murton and Kolstrup (2003) warn against the uncritical use of ice-wedge casts to infer palaeotemperatures. The latter question is whether palaeotemperature interpretations of last glacial ice-wedge casts in north-west Europe can reliably be quantified at present in terms of specific threshold values of mean annual air temperature. Sometimes the sand wedge casts are interpreted as indicating colder, drier conditions than ice-wedge casts, but this is not necessarily true as both types can form in the same environment, for example in the modern Arctic Canadian environment and in western Denmark in a Pleistocene context. Local site differences and the availability of water to infill the crack appear to be the main reasons for the different types, rather than any marked climatic difference over small geographical areas.

In the field a continuum exists between sand wedges and ice wedges, and in their review Murton *et al.* (2000) emphasize that sand wedges and sand veins may form in seasonally frozen ground as soil wedges, and even in nonfrozen environments due to desiccation. However, it does seem likely that relict sand wedges, over 2 m in depth, with well-developed lamination, probably do represent the presence of former permafrost, though sand veins and small sand wedges are potentially ambiguous.

Ice wedges that form at a time after the accumulation of the host sediments are termed 'epigenetic' ice wedges, while those that form as the sediment accumulates and grow upwards as successive sedimentary layers become frozen into the top of the permafrost are called 'syngenetic' ice wedges.

21.8 Growth of ground ice and its decay, and the development of pingos, thufurs and palsas

Segregated ice masses grow within permafrost because of liquid water diffusion. Many forms exist, but the most common is horizontal layers, where water can migrate under pressure from unfrozen parts of the soil to the ice mass. Here when the water comes into contact with the segregated ice it freezes, so enlarging the ice mass. The water diffusion that allows the mass of ice to grow can be caused by pressure and temperature gradients, gravity and the movement of ground water under pressure.

When large areas with segregated ground ice degrade, either by climatic change, disruption of the thermal regime by fire, or vegetation and hydrological changes, thermokarst scenery develops. Here there are sinkholes, thaw depressions and ponds, mass movements along rotational slumps and general disruption

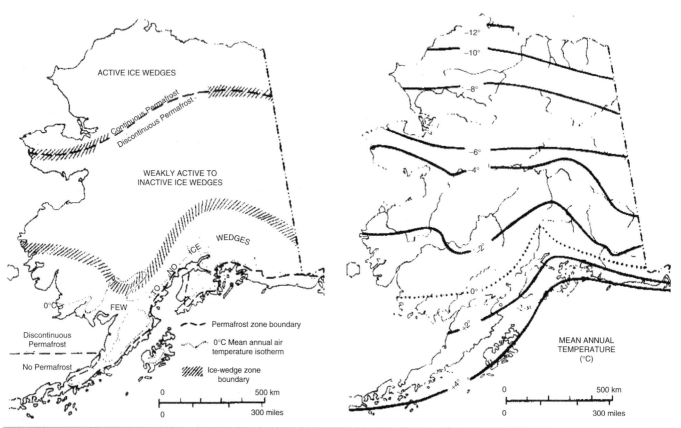

Figure 21.12 Distribution and controls on active ice wedges in Alaska: (a) distribution of active wedges; (b) mean annual air-temperature isotherms (°C); (c) degree-days (°C) of freezing, taken from monthly mean-temperature data. (Source: after Péwé, 1966)

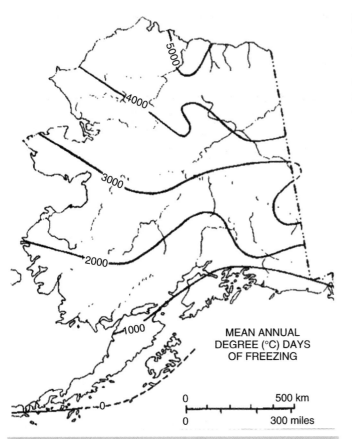

MEAN ANNUAL
DEGREE (°C) DAYS
OF FREEZING

0   500 km

0   300 miles

Figure 21.12 *(continued)*

and collapse of the former permafrost to give a markedly pitted terrain. Thermokarst lakes are particularly prominent landforms on low-relief plains, peninsulas and islands that border the Arctic Ocean, such as in the area close to the mouth of the Mackenzie River.

21.8.1 Pingos

One type of ground ice mass can give rise to a landform known as a pingo. This is an Inuit word for an ice-cored hill. They can be up to 70 m high, with a diameter of up to 1200 m, and are usually circular in shape, some with ice exposed at their summit and along cracks around the dome, and others with summit craters with lakes. There appear to be two types of pingo, with different formative processes, each named after the type locality where they are found: the Mackenzie-Delta, hydrostatic or closed-system pingo and the east-Greenland, hydraulic or open-system pingo (Figure 21.13). However, Gurney (1998) suggests that there is a third category, consisting of polygenetic or unclassified pingos that do not fit neatly into either of the two traditional ones.

All types need the movement of ground water, however, as a result of a pressure gradient, and its concentration as ice to form the mound core. Paradoxically, the-closed system pingo, of which there are up to 1500 present in the Mackenzie Delta, develops initially with a lake (generally deeper than 3 m) where there is

no permafrost beneath it (a talik). This lake gradually fills in with sediment, as all lakes are ephemeral landforms, and invading permafrost isolates the remaining water in the lake's sediments (Figure 21.14). Mackay (1979) realized that often it was not the slow filling of the lake with sediments that was important but its catastrophic drainage, brought about by, for example, the extension of a drainage channel along an ice-wedge network intersecting the lake shore. Continued inward and downward freezing of the now filled-in lake generates enough pressure to move pore water upwards. This injected pore water begins to freeze, forming a segregated ice mass at the core of the developing dome, with maximum rates of vertical growth in young pingos as high as 1.5 m per year; see for example the Porsild pingo. The contained water is expelled ahead of the freezing front and takes the path of least resistance, doming up the relatively thin permafrost below the former lake's centre. With increasing age, the growth rate decreases. Some of the largest pingos in the Mackenzie Delta, such as the Ibyuk pingo, are over 1000 years old and grow at only 2.3 cm per year (Mackay, 1986).

The ice in the pingo core was traditionally thought of as injection ice, frozen from a pool of water injected under pressure. However, it is now thought that the bulk of pingo growth is through ice segregation, although it could be that in the early stages of growth injection ice dominates. As the ground is heaved by the growth of the segregated ice, the tensile stress eventually causes it to crack, exposing the ice core and leading to the degradation of the pingo, ice-melting and the formation of a lake in summer. It is now thought that several stages of closed-system pingo growth can be recognized:

- In the pore-ice stage the ice core does not form, but as the pore ice freezes, the entire lake bottom heaves slowly upwards.

- In the segregated-ice stage, when the pore-water pressure equals or exceeds the overburden pressure, segregated ice tends to form and pingo growth starts.

- The injected-ice stage occurs where pore-water pressures continue to increase and the overburden yields, and water is injected faster than it can freeze.

- Finally, if water continues to be injected faster than it can freeze, the pingo will rupture.

Mackay (1979) realized that not all of the pingos in the Tuktoyaktak peninsula had formed in a truly closed system as there were examples of through-taliks, meaning that ground water could enter the system from beneath adjacent lakes other than the one which drained to form the pingo. Hence he suggested the terms 'hydrostatic' and 'hydraulic' for the traditional closed- and open-system pingos.

East-Greenland or open-system pingos usually develop at the base of a slope, where an intrusion of ground ice is fed by artesian ground-water flow down the slope. They are the result of high hydrological potential, creating high artesian pressure, due to water originating in the upland slopes. The ice develops as a dome

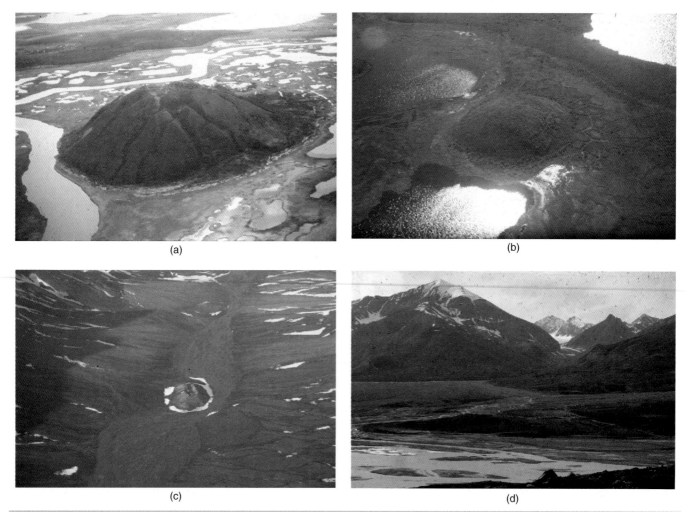

Figure 21.13 Pingos: (a)-(b) Mackenzie-Delta type c) Adventdalen, Spitsbergen, d) Skeldal, East Greenland type.

where water escapes from a confined aquifer underlying discontinuous permafrost (Figure 21.15), with this water continually feeding the dome's development. Eventually cracking develops (as in the previous example), a crater lake forms and in time, with a change in climate, both types degrade completely.

The problem here is that in east Greenland and Spitsbergen the permafrost is not discontinuous, although this type of mechanism works well in Alaska. Here the largest concentration of open-system pingos is found in the Yukon-Tanana uplands in central Alaska and in the Yukon, Canada. These are found in narrow upland valleys, particularly on the lower parts of south- and south-east-facing slopes, but never on north-facing slopes or in the broader river valleys. This is likely to be because there are more opportunities for surface water to enter the ground in the permafrost-free zones of the upper slopes of south-facing valleys than on north-facing slopes that are totally frozen. There has to be restricted ground-water flow too, since a large flow rate would prevent freezing of the water supply. Another controlling factor is that the subsurface temperature is close to 0 °C in order to provide the minimum tensile strength of frozen ground and

prevent premature freezing of the ground water. Pingos can occur as isolated landforms but they are usually found as small groups within the same area and it is not unusual for a new pingo to develop inside the crater, or on the side of an older one. Many are oval or oblong in form and a high percentage are ruptured to varying degrees.

So recharge by slope ground water in a discontinuous permafrost zone explains the Alaskan examples but not the east Greenland or Spitsbergen ones. However, Worsley and Gurney (1996) suggested that for east Greenland, where there is a high concentration of hydraulic pingos, it was higher geothermal heat flows, a residual effect of Tertiary intrusive volcanism, that caused subsequent ground-water upwelling along subsurface structural weaknesses, so that their location is controlled by subsurface geology.

Pingo scars can be recognized by irregular ramparts that mark the place where the sediment around the ground-ice dome collapsed and by a central depression that is the former position of the ground-ice dome and the later crater lake, which has gradually been filled in with sediment and vegetation (Figure 21.16).

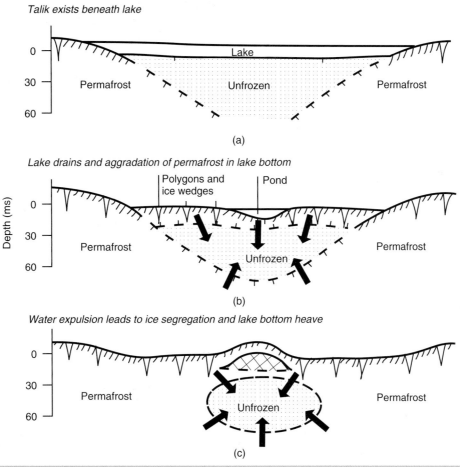

Figure 21.14 Mechanism for closed-system pingo formation. (Source: after French, 1996)

21.8.2 Thufurs

Much smaller, circular or oval ground ice domes or Earth hummocks can also develop, which are called thufurs (an Icelandic term). They occur on level or near-level areas with imperfect drainage and their size varies within fairly narrow limits, between 20 cm and 1 m in height and up to 1.5 m across. Examples from Iceland and the northern Pennines are shown in Figure 21.17(a) and (b).

They usually occur in distinct fields and are closely spaced. Internally their structure is characterized by disrupted horizons and they are formed in fine-grained soils. Externally some hummocks may be split across the apex, or their organic mat may be ruptured near the base. This indicates internal pressure and occasionally trees growing on thufurs are tilted by the heaving of the hummocks. They are not restricted to permafrost terrain.

The origin of these landforms is not known for certain but they do require a heave-sensitive soil, plentiful soil moisture and seasonal frost penetration (Schunke and Zoltai, 1988) and if any of these controls is missing they do not form. The initiation is difficult to understand but may be related to the movement of moisture towards a freezing front, where it carries some fine-grained soil particles. Cellular centres often develop in relatively homogeneous, fine soils and clay-sized particles may be concentrated in such centres by water movement associated with freezing (Figure 21.18). Another mechanism for their initiation may be random development of frost-heaved spots, which may remain without insulating vegetation for many years. These areas have different thermal regimes from the surrounding vegetated areas and may serve as the focus for hummock development. Once an embryonic thufur has been initiated, the vegetation will develop differentially on the mounds compared with the intervening areas. The hummocks are better drained; insulating vegetation develops on them rather than in the moist troughs and the drier mounds will lose heat more slowly. In the moister areas the freezing front penetrates faster and deeper than under the mound, which sets up lateral pressures towards the mound centre, displacing more soil and eventually forming the Earth hummocks.

The permafrost table, if present, acts as an impervious layer to water and as a base that can withstand pressures associated with cryoturbation. In nonpermafrost areas often a finer, impervious layer which holds the moisture is associated with the local bedrock geology, as in the northern Pennines example, where a thin shale bed acts in this way. A review of the genesis and environmental significance of thufurs is given in Grab (2005).

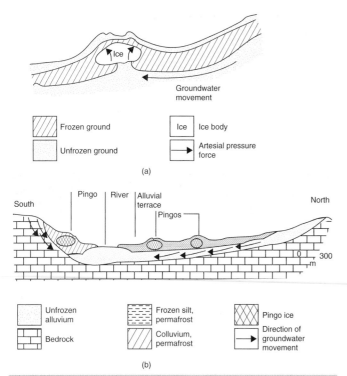

(a)

Frozen ground

Unfrozen ground

Ice   Ice body

Artesial pressure force

(a)

South   Pingo   River   Alluvial terrace   North

Pingos

0   300
m

Unfrozen alluvium

Bedrock

Frozen silt, permafrost

Colluvium, permafrost

Pingo ice

Direction of groundwater movement

(b)

Figure 21.15 Mechanisms for open-system pingo formation. (Source: after French, 1996)

21.8.3 Palsas

Palsas are low permafrost mounds, between 1 and 7 m high, 10 and 30 m wide and 15 and 150 m long, which have cores of layered segregation ice and peat. They are most common at the southern margin of the discontinuous permafrost zone and are believed to form when areas of reduced snow cover allow frost to penetrate more deeply into an unfrozen peat bog. This freezes the water in the peat, forming an initial ice layer, which then grows in size over time as water migrates from unfrozen parts of the peat to the surface of the growing ice mass. As the dome develops the hydrosere vegetation growing on the peat is pushed up above the water table. This vegetation then dies off, resulting in an easily-recognizable bare brown peat surface on top of the mound.

The origins and criteria for recognition of palsas have been discussed in detail by Seppälä (1988) and Nelson *et al.* (1992). There appears to be a continuum between cryogenic mounds in peat (palsas) and similar mounds formed in mineral soils (mineral palsas, or 'lithalsas' in Pissart (2000)). The combination of factors which locates lithalsas remains unclear, but frost-susceptible soils in a discontinuous permafrost with a high local water table appear necessary and it is assumed that localized variation in thermal regime associated with microtopography, snow thickness and vegetation cover leads to localized zones of preferential ice segregation, mound formation and the preservation of the frozen, ice-rich mound core.

These landforms often form in large fields with many mounds and may go through cycles of formation and collapse associated

(a)

(b)

(c)

Figure 21.16 Pingo scars and depressions: (a) Llangurig, mid-Wales; (b) Wicklows, southern Ireland; (c) Skeldal, East Greenland

with slight changes in the local conditions. The result can be several generations of lithalsas at a given location, with later thaw ponds and ramparts disturbing the traces of earlier landforms (Matthews *et al.*, 1997). Pissart (2000) reviewed the likely thermal conditions for these landforms in

(a)

(b)

Figure 21.17 Thufurs: (a) Iceland; (b) northern Pennines, Great Dun Fell

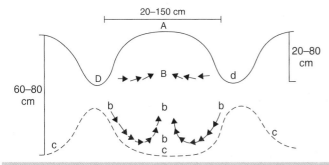

Figure 21.18 A model of cellular moisture and fine-sediment movement in an Earth hummock. (Source: after Schunke and Zoltai, 1988)

Quebec and suggested that mean annual temperatures from −7 to −4 °C, warmest-month temperatures from +8.5 to +11.5 °C and coldest-month temperatures from −25 to −23 °C were necessary. In Scandinavia however, somewhat warmer conditions

support lithalsa formation. Hence a relatively narrow range of temperatures was suggested for their formation:

- Mean warmest month no higher than +10 °C.

- Mean annual temperatures low enough to support discontinuous permafrost, between about −6 and −4 °C.

21.9 Processes associated with snowbanks (nivation processes)

Nivation is a situation in which there is local erosion of a hillside as the result of the intensification of a set of processes at the margins of and underneath snow patches or snowbanks that remain through the summer season. Periglacial environments commonly have less snowfall than warmer climates, especially temperate mountains, but the duration of snow cover is much longer and hence the snow can have a greater geomorphic significance. These processes are illustrated in Figure 21.19; they are frost shattering, mass wasting and the sheetflow or rill activity of melt water.

A pre-existing hollow is needed for the residual snow to accumulate in initially. During the summer melt water percolates down the back and freezes at certain times. So the back and base of the snowbank theoretically come under the action of freeze–thaw cycles. Debris which is loosened by such processes is fairly mobile at the end of the summer because it is lubricated by the snowmelt and rill activity, and gelifluction processes transport sediment away from the front of the snowbank. Gardner (1969) showed that the effect of snow patches on temperatures nearby was to reduce the maximum, minimum and daily mean temperatures; that freezing and thawing occurred more frequently in a restricted zone at the edge of the snow patches than on the rock faces above; and that exposure and aspect were important in controlling freeze–thaw activity, as in exposures with a southerly aspect snowmelt begins in April, but where there is a northerly aspect it is not attained until mid-June or early July. This proves that in summer at least there are temperature conditions around snowbanks that provide a suitable environment for intense frost weathering.

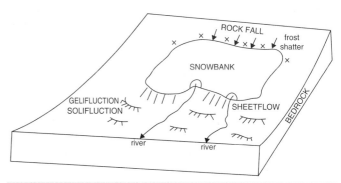

Figure 21.19 Nivation processes associated with snowbanks

Figure 21.20 Nivation cirque, Ystwyth valley, mid-Wales

The net effect of nivation is to produce a nivation hollow with the passage of time and eventually a nivation cirque. Lewis (1939) believed that nivation processes were an essential prerequisite for the development of true corrie landforms associated with ice erosion and periglacial processes in corrie glacier basins. Examples of nivation cirques with protalus ramparts, as opposed to true moraines, can be found in the Ystwyth valley, near Aberystwyth (Figure 21.20). Sometimes it is difficult to differentiate between nivation and glacial landforms, but where there are dominant snow processes there has been no overdeepening of the basin, and the sediments in the pronival or proglacial areas can be used to differentiate the operative processes too (see Section 21.9.1).

The controlling factors in the formation of nivation landforms are a suitable topography with a pre-existing hollow; the position of the orographic snowline; the lithology of the local bedrock as it has been suggested that nivation terraces with coarse debris can be produced on gabbros; nivation benches on sandstone, with finer-grained debris; and small landforms like hollows, ledges and semi-circles in shale, as these break down easily into fine sand and silt.

However, the concept of nivation has been the cause of some confusion in the periglacial literature and in fact modern use of the term is not recommended by Thorn (1988). The main reason for this is that the causal relationship between snowbanks and landforms via the family of processes linked to the term 'nivation' as described above does not always stand up to rigorous analysis. Field studies on the geomorphic effects of snow and the lack of effective freeze–thaw cycles beneath snowbanks led Hall (1980), Thorn and Hall (1980) and Thorn (1988) to question the concept, despite the earlier results of Gardner (1969). The integration of weathering and transport processes in a single term, plus their variable interaction and the difficulty of defining how the intensification of the processes takes place, precludes the easy definition of nivation as a process term. As a landform genesis term it is rarely possible to work out the degree to which nivation has modified the location of a modern snowpatch, and hence the operational definition of a nivation landform is also rarely possible. Yet despite these criticisms, at a simple conceptual

level the family of processes grouped together under the term 'nivation' explain the landforms associated with snowpatches.

21.9.1 Formation of protalus ramparts in upland Britain

A protalus rampart is a ridge of coarse sediment, usually located at the foot of a talus slope, that formed through the accumulation of debris by glissading, bouncing and rolling to the downslope margin of a snowbank or firn field. Often the relict forms have been confused with moraines, landslide deposits, avalanche deposits and protalus rock glaciers. The debris predominantly comes from rockfall on the free face but there can also be contributions from debris flows, snow avalanches and till transported downslope.

In Britain there are currently no perennial snowbeds occurring below rockwalls but landforms interpreted as relict protalus ramparts have been reported widely in upland Wales, the Lake District and the Scottish Highlands. It is thought that these examples formed during the Loch Lomond Stadial of ∼11–10 000 years BP. Ballantyne and Kirkbride (1986) found that rampart development in Scotland and the Lake District has been restricted to locations where there is a marked gradient reduction at the foot of a steep slope overlooked by a cliff, especially the floors of corries and glacial troughs, although some are perched on rock steps high on valley sides. Although there is a concentration in the north-eastern quadrant, there are protalus ramparts with other aspects, including south (Figure 21.21(a)).

Most ramparts are arcuate in planform, with the extremities of the frontal crest curving upslope, although some are linear with approximately straight crests. Few ramparts are over 300 m in length. There is usually a single ramp which bounds a shallow linear depression up to 3 m deep. The distal slopes of the ramps ranges between 34 and 39° but the proximal slopes are much gentler (Figure 21.21(b)). The ramparts show a certain amount of morphometric regularity and there are strong positive linear relationships between rampart width, thickness and distance from the foot of the talus upslope. The relationship between width and thickness is explained in terms of the progressive accumulation of coarse, cohesionless debris and reflects the angle of repose for such debris. The increase in rampart thickness with increasing crest–talus distance has been interpreted as the result of progressive thickening of snowbeds so that the rampart crests moved outwards away from the talus as the ramparts accumulated (Figure 21.22).

Most ramparts are formed of poorly-sorted, coarse, dominantly-angular, openwork clasts, although abundant fines are present at some sites. These fines are likely to have been formed from granular disintegration of some clasts, from supranival wash or avalanches, and from debris flows. Macrofabrics of clast long-axis orientations show both strong transverse and parallel peaks in relation to the ridge axis.

Within individual mountain areas in Britain there is considerable variation in altitude, which reflects local topographic control, particularly the location of rockwalls overlooking suitable sites. However, on a regional scale the altitudinal distribution of protalus ramparts provides an indication of former snowfall patterns.

(a)

(b)

Figure 21.21 Protalus ramparts: (a) below Herdus, Ennerdale, south-facing; (b) below Dead Crags, north-west of Skiddaw, north-facing. Both sites are in the Lake District

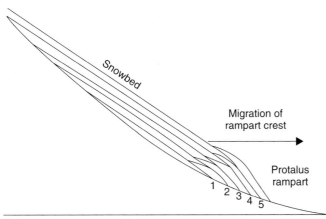

Figure 21.22 Model of protalus rampart development at the foot of a progressively thickening snowbed. (Source: after Ballantyne and Kirkbride, 1986)

Assuming limited spatial variation of stadial palaeotemperatures, ramparts at lower altitudes indicate formerly abundant snowfall,

with both high accumulation and high ablation rates, whereas the restriction of ramparts to high altitudes, as in the Cairngorms, indicates light snowfall, with corresponding limited ablation. In Scotland the lowest altitudes are in the Hebrides and western Scottish Highlands, whereas the highest are in the Cairngorms (Figure 21.23), which reflects the decreasing precipitation (largely as snowfall) during the stadial. Former rockfall rates have also been calculated, as the volume of debris contained in the rampart is equivalent to that lost from the rockwall upslope (Ballantyne and Kirkbride, 1987). The average amount of rockwall retreat implied by the rampart volumes is between 1.14 and 1.51 m.

Thus protalus ramparts indicate association with snowbanks during the Late Glacial Stadial in Britain. However, an individual landform can be the subject of considerable dispute, such as that at the head of Keskadale in the Lake District. Oxford (1985) suggests this landform is a protalus rampart, whereas both Sissons (1980) and Ballantyne and Harris (1994) suggest it is too far from the foot of the talus slope and is a Loch Lomond Stadial end moraine. Similar discussion has taken place regarding the origins of some of the 'moraines' in Cwm Idwal in Snowdonia.

21.10 Cryoplanation or altiplanation processes and their resultant landforms

Altiplanation terraces were first discussed by Russian geomorphologists in the 1930s. They are erosional landforms, often in bedrock, caused by a complex set of frost processes:

1. Those breaking up the bedrock and preparing the loose debris for transport; that is, frost weathering.

2. Those that transport debris, such as supranival processes, deflation and gelifluction. Different workers have emphasized the significance of different processes, but essentially all the above processes are involved in the creation of these landforms.

The terraces are hillside or summit benches, cut in bedrock, that lack a structural geological control and are confined to cold climates, or areas that have suffered Pleistocene cold climates. They are well developed in the mountains of Yakutia in Siberia, in the Hruby Jesenik mountains of Czechoslovakia, Spitsbergen, the northern Urals, the northern Yukon and south-west England. In the latter they have been described from benched hillsides in Exmoor and Dartmoor, particularly on the metamorphic aureole between 300 and 450 m, for example around Cox Tor, where there is some of the most convincing evidence for cryoplanation in Britain (Figure 21.24). Here the steps possess gently-sloping treads between 3 and 8°, whilst the risers slope valleywards and are between 15 and 22°. The Dartmoor examples are small, with widths varying between 10 and 90 m, heights of the risers between 2 and 12 m and lengths up to 800 m. The terrace treads are covered by relict patterned ground (Gerrard, 1988).

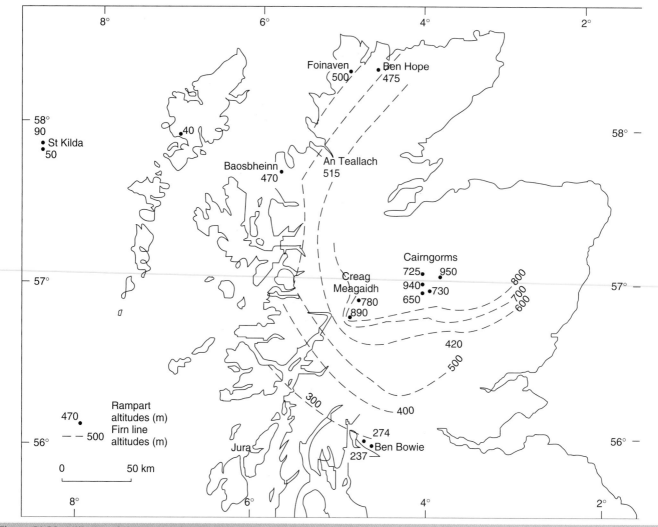

Figure 21.23 Altitudes of the crests of Loch Lomond Stadial protalus ramparts in the Scottish Highlands and reconstructed regional equilibrium-line altitudes of stadial glaciers at their maximum extent. (Source: after Ballantyne and Kirkbride, 1986)

There also appears to be a lithological control and they are found on the interlaminated shales and fine siltstones, shales-limestones-chert, dolerite sills, hornfelses, quartzites and lavas. All these rocks tend to have one common factor, which is that they are well bedded and/or well jointed, and between their well-defined divisional planes they are uniformly hard and dense. Usually the altiplanation terraces have a covering of gelifluction sediments and sometimes patterned ground. So geological structures such as jointing and certain lithological types may help to determine the location and degree of terrace development, but their origin can be independent of such controls. They can range from 10 m to 2–3 km across and can be over 10 km in length, usually with gradients of between 1 and 12°. Much smaller examples from the northern Pennines are shown in Figure 21.25.

The theory of development of these terraces was put forward by Demek (1969), who suggested three stages (Figure 21.26):

1. Where initially nivation processes operate in the region of permanent or temporary snow patches, the freeze–thaw

processes will be more important and more intense than in drier areas. This sediment preparation is transported by gelifluction and melt-water processes. The material on the upper slopes is transported by supranival processes to the front of the snow patches. So the original irregularity increases, either due to the nivation processes or in a pre-existing hollow. But the unequal occurrence, the number and the extent of the altiplanation terraces, even in the same geological environments, suggests that nivation is significant in their development.

2. The terrace consists of a steep section, representing the frost-riven scarp, between 10 and 40°, and a terrace flat on about a 7° slope. On the frost-riven cliff there are nivation processes, which undermine the cliff foot so that there is a parallel retreat of the slope. The debris falling down the upper part of this steep section is transported on the snow-patch surface and accumulates as a protalus rampart. The terrace surface is where the cliff debris is further sorted and transported by

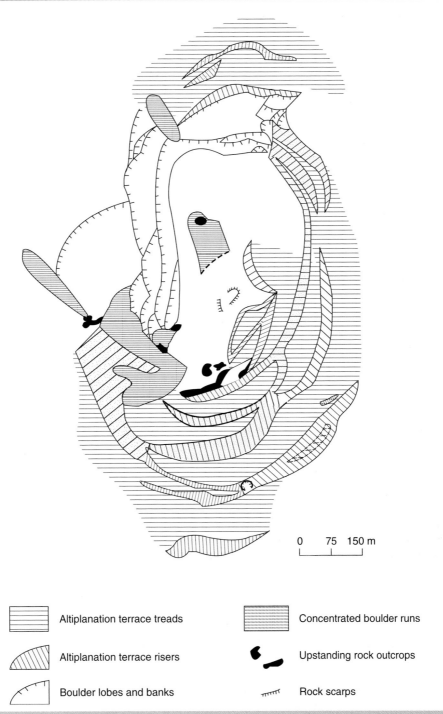

0   75   150 m

Altiplanation terrace treads

Concentrated boulder runs

Altiplanation terrace risers

Upstanding rock outcrops

Boulder lobes and banks

Rock scarps

Figure 21.24 Cryoplanation terraces around Cox Tor, Dartmoor. (Source: after Gerrard, 1988)

a complex set of processes, but everywhere there seems to be evidence of frost sorting. Gelifluction only occurs where there is a sufficient quantity of fines, as on shales, whereas granite terraces are covered by angular block fragments, so the processes prevailing on different terraces depend on local conditions. With the gradual terrace development the direction of removal changes, from the original direction from the foot of the frost-riven scarp towards the terrace edge

to, as the terrace gets wider, sideways removal. Eventually the cliff is worn back.

3. The third stage is characterized by an altiplanation summit flat. The gradient decreases to as low as <2° and the surface is covered with weathered bedrock and polygonal ground. The sediment is finer-grained, but because of the low gradient, gelifluction and snow melt-water transport almost

(a)

(b)

Figure 21.25 Small-scale cryoplanation processes on Great Dun Fell in the northern Pennines: (a) position of late-lying snow, shown in the vegetation contrast and the typical profile; (b) profile of the terrace and effects of snowbank processes. Cross Fell can be seen in the background

stop. Deflation becomes more important due to a higher percentage of finer sediment.

The development of these terraces is helped by the presence of permafrost, but they can develop without it. Climate and topography are the important controls on terrace development, but the time element is considered in the larger ones to be in the order of at least tens of thousands of years. Jointing, especially vertical joints, is important as it promotes the parallel development of the frost-riven scarps, and it is possible to find terraces developed at the contact of two rock types, for example sandstone and shale, with the terrace on the less-resistant rock and the scarp on the more-resistant one. However, most terraces have been found to be independent of geological structure, although in some areas, like the northern Pennines, where there are alternations of limestone, sandstone and shale with differential resistances to erosion, there may be problems

in distinguishing whether altiplanation processes or lithological control played a bigger role in the landscape's development. They seldom appear on granites or syenites and seem to develop most often on slopes consisting of quartzite, phyllite or sandstone.

Yet despite the discussions above, the significance of cryoplanation terraces is highly debateable, according to Büdel (1982), Thorn (1988) and French (1996), although the existence of the landforms is not open to doubt. The problem is the almost total absence of any quantitative process rate data. No process studies have yet demonstrated that they form under contemporary cold-climate conditions; most appear to be inactive or relict today and their age is uncertain. French and Harry (1992), for example, hypothesize that in cold, semi-arid areas, like the northern Yukon, they are effective waste-debris transportation slopes only under certain conditions. In Figure 21.27 it can be seen that they are active during periods of intense frost wedging and high sediment removal and are inactive whenever moisture levels fall below a critical level. This combination of temperature and humidity probably restricts activity to the transitory periods marking the beginning and end of cold-climate fluctuations. Despite the observations that they are erosional in nature, for example where they cut across dip directions and attitudes, there seems to be a real possibility that many landforms identified as cryoplanation terraces are lithological and/or structural benches. Again, despite these reservations, as with nivation, the concept appears useful at a simple level in understanding periglacial slope development.

21.11 The development of tors

Close to the end stage of altiplanation terrace development, upstanding rock masses on the summits can be found as the frost-weathering processes erode back into the hillsides and sediments are carried away. These are tor landforms and it might be thought that a simple periglacial origin best explains them, but as we will see there has been much discussion and controversy related to their formation. There are two major conflicting hypotheses for tors in general and even the same ones have been explained as the result of periglacial frost action or of bedrock residuals formed after the removal of a deeply weathered mantle. Hence a definition of a tor must have no genetic connotation if possible and it has been suggested that a tor is a bedrock residual, with isolated free faces on all sides, the result of differential weathering and mass slope wasting.

They are probably found worldwide but the best known are from the uplands of Central Europe, like the Sudeten Mountains and the Bohemian Massif; from Australasia and Antarctica; and from Western Europe, particularly on Dartmoor and the Pennines (Figure 21.28(a)), but also in other parts of Britain, like the Stiperstones (Figure 21.28(b)), the Prescelli Hills of Pembrokeshire, the Cairngorms and Caithness (the Morven, Smean and Maiden Pap), and in Ireland in the Wicklows and the Galtees. With reference to Britain, it appears that one control is that they

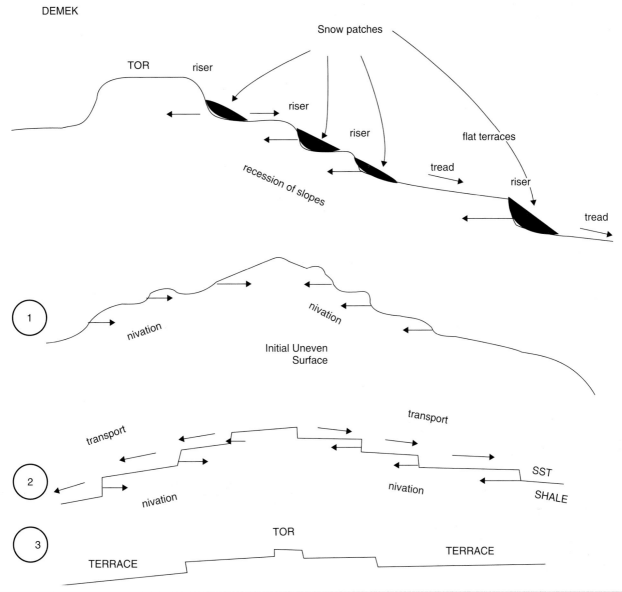

Figure 21.26 Development of cryoplanation terraces. (Source: after Demek, 1969)

are not found in the most heavily glaciated areas and sometimes they are located just outside the ice limits. Topographically they appear in three main locations:

- Summit tors.

- Break-of-slope tors, occurring on the convexities at the top of the valley walls and occasionally on the very edge of the plateau.

- Valley-side tors, located on valley walls, where they are associated with a shattered free face and block talus accumulations.

The interpretation of their origins has shown that similar landforms and sometimes even the same landform can be interpreted in very different ways. Here the climatic significance is in dispute, depending on whether you believe the work of Linton (1955, 1964) or of Palmer and his coworkers (Palmer, 1956; Palmer and Neilson, 1962; Palmer and Radley, 1961). Within these ideas there are both two- and one-cycle theories for tor development.

Linton (1955), originally using Dartmoor examples, although later extrapolating these to other parts of Britain, believed that tors were formed by Tertiary subsurface rotting of rock by acidulated ground water percolating along joints and from them deep into the body of the rock. This deep weathering produced a subsurface rounding, leading to what were called 'corestones', or locally 'woolsacks', under a warm temperate or tropical climatic regime. The pattern of development was largely controlled by structure in the rock and the joint spacing controlled the tor locations (Figure 21.29). The second phase was the exhumation

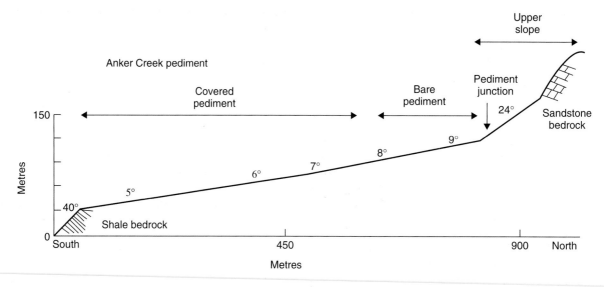

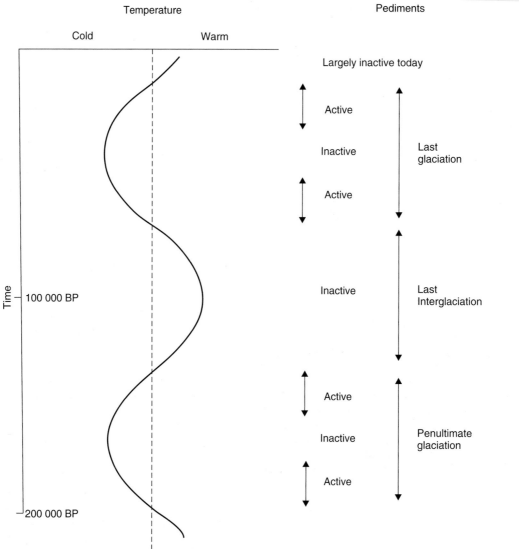

Figure 21.27 Schematic diagram showing the possible relationship between climatic fluctuations and enhanced pediment activity in the Barn Mountains, northern Yukon. (Source: after French and Harry, 1992)

(a)

(b)

Figure 21.28 Tors: (a) Brimham Rocks, central Pennines; (b) Stiperstones, Shropshire

of the fine-grained products of rock decay and their transport downslope to leave the tor isolated. Linton produced a definition of a tor based on this process: it was a residual mass of bedrock produced below surface level by a phase of profound rock rotting effected by ground water and guided by joint systems, followed by a phase of mechanical stripping of the incoherent products of chemical action. So he considered that this type of process resulted in a palaeotropical tor analogous – but on a smaller scale – to the inselbergs and castle koppies of Central Africa and other parts of the tropics. Indeed, at one stage it was suggested that the Dartmoor tors were inselbergs. This sort of preglacial, probably pre-Pleistocene deep weathering, followed by a partial or complete stripping of the buried weathered mantle, was a view long held to explain many tors on the Hercynian massifs of Europe and in Australasia. So this hypothesis for tor formation was not an original view put forward by Linton but he did develop it into a coherent idea. Linton thought that the present-day growan or weathered rock on the Dartmoor granite was a residue of this former warmer deep weathering: an important point that will be returned to later.

Linton (1964) suggested that tors were really caused by the frost activity working on material that had been thoroughly prepared by some antecedent period of deep ground-water weathering. However, these effects of periglacial processes were essentially limited to a removal or redistribution of loose debris already prepared and the exhuming of features already in existence below the surface.

Ford (1967) in Charnwood Forest in Leicestershire followed this explanation by Linton and found Tertiary deep weathering, up to 15 m deep, on granitic and dioritic rocks, but never on lavas or metamorphosed volcanic ashes. This deep weathering was followed by a phase of plucking by a Pleistocene ice sheet to clear away much of the weathered debris and finally frost processes and gelifluction to give the tor-like residuals.

In the Polish Sudeten Mountains there was also convincing evidence, according to Jahn (1974), that summit tors were formed by deep weathering of granite, chiefly during warm interglacials, and by exhumation under periglacial conditions during the glacial phases. On the other hand, in the adjacent Hruby Jesenik Mountains, Demek (1964) described how tors and castle koppies of both a deep-weathered (two-cycle) origin and a periglacial (one-cycle) origin existed in the same area. According to this idea, exhumation of the deep-weathered tors had already begun at the end of the Tertiary and continued into the Pleistocene, at which time the parallel retreat of frost-riven cliffs caused the isolation of other tors and the formation of altiplanation terraces, as described earlier.

Here it should be emphasized that a number of processes could operate to produce what looks like the same landform. This can be seen in Antarctica, where both rounded and angular tors are developed side by side in dolerite (Derbyshire, 1972). The angular forms occur in depressions, where snow patches allow effective frost shattering, whilst the rounded tors occur in more exposed locations where snow patches are absent. Here long continued and slow chemical weathering and exfoliation occur. So we have evidence here for tors of different form adjacent to one another, developing in the same climatic regime through different processes due to the microtopography.

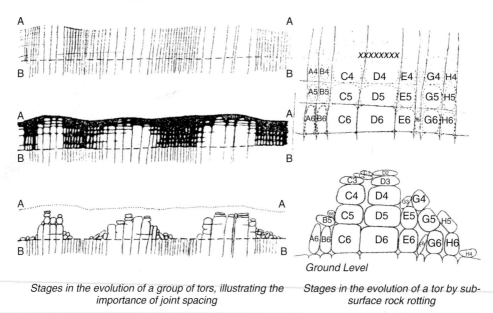

Stages in the evolution of a group of tors, illustrating the importance of joint spacing   *Stages in the evolution of a tor by sub-surface rock rotting*

Figure 21.29 Illustration of Linton's deep rock rotting hypothesis for tor formation

The simple, one-stage periglacial hypothesis of Palmer (1956), Palmer and Radley (1961) and Palmer and Neilson (1962) attributes the Pennine and North Yorkshire Moors tors to fossil landforms formed in a climatic regime very different to the present-day cool, humid climate. These tors are found in a zone peripheral to the last ice phase, which suggests that the processes responsible for their formation operated in a periglacial climate on the well-jointed Millstone Grit and underlying sandstones in the Pennines and on the Corallian Limestones in the North Yorkshire Moors. It is also seen that their angular, shattered form is similar to the Alaskan, Arctic examples and that there is a spatial coincidence with definite periglacial features and deposits. The Dartmoor tors were then looked at, and the authors suggest that they occur on the most resistant part of the granite and that there is no decayed rock from the tor joints. There is decayed granite in the valley bottoms but this is why there are valleys here as they have been preferentially eroded in these rocks. Where the rotten granite extends over the hills, the ridges are devoid of tors. The decomposition of the granite is suggested as a process that was part of the granite mineralization, as a pneumatolytic process which was probably Lower Carboniferous or Permian in age.

The authors suggest that the tors are palaeo-Arctic landforms created by a one-stage hypothesis involves frost weathering and the simultaneous removal of the regolith. This hypothesis ivolves: (a) loss by gelifluction of soil from the tops and hill slopes to give rounded hilltops; (b) frost action on the exposed granite concentrated on the joints to give blockfields and tors; and (c) termination of the cycle before it was fully completed in the case of those hilltops which retained tors. So the process is closely linked to the slope evolution through the altiplanation process, as described earlier. The tors represent the remnants of a frost-riven bedrock outcrop surrounded by an altiplanation terrace, across which the frost-shattered debris is transported by gelifluction and other mass-wasting processes.

It was pointed out that there were some problems with Linton's hypothesis; for example, if tors were formed during the Tertiary period they would be unlikely to retain their morphology through glacials and interglacials. It could be argued here that they were stripped of their weathered debris late in their history, but if they were not exhumed until the last glacial phase then this would be difficult to accept as they are mainly positioned around the tops of hillslopes, which should have allowed their early exhumation. Moreover, deep weathering could not take place in the last interglacial as the climate was not tropical and it didn't last long enough for 17 m of deep weathering to take place, and if deep soils were formed then they should still be visible somewhere today.

Caine (1967) in Tasmania suggested however that the two major hypotheses need not be incompatible. From the Ben Lomond area some tors have developed since the mountain was glaciated in the last phase of the Pleistocene. There has been a process of block removal and toppling from the upstanding bedrock residuals which survived the glaciation. This is proof that frost action is capable of at least remodelling pre-existing landforms to those that we now see as tors. There is also evidence from one particular tor, known as the Watchtower, to prove that the decomposition of dolerite along joints or fault-shattered zones can lead to the formation of tors. It may not give rounded forms in the dolerite but can give comparable angular forms, indistinguishable in terms of morphology from those due to post-glacial frost processes.

The general conclusion from this is that both hypotheses of tor formation are at least partially correct. In Tasmania it appears that the summit and break of slope tors are the result of subsurface decomposition and that they have been exhumed, whilst the valley-side tors are due to post-glacial frost processes. This does show the inadequate nature of hypotheses based solely on the evidence from the general nature of tor distribution, tor morphology and the nature of the blocks around them. Work on the actual sediments associated with tors seems to be one way ahead which might give answers to tor formation.

This was attempted by Eden and Green (1971), who threw some light on the formation of the Dartmoor tors. They suggested a mechanical weathering origin for the growan for the following reasons: it had a relatively low silt and clay percentage; it had a high feldspar residue, which suggested only very limited chemical weathering; it was completely distinct from the pneumatolytic breakdown products (china clay); and it was underlain by solid rock and therefore an atmospheric process had caused it, rather than the other way around. Linton had clearly talked about tropical weathering, but the growan bears little compositional resemblance to weathered granite that has been described from the humid tropics. For example, granite from the Volzberg region of Surinam shows a clayey weathering residue which has 30–55% clay and a sand fraction that is almost entirely made of quartz. In Guayana there is 15–25% clay and the sand is almost devoid of feldspar in weathered granite areas.

Eden and Green convincingly demonstrated that chemical weathering on Dartmoor was far less than that in tropical environments, and on Dartmoor too augering and seismic profiles do not suggest that there is a substantial layer of deeply-weathered growan on the supposed Tertiary summit planation surfaces. Further relevant evidence comes from Hall's work (1985, 1986) concerning the age of the granite saprolites and the conditions under which they formed on the deeply-weathered crystalline rocks of north-eastern Scotland. He showed that the saprolite consisted of two forms: (a) a widespread sandy grus, which like the Dartmoor growan displays limited mineral alteration and reflects prolonged weathering under humid temperate conditions in the Pliocene and Early Pleistocene; (b) a much less widespread, clay-rich grus with advanced mineral alteration, which indicates prolonged weathering under a pre-Pliocene subtropical climate.

Doornkamp (1974), using a scanning electron microscope to look at the surface textures of detrital quartz from Dartmoor, found that there was very little evidence of any intense chemical weathering, except for one striking exception from a sample taken from the Two Bridges Quarry. This was one of Linton's classic sites for unexhumed tors. In the periglacial head deposits the grains showed signs of mechanical breakdown, whereas the growan, which had been generally recognized as a weathering product of the granite, with some chemical weathering playing a role in its formation, showed surface textures similar to the head. The deposits that showed the most chemical alteration were those exposed in the incipient tors in the Two Bridges Quarry, which had surface textures remarkably like those normally to be found on grains subject to chemical weathering in a hotter, more humid climate than that found on Dartmoor today. This supports Linton's ideas. However, there was no evidence of any shallow surface deposits showing any sign of significant chemical weathering. It seems probable that during the Pleistocene most of any pre-existing chemically-weathered mantle was removed, or that the deep chemical weathering of the type described by Linton was localized.

In conclusion, tors are variable in size and shape because the rock type and the geological structure is variable. They are of different ages too. Some are thought to be young and short-lived, like valley-side tors, whilst summit tors could be of great age. Tors seem to be polygenetic and for any individual tor there has to be a detailed study of the weathering sediments in its surrounding area to try and understand the stages of its formation. It seems unlikely that a single hypothesis for tor formation is applicable to all tors and therefore both Linton and Palmer may be correct in part about the origin of the Dartmoor tors. It seems possible that both exhumation and modification of two-cycle tors and the formation of one-cycle tors could have occurred in different parts of Dartmoor, depending on the localization of the deep-weathering processes. Hillslope tors, which occur on the upper parts of slopes and in areas where there is little evidence for deep weathering, such as northern England and Bohemia, are less ambiguous. They seem best explained by a one-stage development of slope retreat caused by frost action, snowbank processes and gelifluction. Even if a one-cycle periglacial origin seems the more likely it is nonetheless unwise to immediately assume that tors represent intense frost action without a weathering study of the sediments. Tors cannot be used as a diagnostic indicator of Pleistocene frost action without a careful investigation of the morphology and weathering patterns of the associated terrain.

21.12 Slope processes associated with the short summer melt season

Downslope soil movement in periglacial environments with a short summer melt season is caused by two processes: frost creep and gelifluction. Frost creep is the ratchet-like movement of particles as the result of frost heaving of the ground and subsequent settlement on thawing of the ice. The heaving is predominantly normal to the slope and the settling more nearly vertical. It tends to be at right angles to the slope because this is the major cooling surface and the direction of ice crystal growth, and heaving is normal to this cooling surface. Soil creep can also be the result of expansion and contraction of the soil because of alternate wetting and drying, which may contribute to creep in periglacial areas where the soil is non-frost-susceptible (that is, not silts). However, frost creep is very difficult to isolate and measure because other processes will be taking place in the soil, but it is thought to be very variable in scale. After several cycles of frost creep though the amount of movement varies directly with the number of soil-freezing cycles, the tangent of the slope angle, the amount of moisture for freezing and the frost-susceptibility of the soil. In general the process decreases rapidly with depth in the soil profile. Other things being equal, frost creep should be greater in subpolar or even temperate climates than in high polar climates where there are fewer freeze–thaw cycles and the depth of thawing is more shallow.

Much more obvious as a periglacial slope process is gelifluction, which is the periglacial equivalent of solifluction, originally defined as the slow flowing of masses of waste saturated with water from higher to lower ground. In a soil profile it decreases with depth and there is obviously a dependency on moisture being available. It is very prominent in frost climates because of the role of the permafrost table in stopping water movement downwards and so aiding soil saturation in the surface layers, and also because of the importance of thawing ice and snow in the summer to the provision of water for saturation. The deeper thawing of ice-rich permafrost in the modern Arctic is commonly associated with release of ground ice slumps in which extensive mudflows are fed by melting ice-rich arcuate headscarps. More highly-plastic clays can suffer pre-failure thaw-induced creep but are especially vulnerable to shallow landsliding along discrete shear surfaces as active-layer detachment slides (Harris and Lewkowicz, 2000).

The overriding importance of moisture can be seen from an example from the Mesters Vig area of East Greenland. Here the rates of near-surface soil movement on a 10–14° slope ranged from a minimum of 0.6 cm per year in relatively dry areas to a maximum of 6.0 cm per year in wet zones. It is more important than the binding effect of vegetation or the gradient effect. Again this can be seen from Mesters Vig, where the highest movement was in an experimental site in the best-vegetated sector, where the slope was less than in the drier parts. The process has been noted on slopes as low as 1° but it probably only takes place to a significant extent where the water content of the soil approximates to or exceeds the Atterberg Liquid Limit (the percentage of moisture at which solids change to fluids), the value at which the soil would have little shear strength. The slope is also important as a control, as the flow depends on the gravity component, which, acting parallel to the slope, increases as the sine of the slope angle. Observations indicate that there is a straight-line relationship between the sine and the rate of gelifluction and soil creep in continuously wet areas. The grain size of the sediment is important too, because the high porosity and permeability of sand and gravel aids good drainage and hence does not favour saturated flows, except where pore-water pressure reduces intergranular pressures. On the other hand, fines tend to remain wet longer than coarser-grained sediment, and silt in particular is subject to flow. This is because it lacks the cohesion of clay and the Atterberg Liquid limit is less in silt than in clay, so less moisture is required for flow. Also, in cold climates mechanical weathering tends to predominate over chemical processes, so silt tends to predominate over clay-sized particles. Gelifluction therefore is a function of the depth of freezing and soil depth, the sine of the slope, the amount of water in the soil and the grain size of the sediment.

During both permafrost degradation and seasonal active-layer thawing, excess pore-water pressures are generated during the consolidation process, reducing effective stress within the soil, and hence its frictional strength. In both the active layer and permafrost, thawing is predominantly from the surface downwards, so that there is downward drainage impedance by the still-frozen substrata. Thus thaw softening is a widespread feature and on a slope usually leads to gelifluction, landslides and mudflows.

(a)

(b)

Figure 21.30 Gelifluction landforms: (a) terrace, Grasmoor, Lake District; (b) Smooth topography caused by gelifluction, Caldbeck Fells, Northern Lake District, UK.

21.12.1 Landforms produced by gelifluction

The landforms produced by the gelifluction process are very varied and there have been many different types of classification. Washburn (1970) suggested:

1. Sheets where there is a smooth surface but which in places are bench-like or have a lobate lower margin.

2. Benches or terraces (Figure 21.30).

3. Lobes which have a tongue-like shape.

4. Streams which have a pronounced linear form at right angles to the contours.

King (1972), working in the Cairngorms, divided them into two major categories: stone-banked and vegetation-covered landforms. The former were characteristic of higher altitudes, at greater slope angles, were more sorted, not so strongly orientated and were thought to have formed by viscous flow. Joint-controlled ledges decelerate the lobes so that they tend to come to rest on these ledges. The stone-banked gelifluction lobes were suggested to be active during the Little Ice Age (1500–1800 AD), whilst some above 900 m are still active today. The vegetated ones though are former stone-banked lobes which ceased movement at the beginning of the post-glacial period and then became vegetated.

Williams (1959), working in the Rondane in Norway, suggested that there were three types:

1. Single terraces at the lower edge of late-lying snow patches. The frontal height is up to 1 m and the slope behind the terrace is concave, which is where the late-lying snow lies. Melt water is thought to be the most important control, partly because it increases the weight on the soil, but mainly because of its effect on soil strength. An increase in water pressure in the soil pores raises the ground-water level and decreases the soil strength. Hence the soil in these terraces becomes saturated during early summer and gelifluction results.

2. Soil tongues, restricted to fairly steep slopes and represented by frontal banks up to 1 m high, with a lobate front.

3. Terraces on slight slopes, associated with frost heaving in the soils, which results from the increase in water content on freezing in fine-grained soils. Most workers on gelifluction have emphasized that it is the melting of the ice layers in the soil that results in the water excess, especially if this occurs at a time of snow melting.

As they were localized in distribution in the Rondane, their absence could be due to the absence of frost-heaving conditions, as water must be available. Ground water must be near enough to the freezing plane for capillarity to move it there, and it also depends often on the bedrock type. For example, where there is phyllite, schist and shale there are fine grain sizes and hence frost heave, but where there are sandstones there is no frost heave since the break-up products of these rocks are permeable. Often gelifluction is associated with springs, which are dependable water sources. This process can result in a sheet of gelifluction sediment, which can smooth the topography (Figure 21.30(a)). Box 21.2 in the companion web site discusses detailed field and laboratory measurements of gelifluction.

21.12.2 Ploughing blocks or boulders

These small-scale landforms are only found in areas showing evidence of active gelifluction, on frost-susceptible soils with low plastic and liquid limits. 'Ploughing block' is a term used by Tufnell (1969–1976) to describe those blocky components of periglacial slope movement which travel faster than their surroundings and during movement rotate to adopt an alignment of least resistance. Such large clasts are able to move downslope under gravity and form distinctive mounds of ploughed sediment in advance of the block, with a marked track upslope behind the block. They indicate differential slope movement, with a mean annual rate of movement of 1–5 cm per year, with a maximum in the spring. The smaller the block, the less likely is it to possess enough weight to form an associated mound and depression, and in the northern Pennines the smallest block had a long axis of 14 cm. The larger the block though, the greater the amount of frost action needed to move it, with again the largest in the northern Pennines being 2.5 m long and 1.4 m wide. These landforms usually occur on grassy slopes where there is sufficient moisture and are rarely found in areas which lack vegetation. This is because the vegetation binds the slope, so retarding its movement. Blocks which stick above the surface can at least partly counteract this retardation and are able to move faster than the surrounding slope. A relatively thick slope regolith is needed too, with contrasting grain sizes between blocks and fines, and there should not be too many bedrock outcrops.

The possible causes of ploughing block movement are illustrated in Figure 21.31. Ballantyne (2001) found ice lenses under the base of boulders in winter and emergent shear planes in the frontal ridges. The differential thermal conductivity of the boulder and the surrounding soil implies that the seasonal freezing plane descends more rapidly through the boulder, causing ice lens growth at the boulder base. During the thaw the plane reaches the base of the boulder whilst the surrounding soil is still frozen. The suggestion is that movement results from trapping of excess water under boulders during thaw, elevation of sub-boulder pore-water pressures to above hydrostatic levels and therefore a reduction in shearing resistance at the boulder bases. Movement is initiated by thaw of soil immediately downslope of the block, but this allows trapped water to escape, reducing the pore-water pressures and hence slope stability again.

21.12.3 Sediments created by gelifluction

The sediments produced by gelifluction are widespread in British upland areas and in the lowlands outside the limits of glaciation. The type of sediment is called head, which is a poorly sorted sediment with stones and a fine matrix which often has a crude stratification parallel with the slope (Figure 21.32(a) and (b)). The flow imparts an orientation up and down the local slope and hence there is usually a strong fabric parallel to the direction of movement. The stones are usually angular and there are often other indications of a periglacial climate, such as vertical stones (Figure 21.33) and periglacial soil deformations called involutions based on morphology or cryoturbations based on cryogenic origin. 'Involution' is a general term for distortion of the soil and where involutions are produced by freezing or thawing of the ground they are called cryoturbations. However, there are many types of involution caused by soft sediment deformation so it is important to be able to demonstrate the periglacial origin if they are called cryoturbations. A useful classification of deformation structures has been proposed by Vandenberghe (1988), who recognized six fundamental types (Figure 21.34). Such deformations produced in a periglacial soil are polygenetic and often it is difficult to decide on the processes involved. However, three main methods of formation have been suggested:

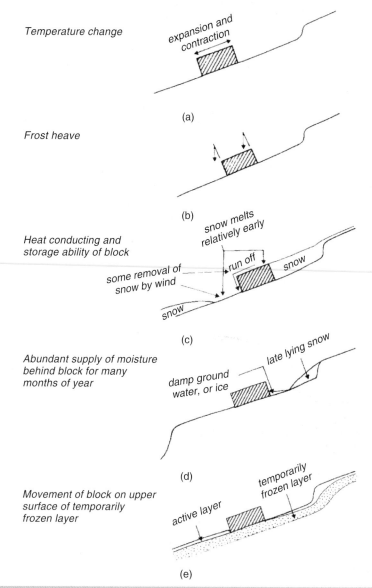

Figure 21.31 Possible causes of ploughing block movement. (Source: after Tufnell, 1972)

- Loading, which is a result of density inversion during thaw of frozen ground. Melting of ice-rich frozen ground may leave a loose, saturated sediment that is liable to liquify as it consolidates. As a result, this low-density sediment at depth is injected upwards into overlying horizons, in many cases producing flame-like structures, whilst the denser sediment above sinks downwards into the liquefied layer. It is likely to produce most of types 2–4 in Figure 21.34.

- Liquefaction of near-surface sediment due to the generation of high pore-water pressures during autumn freezeback. This cryohydrostatic pressure occurs where unfrozen sediment confined between a descending freezing front and an underlying permafrost table may experience raised pore-water pressures and hence deformation.

- Different rates of frost penetration and ice segregation in freezing soil, together with the fact that water in finer sediments freezes at lower temperatures than water in coarser sediments, may generate differential frost-heaving pressures and mass displacement of sediment. Repeated freezing and thawing therefore may lead to cryoturbation structures.

21.13 Cambering and associated structures

Although not widely reported from modern periglacial environments, deep-seated structures such as cambering and valley-bulge structures, interpreted as having formed under permafrost, have

Figure 21.33 Vertical stones, Waterford coast, southern Ireland

(a)

(b)

Figure 21.32 Gelifluction sediment: (a) head, Criccieth, Lleyn Peninsula, west Wales: overlain by an Irish Sea till; (b) head, Waterford coast, southern Ireland

been described in Britain. They are thought to indicate ice-segregation processes during permafrost aggradation at depth and subsequent thaw-consolidation processes caused by thawing at the base of the permafrost (underthaw) during permafrost degradation in the transition from glacial to interglacial stages. Hutchinson (1991) has provided a detailed review of these processes.

Cambering occurs where large-scale valley incision has exposed more-competent caprocks overlying less-competent clay or mudrocks in the valley sides. Caprocks tend to extend down the valley sides as an attenuating drape and extension is accommodated by the development of deep fractures, known as gulls (Figure 21.35), which run parallel to the contours and separate intact caprock blocks. These blocks tend to tilt forward during the extensional phase, increasing apparent dip and producing the 'dip and fault structure'. Cambering is often associated with the development of anticlinal deformation, called valley bulging, within the less-competent strata beneath the valley axis (Figure 21.35). Hutchinson (1991) provided a model for cambering and valley bulging based on estimated displacements in the Lias (Figure 21.36). The stages shown in Table 21.2 were envisaged.

21.14 Wind action in a periglacial climate

In modern high-latitude polar deserts, vegetation is sparse and precipitation is low, which makes wind an important sediment transporter. It is commonly believed too that wind processes played a major role in the mid-latitudes during the Pleistocene, especially during deglacial phases.

Evidence for this comes from wind-modifed pebbles called ventifacts, deflated stone pavements, frosted (matt surface appearance) and rounded sand grains and extensive loess sediments. The reasons for wind being such a dominant factor include the following: there were stronger winds and pressure gradients associated with periglacial areas close to continental ice sheets; there was incomplete vegetation cover; and extensive braided

Figure 21.34 Morphological classification of deformation structures in terms of symmetry, amplitude : wavelength ratio and pattern of occurrence. Type 1 illustrates individual folds of small depth (amplitude) but large wavelength; type 2 fairly regular, symmetrical and intensely convoluted forms with amplitudes of 0.6–2.0 m; type 3 similar to type 2 but with much smaller amplitudes; type 4 solitary 'teardrop' or diapiric forms; type 5 upwards-injected sediment in cracks; type 6 irregular deformation structures. (Source: after Vandenberghe, 1988)

stream deposits occurred, with both coarse and fine sediments at the margins of the ice sheets, and the drying of these surface sediments fostered the development of aeolian periglacial processes.

The wind can move large amounts of fine sediment, especially silts, which do not have the same cohesion as clays, yet are easier to entrain than sands because of the finer grain size. They are also common in a periglacial weathering regime because of the predominance of mechanical break-up of the rocks. The result can be the formation of dunes, thin sheets of wind-blown sediment which over time can collectively build up into large thicknesses of loess sediment. Examples can be found in northern China (200–300 m), the mid-western parts of the United States of America and the cover sands noted in western Europe,

for example in Holland and the Shirdley Hill Sands in south-west Lancashire. However, today wind action in a periglacial climate is of relatively minor importance, is small scale and is localized.

There are nevertheless indirect effects too, such as the part that wind plays in redistributing snow in winter and its role in causing evaporation and latent heat loss from exposed slopes during summer, which can influence the operation of gelifluction and other active-layer processes.

Ventifacts occur where a gravel clast projects above the ground and is abraded by strong winds carrying fine sediment. The pebble eventually becomes grooved and faceted on its wind-exposed side, and is sometimes flipped over by the wind, which then abrades

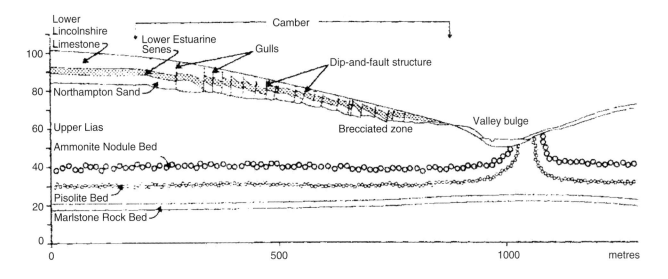

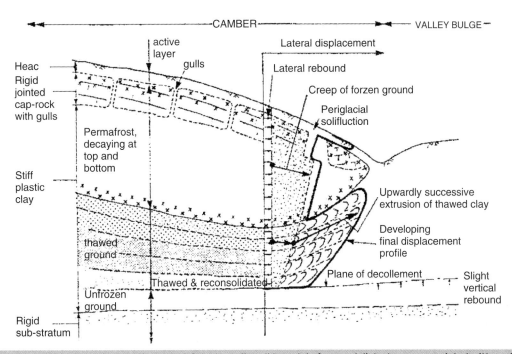

Figure 21.35 Cambering and valley bulging: (a) at Gwashe valley; (b) model of ground disturbance associated with periglacial cambering. (Sources: (a) after Horswill and Horton, 1976; (b) after Hutchinson, 1991)

another face to give a three-sided pebble, or dreikanter. These have been reported from ice-free Peary Land in north-east Greenland and northern Siberia but seem relatively rare in the Canadian Arctic. Fossil Pleistocene examples have been reported from the Cheshire Plain.

Deflation, or the winnowing out of the fines from sediments and their transportation, occurs in unvegetated areas and can produce a lag gravel or desert pavement at the surface and erode shallow basins or depressions. This process also contributes to the niveo-aeolian deposits, which are wind-transported fine sediments incorporated in snowbanks.

Aeolian silt or loess is probably the most widespread result of wind action. It is buff or grey in colour, homogeneous and unstratified, and relatively well sorted. It was formed in essentially dry conditions. This is known because, if wetted, the loess undergoes shrinkage and compaction, proving that it has not been soaked by water or deposited from it. It usually also has calcareous nodules when unweathered and is therefore unleached.

Dune fields and sand sheets today are relatively rare in periglacial regions outside of northern Alaska, parts of the Canadian Arctic and northern Scandinavia. However, there are extensive Pleistocene mid-latitude examples from Europe

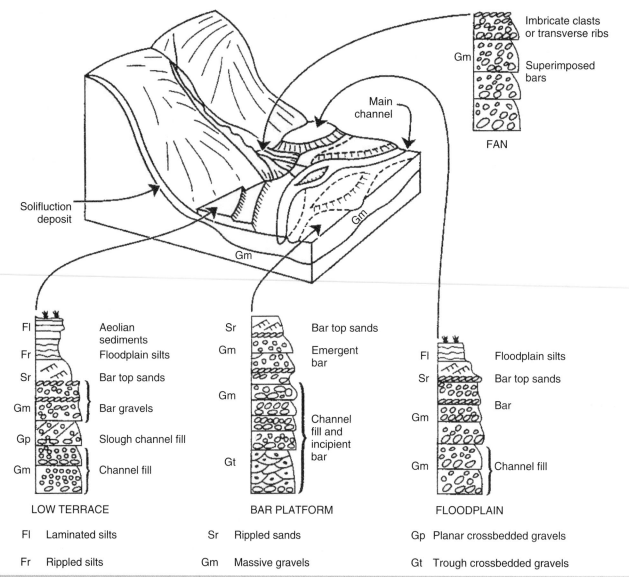

Figure 21.36 Model of periglacial alluviation. (Source: after Bryant, 1982)

and North America. Koster (1988) reviewed the characteristics that indicated a cold, periglacial origin for these aeolian sands. These included: interfingering or conformable relationships with glacial and/or glaciofluvial sediments; evidence for snow melt-water activity and features due to melting and sublimation from niveo-aolian sediments; and incorporation of plant, pollen or insect remains indicative of cold environments and structures of permafrost or seasonal frost origin, such as thermal contraction cracks and ice-wedge casts. However, modern descriptions and analyses of niveo-aolian sand deposits (for example Koster and Dijkmans (1988), from the Great Kobuk Sand Dunes in north-western Alaska) are still relatively rare. Aeolian sands of Pleistocene age often form a mantling deposit, called cover sand, but on occasion they form dunes, as in the Vale of York, where they are up to 6 m high. In Europe and North America they were located immediately outside the ice limits and were usually sourced from unvegetated glacial outwash plains, till plains

and lake shorelines, as in south-west Lancashire, where Wilson *et al.*'s (1981) analysis invoked a source from the present-day Irish Sea basin, with aeolian reworking of glaciofluvial sands. Wind direction can sometimes be inferred from dune morphology and orientation, the dip of the foreset beds and their relationships to known source areas. So, for example, the wind direction in Holland during Late Glacial times was predominantly north-westerly in the Early Dryas period but changed to south-west and westerly directions in the Late Dryas.

21.15 Fluvial processes in a periglacial environment

For much of the year water is stored as snow and ice, but it is released with extremely high discharges during the early part of

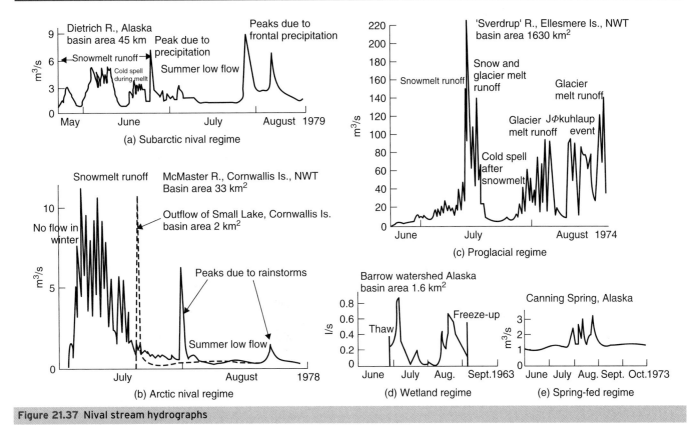

Figure 21.37 Nival stream hydrographs

Table 21.2 Stages of periglacial cambering. (Source: after Hutchinson, 1991)

The following stages are envisaged:

1. Initial incision of the river leading to unloading and valley rebound

2. Permafrost develops, with ice segregation processes increasing ice contents of the frozen clay

3. Valley-ward creep of the frozen clay in response to lateral stresses. This causes extension and initial cambering of caprocks.

4. Permafrost degradation due to large-scale climate warming leads to thaw from the surface downwards, enhanced surficial solifluction, and displacement of caprock valley wards over the thaw softened clays beneath

5. Thawing of the permafrost base due to geothermal heat flux, thaw consolidation and high pore pressures within an effectively confined thawed stratum.

6. Lateral extrusion of the thawed clay at the permafrost base leading to compression along the valley axis and pronounced increase in the valley bulge structure.

the snowmelt season. In the Canadian and Russian Arctic, ice jam floods are typical, where the melt proceeds from the headwaters downstream. Here the flood water passing over ice tends to erode laterally to produce wide, shallow, braided channels. This channel shape is typical because of the fluctuating discharge and highly variable sediment loads, with much sediment available during the melt season. Bank erosion of the frozen sediments involves both hydraulic and thermal erosion, whilst the smaller streams tend to have a diurnal cycle with freezing nightly temperatures.

The lateral thermal erosion produces a thermo-erosional niche. This process is important because it provides much sediment for bedload transport and downstream deposition as the bank collapses. It aids the development of a characteristic flat and braided valley bottom and helps explain the relative efficiency of lateral stream corrosion and subsequent stream migration in permafrost regions. Sometimes the channels can be beaded with deep pools, which develop when a stream passes over a network of ice wedges. This is because the flowing water causes the ice wedges to melt, producing the pools.

As permafrost restricts downward percolation, the basin runoff has a rapid response to snowmelt and any rainfall events. All the different types of runoff regime are dominated by the rapid snowmelt peak in the short winter–summer transition and are called nival (Figure 21.37). After that peak the discharge steadily decreases as less and less snow remains to be melted. The runoff season can be divided into four stages: break-up; the snowmelt or nival flood; late summer; and freeze-back. Large, deep rivers such as the Mackenzie River in northern Canada can maintain their discharge throughout the year; the Mackenzie rises to the south of the continuous permafrost zone and is fed by discharge from the Great Slave Lake. On the other hand, in much smaller catchments the spring snowmelt period is of much greater importance; for example, Umingmak Creek on Ellesmere Island discharges around 95% of its volume in the first two weeks of its annual summer flow (Marsh and Woo, 1981).

Valley asymmetry is thought to be a characteristic of some periglacial stream valleys. In the high Arctic (east Greenland, Siberia, Alaska) the steeper slopes commonly face north and are perennially frozen, whilst the south-facing slopes have an active layer so are lowered by gelifluction processes and there is asymmetric lateral stream corrosion. However, in extreme high-Arctic sites, like Spitsbergen (78° N) and north-west Banks Island (74° N) in the Canadian Arctic, no such regularity exists and instead steeper slopes can face south-west, west, south or east. Pleistocene asymmetric valleys have also been noted throughout Europe and parts of North America, for example in the Chilterns and Downs in England, in Holland, in Belgium and in Germany. In these areas the steeper slopes face towards the west or south-west, unlike the modern high-Arctic areas. The most popular explanation involves differential insolation and freeze–thaw processes operating on exposed south- and west-facing slopes while the colder north and north-east slopes remained frozen.

The characteristics of braided periglacial rivers have been studied by Bryant (1982, 1983) from Spitsbergen and Banks Island in the Canadian Arctic. The Spitsbergen catchments are partly glacierized but the Sachs River on Banks Island has a truly Arctic nival regime. A depositional model to illustrate such regimes is illustrated in Figure 21.36, where two important elements should be noted: first, the accumulation of aeolian deposits on inactive bar surfaces and terraces away from the active channel, and second, the development of ice-wedge polygons on such inactive surfaces. The wind-blown silt is blown around during the summer when river flow is low and is then redeposited as laminated fine sands and silts, forming a cover over the gravel bar surfaces. Ice-wedge growth followed by subsequent casting and burial may result in the formation of intraformational ice-wedge casts that mark migrations of active channels rather than climatic fluctuations.

In general during the spring nival flood the rivers have high competence and high capacity, and most periglacial rivers receive a large supply of coarse sediment through frost-weathering and mass-movement processes down slopes. Therefore these rivers are bedload-dominated. Such river systems occupy wide, shallow channels with rapid deposition of gravel bars during floods. During waning discharge, bars break the surface and the river is braided, and finally during the waning flows progressively finer-grained sediments are deposited so that the bars are composed of stacked, fining-upwards sequences, often truncated by the next season's flood.

In most of the large river valleys south of the last, Devensian ice limit in Britain there are fluvioperiglacial sediments along the lower valley floors, and the Holocene river activity has only trimmed these sediments and deposited a thin, alluvium veneer across them. There may be erratics in the sediments derived from ice further up the catchment but in the modern Arctic there is a continuum between catchments receiving both glacial melt water and snowmelt and those that have no glacial input and are dominated by snowmelt. There have been attempts to quantify the maximum palaeodischarges from some of these Devensian braided rivers in southern England (Cheetham, 1980; Clark and Dixon, 1981). For example, the mean annual flood discharge of the River Kennet during the deposition of the Beenham Terrace was probably between 153 and 418 $m^3\,s^{-1}$, compared to the modern Kennet's mean annual flood discharge of 40 $m^3\,s^{-1}$. For the first and third terraces of the River Blackwater the peak palaeodischarges were between 19 and 37 times the maximum flow recorded for that river today.

21.16 Alluvial fans in a periglacial region

The sedimentological and geomorphological development of alluvial fans has been described from many climatic regimes. However, it would be expected that the seasonal variations and distinct characteristics of the periglacial climate would have a major impact on fan sedimentation and make alluvial fans in these areas distinctive. This has been found to be the case by Catto (1993), who found that the alluvial fans along the eastern escarpment of the Aklavik Range (between the Peel Plateau and the Mackenzie Delta, Northwest Territories) were formed by a combination of subaerial mass movement and fluvial processes. The relative importance of these processes depends on the topographic setting, the type of sediment available and the degree of fan activity. The Mount Goodenough north fan (Figure 21.38) developed primarily through snow-induced debris flow and gelifluction, with fluvial activity playing only a minor role, confined to the fan head. The area is semi-arid, with summer water influx generated largely through snowmelt and augmented locally through degradation of segregated ice during the summer. This type of ice develops preferentially in silty sediment and therefore fans built from silty bedrock are dominated throughout their lengths by debris flows and gelifluction sediments. The aridity of the climate limits the fluvial activity but in itself does not represent the controlling factor in fan morphology and sedimentation, which is the composition and texture of the source material.

In different periglacial climatic regimes the climatic controls can be very different, yet the processes can be similar. For example, Larsson (1982) described the slope evolution in Longyeardalen in Spitsbergen after the rainstorm of 10–11 July 1972. This caused debris flows too. Some mass movements started as slides on plane slopes and continued into flows, creating parallel levées and terminal debris-flow lobes. Other debris flows were associated with pre-existing drainage channels. Small, narrow gullies were deepened by fluvial erosion and in deeply-eroded gullies lateral sediment slumps were removed completely and all the eroded sediment was deposited on older cones. The rainfall intensity

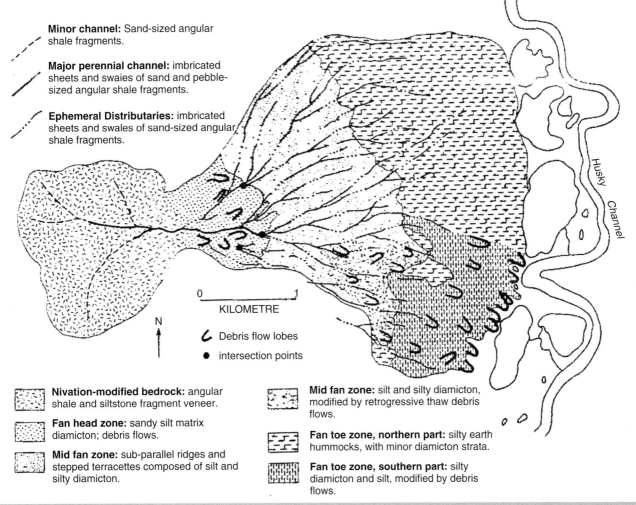

Minor channel: Sand-sized angular shale fragments.

Major perennial channel: imbricated sheets and swaies of sand and pebble-sized angular shale fragments.

Ephemeral Distributaries: imbricated sheets and swales of sand-sized angular shale fragments.

0   1
KILOMETRE

N

⌇ Debris flow lobes

● intersection points

Nivation-modified bedrock: angular shale and siltstone fragment veneer.

Fan head zone: sandy silt matrix diamicton; debris flows.

Mid fan zone: sub-parallel ridges and stepped terracettes composed of silt and silty diamicton.

Mid fan zone: silt and silty diamicton, modified by retrogressive thaw debris flows.

Fan toe zone, northern part: silty earth hummocks, with minor diamicton strata.

Fan toe zone, southern part: silty diamicton and silt, modified by debris flows.

Husky Channel

Figure 21.38 Geomorphological features and sedimentary deposits, Mount Goodenough north fan, Aklavik Range escarpment. (Source: after Catto, 1993)

combined with the existence of an impermeable permafrost table was of great importance as the trigger mechanism, and slope failures appeared when the rainfall intensity reached 2.5 mm per hour, which was abnormally high. Hence these examples show that fluvial erosion and deposition vary in different periglacial regimes, depending on topography, aspect, bedrock control and the rainfall amount and intensity.

The Chelford Sands of eastern Cheshire were deposited in a broad alluvial fan complex sourced from the Pennines (Worsley, 1970). Here fluvial deposition occurred in nival, shallow, braided rivers, but wind action added an aeolian component to the sedimentary succession. The sequence also incorporates organic deposits of Early Devensian age (OIS 4), and numerous intraformational ice-wedge casts indicate the presence of permafrost during fan accumulation (Worsley, 1966). Hence such periglacial structures are of critical importance in the correct interpretation of relict fluvial sediments deposited in a continuous permafrost zone.

21.17 An overview of the importance of periglacial processes in shaping the landscape of upland Britain

Climatic change has affected the importance of particular processes in the past and there have been several cold-climate phases when certain periglacial processes would have been far more prevalent than today. The Little Ice Age of approximately 1500–1800 AD is one example. Pollen Zone 111 (c. 10 300–10 800 BP, the Loch Lomond Stadial), when there was a relatively extensive ice cap in Scotland and small corrie glaciers in most other British upland areas, was much more important, and the range of processes and landforms developed in the British uplands in this period can be seen in Figures 21.39 and 40. During the glacial phases of the Devensian (13 000–26 000 BP) and earlier glacial phases too there would have been nunatak zones in the

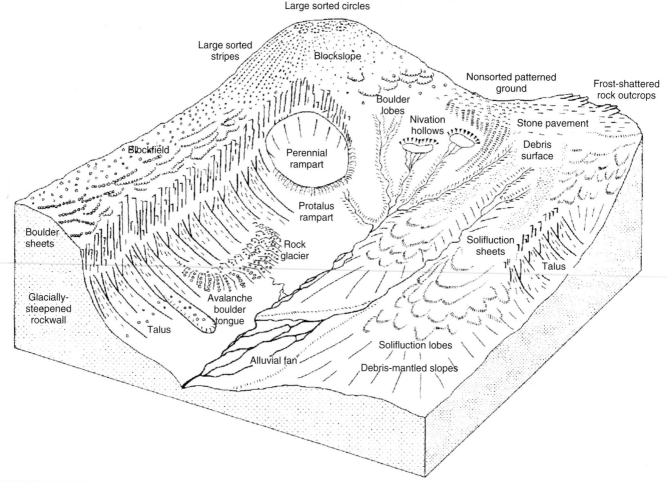

Figure 21.39 The range of Late Devensian periglacial features on Scottish mountains. Most were last active during the Loch Lomond Stadial. (Source: after Ballantyne, 1984)

highest British mountains where periglacial processes were prevalent, and outside the ice sheets there would have been extensive periglacial areas in southern Britain and in the offshore areas of the North and Irish Seas. Some areas of Britain must have been more prone to these periglacial processes, such as Dartmoor and the South Downs, which were always south of the ice sheets, parts of the Pennines adjacent to the ice sheets and even the Cairngorms in eastern Scotland, which, although glaciated, were to the east of the most erosive part of the Highland ice sheet.

In these areas we have already seen that tors and extensive gelifluction sheets developed. Some periglacial processes are still active on the highest ground, for example those illustrated in Figure 21.40. This is because the present climatic conditions on British mountains lie between the extremes of the climatic optimum of the Atlantic period and the Little Ice Age. The climate is strongly maritime, with steep lapse rates, high humidity, strong winds and limited seasonal variations in cloudiness, precipitation and temperature. Weather records are not common, but short records have been obtained from Ben Nevis, the northern Pennines, the Cairngorms, Wales and the Southern Uplands. The

characteristics that are of importance with regard to periglacial processes can be summarized as follows:

- The absence of extreme cold. Screen temperatures are rarely below −10 °C.

- An average of 30−40 air freezing cycles per year.

- An absence of permafrost, but at higher altitudes there may be an annual ground freezing cycle of 2−6 months' length, although the ground can thaw completely at any time during the winter due to incursions of warm maritime air.

- A maximum recorded freezing depth of 0.5 m, though freezing to greater depths is theoretically possible.

- An increase in duration of winter snow-lie approximately linear with altitude. In the Scottish Highlands average snow-lie (over 50% cover) generally exceeds 100 days per year at 600 m and ranges from about 150 to 180 days per year at 900 m. The equivalent figures are 105 days per year for the Cross Fell

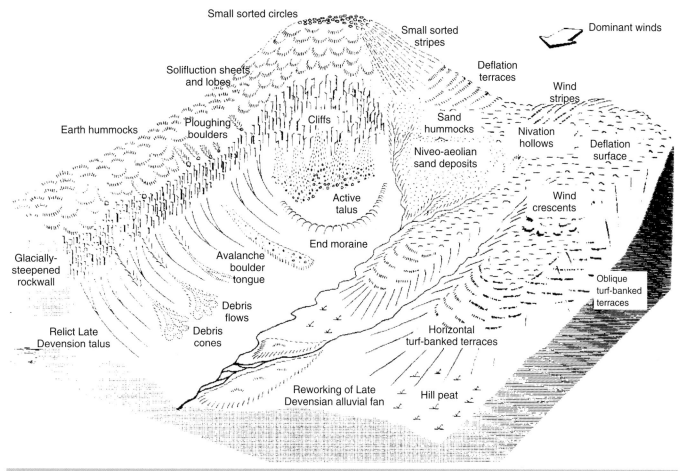

Figure 21.40 The range of active periglacial features on British mountains. (Source: after Ballantyne, 1987)

summit in the northern Pennines and 70 days per year for Snowdon at 900 m. Persistent snow patches can survive for several years before melting, as on Obervatory Gully on Ben Nevis and at 1075 m in a deep corrie on Braeriach in the Cairngorms.

• A strong and persistent wind, which reflects the concentration and acceleration of airflow as it passes over mountain barriers, with averages of 55 km per hour on Ben Nevis and 36 km per hour on Great Dun Fell in the northern Pennines. Mountain storms individually have much higher wind speeds, and on Cairngorm summit for example from 1979 to 1987 the maximum 24 hour wind speed recorded each year was between 93 and 146 km per hour. The strongest gust in the same period ranged from 177 to 275 km per hour.

The British mountains therefore reflect a distinctive maritime, periglacial regime that is very different from those of Arctic, sub-Arctic or Alpine areas, where there are much greater extremes of cold. The periglaciation of upland Britain is considered in detail in Ballantyne and Harris (1994).

There is a current altitudinal zonation of periglacial landforms in the uplands and it follows that there is likely to be a palaeoaltitudinal effect too. However, topography plays a role in the siting of periglacial landforms, so on high mountain summit areas there are blockfields and the sorting of soils by frost. On north-east- and east-facing slopes, aspect allows the favoured preservation of late-lying snow patches and nivation processes, and this too allows a control on valley asymmetry. There are also important drainage controls on landform distribution in the uplands, with, for example, hydraulic, open-system pingo development at the base of upland slopes and thufurs usually located on impervious rock types.

The question of whether rock control is important in shaping how the landscape is modifed by periglacial processes in upland areas can usually be answered in the affirmative. Hard bedrocks give rise to screes from rockfall processes and freeze–thaw break-up of rock on summit plateaux gives rise to blockfields. There is also the importance of jointing in controlling the location of water penetration and hence the freezing of water in the rock, and it is apparent that the effects of joints in granite controlled the development of tors, whichever mechanism for tor development one believes. The bedrock geology can influence the operation of periglacial processes, for example where there is the interbedding of sandstone and thin shales in the Pennines, or where there is an adjacent rock type difference, as around the Dartmoor granite.

The different rock types and their influence on water permeability can be important too. Small-scale fractures and cleavage planes in shale, slate and mudstone allow frost to break up these rocks more easily and allow slope processes like gelifluction to infill the topographic depressions. The scenery is hence made less rugged, as in the Skiddaw Slate country in the Lake District or the Isle of Man, the Silurian rocks of the Howgill Fells and the areas dominated by slates in mid- and north Wales. Not only are there primary periglacial slope sediments in these areas but periglacially-transported glacial till has moved downslope too.

It is likely that the association of frost sorting/freeze–thaw/ snow processes and gelifluction is unique to the upland areas in Britain, compared with most lowland areas. Mechanical processes operated here, as opposed to chemical breakdown of the rocks, and these supplied a lot of sediment to the upland rivers, which helps to partly explain the upland braided river channel pattern.

Periglacial processes operated to supply sediment for glacial processes and helped to shape some glacial processes. So, for example, there is the preparation and break-up of rock prior to glacial transport, the supply of sediment for lateral moraines and the supply of sediment for glacial sediment such as till. Certain periglacial processes operated in conjunction with glacial processes to form corrie backwalls, arêtes and pyramidal peaks, and snow processes (nivation) may be preparatory to corrie development. However, it may also be difficult to distinguish some periglacial landforms from glacial ones, as in the case of large nivation hollows and true corries, and at times it is difficult to distinguish between protalus ramparts and true glacial moraines.

There is no doubt that periglacial processes are one major group of geomorphic processes that are currently shaping the British upland landscape, particularly in the Scottish Highlands and on the mountain tops of the Lake District and Snowdonia. In the more cold-climate phases of the past, these processes would

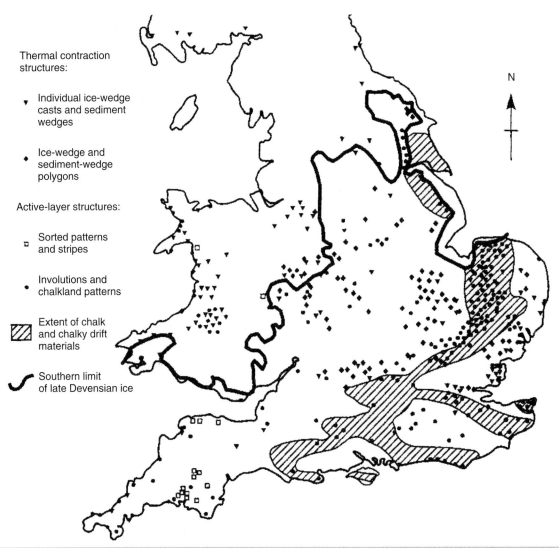

Figure 21.41 Periglacial phenomena and the last glaciation limit in lowland Britain. (Source: after Rose *et al.*, 1985)

have been more extensive and large areas of the British uplands, particularly areas that were not extensively glaciated, would have also suffered from a suite of periglacial processes.

21.18 The periglaciation of lowland Britain

This area generally lies below 400 m in altitude, has few steep slopes, is underlain by sedimentary rocks and is extensively mantled by drift deposits. The area south of the last glaciation is shown in Figure 21.41, but the area south of a line drawn between the Bristol Channel and Essex never experienced glaciations, though it does represent a relict periglacial landscape, having been shaped under climatic conditions much more severe than those of the present day. The relict processes and landforms (some shown in Figure 21.41) that are largely or entirely absent from the uplands include frost cracking and the development of ice-wedge casts, sand wedges, tundra polygons, cryoturbation structures such as involutions, thermokarst landforms and pingo scars (although there are some in mid-Wales, the Isle of Man and in south Cumbria), together with aeolian loess deposits and large-scale slope structures such as cambering. However, some processes and landforms overlap with those in the uplands, in particular frost sorting and the development of patterned ground and gelifluction, which occur in both areas. Again the processes and landform development in this lowland area are discussed extensively in Ballantyne and Harris (1994) and the periglacial legacy in southern England, especially the Chalk landscapes and East Anglia are discussed in French (1996). Box 21.3 in the companion website illustrates Devensian cold stage palaeoenvironmental reconstruction using periglacial and other proxy evidence whilst Box 21.4 illustrates Late PreCambrian and Late Ordovician-early Silurian periglacial structures in West and South Africa.

Exercises

1. Describe the general characteristics of a periglacial environment.

2. Describe the landform–sediment assemblage associated with frost wedging.

3. Explain how frost sorting in soils takes place.

4. What are the characteristics of ice-wedge casts and how do these casts form?

5. Explain the differences between open- and closed-system pingo development.

6. Describe the morphological and sedimentological characteristics of protalus ramparts. How can you distinguish them from glacial moraines?

7. Explain how tors can form and why there has been major controversy as to the climatic significance of such landforms.

8. Why is the active layer significant in the formation of periglacial slope landforms?

9. How have detailed field and laboratory measurements been important in explaining gelifluction processes?

10. Explain the importance of periglacial processes in shaping the landscape in upland Britain.

References

Ballantyne, C.K. 1984. The Late Devensian periglaciation of upland Scotland. *Quaternary Science Reviews.* **3**, 311–343.

Ballantyne, C.K. 1987. The present-day periglaciation of upland Britain. In: Boardman, J. (Ed.). Periglacial Processes and Landforms in Britain and Ireland. Cambridge, Cambridge University Press. pp. 113–126.

Ballantyne, C.K. 2001. Measurement and theory of ploughing block movement. *Permafrost and Periglacial Processes.* **12**, 267–288.

Ballantyne, C.K. and Harris, C. 1994. The Periglaciation of Great Britain. Cambridge, Cambridge University Press.

Ballantyne, C.K. and Kirkbride, M.P. 1986. The characteristics and significance of some lateglacial protalus ramparts in upland Britain. *Earth Surface Processes and Landforms.* **11**, 659–671.

Ballantyne, C.K. and Kirkbride, M.P. 1987. Rockfall activity in upland Britain during the Loch Lomond Stadial. *Geographical Journal.* **153**, 86–92.

Brown, R.J.E. 1970. Permafrost in Canada. Toronto, University of Toronto Press.

Bryant, I.D. 1982. Periglacial river systems: Ancient and modern. Unpublished PhD thesis. University of Reading.

Bryant, I.D. 1983. The utilization of arctic river analogue studies in the interpretation of periglacial river sediments from southern Britain. In: Gregory K.J. (Ed.). Background to Paleohydrology. Chichester, John Wiley & Sons, Inc. pp. 413–431.

Büdel, J. 1982. Climatic Geomorphology. Princeton, NJ, Princeton University Press.

Burn, C.R. 1990. Implications for palaeoenvironmental reconstruction of recent ice-wedge development at Mayo, Yukon Territory. *Permafrost and Periglacial Processes.* **1**, 3–14.

Caine, N. 1967. The tors of Ben Lomond, Tasmania. *Zeitschrift für Geomorphologie.* **11**, 418–429.

Catto, N.R. 1993. Morphology and development of an alluvial fan in a permafrost region, Aklavik Range, Canada. *Geografiska Annaler*. **75**, 83–93.

Cheetham, H. 1980. Late Quaternary palaeohydrology: The Kennet Valley case study. In: Jones, D.K.C. (Ed.). Shaping of Southern England. London, Academic Press. pp. 203–223.

Clark, M.R. and Dixon, A.J. 1981. The Pleistocene braided river deposits in the Blackwater Valley area of Berkshire and Hampshire, England. *Proceedings of the Geologists' Association*. **92**, 139–157.

Corte, A.E. 1966. Particle sorting by repeated freezing and thawing. *Biuletyn Peryglacjalny*. **15**, 175–240.

Demek, J. 1964. Castle koppies and tors in the Bohemian Highland (Czechoslovakia). *Biuletyn Peryglacjalny*. **14**, 195–216.

Demek, J. 1969. Cryogene processes and the development of cryoplanation terraces. *Biuletyn Peryglacjalny*. **18**, 115–125.

Derbyshire, E. 1972. Tors, rock weathering and climate in southern Victoria Land, Antarctica. In: Polar Geomorphology. Institute of British Geographers, Special Publication No. 4. pp. 93–105.

Doornkamp, J.C. 1974. Tropical weathering and the ultra-microscopic characteristics of regolith quartz on Dartmoor. *Geografiska Annaler*. **56**, 73–82.

Douglas, G.R. 1980. Magnitude frequency study of rockfall in Co. Antrim, N-Ireland. *Earth Surface Processes and Landforms*. **5**, 123–129.

Eden, M.J. and Green, C.P. 1971. Some aspects of granite weathering and tor formation, Dartmoor, England. *Geografiska Annaler*. **53**, 92–99.

Eichler, H. 1981. Rock temperatures and insolation weathering in the high Arctic: Oobloyak Bay, northern Ellesmere Island, N.W.T., Canada. In: Results of the Heidelberg Ellesmere Island Expedition 69. Heidelberg Geographisches Institut Universitat Heidelberg Arbeiten (1981). pp. 441–464.

Ford, T.D. 1967. Deep weathering, glaciation and tor formation in Charnwood Forest, Leicestershire. *Mercian Geologist*. **2**, 3–14.

French, H.M. 1996. The Periglacial Environment. 2nd edition. Harlow, Addison Wesley Longman.

French, H.M. 2000. Does Lozinski's periglacial realm exist today? A discussion relevant to modern usage of the term 'periglacial'. *Permafrost and Periglacial Processes*. **11**, 35–42.

French, H.M. and Harry, D.G. 1992. Pediments and cold-climate conditions, Barn Mountains, unglaciated northern Yukon, Canada. *Geografiska Annaler*. **74**, 145–157.

Gardner, J. 1969. Snowpatches: Their influence on mountain wall temperatures and the geomorphic implications. *Geografiska Annaler*. **51**, 114–120.

Gerrard, A.J. 1988. Periglacial modification of the Cox Tor–Staples Tors area of western Dartmoor, England. *Physical Geography*. **9**, 280–300.

Grab, S. 2005. Aspects of the geomorphology, genesis and environmental significance of Earth hummocks (thúfur, pounus): Miniature cryogenic mounds. *Progress in Physical Geography*. **29**, 139–155.

Gurney, S.D. 1998. Aspects of the genesis and geomorphology of pingos: Perennial permafrost mounds. *Progress in Physical Geography*. **22**, 307–324.

Hall, A.M. 1985. Cenozoic weathering covers in Buchan, Scotland, and their significance. *Nature*. **325**, 392–395.

Hall, A.M. 1986. Deep weathering patterns in north-east Scotland and their geomorphological significance. *Zeitschrift für Geomorphologie*. **30**, 407–422.

Hall, K. 1980. Freeze–thaw activity at a nivation site in northern Norway. *Arctic and Alpine Research*. **12**, 183–194.

Hallet, B.S., Walder, J.S. and Stubbs, C.W. 1991. Weathering by segregation ice growth in microcracks at sustained subzero temperatures: Verification from an experimental study using acoustic emissions. *Permafrost and Periglacial Processes*. **2**, 283–300.

Harris, C. and Lewkowicz, A.G. 2000. An analysis of the stability of thawing slopes, Ellesmere Island, Nunavut, Canada. *Canadian Geotechnical Journal*. **37**, 449–462.

Harry, D.G. and Gozdik, J.S. 1988. Ice wedges: Growth, thaw transformation, and paleoenvironmental significance. *Journal of Quaternary Science*. **3**, 39–55.

Horswill, P. and Horton, A. 1976. Cambering and valley bulging in the Gwash valley at Empingham, Rutland. *Philosophical Transactions of the Royal Society*. **A283**, 451–461.

Hutchinson, J.N. 1991. Periglacial slope processes. In: Forster, A., Culshaw, M.G., Cripps, J.C., Little, J.A. and Moon, C.F. (Eds). Quaternary Engineering Geology. Geological Society, Engineering Geology Special Publication 7. pp. 283–331.

Jahn, A. 1974. Granite tors in the Sudeten Mountains. Institute of British Geographers, Special Publication 7. pp. 53–61.

Karte, J. 1979. Raumliche abgrenzung und regionale differenzierung des periglaziärs. *Bochumer Geographische Arbeiten*. **35**.

King, R.B. 1972. Lobes in the Cairngorm Mountains, Scotland. *Biuletyn Peryglacjalny*. **21**, 153–167.

Koster, E.A. 1988. Ancient and modern cold-climate aeolian sand deposition: A review. *Journal of Quaternary Science*. **3**, 69–83.

Koster, E.A. and Dijkmans, J.W.A. 1988. Niveo-aeolian deposits and denivation forms, with special reference to the Great Kobuk Sand Dunes, Northwestern Alaska. *Earth Surface Processes and Landforms*. **13**, 153–170.

Lachenbruch, A. 1962. Mechanics of thermal contraction cracks and ice-wedge polygons in permafrost. Geological Society of America, Special Paper 70.

Larsson, S. 1982. Geomorphological effects on the slopes of Longyear valley, Spitsbergen, after a heavy rainstorm in July 1972. *Geografiska Annaler*. **64**, 105–125.

Lewis, W.V. 1939. Snow-patch erosion in Iceland. *Geographical Journal*. **94**, 153–161.

Linton, D.L. 1955. The problem of tors. *Geographical Journal*. **121**, 470–487.

Linton, D.L. 1964. The origin of the Pennine tors: An essay in analysis. *Zeitschrift für Geomorphologie*. **8**, 5–23.

Lozinski, W. von. 1909. Über die mechanische Verwitterung der Sandsteine im gemässigten Klima. *Bulletin International de l' Académie des Sciences de Cracovie class des Sciences Mathématique et Naturelles*. **1**, 1–25.

Lozinski. W. von. 1912. Die perig;aziale Fazies der mechanischen Verwitterung. Comptes rendus, X1 Congrès Internationale Géologie Stockholm 1910. pp. 1039–1053.

Mackay, J.R. 1979. Pingos of the Tuktoyaktak Peninsula area, Northwest Territories. *Géographie physique et Quaternaire*. **33**, 3–61.

Mackay, J.R. 1986. Growth of Ibyuk Pingo, western Arctic coast, Canada and some implications for environmental reconstructions. *Quaternary Research*. **26**, 68–80.

Mackay, J.R. 1989. Ice-wedge cracks, western Arctic coast. *Canadian Geographer*. **33**, 365–368.

Mackay, J.R. 1992. The frequency of ice-wedge cracking (1967–1987) at Garry Island, western Arctic coast, Canada. *Canadian Journal of Earth Sciences*. **29**, 236–248.

Marsh, P. and Woo, M. 1981. Snowmelt, glacier melt and high arctic streamflow regimes. *Canadian Journal of Earth Sciences*. **18**, 1380–1384.

Matthews, J.A., Dahl, S.O., Berrisford, M.S. and Nesje, A. 1997. Cyclic development and thermokarstic degradation of palsas in the Mid Alpine zone at Lierpullan, Dovrefjell, Southern Norway. *Permafrost and Periglacial Processes*. **8**, 107–122.

Mellor, M. 1970. Phase composition of pore water in cold rocks. United States Army Corps of Engineers, Cold Regions Research and Engineering Laboratory Research Report 292.

Murton, J.B. and Kolstrup, E. 2003. Ice-wedge casts as indicators of palaeotemperatures: Precise proxy or wishful thinking? *Progress in Physical Geography*. **27**, 155–170.

Murton, J.B., Worsley, P. and Gozdzik, J. 2000. Sand veins and wedges in cold aeolian environments. *Quaternary Science Reviews*. **19**, 899–922.

Nelson, F.E., Hinkel, K.M. and Outcalt, S.I. 1992. Palsa-scale frost mounds. In: Dixon, J.C., and Abrahams, A.D. (Eds). Periglacial Geomorphology. Binghamton Symposium in Geomorphology: International Series No. 22. Chichester, John Wiley & Sons, Inc. pp. 305–325.

Oxford, S.P. 1985. Protalus ramparts, protalus rock glaciers and soliflucted till in the northwest part of the English Lake District. In: Boardman, J. (Ed.). Field Guide to the Periglacial Landforms of Northern England. Cambridge, Quaternary Research Association. pp. 38–46.

Palmer, J.A. 1956. Tor formation at the Bridestones in NE Yorkshire and its significance in relation to problems of valley-side development and regional deglaciation. *Transactions of the Institute of British Geographers*. **22**, 55–71.

Palmer, J.A. and Neilson, R.A. 1962. The origin of granite tors on Dartmoor, Devonshire. *Proceedings of the Yorkshire Geological Society*. **33**, 315–339.

Palmer, J.A. and Radley, J. 1961. Gritstone tors of the English Pennines. *Zeitschrift für Geomorphologie*. **5**, 37–52.

Péwé, T.L. 1966. Ice-wedges in Alaska: Classification, distribution, and climatic significance. In: Permafrost International Conference, Lafayette, Indian, 1963. Proceedings of the National Academy of Sciences. National Research Council Publication 1287. pp. 76–81.

Péwé, T.L. 1991. Permafrost. In: Kiersch, G.A. (Ed.). The heritage of engineering geology: The first hundred years. Geological Society of America Centennial Special Volume 3. pp. 277–298.

Pissart, A. 2000. Remnants of lithalsas of the Hautes Fagnes, Belgium: A summary of present-day knowledge. *Permafrost and Periglacial Processes*. **11**, 327–355.

Potts, A.S. 1970. Frost action in rocks: Some experimental data. *Transactions of the Institute of British Geographers*. **49**, 109–124.

Rose, J., Boardman, J., Kemp, R.A. and Whiteman, C.A. 1985. Palaeosols and the interpretation of the British Quaternary stratigraphy. In: Richards, K.S., Arnett, R.R. and Ellis, S. (Eds.). Geomorphology and Soils. London, George Allen and Unwin. pp. 348–375.

Schunke, E. and Zoltai, S.C. 1988. Earth hummocks (thufur). In: Clark, M.J. (Ed.). Advances in Periglacial Geomorphology. Chichester, John Wiley & Sons, Inc. pp. 231–245.

Seppälä, M. 1988. Palsas and related forms. In: Clark, M.J. (Ed.). Advances in Periglacial Geomorphology. Chichester, John Wiley & Sons, Inc. pp. 247–278.

Sissons, J.B. 1980. The Loch Lomond Advance in the Lake District, northern England. *Transactions of the Royal Society of Edinburgh: Earth Sciences*. **71**, 13–27.

Smith, M.W. and Riseborough, D.W. 2002. Climate and the limits of permafrost: A zonal analysis. *Permafrost and Periglacial Processes*. **13**, 1–15.

Taber, S. 1929. Frost heaving. *Journal of Geology*. **37**, 428–461.

Thorn, C.E. 1988. Nivation: A geomorphic chimera. In: Clark, M.J. (Ed.). Advances in Periglacial Geomorphology. Chichester, John Wiley & Sons, Inc. pp. 3–33.

Thorn, C.E. and Hall, K. 1980. Nivation: An Arctic–Alpine comparison and reappraisal. *Journal of Glaciology*. **25**, 109–124.

Tufnell, L. 1969. The range of periglacial phenomena in Northern England. *Biuletyn Peryglacjalny*. **19**, 291–323.

Tufnell, L. 1972. Ploughing blocks with special reference to north-west England. *Biuletyn Peryglacjalny*. **21**, 237–270.

Tufnell, L. 1976. Ploughing block movements on the Moor House Reserve (England), 1965–1975. *Biuletyn Peyglacjalny*. **26**, 313–317.

Vandenberghe, J. 1988. Cryoturbations. In: Clark, M.J. (Ed.). Advances in Periglacial Geomorphology. Chichester, John Wiley & Sons, Inc. pp. 179–198.

Walder, J.S. and Hallet, B. 1985. A theoretical model of the fracture of rock during freezing. *Bulletin of the Geological Society of America*. **96**, 336–346.

Walder, J.S. and Hallet, B. 1986. The physical basis of frost weathering: Toward a more fundamental and unified perspective. *Arctic and Alpine Research*. **18**, 27–32.

Washburn, A.L. 1969. Patterned ground in the Mesters Vig district, Northeast Greenland. *Biuletyn Peryglacjalny*. **18**, 259–330.

Washburn, A.L. 1970. An approach to a genetic classification of patterned ground. *Acta Geographica Lodziensia*. **24**, 437–446.

Washburn, A.L. 1973. Periglacial Processes and Environments. London, Arnold.

Williams, P.J. 1959. An investigation into processes occurring in solifluction. *American Journal of Science*. **257**, 42–58.

Wilson, P., Bateman, R.M. and Catt, J.A. 1981. Petrology, origin and environment of deposition of the Shirdley Hill Sand of southwest Lancashire, England. *Proceedings of the Geologists' Association*. **92**, 211–229.

Worsley, P. 1966. Some Weichselian fossil frost wedges from East Cheshire. *Mercian Geologist*. **1**, 357–365.

Worsley, P. 1970. The Cheshire–Shropshire Lowlands. In: Lewis, C. A. (Ed.). The Glaciations of Wales. London, Longman. pp. 83–106.

Worsley, P. and Gurney, S.D. 1996. Geomorphology and hydrogeological significance of the Holocene pingos in the Karup Valley area, Traill Island, northern East Greenland. *Journal of Quaternary Science*. **11**, 249–262.

22 Aeolian (Wind) Processes and Landform-Sediment Assemblages

Rub Al Khali, seif dunes, arrow-like points upwind. Source: http://www.rst.gsfc.nasa.gov

Earth Environments   David Huddart and Tim Stott
© 2010 John Wiley & Sons, Ltd

Learning Outcomes

After reading this chapter and completing the exercises you will be able to:

➤ Understand the ways in which wind has shaped the Earth's surface by a range of erosional and depositional processes.

➤ Understand how wind systems are formed and the physical processes of wind transport.

➤ Understand the complexity of aeolian sand deposition and the range of bedforms produced.

➤ Describe the characteristics of dune sediments, including dust and loess.

➤ Understand how erosional aeolian landforms are created.

➤ Understand how aeolian rocks can be recognized in the geological record.

Figure 22.1 Sahara Desert satellite image. (Source: http://www. upload.wikipedis.org/wikipedia/commons/a/a7/sahara_satellite. hires.jpg)

22.1 Introduction

In studying aeolian processes the geomorphologist is trying to understand the ways in which the wind has shaped the Earth's surface by erosion, transport and deposition. The scale of these processes can vary from the entrainment of dust or individual sand particles to the deposition of the world's major sand seas (ergs). There is an applied aspect to this too as wherever protective vegetation is removed, especially by the effects of humanity's agriculture, the finer fraction of the soils can become susceptible to soil erosion by wind. Aeolian processes are also important as sedimentary agents in the geological record, as well as currently, because large quantities of sand can be wind-transported in continental deserts; beach sand supplied by longshore drift can be transported inland as coastal dune systems; and the finer-grained fraction, particularly silt, can be carried in suspension as dust storms away from the weathering source areas, and all of this can supply much of the deep-ocean, pelagic sediment. On outwash plains in proglacial environments and in periglacial regions, wind action can abrade rocks, build dunes and deposit thick loess sequences (up to 300 m thick in China) and thinner cover sands far from the glacial source area.

Aeolian processes are also thought to compete with rivers in the amounts of sediment transported, and in certain phases of the Pleistocene they probably played an even greater role. However, today arid-zone aeolian sands are volumetrically greater than the rest, and about a third from the Earth's present-day land area can be classified as semi-arid or arid, although there is discussion as to the exact percentage and the criteria for the definition of drylands (Thomas, 1997). This zone includes the modern desert systems, which occupy about 20% of the arid areas, although bedrock predominates. For example, only 10–20% of the Sahara is sand-covered (Figure 22.1) and in the Arabian peninsula the cover is about one third of the land area. Nevertheless, sand accumulation is a very important and distinctive desert landform, with the Grand Erg Occidental of Algeria occupying an area greater than that of England and the Rub 'al Khali occupying 560 000 km^2 in Arabia.

Hence if the world's deserts today are comparable to those of the geological past then extensive sand deposition will have existed. It is likely that these were more widespread before the colonization of dry land by hardy plant life. However, most aeolian sands are not derived directly from weathered bedrock: they come from fluvial sediments, coastal sands and, in Western Australia, from the highly-weathered upper part of lateritic profiles. Nevertheless, although they inherit some characteristics from the parent material, the characteristics of aeolian transport are imposed on the sands. Wind transport can generally cause an increase in sorting and a reduction in mean grain size in the direction of sand transport.

Desert dune and evaporite sediments in the geological record show that the Earth's aridity goes back into the Pre-Cambrian, with an early phase of dune sediments preserved in the 1.8 billion-year-old Hornby Bay Group in the Northwest Territories of Canada. However, differing plate tectonic configurations through geological time, mountain building and consequent climatic change have caused major changes in the extent and location of arid zones on the Earth. The oldest and most persistent current arid zone is thought to be the Namib, dating back to the Cretaceous at 80 million years, whilst many other deserts seem to date to the Tertiary. For example, deep-ocean core evidence off West Africa suggests aeolian sediment at 38 million years in the Oligocene, whilst other deserts, for example in Australia and the Atacama, seem to date from the Miocene. Quartz sand dominates because many desert sedimentary rocks are sandstones, which is attributed to the long geological periods of these areas, with a continental character and the continual loss of fine sediment during that time.

22.2 Current controls on wind systems

Wind results from differences in air pressure from region to region across the Earth. It generally moves the air from areas of high pressure to those of low pressure. The result is a global air circulation (see Chapter 5). The biggest of the tropical and subtropical deserts are dominated by the subtropical high-pressure systems and can be subject to relatively low-energy wind systems associated with the centres of these highs, but other areas face the relatively high-energy tropical easterly trade wind belts, which blow around the highs over half the planet. The intertropical convergence zone. which separates equatorial wet air from the dry air of the tropical and subtropical deserts, moves seasonally and so the position and strength of the subtropical high-pressure wind systems vary. The trade winds are strongest on land in winter when the highs are best developed, but these areas of tropical and subtropical desert are also affected by the monsoon drawn in towards the continents in summer. The mid-latitude deserts in each hemisphere are affected by the Ferrel westerlies, and these winds are most active in the winter and spring.

So the large-scale wind patterns are the reason for the large-scale patterns of aeolian activity. Superimposed on this global wind system are local winds caused by topography and the influence of the coast. Aeolian processes are also important in the polar regions, and mean wind speeds in Antarctica have been recorded at 2.2 m per second, whilst maximum speeds in excess of 80 m per second have been measured. In the Arctic Ocean up to 10% of the sediment is estimated to be of aeolian origin. However, the global wind systems have changed over geological time and there must have been both major and minor wind-belt shifts and changes in wind speeds. For example, the movement of ice sheets towards the equator in the Pleistocene caused climate changes with windier phases, as at the last glacial maximum, when more sand and dust was transported than is currently.

22.3 Sediment entrainment and processes of sand movement

Wind erosion and sediment movement are related to the properties of wind near the ground surface and the properties of the surface material, and these two sets of properties are interdependent. They are a function of both the erosivity of the wind (its power) and the erodibility of the surface (the grain characteristics holding them in place). Generally where erosivity is greater than erodibility, sediment movement and erosion will occur.

As air moves over the Earth's surface there is frictional retardation and a velocity profile develops in the boundary layer. At the base there is a zero velocity in a very thin layer, which changes to free stream velocity at a height beyond the effects of surface friction. Depending on whether the flow is laminar or turbulent the structure of this velocity profile varies (Figure 22.2), but airflow is nearly always turbulent, with mixing between layers caused by gusting and eddies, and higher shear stresses develop. This is because air has a low viscosity and the boundary-layer depths are usually high. Laminar flow characteristics only develop in very viscous, slow or thin flows.

Under normal atmospheric conditions on unvegetated, flat surfaces, where intense solar heating is absent, the turbulent velocity profile plots as a straight line on semi-log graph paper. The gradient is the result of the surface roughness producing a drag on the overlying airflow. If the gradient of the velocity profile is known, the shear stress at the surface can be determined. To do this the shear velocity ($u*$), which is proportional to the velocity profile gradient, can be calculated from two velocities at known heights. This shear velocity is related to the actual surface shear stress by the following expression:

$$u* = \sqrt{\tau/\rho}$$

where τ and ρ refer to the shear stress and the density, respectively.

The turbulent velocity profile does not reach the surface in Figure 22.2 and very close to the surface the velocity is zero.

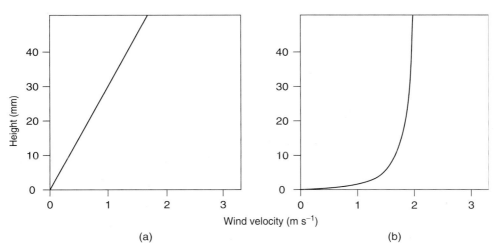

(a)   (b)

Wind velocity (m s^{-1})

Figure 22.2 Velocity profiles close to the surface in laminar and turbulent flow. (Source: after Bagnold, 1941)

The height of the zero-velocity region is called the aerodynamic roughness length (z_0) and is a function of surface roughness: over a rougher surface this parameter is larger, the velocity profile gradients become steeper and shear velocities will increase, with sediment transport likely to rise. As the wind shear increases, the loose sand grains are subject to increasing stress and eventually move when the forces acting to move a stationary grain overcome the forces resisting sediment movement. The critical or threshold velocity is the velocity at which grains just start to move. It varies with increasing grain size, but only for grains over 0.06 mm. Below this value, threshold velocities rise again (Figure 22.2). This is because the sediments of small grain size are more difficult to move due to increasing interparticle cohesion, with weak chemical bonds, and there is greater moisture retention and lower values of surface roughness.

The exact mechanism for entrainment at this critical velocity has been much discussed but possibilities include vertical movements involved in turbulence, which could lift particles from the surface; rotary Magnus effects (lift induced by spin); and electric charges during dust storms. Bagnold (1941), however, suggests that these processes could only lift particles smaller than sand.

In Figure 22.3 the relevant forces are shown for the entrainment of dry bare sand, and the grains are subject to three forces of movement. Lift can be a result of the air flowing directly over the grain, forming a region of low pressure, with relatively high pressure beneath the grain. There is a tendency therefore for the grain to be sucked into the flow. The surface drag of wind is the shear stress on the grain provided by the velocity profile and causes the entrainment of sand-sized grains, along with the form drag, where there are pressure differences around the grain. When these forces overcome the forces of grain cohesion, packing and weight, the grain tends to shake in place and then liftoff occurs, spinning into the airflow. When the movement of particles starts, the bed is bombarded with grains and this initiates further movement, so that sediment movement can be maintained at lower velocities than would be needed to initiate it. There is

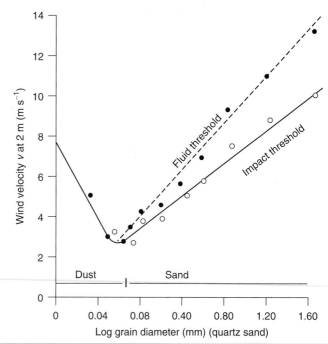

Figure 22.4 Variation of threshold velocity with grain size. (Source: after Bagnold, 1941)

therefore a new threshold value, called the impact or dynamic threshold (Figure 22.4). Of course, whilst these relationships are satisfactory for loose, dry, flat, homogeneous surfaces, the critical thresholds for movement vary according to factors like surface crusting, moisture variations, different grain packing and grain shapes, and vegetation cover (Livingstone and Warren, 1996).

Wind is able to move sediment in a similar way to flowing water, either as bed movement or in suspension, but since air is a much less dense fluid it is only able to carry a much smaller sediment grain size for the same flow velocity. Also, wind does not have to flow in channels or downslope, since it is controlled by global differences in atmospheric pressure. So wind erosion and transport is very broad in its effects and does not show much tendency to concentrate its influence.

22.4 Processes of wind transport

Once the grains have been entrained in response to wind shear they can be carried by the wind in several ways: by saltation, rolling, reptation, suspension and surface creep, principally dependent upon the sediment grain size. The prominence of a fast-moving saltation curtain is the main difference between air and fluvial transport. Figure 22.5 shows the relationships between grain diameter, shear velocity and mode of sediment transport. There is a clear distinction between dust, loess and dune sands.

22.4.1 Saltation

This consists of a curtain of sand grains, between about 0.06 and 0.5 mm in diameter, which as they are ejected from the bed

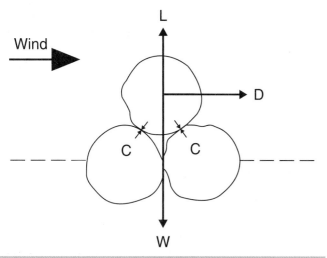

Figure 22.3 Forces exerted on a grain by the wind. L, lift; D, drag; W, weight; C, interparticle cohesion

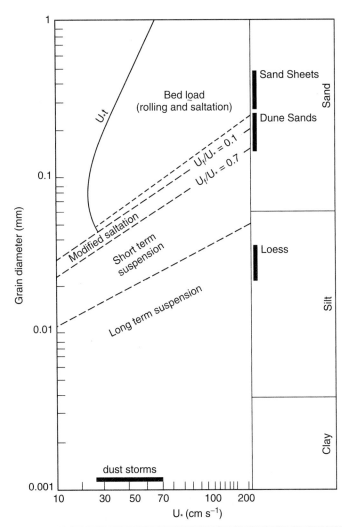

Figure 22.5 Relationships between grain diameter, shear velocity and mode of sediment transport. This shows the distinction between the suspension of dust-sized sediment and the bedload transport of sand-sized sediment. U_f is the particle fall velocity. (Source: after Tsoar and Pye, 1987)

grains descend at angles between 10 and 50°. Saltation stops for an individual grain because when the shear stress decreases, the grain finds a hole in the bed and cannot move out. Therefore saltation is not a continuous process but a discontinuous one, because in turbulent flow both the lift and drag forces fluctuate in magnitude, direction and their point of application. The bed is subject to randomly varying bursts of velocity and shear stress. It is the key process in transport because once sediment movement has begun, it powers all the others.

22.4.2 Rolling and reptation

Rolling is a minor mechanism and is rare because even at high velocities the small mass density of air prevents very high tractive forces from developing. However, slow-moving surface creep can occur when the impact of a moving sand grain is insufficient to propel a stationary grain into the air but nudges it forward along the bed.

By impact, a saltating grain can shift a grain 200 times its own weight and 6 times its diameter. This accounts for about 25% of the total transport. It is to be expected that a high percentage of the coarser grains in a load of mixed sizes will travel in the surface creep zone, though individual grains will alternate between creep and saltation as conditions change from one impact to the next.

There seems to be an important transitional state between creep and saltation, where the low-hopping of several grains occurs consequent upon the high-velocity impact of a single saltating grain. This has been called reptation (Anderson and Haff, 1988). On impact and subsequent rebound a saltating grain may lose 40% of its velocity. This energy, given to the bed, results in the ejection, or 'splashing', of perhaps 10 other grains, with velocities of approximately 10% of the impact velocity, often too low to enter into saltation. Hence each grain takes a single hop, the majority in the downwind direction.

There are other near-surface processes which either help or hinder the entrainment process. For example, bombardment both consolidates the surface, making entrainment harder, and elevates some grains to positions where they are more vulnerable to later dislodgment. There is also preferential movement of coarse grains up to the surface when sand composed of mixed sizes is shaken by bombardment (Sarre and Chancey, 1990). This process is likely to extend no more than five grain diameters beneath the surface and occurs very quickly. It seems to be caused by a compressional–dilational wave which radiates from the impact point of a saltating grain.

22.4.3 Suspension

This is a simple extension of saltation in which the ratio of lift force to submerged weight on grains is so great that the grains remain suspended above the fluid boundary for as long as the flow conditions prevail, which can be for many days in the atmosphere. Suspension is important in dust and loess sedimentation and only affects small grains under 0.06 mm in diameter. The critical shear

into the air gather horizontal momentum from the higher-speed airflow above the bed, and descend to impact the bed, then continue leaping downwind (Figure 22.6). Mean launch angles are between 30 and 50° downwind, although some grains can even bounce back against the flow. There is usually a dense lower layer, with a few lighter grains travelling above the lower zone.

Bagnold (1941) found that sand bounced to a maximum height of 2 m above bouldery areas, but this is exceptional and maximum heights of only 9 cm occur above sand. Since the momentum of the saltating grains is about 2000 times greater than that of the air they displace, the velocity distribution near the ground is entirely controlled by the saltating grains and the surface roughness plays no part. In water the saltation height is usually only a few grain diameters, rather than the few thousand in air.

Saltation hop lengths are about 12–15 times the height of bounce but there is a wide distribution and the majority of

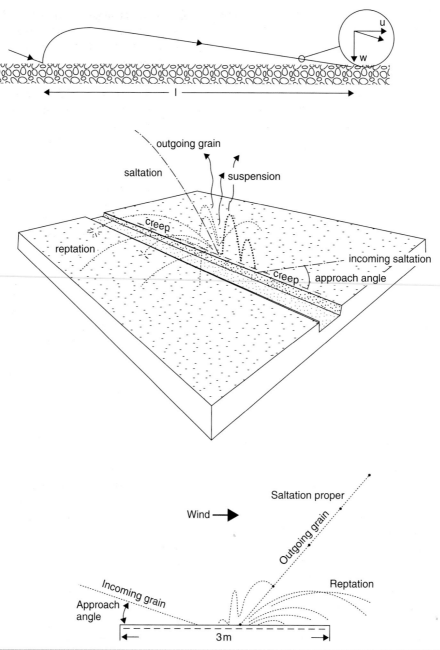

Figure 22.6 Path of a single saltating grain

velocity necessary to initiate suspension is given by the following equation:

$$u_{*c} = 1/7\,\omega$$

where ω is the terminal fall velocity of the grains.

22.5 Aeolian bedforms

Over 99% of arid-zone aeolian sand deposits are found in extensive sand seas, or ergs (Wilson, 1973), which are the largest landform units in aeolian deposition. They are composed of different types of dune, caused by variations in sand supply, vegetation cover and wind regime.

A dune is an aeolian accumulation of sand-sized sediment deposited by the wind and shaped into a bedform by deflation and deposition. Superimposed on all other types of sand bedform can be the smallest bedform, the ripple. Several assertions about the development of aeolian bedforms can be made:

- An initial irregularity is needed on the bed to act as a nucleus before the pattern can develop. These irregularities can be foreign bodies on the plain, like vegetation clumps, oil drums or dead camels; differential erosion of the inhomogenous bedrock; remnants of previous bedforms (this is probably the most common cause); or produced by chance action of eddies in the airflow.

The bedforms can occur spontaneously and are not dependent on pre-existing rhythms on or within the bed. For example, in the Algerian Erg Oriental there are 50 000 draa peaks and in any given area they are spaced very regularly. The hypothesis that each of these peaks has a rock core, or that each draa line parallels an old river bed, is very improbable.

- There is interaction between the air flow and the bedform. As the bedform projects into the airflow it stops uniform motion in a straight line and there is a continuous two-way interaction between the form and flow, so that if one changes the other must also.

- There is bedform migration, as in general airflow accelerates up wind-facing slopes and decelerates down lee slopes. As transport power is directly related to flow velocity, a wind blowing over loose sand must be eroding as it accelerates up the stoss slope and depositing on the lee slope. The effect of this is that the ridge advances downwind. This process occurs even though the body of the ridge is absolutely stationary.

- There are three possibilities for the growth and development of a bedform under constantly applied conditions: indefinite continued growth; cyclic development, with repeated growth and decay; and growth to a dynamic equilibrium, with no further shape or size change. There seems to be no evidence for the first. Most bedforms belong to one of three size groups and forms in between do not grow into one another. There are few bedforms with wavelengths between the three modal groups. Each of the groups has an upper size limit: ripples about 30 cm; dunes 500 m; and draas 5 km. Time is not a factor in limiting their growth so there has to be some other reason why the growth is limited by a dynamic interaction, producing either equilibrium or cyclic development. If there were cyclic development there should be other bedforms associated with different sizes and shapes to form a continuous series, where one represents growth and another decay. However, this does not seem to be the case and therefore it seems that aeolian bedforms must all reach a dynamic equilibrium with the causal conditions, if the latter remain constant for long enough.

- There are universally-distinct bedform groups, which are formed by distinct mechanisms.

- Bedform development takes place when mobile grains come to rest in a sheltered place on the ground surface, through accretion or tractional deposition. It also occurs when wind velocities decrease, causing a reduction in shear stress and a fall in transportational potential, termed grain-fall deposition, and when grains come to rest after rolling down a slope under the effects of gravity, called avalanching or grain-flow deposition.

22.5.1 Ripples

Most ripples are aligned roughly at right angles to the wind-flow direction. Here the profile is slightly asymmetric (gentle windward slopes of between 8 and 13° and lee slopes up to 30°), the wavelength is seldom under 2 cm and rarely over 15 cm, and the heights are between 0.4 and 1.2 cm. Bagnold (1941) noted that the length of a characteristic path of saltating grains and the ripple wavelength were equal, which gave rise to regularly-spaced ripples (the so-called impact or ballistic hypothesis for ripple formation). If there is an increase in wind velocity, or an increase in grain size, then the ripple wavelength increases.

When movement starts over a sand bed, any irregularity will be emphasized, so in Figure 22.7 more grains will land on section AB than on BC, which is sheltered by the rise at B. This bombardment will release more grains from AB than from BC. These will land downwind at a distance approximately equal to the flight-path length. Hence there will be accumulation in zone BC and erosion in AB, and a forward movement of the slope. The ripples have an equilibrium height, because if there were any further growth upward then they would raise their crests into zones of fast flow, where the sand would quickly be eroded.

This simple model of ripple growth only holds for unimodal sands and it is more usual to find sands of mixed grain size, with the coarser grains moving as surface creep. When these zones reach the ripple summit they cannot move on because there are few grain bombardments in the shadow zone and hence there is a concentration of coarser grains at the crest. However, there are many doubts about this ballistic hypothesis. For example, so-called 'aerodynamic ripples' were described in fine sand at high velocities where the wavelengths were much longer than the saltation path length; transverse ripples were noted that met in Y-junctions, which was hard to explain in the ballistic hypothesis;

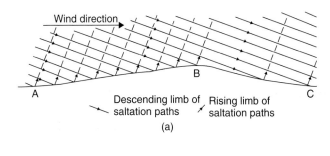

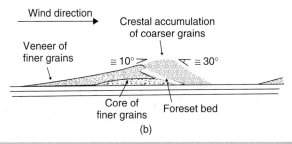

Figure 22.7 Ripple growth and movement

and there is a lot of evidence to suggest that ripple wavelength does not correspond to saltation length.

Today the best-developed model for ripple formation is that of Anderson (1987). From an initial irregularity on the bed, perturbations are generated in the population of reptating grains, and the model gives repeated ripples after about 5000 saltation impacts. The modelled ripples have a strong peak in wavelength in the order of six mean reptation lengths. The ripples grow in wavelength and increase in size as small ones collide and merge. This hypothesis was supported by the experiments of Willets and Rice (1989), who found that as ripples grew, the reptation hop lengths grew on upwind slopes and decreased on the lee, producing rhythmic fluctuations and asymmetric ripple shapes.

Later a model was developed which seemed to show how coarse grains accumulate on ripple summits. This model incorporates two grains sizes, which are fired at the bed, one by one, in a wind field that is compressed over the ripple crests. It shows that small initial undulations on the bed grow under this regime and become well-developed, regularly-spaced ripples after 20 million impacts, or only several minutes of real time. The model also shows that the ripples become coarser at their crests. The impacts on the ripple windward side ejected more fine than coarse grains, so coarsening the surface. These coarse grains could not escape the windward slope because of the lack of impacts in the ripple lee.

22.5.2 Granule ripples

These have much greater wavelengths than normal ripples (0.5–20 m), with heights between 2.5 and 15 cm. Prominent in the composition is 10–20% coarse grains, with 86% of the crest grains over 2 mm in size. These ripples are often nearly symmetrical in cross-section. Bagnold (1941) thought they were the result of long-continued deflation, where all the fine sand was removed from the surface layers so that only coarse grains were left in motion. Continued movement of the creep load was maintained by a supply of relatively fine saltating grains from upwind. Since the coarse grains cannot move beyond the ripple

summits, the ripple's upward growth would continue indefinitely, although it would be very slow. Therefore he thought that they were very old. However, Sharp (1963) found that they appeared in a short time in California. He attributed their greater wavelength to the lower angle of incidence of grain paths at high wind speed, maintaining that they were basically similar to other ripples and that they had an equilibrium height.

22.5.3 Plane beds

These are found in three situations, although the area covered is very small: (a) when wind velocity is very great, though this is rare in nature; (b) when sand is falling out of the wind onto a gentle slope; and (c) when the secondary coarse mode becomes very prominent and the coarse grains rapidly seal off the surface as a lag deposit, preventing any ripple growth.

22.5.4 Dunes

These are generally associated with gentle dips, behind small chance irregularities, or where convergent secondary flows occur. This is because these features will lower the shear velocity and so shear stress and lead to deposition. Once the sand is deposited it will also change the pattern of drag velocity. Generally there appears to be an equilibrium height for dunes in a given area (Figure 22.8). As a dune grows, deposition on the lee slope becomes relatively closer to the summit and the lee slopes become oversteepened. When the angle of repose for sand is reached, avalanching occurs and a slip face is formed at about 33°, compared to the 10–15° slopes on dune backslopes (windward slopes).

Dunes advance whilst conserving their shape both in plan and in cross-section, but the actual movement speed depends on local conditions. They commonly occur in groups in which the dune spacing is very regular (e.g. Figures 22.9–22.13), which is attributed to some regularity in the turbulence patterns in

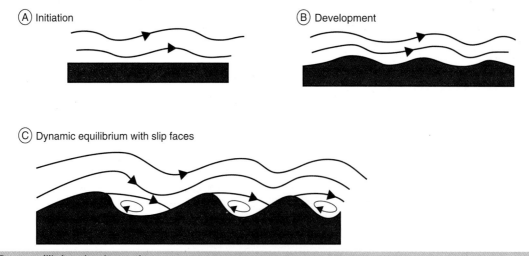

(A) Initiation

(B) Development

(C) Dynamic equilibrium with slip faces

Figure 22.8 Dune equilibrium development

Figure 22.9 Colorado's Great Sand Dunes, which are 230 m in height at the base of the Sangre de Cristo Range

Figure 22.10 Dunes in Sossusvlei region, Namib-Naukluft National Park, Namib Desert. A large slip face has developed. (Source: Luca Galuzzi, http://www.galuzzi.it)

(a)

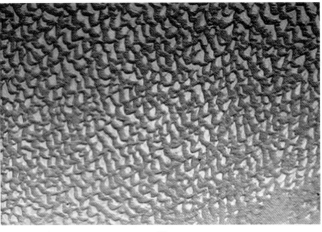

(b)

Figure 22.12 (a) Seif dunes, where the Namib Desert meets the Atlantic Ocean. (b) Interlocked seif dunes, Saudi Arabia. (Source: NASA)

Figure 22.11 Seif dunes. Aerial oblique photo, Arizona, Cocomino County, Moenkapi Plateau, Adeii Eechii Cliffs, Navajo Indian Reservation. The dunes are 2–10 m high, 60 m–several km long and 90 m apart. (Source: http://www.agc.army.mil/research/products_guide/images/p161.gif; 16 December 2005)

the wind. The first dune to form may either fix a whole eddy pattern and so give rise to a regular repetition of forms downwind, or actually initiate eddy motion in the wind. The dune patterns are usually very complex because of secondary flows in the wind, vegetation changes, sand grain size, topography and palaeobedforms. Thus they are not simple in nature, but a classification is given in Figure 22.14.

It is common to describe dunes as either transverse or longitudinal, but often they do not conform to this simple classification and oblique forms are very common. The morphological dune classification is based on their relationship to the formative winds and on the number of slip faces present (McKee, 1979). Linear or longitudinal dunes have long axes which form parallel to the

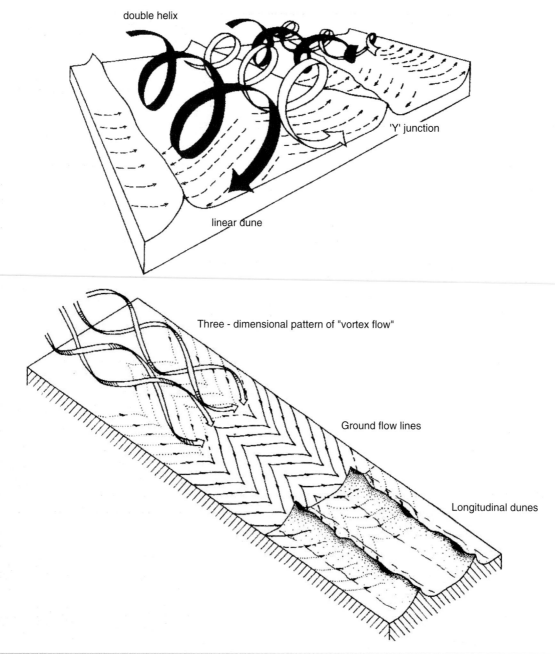

Figure 22.13 Development of spiral vortices (Taylor–Görtler flow) and flowlines. (Source: after Cooke and Warren, 1973)

resultant sand drift direction. Transverse or crescentic forms are aligned normal to the resultant sand drift direction, whilst star dunes have no less than three arms radiating from a central peak and develop in complex multidirectional wind regimes.

However, there are considerable variations in the general dune type form, so this classification is only a start. For example, three broad categories of transverse dune are recognized, whilst a detailed analysis of the south-west Kalahari dunefield shows considerable variation in form and pattern within the general category of 'linear dune' (Bullard *et al.*, 1995). A resultant sand drift direction can be derived from a range of overall wind regimes, and

other factors apart from wind regime can be important for the development of some dune types, such as zibar (Thomas, 1997, p. 390), climbing and falling dunes, echo dunes (Thomas, 1997, pp. 392–393) and parabolic dunes. For example, a second important control is the amount of sand available for dune-building.

Transverse dunes

These develop in wind regimes that show the least directional variability (Figure 22.15). In this dune type, as the bedform grows

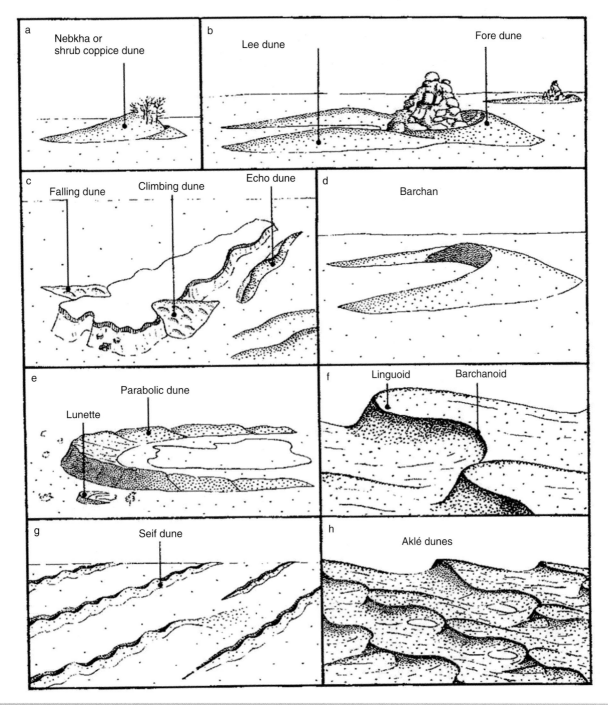

Figure 22.14 Classification of aeolian dunes. (Source: after Cooke and Warren, 1973)

the flow is affected by the ridge projecting into the flow in three main ways:

1. Shear velocity and sand transport increase on the windward side of the ridge and decrease on the leeward side.

2. If the crest angle between the windward and leeward slopes becomes too sharp, flow separation occurs and a roller vortex, with reverse flow, appears.

3. The perturbation, whether wave or vortex, to the lee of each ridge extends further downstream with increasing ridge height.

These changes in flow pattern react simultaneously on the bed, so that there is a continuous interaction. Corresponding effects on the bed are that as the shear increases on the windward slope, vertical growth of the ridge slows down. When the maximum sand discharge arrives at the ridge crest, the whole of the windward

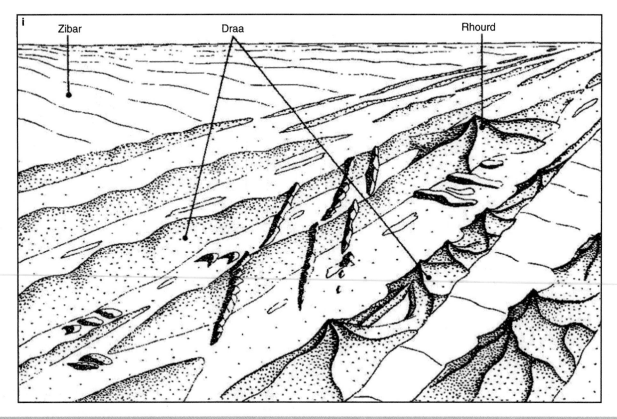

i   Zibar   Draa   Rhourd

Figure 22.14 *(continued)*

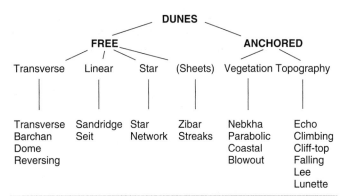

Figure 22.15 Dune classification after Livingstone and Warren (1996)

slope is eroded and the leeward slope is depositional; that is, upward growth has stopped. When flow separation takes place, the saltating grains are unable to cross the zone of reverse flow and they are trapped. There is rapid deposition and steepening of the lee slope just below the crest, and this continues until the slope just reaches the angle of repose. A slip face then forms and

the forward movement of sand across the vortex continues by avalanching.

Stretching of the perturbation downwind increases the bedform's wavelength. Eventually the two-way interaction leads to a dynamic equilibrium and the transverse dune moves forward at constant speed, without changing size or shape.

Barchan dunes

These develop where sand throughflow rates are high in an area, or where sand supply is limited, for example on desert stone pavements. They are usually small and simple dunes, with a maximum height about one tenth of the width. Megabarchans are compound forms which consist of superimposed barchans of different sizes and have been recorded in some dunefields. The morphology is shown in Figures 22.17 and 22.18.

Their development is simple, as when sand saltating over a stony desert encounters a sand patch, the rate of forward movement is slowed because saltation paths are shorter on soft surfaces. As the rate of grains arriving from upwind exceeds the rate of grain despatch on the downwind side, sand accumulation and dune development occur on the patch. Deposition will

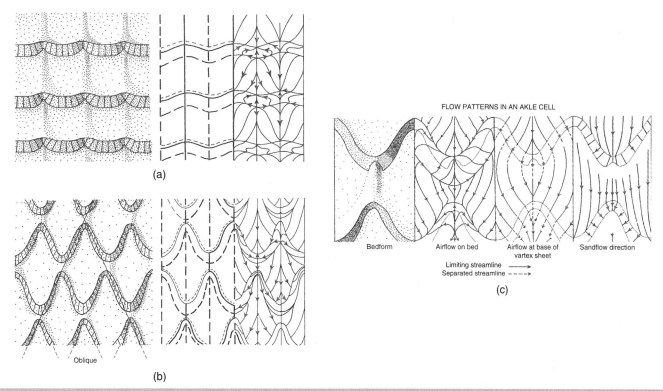

(a)

Oblique

(b)

FLOW PATTERNS IN AN AKLE CELL

Bedform   Airflow on bed   Airflow at base of   Sandflow direction
                                  vortex sheet

Limiting streamline   ——————▶
Separated streamline   – – – ▶

(c)

Figure 22.16 Akle dune patterns resulting from combined two dimensional and three-dimensional flow patterns: a) with in-phase vortices b) with laterally displaced vortices and c) details of flow patterns over akle ridges. (Adapted from Cooke and Warren 1973)

be greatest on the lee side, where wind velocities are lower, flow patterns are divergent and a slip face develops. Sand accumulation is lowest at the edges of the original patch, where saltation rates are greatest, so the dune and slip face become highest in the centre. This leads to the development of the barchan horns, or arms, which trail downwind. The slip face traps the amount of sand regardless of position, but moves forward faster at the edges where the dune is lowest, creating a concave profile (Figure 22.10). Rates of movement have been measured at up to 40 m per year for small barchans, but as the volume of sand contained in a dune increases, the movement rate declines, especially in the case of isolated dunes moving over flat surfaces.

Many linear dunes

These are simple forms, occurring in extensive dunefields, with individual dunes up to 20 m high and up to 1 km apart, although they can reach considerable lengths and heights of up to 200 m. There is a great degree of morphological variation but there are two broad classes: seif dunes (Figure 22.11) and linear ridges (Figure 22.19). Seif dunes have a sharp but sinuous crest, with narrow reversing slip faces, whereas linear dunes are straighter, have convex asymmetrical crests and are generally more subdued

forms. For many years it was thought that probably all linear, longitudinal dune bedforms were initiated by regular spiral vortices, in unimodal wind regimes, with axes parallel to the flow (Glennie, 1970; Wilson, 1972; Warren, 1979) (see Figure 22.13). However, no direct relationship has ever been demonstrated between large vortices and dune development. Some of these spiral vortices were though to pre-exist in the flow, whilst others were stimulated by bed irregularities.

Undulations parallel to the flow develop in response to the local patterns of deposition and erosion. The growth of ridges affects the flow in two ways:

1. Saltating grains, due to gravity, have intermittent contact with the bed. As they hit the sloping sides of the ridge they are deflected sideways and downwards towards the trough.

2. As the crest zones project into the flow the shear velocity and therefore the sand transport rate along them increases.

Both these effects oppose ridge growth. The first produces grain motion from ridges to troughs and the second opposes the lateral velocity gradients and the sand flow due to them. Eventually a

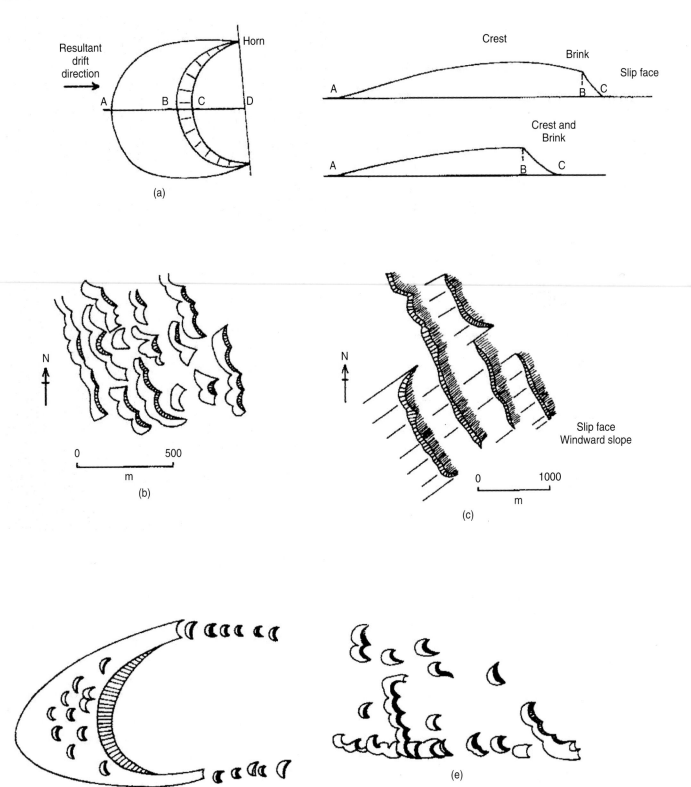

Figure 22.17 Barchan dunes

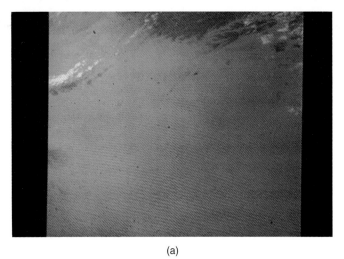

(a)

Figure 22.18 Barchan dunes. Reproduced courtesy of www. freenaturepictures.com. (Source: http://www.freenature.pictures. com/assets/images.htm)

dynamic equilibrium is reached, where the opposing tendencies balance one another out. The mean sand flow at all points now parallels the axes of the undulations, even though spiral airflow still persists. Although there is sand movement over the entire surface, the bedform neither grows nor migrates.

Aklé

The simplest pattern found in dunes is a network known as aklé in the Western Sahara. It is also found in some draas. This pattern is far more common than barchans. Aklé patterns seem to need relatively unidirectional winds and a considerable sand supply. The simplest unit of the pattern is a sinuous ridge, transverse to the wind direction, which is made up of crescentic sections, alternately facing into and away from the wind regularly along the ridge. Those facing away are barchanoid, whilst those facing into the wind are linguoid (Figure 22.16).

It is assumed that the aklé pattern develops where there are fast and slow lanes of secondary flow parallel to the mean flow. Fast lanes (divergent) push the linguoid sections forward, leaving the barchanoid sections in the slow (convergent) lanes. These slow and fast lanes are simply parallel filaments of the wind but it seems unlikely that they could travel side by side without secondary flows from fast to slow lanes, and if continuity is to be maintained this would need the return flow at some level above the ground (see the Taylor-Görtler vortices of Figure 22.13, where there is three-dimensional flow with rotating eddies parallel to the flow).

These flow patterns, however, do not explain one common characteristic of aklé, which is that the barchanoid sections in one ridge are followed downwind by a linguoid section in the next. These two sections enclose a hollow, which is often bounded by a small longitudinal ridge. An explanation of the displacement of barchanoid and linguoid sections in aklé, of uneven elongation of barchan arms and of crossing patterns in sand seas, can

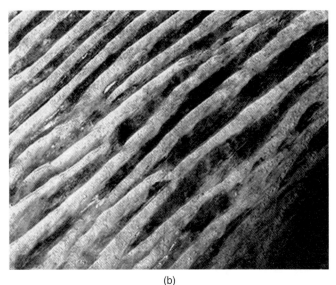

(b)

Figure 22.19 (a) Linear dunes. (b) Linear dunes, Empty Quarter Arabia. (Source: NASA)

be attempted by considering the actual patterns of combined transverse and longitudinal secondary flow and the ways in which these interact with the bedform itself. Simple combinations of transverse and longitudinal flow elements, including return flow of the lee eddy (roller vortex), give a rectangular reticule (Figure 22.16(c)). This form is unusual because of the secondary patterns of return flow in the dune corridors.

It is more usual to find that the longitudinal elements are displaced sideways by half a wavelength, and when this type of flow pattern is well developed the linguoid sections are pushed well forward. The result of this type of flow is to produce dune-ridge elements that are oblique to the flow. This also explains the barchan geometry, because the barchan is probably initiated in one of the convergent nodes in the flow pattern. As these are regularly spaced it will tend to have a regular size. Barchans with one arm longer than the other can develop almost by chance, as once one arm has developed it may extend downwind oblique to

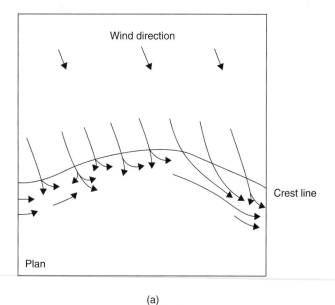

(a)

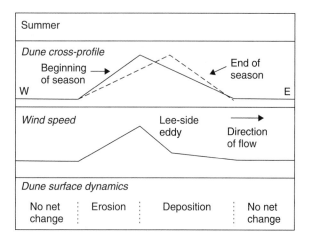

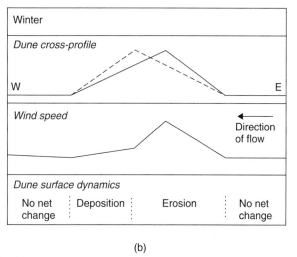

(b)

Figure 22.20 (a) Tsoar's (1978) model of seif dune development. (b) Livingstone's (1986) model of linear dune development

the flow. Its growth might be maintained by the spiral eddy in the lee and by the fact that it will trap most of the sand coming from upwind. Barchans with both arms elongated can be explained as having developed both of the oblique elements and probably occur in regular, unidirectional winds.

However, studies of actual windflow and sand movement over linear dunes in the field (Tsoar, 1978, 1983; Livingstone, 1986) have provided justifiable grounds for the rejection of much of the above elegant hypothesis for the formation of linear dune elements. It has been shown that linear dune development occurs in bidirectional wind regimes, without the formation of helical vortices. Instead, the protrusion of the linear dune itself into the boundary layer creates windflow modifications, which play a central role in dune dynamics. Tsoar (1978, 1983) observed that winds passing obliquely across linear dune crests are deflected through the effect of flow separation at the crest, to blow parallel to the dune axis on the lee side. Hence sand transported across the crest does not leave the dune but is carried down its length. As seif dune crests meander, the angle of wind approach fluctuates along the length of the dune, which affects whether the lee-side flow is erosional or depositional (Figure 22.20). As formative winds are bidirectional, either seasonally or daily, zones of net erosion and net deposition develop on both flanks of the dune. Consequently crestal peaks and saddles develop along the dune and advance down-dune in the resultant drift direction. The internal dune structure reflects this process (Figure 22.21).

Livingstone (1986) investigated the development of a large double-crested linear dune in the Namib Sand Sea, identifying atmospheric flow-line compression in the dune crestal zone as the most important factor for sand movement. He considered that wind velocity changes caused by the protrusion of the dune into the atmosphere created net erosion on the windward side and net deposition to the crest lee, with zones of erosion and deposition alternating seasonally under the influence of winds from each of the modal directions. Interdune areas were zones of sand transport which did not experience net erosion or net deposition. The age and dynamics of linear dunes in the Namib Desert are discussed in Box 22.1, see in the companion website.

The question of how linear dunes are initiated is still a problem. Some workers consider that they could result from the elongation of one arm of a barchan, whereas others believe development could be started through the blowing out of the nose of a parabolic dune (see Section 24.10). Tsoar (1978) suggests an alternative explanation based on the association of coarse sand zibar and linear dunes in the Ténéré and Sinai deserts. Here saltation rates are high across the compact zibar surfaces, slowing down as particles move off this surface and creating favourable conditions for sand accumulation. Linear dunes then extend downwind from the zibar, where this occurs in environments with bimodal winds. However, where other dune types are virtually absent, this explanation is not so convincing.

Star dunes

These contain greater volumes of sand than other dune types and occur in environments with great variability in sand-transporting

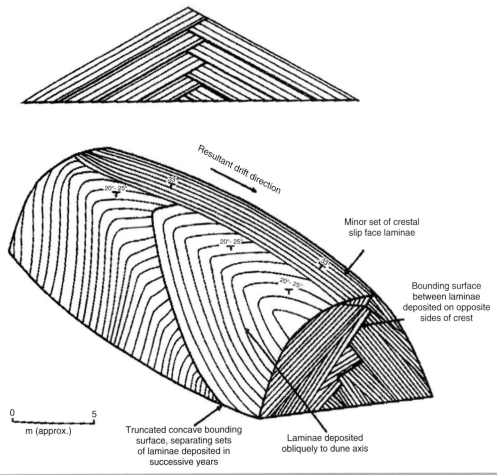

Figure 22.21 Internal structure of linear dunes. (Source: after Tsoar, 1982)

wind directions. Nielson and Kocurek (1986) have shown that star dunes can develop where seasonal variations in dominant wind directions cause other dune types to merge and modify. If one wind direction has overall dominance, the incipient star form can be destroyed as the growth of one arm is favoured, but if the incipient form is big enough, secondary flows created over the other arms can counteract this. This leads to their preservation and the ultimate growth of a star dune.

As star dunes grow in size they increasingly modify their own wind environments, which leads to both the maintenance and growth of the overall star form and the development of other dune forms on their flanks, and hence many star dunes have a complex form (Figures 22.22 and 22.23). Many of the world's largest dunes are of this type, with star dunes up to 400 m high reported. They can occur in linked chains, as in the Gran Desierto in Mexico (Lancaster, 1995), and imbalances in the sand-transporting potential of different components of the overall wind regime can lead to star dune migration in one overall direction.

Zibar

These are dunes with no slip faces that are formed from coarse sand and have hard surfaces. They are widespread in some deserts

and in total area are more important than barchans or star dunes. Most are of low relief and they are common upwind of sand seas in zones from which finer sediment has been winnowed. Most are straight in planform and transverse to the wind, although linear and parabolic forms have been described. In general very little is known of their dynamics.

Dunes associated with obstacles

The dune types described so far have all been free to migrate, but some are anchored around obstacles such as bushes and topographic obstructions. These types are associated with flow separation and acceleration around these obstacles.

Vegetated sand mounds, or nebka, form where vegetation traps sand. They go through a series of stages as the vegetation develops and the ridge grows, especially where the plants undergo adaptations to grow through accumulating sand. They occur on the British and Dutch coasts, where marram grass is the primary colonizer. Here coastal dune development goes through various stages, from embryo, through foredunes to large vegetated dune systems along the coast.

Blowouts are eroded into some of these systems as bare hollows caused by deflation and are usually parallel to the prevailing

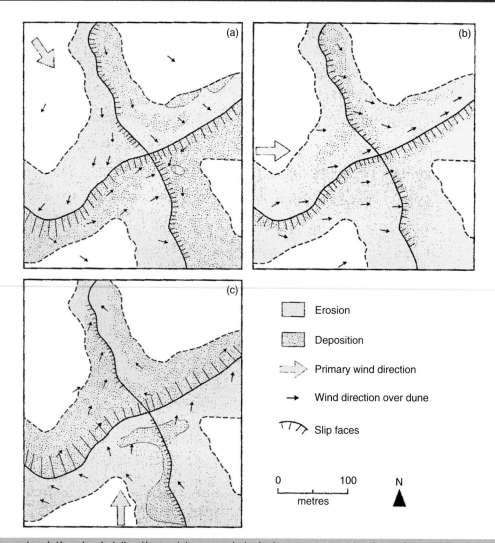

Erosion

Deposition

Primary wind direction

Wind direction over dune

Slip faces

0   100

metres

N

Figure 22.22 Seasonal variations in wind direction and dune morphological response on a star dune. (Source: after Lancaster, 1988, 1995)

wind direction (Figure 22.24). They are caused by vegetation disturbance by animals or humans and by extremely high winds, and in some coastal systems they can be a symptom of a negative sand budget, where the dune is no longer supplied from the beach. They are often associated with parabolic dunes that have a U- or V-shaped planform, in which the arms point upwind (see the development of a parabolic dune in Figure 22.25). They are also characteristic of the vegetated desert margins and vegetated cold-climate dunes, as in Canada and the central United States.

Parabolic dunes develop from blowouts. They grow in size as they feed on sand produced by erosion of the blowout. Eventually the supply of sand decreases as either a firm base or the water table is exposed in the deflation hollow. The vegetation protects the less mobile arms against the wind action, which allows the central section to move downwind. Coastal dunes are discussed further in Section 24.11 and in Livingstone and Warren (1996, pp. 93–100).

Lunettes and clay dunes

These are a distinctive, half-Moon-shaped dune form usually associated with the shores of salt pans. Some are sandy and have the characteristics of free dunes or vegetated coastal dunes, depending on the local microclimate and disturbance. Others have clay contents up to 77%, which also have high salt concentrations. These types rapidly immobilize as the clay aggregates disperse and become cohesive as rainfall percolates through them.

Clay dunes can be created when pellets of aggregated clay move by creep or saltation as if they were sand grains. As they cannot travel far before they become disaggregated by mechanical bombardment, they form a dune close to the source, or are dispersed to form a loess-like sheet. Saline depressions are the most common source of clay, but it can also come from river deposits, especially shortly after floods, or from bare soil.

classified as dunefields. Few sand seas are covered continuously with sand and most include interdune areas of bare rock, soil or lake and river deposits.

However, other large areas of aeolian sand, termed sand sheets or streaks, have no recognizable dune forms and sometimes are of greater total area than the dune accumulations. The controls on these sheets include a coarse sand surface layer that acts as an armour, preventing sand mobilization as dunes; evenly spaced vegetation cover; a high ground-water table; periodic or seasonal flooding; and the development of surface crusts (Kocurek and Neilson, 1986). All of these factors inhibit the development of dunes and slip faces, but it is also possible that some sand sheets are the erosional remnants of previous sand seas (Fryberger et al., 1984). Mapping of these larger-scale aeolian sand accumulations is now possible because of satellite imagery and work has shown that whilst some are close to the sources of their sand, others are far away. Wasson and Hyde (1983) simplified the classification of dunes based on wind-direction variability and sand supply (Figure 22.26).

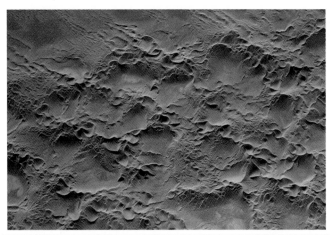

Figure 22.23 Star dunes, Issaoune Erg, Algeria, 16 January 2005. (Source: NASA)

Figure 22.24 Blowout in coastal dunes, Sefton coast.

In salt lakes the clay pellets are created by clay bonding, by salt-cementation, or by algal aggregation, and all of these processes are encouraged by seasonal wetting and drying. In the dry season there must be a wind that is strong enough to entrain the pellets and constant enough to carry them in one direction. This is then followed by a wet season in which the lake refills and the dune is cemented.

So clay dunes are the result of a semi-arid climate, and in many cases their origin dates back to the Late Pleistocene.

22.5.5 Sand seas, dunefields and sand streaks (sheets)

Sand dunes rarely exist in isolation and most cluster in sand seas or dunefields. Cooke et al. (1993) suggest sand seas have a lower size limit of 30 000 km^2, while accumulations smaller than this are

22.6 Dune and aeolian sediments

The vast majority of dune sands are composed of quartz, since it is common in quartz rocks and is chemically stable. Whilst traditionally dune sands have been thought to be rounded, Goudie and Watson (1981) suggest that this has been greatly exaggerated. Most grains are sub-rounded and sub-angularity is not uncommon.

When seen under a microscope, most aeolian sand grains have a surface texture that appears 'frosted'. This is thought to be because of two processes: chemical solution and reprecipitation of silica, and mechanical abrasion. Not only do most grains experience both processes but they are also mutually reinforcing, as solution is more effective on newly-abraded grain surfaces and on small particles that are eroded by abrasion, and abrasion is more effective on grains that have suffered some solution and reprecipitation.

Some surface textural features appear to be caused by transport attrition, such as 'upturned plates' (Kaldi et al., 1978) and 'crescent-shaped impact features' (Le Ribault, 1978). The wind is highly size-selective and aeolian sands are frequently finer, better sorted (always moderately to well sorted) and less positively skewed (coarse) than glacial, fluvial or marine deposits. (The typical size characteristics of aeolian sands are shown in Figure 22.27.) There also appear to be patterns in the variation of grain size on the scale of sand seas and dunefields. These show increased fining, better sorting and decreasing skewedness in the direction of sand transport.

In terms of sedimentary structures it has proven difficult to obtain modern information due to problems in cutting through active dunes and supporting sections in dry sand, but

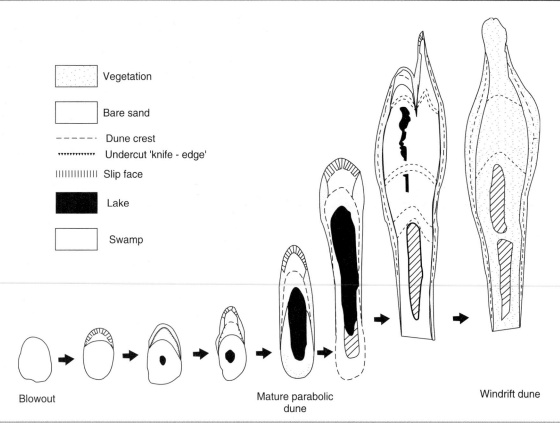

Figure 22.25 Development of parabolic dunes anchored by vegetation. (Source: after Pye, 1982)

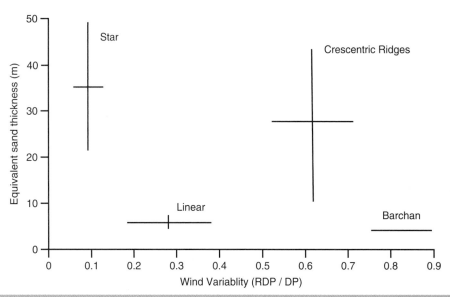

Figure 22.26 Four major dune types, based on wind-direction variability and sand supply. (Source: after Wasson and Hyde, 1983)

ground-penetrating radar can provide some evidence. The great majority of sand-dune deposits consist of cross-bedding as active transverse dunes migrate downwind and sand is eroded from the windward slope and deposited in grainfall and sandflow deposits on the lee slope in the zone of flow separation. The windward slopes are usually capped by a thin layer of topset beds, which are tractional deposits with thin, closely-packed laminae, which can also include small scale ripples. McKee (1979) reported natural sections in White Sands National Monument, New Mexico, from a range of dune types.

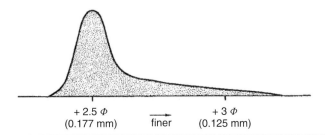

+ 2.5 Φ
(0.177 mm)   finer   + 3 Φ
(0.125 mm)

Figure 22.27 Typical size characteristics of common aeolian sands

22.7 Dust and loess deposition

22.7.1 Dust

Dust is sediment that travels in suspension in the wind, following the air turbulence. This is unlike sand, which falls back to Earth after it has been entrained. It is a very important component of the current ocean basins (Figure 22.28) and in the past it was even more important, especially in periods when there was less river erosion than today. Dust is important in the creation of the world's loess deposits, it is a palaeoenvironmental indicator in deep sea and ice cores and it also has importance in applied aspects of geomorphology, such as soil erosion and other environmental problems in arid and semi-arid lands. It is composed of common rock and soil minerals, salts, ash and biogenic materials, such as pollen, spores and diatoms. The threshold wind speed for dust entrainment is in theory inversely related to size, but in reality the process of entraining dust is not simple as most is created by attrition in saltation and suspension. However, the controls on dust entrainment are:

1. Wind speed and its duration.

2. Lift, which is much more effective for dust than for sand. It is likely that turbulence in dust devils, thunderstorms (for example, dust storms called haboobs in Sudan) and cold fronts is important in dust entrainment, but locally there might be little correlation between wind speed and lift and dust in suspension since much of it will have entrained in other locations and at other times.

3. 'Erodibility controls' related to the character of the surface, such as moisture content, and other soil characteristics, such as clay mineralogy, salt type/content and the presence of organic matter.

4. The amount of bombardment by saltating sand.

5. Vegetation and its seasonal pattern. For example, in the West African Sahel the seasonal minimum of dustiness comes near the end of the rainy season, when the vegetation has fully responded to the rains.

In transport, dust is generally size-sorted by height and the grain size decreases with distance from the source. There is size

Figure 22.28 Dust storms off Alaska, November 2005

bimodality: coarse grains between 10 and 200 microns and fine grains between 1 and 20 microns. It is thought that most of the fine dust is created by the attrition of coarse grains. This attrition comes from the breaking down of fine-grained aggregates, or by blasting fine clays off the surfaces of quartz grains. A prolific sediment source for dust is the dry lake environment (playas and sabkhas), where most are at the distal end of river systems, contain large sediment quantities of appropriate size, have their surfaces kept clear of vegetation by occasional flooding and salinity, and the sediments are broken up, ready for entrainment, by salt crystallization (the Lake Chad basin in West Africa and the Lake Eyre basin in Australia are two such sediment sources that have been shown to be important). Most of this sediment is not contemporary and is inherited from weathering processes in the Late Pleistocene and Early Holocene.

Another source is river alluvium, especially that deposited in the distal section of river valleys, for example the Mississippi and Missouri valleys in the United States. Volcanic ash sources are also important and the finest dust is often pumped as high as the stratosphere.

There has been much discussion about the primary sources of dust and it is now realized that not only are there various contemporary sources but that much of the dust has an inheritance and can have gone through many thousands of transport and deposition events.

22.7.2 Loess

It is generally agreed that loess has been formed from sediment which has been transported by wind, but there has been much dispute as to the origin of the grains from which it is composed. It is an unconsolidated silt or silty sand, commonly buff in colour, characterized by a lack of stratification and its ability to stand in a vertical slope. It is generally porous, highly calcareous and has vertical capillary structures, resembling jointing. Mineralogically it is made up of quartz, with smaller amounts of clay minerals, feldspars, micas, hornblende and pyroxene. Carbonate minerals can form up to 40%. In terms of particle size characteristics, it has a distinct maximum in the silt/fine sand category, with a modal diameter of 20–50 microns. It has 70–95% under 0.06 mm (60 microns) and 97–99% below 0.2 mm (200 microns), with a high degree of sorting. It occurs as extensive, nearly uniform wind-laid sheets of sediment. Usually it has been divided into two kinds:

1. Periglacial loess, deflated from outwash plains, freshly exposed till and rocky tundra surfaces.

2. Continental or desert loess.

However, in some parts of the world there has been much discussion as to its origins, as summarized in Livingstone and Warren (1996, pp. 52–54). Smalley (1966) showed that the fine particles necessary for loess can be produced by the crushing action of glaciers on granite bedrock and it can be seen from the particle size analysis of till that there is always a fine matrix. Wind can transport this fraction easily in suspension and it will only come out of suspension when the fall velocity is greater than the buoyed-up weight of the grains. The time required to fall a measured distance increases markedly when the particle decreases under 200 microns, indicating that particles of this size and smaller will tend to be transported in suspension. Sediments which are deposited after long distance transportation by aeolian action will therefore consist of particles under 200 microns and the expected modal diameter will be 80 microns or less, with the vast percentage between 10 and 50 microns. The 80 micron size is lifted most efficiently but smaller sizes are transported more efficiently.

After deposition the characteristic vertical structure develops. This is related to the cohesive nature of the sediment. The deposits are stabilized by plants which form retaining root systems, but the loess owes its basic structural stability to its small grain size. After deposition the loess undergoes a slight but uniform contraction on drying and randomly-distributed stress centres in the plane contracting system determine the position of the crack net elements. A random network is formed, with a pentagonal or hexagonal series of facets. Thus the history of a glacially-derived loess is as follows:

1. Formation of quartz grains by glacial crushing of granitic bedrock. The material produced falls into two distinct populations, a sand size and a silt size, with little in the intervening grades.

2. Transport of the debris by the glacier.

3. Deposition of mixed debris as the glacier melts.

4. Transport by the wind and the mixed nature of the glacial sediment, allowing the fine grains to be lifted by the wind.

5. Deposition again, with the relatively high cohesive forces associated with the fine sediment enabling the loess to become stable and the vertical structure to develop.

The size distribution of loess formed by either glacial grinding or aeolian abrasion is shown in Figure 22.29.

Each stage is important and affects the final nature of the loess. Whilst the glacial grinding mechanism was supported, it has been suggested that frost action was an important contributory factor, working on the same parent material as the ice. Pye (1987) did not believe that these processes could provide sufficient parent dust for loess.

The question is, are there any major loess deposits that are not composed of glacial loess? In Europe, North America, Argentina and New Zealand loess is found in areas that have been extensively glaciated, or else it is blown in from glaciated areas, as in the southern United States, such as in northern and central Mississippi. However, the Chinese loess may be associated with the glaciation of the Tibetan Plateau, or else with the Gobi and Ordos deserts. Most is now considered to be of desert rather than glacial origin (Zhang et al., 1991).

Some doubts about the reality of desert-derived loess have been raised and the question that is often put forward is: how are small-enough grains produced in deserts? Do grain-to-grain contacts during saltation produce loess-sized grains? Experimental aeolian abrasion did not produce sand grains in the size range 20–50 microns, although it has been shown that 1–2 mm sand grains can be reduced considerably by transport in a 40 mph wind; after about 75 km of transport they were under 50 microns.

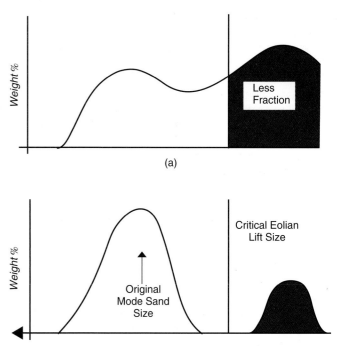

Figure 22.29 Size distribution of loess formed from: (a) glacial grinding; (b) aeolian abrasion

It has been suggested that the size reduction of quartz from the sand mode of 500 microns to the loess mode of 50 microns could be accounted for by a thermal process. Originally when the quartz grains are supplied into the sedimentary system they are sand-sized and ultimately determined by the dimensions of quartz grains in igneous rocks. However, Bagnold (1941) stated that there was little or no evidence for temperature splitting of rock fragments smaller than 1 cm in diameter, which is 20 times the sand mode diameter. Some workers have suggested that insolation weathering in arid and semi-arid climates can produce particles with sizes of the order of 10–100 microns. Nevertheless, the bimodal character of desert grains suggests that thermal fracturing is not significant, since it should yield a continuous range of grain sizes.

On a map of world loess deposits some are close to sandy deserts, yet there is an apparent absence around the Sahara and the American deserts and few are recognized in Australia. McTainsh (1987) does now recognize major terrestrial deposits in the West African Sahel, Tunisia and southern Israel as desert-derived loess deposits, however.

The most likely loess deposit of desert origin appears to be that of the Negev. The deposits of the Beer Sheba basin cover an area of about 1600 km^2 and mineralogical and faunal evidence suggests that this material has been derived mainly from the weathering residues of calcareous rocks in the Sinai Desert. The distance covered during transport was small and deposition took place at the desert margins, where the particles were trapped by steppe-like vegetation. The texture becomes finer away from Sinai. Thus there has been some doubt about the validity of desert processes in producing the required grain sizes, which now seems resolved.

Grain-to-grain contacts by attrition now seem to be the most likely process, but this is difficult to prove. While the experimental evidence appears strong (Whalley *et al.*, 1987; Smith *et al.*, 1991), there are questions as to the reality of the simulations. However, the distinctive clay mineral palygorskite, derived only at very high pH conditions, as found in salt lakes, and therefore of undoubtedly desert origin, can make up between 20 and 70% of the clay-mineral fraction of Tunisian loesses (Coudé-Gaussen, 1987). Dust-sized grains might be formed from salt weathering of quartz, but perhaps only where the quartz has microfractures and silt particles have been produced experimentally by the detachment of siliceous cements or overgrowths on sedimentary grains (Smith *et al.*, 1987).

It has been suggested that the grain-size distribution curves should differ between glacial and desert loess. The small particle product of the desert grain interaction lies within the most easily transported in suspension by the wind. The small particles produced by glacial grinding have to be separated by the wind from a deposit with a continuous range of particle sizes. Hence the sediment sorted from glacial deposits has a larger mean size than the fine material formed in deserts. For desert loess the following stages were suggested:

1. Production of quartz sand from granite and other rocks

2. Transport

3. Deposition elsewhere in the desert area

4. Wind movement of the grains by saltation. When impacts involve more than a certain critical kinetic energy, small-scale fracturing results, giving rise to small quartz chips.

5. Transport in suspension of these chips by the wind that formed them.

6. Deposition beyond the desert margins.

In loess the sediment is usually diagenetically changed by carbonate cementation and many have been reworked by streams after primary deposition. In fact, the need for vegetation to trap the sediment means that many of the landscapes where loess is preserved once had runoff to allow active slope and river processes. Thus in many areas loess was reworked and redeposited at the surface by runoff and a complex landscape developed, which was composed of the accumulating loess and contemporaneous river-cut valleys with terraces and landslides. Loess can be vertical when dry, but when wet the shear strength is greatly reduced, making the sediment prone to subsidence, flow and sliding, and slope stability problems.

The high-latitude loesses are extremely valuable as they have preserved the best terrestrial climatic change record over the Pleistocene in terms of detail and completeness (Kukla, 1987; Kukla *et al.*, 1988). The loess records go back as far as the ocean-core records and the high-resolution magnetic stratigraphy for this record in north-western China extends back to 700 000 years ago, whilst the basal loess from central China is thought to date from about 2.5 million years ago.

Soil horizons developed in the warmer, wetter and less dusty interludes. In the 2.5 million-year history of the Baoji section there have been 37 major cold–warm cycles. The deposition of these loesses intensified in the Pleistocene, with the thicker units lying towards the sequence top, corroborating a picture of intensifying global aridity. This has been further intensified by the aridification caused by the uplift of the Tibetan Plateau during the Quaternary.

In the deep-ocean sediments, dust accumulation provides a useful indicator of past global wind systems and continental environmental change. Detailed core analysis suggests that the northern-hemisphere dominance of dust sources has characterized most of the Cenozoic. This hemisphere was more arid during Pleistocene glacial maxima, with three to five times as much dust transported to the oceans during glacial stages as in interglacials. These kinds of results confirm those found from ice cores and loess profiles.

22.8 Wind erosion landforms

The sediment deposited in dunes or as loess sheets was originally eroded from somewhere, mainly by the wind. Therefore this section is concerned with those landforms eroded by the wind and with the features created by windblown sand bombardment.

Deflation covers the processes of net removal of loose sediment from an area, whilst abrasion is the wearing-down of more cohesive rock by bombardment with wind-transported grains. The most favourable dust-producing surfaces are areas of bare, loose sediment with substantial quantities of sand and silt, but little clay, as the latter provides too much cohesion and the absence of sand reduces the effects of bombardment. However, deflation can rarely proceed very far as there are usually inhomogeneities in the eroding material, or in the surface topography, and these sites form the nuclei of residual features.

22.8.1 Pans, stone pavements and yardangs

The most important deflational landforms are pans and stone pavements, but deflation is also important in the formation of some yardangs. Deflation usually works side by side with other processes, such as aeolian abrasion. The latter is where sand-sized sediment in saltation imparts considerable impact energy on other grains, or rock surfaces. When the object of this bombardment is an individual gravel clast it is called a ventifact, whilst landforms carved from bedrock by the grain impact are called yardangs.

22.8.2 Ventifacts

These are usually pebble-sized, although occasionally they can be boulder-sized. Their characteristic feature is smooth faceting of the surface, but many show pits, flutes or grooves rather than polished surfaces. Those ventifacts with smooth, wind-faceted faces usually have two or three, although sometimes there can be more. The number of edges is indicated by calling them einkanter, zweikanter or dreikanter, meaning one-, two- and three-edged

respectively. These faceted ventifacts have been repeatedly swivelled and overturned. Breed *et al.* (1989) concluded that most ventifact shapes and textures can be explained by impact-face sandblasting, supplemented by fine-particle abrasion of all surfaces by subsidiary wind currents.

22.8.3 Yardangs

These landforms are cut from bedrock and appear in all the major deserts, except in Australia. They are one of the very few landforms that are unique to deserts.

There seem to be two classes, distinguished by size: the smaller one, under 100 m long (Figure 22.30), and ridges or mega-yardangs (Figures 22.31 and 22.32). The shape of the smaller

Figure 22.30 Cuatrocienegas cemented dunes, Coahuila, Northern Mexico.

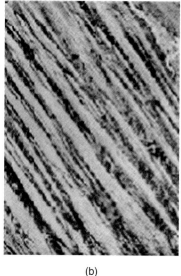

(a)   (b)

Figure 22.31 Yardangs of the Lut Desert of Iran. Views from: (a) low-altitude photograph; (b) high-altitude photograph; (c) Landsat. These yardangs are among the largest on Earth, with almost 100 m of relief. (Sources: (a) J.T. Daniels; (b) US Air Force)

(c)

Figure 22.31 *(continued)*

Figure 22.32 Large yardang. (Source: htto://www.ux1,eiu.edu/~jpstimac/1300/yardang.gif)

yardangs is likened to that of an inverted ship's hull, oriented in the direction of the prevailing wind, and they often occur in closely-packed arrays. They may have concave downwards-tapering or convex and bulbous bows. The top is sometimes flat. The highest and widest part of the landform is generally about one third of the way between the bow and the stern in a well-streamlined yardang. The downwind ends of the yardangs are characterized by gently tapering bedrock surfaces or elongated sand tails.

Halimov and Fezer (1989) expressed scepticism about a general idealized description of streamlined yardangs and described eight types from the Qaidam depression in China (mesas, saw-tooth crests, cones, pyramids, very long ridges, hogbacks, whalebacks and low, streamlined whalebacks). This underlines the variety of forms, although they believed that these represented an evolutionary sequence.

The mega-yardangs, or ridges, have been reported from the Sahara, Egypt, Iran and Peru, and are much bigger landforms. Satellite imagery has shown that they are extensive, covering hundreds of square kilometres, although individual ridges may extend only a few kilometres. An obvious prerequisite for yardangs of any scale is an erodible substrate, but this can range from Cenozoic aeolian, fluvial, lacustrine sediments to sandstones and limestones. Most yardangs occur in unidirectional wind regimes and they are usually parallel to these winds. Abrasion contributes to the undercutting at the bows and flanks, but the aerodynamic shape is a result of deflation and particle dislodgement can be a result of weathering rather than abrasion (McCauley *et al.*, 1977).

22.8.4 Zeugen

An associated landform is the zeugen, which consists of perched rocks created by the abrasion of material in the zone a few tens of centimetres above the ground in which sand transport by saltation takes place. Although early reports from deserts mentioned these perched or mushroom rocks, they are not widespread and could be as much a consequence of rock weathering in the capillary fringe close to the ground as of the action of saltating sand.

22.8.5 Pans

These are shallow, closed depressions, periodically water-filled, which are widely believed to be the result of deflation in semi-arid regions. They are variable in size and have a wide variety of plan-forms, although most are smoothly rounded. Densities greater than $100/100 \text{ km}^2$ have been reported in southern Africa and they are commonly developed on soft sediment susceptible to erosion. Most are associated with a lunette dune on their downwind margin, which is an indicator of deflation. However, other processes must also occur. The sediment on the dry lake floor must be loosened, probably by salt weathering, the salts coming from salt influx in drainage water. None of these processes could take place without an initial closed depression in which water could collect, and these can be formed by processes such as dunes damming water courses, or even the gentle tectonic disruption of a drainage pattern. The smooth form must be the result of wave action.

22.8.6 Stone or desert pavements

These are armoured surfaces formed from intricate mosaics of coarse angular or rounded particles, usually only one or two stones thick, set on or in deposits of sand, silt or clay (Cooke, 1970). They are called *hammada, reg* or *serir* in Arabic and occur widely where there is little vegetation. The deflation hypothesis relies on the grain-size selectivity of wind transport. The wind deflates the finer sediment, leaving behind an increasing concentration through

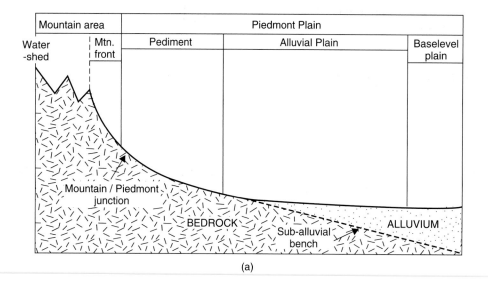

(a)

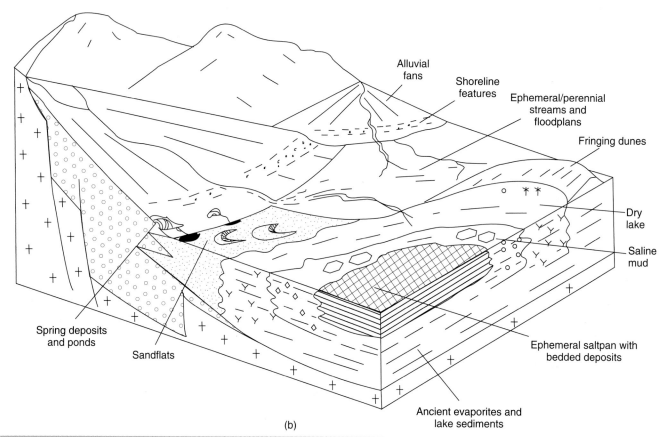

(b)

Figure 22.33 (a) Desert profile from mountain to plain. (b) Range of depositional sub-environments that can occur in closed arid-zone playa basins

time of the coarser sediment to form a lag. This armour then protects the underlying fine sediment from further wind erosion.

22.8.7 Wind-eroded depressions and plains

Although the first geomorphologists to visit deserts thought that deflation could explain the origin of much of the landscape, there was a reversion to the idea that fluvial erosion was the key to desert landscapes, See for example Cotton (1947). However, the role of wind erosion today cannot be dismissed so lightly and for some large enclosed basins in the northern Sahara, such as the Qattara depression (192 000 km^2), researchers have suggested a predominantly deflational origin.

Albritton *et al.* (1990) provided a synthesis of previous evidence together with their own model, which includes fluvial, karst and mass wasting processes, as well as deflation. However, these large depressions are all of considerable antiquity and it would be no surprise to find a variety of processes operational and contributing to their formation over a period of marked climatic change.

Deflation plains were also suggested initially (Keyes, 1909) and there seems little doubt that there are some plains in which wind erosion played a major formative role. For example, Breed *et al.* (1987) described a model for the slow formation of the Tertiary fluvial landscape during the drying up of the south-western Egypt and north-western Sudan landscape during the Quaternary. The desert profile from mountain to plain is shown in Figure 22.33(a), while the potential range of depositional sub-environments that can occur in closed arid-zone playa basins is illustrated in Figure 22.33(b). Box 22.2 in the companion website illustrates a case study from the Triassic of the Cheshire Basin, United Kingdom.

Exercises

1. How are sand grains entrained by wind blowing over a sandy surface?

2. What are the characteristics of granule ripples and how do they form?

3. How do transverse and barchan dunes develop?

4. What is the distinction between saltation and reptation as aeolian sand transports?

5. Evaluate the 'spiral vortex' mechanism for linear, longitudinal bedform development in sandy deserts.

6. What are the morphological characteristics of star dunes and how does this type of dune develop?

7. What are the sedimentary characteristics of aeolian sands?

8. Evaluate whether all loesses are initially derived from a glacial source.

9. Describe the morphology of yardangs and their processes of formation.

10. Discuss the importance of the water table in the formation of the Delamere Member sediments in the mid-Triassic Helsby sandstone formation of the Cheshire Basin, UK.

References

Albritton, C.C., Brooks, J.E., Issawi, B. and Swedan, A. 1990. Origin of the Qattara Depression, Egypt. *Bulletin of the Geological Society of America*. **102**, 952–960.

Anderson, R.S. 1987. A theoretical model for aeolian impact ripples. *Sedimentology*. **34**, 943–956.

Anderson, R.S. and Haff, P. K. 1988. Simulation of eolian saltation. *Science*. **241**, 820–823.

Bagnold, R.A. 1941. The Physics of Blown Sand and Desert Dunes. London, Methuen.

Breed, C.S., McCauley, J.F. and Davis, P.A. 1987. Sand sheets of the eastern Sahara and ripple blankets on Mars. In: Frostick, L.E. and Reid, I. (Eds). Desert Sediments: Ancient and Modern. Oxford, Blackwell Scientific. pp. 337–360.

Breed, C.S., McCauley, J.F. and Whitney, M.I. 1989. Wind erosion forms. In: Thomas, D.S.G. (Ed.). Arid Zone Geomorphology. Belhaven Press. pp. 284–307.

Bullard, J.E., Thomas, D.S.G., Livingstone, I. and Wiggs, G. 1995. Analysis of linear sand dune morphological variability, southwestern Kalahari Desert. *Geomorphology*, **11**, 189–203.

Cooke, R.U. 1970. Stone Pavements in deserts. *Annals of the Association of American Geographers*. **60**, 560–577.

Cooke, R.U. and Warren, A. 1973. Geomorphology in Deserts. London, Batsford.

Cooke, R.U., Warren, A. and Goudie, A.S. 1993. Desert Geomorphology. London, UCL Press.

Cotton, C.A. 1947. Climatic Accidents in Landscape-making. Christchurch, Whitcombe and Tombs.

Coudé-Gaussen, G. 1987. The pre-Saharan loess: Sedimentological characterisation and palaeoclimatological significance. *Geojournal*. **15**, 177–183.

Fryberger, S.G. 1979. Dune form and wind regime. *US Geological Survey*. **1052**, 137–169.

Fryberger, S.G., Al-Sari, A.M., Clisham, T.J., Rizvi, S.A.R. and Al-Hinai, K.G. 1984. Wind sedimentation in the Jafurah sand sea, Saudi Arabia. *Sedimentology*. **31**, 413–431.

Glennie, K.W. 1970. Desert Sedimentary Environments. Developments in Sedimentology 14. Amsterdam, Elsevier.

Goudie, A.S. and Watson, A. 1981. The shape of desert sand dune grains. *Journal of Arid Environments*. **4**, 185–190.

Halimov, M. and Fezer, F. 1989. Eight yardang types in central Asia. *Zeitchrift für Geomorphologie*. **33**, 205–217.

Kaldi, J., Krinsley, D.H. and Lawson, D. 1978. Experimentally produced aeolian surface textures on quartz sand grains from various environments. In: Whalley, W.B. (Ed.). Scanning Electron Microscopy in the Study of Sediments. Norwich, Geobooks. pp. 261–274.

Keyes, C.R. 1909. Base level of eolian erosion. *Journal of Geology*. **17**, 659–663.

Kocurek, G. and Nielson, J. 1986. Conditions favourable for the formation of warm-climate eolian sand sheets. *Sedimentology*. **33**, 795–816.

Kukla, G. 1987. Loess stratigraphy in central China. *Quaternary Science Reviews*. **6**, 191–219.

Kukla, G., Heller, F., Ming, L.X. and An, Z.S. 1988. Pleistocene climates in China dated by magnetic susceptibility. *Geology*. **16**, 811–814.

Lancaster, N. 1988. Development of linear dunes in the southwestern Kalahari, southern Africa. *Journal of Arid Environments*. **14**, 233–244.

Lancaster, N. 1995. Geomorphology of Desert Dunes. London, Routledge.

Le Ribault, L. 1978. The exoscopy of quartz sand grains. Whalley, W.B. (Ed.). Scanning Electron Microscopy in the Study of Sediments. Norwich, Geobooks. pp. 319–327.

Livingstone, I. 1986. Geomorphological significance of wind flow patterns over a Namib linear dune. In: Nickling, W.G. (Ed.). Aeolian Geomorphology. Boston, Allen and Unwin. pp. 97–112.

Livingstone, I. and Warren, A. 1996. Aeolian Geomorphology. London, Longman.

McCauley, J.F., Grolier, M.J. and Breed, C.S. 1977. Yardangs. In: Doehring. D.O. (Ed.). Geomorphology in Arid Regions. Proceedings of the 8th Annual Geomorphology Symposium. pp. 233–269.

McKee, E.D. 1979. Global Sand Seas. US Geological Survey Professional Paper 1052.

McTainsh, H. 1987. Desert loess in northern Nigeria. *Zeitchrift für Geomorphologie*. **31**, 145–165.

Nielson, J. and Kocurek, G. 1986. Climbing zibars of the Algodones. *Sedimentary Geology*. **48**, 313–348.

Pye, K. 1982. Morphological development of coastal dunes in a humid tropical environment, Cape Bedford and Cape Flattery, North Queensland. *Geografiska Annaler*. **64A**, 212–227.

Pye, K. 1987. Aeolian Dust and Dust Deposits. London, Academic Press.

Sarre, R.D. and Chancey, C.C. 1990. Size segregation during aeolian saltation on sand dunes. *Sedimentology*. **37**, 357–365.

Sharp, R.P. 1963. Wind ripples. *Journal of Geology*. **71**, 617–636.

Smalley, I.J. 1966. The properties of glacial loess and the formation of loess deposits. *Journal of Sedimentary Petrology*. **36**, 669–670.

Smith, B.J., McGreevy, J.P. and Whalley, W.B. 1987. Silt production by weathering of a sandstone under hot, arid conditions: An experimental study. *Journal of Arid Environments*. **12**, 199–214.

Smith, B.J., Wright, J.S. and Whalley, W.B. 1991. Simulated aeolian abrasion of Pannonian sands and its implications for the origin of Hungarian loess. *Earth Surface Processes and Landforms*. **16**, 745–752.

Thomas, D.S.G. (Ed.). 1997. Arid Zone Geomorphology: Process, Form and Change in Drylands. 2nd edition. Chichester, John Wiley & Sons, Inc.

Tsoar, H. 1978. The Dynamics of Longitudinal Dunes. Final Technical Report. European Research Office, US Army.

Tsoar, H. 1982. Internal structure and surface geometry of longitudinal (seif) dunes. *Journal of Sedimentary Petrology*. **52**, 823–831.

Tsoar, H. 1983. Dynamic processes acting on a longitudinal (seif) dune. *Sedimentology*. **30**, 567–578.

Tsoar, H. and Pye, K. 1987. Dust transport and the question of desert loess formation. *Sedimentology*. **34**, 139–153.

Warren, A. 1979. Aeolian processes. In: Embleton, C. and Thornes, J. (Eds). *Process in Geomorphology*. Sevenoaks, Arnold. pp. 325–351.

Wasson, R.J. and Hyde, R. 1983. Factors determining desert dune type. *Nature*. **304**, 337–339.

Whalley, B., Smith, B.J., McAlister, J.J. and Edwards, A.J. 1987. Aeolian abrasion of quartz particles and the production of silt-size fragments: Preliminary results. In: Frostick, L.E. and Reid, I. (Eds). Desert Sediments: Ancient and Modern. Oxford, Blackwell Scientific. pp. 129–138.

Willets, B.B. and Rice, M.A. 1989. Collision of quartz grains with a sand bed: The influence of incident angle. *Earth Surface Processes and Landforms*. **13**, 717–728.

Wilson, I.G. 1972. Aeolian bedforms: Their development and origins. *Sedimentology*. **19**, 173–210.

Wilson, I.G. 1973. Ergs. *Sedimentary Geology*. **10**, 77–106.

Zhang, L., Dai, X. and Shi, Z. 1991. The sources of loess material and the formation of the Loess Plateau in China. *Catena*. **20**(S), 1–14.

Further reading

Goudie, A.S., Livingstone, I. and Stokes, S. 1999. Aeolian Environments, Sediments and Landforms. Chichester, John Wiley & Sons, Inc.

Thomas, D.S.G. 1989. Arid Zone Geomorphology. London, Belhaven Press.

Warren, A., Gill, T.E. and Stout, J.E. 2009. Bibliography of Aeolian Research, 1646–2009. http://www.lbk.ars.usda.gov/wewc/biblio/bar.htm.

SECTION V
Principles of Ecology and Biogeography

23 Principles of Ecology and Biogeography

Desert Ecosystem from the Chihuahan Desert in the Cuatrocienegas Basin, Northern Mexico:
psammosere dune sands blown from playa lake in the background

Earth Environments   David Huddart and Tim Stott
© 2010 John Wiley & Sons, Ltd

Learning Outcomes

After reading this chapter and completing the exercises you will be able to:

➤ Understand the differences between biogeography and ecology.

➤ Understand why organisms live where they do.

➤ Understand that materials essential for the maintenance of life are recycled in natural ecosystems and the ways in which the recycling processes work.

➤ Understand the principles related to energy flow in ecosystems and their impact upon the form of natural communities.

➤ Understand why vegetation succession occurs and the controls on the types of climax vegetation.

23.1 Introduction

Biogeography is a huge, diverse area of geography that has developed from several different perspectives. The 'bio' comes from biology, where traditionally the history and geography of plants and animals is covered, along with subdisciplines like biochemistry, cell biology, physiology, histology and ecology. Thus historical biogeography covers the long-term evolution of life and the influence of such factors as plate tectonics, global climatic change and sea-level changes. Biologists Charles Darwin and Alfred Wallace laid the foundations of this area in the nineteenth century.

The science of ecology, on the other hand, studies ecosystems and communities. It developed as an independent area of study early in the twentieth century but gradually infiltrated traditional biogeography. It has been suggested that ecology is the scientific study of the interactions that determine the distribution and abundance of organisms.

There are both analytical and ecological branches of biogeography. Analytical biogeography studies where organisms live today and how they disperse, whereas ecological biogeography looks at the relationships between life and the complex of environmental factors, like climate and soils. It originally considered mainly present-day conditions but it has gradually edged back through the Holocene and into the Pleistocene.

Biogeography is an important aspect of physical geography and environmental science, and it includes the human impact on ecosystems too. It considers ecological principles and biogeographical concepts in relation to ecosystem management and it even impinges on environmentalism and is concerned with a discussion of environmental and ethical issues in topics such as species exploitation, environmental degradation and biodiversity.

So biogeography covers the geography, ecology and history of life: where it lives, how it lives there and how it came to live there. In other words, it is concerned with the analysis and explanation of patterns of distribution, and with the understanding of changes in distribution that have taken place in the past and are taking place today. It is therefore an interdisciplinary subject and a forum for data, principles and concepts from the natural and Earth sciences which are relevant to the understanding of past and present distributional patterns of organisms over the biosphere.

23.2 Why do organisms live where they do?

Often biogeographers ask a misleadingly simple question: why do organisms live where they do? The case of the marsupials (pouched mammals) who live in Australia and South America but not in Europe, Asia, Africa and Antarctica is a good example of a complex reason for animal distributions. In this case plate tectonics help explain this disjunct pattern but there are many reasons why different regions of the planet have distinct assemblages of plants and animals.

The Irish flora illustrates this well, as there are many peculiar distributions. Some plants are confined to the east coast, particularly the north-east, such as *Geranium pratense*, *Geranium sylvaticum* and *Carex pauciflora*, which are all confined to Antrim. Others are confined to the south-east coast, for example *Juncus acutus*, *Trifolium glomeratum* and *Asparagus maritimus*. It seems likely that these species are the very latest arrivals to the island, which just made it before sea level rose sufficiently to cut Ireland off from Britain during the mid-Holocene.

There are several discontinuous ranges, which are probably very ancient, and it is extremely difficult to suggest their origins. So for example, *Hydrilla verticillata* is found in Australia, Madagascar, Central Africa, Russia, Esthwaite Water in the Lake District and Renvyle Lough in western Galway, Ireland. *Inula salicina* is widespread in Europe, absent from Britain and found in Ireland only on the shores of Lough Derg. On the other hand, the Lusitanian flora of western and south-western Ireland, which has 16 species, including the strawberry tree (*Arbutus unedo*), four heaths (*Erica mediterranea*, *Erica mackaina*, *Erica vagans* and St Dabeoc's Heath, *Dabeocia cantabrica*), Lusitanian butterwort (*Pinguicula lusitanica)* and the close-flowered orchid (*Neotonea intacta*), can be explained more readily. This flora is also found in Europe in the Spanish peninsula, south-west France and south-west England. The group as a whole is very varied (with rock plants, bogs, herbs and moors), although distinctly calcifuge in preference. It has been suggested that the reason for its presence is climatic, with the controlling influence being the January isotherm of $5.5°C$. It either survived on the mild, western coasts throughout the Pleistocene, from interglacial periods, or else it entered during a phase of lowered sea level in the early postglacial. What is certain is that these species must have migrated as far north as Ireland either when the climate was favourable or when the coastline was continuous in a way in which it is not now.

However, during the Hoxnian interglacial (the interglacial before the last one) at Gort in western Ireland there was a flora which included *Rhododendron ponticum* and the Lusitanian flora plants. This might suggest that this characteristic flora was already in existence then and is not a recent arrival, although there may have been dying out and recolonization. Today it does not show any signs of spreading, for it has been observed in Kerry for 40 years, and in fact there has been a decrease in some species: for example the strawberry tree is only known from place names in Mayo and Galway; there is a single patch of *Erica vagans* in Fermanagh; and the last sighting of an annual spurge (*Eurphorbia peplis*) was in 1839 at Garraris Cove, Waterford. It is unlikely that they could have survived in Ireland during the last glacial phase, due to the cold periglacial climates south of the ice sheets, and in fact the simplest explanation is that they remigrated to Ireland about 7500 years ago when the climate was wetter, with a warmer winter. Later they are likely to have become more restricted in range with increasing winter coldness and all areas that were not oceanic would have lost this group of plants.

Another example shows a very patchy and discontinuous distribution, as it is found along the western coast of Ireland, around Lough Neagh in Fermanagh and in the Wicklows, but also in the Scottish Isles and North America. It is not found in Europe south of Ireland. This so-called North American element of the flora includes two orchids (*Spiranthes gemmipara* in western Cork and Kerry and *Spiranthes romanzoffiana* from Lough Neagh and the River Bann); the blue-eyed grass (*Sisyrhinchium angustifolium*), a rush species (*Juncus tenuis*), a pondweed (*Najas flexilis*) from the west-coast lakes from Kerry to Donegal) and the pipewort (*Eriocaulon septangularia*) from Kerry to Donegal. Some botanists consider that this group was a preglacial survival and not a recolonization, comprising the last relics of a group that existed on both sides of the Atlantic in preglacial times. However, the pipewort has been found at the Gortian interglacial site and *Najas flexilis* was abundant and widespread throughout western Europe during the mid-Holocene and therefore no preglacial survival is required.

The accidental dispersal hypothesis (by currents, winds and birds) has sometimes been put forward to explain problem distributions. This was the case for this North American element when an Irish lighthouse keeper called Sullivan suggested that the Greenland white-footed goose was a possible dispersal agent. This bird breeds in north-west Greenland and overwinters in Ireland. Some of the Greenland population may also winter in the region around the St Lawrence mouth in North America and they frequent lake margins. It so happens that the Irish North American floral element is all freshwater species, or occurs in Ireland in lake-margin habitats. Some of the species are also known to be dispersed by water fowl in America, and one, the blue-eyed grass, grows in north-west Greenland, within the breeding ground of the goose. Nevertheless, the digestion of birds is very rapid and the seeds are often crushed first, and therefore extended transport is unlikely. They could stick on the birds' feet, but this seems unlikely too.

The Arctic–Alpine element in the Irish flora has a peculiar distribution too, for the Irish climate, with its high precipitation, low summer and high winter temperatures and windiness, seems to have had an effect on the survival of this group. Many members do not seem to have any desire for a mountain habitat and do not today show much tendency to retreat to the lower temperatures offered by the mountains. Of the 26 species, 11 occur at sea level and 10 more under 330 m, and at 600 m there is a marked diminution. The Sligo–Leitrim plateau, between 300 and 500 m, is the main area, with 19 species; the Burren limestone in Galway is another. There is no difficulty in explaining this element as it is likely to have migrated during cold phases of the Pleistocene and survived in mountain refuges, south of the ice, or offshore, as sea levels were considerably lower. Two species show an endemic distribution, only occurring in Ireland: *Arenaria ciliata hibernica* on Ben Bulben in Sligo and *Salix hartii* on Aranmore Island in Donegal. The distribution in the United Kingdom of an Arctic–Alpine species, such as the purple saxifrage (*Saxifraga oppositifolia*) or the dwarf birch (*Betula nana*) and dwarf willow (*Salix herbacea*) is different, with high mountain locations being prevalent (Figure 23.1).

Generally there are two groups of reasons to explain why organisms live where they do – ecological and historical biogeographical reasons – and the Irish flora examples illustrate these concepts well.

The ecological explanation includes several interrelated ideas. First, each species has a characteristic life history, reproduction rate, behaviour and means of dispersal, and these traits affect a particular population's response to the environment in which it lives. Second, the response of a population to its physical surroundings (the abiotic environment), including such factors as temperature, light, soil, bedrock geology, fire, water and oxygen levels, and its living surroundings (the biotic environment), including such factors as competing species, parasites, predators, diseases and human influences. Each species can only tolerate a certain range of environmental factors: it can only live where these factors lie within its tolerance limits. If one factor is outside this range, the species cannot survive. This is known as the limiting factor. Some species are tolerant of a wide range of conditions and may be found over large areas of the world (eurytypic species), whereas others are very intolerant and may be restricted to small areas by the effects of one factor (stentypic species).

The historical explanation includes two basic ideas. The first is that a species originated in one particular place and then spread to other regions if it could and was willing to do so. The second is the importance of both geological change, such as plate movements or sea-level rise, and climatic change, which can split a single population into two or more isolated groups.

An example here is the modern tapir, which is a close relative of the horse and rhinoceros and shows a broken distribution. There are four living species, one in south-west Asia and three in Central and South America (Figure 23.2(a)). At first sight it is difficult to understand how such close living species can live in such distant parts of the world. However, the fossil record is instructive here and members of the tapir family were once more widely distributed than at present; they are known to have lived in North America and Eurasia. The oldest fossils come from Europe and it would seem likely that the tapirs evolved here

(a)

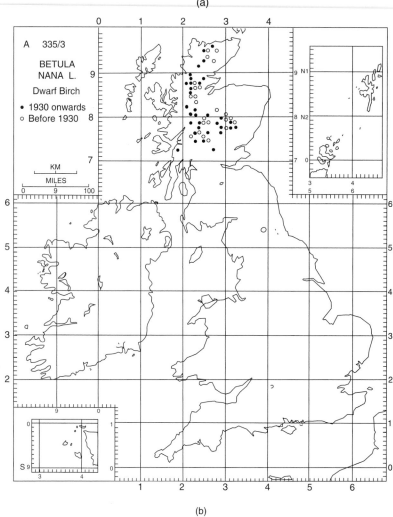

(b)

Figure 23.1 (a) Purple saxifrage (Saxifraga oppositifolia). (b) and (c) Distribution of dwarf birch (b) and dwarf willow (c) in the British Isles today

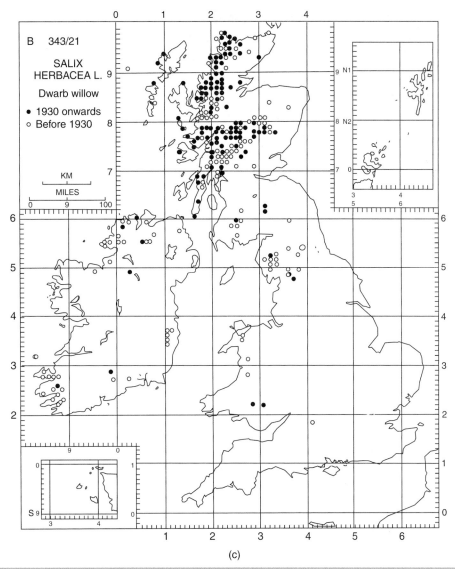

B   343/21

SALIX
HERBACEA L.
Dwarb willow

• 1930 onwards
○ Before 1930

(c)

Figure 23.1 *(continued)*

(their centre of origin) and then dispersed east and west, and that those that went north-east reached North and South America. Subsequently, probably owing to climatic change, the tapirs in North America and the Eurasian homeland became extinct, while the survivors at the tropical edges of the distribution produced the present species.

This is fine as a model until a geologist finds an older fossil tapir from another part of the world, because we are always faced with an incomplete fossil record. Thus there are many examples of evolutionary relict species, such as the dwarf mammoth species on Wrangell Island in the Siberian Arctic, which survived until 6500 BP while most other mammoths in the world were extinct by 12 000 BP, or a smaller variety of the giant Irish deer, which survived in the Isle of Man and south-west Scotland apparently 2000 years longer than on the Irish mainland at the end of the Pleistocene and into the Holocene (it is possible that the grassland

habitat on which the deer depended survived in these areas later than in Ireland after the migration of post-glacial trees).

The climatic relicts like the Arctic–Alpine species discussed earlier are common, and just one additional animal species will be given here. The dung beetle (*Aphodius holdereri*) is today restricted to the high Tibetan plateau, between 3 and 5 km, with its southern limit at the northern slope of the Himalayas. In the period between 25 000 and 40 000 BP in the mid-Devensian interstadials in the British Isles it was widespread, but since then climatic change has severely restricted the availability of suitable habitats for its survival.

23.3 Components of ecosystems

All life forms require the Sun's energy for their existence; this is true whether we are looking at terrestrial or marine life. The

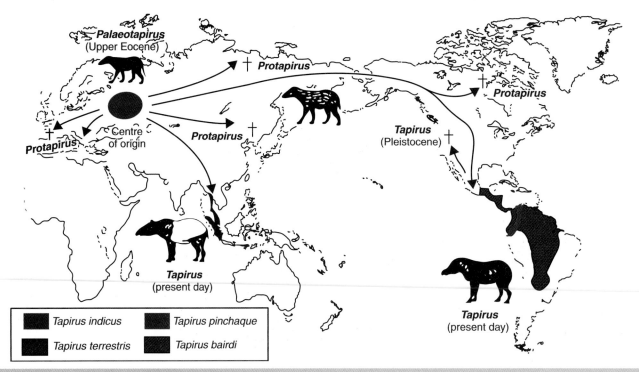

Figure 23.2 (a) The distribution of the living species of tapir

major exception is the specialized life that exists in the 'smokers' in the mid-ocean ridge systems, which exist without sunlight.

Ecology is all about the study of interrelationships within an ecosystems. An ecosystem is the whole complex of plant and animal populations that are interlinked in their living through the effects they exert on one another, coupled with the sum total of all environmental factors that influence any one or all of the component organisms. For example, Mull Woods, Croxteth, a small wood close to Liverpool, is made up of several different components (Figure 23.3 and 23.4(a)): the mature trees, the understory of coppiced woodland, the rhododendron scrub woodland, the ground vegetation such as brambles and the animals that live within the undergrowth and trees. All these are closely associated with one another.

Many of the animals in this wood derive their food from the plants, but others are predators that live off other animals. Some plants are parasitic and live off other plants, for example on the decaying woodland trees. Further, the plants obtain their energy from the soil, the Sun and the atmosphere: water and nutrients are taken from the soil, vital elements are taken from the atmosphere and the whole process is powered by solar energy. So this woodland includes not only the trees, the vegetation and the surface animals, but the soil and its organisms and the immediate atmospheric envelope with which all the rest interacts.

Any ecosystem has two parts:

1. The non-living or abiotic environmental factors. These include the physical factors such as winds, moisture, temperature and solar radiation, the inorganic elements and compounds, and the organic compounds that are the by-products of organism activity or death.

2. The biotic components, such as plants, animals and microbes, which interact in a fundamentally energy-dependent manner.

Ecosystems can be any size but they usually cover several distinctive biotic associations and form a complete and self-contained

(a)

Figure 23.3 (a) Mull Woods, Croxteth, Liverpool. (b) Generalized forest ecosystem

TREE LAYER

SHRUB LAYER

FIELD LAYER
GROUND LAYER
LITTER LAYER
TOPSOIL

SUBSOIL

REGOLITH

(b)

Figure 23.3 (*continued*)

unit. For example, the woodland ecosystem of the northern coniferous forests has islands of land that are not forested within it. Similarly, we could discuss the moorland ecosystem of Upper Teesdale, which would include distinctive plant associations on the Whin Sill dolerite, the limestone and metamorphosed limestones, and the drift-covered, acid moorlands. It is also possible to describe a woodland ecosystem without reference to a particular place.

An ecosystem is usually self-sufficient in resources and self-perpetuating, although there are usually broad transitional zones, or ecotones, between ecosystems, rather than sharp dividing lines. However, just as no organism is completely self-sufficient, neither is an ecosystem a completely discrete entity delimited sharply from all other ecosystems; rather it is complexly influenced by any surrounding and connecting ecosystems.

Ecosystem ecology emphasizes the movements of energy and chemical elements (nutrients). Major concerns include how much and at what rate energy and nutrients are stored and transferred between the components of a given ecosystem. The cycling of matter and energy is therefore central to how any ecosystem works. The fundamental driving force for an ecosystem is solar energy or radiation, but this is only one of the many inputs (Figures 23.5

and 23.6). Rainfall supplies water and minerals, with much of it being stored in the soil for later use. Weathering breaks down rocks and supplies inputs to the soil. Organisms can also enter the ecosystem: migrating animals can move in and winds can transport seeds. Humans provide many inputs too, such as the addition of fertilizers to soils, or of pesticides for the control of crop pests.

The outputs can also be seen in Figure 23.5. Soil water flows away, animals migrate, wind carries leaves, gases escape back to the atmosphere, and humans crop the plants and timber. Such an ecosystem is an open system, receiving and losing energy and matter across its boundaries.

To achieve balance or equilibrium in the system there are self-regulatory controls by which it can respond to changes. These can be achieved through various feedback mechanisms. Negative feedback has a stabilizing ecosystem-perpetuating influence in that it prevents major changes in any one component. An example of this is that an animal population within a particular ecosystem may increase, resulting in the depletion of the food supply. Some of the animals will subsequently die because of the food shortages and the animal population will be reduced, ultimately restoring the original ecosystem characteristics. Positive feedback is damaging to the ecosystem equilibrium as when it occurs, one

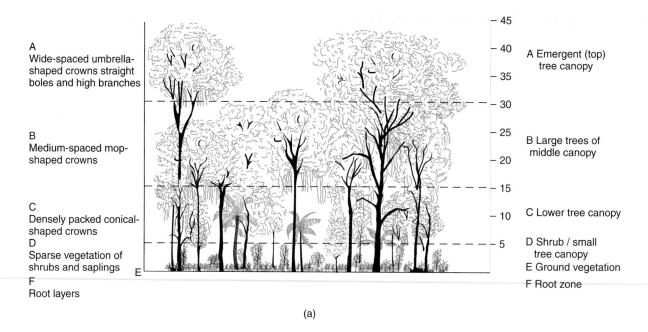

A
Wide-spaced umbrella-shaped crowns straight boles and high branches

B
Medium-spaced mop-shaped crowns

C
Densely packed conical-shaped crowns
D
Sparse vegetation of shrubs and saplings
E
F
Root layers

A Emergent (top) tree canopy

B Large trees of middle canopy

C Lower tree canopy

D Shrub / small tree canopy
E Ground vegetation
F Root zone

(a)

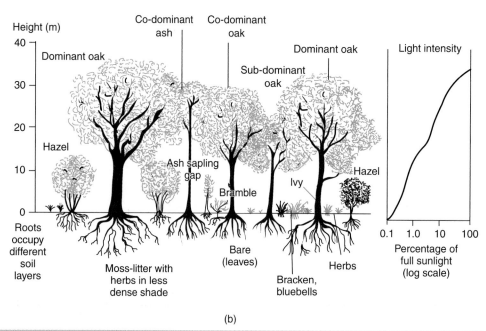

(b)

Figure 23.4 Major elements in: (a) an evergreen, tropical rainforest ecosystem; (b) a temperate deciduous forest. (Sources: (a) after Richards, 1979; (b) after Cousens, 1974)

component becomes dominant, normally at the expense of others. An example of this is fire, which is an important mechanism in the maintenance heathland communities in many temperate parts of the world. If it is eliminated, the ecosystem will change, be invaded by non-heathland species and eventually develop into scrub woodland.

Occasionally a threshold is crossed that disturbs the dynamic equilibrium, which is the ideal state in ecosystems. Impetus for change becomes so great that control within the ecosystem cannot accommodate it and a state of imbalance occurs. The point of no return is reached and a new ecosystem is established. Large-scale clearance of tropical rain forest is a good example of this, resulting in nutrient depletion, soil erosion and an overall degradation of the ecosystem. Examples can be found in many parts of the Amazon basin, Malaysia and Indonesia.

23.4 Energy flow in ecosystems

All ecosystem organisms require energy to move, grow, reproduce and carry out bodily functions, and in most ecosystems this energy comes from sunlight. The biological parts of an ecosystem can be divided into several energy levels: producers, consumers and decomposers (Figure 23.7). However, the producers, which are all the green plants, use part of the solar energy for photosynthesis (energy fixation).

The amount of solar radiation available to plants varies with geographical location, although much of the energy entering the Earth's atmosphere is lost ($\sim$15.3 $\times$ 10^8 cal/m^2 per year). For example, in Britain this solar radiation is only $\sim$2.5 $\times$ 10^8, whilst in Michigan it is $\sim$4.7 $\times$ 10^8 and in Georgia it is $\sim$6.0 $\times$ 10^8. In fact, about 95–99% of this energy is immediately lost from the plant in the form of sensible heat and heat of evaporation. Only the remaining 1–5% is available for photosynthesis.

This energy, once received by an organism or an ecosystem, passes through a series of transformations, mostly from one form of chemical energy to another. With each transformation some energy is converted to heat and ultimately radiated away. This is therefore a one-way flow which diminishes in amount until all the energy has been lost as waste heat.

This energy is initially used in the photosynthetic process, whereby CO$_2$ from the atmosphere is converted into energy-rich carbon compounds, the carbohydrates. Oxygen is liberated into the atmosphere as a by-product. Thus, this process produces organic matter – that is, food for the next level in the system – and oxygen, which all living things require. The producers fulfil this crucial role. As the energy incorporated is subsequently synthesized into other molecules that serve the nutritional requirements of the producers' own growth and metabolism, it is possible to speak of the producers as being self-feeding, or autotrophic.

The autotrophs manufacture their own food from inorganic substances using light. They require water, CO$_2$ and oxygen, and 16 inorganic nutrients. Their biochemical systems are capable of producing the biological macromolecules needed for life entirely from these inorganic substances.

The overall effect of photosynthesis is to unite the hydrogen atoms of water with the atoms of CO$_2$ to form carbohydrates and release oxygen. A simplified equation for this is:

$$6CO_2 + 6H_2O + \text{light energy}$$
$$= C_6H_{12}O_6 \text{ (organic sugar)} + 6O_2$$

However, this is a complex process which involves two forms of chlorophyll, a green pigment particularly adapted to absorbing light energy and releasing it in a useable form.

Respiration (the degradation of chemical energy to heat) is the complimentary process to photosynthesis, in which carbohydrate is broken down and combined with oxygen to give CO$_2$ and water. Hence solar energy is transformed to chemical energy by plants, which conforms to the laws of thermodynamics:

solar energy assimilated by plants

= chemical energy of plant growth + heat energy of respiration

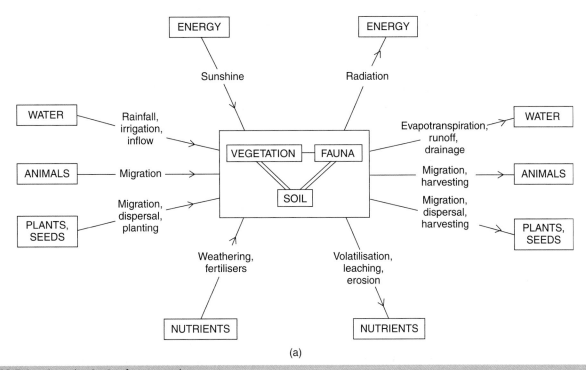

(a)

Figure 23.5 Inputs and outputs of an ecosystem

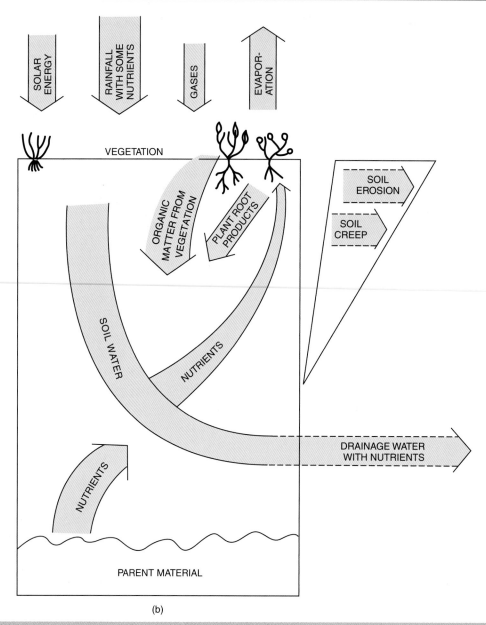

(b)

Figure 23.5 *(continued)*

So the amount of energy that ultimately leaves the ecosystem is equivalent to that entering it, but it does so in a transformed state. It is in conformity with the second law of thermodynamics that energy is degraded from the free or high-grade form (equivalent to work) to the low-grade form (equivalent to heat).

Figure 23.8(a) illustrates the autotroph as an energy system. F1 is the flow of light energy from the Sun, which provides the power source. Only about half of this available energy is absorbed in a unit of time, and the rest is reflected (F2). Of this absorbed energy flow (F3), most is absorbed by molecules other than chlorophylls and so is quickly re-radiated away as heat (F4). Some, however, is stored as chemical energy in the form of carbohydrate. This is labile storage in a form which can be readily broken down. This flow into labile storage (F5) is the primary productivity. Flow of energy through the respiration process is shown in Figure 23.8(b), where the power source is the chemical energy of labile storage. The flow of energy from this source (F6) is partly lost by the inefficiency of respiration (F7) but the remainder (F8) is used to maintain and increase the total energy stored in the biomass.

From the point of view of photosynthesis, the autotroph has one input (solar energy) and three outputs:

1. Flow of energy unused in photosynthesis, which is released to the surroundings.

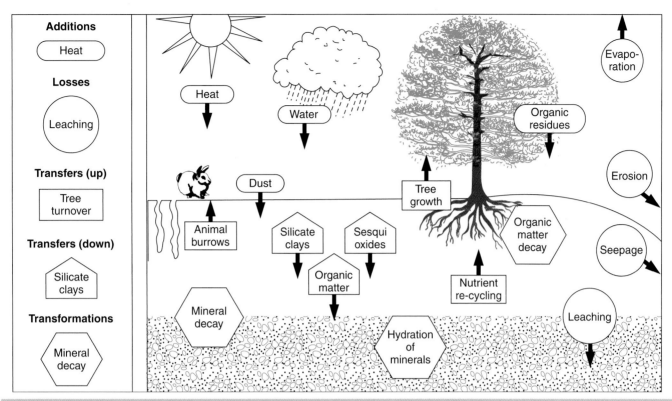

Figure 23.6 A soil system showing the additions, losses, transfers and transformations of material

2. Energy lost as heat in the conversion of photosynthesis, which is released to the surroundings.

3. Energy lost by respiration in converting one form of chemical energy to another.

Some autotrophs use material with a low potential chemical energy as their energy source. They are said to be chemo-autotrophic or chemosynthetic. For example, the energy used by *Beggiatoa*, a blue-green algae, in synthesizing organic materials is produced by oxidation processes. It is able to grow in the complete absence of organic substances. However, this is insignificant when compared with green plants as a source of food for the next stage.

Primary productivity is the amount of chemical energy of all forms produced by autotrophs from inputs of light and inorganic matter. The values vary in different ecosystems (Table 23.1). In areas of low biological activity, such as deserts or the Arctic tundra, primary productivity may be as little as one hundredth of that in tropical rain forests or coral reefs. However, as it is expressed on an areal basis, low productivity does not necessarily mean that the plants are less productive, just that there are fewer plants.

Gross primary production is the total amount of chemical energy stored by autotrophs per unit area per unit of time. It does not represent the food potentially available to consumers, since the producers have to perform work in synthesizing the

organic matter and some energy is lost in respiration. Therefore, gross primary production respiration represents the food energy potentially available. This is called the net primary production and is about 80–90% of the gross primary production.

This is linked with the concept of photosynthetic efficiency. Green plants are the main agents for the manufacture of the high-energy materials necessary for the maintenance of life. It is therefore important to know the efficiency with which various plants convert the Sun's radiant energy into chemical energy of plant protoplasm. Photosynthetic efficiency is the net productivity divided by the total amount of visible light reaching the plant. The values are usually very low, in the order of 1–5%, and values for productivity are normally based on a year. As might be expected, the most productive plant communities occur in the tropics, where rain forest and perennials under intensive cultivation may produce 50–80 million tons per hectare, whereas in temperate regions reedswamps may produce 30–45 million tons per hectare.

A factor of significance when comparing production at low and high latitudes is the predominance of different biochemical pathways in different environments. The Calvin cycle is the classic description of photosynthesis, where carbon dioxide is fixed initially as a three-carbon compound, phosphocleric acid, in a reaction catalysed by the enzyme ribulose biphosphate carboxylase. Plants with this pathway are called C_3 plants. This is in contrast with C_4 plants, which have a four-carbon compound as the first product. C_4 carbon fixation is catalysed by phosphoenol pyruvate carboxylate to produce oxaloacetic acid. This pathway

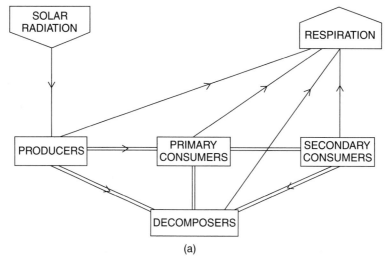

(a)

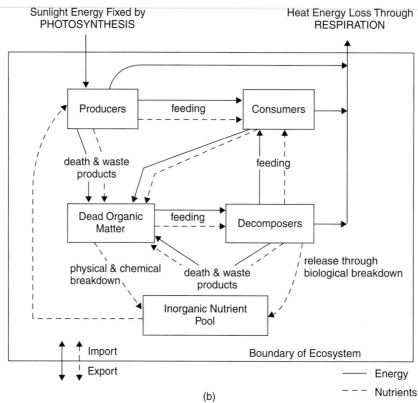

(b)

Figure 23.7 (a) Biological parts of an ecosystem (energy levels). (b) Patterns of energy and nutrient exchange in an ecosystem

is only found in flowering plants and in less than 20 out of 300 families.

A third biochemical pathway, crassulacean acid metabolism (CAM) is found in a few specialized plants that undergo extreme water stress. To conserve water, the stomata remain closed during the day, and at night carbon fixation occurs by the C_4 route to produce malic acid, which is stored in the cell vacuoles. During daylight the malic acid is decaroxylated and carbon is refixed by the Calvin cycle. The generalizations that can be made in the three systems of carbon fixation are that:

1. C_4 plants are the most efficient photosynthetically in optimum conditions and are adapted to hot, bright and fairly dry climates.

2. C_3 plants are adapted to cool, moist and shady environments.

3. CAM plants are adapted to very dry environments.

The next energy level is the primary consumer and all herbivorous animals come into this category, including caterpillars.

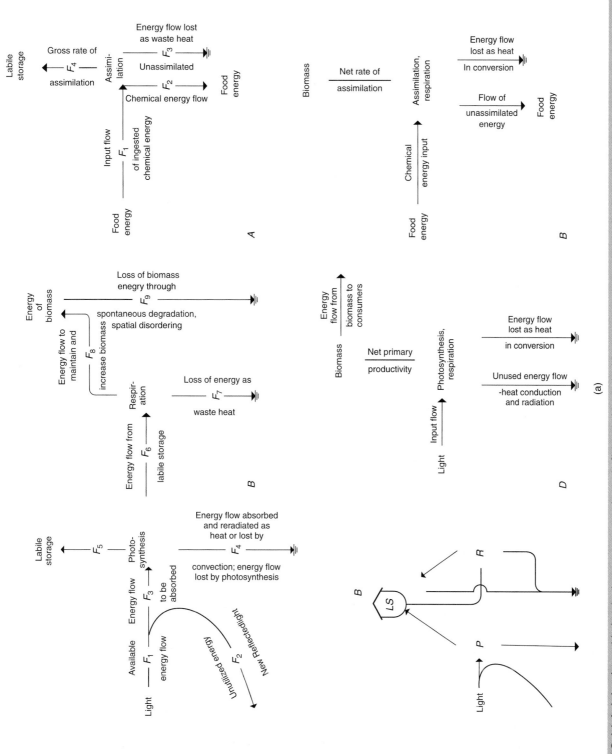

Figure 23.8 (a) Autotrophs and heterotrophs as energy systems. Left-hand side: Autotroph as an energy system: A, photosynthesis; B, respiration; C, combined diagram of photsynthesis and respiration. Right-hand side: Simplified representation of the heterotroph as an energy-flow system; D, simplified representation of the autotrophs as an energy-flow system. (b) Fate of energy incorporated by autotrophs in Cedar Bog Lake, Minnesota, in gm calories per cm² per year. (Sources: (a) after Odum, 1971; (b) after Lindeman, 1942)

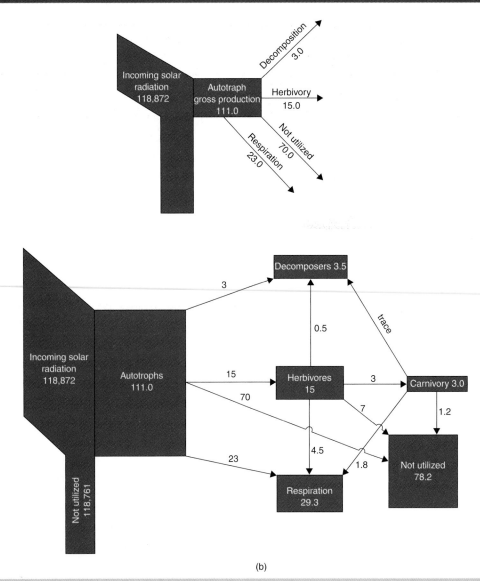

(b)

Figure 23.8 (*continued*)

After this there is the secondary consumer level, the carnivores. The secondary consumer level is sometimes thought of in terms of any animal living off herbivores, with tertiary consumers being carnivores that feed on other carnivores. Thus the secondary consumer category includes spiders, snakes, lizards, frogs, eagles, hawks and fish-eating sea birds, as well as all the obvious ones.

The fourth level comprises the decomposers, which live on dead plant and animal material. Their function is to break down the complex organic compounds into simple compounds and elements, which are returned to the soil and become available as raw materials again. This level is divided into two groups:

1. Small detritus-feeding animals like soil mites, millipedes and worms.

2. Bacteria and fungi of decay.

The relative importance of these groups is difficult to establish, although it is known that on occasions microorganisms have been responsible for transmitting as much as 90% of the energy flow through such a system.

Humans are a consumer of both plant and animal material, as are animals such as bears. These omnivorous species do not fit easily into the above scheme.

Scavengers are carnivores. Any organism whose nutritional needs are met by feeding on other organisms is referred to as heterotrophic, or other-feeding.

In almost all ecosystems chemical energy is stored and used by heterotrophs higher up the food chain. This intake of chemical energy and its conversion to useful forms is the process of assimilation. A portion (F2) of the total chemical energy ingested (F1) by heterotrophs is unused and excreted as faecal matter. There is a loss of energy through work (F3) and a flow of energy into labile storage (F4) (Figure 23.8).

Table 23.1 Net primary productivity and total biomass from selected ecosystems. (Source: Whittaker and Likens, 1975)

Ecosystem	Net primary productivity (t ha^{-1}y^{-1})		Biomass (t ha^{-1})		Area 10^6 km^2
	Range	Mean	Range	Mean	
Tropical rain forest	10–35	22	60–800	450	17
Temperate deciduous forest	6–25	12	60–600	300	7
Temperate evergreen forest	6–25	13	60–2000	350	5
Boreal forest	4–20	8	60–400	200	12
Savanna grassland	2–20	9	2–150	40	15
Temperate grassland	2–15	6	2–50	16	9
Tundra	0.1–4	1.4	1–30	6	8
Semi-desert	0.1–2.5	0.9	1–40	7	18
Desert	0–0.1	0.03	0–2	0.2	24
Arable land	1–40	6.5	4–120	10	14
Lakes and streams	1–15	4	0–1	0.2	2
Estuaries	2–40	15	0.1–40	10	1.4
Upwelling ocean zones	4–10	5	0.05–1	0.2	0.4
Continental shelves	2–6	3.6	0.01–0.4	0.01	26.6
Open ocean	0.02–4	1.25	0–0.05	0.03	332

It is clear that there must be a dynamic equilibrium between net primary production by plants and the amount of food assimilated by heterotrophs. Individual heterotrophs do not assimilate all the food they consume: in herbivores as much as 90% may pass out of the body, whereas in omnivores this is between 30 and 75%. So we see that for a heterotroph:

chemical energy eaten = chemical energy assimilated

+ chemical energy of the faeces produced

and:

chemical energy assimilated = chemical energy of growth,

including production of young and excretory products

+ heat energy of respiration

The excretory products serve as food material for other heterotrophs and at each stage heat is used and lost to the system. The end result conforms with the laws of thermodynamics in that the solar energy entering the system equals the heat energy leaving it.

Losses through respiration are important and in any ecosystem the losses by individual species should be investigated. It has been shown that different species have different respiration rates since some species are more active than others. For example, the activity of adult oribatid mites varied between the 46 different species. Life stages within a single species also vary and when measurements are made it is necessary to measure the respiratory rate over a period exceeding 24 hours, which counteracts the differential respiratory rates due to either diurnal or nocturnal activity.

23.5 Food chains and webs

The exact sequence of who eats what is called a food chain, if it is fairly simple, or a food web if complex. In nature a food chain is usually part of a large number of other food chains and there are very complex relationships between the components. Examples of simple food chains are:

1. leaves → worms → blackbird → sparrowhawk

2. frog → snake → owl

3. leaves → worms → blackbird → sparrowhawk → frog → snake → owl

4. on a British chalk grassland: grass → rabbit → human

In all these cases there are plants, herbivores and carnivores, and primary producers, primary consumers and secondary consumers, as can be seen in Figure 23.9.

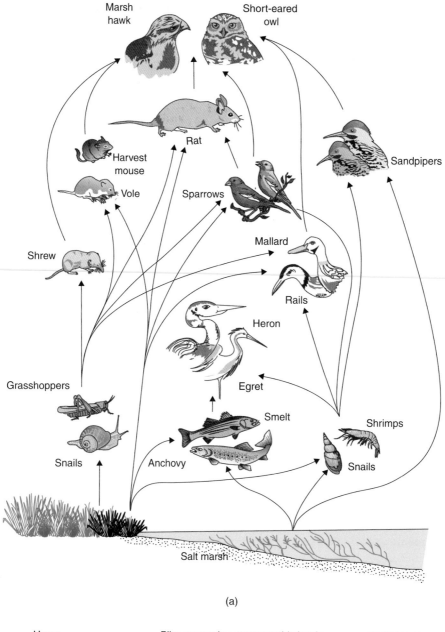

(a)

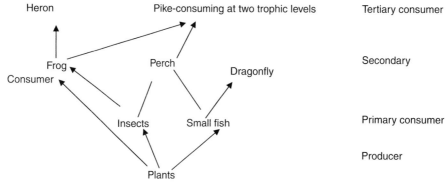

(b)

Figure 23.9 Simple food chain and food web, from a salt marsh in San Francisco Bay. (Source: after Smith, 1966)

Each step in the transfer of food within a chain is called a trophic level and at every one potential energy is lost through the dissipation of heat. Energy flow is unidirectional and noncyclic.

Take another example, the diet of the herring (*Chaetoceros decipiens*). Here there could be a simple, linear food chain from a diatom to a copepod (such as the small crustacean *Calamus*) to the herring, but really this is part of a much greater food web containing a series of organisms that are all partially dependent on one another.

Looking at the diet of a sparrowhawk (80% rodents, 15% small birds, 5% insects), it is clear that there must be more mice in the ecosystem than sparrowhawks. To emphasize this point, in the marine ecosystem for every 1000 lb of plant plankton there is a 100 lb increase in animal plankton. For every 100 lb increase in animal plankton there is only a 10 lb increase in fish, and by eating 10 lb of fish, humans will increase by 1 lb.

Another example can be taken from the Kruger National Park in South Africa, where in the 1958–1959 season leopards killed 420 herbivores, of which 320 were impala, while cheetahs killed 65 herbivores, of which 47 were impala. This illustrates the large number of impalas involved. In this ecosystem there must be a balance of numbers at each level because each species must maintain its population and any losses must be replaced by new births. The biomass of impala is related to the food production of the vegetation. If there were too many impala they would eat the vegetation reserves, adequate regeneration of the vegetation would be prevented and overgrazing would result. This would mean that the ecosystem would be out of balance. In a balanced ecosystem, all levels are adjusted to the production available.

Food chains can be differentiated into several categories. For example, there can be:

1. Predator chains, in which energy moves from small organisms to larger animals.

2. Parasite chains, in which smaller organisms prey on larger ones.

3. Saprophyte chains, where energy is removed from dead material by a series of microorganisms.

On the other hand, two major types of food chain can exist separately, or in conjunction with one another:

1. A grazing food chain, where there is a fairly direct and rapid energy transfer from living plants, through grazing herbivores to carnivores.

2. A detritus food chain, where energy is transmitted much more slowly through the decomposed elements of dead plant and animal material to other organisms which feed on them.

In the marine ecosystem the grazing chain is the major pathway of energy flow, whereas in the forest ecosystem the detritus chain is the more important. The two food chains are not completely isolated from one another; for example, carcasses and animal faeces that were once part of the grazing food chain become incorporated in the detritus chain. In many ecosystems the detritus food chain is of greater importance in terms of energy flow than the grazing chain. For example, in Silver Springs, Florida (Figure 23.10), where the heterotroph food supply was initially present as living plants, there was a greater flow through the decomposers (5060 cal/m^2 per year) than through the grazers (3368 cal/m^2 per year). It is considered that microorganisms can account for as much as 90% of the energy flow through an ecosystem.

As all organisms dissipate energy through respiration, the production of a trophic level is always less than that of the preceding level. This can be illustrated by an ecological pyramid,

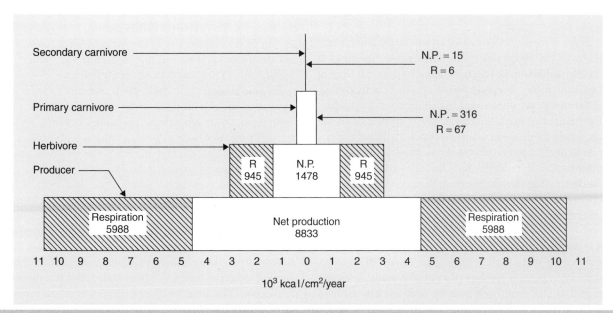

Figure 23.10 Ecological pyramid: example from Silver Springs, Florida, in cal/m^2 per year. (Source: Odum, 1957)

Numbers

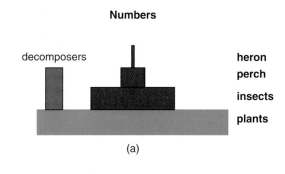

(a)

Biomass

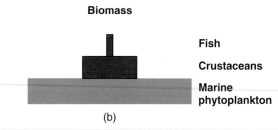

(b)

Figure 23.11 (a) Pyramid of numbers. (b) Pyramid of biomass

with plant production at the broad base and production of top-level consumers at the apex (Figure 23.10 and 23.11(a)). The animals at the base of a food chain are relatively abundant, while those at the top are relatively few in number, and there is a progressive decrease in between the two extremes.

This sort of analysis allows an ecologist to state the number of herbivores supplied by a given number of plants, for example, but in comparing ecosystems it is useless. This problem is partly overcome by using the weight of the organisms, or biomass, rather than the numbers. This is called the pyramid of biomass (Figure 23.11(b)). It only samples the quantity of organic material present at any one moment – that is, the standing crop – and gives us no idea of the productivity.

A third type of ecological pyramid, an energy pyramid, overcomes many of the objections to the former two types and is defined as the total amount of energy utilized by different feeding types in a square metre over a set period of time (e.g. one year). Studying the energy provides a unifying concept, as the productivity of an individual organism can be given, or that of all the organisms within the ecosystem, and the productivities of regions as different as the tropical rainforest, Arctic tundra and sea floor can be compared.

The fact that energy flow decreases up the food chain or web of an ecosystem means that the number of trophic levels is ultimately limited, usually to five. An ecosystem is actually an energy system distributed over space, and as we go up the food chain less and less energy is available in a unit of area. Consequently the organisms feeding at the highest levels must range over a wider area. Eventually a point is reached at which the available energy is spread so thin that more work must be done to obtain it than is gained by it. At this point no organisms can be supported and the limits of the food chain are reached.

If an organism feeds at more than one level, a larger population can be supported by feeding at the lower level than the upper one. This results from the idea that only 10–15% of the available energy can be passed up the grazing chain without damaging the supplying level through overgrazing.

Energy flows in ecosystems are of great importance to humans as many of our agricultural activities serve to change, or rechannel, energy flows to maximize primary productivity. For example, pesticides alter the energy flows by reducing the number of primary consumers, while bringing food to animals saves them from expending energy in searching for it.

23.6 Pathways of mineral matter (biogeochemical cycling)

In the process of converting solar energy into chemical compounds by photosynthesis, the green plants also incorporate into their protoplasm a variety of inorganic elements and compounds. Among the important ones are the direct components of photosynthesis, carbon dioxide and water, and those that are critical to protoplasmic synthesis: nitrogen, sulphur, magnesium, calcium and phosphorus. As the green plant is grazed on by herbivores not only chemical energy in the form of carbohydrates, fats and proteins is transferred to the herbivores, but a lot of nutrients as well.

This process continues throughout the system to the decomposers. There is no diminution of the nutrients in the cycle and at certain points along the chain there may be a concentration, as in the case many years ago of mercury in tuna fish. After the decomposers' activity the nutrients are released to the environment for available use. There is re-cycling. This interaction between the physicochemical environment and the plants and animals lies at the core of ecosystem dynamics. Biogeochemical cycling illustrates the idea that interchanges of elements and compounds occur between both the biotic and abiotic ecosystem components. There is some loss as carbon, hydrogen and oxygen are lost in decomposition in the form of gases, carbon dioxide and water vapour. However, there are unlimited amounts of these components so this loss does not really matter.

Biogeochemical cycles have pools and flux rates. A pool is a quantity of any chemical substance in either the biotic or abiotic component of an ecosystem. Exchanges between pools occur through processes such as photosynthesis and erosion and the rate at which this exchange occurs is the flux rate; that is, the quantity of material passing from one pool to another.

In grassland ecosystems where there is a high biomass of consumers, the bulk of the nutrients follow a pathway from producer to consumer. In other ecosystems the balance is different, for example in forests there is generally a comparatively small biomass of consumers and the general nutrient flow path is directed from the producers to the decomposers. For example, the total weight of leaves in a temperate forest is around 5000 kg per hectare. The total weight of leaves which are processed by consumers is fairly small, for example aphids eat about 70 kg per hectare and voles perhaps 500 kg per hectare.

The actual quantities of elements used depends on the species present. Take the example of the calcium content of leaves (percentage of dry weight) from different tree species and some other plants: beech has 2.46%, oak 1.7%, pine 0.99%, bracken 0.83%, whilst heather has only 0.44%. In a mixed oak woodland the calcium use taken up by the trees is 102 kg per hectare per year, 16 kg per hectare per year is retained and 86 kg per hectare per year is released by leaf fall and returns to the soil. This means that about 10% of the total input of calcium every year needs to come from soil-weathering.

One of the most complete studies of major elements has been from tropical montane rainforest ecosystems. Nutrients are stored in the soil, the living biomass and litter layers (kg per hectare), and the fluxes are uptake (not quantified), litterfall and inputs in rainfall. As the rain passes through the plant canopy it leaches nutrients from the plants, adding to those in the rainfall to give the total amount of throughfall. Nutrients can be lost to the ecosystem by leaching and runoff from the soil, but these have not been quantified. Similar systems can be seen in Figure 23.4.

Of the essential nutrients for plant growth, nitrogen is the most abundant, for it is contained within each amino acid building block of protein. Phosphorus is used in the skeletal structure of vertebrates and is also vital to processes of biochemical synthesis at the cellular level in both plants and animals. Calcium and potassium both play a role in regulating the osmotic pressure in cells and calcium is precipitated as carbonate or phosphatic compounds in external and internal skeletons. Iron, magnesium and manganese are important constituents of enzymes as well as such macromolecules as chlorophyll. Sulphur is a component of proteins and is necessary in some enzymes as well, whereas sodium and chlorine are required by animals in reasonable amounts but seem unused by all but a few plant species.

The micronutrients, or trace elements (such as copper, zinc, manganese, iron, boron, molybdenum and cobalt) are elements required by organisms only in small amounts but are just as vital. Many of these play specialized roles in enzymes, vitamins and other molecules involved in biochemical reactions ranging from photosynthesis to nitrogen metabolism. Other elements, such as silicon and aluminum, are present in plant tissues but are probably not essential for growth. Table 23.2 summarizes the function in plants, the amounts needed and the nutrient sources. The nutrient cycle in the soil, with the main inputs, outputs and pathways for transfer of nutrients, is shown in Figure 23.12.

23.6.1 The carbon cycle (Figure 23.13)

This is probably the simplest of the nutrient cycles and is essentially a perfect cycle in that carbon is returned to the environment about as quickly as it is removed. The basic movement of carbon is from the atmospheric reservoir to producers, to consumers, from both these groups to the decomposers, and then back to the reservoir. The atmosphere has a concentration of about 0.030−0.04% carbon dioxide. As between 4 and 9×10^{13} kg of carbon is fixed annually in photosynthesis, along with a much smaller amount of direct fixation by marine invertebrates, it is clear that there is either a great mobility of carbon or an additional reservoir, or both. The additional reservoir is the ocean, which is estimated to contain over 50 times as much as the atmosphere.

Respiratory activity in the producers and consumers accounts for the return of a considerable amount of the biologically-fixed carbon to the atmosphere as gaseous carbon dioxide. However, the largest return is made by the respiratory activity of decomposers in their processing of waste materials and the dead remains of other trophic levels. There is a geological component too, which involves the deposition of plant material (peat, coal, oil) and such animal remains as mollusc shells and protozoan tests as carbonate rocks. We can see here that at times in the past net primary production must have exceeded community respiration and that carbon-rich organic matter accumulated in ecosystems. This process occurs today in a minor way in upland and lowland bogs, where peat is accumulating.

Where rocks are uplifted through tectonic processes carbonate rocks then occur on land and add to the soil nutrients by weathering. However, such transfers are small on a global scale compared with the exchanges between the biota and the atmosphere. In addition a number of aquatic plants occurring in alkaline waters release calcium carbonate as a by-product of photosynthetic assimilation. Upon weathering and dissolution of carbonate rocks, and the combustion of fossil fuels and volcanic activity involving deposits of fossil fuel and carbonate rocks, atmospheric carbon is returned to the reservoir. Some simple effects of humanity's activities are illustrated in Figure 23.13(b). Changes in carbon storage are discussed in Box 23.1 (see in the companion website).

23.6.2 The nitrogen cycle (Figure 23.14)

Nitrogen is converted from one form to another by the biochemical processes summarized in Table 23.2. Within the active nitrogen pools it exists in two forms: as inorganic nitrogen (nitrate and nitrite) and as organic nitrogen (amino acids and other nitrogen-containing compounds). By far the largest active nitrogen pool is that in decaying organic matter.

Nitrogen in storage pools occurs in two forms. One is a relatively inert molecular nitrogen (N_2) and the other is nitrogen in rock-forming minerals. More than five times as much nitrogen is stored in rocks as is found in the atmosphere.

There are three processes of nitrogen fixation: biological fixation, industrial fixation and atmospheric fixation. Plants obtain most of their nitrogen from the soil as nitrate or ammonium ions, with 60% occurring in agroecosystems and most of the remainder in forests, while other terrestrial ecosystems account for 7% of total terrestrial nitrogen fixation.

Industrial fixation includes the production of nitrogen fertilizers and the oxidation of nitrogen during fossil-fuel combustion. For example, in the combustion of fossil fuels N_2 and oxygen are brought together at high temperatures. Besides the carbon dioxide and water produced by the fuel itself, significant amounts of NO and NO_2 gases are often produced as an undesirable by-product. As they are released to the atmosphere these two gases soon

Table 23.2 Nutrient sources and their function in plants. (Source: after Briggs and Smithson, 1985)

Nutrient	Amount needed (mg/litre)	Function in plants	Sources
Oxygen	–	Respiratory processes	Atmosphere
Carbon	–	Carbon used in photosynthesis	Atmosphere
Nitrogen	15	Protein synthesis, formation of nucleic acids, amino acids, vitamins	Soil, as ammonium, nitrite or nitrate; often fixed by bacteria in plant roots
Sulphur	1	Protein and vitamin synthesis	Soil, as sulphates, gypsum or pyrites; as hydrogen sulphide in waterlogged soils
Calcium	3	Metabolic processes; vital constituent of cell membranes	Soil, from limestone, feldspars, augite, gypsum
Potassium	5	Protein synthesis and transphosphorylation (conversion of sugar to phosphate	Soil, from feldspars, micas and clay minerals
Magnesium	1	Constituent of chlorophyll; enzymic reactions	Soil, from dolomitic limestone, montmorillonite clays, biotite, augite, hornblende
Phosphorus	2	Component of many organic molecules; major source of energy through conversion of ATP to ADP	Iron, aluminium and calcium phosphates in soil; dissolved phosphates in soil solution
Iron	0.1	Oxidation-reduction processes in respiration	Iron oxides, sulphates and silicates; often chelated with organic acids
Manganese	0.01	Small quantities used in enzymic reactions	Iron-magnesian minerals
Copper	0.0003	Respiratory metabolism	Igneous vein minerals
Zinc	0.001	Enzymic reactions	Igneous vein minerals
Boron	0.05	Cell division during growth	Soluble borates (mainly from marine sources)
Molybdenum	0.0001	Nitrogen fixation and assimilation	Igneous vein minerals
Cobalt	0.00001	N fixation in root nodules	Igneous vein minerals
Sodium	0.05	Unknown	Sodium chlorides; sea spray
Silica	0.0001	Unknown	Silicate minerals
Chloride	0.05	Regulates osmotic pressure; balances cation concentration in cells and sap	Dissolved in rainwater entering soil

combine with water to produce nitrous acid (HNO_2) and nitric acid (HNO_3). This reaches the soil or the ocean by precipitation. Atmospheric fixation on the other hand only occurs during lightning strikes, when the high temperatures and pressures act to allow oxidation of N_2 to NO and NO_2, which eventually reach the Earth as nitrate and nitrite. Overall it accounts for about 2% of global nitrogen assimilation, the rest being cycled in nongaseous forms. However, a lot of the nitrogen in soils and biota originated in the atmosphere and has accumulated over the millions of years in which nitrogen fixation has been taking place. Volcanic processes have contributed another small addition over time as outgassing of NO and NO_2.

It is important to note that at present nitrogen fixation far exceeds denitrification and so useable nitrogen is accumulating at the surface. This is produced almost entirely by human activities in industrial fixation and the widespread cultivation of nitrogen-fixing food plants. Most of this excess nitrogen flows into rivers, lakes and oceans, where it causes a pollution problem by allowing the growth of algae and other phytoplankton, often to the detriment of desirable aquatic organisms, and contributes to the process of eutrophication, in which there is a reduction of the oxygen content to low levels. The simple effects of humanity's activities on the nitrogen cycle are illustrated in Figure 23.14(b).

23.6.3 The sulphur cycle (Figure 23.15)

The storage pool of sulphur is the sedimentary rocks. Of the active pools, the oceans contain the largest amount, mostly in the form of sulphate ions. Many of the other active pools are small, such as the decaying organic matter, the land plants and the marine

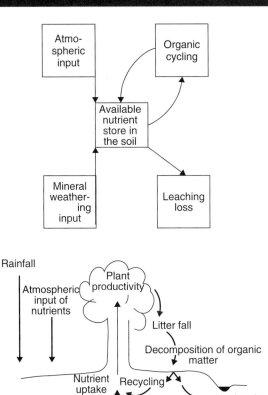

Figure 23.12 The nutrient cycle in soil

is in the form of sulphur dioxide or hydrogen dioxide and is eventually dissolved in water droplets, and precipitation returns some of this sulphur to the oceans. However, a large proportion falls out over land and so reverses the seaward flow of sulphate in ground and surface runoff. Direct plant uptake also removes a substantial part of the sulphur from the atmosphere, converting it to biologically-useful forms.

In addition to movements through the atmosphere, sulphur moves through the crust as sulphate and becomes incorporated into sedimentary rocks under the sea. As sulphur-containing rocks weather, SO_4 ions are released to the terrestrial portion of the biosphere, but movement through this section of the cycle is slow. Currently the amount of sulphur is increasing on land and in the oceans due to the burning of fossil fuel. The implications of this for the oceans should be small because of the vast size of the active pool of dissolved marine sulphates, but on land the story is different as pHs are lowered. The ecological effects are potentially many and complex, and include changes in the leaching rates of nutrients from plant foliage, changes in the leaching rates of soil nutrients, acidification of lakes and rivers, the effects on metabolism of organisms and the corrosion of buildings and other structures.

23.6.4 The phosphorus cycle (Figure 23.16)

Phosphorus is derived from the weathering of phosphatic minerals, such as apatite and some of the clay minerals, and is often a limiting factor to growth, mainly because it lacks an atmospheric pool and its biogeochemical cycle is linked closely with longer-term geological cycles. It occurs in the soil in three main forms:

1. Inorganic phosphate in the original minerals or compounds coating the mineral grains.

plants, with the active sulphur pool in the atmosphere being the smallest of all. However, the small size of this pool compared to the much larger yearly flows in and out of it shows that sulphur turnover in the atmosphere is very rapid.

There are four major flows to the atmosphere: bacterial emission, fossil-fuel burning, blowing of sea salts and volcanoes. Most

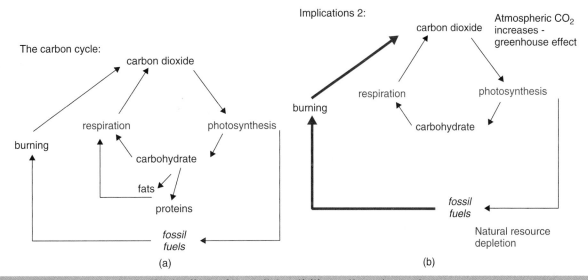

(a)   (b)

Figure 23.13 (a) The carbon cycle. (b) Simple effects of humanity's activities on the carbon cycle

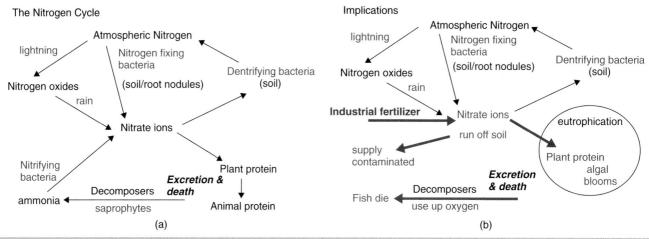

Figure 23.14 (a) The nitrogen cycle. (b) Simple effects of humanity's activities on the nitrogen cycle

Sulphur Cycle

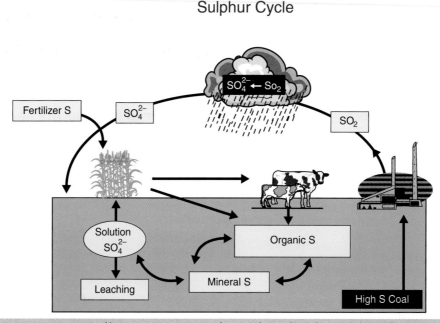

Figure 23.15 The sulfur cycle. (Source: http://www.omafra.gov.on.ca/english/crops/field/croptalk/2006/ct.o606a2.htm)

2. Organic phosphate bound up within the plant debris and soil fauna.

3. Soluble phosphate in soil solutions.

Generally phosphorus is highly insoluble, so quantitatively the last form is least important, although it is mainly from here that plants obtain their phosphates. So it is crucial that there is a constant phosphorus turnover between the other two forms and the soluble phosphorous. Phosphorus is taken up by plants in the phosphate form and is often aided by soil organisms. It can be returned directly to the soil through leaf fall or organic-matter decomposition, or it can pass along the food web. In the latter case much phosphorus is returned through animal waste, such as dung, and eventually, after a much longer time period,

through animal bone decay. Phosphates are also returned to the soil through fertilizers of various types. For example, nutrients washed from the land are carried to the ocean, absorbed by fish, eaten by seabirds and deposited as guano which is then collected by humans and used as fertilizer.

The phosphorus that is returned to the soil in an organic form is slowly released during the decay of organic matter. Bacteria play an important part in this process as they mineralize the phosphorus and convert it into inorganic forms. These may then be taken up by plants, or can combine with other elements, such as iron or calcium, to form the insoluble phosphate compounds in which they are stored. Small amounts are also adsorbed onto colloids. Some simple effects of humanity's activities on the phosphorous cycle are illustrated in Figure 23.16(b).

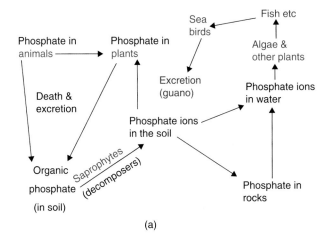

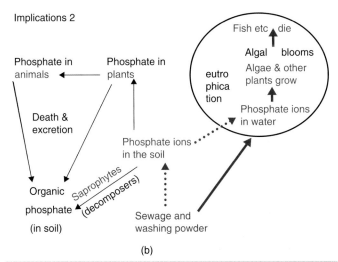

Figure 23.16 (a) The phosphorous cycle. (b) Simple effects of humanity's activities on the phosphorous cycle

23.6.5 The potassium cycle (Figure 23.17)

The main natural potassium sources are the potassium feldspars and micas, which account for around 90–98% of the total potassium in ecosystems. These minerals slowly weather, producing potassium-containing clay minerals such as illite and vermiculite and release potassium into the soil. However, most of the potassium is held in the clay minerals, within the clay lattice, and is not readily available to plants, although in the longer term it is slowly released by the gradual breakdown of the clay. A small proportion is held in an exchangeable form on the surface of the clay minerals and this is released by exchange with other cations in the soil solution. This source is available to plants, and is replenished by potassium from the clay lattice.

At any particular time around 10% of the readily-available potassium in the soil is dissolved in solution and so is directly available for plants. Some of this is carried away by leachates and reaches rivers and the ocean; some is lost by soil erosion, too. Losses by leaching are greatly affected by soil pH; almost four

times more potassium can be lost from a soil with pH 4.5–5.0 than from a neutral soil (pH 7.0). Some potassium is lost by harvesting and stored in humans, from where it passes into the sewage system, but much is returned to the soil in plant residues, animal wastes and manures. These, with inorganic fertilizers, provide the main sources of soil potassium. In the soil these inputs are released by decomposition and chemical reactions, then recycled, leached, adsorbed or fixed within the clays.

Biogeochemical cycling is crucial as, through this process, life-giving nutrients are circulated and all the ecosystem components are tied closely together, including the hydrological and landscape systems. Water is one of the main agents of nutrient cycling, acting in all kinds of ways: transfer of nutrients to plants, washing of nutrients from leaves and animal wastes, and decomposition of organic matter. It also helps initially in the weathering of minerals and it transfers the nutrients from the ecosystem, too.

23.7 Vegetation succession and climaxes

A plant community is one in which there is a complex of species. It has a definite structure and is composed ecologically and phenologically of different elements in time and space. Yet despite its dynamic nature it forms a persevering system which depicts botanically the physicogeographical relationships and the history of the region in which it is located.

A unity of life is impressed upon an aggregate of plants by their living under common conditions in a common habitat. So here there is an analogy with a human community, which shows interdependence between its members and an overall dependence on the common habitat. However, plant communities differ from one another in the species present and in the life forms present, whether they are on a world scale or more locally in a particular region. In some communities the plants are in physical, lateral contact, so that they form a closed community; when there is a space that can be colonized, it is known as an open community. Although areas of mature vegetation may appear to have gaps, often the root systems are in contact, leaving no room. Within a community certain species become dominant, and these species frequently control the ecological conditions that affect the others; for example, the tallest species overshadow the rest.

When there is more than one dominant species they usually exhibit the same life form, so that in a forest the typical life form is the tall tree trunk. However, it is often possible to distinguish a number of strata, particularly in forest communities, which is known as layering. It may be possible to distinguish the canopy, shrub, herb and moss layers. Each layer has one or more dominant species but sometimes the layering is incomplete, especially in forests in which the larger trees form such a thick, dense canopy that they shade the lower layers, which may be missing or poorly developed. The layering is only developed when there has been time for vegetation to work its way into these communities. Where there has been human interference it may not be possible to recognize the layers. For example, the shrub layer may have

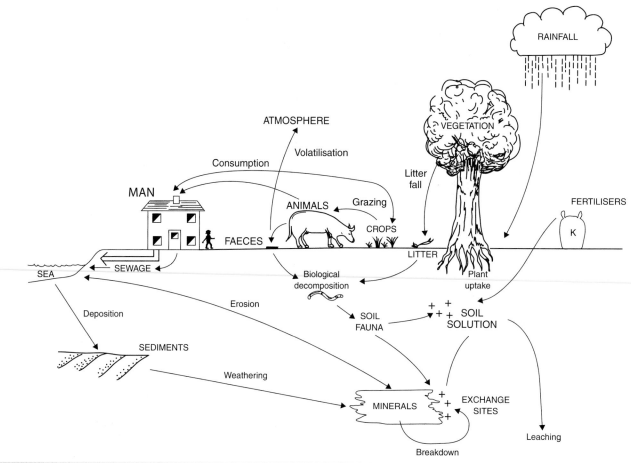

Figure 23.17 The potassium cycle

been cleared for firewood, or reduced by animal grazing. When plant communities cannot be recognized or easily defined there is a tension belt featuring a transition zone, which exhibits blending of the components of each community. An ecotone occupies such a habitat between adjacent communities.

A plant succession is where there is colonization of a given habitat both in time and space as it builds out. It is any unidirectional change that can be detected by the proportions of species in a stand, or the complete replacement of one community by another. Time is important in succession because on any given area subject to a certain climate and soil the natural vegetation is different according to its age. The succession typically begins on a bare habitat soon after it has been created, such as a new lava flow, mud flat, shingle beach, mudflow or scree and the seral successions that develop are called priseres (primary successions) (Figure 23.18). The first group of plants to colonize the new area is the pioneer species. Once they have established themselves the habitat will become modified, because, for example, the plants give shelter and humus as they die and break down. Competition comes into play and there is survival of the fittest as species fight for soil moisture and light. Thus a series of communities is initiated in turn until the area comes under the control of plants that seem capable of perpetuating themselves indefinitely. These plants form the climatic climax community that ends the

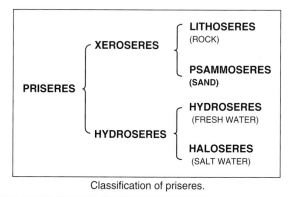

Classification of priseres.

Figure 23.18 Classification of priseres

succession. Here there is relative stability in the community, although there are some fluctuations around a stable, constant, mean condition. This means there is a dynamic equilibrium or steady-state condition in the community. Each community of the succession is marked by several stages:

1. **Pioneer or invasion stage:** Which species colonize depends on the proximity of potential seed parents and factors such

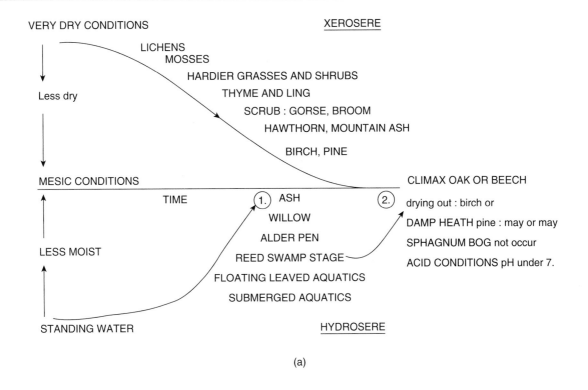

VERY DRY CONDITIONS   XEROSERE

LICHENS
MOSSES
HARDIER GRASSES AND SHRUBS
THYME AND LING
SCRUB : GORSE, BROOM
HAWTHORN, MOUNTAIN ASH
BIRCH, PINE

Less dry

MESIC CONDITIONS   CLIMAX OAK OR BEECH

TIME   ① ASH   ② drying out : birch or

WILLOW   DAMP HEATH pine : may or may

ALDER PEN   SPHAGNUM BOG not occur

REED SWAMP STAGE   ACID CONDITIONS pH under 7.

LESS MOIST

FLOATING LEAVED AQUATICS
SUBMERGED AQUATICS

STANDING WATER   HYDROSERE

(a)

PIONEER STAGE

Pioneer species slowly invade; increased weathering; slow organic matter accumulation.

BUILDING STAGE

Early colonising species become established. shading out pioneers; increased shade, organic matter accumulation; nutrient cycling more rapid.

MATURE (CLIMAX) STAGE

Late colonising species become established, shading out some early colonisers; active cycling of nutrients; intense competition for light, water, nutrients; multi-layered vegetation cover.

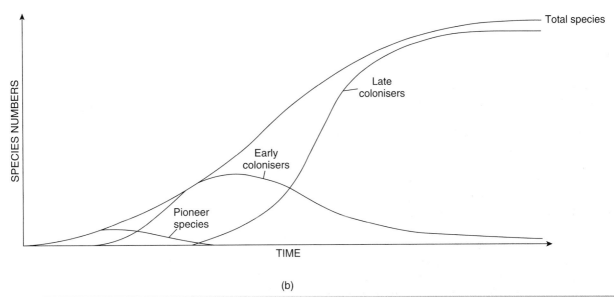

(b)

Figure 23.19 (a) Convergence to a single climax community. (b) Changes involved in seral succession. (c) Possible stages from Xerosere and Hydrosere with time. (Source: after Briggs and Smithson, 1985)

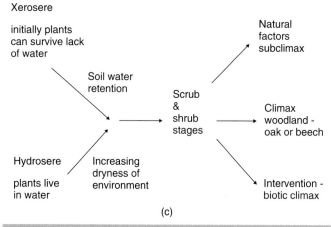

Figure 23.19 *(continued)*

as the methods of seed dispersal, whether by wind, animals or humans.

2. **Ecesis:** This involves successful seed germination, successful plant growth to maturity, and the flowering and setting of seed so that the community is perpetuated.

3. **Reaction:** Changes occur on the habitat, for example the addition of organic matter to the soil, changes in soil nutrients and water supply and changes in the microclimate.

4. **Stabilization:** The climax community is reached and the principal life form dominates, usually trees.

In any given area there is a degree of convergence to a single climax community amongst the successions, so that rock surfaces, ponds and dune successions may all end in forest. This is the convergence to a single climax community, which is determined by the climate of the region (Figure 23.19(a)). So, for example, a xerosere and a hydrosere can converge to produce a medial soil, with sufficient water and aeration to support the highest life form for a plant. These are mesophytes, or mesophilous trees, and the conditions for their occurrence are known as mesic conditions. In Figure 23.19(b) a general outline of the changes involved in a seral succession is illustrated, while Figure 23.19(c) illustrates how initial bare habitats can converge to produce similar climax vegetation even though the original habitats and different stages are so different. However, the convergence is only partial and major differences remain among the climax communities which develop in different habitats. Hence an area will contain a number of climax communities, forming a mosaic.

All the temporary communities in the sequence are collectively called a sere or chronosequence where there is a building-up in time, or a zonation where there is a horizontal succession in space. A number of trends, or progressive developments, can generally be recognized underlying most successional processes. A general outline of the changes involved is given below:

1. Progressive development of the soil, with increasing depth, increased organic content and an increased differentiation of horizons towards the mature soil of the final community.

2. An increase in height and massiveness and a differentiation into strata of the plant community,

3. Productivity, or the rate of formation of organic matter per unit area, in the community increases with increasing development of the soil and of the community structure and there is an increased utilization of the environmental resources. The microclimate is increasingly determined by the characteristics of the community as the plant height and density increase.

4. Species diversity increases from the simple communities of the early part of the succession to the richer communities of the late stage.

5. Populations rise and fall and replace one another across the passage of time. The rate of this replacement in many cases slows through the course of succession as smaller and shorter-lived species are replaced by larger and longer-lived ones.

6. Relative stability of the communities consequently increases. The early stages are usually unstable, with populations rapidly replacing one another. The final community is usually stable, dominated by longer-lived plants which maintain their populations, with the community composition no longer changing directionally. Good examples of succession can be seen on volcanoes, as in Figure 23.20 and earlier in Figure 13.13.

It is possible to recognize primary and secondary seral successions. The characters of the initial bare areas have a very important bearing on the kind of community which invades them, as well as on the rate at which the sere advances. There are primary bare areas created after deglaciation by erosion or after volcanic activity, and the successions that begin on such areas are called primary seres or priseres. They are characterized by having no break in the succession from the bare area to the climax.

Secondary bare areas are those that result from destruction of pre-existing vegetation by, for example, fire, cultivation, recreational trampling or grazing pressure. Any community sequences which develop in such cases are called secondary seres. The original destruction often does not destroy all the pre-existing vegetation and residual species may be mingled with invaders. In secondary successions the early seral communities benefit tremendously from the fact that the soil has been occupied by previous plants. It already contains humus with nitrogen, which supports an active biota; it contains a supply of dormant seeds and spores; and it is permeated by worm and root channels that promote aeration and allow water to enter easily. As the roots and litter of the devastated community decay, nutrients may become

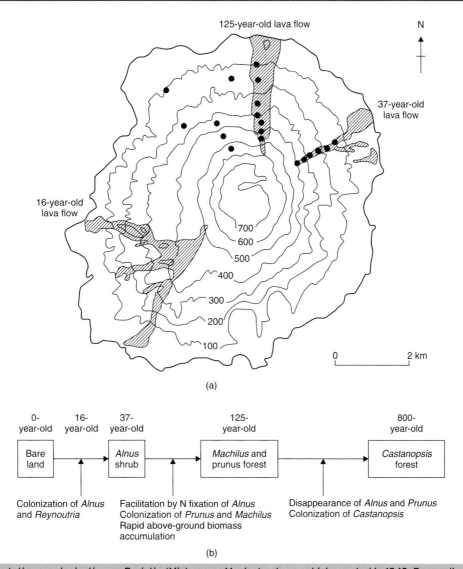

125-year-old lava flow

N

37-year-old
lava flow

16-year-old
lava flow

700
600
500
400
300
200
100

0   2 km

(a)

0- year-old	16- year-old	37- year-old	125- year-old	800- year-old
Bare land	→	*Alnus* shrub	→ *Machilus* and prunus forest	→ *Castanopsis* forest

Colonization of *Alnus*
and *Reynoutria*

Facilitation by N fixation of *Alnus*
Colonization of *Prunus* and *Machilus*
Rapid above-ground biomass
accumulation

Disappearance of *Alnus* and *Prunus*
Colonization of *Castanopsis*

(b)

Figure 23.20 (a) Vegetation recolonisation on Parictin (Michoacan, Mexico) volcano which erupted in 1943. Fumarolic activity can stille be seen on the left and the lava flows are now beginning to be colonized by trees. (b) Vegetation colonization on the Paricutin lava flows, with a Late Pleistocene, densely wooded cinder cone in the background as a contrast. (Source: after Kamijo *et al.*, 2002)

more plentiful than previously and there is a consequent increase in the environmental potential. This is unlike the primary succession, where the soil is nearly sterile and structureless and certain nutrients may not be present. Therefore the communities usually differ to a marked degree.

Fortunately there exist relatively few fundamental patterns of vegetational development, despite the diversity of expression of each pattern. The pioneer communities can be grouped into two broad categories:

1. Those adapted to a shortage of water, characterized by xerophyte plants (xeroseres).

2. Those adapted to an excess of water, characterized by hydrophyte plants (hydroseres).

23.7.1 Xeroseres

These seral successions either develop on bare rock surfaces or in rock crevices, when they are called lithoseres or psammoseres, which develop on the beach to sand dune environments.

Lithoseres

On bare rock there are very severe environmental conditions, yet in all but the driest areas plants colonize them. Crustose lichens are the most common pioneers but club moss (*Selginella*) and mosses such as *Bryum* and *Grimmia* are also abundant. Lichens and certain mosses have several significant ecological adaptations, which allow them to colonize rock surfaces. Not only can some become established without soil but they can endure long periods without moisture, to the point of becoming brittle. This adaptation is crucial as rock surfaces can only hold water for very short periods following rain. Where the rock surfaces are freely exposed to insolation, long drought periods, separated by brief intervals during which water is available, are the norm. As photosynthesis is confined to these brief and widely separated intervals, lichens which grow on exposed rock surfaces make very slow and sporadic growth. The lichens tend to disintegrate the rock surface mechanically through the wedge-like action of their rhizoids, and they also weaken the rock by secreting corrosive carbonic acid. A loose debris film may accumulate and in some cases the pioneer lichens are replaced by one or more subsequent lichen communities. The simple stages in development of a lithosere are shown in Figure 23.21(a).

In rock crevices there is an entirely different story. Wind-carried debris fills a rock fracture as soon as it opens and keeps it filled as it is widened by weathering. Therefore, from the start soil conditions are distinctly better and community development is not as slow as on the bare rock surface. Not only does the crevice sere begin with a higher type of plant, for example mosses and sometimes even herbs, but the shoots of these taller pioneers project from the crevice to intercept debris that is being wind-moved or is under the action of gravity, eventually building up the soil above the level of the crevice.

On gently-sloping and unfractured rock surfaces, randomly located pioneers may provide foci for development, which spread out centrifugally as the mat thickens. The most advanced stage is centrally located in each expanding island. As the rock surface disappears beneath a coalescing vegetation cover and soil, other communities may enter while the rooting mechanism thickens. The microclimate of a rock outcrop frequently sets definite limits to the extent of vegetation encroachment. The main controlling factor in any xerosere is the initial degree of dryness of the habitat and the relative rapidity with which this is ameliorated. Drought tends to be extreme because of the lack of a rooting medium that retains enough water to satisfy transpiration demands between storms, and the rate of vegetation development is limited by the rate of accumulation of such a water-retaining medium.

The sequences are usually from lichens/mosses/herbs to trees, if the climate and the parent material permits. However, the shift in dominance rarely eliminates the life forms that were characteristic of the preceding stages. Once the *Rhacomitrium* mosses become established they tend to form a mat which traps any wind-blown silt. Grasses can then become established and root. As they decay they provide humus, which eventually leads to a soil depth sufficient to support higher plants, such as ling heather or thyme. With deeper soils moisture is less of a problem and eventually gorse, scrub and even trees become established in the less exposed areas. This primary sere is very slow to develop and in all but the wettest climates many centuries are required for the completion of a sere.

Other examples of primary colonization come from volcanic islands such as Surtsey, which was created from the Mid-Atlantic Ridge off Iceland in 1963 and has been studied intensively by ecologists, or from areas where glacier recession has provided fresh space for vegetation to colonize. A classic case of the latter can be taken from work in Glacier Bay, Alaska. Here there were comparative observations of sites that had been exposed at various times due to deglaciation over 19 years in permanent quadrats. The results showed that in the vegetational development, three principal stages could be recognized.

The pioneer stage was dominated by herbaceous and mat-forming species, including *Rhacomitrium* mosses, the horsetail, broad-leaved willow herb and mountain avens (*Dryas*). The next stage consisted of thickets of *Dryas* and dwarf willows, with the establishment of tree willows, alder and the cottonwood. A forest stage was the final development, dominated initially by Sitka spruce and later by a mixture of spruce and hemlock species. There was a steady increase in the density of herbs and in the size and cover of mat-forming species, but despite the presence of the apparent stages, the individuals of all stages, including the forest trees, appeared throughout. Such changes can be explained in terms of seed banks, dispersal rates and varying growth and reproductive rates. So species with especially mobile propagules, such as the seeds of the willow herb or moss spores, arrived in greatest numbers and therefore formed the majority of the initial colonization. The later shrub invaders progressively outshaded the earlier vegetation and were later outshaded themselves by the conifers.

Psammoseres

These develop on sand dunes at the back of sandy beach fore-shores, where sand is blown inland by the wind. The size and shape of the dunes that develop depend on the extent of the sand supply, the strength and duration of onshore winds and the shore profile. The plants that colonize these loose, dry sand dunes have to cope with several problems as pioneers: lack of water, sand mobility, salt spray, wind breakage, sand abrasion from sand carried by the wind and worsening aeration about their root systems due to sand deposition.

Dune formation usually begins at the foreshore drift limit, where sand accumulation takes place around flotsam and jetsam and the pioneer plant species. The plants are scattered and there is no development of closed vegetation. Most of the pioneer species are annuals, such as sea rocket, saltwort, sea sandwort and shore orache, but the sand couch grass (*Agropyron*) in the UK is also important as an initial colonizer. These plants grow at the upper tidal limit but cannot endure prolonged sea-water immersion and the habitat is naturally very unstable from reworking by winds and waves at high tide.

The main pioneer is the sand couch grass, which prepares the way for the main dune builder, the much more vigorous marram (*Ammophila arenaria*). The sand couch grass is capable of sending up aerial shoots as soon as it is covered by fresh, wind-transported sand and it can also withstand periodic inundation by salt spray. It extends the initial stage of dune formation, the embryo or foredunes (Figures 23.21(b) and 23.22), because it possesses extensive underground rhizomes, which send up new aerial shoots. These shoots help in the build-up of more sand and the roots help to bind the sand that is already deposited.

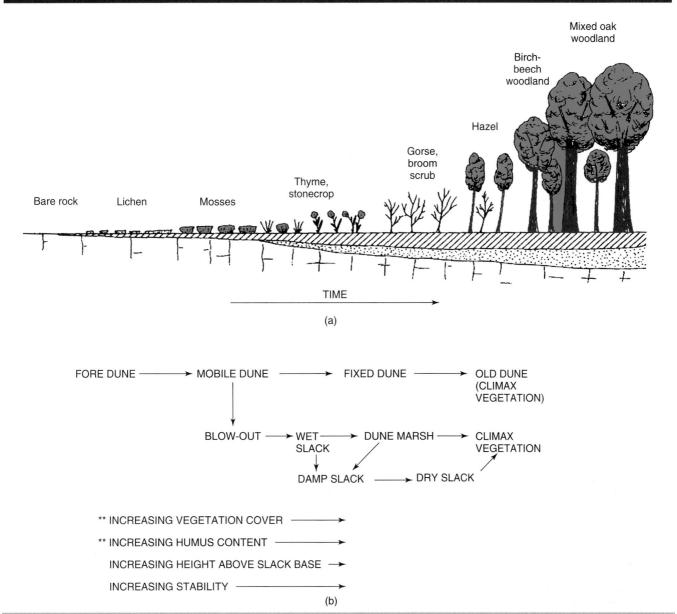

Figure 23.21 (a) Simple stages in lithosere to mixed oak woodland. (b) Dune succession with the effects of a blowout. (Source: (a) after Briggs and Smithson, 1985)

Once marram grass is established it grows very rapidly and fixes the foredunes, and each tuft of marram produces its own small dune with the help of an extensive horizontal- and vertical-spreading root system. It is better able to cope with the drier conditions on higher ridges, and has tough, rolled leaves that help it cope with the sand blasting and desiccating winds. In a few localities, such as parts of Lancashire and Ireland, the sand couch grass forms extensive communities, almost to the exclusion of marram, and few other plants are found.

Behind the foredunes, the main mobile dunes develop where sand is blown off the beach onto the seaward slope. The sand is fixed there, so that the main dunes appear to move; that is, they prograde. These dunes can be up to 30 m high. This stage, where there is a major increase in dune height, is often called the yellow

dune stage, or the partially-fixed stage. It is not until this stage is reached that marram becomes dominant, because it cannot tolerate over 2% salt in the soil. This species reacts vigorously to sand covering by producing new shoots.

Other species that are associated with marram at this stage are lyme grass, sea sandwort, sea holly, sea spurge, sea bindweed, sand sedge, sand fescue and sea convolvulus. In this stage there is still sand mobility, as there are open patches of sand that are not vegetation-covered.

In the landward stages of this marram phase, particularly in depressions where the sand is fairly stable, mosses make their appearance and help to stabilize the sand. Plants from the interior hinterland begin to colonize the inner sides of the dunes: usually weeds, such as groundsel, ragwort, colt's foot, rose bay willow

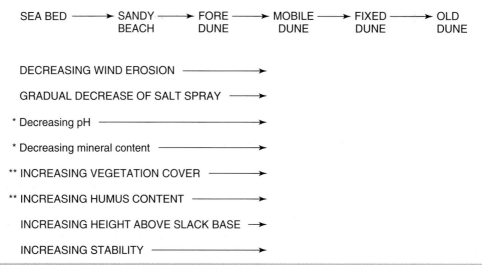

Figure 23.22 Dune seral relationships

herb, thistles, hawkweed, curled dock, dandelion and scarlet pimpernel. These plants grow in a naturally disturbed habitat. Common restharrow, dune fescue, thyme-leaved sandwort, sand cat's tail, spreading meadow-grass and wild pansy are found in this stage too.

Eventually this leads to the stage in which there is complete vegetation cover, which is called grey dunes, or the fixed-dune stage. The only open sand here occurs where there are paths through the dunes. The species involved in the vegetation cover are extremely varied and mosses become more important if the humidity is high enough. Different types depend on the sand type but one of the most common is *Tortula*. On highly calcareous shell sand, the species *Camptothecium* is also frequently found.

Lichens also appear on some dune areas following the mosses, especially species of the genus *Cladonia*, and it is these species that give the characteristic grey colour to the dunes, along with the development of humus formation.

Marram tends to have died out, likely because of the lack of aeration around the roots as the vegetation cover grows. Other species which began as pioneers are also beginning to be choked out, such as sea sandwort, bindweed and sea spurge. Sand sedge and a grass, sand fescue, take over as dominants, whilst common bent, sweet vernal grass, sheep's sorrel, lady's bedstraw, dewberry, common stork's-bill, Portland spurge, early forget-me-not, restharrow and heath-grass are also present.

Ultimately the dunes are colonized by an inland flora and 75% of the species will be from inland. If grazing is at all severe, the maritime species will disappear. The soil shows an increase in the percentage of humus with time and the calcium carbonate, which was originally present in shell sand, is rapidly leached, with the result that the dunes become more acid and a dune heath vegetation develops. This is dominated by species like heather, gorse, broom, *Agrostis* species, mat grass, sweet vernal grass and sedge, and wavy hair grass. Herbs such as tormentil, harebell and sheep's sorrel can be found in the drier areas, whilst in the damper hollows heath rush and common sedge grow.

Several possibilities are available for the final stage of dune development:

1. Grassland, especially on calcareous dunes where there is grazing by rabbits or domestic animals, or where they are converted to golf courses. This stage has occurred in the majority of British dunes systems.

2. Calluna heath on noncalcareous dunes, particularly on the west and south coast of Britain, such as at Walney Island (off the Furness peninsula), the Formby–Ainsdale–Birkdale dunes in south-west Lancashire, Newborough Warren in Anglesey and Dorset.

3. Forest, although natural forest hardly occurs in British dunes (for example, only small patches of silver birch on the Ainsdale dunes are thought to be natural), probably because of the lack of suitable seed parents, or the restraining influence of humans. Trees are found in north Holland. Here the final dune colonization stage is a low birch woodland, with sea buckthorn and privet into which oak might eventually enter. The most successful dune stabilizers are several varieties of pine and birch.

4. Bracken will invade the noncalcareous dune pasture if given an opportunity and, if not checked, a closed bracken community can result.

5. Dune scrub can be regarded as the forerunner to indigenous forest but in Britain is not common. It consists of any of the spiny shrubs that can grow on light sandy soils, such as Burnet rose, bramble, gorse, blackthorn, wild rose, sea buckthorn, birch, hawthorn and elder.

At any stage in the development of this dune plant cover, sand may be set in motion again if the vegetation cover is broken. In fact, the usual history is one of alternate break-up and restabilization of the dune complex as a whole, with no dune retaining its identity for long. Many factors have been known to break up the vegetation, such as fire, drought, overgrazing by rabbits or livestock, livestock paths and recreational trampling pressure from people, off-road vehicles, scramble bikes and horse-riding. Once this has occurred, wind erosion gets started and blow-outs are created, which quickly destroy the soil.

Erosion may get as far as the capillary fringe, which lies just above the local water table. This allows flat, moist areas to develop as dune slacks, which can have standing water during all or parts of the year and are commonly interspersed in the moving dunes. In these dune slacks a hydrosere succession develops, with pond and aquatic plants, or if there is only standing water during winter or the soil is marshy, several rush varieties, such as the jointed rush and the common spike-rush, a variety of marsh plants, and the moss genus *Hypnum*. In the incipient slacks, herbs, such as the early marsh orchid, yellow wort and seaside centuaries are common.

On the west coast of Britain, for example at Ainsdale on the south-west Lancashire coast, the vegetation is frequently of a fen type, dominated by sedges and rushes. Typical plants of the wet slacks are the common sedge, brown sedge, sea rush, creeping bent, water mint, purple loosestrife, marsh pennywort and water dock, but in places the slacks are colonized by a carpet of creeping willow (*Salix repens*). This species cannot survive permanently-waterlogged habitats and its lower limit is set by the degree of waterlogging. Dry slacks only exceptionally have standing water

in them and share much vegetation with the more humid parts of the dune ridges, such as common bird's foot trefoil, heath speedwell, yellow rattle and common century. At Ainsdale the rare dune helleborine *Epipactus dunensis* is widespread in dry slacks. In the acidic dunes the slacks eventually carry vegetation dominated by *Salix atrocinerea* and *Betula pubescens*, into which other species may enter, such as the yellow flag iris.

It is generally thought that evolution from bare sand to dune heath would take about 300 years. The dune seral relationships in a generalized dune system are illustrated in Figure 23.22. See in the companion website for Box 23.2 Case study of the Gibraltar Point sand-dune system, Skegness, Lincolnshire.

23.7.2 Hydroseres (including haloseres)

There are two main seral successions, depending on whether there is fresh water (hydroseres, in ponds, lakes, fens and bogs) or saline water (haloseres, on salt marshes).

Hydroseres

The important factors in this case are the depth of water and the lack of aeration, but the overall trend of development is upwards and into the water through the continual building of decaying vegetation until the initial water body is completely filled (Figure 23.24(a) and (b)).

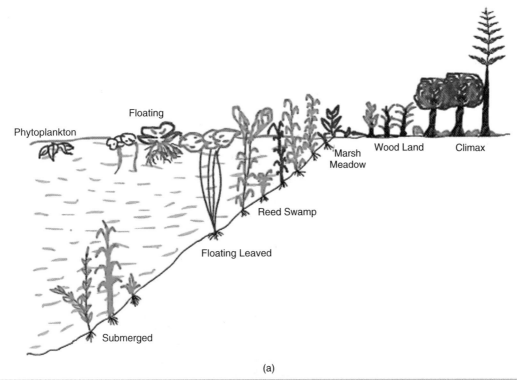

Floating

Phytoplankton

Marsh Meadow

Wood Land

Climax

Reed Swamp

Floating Leaved

Submerged

(a)

Figure 23.24 (a) Development of hydroseres. (b) Hydroseral development in lake basins, Danes Moss, Cheshire. (c) Hydroseral development in a small pond, Mull Wood, Croxteth, Merseyside. (d) A typical hydrosere developed in a glacial kettlehole in northern England generally (Source: Briggs and Smithson, 1985) and in Sweet Mere, Ellesmere in particular (Source: after Cousens, 1974)

Trend of succession in a Hydrosere

Pioneer Community	Seral Communities					Climax Community
1	2	3	4	5	6	7
Phytoplankton stage	Rooted-submerged stage	Rooted-floating stage	Reed-swamp stage	Sedge-meadow stage	Woodland stage	Forest stage
e.g	*e.g*	*e.g*	*e.g*	*e.g*	*e.g*	*e.g*
Diatoms, Bacteria, Cyoanobacteria, Green algae	Elodae, Hydrilla, Vallisneria, Ultricularia, Potamogeton	Nymphaea Trapa, Azolla, Lemna, Wolffia, Pistia	Sagittaria, Phragmites, Typha, Scirpus	Juncus, Cyperus, Eleocharis. Carex	Salix, Populas, Alnus, Cornus	Temperate mixed Tropical rain tropical Deciduous

⟶ General trend of succession ⟶

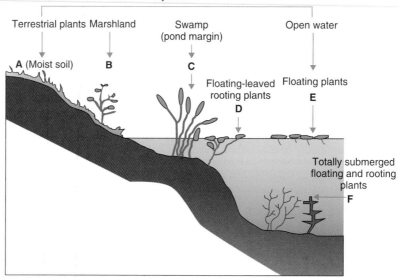

(b)

Figure 23.24 (*continued*)

(c)

Figure 23.24 *(continued)*

Early on the dominant environmental factor controlling hydroseral development is the depth of water in relation to light availability on the bottom, but later the nutrition, aeration and temperature of the rooting medium tend to take precedence. The first colonizers are plankton and algae. They have a limited influence but sometimes when covering a large percentage of the water surface they can restrict the development of communities on the bottom by intercepting light. These are soon followed by the submerged anchored aquatics, such as the water crowfoot, hornwort and milfoil, and because they are the first macrophytes to advance over the bottom they usually make up the innermost zone of vascular plants in pond or lake-margin vegetation, extending to a depth of about 7 m.

After the decay of this vegetation the water becomes shallower and the floating-leaved, anchored species will appear, such as the water lily, duckweed, frogbit and pondweed. Frequently these form a zone in which the water is no deeper than about 3 m. These plants shade out many of the submerged aquatics and the decay of vegetation gradually raises the level of the water body floor.

The next stage is represented by species that are rooted on the lake floor but have aerial leaves (the emerged, anchored hydrophytes). The species here are reeds (*Phragmites*), sedges (*Carex*), rushes, yellow iris and water plantain. Transpiration takes place through the aerial leaves, which aids in the lowering of the water level, and this stage is known as a reed swamp. The water is generally no deeper than 1 m.

As the water becomes shallower the plants in this stage are able to encroach on the floating-leaved plants by sending shoots up into the air between their blades. Gradually the submerged sediment surface appears above the water and marsh plants such as water mint form marginal mats. Here both light and oxygen content favour maximum productivity. A narrow belt of vascular plants and mosses can expand towards the centre of the water body at a relatively rapid rate. These plants form a tangled peat

accumulation mat, which either remains anchored to the bottom and fills the basin as it advances, or pushes out over the water surface as a floating shelf that is attached only to the shore.

Expanding in a centripetal manner, a floating mat overrides communities of emergent and submerged, anchored plants. In the opposite direction, as the older, shoreward edge of the mat becomes thicker and firmer with age, it builds up over the water table and becomes a suitable habitat for invasion by communities that are dominated by shrubs and trees, each of which in turn demands a greater degree of firmness and drainage of the peat. This eventually leads to terrestrial vegetation, which can go one of two ways, depending on the pH of the water. If the water is base-rich, fen vegetation will occur, where grasses and sedges will pass into a carr woodland with alder and willow. This dries out completely into ash woods and eventually an oak climax. On the other hand, if the water tends towards acidity, the succession may be deflected at the reed-swamp stage and sphagnum moss can come in. The vegetation will tend to remain as a bog until it dries out due to climatic change or a drop in the water table. It is then colonized by heath, then birch and pine woodland. Generally the soils developed are too acid for the bog to pass beyond the heath stage. When this is in equilibrium it is an example of an edaphic complex, which is held at this stage by local soil peculiarities.

Lakes and ponds that result from deglaciation tend to be oligotrophic when first formed as they have few of the essential plant nutrients and so support a very sparse flora, and they also have high oxygen content. This is illustrated in **Figure 23.23(d)** in a generalized form from glacial kettleholes in northern England, such as at Blelham Bog, Cumbria (Huddart, 2002) or The Bog, Roos (Innes, 2002). An oligotrophic condition is also favoured by deepness of the water body, which restricts the development of highly-productive, rooted vegetation, or by wave action or stoniness, sufficient to inhibit marginal vegetation. The gradual accumulation and retention of nutrients by aquatic ecosystems commonly changes oligotrophic water to eutrophic where nutrients are abundant and there is low oxygen content, due to active decay processes.

Mires include all aquatic ecosystems with wet peat serving as the rooting medium for plants, such as the two types mentioned above: fens and bogs. Fens are distinguished by their seedlings having available to them water that has been in contact with mineral soil and consequently has a much better nutrient supply than is contained in rainwater. Cyperacae, grasses or reeds are usually dominant in the early stages of fen succession, with non-ericaceous shrubs and trees invading later (the carr vegetation of the British ecological literature). Mosses, when abundant, seldom include sphagnum. The surface of the peat is flat or slightly concave. Fens occur in basins, or on slopes with seepage, in areas of generally low to moderate rainfall and where the pH is over 5.

Bogs (ombrogenous mires or mosses) include mires in which seedlings have available to them scarcely more nutrients than are concentrated in rainwater. They generally build up a surface well above the water table and occur in areas with high rainfall. Generally they are limited to wet and cold climates and sphagnum moss is usually conspicuous, forming closely-spaced hummocks. Bogs are normally about 5 m high and accompanied by ericaceous

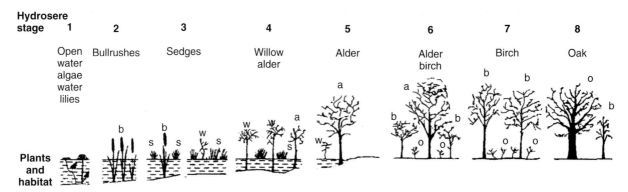

Habitat description	Reed swamp	Marsh or fen	Open wooded fen	Closed wooded fen	Woodland		
Habitat Processes	Accelerated deposition of silt and clay. Floating raft of organic matter forms ——→ thickens		Raft now a mat resting on mineral soil	Black mineral soil revealed in patches. Earthworms	Ground level now above water table; oak seedlings	Birch canopy forms; oak saplings	Oak grows through and then over the birch
pH level	–	–	7.3		4.3	3.7	–
Number of species of plant	6	10	14	26	18	14	10

(d)

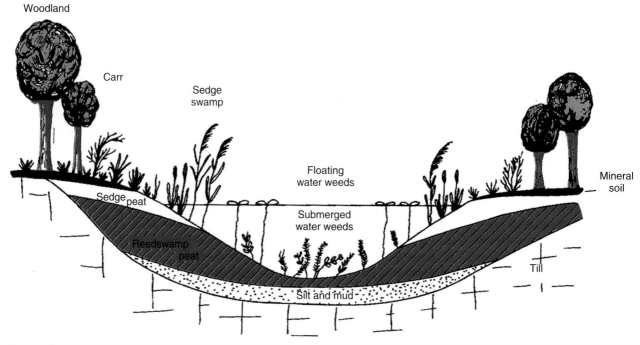

Figure 23.24 (continued)

shrubs and even conifers in places. Floristically they are more impoverished than a fen and the pH is 4.5 or under. There are three basic types:

1. **Blanket bog:** Initiated in locally wet places, from where it spreads to cover all the topography, except steep slopes. It is found in wet, foggy climates, along cool temperate coasts.

2. **Raised bog:** Confined to areas of wet basins as a result of only moderate precipitation. The surface becomes convex.

3. **Flat bog:** Confined to the limits of a wet basin. It remains flat and is found in a drier climate than a raised bog.

Haloseres or salt-marsh successions

These successions start from a salt-impregnated soil on the coast and can be found where any of the following physiographic conditions occur:

- Estuaries, such as at the Humber, Solway Firth or Ribble.

- In the shelter of spits, such as at Blakeney on the north Norfolk coast.

- In the shelter of offshore barrier islands, such as on the north Dutch coast adjacent to the Wadden Sea.

- Where there are relatively protected bays with shallow water, such as around Morecambe Bay and the Wash.

The extent of salt marsh in any area depends on the slope of the land and the amount of silt in suspension in the water. The zonation that is found indicates the different lengths of time for which an area is covered by the tide and represents the successive stages in the sere which proceed as the marsh gradually encroaches on the open water. The raising of the soil level is caused by the trapping of the silt by vegetation during the ebb and flood tides. The sea water leaves a saline sediment with a 3% salinity, but this rises to 7% salt concentration when the water evaporates. The effects of the inundation are many but include:

- A reduction in the supply of carbon dioxide, which reduces photosynthesis.

- A reduction of the supply of sunlight, caused by silt in the water.

- Disturbing of the environment of the plant's roots.

- A reduction in the supply of oxygen, which affects respiration.

- Affects on metabolic plant processes, caused by the water salinity.

- Tides acting as a mechanism for seed dispersal.

- Prevention of successful germination of seedlings by uprooting by currents.

The highest levels of the salt marsh are only covered by the highest spring tides and are affected more by sub-aerial agencies, and the salt is more easily leached out by rainfall. Lower down the salt marsh, where there is a daily covering by the tides the higher concentration of salt leads to problems for plant colonization and only plants that are especially adapted to these conditions can grow (the halophytes). However, there are some plants, called semi-halophytes, which have a wide range of tolerance, such as sea aster and sea spurrey. There are also glycophytes, for which low salinity is essential, such as rushes in the upper marsh.

Plants living on salt marshes have to have special adaptations, especially in the lower marsh, to cope with the prevailing conditions, like the salt concentrations and currents. The seeds of the salt marsh species are dispersed by high tides during the autumn equinox and the winter. However, the germination of seedlings in spring is inhibited by tidal inundation and the lower limit of pioneer plant germination is set by the erosion of the seedlings by the currents. There is a sharp rise in the disappearance and mortality rate of young plants at the level at which there is daily flooding; that is, below HWMNT (high water mark neap tides). Glasswort seedlings need to be left uncovered for two to three days, which only occurs above HWMNT. In that period the roots can penetrate the sediment to a sufficient depth to anchor it when the salt marsh is next flooded and so prevent it being eroded. The seeds of the sea aster require five days before they are properly anchored in the sediment; therefore this period represents the minimum required exposure time between tidal inundations.

Sea water has an osmotic pressure (the suction pressure, which develops between roots and soil water when they are in contact and allows the plants to obtain water and soluble salts from the sediment) of about 20 atmospheres due to its high salt concentration. The periodic inundation of the salt marsh affects the salinity levels of the sediment water and creates high osmotic pressures. An average plant with an osmotic pressure of 10 atmospheres would quickly wilt if grown on a salt marsh because water and salts would be sucked from the plant into the soil. Halophytes are adapted however and their average osmotic pressure value is 40 atmospheres. By exerting a greater osmotic pressure than that of the soil water in the salt marsh, halophytes can overcome the pressure exerted by the saline sediment water in which they have to grow.

In summer and during periods of neap tides, salt marsh vegetation is often subject to drought. In order to combat the danger of drying out, many halophytes have adaptations to check transpiration. These include:

- Thick cuticles or the secretion of wax on the leaf surface, which confine evaporation to the stomata (for example, Danish scurvy grass and orache species).

- Stomata that are protected in grooves or hollows in the plant's leaves, or further protected when the leaves roll up in dry weather, trapping a layer of moist air in the leaf and so reducing evaporation from the stomata. When water is plentiful the leaf blade opens out, exposing a greater surface area for transpiration.

- Hairs on the leaf surfaces, which maintain a damp atmosphere around the stomata because moisture tends to condense on the hairs.

- A silvery or mealy foliage surface, which aids the retention of moist air near the leaf surface and reflects much of the Sun's heat. For example, the sea purslane has silvery, scale-like hairs covering its leaves.

- A reduction in leaf surface, which discourages transpiration. The glasswort is an excellent example of this as it only has rudimentary leaves (Figure 23.25(a)), just visible as tiny scales attached to the joints of the stems.

Salt marsh plants can be grouped into three types according to the manner in which they adapt to the saline environment. There are succulents, in which the salt uptake is compensated for by water uptake, though at the end of the season the succulent portion may dry up and be sloughed off, as in the glasswort. Then there are salt-excreting forms with special glands that remove salt to the exterior, such as the cord grass or sea purslane. Finally, there are plants which show deciduous leaf death when the leaves contain too much accumulated salt. Their death and decay at the end of the season removes salt from the plants. Examples include the sea aster.

The majority of halophytes possess relatively deep root systems and generally they are markedly woody. These roots enable them to secure a firm anchorage in the unstable mud and allow them to derive their main water supply from regions in which the salt concentration is less variable than in the surface zones. Individual species have the following adaptations to the unstable salt marsh:

- Cord grass (*Spartina*) has an extensive root system so that it becomes firmly anchored in mud and can easily resist strong tidal currents. This allows it to become a pioneer species. It has long creeping rhizomes which bind the soft mud efficiently and its roots are of two kinds (Figure 23.25(b)): there are a few long anchoring roots that grow vertically downwards, which hold the plant firmly in the mud, and a large number of fine absorbing roots, which grow out horizontally just below the surface.

- Small, low-growing plants have extensive creeping roots. An example is the sea sandwort.

- Red fescue produces horizontally-creeping roots just below the sediment surface, which stabilize the salt marsh surface.

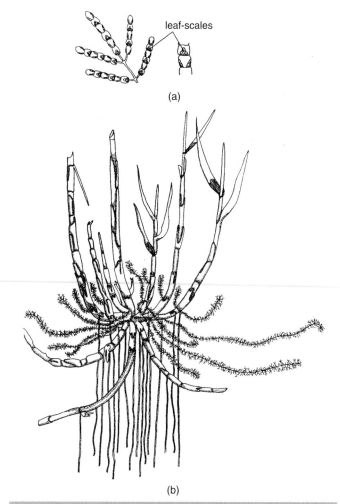

leaf-scales

(a)

(b)

Figure 23.25 Adaptations of salt marsh plants: (a) rudimentary leaves of the glasswort; (b) root systems of the cord grass (*Spartina*)

The plant communities present are as follows:

1. A pioneer stage, in which there is a bare mud bank. The pioneers are minute green algae which form an algal mat; they are mainly species of *Vaucheria* on the finer flats and *Enteromorpha* and *Cladomorpha* on the more sandy flats.

2. The next stage is where eel grass (*Zostera marina* and *Z. nana*) come in, which give more stability to the mud. In this stage silting immediately starts. Eel grass beds were once much more extensive in Britain than they are today, but they were decimated by disease following the 1930s, though many have since recovered.

3. At a higher level, but still in areas that are submerged for 90% of the day, glasswort (*Salicornia*) appears. This is an annual plant and therefore its mud-binding effects are seasonal. Along with this plant come the sea aster, the annual seablite

and marsh samphire. These all help fix small mud islands, leading to an integrated channel system developing in the mud flats. The creeks deepen through tidal scour and another face is created, which is colonized by sea purslane.

4. The high marsh stage stands above all but the highest spring tides and is very important. In this zone there is rapid change from an open community to a closed one. Sea thrift and sea lavender come in here.

5. The next stage is the colonization of the high marsh by rushes, reeds and some grasses, such as red fescue. This can result in a fine turf which encourages rough grazing, called salting, as around Morecambe Bay and the Ribble estuary.

If the succession were left it would pass to scrub, with species like gorse, and to woodland in sheltered locations.

One major exception occurs in areas in which rice or cord grass (*Spartina*) has been introduced. This arrived in Britain in 1870 and was planted to stabilize mud flats in Southampton Water. It flourished rapidly and spread as far as the Dee Estuary on the west coast. It bypasses all the stages in the succession and can form a pure stand right from the pioneer stage.

In general in Britain there is the following succession: algal community → low marsh (*Salicornia*) → high marsh → highest marsh (rushes) → saltings, which are generally reclaimed and held by grazing. However, in detail the succession varies from coast to coast in the British Isles due to differences in the proportion of sand and mud present. The muddiest coasts are in the south and east of the country, derived from the newer sedimentary rocks. The coast becomes progressively sandier towards the north and west.

This is an important distinction because the sea manna grass (*Puccinellia*) dislikes soft, sloppy mud and hence is relatively uncommon in East Anglia and along the south coast but becomes one of the dominants on the north and north-west sandy coasts.

The regional variations in Britain in the halosere successions are illustrated in Figure 23.26. See in the companion website for Box 23.3 Regional variations in the halosere successions in the United Kingdom.

23.7.3 Plagioclimax and subclimax conditions

If a sere is unable to reach a climax it is known as a sub-sere. These can be formed in two ways:

1. Through a shorter sere, which began not on a bare habitat but in a partially-destroyed climax.

2. Through an ordinary sere which has had its progress arrested either by the biotic factor or by periodic natural calamities, such as volcanoes, hurricanes and avalanches. For example, tropical rain forest vegetation is usually virgin forest, which is

the climatic climax, but once the biotic factor appears, such as tree felling, grazing and trampling of livestock, and burning and ploughing, secondary vegetation soon appears.

Whenever the vegetation is held permanently below the climax it is called a sub-climax (Figure 23.29). Arresting factors are of three types:

1. **Physiographic:** This is closely linked with the edaphic factor. For example, on steep slopes soil instability may be the main factor, while shaded slopes might be too cold. Sunny slopes on the other hand may be too hot, sea coasts too exposed and therefore too windy, ridge tops may be too dry and too far from the water table, and hollows may be left too wet.

2. **Edaphic:** Especially with regard to the amount of soil moisture present. For example, in England and Wales very wet soils with a high water table will not go to natural oak–beech woodland but go to a fen. Freely-drained soils will be too dry and the water table will be too low, so that deciduous woodland cannot fully develop and a sub-climax vegetation of shallow-rooted herbs will exist instead.

3. **Biotic:** Where there is artificial interference by humans and animals. For example, the hay meadow and rough grazing sub-climax vegetation will, if the edaphic and physiographic factors are suitable, return to oak–beech climax woodland in 200 years in southern Britain. Heathland is also a type of sub-climax that has been successively arrested by pasturing and burning, but in addition here there is an adverse edaphic factor, with acid soils and free drainage, which creates a water deficiency. The constant intervention of humans is not an arresting factor but merely a deflecting one and is periodic rather than permanent. It is known as a plagioclimax. This can vary in form according to the precise form of the intervention but can include the effects of domesticated animals, rabbits and fire.

We have seen that communities change with time through succession to a climax (Figure 23.30) and that once a community achieves equilibrium it tends to stay that way unless something upsets it. However, it should be noted that communities are stable, not static. The population dynamics of the assembled animal and plants groups can change in any ecosystem. The population size depends on births, deaths and migration into and out of the ecosystem. These are all affected by predation, parasitism, disease and competition between species.

This brings us to some concepts related to life-history patterns, where there are links made between types of habitat and different types of life-history. Clearly there is a limited amount of energy and other resources available to any species for growth and reproduction. Hence species may have to grow more and reproduce less, or reproduce more and grow less. The potential for a species to multiply rapidly and produce large numbers of

EAST ANGLIA:-

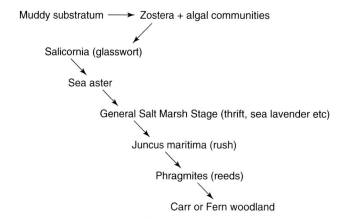

SOUTH COAST :-

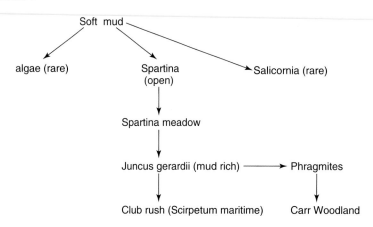

WEST and NORTH COAST MARSHES

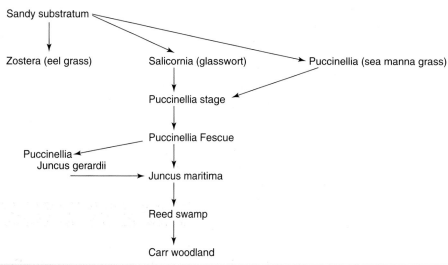

Figure 23.26 Regional variations in the halosere successions in the United Kingdom

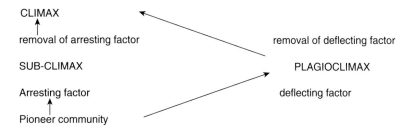

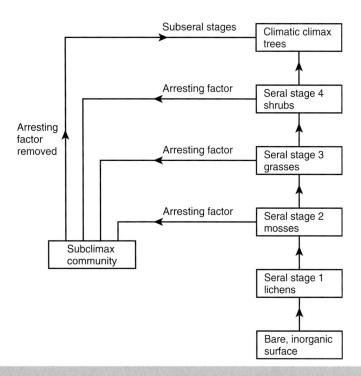

Figure 23.29 Sub-climax vegetation

offspring early in the life cycle is a big advantage in environments that are short-lived, allowing the organisms to move into habitats quickly and exploit any new resources. This is a characteristic of the life cycles of terrestrial organisms that invade disturbed land (say by fire or the opening up of woodland by felling), such as the annual weed species. These have life-cycle properties that are favoured by natural selection in such places. They have been called r-species, and the habitats in which they are likely to be favoured are called r-selecting.

On the other hand, species living in habitats in which there is intense competition for limited resources have different life histories. Those that are successful are those that have garnered the larger share of available resources. Their populations are usually crowded and they have won because they have grown faster or have spent more of their resources on aggression or another activity that has allowed them to compete successfully with others. They are called K-species and the habitats they are likely to favour are called K-selecting.

Trees in a wood are good examples of K-species because they have to compete for light in the canopy and the survivors are those that put resources into early growth to overtop their neighbours. They usually delay reproduction until their branches have an assured position in the leaf canopy. They have a long life, usually with a relatively low allocation to reproduction overall but with large individual seeds. In the open, disturbed habitats the r-selecting plants tend to have a greater reproductive allocation but smaller seeds, earlier reproduction and a shorter life. For a more detailed discussion of concepts related to life histories the reader should refer to Begon *et al.* (2006).

23.8 Concluding remarks

It is clear that human activity is changing the Earth's environments in many ways and that this change is occurring at a speed never before witnessed by *Homo sapiens*, although rapid change has

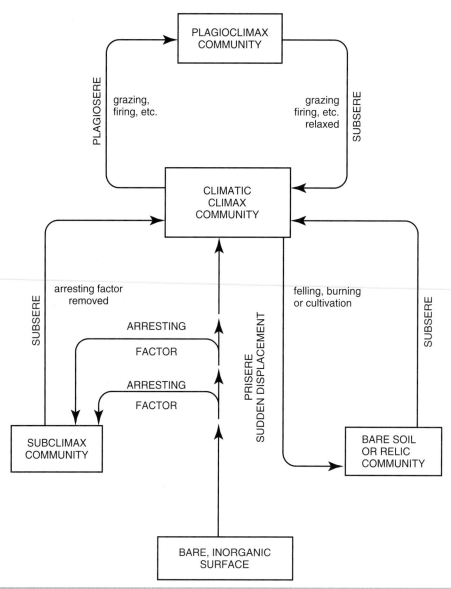

Figure 23.30 Climax vegetation development

occurred many times in the geological record, as we will see in Chapter 29. Hence it is important that all of us have an appreciation and understanding of how the planet's life forms interact with their physical and biological environments, so that we can adapt to any changes that are occurring as quickly as possible in order to try and prevent disastrous problems for the human race and other life forms.

In other words, it is important for everyone to understand their place in the planet's ecology and their links within the ecosphere as fully as possible, much more so than ever before. Hence this chapter has stressed the basic ecological frameworks and concepts so that the reader can understand the interaction between the Earth's life forms and the abiotic and biotic environments.

Exercises

1. Explain the distribution of the Lusitanian, North American and Arctic–Alpine flora in Ireland.

2. Describe the inputs and outputs to ecosystems.

3. Define the terms 'producers', 'consumers' and 'decomposers' and outline their roles in ecosystems.

4. What types of food chains can be differentiated in ecosystems?

5. How will the carbon balance in the Arctic respond to global warming?

6. What are the processes of nitrogen fixation in the nitrogen cycle?

7. Why is biogeochemical cycling important in the Earth's ecosystems?

8. What are the characteristic stages that can be recognized in the development of hydroseres?

9. How do salt marsh plants cope with living in such an environment?

10. Explain what an ecologist means by the terms 'subclimax' and 'plagioclimax'.

References

Begon, M., Townsend, C.R. and Harper, J.L. 2006. Ecology from Individuals to Ecosystems. 4th edition. Oxford, Blackwell Publishing.

Briggs, D. and Smithson, P. 1985. Fundamentals of Physical Geography. London, Hutchinson.

Cousens, J. 1974. An Introduction to Woodland Ecology. Edinburgh, Oliver and Boyd.

Huddart, D. 2002. Blelham Bog. In: Huddart, D. and Glasser, N.F. (Eds). Quaternary of Northern England. Geological Conservation Review Series 25. Peterborough, Joint Nature Conservation Committee. pp. 230–240.

Innes, J. 2002. The Bog, Roos. In: Huddart, D. and Glasser, N.F. (Eds). Quaternary of Northern England. Geological Conservation Review Series 25. Peterborough, Joint Nature Conservation Committee. pp. 429–435.

Kamijo, T., Kitayama, K., Sugawara, A., Urushimichi, S. and Sasai, K. 2002. Primary succession of the warm-temperate broad-leaved forest on a volcanic island, Miyake-jima, Japan. *Folia Geobotanica*. **37**, 71–91.

Lindeman, R. 1942. The trophic dynamic aspect of ecology. *Ecology*. **23**, 399–418.

Odum, H.T. 1957. Trophic structure and productivity of Silver Springs. *Ecological Monographs*. **27**, 55–112.

Odum, H.T. 1971. Environment, Power and Society. New York, Wiley-Interscience.

Richards, P.W. 1979. The Tropical Rain Forest. Cambridge, Cambridge University Press.

Smith, R.L. 1966. Ecology and field biology. New York, Harper and Row.

Whittaker, R.H. and Likens, G.E. 1975. The biosphere and man. In: Leith, H. and Whittaker, R.H. (Eds). Primary Productivity of the Biosphere. New York, Springer-Verlag. pp. 305–328.

Further reading

Anderson, J.M. 1981. Ecology for Environmental Sciences. London, Arnold.

Cotgreave, P. and Forseth, I. 2002. Introductory Ecology. Oxford, Blackwell Science.

Cox, C.B. and Moore, P.D. 1994. Biogeography: An Ecological and Evolutionary Approach. Oxford, Blackwell Scientific.

Ganderton, P. and Coker, P. 2005. Environmental Biogeography. Harlow, Pearson Education.

Huggett, R. 1998. Fundamentals of Biogeography. London, Routledge.

Krebs, C.J. 2001. Ecology. 5th edition. San Francisco, CA, Benjamin Cummings.

Packham, J.R. and Willis, A.J. 1997. Ecology of Sand Dunes, Salt Marshes and Shingle. London, Chapman and Hall.

Ranwell, D.S. 1972. Ecology of Salt Marshes and Sand Dunes. London, Chapman and Hall.

Townsend, C.R., Begon, M. and Harper, J.L. 2003. Essentials of Ecology. 2nd edition. Oxford, Blackwell Publishing.

24 Soil-forming Processes and Products

Severe soil erosion in a wheatfield, near Washington State University.
(Source: Wikipedia Commons, Jack Dykinga, http://www.ars.usda.gov/is/
graphics/photos/k5951-1.html)

Earth Environments   David Huddart and Tim Stott
© 2010 John Wiley & Sons, Ltd

Learning Outcomes

After reading this chapter and completing the exercises you will be able to:

➤ Understand the soil as a system, the controls on soil formation and how soil horizons develop through time.

➤ Be able to describe soil characteristics in the field and the laboratory.

➤ Be able to classify soils based on these characteristics.

➤ Understand how topography and altitude have influenced British soils.

➤ Recognize the value of palaeosols in understanding environmental change in the geological record.

24.1 Introduction

24.1.1 What are soils?

Depending on who uses the term it can have several definitions. A farmer might suggest that a soil is 'what my plants grow in', a geomorphologist might say it is a zone in which rocks are altered by chemical and biological processes, whereas a civil engineer will be much more concerned with the stability of the end product of these processes. Two commonly used definitions are that:

1. Soils are a collection of natural bodies occupying the portions of the Earth's surface that support plants and have properties due to the interaction of climate and living matter on parent material.

2. Soils are a product of the soil parent material reacting to its environmental conditions through various chemical and biological processes, taking place within the soil profile over time.

The emphasis in these definitions is that the environmental conditions influence chemical and biological processes which take place at the Earth's surface over periods of time. Soils are chiefly an admixture of mineral grains and finely-divided organic matter, so for example a good surface silty loam (see definition in Table 24.1) contains 5% organic material derived from the decomposition of plants, 45% mineral matter derived from the breakdown of rocks, 25% soil water enriched with dissolved solids and 25% soil air enriched with CO_2 from the respiration of soil animals and plant roots. The lower horizons have less organic matter and more water and minerals. This is therefore a three-phase system with solid, liquid and gas phases, where the solid phase is the biggest. It is an amalgam in which the solids are loosely packed and the voids are filled with air and water.

24.1.2 Why is the study of soil important?

Soil plays a pivotal role as a resource in agriculture, as a crucial part of our environment and how we develop the land. It played an important role in controlling where initial settlement occurred and prehistoric agriculturalists' ability to grow good crops. This is still the case today as the human race requires ever more agricultural land for food as the population rises. This is despite the increases in crop yields and genetically-modified strains for specific locations. Virtually all our food is grown in soil, as is that which we feed our animals. Soil is the foundation of all ecological habitats and therefore of ecological health and biodiversity. The maintenance of healthy soil biological communities means long-term soil sustainability, clean water and clean air.

However, soils can be naturally fragile and erosion can take place relatively easily if we are not careful in their management (see the frontispiece for a case in the USA). An example of can be taken from the British uplands, where the thinness of the soils, the hostile climate and the generally acid parent materials, coupled with the recreational problems caused largely by the trampling pressure of hillwalkers, has caused damage to the soils. There has been footpath erosion and widespread landscape changes have been caused by the management of this problem (for example, artificial surfaces, stone paths, non-native vegetation used to recolonize the trampled soils and a general change in the natural landscape).

Soil is also important because it provides the foundation material for buildings, roads, reservoirs and dams. A civil engineer must make sure the soil properties are safe for such structures so that they will not fail, particularly under specific conditions, such as high pore-water pressures. The soil properties can also be important for structures such as pipelines and cables, which need to be as stable as possible and must not be corroded by waters from the soil.

Soil also plays a role in managing the Earth's water resources and water quality. It can hold water, it allows the movement of water through it and it can filter out pollutants. It regulates the time span between rainfall on the surface and when that rain reaches streams, rivers and aquifers. Sometimes it is expected that soils will reprocess our waste products as rubbish is dumped in landfill sites, but it can only do so to a certain extent.

Soil is so important to the welfare of life on Earth that we must sustain and protect it at all costs.

24.2 Controls on soil formation

Soil is a dynamic medium which tries to achieve a slowly changing equilibrium with its formative environment and is constantly reacting to changes in that environment. Some of the changes can be extremely slow, such as the general climatic amelioration during the Holocene from the last glacial phase, or extremely fast, such as when a farmer drains a field or adds fertilizer to the soil, but all of the many influences on soils are collective and not individual.

Table 24.1 Characteristics of various textural categories

sand	Predominantly (>85%) sand. Well drained, poor nutrient and water rete little, if any, structure. Very high risk of erosion.
loamy sand	Similar to sand but begins to hold together.
sandy loam	A mixture of sand with a little silt and clay. Can form a good structure if sufficient organic matter present. Good water availability, moderate nutrient. Easily cultivated. Moderate risk of erosion too easily.
sandy silt loam	A mixture of sand and silt with very little clay. Will form a reasonable b availability but low nutrient retention. Well-drained with high leaching risk.
silt loam	Predominantly silt with very little sand or clay. Forms a weak structure only moderate nutrient retention. Moderately well-drained with high leaching. Mixture of sand and clay with some silt, forms a strong structure.
sandy clay loam	Reasonably good water availability and nutrient retention. Needs careful management but will give good yields.
clay loam	Equal mix of sand, silt and clay, forms a strong structure and high nutrient retention but imperfect drainage. Low risk will give very good yields.
silty clay loam	Mixture of silt and clay with some sand, forms a good structure that may good water availability, good nutrient retention but imperfect drainage management but will give very good yields.
sandy clay	Mainly clay and sand, forms a strong structure but danger of compaction retention but often with very poor drainage; very low risk of leaching or erosion.
silty clay	Mainly clay and silt, forms a strong structure but danger of compaction very poor drainage; very low risk of leaching or erosion. Very difficult to cultivate.
clay	Mainly clay, forms a strong structure but danger of compaction. Good poor drainage; very low risk of leaching or erosion. Difficult to cultivate.

A soil starts developing with newly-exposed parent material, such as dune sand, a lava flow or salt marsh sediment and then goes through several stages. The prerequisite is breakdown of rock or pre-existing sediment at the Earth's surface by physical and chemical weathering processes. The physical weathering produces mineral particles, which give support for plant roots, whereas the chemical weathering produces two of the three nutrients (P and K) required by plants. The mineral soil does not contain the third nutrient (N), which comes from the atmosphere.

The second stage of development is when seeds arrive and a cover of a few hardy, pioneer plants becomes established. This results in the photosynthetic process: the liberation of O_2 to the atmosphere and CO_2 to the soil. The plants that initially colonize die and decompose, and organic material is carried underground by rain and soil organisms, making the soil more fertile so that it can support a richer flora.

The cycle continues and the fertility rises to a maximum when the climatic climax vegetation exists. A sequence of plant communities is developed at this stage (see Section 23.7). This climatic climax stage is at equilibrium with the environmental conditions and the soils are referred to as equilibrium soils, since they are in balance, or in steady state, with their overall environmental controls.

The soil goes from the early stages to maturity very slowly and gives rise to varying stages in the weathering of minerals in the under-2 mm fraction of soils. Zones in the soil develop and soil horizons are apparent. Mature soils only change very slowly with time. In Table 24.2 it is apparent that the mineral composition changes as soils develop and that there is a sequential set of weathering processes, but this is just one example of the changes that take place as soils develop with time.

A general expression relating the formative soil controls was first suggested in the 1940s:

$$S = C.P.B.R.T$$

where S is soil formation, C is the climatic factor, P is the parent material, B is the biotic factor (vegetation, animals and humans), R is the relief or topographic factor and T is the time over which the factors have operated. These five factors determine the nature of the initial inputs, how they are transformed and how quickly they are moved and lost from the soil. They define the state of the soil system.

The equation was revised to make it more applicable to modern ideas on ecosystems and a discussion of these ideas can be found in Gerrard (1992, pp. 3–5) but a simple approach based on the original ideas is still extremely useful and is adopted here.

24.2.1 Climate

The first soil scientists thought that climate was the all-important control. Now we see that it is indeed important, but not pivotal. It controls the weathering regime by influencing how much water is present and the temperatures that occur in a given area. It is important in controlling the chemical reactions that can take place in the soil and the soil life. Climate influences the type of

Table 24.2 Sequential weathering changes in soil minerals in the under-2 mm soil fraction. (Source: after Jackson *et al.*, 1948)

Stage	Type mineral	Soil characteristics
Early weathering stages		
1	Gypsum	These minerals occur in the silt and clay fractions of young soils all over the world, and in soils of arid regions where lack of water inhibits chemical weathering and leaching.
2	Calcite	
3	Hornblende	
4	Biotite	
5	Albite	
Intermediate weathering stages		
6	Quartz	Soils found mainly in the temperate regions of the world, frequently on parent materials of glacial or periglacial origin: generally fertile, with grass or forest as the natural vegetation.
7	Muscovite (also illite)	
8	Vermiculite and mixed layer minerals	
9	Montmorillonite	
Advanced weathering stages		
10	Kaolinite	The clay fractions of many highly weathered soils on old land surfaces of humid and hot intertropical regions are dominated by these minerals: often of low fertility.
11	Gibbsite	
12	Hematite (also goethite)	
13	Anatase	

soil water processes that can take place, such as eluviation and leaching, and in arid areas water can actually move up through the soil profile.

24.2.2 Biota

The presence of soil life is critical to soil development and the biota have two important functions. First, they provide organic matter, especially humus, which influences the solubility of the soil minerals. Second, the soil animals mix soil particles and help aerate minerals and promote water movement through channels and burrows. Microflora (algae, bacteria and fungi) decompose organic matter and release many of the nutrients that otherwise would be locked up in the organic compounds.

For example, the depth of roots can be important: hardwoods can reach deep lime, which can be recirculated to the surface, giving the soil a pH of 7, whereas nearby grass cannot reach the lime, leading to a soil pH as low as 5.5. Meanwhile, woodland trees can intercept rainfall, decreasing the runoff velocity, and erosion and transpiration from the leaf surfaces can subsequently lead to drying of the soil.

24.2.3 Relief

The relief has indirect effects on soil development via other processes (like the biota, drainage and microorganisms) such as soil movement and how much incoming solar radiation is received. There are three components to the relief factor: morphology, aspect and altitude. The latter two are passive in the sense that their influence is via climate, whereas the morphology has a direct effect. It can produce a clinosequence or toposequence depending on the slope angle, which affects the soil water drainage, although this is also controlled by the soil texture and the parent material. As the slope angle increases, so does the runoff and erosion. The result can be that soil profiles occur in a repeated manner and are geographically associated with relief.

A soil catena (derived from the Latin for 'chain') is produced by different drainage conditions, differential transport of eroded materials and leaching, translocation and redeposition of mobile chemical constituents. An illustration of this can be seen in Figure 24.1, where it should be noted that the influence of bedrock control also helps catena development.

It is in the recognition of the systematic relationships between soils, landforms and geomorphological processes that the catena concept is of great importance. Such studies show that soil development is integral to the overall evolution of the landscape and that soils cannot be studied in isolation from the geomorphological systems in which they are located (see further discussion in Gerrard, 1992, Chapter 3).

Aspect is important too in that it affects the amount of incoming solar radiation, so that different soils will form on sun-facing slopes versus shady slopes (Figure 24.2). For example, south- and west-facing slopes are warmer and drier than north- and east-facing slopes in the northern hemisphere, so the growth potential varies significantly. This growth potential is estimated to be 33% higher on south-facing slopes than on north-facing slopes, while west-facing slopes have a 24% advantage and east-facing slopes a 7% advantage over north-facing slopes.

Altitude affects the climate in that there is higher rainfall, lower temperature, increasing cloudiness and lower solar radiation as altitude increases, which all lead to an excess of precipitation over evaporation. Leaching is high where parent materials are freely drained and water-logging is common where they are not. The result is a reduction in biological activity, a slower breakdown of plant debris and the development of peat.

24.2.4 Parent materials

These can be organic, unconsolidated or lithified. Weathering can take place away from the site where soil parent material was formed, for example in glacial drift or parent materials derived from river sediments. Initially the parent material controls many of the physical and chemical properties of the soil but as time progresses it becomes of less importance and the soil becomes less related to it. In general, fine-grained parent materials, such as some glacial tills from the Irish Sea basin and shale, slate or mudstone bedrocks, produce slow-draining, moisture-retentive, waterlogged or gleyed soils. Coarse-grained parent

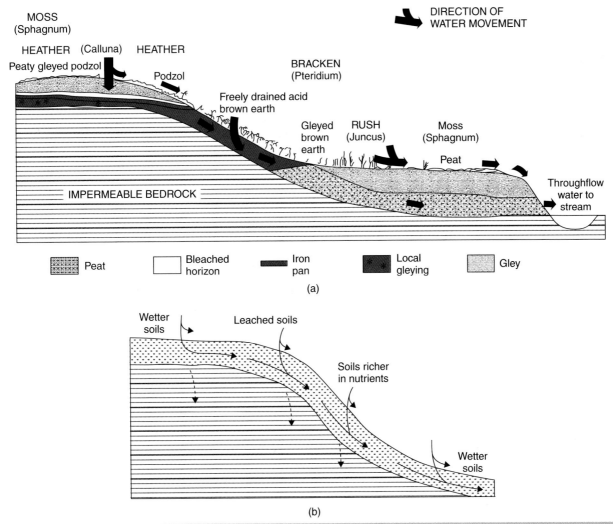

Figure 24.1 Soil catena development: (a) vegetation, soil, nutrient and water movement; (b) nutrient movement in relation to soils on hillslopes. (Source: after Courtney and Trudgill, 1976)

materials, such as fossil screes, periglacial deposits, sands and gravels, granite, sandstone or quartzites produce freely-draining soils where podsolization is an effective process through leaching. The base content of the parent material is important too as parent materials with high base contents, such as limestone, marble, basalt, dolerite and gabbro, generally have higher populations of soil microorganisms and higher rates of biological activity. The result is that plant breakdown and humification are quicker in these soils.

24.2.5 Time

This is an important control on soil development. Early stages of soil development are devoid of horizons and are sometimes called raw soils. They may be reworked by erosion, as in dune soils, so that time returns to zero, but with maturity horizons develop. In the tropics the time taken to achieve maturity may be only a few decades, whereas in Britain dune soils develop from the foreshore

to dune heath over 300 years and in British heathlands the soil assumed its present characteristics as a result of processes which began about 4000–5000 years ago. There is a major contrast between azonal soils on bedrock and a mature, many-horizoned soil (Figure 24.3).

24.3 Soils as systems

A systems approach to the study of soil development is useful. The soil or the system can have inputs, outputs and internal processes, as we have seen from Chapter 23. The soil can be looked at as an open system, as materials and energy are gained and lost at its boundaries. The inputs or additions to the soil are: nutrients from weathered parent material; nutrients from the atmosphere from water, gases and windborne particles and nutrient cycling by plants and animals; respiration by soil animals; and excretions from plant roots and organic matter. The outputs

SHADED SLOPES

Colder soils
Wetter soils
Restricted soil fauna
Surface accumulation
of acid organic
matter

SUN-FACING SLOPES

Warmer soils
Drier soils
Varied soil fauna
Organic matter
incorporated

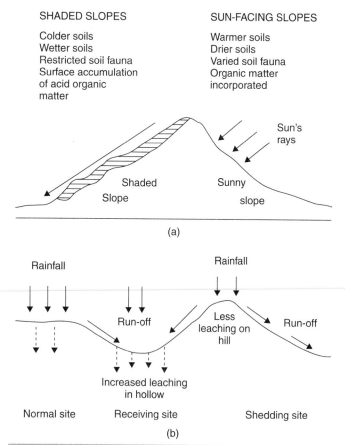

(a)

(b)

Figure 24.2 Effects of aspect and relief: (a) soils on shady and sunny slopes; (b) normal, receiving and shedding sites. (Source: after Bridges, 1970)

Figure 24.3 Mature, many-horizoned humus iron podzol, screen formation, Wexford, southern Ireland

to the system are: nutrients taken up by the plants; nutrient losses into mobile soil water, overland flow and throughflow and groundwater percolation; soil erosion and soil creep, which result in losses of soil downslope; and evaporation and transpiration from the soil surface. These inputs and outputs interact with the nutrient store in the soil.

The relationships described above are greatly simplified. In nature they are much more complex, so for example nutrients lost to plants may return as leaf litter the following autumn and therefore there is nutrient recycling. However, if the vegetation is cropped, nutrients become depleted and fertilizers and animal manures have to be used. Transfers in the system consist of both upward and downward movements of organic and mineral matter by soil fauna or water and there can be transformations resulting in change of both composition and form in the inputs.

24.4 Soil profile development

24.4.1 Nomenclature and horizons

The profile is the basic soil unit for description and classification and its division into horizons is fundamental. The sequence and

type of horizons helps to define the soil type. This horizonization is due to the concentration of particles, grain size, chemicals such as iron and aluminium, and organic matter into specific soil zones. The horizons are horizontal layers which are not sedimentary in origin but rather the expression of the soil-forming processes caused by internal rearrangements of the soil constituents. They are differentiated in detail by several variables, such as colour, texture, structure, stoniness, acidity and mineral content.

The organic layers are almost always at the surface, but in humid regions there can be downwashing of solutes and colloids and redeposition further down the profile, whilst in arid regions there can be the reverse, with upward movement and deposition. These processes are called eluvial where material is lost and illuvial where material is gained.

The old system of horizon description was to divide the profile into A, B and C, where A is the surface organic horizon with some mineral matter, down to the furrow depth; B is the mineral horizon, which is weathered with minimal organic matter and has a complex formation; and C is the parent material or unweathered bedrock. These divisions were also called the surface soil, the subsoil and the substratum. They were not particularly good sub-divisions as they used preconceived labels; today we use symbols for horizons containing particular diagnostic properties, irrespective of where they occur in the vertical profile sequence. These horizon symbols are illustrated in Table 24.3; a full designation can be found in Avery (1980).

Different environmental conditions with regard to the rainfall balance input and soil drainage output are important in controlling the soil development, as can be seen from Figure 24.4. In Figure 24.4(a), A's rainfall input is equal to the drainage output, drainage is good but the input is small, whereas in B inputs equal outputs but the input is large. In C the output cannot keep pace with the input rate and drainage is poor, and the inputs can be high or low. This results in several major soil-profile processes:

1. Leaching, where there is downwashing of soluble material.

2. Clay translocation, or the downwashing of clays (lessivage).

3. Cheluviation and podsolization, where there is downwashing of most of the soil solutes to leave a bleached, white or grey upper soil mineral horizon.

Table 24.3 Horizon symbols. (Source: Avery, 1980)

Organic and organic-mineral surface horizons

L	Undecomposed litter.
F	Partially decomposed litter.
H	Well decomposed humus layer, low in mineral content.
A	Mixed mineral – organic layer.
Ap	Ploughed layer of cultivated soils.
Ah	A humose horizon.
Ag	A horizon with rusty mottling subject to periodic waterlogging.

Mineral sub-surface horizons

E	Eluvial horizon, depleted of clay and/or sesquioxides.
Ea	Bleached ash-like horizon in podzolised soils.
Eb	Brown (paler when dry) friable weakly structured horizon depleted of clay.
B	Altered horizon distinguished from the overlying A or E horizons and the underlying less altered C, by colour and/or structure, or by illuvial concentrations of the following materials denoted by suffixes.
t	Illuvial clay, characteristic of brown Earths (*sols lessivés*).
h	Illuvial humus, characteristic of podzols.
fs	Illuvial iron and aluminium (sesquioxides)
C	A horizon that is little altered, except by gleying and is either like or unlike the material in which overlying horizons have developed. (Where two or more distinct depositional horizons occur in the lower part of the profile they are designated C1, C2, etc.)
Bg Eg Cg	Mottled (gleyed) horizon subject to waterlogging; where gleying is only weakly expressed the suffix (g) is used.
Bfeg Cfeg	In gley soils, horizons of maximum iron accumulation.
A/C B/C	Horizons of transitional or intermediate character.

Horizons in soils derived from peat

Aop	Ploughed layer of peaty soil or peat.
Bo	Decomposed sub-surface horizon distinguished by its structure.
Co	Little altered peat substratum.

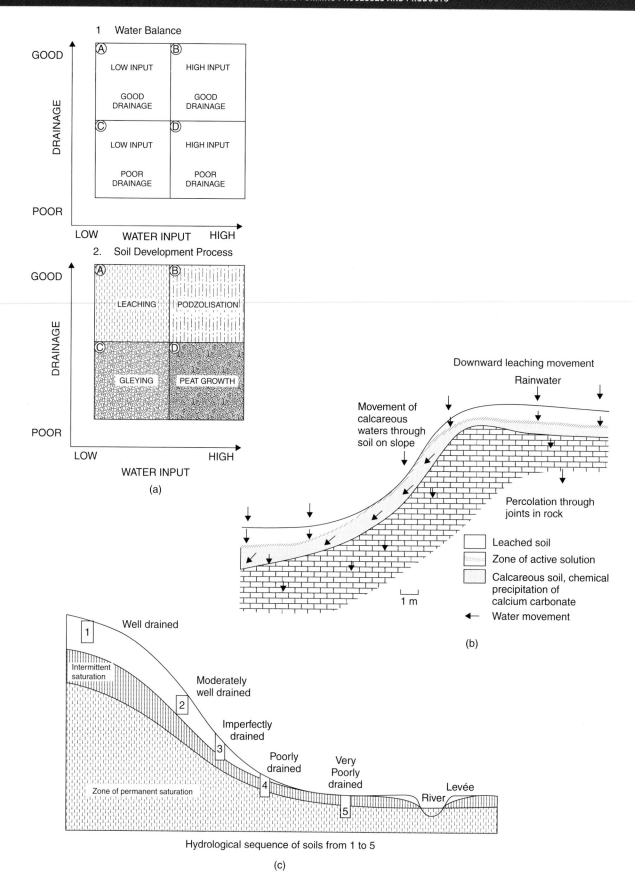

Figure 24.4 Soil development and drainage conditions: (a) water balance and soil-development processes; (b) slope and valley bottom showing a hydrological soil sequence; (c) leaching and slope on limestone. Courtney and Trudgill, 1976) (Source: after courtney and Trudgill, 1976)

4. Gleying, which occurs in waterlogged soils, where alternate phases of oxidizing and reducing conditions are present according to the soil water level.

24.4.2 Processes of translocation (eluviation)

Leaching

This process involves the movement of chemical compounds as soluble ions along with the percolation of soil water during its drainage through the soil. Strictly speaking it can involve the leaching of calcium carbonate (decalcification), chelates (cheluviation) and clays (lessivage). Rain can pick up atmospheric substances and form vegetation. This makes it slightly acidic and in many cases this rain is a weak carbonic acid due to the atmospheric carbon dioxide. This can dissolve soluble minerals and transport them through the soil and it can also remove nutrients attached to the surfaces of clay and organic matter by cation exchange and wash these through the soil too.

Cheluviation

This process is caused by soluble organic complexes picked up by rain water from the surface soil layers, which translocate soil metal cations, particularly iron and aluminum oxides. These complexes are highly mobile and are easily washed through the soil to precipitate the less-soluble mineral material at depth as a hard, distinct layer (ironpan), whilst the most soluble can be removed from the profile completely. The precipitation may be a response to changes in the degree of aeration and the quantity of organic substances lower down the profile.

A summary of two groups of theories related to the podzolization process can be seen in a case study on podzolization taken from De Corninck (1980) and Farmer (1982). Two main groups of organic matter are suggested as being responsible complexing agents: organic compounds, such as polyphenols, washed directly from plant foliage and litter; and condensed humic and fulvic acids and the organic end products of their decomposition. The extensive literature relating to this process is discussed clearly and in much detail in Ross (1989, pp. 112–119).

Clay translocation or lessivage

This process involves the removal of individual clay particles or the dissolution of clay minerals in the upper soil layers and their transport in suspension through the profile. The clay is washed down root channels, worm holes, cracks and pore spaces and is deposited further down the profile. Here it tends to form thin clay skins (cutans) around larger soil grains and along the pore walls.

The evidence for this originally came from micromorphological studies of soil thin sections. In freely-drained soils the B horizons, which show clay accumulation, are called 'argillic'. It might be expected that there would be a correlation between the amount of water percolation through soils and the amount of clay translocation but this process is not always found in the wettest climates.

Other factors are clearly important. As the clay is largely moved in suspension, it must first be suspended and then kept in a dispersed state. Hence the stability of soil colloids may be an important control on the transport of clays. Increased mobility has been found to be associated with increased solution pH over the range 5.5–8.5 and with increased organic matter. As pH rises, hydroxyl groups at clay crystal edges dissociate, increasing the colloid negative charge and its cation-exchange capacity. This is the small but significant pH-dependent cation-exchange capacity of mineral soil colloids. As organic matter has a much higher pH-dependent cation-exchange capacity through the dissociation of carboxyl and phenyl groups, it might be expected that organic colloids will be even more mobile than mineral colloids when solution pH is increased.

Other material in the soil like silt, iron oxides and organic matter can move in the same manner. The latter is important and, although the organic matter may move in solution as well, it can be reprecipitated further down the profile. Here this redeposition can be controlled by small changes in soil wetness, acidity and aeration so that distinct layers of humus can accumulate in the subsoil away from the soil surface layers where they are normally found.

24.4.3 Other soil-forming processes

Gleying

This process operates from the effects of oxidation and reduction processes and involves little redistribution of the soil materials, where the soil water output is not rapid. During periods of waterlogging there are anaerobic conditions because the soil pore spaces become water-filled and air cannot enter the soil. This means that the oxygen that normally acts as a sink for electrons released by plants and soil organisms is lacking and other compounds, like iron and aluminum oxides, accept these electrons, resulting in chemical reduction of these compounds.

Ferric iron is converted to the ferrous form, which is characteristically grey or green in colour. Other compounds can be reduced in a similar way and the result is that the soil becomes duller in colour and many of the reduced compounds, typically more soluble, can be washed from the soil. On the other hand, during drier periods air enters the soil and a process reversal occurs as oxygen accepts the electrons and combines with the reduced materials to form higher-valency compounds. The ferrous iron is converted back to its ferric state, which is typically red or yellow in colour. The soil then becomes brighter.

As there are often fluctuations in the water table, alternate phases of reduction and oxidation are common in some soils. This results in the variation of soil colour between grey and red and in a mottling effect. Oxidation is usually well developed along the larger pore spaces (structural cracks, root and earthworm channels) because of the better air access along these.

Calcification

The processes discussed so far all involve eluvial or loss processes. However, it is possible in some climatic regimes, such as low-rainfall continental interiors, particularly steppe and prairie areas, that there is limited output of both water and soluble materials from the base of the soil profile and also of the movement and deposition of soluble salts within the profile. Under moderate rainfall, calcium carbonate is translocated from the upper profile down, where it is reprecipitated in an accumulation zone, called the calcic horizon (B_{ca}). This surface leaching can result in slightly acidic surface horizons.

The deposition of calcium carbonate can clearly be seen, as the redeposited calcium carbonate infills soil pores and cracks to give the B horizon a speckled white appearance. This transfer of calcium carbonate and its deposition is known as the calcification process and its characteristics are shown in Figure 24.5.

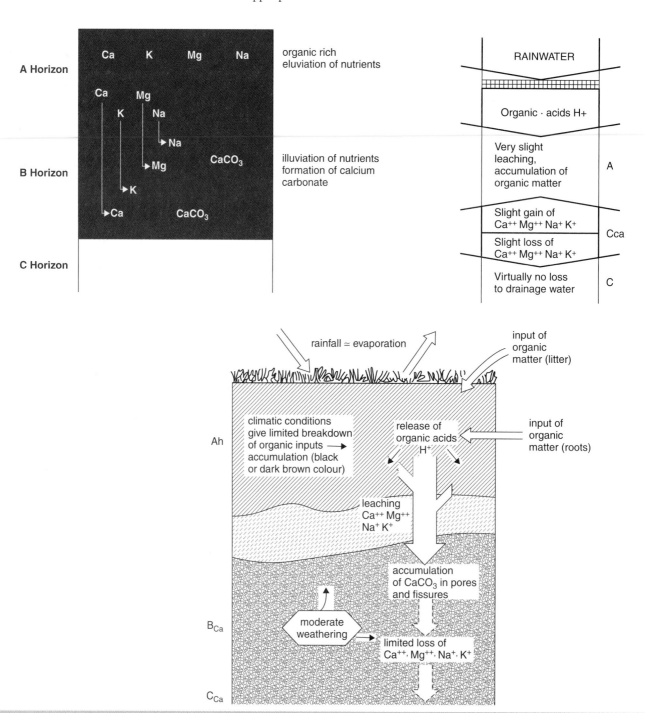

Figure 24.5 The calcification process

Salinization

In fully arid climates there is only limited downward translocation of soil materials because of the low and irregular rainfall and there can be even upward movement of water and soluble salts (illuvial processes) wherever a water table occurs near the soil surface. Where salts are moved upwards and accumulate, the process is known as salinization. This leads to enrichment in the upper part of the soil, with reprecipitated, soluble salts following the output of water by evaporation as a salic horizon or salt pan (Figure 24.6). There is a high salt concentration in the surface horizons and often there are surface salt encrustations, which give the surface an irregular surface form. These are also known as solonchaks or white alkali soils and they show little or no change down the soil profile, indicating that the parent material is little affected by the soil-forming processes. Most saline soils are only slightly alkaline in reaction and the salts present are predominantly sodium and calcium sulphates and chlorides, although nitrates and magnesium may occur in some places.

Ferrallitization and ferrugination

Most of the processes discussed so far operate throughout the world's temperate regions, the Mediterranean and arid regions of the tropics. However, in much of the tropics soil development has continued for a long period of time and it was not interrupted by glacials as in many of the temperate regions. Over a long time span and in the intensive climatic regimes of high rainfall and high temperatures, some soil-forming processes may be restricted to the tropics.

Ferrallitization (Figure 24.7) combines intense transformations of soil material by weathering with outputs from the soil in the form of solutions, including all salts and cations and much of the silica. The residue that is left behind consists predominantly of quartz, some clay minerals, and iron and aluminum sesquioxides. These soils are usually very acid, with low base status and high levels of exchangeable aluminum. In certain situations there can be a net accumulation of sesquioxides by lateral inputs, or by vertical transfers from materials at depth through water-table fluctuations. Iron-rich nodules may occur in the zone of alternate wetting and drying to form a laterite layer.

Ferrugination (Figure 24.8) combines rather less intense chemical weathering and transformation with complete removal of the soluble salts, a moderate degree of cation leaching and some silica. The distinctive feature is the red colour, as the released iron from the minerals during transformation is transferred to the ped surfaces, where it is dehydrated. This process of reddening (rubification) does not require much iron in the soil.

Biological processes

It is clear that various biological processes occur in the redistribution of soil-profile materials. For example, plants can cause upward translocation of compounds as they extract water from

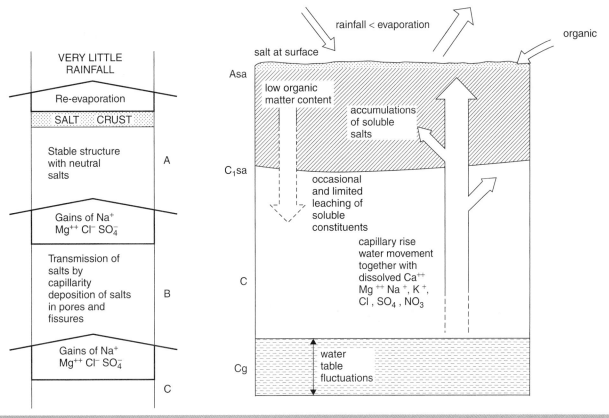

Figure 24.6 The salinization process

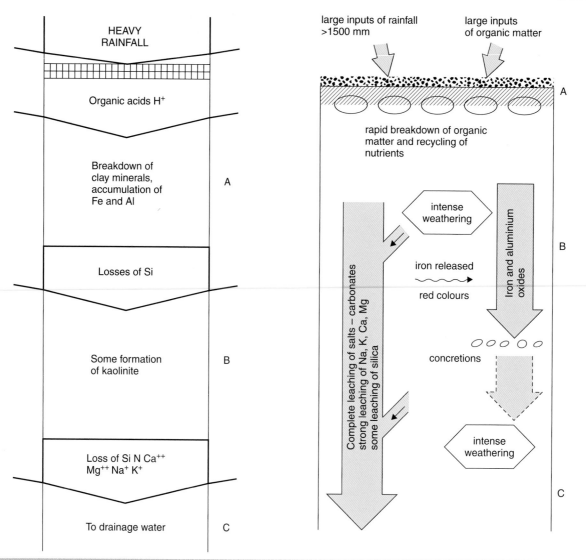

Figure 24.7 The ferrallitization process

the lower soil levels and so draw the dissolved materials in the water to the surface. There is physical mixing caused by development of ice crystals, burrowing of animals, earthworm tunnelling and the smaller-scale activity of microorganisms, which overturn the soil and increase the rate of soil-forming processes. Similarly, root growth, uprooting of trees by wind or animals and ploughing by humans all cause mixing of the soil.

24.5 Soil properties

24.5.1 Soil texture and grain size

Soil can consist of mineral particles of any size from stones to clay, but it is usually a mixture of different sizes. The smaller the particles, the more active they are in the soil profile and the more easily they can be moved. The grain size can be measured in the field using a hand-texturing technique, or more accurately in the laboratory by sieving, by a sedimentation tube technique measuring the rate of fall through water or by a Coulter Counter or sedigraph. The relative proportions of the different sizes in a soil sample give the soil its texture, with the exclusion of gravel (stones) and organic matter. The constituent parts can be referred to different grain-size scales, such as the ones below.

The Atterberg Scale

gravel 0.05–5 mm

coarse sand 2–0.2 mm

fine sand 0.2–0.02 mm

silt 0.02–0.002 mm

clay under 0.002 mm.

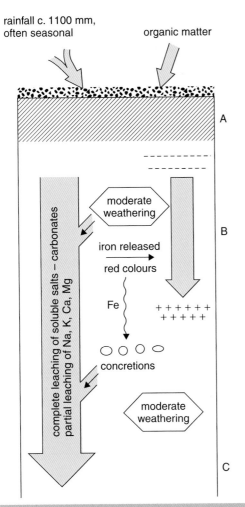

Figure 24.8 The ferrugination process

United States Department of Agriculture 'American' Scale

gravel 75–2 mm

coarse–medium sand 2–0.25 mm

fine sand 0.25–0.05 mm

silt 0.05–0.002 mm

clay under 0.002 mm.

The various size groups are called fractions and most soils have a variety of these:

1. **Coarse, including sand:** Often rounded and flat. Unless they are coated with fines, they have no plasticity and poor water retention.

2. **Fines:** Silt is irregular and fragmental, with some plasticity, cohesion and absorption. Clay is composed of clay minerals like kaolinite, mica, vermiculite, montmorillonite, chlorite

and allophane. There are also plates or flakes, often with laminated water between the plates, with a surface area much greater than that of coarse sand, for example.

The mineral composition of the different fractions varies. In the coarser fraction silica predominates, often with feldspars and mica. Using a textural triangle a limited number of categories, described as the main textural classes, are shown in Figure 24.9.

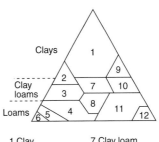

1 Clay
2 Sandy clay
3 Sandy clay loam
4 Sandy loam
5 Loamy sand
6 Sand
7 Clay loam
8 Loam
9 Silty clay
10 Silty clay loam
11 Silt loam
12 Silt

LIMITING VALUES FOR SOIL TEXTURE CLASSES

Sand	Material containing at least 85% sand, provided that the percentage of silt plus 1.5 times the percentage of clay shall not exceed 15.
Loamy sand	Material containing not more than 90% nor less than 70% sand, together with a percentage of silt plus 1.5 times the percentage of clay not less than 15 at the upper sand limit, or a percentage of clay to not exceed 30 at the lower sand limit.
Sandy loam	Material containing either less than 20% of clay with the percentage of silt plus twice the percentage of clay exceeding 30, or having between 43 and 52% of sand with less than 7% of clay and less than 50% of silt.
Loam	Material containing between 7 and 27% of clay, 28 to 50% of silt with less than 52% of sand.
Silt loam	Material containing either more than 50% of silt together with between 12 and 27% of clay, or which has between 50 and 80% of silt with less than 12% of clay.
Silt	Material that contains more than 80% of silt with less than 12% of clay.
Sandy clay loam	Material containing between 20 and 35% of clay, with less than 28% of silt and more than 45% of sand.
Clay loam	Material that contains between 27 and 40% of clay and between 20 and 45% of sand.
Silty clay loam	Material that contains between 27 and 40% of clay and less than 20% of sand.
Sandy clay	Material that contains 35% or more of clay together with 45% or more of sand.
Silty clay	Material that contains 40% or more of clay together with 40% or more of silt.
Clay	Material that contains more than 40% of clay together with less than 45% of sand and less than 40% of silt.

Figure 24.9 Main textural classes

Table 24.4 Soil organisms in topsoil. (Source: after Jones, 1997)

Organisms	number per teaspoonful	weight (kg) per hectare
Bacteria	90 000 000	1-2000
Fungi	200 000	2-5000
Algae	30 000	~500
Protozoa	5000	100-500
Nematodes	30	~200
Earthworms	< 1	0-2000
Other soil animals	< 1	0-500
Total	~91 million	~6000

The characteristics of the various categories are given in Table 24.4; these are taken from Conway (2009).

24.5.2 Soil structure

The soil structure, or tilth, is where there is a grouping of particles into aggregates. Thus it describes how the components are joined together or organized, especially in medium-textured soils. The lumps into which the soil disintegrates in water are called peds, and soils often have a different structure in different horizons. They are important because they can affect water movement, heat transfer, aeration, density and porosity. The following terms are used in assessing the distinctiveness of structure:

1. **Structureless or massive:** Here there are no observable peds. The structure can be massive if coherent or there can be single grains if it is noncoherent, as in salty soils, where the salts change the physical state of the colloids so that they cannot bind the particles together.

2. **Weak:** Here there are indistinct peds and there is much unaggregated material when disturbed.

3. **Moderate:** Here there are well-formed peds and there is little unaggregated material when disturbed.

4. **Strong:** Here the peds are distinct in places and remain aggregated when disturbed.

The structure can develop from the single-grain state or from the massive state and these define the boundaries of common structures shown in Figure 24.10 and Table 24.5. Each type can be divided into size classes from very fine or very thin to very coarse or very thick.

The genesis of any structure is often complicated and obscure but the factors that control it include the nature and origin of the parent material, the soil-forming conditions (especially the clay/humus conditions) and the presence of salts in arid areas. The mechanisms are really any processes that develop lines of weakness, shift particles and force contacts, such as wetting and drying, freezing and thawing, roots/animals/tillage, cation exchange and flocculation. Chemical cements, clay particles and organic matter have a cohesion due to electrochemical bonding. The structural stability can change through processes such as dissolving in rain and ploughing. In general, the larger the aggregates, the lower their stability.

24.5.3 Soil types based on texture and structure

Soils can be described on the basis of the relative proportions of the grain sizes and their structure as light/heavy, sandy/loamy or clayey. A light soil has a large proportion of coarse particles, hence large air spaces, and is light in terms of ploughing. If it has over 60% sand it is termed sandy; under 30% clay and lots of sand, it is loamy; over 30% clay and under 50% sand, it is clayey. The characteristics of these three soil types are as follows:

1. **Sandy:** Light, too permeable, rapidly dries out, low water storage, poor mineral nutrients, highly aerated and therefore rapid oxidation and rapid heating. These are known as 'hungry' soils. An organic binder and water have to be added in order for them to be a good growing medium.

2. **Clayey:** Heavy, impermeable or poor permeability, chemically rich in nutrients but often inaccessible due to poor aeration, liable to waterlogging. Clay particles swell when wet and the whole soil may be reduced to a structureless slush, when it can be described as 'puddled'. These soils form clods when dry; if they are ploughed when wet, the structure is destroyed. They are cold soils, especially in the spring.

3. **Loamy:** These soils have the optimum combination of permeability and water retention and they also aggregate easily to facilitate root holding.

24.5.4 Soil organic matter

This is derived from decaying leaf litter and faecal and other material from animals. As it develops with time, it loses its structure. Humification is the process of this breakdown, and humus the result. It is a resistant complex of amorphous and colloidal substances and is almost jelly-like in appearance. It is what results when all the flora and fauna have eaten and digested one another and can get nothing further out. Hence it is simultaneously of plant, animal and microorganism origin and is always in a state of flux. It is noncrystalline and dark in colour, and the process releases nutrients needed for plant growth. Its porosity

Size class	Type of structure	Description of aggregates	Appearance of aggregates (peds)	Common horizon location
1 mm– 6 mm	Crumb	Small-fairly porous spheres: not joined to other aggregates		A-horizon
1 mm– 10 mm	Platy	Like plates: often overlapping which hinders water passing through		Plough pan Ea-horizon in forest and claypan soils
5 mm– 75 mm	Blocky	Like blocks: easily fit closely together: often break into smaller blocks		B-horizon: heavy clays, e.g. London Clays
10 mm– over 100 mm	Prismatic	Column-like prisms: easily fit closely together: sometimes break into smaller blocks		B/C - horizon: heavy clays, e.g. Carboniferous clays
10 mm– over 100 mm	Columnar	Like columns with rounded caps: easily fit closely together		B/C - horizon in alkali soils

Figure 24.10 Common soil structures

is between 80 and 90%. It has a low plasticity and cohesion and so ameliorates the effects of mineral clays. In the soil, three layers of distinct organic matter can often be recognized (Figure 24.11):

1. **L:** Leaf litter, where leaves and other plant remains can still be recognized.

2. **F:** Fermentation or humification layer, where decay is active.

3. **H:** Humus layer, where the plant remains are unrecognizable.

When soils from different areas are compared, three distinct types of humus can be recognized:

1. **Mull:** Soft, blackish matter, crumbly when dry, rich in nutrients, and found where the soil is not too acidic, like in lowland hardwoods and fertile grasslands, where there is free drainage and good aeration. It rapidly decomposes and may combine with clay particles to form clay–humus complexes.

2. **Mor:** Raw, fibrous, acid humus, poor in nutrients. Found in upland heath, coniferous woodland or bog. Decomposes slowly.

3. **Moder:** An intermediate form, both in terms of its rate of decomposition and the development of clay–humus complexes. There are distinct layers, as in mor, but it is further decomposed like mull, and is more readily incorporated into the soil than mor.

The importance of humus is that it holds the nutrients in reserve and it can affect the soil structure. It consists of four main constituents: carbohydrates, proteins, cellulose and lignin, which all vary in their resistance to decomposition, with carbohydrates being the least resistant. These decay most rapidly. Different plants decay at different rates: for example, pines, firs and heather decay very slowly, whereas most grasses, herbs and the leaves of broad-leaved trees decompose more rapidly. During decomposition the organic matter changes to a chemically more simple form.

Table 24.5 Terms used in the structural description of soils

Structureless:	no observable peds, massive if coherent and single grain if non-coherent.
Weak:	indistinct peds: when disturbed breaks into much unaggregated material.
Moderate:	well formed peds; little unaggregated material when disturbed.
Strong:	peds distinct in places; remains aggregated when disturbed.

The basic types of *peds* are defined as follows:

Platy:	vertical axis much shorter than horizontal.
Prismatic:	vertical axis longer than horizontal, vertical faces well defined, vertices usually angular.
Blocky:	dimensions of peds are of the same order, enclosed by plane or curved surfaces that are casts or moulds formed by faces of adjacent peds; subdivided into *angular* and *subangular*.
Granular:	small, subrounded or irregular, peds without distinct edges or faces, usually hard and relatively non-porous.
Crumb:	soft porous granular aggregates.

Each type is divided into size classes, from very fine or very thin to very coarse or very thick.

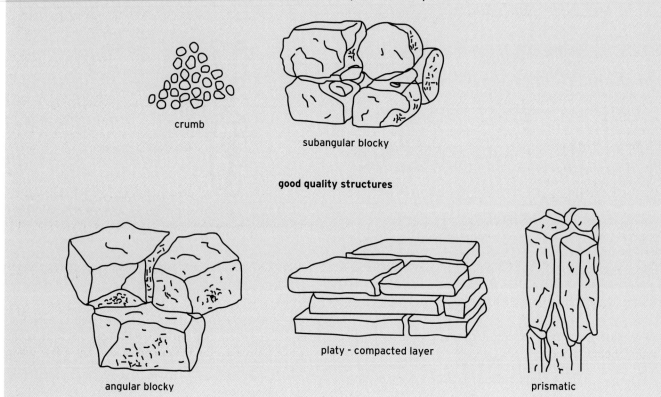

crumb

subangular blocky

good quality structures

angular blocky

platy - compacted layer

prismatic

When there is much organic matter it becomes difficult and less useful to assess the particle-size class of the mineral fraction. In such situations the materials are either classed according to their estimated organic matter content or checked against laboratory determinations of organic carbon. Soils can be termed 'humose' or 'peaty' according to the relationship between percentage clay in the mineral fraction and the percentage organic carbon. The degree of humification can be described according to the classification illustrated in Table 24.6.

24.5.5 Soil chemistry

There are several aspects of soil chemistry, which are important in understanding how soil processes operate.

Colloids and cation-exchange capacity

Colloids are particles under 0.001 mm. In the colloidal state there is a two-phase system, where one material is dispersed through

(a)

(b)

Figure 24.11 Soil organic matter: (a) organic layer in the soil-surface, litter (L) layer of Holm oak leaves in the Puig Mayor region of Majorca; (b) litter (L), fermentation (F) and humus (H) layer in Holm oak woodland, Majorca

Table 24.6 Degree of humification

Degree of humification

A. Slightly humified peat

H1 — Completely non-humified and free of dy*. Yields only colourless water when squeezed.

*dy: brown or yellowish brown flocculent material that can be squeezed out with the water from peats.

H2 — More or less unhumified and free of dy. Yields yellowish brown water when squeezed.

H3 — Very slightly humified with a small amount of dy. Yields muddy water when squeezed but the peat itself does not pass through the fingers.

H4 — Slightly humified peat and weakly dy-charged. Yields very muddy water when squeezed. Residue slightly plastic.

H5 — Humified peat with a large amount of dy. Plant structure still evident. Yields very muddy water when squeezed and some peat escapes through the fingers. Residue quite plastic.

B. Medium humified peat

H6, H7. — Well humified peat and strongly dy-charged. Very few visible plant remains. Up to 2/3 of mass passes through the fingers on squeezing. Residue consists mainly of root fibres and wood, and is strongly plastic.

C. Strongly humified peat

H8–H10. — Very strongly humified and almost completely dy-charged. No plant structures visible; 2/3 to the whole of the mass squeezes through fingers.

K^+/Na^+, whereas in arid regions there are Ca^{++}/Mg^{++}, Na^+/K^+ and H^+. Different clay minerals behave differently:

1. **Kaolinite:** This has a fixed lattice structure with no expansion on wetting; that is, no adsorption between plates. Here the active surface is only the exterior and it has a low cation adsorption.

2. **Montmorillinite:** This has a weak lattice, with much smaller units that can expand and allow much swelling (adsorption 10–15 times that of kaolinite).

3. **Mica (illite):** This is comparable to montmorillinite, but with bigger particles and therefore less adsorption.

4. **Chlorite (vermiculite):** This is similar to montmorillinite and mica. Vermiculite fixes K well.

In organic colloids in humus the cation adsorption capacities are far greater than in clays. The order of adhesion of cations if present in equivalent quantities is Al-Ca-Mg-K-Na, and H is also

another (analogies being milk, cheese and clouds). In soils most clays and humus come into this category. They are important because they store nutrients and water, affect the plasticity (which can encourage a rapid change in structure), affect the soil cohesion as the water in a clay gel is reduced (so that the colloids stick together), influence the swelling and contraction due to the water uptake and expulsion, affect the flocculation or dispersion of soil particles and affect the leaching and the soil structure. Mineral colloids have a platy form and a huge surface area and are generally silicate clays, which carry a negative charge, so they attract cations, which have a positive charge. This gives an ionic or 'electric' double layer with an inner layer, an anion and an outer layer, with loosely-held cations. Here the adsorbed ions are attached to the surface. There is also much water held at the colloid surface and the water molecules between the plates of the clay micelles. Under natural conditions certain cations predominate, so that in temperate/humid areas there are H^+/Ca^{++}, Mg^{++} and tightly held. In wet/organic areas H + Al can be dominant as the other metal ions are leached and the result is a low pH. In arid zones the metallic ions dominate and the pH is 7+.

Questions that need to be considered are: How do nutrients become detached from clays and move into the plant roots? and Why are water soluble fertilizers not washed out of the soil before plants can use them?

The answers are related to cation exchange and storage on the colloid surfaces. This is one of the most important processes as it is the main mechanism of plant nutrition. Colloids can exchange one ion for another in solution in soil water. When a plant rootlet comes into contact with a clay particle with adsorbed cations the process of cation exchange takes place and the root gives out H ions in exchange for nutrient ions. Hence the cation-exchange capacity of a soil depends on the capacity of colloids (clays/humus) to exchange ions for those in the soil solution. Calcium, magnesium, potassium and sodium migrate from the swarm of cations adsorbed on to the clay surface and move to the plant root, with their places taken by hydrogen ions. The nutrient ions are translocated in the water-transporting, xylem plant tissue to the stems and leaves. The hydrogen ions substituted on the clay surface can be used to weather soil minerals, releasing further nutrient cations which can be adsorbed on to the clay surface.

The CEC is measured in milliequivalents (me) per 100 g of soils, which equals 1 mg of H, or the amount of any other ion that will combine with or displace it. For example, clay of 1 me can adsorb and hold 1 mg of H or its equivalent. The CEC me of other ions is the atomic weight divided by the ionic charge, giving the adsorption in mg/gm. Different colloid types vary, so that humus has a figure of 200, montmorillonite 100 and kaolin 4, and therefore there are differences in CEC in different soils. Most soils contain a mixture of cations, so that a normally-cultivated soil would have Ca 80–90%, Mg 5–10%, Na 0–5% and K 2–3%. Hence the point of liming soils is to keep up the proportion of Ca

ions for the optimum plant growth as they tend to decline as Ca exchanges with other ions in percolating soil water.

Soil pH

The soil acidity, reaction or pH is a measure of the concentration of the free H ions in soil water. The scale can be seen in Figure 24.12. As an acid is a compound which dissociates in water to give hydrogen ions, soil pH is a measure of the soil-water acidity. Here the pH is the logarithm of the reciprocal of the molar concentration of the hydrogen ions in the soil water. As the scale is logarithmic, each time the hydrogen ion concentration shows a tenfold change, the pH alters by one unit on the scale. These H ions are provided by the dissociation of water, by root activity and by weathering processes. As the concentration of hydrogen ions increases, the soil becomes more acid. Values of 3 can be found on moors and heaths, whereas alkaline values can only be found when there is free calcium carbonate present. A pH of 10–11 can only be found in arid climates. Most crops prefer acid soils (6–6.5) and low yields are given at pH below 5, with the lower tolerance limit for wheat 5.1, potatoes 4, oats 4.2 and beet 4.9. The importance of pH to soil processes and fertility is that it influences the solubility of many nutrients.

Soil calcium carbonate content

The calcium carbonate content of a soil affects its pH and hence many other properties. In particular, it affects soil structure, as calcium ions act as bridges between the colloids and calcium carbonate is a strong cementing agent. As with pH, the calcium carbonate content is influenced by leaching; as the leaching rate increases, so the depth to calcium carbonate in the soil increases too.

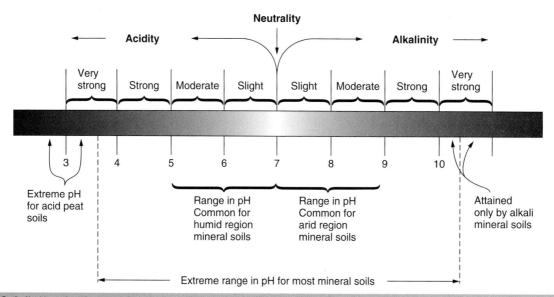

Figure 24.12 Soil pH scale. (Source: after Curtis et al., 1976)

Soil nutrients

Nutrients are important in soils because they are the chemical elements necessary for plant growth. The active ingredients are dissolved in soil water or held on colloids and the absorption of nutrients from the soil is through the plant roots. When plants die there is a return of nutrients to the soil. Decomposition of plant residues allows the nutrients to be reused. There are several types of nutrient:

1. Nutrients in solution, which are freely available to plants but can be washed out of the soil.

2. Nutrients attached to the clay–humus complexes, which are available to plants but are not easily washed out of the soil.

3. Nutrients stored in the mineral grains, which are unavailable to plants, unless released by weathering or bacterial/fungal attack.

The chief nutrients are N, K and P (primary nutrients), with Ca, Mg, S, Fe and Zn the secondary nutrients. For example:

1. Nitrogen is made available by biological means and is nearly all combined in the humus, so is unavailable. It is usable only in the nitrate form.

2. Phosphorus is taken up as phosphates. No matter how these are added to the soil, they quickly become insoluble, fixed and unavailable. Phosphorus is present in soils in various forms, such as insoluble apatite ($CaPO_4$), insoluble Fe and Al phosphates, slightly soluble dicalcium phosphate, exchange ions on colloids and in organic compounds. Only the last three are useful to plants, although some plants, like lupins, can extract phosphorus from insoluble compounds.

3. Potassium is usually present in abundance but most is rigidly held or easily leached. Most is derived from colloidal ions, although some K-containing minerals, such as muscovite mica, weather readily and may provide a steady source.

4. The trace elements, such as manganese, copper and boron, are essential to plant growth in small quantities. They are toxic in too large a quantity, and are sometimes not present at all in alkaline soils.

24.5.6 Soil water

Moisture conditions, the amount of water available and the way in which water flows through the soil are important to many soil processes. Fundamental to how water behaves in a soil is the pore-size distribution, which depends on its texture and structure and determines how tightly soil water is held. Small pores, as in clays, exert high tension and therefore make release difficult, whereas large ones, as in coarse sands, have low tension and allow easy water movement. All pores smaller than 60 μm diameter will hold water, but they are divided into those with available water capacity (between 0.01 and 60 μm diameter), which can be extracted by plant roots, and those with unavailable water capacity (smaller than 0.01 μm diameter), which hold water too strongly for plants to use.

The relationship between water content in a soil and tension can be shown using a moisture-release curve (Figure 24.13). This

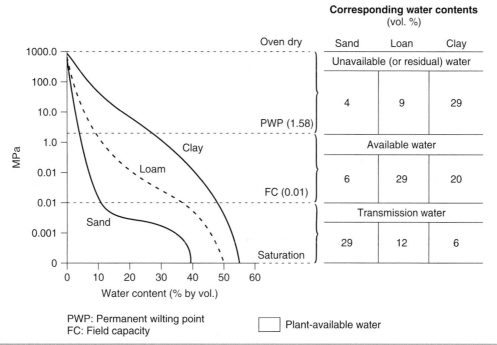

Figure 24.13 Relationship between water content and tension using a moisture-release curve

shows three soils with different textures. The amount of plant-available water lies between the field capacity and the wilting point, where the water is held under a great deal of tension. It is a measurement of how much water a soil can keep for plant growth under nonsaturated conditions.

The moisture held at field capacity is of two types. Part is held loosely in the soil and can be used by plants. This is the available as capillary water. The rest is not available as it is held against the surfaces of the soil particles by very strong forces. This is called hygroscopic water. After plants have used all the available soil capillary water, they continue to lose water through evaporation and transpiration and the result is too much water loss and wilting. At this stage the soil wilting point is reached; this roughly corresponds to the lower limit of available water. Using the soil texture, various hydraulic soil properties, such as porosity, wilting point, field capacity and water-holding capacity, can be calculated at www.bsyse.wsu.edu/saxton/soilwater/soilwater.htm. (see discussion in Saxton et al., 1986).

24.5.7 Soil organisms

Soils contain a great range of organisms in huge numbers. This can be seen in Table 24.4. The total weight of living microorganisms in soil is about 2% of the total weight of organic matter, so that a soil with 5% organic matter will contain about 2.5 tonnes of microorganisms per hectare, although it has been estimated that in typical grassland topsoil there are about 6 tonnes per hectare. All play a role in the decomposition of dead plants, animal waste and other dead organisms. The macro-animals include the herbivores, such as slugs, which only eat living plant issue; detritivores, such as mites, beetles and woodlice, which live mainly on decayed plant tissues; and predators, such as moles, spiders and many insects, which live mainly on other animals. Soil flora are far more numerous and provide a much greater biomass than soil fauna, and they themselves are dominated by the microflora (see Table 24.4). The pH has a major control on the number of bacteria and fungi, as can be seen from Table 24.7. A good description of soil organisms can be found in Thompson et al. (1986).

Table 24.7 Soil flora and the effect of soil pH on the number of bacteria and fungi

Soil pH	Bacteria (millions/g)	Fungi (thousands/g)
7.5	95	180
7.2	58	190
6.9	57	235
4.7	41	966
3.7	3	280
3.4	1	200

24.6 Soil description in the field

24.6.1 Describe the site characteristics

General details, initially including location, position (grid reference), date, the weather for the last few days and the observer's name.

Describe the following characteristics

1. regional relief (e.g. a major landform unit such as a coastal plain or dunes)

2. elevation (height above sea level) and local relief such as a valley side or valley floor

3. slope (angle measured and description of slope, such as concave, convex, top or base of slope)

4. aspect (the direction the slope faces in degrees)

5. microrelief on the slope (e.g. ridges, furrows, ditches, rock outcrops)

6. site drainage: does it shed water or receive it, is it normal or flooded?

7. parent material (rock type, age, was it transported like a glacial sediment, fluvial gravel or glaciofluvial sand?)

8. vegetation: type and % cover

9. land use: both past and present day.

24.6.2 Dig the soil pit

Pit size

1. Should be about 1.0 m × 0.7 m and should be down to about 1.0 m depth, or until the parent material is reached.

2. Clean up the face that gets the most light during the observation period.

3. After observation, replace all the soil and tamp it well down, keeping the topmost sod until last.

24.6.3 Profile description

Profile drainage (Table 24.8)

1. **Excessively drained:** Soil is loose, powdery, fluffy, few earthworms, much organic matter on the surface.

Table 24.8 Classification of soil profile drainage from gley morphology

Drainage class	Drainage conditions	Gley morphology
Free	No waterlogging above 80 cm except during first four days after heavy rainfall	No motting in upper 80 cm
Moderate	Waterlogging above 60 cm for less than one month	Distinct mottling below 60 cm
Imperfect	Waterlogging above 60 cm for more than 1 month continuously	Distinct mottling 60 cm; faint mottling 30–60 cm
Poor	Waterlogging for long periods during the summer	Distinct mottling below 30 cm
Very poor	Almost continuous waterlogging	Grey colours in and below the topsoil

2. **Freely drained:** Profile has no sharp colour boundaries, no mottling in top 75 cm.

3. **Poorly drained:** Dominant grey gleyed zone or layer at less than 75 cm depth.

4. **Imperfectly drained:** Mottling along root channels.

5. **Very poorly drained:** Grey, may be peaty at the surface.

Profile description

Each horizon should be described from the surface downwards under the following headings:

Depth of horizon and thickness in cm

Zero is top of A horizon; clarity of horizons: clear/sharp/merging boundaries; regularity of the boundaries: smooth, wavy, irregular.

Moisture

1. **Dry:** Darker when water added.

2. **Slightly moist:** Can be rolled into clods.

3. **Moist:** Does not immediately wet the fingers.

4. **Wet:** Sticks or wets the fingers when moulded.

This can also be measured easily in the laboratory by bagging up samples, weighing, oven-drying and then reweighing for each horizon.

Colour

Identify using the Munsell Soil Colour Chart, which contains carefully-matched colour chips, both the hue name (top of the chart), value (vertical) and chroma (horizontal). To check the colour the soil is first moistened to saturation, as dry soil would have a different colour, and the soil is then matched to the colour in the book (for example, light reddish brown is 5YR 6/4). Look at both the soil matrix and any mottling. If colour charts are not available, visual colour descriptors can be used.

Texture (Table 24.9)

Carry out the hand-texturing procedure for each horizon: moisten to maximum stickiness, rub between thumb and forefinger and assess which textural category is present:

1. **Loamy sand:** Loose individual grains can be felt.

2. **Loam:** Slightly gritty but smooth.

3. **Clay loam:** Fine, will form ribbons, does not crumble when wet.

Samples may be taken for laboratory analysis if required.

Organic matter

Its position in the profile and the decomposition state it has reached should be described and the degree of humification should be worked out (Table 24.6). If there is more than 15% organic carbon the horizon is humose and the mineral texture procedure is not used. The broad subdivisions shown in Table 24.10 are noted.

In practice it is easier and much more accurate to collect samples for loss on ignition in the laboratory using a furnace, where all the organic matter is burnt off at high temperatures.

Stoniness

Estimate visually by volume:

- under 1%: stoneless
- 1–5%: slightly stony
- 5–20%: stony
- 20–50%: very stony
- 50–75%: extremely stony
- over 75%: stone dominant.

Also describe:

- Quantity: many/none/few.
- Size: gravel, cobbles, boulders.

Table 24.9 Hand-texturing of soils

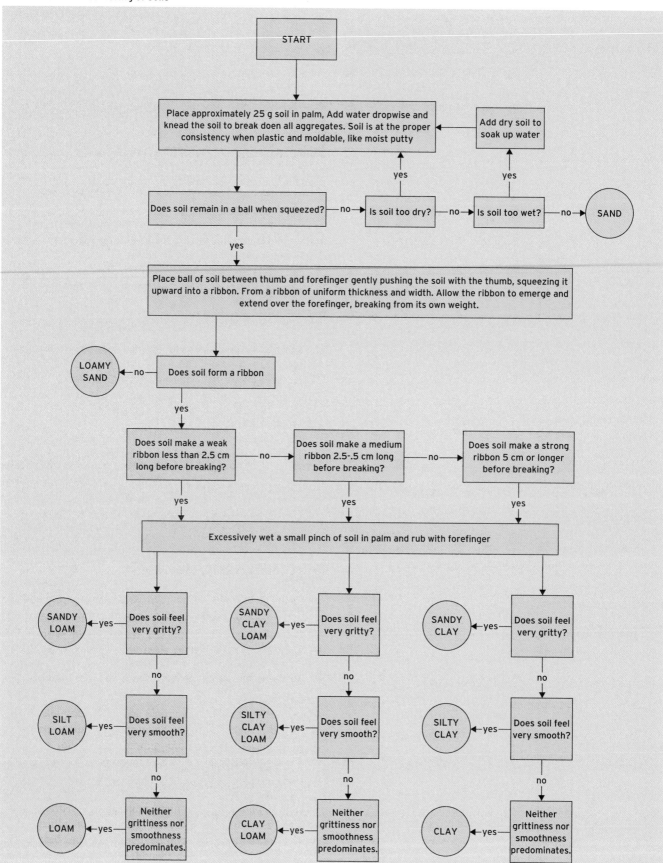

Table 24.10 Organic matter subdivisions

Descriptive term	Organic carbon (%)	Organic matter (%)
peat	>25	>43
peaty soil	15-25	26-43
humose mineral soil	7.5-15	13-26

- Shape: angular, rounded, flat, etc.
- Lithology.

Structure
Describe the size and shape of the peds. See Table 24.5 for the terms used and descriptors.

Porosity and permeability
Describe and quantify in general terms the degree and frequency of small holes and cracks for aeration (voids, fissures, cracks, root/worm channels).

Roots
Quantity and size should be noted.

Fauna
Describe any soil fauna seen and where.

Chemical characteristics
1. **pH:** use the testing kit.

2. **Calcareous or not:** Use dilute HCl on a small soil sample and observe the reaction:

   under 0.1%: no reaction

   0.1–0.5%: faint spitting audible

   0.5–1.0%: spitting audible

   1.0–2.0%: clearly audible, slight reaction visible

   2.0–5.0%: easily audible and visible

   5.0–10.0%: vigorous effervescence

   Samples can be taken for more accurate laboratory determination of calcium carbonate content.

3. **Iron:** Red or grey deposits may be present.

4. **Silica:** White powder may be present, which does not react with HCl.

Once the soil has been described in the field it should be possible to classify it into a broad category.

Further detail on field description of soils can be found in Hodgson (1993) and Trudgill (1989). In the latter some introductory and more advanced keys for soil-type identification are provided for field use in Britain.

24.7 Key soil types, with a description and typical profile

In the British Isles the sequence of soil changes in response to increased rainfall input but with good drainage is as follows:

1. Acid brown Earth, with a low to moderate input.

2. Brown Earth, with moderate rainfall.

3. Grey brown podzolic soil, with moderate to high rainfall.

4. Podzol, with high rainfall.

5. Peaty gleyed podzol. Groundwater gleys and peats are present where the drainage is poor and brown calcareous soils and rendzinas are found on limestone parent material.

24.7.1 Acid brown Earth

The A horizon is capped by a mull or moder humus and the B horizons show little sign of differentiation. However, they may be slightly lighter in colour in the upper horizons due to the removal of solutes. No clay and very little iron has been carried down the profile, with only the most soluble elements being actively removed, such as Ca and Mg. The pH is generally about 5.

24.7.2 Brown Earth

This soil (Figure 24.14) has a strongly leached layer beneath a mull or moder humus layer, but the main characteristic is that there is clay movement through the profile, as a jelly-like solution. This is deposited as a skin or cutan but the clay is not broken down. The lower B horizon is therefore rich in clay and can be distinguished by its texture. Some iron is also moved down.

24.7.3 Grey brown podzolic

This soil (Figure 24.15) is found under a mull or moder humus, but is not particularly acid. There is a lighter horizon below the A, which has physically lost some clay by downwashing to give a clay illuvial B horizon, where the clay has increased by 2–3 times. The clay is not broken down chemically and forms a layer in the B horizon.

Figure 24.14 Brown Earth from a Burren field, western Ireland

24.7.4 Podzol

This soil (Figure 24.16) is recognized by sharply-contrasting horizons in areas of high rainfall on sites with good drainage and porous parent materials, especially in cold, humid areas. It is usually found beneath heathland, conifers or other acid vegetation, where litter accumulates due to the slow decay, and at the surface there is a very acid humus (mor). Rainwater is acidic and dissolves plant substances that complex with the soil Fe/Al, and this moves downwards in solution. There is a breakdown of the clay minerals, and hence silica is left from the aluminosilicates in the surface layer as a bleached-grey, eluvial horizon. There is deposition lower down the profile as an illuvial horizon, forming a hard red-brown or dark ironpan. The processes related to podzolization are shown in Figure 24.17.

24.7.5 Peaty gleyed podzol

High rainfall means leaching and gleying, so a podzol may be formed by leaching, but very high rainfall, especially in winter, may lead to the mottling of some of the soil horizons. A peaty gleyed podzol can show mottling below the ironpan, and in some places the ironpan may become so well developed as to be impermeable to water; mottling may then occur above the ironpan. Usually there is a very thick, peaty A horizon. The simpler podzols without gleying are referred to as humus–iron podzols.

24.7.6 Gley

Although many soils can show signs of gleying (Figure 24.18) a true gley is usually found in thick clay deposits, where the

Figure 24.15 Grey brown podzolic profile

lower part of the profile is an unweathered clay. The upper horizons are mottled and merge with organic horizons. The upper parts of the profile tend to be redder/browner and the lower parts pale, green/blue-grey because of water-table fluctuations. It is possible to get surface-water gleys (Figure 24.19), formed by an impervious horizon within the soil, and groundwater gleys, where water backs up from an impervious layer below the soil. Organic decomposition is slow and peat may accumulate.

24.7.7 Organic soils or peats

These soils accumulate in bogs with very poor soil drainage. In uplands there is moor peat, which is acid caused by high rainfall and the leaching of all bases, and in the lowlands, fen peat, which is neutral or alkaline. Figure 24.20 illustrates peat development and representative profiles of organic soils from various types of peat-accumulating situation in Lancashire.

(a)

(b)

(c)

Figure 24.16 (a) Peaty, ironpan podzol, Broad Law, Ireland. (b) Humic podzol, Malaysia (Tropohumod). (c) Podzol, Sweden (Cryothod)

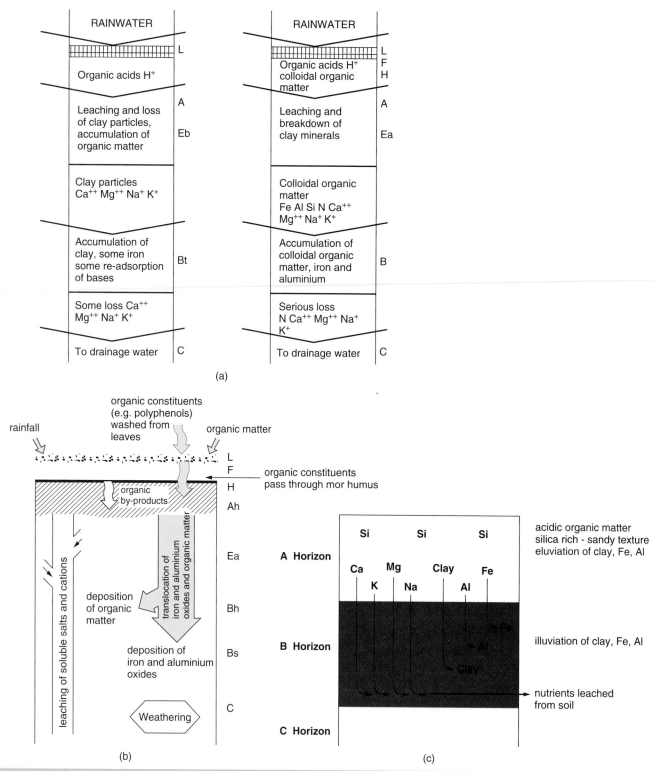

Figure 24.17 Processes related to podzolization: (a) leaching; (b) podzolization; (c) diagrammatic view of the podzolization process

Figure 24.18 Gleys

Figure 24.19 Surface-water gley, Calary, County Wicklow, Ireland

24.7.8 Calcareous soils

These soils are little affected by rainfall and leaching and they owe their presence to a limestone parent material in an area of low rainfall. They do not occur on limestone that does not break up easily or where the rainfall is high, as the calcium is leached out easily and is not well replenished by a weathering input.

Soil where the mineral input is limited is much thinner and soil development is usually limited to the accumulation of organic matter, although some mineral matter is also incorporated into the organics. These soils do not possess a mineral B horizon. For example, on hard limestone a rendzina (USDA Rendoll) (Figures 24.21 and 24.22) forms with a humus of the mull-type, a high proportion of free carbonates and so base saturation of the A horizon. There is a relatively high level of soil-organism activity and a stable organic matter, which directly overlies the parent material, usually without a B horizon. On hard acid rocks a ranker forms, with mor-type humus immediately on top of bedrock (Figure 24.21). This often forms on steep slopes where there is continual loss of material by lateral creep.

24.8 Podsolization: theories

Theories of podsolization largely fall in two groups:

1. **The organic theory:** In which translocation and deposition of Fe and Al occur as organic complexes (see Table 24.11(a), from De Corninck, 1980).

2. **The inorganic theory:** In which the translocation and deposition of Fe and Al occur as inorganic compounds (see Table 24.11(b), from Farmer, 1982).

In the organic theory the traditional view was that humic and fulvic acids were responsible for complexing and mobilizing Fe and Al. Subsequent modifications included complexation by soluble organic materials in leaf leachates and the reasons for the immobilization of organomineral complexes at depth.

In the inorganic theory imogolite-like compounds have been identified in the podzol B horizons, which suggest that the translocation of soil Fe and Al also occurs in inorganic compounds, such as gel-like aluminosilicates.

'Podsolization' is therefore used to describe three component processes:

1. *In situ* **weathering:** The hydrolysis of soil minerals to produce hydroxy ions of Fe and Al.

2. **Translocation:** The eluviation of soluble and chelated hydroxy Fe and Al ions and fine clay particles from surface mineral horizons.

3. **Deposition:** The illuviation of hydroxy Fe and Al ions and fine clay particles in subsurface mineral horizons.

Since the pioneering work of Bloomfield (1953–55), soluble organic compounds derived from live plant leaves, leaf litter and soil organic matter have been implicated in soil-translocation processes under forest and heathland vegetation. There is debate about the relative importance of mobilization by simple organic compounds, such as polyphenols, derived from leaf litter and litter washings, and the much more complex humic and fulvic acids, derived from soil organic matter. There is also a lot of discussion related to the mechanisms of deposition responsible for the iluviated podsolic B horizon.

The traditional theory for the deposition of the chelated Fe and Al cations is that of reduced solubility of the organometallic complex with higher pH at depth and as biodegradation reduces the ratio of organic matter to metal in the complex. The discovery of imoglite-type allophanes in podzolic B horizons has raised

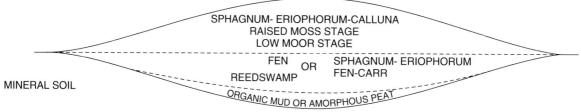

Generalized development of basin peat to raised mess stage

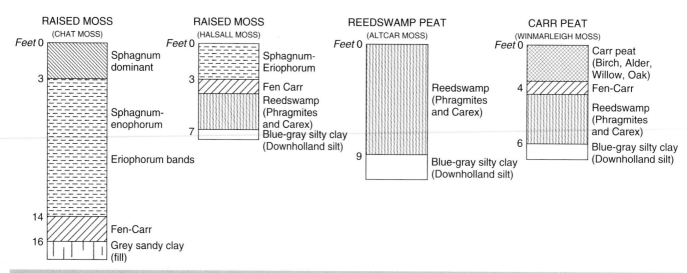

Figure 24.20 Distribution of soils in the Ainsdale and Formby Associations. (Source: Hall and Folland, 1967)

questions about whether organic compounds play any important role in Fe and Al translocation in podzolic profiles.

24.9 Soil classification

Soils vary continuously across landscapes and this spatial variability affects all soil properties. So why do we need to classify soils? There are two main reasons:

1. To allow generalizations to be made about information so that it can be transmitted easily between individuals who are studying the soil. In other words, it saves time and effort because a single term can be used to summarize whole soil profiles.

2. To allow theory construction. For example, for agricultural purposes it is helpful if one can make predictive statements about how a group of soils will respond to a fertilizer. In other words, it aids agriculture and makes possible better plans for agricultural management.

Therefore the aim is to group soils in such a way that each group has a minimum diversity within it and a maximum diversity between groups. Although classification can be an arbitrary process, it is useful, but the methods of classification depend on

the country, usage and approach. As such, classification does not increase knowledge, but it may show where the gaps are. In early classificatory systems soils were grouped if they formed under similar conditions, but modern systems have placed more emphasis on classifying soils using observable and measurable properties. Thus there are various types of classification, discussed below.

24.9.1 The genetic method or Dokuchaeiv approach

This is a Russian classification (Table 24.12) which has been adapted and modified many times elsewhere. It is based on large-scale variation and the relationship that was discovered between soil, natural vegetation and climate. It emphasizes that soils are dynamic and can reach a mature state of equilibrium. It starts with a big division into three orders:

1. **Zonal:** Soils are mainly the result of climatic or biological factors and were correlated with vegetation zones.

2. **Intrazonal:** Soils that have properties more related to local conditions, for example drainage, topography or the parent material. These local factors override the zonal factors, such as where soils develop on limestone, for example.

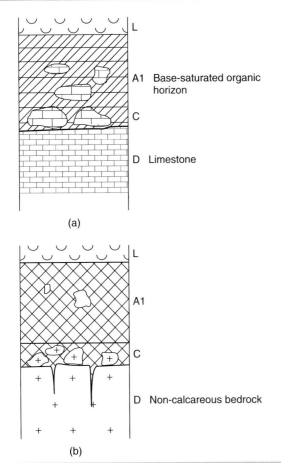

(a)

(b)

Figure 24.21 (a) Rendzina profile (USDA Rendoll). (b) Ranker profile (USDA Inceptisol)

Figure 24.22 Rendzina near Boyle, Ireland

3. **Azonal:** Immature, unconsolidated soils without horizons in which the erosional and depositional processes dominate the more usual soil-forming processes, as in alluvium or sand dunes.

Table 24.11 Podzolization theories: (a) organic theory; (b) inorganic theory

(a)	Organic theory of podzol formation (from De Corninck, 1980)
Stage	Development process
(1)	Mobile organic substances are formed during the decomposition of surface litter and soil organic matter.
(2)	If there are sufficient polyvalent cations (Al and Fe) at the top of the mineral soil profile, the mobile organic substances are immobilised immediately and no downward migration occurs.
(3)	If insufficient amounts of Al and/or Fe are available to completely immobilise the mobile organic matter, these cations are complexed by the mobile organic matter and transported downwards.
(4)	Immobilisation of organomineral complexes may occur at depth due to (a) supplementary fixation by cations (b) desiccation (c) on arrival at a level with different ionic concentration

(b)	Inorganic theory of podzol development* (Farmer, 1982)
Stage	Development process
(1)	Early stage of podzol formation, before development of A/E horizon, mobile Al, Fe and Ca are relatively abundant in the A horizon, and any fulvic acid liberated by organic matter decomposition will be precipitated *in situ* as an insoluble salt.
(2)	Once A/E horizon develops, fulvic acid becomes excess of the A horizon capacity and passes down the profile. This fulvic acid can carry only small amounts of complexed Al and Fe, but it will attack imogolite and proto-imogolite already deposited in the B horizon; this liberates silica and forms an insoluble Al fulvate *in situ*. This mechanism can account for the absence of imogolite-type minerals in Bh horizons.
(3)	Thin ironpan (Bf) horizons can provide effective barriers to the passage of fulvic acids - so in profiles with ironpans, imogolites persist up to the ironpan.
(4)	Finally, downward migrating organic matter is absorbed on imogolite-like material of B_2 horizons.

The zonal soils were further divided into 10 suborders on the basis of colour and vegetation, while the intrazonal soils were divided according to drainage, chemistry and parent material. There were no suborders within the azonal group. Then there was a subdivision into great groups, but the scheme was not taken beyond this level to a series or type unit.

The deficiencies of this type of classification were seen to be as follows:

1. The characteristics of soils that have been greatly modified by humanity were not considered.

Table 24.12 Dokuchaeiv soil classification

ORDER	SUBORDER	GREAT GROUPS
1. BOREAL	cold zone soils	tundra soils
2. TAIGA	light col.podzolics soils of the forests	podzols brown poizols/grey-brown podzo red-yellow podzols grey podzols
3. SUBTROPIC TROPICAL	warm temperate and tropical forests	laterite or red soils
4. FOREST/STEPPE	forest/grassland transition	degraded chernozems non-calcic browns
5. STEPPE	dark soils of the subhumid grasslands	praire soils chernozems
6. DESERT/STEPPE TRANSITION		chestnut brown soils
7. DESERT	light coloured soils of arid areas	sierozems/red desert soils
All the above are ZONAL		
8. WET AREA	hydromorphic soils of bogs, swamps, marshes.	bog soils/planosols/gleys
9. SECONDARY ALKALINE	halomorphic soils	Solonchak/solonetz/solod
10. CARBONATE	calcimorphic soils	brown forest soils rendzinas
All the above are INTRA-ZONAL		
ZONAL	no suborders	lithosols/regosols/alluvium

2. There was no provision made for multiple soil profiles due to climatic change.

3. There was no account taken of what zonal soils that did not develop under ideal conditions of horizontal topography and good drainage were like.

4. The influence of climate and vegetation on soil formation, although important, was overstressed.

24.9.2 The USDA (7th Approximation) Classification

This was designed for use throughout the world and was based on the morphological characteristics of the profile, and especially on horizon formation and sequences (genetic properties). It retains some features of the genetic approach, for example it recognizes degrees of maturity, zonality and intrazonality, but not in a direct form. If other things are equal, properties affecting plant growth are used. It uses names based on zoological and biological roots and is a hierarchical, soil taxonomic approach (see the summary in Table 24.13).

The advantages of this classification are that it classifies soils rather than soil-forming processes; it deals with soils as they are today and not their genesis; and it is relatively objective, as it measures the field properties of soils and classifies according to these. It uses international nomenclature based on a system of diagnostic horizons which have a fixed range of chemical and physical properties with respect to moisture, texture, structure, temperature, pH, organic matter, depth, colour, base saturation, clay content, Fe/Al content and salts present. So the soils are classified according to which horizons they contain and in what order.

The classification is hierarchical, with the orders based on broad climatic groups, through orders, suborders, great group, subgroup, family, series, type and phase. The root name of an order is preserved and subdivisions add new prefixes or suffixes to indicate their properties.

This classification has been criticized for being too inflexible since it does not allow for the natural soil variation found across the landscape. It is also rigid in that the absence of a single diagnostic characteristic is enough to exclude a soil from a particular group. There is also a need for laboratory measurements in addition to field observations before a soil can be classified, which increases the time and cost of a survey. Attempts were made to relate British soils to this classification but it became generally apparent that a different system was necessary.

24.9.3 British Classification

The original soil classification for England and Wales was developed in 1940 when six main soil groups were recognized (brown Earth, podsol, gley, calcareous soils and alluvium), which were closely related to the European soil classifications. This was superseded by the 1973 New British Classification (Table 24.14), which has been used in all Soil Survey publications since 1974. The basic soil unit is the soil profile and the classification was based on observable field properties. The soils were divided into major

Table 24.13 USDA (7th Approximation) Soil Classification

Summary of USDA Soil Survey staff classification (1960)

(7th Approximation) (After Bridges, 1970)

1. *Entisols.* Weakly developed (generally azonal) soils — Aquents
   - with features of gleying — Arents
   - with strong artificial disturbance — Fluvents
   - on alluvial deposits — Psamments
   - with sand or loamy sand texture — Orthents
   - other Entisols (e.g. lithosols, some regosols)

2. *Vertisols.* Cracking clay soils — Uderts
   - usually moist — Usterts
   - dry for short periods — Xererts
   - dry for a long period — Torrerts
   - usually dry

3. *Inceptisols.* Moderately developed soils, not in other orders — Aquepts
   - with features of gleying — Andepts
   - on volcanic ash — Tropepts
   - in a tropical climate — Umbrepts
   - with an umbric epipedon — Ochrepts
   - other Incepitsols (e.g. most brown Earths)

4. *Aridisols.* Semi-desert and desert soils — Argids
   - with an argillic horizon — Orthids
   - other soils of dry areas (e.g. grey desert soils)

5. *Mollisols.* Soils of high base status with a dark A horizon — Albolls
   - with albic and argillic horizons — Aquolls
   - with features of gleying — Rendolls
   - on highly calcareous parent materials — Borolls
   - others in cold climates — Udolls
   - others in humid climates — Ustolls
   - others in sub-humid climates — Xerolls
   - others in sub-arid climates

6. *Spodosols.* Soils with a spodic horizon (e.g. podzols) — Aquods
   - with features of gleying — Ferrods
   - with little humus in spodic horizon — Humods
   - with little iron in spodic horizon — Orthods
   - with iron and humus

7. *Alfisols.* Soils with an argillic horizon and moderate to high base content
   - with features of gleying — Aqualfs
   - others in cold climates — Boralfs
   - others in humid climates — Udalfs

(continued overleaf)

Table 24.13 *(continued)*

– others in sub-humid climates	Ustalfs
– others in sub-arid climates	Xeralfs
8. *Ultisols.* Soils with an argillic horizon and low base content	
– with features of gleying	Aquults
– with a humose A horizon	Humults
– in humid climates	Udults
– others in sub-humid climates	Ustults
– others in sub-arid climates	Xerults
9. *Oxisols.* Soils with an oxic horizon or with plinthite near the surface	
– with features of gleying	Aquox
– with a humose A horizon	Humox
– others in humid climates	Orthox
– others in drier climates	Ustox
10. *Histosols.* Organic soils (sub-orders not finalised)	

Table 24.14 New England and Wales Soil Classification. (Source: after Avery, 1973)

Major group	Group	Sub-group
Lithomorphic (A/C) soils Normally well-drained soils with distinct, humose or organic topsoil and bedrock or little altered unconsolidated material at 30 cm or less	*Rankers* With non-calcareous topsoil over bedrock (including massive limestone) or non-calcareous unconsolidated material (excluding sand)	Humic ranker Grey ranker Brown ranker Podzolic ranker Stagnogleyic (fragic) ranker
	Sand-rankers In non-calcareous, sandy material	Typical sand-ranker Podzolic sand-ranker Gleyic sand-ranker
	Ranker-like alluvial soils In non-calcareous recent alluvium (usually coarse textured)	Typical ranker-like alluvial soil Gleyic ranker-like alluvial soil
	Rendzinas Over extremely calcareous non-alluvial material, fragmentary limestone or chalk	Humic rendzina Grey rendzina Brown rendzina Colluvial rendzina Gleyic rendzina Humic gleyic rendzina

Table 24.14 (continued)

Major group	Group	Sub-group
	Pararendzinas Over moderately calcareous non-alluvial (excluding sand) material	Typical pararendzina Humic pararendzina Colluvial pararendzina Stagnogleyic pararendzina Gleyic pararendzina
	Sand-pararendzinas In calcareous sandy material	Typical sand-pararendzina
	Argillic brown Earths Loamy or loamy over clayey, with sub-surface horizon of clay accumulation, normally brown or reddish	Typical argillic brown Earth Stagnogleyic argillic brown Earth Gleyic argillic brown Earth
	Paleo-argillic brown Earths Loamy or clayey with strong brown to red sub-surface horizon of clay accumulation attributable to pedogenic alteration before the last glacial period	Typical paleo-argillic brown Earth Stagnogleyic paleo-argillic brown Earth
Podzolic soils Well drained to poorly drained soils with black, dark brown or ochreous sub-surface (B) horizon in which aluminium and/or iron have accumulated in amorphous forms associated with organic matter. An overlying bleached horizon, a peaty topsoil, or both, may or may not be present	*Brown podzolic soils* Loamy or sandy, normally well drained, with a dark brown or ocheous friable sub-surface horizon and no overlying bleached horizon or peaty topsoil	Typical brown podzolic soil Humic brown podzolic soil Paleo-argillic brown podzolic soil Stagnogleyic brown podzolic soil Gleyic brown podzolic soil
	Gley-podzols With dark brown or black sub-surface horizon over a grey or mottled (gleyed) horizon affected by fluctuating ground-water or impeded drainage. A bleached horizon, a peaty topsoil or both may be present	Typical (humus) gleypodzol Humo-ferric gley-podzol Stagnogley-podzol Humic (peaty) gley-podzol
	Podzols (sensu stricto) Sandy or coarse loamy, normally well drained, with a bleached horizon and/or dark brown or black sub-surface horizon enriched in humus and no immediately underlying grey or mottled (gleyed) horizon or peaty topsoil	Typical (humo-ferric) podzol Humus podzol Ferric podzol Paleo-argillic podzol Ferri-humic podzol
	Stagnopodzols With peaty topsoil, periodically wet (gleyed) bleached horizon, or both, over a thin ironpan and/or a brown or ochreous relatively friable subsurface horizon	Ironpan stagnopodzol Humus-ironpan stagnopodzol Hardpan stagnopodzol Ferric stagnopodzol

(continued overleaf)

Table 24.14 (continued)

Major group	Group	Sub-group
Pelosols Slowly permeable nonalluvial clayey soils that crack deeply in dry seasons with brown, greyish or reddish blocky or prismatic sub-surface horizon, usually slightly mottled	*Calcareous pelosols* With calcareous sub-surface horizon	Typical calcareous pelosol
	Argillic pelosols With sub-surface horizon of clay accumulation, normally non-calcareous	Typical argillic pelosol
	Non-calcareous pelosols Without argillic horizon	Typical non-calcareous pelosol
Gley soils With distinct, humose or peaty topsoil and grey or grey-and-brown mottled (gleyed) sub-surface horizon altered by reduction, or reduction and segregation, of iron caused by periodic or permanent saturation by water in the presence of organic matter. Horizons characteristic of podzolic soils are absent	1. Gley soils without a humose or peaty top-soil, seasonally wet in the absence of effective artificial drainage.	
	Alluvial gley soils In loamy or clayey recent alluvium affected by fluctuating ground-water	Typical (non-calcareous) alluvial gley soil Calcareous alluvial gley soil Pelo-alluvial gley soil Pelo-calcareous alluvial gley soil Sulphuric alluvial gley soil
	Sandy gley soils Sandy, permeable, affected by fluctuating ground-water	Typical (non-calcareous) sandy gley soil Calcareous sandy gley soil
	Cambic gley soils Loamy or clayey, non-alluvial, with a relatively permeable substratum affected by fluctuating ground-water	Typical (non-calcareous) cambic gley soil Calcaro-cambic gley soil Pelo-cambic gley soil
	Argillic gley soils Loamy or loamy over clayey, with a sub-surface horizon of clay accumulation and a relatively permeable substratum affected by fluctuating ground-water	Typical argillic gley soil Sandy-argillic gley soil
	Stagnogley soils Non-calcareous, non-alluvial, with loamy or clayey, relatively impermeable sub-surface horizon or substratum that impedes drainage. 2. Gley soils with a humose or peaty topsoil normally wet for most of the year in the absence of effective artificial drainage	Typical stagnogley soil Pelo-stagnogley soil Cambic stagnogley soil Paleo-argillic stagnogley soil Sandy stagnogley soil

Table 24.14 (*continued*)

Major group	Group	Sub-group
	Humic-alluvial gley soils	
	In loamy or clayey recent alluvium	Typical (non-calcareous) humic-alluvial gley soil Calcareous humic-alluvial gley soil
	Humic-sandy gley soils	
	Sandy, permeable, affected by high ground-water	Typical humic-sandy gley soil
	Humic gley soils (sensu stricto)	
	Loamy or clayey, nonalluvial, affected by high ground-water	Typical (non-calcareous) humic gley soil Calcareous humic gley soil Argillic humic gley soil
	Stagnohumic gley soils	
	Non-calcareous, with loamy or clayey, relatively impermeable sub-surface horizon or substratum that impedes drainage	Cambic stagnohumic gley soil Argillic stagnohumic gley soil Paleo-argillic stagnohumic gley soil Sandy stagnohumic gley soil
Man-made soils	*Man-made humic soils*	
With thick man-made topsoil or disturbed soil (including material recognisably derived from pedogenic horizons) more than 40 cm thick	With thick man-made topsoil	Sandy man-made humus soil Earthy man-made humus soil
	Disturbed soils	
	Without thick man-made topsoil	
Peat soils	*Raw peat soils*	
With a dominantly organic layer at least 40 cm thick formed under wet conditions and starting at the surface or within 30 cm depth	Permanently waterlogged and/or contain more than 15 per cent recognisable plant remains within the upper 20 cm	Raw oligo-fibrous peat soil Raw eu-fibrous peat soil Raw oligo-amorphous peat soil Raw eutro-amorphous peat soil
	Earthy peat soils	
	With relatively firm (drained) topsoil, normally black, containing few recognisable plant remains	Earthy oligo-fibrous peat soil Earthy eu-fibrous peat soil Earthy oligo-amorphous peat soil Earthy eutro-amorphous peat soil Earthy sulphuric peat soil

groups and subgroups, where the major groups were distinguished by two main factors:

1. The composition of the soil within specified depth limits.

2. The presence or absence of diagnostic horizons, generally reflecting the degree or kind of alteration of the original material.

The distinguishing criteria should be recognizable in the field and laboratory information should not be required to back these up. The criteria used can usually be related to other environmental attributes, such as landforms, geology, vegetation and climate. As the classification was conceived largely for helping land-use decisions, soils should be classified as far as possible on properties that affect land-use capability.

24.9.4 Food and Agriculture Organization soil reference groups (1998)

In 1961 a joint scientific advisory panel from FAO, UNESCO and the International Society of Soil Science started to organize a project whose main aim was to catalogue the world's soil resources and put in place a generally acceptable soil classification framework. It was not intended to replace national schemes but to formulate a set of broad classes that could be used to classify the world's main soils. The first phase was completed by 1974 and the latest amendments were made in 1998, with the creation of the World Reference Base for Soil Resources (Table 24.15). This

provides a way of correlating the many national soil classification systems. There are two levels of classification, consisting of 30 reference soil groups and 170 subunits.

24.10 Regional and local soil distribution

24.10.1 Zonal soils

From our discussion of the soil-forming processes and controls it seems there could be an infinite variety of soils produced by the

Table 24.15 FAO soil reference groups

Soil group	Description
Histosols	Soils that have a peat layer or 'H' horizon >40 cm deep
Cryosols	Soils that have permanently frozen horizons (below 0 °C for two or more years) within 100 cm of the soil surface
Anthrosols	Soils that have horizons strongly influenced by human activity such as cultivation (Hortic horizon), irrigation (Irragic horizon) or manure incorporation (Terric horizon)
Leptosols	Soils with hard rock within 25 cm of the surface or which overlie materials with >40% calcium carbonate within 25 cm of surface or contain <10% fine Earth to a depth of 75 cm or more (weakly developed soils)
Vertisols	Soil that has a vertic B horizon (a clay-rich, self-turning horizon, >30% clay) >50 cm deep and within 100 cm of the soil surface
Fluvisols	Soil having formed on recent alluvial deposits, having a fluvic horizon (a dark-coloured horizon, usually resulting from pyroclastic deposits) within 25 cm of the surface, continuing to depths greater than 50 cm
Solonchaks	Soils that have surface or shallow subsurface horizons >15 cm deep that are enriched with soluble salts. Common in soils forming from recent alluvial deposits (soils of salty areas)
Gleysols	Soils which have evidence of gleying within 50 cm of the surface
Andosols	Soils that have vitric horizons (>10% volcanic glass or other volcanic material) >30 cm deep or andic horizons (weathered pyroclastic deposits) within 25 cm of the soil surface (volcanic soils)
Podzols	Soils that have a spodic B horizon (subsurface horizons containing illuvial organic matter and or aluminium and iron). Bleached surface horizons and an iron pan at depth are also usually present
Plinthosols	Soils with a plinthic horizon (iron-rich horizon >15 cm with >25% plinthite) within 50 cm of the surface which harden when exposed
Ferralsols	Ferric B horizon (distinctive red mottling >15 cm deep), highly weathered with high concentrations of iron and aluminium
Solonetz	Soils with a natric B horizon (a dark-coloured horizon >7.5 cm deep with clay enrichment and a high concentration of exchangeable sodium)
Planosols	Soils that have E horizon resulting from prolonged exposure to stagnant water within 100 cm of surface, often marked with an abrupt change in textural properties
Chernozems	Soils that have a mollic A horizon (dark-coloured, well-structured surface horizon with a high base saturation) to a depth of at least 20 cm
Kastanozems	Soils with a mollic A horizon >20 cm depth coupled with concentrations of calcium compounds within 100 cm of the soil surface
Phaeozems	All other soils which have a mollic A horizon
Gypsisols	Soils that have gypsic (concentrations of calcium sulphate) horizons within 100 cm of the surface or concentrations >15% gypsum over 100 cm
Durisols	Soils having a duric horizon (cemented silica) within 100 cm of the surface

Table 24.15 (*continued*)

Soil group	Description
Calcisols	Soils that have calcic horizons (discontinuous concentrations >15% of calcium carbonate) within 125 cm of the soil surface
Albeluvisols	Soils that have an argic horizon (B horizon showing signs of clay enrichment) with an irregular upper boundary (sometimes referred to as 'tonguing')
Alisols	Soils that have an argic B horizon with a CEC >24cmol$_c$/kg clay and a base saturation <50% within 100 cm of soil surface (soils with high concentrations of aluminium)
Nitisols	Soils that have a nitric B horizon (a clay-rich horizon with >30% clay mainly consisting of 1:1 minerals) >30 cm deep with a CEC <36 cmol$_c$/kg with no evidence of clay lessivage detected using thin sections within 100 cm of the surface
Acrisols	Soils that have an argic B horizon with a low CEC(<24 cmol$_c$/kg). These soils are characterized by acidity
Luvisols	Soils with an argic B horizon with a CEC >24 cmol$_c$/kg with illuvial accumulations of clay
Lixisols	All other soils having argic B horizons within 100–200 cm of the surface
Umbrisols	Soils having umbric horizons (thick, dark-coloured base, poor surface horizons)
Cambisols	Soils with a cambic (evidence of change or alteration) or mollic horizon over a subsoil with a base saturation <50% in the top 100 cm or an andic, vertic or vitric horizon starting between 25 cm and 100 cm depth
Arenosols	Weakly developed coarse-textured soils
Regosols	All other soils

interaction of the various parameters. However, as can be seen from Dokuchaeiv's approach to soil classification, there is a very close association between vegetation, climate and soil type. Each climatic zone has a characteristic vegetation and soil; here zonal soils are produced (Figure 24.23).

Four examples of these zonal soils are described below: Arctic, tundra soils; chernozems; chestnut-brown soils; and the brown and red Mediterranean soils. Tropical rain forest and savanna soils are discussed in Chapter 25.

Arctic, tundra soils

Tundra soils are poorly developed, with the dominant factors in their formation being the low temperatures and the presence of an impermeable substratum. Generally they have poorly-defined horizons due to the mixing of the upper soil by frost processes (cryoturbation). This creates involutions and warping of the soil. However, there is often a layer of surface peat since the vegetation does not break down quickly in the cold temperatures, but where frost action is long continued and occurs to depth, peat can be buried and organic matter can be forced deep into the mineral material.

The thickness of the mixed zone is controlled by the local permafrost table, which causes a block to soil drainage during the short summer. The blue and grey colours that are predominant in the mineral horizons are caused by these prevailing waterlogged conditions. These maintain iron in its reduced state so that the ferric oxide form, which is so common in the soils of lower latitudes, is generally missing, although mottling can occur. Organic matter streaks can also occur. Physical weathering dominates over chemical processes and as a result nutrients are lacking.

Soil fauna too is limited. Hence the shortness of the growing season only allows slow-growing perennials of dwarf habit, which leave peaty, carbonaceous organic matter whose breakdown is severely restricted by the inactivity of the few soil organisms in conditions of low temperatures and poor aeration. This poor aeration is due to the lack of evaporation, the abundance of water in the short summer melt season and the restricted drainage. The result is gleying, although in places on higher, gravelly ridges where there are steeper slopes, better-drained soils occur, usually with moisture contents below field capacity A (referred to as Arctic brown soils). Here oxidation rather than reduction can occur and there can be leaching of soluble nutrients. These types of soil are more stable and less prone to frost heaving; that is, they are less frost-susceptible, although variation on slopes can occur because of stone-stripe development. These soils are likely to have developed horizons, but the area of freely-drained soil is limited.

In waterlogged, low-lying depressions, thick layers of peat dominate and the whole active layer may be in this material. The great variability in water availability, the depth of freezing processes, depth of thaw, snow cover and vegetation thickness, as well as a great microrelief, can result in an extremely complex local soil distribution. However, soils with a silty or loamy texture are common due to the dominant physical break-up of any bedrock. There is often a zone of organic-matter accumulation immediately above and below the permafrost table, especially where the soils are shallow and the permafrost table is close to the surface. This may be an illuvial humic horizon and can be caused by several processes, such as downward flow of colloids and dissolved humus, although the layer might be fossil and

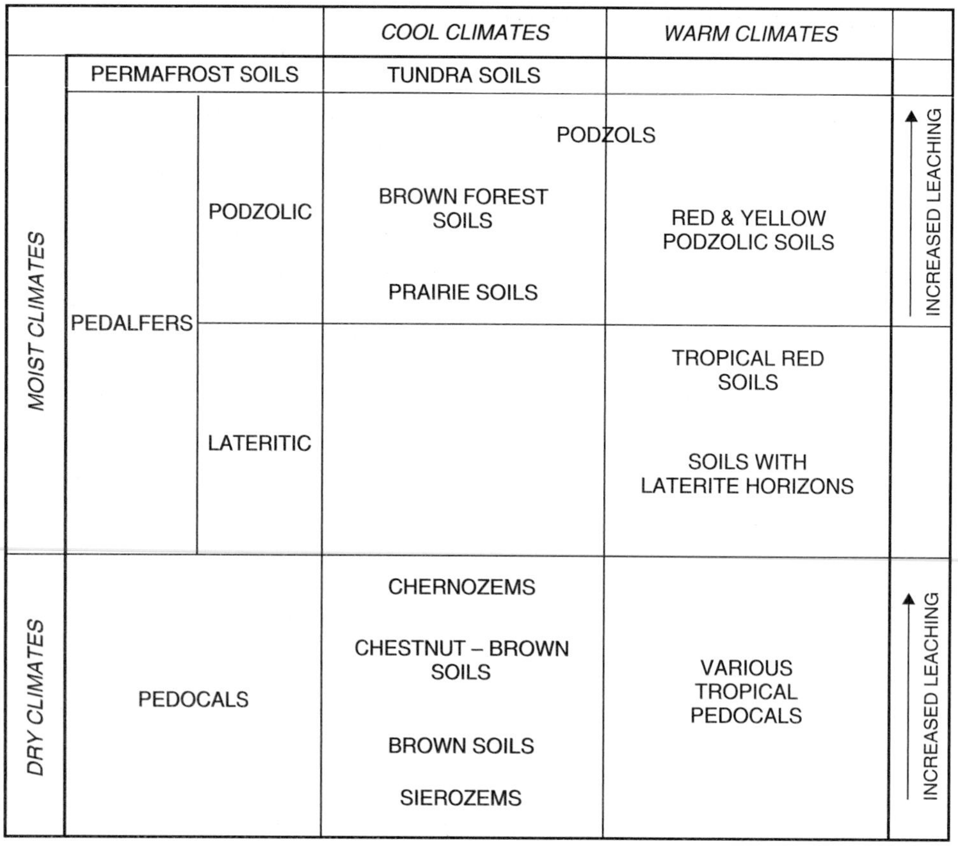

			COOL CLIMATES	WARM CLIMATES	
MOIST CLIMATES	PEDALFERS	PERMAFROST SOILS	TUNDRA SOILS		
		PODZOLIC	PODZOLS		INCREASED LEACHING
			BROWN FOREST SOILS	RED & YELLOW PODZOLIC SOILS	
			PRAIRIE SOILS		
		LATERITIC		TROPICAL RED SOILS	
				SOILS WITH LATERITE HORIZONS	
DRY CLIMATES	PEDOCALS		CHERNOZEMS		INCREASED LEACHING
			CHESTNUT – BROWN SOILS	VARIOUS TROPICAL PEDOCALS	
			BROWN SOILS		
			SIEROZEMS		

Figure 24.23 Zonal soils. (Source: after Eyre, 1963)

related to a warmer climatic phase. It has even been suggested that frost stirring and gelifluction could be responsible.

Arctic soils and their fossil equivalents also show a characteristic micromorphology. The soils can show four major types:

1. **Platy or lenticular structure:** This is caused by migration of soil water during freezing, which causes dehydration of the soil immediately below the freezing front. Frost-heave pressures can lead to compaction of unfrozen soil and the formation of dense, platy peds separated by large, laterally-extensive planar voids. The latter are thought to be the locations of former ice lenses and, because of the pressures created during ice lens formation, often contain thin stress coatings of fines.

2. **Skeletal grain coatings:** Plasma coatings on the surfaces of skeletal grains are attributed to the thixotropic behaviour of the soil during thawing and are often associated with patterned ground. Wetting and drying associated with the freezing and thawing may also be the cause, but it is difficult to see how fines can be transported downwards easily in permafrost soils. Similar grain coatings could be caused by gelifluction and they have been described as smooth-surfaced and streamlined coatings from Okstindan (northern Norway). Here it was thought that rotation of the grains in

the gelifluction layer led to the coatings and to plastering of orientated silt on to the grain surfaces.

3. **Reorientation of skeletal grains:** Skeletal grains can become reorientated through flow in geliflucted soils in a downslope orientation, or they can be vertically orientated due to the differential movement of coarse and fine sediment. This can occur as the freezing front moves down through the soil, with the result that the coarse sediment moves upwards to the surface and fines moves down in advance of the freezing front. This results in the vertical orientation of both stones and sand grains.

4. **Vesicular voids:** Many Arctic soils show large, bubble-like pores or vesicles, up to 2–3 mm in diameter, and they seem especially characteristic of wet soils. Several explanations have been suggested, such as wetting, expulsion of air during freezing of wet soils and thixotropic flow, which may generate air bubbles. It has been possible to illustrate that soil vesicles from the centre of sorted circles and from geliflucted till could be caused by thixotropic behaviour during thaw consolidation. As vesicles can also be caused by puddling a wet soil, freezing need not be present, although liquefaction during rapid thaw consolidation is probably significant.

There has been great interest in Arctic tundra soils, initially because of American and Russian strategic needs during the Cold War, such as the need to develop communications networks in the Arctic, particularly a stable road system and airfields, which resulted in the overcoming of engineering problems caused by these soils. There was also a need to extract minerals from these areas. In particular, a series of problems were caused by the transmission of oil and gas in pipelines, for example the Trans-Alaska Pipeline from the north slopes of Alaska to the Gulf of Alaska through zones of both continuous and discontinuous permafrost.

Chernozems (USDA mollisol, suborder boroll)

These dark soils are characteristic of the sub-humid grasslands of the steppes of Russia and the prairies of North America. Typically this soil is so brown that it appears black when exposed fresh in the field. This is caused by the mull humus content, which is caused by the combination of the physical nature of grass humus, the high base-status and the high soil temperatures during the summer.

The grasses produce large amounts of organic matter, both by excreting water-soluble compounds from their roots and by their decomposition *in situ*, all aided by the earthworm activity, which is intense. Also, the climatic seasonality between wet and dry seasons not only increases the decomposition rate of their most unstable components but also stabilizes other humic compounds against degradation. It also immobilizes clay and iron oxides within aggregates and prevents the formation of B_t horizons. However, despite the even spread of this humus throughout the chernozem profile, the average chernozem has only 8–10% organic content.

Evaporation, with consequent upward soil water movement, predominates and only rarely is precipitation so prolonged, or snow-melt so great, that percolation is beyond the root range. Thus means that the most soluble potassium and sodium salts are leached but the less-soluble calcium salts, particularly calcium carbonate, accumulate. These soils have a pH of > 7 and because of this the clay compounds have no tendency to dissociate and the clay minerals are completely stable. Thus the chernozem profile is composed of a thick, dark layer, often up to 80–100 cm thick, with a well-developed crumb-granular structure, which is darker than the lower layer. Hence there is no distinction between eluviated and illuviated horizons.

There is usually a marked upper A horizon where the yet undecomposed dead leaves and leaf bases of the grasses have accumulated because of the restricted breakdown rate of the organic matter caused by the relatively cold winters and relatively dry summers. This grades downward into a predominantly organic, almost structureless horizon, which is rapidly mixed downwards by the plentiful soil fauna. The former presence of small invertebrates is shown by their infilled burrows, known as krotovinas in Russia. Sometimes the A horizon gives way sharply to the underlying C horizon of partially-weathered parent material, but usually there is development of a thick calcium carbonate zone, which occurs in the form of irregular concretions or nodules. Less frequent carbonate concretions occur irregularly throughout the lower part of the organic horizon, often as fine tubes or filaments, the remains of the burrows of the soil fauna. As air can penetrate downwards freely in these burrows, evaporation and redeposition of lime are common.

Below the zone of calcium carbonate accumulation there can be zones of deposition and accumulation of other salts, such as gypsum. As the clay minerals are montmorillonite, the exchange capacity and soil fertility are high in chernozems. These soils have generally developed on loess or loess-like sediments as the parent material, outside the areas covered by the Pleistocene ice sheets where there was a tundra climate. Strong winds winnowed fines from the glacial sediments, and transported and deposited them to the south as the parent material for the chernozem soils.

Chestnut-brown soils (USDA mollisol, suborder xeroll)

These soils occur where the climate is drier at the steppe–desert transition, where the vegetation becomes sparser and more xerophytic, but they are closely related to the chernozems. They are paler in colour because the more arid climate only allows a reduced annual production of vegetation per unit area, less luxurious grass, less-dense root mat and a shorter root system, and hence on average only 3–5% organic matter (mull humus) in these soils. Both the depth to which organic matter reaches (usually under 25 cm) and its content at each depth in the soil are less than in the chernozems. There is little or no removal of soluble products from the soil, but the depth of calcium carbonate and gypsum deposition is shallower, usually between 40 and 50 cm from the surface.

Bridges (1970) describes the soil pattern from the southern Ukraine, where there is a close relationship with the micro-relief. Here broad, shallow depressions occur, which receive water after the spring snow melt. Due to the movement of water through the soil to these depressions, there is a tendency for soluble salts to accumulate, but at the same time the additional water can cause leaching of the salts to greater depths in the wetter season. However, in the drier season there is migration of the salts upwards and accumulation near the surface, which has led to a whole range of solonetz-type soils and solods (Figure 24.24).

Brown and red Mediterranean soils

The vegetation in the Mediterranean was originally an evergreen forest of broad-leaved and coniferous trees, dominated by oak and pine species. However, a long-continued human intervention has removed a lot of this climax woodland through burning, grazing and lumbering. In these areas a secondary growth known as maquis on noncalcareous soils and garrigue on calcareous soils has developed as subclimax vegetation types, and soil erosion is common.

Maquis is dominated by evergreen shrubs, with spiny or leathery drought-resistant foliage (for example, wild olive, carob, lentisk, cistus and arbutus). Garrigue on the other hand has even lower growing plants, particularly herbs and prickly shrubs.

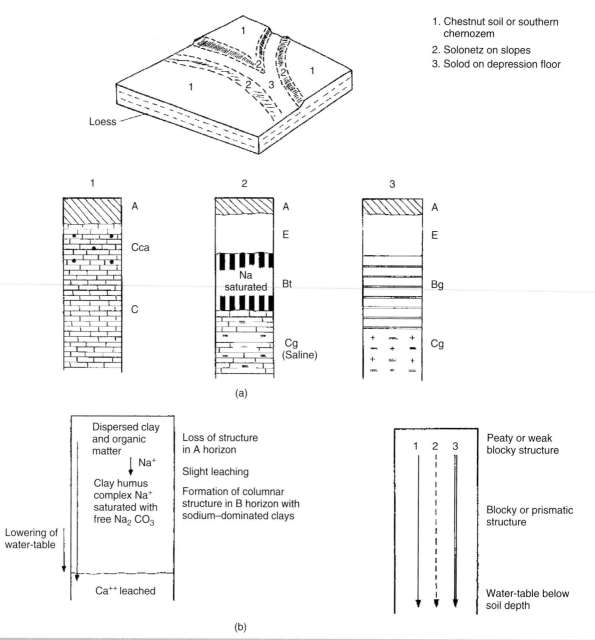

Figure 24.24 Soils of the southern Ukraine and the formation processes of solonetz and solod soils: (a) three soil types; (b) solonetz profile; (c) solod profile. (Source: after Bridges, 1970)

Nevertheless, there is likely to have been some climatic climax maquis, particularly of the wild olive–lentisk association, on wind-swept areas where the rainfall was very low and drainage free.

Brown Mediterranean soils developed beneath this mixed woodland, characterized by a brown colour and a friable humus-rich A horizon overlying a denser and less friable B horizon. Where these soils have developed on calcareous parent material, the upper horizons have been decalcified and clay movement has occurred. In the lower part of the iluvial horizon and on the linings of fissures in the C horizon, clay has been redeposited in calcareous conditions as the soil dries during the hot, dry summer.

At the same time siliceous–iron complexes are precipitated to give the rich red colour of the lower horizons of these soils. This is a process of slight ferralitization, due to the presence of iron sesquioxides, one of the products of clay decomposition. Some of the red Mediterranean soils have resulted from soil processes in the eroded remnants of these brown soils.

24.10.2 Intrazonal upland soils in Britain

When a landscape is traversed there are local soil changes based on the topography, aspect, small-scale bedrock and drift controls, and microclimate. This is especially so when the traverse is made from a lowland area to the uplands, where in general the upland soils are shallow, rocky, acidic, infertile, immature and heterogeneous compared to those in the lowlands. In the uplands the soils can be azonal where they are developing on surfaces too young for all but the early stages of soil development, and often intrazonal where they are controlled by local conditions of drainage, rock type and slope. There are no mature zonal soils as can there are in the lowlands. If we apply the general relationships that we discussed earlier, where S = C.P.R.B.T, we can see the general upland soil characteristics in Britain.

Climate

Soils are affected by: lower temperatures; higher precipitation; increased cloudiness and hence lower solar radiation; stronger winds, which cause increased evaporation and desiccation on steep hillslopes during dry spells, especially on the south-facing slopes, and can act as an eroding agent for dried-out soils, especially peats; currently-active periglacial processes involving frost action, disturbance and frost heave; waterlogging from poor aeration; increased soil acidity; higher leaching; limited biological activity, with slow decay of organic matter, which leads to the build-up of organic mor horizons or peat; and rare soil fauna, caused by the climate, where 5.5 °C is the limit for biological activity.

However, some upland soils are a product of climatic conditions that existed in the past and may have inherited structures from these past climatic phases. For example, many periglacial structures remain in soils from Pollen Zone 111 in the Late Glacial period. Climatic change in the current post-glacial period has created drier, colder and wetter conditions in our uplands at different times. For example, there is evidence for much higher agriculture and settlement in our uplands during the Medieval warm period, whilst the climate was much colder in the Little Ice Age, with greater snowfall in the uplands, and hence these climatic controls would have influenced soil processes and development.

Parent material

In the uplands solid bedrock can be the parent material for soil development, or unconsolidated sediment, such as extensive glacial and periglacial sediments, slope deposits like screes and alluvial fans and cones. The soft, weak, sedimentary rocks like shale and mudstone generally break down more rapidly than the hard, resistant igneous or metamorphic rocks. Generally hard, crystalline rocks predominate in mountains because of the orogenic processes, but there is also much sediment that was originally created in geosynclinal, deep-sea basins, and this has been converted to slates.

Slates are fine-grained, with some weathering to clays, and as a result are slow-draining and moisture-retentive, with water-logging and gleying taking place. In the Pennines there is much coarser sandstone and gritstone from Carboniferous fluvial and submarine deltas and here podzolization through leaching is common because of the good drainage, both vertically and laterally in

the soils. The limestones in the Pennines and Peak District have larger numbers of soil organisms and generally a higher biological activity, with the result that organic-matter breakdown is faster and there is quicker humification. As drainage is affected directly by the grain size there is often a great deal of variability because of rock-type change and local slope factors. Aspect also affects the dryness and warmth of upland soils. Many of the upland igneous and metamorphic rocks are acidic, such as the granites, gneisses and rhyolites, whilst many of the sedimentary rocks are basic, or at least have calcareous cement. Some metamorphic rocks are basic too, such as some of the Scottish Highland schists, whilst some metamorphics are acidic, such as the quartzites. The effects of the parent material are generally felt during the early stages of soil development and diminish with age. Hence the influence of parent material is a very important control on soil development in mountains.

Topography or the relief factor

Included in this category are: aspect; slope steepness, which affects the transport of material and subjects the soils to continual disturbance; and drainage, with slopes generally shedding water. Plateau surfaces or topographic flats may be caused by a geological control like a shale bed, which may allow water to accumulate at the surface. Soil catenas develop in the uplands because slopes are very common (Figures 24.25 and 24.26).

Biotic factor

The vegetation type controls the amount and type of organic matter, with grasses producing in general more basic soils and conifers having more silica in the needles, producing more acidic soils. Some of the evergreen species are slow to decay. The result is that, except on limestones, soil fauna is rarer in the uplands.

Humans too have had a major affect on the upland soils, largely through the destruction of the natural vegetation and the creation of non-natural, human-controlled vegetation types, such as the heather moorlands. We can trace the modification of the upland soils back to the Mesolithic, where at the woodland limit fire modification of the woodland was carried out to encourage big game animals to graze and hence to make for easier hunting in clearings in the less-dense upland woods. Leaching was thus initiated and the soils never recovered at this altitude. Lower upland woodland was affected by bronze-age pastoralism and it was then that the upland vegetation first took on similarities with the present day, with large areas dominated by grasslands. Much later the upland valley floors, which were densely wooded, were cleared by Viking settlers in places such as the Lake District. Acidic pollution and precipitation since the Industrial Revolution has also affected upland soils in the Pennines and Peak District.

Time factor

This is also important in the development of upland soils. They can quickly go back to time zero due to erosion caused by slope movement or catastrophic rainfall events, which can cause peat

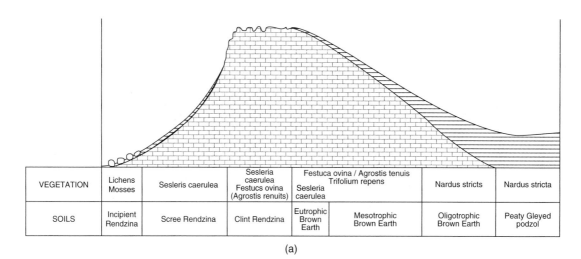

VEGETATION	Lichens Mosses	Sesleris caerulea	Sesleria caerulea Festucs ovina (Agrostis renuits)	Festuca ovina / Agrostis tenuis Trifolium repens		Nardus stricts	Nardus stricta
				Sesleria caerulea			
SOILS	Incipient Rendzina	Scree Rendzina	Clint Rendzina	Eutrophic Brown Earth	Mesotrophic Brown Earth	Oligotrophic Brown Earth	Peaty Gleyed podzol

(a)

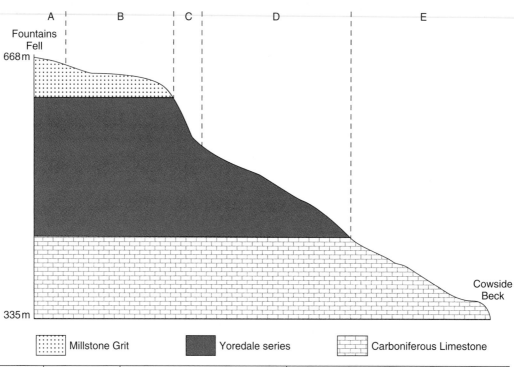

Millstone Grit   Yoredale series   Carboniferous Limestone

Zone	Elevation	Landscape/soil parent material	Soil types
A	654 m.+	Summit area of the Millstone Grit plateau	Peat, Podzols (Peaty Podzols*, Peaty Gleys and Peat Rankers)
B	594–654 m.	Lower slopes of Millstone Grit plateau	Peat, Peat Rankers (Podzols, Peaty Gleys)
C	562–594 m.	Steep shale slopes	Podzolized Brown Earths, Mesotrophic Brown Earths, Surface Water Gleys (Peat)
D	439–562 m.	Boulder clay of the middle slope	Peaty, Peaty Gleys, Peaty Gleyed Podzols (Surface Water Gleys)
E	335–437 m.	Lower limestone slopes	Rendzinas, Mesotrophic and Oligotrophic Brown Earths, Peaty Gleyed Podzols

*Soils shown in brackets occupy a very small area of each zone.

(b)

Figure 24.25 The relationship between geology, vegetation and soil type in the north-west Pennines: (a) Ingleborough; (b) Fountains Fell; (c) soil development on limestone; (d) sequence of soils with increasing drift cover

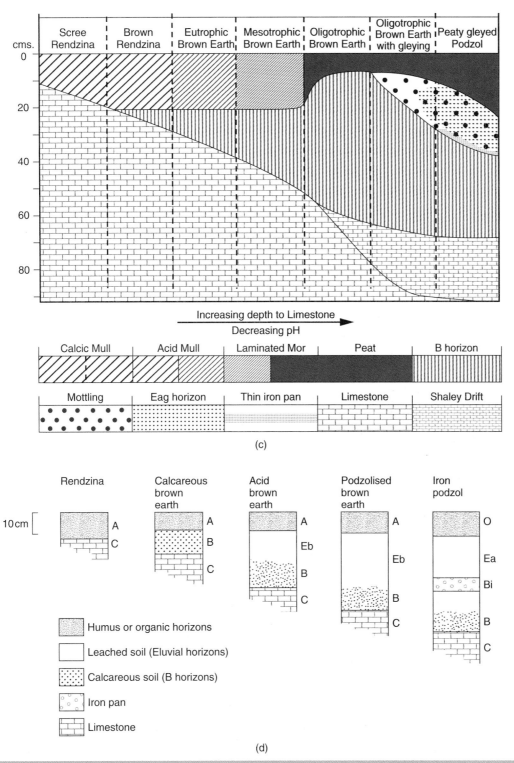

Figure 24.25 (continued)

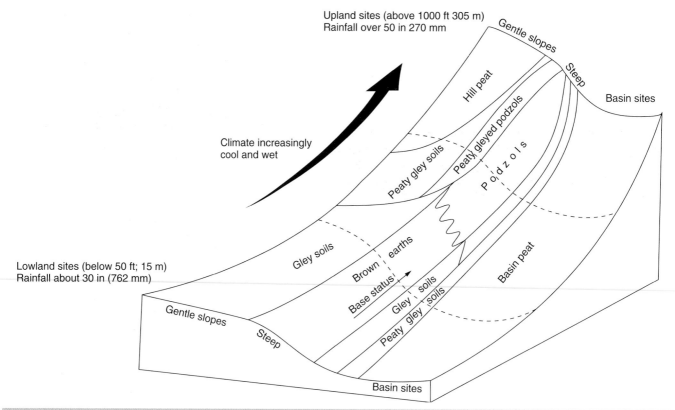

Upland sites (above 1000 ft 305 m)
Rainfall over 50 in 270 mm

Climate increasingly
cool and wet

Lowland sites (below 50 ft; 15 m)
Rainfall about 30 in (762 mm)

Figure 24.26 Soil catenas and soil types in the British Uplands

slides, flooding or normal river erosion at the base of the slopes and in the valleys. There has been much climatic change in the uplands too, which has modified the soil development at different periods of time in the post-glacial.

Upland soils are important because as altitude increases, soils become thinner, coarser, generally more acidic, more unstable, more infertile and more immature. Hence their potential is limited as the soil productivity is low due to their being frozen, often with low temperatures, below the limit for biological activity, and with a decreasing decomposition of organic matter, which takes nutrients out of circulation. The soil's stability is low too once the vegetation cover is broken and soil erosion can take place extensively, caused by animals and by recreationalists. This can be seen in all the popular British upland hill-walking areas and particularly on peaty soils, such as on the Pennine Way in areas like Kinder Scout. This limited soil potential affects the limits of permanent settlement in the uplands, the limits of agriculture and the use of the land for forestry, where only coniferous plantations can really grow, due to soil changes since the destruction of the deciduous wildwood. This type of forest monoculture also perpetuates the soil acidity.

During the 1939–1945 war much marginal land came under cultivation, but much of this has reverted to an uncultivated status once again. Nevertheless, soil improvements can take place in the uplands through drainage of wetlands and poorly-drained soils on plateau areas as well as improvements to hill pasture for grazing by the reseeding of the grasslands. As the sheep tend to eat

the more palatable grasses, like the *Festuca* and *Agrostis* species, the less palatable, like *Nardus*, become dominant. Care has to be taken to reseed with native species and some mistakes have been made in the management of recreation damage during the building of the ski lifts and ski runs close to Aviemore and in the restoration of soils on the summit plateau of Whernside on the Three Peaks Walk in the Pennines. In both cases the bright green grass of the non-native species was obvious.

The two main factors which inhibit growth are the soil wetness and the acidity, even in limestone areas. These can be managed by drainage and liming, but other factors, like the soil stoniness and the persistence of bracken rhizomes, can be difficult to eradicate. Hence the upland soil is one of the main controls on what can grow and what the landscape looks like.

24.10.3 The relationship between geology, vegetation and soils: an example from the north-west Pennines

In the Pennines there is a close relationship between the bedrock geology, the vegetation and the soil types developed (Figure 24.25(a)). On the bare limestone, or limestone with only a thin drift cover, rendzinas have developed neutral or just acidic brown Earths and basic organic soils (histosols). On the thicker drift of the Yoredale Series rocks (formed of cycles of shale, sandstone and thin limestones), above the Great

Scar of Carboniferous Limestone (Figure 24.25(a)), acid brown Earths, podzols and gleys have developed, depending on the local slopes and drainage conditions. On the summit areas of the Millstone Grit peats (acid histosols), peaty gleyed podzols and peat rankers have developed. The characteristics of these soil types are summarized in Figure 24.25(c) and (d).

24.11 The development of dune soils: an example from the Sefton coast

On the Sefton coast there has been a long history of dune soil study, with the pioneering work of Salisbury (1925), the Soil Survey monographs of Hall and Folland (1967, 1970) and the review of James (1993). The chief soil-forming processes for the various ecological zones are given in Table 24.16. There are general hypothetical trends in the soils with distance inland from the coast:

- Progressive leaching of carbonates with age (although there is the addition of $CaCO_3$ from land snails throughout the dunes).

- Change from alkaline to acid with time.

- Progressive increase in organic content with age.

- Reflection of the associated vegetational changes from marram, through various dune stages and stabilized dunes with pinewoods, to dune heathland.

The changes are thought to be on a time scale of no more than 300 years, which is part of the historical phase of dune development on this coast, although dune formation has taken place over the last 5000 years. The soil classification and profiles for the types of soil present are illustrated in Figure 24.27 and figures for organic matter and pH trends are given in Table 24.17. The characteristics of the soil types (the Ainsdale Association in Hall and Folland (1970), Figure 24.20) are as follows:

1. **Azonal raw sands and sand pararendzinas or sandy regosols:** Newly deposited sand at the back of the beach is alkaline (pH near 9), calcareous (% $CaCO_3$ between 4 and 8) and contains under 1% organic matter. The pararendzinas have a distinct A horizon, though it is only about 10 cm thick; there is weak soil development because of dune erosion and there are buried A horizons because of sand deposition. Buried organic layers are common because of wind erosion and deposition in the dunes (Figure 24.28).

2. **Acid sand and micropodzols:** These are associated with the pinewoods planted since 1887 as a technique for managing dune stability. The chief factor is the decay of acid needle litter and the pH falls to a low of 3.5 at the base of the thin humus horizon and remains very low throughout the profiles. Micropodzols have a narrow, bleached, grey-sandy, eluviated layer and a weakly-developed rusty-mottling B horizon caused by the iron mobilization and leaching of the iron during podzolization. Where dune heathland is found, for example north of Freshfield heather, broom and acid grasses invade the dune flats and a mor organic layer is associated with a thicker F and H horizons at the surface and a bleached Ea horizon with a podzolic B horizon.

3. **Peaty gleys:** These are associated with dune slacks and the chief controlling process here is the fluctuating ground-water table. Two main types of profile exist, with an oxidized

Table 24.16 Organic matter and pH trends at four British dune sites

Quantity	Site	Dune age (years)		
		0	50	100
Chief soil-forming processes in the Sefton duneland	Blakeney, Norfolk (0-10 cm)[a]	0.2	0.4	0.7
	Southport, Merseyside (0-10 cm)[a]	0.5	0.9	4.0
	Studland, Dorset (0-5 cm)[b]	0.2	0.6	2.5
	Holkham, Norfolk (0-15 cm, excluding any superficial organic layer)[c]	0.2	0.6	1.7
pH (H_2O)	Blakeney, Norfolk (0-10 cm)[a] (average initial $CaCO_3$ content 0.3%)	7.2	7.0	6.4
	Southport, Merseyside (0-10 cm)[a] (average initial $CaCO_3$ content 2%	8.2	7.8	7.2
	Studland, Dorset (0-5 cm)[b] (average initial $CaCO_3$ content <0.1%)	7.0	5.2	4.3
	Holkham, Norfolk (0-15 cm)[c] (average initial $CaCO_3$ content c. 2%	8.8	6.5	5.5

[a] Salisbury (1922, 1925, 1952)
[b] Wilson (1960)
[c] Ball and Williams (1974)

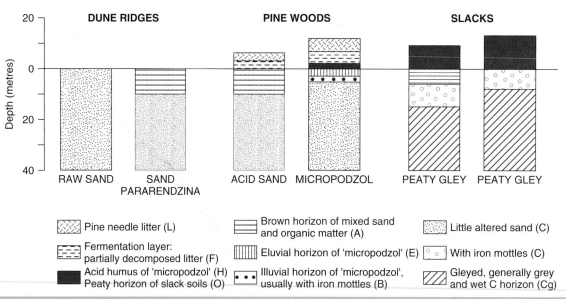

Figure 24.27 Soil types in the Sefton coast dune system. (Source: after James, 1993)

Table 24.17 Chief soil-forming processes in the Sefton duneland

Dune ridges

1. Deposition and erosion of sand.

2. Organic matter incorporation and mineralisation.

3. Cation adsorption on humus.

4. Decalcification.

5. Acidification.

6. Mineral weathering.

7. Leaching.

8. Atmospheric deposition of solutes and solids.

Pinewoods

1. Pine needle litter accumulation.

2. Acid humus formation.

3. Acidification.

4. Weak podzolization.

Dune slacks

1. Accumulations of peaty organic matter.

2. Gleying: chemical reduction of oxides (e.g. of iron) in oxygen-deficient, wet mineral soil.

mineral horizon occurring immediately below the organic surface horizon in the drier soils where the water table is relatively deep, and an absence of such a horizon where the water table is closer to the surface. As in the dunes, there are significant decreases in the pH, depth and exchangeable sodium content of the organic topsoil with distance from the sea in slack soils.

24.12 The development of woodland soils in Delamere Forest

Delamere Forest is situated in the Cheshire Basin between the North Wales hills and the Pennines and the bedrock is composed of a range of Triassic rocks (mudstones, siltstones and

(a)

(b)

Figure 24.28 (a) Micropodzol in the Sefton dunes. (b) Buried organic layers in the Sefton dune system

freely-draining sands/gravels, since the annual rainfall is not particularly great. Humus is moved through the profile, collects iron and is subsequently relocated as an ironpan lower down. There is often a white zone marking leaching in the upper profile, particularly in the humus iron podzols and peaty podzols towards the base of the slopes. Podsols are the most common soil type at Delamere because of the sandy soils and freely-draining slopes.

At the base of these catenas there are examples of surface-water gleys and peats where the soil is waterlogged for most of the year; this waterlogging is usually a result of silts/clays in the glacial sediments. Where the water table fluctuates towards the margins of the kettleholes there are examples of ground-water gleys with a mottled grey horizon below slowly decomposing peat.

24.13 Intrazonal soils caused by topographic change

A slope between a plateau and a valley bottom causes changes in soil drainage and hence changes in soil type over often short distances. However, the underlying bedrock geology is important too and soils on impermeable parent materials are different to those on permeable parent materials on a similar slope and slope angle. The resultant relationships between soils, relief and climate in uplands can be complicated. Even small-scale landforms such as gelifluction terraces can cause different types of soil to form, according to the micro-topography (Figure 24.26).

24.14 Palaeosols

Many archaeological and geological sedimentary successions contain features which indicate former soil development. Hence a palaeosol is a fossil soil that has developed on the land surface of the past. There are two types: buried palaeosols, which have been covered by younger sediments, and relict palaeosols, which occur on land surfaces that have never been buried. There is also a category called the exhumed palaeosol, which was formerly buried but has now been exposed by erosion.

Palaeosols are important because they usually formed under markedly different environmental conditions from those currently prevailing at a particular location. They can be recognized through many characteristic criteria related to soil-forming processes, such as the presence of root/rootlet horizons, phytoliths, calcareous nodules, typical soil profiles, structural features, minerological characteristics such as clay minerals like kaolinite, various major and trace elements, and microstructures. By relating the fossil soil to these environmental characteristics of present-day soils, deductions can be made about the environmental conditions at the time of the palaeosol's formation, giving valuable insights into specific environmental conditions and processes. Buried soils in a stratigraphic sequence are valuable indicators of an interval of nondeposition and mark the presence of a stabilized landscape surface, often within cycles of deposition and erosion,

sandstones), which are overlain by glacial sediments from the latest phase of the Devensian glacial. These silt, sand and gravel sediments form a hummocky terminal moraine across this part of Cheshire, with many ice-block melt depressions (kettleholes), such as Hatch Mere, Flaxmere and many smaller ones. To the south are proglacial sands and gravels.

The controls on the soil development in this area include this geological history and the previous long land-use history; the land was once deciduous wildwood, with clearances made for agriculture (place names ending in -ley indicate Anglo-Saxon clearings in the wood, e.g. Manley, Dodsley, Kingsley); it has been used as a royal hunting forest and has been part of the Forestry Commission since 1919, with largely coniferous woods and deciduous amenity planting along the roads through the forest.

The soils are usually controlled by the topography and there are excellent examples of soil catenas down the kettlehole slopes. Beneath the deciduous woodland strips there are acid brown Earths, many of which are slightly podsolized because of the

which in an archaeological context may be an area of human occupation and artefact accumulation.

Palaeosols are created by rapid burial of the former land surface by processes such as wind-blown loess, volcanic ash deposition and fluvial, overbank, silt deposition. They are buried deep enough to be unaffected by later (current) soil-forming processes. The United States Department of Agriculture (1987) classifies a buried soil as one covered by a mantle of new material that is 50 cm thick or more, or between 30 and 50 cm thick if it is at least half of the diagnostic horizons preserved in the palaeosol.

However, although it is common to be able to relate the fossil characteristics to modern soil-forming processes, depending on the length of geological time, certain attributes of the original soil development may have been lost or altered. Usually some secondary changes will have occurred, such as deposition of manganese, iron or calcium carbonate, but these are often minor in nature. Also, the vegetation and climate patterns cannot be assumed to have been present in the past and hence different conditions throughout geological time can create palaeosols with different characteristics to present-day soils.

As the classification of present-day soils is based partly on characteristics that do not survive to the geological record, and because certain climatic conditions must be known for a proper classification, Earth scientists have developed a classification system especially for palaeosols that focuses on the features that are likely to be preserved and attempts to minimize the use of interpretation. Some authors, in describing the Lower Carboniferous calcrete palaeosols from South Wales, have used the terms 'pedoderm' for a mappable unit of soil, either entire or truncated, whose physical and stratigraphic characteristics permit consistent recognition and mapping, and 'pedocomplex' for a sequence of soils that lie in close vertical succession but do not overlap.

Palaeosols are discussed in great detail in Reinhardt and Sigleo (1988), Retallack (1983, 1990), Wright (1982–1992) and Kemp (2001). The techniques for identifying palaeosols are discussed in Catt (1986) and Gerrard (1992), but it is usual for both field and laboratory techniques to be used, and identification of the processes that have occurred is rarely easy. See the companion website for Box 24.1 Soil erosion related to Mayan deforestation.

Exercises

1. What is a soil catena?

2. Illustrate how parent materials can influence soil development.

3. What are the inputs and outputs to soil systems?

4. Explain the processes taking place during eluviation or translocation.

5. Describe the characteristics of gleyed soils.

6. Why are there three distinct types of humus?

7. Define the cation-exchange capacity of a soil. Why is it important?

8. What are the important soil nutrients required for plant growth?

9. What are the characteristics of podzolic soils?

10. Describe the soil types associated with sand dunes and the changes that take place with time in such soils.

References

Avery, B.W. 1973. Soil classification in the Soil Survey of England and Wales. *Journal of Soil Science*. **24**, 324–338.

Avery, B.W. 1980. Soil classification for England and Wales. Soil Survey Technical Monograph 14. Harpenden, Soil Survey of England and Wales.

Bloomfield, C. 1953–1955. A study of podzolisation, I–IV. *Journal of Soil Science*. **4**, 5–16, 17–23. **5**, 39–45, 46–49, 50–56.

Bridges, E.M. 1970. World Soils. London and New York, Cambridge University Press.

Courtney, F.M. and Trudgill, S.T. 1993. The Soil: An Introduction to Soil Study. London, Hodder and Stoughton.

Catt, J.A. 1986. Soils and Quaternary Geology: A Handbook for Field Scientists. Oxford, Clarendon Press.

Crampton, C.B. and Taylor, J.A. 1967. Solifluction terraces in South Wales. *Biuletyn Peryglacjalny*. **16**, 15–36.

Conway, J. 2009. Soils of Wales. Available online at: http://www.rac.ac.uk/?_id=3723.

Curtis, L.F., Courtney, F.M. and Trudgill, S.T. 1976. Soils in the British Isles, Longman London.

De Corninck, F. 1980. Major mechanisms in formation of spodic horizons. *Geoderma*. **24**, 101–128.

Eyre, S.R. 1963. Vegetation and Soils: A World Picture. Arnold, London.

Farmer, V.C. 1982. Significance of the presence of allophane and imogolite in podzol B_s horizons for podzolisation mechanisms: A review. *Soil Science and Plant Nutrition*. **28**, 571–578.

Gerrard, J. 1992. Soil Geomorphology: An Integration of Pedology and Geomorphology. London, Chapman and Hall.

Hall, B.R. and Folland, C.J. 1967. Soils of the south-west Lancashire coastal plain. Sheets 74 and 83. Memoir of the Soil Survey of Great Britain. Harpenden, Soil Survey.

Hall, B.R. and Folland, C.J. 1970. Soils of Lancashire. Bulletin of the Soil Survey of Great Britain 5. Harpenden, Soil Survey.

Hodgson, J.M. 1993. Soil Survey Field Handbook. Technical Monograph 5. Harpenden, Soil Survey.

Jackson, M.L., Tyler, S.A., Willis, A.L., Bourbean, G.A. and Pennington, R.P. 1948. Weathering sequence of clay-size minerals in soils and sediments. Fundamental generalizations. *Journal of Physical Colloidal Chemistry*. **52**, 1237–1260.

James, P.A. 1993. Soils and nutrient cycling. In: Atkinson, D. and Houston, J.A. (Eds). The Sand Dunes of the Sefton Coast. Liverpool Museum. pp. 47–54.

Jones, C. 1997. quoted from The Soils of Wales: an online textbook and field guide by Dr.John.

Kemp, R.A. 2001. Pedogenic modification of loess: Significance for palaeoclimatic reconstructions. *Earth Science Reviews*. **54**, 145–156.

Reinhardt, J. and Sigleo, W.R. 1988. Paleosols and weathering through geologic time. Geological Society of America Special Paper 216.

Retallack, G. 1983. A paleopedological approach to the interpretation of terrestrial sedimentary rocks: the mid-Tertiary fossil soils of Badlands National Park. South Dakota, Bulletin of the Geological Society of America. **94**, 823–840.

Retallack, G.J. 1990. Soils of the Past: An Introduction to Paleopedology. Boston, Unwin Hyman.

Ross, S. 1989. Soil Processes: A Systematic Approach. London and New York, Routledge.

Salisbury, E.J.. 1925. Note on the edaphic succession in sand dune soils with special reference to the time factor. *Journal of Ecology*. **13**, 322–328.

Saxton, K. *et al.* 1986. Estimating generalized soil-water characteristics from texture. *Soil Science Society of America Journal*. **50**, 1031–1036.

Thompson, R.D., Mannion, A.M., Mitchell, C.W., Parry, M. and Townshend, J.R.G. 1986. Processes in Physical Geography. London and New York, Longman.

Trudgill, S. 1989. Soil types: A field identification guide. *Field Studies*. **7**, 337–363.

United States Department of Agriculture. 1987. Keys to Soil Taxonomy. Technical Monograph 6. Ithaca, NY.

Wright, V.P. 1982. Calcrete palaeosols from the Lower Carboniferous Llanelly Formation, South Wales. *Sedimentary Geology*. **33**, 1–33.

Wright, V.P. 1986. Paleosols: Their Recognition and Interpretation. Oxford, Blackwell Scientific.

Wright, V.P. 1992. Paleosol recognition: A guide to early diagenesis in terrestrial settings. In: Wolf, K.H. and Chilingarian, G.V. (Eds). Diagenesis 111. Developments in Sedimentology 47. Rotterdam, Elsevier. pp. 591–619.

Further reading

Ashman, M.R. and Puri, G. 2002. Essential Soil Science: A Clear and Concise Introduction to Soil Science. Oxford, Blackwell Scientific.

http://www.irim.com/ssm/home.htm Online soils book. An American bias, but still a useful site.

http://www.agroforester.com/overstory/overstory81.html. Soil food web.

25 World Ecosystems

Chihahuan Desert in northern Mexico
Shoreline ecosystem of Lake Turkana, northern Kenya

Earth Environments   David Huddart and Tim Stott
© 2010 John Wiley & Sons, Ltd

Learning Outcomes

After reading this chapter and completing the exercises you will be able to:

➤ Understand that there are marked world changes in vegetation and soils in terrestrial ecosystems and how these can be subdivided.

➤ Understand the ecozone concept and have a detailed knowledge of three world examples: the tundra ecozone, the tropical (equatorial rain forest) ecozone and the seasonal tropics (savanna) ecozone.

➤ Understand the potential effects of global warming on the world's ecozones.

25.1 Introduction

When the surface of the Earth is traversed it is clear that there are marked changes in both vegetation and soils in the terrestrial ecosystems and that there are major differences too in aquatic ecosystems. The classification of these ecosystems has been undertaken in various ways; for example, terrestrial biomes have been classified using a life zone system, which is based on two major assumptions: first, that mature stable plant formations represent physiognomically-discrete vegetation types that are recognizable on a worldwide basis: plant formations have a distinctive form and structure; and second, that the geographical boundaries of such vegetation groups correspond closely with the climatic zone boundaries that we covered in Chapter 7. So there are distinctive vegetation formations, or floral provinces, which have a uniform structure, appearance and composition. These can be grouped into larger classes called formation types. On a world scale they can be presented in a hypothetical continent, as in Figure 25.1.

Alternative methods of subdivision, for example using the energy characteristics, have been suggested, but because climate has a major role in determining the ecosystem and land use, the major terrestrial biomes have been classified according to Figure 25.2 in this chapter. A biome is defined as a major regional community of plants and animals with similar life forms and environmental conditions. It is ecologically integrated and uniform. It is the largest biogeographical unit and is named after the dominant form of life, such as tropical rain forest or coral reef. The dominant life forms are usually conspicuous plants, but a biome represents the soil, the vegetation, the animals and their physical, abiotic environment too. The major biomes and formation classes are illustrated in Table 25.1.

A single biome can be scattered widely over the surface of the Earth, but due to similar natural selection pressures, the species in different parts of a biome may converge in their appearance and behaviour. This can occur even when they do not share the same ancestors. So we can see that this convergent evolution, where there is independent development of similarity between species as a result of their having similar ecological roles and selection pressures, allows plants of different families which superficially resemble one another to be found in African and North American deserts.

The climate is important in determining where a particular biome is located and as there are longitudinal climatic patterns, so there are longitudinal patterns of biome distribution. However, the biomes grade into one another in transitional zones and do not have sharp boundaries, which makes accurate subdivision on a map difficult. Nevertheless if maps of world soils and vegetation are compared there is a broad similarity between them, which reflects the climatic control on both.

It is also useful to consider the ecozone concept. Ecozones are world divisions which have their own characteristic interplay of climatic factors, morphodynamics, soil-forming processes, living conditions for plants and animals and production potentials for agriculture and forestry. They could be defined as a geozonal ecosystem. There are many problems in trying to subdivide the world into only a few major ecozones: there is a wide range of small-scale variations in environmental conditions existing everywhere on the Earth, which can only be fitted into an ecozonal classification by applying many constraints and by accepting considerable blurring of the data as a result. Many geographical elements also do not have distinct boundaries and changeover is gradual.

There is also the problem of many phenomena on the Earth having developed over long time periods; their present form is the result of various environmental conditions to which a particular area was subjected across geological time. This is particularly the case in landform and soil development. As a result, ecozonal boundaries must be drawn arbitrarily to a certain extent and must be applicable for only a certain number of geographic characteristics, especially climatic thresholds. However, the variation in conditions in each of the zones is understood to be large.

Nevertheless, highly significant common elements do exist within each ecozone. These can be characterized by only the average conditions that predominate in them and any quantitative data from an ecozone can be viewed only as a guideline to make the global differences between ecozones clearer.

Schultz (1995) only considered the terrestrial world regions and ignored the marine environment when he divided the Earth's land surface into nine ecozones (Figure 25.3). Table 25.2 illustrates his ecozone comparisons according to selected features. The same approach is adopted here and the marine ecozones are disregarded. Three terrestrial ecozones will be considered in more detail in this chapter as examples of this concept: the tundra, the tropical rain forest and the savanna (seasonal tropics). Each is summarized in the form of a similar diagram structure in order to emphasize their interrelated characteristics.

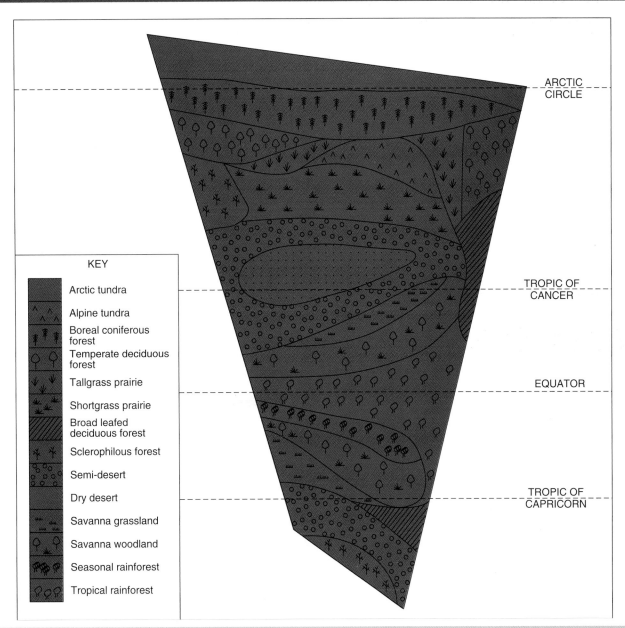

Figure 25.1 Vegetation types in a hypothetical continent. (Source: after Briggs and Smithson, 1985)

KEY

Arctic tundra

Alpine tundra

Boreal coniferous forest

Temperate deciduous forest

Tallgrass prairie

Shortgrass prairie

Broad leafed deciduous forest

Sclerophilous forest

Semi-desert

Dry desert

Savanna grassland

Savanna woodland

Seasonal rainforest

Tropical rainforest

ARCTIC CIRCLE

TROPIC OF CANCER

EQUATOR

TROPIC OF CAPRICORN

25.2 The tundra ecozone

The word 'tundra' is derived from the Finnish word '*tunturia*', which means treeless plain, and it is no surprise when one considers its distribution that this ecozone is characterized by cold temperatures, little precipitation, poor nutrients supplies, a short growing season and frost processes. We considered the latter in detail in Chapter 21 and there is no need to repeat the discussion.

The tundra's general characteristics are: an extremely cold climate; low biotic diversity; a simple vegetation structure, but with temperatures high enough for a continuous plant cover to develop; drainage limitations caused by the permafrost table close to the surface; a short growing season and reproductive time period; energy and nutrients in the form of dead organic material; and large population oscillations. In the northern hemisphere the tundra covers nearly 5 million km^2, encircling the North Pole and extending south to the taiga coniferous forests, whereas in the southern hemisphere it is almost absent. There is also a much smaller Alpine tundra, which is found high on mountains above the treeline. Even in Britain the Cairngorm High Plateau has many tundra characteristics.

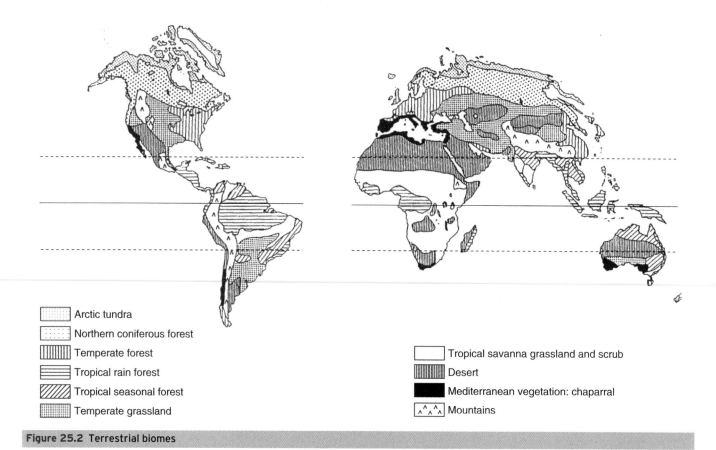

Arctic tundra

Northern coniferous forest

Temperate forest

Tropical rain forest

Tropical seasonal forest

Temperate grassland

Tropical savanna grassland and scrub

Desert

Mediterranean vegetation: chaparral

Mountains

Figure 25.2 Terrestrial biomes

Table 25.1 Major biomes and formation classes

Biome	Formation-type	Climate*	Soils†
Forest	Tropical rain forest	Tropical rainy (Af)	Oxisols, ultisols
	Seasonal rain forest	Tropical monsoon (Am)	Oxisols, ultisols, ustalfs, vertisols
	Broad-leafed evergreen	Moist warm temperate (Cfa)	Udalfs, udults
	Sclerophyllous	Mediterranean (Csa)	Xeralfs, xerolls, xeralts
	Temperate deciduous	Warm temperate (C)	Udalfs, boralfs, udolls
	Boreal coniferous	Cold boreal (Dfb, Dfc)	Spodosols, boralfs, histosols, cryaquepts
Savanna	Savanna woodland	Seasonal tropical (Aw)	Ustalfs, ultisols, oxisols, vertisols
	Savanna grassland	Dry steppe (BSh)	Ustalfs, ultisols, oxisols, vertisols
Grassland	Tall-grass prairie	Moist warm temperate (Cfa)	Udolls
	Short-grass prairie (steppe)	Cold steppe (Bsk)	Borolls, ustolls, xerolls, aridisols
Desert	Semi-desert scrub	Hot dry desert (BWh)	Aridisols, psamments
	Dry desert	Hot dry desert (BWh)	Aridisols, psamments
Tundra	Alpine tundra	Mountain (H)	Mountain soils
	Arctic tundra	Tundra (E)	Cryaquepts, cryorthents

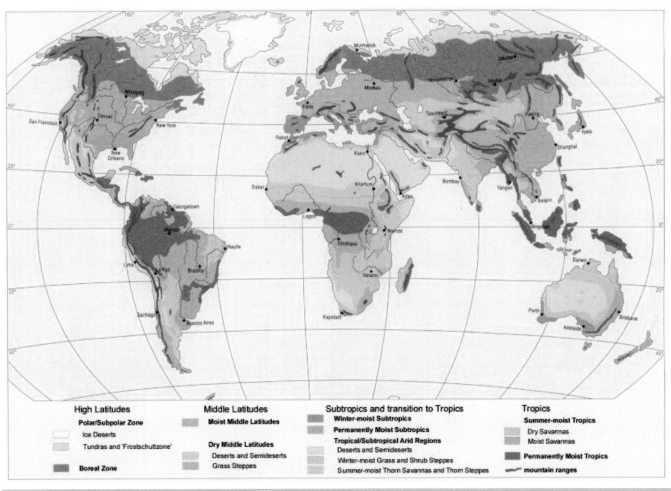

Figure 25.3 Ecozones. (Source: after Schultz, 1995)

25.2.1 Climate of the tundra

This ecozone has one of the most extreme environments for animal and plant life and this extreme condition of existence imposes constraints on organic activity. The crucial external variable controlling both the ecosystem and the landscape system is the radiant energy input, for there is a massive decrease in incident radiant energy received at the surface near the poles due to the low angle of the Sun at these high latitudes. There is also a pronounced annual periodicity in solar radiation received at the surface and the climate is controlled by this annual solar radiation cycle. The radiation balance is only positive during the short period from April to September in the northern hemisphere, but during this period the daily radiation is relatively high. At the peak period from May to July in the north, total daily radiation is comparable to that received by the tropics, although the intensity is only half as great since it is spread over 24 hours. Towards the poles the diurnal variations in incidence of solar radiation decrease progressively and the daily change from day to night is replaced by the semi-annual change from polar night to polar day. In line with this, the diurnal variations of the air temperatures diminish so that a thermal and solar seasonal climate predominates. There is a negative energy balance because there is a net and continual loss of heat by back radiation to space, which produces clear cloudless skies and hence a relatively dry and very cold climate.

Air temperatures can be as low as −60 °C and winters are long and cold, with an average winter temperature of −34 °C. The annual range of temperature can be over 50 °C in Siberia but the daily range is small by contrast in all areas. However, the average summer temperatures are between 3 and 12 °C, with at the most three months having a mean temperature above 5 °C, the threshold for plant growth. This allows a growing season of usually between 50 and 60 days and it is this crucial period which allows this ecozone to sustain life. This short Arctic summer has long sunny days, with daylight close to 24 hours. Hence there are appreciable quantities of radiant energy (2700–3500 kW m^2 per day). However, the air temperature still remains low during the short summer season and much of the energy is absorbed as the latent heat of the change of state from ice to water.

The summer heating of both the land and the atmosphere progresses very slowly and its optimum also remains far behind

Table 25.2 Ecozone comparisons according to selected, quantifiable characteristics. (Source: after Schultz, 1995)

Symbol key: ● Very high · ◕ High · ◐ Medium · ◔ Small · ○ Very small or no value. Where a cell contains two symbols, they correspond to the two sub-zones (e.g. 4.1/4.2 or 5.1/5.2).

Feature[1]	1. Polar/Subpolar Zone (1.1 Ice desert / 1.2 Tundra and frost debris zone)	2. Boreal Zone	3. Humid Mid-Latitudes	4. Arid Mid-Latitudes (4.1 Grass Steppes / 4.2 Deserts and semi-deserts)	5. Tropical/Subtrop. Arid Lands (5.1 Thornscrub savannas and subtrop. steppes / 5.2 Deserts and semi-deserts)	6. Mediterranean-Type Subtropics	7. Seasonal Tropics	8. Humid Subtropics	9. Humid Tropics
Annual precipitation (P)	○	◔	◐	◔ ○	◔ ○	◐	◔	◕	●
Mean annual temperature	○	◔	◐	◐	◕	◕	●	◕	●
Potential annual evapotranspiration	○	◔	◐	◕	●	◕	◕	◐	◕
Runoff – Amount (Q)	◔	◐	◐	○	○	◔	◐	◕	◕
Runoff – Coefficient (Q/P)	●	◕	◐	○	○	◔	◔	●	●
Annual global radiation	○	◔	◐	◕	●	●	●	◕	◕
Length of growing season	○ ◔	◐	◕	◔ ○	◔ ○	◕	◕	●	●
Incoming short-wave radiation during growing season	○	◔	◐	◔	◔	◔	◕	●	●
Mean temperature during growing season	○	◔	◔	◐	◕	◔	●	◕	●
Phyto-mass – Total	◔	◐	◕	◔ ○	◔ ○	◔	◐	●	●
Phyto-mass – Root/shoot ratio	◕	◔	◔	◕ ●	◐ ●	◐	◐	○	○
Leaf area index	◔	◕	◐	◐ ○	◔ ○	◔	◐	◕	●
Primary production	○	◔	◐	◐ ○	◔ ○	◔	◕	◕	●
Litter accumulation	◕	●	◐	◔ ○	◔ ○	◐	◔	◐	◔
Dead soil organic matter	●	◔	◕	● ○	○	◐	◔	◐	◔
Decay period of organic wastes	●	◕	◐	◔	◔	◕	◔	◔	○

1)
● Very high (value)   ◐ Medium (value)   ○ Very small or no (value)
◕ High (value)   ◔ Small (value)   Without symbol = inapplicable

2)
For thornscrub savanna only

that of other ecozones. This is because of the long-lasting snow cover and the very high soil water content after snow melt. The snow cover reflects most of the Sun's rays and only 10–20% of the radiation energy is absorbed to warm the snow. Once the snow cover begins to melt, the amount of radiation that is absorbed rises sharply and the albedo drops from 80–90% to 10–20%, while the radiation balance triples and can even be positive at midnight. After the snow has melted the radiation balance continues to be positive but despite this the soil and air still warm up only slightly as the active layer in the soil is waterlogged and a large percentage (up to 50%) of the radiation absorbed is used to evaporate this water and is thus lost as latent heat. Hence the heat budget to warm up the atmosphere over the tundra is below that which might be expected and as a rule the root zone of the plant cover does not thaw out throughout its depth until a few weeks after the snow has melted. Plant growth cannot really begin until this has occurred. Hence most of the solar energy entering this ecozone remains unused for plant production.

Surface water, soil moisture and ground water as ice are unavailable to organisms during the winter. Rainfall varies in different parts of the tundra but is generally low, with yearly precipitation figures between 150 and 250 mm. This is not because of the infrequency of precipitation events but rather their temperature-related low intensities. However, because potential evaporation is also very low, the climate is humid. The winter snow cover is little more than 20–30 cm thick and exposure to wind, not to insolation, is the main factor behind the variations in thickness and snow cover duration that occur frequently over short distances. The plants have a low albedo, which enables them to absorb more heat than the surrounding air. This is believed to speed up the physiological activity such as photosynthesis and respiration but the intense physiological drought, which is often accentuated by the cold, desiccating, strong winds, is the greatest limiting factor in the tundra. The interactions between environmental conditions and vegetation are illustrated in Figure 25.4.

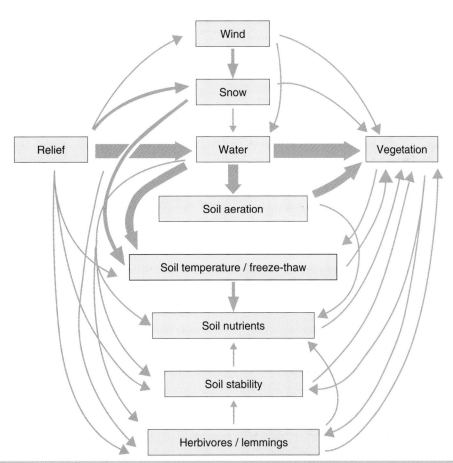

Figure 25.4 Interactions between environmental conditions and vegetation in the tundra. (Source: after Webber, 1978). The major controls are indicated by thicker arrows. Local differences in tundra vegetation are mainly relief-related because the relief affects the snow distribution in winter and melt-water flow in summer and so dictates dry, moist or wet conditions and also aeration and temperature

25.2.2 Tundra soil system

The soil system is extreme too, with immature A/C profiles of largely undecayed organic material overlying the unweathered parent material. Minerals remain unweathered because of the low temperatures and the existence of permafrost. Horizonation is poorly developed, due largely to frost processes in the active layer above the permafrost, which result in involutions and ice wedges, the development of patterned ground indicative of frost sorting and the movement of the active layer as gelifluction in summer. This is reflected in a great deal of soil instability.

The thawing of the active layer in summer forms a saturated and waterlogged mineral soil profile (gleysols) and organic decomposition is limited. Hence peat is a sink for both energy and nutrients and acts as a buffer damping climatic oscillations, immobilizing or regulating the flow of carbon dioxide and water.

25.2.3 Tundra ecosystem structure

In terms of ecosystem structure, the life form spectra are dominated by low-growing herbaceous and largely perennial plants. Low shrubs, sedges, grasses, mosses and lichens are typical. The following plant habits are characteristic at low temperatures and low light intensities: the rosette form, with the aerial shoots dying back in winter; the cushion herb; the hemicryptophyte habit of the grasses and sedges; and geophytes (plants with their perennating bud below the soil surface). When they are present, woody plants usually have a prostrate habit as conditions at the surface are often very different and less extreme than at 1 m above it (Figure 25.5). Hence microrelief becomes very important to plant growth and soil surfaces are warmer than the air wherever snow protection provides insulation. Owing to the soil's higher thermal conductivity, temperatures are substantially higher near the soil surface in summer. Asexual and vegetative reproduction is

Figure 25.5 Prostrate habit of tundra plants: Arctic or dwarf willow (*Salix herbacea*) from the Mesters Veg region of east Greenland

very common, where budding and division rather than flowering and the production of seed are prevalent. Plants can carry out photosynthesis. The primary adaptation of the vegetation is to the extreme environment and not in response to competition for space or resources. So the tundra is structurally a simple ecosystem, both vertically and horizontally. It is unsaturated in the sense that bare areas are common.

25.2.4 Tundra diversity

The tundra lacks both floral and faunal diversity. Of the approximately 250 000 flowering plants only about 900 are found in the circumpolar Arctic flora. In tropical Brazil alone the figure would be 40 000 species, but for example Alaska has only 440 species of plant, of which 250 are in the south of the state, dropping to 100 at the north coast at Point Barrow. The animals, including herbivorous mammals like lemmings, voles, caribou and Arctic hares; carnivorous mammals like Arctic foxes, wolves and polar bears; migratory birds such as ravens, snow buntings, owls, geese and sandpipers; insects, which are abundant in the short summer, like mosquitoes, blackflies and deerflies; and marine mammals and fish are all adapted to cope with the long, cold winters and are able to breed and raise their young quickly.

Animals often have additional insulation from fat and although hibernation during the winter might seem a likely strategy because food is not abundant, it is almost unknown. Another alternative would be winter migration to the south. Most large mammals usually migrate south, including the reindeer, followed by the wolves and wolverines, and most birds do the same. However, a few species overwinter in the tundra, such as the ptarmigan and the Arctic hare, and since these herbivores do not leave some carnivores stay too, like the Arctic fox and snowy owls. Winter coats and plumage are often white. Due to the constant immigration and emigration, the populations continually oscillate. The most famous example of this is the periodic lemming population crashes.

In Alaska a close correlation was found between the availability of plant nutrients (nitrogen, potassium and phosphorus) and the massive increase in the lemming population every three to seven years (Schultz, 1969). However, there is no generally applicable explanation for this. At Point Barrow, Alaska, every four to six years the lemming population increases to a density of 200 individuals per hectare or higher, and in between these peaks it drops to less than one lemming per hectare (Figure 25.6), where the lemming population includes the collared lemmings (*Dicrostonyx*) and the true lemming (*Lemmus*). The extreme fluctuations are a result of the yearly change in food supply, which depends on the plant growth conditions and hence on the weather, the density of the herbivore population and the number of predators, which with a slight delay also follows the herbivore density.

Large groups of amphibians and reptiles are rarely if ever found in this ecozone. The number of species is again low for the tundra, for example only 70 species of birds out of a worldwide

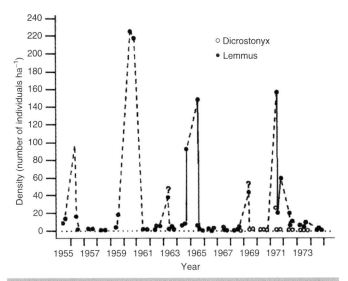

Figure 25.6 Variations in density of the brown lemming population at Point Barrow, Alaska. *Dicrostonyx*: collared lemming, *Lemmus*: true lemming. (Source: after Batzli, 1981)

total of 8600 and 23 species of mammal out of a world total of 3200. On the other hand, many species of herbivores are present in large numbers.

25.2.5 Tundra plant communities

The tundra plant communities can be subdivided into several categories:

1. **Grass sedge tundra:** Grasses like bents, fescues, foxtails; sedges; flowering plants like *Saxifraga*, *Dryas*, *Primula* and *Potentilla*; variants with mosses and lichens as dominants; and occasionally there can be creeping shrubs growing prostrate through the moss or lichen mat, between the grasses and sedges.

2. **Tundra heath:** Dwarf trees like *Betula nana*, *Salix herbacea*, Arctic bell heather and *Arctostaphyllos*.

3. **Tundra scrub:** Plants like the Labrador tea, Lapland rosebay, *Salix* species, *Alnus* species, birch. Scrub depends on shelter and is absent in the High and Middle Arctic.

4. **Bogs and marsh:** Mainly acid sphagnum bog and mire with sedges and cotton grass, but with tussocks and hummocks invaded by dwarf shrubs.

5. **Fellfields and Arctic barrens:** These areas are virtually devoid of vegetation, except for crustose and foliose lichens and isolated vascular plants. This community dominates the High Arctic and is common on many wind-exposed sites, where there is little snow cover.

With the north–south change in vegetation and soils, the Arctic can be divided into three circumpolar zones: the Low Arctic tundra, with over 80% plant cover; the High Arctic tundra, with between 10 and 80% cover; and the polar desert, where there is under 10% cover.

In terms of tundra vegetation dynamics, the norm is continuous oscillations, invasions and retreats of the vegetation and many areas undergo a kind of perpetual readjustment. There appears to be no long-term tendency towards equilibrium, which would be expected in climax vegetation. Each patch of the vegetation mosaic seems to undergo a short-term succession and each area's stability is limited in extent both in time and space, particularly by permafrost interactions.

During the growing season photosynthesis takes place all the time, and with the exception of cloudy nights and the time near the end of the growing season, the gross primary production is higher than respiration loss. However, this is not able to counterbalance the disadvantages of the short and cool growing season or the soil problems. Hence the above ground productivity of the tundra is low, with less than 1 g dry matter production per m^2 per day, or from 10 to 900 kg per hectare per year. This is in contrast with the figures for tropical rain forests of from 28 000–100 000 per hectare per year. The photosynthetic efficiency is relatively low as well.

The low productivity means that it takes time to regenerate following destruction. It has a low resilience and this ecozone is sensitive. There is a much smaller biomass of producer organisms in the tundra and it has an extremely simple vertical arrangement, reflecting the short growing season. This primary production is not sufficient to support animal life if only small areas of tundra are considered, so the large herbivores and carnivores are dependent on the productivity of vast areas of tundra and have adapted a migratory life style. Small herbivores feed and live in the vegetation mat eating the roots, rhizomes and bulbs. Their grazing pressure is at least partly responsible for the population fluctuation associated with animals like the lemming and the smaller carnivores dependent on them such as the Arctic fox and snowy owl.

Even worse than the primary production figures for the tundra are the disadvantages associated with the decomposition of dead organic material (Figure 25.7).Compared with the tropical rain forest, the primary production is 10 times lower, but the decomposition takes 100–1000 times longer. A deep litter layers and high humus content are therefore characteristic of most tundra ecosystems.

Decomposition is inhibited by the low temperatures and the acidic and usually waterlogged conditions, which mean an oxygen-poor environment. The accumulation of humus means that most of the nutrients are present in a form unavailable to plants. This has a negative effect, particularly on the nitrogen supply, because legumes with nodule-forming rhizobia in their roots are rarely found and the fixation of N$_2$ by bacteria living freely in the soil seems to be severely impeded by the cold, acidic conditions. The slow decomposition of litter can also limit the availability of phosphorus and other nutrients to plants, to the point that growth is restricted. Due to the pH usually being lower

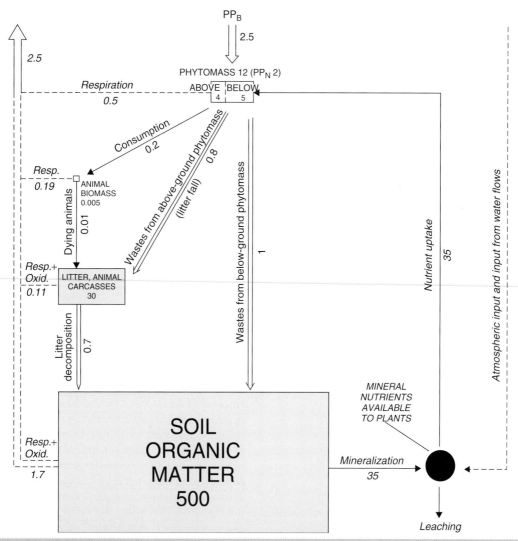

Figure 25.7 Summary of model for dwarf shrub tundra on a gelic gleysol. The huge reserves of organic matter in the soil, mainly as peat, should be noted compared with the small amount of phytomass. (Source: after Schultz, 1995)

than 4.5, inorganic phosphorus is mainly present in the form of relatively insoluble Fe-P or Al-P complexes.

25.2.6 Conclusion

In the tundra ecozone the biological structure and diversity is kept simple by the powerful, limiting climatic factors. This simple community cannot damp down and buffer itself against fluctuation and instability in the external environment, nor can it control by interaction fluctuations in the population of individual organisms. There is a low biological productivity, a simple pathway of energy transfer and a slow turnover in nutrients, all of which set further limits to the development of organic structure.

The ecozone's inherent instability means that the attainment of a mature, stable equilibrium state, the climax by progressive successional stages, is not possible, and at best it has been regarded as a mixed polyclimax. In the tundra model it can be assumed that an equilibrium will never be reached and that as primary

production is always larger than the rate of decomposition, the reserves of dead organic matter will go on increasing. This ecozone is the only one in which excess organic matter is produced on an ongoing basis; all other zonal ecosystems have a long-term balanced input–output cycle of matter. The summary diagram for the tundra ecozone is illustrated in Figure 25.8.

25.3 The tropical (equatorial) rain forest, or humid tropics *sensu stricto*, ecozone

The distribution of the tropical rain forest can be seen in Figures 25.1 and 25.2. Its global distribution is closely tied to the warm, moist climates that occur near the equator. Where a dry season of more than 2–3 months occurs, it passes into a moist

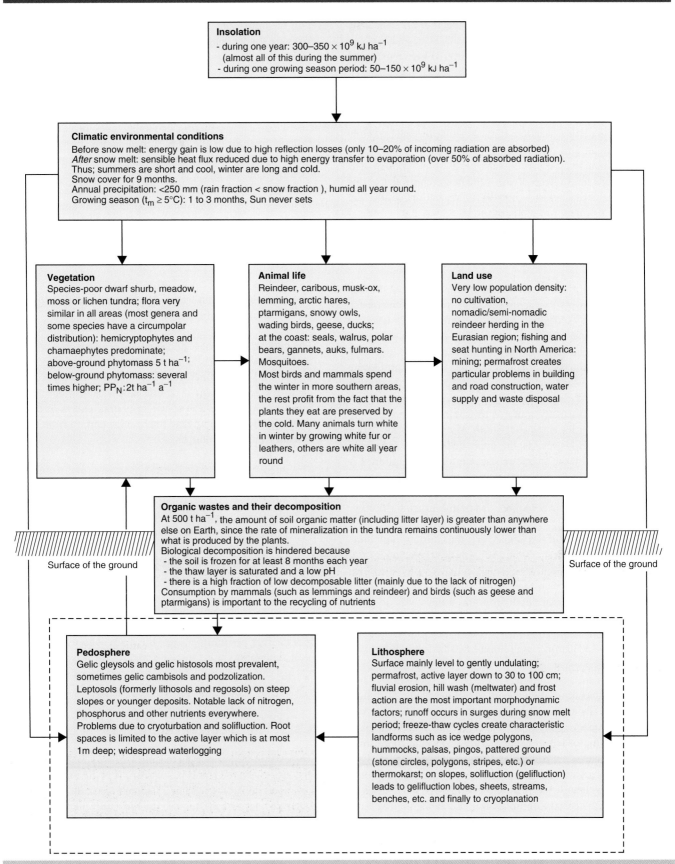

Insolation
- during one year: $300–350 \times 10^9$ kJ ha^{-1}
  (almost all of this during the summer)
- during one growing season period: $50–150 \times 10^9$ kJ ha^{-1}

Climatic environmental conditions
Before snow melt: energy gain is low due to high reflection losses (only 10–20% of incoming radiation are absorbed)
After snow melt: sensible heat flux reduced due to high energy transfer to evaporation (over 50% of absorbed radiation).
Thus; summers are short and cool, winter are long and cold.
Snow cover for 9 months.
Annual precipitation: <250 mm (rain fraction < snow fraction), humid all year round.
Growing season ($t_m \geq 5°C$): 1 to 3 months, Sun never sets

Vegetation
Species-poor dwarf shrub, meadow, moss or lichen tundra; flora very similar in all areas (most genera and some species have a circumpolar distribution): hemicryptophytes and chamaephytes predominate; above-ground phytomass 5 t ha^{-1}; below-ground phytomass: several times higher; PP$_N$:2t ha^{-1} a^{-1}

Animal life
Reindeer, caribous, musk-ox, lemming, arctic hares, ptarmigans, snowy owls, wading birds, geese, ducks; at the coast: seals, walrus, polar bears, gannets, auks, fulmars. Mosquitoes.
Most birds and mammals spend the winter in more southern areas, the rest profit from the fact that the plants they eat are preserved by the cold. Many animals turn white in winter by growing white fur or leathers, others are white all year round

Land use
Very low population density: no cultivation, nomadic/semi-nomadic reindeer herding in the Eurasian region; fishing and seat hunting in North America: mining; permafrost creates particular problems in building and road construction, water supply and waste disposal

Organic wastes and their decomposition
At 500 t ha^{-1}, the amount of soil organic matter (including litter layer) is greater than anywhere else on Earth, since the rate of mineralization in the tundra remains continuously lower than what is produced by the plants.
Biological decomposition is hindered because
- the soil is frozen for at least 8 months each year
- the thaw layer is saturated and a low pH
- there is a high fraction of low decomposable litter (mainly due to the lack of nitrogen)
Consumption by mammals (such as lemmings and reindeer) and birds (such as geese and ptarmigans) is important to the recycling of nutrients

Surface of the ground

Surface of the ground

Pedosphere
Gelic gleysols and gelic histosols most prevalent, sometimes gelic cambisols and podzolization.
Leptosols (formerly lithosols and regosols) on steep slopes or younger deposits. Notable lack of nitrogen, phosphorus and other nutrients everywhere.
Problems due to cryoturbation and solifluction. Root spaces is limited to the active layer which is at most 1m deep; widespread waterlogging

Lithosphere
Surface mainly level to gently undulating; permafrost, active layer down to 30 to 100 cm; fluvial erosion, hill wash (meltwater) and frost action are the most important morphodynamic factors; runoff occurs in surges during snow melt period; freeze-thaw cycles create characteristic landforms such as ice wedge polygons, hummocks, palsas, pingos, pattered ground (stone circles, polygons, stripes, etc.) or thermokarst; on slopes, solifluction (gelifluction) leads to gelifluction lobes, sheets, streams, benches, etc. and finally to cryoplanation

Figure 25.8 Tundra ecozone. (Source: after Schultz, 1995)

savanna region. The two climatic zones together are called the moist or humid tropics (Schultz, 1995).

The tropical rain forest biome once covered between 14 and 17% of the Earth's land surface but it has been decreasing at an alarming rate over the last few decades to a current figure of about 6% and some expert opinion suggests it could be consumed completely in the next 40–50 years.

A vertical gradation to colder-climate vegetation occurs rapidly on tropical mountains and above a few hundred metres of altitude the forest character starts to change, until above 2000 m it is called montane forest. The climate is the major controller of this ecozone's characteristics as it controls the vegetation, weathering and soil development, productivity and land use.

25.3.1 Climate of the humid tropics

Close to the equator there is an intense energy input from the Sun throughout the year, producing the inter-tropical convergence zone, which is a convection zone of rising air that loses its moisture as frequent, intense rainstorms. In those equatorial regions that are relatively isolated from the ocean winds that carry water vapour inland there are breaks in the rain forest and to the north and south away from the equator there is a general rainfall decrease as the influence of the ITCZ becomes weaker.

The region is normally confined to within 5° N and S of the equator and includes equatorial South America, central Africa and Indonesia. However, as the areal delimitation is usually also demarcated by the characteristic rain forest vegetation it is often difficult to carry out, especially where regions border on the adjacent humid subtropics, as in south-eastern Brazil and Vietnam, and in areas where there has been human destruction of the forest.

The characteristics of the climate are: the constant high temperatures, usually between 25 and 27 °C (always exceeding 18 °C); the absence of low temperatures; a small annual temperature range (±5 °C) which is less than the diurnal range (6–11 °C); limited seasonality with an absence of dry months, usually with no more than one month with under 50 mm precipitation; regular, high and intense rainfall (mean annual levels of 1500 mm, frequently values of 2000–3000 mm); high humidity, sultriness and unstable air conditions; and constant day length, with 12 hours a day because of the vertical Sun's rays. The high cloudiness and high atmospheric humidity explain why the amount of diffuse atmospheric radiation is up to 40%.

By far the greatest amount of the received solar energy is consumed in the process of evaporating water and more than 1000 mm of water is evaporated and transpired per annum. This is significantly greater than in any other ecozone. The climate is uniform throughout the year, and because precipitation is in excess of potential evapotranspiration, this allows a luxuriant vegetation development and plant growth that continues all year round. The annual periodicity which is obvious in all other zonal plant formations is lacking in the tropical rain forest and in most species the different developmental stages occur at varying times, even among members of the same species, and occasionally

even in the individual branches of the same tree. Hence seasonal changes are totally missing.

25.3.2 Weathering regime

Despite some climatic change in the Quaternary, extremely intense chemical weathering generally takes place, largely by hydrolysis, because of the high soil moisture contents all year round, the consistently high temperatures and the usually high soil acidity. This strong chemical weathering produces extremely thick zones of saprolite (decomposed rock), often tens of metres thick (over 100 m on crystalline rocks in Brazil) and practically no silicates are left from the parent rock, with clay minerals the result. Bare rock outcrops are relatively rare and short-term disintegration and rock decomposition are more rapid than transport. This is because of:

- The generally high infiltration rate and the fact that moisture is not a limit to weathering.

- The high ground surface and soil temperatures, as a temperature increase of 10 °C increases the rate of chemical reactions by two and a half times. At a depth of about 1 m, soil temperatures are constant at about 2–4 °C above the average free air temperature due to the insulating effects of the layered vegetation canopies and the heat generation from rapidly decomposing dead biomass.

- The intensity of the biochemical processes, due to the great amount and rapid supply of litter, which releases organic acids and complexing and chelating agents. The decomposition of humus promotes high carbon dioxide levels, on average five times greater than those of temperate soils.

- A long time period with a stable biome that has adjusted to relatively little climatic change.

There is extensive leaching, including the quartz, acidic soil water, which allows iron oxides to be very mobile; and because there are high discharges per unit per area, there are considerably greater tropical denudation rates. Surface and near-surface rock solution results in a variety of solution microforms on rock such as granite and basalt, as well as limestone (see Section 15.5.2).

25.3.3 The tropical rain forest

The vegetation is the most luxuriant, complex and diverse terrestrial ecosystem the world has known. It is the mature, stable, climax vegetation of the equatorial tropical lowlands, where there is no marked dry season. It is also the most ancient ecosystem, as despite contracting and expanding in the Quaternary, responding to climatic change, in its heartland it has existed with little change since the beginning of the Tertiary. Although there are regional variations, the tropical rain forests have a large number of common characteristics, as outlined below.

Structural organization

The tropical rain forest has a complex vertical and horizontal structure, with the traditional view defining between three and five storeys or tree strata (Figure 25.9). However, there is a debate as to whether this is a valid view or an illusion created by the spatial mosaic of patches of very differently aged trees. Each of the storeys shows a marked decline in the amount of light available compared to that above, and as a result the living conditions are significantly different for both plants and animals in each. The percentage of photosynthetically-active radiation declines in inverse proportion to the height; for example measurements were 53% in a clearing and 18% in the forest shade.

The species diversity is very high and often the number of species per hectare can reach over 100. Hence there is a lack of dominance of any one species, coupled with low population densities of all species, which gives rise to the horizontal complexity. There are large numbers of dependent life forms, especially the epiphytes (Figure 25.10(a)) and climbers/lianas. Often more than 70% are tree species, mostly evergreen, and the shrub and

ground strata are limited by light getting through the tree canopy, although along river valleys there are the so-called gallery forests.

The layer of ground herbaceous plants may be denser for short time periods where trees have fallen over and more light can reach the forest floor. The tree seedlings make use of these clearings and the survival of young plants depends on how quickly they succeed in gaining height and can thus derive more benefit from the light than their competitors. Hence the ability to grow rapidly in length is a characteristic of many species, and bamboo for example can grow at a rate of 0.5 m per day. Elsewhere saprophytes and parasites are important in the reduced-light areas.

Convergent evolution is evident in the physiognomic characteristics of the trees. For example, all are evergreen hardwoods, with broadleaves, and many have long drip tips. The function of the drip tip is uncertain, although they will accelerate rainwater runoff and facilitate the gas exchange required for photosynthesis and respiration. The uncertainty comes from the facts that many drip tips are turned upwards and that they are lacking in many of the wettest forests. Usually smaller, laurel-like leaf forms, with facilities for reducing transpiration loss such as a thick cuticle

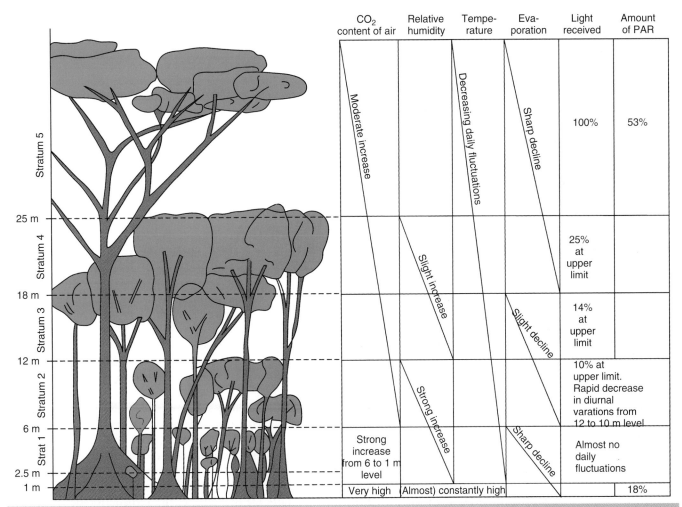

Figure 25.9 Structural organization of vegetation in a tropical rain forest and associated ecological conditions, taking an example from the Ivory Coast. (Source: after Bougeron, 1983)

(a)

(b)

Figure 25.10 (a) Epiphytes in tropical rain forest. (b) Buttresses in tropical rain forest

from several hundred to around 1000. The growth habit is a tall, unbranched, slender-to-columnar stem, frequently broadened with buttresses or stilt roots. The bark is thin, smooth and light-coloured and the roots are shallow and spread out just below the surface, which is optimal for the uptake of nutrients. However, the shallow roots do not provide sufficient anchorage and are given support by the buttresses (Figure 25.10(b)), which may also support respiration processes, although they do not extend below the surface but rather form a series of comb-like rows of shallow roots. Many trees also have aerial roots for respiration as the root respiration in the soil is impeded by frequent waterlogging, the high oxygen consumption of decomposers and low gas exchange with the outer atmosphere.

Cauliflory, the growth of flowers and fruit on leafless stems (e.g. the cacao tree), although not common, is much more frequent than in any other ecozone. The trees are often pollinated by insects, birds (such as the hummingbirds in South America) and bats, and some animals help distribute seeds. The canopy height reaches 30–40 m, with some crowns above this to 70–80 m. This canopy is uneven due to the tree height variations. There is a similar convergence in other life forms, such as the lianas and epiphytes (mosses, ferns, orchids and other flowering plants).

Functional organization

Each trophic level is occupied by a very large number of species, but each of these is adapted to highly-specialized ecological niches which are narrowly defined, and hence competition is avoided. The species are mainly K-selected stress-tolerators, which are characteristic of climax communities. There is a highly integrated community structure in terms of the interrelationships between organisms and the complexity of the functional pathways, such as food webs.

The climate and microclimate within the forest are largely created by the forest itself. The multilayered forest structure explains why the leaf area index is higher than for any other zonal plant formation. The nutrient reserves are concentrated in the living biomass and hence the forest is largely independent of mineral soil because of the constant, efficient and rapid nutrient turnover. Overall the very effective energy filter of the complex canopy, plus the high sunlight and general microclimate, all lead to a very high productivity, with many efficient C4-pathway photosynthesis plants.

Most of the measurements of phytomass are in the order of 300–650 tonnes per hectare, with the average just below 500 tonnes per hectare. In each case between 75 and 90% of the phytomass is above ground and 90% is in the form of living tree wood. Only about 2–3% is accounted for by the tree leaf mass. There is general agreement that the primary production of this ecozone exceeds that of any other.

Despite the considerable supply of leaf litter, generally the forest floor is not covered by an unbroken litter layer. This is partly because litter from the upper storeys gets caught up in the lower canopy layers where it is available to epiphytes, but the major reason is that litter is rapidly decomposed on the ground, usually within a few months. In addition to termites, leaf-cutter

and a waxy layer, predominate in the upper canopy layer where the leaves may be exposed to much direct sunlight and strong wind each day, so that for a while a considerable drought may exist and the plants will need these xeromorphic Sun leaves. In the more protected stem region, the leaves are softer, larger and a darker green in colour (hygromorphic shade leaves) and no bud scales are developed. These large-leaf areas favour photosynthesis in areas of lower light.

Problems related to transpiration from the vapour saturation of the air are often solved with the help of water pores (hydathodes), through which water droplets can be actively excreted, and sometimes the stomata are raised above the leaf surface to facilitate transpiration. They show a wide range of adaptations to the constant high humidity and can be classified as hygrophytes, or moisture-loving plants. The luxuriance of the vegetation is great and the tree density (number of trees per hectare) ranges

ants and some earthworms, fungi play the major role in litter breakdown.

It is characteristic of the rain forest ecosystem that the nutrient turnovers take place within what is largely a closed system. This is because otherwise the extremely luxuriant vegetation development on many of the highly nutrient-deficient soils could not be explained. An obvious indication of this closed cycle is that river water usually contains very low concentrations of dissolved and suspended matter and in many cases the soil fertility declines rapidly once the forest has been cleared, and the resupply of nutrients is thus stopped.

The losses by leaching are small because, although the soils thicknesses are great, most nutrients are used in the upper part of the soil. Studies of the amounts and percentages of the nutrient elements supplied to the soil in the Ivory Coast and New Guinea show that the inputs of nutrients through precipitation are significant. These compensate for the low leaching loss. However, the major inputs occur via the two most important nutrient recycling pathways: the plant wastes and throughfall from crown wash/stemflow. The latter is important for potassium supply,

and twice as much of this is returned to the soil via throughfall than via litter fall. Hence potassium circulates much faster than calcium and throughfall also plays a major role in the return of magnesium. For all other nutrients, litter fall is more important. A simplified model for a tropical rain forest is illustrated in Figure 25.11.

So the vegetation is a highly complex, diverse forest community that is very well integrated with a high level of internal organization and interdependence. In the absence of external disturbance, it appears to be inherently stable. It has a resistance, which is the degree to which a system is buffered against change. It has a capacity to resist change or to accommodate changes in its external environment: it functions largely independently of fluctuations in its external environment, patterns of change or levels of disturbance, which the forest long ago learned to accommodate.

It is also a resilient ecosystem that has a robustness. This is made up of two components: its amplitude or lability and its elasticity or recovery. The tropical rain forest in fact proves to have limited lability, or a limited tolerance for amplitude or disturbance. In the face of major disturbance it has limited

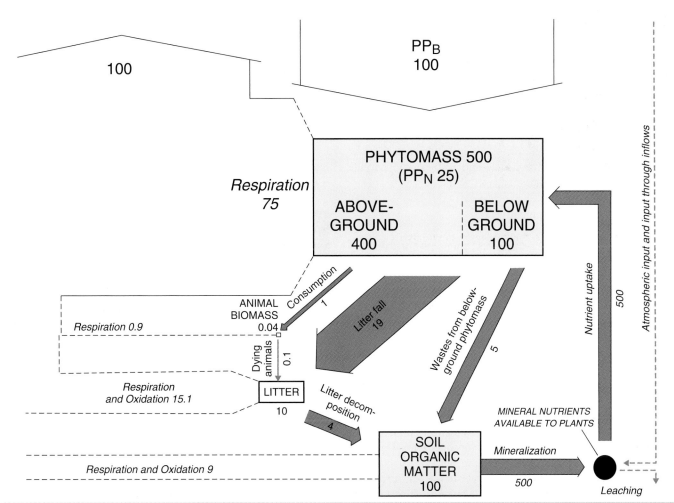

Figure 25.11 Simplified ecosystem model of a tropical rain forest. Note the large amount of phytomass and the small amount of litter in the soil, the high turnovers of energy and minerals and the low amounts of exchangeable nutrients in the soil. (Source: adapted from Schultz, 1995)

elasticity and its trajectory of recovery is slow and incomplete. This means that once a disturbance crosses a certain threshold the forest is not resilient, and it cannot recover once the disturbance has ceased to anything like its former complexity and diversity.

Animal life

Many tropical rain forests appear to be uninhabited since the vegetation absorbs all noise, but the greatest number of species anywhere on the planet exists in this ecozone. Most animal species are represented by just a few individuals, which are distributed among a large number of ecological niches, with many in the upper storeys. The animal communities are stratified and ground-dwellers are rare because the sparse ground flora offers a food base to only a few herbivores, such as elephants, okapi and tapirs. The animal groups that contain particularly large numbers include the reptiles and amphibians, because of the uniform, warm–humid conditions.

Soils

The typical soils are ferralsols (identical with the Oxisols of the USDA 1975 Classification), which show a high degree of weathering and are typical of old land surfaces that have formed over long time periods. The term 'ferrallitization' describes the process by which the parent material is changed into a soil consisting of highly-weathered material rich in secondary hydrous oxides of iron and aluminum (the sesquioxides), which is nearly devoid of bases and primary silicates but can contain large amounts of clay minerals, especially kaolinite. As a consequence of the deep chemical weathering the soil profile is uniform in colour (bright yellow to deep red) and texture, and has the following other characteristics:

- extremely thick development

- no clay illuviation and no clay skins (immobility of clay)

- sandy loam or finer texture

- a clay fraction composed of kaolinite, iron oxides and aluminum oxides, and a sand and silt fraction consisting almost entirely of quartz

- low cation-exchange capacity

- poorly-developed structure

- weakly-developed horizonation with merging boundaries

- highly acid soils

- variable agricultural potential, usually moderately low to very low.

Tropical rain forest vulnerability

Tropical rain forests once occupied far greater percentages of the planet's surface than they do currently and some of the deforested areas are vast. For example, it is estimated that in Brazil 5.4 million acres have been destroyed per year, averaged for the period 1979–1990, and in Costa Rica the area of forest was reduced from 67% primary forest in 1940 to 17% by 1983. In parts of the south-eastern coast of Brazil the Atlantic rain forest has been almost totally destroyed, with primary forest occupying less than 2% of its original area. This has taken place because of human population increase and because the value of the land was perceived as lying only in its timber by short-sighted governments, multinational logging companies and landowners.

Commercial timber includes teak, mahogony and rosewood, which have generated big profits for the companies and are a large economic resource for often impoverished governments. Traditional slash and burn/shifting subsistence farming has been displaced in some places by cattle ranching and in others by mining, or the development of hydro-electric schemes on the rivers. This loss of tropical rain forests is believed to have a devastating world impact as biologically they are so diverse. It is estimated that less than 1% of the huge biodiversity has been scientifically studied for possible uses by man and many species are probably lost every day, becoming extinct before they have been identified, let alone evaluated. It is also estimated that carbon dioxide entering the atmosphere from fires associated with the destruction of the rain forest and from rotting vegetation may lead to a further increase in global warming and more climatic change, and there is accelerated soil erosion after clearance.

The summary model for the tropical rain forest ecozone is illustrated in Figure 25.12.

25.4 The seasonal tropics or savanna ecozone

The savanna biome covers approximately 20% of the planet's land area and includes many of the subtropical regions adjacent to the rain forest. Although grassland may dominate, it is an open woodland in many areas, with widely-spaced and scrubby trees (see the types in Figure 25.13). As the vegetation is non-uniform in floristic and physiognomical terms, both on a macro- and microscale, it is sometimes difficult to determine the outer margin of the savanna ecozone.

The formation of this ecozone has generated some dispute as it seems that climatic factors alone cannot account for its characteristics. Although these areas have a marked dry season and a variable precipitation, often the climate could sustain a much more luxuriant and diverse flora than it does. A possible reason for this is that the savanna represents a plagioclimax that has been severely affected by man, especially through fire. If it assumed that the savanna is characterized primarily by a strict

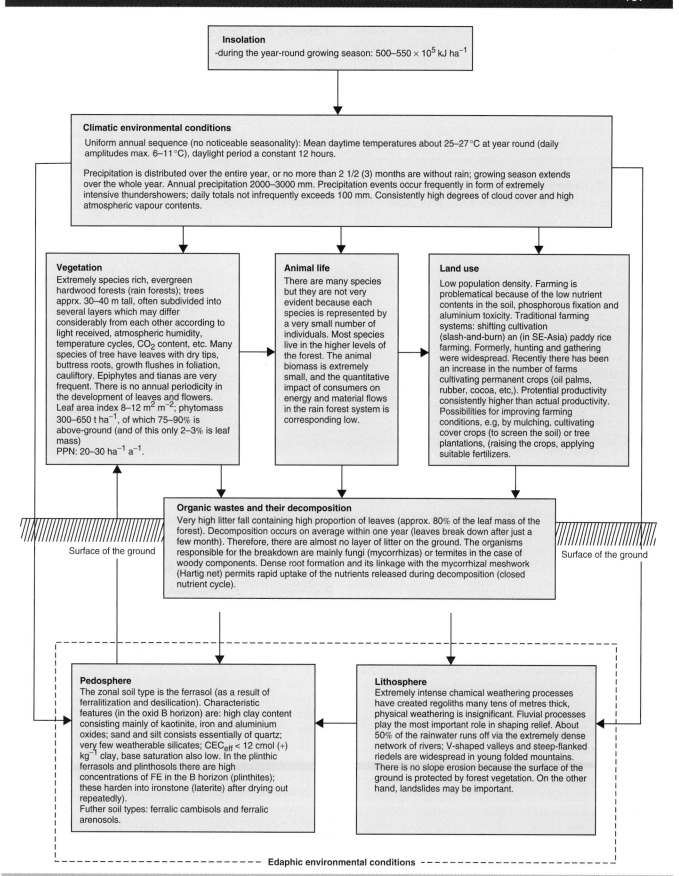

Insolation
-during the year-round growing season: 500–550 × 10⁵ kJ ha⁻¹

Climatic environmental conditions

Uniform annual sequence (no noticeable seasonality): Mean daytime temperatures about 25–27°C at year round (daily amplitudes max. 6–11°C), daylight period a constant 12 hours.

Precipitation is distributed over the entire year, or no more than 2 1/2 (3) months are without rain; growing season extends over the whole year. Annual precipitation 2000–3000 mm. Precipitation events occur frequently in form of extremely intensive thundershowers; daily totals not infrequently exceeds 100 mm. Consistently high degrees of cloud cover and high atmospheric vapour contents.

Vegetation
Extremely species rich, evergreen hardwood forests (rain forests); trees apprx. 30–40 m tall, often subdivided into several layers which may differ considerably from each other according to light received, atmospheric humidity, temperature cycles, CO₂ content, etc. Many species of tree have leaves with dry tips, buttress roots, growth flushes in foliation, cauliftory. Epiphytes and tianas are very frequent. There is no annual periodicity in the development of leaves and flowers. Leaf area index 8–12 m² m⁻²; phytomass 300–650 t ha⁻¹, of which 75–90% is above-ground (and of this only 2–3% is leaf mass)
PPN: 20–30 ha⁻¹ a⁻¹.

Animal life
There are many species but they are not very evident because each species is represented by a very small number of individuals. Most species live in the higher levels of the forest. The animal biomass is extremely small, and the quantitative impact of consumers on energy and material flows in the rain forest system is corresponding low.

Land use
Low population density. Farming is problematical because of the low nutrient contents in the soil, phosphorous fixation and aluminium toxicity. Traditional farming systems: shifting cultivation (slash-and-burn) an (in SE-Asia) paddy rice farming. Formerly, hunting and gathering were widespread. Recently there has been an increase in the number of farms cultivating permanent crops (oil palms, rubber, cocoa, etc,). Protential productivity consistently higher than actual productivity. Possibilities for improving farming conditions, e.g, by mulching, cultivating cover crops (to screen the soil) or tree plantations, (raising the crops, applying suitable fertilizers.

Organic wastes and their decomposition
Very high litter fall containing high proportion of leaves (approx. 80% of the leaf mass of the forest). Decomposition occurs on average within one year (leaves break down after just a few month). Therefore, there are almost no layer of litter on the ground. The organisms responsible for the breakdown are mainly fungi (mycorrhizas) or termites in the case of woody components. Dense root formation and its linkage with the mycorrhizal meshwork (Hartig net) permits rapid uptake of the nutrients released during decomposition (closed nutrient cycle).

Surface of the ground

Surface of the ground

Pedosphere
The zonal soil type is the ferrasol (as a result of ferralitization and desilication). Characteristic features (in the oxid B horizon) are: high clay content consisting mainly of kaotinite, iron and aluminium oxides; sand and silt consists essentially of quartz; very few weatherable silicates; CEC_eff < 12 cmol (+) kg⁻¹ clay, base saturation also low. In the plinthic ferrasols and plinthosols there are high concentrations of FE in the B horizon (plinthites); these harden into ironstone (laterite) after drying out repeatedly).
Futher soil types: ferralic cambisols and ferralic arenosols.

Lithosphere
Extremely intense chamical weathering processes have created regoliths many tens of metres thick, physical weathering is insignificant. Fluvial processes play the most important role in shaping relief. About 50% of the rainwater runs off via the extremely dense network of rivers; V-shaped valleys and steep-flanked riedels are widespread in young folded mountains. There is no slope erosion because the surface of the ground is protected by forest vegetation. On the other hand, landslides may be important.

- - - Edaphic environmental conditions - - -

Figure 25.12 Tropical rain forest ecozone. (Source: after Schultz, 1995)

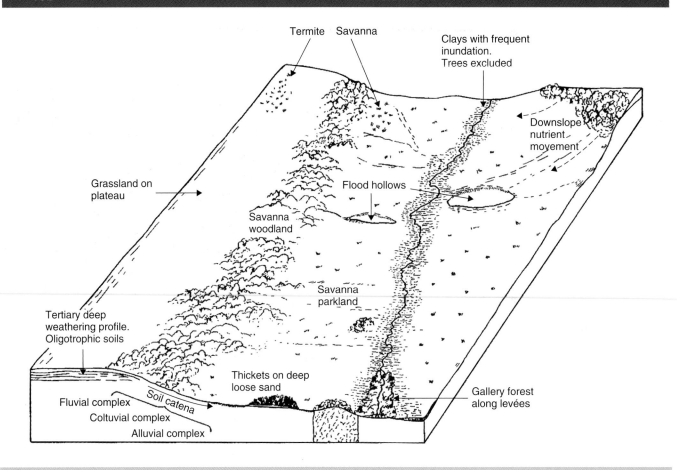

Figure 25.13 Environmental relationships in a savanna area. (Source: after Collinson, 1977)

seasonality (as a rule a summer rainy season and a winter dry season) and vegetation consisting of a continuous grass layer and a discontinuous tree or shrub layer, then it is found that this also applies to some regions that extend northwards beyond the tropical boundary. Schultz (1995) argues a case for not including the Sahel-type, thorn savanna regions in this ecozone.

25.4.1 Climate of the seasonal tropics

Major large-scale differences in the savanna regions are linked with differences in humidity that result from annual rainy periods of unequal length. Hence, the dry savannas have between 4.5 and 7 rainy months and can be differentiated from the moist savannas, which have approximately 7–9.5 rainy months.

This of course is controlled by the climate and the seasonal tropics fall between the equatorial low-pressure trough and the subtropical/tropical belts of high-pressure cells. This difference between two fairly stable pressure systems generates the constant trade winds; during any one year these pressure and wind systems are displaced as the height of the Sun changes and the high-pressure belt is broken up in the respective summer hemisphere by continental thermal lows. Periods in which strong trade winds are blowing are also periods with little or no precipitation (when the Sun is at a low angle). The reverse happens as the height

of the Sun increases during the summer months: the distance between the tropical high-pressure zones and the equatorial low-pressure trough increases once more, and in the high-pressure belt thermal lows form above all the continents. As a result the pressure gradient towards the equatorial low-pressure trough decreases significantly, the trade winds die down and greatly-modified circulation patterns can result over the continents. These include highs reaching convective air movements, which generate strong rain showers.

Nearer the equator this precipitation is distributed between two maxima in the spring and autumn (as in Ethiopia and in the Llanos of Venezuela), according to the two zenithal positions of the Sun, which occur a long time apart here. With increasing distance from the equator, the rain maxima come closer together (as the time between the maximum declinations of the Sun decreases) to form a twin-peaked and finally a single rainy season, which alternates with a rainless dry season. At the same time the precipitation period becomes shorter and less rain falls. With this decreasing precipitation, the variability increases.

The radiation budget is positive all year round, although it is significantly weaker in the winter season. The day length deviates by no more than an hour from 12 hours, resulting in a moderately balanced temperature cycle, with mean seasonal temperature differences lower than the daily differences. At sea level all the monthly averages are above 18 °C, the maximum temperatures occur immediately before the start of the rainy seasons and a bigger temperature drop occurs at the start of the dry season.

25.4.2 Weathering regime

Strong chemical weathering characterizes this ecozone in the usually permanently-wet subsoil. As a result a deep, clay-rich soil develops and the rock is decomposed to a considerable depth. The landform characteristics include some of the following: the great effectiveness of sheet flows from overland flow is correlated with the high precipitation intensity; peneplains and inselbergs are typical landforms; river runoff is usually restricted to certain periods of the year and after that the beds are dry and runoff is episodic.

25.4.3 Soils

The soils of the savanna zone are variable but include oxisols, ultisols, vertisols and alfisols. Their distribution is related to climatic, geomorphological and geological conditions. Slope processes are active, for the vegetation is often not sufficient to prevent erosion and downwashing of nutrients. Thus marked slope catenas, with a gradation from shallow stony soils to deeper, less well-drained, base-rich alluvial soils, develop on hillslopes. Kaolinite forms as the dominant clay mineral, and the extent of oxide formation as sesquioxides is high. When the iron is oxidized, red hematite is produced to give the widespread red soil colour. Due to the intensive chemical weathering, most tropical soils only contain small amounts of weatherable silicates, or even none.

The vertisol is typical of grass-covered, poorly-drained plains and depressions consisting of clay-rich (usually calcium carbonate-containing) weathering sediments. It is dark grey to black or brown in colour, with at least 30% clay occurring in areas with at least 200–300 mm precipitation per annum and 3–9 dry months. During the rainy season it has a dense, viscous-plastic consistency and high swelling pressures are generated but during the dry season shrinkage cracks develop. As a result movement called hydroturbation leads to deep mixing and homogenization of the soil and evidence of such movement from shiny sliding surfaces or slickensides can be seen on the soil aggregates. At the surface of the soil the hydroturbation is evident in the form of gentle undulations in the microrelief, known as gilgai (Figure 25.14). The pronounced swelling and shrinkage characteristics are linked with the generally high percentages of swellable clay minerals, especially smectite. These clay minerals have a high cation-exchange capacity, carbonate precipitation can occur in the profile and, despite the dark colour, the humus content is under 3%. Due to their considerable nutrient content, vertisols have a very high production potential, but there are problems, such as the difficulty of working in both the wet and dry states. Many vertisol regions are therefore used as grazing land.

Figure 25.14 Undulations of the microrelief or gilgai on silty clay soils developed on lake sediments, east of Lake Turkana, northern Kenya

25.4.4 Vegetation characteristics

Although there is a diversity of different vegetation types, there is one common characteristic that is not found elsewhere in any ecozone and that is the mixed combination of trees and grasses. The regional variations are based mainly on the varying degrees of tree cover, which can range from almost treeless grassland to closure of the tree stand canopy. The grass cover is always closed however and the predominant species of grasses belong to the C4 plants, which are regarded as more efficient than C3 plants under the conditions of strong solar radiation, high temperatures and high evaporation rates characteristic of the savanna. Figure 25.15 shows a classification of the savannas according to the physignomic and floristic criteria and the environmental conditions (soils and precipitation).

The dry season, lasting between three and seven months each year, is the limiting factor for plant growth. Generally trees shed their leaves as a response to drought and the grasses and herbaceous plants react by allowing their above ground shoots to die off. During the rainy season infiltrating water increases the soil water to a great depth and field capacity is exceeded. The result is that a great deal of water is available to the savanna vegetation even after the rainy season ends. Despite this moisture seasonality, the individual development phases of the plants are by no means synchronous; for example, many trees blossom in the late dry season, shortly before or simultaneously with the development of foliage, which also usually starts before the new rainy season. The impulse for triggering these events is probably not moisture but correlations with radiation and temperature conditions. In the case of plants that bloom in the rainy season, the reproductive phases may occur at the beginning, in the middle or at the end of the wet period. The date of shedding of leaves depends on the soil-water reserves available; because the deeper-rooted, woody plants have access to this water for longer, the leaf fall can occur several weeks after the herbaceous plants have dried out.

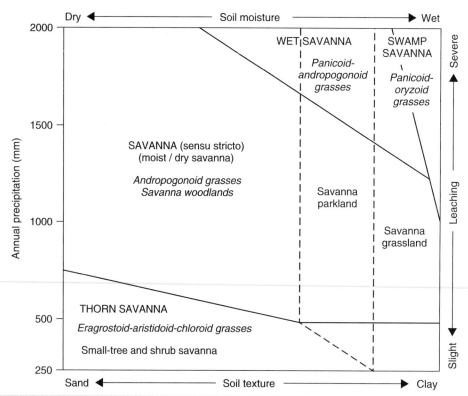

Figure 25.15 Classification of savannas. (Source: after Johnson and Tothill, 1985)

In the moist savanna evergreen trees and shrubs gradually increase in number until forests are formed. The leaves of these evergreens show a scleromorphic structure, where the leaves are hard, stiff and leathery, due to the fact that they contain large amounts of supporting tissue. As a result, even when they suffer great water losses, they do not wilt. These evergreen forests frequently occur along rivers as gallery forests.

The shade from the trees influences the grass layer, for example in the wetter areas this shade impedes grass growth. Hence mostly shade-tolerant, low-production herbs occur in the place of grasses and they are usually dominated by only a few species. In drier areas the shade from trees can favour highly productive grass growth.

Many of the trees are xerophytes and are adapted to survive the dry period. They are also deep-rooted and have flattened crowns. Some shed their leaves during the dry season to reduce transpiration. Some are stunted and can be overtopped by the tall grasses. Browsing by some of the savanna animals is a major constraint on tree growth and survival, and some plants have developed thorns as a protection against this.

The above ground grass mass is characterized by wide seasonal variations, and considerable fluctuations occur from year to year as the grass mass production is much higher during heavy rainfall periods. However, in general there is a linear correlation, although it depends on the composition of species, the soil quality and the rainy-season length.

As the nutrient supply in the soils is poor, it is important to maintain and make optimal use of what is available. The grasses do this through their dense root systems, which are closely linked with microorganisms, and by forming high root : shoot ratios. The rapid decomposition of dead organic matter which is characteristic of all savanna regions is beneficial in view of the low nutrient reserves in the soil. The material is broken down by fire and termites because most herbivores cannot use dead grasses/leaves. They contribute to the turnover of matter and energy essentially by consuming living plants, and in the case of large mammals by trampling, which causes mechanical size reduction.

25.4.5 Animals

This ecozone is characterized by a rich insect and spider fauna, but compared to the tropical rain forest the fauna is impoverished, though large numbers of the individuals of these species are found. Grasshoppers and locusts are the most important primary consumers, whilst in the rainy season caterpillars are prevalent. Amongst the secondary consumers, spiders and ants are dominants. An example of the density of the herb-layer arthropods in different stages of an arrested savanna-to-woodland succession in Upper Shaba, Zaire is illustrated in Table 25.3.

The macrofauna on the other hand has developed differently in individual continents and there are major differences in the species, the animal densities, the life forms represented and their functions in the savanna ecosystem. Many of the African savannas are extremely game-rich, including herbivorous antelopes,

Table 25.3 Types and densities of herb-layer arthropods in different stages of an arrested savanna-to-woodland succession in Upper Shaba, Zaire

		SAVANNA	MIOMBO WOODLAND	
			BURNED	UNBURNED
HERBIVORES				
Chewers:	Short-horn grasshoppers*	342	124	118
	Long-horn grasshoppers*	80	29	29
	Crickets	66	190	164
	Caterpillars (larvae of moths and butterflies)	52	34	50
Suckers	Hompoteran bugs	69	71	58
CARNIVORES				
	Spiders	176	257	308
	Mantids	57	29	37
TAXA WITH MIXED FEEDING HABITS				
	Adult beetles	283	74	87
	Cockroaches	109	123	203
	Heteropteran bugs	60	86	208
Totals		1294	1017	1262

*The short- and long-horn grasshoppers were the predominant groups within the Acridoidea and Tettigonoidea, respectively, which were each counted as a whole.

gazelles, elephants, giraffes, hippopotami, rhinoceros, warthogs and ostriches, as well as carnivorous lions, leopards and hyenas. The animals show ecological segregation and occupy specific niches (Figure 25.13).

This can be seen when closely-related ungulate species are examined. Segregation through difference in size is common, for example extremely small dikdiks can penetrate into thickets and select smaller browse items than giraffes. Similarly, small muzzled gazelle can select specific plant parts from a grass sward, unlike broad-muzzled wildebeest. Other differences include habitat and food preference, but these do not eliminate all overlap, particularly between medium-size grazers. For example, topi, wildebeest and zebra are commonly found grazing in the same region. They feed almost exclusively on grasses but choose differently between leaf, sheath and stem, particularly during the dry season. One adaptation is to migrate for better forage and a good example of this is the wildebeest migration in the Serengeti-Mara region of Tanzania and Kenya.

Indian savannas were originally similar to those of Africa but always contained fewer species, yet in South America there are few large herbivorous mammals. In Australia various marsupials are the primary consumers and many occupy similar niches to those filled by certain species of mammal in the other continents. Many savanna-dwelling animals live in large herds, and as well as grasses, trees and shrubs form important food sources, especially during the dry season. Many of the ungulates migrate with the rain. Studies have confirmed the view that the savanna animal grazing does not damage the primary production and actually stimulates increased production.

Termites are a characteristic component of the savanna ecozone and many species build above ground structures, up to several metres high, for their nests, which are often major features of the savanna landscape. They transport fine grain sizes up to fine sand from depths of 1.5 m and the amount shifted can be considerable; for example, calculations in western Nigeria and Zaire have shown that if all the mounds were flattened, the soil surface would be covered by a layer of sediment 20–30 cm thick. Most species live exclusively on dead organic matter and their density increases with the percentage of this matter in the same soils. They play an extremely important role in breaking down this matter as in the most productive savannas they are present at densities of over 100 million per hectare. They make the plant matter more readily available for decomposition by other organisms and they also eat growing plants, especially during the dry season.

25.4.6 Savanna fires

Fires in the dry season are characteristic and often spread from grassland to the tree and shrub layer. Over several years probably all areas are burnt. Fires can be caused by lightning strikes, but much more commonly by man. The custom of setting fires goes back tens of thousands of years and is used to maintain and improve grassland, to clear land for cultivation and provide ash for fertilizer and to keep away wild animals and destroy vermin like snakes.

These fires have a far-reaching effect on the savanna. For example, they are important selection factors for the flora and fauna, they determine the vegetation and influence the heat and water budgets of the soil, the plant cover and the layer of air near to the ground, and they modify the energy and matter turnover in the system. Various degradation or secondary succession stages occur, depending on the burning frequency, the time when burning takes place and the intervals between fires. These different seral stages are joined together in a vegetation mosaic which changes with time, but the overall mosaic can be regarded as a fire-climax formation. As a result of the many fires, the dominant tree species – the pyrophytes – have some fire resistance. Once burning has stopped there is a grass growth impulse, which happens for many reasons, including more favourable net photosynthesis than before and extra minerals in the soil.

25.4.7 Savanna ecosystems

The savanna ecosystems differ markedly from each other as regards reserves and turnover rates, depending on the humidity (moist or dry savannas), tree density (dry forests, semi-evergreen humid forests, tree savannas, grass savannas), the significance of animal browsers and the frequency and intensity of fire. Therefore there is no simple ecosystem model.

However, we will discuss here one moist savanna example taken from the Lamto savanna in the Ivory Coast. The Lamto savanna consists of a grass layer 1.5–2.5 m high, with at some locations open woodland and at others dense tree stands. Various savanna types can be recognized, depending on the species present and the soil types. The most important producers are the grasses and because of the favourable humidity conditions (on average 1250 mm of annual rainfall) the primary production is very high at 27 tonnes per hectare per years. The grass burns off each year and as a result 6–8 tonnes of dry matter is destroyed per hectare.

The remainder is consumed by herbivores, the most important of which are grasshoppers, termites, caterpillars and ants.

Only a very small percentage of the food taken up by the primary consumers is assimilated (~13%) and hence the secondary production by the primary consumers is remarkably low at only about 1% of the ingested food energy, although their respiration losses are relatively small. Most of the food intake is defecated again virtually unchanged and together with the smaller amount of primary production, which neither burns nor is consumed by animals, is available to the decomposers to be further broke down. This process accounts for over 60% of the primary production.

In the case of the secondary consumers (carnivores) there are two trophic levels. The fact that the energy taken up by the carnivores is greater than the energy stored in the herbivores is probably because many of the secondary consumers feed partially on the faeces of the herbivores and also consume detritivores. The low respiration losses of the carnivores as well as their high digestion efficiency explain the high degree of gross production

Table 25.4 Effects of the the prevention of fires in savannas in Olokemeji Forest Reserve, south-western Nigeria. Plots were cleared in 1929 and subsequently subjected to different fire treatments. The results are from a survey in 1957 and show that complete protection causes reversion to climatic climax forest, while early burns (less intense, at the beginning of the dry season) allow development of a wooded savanna with fire-resistant clumps of trees, while late burns (more intense, at the end of the dry season) allow only a few fire-resistant trees to grow. (Source: after Hopkins, 1974)

	NO BURN	EARLY BURN	LATE BURN
Vegetation	Trees + shrubs, few grasses, forest succession	Savanna with clumps of closed canopy woodland	Grass savanna with scattered trees
Number of trees in 0.17 ha	433	163	98
Number of fire-sensitive trees in 0.17 ha	279	11	0

(a)

(b)

Figure 25.16 Savanna ecosystem types in northern Kenya, east of Lake Turkana (a) Dry, wind-eroded surface and thorn scrub vegetation during the dry season. A jackal is seen in the foreground. (b) Savanna grassland with isolated thorny shrubs. A warthog group is seen in the middle distance. (c) Mixture of grassland and trees, with thorny shrubs. Thompson gazelle are seen in the middle distance. The grasslands are grazed by large herds of zebra and both small and large gazelle species

(c)

Figure 25.16 (*continued*)

efficiency at both levels, and about a quarter of the food intake goes into the production of the secondary consumers. Important secondary consumers include some species of birds, lizards, mantises, spiders and ants.

In fact, most savannas are not climatic climaxes and they would often develop a different vegetation cover if fire and grazing were excluded. Possible effects of the prevention of fires (and by analogy, of the removal of herbivores) are shown in Table 25.4. Illustrations of the variability of ecosystems east of Lake Turkana in northern Kenya are shown in Figure 25.16.

The summary model for the savanna ecozone is illustrated in Figure 25.17.

25.5 Potential effects of global warming on the world's ecozones

It is clear that at regional to global scales there is going to be a series of marked impacts of global warming on the world's ecozones. The Intergovernmental Panel on Climatic Change has made extensive use of a MAPSS model output in regional and global assessments of climatic change on vegetation.

With future global warming, forest boundaries that are limited by cold temperature will shift toward the poles and upward in height. However, boundaries that are limited by water, for example transitions between closed forest and open savanna, could shift toward either drier or wetter conditions, depending on several factors. Increased temperatures will tend to cause drought-induced declines; regional increases in precipitation could mitigate the negative effects of increased temperatures; and elevated carbon dioxide concentration could increase the water-use efficiency of trees and also mitigate the negative impacts of elevated temperatures. In early stages of global warming, benefits from this elevated carbon dioxide concentration could dominate the negative effects of increased temperatures, resulting in a broad scale increase in productivity and density of most forests world-wide. However, these benefits could possibly be saturated within a few decades while temperatures continue to rise.

Eventually, maybe towards the end of this century, the negative effects from elevated temperatures would become dominant and forests could begin a major drought-induced decline. In North America, for example, temperate coniferous forests will expand in importance in Alaska and south-west Canada. Tundra and taiga regions of Alaska and Canada will be much reduced in area, with expansions of both boreal and temperate coniferous forests into those areas. Forests in Washington and Oregon could initially expand in area and density over the next few decades but these gains could be more than offset by later losses in both forest area and density. Drier forest types could expand further north in the Coast Ranges and fires could become more prevalent west of the Cascades in the mid to late century. If forests in the north-west initially responded with increased productivity and density, they would sequester more carbon, acting as a negative feedback to further global warming. However, increased forest densities would result in more tree water use and there would be less water for irrigation and domestic uses. Under continued warming, if forests began to decline, they would use less water, which in combination with less snow formation and earlier melt could result in increased winter flows and possible flooding.

Climatic change, both natural and man-induced, will undoubtedly create change in the world's ecozones, but because there are complex, interrelated variables controlling these likely changes, the impact is difficult to predict with any great accuracy at this moment in time. For example, the permafrost zone may start to melt three times faster than we originally thought because of the speed at which the Arctic sea ice is melting. Studies have shown that the effects of sea ice loss, which reached an all time record in the summer of 2007, extend far inland to areas where the ground is normally frozen all year round (as much as 1800 km). The smaller the area of sea ice, the less sunlight is reflected and the more heat is absorbed. The result will be rapid Arctic land warming and permafrost thaw. In September 2007 the Arctic sea ice shrank to 30% of its average extent for that time of year. Also, air temperatures over the Arctic rose by about 2 °C above the long-term average for the period 1978–2006.

Climate is a complex change agent. If global warming continues as projected, its effects will depend not only on the direct impacts on land and water resources, but also on how technology, economies and societies change over time (see further discussion in Chapter 31). In the Arctic there are unresolved questions related to how much accelerated coastal erosion there will be, how much methane emission will occur and whether faster shrub encroachment will happen in the tundra regions if sea ice continues to retreat rapidly as predicted. Undoubtedly there will be widespread changes across the Arctic, not just in terms of climate, but on the local communities through damage to roads, buildings and other types of infrastructure as the frozen ground thaws and destabilizes. See the companion website for Box 25.1 Impact of global warming on the Mackenzie Basin, north-western Arctic Canada.

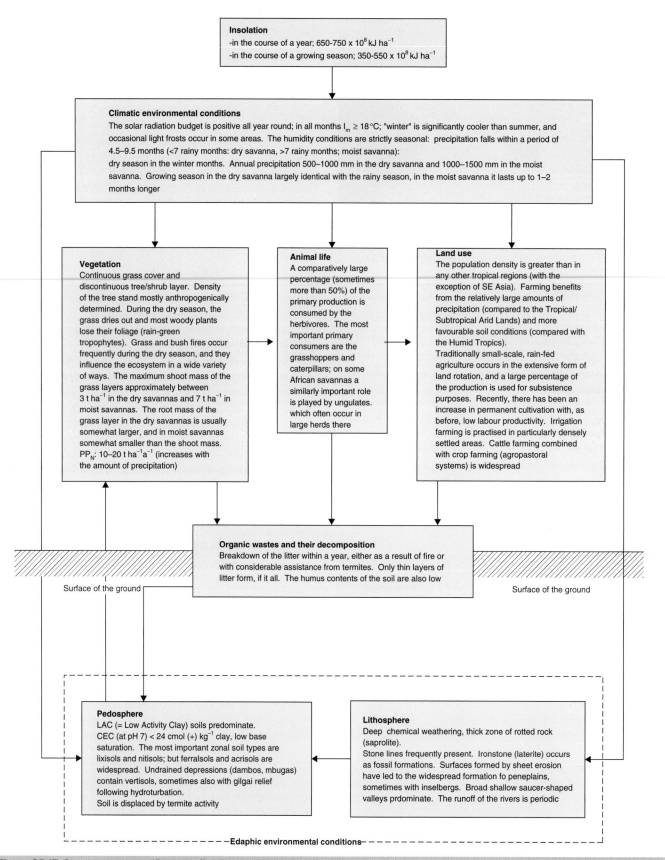

Figure 25.17 Savanna ecozone. (Source: after Schultz, 1995)

Exercises

1. Define (a) a biome and (b) the ecozone concept. What are the differences between the two terms?

2. What are the main biological characteristics of the plants that live in the tundra? Describe the categories of plant communities in this ecozone.

3. Explain why the lemming population increases and decreases on an apparently regular basis in the Arctic.

4. What are the common biological characteristics of all tropical rain forests?

5. Why is the savanna considered to be a plagioclimax?

6. What are the biological characteristics of vegetation in savanna regions?

7. Why is the tropical rain forest so vulnerable?

8. Describe the complex vertical structure of a typical tropical rain forest.

9. What are the climate characteristics of the humid tropics?

10. What are the disadvantages associated with the decomposition of organic matter in the tundra ecozone?

References

Batzli, G.O. 1981. Populations and energetics of small mammals in the tundra ecosystem. In: Bliss, L.C., Heal, O.W. and Moore, J.J. (Eds). Tundra Ecosystems: A Comparative Analysis. International Biological Programme 25. Cambridge, Cambridge University Press. pp. 153–160.

Briggs, D. and Smithson, P. 1985. Fundamentals of Physical Geography. London, Hutchinson.

Brown, R.J.E. 1968. Occurrence of permafrost in Canadian peatlands. *Proceedings of the 3rd International Peat Congress, Quebec.* 174–181.

Collinson, A.S. 1977. Introduction to World Vegetation. George Allen and Unwin.

Hopkins, B. 1974. Forest and Savanna. London, Heineman.

Johnson, R.W. and Tothill, J.C. 1985. Definition and broad geographic outline of savanna lands. In: Tothill, J.C. and Mott, J.J. (Eds). Ecology and Management of the World's Savannas. Canberra, Australia Academy of Science. pp. 1–13.

Schultz, A.M. 1969. A study of an ecosystem: the arctic tundra. In: Van Dyne, G.M. (Ed.). The Ecosystem Concept in Natural Resource Management. New York, Academic Press. pp. 77–93.

Schultz, J. 1995. The Ecozones of the World. Berlin, Springer-Verlag.

Further reading

Golley, F.B. 1983. Tropical Rain Forest Ecosystems. Ecosystems of the World 14A. Amsterdam, Elsevier.

Harris, D.R. (Ed.). 1980. Human Ecology in Savanna Environments. London, Academic Press.

Jacobs, M. 1988. The Tropical Rainforests: A First Encounter. Berlin, Springer-Verlag.

Mabberley, D.J. 1983. Tropical Rain Forest Ecology. London, Blackwells.

Walter, H. 1971. Ecology of Tropical and Subtropical Vegetation. Edinburgh, Oliver and Boyd.

SECTION VI
Global Environmental Change: Past, Present and Future

26 The Earth as a Planet: Geological Evolution and Change

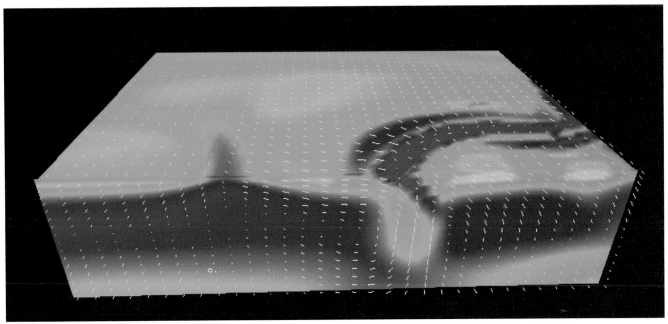

Plate generation and subduction. Source: Tackley, P. in
http://ct.gsfc.nasa.gov/olson.finalreport/final_report.html

Earth Environments   David Huddart and Tim Stott
© 2010 John Wiley & Sons, Ltd

Learning Outcomes

After reading this chapter and completing the exercises you will be able to:

➤ Understand how unique the Earth is as a planet.

➤ Understand what we know about the early Earth and how we know it.

➤ Understand how the Earth's core formed and evolved.

➤ Describe the evidence for a molten magma ocean.

➤ Understand how the Earth's mantle and continental crust formed and evolved.

26.1 Introduction

The first five sections of this book established the framework for a holistic study of the Earth, with an introduction to Earth systems, and described the processes, patterns and Earth materials in the atmospheric and ocean systems, the endogenic and exogenic geological systems and the ecological and soil systems. This was mainly a study of contemporary or Quaternary Earth systems, although in places reference was made to geological case studies.

It is inevitable that an important set of antecedent processes must be considered when looking at what is going on in the Earth today. In whatever Earth system is under consideration, a set of past processes will have an influence on today's environments and on the future development of the Earth. Although we dealt with each Earth system in a separate section of the book, the close interrelationships between these systems was stressed in the introductory section, is obvious in many chapters and will be reinforced in this final section.

In this concluding section we will initially consider the Earth as a planet and its geological evolution and change. It is not possible to cover the whole of the Earth's geological history in detail in a book that aims to be wide ranging, and so we will only discuss an outline early evolution in this chapter, before considering changes in the evolution of the atmosphere and ocean systems in Chapters 27 and 28. This will provide the basis for a consideration of the Earth's biosphere evolution in Chapter 29 and for a series of case studies of geological change, particularly icehouse and greenhouse phases from the Proterozoic through to the Quaternary and Holocene, in Chapter 30. This will lead logically on to a consideration of future climatic change and the future of the Earth's environmental systems in the final chapter, Chapter 31.

Before we look at early Earth history we need to ask a fundamental question with regard to the planet Earth.

26.2 How unique is the Earth as a planet?

Planetary exploration of our solar system and beyond has not yet discovered any other life forms, although the scientific search to answer the fundamental question for mankind – are we unique life forms? – goes on. The answer seems likely to be yes, simply because it appears from what is known about the planets in our solar system, and those discovered around other stars, that the Earth is unique as a planet. Conditions on Earth make it suitable for higher life forms, although it is likely that some of the lower life forms would be able to survive on other planets and indeed may even have an extra-terrestrial source (see Chapter 29). Nonetheless, the Earth seems just perfect for life. It is difficult to understand why this is so but it is likely to just be chance. Condie (2005, pp. 8–9) offers the following analysis:

- The Earth's near circular orbit results in a more or less constant heat amount from our Sun. If the orbit were more elliptical, the Earth would freeze over in the winter and heat up to high temperatures in the summer. Higher life forms could not survive if that were the case.

- If the Earth were much larger, the force of gravity would be too strong for higher life forms to exist.

- If the Earth were much smaller, water and oxygen would escape from the atmosphere and higher life forms could not survive.

- If the Earth were only 5% closer to the Sun, the oceans would evaporate and greenhouse gases would cause a rise in surface temperature that would be too high for any life forms.

- If the Earth were only 5% further away from the Sun, the oceans would freeze over and photosynthesis would be reduced greatly, which would lead to a decrease in atmospheric oxygen. Higher relief forms would not be able to survive.

- If plate tectonics did not exist on the Earth, there would be no continents and so terrestrial higher life forms would not be able to exist. The Earth appears to be the only planet we have discovered so far that has plate tectonics.

- If the Earth did not have a magnetic field of just the right strength then lethal cosmic rays would kill most life forms on the planet.

- If the Earth did not have an atmospheric ozone layer to filter out the Sun's harmful UV radiation, higher life forms could not exist.

- If the Earth's axial tilt were either greater or smaller, surface temperature differences would be too extreme for life to exist. Without the Moon, the Earth's spin axis would wobble too much to support life.

- Without Jupiter's huge gravity field, the Earth would be bombarded by meteorites and comets and higher life forms would not survive.

- The diversification of mammals and hence humans only took place because of the asteroid collision 65 million years ago. Evolution on the Earth may have been subject to many chance events.

If there were changes in any of these characteristics then we would not exist: it is as simple as that. It looks like chance that all these characteristics came together on one planet as they did.

26.3 What do we really know about the early Earth?

It is safe to say that the Earth is not the planet that it once was. Since it was first formed it has changed quite markedly, and it continues to change today. The first 2 billion years of its 4.6 billion-year history (see Figure 26.1) are relatively poorly understood compared with later geological times. It is difficult to understand this period because the rock record is incomplete, particularly for the first 750 million years of the planet's history. Nothing much of the rock record is preserved before 3.8 billion years ago as the planet is a dynamic system that has destroyed any evidence of its earliest phases. Early in its history the Earth would have had major impacts from asteroids and comets, like the Moon and our planetary neighbours, but none of this is preserved on Earth.

It is likely too that the early geological history has a preservation bias. Rollinson (2007, p. 27) asks the following question of the

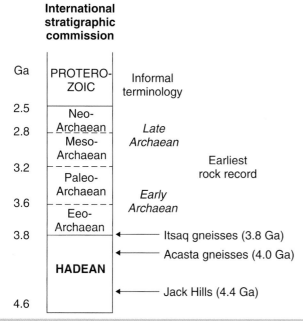

Figure 26.1 A geological time scale for the early part of the Earth's history

Archaean rock record: 'Is what we see now, representative of what was there then?' The answer has got to be 'no' because given the Earth's erosional processes, those rocks that formed at the Earth's surface in the Archaean are more likely to have been destroyed than any deep crustal, plutonic rocks. The relative rarity of Archaean andesites demonstrates this, as typically this rock type today forms subaerial strato-volcanoes that are actively eroded and therefore likely not to be preserved. However, it is normal for us to believe in the principle of uniformitarianism, where the physical and chemical processes which operate today operated in the past and are used to explain the events of the rock record.

There are some caveats to this principle when dealing with the early Earth. The earliest phase of Earth history was very different because it was subject to planetary processes such as impacting, accretion and the related melting, which are not part of uniformitarian principles. There are rock types which are unique to the Archaean, implying some differences from today. These rock types include ultramafic komatiites, banded ironstone formations in greenstone belts and calcic anorthosites, typically found in the deeper crust.

Rollinson (2007) argues too that many of the Earth's main reservoirs change gradually over time and that these changes are not cyclical but unidirectional, on a time scale of billions of years. This is known as secular change and is seen for example in the chemical evolution of the atmosphere, the oceans and probably in the mantle too. Nevertheless, it is clear that many Archaean rocks formed under conditions similar to those found in modern sedimentary environments.

26.4 The early geological record

There are few terrestrial geological samples older than 3.8 billion years and so knowledge of the Hadean is dependent on a small amount of physical evidence, added to by inferences from meteorites and planetary materials and deductions based upon the distribution of some radiogenic isotopes in the Earth's mantle. In contrast, Archaean rocks are found in most continents in areas known as cratons, or 'shield areas' (Figure 26.2), the term implying an area of continental crust usually composed of crystalline rocks that have been stable for a very long time and have a thick mantle lithosphere and low heat flow. The conventional wisdom is that they are the stable core of continents: that since their formation they have been inert, with later-formed continental crust accreting upon them. However, as we can see from Box 26.1 (see the companion website), there is evidence that these cratons have experienced episodic rejuvenation events throughout their history.

These Archaean rocks can be divided into three types of lithological association: greenstone belts, Late Archaean sedimentary basins and granite-gneiss terranes. The first two types have relatively low-metamorphic-grade rocks and represent Archaean upper crust, whilst the latter belong generally to the deeper, continental crust. Commonly all these rocks have long deformation histories and are metamorphosed and altered, and working

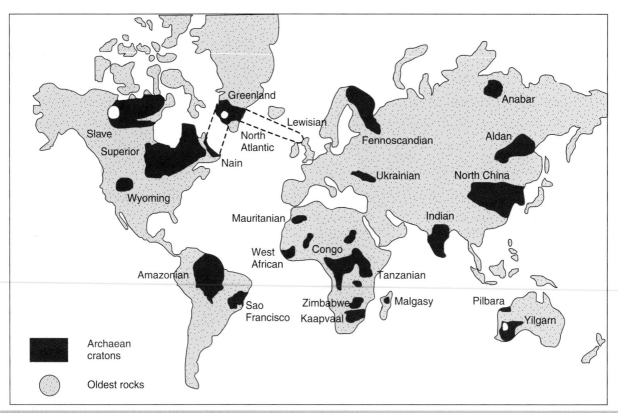

Figure 26.2 The distribution of Archaean rocks on Earth, showing the main cratons and the areas with the oldest known rocks. (Source: after Rollinson, 2007)

out their original character requires patient fieldwork and much laboratory geochemistry.

26.4.1 Archaean greenstone belts

These are sequences of volcanic and sedimentary rocks occurring in sublinear belts up to hundreds of kilometres long and with varying deformations. Their name is derived from the colour of the lightly metamorphosed basalts. Within individual cratons the ages can vary by as much as 1 billion years, but normally the time span is much shorter (a few hundred million years), and they may incorporate large time gaps in the sequences. When these age differences are associated with differing lithological associations it suggests that individual greenstone belts may be composed of accreted fragments that formed in very different tectonic settings.

Eriksson *et al.* (1994) subdivided greenstone belts into seven dominant lithological associations (Table 26.1), which they used to define the different tectonic environments, which between 3.8 and 2.5 billion years ago were similar to those of modern Earth. Basalts dominate the volcanic rock types, usually as pillow lavas metamorphosed to chloritic schists. Precise zircon geochronological dating, igneous geochemistry and structural mapping in greenstone belts have demonstrated that 'modern' tectonic processes, such as arc accretion, were taking place 2.7 billion years ago.

26.4.2 Late Archaean sedimentary basins

These sedimentary basins are very different from greenstone belts: they are not strongly deformed, they do not have an elongate shape and they do not feature abundant basalts. The unconformable contacts between these basins and greenstone belts and gneisses of the underlying craton show that they formed after the craton became stable, or during the later stages of craton formation. A good example is the Witwatersrand Basin in South Africa (3.1–2.7 billion years old), composed of a 7 km-thick sequence of terrestrial sediments and volcanic rocks. These basins can provide information related to the processes of epicontinental basin development, and the rocks can give us information about the environment:

- The presence of detrital grains of pyrite and uraninite, which are normally oxidized in the modern weathering regime, suggests that the atmosphere was low in oxygen (Frimmel, 2005).

- Torroidal-shaped gold grains and faceted pebbles indicate that aeolian deflation was affecting the surrounding land as part of the gold concentration process.

- There are glacial-origin diamicts and palaesols in the Pongola Supergroup rocks of Swaziland.

Table 26.1 Key lithological associations recognized in Archaean greenstone belts

Tectonic setting	Principal lithologies	Depositional system	Basin configuration	Tectonic elements
Volcanic basin	Mafic/ultramafic volcanic rocks	Individual volcanoes; shallow water, chemical-biogenic, evaporitic;	Not preserved	May be thrust bound at site of emplacement
Arc – Forearc	Conglomerate-sandstone-mudrock/turbidite/ironstone/chert/ tuff	Fan delta and submarine fan or ramp	Elongate between volcano plutonic and metasedimen-tary/ophiolitic elements	Bounded by thrust complex on one side; may be deformed into a lithotectonic complex
Arc – Interarc	Calc–alkaline volcanic/plutonic/mafic volcanic; pyroclastic	Individual volcanoes; subaerial and subaqueous fans; chemical-biogenic, evaporitic, and pelagic	Outer margins pass to forearc and backarc elements	Local structures around tectonic elements
Cratonic extensional	Sedimentary and bimodal mafic/ felsic volcanics	Fluvial and fan delta or submarine fan; shallow to deep marine shelf; subaerial and subaqueous volcanic centers	Elongate between cratonic margins or active and remnant arcs	Fixed position listric or back-stepping normal faults; intrabasinal transfer faults
Stable Shelf	Chemical-plagic +/– siliclastic	Chemical and siliciclastic platform or fluviodeltaic; tidal and delta channels	Regionally or interregionally tabular	Not seen; thermal relaxation and eustatic change
Compressional foreland	Conglomerate-sandstone-mudrock/turbidite/ironstone/ volcanics	Alluvial fan, fluviodeltaic; shallow marine; submarine fan or submarine ramp	Elongate, wedge shaped against, between or on thrust sheets	Extra- and intra basinal thrust faults; may be divided into separate compartments
Collisional Graben	Breccia-conglomerate-sandstone-mudrock/mafic–felsic volcanic	Alluvial fan, fluvial, fan delta; shallow to deep marine; lacustrine; in volcanic basins subaerial to subaqueous fan	Elongate, wedge shaped between bounding faults; enclosed or open to marine basin	Longitudinal, strike-slip or oblique slip faults and various intra basinal faults

- The gold is often associated with bituminous carbon, and whilst this was originally thought to be of algal mat origin, it is now believed to have originated as immature algal kerogen that was subsequently mobilized by hydrothermal fluids to its present site.

26.4.3 Archaean granite-gneiss terranes

Granitic gneiss is the most abundant rock type in the cratons, but it occurs in many forms, for example as undeformed plutons and highly deformed banded gneisses, some at granulite grade. Some gneisses may form the basement to greenstone belts, but others can intrude these rocks. They are also variable in composition, but many belong to the magma suite tonalite-trondhjemite-granodiorite (TTG). Examples include the Lewisian Gneiss in the North Atlantic craton and the Zimbabwe craton in southern Africa.

Granite-gneiss terranes of the upper and middle crust are frequently associated with greenstone belts but the granulite

gneiss belts of the lower crust have a slightly different rock assemblage and contain lenses of basalt, a variety of sedimentary rock types and layered ultramafic/mafic anorthositic intrusions. However, they could be fragments of greenstone belts, folded deep into the crust. These rocks provide us with information about how continental crust formed and its development through time.

26.4.4 Where are the oldest rocks on Earth?

Rocks any older than 3.8 billion years are very rare on Earth and have only been found at a few localities. The Gothåbsfjord area of west Greenland is the most extensive ($2000 \, km^2$) and embraces the Itsaq gneisses, with a number of geological units between 3.6 and 3.9 billion years. One of the most important areas within this zone is the Isua Greenstone belt, which is composed of a sequence of metabasalts, ultramafic rocks, clastic and chemical sediments with general ages between 3.7 and 3.8 billion years, although there are mineral inclusions in pyrite grains from within the iron formation which give ages of 3.86 billion years, and one population gave ages of 4.31 billion years (Smith *et al.*, 2005). This greenstone belt appears to be a series of stacked thrust sheets of different provenances and ages but it does contain a record of the Earth's oldest mafic magmatism, possible evidence of early life and chemical and clastic sediments which contain information about the early atmosphere and oceans.

The Acasta Gneisses of north-west Canada are on the western margin of the Slave craton and contain zircons with crystallization ages of 3.96 billion years and even older inherited cores with ages up to 4.2 billion years. They comprise a heterogeneous assemblage of strongly foliated and banded tonalitic and granodioritic gneisses, which include lenses of amphibolite and metasedimentary and ultramafic inclusions. They preserve a complex history of intrusion and deformation.

However, the oldest known terrestrial materials are the 4.1–4.404 billion-year-old detrital zircons from the Jack Hills (Mount Narryer area) of Western Australia, which come from a sequence of highly metamorphosed quartzites, conglomerates and pelites. Zircon xenocrysts from felsic gneisses with ages of up to 4.18 billion years have been found in the same region. Together these indicate that the rocks of this part of the Pilbara craton, generally with ages between 3.3 and 3.7 billion years, were derived from substantially older rocks which now appear to have been destroyed. Some of the zircon crystals show a number of discrete age events within the same grain, indicating that they have had complex thermal histories. These histories likely indicate the reworking and metamorphism of already-formed crust, which also means that the crust was sufficiently thick for metamorphism to take place and so represents a continental, rather than an oceanic environment (Peck *et al.*, 2001).

26.5 The first Earth system

We know that the Earth experienced many extreme changes in the first 100 million years of its life that were very different to any in its recent geological history. We can pose some fundamental scientific questions: When did the Earth form? How different is it to other planets? What was the composition of the earliest atmosphere and oceans? Did they come from inside the Earth or were they acquired from somewhere else within the solar system? When was the first crust formed, what was it like and has any of it been preserved? Was the mantle hotter early in the Earth's history? How did subduction originate?

This last question is discussed in Box 26.2 (see the companion website), but of course this is just an idea that incorporates some of the characteristics of early Earth into a hypothesis that needs further research.

It seems likely that the Earth formed according to the standard model of planetary accretion, discussed in Chapter 11 and in detail in Rollinson (2007, Chapter 2). We know the following of the earliest Earth system:

- It had a chondritic bulk composition, because the primitive material from the solar nebula is thought to be identical to chondritic meteorites.

- It experienced at least one giant impact. The Moon formed through one of these.

- This melting led to the formation of a terrestrial magma ocean during the late accretion stage.

- The extensive melting may also have facilitated core formation. It is probable that during its magma-ocean stage the Earth differentiated into an outer silicate mantle and an inner metallic core.

Several parts of the early Earth system seem linked to either one or many giant impacting events: the formation of the Moon, the existence of a magma ocean and the separation of the core. The model for the Moon's origin suggests that a planetisimal the mass of Mars (~15% the mass of the Earth) collided with the proto-Earth at a time after core formation. This generated a huge amount of thermal energy and both the impactor and part of the Earth were vaporized; some of this material coalesced around the Earth to form the Moon, and it is widely believed that both the Earth and the Moon went through magma-ocean stages. Another result of this impact was that 30% of the Earth's mass was raised to temperatures greater than 7000 K, so that the Earth was surrounded by a silicate vapour atmosphere and much of it was in a molten state. It is possible that if the impactor had a metallic core of its own, this was added to the Earth's core immediately following the impact. See the companion website for Box 26.2 A hypothesis for the origin of subduction on early earth.

26.6 How did the Earth's core form?

We have seen that there is geophysical evidence for a liquid, Fe-Ni-S-alloy outer core and a solid, Fe-Ni-alloy inner core. Temperatures are thought to be around 3500–4000 K at the top of the core, rising to 5000–6000 K at the Earth's centre. At present the inner core continues to grow at the outer core's expense. This generates heat in the modern Earth.

The Earth's magnetic field is thought to be driven by a geomagnetic dynamo, governed by thermal convection in the outer core. The origin of this convection is not well understood but it may be radioactive heating, the cooling of the core or the latent heat caused by crystallization at the inner/outer-core boundary. It has been found that there would be a core density deficit if the outer core were made up of a Fe-Ni alloy, so there must be an extra dilutant which can form an alloy with Fe at core pressures and this must match the known seismic properties and density of the outer core. The main suggestions have been O, Si and S, with the latter the most likely.

When and how did the core form? It is thought to have been during the early stages of accretion, maybe as early as 10 million years after the solar system's formation. The main constraint on its formation must have been that the Earth needed to be sufficiently molten for metal droplets to separate and sink through a silicate melt. The most important heat source during the early stages of planetary accretion was the decay of short-lived isotopes such as ^{26}Al, but later impact melting became a more important thermal energy source, which led to the creation of magma oceans.

The formation of the Moon is likely to have led to widespread melting about 30 million years ago, which would have resulted in homogenization of discrete metal and silicate phases that were present in the initial accretionary components and in the formation of a magma ocean where the silicate and metal phases eventually separated, leading to the formation of the Earth's core and mantle.

There have been several models to explain core formation:

- It began when the Earth was only 10% formed and took place in a shallow magma ocean where protoplanets had already formed cores. On impacting, these merged without fully equilibrating with the silicate mantle.

- There was an initial major accretion event during which the core formed and siderophile elements (those elements that have a chemical affinity for metallic iron) were removed from the mantle to the core. This was followed by a later stage of accretion during which more oxidizing material was added to the Earth (about 7% of the Earth's mass), which contributed to the moderately siderophile part of the mantle. A still-later part added about 1% of the Earth's mass.

- The current favourite explanation is that silicate–metal equilibration took place at high pressures in a deep magma ocean. Experimental studies suggest that this ocean was 1000 km deep where the silicate–metal separation took place and that liquid metal percolated through a silicate magma ocean as droplets of 'metal rainfall'. These droplets accumulated at depth into a metal layer and descended deeper as diapirs (Figure 26.4).

- A further alternative is that the Earth's core grew gradually through the coalescing of the metal cores of planetisimals without fully equilibriating with the Earth's mantle.

26.6.1 Was there really a molten magma ocean?

There is a lot of circumstantial evidence for a magma ocean from the origin of the core and the formation of the Moon but it has been hard to find real direct geochemical evidence in the Earth's mantle to give evidence that such an ocean ever existed (although see Box 26.1, see the companion website). However, it seems likely that the early Earth was extremely hot and it has been suggested that from a theoretical viewpoint it is difficult to accrete the Earth without a magma ocean. Some likely magma ocean scenarios are illustrated in Table 26.2.

There seem to be two major controls on the formation: the thermal input and the melting temperature of the planetary material. The thermal controls are the presence of a blanketing atmosphere and the rate of delivery and size of the impactors. The blanketing atmosphere could be formed from gases of the solar nebula, outgassed steam from the Earth or as a result of the impacting event. If in fact the surface temperatures are over 2100 K then melting will extend deep into the lower mantle. The melting temperature of the proto-mantle depends on its initial composition. Most experimental models assume a peridotitic composition with liquidus temperatures of about 2500 °C at 25 GPa but there are alternatives which might lower the melting temperature by several hundred degrees.

In this earliest Earth system the interactions were strongest between the different Earth reservoirs, with intense, dynamic interactions between the core, mantle, proto-ocean and atmosphere. There is also likely to have been an early basaltic crust, which has long been lost to recycling into the mantle. In the next section we will look at how the mantle is thought to have evolved. See the companion website for Box 26.3 Geochemical clues to the former magma ocean.

26.7 Evolution of the Earth's mantle

26.7.1 What do we know about the mantle?

The mantle is the most important reservoir in the Earth's system. It is generally thought that the mantle is composed of peridotite and is significantly depleted in silica relative to primitive chondrites. We know from seismic studies that the mantle is layered and is divided into an upper and lower mantle, separated by a transition zone at 400–600 km depth (Figure 26.5). Above this the mantle is dominated by olivine and orthopyroxene, with minor garnet and clinopyroxene. The lower mantle is composed of phases of Mg- and Ca-perovskite and magnesiowustite. The seismic velocity contrasts between the two layers probably reflect phase transformations between the two and are not related to differences in bulk chemical composition.

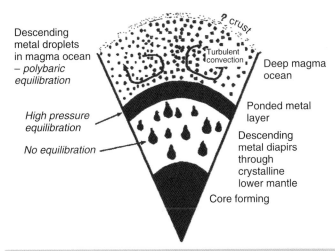

Figure 26.4 A generalized model for silicate–metal separation in a deep, turbulent, convecting terrestrial magma ocean. Descending metal droplets equilibrate with silicate melt and pond at the base of the magma ocean. Diapirs of metal descend to the growing core through a crystalline or partially-molten lower mantle. (Source: after Rubie et al., 2003)

Table 26.2 Magma ocean scenarios

Scenario	Deep or shallow magma ocean	Differentiation
1. The Earth forms in a solar nebula, blanketing effect is provided by solar gases	Deep and shallow	Shallow turbulent magma ocean with no fractionation.
		Differentiation takes place at lower mantle pressures and is affected by high-pressure minerals such as Mg- and Ca-perovskite
2. The Earth forms in "gas-free" space, blanketing effect of a steam atmosphere	Shallow	Upper mantle differentiation.
		Differentiated upper mantle material is buried in the lower mantle during the accretionary growth of the Earth
3. Giant impact	Deep and shallow	Chemical differentiation in the lower mantle is less likely due to rapid cooling, unless there is a thick transient atmosphere.
		The upper mantle is differentiated

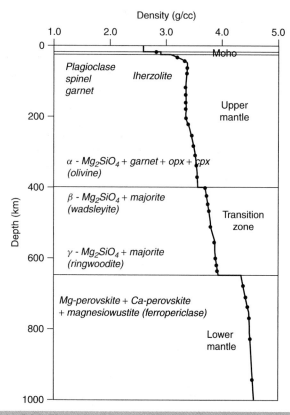

Figure 26.5 The main mineral phases and changes in density with depth for the top 1000 km of the Earth's mantle. (Source: after Rollinson, 2007)

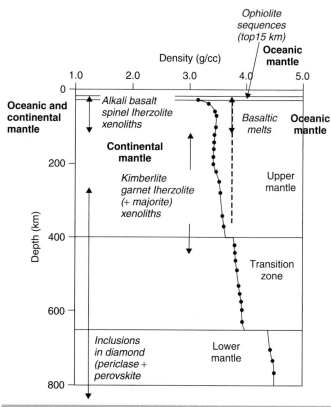

Figure 26.6 The natures of the upper and lower mantles. (Source: after Rollinson, 2007)

The upper part of the mantle (Figure 26.6) is composed of the mineral assemblage olivine (60%), orthopyroxene (30%) and clinopyroxene (5%) as well as an aluminous mineral (5%). At the shallowest depths this aluminous phase is plagioclase, but this changes to spinel at around 30 km (pressure of 0.1 GPa) and then to garnet at around 75 km (0.2 GPa). Between a depth of around 300 and 460 km pyroxene begins to dissolve in garnet to form the higher-density phase majorite garnet and at about 400 km (14 GPa) olivine also undergoes a change to the more dense wadsleyite. At around 500 km olivine undergoes a further phase change to ringwoodite.

At 660 km there is marked increase in seismic velocity and mantle density. Here the ringwoodite dissociates to magnesian-perovskite (Mg, Fe) SiO_3 and magnesiowustite or ferroperi-clase, and the majorite-garnet breaks down to form the phase calcium-perovskite ($CaSiO_3$). Recently a new silicate structure has been found in conditions corresponding to a depth of 2700 km below the Earth's surface. It is important because it has changed our understanding of the boundary between the core and the mantle (see Box 26.3, see the companion website).

26.7.2 Can we see mantle rocks on the Earth's surface?

In places ophiolite complexes which represent obducted ocean floor fragments are now exposed as thrust slices on continental crust, and mantle rocks are frequently present as part of these successions, underlying basaltic pillow lavas, sheeted dykes and layered gabbros. For example, in the Oman ophiolite over 10 000 km^2 of mantle rocks are exposed. However, only the uppermost 10–15 km of the mantle is represented.

The complexes have been divided into two types: the harzburgite-type, thought to represent oceanic crust created at a fast-spreading ridge and a mantle sequence that has experienced a high degree of partial melting, and the lherzolite-type, which represents crust formed at a slower spreading ridge and a mantle section which experienced a lower degree of partial melting. In places there are orogenic lherzolites, thought to represent slices of mantle peridotite emplaced on to or in to the continental crust during continental collision.

Mantle xenoliths on the other hand can bring unmodified materials from much deeper in the mantle to the surface through volcanism. These are represented by a variety of types:

- Kimberlites sample the deep mantle beneath ancient continental crust. These include ecologite and a peridotite group which is made up of garnet lherzolite, garnet harzburgite and dunite. The depth of origin and formation temperature can be calculated using mineral–chemical equilibria between the co-existing xenolith phases.

- Alkali basalt xenoliths are most commonly spinel-bearing peridotites, but there are other xenoliths formed in the upper part of the spinel lherzolite field and these represent disrupted mantle fragments permeated by melts. Other types represent potassic, ultramafic magmas from the Ontong Java oceanic plateau, which come from much deeper in the base of the transition zone.

- Diamond inclusions include materials that have been brought to the surface without re-equilibration. Majorite garnet inclusions which indicate a derivation from 250–440 km have been found but the deepest mantle inclusions include the phases periclase, ferropericlase and Mg- and Ca-perovskite, which come from a range of depths down to 1700 km and indicate the depth of origin for some kimberlites.

Basalt is the melting product of mantle peridotite. Therefore, even though the Earth's mantle is not really accessible, its internal composition can be seen from its melt products, and because these basalt rocks are preserved in continental crust as old as 3.6–3.9 Ga. billion years, they also provide a view of the Earth's mantle across much of geological time. We can see variations in mineralogy with depth, and the chemical composition, in particular the trace element and isotopic characterization, has revealed a number of different basalt types from different chemical domains within the mantle. This shows there is chemical heterogeneity present in the mantle.

As we go into the mantle it is clear that our knowledge decreases with depth and our only detailed information is from the upper mantle. For example, we do have information about the composition of the upper mantle from the ophiolites and the xenoliths. From these we know that the mantle is ultramafic and that the upper mantle is peridotite. Other data come from seismic and density data and peridotite melting experiments.

The rock types present are illustrated in Figure 26.7 and expressed as proportions of olivine, orthopyroxene and clinopyroxene; all are found as xenoliths and in mantle ophiolites. Two fundamental types of mantle rock are those that are classified as relatively enriched in the elements Ca, Al, Ti and Na and those that are not. Lherzolites are enriched peridotites and are not thought to have melted; that is, a basaltic melt has not been taken away. On the other hand, mantle that has experienced melt extraction is known as 'depleted' mantle and has a lower concentration of Ca, Al, Ti and Na. The rock types here are harzburgite and dunite. The relationship between these rocks is illustrated in Figure 26.7.

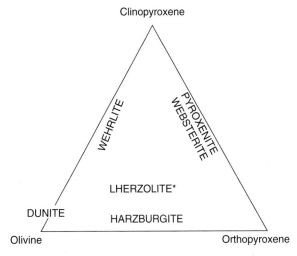

Figure 26.7 Classification of ultramafic igneous rocks, illustrating the main upper-mantle rock types

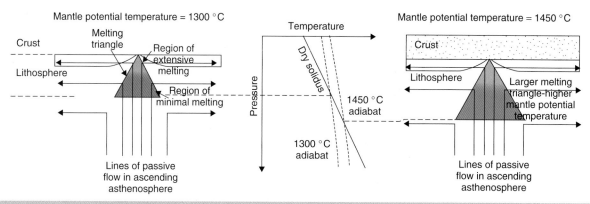

Figure 26.8 Model for mantle melting beneath a mid-ocean ridge. The pressure–temperature diagram in the middle shows the intersection of mantle adiabats for 1300 °C and 1450 °C with the dry mantle solidus. The right-hand-side diagram shows the melting triangles for mantle potential temperatures of 1450 °C, whilst the left-hand side illustrates those for 1300 °C. The higher the mantle potential temperature, the deeper and more extensive the melting, resulting in a thicker ocean crust and lithosphere. (Source: after Asimow and Langmuir, 2003)

The lower mantle is divided from the outer core by the D″ layer, which is a hot thermal boundary layer with an enigmatic composition. It is variable in thickness, contains discrete discontinuities and is thought to contain lateral chemical heterogeneities, which make it significantly different in composition to the overlying mantle. Although various origins have been suggested, such as the final location for subducted slabs, or part of a very old, original accretion, it is agreed that it plays an important role in mantle plume generation.

26.7.3 Mantle melting and mantle convection

Melting in the mantle is important because it provides a mechanism of heat loss for the Earth and supplies the volcanic basaltic melts, especially at mid-ocean-ridge systems (see Section 13.3.2). The model for mantle melting at this location is shown in Figure 26.8. Initially melt will move within the melting zone as small droplets, which will tend to migrate upwards from the melting and collect together to move as larger drops and melt patches. As the melt reaches the surface it will tend to move along channelways in the mantle whose orientation is related to mantle flow and upwelling beneath a spreading ridge. Kelemen *et al.* (2000) suggested a pattern of melt channels like that in Figure 26.9, in which a large number of small melt channels coalesce to produce a smaller number of wider channels which focus melt beneath a mid-ocean ridge.

Originally it was believed that the lower and upper mantle were distinct geochemical sources, with minimal mass transfer between the two. Hence it was thought that the mantle convected as two separate layers (Figure 26.10(a)). This is now known to be incorrect as seismic tomographic studies illustrate that subducting, lithospheric slabs penetrate through the mantle transition zone deep into the lower mantle. Hence there is strong case for whole-mantle convection, with significant mass transfer from the upper to the lower mantle (Figure 26.10(b)). Return flow from the lower to the upper mantle may be responsible for 'superplumes' beneath the Pacific Ocean and in east Africa.

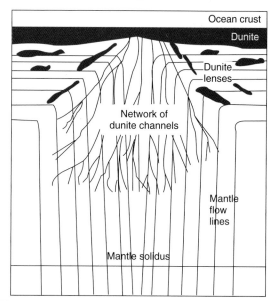

Figure 26.9 Network of coalescing dunite channels in an adiabatically-upwelling mantle beneath a fast-spreading mid-ocean ridge. (Source: after Kelemen *et al.*, 2000)

26.7.4 Mantle plumes

This brings us on to the idea of mantle plumes, the mechanism by which large volumes of basaltic melt are brought rapidly to the surface. These have been invoked to explain hot-spot volcanism, such as the linear, time-transgressive Emperor–Hawaii island chain, volcanic islands like Iceland, the origin of large igneous provinces with flood basalts and volcanic activity in the Archaean greenstone belts. The conventional mantle plume originates in the thin D″ layer of the lower mantle (the evidence for this is summarized in Condie (2005, pp. 144–145)). It represents a narrow (100 km diameter) vertical column of hot mantle which,

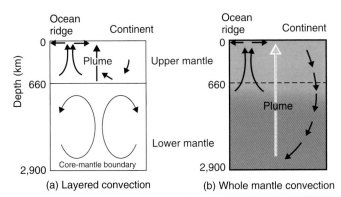

Figure 26.10 Diagram showing the two different models of mantle convection. (a) A two-layered mantle in which the upper mantle is chemically isolated from the primitive, lower mantle. The plume source here is in the upper mantle and subducting slabs do not penetrate the 660 km upper mantle–lower mantle boundary. (b) The whole-mantle model, in which there is mass exchange between the upper and lower mantles. Plumes here are sourced in the D″ layer, at the core–mantle boundary, and subduction penetrates the 660 km discontinuity. (Source: after Rollinson, 2007)

because of its buoyancy, rises and cuts across the entire mantle thickness. Where it hits rigid lithosphere, the hot mantle spreads out beneath it as a mushroom-shaped plume head several hundreds of kilometres across (as can be seen in Figure 26.11 from numerical modelling). Where there is voluminous volcanism, basalts are the main rock type but there are also picrites, which represent a higher temperature melt of basalt. The plumes appear to behave independently of plate tectonics and some authors, such as Condie (2005), have suggested that mantle plumes have developed episodically through geological time and may be responsible for the major growth stages of the continents. Nevertheless, there are several scientific arguments against the plume hypothesis:

1. Mantle tomography has not been able to image narrow, vertical structures cutting across the whole mantle that would represent plume stems.

2. The high temperatures needed by mantle plumes are not supported by the expected positive heat-flow anomalies.

3. The temperature contrast between plumes and the normal mantle is not as extreme as that claimed by the advocates of mantle plumes, because lateral temperature fluctuations in the upper mantle are greater than has previously been recognized.

4. Physical models of the mantle suggest plumes are impossible because of the high pressures at depth, which suppress the buoyancy of the material and prevent the formation of narrow plumes.

5. The geochemical arguments used in support of mantle plumes are ambiguous and do not require the deep and shallow sources proposed by plume theorists.

This is the place to summarize what we know about supercontinents and debate their relationship to mantle plume events. Supercontinents have aggregated and dispersed several time in the Earth's history, although the geological record is only well documented for the last two cycles: Gondwana-Pangea and Rodinia. It is agreed though that the continental masses are constantly being reconfigured. Continental collision makes fewer and larger continents, whilst rifting makes more and smaller continents. The last supercontinent, Pangaea, formed about 300 million years ago. The previous supercontinent, Pannotia, or Greater Gondwanaland, formed about 600 million years ago, and its dispersal formed the fragments that ultimately collided to form Pangea.

Beyond this the time span between supercontinents becomes more irregular. For example, the supercontinent before Gondwanaland, Rodinia, existed ~1.1 billion to ~750 million years ago – a mere 150 million years before Gondwanaland. The supercontinent before that was Columbia, ~1.8–1.5 billion years ago; then Kenorland, ~2.7–2.1 billion years ago; Ur, ~3 billion years ago; and Vaalbara, ~3.6–2.8 billion years ago. Generally one complete supercontinent cycle is thought to take 300–500 million years to occur.

However, the timing of break-up and dispersal in the last 1 billion years does not support a simple cycle in which a break-up phase is always followed by a growth phase, the growth phase by a stasis phase and the stasis phase by another break-up.

Mantle plume events are relatively short-lived mantle events ($\leq$100 million years) in which many mantle plumes bombard the base of the lithosphere. A fundamental question is whether there is a relationship between an apparent episodic growth of continental crust, the supercontinent cycle and mantle plume events. It seems possible to identify mantle plume events in the Pre-Cambrian by the distribution of komatiites, flood basalts, mafic dyke swarms and layered mafic intrusions. When these are dated, it appears that there have been many plume-head events in the last 3.5 billion years, with no plume-free intervals greater than around 200 million years (e.g. Prokoph et al., 2004). The frequency of events does depend on which database is used, but whichever is chosen, the activity seems strongly episodic (e.g. Figure 26.12). The evidence for the various major events is reviewed by Condie (2005, pp. 332–347).

Most models for episodic continent growth involve the catastrophic sinking of slabs through the 660 km seismic discontinuity in the upper mantle, which upon arrival at the D′ layer above the core initiate a mantle plume event. From seismic tomographic evidence the descending slabs currently sink into the lower mantle, but this may not have been the case when the Earth was hotter. There may have been layered convection throughout most of the Archaean, which would have prevented the occurrence of 'slab avalanches' (as they have been called), and it may not have been until the Late Archaean, when the 660 km discontinuity became weaker, that slabs fell through to the lower mantle. As the temperature and the Rayleigh number decreased with time, slabs could more easily penetrate the 660 km discontinuity, leading eventually to whole-mantle convection.

Another possible cause of mantle plume events is the effect of impacts (see Box 26.5, see the companion website). It has also

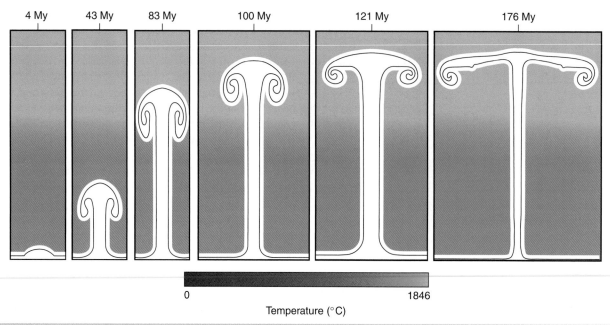

| 4 My | 43 My | 83 My | 100 My | 121 My | 176 My |

Temperature (°C)

0   1846

Figure 26.11 Modelling of the rise and growth of a mantle plume from the D″ thermal-boundary layer. (Source: after Rollinson, 2007)

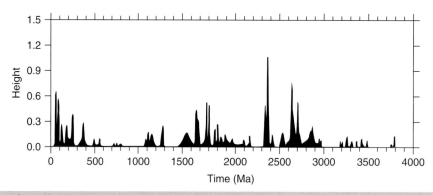

Figure 26.12 Distribution of mantle plume events deduced from time-series analysis of plume proxies

been suggested that large-scale melting can take place without the involvement of mantle plumes (Box 26.6, see the companion website).

This brings us back to a consideration of the possible relationship between mantle plume events and the supercontinent cycle (See Figure 26.13). The timing of events within this cycle is constrained by data from the isotopic ages of juvenile continental crust and from computer simulations of mantle processes. Beginning with a supercontinent, the latter models suggest that it takes an average of 200–400 million years for shielding of a large supercontinent to cause a mantle upwelling beneath it. This upwelling breaks up the land mass over a 200 million-year period and could be the trigger for a mantle plume event by producing a slab avalanche through the 660 km discontinuity. The time from the production of the slab avalanche to the production of the juvenile crust is likely to be under 100 million years. This suggests the link is between slab avalanches and supercontinent formation rather than break-up.

If supercontinents are really cyclic, it seems that they are not periodic, especially as break-up and aggregation of supercontinents overlap by 50–100 million years. As Condie (1998) has suggested, the duration of supercontinent formation and the total life span of supercontinents decrease with time. For example, the duration of the first well-documented supercontinent (1.9 billion years ago) was at least 500 million years, while Rodinia existed for about 300 million years and Gondwana and Pangea lasted only 100–200 million years. There is a corresponding decrease in the volume of juvenile continental crust produced during each cycle: 40% of the present juvenile continental crust was created in the formation of supercratons of a supercontinent at 2.7 billion years; 30% during the 1.9 million-year supercontinent; and only 12% during the formation of Rodinia. The likely reason for this is that the 660 km discontinuity decreased in strength with time: it became more permeable to descending slabs with falling mantle temperatures and as a result slabs would begin to sink through the boundary, with fewer accumulating at the boundary. When a slab avalanche occurred it was smaller, the

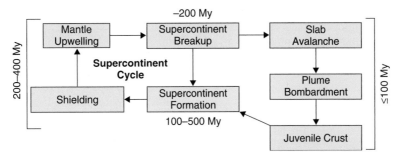

Figure 26.13 The possible relationship of the supercontinent cycle to mantle plume events. (Source: after Condie, 1998, 2000)

number of mantle plumes decreased and hence the juvenile crust volume associated with any plumes decreased. If this overall model is correct, the decreasing aggregation time of supercontinents with time is likely to reflect mantle cooling as radiogenic heat decreased. See the companion website for Box 26.5 Effect of terrestrial impacts on mantle plume events and Box 26.6 Global warming of the mantle at the origin of flood basalts over supercontinents.

26.7.5 The mantle geochemical system

The main inputs and outputs of the mantle are summarized in Figure 26.14. The largest input is oceanic lithosphere, which is strongly depleted in water-soluble elements through the subduction process, maybe down as far as the D″ layer. There is a smaller percentage in the form of subducted sediment and a significant flux of water into the mantle through the subduction zones, although there is debate as to how far it can travel. The main outputs are the basaltic melts at oceanic ridges and ocean islands in arcs and beneath continents, and the subtle chemical variations in the melt types are a combined effect of the different mantle sources and the different melting processes that take place. There is a volatile output, dissolved in basaltic melts to the atmosphere and oceans. As part of the longer-term evolution of the Earth, the reprocessing of oceanic crust during subduction leads to the formation of continental crust and hence the continental crust

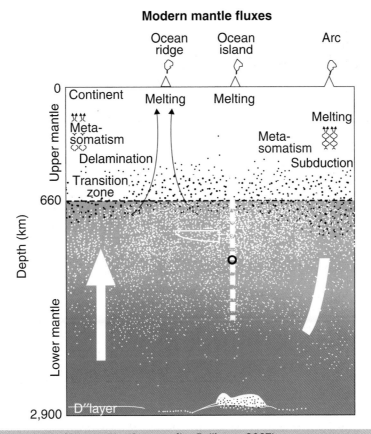

Figure 26.14 The main fluxes into and out of the mantle. (Source: after Rollinson, 2007)

can be looked on as an output from the mantle, but as part of a two-stage process. Of course, not all of the melts in the mantle reach the surface but they may consolidate in the mantle in a position away from the source. This contributes to the mantle metasomatism process.

It has been estimated that the whole mantle could have been replaced over the past 2–3 billion years based on the mass flow out of the mantle through basal melting and into the mantle through subduction. Although there is much variation in the figures produced by scientists, it is clear that a significant portion of the Earth's mantle has been reprocessed during its history.

26.7.6 What was the Earth's earliest mantle like?

The direct sampling of Archaean mantle is rare and geophysical evidence is impossible to obtain, so the major evidence comes from the Archaean mafic and ultramafic magmas. For example, the composition of Arachaean tholeiites was compared with that of modern ocean-ridge basalts and it was found that the Archaean tholeiites had higher silicon and iron oxide and lower, incompatible trace element concentrations. Either they formed from a different source or they are the product of a different melting process.

Arndt *et al.* (1997) interpreted the trace element concentrations and high Si, Fe, Ni and Cr to imply a high degree of partial melting. This would suggest that the Archaean tholeiites were produced by hotter, deeper mantle melting; that is, there was a thicker melt column. Calculations based on radioactive heat production also show that the Archaean mantle was hotter than the modern mantle. These high temperatures were calculated from the common, Archaean, ultramafic lavas known as komatiites, first recognized in the 1970s. The general view is that they were formed by high-temperature melts at around 1600 °C, rising from a depth range of 300–700 km.

However, the proposal that komatiites were the product of cooler, wet mantle melting (Box 26.7, see the companion website) weakens the argument for a very hot Archaean mantle, which was always a fundamental assumption about the Archaean Earth. Maybe there are good reasons now to argue that the temperatures of the Earth's mantle has only declined by 100–200 °C since the Mid–Late Archaean.

26.7.7 Models for the evolution of the mantle

There are two major models for the structure of the mantle. One believes the mantle is chemically stratified (a layer-cake model) whilst the other sees it as heterogeneous, with blobs of enriched mantle randomly distributed throughout a matrix of depleted mantle. An example of the first type of model is the lava-lamp model (Figure 26.15); an example of the second is the plum-pudding model (Figure 26.16). A layered mantle might be explained by either the deeply-subducted ocean crust now located

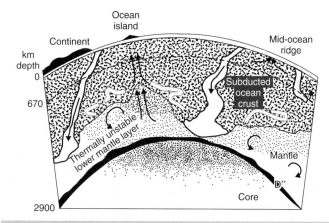

Figure 26.15 The lava-lamp model for the Earth's mantle. (Source: after Kellogg *et al.*, 1999)

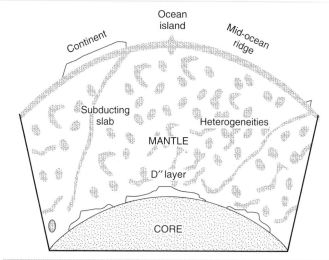

Figure 26.16 The plum-pudding model for the Earth's mantle. Subducted ocean lithosphere becomes mixed by convective stirring into the mantle and so introduces chemical heterogeneities. (Source: after Helffrich and Wood, 2001)

in the lower mantle or the date of the primary differentiation of the Earth in a magma ocean. The heterogenous model for the mantle is thought to have been produced through the incomplete mixing of deeply-subducted oceanic lithosphere over geological time. Further details related to the various models are given by Rollinson (2007, pp. 124–131).

We have seen that the mantle records a history of the extraction and recycling of both basaltic and continental crust. Due to the slow mixing and diffusion rates, mantle compositional heterogeneities produced by these processes may be preserved for over 1 billion years, so significant parts of the Earth's history can be observed in recent mantle melts. It seems likely that there is no preservation of a primitive, undifferentiated mantle as we have already seen that Nd-isotopes show the mantle experienced a major differentiation event early on, perhaps as early as 30 million years after the formation of the solar system. Here a Fe-rich basaltic crust formed on a magma ocean and was subsequently buried in the lower mantle. Oceanic basalt extraction and recycling can be traced back as far as 4.2 billion years ago from Os-isotopes (osmium), but the extraction of the continental crust on a large scale only began about 3 billion years ago and it is thought that continent-derived sediment recycling back into the mantle did not begin until around 2 billion years ago.

26.8 Evolution of the continental crust

We introduced ideas related to the processes, structure and composition of the continental crust in Chapters 12 and 13 but in this section we will look at the origin of the continental crust in more detail and consider what the characteristics of the primitive crust were and how the crust originated in the Archaean.

The Earth is the only planet in our solar system with a well-developed, felsic continental crust, which, although thin relative to other layers, had been progressively taken out of the Earth's mantle over the last 4 billion years. In fact, the reservoir of depleted mantle from which the crust was created has grown in volume as the continental crust has grown. The continental crust is ancient in parts and geologically recent in others, and is composed of a wide range of different rock types.

26.8.1 Crustal growth at destructive plate boundaries

The main site for modern crustal growth is in island arcs, in association with destructive plate boundaries. These are the sites at which new crust is created, through partial melting in the

mantle wedge, and where the main flux from mantle to crust is basaltic. They are also the locations where crust is recycled back into the mantle as subducted sediment. The general model is illustrated in Figure 26.17. The process is as follows:

1. Amphibolite, formed from hydrated ocean floor basalt in the upper layer of a subducting slab, begins to dehydrate at around 50–60 km depth, releasing fluids into the overlying mantle wedge.

2. The hydration of the mantle wedge results in the formation of amphibole in mantle peridotite. This amphibole-peridotite is dragged deeper into the mantle, to around 110 km, through viscous coupling to the subducting slab. Here the amphibole breaks down and releases aqueous fluid into the overlying mantle, which causes partial melting in the mantle wedge.

3. The partial melt zone migrates upwards as a diapir into a hotter region of the wedge, promoting further melting.

4. The diapir is dragged into the corner of the wedge by convective flow within the wedge, where the melt is focused.

The most robust crustal growth models suggest that continental crust grew progressively with time, possibly in an episodic fashion, which indicates phases of intense crustal growth at certain time periods during the Earth's history. There was certainly rapid growth during the Late Archaean.

Moreover, the composition of the continental crust during this early phase was different, illustrating the fact that the composition of the magmas was different. In the Archaean the main flux was magmas of the Si-Na rich tonalite-trondhjemite-granodiorite

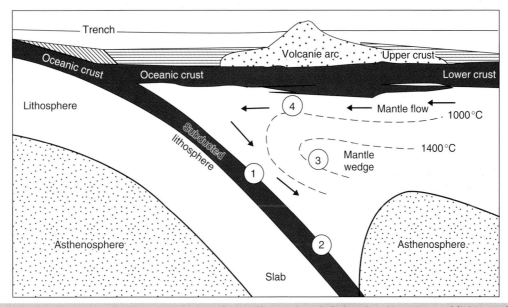

Figure 26.17 Model for the formation of new crust in an oceanic arc. The numbered circles show the process of fluid transfer from the subducted oceanic crust to mantle and melting within the mantle wedge. (Source: after Rollinson, 2007)

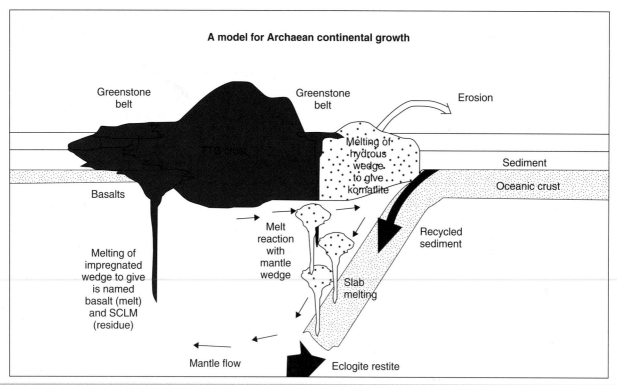

Figure 26.18 Model to show the different probable contributions to Archaean continental growth. (Source: after Rollinson, 2007)

association. It has been shown from experimental and geochemical studies that these magmas are likely to have formed by the partial melting of hydrous basalt.

Without water, magmas of TTG compositions cannot form. It appears that the most efficient method for generating the large volumes of these magmas needed to make the new crust is the partial melting of a wet basalt slab in a subduction zone. There is a modern analogue in the relatively rare, high silica adakite magmatic suite (representing a rock series with a composition range from andesite to rhyolite, typically with SiO_2 over 56% by weight, N_2O over 3.5 but under 7.5% by weight and low K_2O. The series isnnamed after an island in the Aleutian Islands, where it was first described.). This comprises slab melts which have interacted with the mantle wedge on their way to the crust. The complement to the TTG magma production is the formation of a melt-depleted, ecologitic slab residue, which is returned to the mantle. The best estimate for an average, time-integrated continental crust composition is that it is of andesitic composition. This is a problem as in modern arc systems the main flux from mantle to crust is basaltic in composition. To try and explain this situation, several solutions have been suggested:

1. A process of crustal delamination, where ultramafic material is removed from the roots of continents and returned to the mantle.

2. The composition of the continents is modified by weathering and selective element recycling.

3. As there are difficulties with both these ideas it has been suggested that there is a changing composition with time, so that the present andesitic composition is a time-averaged composition between the present basaltic flux and a more felsic TTG flux during the Archaean.

The models of Archaean crust generation imply that subduction has been operating since at least about 3.5 billion years ago. A summary of the different possible contributions to Archaean crustal growth is given in Figure 26.18. However, the big question is about the relative contribution that plume magmatism has made to the creation of felsic continental crust.

Exercises

1. Why do you think the Earth seems perfect for the evolution of life?

2. What evidence is there for parts of the early Earth system being linked to giant impacting events?

3. When and how did the Earth's core form?

4. What evidence is there for an early molten magma ocean?

5. What evidence can we see at the Earth's surface for the composition of the mantle?

6. What are the arguments for and against mantle plumes?

7. Describe the inputs and outputs to the Earth's mantle system.

8. Describe the two major models for the evolution of the mantle.

9. What are komatiites and how do they form?

10. What characteristics of the system are required for the general model of crustal growth?

References

Arndt, N.T., Albarede, F. and Nisbet, E.G. 1997. Mafic and ultramafic magmatism. In: De Wit, M.J. and Ashwaal, L.D. (Eds). Greenstone Belts. Oxford, Clarendon Press. pp. 233–254.

Asimow, P.D. and Langmuir, C.H. 2003. The importance of water to oceanic mantle melting regimes. *Nature*. **362**, 144–146.

Condie, K.C. 1998. Episodic continental growth and supercontinents: A mantle avalanche connection. *Earth and Planetary Science Letters*. **163**, 97–108.

Condie, K.C. 2000. Episodic continental growth models: Afterthoughts and extensions. *Tectonophysics*. **322**, 153–162.

Condie, K.C. 2005. Earth as an Evolving Planetary System. San Diego and London, Academic Press, Elsevier.

Eriksson, K.A., Krapez, B. and Fralick, P.W. 1994. Sedimentology of Archaean greenstone belts: Signatures of tectonic evolution. *Earth Science Reviews*. **37**, 1–88.

Frimmel, H.E. 2005. Archaean atmospheric evolution: Evidence from the Witwatersrand goldfields, South Africa. *Earth Science Reviews*. **70**, 1–46.

Helffrich, G.R. and Wood, B.J. 2001. The Earth's mantle. *Nature*. **412**, 501–507.

Kelemen, P.B., Braun, M. and Hirth, G. 2000. Spatial distribution of melt conduits in the mantle beneath oceanic spreading ridges: observations from the Ingalls and Oman ophiolites. Geochemical and Geophysical Systems 1, Paper 1999GC000012.

Kellogg, L.H., Hager, B.H. and van der Hilst, R.D. 1999. Compositional stratification in the deep mantle. *Science*. **283**, 1881–1884.

Peck, W.H., Valley, J.W., Wilde, S.A. and Graham, C. 2001. Oxygen isotope ratios and rare Earth elements in 3.3 to 4.4 Ga zircons: Ion microprobe evidence for high $\partial^{18}O$ continental crust and oceans in the early Earth. *Geochimica Cosmochimica Acta*. **65**, 4215–4229.

Prokoph, A., Ernst, R.E. and Buchan, K.L. 2004. Time-series analysis of large igneous provinces: 3500 Ma to present. *Journal of Geology*. **112**, 1–22.

Rubie, D.C., Melosh, H.J., Reid, J.E., Liebske, C. and Righter, K. 2003. Mechanisms of metal-silicate equilibration in the terrestrial magma ocean. *Earth and Planetary Science Letters*. **205**, 239–255.

Rollinson, H. 2007. Early Earth Systems: A Geochemical Approach. Oxford, Blackwell Publishing.

Smith, P.E., Evensen, N.M. York, D., 2005. Oldest reliable terrestrial ^{40}Ar-^{39}Ar age from pyrite crystals at Isua. West Greenland. Geophysical Research Letters 32, L21318/doi 10.1029/2005 GLO 22640.

Further reading

http://www.classzone.com/books/earth_science/terc/content/visualisations/es0805/page01.cfm..

27 Atmospheric Evolution and Climate Change

Large meltwater channel on Nordenskiold glacier, Svalbard. Photo: T. Stott

Learning Outcomes

After reading this chapter and completing the exercises you will be able to:

➤ Describe the Earth's early atmosphere and the effect of volcanoes.

➤ Understand how and why oxygen levels in the Earth's atmosphere have risen.

➤ Describe the effects of humans on the balance of gases in the Earth's atmosphere.

➤ Understand the differences between 'predictions', 'scenarios' and 'projections' as terms used to characterize future climate changes.

➤ Understand the role and main findings of the Intergovernmental Panel for Climate Change (IPCC).

➤ Explain why a number of climate prediction models and a range of scenarios have been used by the IPCC to predict future climate.

➤ Describe the range of likely impacts of global warming predicted for the twenty-first century.

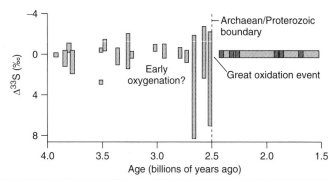

Figure 27.1 Range of MIF of sulphur over time. The great oxidation event occurred ~2.45 billion years ago, and an early, failed, oxygenation event might have occurred around 3.2 billion years ago, but this is hotly debated. The degree of MIF is indicated by $\Delta^{33}S$, which is the parts per thousand (‰) deviation of the standardized $^{33}S/^{32}S$ ratio from the value predicted from the $^{34}S/^{32}S$ ratio and mass-dependent fractionation. The range of values from samples of a given age is shown by vertical bars

27.1 Evolution of the Earth's atmosphere

As we saw in Chapter 9, in the 4.6 billion years since the Earth's formation there have been some traumatic upheavals which have affected the evolution of the atmosphere. Scientists believe that the Earth's atmosphere contained only hydrogen and helium at the time of its formation. Some of the influences on the evolution of the atmosphere include: changes in solar luminosity, volcanic activity, the formation of the oceans, mountain building and continental drift, and the origin of life on the planet.

This section presents case studies on the rise of atmospheric oxygen and the formation of supercontinents, which is linked to increases in atmospheric oxygen.

27.1.1 The rise of atmospheric oxygen

Clues from ancient rocks are helping to produce a coherent picture of how the Earth's atmosphere changed from one that was almost devoid of oxygen to one that is one-fifth oxygen. A battery of geological indicators suggest a shift from an anoxic atmosphere sometime between 2.5 and 2.0 billion years ago, known as the great oxidation event. The most compelling evidence for this is the absence in older stratigraphic units of 'red beds', sedimentary rocks stained by red iron oxide. Instead, an abundance of lithified ancient soils that lost their iron during weathering have

been found, reflecting the absence of oxygen in the weathering environment.

The 'smoking gun' for the rise of atmospheric oxygen was discovered and reported in 2000, when rocks older than about 2.45 billion years were found to contain a large degree of mass-independent fractionation (MIF) of sulphur isotopes; rocks younger than 2.32 billion years show essentially none (Figure 27.1).

Some workers have argued that the contraction of the spread in MIF values at ~2.45 billion years (Figure 27.1) was the direct result of a collapse in atmospheric methane levels. The loss of 'greenhouse warming' is then invoked to explain the ensuing first major glaciations in Earth's history, perhaps of 'Snowball Earth' proportions, with ice extending to the tropics.

With regard to the pursuit of reconstructing ancient oxygen levels on Earth, most geological indicators imply only presence or absence. MIF disappears when oxygen levels reach 0.001% of the present atmospheric level (PAL) and iron is retained in ancient lithified soils when oxygen is at 1% of its PAL. Persistent anoxia of the oceans in the Proterozoic (from 1.8 to 0.5 billion years ago) is argued to require oxygen levels below 40% of PAL. Fire is sustained only above about 60% PAL, and the more or less continuous geological record of charcoal over the past 450 million years sets this as a lower limit for atmospheric oxygen since the advent of forests on Earth. The interesting exception is the Middle to Late Devonian, ~380 million years ago, which shows a charcoal gap coincident with widespread evidence for marine anoxia. The other available redox indicators are from marine sediments, requiring that internal ocean processes that affect deep-ocean oxygen levels be untangled before inferences about atmospheric oxygen levels can be made.

New approaches to this problem are based on the effect that oxygen has on carbon-isotope fractionation and on the physiological effects of, adaptations to and defences against oxygen of plants and animals.

27.1.2 Formation of supercontinents

Atmospheric oxygen concentrations in the Earth's atmosphere rose from low (almost negligible) levels in the Archaean to about 21% today. This increase is thought to have occurred in six steps: 2.65, 2.45, 1.8, 0.6, 0.3 and 0.04 billion years ago, with a possible seventh identified at 1.2 billion years ago. Campbell and Allen (2008) show that the timing of these steps correlates with the amalgamation of the Earth's land masses into supercontinents (see Chapter 28). They suggest that the continent–continent collisions required to form supercontinents produced super-mountains, which eroded quickly and released large quantities of nutrients such as iron and phosphorus into the oceans. This led to an explosion of algae and cyanobacteria, a marked increase in photosynthesis and therefore production of O_2. Enhanced sedimentation during these periods promoted the burial of a high proportion of organic carbon and pyrite, which prevented their reacting with free oxygen. This meant that the atmospheric increases in O_2 from photosynthesis could be sustained.

27.2 Future climate change

The issue of the future of the Earth's climate is now on most people's minds. Barely a news broadcast or bulletin goes by without some mention or a new prediction or finding which may affect our understanding of our future climate.

27.2.1 Predictions, scenarios and projections

A prediction is a statement that something will happen in the future, based on known facts at the time the prediction is made and assumptions about the processes that will lead to change. Predictions are rarely certain; for example, a weather forecaster might predict that there will be a 60% probability of rain tomorrow.

A scenario is a plausible description of some future state, with no statement of probability. Scenarios are alternative pictures of what the future might be like, given that a certain set of decisions are taken. They may be used to assess the consequences of a particular decision, policy or strategy. For example, what would happen to a particular riverside community if a flood of a certain magnitude occurred?

Projections are sets of future conditions, or consequences, based on explicit assumptions, such as scenarios. For example, if the aforementioned flood happened, how many houses would get flooded, how many people would be affected, what would be the depth of the water in certain locations and how long would it take for the water to subside? A projection would answer such questions as specifically as possible.

27.2.2 Climate models

Climatic models have been used to investigate the effects of known or highly likely changes in the future. For models to be successful they need to take account, at the very least, of the feedbacks we considered in Chapter 10: water vapour, clouds, ice-albedo and ocean circulation. They must take account of variations in solar radiation and Milankovitch-type orbital oscillations, and changes in greenhouse gases will need to feature strongly. Because of the very different scales of operation it is difficult to account for astronomic, oceanic and atmospheric factors in the same model.

Our uncertainty about future climate is based on many different forcing factors, discussed earlier in this chapter. Some change in a linear way while others are nonlinear; they operate at different time scales; some may reach thresholds; and all are superimposed upon one another. It is therefore just about impossible to predict how long any particular trend will last, because of the complex interactions between all parts of the climate system.

Despite, or because of, the difficulties and challenges of modelling climate change, the Intergovernmental Panel for Climate Change (IPCC), a scientific intergovernmental body, was set up in the 1980s by the World Meteorological Organization (WMO) and the United Nations Environment Programme (UNEP). It was established to provide decision makers and other interested parties with an objective source of information about climate change. The IPCC does not conduct any research, nor does it monitor climate-related data or parameters. Its role is to assess on a comprehensive, objective, open and transparent basis the latest scientific, technical and socioeconomic literature from across the world that is relevant to the understanding of the risk of human-induced climate change, its observed and projected impacts and the options for adaptation and mitigation. The IPCC has reported four times since 1990, drawing together hundreds of climate models developed by around 3000 scientists. See the companion website for Box 27.1 The future of Atlantic hurricanes.

27.2.3 Emissions scenarios used by the IPCC

In order to provide advice to policy makers on the consequences of anthropogenically-caused climate change in the twenty-first century, the IPCC commissioned a range of scenarios for greenhouse gas and sulphate aerosol emissions up to the year 2100. The scenarios, developed by a panel of authors, with wide consultation of experts, were reported in the Special Report on Emissions Scenarios (SRES), published in 2000. They were intended to feed into the projections of climate change in the Third Assessment Report (2001) and have been retained for use in the Fourth Assessment Report (2007). Future emissions of greenhouse gases are the product of complex interacting systems driven by population change, socioeconomic development and technological advance, all of which are uncertain. In all there are 35 scenarios, containing data on all gases required to force climate models (the main ones being carbon dioxide, methane, nitrous oxide and sulphur dioxide), based on four different 'storylines' of how the future might develop (Figure 27.3).

The models calculate accumulated emissions expressed in units of thousands of millions of tonnes of carbon equivalent (GtC) or carbon dioxide ($GtCO_2$-eq.). The IPCCs Fourth Assessment

Box 3: The SRES emissions scenarios

A1. The A1 storyline and group of related scenarios describe a future world of very rapid economic growth, global population that peaks in mid-century and declines thereafter, and the rapid introduction of new and more efficient technologies. Major underlying themes are convergence among regions, capacity building and increased cultural and social interactions, with a substantial reduction in regional differences in per capita income. The A1 group scenario is split into three groups that describe alternative directions of technological change in the energy system. The three A1 groups are distinguished by their technological emphasis: fossil intensive (A1FI), non-fossil energy sources (A1T), or a balance across all sources (A1B) (where balanced is defined as not relying too heavily on one particular energy source, on the assumption that similar improvement rates apply to all energy supply and end-use technologies).

A2. The A2 storyline and group of scenarios describe a very heterogeneous world. The underlying theme is self-reliance and preservation of local identities. Fertility patterns across regions converge very slowly, which results in continuously increasing population. Economic development is primarily oriented to particular regions and per capita economic growth and technological change more fragmented and slower than other storylines.

B1. The B1 storyline and group of scenarios describe a convergent world with the same global population that peaks in mid-century and declines thereafter, as in the A1 storyline but with rapid change in economic structures toward a service and information economy, with reductions in material intensity and the introduction of clean and resource-efficient technologies. The emphasis is on global solutions to economic, social and environmental sustainability, including improved equity, but without additional climate initiatives.

B2. The B2 storyline and group of scenarios describe a world in which the emphasis is on local solutions to economic, social and environmental sustainability. It is a world with continuously increasing global population, at a rate lower than A2, intermediate levels of economic development, and less rapid and more diverse technological change than in the A1 and B1 storylines. While the scenario is also oriented toward environmental protection and social equity, it focuses on local and regional levels.

Source: IPCC 2001 a, Box 4 of Technical Summary

Figure 27.3 The IPCC Special Report on Emissions Scenarios (SRES), published in 2000

Report (2007) concludes that 'there is high agreement and much evidence that with current climate change mitigation policies and related sustainable development practices, global greenhouse gas emissions will continue to grow over the next few decades'.

In 2000 the IPCC Special Report on Emission Scenarios projected an increase in global greenhouse gas emissions of 25–90% (CO_2-eq) between 2000 and 2030 (Figure 27.4), with fossil fuels maintaining their dominant position in the global energy mix to 2030 and beyond.

For the next two decades a warming of about $0.2\,°C$ per decade is projected for a range of SRES emission scenarios. Even if the concentrations of all greenhouse gases and aerosols were kept constant at year 2000 levels, a further warming of about $0.1\,°C$ per decade would be expected. Afterwards, temperature projections increasingly depend on specific emission scenarios.

In terms of regional differences in the effects of global warming, Figure 27.5 shows that warming will be greatest over land and at most high northern latitudes, and least over southern ocean and parts of the northern Atlantic Ocean, continuing recent observed trends.

Model predictions also include the contraction of snow-cover area, increases in thaw depth over most permafrost regions and decrease in sea-ice extent; in some projections using SRES scenarios, Arctic late-summer sea ice disappears almost entirely

by the latter part of the twenty-first century. They predict a very likely increase in the frequency of hot extremes, heat waves and heavy precipitation; likely increases in tropical cyclone intensity; a poleward shift of extra-tropical storm tracks with consequent changes in wind, precipitation and temperature patterns; very likely precipitation increases in high latitudes and likely decreases in most subtropical land regions. Just one example of these changes is the 6 February 2008 series of tornadoes that swept across the southern USA, killing over 48 people, which seem to have been triggered by unusually warm temperatures in the region.

There is also high confidence that by mid-century, annual river runoff and water availability will increase at high latitudes (and in some tropical wet areas) and decrease in some dry regions in the mid-latitudes and tropics. There is similarly high confidence that many semi-arid areas (e.g. the Mediterranean basin, the western United States, southern Africa and north-east Brazil) will suffer a decrease in water resources due to climate change.

27.2.4 Impacts and implications of global warming

The impacts and implications of global warming predicted in the various scenarios are so far-reaching and wide-ranging that it

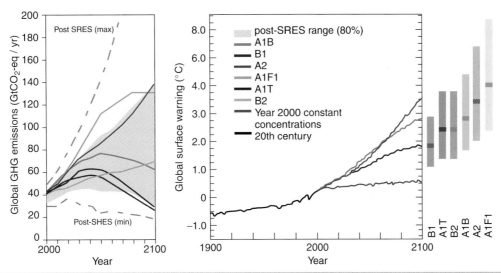

Figure 27.4 Left: Global greenhouse gas emissions (in CO_2-eq) in the absence of climate policies: six illustrative SRES marker scenarios (coloured lines) and the 80th percentile range of recent scenarios published since SRES (grey shaded area). Dashed lines show the full range of post-SRES scenarios. The emissions cover CO_2, CH_4, N_2O and F gases. Right: Solid lines are multimodel global averages of surface warming for scenarios A2, A1B and B1 (Figure 27.3), shown as continuations of the twentieth-century simulations. These projections also take into account emissions of short-lived greenhouse gases and aerosols. The pink line is not a scenario, but shows Atmosphere-Ocean General Circulation Model (AOGCM) simulations, where atmospheric concentrations are held constant at year 2000 values. The bars at the right of the figure indicate the best estimate (solid line within each bar) and the likely range assessed for the six SRES marker scenarios in 2090–2099. All temperatures are relative to the period 1980–1999. (Source: IPCC AR4, 2007, Climate Change 2007: Synthesis Report, Figures 3.1 and 3.2)

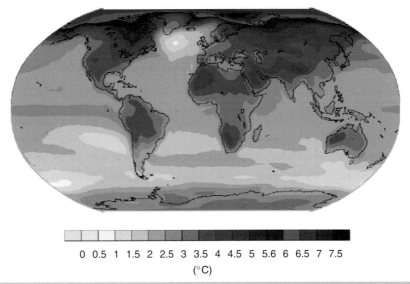

Figure 27.5 Projected surface-temperature changes for the late twenty-first century (2090–2099). The map shows the multi-AOGCM average projection for the A1B SRES scenario. All temperatures are relative to the period 1980–1999. (Source: IPCC AR4, 2007, Climate Change 2007: Synthesis Report, Figure 3.2)

is difficult to know where to begin what must be a relatively short discussion here. As climate changes, so all the environmental elements dependent upon climate will also change. Perhaps Figure 27.6 is a useful summary on which to focus any discussion.

The most obvious and immediate effects are likely to be on ecosystems: the plant and animal kingdoms. If the change happens gradually and with no human interference, some species may migrate to more suitable areas and other species will move in to replace them. This has happened many times in the last 2 million years. However, the main difference is that not only is the change happening quickly, there is a vast and growing human population which will require food and water. Plant and animal species might come second if there were a fight for survival.

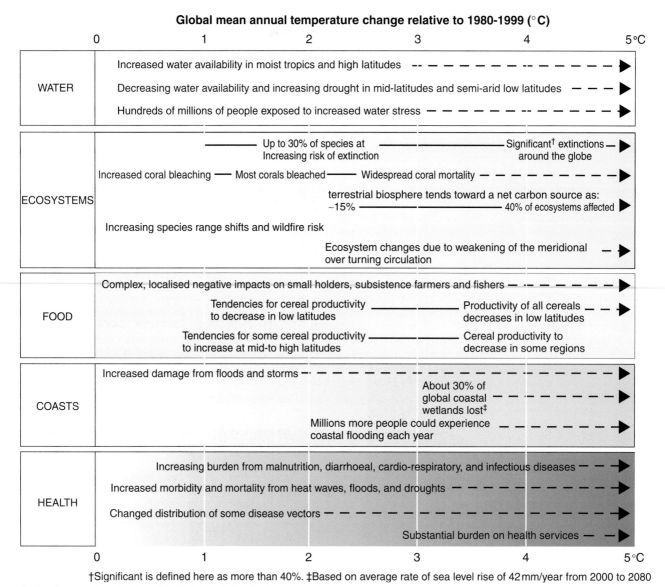

Global mean annual temperature change relative to 1980-1999 (°C)

†Significant is defined here as more than 40%. ‡Based on average rate of sea level rise of 42mm/year from 2000 to 2080

Figure 27.6 Examples of impacts associated with projected global average surface warming. Examples suggest global impacts projected for climate changes (including sea-level and atmospheric CO_2 where relevant) associated with different amounts of increase in global average surface temperature in the twenty-first century. The black lines link impacts and broken-line arrows indicate impacts continuing with increasing temperature. Entries are placed so that the left-hand text indicates the approximate level of warming that is associated with the onset of a given impact. Quantitative entries for water scarcity and flooding represent the additional impacts of climate change relative to the conditions projected across the range of SRES scenarios A1FI, A2, B1 and B2. Adaptation to climate change is not included. Confidence levels for all statements are high. (Source: IPCC AR4, 2007, Climate Change 2007: Synthesis Report, Figure 3.6)

It is likely that some pests and diseases will proliferate in a warmer world. Many pests currently have their geographical range and/or physiology restricted by winter cold, so warmer winters might allow expansion of their habitat. The mosquito, for example, could re-establish itself in southern Britain, and certain crop diseases could become more prevalent, and if not controlled could affect crops and therefore food supplies.

The likely global redistribution of water could leave hundreds of millions more people without sufficient drinking water. The effects of increased incidence of drought on human life in the Sahel region has already given a clear insight into what water shortages in some regions can mean. There may be increased precipitation in some areas and decreases in others; the precise changes are very difficult to predict. However, the effect of increased evaporation through higher temperatures would be decreased soil moisture levels, which would certainly affect crops, and climax ecosystems might change; for example, rain forest could convert to savanna grassland. A change in moisture regime would have geomorphological impacts affecting river levels, erosion and sediment transport, and channel form. Increased storminess would generate more floods, which could be associated with channel erosion, slope failures and landslides, and generally more variable river regimes with all the predictable consequences for people who live along or near to rivers. Permafrost areas would melt further, leading to greater microbial activity, and as organic matter began to decompose more rapidly, methane gas would be released, contributing more of its powerful greenhouse effects to the atmosphere.

If there were an increase in global mean temperature, the oceans would get warmer and expand, causing sea level to rise. Warmer temperatures would also be highly likely to cause glaciers and ice caps to melt, adding more water to the oceans and increasing the expansion already mentioned.

The combination of these processes is estimated to cause sea levels to rise by between 0.3 and 1 m by 2100, though some models and scenarios estimate it to be considerably more. Large areas of low-lying coastal wetlands would be flooded, farmland could be contaminated by salt water and hundreds of millions of people might be displaced. Coastal erosion might increase in some areas, posing a threat to land and property (see Chapter 19 for some examples).

Taking all these environmental impacts together, the predicted rapid increase in global mean temperature over the twenty-first century is likely to have some dramatic consequences.

27.2.5 Mitigating human-induced global warming

When considering a problem as vast as global warming, people often feel overwhelmed and powerless – that their individual efforts will make little difference. However, most people would agree that we need to resist that response, since the solution requires individuals to take responsibility. This can happen by educating people about the problem and the potential solutions, by doing our part to minimize our use and waste of resources, by becoming more politically active and demanding change.

You can start by visiting a Web site such as http://www.climatecrisis.net, where you can use an interactive energy meter to calculate your impact or 'carbon footprint' so that you can evaluate which areas of your life produce the greatest emissions, and therefore work towards reducing those emissions to live a carbon-neutral lifestyle. Here are some areas where individuals can make an impact:

Save energy at home

1. Choose energy-efficient lighting and appliances, and properly operate and maintain them.

2. Heat and cool your house efficiently; insulate your house.

3. Get a home energy audit carried out.

4. Conserve hot water.

5. Reduce standby power waste.

6. Improve the efficiency of your home office.

7. Switch to a green power supplier.

Get around on less

1. Reduce the number of miles you drive by walking, cycling, car pooling or taking public transport wherever possible.

2. Drive more efficiently; reduce your speed and acceleration.

3. Make your next vehicle purchase a more efficient one: consider alternative-fuel cars, fuel-cell vehicles and hybrids.

4. Work from home if possible by using the telephone/Internet.

5. Reduce air travel.

Consume less and conserve more

1. Buy less: consider borrowing or renting things, or buy second hand.

2. Buy things that last.

3. Pre-cycle: reduce waste before you buy: choose produce without excessive packaging.

4. Recycle paper, glass, steel, aluminum and plastic wherever possible.

5. Obtain and use reusable grocery bags.

6. Compost your kitchen waste.

7. Carry your own refillable bottle for water or other drinks.

8. Modify your diet to include less meat.

9. Buy locally-produced food.

10. Purchase offsets to neutralize your remaining emissions.

Be a catalyst for change

1. Learn about climate change.

2. Let others know.

3. Encourage your school, organization or business to reduce emissions in the workplace.

4. Vote with your money: buy only brands and from stores which act in an environmentally-responsible manner.

5. Consider the impact of your investments: invest in companies, products and projects that responsibly address climate change.

6. Take political action: responsibly lobby your local politician.

7. Support an environmental group.

See the companion website for Box 27.2 Formation of supercontinents.

Exercises

1. Approximately when did the shift in the Earth's atmospheric oxygen levels known as the great oxidation event take place?

2. In what way do scientists think that the formation of supercontinents might be linked to increases in atmospheric oxygen?

3. Outline the differences between 'predictions', 'scenarios' and 'projections'.

4. For climate models to be successful, what types of feedback do they need to take account of?

5. For a range of IPCC SRES emission scenarios, what is the projected increase in mean global temperature per decade over the next two decades?

6. If the concentrations of all greenhouse gases and aerosols had been kept constant at year 2000 levels, what would be the expected increase in mean global temperature per decade over the next two decades?

7. Comment on what model predictions are for: snow cover; thaw depth; sea-ice extent; hot extremes; heat waves; precipitation; tropical cyclone intensity; river runoff; semi-arid regions.

8. Give some examples of what IPCC models predict might happen to ecosystems, food and human health.

9. Outline four main areas in which the average person could change their lifestyle to help mitigate the effects of climate change.

10. Explain the concept of 'stabilization wedges' (as put forward by Pacala and Socolow) as a means of tackling global warming.

11. Describe the three broad approaches to carbon capture and storage (CCS).

References

Campbell, I.H. and Allen, C.M. 2008. Formation of supercontinents linked to increases in atmospheric oxygen. *Nature Geoscience*. **1**, 554–557.

Further reading

Gore, A. 2006. An Inconvenient Truth: The Planetary Emergency of Global Warming and What We Can Do About It. London, Bloomsbury Publishing.

Henderson-Sellers, A. and Robinson, P.J. 1999. Contemporary Climatology. London, Longman.

Henderson-Sellers, A. 1985. The evolution of the Earth's atmosphere. In: Fifield, R. (Ed.). The Making of the Earth. Blackwell and New Scientist. pp. 79–86.

Houghton, J. 2004. Global Warming: The Complete Briefing. 3rd edition. Cambridge, Cambridge University Press.

Kennett, J.P., Cannariato, K.G., Hendy, I.L. and Behl, R.J. 2003. Methane Hydrates in Quaternary Climate Change: The Clathrate Gun Hypothesis. American Geophysical Union Special Publications Series Volume 54. Washington, DC.

Kump, L.R. 2008. The rise of atmospheric oxygen. *Nature*. **451**, 277–278.

Pittock, A.B. 2007. Climate Change: Turning Up the Heat. Collingwood, Victoria, CSIRO Publishing.

28 Change in Ocean Circulation and the Hydrosphere

Sunset at Noosa Head, Sunshine Coast, Queensland, Australia. Photo: T. Stott

Earth Environments   David Huddart and Tim Stott
© 2010 John Wiley & Sons, Ltd

Learning Outcomes

After reading this chapter and completing the exercises you will be able to:

➤ Understand the relationship between sea level and the supercontinent cycle.

➤ Understand the supercontinent cycle and its relationship with climate.

➤ Explain the relationships between global temperature, ice ages and sea-level change.

➤ Describe the mechanism and potential effects of a weakening of the Atlantic Heat Conveyor on climate of the North Atlantic Region.

28.1 Introduction

As we noted in Chapter 8, the oceans are a major influence on the planet's climate. They cover 71% of the Earth's surface and store 97% of its water by volume. Ocean circulation is responsible for about half of the transport of energy from the tropics to the poles, the atmosphere transferring the other half. In Chapter 3 we saw that the atmospheric circulation is driven by the heating of surface waters in the tropics, and cooling at the poles. Cold surface currents travel equatorwards and warm surface currents travel polewards. Knowledge of the oceans, ocean circulation and how circulation responds to climate change is important for a number of reasons:

1. Oceans are capable of transporting heat from equatorial regions polewards.

2. Ocean thermal conditions are one of several factors affecting the growth and melting of sea ice.

3. Ocean thermal conditions affect the evaporation rate, thereby controlling atmospheric moisture.

4. Ocean temperatures are a control on the rate of ice-shelf melting and therefore may play an important role in terms of ice-sheet mass balance.

5. Eustatic sea-level change is a direct control on the growth and decay of marine ice sheets, which play a vital role in moderating the Earth's climate.

It is therefore important that we have some understanding of the patterns of ocean circulation, particularly as these patterns, and the sea level itself, could change in a warmer future, with important implications for the Earth's climate and for people living in low-lying coastal areas.

28.2 Sea-level change and the supercontinental cycle

28.2.1 The supercontinent cycle

The supercontinent cycle describes the quasi-periodic aggregation and dispersal of the Earth's continental crust. One complete supercontinent cycle is said to take 300–500 million years to occur. Continental collision makes fewer and larger continents, while rifting makes more and smaller continents. The last supercontinent, Pangaea, formed about 300 million years ago. The previous supercontinent, Pannotia, formed about 600 million years ago, and its dispersal formed the fragments that ultimately collided to form Pangaea.

Further back in the Earth's history the time between supercontinents becomes more irregular. For example, the supercontinent before Gondwanaland, Rodinia, existed ~1.1 billion to ~750 million years ago – some 150 million years before Pannotia. The supercontinent before that was Columbia: ~1.8–1.5 billion years ago, and before that was Kenorland: ~2.7–2.1 billion years ago. The first continents were Ur (~3 billion years ago) and Vaalbara (~3.6–2.8 billion years ago).

The hypothetical supercontinent cycle in some ways complements the Wilson cycle, named after plate tectonics pioneer J. Tuzo Wilson, which describes the periodic opening and closing of ocean basins (Figure 28.1).

Because the oldest sea floor is only 170 million years old, whereas the oldest bit of continental crust goes back to 4 billion years or more, it makes sense to emphasize the much longer record of the planetary pulse that is recorded in the continents.

It is known that sea level is low when the continents are together and is high when they are apart. Thus sea level was low at the time of the formation of Pangaea (Permian) and Pannotia, rising rapidly to maxima during the Ordovician and Cretaceous, when the continents were dispersed. Sea level is controlled by the age of the sea floor. Oceanic crust lies at a depth (d) that is a simple function of its age (t):

$$d(t) = 2500 + (350t^{1/2})$$

where d is in metres and t is in millions of years. So newly-formed crust at the mid-ocean ridges lies at about 2500 m depth, whereas 100 million-year-old sea floor lies at a depth of about 6000 m.

In the same way that the water level in a bathtub is controlled by the size of the person in the bath, sea level is controlled by the depth of the sea floor (ignoring complications resulting from glacial-ice-sheet and temperature fluctuation). The relationship between sea-floor depth and sea level can be expressed as follows:

The mass(M)of water on the Earth is a constant = K1

where:

$$K1 = M(sea\ water) + M(fresh\ water) + M(ice)$$
$$+ M(atmospheric\ water)$$

STAGE	MOTION	PHYSIOGRAPHY	EXAMPLE
EMBRYONIC 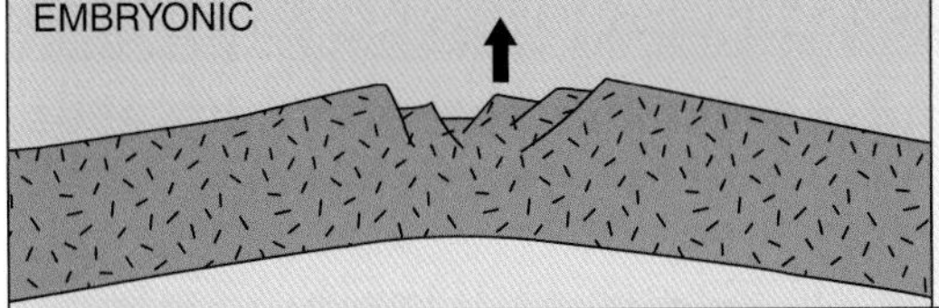	Uplift	Complex system of linear rift valleys on continent	East African rift valleys
JUVENILE 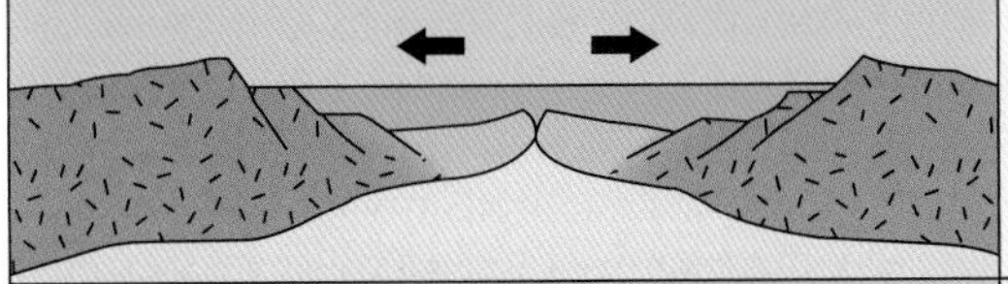	Divergence (spreading)	Narrow seas with matching coasts	Red Sea
MATURE 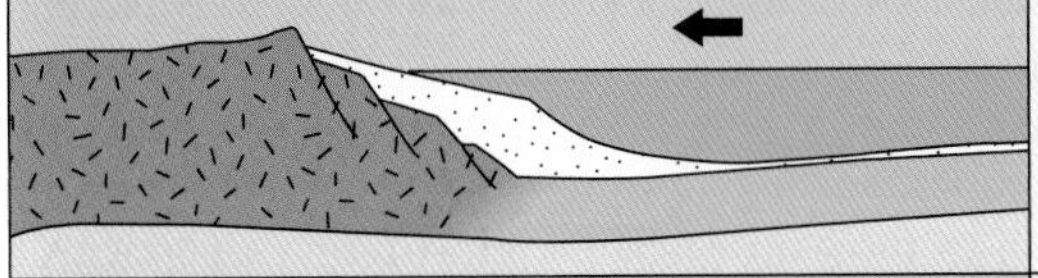	Divergence (spreading)	Ocean basin with continental margins	Atlantic, Indian, and Arctic Oceans
DECLINING 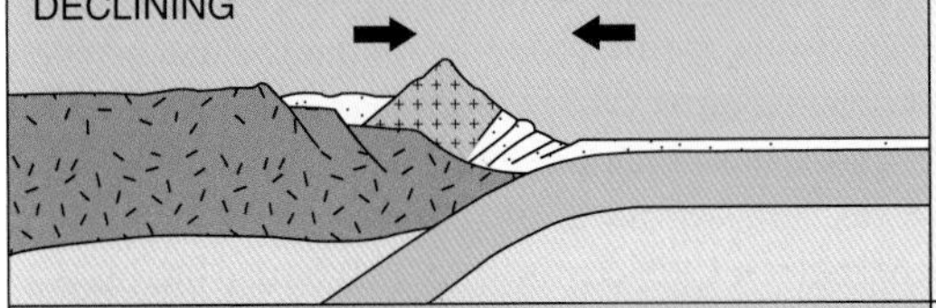	Convergence (subduction)	Island arcs and trenches around basin edge	Pacific Ocean
TERMINAL 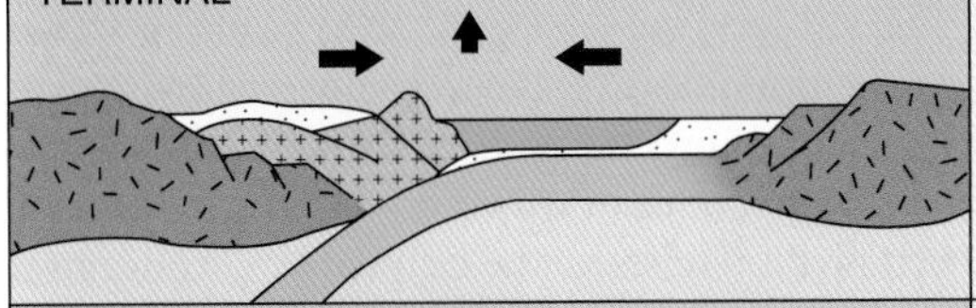	Convergence (collision) and uplift	Narrow, irregular seas with young mountains	Mediterranean Sea
SUTURING 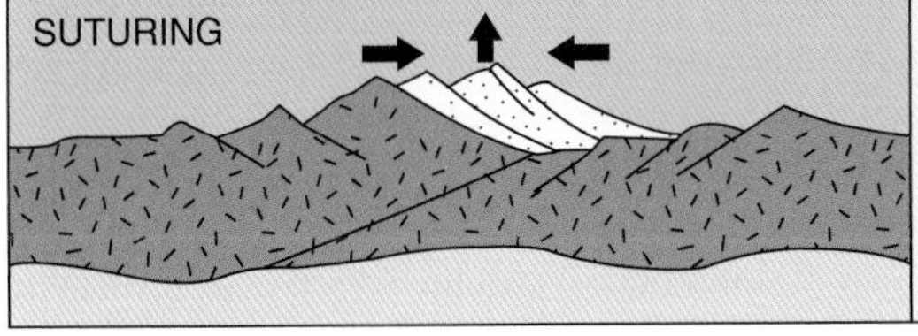	Convergence and uplift	Young to mature mountain belts	Himalayas

Figure 28.1 The Wilson cycle, describing ocean-basin development in six stages. (Source: Dutch *et al.*, 1998, p. 279)

We can neglect M (fresh water) + M (atmospheric water), so:

$$K1 = M(\text{sea water}) + M(\text{ice})$$

Consider the ice-free world:

$$V(\text{sea water}) = K1/(\text{mean density of sea water})$$

This volume fills the ocean basins to a depth determined by $A \times d'$, where A = area of the ocean basins and d' = mean depth of the ocean basins. d' is determined by the mean age of the seafloor. A can change when continents rift (stretching the continents decreases A and raises sea level) or as a result of continental collision (compressing the continents leads to an increase in A and lowers sea level). Increasing sea level will flood the continents, while decreasing sea level will expose continental shelves. Since the continental shelf has a very low slope (see Chapter 8), a small increase in sea level will result in a large change in the percentage area of continents flooded.

If the world ocean on average is young, the sea floor will be relatively shallow, sea level will be high and more of the continents will be flooded. If the world ocean is on average old, sea floor will be relatively deep, sea level will be low and more of the continents will be exposed. There is thus a relatively simple relationship between the supercontinent cycle and the mean age of the sea floor: a supercontinent means lots of old sea floor, resulting in low sea level, whereas dispersed continents means lots of young sea floor, causing a high sea level.

There may also be a climatic effect of the supercontinent cycle, which could amplify this further: a supercontinent will give rise to a continental climate that is more likely to favour continental glaciation, resulting in an even lower sea level. On the other hand, dispersed continents are more likely to experience a maritime climate, meaning that continental glaciation is unlikely and sea level is not lowered by this mechanism.

There is a progression of tectonic regimes that accompanies the supercontinent cycle. During break-up of the supercontinent, rifting environments dominate. These are followed by passive margin environments, while sea-floor spreading continues and the oceans grow. These in turn are followed by the development of collisional environments, which become increasingly important with time. The first collisions are between continents and island arcs, but these lead ultimately to continent–continent collisions. This is the situation that was observed during the Palaeozoic supercontinent cycle and is being observed for the Mesozoic–Cenozoic supercontinent cycle, which is still in progress.

28.2.2 The supercontinent cycle and its relationship with climate

There are two types of global Earth climate: icehouse and greenhouse. Icehouse is characterized by frequent continental glaciations and severe desert environments. We are now in the icehouse phase, but moving quickly towards greenhouse. Greenhouse is characterized by warm climates. Both reflect the supercontinent cycle. In the icehouse climate, continents move together, sea level is low due to lack of sea-floor production, climate is cooler and more arid, and the formation of supercontinents begins.

In a greenhouse climate, on the other hand, continents are dispersed, sea level is high and there is a high level of sea-floor spreading, with relatively large amounts of CO_2 production at oceanic rifting zones. The global climate is warm and humid and is associated with calcite seas.

While this chapter is about change in the global oceans, it is useful to reconsider the changes in global mean temperature over time (Figure 28.2), since temperature, ice ages and sea level are closely linked.

In terms of geological time, Figure 28.2 shows some very abrupt changes, for example between 140 000 and 125 000 years ago (Figure 28.2(c)) and between 20 000 years ago and the present (Figure 28.3(d)). In the latter example, the climate changed from one supporting huge mid-latitude ice sheets to one supporting none of theses, and even the ice caps and mountain glaciers that remain are melting year on year. The effect of these variations in global mean temperature on sea level in the past 120 000 years are

seen in Figure 28.2, where icehouse and greenhouse phases can be identified.

28.2.3 Global temperature, ice ages and sea-level change

Sea-level reduction is caused by an increase in global ice volume, transferring water from oceans to the continents. A lowering of sea level causes marine ice sheets to grow more, which causes more sea-level fall – a simple positive feedback process. A drop in sea level is called a regression, whereas a rise is called a transgression. Marine ice sheets cannot start sea-level fall, which is caused by the growth of terrestrial ice masses. Since these are formed when surface temperature is reduced, the main cause of ice ages, and therefore change in sea level, is related to climate change.

Although a long-term component (200–400 million years) in sea-level change during the Phanerozoic seems to be agreed upon by most investigators, short-term components and amplitudes of variation still remain uncertain and controversial. Major transgressions are recorded in the early Palaeozoic and the late Cretaceous (Figure 28.3). Estimates of the amplitude of the Cretaceous sea-level rise range from 100 to 350 m above the present sea level, and estimates of the early Palaeozoic rise are usually tens of metres. Cyclicity in sea-level change has been proposed by many investigators, with periods in the range 10–80 million years. The Vail sea-level curve (Figure 28.3) has been widely accepted, although amplitudes are subject to uncertainty.

Figure 28.4 shows two independent sea-level estimates (metres below present), both of which are subject to some inaccuracies. The first method relies on the fact that $\delta^{18}O$ concentration within an ice sheet is different to that in an open ocean. The $\delta^{18}O$ concentration within an open ocean will depend on global ice volume and is recorded in deep-sea sediment cores that are rich in benthic foraminifera. Sea-level change can therefore be estimated through benthic oxygen isotope data.

A second method of eustatic sea-level reconstruction uses the ^{14}C dating of low-latitude coral reefs; the curve estimated from raised marine terrace studies from New Guinea is shown in Figure 28.4. This curve suggests that the global sea level at the late glacial maximum was about 120 m lower than at present.

Several factors can cause short-term changes in sea level. Tectonic controls are important, particularly continental collisions, which operate on time scales of 1–100 million years, and produce rate change of less than 1 cm per thousand years. Then there are glacial controls that operate on time scales of 10 000 to 100 million years, which produce rate changes of up to 10 m per thousand years. With the present ice cover on the Earth, the potential amplitude for sea-level rise if all ice were to melt

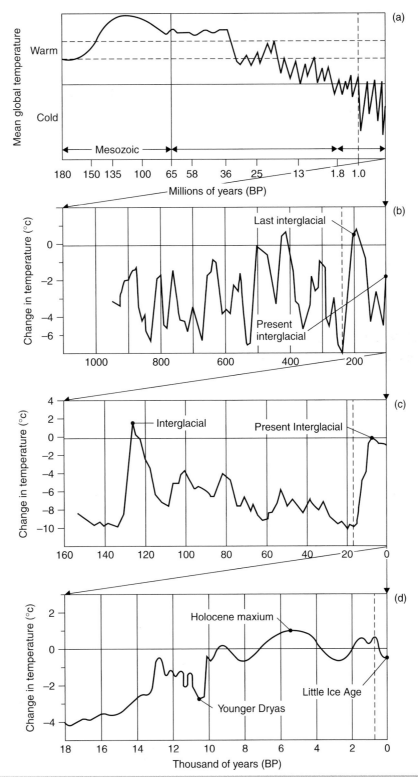

Figure 28.2 Global temperature variation since: (a) 180 000 000 years ago; (b) 1 000 000 years ago; (c) 160 000 years ago; (d) 18 000 years ago. The Quaternary period from 1.8 million years ago to the present day can be split into two subdivisions, the Holocene (representing the last 10 000 years) and the Pleistocene (between 1.8 million and 10 000 years ago). Last glacial period between 130 000 and 10 000 years ago can be termed the 'Late Quaternary'. However, it is also referred to by formal chrono-stratigraphic names such as 'the Weichselian' in northern Europe, 'Devensian' in the UK, 'Würm' in central Europe and 'the Wisconsin' in North America. The last glacial maximum (LGM) was recorded at about 21 000 years ago. This glacial episode is called (in northern Europe) the Late Weichselian between about 40 000 and 10 000 years ago. The last ice sheets had disintegrated by 10 000–7000 years ago. (Source: Siegert, 2001, p. 2). Reprinted from Ice sheets and Late Quaternary environmental change / Martin J. Siegert, 2001, p. 2, with permission from John Wiley & Sons Ltd.

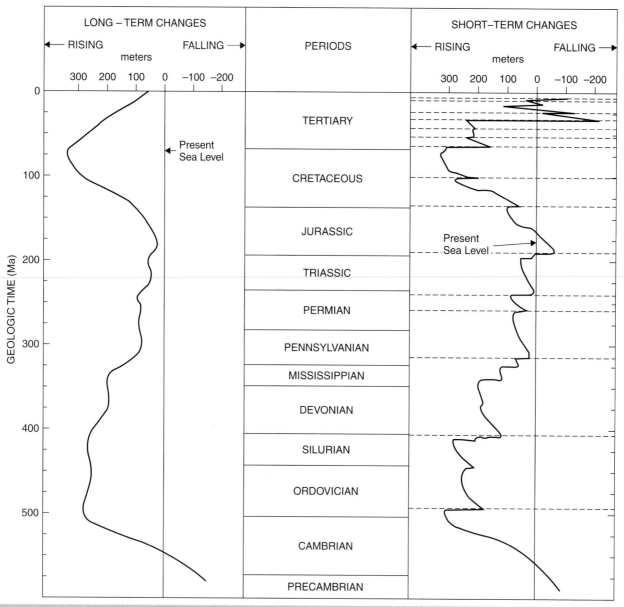

Figure 28.3 Long- and short-term changes in sea level during the Phanerozoic (Source: Condie, 2005, p. 200)

is 200 m. The short-term drops in sea level 500, 400 and 350 million years ago (Figure 28.2) probably reflect glaciations at these times. The rise of the Himalayas in the Tertiary period produced a drop in sea level of around 50 m. Also, waves and tides, storm winds and hurricanes, tsunamis, melting methane hydrates (clathrates) and catastrophic sediment slumps can affect tidal range.

Long-term changes in sea level are related to (a) the rates of seafloor spreading and the volume of ocean ridge systems; (b) the characteristics of subduction; (c) the motion of the continents; and (d) the supercontinental insulation of the mantle: as the mantle warms beneath a supercontinent, the continental plate rises isostatically, with a corresponding fall in sea level.

28.3 Ocean circulation in a warming climate

The characteristics of ocean circulation were discussed in Chapter 8. Climate models discussed in Chapters 10 and 27 predict that the ocean's circulation will weaken in response to global warming, (See the companion website for Box 28.1 Ocean circulation in a warming climate.) but the warming at the end of the last ice age suggests a different outcome. See the companion website for Box 28.2 Sea-level rise: what makes accurate prediction so difficult? and Box 28.3 Rapid climate change: daily measurements of the Atlantic overturning circulation.

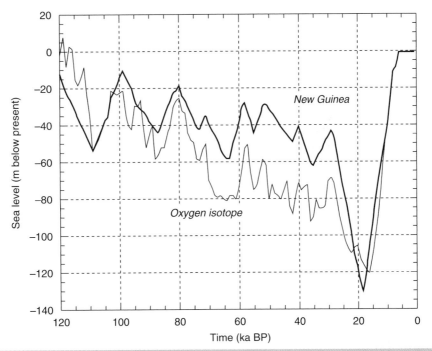

Figure 28.4 Records of global sea-level change over the past 120 000 years, from New Guinea terraces and sea-floor-sediment δ^{18}O records. (Source: Siegert, 2001, p. 4). Reprinted from *Ice sheets and Late Quaternary environmental change* / Martin J. Siegert, 2001, p. 4, with permission from John Wiley & Sons Ltd.

Exercises

1. How long does it take for one complete supercontinent cycle years to occur?

2. Explain the difference between the two types of global Earth climate: icehouse and greenhouse.

3. When the continents are together, is sea level high or low?

4. In terms of sea-level change, explain the difference between regression and transgression.

5. Describe the factors which cause short-term changes in sea level.

6. Describe the factors which cause long-term changes in sea level.

7. Outline the likely changes in ocean circulation in a warming climate.

References

Condie, K.C. 2005. Earth as an Evolving Planetary System. London, Elsevier.

Dutch, S.I., Monroe, J.S. and Moran, J.M. 1998. Earth Science, Boston, International Thomson Publishing.

Siegert, M.J. 2001. Ice Sheets and Late Quaternary Environmental Change. Chichester, John Wiley & Sons, Inc.

Further reading

Gurnis, M. 1988. Large-scale mantle convection and the aggregation and dispersal of supercontinents. *Nature.* **332**, 695–699.

Murphy, J.B. and Nance, R.D. 1992. Supercontinents and the origin of mountain belts. *Scientific American.* **266**(4), 84–91.

Nance, R.D., Worsley, T.R. and Moody, J.B. 1988. The supercontinent cycle. *Scientific American.* **259**(1), 72–79.

Sverdrup, K.A., Duxbury, A.C. and Duxbury, A.B. 2003. An Introduction to the World's Oceans. London, McGraw-Hill.

Thurman, H.V. 2001. Introductory Oceanography. Upper Saddle River, NJ, Prentice Hall.

29 Biosphere Evolution and Change in the Biosphere

A reconstruction of Tiktaalik roseae. (Source: Wikimedia Commons)`

Earth Environments   David Huddart and Tim Stott
© 2010 John Wiley & Sons, Ltd

Learning Outcomes

After reading this chapter and completing the exercises you will be able to:

➤ **Understand how evolution occurs in the fossil record.**

➤ **Understand how life on Earth may have originated.**

➤ **Describe the outline stages of the Earth's biospheric evolution, including the earliest life on Earth, the eukaryotes and the diversification of life, the evolving metazoans and the Ediacaran fauna, the Cambrian explosion of life, the colonization of the land, the reptilian radiation, the dinosaurs and their predecessors and the origin of mammals.**

➤ **Understand that mass extinctions have occurred which have had major effects on the biosphere's evolution and why these might have occurred.**

29.1 Introduction

The origins of life on Earth and the key stages of its evolution will be covered in this chapter. In this evolution of the planetary biosphere there are close links with other sections of this book, from the major geological processes like plate tectonics and volcanic activity, to the evolution of the atmosphere and major sea-level changes.

Much is unknown about the development of life on this planet simply because of the paucity of the fossil record, which reflects only a fraction of the life that has existed. This is especially so on land, where there is generally poor fossil preservation potential, and in animal groups such as the insects, where even the modern fauna is poorly studied and the species are vast in number. The record from the marine environment is much better but even here it is far from complete. The story is also extremely complex and for a fuller understanding the reader must turn to one or more of the specialist texts that are recommended later.

Absolutely crucial to any understanding is the idea that fossils and their study in palaeontology give a special perspective into geological time and insights into how evolution takes place. Through the fossil record, the history of life across time and the relationship between evolving lineages can be better understood, and evolutionary patterns can be seen more clearly.

It is also apparent that rates of evolution have varied through geological time: occasionally the change has been rapid, whilst at other periods there seems to have been little evolutionary modification. So it appears that we have rapid bursts of change in evolution, where there is diversification from a single stock, with its descendents becoming adapted to life in new habitats. This adaptive radiation is an important pattern revealed by the study of fossils.

29.2 Mechanisms of evolution in the fossil record

The two main modes of evolution are termed microevolution and macroevolution. The former refers to small-scale changes that take place within species, and especially to the transformation from an existing species to a new one. Macroevolution concerns evolution above the species level, including the origins of major groups, such as families or classes, and adaptive radiations.

In microevolution there are changing interactions between the organism and its environment, gradual descent with modification under the processes of classic Darwinian natural selection, where individuals with character combinations suited to survival are favoured (see the classic example of the evolution of the horse in Box 29.1, see the companion website). Gradual change takes place through the accumulation of small variations from one generation to the next.

In the 1950s it was realized that species could arise by lineage splitting, or allopatric speciation (cladogenesis), in small isolated communities at the fringe of the geographical range of a species, or by transformation of the whole population through intermediate stages in an otherwise stable environment (anagenesis, or sympatric speciation). Together, these types of speciation are referred to as the process of phyletic gradualism (Figure 29.1), which implies a slow, even methodical change in the whole gene pool.

It is unfortunate that there are very few cases in which fossil populations show a gradual change through time, with a series of graded intermediate forms. It has been argued that this is due to the incompleteness and bias of the fossil geological record. However, it has also been suggested that the rock record can be taken at face value and that evolutionary change was not slow and steady but took place only in punctuations, or short-lived bursts, in which a new species evolved from a parent one, often with relatively large morphological change, and then lived without much change until extinction. All the evidence from Devonian trilobites from eastern North America and Pleistocene land snails in Bermuda suggests the abrupt origins of new species, followed by stasis (standing still). This model of evolutionary change has been called punctuated equilibrium (Figure 29.1).

Rapid changes might take place through geographic speciation, or allopatric speciation, where a small part of a homogeneous breeding population in one area moves and becomes geographically isolated, for example as part of plate movement. Whereas gene flow is achieved throughout the larger population by frequent interbreeding, maintaining stability, natural selection or genetic drift in the small population will modify it quite rapidly, especially if it is adapting to a new environment. The new population will carry only a small part of the total gene pool of the parent population (Mayr's founder principle) and this will be reinforced by mutations. The result can eventually be a new species, reproductively distinct from the initial one.

It seems that there are now a number of documented cases showing both types of evolutionary change: long-term stasis and the speciation by abrupt morphological leaps. Both appear to be

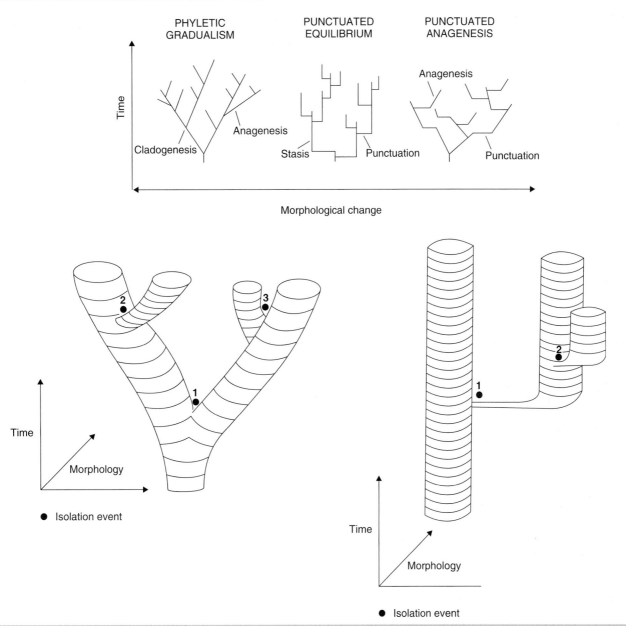

Figure 29.1 Modes of evolutionary change. In phyletic gradualism (right), evolution proceeds through either gradual morphological change (anagenesis) or lineage splitting through the isolation of peripheral populations (cladogenesis). In punctuated equilibrium (left), evolution proceeds through rapid morphological change after isolation of peripheral populations. No gradual changes occur in between these events. (Source: Doyle, 1996)

possible. For example, work on Lower Ordovician trilobites from Spitsbergen shows many olenid, benthic trilobites with a clear stepwise evolution with the abrupt origin of new species, followed by stasis. In the same beds the large-eyed trilobite *Carolinites* is found, which was pelagic and swam in the upper waters of the sea. It showed a continuous and gradual change through the same succession, demonstrating that the two competing theories are in fact complementary.

In reviewing examples of change in fossil species, Doyle (1996, pp. 80–84) suggests that evolution may proceed as punctuated anagenesis (punctuated gradualism), with gradual change, rather than stasis, occurring between rapid bursts of punctual change (Figure 29.1).

These processes are related to how species change, but how do major plant and animal groups form? This is the domain of macroevolution, which deals with the origin and evolution of higher taxonomic groups and reflects the overall change in the fossil record through time. In macroevolution the unit of selection is the species and the source of variation is rapid speciation events and species selection acts on those species with favourable character combinations. Here after a geologically brief period of change the species have their characteristics defined from the beginning. Usually there is then little change, although over long time periods there is often a trend, such as the increased complexity of ammonite sutures (traces of the septa on the wall of the shell).

In the fossil record, as we will see later, there are sudden changes indicating an initial rapid evolutionary period. These changes can be relatively small at lower taxonomic levels, but much larger ones result in the origin of higher taxa: orders, classes and phyla. The links between higher taxa are obscure and poorly represented in the fossil record. For example, linking or transitional forms appear to be absent in the rapid diversification of the early Cambrian life. Many new structural plans appear rapidly in the record; these allowed great evolutionary diversification, such as the derivation of the graptolite from monograptids to the diplograptids. Such changes are now understood in terms of changes of regulatory genes, which control the operation of structural genes.

Evolutionary rates within evolving groups are not constant and for most fossil groups there is a general pattern. Evolutionary change is much more rapid in the initial stages of the life of a higher taxon than later on. When a new successful animal group occurs for the first time, with key adaptive innovations, there is a rapid evolutionary burst as new taxa proliferate rapidly and expand their range quickly, and new groups adapt to different environments. This process is known as adaptive radiation, in which there is a high production of species and genera, and sometimes families and orders, and will be illustrated later in this section.

The initial impetus for an adaptive radiation is commonly a new structural plan. In marine organisms these are often linked with marine transgressions. Here new space is created and new environments are generated, which allow opportunistic groups to arise and colonize. Then the rate of species production slows down and the species tend to survive longer. Phyletic micro-evolutionary change allows adaptive fine-tuning and if a particular species becomes extinct it is likely to be replaced by a new one from a related stock, which will become adapted to its environment in the same way.

If the origin of new species more or less equates with the extinction of older ones then the lineage will survive and remain vigorous. Living fossils, like the coelocanth, are those species that survive in the virtual absence of evolutionary change; the deep sea, which is a quiet and stable environment, is a good habitat for such populations. But usually it seems to be the case that all taxa become extinct. Throughout the history of life individual taxa have originated, flourished for a period of time and then become extinct. The end of individual species has left little trace on the pattern of evolution, but much more important are the periods of mass extinction. More than any other factor, these have controlled the whole pattern of life's evolution on this planet. Conditions have been reset in new and different combinations, but as we will see, the controls on such changes are potentially many and often it is difficult to unravel the potential complexity. They include sea-level change, climatic change, vulcanicity, extraterrestrial impacts and, in the case of the latest event, the effects of humans, but in reality many of these factors may be interrelated.

The pattern of extinction and radiation is illustrated in Figure 29.2, which shows the diversity of life at different periods in the Earth's history. The clusters of evolving groups of animals and plants, or clades, show rapid reductions of diversity (extinctions),

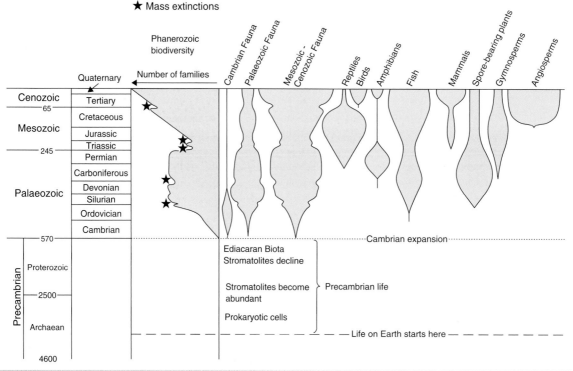

Figure 29.2 Changes in the biosphere through time. (Source: Doyle, 1996)

whilst radiations are shown by rapid increases in faunal diversity over a brief time interval. See the companion website for Box 29.1 Case study of an evolutionary trend: horse evolution.

We now need to consider some of the major events in an outline history of the biosphere. The first stage is to try and answer the question: How did life originate on planet Earth?

29.3 The origins of life

In the early stages of the Earth's evolution, until about 3.9 billion years ago, the planet was bombarded by large meteorites, which would have been able to vaporize any liquid on the surface and even to have melted the solid crust. We must look for the evolution of life on the planet at about this time period, as the first sedimentary rocks are dated to about 3.8 billion years.

These rocks show that there was surface liquid water and an atmosphere that was reducing and lacked free oxygen. There is much debate as to the actual composition but it may have consisted of methane (CH_4), ammonia (NH_3), water and hydrogen, although alternative ideas suggest nitrogen, carbon dioxide and monoxide, water and various other mixtures. Other important chemicals included formaldehyde (CH_2O) and cyanide (CHN) as the basic building blocks of more complex organic molecules, although experimental work has suggested that many simple organic molecules could have been synthesized abiotically from any of these gaseous mixtures, provided that there was a suitable energy source for the reactions.

It is clear though that life evolved in an environment that would have killed us instantly. Amino acids, some of the nucleotide bases and carbohydrates have been produced experimentally and these are vital to the origins of life. In 1953 Miller made a model of the Pre-Cambrian ocean in a laboratory glass vessel. He exposed a mixture of water, nitrogen and carbon monoxide to electrical sparks, to imitate lightning, and found that after a few days there was a brownish sludge in the bottle, which contained sugars, amino acids and nucleotides. Further experiments led to the production of polypeptides, polysaccharides and even large organic molecules. Other experiments even produced cell-like structures, in which a soup of organic molecules became enclosed in a membrane. Protocells seemed to feed and divide, but they did not survive for long. Nevertheless, the crucial step from molecules in solution to living cells has not been achieved in the laboratory.

Not all scientists accept an organic origin for life and there is an inorganic model in which it is proposed that the first genetic material arose in association with clay minerals on the early Earth and that organic molecules became involved only later (see Benton and Harper, 1997, p. 65). However, some simple organic molecules have been detected in interstellar space and complex ones have been found in meteorites, thus raising the idea of an extraterrestrial source of life, brought in by comets and meteorites (see Box 29.2, see the companion website). On Earth the energy sources could have been electrical discharges, ultraviolet light and ionizing radiation, hot springs, lava or solar heat.

We can suggest that eventually the Earth's surface was covered by oceans of some kind, which contained, in Haldane's classic phrase, 'a hot dilute primitive soup' of organic material, or Darwin's 'warm pond'. Most theories suggest that the location for the evolution of life might include tropical lagoons or cold volcanic islands, although there have been suggestions that life began not on the surface but deep down mid-ocean-ridge systems at the hot smokers.

The further stages are not as clear, as the amino acid molecules had to be brought together and polymerized as proteins which could replicate themselves through DNA. The origin of the primitive cell was essential for further evolution as the cell membrane acts as a buffer between the living material and the outer environment.

So life went through a stage of chemical evolution on a lifeless planet. The necessary conditions must have included those in Table 29.1, while the energy sources available are illustrated in Figure 29.5. However, it has been suggested that early volcanic gas had the chemical credentials to spawn basic proteins. Carbonyl sulphide (COS) may have been instrumental in stringing together the first molecular building blocks and this may answer one of the

Table 29.1 Necessary conditions for the evolution of life from nonliving chemicals. (Source: Cowen, 2000)

NECESSARY CONDITIONS FOR THE EVOLUTION OF LIFE FROM NON-LIVING CHEMICALS

Energy is needed to form complex organic molecules. Some laboratories have used electrical discharges to simulate lighting on the early Earth; others use high-energy particles from cyclotron in place of radioactivity from rocks and cosmic pays heat for volcanic activity, shock waves or laser beams for meteorite impacts, or lamps for solar UV radiation. All these energy sources were present on the early Earth.

Protection. Continued energy input (especially heat) will destroy any complex organic molecules that form in reactions, so after they form they must quickly be protected from strong radiation. Laboratory experiments are often designed to allow organic molecules to drop into cold water away from the energy source. On the early Earth, molecules may have been protected under shallow water, in sheltered tide pools or rock crevices, under rocks, ice, or particles of sediment.

Concentration. All chemical reactions run better at high concentration, but almost all reactions leading toward life give low yields in the laboratory. Life is water-based, yet too much water dilutes chemicals so that they react slowly. Some process must have concentrated chemicals on the early Earth. Evaporation is one, and there are others.

Catalysis. Catalytic converters in the exhaust systems of cars contain platinum as a catalyst that encourages the breakdown of pollutants. An organic substance that works as a catalyst is called an **enzyme**. All the reactions inside our cells and our bodies are aided by enzymes, which are necessary even in the simplest possible living cell. Suitable catalysts may have encouraged difficult reactions on the early Earth, even at low energy levels and low concentrations. Later, the last stages leading toward life may have been aided by catalysts trapped on or inside membranes.

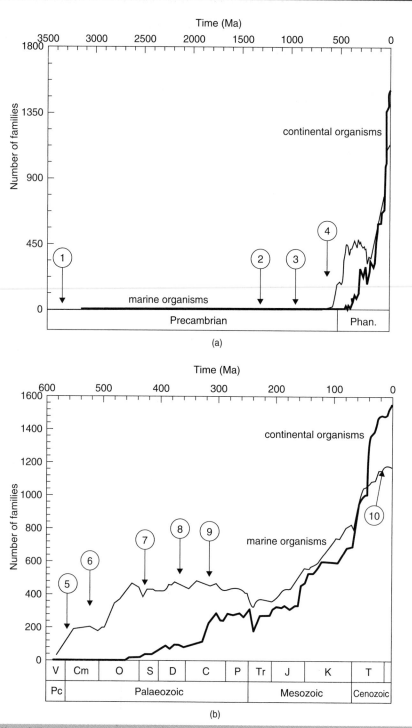

Figure 29.5 The diversification of life, plotted for (a) the whole of the past 4000 million years and (b) the Phanerozoic. 10 major biological advances are illustrated (see the text). (Source: after Benton and Harper, 1997)

difficult questions surrounding the origins of life: How did the first complex biological molecules appear, given that there were no organisms around to produce them? Volcanic gas may have been responsible for creating the first rudimentary proteins. Amino acids have been exposed to COS in the laboratory and peptide chains (the strands from which proteins are woven) appeared very

quickly and could be created through several chemical processes, such as oxidation, alkylation and metal catalysis.

COS makes up only about 0.1% of volcanic gases today but the prebiotic atmosphere probably had a significant percentage. This means that volcanoes could have been the cradles for life and the reactions found in the laboratory could have taken place in lakes close to volcanoes, or in areas of underwater volcanism associated with deep vents (hot smokers). This chemical agent was capable of stringing together amino acids but there are problems related to how amino acids appeared in the first place: were they formed on Earth or brought in from space?

It is clear that the palaeontological record cannot provide any information on these processes as direct information on such important components to the origin of life, such as the terrestrial atmospheric composition, the ocean temperature and pH values, is unknown. This is why the experimental approach has proved so valuable. Further ideas on how the first living cells evolved are covered in Cowen (2000, pp. 6–11) and a detailed discussion on the origins of life is given by Lazcano (2003) and Rollinson (2007, Chapter 6), but what is clear is that the primitive life forms would have been very different to the simplest cells that are alive today as these are very complex, having evolved through many billions of generations. After this initial stage, however, there is much more direct evidence for the evolution of the biosphere from the fossil record. See the companion website for Box 29.2 Meteorite impacts and the origins of life.

The question of how life originated and what processes were involved is one of the most fundamental to be posed but one of the most difficult to answer. In the next section the overall history is briefly considered, before subsequent sections deal with the major events.

29.4 An outline history of the Earth's biospheric evolution

Palaeontology has given clues as to how life on Earth has reached its present diversity but what can be seen in any overview is that many major events in life's history can be identified. Benton and Harper (1997, pp. 306–310) present one such overview, which shows that there were major adaptations which enabled substantial increases in diversity to occur (Figure 29.5):

1. The origin of life took place. There was a synthesis of organic molecules in the Pre-Cambrian oceans. The first living organisms were bacteria, which were prokaryotic cells that were small and lacked a nucleus. These initial life forms operated without oxygen, but they caused a significant change: the raising of oxygen levels.

2. Large eukaryote cells developed, with organelles and membrane-bounded nuclei containing the chromosomes. They reproduced sexually. This mixing of genetic material during sexual reproduction opened the door to the exchange of genetic material, mutation and recombination, and the development of variation in populations, which are the basic materials for evolution.

3. Multicellular organisms (metazoans) developed. These were clusters of eukaryote cells organized into different tissue types and organs, in which different parts of the organism were responsible for particular functions and tasks.

4. During the Cambrian a wide variety of mineralized skeletons appeared, which provided protection, support and areas for the attachment of muscles. Predator pressure may have been the reason for the development of these hard parts, but marked changes in ocean chemistry during the early Cambrian marine transgression may have forced organisms to take in and excrete large quantities of minerals. Enhanced oxygen levels in the world's oceans made precipitation of minerals a lot easier.

5. There is evidence of much predation, which must have formed a new part of the ecosystem in which there was a rapid diversification of armoured and protective strategies and the development of predator–prey systems.

6. Biological reefs developed. Throughout the Phanerozoic, from the early Cambrian through to the Tertiary, the main reef builders changed, with major changes punctuated by mass extinctions.

7. The colonization of the land added major new environments for life. Although soils have been reported from the Pre-Cambrian, colonization by microbes in the hot, anaerobic and irradiated landscape was unlikely. By the Mid-Ordovician, soils were coated by a green scum of cyanobacteria, and by the Silurian, small vascular plants with stomata, a waxy covering and trilete spores were established. This development of plant growth stabilized the land, cut down erosion and had a major effect on the physical nature of the planet. Colonizing invertebrates were faced with problems of dehydration, respiration and support, but they developed waterproof skins, lungs and skeletal support, and by the early Devonian the land was colonized by myriopods, insects and arachnids. By the Carboniferous, oligochaete worms, scorpions and gastropods were present. Vertebrates moved onto land during the Devonian to exploit the new sources of plant and invertebrate food. Full terrestrialization occurred with the reptiles in the Mid-Carboniferous, when the amniotic egg developed. The early reptiles gave rise to the mammal-like reptiles in the Permian and Triassic. The dinosaurs arose from another group of early reptiles, the thecodonts, in the Mesozoic. By the end of the Mesozoic most of the land, sea and flying reptiles had been killed off in the End-Cretaceous extinction. These were replaced by a radiation of mammals, derived from the mammal-like reptiles, and of birds, derived from the dinosaurs.

8. The major development of forests occurred during the Carboniferous, although trees had already developed during the Devonian. This created a vertically-tiered range of new habitats and led to a dramatic burst in the diversification of vascular land plants, as well as a radiation of insects and their predators, such as spiders, amphibians and reptiles.

9. The next major expansion of life was into the air with the development of flight. This was the reason for the great

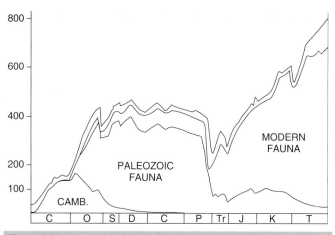

Figure 29.7 Sepkoski's three great faunas

success of the insects. There has been development of flight in the vertebrates at least 30 times, from the gliding diapsid reptiles of the Late Permian, through the pterosaurs which dominated Mesozoic air space, to birds and bats and gliding marsupial and placental mammals.

10. The final stage was the development of the extremely successful human species and of consciousness (the ability to think about one's existence, about the past and the future), which allowed us to plan. Success in evolutionary terms involves a species' abundance, its dominance of the physical and biotic environment (and its manipulation of it in major ways) and its longevity. We can certainly claim the first two characteristics, but we are not certain about the last one just yet.

Another way of looking at the faunal evolution is to consider the diversity of life and how it changed through time. The general pattern has been known since Phillips defined the Palaeozoic, Mesozoic and Cenozoic eras in 1860, before the fossil record was well known, and the general trends can be seen in the Sepkoski curves (Sepkoski, 1984 and Figure 29.7), which illustrate the diversity of marine families through time. These curves show clear trends: few families existed in Ediacaran times but the Cambrian saw a dramatic increase to around 200; there was a new, dramatic rise at the beginning of the Ordovician, which raised the total to over 400, a number that was relatively stable through the Palaeozoic; in the Late Permian there was a dramatic drop to 200 families, but a steady rise in the Triassic continued to the present, with over 800 families, with a short reversal at the end of the Cretaceous.

The positions of four mass extinctions are shown in Figure 29.8, but there have been others. Diversity is the net product of the species originations minus mass extinctions. Sepkoski noted three great divisions of marine life through time: the Cambrian, Palaeozoic and Modern faunas (Mesozoic–Cenozoic Faunas) (Figure 29.7). The histories of these three marine faunas are illustrated in Figure 29.8.

The Cambrian faunal increase in diversity was dominated by trilobites (77%), which were mainly surface-digging deposit feeders, but after a Late Cambrian diversity peak, the Cambrian fauna declined in diversity through the Ordovician and afterwards. The Palaeozoic fauna, dominated by articulate brachiopods and other suspension-feeding animals, was responsible for the great diversity rise in the Ordovician and slowly declined thereafter. The Palaeozoic fauna suffered markedly in the Late Permian extinction and its recovery afterwards was insignificant compared with the dramatic diversification of the Modern fauna. This is rich in

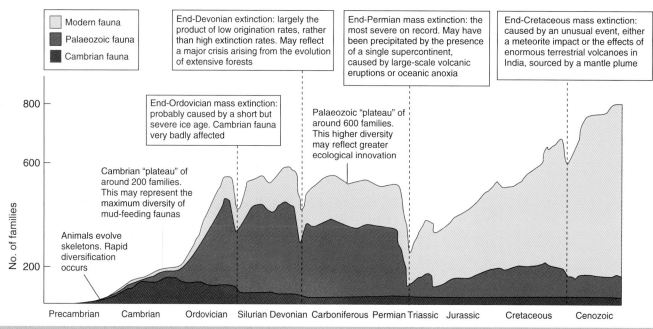

Figure 29.8 History of the three great faunas. (Source: after Cowen, 2000)

molluscs and is dominated by swimming predators such as fish and cephalopods; strongly burrowing groups such as most bivalves, crustaceans and echinoids; and versatile opportunists such as crustaceans and gastropods. The overall trend has therefore been to add new ways of life, or guilds, to marine faunas through time, and each of the three faunas exploited an increased number of potential niches available on the shallow marine shelves. They did this physically by growing above and below the sea floor and also biologically by introducing feeding-method innovations and increasing predation.

Abiotic effects seem to have influenced the shape of the diversity curves. For example, there is an apparent relationship between the formation of the Pangaea supercontinent and the End-Permian mass extinction. Major continental break-ups such as those of Pangaea during the early Mesozoic and Rodinia during the Cambrian seem to have resulted in increased adaptive radiation. This could be because continents were becoming isolated, causing biota to evolve independently and therefore increasing species diversity. Conversely, as plates collide and supercontinents come together the range of habitats declines and species diversity decreases.

However, plate-tectonic events cannot be the sole explanation because if they were the only control on diversity then much the same groups of animals would rise and fall with plate configurations. Instead there are dramatic changes in the different animal groups that succeed one another through time. Also, the overall increase in diversity is not predicted on plate tectonic grounds and there are too many mass extinctions in the fossil record for the plate-tectonic hypothesis to be the full explanation.

So even if plate-tectonic factors set the planet up for an extinction there does need to be a separate cause or causes to explain the extinctions themselves. For example, changes in climate from icehouse to greenhouse situations may have had an effect on species diversity, though the relationship is not clear-cut. The abrupt transition from greenhouse to icehouse in the Late Ordovician may have brought about the End-Ordovician mass extinction, yet other similar transitions such as that in the Carboniferous appear to have had no effect.

The development of the terrestrial ecosystem, dominated by plants, was an extremely important part of the Phanerozoic life story. This flora was associated with the first land colonization in the Ordovician–Silurian. Here there were the adaptive advantages of a rigid stem and later tissue that was capable of carrying nutrients through the plant, which allowed plants to leave the aquatic environment for the first time. Continued evolution of the woody and root tissue of plants saw the development of the lycopod/sphenopsid coal forest and the tree ferns in the later Palaeozoic. These seedless trees of the Carboniferous were replaced by the widespread Mesozoic gymnosperm floral radiation with the development of a reproductive system that was not dependent on water, as it was in the coal forests. Reproduction was through seed that was fertilized on the plant, and, as in the modern conifers today, they had male (with pollen) and female (seed) cones. The dominance of the gymnosperms was reduced with the later Mesozoic angiosperm radiation (flowering plants) in the Cretaceous.

Today these plants dominate the land, with 96% of the present-day flora. Their success is associated with their improvement of the seed-bearing reproductive system and of dispersal mechanisms associated with insects, and their ability to colonize habitats unavailable to other plants.

Another important question when considering the diversity of life is: How far is the fossil record complete? Most organisms are incorporated into rocks by chance and it is the time interval between death and burial which is crucial to the likelihood of preservation. During this period there is the most chance for potential fossils to be destroyed. Rapid burial, excluding oxygen and potential scavengers, is the preservation key. The fossil record is therefore bound to be incomplete and not all organisms will be preserved. For example, organisms living on land are less likely to be preserved than aquatic organisms, and soft-bodied organisms, or fragile ones, are less likely to be preserved that organisms which have hard parts or are more robust.

The fossil record of shelly organisms living in an aquatic environment is very good, comparing favourably with the diversity of modern species. This is particularly true of the *Brachiopoda*, which are a major group of shelly organisms more abundant in the geological past than the present day.

On the other hand, the most diverse group of animals living today is the insects. Their fossil record is poor because they are mostly terrestrial and they have very fragile jointed limbs and appendages. Both this group and worms are under-represented in the fossil record. During the process of fossilization there is loss of data about the organism and its life habit. The fossil record therefore has biases to organisms that have tissue resistant to decay, to marine organisms and to organisms living in low-energy environments.

However, it is likely that the idea that the fossil record is incomplete and often inadequate to demonstrate the diversity and evolution of the Earth's life has been overstated at times. This record is an extremely important data source on the nature and evolution of life and it is as reliable as any other Earth line of evidence in science. For example, in about 95% of cases the sequence of fossil species as found in the fossil record reflects the order in which they actually evolved; the sequence of fossils provides evidence of the evolution pattern but not the mechanisms; the sequence of fossils provides the most rigorous test of evolutionary hypotheses derived from comparisons of their morphology; and the fossil record is a rigorous test of evolutionary hypotheses. See the companion website for Box 29.3 Cycles in fossil diversity.

29.4.1 Earliest life in the fossil record

The first organisms on Earth operated in the absence of oxygen, had anaerobic metabolisms and go back at least 3.5 billion years. They include rare structures identified as possible stromatolites from various parts of the world. These stromatilitic structures were presumably constructed, as today, by cyanobacteria and other prokaryotes (see Figure 29.10 for a modern example).

Cyanobacteria live in shallow seas and require good light for photosynthesis. They form thin mats on the sea floor in

Figure 29.10 Stromatolites from the modern freshwater pozas (pools), Cutarocienegas, Coahuila, Northern Mexico

order to maximize their sunlight intake, but occasionally they are overwhelmed by sediment, migrate towards the light again and recolonize the top of the sediment. This process may recur, giving extensive layered structures, which are preserved.

Prokaryotes are a paraphyletic assemblage of bacteria that today occupy a range of habitats, from hot springs, Antarctic snowfields, human digestive systems to hundreds of metres into the crust. There are two main sub-groups: the Kingdom Eubacteria includes the cyanobacteria (blue-green algae) and most groups commonly called 'bacteria'; the Kingdom Archaebacteria consists of the Halobacteria (salt-digesters), Methanobacteria (methane producers) and Eucytes (heat-loving, sulphur-metabolizing bacteria). Today insights from molecular phylogenetics suggest a tree of life with three domains, and the geological evidence from isotopic and biomarker evidence suggests that the bacteria were in existence by 3.5 billion years ago and that the Eucarya had separated from the Archaea before 2.7 billion years ago. Photosynthesizing cyanobacteria were in existence by or before 2.7 billion years ago, several hundred million years before the rise of atmospheric oxygen. The problems with recognizing the earliest life forms can be seen in Box 29.4 (see the companion website).

Rare single-celled fossils are known from Australian and African cherts, dated to about 3.5 billion years ago, and include spherical and filament-like chains of cells, for example the 11 species of bacteria and cyanobacteria from the Apex chert in Western Australia, which are between 10 and 90 microns in length. Unusual carbon spheres from 3.8 billion-year-old rocks at Isua in Greenland have been interpreted by some as the remains of simple cells, but doubts remain. Still, isotopic evidence suggests that life evolved the ability to produce complicated biochemical pathways early in its history. Modern photosynthesizing organisms draw carbon dioxide from the atmosphere and preferentially extract more of the lighter stable isotope of carbon, ^{12}C. Hence a light carbon isotope signature is evidence for photosynthesizing organisms. This may be present in the Isua rocks (see Box 29.4,

see the companion website) and is certainly common in rocks less than 3.7 billion years old.

In the Warrawoona Group (north-western Australia) and the Swaziland Supergroup in southern Africa there are five lines of evidence to support the presence of an active ecosystem 3.4 billion years ago: populations of spheroidal organic microspheres (Archaeosphaeroides) embedded in cherts with fine-scale organic laminae; isolated organic filaments; stromatolitic sediments; an organic geochemical signal of negative δ^{13}C values; and absolute buried carbon abundances (average 0.46% by weight), which are comparable to those found in modern sedimentary basins (where there is a range from 0.18 to 0.76%). All this data suggests that cellular life on Earth evolved between 3.9 and 3.4 billion years ago and it seems that at least 12 major groups of prokaryotes arose in the Pre-Cambrian and survived to the present day. These are still the most numerous life form on the planet. However, in light of the debates above, some of these sites, and others reviewed in Rollinson (2007), may need future discussion. See the companion website for Box 29.4 Searching for the oldest life in western Greenland.

29.4.2 Eukaryotes and the diversification of life

The generally larger eukaryotes, as single-celled or multicellular organisms, evolved later and there has been intense debate as to the origin of this form of life. All living organisms other than prokaryotes and viruses are eukaryotes. They are distinguished from prokaryotes by having a nucleus that contains their DNA in chromosomes and cell organelles. The latter are specialized structures which perform key functions, such as mitochondria for energy transfer, cilia or flagella for movement, and chloroplasts in plants for photosynthesis. They are a much higher grade of organization than prokaryotes as the latter have no nucleus and only a single strand of DNA.

The origin of eukaryotes was through a process called endosymbiosis. Here individual prokaryotic organisms were incorporated into eukaryotic cells to produce independent entities whose functions were carried out by specialized organelles derived from the absorbed organisms. The mitochondria and chloroplasts of eukaryotic cells have an independent genetic code, reflecting their origins as independent organisms. In addition, some modern prokaryotes have a marked resemblance to the organelles of eukaryote cells. The eukaryotic cell is a community of microorganisms working together in a 'marriage of convenience'.

Most single-celled eukaryotes are relatively large, between 10 and 100 microns, but diversification of these organisms could only have taken place after the oceans and the atmosphere had become oxygenated. It seems likely that gaseous oxygen began to accumulate in the Earth's atmosphere around 2 billion years ago, and this might be the time in which eukaryotes began to spread geographically and to evolve rapidly as they encountered new environments. However, the oldest evidence for eukaryotes comes from chemical fossils 2.7 billion years old, whilst the oldest body fossils are the acritarchs, which are thought to be the resting stages of eukaryotic planktonic algae. The oldest of these are 2.5 billion years old and they appear to have undergone a major radiation about 1 billion years ago, when they become common in the fossil record.

True multicelled organisms only arose among the eukaryotes. One of the oldest is a bangiophyte red alga preserved in silicified carbonates of the Hunting Formation (eastern Canada), dated to 1260–950 million years. It is a simple strand of 50 micron-wide cells, but in the slightly later Lakhanda Group of eastern Siberia (1000–900 million years) new modes of life became possible and up to six metaphyte species have been found, as well as a colonial form that creates networks rather like a slime mould. The later still (750 million years) Svanbergfjellet Formation of Spitsbergen has yielded various multicellular green algae.

Thus the Late Pre-Cambrian ocean harboured a dynamic and vital ecosystem of autotrophs and filter feeders, but with very short food chains in various facies: chert-carbonate facies, such as in the Gunflint Chert (southern Ontario), stromatolitic facies and a shale-siltstone facies, yielding acritarchs. Its collapse was abrupt and its successor, the Cambrian radiation of animal phyla in a new, competitive and multitrophic ecosystem, was a major event in the history of life. See the companion website for Box 29.5 The origin of oxygen in the earth's atmosphere.

29.4.3 Evolving metazoans and the Ediacaran fauna

Animals, all of which are multicellular, may have begun as clustered aggregates of protozoans, such as foraminifera or radiolaria, which remained together after cell division. One line led to the sponges and another to the metazoans, in which different parts of the body became specialized for different functions and in which tissues developed. Once these multicellular animals were established, metazoan life underwent successive radiations.

Classification of the metazoa is a key element in deducing knowledge about their origin and early history, and a generally-agreed classification for all animals (Figure 29.13) points to various early events in their evolution and the order in which they occurred. However, before 1947 almost nothing was known of Pre-Cambrian metazoan life, except for the stromatolites. Then a rich and well-preserved fauna of frond- and disc-shaped, soft-bodied animals was found in the Ediacara Hills of southern Australia, which was the first acceptable record of life from the Pre-Cambrian. Thirty genera are now known and traditionally they have been considered early representatives of modern phlya, especially Cnidaria (jellyfishes and soft corals), with over 100 species, along with animal tracks and burrows. They have since been found almost worldwide, distributed in rocks just above tillites of the last Pre-Cambrian glaciation up to the Pre-Cambrian boundary and possibly beyond. Most are probably between 564 and 543 million years old and come from marine, shallow-water sediments. The fossil assemblages probably represent ecosystems as they could not have been transported any distance due to their fragility.

Structurally three groups of organisms seem to be present in the fauna: those with radial symmetry, which may be related to jellyfish or corals, an example being *Cyclomedusa*, an organism with concentric rings; those with bilateral symmetry, possibly related to worms or arthropods, such as *Spriggina*; and those with an unusual symmetry that may not be represented in living groups, such as *Tribrachidium*. Very unusual symmetries like that in *Dickinsonia*, which is like no living organism, may imply that some of the Ediacaran organisms should be classified in a group of their own, separate from all living metazoans. Seilacher (1989)

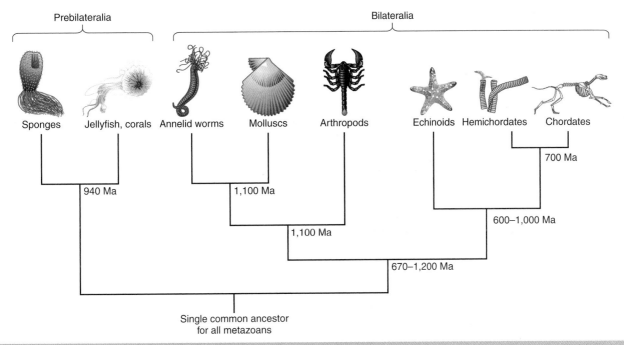

Figure 29.13 Classification of multicellular animals. (Source: after Milsom and Rigby, 2004)

suggested that they were a failed evolutionary experiment, quite unrelated to any modern organisms, and categorized all of them to the Vendozoa. He made his case based on the distinct structural morphology, but this suggestion has generated much debate.

The fauna ranged through a time span of about 40 million years and the species diversity increased through time, reaching its maximum just before the Pre-Cambrian–Cambrian boundary. There are some possible Ediacaran survivors into the Middle Cambrian, such as the *Thaumaptilon* from the Burgess Shale, which closely resembles *Charniodiscus*, forming a minor component of the much more diverse fauna. Most of the animals were probably benthic, but a few were planktonic. All lacked a gut and they probably lived by absorbing food and undergoing gaseous exchange directly across their body surfaces, like simple flatworms do today. However, it has also been suggested that the quilted segments of some Ediacaran animals were used as algal farms, and even that some were lichens and not animals at all. It is likely that they died out because of the rise of predators and scavengers, which together with an increase in atmospheric oxygen may have prevented the routine preservation of soft-bodied organisms.

The Pre-Cambrian–Cambrian transition marked one of the biggest faunal turnovers in the geological record, with a significant move from soft-bodied animals, possibly photo-autrotrophic or chemo-autotrophic animals, to heterotrophs, which relied on a variety of nutrient-gathering strategies.

Photo-autotrophy involves a symbiotic relationship between photosynthetic algae or their chloroplasts and animals. The algae receive protection from the host's tissues and release nutrients to and remove waste products from the host. For such photosymbiosis to work, a large part of the host's body must be exposed to sunlight so that the algae can photosynthesize.

Chemo-autotrophy is where there is direct uptake of energy-supplying nutrients from sea water and can involve symbiotic relationships with chemosynthetic bacteria, although some animals could probably absorb directly dissolved nutrients without bacterial aid. The high surface area of the flattened or sac-shaped bodies of many of the Ediacaran fauna would have allowed either efficient uptake of sea-water nutrients or greater light absorption for photosynthesizing, endosymbiotic algae. We will deal with the heterotrophic expansion in the next section.

29.4.4 The Cambrian explosion of life

The Cambrian evolutionary radiation initiated the first complex heterotrophic, metazoan communities and established new phyla and classes, which seem to have been phased through 30–40 million years ago, but it was a crucial point in the development of life on Earth. Predators appear to have been important, as skeletons first appeared in a number of higher-animal taxa as a direct response to predator pressure. The predators may have wiped out the Ediacaran fauna.

Several lines of evidence suggest that there was intense predation pressure during the Early Cambrian: there are specimens of damaged (boreholes) and sometimes healed prey (e.g. trilobite fossils with bites out of the carapace); there are fossils of predators and the development of anti-predatory adaptations (deep burrows; spines on trilobites; multi-element spiny mail coats or sclerites; utilization of chemotoxins to discourage predation, known from numerous punctae running through the shelly walls of some Early Cambrian brachiopods that could have delivered chemical deterrents); mineralized hard parts necessary for protection were co-opted for other functions, such as support and feeding, and allowed the evolution of new body plans that were dependent on skeletons for their function: for example, brachiopods cannot operate their efficient, internal, filter-feeding currents without the aid of a bivalved shell.

Most of the major body plans of animals evolved and each phylum is radically different from any of the others so they must have had different evolutionary paths. Even the hard parts evolved in very different ways; for example, sponges had an internal skeleton of fine silica needles whilst brachiopods evolved an external shell of calcium carbonate. Arthropods evolved chitin, but different groups of arthropods impregnated the chitin with either calcium carbonate or calcium phosphate.

The Cambrian fauna were dominated by trilobites and other arthropods, brachiopods, hyolithids (an extinct molluscan group) and the earliest echinoderms. The fauna was based mainly on mud-eating and most of the animals lived on the sediment surface. However, a much higher diversity of animals and niches is recorded by some locations with exceptional preservation, such as the Burgess Shale, which probably means the traditional picture is too simple. There must have been an equivalent radiation of soft-bodied animals at the phylum level and although most traces have been lost, there are a few isolated occurrences, like the soft-bodied, <10 cm-diameter organisms found on a single bedding plane in the Scottish Highlands, which may have echinoderm affinities, but is most likely an unknown phylum.

One characteristic of the Early Cambrian fauna is many short-lived, often geographically isolated and frequently bizarre forms which cannot be related to any other groups and are now extinct. It was a period in which many independent lineages originated but only a few survived and there were many more body plans in this period than afterwards. One common characteristic of the successful Cambrian animals was a larger body size, which suggests that the expansion was driven by worldwide ecological change.

One obvious question to try and answer is: Why did this expansion happen? Yet despite a synthesis of the physical and biological world across the Pre-Cambrian–Cambrian boundary we still do not really know the answer. It appears that during the later stages of the Pre-Cambrian (Vendian) and the Early Cambrian there were major changes in the atmosphere, ocean chemistry, plate tectonic activity and climate which must have made a major contribution to the expansion. There was a slow build-up of free atmospheric oxygen because of photosynthesis, which evidently attained a sufficient level to support animal life, and there was a decrease in CO_2 in the atmosphere. Other factors include a worldwide series of marine transgressions, the first following the widespread Varangian glaciation, which successively flooded the continental shelves and opened up many shallow-water ecological niches. The availability of extra plankton may be related to major periods of ocean upwelling and possibly to the synchronous phosphorite deposits on a global scale around the boundary. It seems likely that during the Late Pre-Cambrian, phosphorous-rich sediments accumulated on the deep-ocean floor, as this element was depleted from the surface waters. During a period of enhanced oceanic overturn in the Early Cambrian, probably associated with the marine transgressions, phosphorous was brought up to the surface water to give increased nutrient levels in the photic zone. This could have been a trigger for the sudden diversification of living organisms. Many of the fauna at this period have shells of calcium phosphate. The spread of nutrient-enriched waters over low-latitude shelves, as recorded in the C-isotope fluctuations, may well be linked to a global temperature rise as the Earth emerged from the Late Pre-Cambrian icehouse to the greenhouse effect of the later Cambrian. The uptake of calcium nutrients also allowed the further development of skeletons.

There has also been a suggestion that the evolution of predation triggered the Cambrian radiation. Predators could have caused diversification among prey as those preyed upon evolved large sizes, hard coverings from any available biochemical substance, the use of powerful toxins, changes in lifestyle and behaviour, or any combination of these, in order to become more predator-proof. At the same time the new predators would have had to develop more ways of attacking their prey and these overall changes could have caused a significant burst of evolutionary change as positive feedback mechanisms came into play.

Although we can see the evidence for these characteristics in the fossils, this does not explain the timing of the Cambrian explosion and it is thought that the development of predation alone cannot explain the rich variety of skeletons in the fossil record. In fact, although we have called this event an explosion, which implies a sudden geological event, some palaeontologists prefer to see it as marking the time period in which animals first appear as shelly fossils, and not necessarily as when they evolved.

So there are two ways of looking at this event: either it was the nearly-simultaneous evolution of a large group of animal phyla (the 'bang' hypothesis) or Cambrian animal phyla had long Pre-Cambrian histories that were not recorded as fossils (the 'whimper' hypothesis). However, of the phyla that existed in the Cambrian, only very few show any evidence of a Pre-Cambrian ancestry and so it is thought that the genesis of most animal phyla must date to the Cambrian and not before. On the other hand, we know that metazoans originated well before the Cambrian boundary but their potential for adaptive radiation and their ecological impact were not fully realized until this time. The fauna that did develop however did not dominate the marine environment for long and its evolutionary history is short in geological terms. See the companion website for Box 29.6 The Burgess Shale fauna as a Cambrian case study of a lagerstätten.

29.4.5 Colonization of the land

This process was likely to have been gradual, occurring throughout the Palaeozoic and perhaps even originating in the Pre-Cambrian. As the atmosphere evolved, land surfaces were changed by mats of microbes and thallophytes, which produced a suitable surface for the colonization by larger plants and animals. Bacteria probably invaded fresh water and damp areas early in the Pre-Cambrian and lichens were likely contributing to soil formation 1 billion years ago. However, it was during the Palaeozoic that larger animals and plants moved from the sea on to land and hence radically increased the Earth's living space and changed the characteristics of the planet, such as its weathering cycles, the composition of the atmosphere and the climate.

During the Ordovician, plants began to migrate on to land, and the earliest spores, which probably belonged to land plants (like the spores of modern liverworts), come from Middle Ordovician rocks.

By the Silurian there was a widespread low-lying plant cover wherever there were wet conditions. However, there are no fossils of plant parts other than spores before the Middle Silurian. Molecular evidence in fact suggests that all land plants today are descendents of liverworts. The Late Silurian rocks contain the first well-preserved land plants, which probably included vascular plants like *Cooksonia*. This was only a few centimetres tall but had a simple structure of thin, evenly-branched stems, with sporangia at the tips and no leaves. However, it also had central structures that were probably xylem, rather than simple conducting strands.

By the Devonian period, plants were much more diverse, had evolved into drier and higher locations and were beginning to form the first forests. Seed-bearing plants had evolved by the Late Devonian and could reproduce away from water. By the end of the Devonian all the major innovations of land plants, except flowers and fruit, had evolved. Forests of seed-bearing trees and lycopods (ancestors of living club mosses) had appeared, with understoreys of ferns and smaller plants. The evolution of the seed seems to have been the basis for competitive success in the Early Carboniferous and these plants invaded drier habitats, with seed dispersal by wind, rather than water, becoming important by the Late Devonian. The increasing success of land plants must have produced large amounts of rotting plant debris in swamps, rivers and lakes, which led to very low oxygen levels in the tropics. At the same time, the increasing photosynthesis by land plants drew down atmospheric carbon dioxide and increased atmospheric oxygen. This must have helped to encourage air breathing among contemporary fresh-water arthropods and fish.

Animals soon followed through the Late Silurian (trace fossils of arthropod footprints), Devonian and Carboniferous. The earliest were small arthropods (mites and springtails) that ate organic debris from the Early Devonian, followed by other arthropods that ate them, such as primitive spiders. Animals like millipedes, spiders, mites and other primitive arthropods made the transition and insects evolved on land during the Carboniferous. The early leaf litter of the forests were colonized by worms, slugs and snails, and following these prey animals came predatory tetrapods.

As early as the Silurian, lobe-finned fish were capable of limited movement on land; they may have used this to escape marine predators or drying-out pools. Air breathing probably evolved in many groups of early fish that lived in environments where oxygen was seasonally low, in shallow lagoons along the shore and in tropical rivers, deltas and lakes.

The Devonian period saw the first amphibians, which were mainly fish eaters that spent most of their time in the water, but their relatives soon began to exploit the forest floor resources. By the Late Devonian a group of rhipidistians, or Tetropodomorpha (lobefins and ancestors of all amphibians and all other land vertebrates) had evolved nostrils, which are now used in all tetrapods for air breathing, and the choana, which allows air breathing through the nostrils, and they must have evolved lungs at roughly the same time as the lungfish did. The dorsal lobe fins in the rhipidistians, such as *Eusthenopteron*, which evolved towards an amphibious life, were attached to the spine just as firmly as the ventral fins and played just as important a role in swimming, but the ventral fins evolved to become limbs because they happened to be placed where they interacted with the land. Cladograms of lobefin fish illustrate where these rhipidistians are located (Figure 29.20), but only one line took the next evolutionary path to become tetropod ancestors (four-legged land vertebrates). These were the osteolepiforms, which had skull bones and the patterns of bones in the lobe fins and the general size, shape and geographic distribution of the early amphibians.

Definite early amphibians are known from the late Devonian of Scotland and Greenland, such as *Elginerpeton* (the oldest) and other groups. They showed increasing adaptation towards effective movement on land and radiated fast into a variety of sizes, shapes and ways of life that we now recognize as characteristic of amphibians and eventually reptiles. Compared with the classic scene of *Eusthenopteron* lurching between ephemeral pools, the Devonian habitats now appear more aquatic and the groups show that they were restricted to vegetation-choked active channels in meandering river systems or shallow, warm, epicontinental seas. Many were suited to only occasional terrestrial excursions and surprisingly many of the features characteristic of tetrapods probably evolved before terrestrialization.

Daeschler *et al.* (2006) and Shubin *et al.* (2006) produced evidence of a 375 million-year-old fish that seemed to be an evolutionary intermediate between fish and the first early amphibians, as reported in Box 29.7 (see the companion website). This is important because transitional fossils tell us how much a new species resembled and differed from its nearest neighbours on the phyological tree.

The physical challenges to life on land are summarized in Table 29.2. Organisms needed much stronger supports to stay upright in air than they did in the water. Plants developed tissues such as lignin and cutin whilst animals strengthened their preadapted hard cutiles (in the case of arthropods) or internal skeletons (in the case of vertebrates).

Maintaining an internal watery environment is also important for all land dwellers and some organisms have only partially solved this, so that they live in areas that are wet, or at least protect them from drying out, such as in leaf litter and soils. Most large land dwellers have developed a waterproof membrane to keep their water inside, so that plants have a waxy cuticle on their leaves and animals have waterproof cuticles or skin that lessens the risk of

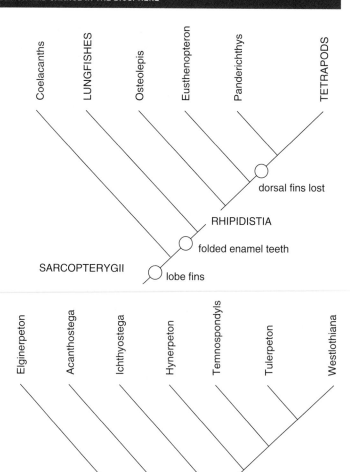

Figure 29.20 Cladograms of lobefin fish and some early tetropods. (Source: after Cowen, 2000)

drying out. Animals have also developed modified metabolisms that lose less water during digestion than do aquatic organisms.

A problem that arises with this waterproofing is the difficulty of exchanging gas with the atmosphere. Plants need to extract carbon dioxide from the air and expel oxygen during photosynthesis. During respiration, both plants and animals need to extract oxygen from the air and expel carbon dioxide. Most plants carry this out through stomata, which are small holes in the waxy cuticle that are opened and closed by a pair of guard cells. Small animals such as spiders can rely on diffusion along a tube network within the body, but larger animals exchange gases with the atmosphere using specialized organs, such as the vertebrate lung.

The refractive index of light is lower in air than in water and sound transmission is different too, so vision and hearing have to be modified in land animals.

Table 29.2 Challenges to life on land: adaptations. (Source: after Milsom and Rigby, 2004)

Challenge	Plants	Animals
Physical support	Woody tissues, lignin, and cutin (which may have evolved from precursor molecules involved in protection from UV radiation)	Arthropod hard cuticle and tetrapod skeletal strengthening. To some degree animals such as these are preadapted to live on land
Retention of water	Development of waxy cuticles, and behavioral methods of retaining water	Waterproof cuticle, water-retaining skin, behavioral methods, and modified excretion
Gas exchange	Via stomata in advanced plants	By diffusion if small, or via specialized organs such as lungs
Reproduction	Spores, protective seed pods, and internal fertilization	Waterproof eggs and internal fertilization

A good review of the physiological considerations related to the early colonization of the land is given in Selden and Edwards (1989). See the companion website for Box 29.7 *Tiktaalik*: a firm step from water to land?

29.4.6 Reptilian radiation

The first reptiles evolved in the warm, humid, tropical regions along the southern shores of Euroamerica and life away from the swamps and forests demanded adaptations for dealing with seasonality. In this section we will show how reptiles became the dominant large animal in the Permian.

One of the oldest known reptiles is *Hylonomus*, from the Mid-Carboniferous of Canada and although it looked little different from some of the contemporary amphibians, it shows several clear reptilian characteristics, such as a high skull, evidence for additional jaw muscles and an astragalus bone in the ankle. *Westlothiana* and *Casineria* from East Kirkton in Scotland, only about 20 cm long, are likely to be even older, Early Carboniferous reptiles.

The key development for reptilian success is the evolution of the cleidoic (closed) egg. This is not known from these fossils, but it was the crucial development. Here amniotes (reptiles, birds and mammals) broke with aquatic reproduction and enclosed their eggs within a tough, semi-permeable, hard shell. The developing embryo was protected from the outside world, there was no need to lay eggs in water and there was no larval development stage.

The reptiles radiated during the Late Carboniferous; the three main amniote lines that became established can be distinguished by the pattern of openings in the sides of their skulls. These diagnose the key clades among amniotes.

The oldest anapsids were small insect-eaters, but others include Permian and Triassic reptiles and turtles, which appeared in the Late Triassic. The anapsid skulls, with no openings behind the eye, inherited this primitive characteristic from the fish and amphibians.

The synapsids include the mammal-like reptiles and the mammals, and the oldest, the pelycosaurs, are known from the Late Carboniferous and Early Permian. They were small-to-medium insectivores with powerful skulls and sharp, flesh-piercing teeth. Some later animals, such as *Dimetrodon*, had large sails supported on vertical spines growing from the vertebrae, used to control body temperature, both to warm them up more easily and to cool them down. Over 50% of Late Carboniferous reptiles and over 75% of Early Permian reptiles were pelycosaurs. This group also evolved into some forms that were plant-eaters (caseids and edaphosaurs), which were the first herbivorous land vertebrates.

Mammal-like reptiles radiated dramatically in the Late Permian as a new clade, the Therapsida. Some were carnivores, such as the gorgonopsians, with large wolf-like bodies and massive sabre teeth, whilst dicynodonts were successful herbivores and some of the first animals to have a complex chewing cycle that allowed them to eat a wide variety of plant foods. Most lived at higher latitudes and we can be sure that their metabolism was not like that of living reptiles. They had sprawling forelimbs and did not move efficiently compared with later reptiles and mammals. Unlike other reptiles they had short, compact, stocky bodies, with short tails, which were good adaptations for conserving body heat, if not for generating it. They may have had hair or thick hides for conserving heat but we cannot detect that from their fossils. All these characteristics suggest they did not have a large energy budget. They may have had some form of internal temperature control, but it would have been nowhere near as good as that of living mammals. Most of these groups were wiped out at the end of the Permian.

The cynodonts were an important Triassic synapsid group. The early Triassic form *Thrinaxodon* was dog-like and had numerous small canals in its skull, indicating small nerves that served the roots of sensory whiskers. This means hair as insulation and temperature control. The cynodonts evolved along several lines in the Triassic but gave rise to mammals such as *Megazostrodon*.

The third major amniote group, the diapsids, initially comprised small-to-medium carnivores and was never as important as the synapsids until the Permian extinction event which had such an important effect on therapsid communities. Small and large carnivores such as *Erythrosuchus*, one of the first archosaurs, appeared: a group later to include the dinosaurs, pterosaurs, crocodiles and birds.

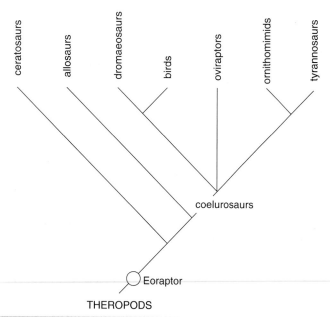

Figure 29.27 Phylogram of the ornithischian dinosaurs. (Source: after Cowen, 2000)

Further detail related to the reptilian radiation and in particular the development of locomotion, herbivory and thermoregulation in reptile evolution is given in Cowen (2000, Chapter 10).

29.4.7 Dinosaurs and their predecessors

Introduction

Dinosaurs dominated terrestrial communities for 100 million years and it was only after they disappeared at the end of the Cretaceous that mammals became dominant. It seems likely therefore that the dinosaurs were competitively superior to the mammals, which confined the mammals to small body size and ecological insignificance for a very long time.

The dinosaurs had appeared at about the same time as the first mammals in the Late Triassic of Gondwanaland. The earliest examples were small, bipedal carnivores, but by the end of the Triassic all four major dinosaur groups had evolved. They had descended from small, bipedal thecodonts, like *Euparkeria*, which was only about a metre long, very lightly built, with a long strong tail to give balance in running and a long and light skull with many long, sharp, stabbing teeth.

Theropods were all bipedal, retaining the body plan of the ancestral form. However, some of the most advanced dinosaurs were theropods, as are many living birds. Some were extremely small, like *Eoraptor* (Late Triassic of Argentina), which had a skull only about 2.5 cm long but was a fast running carnivore, with sharp teeth and grasping claws on its forelimbs (Figure 29.27). *Coelophysis* (Late Triassic, North America) was much bigger at 2.5 m long, although it was still lightly built and adapted for fast running. The bones of its skeleton were fused into stronger units than the earlier theropods so it is placed into the first of the derived theropod groups, the ceratosaurs (Figure 29.27).

During the Triassic some diapsids became large carnivores, some specialized fish-eaters, others specialized grubbing herbivores and still others small, two-limbed, fast-moving insectivores (which evolved into the crocodiles and dinosaurs). Some became proficient flyers and evolved into pterosaurs, others such as the nothosaurs and ichthyosaurs became fully-adapted marine reptiles.

Another mass extinction event near the beginning of the Triassic (~225 million years ago) set the new age of diapsids in train as most of the synapsids and various basal archosaur groups died out. Many new kinds of terrestrial tetrapod then diversified.

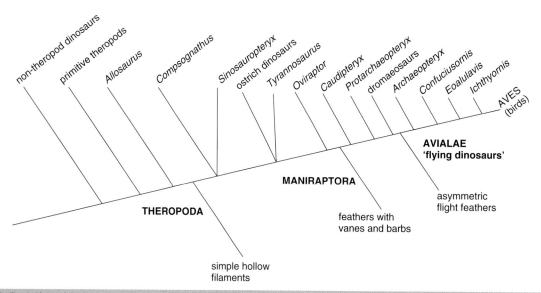

Figure 29.28 Evolution of Theropoda, Maniraptora and Avialae. (Source: after Milner, 2002)

More advanced theropods separated early into the allosaurs, which were massive, powerful predators, and the lighter, more agile coelurosaurs. Within this lineage the small, agile carnivores included the birds and the dromaeosurs (including *Velociraptor* and *Deinonychus*). As more fossils are collected from this lineage it becomes increasingly difficult to draw the line between dinosaurs and birds. For example, *Unenlagia* from Argentina had arms that were very bird-like in the way they folded, and like birds, but unlike other theropods, its shoulder point faced sideways. Researchers have identified more than 100 characteristics that link these dinosaurs and modern birds, including a wishbone, swivelling wrists and the presence of three forward-pointing toes. There are also reports from Liaoning Province (China) that *Meilong* ('soundly sleeping dragon'), a juvenile animal, slept like a bird, with its head tucked between forearm and trunk and its tail encircling its body, presumably to conserve heat (it was warm blooded). Only birds with long, flexible necks tuck their heads behind a forelimb to sleep, a posture that helps to keep them warm.

The other two lineages of coelurosaurs are distinct groups: Oviraptors, nest-building dinosaurs from Mongolia, and Ornithomimids, the so-called ostrich dinosaurs. The latter might have been the most intelligent of the dinosaurs and evolved a plan like a living ostrich, but they had long arms and slim, dextrous fingers instead of wings, and large eyes but no teeth. They had feet well adapted to running and are related to the Tyrannosaurs.

Several lineages of theropods evolved to great size, known informally as carnosaurs, though they are not a clade and evolved at least four times: in Jurassic allosaurs and in three Cretaceous groups, tyrannosaurs in North America, *Gigantosaurus* from Argentina and *Carcharodontosaurus* from North Africa. Each stood at least 6 m high, with a total length of between 12 and 14 m and a weight of 6–8 tonnes. They must all have relied on massive impact from the head for killing, aided by huge stabbing teeth and forelimbs with slashing claws. The earliest Ornithischians were the first herbivorous dinosaurs and they radiated widely. They ate rather coarse, low-calorie vegetation and many were medium-sized animals. Small bipedal ornithischians, such as *Lesothosaurus* and the heterodontosaurids, gave rise at various times to derived groups that were much heavier (some weighing over five tonnes). The armoured dinosaurs form one derived clade, which includes the stegosaurs and ankylosaurs, while the horned dinosaurs or ceratopsians form another, but most of the larger ornithischians were ornithopods, including iguanodonts and the 'duck-bill' dinosaurs, the hadrosaurs.

Some early dinosaurs evolved to become very large, heavy, quadrupedal vegetarians, with broad feet and strong, pillar-like limbs. These sauropodomorphs had an early radiation as prosauropods from the Late Triassic into the Early Jurassic, when they were typically 6–10 m long browsing herbivores. They had particularly long, lightly-built necks and heads and light forelimbs, and they seem to have been adapted to browsing high in vegetation (e.g. *Plateosaurus*). They had a later radiation as the sauropods, which were the largest land animals that ever lived, between 50 and 80 tonnes in body weight, with *Argentinosaurus* close to 100 tonnes. *Apatosaurus* was 20 m long and weighed 30 tonnes. Other sauropods included *Diplodocus, Brachiosaurus, Supersaurus* and *Seismosaurus*. All were quadripedal herbivores, with very small heads and very long necks. Their tails were also long but their bodies were massive, with powerful load-bearing limb bones and pelvises.

Dinosaur ecology and ethology

Vegetarian dinosaurs must have had bacteria in their guts to help them break down cellulose. Large animals also have slower metabolic rates than small ones and for vegetarians this means a slower movement of food through the gut and more fermentation time. Vegetarians usually grind their food well so that it can be digested faster. This was achieved in two different ways in the two major groups of vegetarian dinosaurs, the ornithischians and the sauropods.

The saurapods had very small heads for their size and small, weak teeth and it has sometimes been thought that this indicates a soft diet. However, they probably had a small head because it sat at the end of a very long neck, like a giraffe. In both animals the food is gathered by the mouth and teeth, then swallowed and macerated later. Giraffes are ruminants and boluses of food are regurgitated and chewed with powerful molar teeth. The sauropods probably used a different system for grinding food and they had enormous powerful gizzards in which food was ground up between stones that they had swallowed (the so-called fossil gastroliths). This characteristic is also seen in birds like chickens and the fossil moa.

Ornithischians on the other hand generally had lots of massive grinding teeth, especially the hadrosaurs and ceratopsians, and they would have chewed their food thoroughly. Food gathering and processing was no problem to these vegetarian dinosaurs and some reached enormous size.

In terms of behaviour, some dinosaur trackways illustrate stampeding effects, which must mean that at least some dinosaurs gathered in herds and responded immediately and instinctively to dangerous predators. Other behaviour indicators are preserved in the skeletons, such as that of the ornithischian dinosaur *Pachycephalosaurus*, which had a dome-shaped bone area on the top of its skull. This was not solid but air-filled and porous, with an internal structure like those found in sheep and goats, which fight by ramming their heads together. It might have fought like that, or perhaps it butted opponents from the side rather than head-to-head. Similarly, *Triceratops* and other ceratopsians had air spaces between the horns and the brain case, which may also indicate that they competed by direct head-to-head impact.

Some hadrosaurs had huge crests on their heads, such as *Parasaurolophus*, which were are not solid but contained tubes running upward from the nostril and back down into the roof of the mouth. Only large males had these crests; females had smaller ones and juveniles none at all, so it is not likely that they evolved for additional respiration or thermoregulation. Perhaps these dinosaurs could produce different sounds to go with different visual signals provided by the crests. Thus the hadrosaurs may have had a sophisticated social system.

There have also been major finds of fossilized dinosaur eggs and nests from all three major dinosaur groups (in Montana, Alberta, Mongolia, Spain, Argentina, Argentina and China). The nests were carefully constructed mud bowls lined with vegetation and in each one the eggs were laid or arranged in a neat pattern so that they would not roll around. Some nests had protective barriers around their edges. Many are clustered together, suggesting that they were in communal nesting colonies, and in most instances a stratigraphic horizon yields numerous clutches, which appear as conspecific with one another. Some of these associations extend over $1 km^2$. The presence of large numbers of adults improved the safety of both the eggs and hatchlings. There is good evidence for long-term parental care by dinosaurs too, and one nest in Montana contained 15 baby duckbilled dinosaurs that were not new hatchlings since they were about two times too large for the eggshells nearby and their teeth had wear marks. When they died, were buried and became fossilized, they had an adult close by. In Mongolia three of seven known oviraptor nests had adults on or near them, which are likely to have been brooding the eggs.

Based on fossil assemblages, numerous hadrosaur taxa and certain ceratopsian taxa travelled in large, monospecific aggregations that died together in either flood or drought. Fossil evidence supports the possibility of pack hunting behaviour in dromaeosaurs when they were attacking prey significantly bigger than themselves. For example, *Deinonychus* teeth and body fossils are frequently found in association with a single *Tenontosaurus*, an orithischian at least twice as big as them. It is envisaged that multiple *Deinonychus* leaped at the tenontosaur, employing a forelimb attack.

It has been found in China that coelurosaurian theropods were feathered and it has been suggested that feathers on their hands increased manoeuvrability whilst running, cornering and braking and control during leaps. When attacking smaller prey, feathers could be used as 'blinders', making their hands and arms appear larger and hence discouraging their prey from dodging to either side at the close of a strike. In a pack-hunting situation, this same technique could be employed in herding behaviour. Feathers, it appears, did not evolve originally for flight, and the Chinese theropods all have down, probably as insulation to keep their body at an even temperature. There have even been reports of a feathered predecessor of *Tyrannosaurus rex* in China, *Dilong paradoxus*, which probably used the protofeathers for insulation (Xu *et al.*, 2004). By the time *T. rex* evolved 65 million years later it had probably lost the featherlike characteristics of its predecessor as with its much larger size it would have had more difficulty losing heat than keeping it. Nevertheless, *Tyrannosaurus* chicks may have had a downy cover. However, feathers are more complex to grow, more difficult to maintain in good condition, more liable to damage and more difficult to replace than fur and there is no apparent reason for the evolution of feathers rather than fur. Cowen (2000, p. 206) suggests in fact that feathers evolved for display. Hence dinosaurs should therefore be compared with living mammals and birds in running a complex society, rather than living reptiles.

Mammals and birds are warm-blooded, with high metabolic rates, and they are much more intelligent than other living vertebrates. So the question has often been asked, 'Were dinosaurs warm-blooded?' Also, 'Were all dinosaurs alike in their physiology?' Based on many kinds of evidence it has been suggested that dinosaurs had a unique type of physiology: small theropods (and possibly others) had down feathers and thermoregulation; dinosaurs did not have a high resting metabolic rate (they had no complex nasal passages, called turbinates, which are found in living mammals and birds); they could breathe whilst they ran and had an erect gait, and so had an aptitude for fast sustained running, noted in footprints. As there are no living animals with this combination of features, it has been suggested that they could turn their metabolic rate up or down, possibly with the seasons.

Evolution of birds

Two of the theropod dinosaurian innovations, feathers and brooding, led to a completely new group of dinosaurs: the birds. Since T.H. Huxley hypothesized a theropod origin for birds in 1869 a large quantity of fossil evidence has been found to support this dinosaurian ancestry, and this evidence is now indisputable. There are many transitional fossils between theropods and the first birds (Figure 29.28) and feathers have been found on several theropods, which were ground-running animals.

The evidence comes from a variety of sources:

1. **Bone characteristics:** The presence of air spaces connected to the ear region (rostral, dorsal and caudal tympanic recesses); ventral processes on cervicothoracic vertebrae; ventral segments of thoracic ribs; forelimbs that are more than half the length of hindlimbs; a semi-lunate carpal bone allowing the swivel-like movements of the wrist; and clavicles fused into a wishbone. There are also skeletal features correlated to the bird system of lung structure and ventilation.

2. **Behaviour:** Brooding on their own nests (for example, oviraptors).

3. **Microstructure of the eggs:** The basic ornithoid type has shell units with at least two distinctly separated microcrystalline zones. The innermost zone is formed from a calcite crystalline structure that radiates from an organic core, whilst the external zone has a scaly microstructure. In birds this external zone may grade into or be completely separated from a third, outermost zone. No other reptilian egg is known to have these microstructural characteristics.

4. **Feathers:** Particularly the Chinese feathered dinosaurs, such as *Sinosaurpteryx, Protarchaeopteryx, Caudipteryx* and *Sinorthosaurus*.

5. **Sleeping posture:** *Meilong*, a fossil juvenile troodontid that perished in its sleep around 130 million years ago in the Yixian Formation (Liaoning Province), slept like a bird, in a stereotypical sleeping or resting position, with its head tucked between its forearm and trunk and its tail encircling its body, probably to conserve heat. Only birds with long

flexible necks tuck their heads behind a forelimb to sleep, a posture that helps keep them warm. This suggested to Xu and Norell (2004) that many bird-like features occurred early in dinosaurian evolution.

Living birds are warm-blooded, with efficient thermoregulation that maintains body temperature; they breathe more efficiently then mammals, pumping air through their lungs rather than in and out; they have better vision than any other animals; and they can fly better, faster and further than any other animals, an ability that requires complex energy supply systems, sensing devices and control systems. The earliest known bird is *Archaeopteryx*, from the Upper Jurassic Solnhofen Limestone in Germany. It had feathers on its wings and tail but it looked like a small theropod dinosaur, had a theropod pelvis, and had a long bony tail, clawed fingers and a jaw full of teeth, which are all theropod characteristics. It lacked many features of living birds, with no breastbone and no hole through the shoulder joint through which to pass the large tendon required for the rapid, powerful, twisting wing upstroke in living birds.

How did powered flight evolve in birds?

There appear to be several hypotheses (Cowen, 2000):

1. **The arboreal hypothesis:** Small dinosaurs were climbers and ancestral forms jumped out of trees from a height. This was long favoured but now must be reconsidered in the light of the Chinese evidence. With long, erect limbs, a comparatively short trunk and bidpedal locomotion, *Archaeopteryx* and the feathered Chinese theropods are exactly the opposite in body plan of all living animals that jump and glide between trees. There is not much in the bird ancestry to suggest any arboreal adaptations. Aerodynamically the transition from gliding to flapping is difficult and the flapping of proto-wings and any feathers at all on the wings or tail serve only to increase drag. The smallest dromaeosaur from Liaoning, *Microraptor*, provides a test of the arboreal habits of these animals. It was only 47 mm long, was feathered, had slender, very curved claws on its toes and fingers, and the degree of its claw curvature fell within the range of modern climbers. It has been suggested that it was a tree-dweller not a ground-runner and so could have been a stage in the development of flight.

2. **The cursorial hypothesis:** Certain adaptations of ground-dwelling theropods provided some of the anatomy and behaviour necessary for flight, such as the lengthening of the forearms, especially the hands, placing long, strong feathers in those areas and evolving powerful arm movements. However, a running theropod that flapped its arms would increase its drag, while some weighed 20–30 kg and are thought to have been too heavy to have flown.

3. **The running raptor hypothesis:** This was a mechanically-sounder idea that included behaviour involving strong, synchronized arm strokes and the evolution of strong pectoral muscles. Here it was suggested that the proto-bird hunted by running fast and leaping after any flying or jumping insects it disturbed, or by using its sweeping arm as an insect trap. However, this proposal would consume much energy and no predator today runs at high speed to flush out insects it can leap after. Also, the speed required would have been far too great for the suggested proto-bird. Experts on aerodynamics are sceptical that flight developed this way.

4. **The display and fighting hypothesis:** Theropods had long, strong display feathers on the arms and tail. Successful display was increased by lengthening the arms, especially the hand, and by actively waving them, perhaps by flapping them rapidly and vigorously. This would also have encouraged the evolution of powerful pectoral muscles. So strong wing flapping is a simple extension of display flapping, encouraged by fighting behaviour. Here powerful flapping used to deliver forearm smashes could have lifted the proto-bird off the ground. But the proto-birds probably gained flight behaviour, anatomy and experience at low ground speed and low height. Once lift-off was achieved, flapping flight probably quickly followed.

There has also been a suggestion that the evolution of flight involved a four-wing stage (Box 29.9, see the companion website). However, the first generally accepted proto-bird, *Archaeopteryx*, probably hardly flew at all as it did not have adaptations for sustained flapping flight. In fact, this animal evolved structures that were active deterrents to flight. For example, its tail was long and bony, with long feathers that added a lot of drag and little lift, and it must have spent its life as a fast-running, displaying bird, hunting for small prey.

Nonetheless, after *Archaeopteryx* bird evolution was rapid. Early Cretaceous rocks in all the northern continents and Australia show a radiation of birds and fossils that demonstrate they had dramatically improved their flying and perching ability. Most were small and light, with lighter bones than *Archaepteryx* and genuine beaks. The Liaoning, China, deposits show that *Confuciousornis* and *Changchengornis* (122–124 million years ago) had toothless jaws and beaks. The former lived in colonies around forested lake margins and was more bird-like than *Archaeopteryx*, with modern flight apparatus, although it retained fully functional grasping claws on all but the index finger, which supported the flight feathers. Its wings were very long, with exceptionally long flight feathers to increase lift, and its body was compact and streamlined, with an enlarged breastbone to anchor flight muscles and strut-like bones to brace the shoulder to the chest. The tail was much reduced, well on the way to forming a pygostyle (the tight knot of bones that form the tail stump, or parson's nose, and control the tail feathers in modern birds). By 70 million years ago, in the Late Cretaceous, fully modern birds had begun to appear.

When the dinosaurs died out at the end of the Cretaceous they left ecological niches on the ground associated with larger body sizes. The two species that took advantage were the mammals that became large herbivores (see Section 29.4.8) and the birds that became the dominant land predators in some regions in the Paleocene. These evolved to become flightless, terrestrial bipeds once again. Large (2 m tall), flightless birds called diatrymas lived in the northern hemisphere. They had massive legs, vicious claws and huge, powerful beaks. Slightly later carnivorous birds with similar adaptations, called the phorusrhacids, which were up to 2.5 m tall, dominated the grassy plains of South America until the Eocene.

The southern continents today have a number of large flightless birds, such as the ostrich, cassowary, rhea and emu, but several others have become extinct, such as the moas of New Zealand, which reached over 3 m tall. The elephant bird (*Aepyornis*) of Madagascar was living until quite recently. These birds were much heavier than the birds that continued flying, such as the teratorns in South and North America, which reached a wingspan of 7.5 m and a body weight of around 75 kg, and the pelagornithids, gigantic marine birds with a wing span of 6 m and a weight of around 40 kg. See the companion website for Box 29.8 Were dinosaurs able to swim? and Box 29.9 Four-winged dinosaurs from China.

29.4.8　The origin of mammals

Mammals are animals that have evolved a very good ability to regulate their body temperature and internal chemical composition and to live an active life in the spatially heterogeneous and fluctuating terrestrial environment in which they occur. They suckle their young, have live births, possess acute hearing at high frequency and are warm-blooded. Their temperature is maintained constant with the help of devices for controlling the conductivity of the body surface, such as variable insulation and rate of blood flow through the skin. Their elevated metabolic rate is achieved by a high rate of food collection, made possible by sophisticated locomotory, sensory and central nervous systems. Their high rate of food ingestion is achieved by the complex dentition and jaw musculature. Elaboration of their rate of gas exchange is achieved by the enlarged lung, the diaphragm, the modified locomotory gait (which allows higher levels of ventilation whilst active) and the rapid rate of blood flow and gas exchange at the tissues that is possible with complete double circulation. Their high blood pressure is important for the high ultrafiltration rate of the kidneys and consequently for the rapid and precise regulation of the body fluids. They also have enlarged and specialized brains. All of these characteristics are interrelated and all contribute to the very high level of overall regulatory ability (homeostasis) typical of mammals.

The first stage in mammalian development was the emergence of the mammal-like reptiles (synapsids) that occurred from the Late Carboniferous onwards; by the Late Triassic and earliest Jurassic, skeletons of small animals, such as *Megazostrodon*, are found with nearly all the skeletal characteristics of the Mammalia. Over around 100 million years a complex pattern of originations and extinctions of synapsid taxa occurred, with an erratic trend through time for animals with an increasing number of mammalian characters.

The evolutionary trend culminated in a sequence of relatively small carnivores that evolved increasingly mammalian characteristics of different structures and body systems more or less simultaneously. There is no indication of a single key adaptation leading to the rapid evolution of mammals and the rate of evolution of the morphology was not constant. In the cynodonts at least six groups of carnivorous and herbivorous taxa evolved some mammalian characteristics during the Triassic and a small-bodied carnivorous cynodont group probably evolved into the first true mammals in the very Late Triassic. It appears that they were ecologically squeezed between the first dinosaurs, which were fast and sustained runners, and small lizard-like reptiles, which had a low resting metabolic rate fed by solar energy. As a result they escaped extinction by evolving into a habitat suitable for small warm-blooded mammals and nothing else: the night. Burrowing in the dark, they were invading a habitat that required much greater sensitivity to hearing, smell and touch. This may have selected the relatively large complex brain that allowed them to emerge as the dominant large land animals after the extinction of the dinosaurs.

However, the first mammals were very small and shrew-like (but see Box 29.10, see the companion website for a different story) and the family Morganucodontidae is probably ancestral to many Mesozoic mammals that are now extinct. These were small (only about 10 cm long) carnivores, eating worms, grubs and insects. The skeleton shows that they were agile climbers and jumpers, their neck was very flexible as in living mammals and their teeth were fully differentiated into mammalian types. As in most mammals today the front teeth were replaced once and there was only one set of molars, with double roots.

Mammals increased in diversity through the Cretaceous, but not spectacularly. Their moderate diversity increase can be correlated with the evolution of the flowering plants and the insects that diversified with them. However, after the Cretaceous extinctions and the dinosaur disappearance mammals radiated into all body sizes and into many habitats. They must have been able to compete successfully in open terrestrial habitats for the first time.

The Cenozoic Era shows this explosive mammalian radiation, when their diversity rose from 8 to 70 families, as well as the dominance of flowering plants, insects and birds. But was it real explosive radiation or had the different groups of mammals already diverged genetically, only to be ecologically released after the End-Cretaceous extinction? It is difficult to look more carefully at Cretaceous mammals to try and find advanced characteristics among them because the fossil record is so poor.

It might be possible to use genetics and molecular clocks, where molecular changes might allow us to determine the times of divergence of living animals without ever having to look at their fossil ancestors (using proteins, specific genes and DNA sequences from the nucleus or from mitochondria). It is clear however that the clocks run at different speeds within different animal clades, dates sometimes disagree with the fossil record and there are many other problems (see for example Cowen, 2002, pp. 301–303).

What can be seen is that marsupials and monotremes always fall outside the groups that form the placental mammals, but that there are some surprising insights into mammalian radiation that were not discovered by standard morphological comparison. For example, a large group of African mammals forms a clade separate from the other placentals: this includes elephants, sea cows, hyraxes, aardvarks, elephant shrews and golden moles, which illustrates a great array of different body plans, sizes and ecologies. It also implies that Africa became isolated in Cretaceous times, with a set of early placentals that evolved to fill all these ecological roles, separate from evolution on other continents. Some of the results seem counter-intuitive, for example where a clade links carnivores with perissodactyls (horses, rhinos and tapirs).

Molecular evidence also suggests that rodents may not be a true clade but an artificial assemblage of gnawing mammals that evolved much the same set of teeth, habits and ecology. It is clear that there needs to be much more analysis of the existing data and further research, but a 'relaxed' molecular clock has been created that allows for different mutation rates in different groups of species. This uses 36 diverse living species to create an evolutionary tree that includes all the major groups of organism, then ties it to the fossil record at six points. The rest of the tree has to fit with these six knowns. Then over 100 proteins are looked at in each of the 36 living species; these proteins are all essential molecules for life and their amino acid sequences remain remarkably unchanged during evolution. However, over long time periods small mutations in the organisms' DNA make the protein sequences drift apart. The differences between the species can be used to estimate how fast the mutation rate is in each group. A computer model is used to fit the different mutation rates to their trees, together with the dates from the six fossils. The resulting family tree overall fits well with the fossil record.

In the Paleocene the mammals were generally primitive in their structure but included recognizable ancestors of many living groups. However, South America was already geographically isolated and ancestors of its peculiar fauna can be recognized there. The condylarths were the dominant group. They were generalized, rapidly-evolving early ungulates and most were herbivores of varying size. The arctocyonids had low, long skulls with canines and primitive molars and were probably racoon-like omnivores. Other mammals continued to occupy the same small-bodied guilds that Mesozoic mammals had occupied for a long time period and even today 90% of all mammal species weigh less than 5 kg.

After a marked turnover at the end of the epoch many new groups appeared in the Eocene that continue to the present day. This turnover was partly due to climatic change and partly to free migration of mammals across the northern continents of Eurasia and North America. Many modern groups of mammals appeared early in the Eocene, including rodents, advanced primates and modern artiodactyls and perissodactyls. Many different herbivores of all sizes evolved. Larger mammals seem to have been omnivores or herbivores but did not evolve quickly into large carnivores and early examples were only the size of hyenas and coyotes. Most were creodonts, including ambush predators and runners.

At the end of the Eocene the larger carnivorous groups died out and miacids (small carnivorous mammals) underwent a radiation that produced all the modern carnivore types. However, there appears to be no obvious key innovation behind this

radiation. They must have succeeded because their larger competitors disappeared for unknown reasons at the end of the Eocene. This is an example of guild replacement, not displacement.

During the Oligocene, as Antarctica became isolated and refrigeration set in, the climate began to cool on a global scale. Stresses were set up on the ecosystems of the various continents at different times but by the Miocene the cold increased and Antarctica's ice cap grew, affecting the world's climate. Vegetation changed, creating more open country with savanna grasslands. Mammals responded to this. The climatic and geographical changes allowed exchanges of mammals between continents, often in pulses as opportunities arose.

The animals that were successful on the open plains, with only scattered woodlands, were the grazers or browsers and deer and antelope evolved to great diversity. Their teeth often evolved to become very long for their height or hypsodont, with greatly increased enamel surfaces. Elephants, rodents, horses, camels and rhinos all evolved jaws and teeth with adaptations for better grinding. The grazer chewed on tough fibres and resisted the abrasion from tiny silica fragments (phytoliths). Savanna animals also became taller and longer-legged, well adapted for running fast. By the end of the Miocene the mammalian faunas were essentially modern. See the companion website for Box 29.10 Large carnivorous mammal-like badgers from Liaoning, China and Box 29.11 The evolutionary history of mammals.

29.5 Mass extinctions and catastrophes in the history of life on Earth

Extinction affects life on this planet all the time. It is the expected fate of a species, rather than a rarity, and more than 99% of all species that ever lived are extinct today. Sometimes extensive losses of whole populations, clades or groups that share particular morphologies, such as large body sizes, or functional attributes, such as feeding mechanisms, can occur. Whole ecosystems can collapse, and particularly in the face of rapid human-induced species loss, the difficulty is to understand the biological attributes that govern a species' vulnerability and the consequences of its extinction.

There have been periods in which there has been extremely rapid extinction, just as there have been periods with rapid evolutionary radiation. We have made reference to these on several occasions so far in this chapter. These dramatic drops in diversity are called mass extinctions, where there is extinction of a significant component of the global fauna and/or flora, and they are sudden in geological terms. A plot of family diversity through time (Figure 29.30) shows these crisis periods, which are listed in Table 29.3.

These extinctions were followed by recovery, but the larger the extinction, the longer the recovery time and the more the global ecosystem changes across the event. There were probably mass extinctions in the Pre-Cambrian too; for example, there was a major loss in acritarch diversity about 650 million years

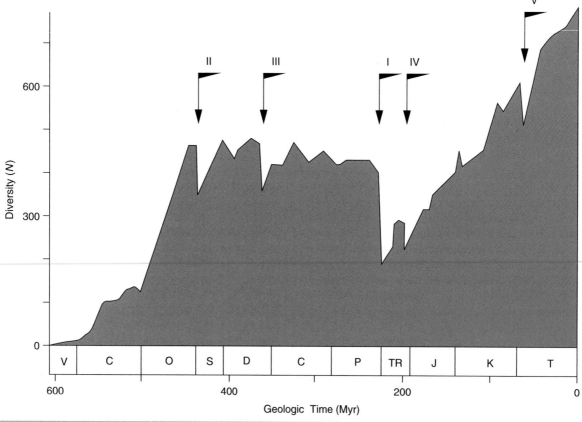

Figure 29.30 Characteristics of the five major mass extinctions. (Source: after Hallam, 2004)

Table 29.3 Magnitudes of the major crises in the history of life on Earth. For each event the percentage of families of marine animals which perished is given

Mass extinction	Marine families eliminated (%)
I. Late Permian	50
II. Late Ordovician	22
III. Late Devonian	21
IV. Late Triassic	20
V. Late Cretaceous	15

ago when about 70% of the flora died out, and the Ediacaran biota also appears to have vanished abruptly in the latest Pre-Cambrian.

Mass extinctions occur in the geological record where there are relatively sudden, near-simultaneous disappearances of a large percentage of the living species, affecting a great variety of organisms. They have been recognized at several well-documented periods of geological time: at the end of the Ordovician (439 million years ago), the Late Devonian (365 million years ago), the end of the Permian (250 million years ago), the end of the

Triassic (201 million years ago) and the end of the Cretaceous (65 million years ago). In these major extinctions at least 75% of marine life and many non-marine species died out. The characteristics in terms of species dying out are illustrated in Table 29.3.

There were also another 20 smaller-scale extinctions, such as that at the end of the Pleistocene, when many mega-mammals died out. This event was caused by either climatic change or the effects of humanity's hunting, or a combination, but it is thought likely that many mass extinctions have been caused by catastrophic events such as bolide (comet or asteroid) impacts or mega-eruptions, or at least that these have made a major contribution, with both greenhouse effects and volcanic/nuclear winters postulated. However, some have argued that volcanic eruptions did not have major effects and it is suggested that volcanic gases, apart from carbon dioxide, only really affect the weather, rather than the climate.

These mass extinctions have caused great controversy in the geological literature and have often polarized discussion. Some workers have suggested a periodicity to the extinctions, for example every 32 million years in the past 250 million years, based on periodic bursts of extinction in ammonites and foraminifera during the Mesozoic and Cenozoic, or a 26 million-year periodicity since the Late Permian. This might suggest a single causal mechanism for the periodicity rather than independent or unique causes. However, there seem to be problems with the original suggestions of periodic extinctions as there is a lack of correspondence between predicted and observed extinctions in the geological record, as well as problems with imprecise and variable dating of the time scale used in computing the temporal spacing of extinction events. In other words, this periodicity probably does not really exist.

Yet the mass extinctions do have to be explained by global processes as they were global phenomena. There has been much discussion and debate for every one of the mass extinction periods involving some or all of the following processes: plate tectonic changes, although usually these are thought to occur at too slow a rate; failure of the normal ocean circulation, which would affect ocean chemistry enough to cause global changes in climate and atmosphere; rapid change in sea level, affecting the global ecology and climate; an enormous volcanic eruption; and an extraterrestrial impact, from either an asteroid or a comet. In the latter case, an impact big enough to cause a global effect would leave behind physical evidence in the form of a sharply-defined layer of the element iridium, and/or impact-generated glass spherules or tektites, and/or quartz crystals with shock marks in the rocks associated with the extinction.

29.5.1 Extinction in the Late Ordovician period

This extinction event seems to be linked with major climatic change associated with global cooling. The first pulse of extinction occurred as a glacial period began (in the high palaeolatitudes of the Gondwana supercontinent) and the second occurred as it ended. No evidence has been put forward for an impact but the event did coincide with a regression that was quickly followed by a transgression as a result of ice build-up and then melting. Associated with the latter event was an extensive spread of black shales and anoxia. In fact it has also been suggested that this extinction event may have been comparatively minor, or certainly fairly gradual and multiple.

29.5.2 Extinction in the Late Devonian period

An extinction event took place, probably in several steps, between the last two stages of the Devonian, the Frasnian and the Famennian. There were major global coral reef extinctions, and their associated fauna and other groups of plants and animals were affected as well. Iridium spikes, shocked quartz and glass spherulites have been reported from China and Western Europe but there are also indications of changes in climate, sea level and ocean chemistry at the same time. Carbon isotope shifts indicate that global organic productivity was rapidly changing before the boundary. The geological evidence suggests several closely-spaced but medium-sized impacts over a couple of million years and it has been suggested that the global ecosystem was already stressed when an impact occurred.

29.5.3 Extinction in the Permo-Triassic period

This extinction 250 million years ago defines the end of the Palaeozoic Era and the beginning of the Mesozoic when an estimated 57% of all families and 95% of all marine animals became extinct. It was a rapid event, probably occurring over a million years and possibly much faster, and although it was much more serious in the oceans, it affected terrestrial ecosystems too. There is no evidence for an impact at the Permian–Triassic boundary and the plate collisions that formed Pangaea in the Permian would account for a major drop in diversity but not a mass extinction. However, the Earth was affected by plume activity at this boundary, when heat rising created a flat head of magma that could have been up to 1000 km across and 100 km thick, which created flood basalts over the surface to form the Siberian Traps. These lasted for only about one million years but were the largest and most intense eruptions known on Earth.

For the Siberian Basalts the palaeontological evidence suggests that Permian–Triassic mass extinctions were in the lowest lava sequence. It has been suggested that the volcanism caused the release of large quantities of methane from hydrates that had accumulated on adjacent high-latitude continental shelves, leading to greenhouse warming and a scenario in which the eruption was the primary cause of the extinction. The plume rose to the surface and erupted when a large amount of sulphate aerosol cooled the climate enough to rapidly form ice-caps. This caused a drop in sea level, along with global cooling, early in the eruptive sequence. Finally, as the eruption died away the crust subsided and the aerosols dispersed, making for a rapid end to the volcanically-induced glaciation and another rapid climatic change.

It is possible that the volcanic gases that had built up during the eruption had a greenhouse effect for some time after the eruption ended, taking the Earth from a volcanic glaciation to a volcanic hothouse. If the other effects of a giant eruption are added to this scenario, such as acid rain, ozone depletion, a massive injection of carbon dioxide into the atmosphere, or any combination of these, then all the ingredients are there for a mass extinction. Yet it is by no means certain that a giant eruption would cause a mass extinction: big eruptions like Krakatau (AD 1883) and Toba (75 000 years ago) had no lasting biological trace and some plume eruptions, such as the Jurassic Karroo Basalts and the Miocene Columbia River Basalts, have not been linked with extinctions, although there could be a threshold effect here. If an event is not big enough then it will do nothing but if it is big enough it will do everything.

There may have also been a climatic impact that the volcanic eruption added to. In the Late Permian the continents were grouped together in the supercontinent Pangaea and there was a giant ocean, Panthalassa, which covered 70% of the planet's surface. It appears that normal conditions deteriorated in this ocean around 260 million years ago as the deep water became anoxic. However, the surface waters remained normal, as we can see from abundant radiolarians in chert deposits. Then about

255 million years ago the radiolarians become rare towards the Permian–Triassic boundary and across the boundary there were none at all, suggesting that the ocean became anoxic to the surface. Long after the extinction around 245 million years ago the surface waters became sufficiently oxygenated to support radiolarians again, but the deep waters remained anoxic until about 240 million years ago. There were also rapid changes in carbon isotopes and in sea level across this boundary which cannot be explained by this model.

There have been reports of extreme abundances of fossil fungal cells in land sediments at the Permian–Triassic boundary, where the fungi broke down massive amounts of vegetation that had been catastrophically killed, especially with the extinctions of gymnosperms in Europe and the coal-generating floras of the southern hemisphere. The extinctions may have been caused by another ocean model in which there was a catastrophic overturn of an ocean supersaturated in carbon dioxide. This resulted in a very quick and large degassing that rolled a cloud of dense carbon dioxide over the ocean surface and low-lying coastal areas. The carbon dioxide build-up resulted from the global geography of the Late Permian, where it is speculated there was abnormal ocean circulation that did not include enough downwards transport of oxygenated surface water to keep the deep water oxygenated. With normal respiration and decay of dead organisms, the deep water evolved into an anoxic mass loaded with dissolved carbon dioxide, methane and hydrogen sulphide. Carbon continued to fall to the sea floor from normal surface productivity but it was deposited and buried because there was no dissolved oxygen to oxidize it. As carbon dioxide levels fell in the atmosphere, the Earth and the surface ocean cooled. Finally the surface waters became dense enough to sink, triggering a catastrophe as the carbon dioxide-saturated deep waters were brought to the surface, degassing violently. This event triggered greenhouse heating and a major climatic warming.

It seems likely that the carbon isotope change at the boundary was very short-lived, which suggests a major addition of non-organic carbon to the ocean, rather than a failure in the supply of organic carbon. Evidence suggesting a prolonged greenhouse warming at the Permian–Triassic boundary has also been found in Australia. Several palaeoclimatic indicators suggest that carbon dioxide had a vital role to play. Carbon dioxide in the atmosphere could have been increased by volcanic eruptions, through organic turnover, and it would have been accentuated and prolonged if plants were killed off globally as world floras and oceanic plankton would need to recover before the carbon dioxide could be drawn down out of the atmosphere. So it may be that a complicated story involving plume eruptions and ocean circulation changes are involved in these major extinctions.

Nevertheless, controversy about the cause of this extinction event has reigned, particularly over the claims of a cometary impact. There have been reports that the ratios of noble gases found in sediments in southern China were consistent with the idea that they had originated during a comet or asteroid impact at a location off north-west Australia. Yet many others vociferously dispute these claims and NASA is funding further work to try and duplicate the original results.

29.5.4 Extinctions at the End-Triassic period

The Triassic–Jurassic boundary marks a turnover of both terrestrial and marine groups. It has been suggested that a gigantic eruption took place at around this time, either synchronous with, or slightly postdating the boundary, in a volcanic episode when the first opening up of the plate boundary that split the Atlantic Ocean occurred. In the Newark Basalts of eastern North America the pollen changes that mark the Triassic–Jurassic boundary have been located just below the first lava flows. However, some suggest that the volcanic event took place slightly after the boundary and to complicate the picture shocked quartz has been discovered very near the boundary in northern Italy. In fact, several craters are dated to a Late Triassic period, as if they had been formed from an incoming extraterrestrial mass that fragmented at the last moment, but it is not clear whether they formed simultaneously, or if so whether they were formed at the same time as the extinction. For this period it is not at all clear whether the extinctions were even large and close inspection of the fossil record shows that many groups thought to be affected by this event, such as ammonoids, bivalves and conodonts, were in fact in decline throughout the Late Triassic and that other groups were relatively unaffected or subject to only regional effects. Yet explanations for the biotic turnover have included both gradualistic and catastrophic events.

Regression during the Rhaetian, with consequent habitat loss, is compatible with the disappearance of some marine faunal groups but may be regional and not global in scale and cannot explain the apparent synchronous decline in the terrestrial realm. Gradual widespread aridification of the Pangaean supercontinent could explain a decline in terrestrial diversity during the Late Triassic. Although an impact exactly at the boundary is lacking, there are the Late Triassic impact structures, which could have produced environmental degradation. The eruption of flood basalts would have caused emissions of carbon dioxide and sulphur dioxide and there seems to have been a substantial excursion in the marine carbon-isotope record of both carbonate and organic matter. This suggests a significant disturbance of the global carbon cycle at the boundary. Release of methane hydrates from sea-floor sediments is a possible cause for this isotope excursion but the trigger mechanism and resultant climatic effects are still uncertain.

29.5.5 Extinctions at the Cretaceous-Tertiary boundary

The end of the Mesozoic Era is marked by a major, worldwide mass extinction event that affected almost all the large vertebrates on Earth, on land, at sea and in the air, as all the dinosaurs, plesiosaurs, mosasaurs and pterosaurs became extinct suddenly at around 65 million years ago. Most plankton and many tropical invertebrates, especially reef dwellers, became extinct too, as well as many land plants. However, many groups survived, such as insects, mammals, birds and flowering plants on land, and fish, corals and molluscs in the ocean. There have been many explanations for this event, but two have dominated: a bolide (large asteroid or comet) impact or a giant volcanic eruption.

The scenario is that at the Cretaceous–Tertiary extinction event a small asteroid hit the Earth. Evidence for this came initially from the very large quantities of iridium present in the boundary rocks on land and in the sea. These spikes record a very short, unusual event, which was nonetheless very big as iridium is usually rare on Earth, and it was suggested that an asteroid impact was the cause. The asteroid Baptistina (170 km across) broke up about 160 million years ago in the inner main asteroid belt and fragments produced from this break-up were slowly delivered to orbits from which they could strike the terrestrial planets. This asteroid shower seems to be the most likely source (over 90% probability) for the 10 km wide Chicxulub crater off the Yucatan coast of Mexico (Bottke *et al.*, 2007), although the overall structure is 180 km across (see Figures 29.32 and 29.33 and the frontispiece to this chapter).

A core through the Chicxulub structure hit 380 m of unusual igneous rock, possibly generated by melting together the sedimentary rocks of the region. On top of this igneous rock lies broken rock, probably the largest surviving debris particles that fell back to the crater without melting, and on top of that are normal sediments. Haiti was 800 km from this crater but at the Cretaceous–Tertiary boundary there is a 30 cm layer of glass spherules overlain by a turbidite, submarine landslide layer that contains large rock clasts. Some of these look like shattered ocean crust, but the spherical pieces of black and yellow glass up to 8 mm are tektites (glass beads formed when rocks are instantly melted and splashed out from impact locations in the form of pieces of molten glass, then cooled while spinning through the air). The Beloc tektites, dated precisely to 65 million years ago, formed at 1300 °C from two different rock types. The

black tektites formed from continental volcanic rocks and the yellow ones from evaporite sediments with a high sulphate and carbonate content. This is precisely the mixture of rocks from the Yucatan around Chicxulub and the igneous rocks from the core have a chemistry of a once-molten mixture of the two.

Above the turbidite is a thin red clay layer between 5 and 10 mm thick which contains iridium and shocked quartz. The egg shape of the crater indicates that the asteroid hit at a shallow angle, about 20–30°, pushing out more debris to the north-west than other directions, with an impact energy of around 100 million megatons. In recent years, several other craters of around the same age as Chicxulub have been discovered, all between latitudes 20° and 70 °N. Examples include the Silverpit crater in the United Kingdom and the Boltysh crater in the Ukraine, both much smaller than Chicxulub but likely to have been caused by objects many tens of metres across striking the Earth. This has led to the hypothesis that the Chicxulub impact was only one of several that happened at the same time.

The near coincidence in time of the Deccan flood basalts has potential implications for this event too. Despite the fact that some workers have found a coincidence of flood basalt events with a number of mass extinctions (e.g. the Siberian flood basalt province and the End-Permian extinctions at 250 million years ago) and a possible cause and effect relationship, it is possible that both types of extreme event could be implicated. Flood basalt activity could represent only one of many interrelated geological factors and it has been suggested that the coincidence of these events with large impacts may be necessary in causing major mass extinction. It seems possible that large impacts could trigger or enhance the volcanism. The Deccan flood basalts, where there are over 50 flows in 1200 m of section, are thought to have erupted over about 500 000 years, bracketing the Cretaceous–Tertiary boundary. The impact fallout layer has also been found in sediments interbedded within the lavas, which supports the radiometric dating and palaeomagnetic dating, suggesting that the eruptions began prior to the boundary. There seems therefore to be a major probability that the correlations are correct, especially as there are other examples of flood basalt and lesser faunal and floral changes in the geological record (e.g. the Karroo Basalts might correlate with the boundary between the Early and Middle Jurassic, which is marked by a faunal turnover event).

The processes that have been suggested as the environmental effects of flood basalts and therefore linked to mass extinctions are as follows:

1. A cooling caused by the formation and spread of stratospheric aerosols produced by high fire fountaining and associated convective plumes above them. This would be associated with rich dissolved sulphur taken into the stratosphere. Volcanic aerosols at all altitudes would cause short-term climatic cooling but a flood basalt eruption could release such large quantities of sulphuric acid aerosols that a regional tropospheric aerosol cloud might be maintained, despite aerosol washout. This could be aided by a suppression of convection in the atmosphere by the cooling, which would mean that rainout would be less effective at removing the aerosols.

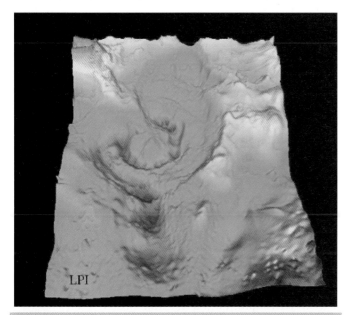

Figure 29.32 Chicxulub impact crater. (Source: courtesy V.L. Sharpton, LPI, http://www.apod.nasa.gov./apod/ap000226.html)

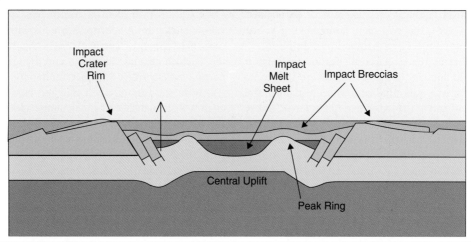

Figure 29.33 Schematic diagram of a buried complex crater with impact breccias overlying impact melt. This is a generalized diagram of how the Chicxulub impact crater may be structured. It is buried beneath several hundred to one thousand metres of sediment, so drilling is required in order to sample rocks from the crater. The Chicxulub Scientific Drilling Project is designed to drill through the overlying sediments, the impact breccias, the impact melt, and into the underlying impact-fractured rock. (Source: http://www.lpl.arizona.edu/SIC/news/chicxulub1.html)

2. Greenhouse warming caused by large carbon dioxide emissions, although recent estimates suggest that the total increase in CO_2 would have been under 200 ppm, leading to a predicted global warming of only 2 °C over the eruption period of the lavas, and it is not likely that such a small and gradual warming would have been an important factor in mass extinctions. Another source of warming could have been from the SO_2 concentrations, which could have led to major severe regional warming.

So the question that is pertinent here is: Is there a relationship between impacts and flood basalts? Large impacts by comets or asteroids might excavate initial holes deep enough to result in decompression upper-mantle melting but there is no firm evidence that impacts can directly cause volcanism, or of local, large impacts at flood basalt locations. However, an indirect connection may be possible as impact energy would cause major lithospheric stresses, which might lead to rapid decompression in the mantle above rising mantle plumes and help induce or increase ongoing volcanism. It does seem though that the Deccan lavas began erupting prior to the Chicxulub impact and therefore that the impact could not have triggered volcanism.

Moreover, other evidence suggests that the mass extinctions at the Cretaceous–Tertiary boundary were probably caused by the impact and not volcanism. For example, there is low iridium (a mineral thought originally to have been brought in from an extraterrestrial source, although volcanic sources have also been found) in the Deccan lavas, in soils and in weathering profiles between flows and at the boundary sections, and the shocked quartz and stishovite cannot be produced at pressures characteristic of volcanic eruptions (although there has been dispute here too about whether these shocked minerals can be formed by volcanic processes). Further, the tektite glass (impact melt glass) found is not compatible with a volcanic origin.

There remains much dispute still about the role of volcanism in the Cretaceous–Tertiary extinctions and it has been suggested that worldwide volcanism coincident with the end of the Cretaceous provides the source of iridium, shocked minerals, the carbon (CO_2) excursion and the attendant greenhouse effect.

What is clear is that both impact effects and volcanic eruptions could have had major global effects on the atmosphere, and it is even possible that it was both together which combined to cause the mass extinctions. The impact scenario of a 10 km asteroid would blow a mass of vaporized rock and steam high above the atmosphere. This would cause a so-called nuclear winter. Because the rocks at Chicxulub contain high sulphur contents, after impact enormous quantities of sulphate aerosols would have been produced in the atmosphere, which would have acted as nucleation points for severe acid rain. This would be sufficient to have suffocated some air breathers, destroy plant foliage and dissolve marine shells. The carbon dioxide balance between air and ocean would have been upset, which would make surface waters barren for tens of years. Among other effects would be the impact of dust, smoke and aerosols in cutting down the sunlight, so that land plants and plankton could not photosynthesize. This would initially cause cool or even freezing temperatures for months, and later after the dust and aerosols had settled the enormous carbon dioxide levels released into the atmosphere by the impact would generate a greenhouse effect that elevated the Earth's temperatures for a thousand years or more.

The most extreme impact scenario suggested has been called the microwave summer, in contrast to the nuclear winter idea. In this scenario some of the material was blasted upwards at a velocity greater than the Earth's escape velocity, although most fell back after a travel time of about one hour emitting thermal radiation as it re-entered the atmosphere. There was a radiation pulse 1000 times more than enough to ignite dry forests, allegedly leaving soot in the Cretaceous–Tertiary boundary sections. However, the geological evidence is that birds, tortoises and mammals, all living

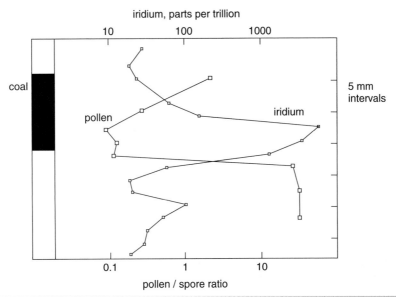

Figure 29.34 Spore and iridium spike at the Cretaceous–Tertiary boundary. (Source: after Cowen, 2000)

on land and breathing air, all survived the Cretaceous–Tertiary boundary event and hence these extreme scenarios are not likely to have happened.

The fossil record is ambiguous when it comes to producing real evidence about the Cretaceous–Tertiary event. Land plants in North America were devastated at the Cretaceous–Tertiary boundary and this is seen where the sediments below the boundary are dominated by angiosperm pollen, but the boundary itself is dominated by fern spore: a spore spike that mirrors precisely the iridium spike and is equally intense and short-lived (Figure 29.34). This could be explained by a short, severe crisis for the land plants, generated by impact or eruption, in which all adult leaves died off for lack of light, in prolonged frost or in acid rain. Ferns are likely to have been the first plants to recolonize, with higher plants doing so much later. In other parts of the world the plants were affected less; for example, in Japan and in the southern hemisphere plants were hardly affected at all. This suggests the effects of the Chicxulub impact had a far greater effect in North America than elsewhere, which is logical and fits with the rock evidence.

Another problem is that a very sudden change in temperature should have caused a catastrophe in the ectothermic reptiles but did not, as crocodilians and turtles were hardly affected at all and lizards were affected only in a minor way. The fact that high-latitude Late Cretaceous dinosaurs lived in Alaska, South Australia and Antarctica, where they would have been well adapted to seasonal variations, periods of darkness and very cool temperatures, makes an impact scenario difficult for such animals. The survival of birds is also difficult to explain as the smaller dinosaurs overlapped with the larger birds in size and in ecological niches as terrestrial bipeds. How did the birds survive and not the dinosaurs? Birds seek food in the open by sight, are small and warm-blooded, with high metabolic rates and small energy stores, and today even a small problem, like a sudden

storm or severe winter, causes high bird mortality. It seems that the extreme impact models must be wrong.

However, an impact or a large volcanic eruption might have caused mass extinction by longer-term climatic changes. Oxygen isotope measurements across the Cretaceous–Tertiary boundary suggest that oceanic temperatures fluctuated markedly in the Late Cretaceous and through the boundary events. Added to this, carbon isotope measurements across the boundary suggest that there were severe, rapid and repeated fluctuations in ocean productivity in the three million years before the final extinction and that productivity and ocean circulation were suppressed for at least several tens of thousands of years after the boundary, perhaps as long as two million years. These changes could have caused marked effects in both marine and terrestrial environments.

It has been suggested that it is climatic change that is the connection between the impact and the mass extinction. The impact upset normal climate, with longer-term effects that lasted much longer than the immediate and direct consequences of the impact. However, there were survivors and hardly any groups of organisms became entirely extinct. But there is still missing information that is required to fully explain the Cretaceous–Tertiary mass extinction. The unusual severity of the Cretaceous–Tertiary extinction, its global scale and the sudden and dramatic biological characteristics, such as the fern spore spike, may have occurred due to a gigantic plume volcanic eruption and an extraterrestrial impact, which occurred when the global ecosystems were particularly vulnerable to an ocean stability disturbance.

29.5.6 End-Pleistocene extinctions: Humans or environmental change?

The Pleistocene fossil record accumulated in a period in which there was a series of glacials and interglacials and extreme climatic change, yet it appears that the extinction rate overall did not rise above background for most groups of organisms (for example, terrestrial mammals, North American amphibians and reptiles, Eastern Pacific mollusca). There were many local extinctions and responses to climatic change because of the ice-sheet movements, yet there is very little evidence for global species extinction.

An event that does stand out as exceptional is the End-Pleistocene extinction of the terrestrial mammal megafauna (where the mean body weights are over 44 kg), although this is not classified as a mass extinction event.

- It appears to have been a global extinction, although the magnitude markedly varied between continents.

- The losses were 33 out of 45 genera in North America, 46 out of 58 genera in South America and 15 out of 16 genera in Australia. The figures were lower for Europe (7 out of 24 genera) and lowest in Africa (7 out of 44 genera).

- The species lost in North America included sabre-toothed cats, mastodon, mammoths, short-faced bear, camels, horses, giant armadillo, glyptodonts and ground sloths. The South American extinctions included gompotheres, ground sloths, glyptodonts, horses and large notoungulates. In Australia a number of giant marsupials died out, including the diprodonts, giant kangaroos and large carnivores like the marsupial lion.

- Animals that became extinct were not replaced.

- Where the extinctions are well dated they were sudden, as in North America and New Zealand.

- Taxa other than megafauna were affected; for example, there was a 75% loss of large mammalian genera in the American tropics, with only 4 of the 23 genera known in the Amazon basin surviving.

- Small vertebrates on islands died out yet the percentage of small mammals that became extinct overall is much lower than that for the larger species.

The causes of these extinctions have generated much debate and many suggestions, with two main contending hypotheses: the effects of climatic change and the effects of human predation. It is generally agreed that these extinctions occurred when there was both major climatic change and an increasing predation from the human populations but their relative importance is contended.

The climatic explanation involves the fact that the transition from a glacial to an interglacial, including major Late-Glacial climatic oscillations, resulted in extensive plant community reorganization in which mosaic vegetational communities were replaced by climatically-zoned vegetation. Herbivores that had evolved to feed on a wide range of plants disappeared first, which led to the extinction of the large carnivores and scavengers that depended on them. The main problem with this hypothesis is

that there were multiple glacial–interglacial transitions throughout the Pleistocene with no global extinctions associated with them. It has been argued that the latest glacial-interglacial transition was fundamentally different climatically but this seems unlikely.

The overkill hypothesis argues that in North America the extinctions correlate well with the arrival of the Clovis culture (~11500–10 500 years ago) and so hunting by humans must have been the main driving force behind them. The evidence shows the diachronous nature of the megafaunal extinctions on a global scale, and it has been argued that the extinctions in different places roughly coincided with the arrival of humans, yet often the synchronicity of the extinctions has not been demonstrated. It has also been suggested that there are very few places where mammal remains occur with human remains or human artefacts, which implies that co-existence was brief.

A third possibility is that the combination of climatic changes and human predation explains the observed patterns. In this case, human predation only became critical once the populations of the large mammals had declined and/or become geographically restricted in response to climatic and vegetational changes. Yet it is not really clear why when the ice-sheet retreat had increased the available habitable area, the large mammals should have declined. Some species survived until much later than the climatic oscillations at the end of the Pleistocene; for example, the mammoth survived on St Paul Island (Pribilofs) in the Bering Seas until around 8000 years ago and on Wrangell Island (north-eastern Siberia) until 4000 years ago (Vantanyan et al., 1993; Guthrie, 2004), and the Irish elk survived in western Siberia until 7700 years ago (Stuart et al., 2004). This suggests that there may be no single explanation for the End-Pleistocene extinctions.

As has been mentioned, the 'ragged nature' of the End-Pleistocene extinctions, with isolated pockets of populations surviving for longer, suggests that extinctions have a complex ecology. Before their extinction, both the Irish elk and the woolly mammoth underwent dramatic shifts in distribution, driven largely by climatic/vegetational changes.

The fossil record, combined with modern data, can provide a deeper understanding of biological extinction and its consequences, which is important in understanding the latest great mass extinction period, caused by humans (sometimes discussed as the sixth mass extinction). The fossil record has locked into it many periods of global warming and cooling episodes that exceeded the ability of species to cope with local conditions, causing the disruption of ecological communities and the colonization of previously sparsely-populated habitats. Extinctions, whether 'normal' or caused by some extraordinary event such as an asteroid impact, are complex but crucial to a fuller understanding of the dynamics of biodiversity on Earth.

Exercises

1. Explain the difference between macro- and microevolution.

2. How do you think life originated on Earth?

3. How far do you think the fossil record is a complete record of the life that has appeared on Earth?

4. Why do you think the Cambrian explosion of life took place and what were the characteristics of this explosion?

5. What were the physical challenges to life on land that the primitive amphibians had to cope with?

6. What evidence do we have for dinosaur behaviour?

7. What evidence do we have that birds are the descendents of dinosaurs?

8. Outline the characteristics of the major mass extinctions in the history of life on Earth.

9. Describe the geological evidence for the 65 million-year-old asteroid impact at the Cretaceous–Tertiary boundary.

10. Discuss whether climate change or humanity was responsible for the megafaunal extinctions at the end of the Pleistocene.

References

Ahlberg, P.E. and Clack, J.A. 2006. Palaeontology: A firm step from water to land. *Nature*. **440**, 747–749.

Benton, M. and Harper, D. 1997. Basic Palaeontology. Harlow, Prentice Hall.

Bottke, W.F., Vorrouhlicky, D. and Nesvorny, D. 2007. An asteroid breakup 160 Myr ago as the probable source of the K/T impactor. *Nature*. **449**, 48–53.

Cowen, R. 2000. History of Life. Oxford, Blackwell Science.

Daeschler, E.B., Shubin, N.H. and Jenkins, F.A., Jr. 2006. A Devonian tetrapod-like fish and the evolution of the tetrapod body plan. *Nature*. **440**, 757–763.

Doyle, P. 1996. Understanding Fossils. Chichester, John Wiley & Sons, Inc.

Guthrie, R.D. 2004. Radiocarbon dating evidence of Holocene mammoths on an Alaskan Bering Sea island. *Nature*. **429**, 476–478.

Hallam, T. 2004. Catastrophes and Lesser Calamities: The Causes of Mass Extinctions. Oxford, Oxford University Press.

Kemp, T.S. 1999. Fossils and Evolution. Oxford, Oxford University Press.

Lazcano, A., 2003. The Never-Ending Story. American Scientist. 5, 91. On the Bookshelf book review.

Melosh, H.J. *et al.* 1990. Ignition of global wildfires at the Cretaceous/Tertiary boundary. *Nature*. **343**, 251–254.

Milner, A. 2002. Dino-Birds. London, The Natural History Museum.

Milsom, C. and Rigby, S. 2004. Fossils at a Glance. Oxford, Blackwell.

Rollinson, H. 2007. Early Earth Systems: A Geochemical Approach. Oxford, Blackwell Scientific.

Shubin, N.H., Daeschler, E.B. and Jenkins, F.A., Jr. 2006. The pectoral fin of *Tiktaalik roseae* and the origin of the tetrapod limb. *Nature*. **440**, 764–771.

Seilacher, A., 1989. Vendoza: Organismic Construction in the Proterozoic Biosphere. Lethaia. **22**, 229–239.

Selden, P.A. and Edwards, D. 1989. Colonisation of the land. In Allen, K. and Briggs, D.E.G. (eds.). Evolution and the Fossil Record, Belhaven Press. London, 122–152.

Sepkoski, J.J. 1984. A kinematic model for Phanerozoic taxonomic diversity, *Paleobiology* **10**, 246–267.

Stuart, A.J., Kointsev, V.E., Higham, T.F.G. and Lister, A.M. 2004. Pleistocene to Holocene extinction dynamics in giant deer and woolly mammoth. *Nature*. **431**, 684–689.

Vantanyan, S.L., Garrut, V.E. and Sher, A.V. 1993. Holocene dwarf mammoths from Wrangel Island in the Siberian Arctic. *Nature*. **282**, 337–340.

Wignall, P.B. 2001. Large igneous provinces and mass extinctions. *Earth-Science Reviews*. **53**, 1–33.

Xu, X., Norell, M.A., Khuang, X., Wang, X., Zhao, Q. and Jia, C. 2004. Basal tyrannosauroids from China and evidence for protofeathers in tyrannosauroids. *Nature*. **431**, 680–684.

Xu, X. and Norell, M.A. 2004. A new troodontid dinosaur from China with avian-like sleeping posture. *Nature*. **431**, 838–841.

Further reading

Allen, K.C. and Briggs, D.E.G. 1989. Evolution and the Fossil Record. London, Belhaven.

Clarkson, E.N.K. 1998. Invertebrate Paleontology and Evolution. Oxford, Blackwell Science.

30 Environmental Change: Greenhouse and Icehouse Earth Phases and Climates Prior to Recent Changes

Spiral of geological time and the associated changes. (Source: NASA)

Earth Environments   David Huddart and Tim Stott
© 2010 John Wiley & Sons, Ltd

Learning Outcomes

After reading this chapter and completing the exercises you will be able to:

➤ Document and understand how the Earth's climate has changed dramatically over geological time by studying specific examples from various phases of the planet's history.

➤ Understand that the climate has oscillated between greenhouse and icehouse conditions and have some knowledge of the evidence on which this finding is based and the processes that are thought to control these changes.

➤ Understand and be able to evaluate the evidence for specific examples: the Proterozoic Snowball Earth; the greenhouse Cretaceous period; the Palaeocene supergreenhouse; the Oligocene and subsequent climatic oscillations and cooling; the Late Cenozoic ice ages; Late Glacial evidence for rapid climatic change; the Medieval Warm period; and the Little Ice Age.

30.1 Introduction

'Greenhouse Earth' describes the state in which the planet's atmosphere contains enough greenhouse gases for ice to be completely absent from its surface, when it is tropical even to the poles. It has been suggested that 'icehouse Earth' phases can also occur, where glacial ice is present in fluctuating amounts and the Earth has the possibility of ice ages and associated glacials, interglacials, stadials and interstadials.

• **Ice age and glacials:** There is an expansion of ice caps to cover a varying percentage of the planet, usually with an expansion from the two polar regions and from high mountains, forming periodically-occurring ice sheets in more temperate latitudes.

• **Interstadial:** A brief warmer phase during an ice age. This is not as warm as an interglacial but can last for several thousand years.

• **Interglacial:** A longer warm period during an ice age, such as today's climate, where there is ice present mainly at the poles and in high mountains. This may last for tens of thousands of years.

• **Stadial:** A brief cooler period during an interglacial when glaciers may reform in high mountains in temperate latitudes.

As we are presently in the middle of an ice age it is essential that we try and learn from the past in order to interpret what might

happen in the future. (See the companion website for Box 30.6 Origins of Mid-Cretaceous high subsurface ocean temperatures.) This is even more important because we are currently altering the climate by our own anthropogenic actions and this is being superimposed on the natural change. We therefore hope that we can learn from the way in which the Earth's climate has changed drastically over geological time and perhaps put the recent global warming into a better perspective.

Whilst it is impossible to reconstruct the Earth's climate for every geological period, examples of the various types and what the controls on these have been are given in this chapter.

30.2 Early glaciations in the Proterozoic phase of the Pre-Cambrian (the Snowball Earth hypothesis)

30.2.1 Introduction: The evidence for a Snowball Earth

It has been suggested that the Earth was covered completely by ice from pole to pole for long stretches of time during part of the Cryogenean period of the Proterozoic Eon (850–630 million years ago; see Figure 30.1 for time scales). Here there were two phases: the youngest, the Marinoan, from about 660 million years ago, estimated to have lasted 6–12 million years, and the Sturtian, later than 746 million years ago. There was a

Neoproterozoic	Ediacaran	630–542 Ma
	Cryogenian	850–630
	Tonian	1000–850
Mesoproterozoic	Stenian	1200–1000
	Ectasian	1400–1200
	Calymmian	1600–1400
Paleoproterozoic	Statherian	1800–1600
	Orosirian	2050–1800
	Rhyacian	2300–2050
	Siderian	2500–2300

Figure 30.1 Divisions of the Proterozoic Eon

much older Snowball Earth too, about 2200 million years ago, known as the Makganyene Formation in South Africa. Although there are sedimentary deposits about 200 million years older than this in central North America and Northern Europe, it is not thought that these represent earlier Snowball Earths.

This 'Snowball Earth hypothesis', as it has come to be popularly called (first coined by Kirschvink in 1992), was developed to account for sedimentary rocks generally regarded as glacial in origin in tropical latitudes and other features of the geological record at this time. The importance of this period is that multicellular evolution began to accelerate after the last glacial ended. However, Snowball Earth is a controversial hypothesis and there are many scientists who oppose the idea that the Earth could be completely frozen, or even the geological evidence on which this is based. This is thus an interesting and important example of how climatic and geological evidence can be interpreted differently by well-respected scientists.

The evidence for global ice cover comes from sediments interpreted as glacial in equatorial latitudes at the time of deposition (Figure 30.2), including striated pavements, diamictites interpreted as tillites, facetted and striated stones and dropstones (see http://snowballearth.org).

Snowball Earth may have been facilitated by an equatorial continental distribution which allowed rapid, unchecked weathering of the continental rocks, absorbing large quantities of carbon

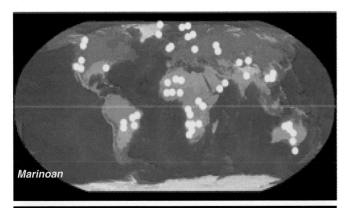

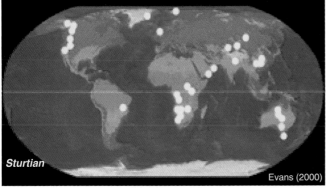

distribution of glacial deposits

Figure 30.2 Global distribution of the Sturtian (~720–700 million years ago) and the Marinoan (~660–635 million years ago) Cryogean glacial and glaciomarine deposits

dioxide from the atmosphere. As this greenhouse gas depleted, so ice accumulated, which further cooled the Earth by reflecting solar energy. This runaway system led to a new ice-covered equilibrium with equatorial temperatures like those of modern Antarctica, at around $-20\,°C$. The global mean temperature is suggested to have been around $-50\,°C$. To break out of this glacial state, large quantities of greenhouse gases emitted mainly from volcanic processes would have had to accumulate over millions of years, but once melting began it could be a rapid process, maybe lasting as little as 1000 years. Then weathering of glacial sediments, by reaction with carbon dioxide and fertilization of ocean photosynthesizers, could eventually have captured enough of the greenhouse gases to initiate another Snowball Earth phase.

The evidence for global ice cover comes from sediments interpreted as glacial in equatorial latitudes at the time of deposition. Geochemical evidence from rocks associated with low-latitude glacial deposits has been interpreted as illustrating a crash in oceanic life during glacial phases, which is consistent with a freezing of the surface oceans. Although the presence of glacial ice cannot be disputed, the idea that the entire Earth was ice-covered is contentious and some scientists have suggested a 'slushball' rather than a 'Snowball'. In a slushball alternative there would be a band of ice-free or ice-thin waters remaining around the equator, allowing a continued hydrological cycle.

This model was required by some scientists in 2007, who devised a computer simulation of the CO_2 which factors in the role of oxygen in the ocean. Progressive cooling, they suggested, would have allowed atmospheric oxygen to spread more deeply into the sea, transforming rich layers of dissolved organic carbon, formerly created by photosynthesis, into CO_2. This would be released back into the atmosphere and warm it up (by the greenhouse effect) sufficiently to create thawing, allowing sea ice and glaciers to shrink before a cooling cycle resumed. This would mean that the Cryogenian would have been a milder, slushier, shorter period, with ice-free seas in the tropics, where sunlight would have generated photosynthesis rather than a deep, long phase of planetary freezing. The scientists called this 'slushball Earth' and it supports others who believe that certain phenomena in the sedimentary record need ice-free locations for ice to move to and also that some sedimentary structures could only form in an open ocean.

Attempts to construct computer models of a Snowball Earth have found it difficult to accommodate global ice cover without major changes in the laws and constants that govern the Earth and in fact recent research using observed geochemical cyclicity in clastic rocks suggests that glacial phases were interrupted by warm phases similar to the Pleistocene.

Snowball Earth must have had profound implications for the history of life on the planet. Whilst many refugia have been suggested worldwide, ice cover would have had a drastic effect on ecosystems dependent on sunlight. However, melting of this global ice cover may have offered new opportunities for diversification and could have driven the rapid evolution which took place at the end of the Cryogenian.

30.2.2 What could initiate Snowball Earth?

The cause of Snowball Earth is suggested to be a lowering of atmospheric greenhouse gases (such as carbon dioxide and methane) through tectonically-mediated rock weathering when the Sun was dimmer than at present. The tropical distribution of continents is necessary to allow the start of Snowball Earth. Tropical continents are more reflective than open ocean and therefore absorb less of the Sun's heat (most of the absorption of solar energy today occurs in tropical oceans). Tropical continents are also subject to more rainfall, which results in increased river discharge and erosion.

When exposed to the air, silicate rocks undergo weathering reactions of the form rock-forming minerals $+ CO_2 +$ water, which result in cations $+$ bicarbonate and silicon dioxide. The released calcium cations react with the dissolved bicarbonate in the ocean to form calcium carbonate as a chemically-precipitated rock. This transfers carbon dioxide from the atmosphere to the Earth's rock record and offsets the carbon dioxide released from volcanoes into the atmosphere. However, a lack of suitable sediments for palaeomagnetic analysis makes precise continental distributions during this period difficult to work out, though some reconstructions point to the polar continents providing a place where ice could nucleate. It has also been suggested that changes in ocean circulation patterns could then have provided the trigger for Snowball Earth.

Another factor that may have contributed to initiation is the introduction of atmospheric free oxygen, which could have reached sufficient amounts to react with methane in the atmosphere, oxidizing it to carbon dioxide (a much weaker greenhouse gas). Also a younger, fainter Sun would have emitted 6% less radiation.

Usually as the Earth gets cooler the weathering reactions slow and so less carbon dioxide is removed from the atmosphere, causing the Earth to warm again as this gas accumulates. But in the Cryogenian period the Earth's continents were in tropical latitudes, making this moderating process less effective as high weathering rates continued on land even as the Earth cooled. This allowed ice to advance beyond the polar areas, and once it reached within $30°$ of the equator positive feedback occurred, with the increased albedo of the ice leading to further cooling and the formation of more ice until the whole planet was ice-covered.

Polar continents have low evaporation rates and so are too dry to allow substantial carbon deposition, restricting the amount of atmospheric carbon dioxide that can be removed from the carbon cycle. A gradual rise of the proportion of the isotope C^{13} relative to C^{12} in sediments predating global glaciation indicates that carbon dioxide drawdown before a Snowball Earth was a slow and continuous process. However, the beginning of a Snowball Earth is always marked by a sharp downturn in the C^{13} value of the sediments, which may be attributed to a crash in biological productivity as a result of the cold temperatures and ice-covered oceans. The global temperatures were so low that the equator was as cold as modern Antarctica. This was maintained by the reflective ice and a lack of heat-retaining clouds, caused by water vapour freezing out of the atmosphere.

Two further phenomena aided the Cryogenian high silicate weathering rates and therefore a cold climate. The first was the break-up of the Rodinia supercontinent, which began about 830 million years ago (Figure 30.3) and lasted for nearly 200 million years. Silicate weathering rates are low when such a supercontinent exists as most land area is far from the ocean and hence dry. When a supercontinent breaks up, formerly dry areas become wetter and weathering rates increase. The second event was the eruption of flood basalt 723 million years ago in Arctic Canada, as basalt weathers fast and is a rich source of Ca_{2+} ions.

The effects of methane changes could also be important, especially as CH_4 levels were thought to be far higher than those of the present day. When O_2 levels rose, CH_4 levels fell accordingly, initiating a loss of greenhouse warming. The most likely source of a sudden rise in O_2 is the evolution of oxygenic photosynthesis. However, preliminary evidence from fossil organic molecules suggests that oxygenic photosynthesis was in existence half a billion years before the first Snowball Earth.

An alternative explanation for the first Snowball Earth phase 2.3 billion years ago has been suggested: when bacteria suddenly developed the ability to break down water and release oxygen this led to a radical change in the biosphere. This oxygen influx destroyed methane in the atmosphere, which had acted like a blanket to keep the Earth warm. In modelling this scenario Kopp *et al.* (2005) say that the Earth's exact position from the Sun was the only thing that saved the planet from a permanent deep-freeze. Before the Snowball Earth the Sun was only 85% as bright as it is now and the atmosphere was loaded with methane. Then along came cyanobacteria, which evolved into the first organism to use water in photosynthesis, releasing oxygen as a by-product. This happened about 2.3 billion years (not as early as 3.8 billion years ago, as some scientists had thought). Computer modelling suggests that most of the atmospheric methane could have been destroyed within 100 000 years, and certainly within several million years global temperatures would have plummeted to $-50°C$ and ice would have been extremely thick at the equator. Most organisms would have died, but many survived (as described later) and adapted to breathe oxygen. Thus life eventually evolved into more complex forms. Eventually the changed biology and chemistry caused CO_2 to build up sufficiently to generate another greenhouse phase.

30.2.3 What could cause a break from global glaciation and the end of a Snowball Earth?

The carbon dioxide levels necessary to unfreeze the Earth have been estimated at 350 times what they are today: about 13% of the atmosphere. Since the Earth was almost completely ice-covered, carbon dioxide could not be withdrawn from the atmosphere by the weathering of siliceous rocks. However, over a period of between 4 and 30 million years enough carbon dioxide and methane accumulated, mainly emitted from volcanoes as plate tectonics continued. Carbon dioxide entered ocean basins through sub-sea volcanoes and vents and subglacial terrestrial volcanoes. This built up due to an imbalance between out-gassing and consumption finally caused enough greenhouse effect to make surface ice melt in the tropics until a zone of ice-free land and water developed, which was darker than the ice and absorbed more energy from the Sun, again initiating a positive feedback effect. On the continents the glacial melting released extensive quantities of glacial deposits, which were eroded and weathered. The resulting sediment supplied to the oceans was high in nutrients such as phosphorus, which combined with the abundance of carbon dioxide to trigger a cycanobacteria population explosion. This caused a relatively fast re-oxygenation of the atmosphere, which may have contributed to the rise of the Ediacaran biota and the subsequent Cambrian explosion of life.

This positive feedback loop melted ice over a short period of geological time, but the replenishment of the atmospheric oxygen and depletion of carbon dioxide levels took further millennia. It has been suggested that carbon dioxide levels fell sufficiently for the Earth to freeze again and a repetition of this cycle occurred until the continents had moved to more polar latitudes.

30.2.4 Arguments for Snowball Earth

A variety of arguments have been made for the Snowball Earth hypothesis from across a range of scientific disciplines, and most have proved controversial and initiated much discussion. These arguments include the following:

1. Palaeomagnetic measurements indicate that some sediments of glacial origin were deposited within $10°$ of the equator, although the accuracy of this reconstruction has been questioned. For example, it is possible that during the Proterozoic the Earth's magnetic field did not did not have a dipolar distribution as it does today (see Box 30.1, in the companion website). A hotter core may have circulated more vigorously and given rise to four, eight or even more poles, in which case palaeomagnetic data would have to be reinterpreted as particles could have aligned to a west pole rather than a north one. However, these data suggest that multiple reversals of the geomagnetic field occurred, indicating that during the Marinoan glaciation the deposits were formed close to the palaeo-equator over at least several hundreds of thousands of years, if not millions. Other data suggest that the carbonate successions formed at low latitudes, indicating that the meridional climate gradient was not reversed as it would have been if the rotation axis had been highly oblique to the ecliptic.

2. Glacial and glaciomarine deposits are found at low latitudes with features such as dropstones, varves and striations, and diamictites interpreted as tillites. However, many of these features could be formed by other processes and many of the key localities for Snowball Earth have been contested vociferously.

3. During the proposed Snowball Earth period there are variations in C^{13} that are rapid and extreme compared to normal observed modern variations and this is consistent with a deep freeze that killed off most or nearly all photosynthetic life in the oceans. Close analysis of the timing of the C^{13} spikes in deposits across the planet suggests four or possibly five glacial events in the later Proterozoic.

4. Banded ironstone formations (BIF) are deposits of layered iron oxide and iron-poor chert. Generally they are very old and are related to the oxidation of the Earth's atmosphere in the Early Proterozoic, when dissolved iron in the oceans came in contact with photosynthetically-produced oxygen and precipitated out as iron oxide. The bands were produced at the tipping point between an anoxic and an oxygenated atmosphere; today, since the atmosphere is oxygen-rich, it is not possible to accumulate enough iron oxide to create a banded deposit. However, the only BIFs after the Early Proterozoic (after 1.8 billion years ago) are associated with the Cryogenian glacial sediments. For such rocks to be deposited there would have to be anoxia in the ocean and so much dissolved iron would have to accumulate before it could meet an oxidant that would precipitate it as ferric oxide. For the oceans to become anoxic there must have been limited gas exchange with the oxygenated atmosphere and it has been argued that the reappearance of BIFs is a result of limited oxygen levels in an ocean sealed by sea ice. However, detractors suggest that the rarity of BIF deposits indicates deposition in inland seas, such as the current Black Sea, which being isolated from oceans may have been stagnant and anoxic at depth, with a sufficient iron input to provide the required conditions for BIF formation.

5. Cap carbonate rocks are found as a sharp transition from Proterozoic glacial deposits to a limestone or dolostone (dolomitic limestone), typically between 3 and 30 m thick. These deposits sometimes occur in sedimentary successions that have no other carbonate rocks, which suggests a remarkable change in ocean chemistry. They have an unusual chemical composition, microbial mounds with a vertical tubular structure, primary and early diagenetic barytes, and giant wave ripples that distinguish them from most standard carbonate rocks. Their formation could be caused by a large influx of positively-charged ions, as would be produced by rapid weathering during the extreme greenhouse following a Snowball Earth event. The $\delta^{13}C$ isotopic signature of these cap carbonates is nearly $-5‰$ and such a low value is usually indicative of absence of life, since photosynthesis usually raises the value. However, the release of methane deposits (from permafrost and submarine gas hydrates) could have lowered the value and counterbalanced the effects of photosynthesis. The explanation for these cap carbonates suggests that at the melting of a Snowball Earth, water dissolves the abundant carbon dioxide from the atmosphere to form carbonic acid, which falls as acid rain. This weathers any exposed silicate and carbonate rocks, releasing calcium, which is washed into the oceans to form the cap carbonates. However, the high carbon dioxide atmospheric concentration causes the oceans to

become acidic and dissolve any carbonates within: a paradox for the formation of cap carbonates. The thickness of some of these deposits is far above what could reasonably be expected in the relatively quick deglaciations, and the explanation is further weakened by the lack of cap carbonates above many clear glacial deposits at a similar time and by the occurrence of similar carbonates within the sequences of proposed glacial origin.

6. There is evidence for deep flooding of previously shallow-water shelves and platforms after the Sturtian and Marinoan melt phases, sustained after isostatic readjustments, reflecting slow tectonic subsidence over millions of years under glacial ice.

7. Isotopes of boron in Namibia suggest that the pH of the oceans dropped dramatically before and after the Marinoan snowball event, which may suggest a large decrease in sea-water pH and indicate a build-up of carbon dioxide in the atmosphere, some of which would dissolve into the oceans to form carbonic acid. However, although the boron variations may be evidence of extreme climatic change, they do not necessarily imply a global glaciation.

8. Iridium is rare on the Earth's surface and the only significant source of this element is cosmic particles that reach Earth. During a Snowball Earth, iridium would accumulate on the ice sheets and when the ice melted the resulting sediment layer would be iridium-rich. Such an iridium anomaly has been discovered at the base of cap carbonate formations in central Africa and has been used to suggest that the glacial episode lasted for at least 3 million years. However, this does not imply a global extent to the glaciation and the anomalies could be explained by the impact of a large, extra-planetary object, such as a comet or meteor.

9. There has been a suggestion of an astronomical origin for Snowball Earths (see Box 30.2, in the companion website). This has little good evidence to support it, although it is an interesting idea.

30.2.5 Arguments against Snowball Earth

There has been much opposition to the Snowball Earth hypothesis since it was first suggested and it still gives rise to controversy. This opposition includes the following:

1. The so-called 'Zipper rift' hypothesis suggests that a Snowball Earth was no different to any other glacial phase in the Earth's history. Eyles and Januszczak (2004) suggest that two pulses of continental 'unzipping' coincided with the glaciated periods. These were: initially the breakup of the Rodinia supercontinent, foming the proto-Pacific ocean; followed by the splitting of the continent Baltica from Laurentia, forming the proto-Atlantic. The associated tectonic uplift would form high plateaus, which could then host glacial ice. Continental rifting, with associated subsidence, tends to produce land-locked seas suitable for the accumulation of BIFs and would produce the space for the fast deposition of sediments without the need for an immense and rapid melting to raise global sea levels.

2. The high-obliquity hypothesis also explains the presence of ice on the equatorial continents, suggesting that the Earth's axial tilt was quite high, close to $60°$, which would place its land in high 'latitudes' (see Box 30.1, in the companion website), although supporting evidence for this is scarce. A less extreme idea is that it was the Earth's magnetic pole that wandered to this inclination, as the magnetic readings which suggested ice-filled continents depend on the magnetic and rotational poles being relatively similar. In either of these two situations, the freeze would be limited to relatively small areas, as is the case today, and severe changes to the Earth's climate are not necessary.

3. Analyses of glacial sedimentary rocks from the Huqf Super-group in Oman have produced evidence of hot–cold cycles in the Cryogenian. Using the chemical index of alteration, the chemical and mineral composition of these rocks was examined. A high alteration index would indicate high rates of chemical weathering of the land surfaces, which would cause rocks to decompose quickly and would be enhanced by humid or warm conditions. On the other hand, a low chemical alteration index would indicate low rates of chemical weathering in cool, dry conditions. Three intervals with evidence of extremely low rates of chemical weathering are present, indicating pulses of cold climate, but these alternate with periods of high chemical weathering, likely to represent interglacial phases with warmer, humid climates. The hydrological cycle and the routing of sediment were active throughout the glacial epoch, which would require substantial open ocean water (Rieu et al., 2007). These climatic cycles with evidence for water negate the idea of a Snowball Earth and are inconsistent with a deeply-frozen planet.

4. In limestones and dolomites from the Precambrian glacial deposits there is a record of the carbon isotopic variation and the results give consistent positive values from different parts of the world. The evidence shows that the ratio of C^{13} to C^{12} was actually higher during the glaciation, indicating the presence of a highly productive marine ecosystem. This ratio dropped only after the ice had melted, suggesting that other influences than those produced in the Snowball Earth hypothesis must have been active. There seems to be no evidence that a global ice sheet impacted on the overall marine productivity.

5. There has been a reinterpretation of some of the 'glacial' rocks as subaqueous mass flows (see Box 30.3, in the companion website), which would mean there were no Snowball Earth-type temperature changes.

6. It has been found by climatic modelling that where simulations of certain climatic forcings have been used to try and reproduce conditions suggested by the geological record, these forcings were not sufficient to reproduce the first-order characteristics of Snowball Earth (Sohl and Chandler, 2006). The key forcings used were a reduction in solar luminosity to 6.19% less than that today, a reduction in carbon dioxide levels to 140 ppm and a cut in ocean heat transport by 50%. However, although this produced a significant overall cooling of the Earth, the forcings were not sufficient to give the annual average temperatures suggested or the degree of ice cover on topical continents needed to initiate or maintain ice sheets.

30.2.6 How did life survive Snowball Earths?

If Snowball Earths occurred, how could life survive them? No Snowball Earths have occurred since the first appearance of bilaterian animals (above sponge grade) in the fossil record about 555 million years ago in Arctic Russia. However, a host of microscopic organisms (both prokaryotes and eurkaryotes) and a few cm-scale organisms (such as the coiled Grypania (1.9 billion years ago), the necklace-like, colonial organism Horodyskia (1.5 billion years ago) and the worm-like Parmia (1.0 billion years ago)) evolved before the final two Snowball Earth phases and survived. Palynofloras were relatively impoverished during the Cryoginian period but the characteristic assemblage of simple, thin-walled microspheres (leiospheres) passed through the Marinoan cold phase without change. The implication is that sunlight and water co-existed on Snowball Earths and that it was possible for life to survive in the following ways:

1. Reservoirs of anaerobic and low-oxygen life powered by chemicals existed in the deep-ocean hydrothermal vents, although photosynthesis was not possible there.

2. Cells and spores lay dormant in the frozen ice and under the ice layer in chemolithotrophic (mineral-metabolizing) ecosystems, theoretically resembling those in modern glacier basal layers, high Alpine and Artic scree permafrost, and especially in areas of volcanism and geothermal activity, where there could be oases of molten water. Cyopreserved bacteria that were alive when 32 000-year-old Arctic ice melted have been reported 400 m under the floor of the ocean.

3. Life existed in pockets of liquid water both within and under ice sheets, similar to Lake Vostock (Antarctica), possibly resembling microbial communities living in the perennially frozen lakes of the Antarctic dry valleys. Photosynthesis can occur under up to 100 m of ice.

4. Diatoms are preserved in brine channels from present-day sea ice, and along with protozoa and foraminifera must be tolerant of hyperoxia since photosynthetic oxygen cannot escape the brine channels.

Organism size and complexity increased considerably after the Snowball Earth glacials, especially the Marinoan, and the development of multicellular organisms may have been the result of increased evolutionary pressures resulting from multiple icehouse–hothouse cycles. However, there is no empirical support for this from the fossil record and fluctuating nutrient levels and rising oxygen may also have played a part in the achievement of multicellularity in microscopic animals.

Currently aspects of this Snowball Earth hypothesis remain controversial and it is being debated under the auspices of the International Geoscience Programme (IGCP) Project 512: Neoproterozoic Ice Ages (http://www.igcp512.com/). A review article by Allen and Etienne (2008) indicates the doubts about this hypothesis of some geologists (see Box 30.4, in the companion website).

30.3 Examples of changes from greenhouse to icehouse climates in the Earth's past

30.3.1 Links between surface temperatures and CO₂ concentrations during the Palaeozoic era

The link between atmospheric CO_2 concentrations and the planet's surface temperatures is at the core of our understanding of environmental change at many periods in the Earth's history. One of the most puzzling times for the understanding of the climatic effects of CO_2 is the Palaeozoic Era, between 543 and 248 million years ago, which was the time in which the diversification of the major classes of large-bodied animal and plant life forms occurred.

Modelled CO_2 concentrations between the Mid-Cambrian and latest Silurian (530–417 million years ago) suggest that the CO_2 levels were between 12 and 17 times higher than they are in the modern atmosphere. These were followed by far lower levels, comparable to the modern atmosphere, during the Carboniferous (354–290 million years ago) (Berner and Kothavala, 2001). The latter levels seem likely as a plausible cause of the extensive Carboniferous glaciation.

However, northern Africa had extensive glaciations during the Late Ordovician and Early Silurian. There are two possible explanations for this: either model reconstructions of atmospheric CO_2 levels are susceptible to large errors or climate can change dramatically independently of atmospheric CO_2 variations. Moreover, Veizer et al. (2000) have reconstructed tropical, shallow marine temperatures during the Palaeozoic by measuring oxygen isotopes in carbonate fossils at low latitudes and the results suggest that the temperatures were within around 5 °C of modern conditions during times of high inferred atmospheric CO_2, such as the Silurian, and also during times of lower inferred atmospheric CO_2, such as the Carboniferous. Veizer et al. suggested that the Phanerozoic temperature trend has four icehouse–greenhouse modes.

This kind of climate reconstruction debate has two uncertainties. First, the geological evidence for the spatial and temporal distribution of sediments and fossils provides qualitative constraints on climate but cannot easily be translated into global temperature and therefore test models of global climate. Second, oxygen-isotope constraints on surface temperatures are vulnerable to artefacts from diagenetic or burial-metamorphic overprints. The result has been uncertainty as to whether the systematic temporal variations in oxygen-isotope compositions of Phanerozoic marine carbonate fossils reflect climatic change, variation in the $\delta^{18}O$ of sea water or post-depositional alteration.

To try and get round these problems, Came *et al.* (2007) present estimates of sea-surface temperatures that were obtained from fossil brachiopod and mollusc shells using the 'carbonate clumped isotope' method – an approach that, unlike the $\delta^{18}O$ method, does not require independent estimates of the isotopic composition of the Palaeozoic ocean. Their results indicate that tropical sea-surface temperatures were significantly higher than today during the Early Silurian period (443–423 million years ago), when carbon dioxide concentrations are thought to have been relatively high, and were broadly similar to today during the Late Carboniferous period (314–300 million years ago), when carbon dioxide concentrations are thought to have been similar to the present-day value. These results are consistent with the proposal that increased atmospheric carbon dioxide concentrations drive or amplify increased global temperatures.

30.3.2 A greenhouse climate in the Middle and Late Cretaceous

Many geologists think that major climatic cycles are related to plate tectonics and every time the continents are pulled apart huge quantities of volcanic water, carbon dioxide and methane are released into the atmosphere and greenhouse climates prevail on Earth. On the other hand, when continents come together and mountain ranges are created, the mountains are eroded, new soils form and remove CO_2 from the atmosphere (Figure 30.4(a)), these soils are also eroded and the CO_2 becomes locked in ocean-floor sediments. So when atmospheric CO_2 is low, glaciation occurs; this has happened in the Late Ordovician about 440 million years ago and in the Permo-Carboniferous at about 310–270 million years ago.

It appears that there are supercycles of greenhouse–icehouse related to these phases of plate tectonics and sea-level changes, as can be seen from Figure 30.4(b). There have also been major changes in CO_2 over time (Box 30.4, see in the companion website), related to the processes of volcanism, sedimentation, erosion, mountain building and the evolution of life. Each change in climate has been moderated by the Earth's heat-regulating mechanism; that is, increased chemical weathering reduces warming trends by taking up more CO_2 from the atmosphere, and similarly it reduces cooling trends by taking out less CO_2. This is because of the relationship between climatic factors, such as temperature and precipitation, and the chemical-weathering rate.

One such phase of greenhouse climate occurred during the Middle and Late Cretaceous when global sea level was more than 100 m higher than today (possibly as much as 200 m higher), which drowned out most coasts and covered the continental interiors (although there were major sea-level transgressions and regressions, with at least six in the Late Cretaceous with a periodicity of around 5.5 million years); the sea-surface temperature was $10-15\,^{\circ}C$ higher than now, the climate was $2-6\,^{\circ}C$ warmer at the equator and $20-60\,^{\circ}C$ warmer at the poles, and many continents were covered by shallow tropical seas with carbonate rock deposition. The Earth had a warm, wet, greenhouse climate, with thick vegetation covering the land areas, and temperatures were high enough to allow vertebrates and forests to exist close to the poles. Oxygen isotope ratios in carbonate sediments recovered from deep-ocean cores provide evidence for the sea temperatures in which the carbonate shells formed and for the amount of water stored in the polar ice caps. The data here indicate that the ocean water was much warmer than it is today, especially in the deep ocean where temperatures were as high as $15\,^{\circ}C$, compared with today's values of around $2\,^{\circ}C$. There was no polar ice. A further discussion of the origin of these high temperatures is discussed in Box 30.5 (see the companion website).

It is thought that atmospheric CO_2 was about 1% when the world's major coal deposits formed during the Carboniferous, from 368 to 248 million years ago, but that from 250 to 120 million years ago the CO_2 content varied greatly, increasing to a peak of 6% CO_2 by 120 million years ago (a factor of $4\times$ the present-day levels). The higher atmospheric CO_2 levels were the cause of warming, with the evidence for these levels coming from a variety of sources:

- Palaeomagnetic evidence that sea floor was spreading faster at that time; these faster rates would have resulted in faster subduction of carbonate sediments, which would have led to increased rates of CO_2 production and increased CO_2 from the mid-ocean ridges.

- Sea levels were higher, which would have meant less land available for weathering of silicate rocks. Coupled with this, the tectonic uplift hypothesis suggests that the mean rate of chemical weathering is strongly affected by the rate of uplift of fresh rock surfaces, which outweighs the influences of climate (Figure 30.4(a)). In the Cretaceous the amount of tectonic uplift was not great and therefore global warming occurred.

- The C-isotope data, where the evidence is that the sediments had less ^{13}C, which is consistent with the idea that CO_2 levels were higher.

It is thought that the equator–pole contrast in temperature was around $20-30\,^{\circ}C$, whereas today it is $50-60\,^{\circ}C$, so the circulation was different and more mixing took place. As there was no polar ice there was a changed albedo in these regions, with incoming heat stored, thus warming them. There is likely to have been a change too in the thermohaline circulation, which probably ran backwards, with dense, highly-saline water forming at low

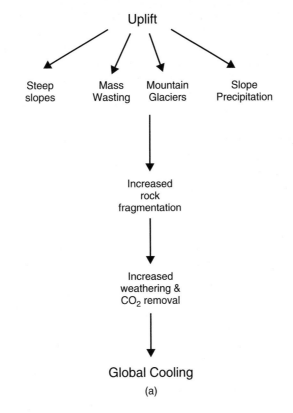

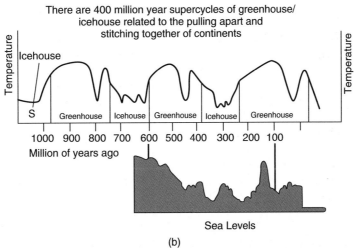

Figure 30.4 (a) Model of plate tectonics, mountain uplift and global cooling. (b) Greenhouse–icehouse supercycles

latitudes. There is evidence for forests at regions which today are tundra or ice-covered, and reptiles and dinosaurs lived at high latitudes. Massive flood basalts mark the break-up points of Pangaea and Gondwana and the Earth's magnetic field, which normally reverses at odd intervals, remained with a constant polarity for over 30 million years (Figure 30.7). It has been suggested that all this was the result of a major period of plume activity rising from deep in the mantle.

For a period prior to the Cretaceous the lower mantle heated up via heat from the core and radioactive decay in the mantle.

Eventually the lower mantle became so hot that it was less dense than the mantle above it. This resulted in blobs of hot, low-density mantle rising from the core–mantle boundary towards the Earth's surface as mantle plumes over tens of millions of years. When the plume heads reached close to the Earth's surface the heat caused widespread melting, the creation of flood basalts and rapid sea-floor spreading. This volcanic-activity increase resulted in additional CO_2 liberation from the mantle into the atmosphere, increasing the greenhouse effect and creating a warming climate. For example, the basalt eruption forming the

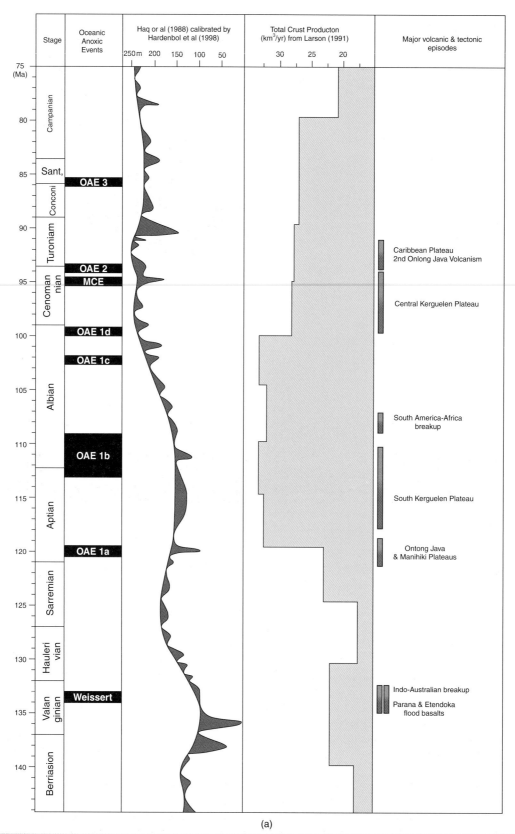

(a)

Figure 30.7 Anoxic black shales or oceanic anoxic events (OAEs): correspondence in the Mid-Cretaceous between a high rate of oceanic crust production, accumulation of abundant black shales and absence of magnetic reversals. (a) Associated sea-level changes and the production rate of the ocean crust. (b) Mantle plume production and the magnetic reversal time scale. (Source: Takashima et al., 2006)

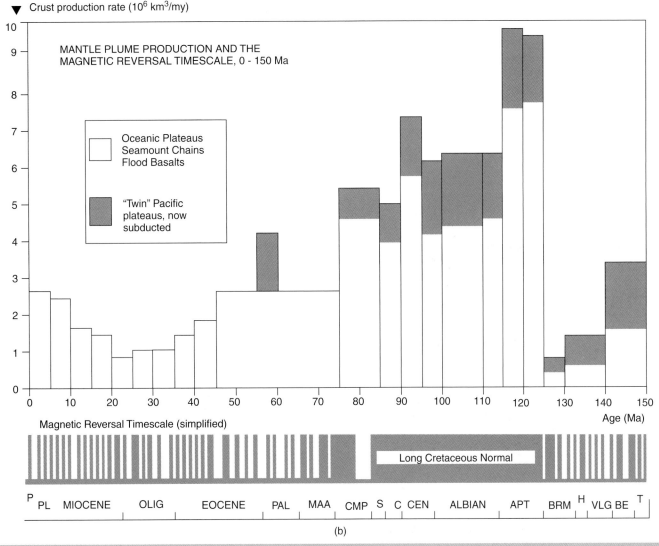

Figure 30.7 (continued)

Ontong–Java plateau in the Western Pacific could have increased carbon dioxide levels in the atmosphere to 4–15 times the present levels, which would have caused a global warming of 3–8 °C (perhaps half the calculated warming for the Mid-Cretaceous). As there are several such Cretaceous plateaux in the Pacific, it would seem that carbon dioxide degassing associated with plumes could account for a large percentage of climatic warming at this time. The release of heat from the core–mantle boundary increased the convection in the Earth's outer core, making the Earth's magnetic field more stable, so accounting for the long period of constant normal polarity.

The break-up and movement of the components of Pangaea and Gondwanaland during the Cretaceous caused expansion of narrow oceans and sea levels stood high during the Early Cretaceous. Extensive marine deposits covered most continents (see Figure 30.8 for North America late in the Early Cretaceous). This rise in sea level resulted from the major expansion of the mid-ocean ridge systems and plume systems mentioned above.

In the Mid-Cretaceous marine muds, rich in organic matter (now black shales), accumulated as a result of poor, sluggish circulation (anoxic bottom waters), with little vertical mixing in the ocean and stagnation of much of the water column (Figure 30.7). Polar seas were too warm for their surface waters to descend and spread oxygen throughout the deep ocean, and the end result was that the low-oxygen zone was increased greatly, expanding into the photic zone. Because of the low temperature gradients, winds blowing over the oceans were weaker on average than they are today, and hence upwelling would have been weaker. The absence of a strong, wind-driven circulation would have added to the general stagnation of the lower part of the Mid-Cretaceous ocean.

In the Late Cretaceous there are indications of changes in the ocean circulation from the oxygen isotope ratios of foraminifera. There was a change toward lighter oxygen in the benthic foraminifera, indicating that temperatures in the deep sea rose from about 14 °C to 20 °C in the Mid-Cretaceous and then fell

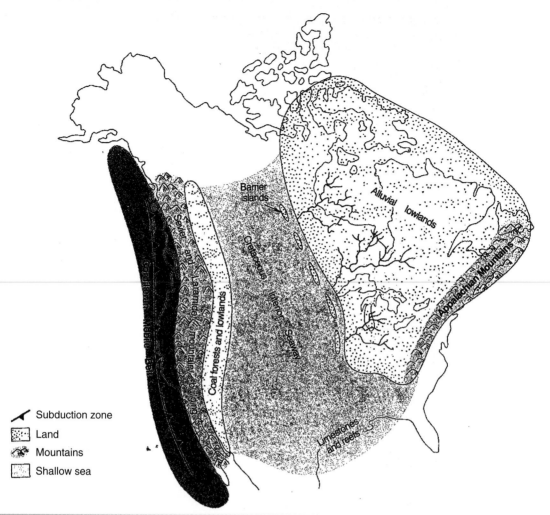

Subduction zone
Land
Mountains
Shallow sea

Figure 30.8 North America late in the Early Cretaceous

again, reaching around 9 °C during the Late Cretaceous. Isotope ratios in planktonic foraminifera indicate that high-latitude surface waters also cooled in the same period, suggesting that the Mid-Cretaceous ocean circulation changed to one in which the high-latitude cooler waters sank to the deep sea, carrying oxygen with them. In the Late Cretaceous calcareous nannoplankton were so abundant in warm seas that the small plates that covered their cells accumulated in huge volumes to form the fine-grained limestone that we know as chalk (Latin *creta*), which rock is made from 75% of this minute skeletal sediment. This chalk is best known from Western Europe but chalk seas formed extensively in other parts of the world too at this time. The general presence of oxygenated conditions is shown by the widespread bottom-dwelling fauna.

Of course, the meteorite impact in the Yucatan Peninsula in Mexico around 65 million years ago that we discussed in Chapter 29 had an impact on the immediate climate of the planet, but this was only a brief disturbance of the super-warm climate of the Late Cretaceous and there does not appear to have been a measurable long-term shift in the Earth's climate.

30.3.3 The Palaeocene supergreenhouse and the changes in the Eocene

The Earth has seen dramatic climatic changes since the Cretaceous (Figure 30.9), as can be seen from the oxygen-isotope record from benthic foraminifera recovered from deep-sea sediment cores (Figure 30.10) and the relationship between fossil smooth-margined and jagged-margined leaves in North America (Figure 30.11). This record indicates that the change from the Cretaceous greenhouse to the Holocene icehouse has been far from smooth, characterized by periods of rapid change interspersed with phases of relative climatic stability (Figure 30.12). The records for planktonic foraminifera suggest that the climatic changes were greatest at high latitudes, with the conditions at the equator 65 million years ago similar to those of the present day. The explanations for such changes are complex and we will discuss some of them in this section.

The traditional explanation involves the positions of the continental masses, as it was considered that the relative locations of the continents played an important role in oceanic circulation and so in heat transport on Earth. At the end of the

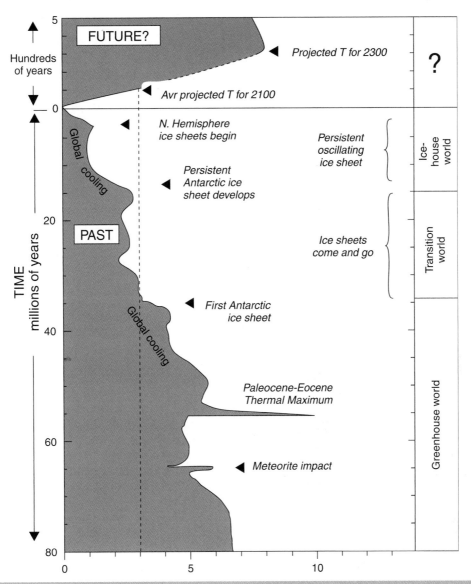

Figure 30.9 Changes in the average global temperatures over the last 80 million years

Cretaceous the continents were distributed. The Tethys Ocean had contracted significantly but there was still open access at both ends and so there was free water flow around the equator. This current was warmer than today's equatorial currents because the water contained within it spent a significantly longer time along the equator before being fed off into the northern and southern gyres. This also meant that the water within the north- and south-flowing currents that diverged from the equator towards the polar regions was warmer and fed heat to these regions.

Another difference was the lack of a circum-polar current in the Cretaceous. Antarctica today is ringed by a cold surface current that feeds cold water towards the equator and acts to insulate the south pole from the low-latitude warm waters. In the Cretaceous, Antarctica was joined with South America and there was only a small gap between it and Australia. These two factors

mean that there is a much greater temperature gradient today between high and low latitudes.

The Early Eocene (around 55–50 million years ago) is thought to be the warmest climate phase of the last 100 million years and this warmth is usually thought to be caused by unusually high atmospheric CO_2 conditions, possibly four times current levels, although there are alternative explanations for the greenhouse warming (see below). In the deep-sea cores there is an abrupt change in foraminiferal oxygen-isotope ratios towards lighter values (Figure 30.14) in both benthic and planktonic species. This indicates that even close to Antarctica both surface and deep-sea waters warmed by several degrees to about 18 °C within less than 3000 years. More than 70% of the deep-sea foraminiferal species became extinct, probably because of this warming, but also because the deep oceans became oxygen-depleted as cool polar waters no longer transported oxygen to the deep sea. The

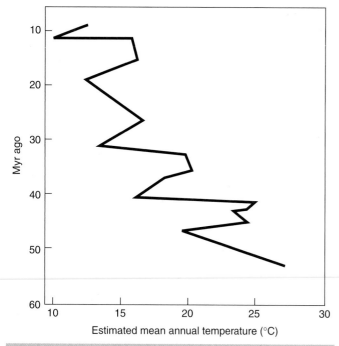

Figure 30.10 Oxygen-isotope record from foraminifera

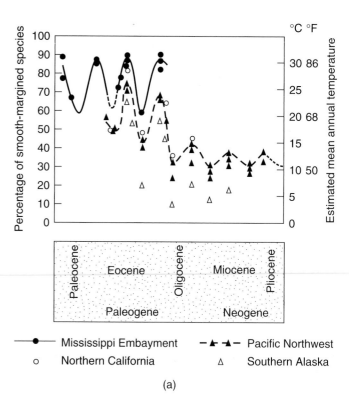

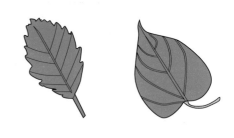

carbon-isotope ratios also shifted quickly to very low levels and similar major shifts towards isotopically-light carbon are seen in the organic matter produced by marine phytoplankton and in terrestrial plant fossils. It is therefore apparent that carbon dioxide used in photosynthesis suddenly became highly enriched in carbon 12.

It has been suggested that this shift towards isotopically-light carbon was so marked that melting of frozen methane hydrates along continental slopes might be a reasonable explanation. There has been much debate as to why this should happen, but if it happened it would have set up a positive feedback because as a greenhouse gas methane is around 30 times more powerful than carbon dioxide. Hence a small amount of greenhouse warming from an initial methane release would have led to further melting of methane hydrates and further warming. The greenhouse warming would have weakened gradually as the methane added to the atmosphere oxidized to carbon dioxide (see Box 30.7, see the companion website). It has even been suggested that natural wildfires developing under more extreme climatic conditions could have had a major impact on the carbon cycle at this time (see Box 30.8, see the companion website).

Some scientists have suggested that this Eocene temperature rise cannot be explained by greenhouse warming caused by an elevated carbon dioxide concentration. For example, evidence from the leaves of fossil ginkgos, where the stomata cells of the leaf surfaces have been compared to the densities of modern ginkgos grown under a variety of carbon dioxide concentrations, indicates that the Early Eocene carbon dioxide concentration was not that much higher than it is today. Computer modelling of levels of atmospheric carbon dioxide shows similar results if these generalizations can be believed. However, greenhouse warming

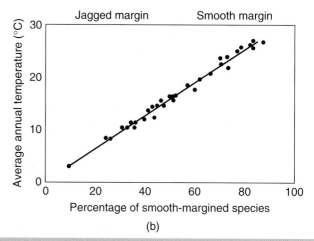

Figure 30.11 Leaves and climate: (a) changes in the percentages of smooth-margined leaves in fossil floras, revealing changes in temperatures during the Eocene and Oligocene; (b) relationship between climate and the shapes of leaves of flowering plants

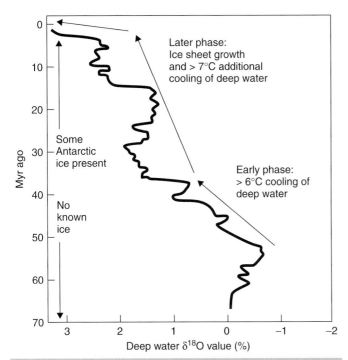

Figure 30.12 Rapid climate changes and phases of climatic stability from the Cretaceous to today

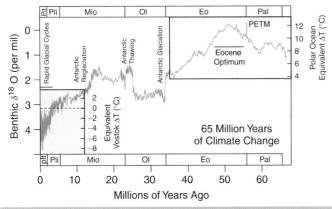

Figure 30.14 Deep-sea core record and abrupt changes in foraminiferal oxygen-isotope records towards lighter values at the end of the Palaeocene Epoch. The climatic change over the last 65 million years, based on the oxygen-isotope measurements ($\delta^{18}O$) on benthic foraminifera, is also shown; this reflects the contribution of local temperature changes on the environment and of changes in the isotopic composition of sea water associated with the growth and retreat of continental ice sheets. (Source: Wikimedia Commons, prepared by Robert R. Rohde and incorporated into the Global Warming Art project; based on data from Zachos *et al.*, 2001)

may still have been stronger during this period than it is today because water vapour is also a greenhouse gas and forests were more extensive and transpired large quantities of water into the atmosphere, which may have resulted in a high water-vapour concentration and substantial greenhouse warming over large areas of the planet.

Antarctica did not have an ice sheet but there may have been regions with Alpine glaciation and a temperate climate that supported palms, ferns and rainforest trees. Subsequently there was a significant but gradual cooling over 20 million years (the so-called 'doubthouse') to an icehouse climate. Some of the evidence for these climatic conditions comes from coring at sites in the Arctic Ocean (e.g. IODP Leg 302). Here Palaeogene sediments totally lack carbonate and thus do not allow the use of conventional palaeothermometers. However, studies from the University of Utrecht suggest that a new organic palaeothermometer based on archaeal membrane lipids will allow the determination of Arctic summer sea-surface temperatures in great detail.

The initial results suggest that the Eocene cooling occurred much earlier and faster than was previously assumed and that summer sea-surface temperatures were extremely high during the Palaeocene–Eocene Thermal Maximum, at around 20–24 °C 55 million years ago, dropping to values as low as around 5 °C 5 million years ago. Here large volumes of cool fresh water appeared in the Arctic, chilling the sea and diluting its saltiness so much that there was a massive growth of the hydropterid fern Azolla in the Arctic surface waters at this time. These figures are in agreement with initial dinoflagellate-based sea-surface temperature trends.

It has been suggested that this dramatic, fast cooling could be related to a substantial drop in CO_2 resulting from the highly efficient production and burial of large quantities of organic carbon in the Arctic Basin. This may have been due to anoxic water column conditions in the stratified basin and the growth of this surface fresh-water fern. At 45 million years ago the first ice started to form in the Arctic Ocean, as evidenced by dropstones from icebergs, and the relative cooling has continued since.

An interesting side issue is the evidence from the Late Eocene impact craters of Chesapeake Bay and Toms Canyon (on the east coast of the United States) (Figure 30.15) and Popigai (northeastern Siberia), which are the result of a comet shower. Based on carbon- and oxygen-isotope records, these impacts may have had a global effect on the carbon cycle and so on global temperatures. It may be that the impending icehouse was delayed by the injection of carbon dioxide into the atmosphere and when the impact effects moderated the temperature, decline was drastic.

However, by the end of the Eocene a well-developed spreading centre had formed between Australia and Antarctica so that there was now vigorous water flow between the Indian and Pacific Oceans, and some geologists suggest this circulation increase was the reason for significant cooling throughout the period from 60 to 40 million years ago (see Stanley, 2005, pp. 458–459). As sea-floor spreading moved Antarctica away from Australia and South America it is thought that an opening of ocean gateways produced a strong circumpolar current in the southern ocean, which thermally isolated the Antarctic continent. Surface waters dropped from 20 °C to as low as 10 °C by the end of the Eocene. The temperatures remained high enough however that there is no evidence of significant glaciation during this period and overall the climate was warm and humid, although this was originally suggested to be the reason for cooling of the continent in the Oligocene so that an ice sheet could grow.

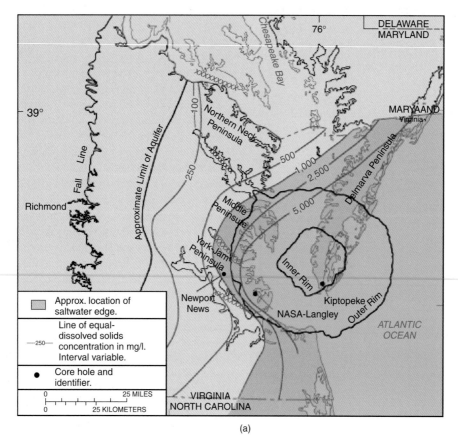

(a)

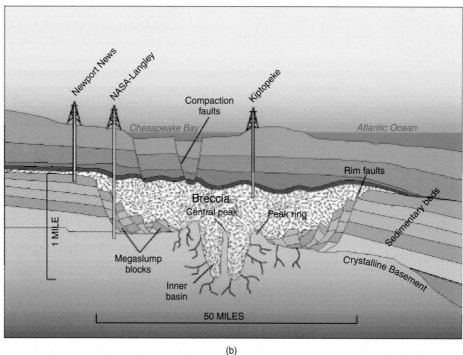

(b)

Figure 30.15 Late Eocene impact craters, Chesapeake Bay and Toms Canyon. (a) Impact site; the smaller Toms Canyon crater was formed by the impact of several small fragments; distribution of sedimentary deposits by the tidal wave from the Toms Canyon impact; dots show where sedimentary cores were obtained. (b) Idealized east–west cross-section of the crater; there is considerable vertical exaggeration. (Source: after Stanley, 2005)

30.3.4 Oligocene Antarctic glaciation and its potential impacts globally (the first major glaciation of the Cenozoic)

Around 34 million years ago at the end of the Eocene there was another plunge in temperatures of 3–4 °C in only 200 000 years, which gave rise to the Oligocene glaciation when Antarctica was changed from a lush, vegetated continent to the ice sheet-covered region that we know today. This climatic step from greenhouse to icehouse has long interested researchers because of the potential insights that may be gained into the factors that drive climate change from one state to another.

The end of the Eocene had been unusually warm but the Oligocene saw the development of a permanent ice sheet in Antarctica up to 10 million km^3 in area and a sea-level drop of 40 m or greater (see Box 30.9, see the companion website) as ocean water was locked up in the continental ice sheet. Evidence for this global cooling comes from an array of fossil, geochemical and sedimentary data, especially at high latitudes; it is rich from sea-floor sediments around the Antarctic continent and from deep-sea records at the equatorial Pacific. The deep-sea data show that the climatic shift occurred in multiple stages (see Figure 30.16) and involved a combination of continental ice-sheet growth and deep-water cooling and a dramatic change in ocean carbonate saturation that can be calibrated on orbital time scales. There is also geochemical and palaeontological evidence from a hemipelagic sequence drilled in Tanzania (Tanzania Drilling Project Sites 11, 12 and 17) with exceptional microfossil preservation, which shows a closely coincident mass extinction of shallow-water carbonate-secreting organisms with an extended phase of ecological disruption in the plankton; for example, the extinction of the foraminifera family Hantkeninidae (Pearson *et al.*, 2008). These records show that there was a phase of very severe global climatic and biotic change that lasted about half a million years, which preceded maximum glacial conditions in the early Oligocene by around 200 000 years.

In Antarctica, sediments on top of the Eocene marine shelf section on Seymour Island (Antarctic Peninsula) include glaciomarine deposits and a terrestrial till, with clasts derived from a variety of rock units on the peninsula (Ivany *et al.*, 2006). These have been dated at or very close to the Eocene–Oligocene boundary. Glacier ice reaching sea level in the northern peninsula at this time indicates a regionally-extensive West Antarctic ice sheet.

There is evidence too that northern continents cooled and dried. The implication is that the cause was global rather than just regional change. Previously in the Cenozoic the only significant ice concentrations had been in the Antarctic highlands and in and around the Arctic Ocean, as we described earlier. Two papers in *Nature* by Dupont-Nivet *et al.* (2007) and Zanazzi *et al.* (2007) and the commentary on them by Bowen (2007) relate to the effects of this major global event. Sedimentary records from the Xining basin in the Tibetan Plateau suggest a drop in atmospheric water which caused cooling and aridification, shown by the disappearance of playa lakes and other palaeoenvironmental and

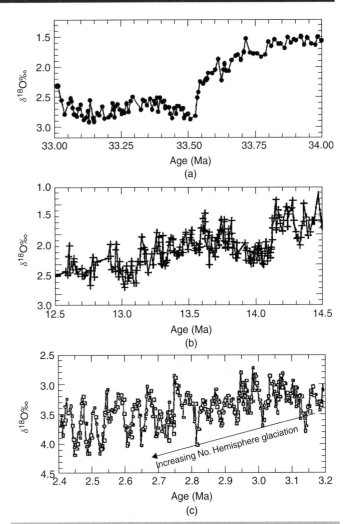

Figure 30.16 High-resolution $\delta^{18}O$ records at: (a) the Eocene–Oligocene boundary; (b) the Middle Miocene; (c) the Late Pliocene

paleontological evidence. This aridification was exactly coincident with the Antarctic cooling, although previously it had been attributed to rapid uplift of the Tibetan Plateau after India collided with Eurasia, raised the Tibetan Plateau and created a rain shadow in the lee of the Himalayas. Hence this new evidence suggested that the Tibetan Plateau climate was influenced by global events.

In North America this cooling event was explored by measuring stable isotopes from fossil teeth and bone to create a proxy temperature record and it was found that there was a large drop in mean annual temperatures of 8.2 °C (a greater fall than in mid-latitude oceans). This climatic transition from warm to cold may explain why there was a faunal turnover in the Early Oligocene and why so many cold-blooded reptiles and amphibians became extinct when the mammals escaped relatively unscathed, as the latter were able to regulate their body temperatures.

The two studies demonstrate the global nature of the climatic and biotic change at this time and strongly challenge the ocean-circulation hypothesis for the Early Oligocene climatic event. They imply that a global forcing agent such as a decline in the atmospheric greenhouse gas concentration (likely to be CO_2 changes) must have played a role in triggering the build-up of the Antarctic ice and was a driver for changes in both the northern and the southern hemispheres, rather than just a change in ocean gateways as we suggested in Section 30.3.3. What is now believed is that a long-term decline in atmospheric CO_2 was the major influence on polar cooling. Computer modelling using the global climate model (GCM) uses ocean and atmospheric circulation, greenhouse gas levels and the past distribution of the continents and oceans to produce an ice sheet. The values for the Earth's past CO_2 levels that have been reconstructed from geological records show a decline between 50 and 34 million years ago from 1200 ppm to around 600 ppm. It is thought that, as in the current global warming, CO_2 has had a major influence on past temperature changes. The opening of the ocean gateways, although important, is now considered to have played a secondary role.

30.3.5 Has the Antarctic ice sheet always been stable?

There is geological evidence to suggest that global sea level has oscillated by 10–40 m (over time periods of 40–100 thousand years) during the last 34 million years. This suggests that the Antarctic ice sheet had extensive periods of instability and a fluctuating volume of up to 80%. This is thought to be the case particularly between 34 and 15 million years ago, when atmospheric CO_2 levels were twice as high and global average temperatures were 2 °C warmer than in the year 1900. A second major cooling took place during the Middle Miocene (see Box 30.10, see the companion website).

There has been widespread debate about how stable the Antarctic ice sheet has been over the last 15 million years. There are scientists known as 'stabilists' who suggest the ice sheets have been cold and have not changed significantly in area/size and there are others known as 'dynamicists' who argue that the oxygen isotope records support unstable, dynamic ice sheets, with changes of up to 50% compared to the present-day volumes. Evidence for the latter view has been provided by marine microfossils found in glacial sediments at a number of places in the Transantarctic Mountains, although the stabilists have argued that numerous wind-transported marine diatoms can be found at the south pole today, so there is continued discussion here. However, the Cape Roberts Project drilled cores 1.5 km deep to provide a record of ice-sheet behaviour during the period between 34 and 17 million years ago, which gave us the first direct evidence that the Antarctic ice sheet had been highly dynamic in a past warmer world, closely matching the deep-sea oxygen-isotope records.

Evidence from the Prydz Bay drill cores (Crawford, 1997) shows that the Antarctic ice sheet has undergone rapid changes in volume in response to climatic change. There was a dynamic, waxing and waning ice sheet through much of the Cenozoic, while the Late Pliocene sedimentary evidence points to a significantly reduced ice mass on the eastern Antarctic continent, allowing the deposition of distal, ice-rafted glaciomarine sediment, which

supports the conclusion of Hambrey et al. (1991) that a warmer climate prevailed in Antarctica during the Late Pliocene.

30.3.6 What evidence is there in the Arctic Ocean for Cenozoic climate change?

Moran et al. (2006) presented a palaeooceanographic record constructed from over 400 m of sediment core from the Lomonosov ridge in the Arctic Ocean. This illustrates a palaeoenvironmental transition from a warm greenhouse world to a colder icehouse, influenced by sea ice and icebergs from the Middle Eocene to the present. The average of sedimentation rates from 1 to 2 cm per thousand years is an order of magnitude higher than previous estimates, where the Arctic had been interpreted as sediment-starved. For the most recent 14 million years there is synchronous cooling of the Arctic, with the expansion of the east Antarctic ice around 14 million years ago and expansion of the Greenland ice around 3.2 million years ago. However, in the earliest Eocene temperatures were estimated to be around 24 °C, and there were major changes in the hydrology of the region during the Palaeocene–Eocene Thermal Maximum (around 55 million years ago).

The first occurrence of ice-rafted debris comes from the Middle Eocene (about 45 million years ago), considerably earlier than was previously believed, and fresh surface waters were present around 49 million years before that, with temperatures around 10 °C. The revised timings of the earliest Arctic cooling events seem to coincide with those from Antarctica, which suggests there was a bipolar symmetry in climate change. These bipolar changes in turn suggest a greater control on global cooling by changes in greenhouse gases, as opposed to tectonic forcing.

Tectonic changes such as the opening of the Drake Passage may modify portions of the Earth's ocean circulation but they initiate climate changes that progress too slowly to induce synchronous global cooling patterns. See the companion website for Box 30.10 From greenhouse to icehouse: evidence from Middle Miocene climate rhythms and extremes of climate.

30.3.7 Summary

From this brief discussion of changes from greenhouse to icehouse climates on our planet it is clear that there are many possible initiators of climate change. Extreme aberrations in global climate can occur through a number of mechanisms, as discussed in Zachos et al. (2001). A major question is whether the Earth's climate is controlled by processes internal or external to the planet.

External forces such as changes in the form of orbital parameters like the Milankovitch cycles can give rise to short-term cyclicity, which we will return to in the next section. There are key questions: Is there any controlling factor to the cyclicity in the climatic patterns that have been recognized? and How does the Earth change from cold to warm and vice versa? The abrupt transition indicates very large changes in the Earth's climate system, which can occur relatively quickly. Other extraterrestrial forcing could operate too, through changes in solar luminosity, the passage of the solar system through clouds of matter, and the impacts of comets, meteorites or asteroids.

There are also diverse terrestrial climate forcing factors, such as changes in atmospheric greenhouse gases, ocean geochemistry and biochemistry, land–sea distributions and their effects on the Earth's overall albedo, ocean circulation, and water-mass formation. The tectonic forces and their rate of action (as related to sea-floor spreading and ocean-crust subduction) can also have major impacts on the Earth's sea level and climate, with many feedback loops, and it is clear that continental fragmentation and assembly at various stages in the Earth's history must have had a major impact on the climate.

We need to be able to learn from these geological examples and realize that even without human influences there have been rapid oscillations, which have had major impacts on plant and animal populations and diversity.

30.4 Late Cenozoic ice ages: rapid climate change in the Quaternary

30.4.1 Introduction

An ice age, as we have seen, is a period of long-term reduction in the Earth's surface and atmospheric temperature which results in expansion of polar ice sheets and Alpine glaciers. There are three main lines of evidence for such periods (these have been discussed briefly in Section 30.2):

- **Geological/geomorphological:** In the form of landforms of glacial erosion and glacial deposition, and the deposition of glacial, glaciofluvial, glaciolacustrine and glaciomarine sediments.

- **Chemical evidence:** From the variations in isotope ratios in fossils present in ocean sediment cores and ice cores. This is because water containing heavier isotopes has a higher rate of evaporation, which decreases with colder conditions, allowing a record of temperature to be constructed.

- **Palaeontological evidence:** During a glacial period, cold-adapted organisms move into lower latitudes and organisms that prefer warmer conditions become extinct or are squeezed into even lower latitudes.

We will discuss the chemical evidence first, before elaborating on why these ice ages occur and why they are interspersed with interglacials between glacial maxima.

30.4.2 Evidence for rapid Quaternary climate change

Oxygen-isotope record in the ocean basins

Evidence for climatic change since the Pliocene in the last 3 million years (the Quaternary) comes from many sources, but one of the principal lines of evidence is the oxygen-isotope record from deep-sea cores (see Figures 30.18 and 30.19).

The argument is that any variations of water stored terrestrially through time, usually in the form of ice, can have a major effect on the mean ocean $\delta^{18}O_{water}$ value and so the marine $\delta^{18}O_{calcite}$ record. Today high-latitude precipitation goes back to the oceans through summer melting of snow and ice. During glacial phases, snow and ice accumulated in the ice sheets and because the difference in ice-sheet and mean ocean values is large, ice-sheet fluctuations are reflected in mean ocean $\delta^{18}O_{water}$ values. This can be seen by the fact that $\delta^{18}O_{ice}$ is -35 to $-40‰$, whilst the mean ocean value of $\delta^{18}O_{water}$ is $\sim0‰$. Hence this relationship can be shown by how the mean $\delta^{18}O_{water}$ value increased during the last glacial maximum (LGM) relative to the present. During the LGM, water stored in continental ice sheets lowered global sea levels by ~120 m, removing approximately 3% of the ocean's volume. This meant that the $\delta^{18}O_{water}$ value increased by 1.2‰ during the LGM relative to the present.

The $\delta^{18}O_{calcite}$ records come from benthonic and planktonic foraminifera fossils in sediment cores in all the world's ocean basins and these show a cyclic variation through the Pleistocene and into the Late Pliocene. The record has been divided up into 100 oxygen-isotope stages, representing 50 glacial–interglacial cycles in the last 2.6 million years (see Figures 30.18 and 30.21). However, the ocean record from these fossils has been used throughout the Cenozoic to show temperature oscillations and changes.

The last of the $\delta^{18}O$ steps was recorded during the late Pliocene from 3.2 to 2.6 million years ago, although it is better characterized as a series of $\delta^{18}O$ cycles with increasing amplitudes and values over this interval (Figures 30.18 and 30.19(a)).

Ice-core isotopic record

Global mean sea-surface temperatures can be worked out from measurements of the isotopic fractionation of oxygen in ice cores and it has been possible to obtain estimates of sea-level air temperature over the past 160 000 years. For example, in Figure 30.22 there are two sets of ice-core isotopic measurements, one from Byrd Station in Antarctica and the other from Camp Century in Greenland. The two measurement sets are correlated and it is possible to see the temperature reduction of the most recent Pleistocene glacial between 60 000 and 15 000 years ago. The warming trend to the present interglacial started around 15 000 years ago. Similarly, the variations in temperature, carbon dioxide and dust from the Vostok ice core have occurred over the last 400 000 years.

Composition of air in ice bubbles

The isotopic record in the above paragraph is complemented by studying the composition of ancient air trapped in air bubbles in the glacier ice. The records for methane and carbon dioxide are summarized in Figure 30.24 and both curves correlate extremely well with the temperature data for the last 180 000 years.

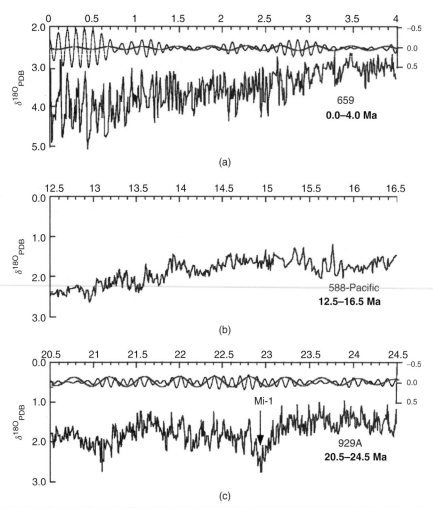

Figure 30.19 Oxygen-isotope record for: (a) the last 4 million years; (b) 16.5-12.5 million years ago; (c) 24.5-20.5 million years ago

30.4.3 Positive and negative feedbacks in glacial periods

Each glacial phase is subject to positive feedback, which makes it more severe, and negative feedback, which mitigates and eventually ends the cold period. Precipitation as ice and snow increases the Earth's albedo and so when the air temperature decreases, ice and snowfields grow, and this continues until equilibrium is reached. At the same time, the forest reduction caused by ice-sheet expansion increases the albedo.

Another hypothesis suggests that an ice-free Arctic Ocean leads to increased snowfall at high latitudes. When low-temperature ice covers the Arctic Ocean there is little evaporation or sublimation and the polar regions are dry in terms of precipitation, which allows the high-latitude snowfall to melt during the summers. An ice-free Arctic Ocean absorbs solar radiation during the long summer days and evaporates more water into the Arctic atmosphere. With higher precipitation, parts of this snow may not melt during the summer and glacial ice can form at lower latitudes, reducing the temperatures and increasing albedo.

Additional fresh water flowing into the North Atlantic during a warming cycle might also reduce the global ocean-water circulation (see Box 30.11, see the companion website). Such a reduction (by reducing the effects of the Gulf Stream) would have a cooling effect on northern Europe, which in turn would lead to increased low-latitude snow retention during the summer. It has also been suggested that during an extensive ice age, glaciers might move through the Gulf of Saint Lawrence, extending into the North Atlantic Ocean to the extent that the Gulf Stream becomes blocked.

30.4.4 Processes which mitigate glacial periods

Ice sheets that form during glaciations cause erosion of the land beneath them. After some time, this will reduce land below sea level and thus diminish the amount of space on which ice sheets can form. This mitigates the albedo feedback, as does the lowering in sea level that accompanies the formation of ice sheets.

Another factor is the increased aridity occurring with glacial maxima, which reduces the precipitation available to maintain

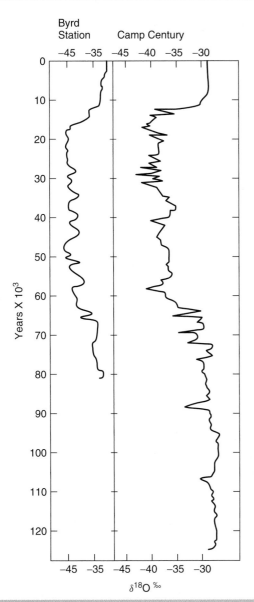

Figure 30.22 Ice-core oxygen-isotope measurements from Greenland and Antarctica

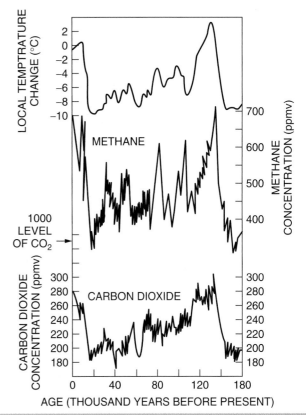

Figure 30.24 Correlation between the ice-core measurements of palaeoclimatic temperatures and the abundances of methane and carbon dioxide

glaciation. The glacial retreat induced by this or any other process can be amplified by similar inverse positive feedbacks to those which affect glacial advances.

30.4.5 Causes of ice ages

The causes of ice ages remain controversial for both the large-scale ice age periods and the smaller ebb and flow of glacial–interglacial periods within an ice age. The consensus is that several factors are important: atmospheric composition (the concentrations of carbon dioxide and methane); changes in the Earth's orbit around the Sun, known as Milankovitch cycles (and possibly the Sun's orbit around the galaxy); the motion of tectonic plates, resulting in changes in the relative location and amount of continental and

oceanic crust on the Earth's surface, which could affect wind and ocean currents; variations in solar output; the orbital dynamics of the Earth–Moon system; and the impact of relatively large meteorites and volcanism, including eruptions of supervolcanoes.

Some of these factors are causally related to each other. For example, changes in the Earth's atmospheric composition, especially the concentrations of greenhouse gases, may alter the climate, while climate change itself can change the atmospheric composition, for example by changing the rate at which weathering removes CO_2. The Tibetan and Colorado Plateaus have been suggested as immense CO_2 'scrubbers' with a capacity to remove enough CO_2 from the global atmosphere to be a significant causal factor of the 40 million-year Cenozoic cooling trend. It has been further claimed that approximately half of their uplift (and CO_2 'scrubbing' capacity) occurred in the past 10 million years.

Changes in the Earth's atmosphere

There is evidence that greenhouse gas levels fall at the start of ice ages and rise during the retreat of the ice sheets, but it is difficult to establish cause and effect. Greenhouse gas levels may also be affected by other factors that have been proposed as the causes of ice ages, such as the movement of continents and volcanism.

William Ruddiman has proposed the early anthropocene hypothesis, according to which the anthropocene era, as some people call the most recent period in the Earth's history, when the activities of the human race first began to have a significant global impact on the Earth's climate and ecosystems, did not begin in the eighteenth century with the advent of the Industrial Era, but 8000 years ago with the intense farming activities of our early agrarian ancestors. It was at that time that atmospheric greenhouse gas concentrations stopped following the periodic pattern of the Milankovitch cycles. In his overdue glaciation hypothesis, Ruddiman claims that an incipient ice age would probably have begun several thousand years ago, but was forestalled by the activities of early farmers.

The position of the continents

The geological record appears to show that ice ages start when the continents are in positions which block or reduce the flow of warm water from the equator to the poles and thus allow ice sheets to form. The ice sheets increase the Earth's reflectivity and thus reduce the absorption of solar radiation. With less radiation absorbed, the atmosphere cools; the cooling allows the ice sheets to grow, which further increases reflectivity in a positive feedback loop. The ice age continues until the reduction in weathering causes an increase in the greenhouse effect.

There are three known configurations of the continents which block or reduce the flow of warm water from the equator to the poles:

- A continent sits on top of a pole, as Antarctica does today.

- A polar sea is almost land-locked, as the Arctic Ocean is today.

- A supercontinent covers most of the equator, as Rodinia did during the Cryogenian period (see Section 30.2).

Since today's Earth has a continent over the south pole and an almost land-locked ocean over the north pole, geologists believe that the Earth will continue to endure glacial periods in the geologically-near future.

Some scientists believe that the Himalayas are a major factor in the current ice age, because these mountains have increased the Earth's total rainfall and therefore the rate at which CO_2 is washed out of the atmosphere, decreasing the greenhouse effect (Ruddiman and Kutzbach, 1991). The Himalayas' formation started about 70 million years ago when the Indo-Australian Plate collided with the Eurasian Plate, and the Himalayas are still rising by about 5 mm per year because the Indo-Australian plate is still moving at 67 mm per year. The history of the Himalayas broadly fits the long-term decrease in the Earth's average temperature since the Mid-Eocene, 40 million years ago.

Another important aspect which contributed to ancient climate regimes is the ocean currents, which are modified by the positions of continents, among other factors. These have the ability to cool (e.g. aiding the creation of Antarctic ice) and to warm (e.g. giving the British Isles a temperate as opposed to a boreal climate). The closing of the Isthmus of Panama about 3 million years ago may have ushered in the present period of strong glaciation over North America by ending the exchange of water between the tropical Atlantic and Pacific Oceans.

Variations in the Earth's orbit (Milankovitch cycles)

The Milankovitch theory involves a set of cyclic variations in characteristics of the Earth's orbit around the Sun, which cause changes in the amount and distribution of solar radiation reaching the Earth's surface (known as solar forcing). This theory was first discussed in the 1920s by a Serbian astronomer, Milankovitch, though initially it was not widely accepted. Since the 1970s, however, data collected have generated broad support for Milankovitch's ideas.

Each cycle has a different length, so at some times different cycles' effects reinforce each other and at other times they (partially) cancel each other out.

Three orbital parameters are especially important in causing ice-sheet waxing and waning:

- Changes in the eccentricity of the Earth's orbit.

- Changes in the tilt of the Earth's axis.

- The precession of the equinoxes.

In combination, these factors influence the amount and distribution (seasonality and location) of solar radiation reaching the Earth. Changes vary with both latitude and season. Because of the different periodicities of variation of the three factors, the composite variations in solar radiation are very complex.

Although the connections are not obvious and direct, changes in the amount of solar radiation are thought to drive the growth and melting of major ice sheets. Over the last 750 000 years, ice sheets have expanded into the midwestern United States at least eight major times, although the timing of the earlier of these advances is not well known.

Eccentricity

This refers to the shape of the Earth's orbit around the Sun. It is not a circle, but rather an ellipse. The shape of the elliptical orbit, which is measured by its eccentricity (Figure 30.25), varies between 1% and 5% through time.

The eccentricity affects the difference in the amount of radiation the Earth's surface receives at aphelion (around 4 July) and at perihelion (around 3 January). The effect of the radiation variation is to change the seasonal contrast in the northern and southern hemispheres. For example, when the orbit is highly elliptical, one hemisphere will have hot summers and cold winters whilst the other hemisphere will have warm summers and cool winters. When the orbit is nearly circular, both hemispheres will have similar seasonal contrasts in temperature.

Although the amount of change in radiation is very small (less than 0.2%), it is thought to be extremely important in the expansion and melting of ice sheets. The eccentricity of the Earth's orbit varies in a periodic manner; there are some variations occurring on a period of 413 000 years and others on a period of approximately 100 000 years.

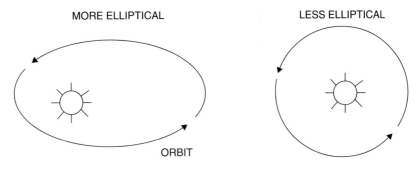

MORE ELLIPTICAL

LESS ELLIPTICAL

ORBIT

PERIODICITY:

100 000 YEARS

Figure 30.25 Eccentricity

Obliquity or axial tilt

The Earth's axis is tilted (inclined) with respect to its orbit around the Sun. Today the tilt is approximately 23.5°. It varies between 21.5° and 24.5° in a periodic manner (Figure 30.26). A graph of the tilt over the last 750 000 years shows that the dominant period of this variation is approximately 41 000 years.

Changes in the tilt of the Earth's axis cause large changes in the seasonal distribution of radiation at high latitudes and in the length of the winter dark period at the poles. Changes in tilt have very little effect on low latitudes. The effects of tilt on the amount of solar radiation reaching the Earth are closely linked to the effects of precession. Variation in these two factors causes radiation changes of up to 15% at high latitudes, and radiation variation of this magnitude greatly influences the growth and melting of ice sheets.

Precession of the equinoxes

The Earth does not have a perfect spin about its axis. There is a change in its orientation, referred to as a slow wobble, which can be likened to a spinning top running down and beginning to wobble back and forth. There is also a turning round of the elliptical orbit of the Earth itself.

Twice a year, at the equinoxes, the Sun is positioned directly over the equator. Currently the equinoxes occur on approximately 21 March and 21 September. However, because the Earth's axis

of rotation wobbles, the timing of the equinoxes changes. The change in the timing of the equinoxes is known as the precession (Figure 30.27).

Although the timing of the equinoxes is not in itself important in determining climate, the timing of the Earth's aphelion and perihelion also changes. Like the timing of the equinoxes, the timing of the aphelion and perihelion is affected by the wobble in the axis of rotation. The change in aphelion and perihelion is important for climate because it affects the seasonal balance of radiation. For example, when perihelion falls in January, the northern-hemisphere winter and southern-hemisphere summer are slightly warmer than the corresponding seasons in the opposite hemispheres.

The aphelion and perihelion change position on the orbit through a cycle of 360°. The cycle has two periods of approximately 19 000 and 23 000 years. Together these combine to produce a generalized periodicity of about 22 000 years. The effects of precession on the amount of solar radiation reaching the Earth are closely linked to the effects of tilt. Variation in these two factors causes radiation changes of up to 15% at high latitudes. Radiation variation of this magnitude greatly influences the growth and melting of ice sheets.

Using these three orbital variations, Milankovitch was able to formulate a mathematical model that calculated latitudinal

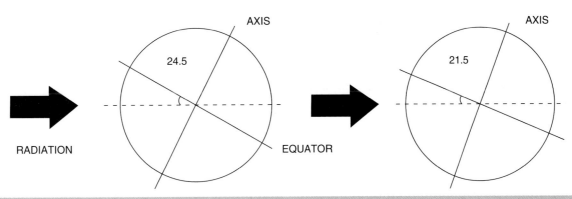

AXIS

24.5

RADIATION

EQUATOR

AXIS

21.5

Figure 30.26 Obliquity or axial tilt

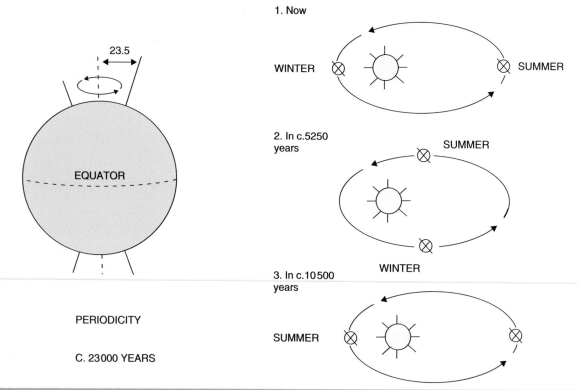

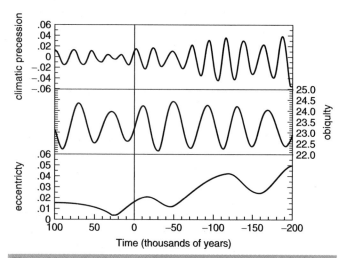

EQUATOR

PERIODICITY

C. 23 000 YEARS

1. Now

WINTER   SUMMER

2. In c.5250 years

SUMMER

WINTER

3. In c.10 500 years

SUMMER

Figure 30.27 Precession of the equinoxes

Figure 30.28 Graph showing the calculated values for 300 000 years of orbital variation

differences in insolation and the corresponding surface temperature for 600 000 years prior to 1800 (correlated in Figure 30.28). His theory concludes that ice caps at the poles increase and decrease in size as a reflection of the insolation at high latitudes. Specifically, he used calculations from 65 °N to develop his model.

These variables are only important because the Earth has an asymmetric landmass distribution, with most of the large continental landmasses, except Antarctica, parts of South America and Africa, located in the northern hemisphere. At times when

the northern-hemisphere summers are coolest (i.e. furthest from the Sun, due to precession and greatest orbital eccentricity) and winters are warmest (maximum tilt), snow can accumulate on and cover broad areas of North America and Europe. At present only precession is in glacial mode, with tilt and eccentricity not favourable to glacials. Even when all the orbital parameters favour glaciation, the increase in winter snowfall and decrease in summer melt is barely enough to trigger glaciation, never mind growing large ice sheets. Hence ice-sheet growth requires the support of positive feedback loops, the most obvious of which is that snow and ice have much lower albedo than the ground and vegetation, and thus ice masses tend to reflect more radiation back to space, cooling the climate and allowing ice sheets to expand.

In 1976 a seminal study based on the use of two deep-sea cores to measure $\delta^{18}O$ showed the existence of periodic signals of 105 000, 41 000, 23 000 and 19 000 years (Hays *et al.*, 1976) and since then many workers have established the existence of the dominant 100 000 cycle in deep-sea sediment cores. Today it is generally agreed that the orbital effect is the primary cause of glacial–interglacial changes.

Milankovitch theory has also helped to stitch together a near-continuous orbital scale chronology for the past 40 million years, allowing much more precise timing of important evolutionary and extinction events and better estimates of the timing and rate of climatic change.

A 23 000-year cycle has been found in deposits of carbonaceous sediments in the Pacific, in cave flowstone and in corals. A 5 million-year record for Lake Baikal of silica-bearing sediments using multiple proxies for climate shows evidence of the 41 000-year cycle throughout the record, but it is particularly strong for the period 1.8–0.8 million years ago. The 23 000-year cycle has been strongest during the last 400 000 years, while the data show that the 100 000-year cycle has been strong only during the last 800 000 years.

Summary

The spectacularly long records that show continuous response to the Milankovitch cycles and the evidence that all over the planet ice adjusted in the same way seems to be overwhelming corroborative evidence for the theory's validity. Despite some remaining problems, the orbital ellipticity, obliquity and precession have clearly influenced glaciation timing and severity. See the companion website for Box 30.12 Northern–hemisphere forcing of climate cycles in Antarctica.

What does the Milankovitch hypothesis say about future climate change?

Orbital changes occur over thousands of years, and the climate system may also take thousands of years to respond to orbital forcing. Theory suggests that the primary driver of ice ages is the total summer radiation received in northern-latitude zones where major ice sheets have formed in the past, near 65 °N.

Past ice ages correlate well to 65 °N summer insolation. Astronomical calculations show that 65 °N summer insolation should increase gradually over the next 25 000 years, and that no 65 °N summer insolation declines sufficient to cause an ice age are expected in the next 50 000–100 000 years.

Problems associated with the Milankovitch hypothesis

The observed periodicities of climate fit very well with the orbital periods, so that this orbital theory has overwhelming support. Nevertheless, there are some problems reconciling theory with observations. For example, there is the 100 000-year problem, where the eccentricity variations have a significantly smaller impact on solar forcing than precession or obliquity and so might be expected to produce the weakest effects. Yet observations show that in the last 1 million years the strongest climate signal has been the 100 000-year cycle. There is also a 400 000-year problem, where the eccentricity variations have a strong cyclicity with this length and should perhaps also be expected to be apparent in the records. The relative absence of this periodicity in the marine isotopic record may be due at least in part to the response of the climate system components involved, especially the carbon cycle. Another problem is referred to as the causality problem, where for example the penultimate interglacial (in marine isotopic stage 5) appears to have begun 10 000 years in advance of the solar forcing hypothesized to have caused it.

The effects of these variations are mainly believed to arise from variations in the intensity of solar radiation on various parts of the planet. Observations show that climate behaviour is much more intense than the calculated variations. Various internal characteristics of climate systems are believed to be sensitive to the insolation changes, causing positive feedback or amplification and negative feedback or damping responses.

Finally, we must mention the transition problem. This refers to the change in frequency of climate variations 1 million years ago. From 1 to 3 million years ago climate had a dominant mode, matching the 41 000-year obliquity cycle, but after that time this changed to a 100 000-year variation, matching eccentricity. It is difficult to understand why this happened.

It is very unlikely that the Milankovitch cycles can initiate or end an ice age. Even when their effects reinforce each other they are not strong enough. The peaks (effects reinforce each other) and troughs (effects cancel each other out) in the cycle are much more regular and much more frequent than the observed ice ages. In contrast, there is strong evidence that the Milankovitch cycles affect the occurrence of glacial and interglacial periods within an ice age.

The present ice ages are the most studied and best understood, particularly over the last 400 000 years, since this is the period covered by ice cores that record atmospheric composition and proxies for temperature and ice volume. Within this period, the match of glacial–interglacial frequencies to the Milankovitch orbital forcing periods is so close that orbital forcing is generally accepted. The combined effects of the changing distance to the Sun, the precession of the Earth's axis and the changing tilt of the Earth's axis redistribute the sunlight received by the Earth. Of particular importance are changes in the tilt of the Earth's axis, which affect the intensity of seasons. For example, the amount of solar influx in July at 65 °N latitude varies by as much as 25% (from 400 W/m^2 to 500 W/m^2). It is widely believed that ice sheets advance when summers become too cool to melt all of the accumulated snowfall from the previous winter. Some workers believe that the strength of the orbital forcing is too small to trigger glaciations, but feedback mechanisms like CO_2 may explain this mismatch.

While Milankovitch forcing predicts that cyclic changes in the Earth's orbital parameters can be expressed in the glaciation record, additional arguments are necessary to explain which cycles are observed to be most important in the timing of glacial–interglacial periods. In particular, during the last 800 000 years the dominant period of glacial–interglacial oscillation has been 100 000 years, which corresponds to changes in the Earth's eccentricity and orbital inclination. Yet this is by far the weakest of the three frequencies predicted by Milankovitch. During the period 3.0–0.8 million years ago, the dominant pattern of glaciation corresponded to the 41 000-year period of changes in the Earth's obliquity. The reasons for the dominance of one frequency over another are poorly understood and an active area of current research, but the answer probably relates to some form of resonance in the Earth's climate system.

The 'traditional' Milankovitch theory struggles to explain the dominance of the 100 000-year cycle over the last eight cycles. Some scientists have pointed out that those calculations are for a two-dimensional orbit of the Earth and that the three-dimensional orbit also has a 100 000-year cycle of orbital inclination. They proposed that these variations in orbital inclination lead to variations in insolation, as the Earth moves in and out of known dust bands in the solar system. Although this is a different mechanism to the traditional view, the 'predicted' periods over the last 400 000 years are nearly the same.

Another worker, William Ruddiman, has suggested a model that explains the 100 000-year cycle by the modulating effect of eccentricity (weak 100 000-year cycle) on precession (23 000-year cycle) combined with greenhouse gas feedbacks in the 41 000- and 23 000-year cycles.

Yet another theory has been advanced by Huybers and Wunsch (2005), who argued that the 41 000-year cycle has always been dominant, but that the Earth has entered a mode of climate behaviour in which only the second or third cycle triggers an ice age. This would imply that the 100 000-year periodicity is really an illusion created by averaging together cycles lasting 80 000 and 120 000 years. This theory is consistent with the existing uncertainties in dating, but is not widely accepted at present.

30.4.6 Variations in the Sun's energy output

There are at least two types of variation in the Sun's energy output:

- In the very long term, astrophysicists believe that the Sun's output increases by about 10% per billion (10^9) years. In about 1 billion years the additional 10% will be enough to cause a runaway greenhouse effect on Earth: rising temperatures produce more water vapour, which is a greenhouse gas (much stronger than CO_2); the temperature rises; more water vapour is produced; and so on.

- In the shorter term, since the Sun is huge, the effects of imbalances and negative feedback processes take a long time to propagate through it, so these processes overshoot and cause further imbalances, and so on. A 'long time' in this context means thousands to millions of years.

The long-term increase in the Sun's output cannot be a cause of ice ages.

The best known shorter-term variations are sunspot cycles, especially the Maunder minimum, which is associated with the coldest part of the Little Ice Age. Like the Milankovitch cycles, sunspot cycles' effects are too weak and too frequent to explain the start and end of ice ages but very probably help to explain temperature variations within them.

30.4.7 Volcanism

It is theoretically possible that undersea volcanoes could end an ice age by causing global warming. One suggested explanation of the Paleocene–Eocene Thermal Maximum is that undersea volcanoes released methane from clathrates and thus caused a large and rapid increase in the greenhouse effect. There appears

to be no geological evidence for such eruptions at the right time, but this does not prove they did not happen (see Box 30.3, see the companion website). It is harder to see how volcanism could cause an ice age, since its cooling effects would have to be stronger than and outlast its warming effects. This would require dust and aerosol clouds, which would stay in the upper atmosphere blocking the Sun for thousands of years, which seems very unlikely. Undersea volcanoes could not produce this effect because the dust and aerosols would be absorbed by the sea before they reached the atmosphere. See the companion website for Box 30.13 Links between Pleistocene glaciation and methanogenesis in sedimentary basins.

30.5 Late Glacial climates and evidence for rapid change

30.5.1 Introduction

The Late Glacial can be defined as the time between the first rise of the temperature curve after the maximum of the Last Glacial (Devensian) and the very rapid rise at the opening of the Holocene (Flandrian). For continental north-western Europe the date of 13 000 BP has been adopted as the beginning of the so-called classic Late Glacial, and the beginning of the Flandrian has been set at 10 000 BP. In Britain, C^{14} dates on the lowest organic deposits in kettleholes in North Wales and north-west England show they were open and accumulating organic sediment ~14 500 years ago, which was much earlier than the rest of Europe. The period covers major change, with the warming up from a glacial stage, through the fluctuations of a Late Glacial period to the rapid climatic improvement at the opening of the present interglacial. The techniques used to obtain a picture of the environment and the vegetation in particular include studies of pollen, soils, sediment geochemistry and stratigraphy from lakes (Figure 30.29), beetle remains, geomorphology, and the ice-core and deep-sea core records (see Figure 30.30). Although our early understanding of this period came from north-western Europe we need to establish whether these rapid climatic changes were apparent throughout the world.

30.5.2 Classic divisions of the Late Glacial

The first evidence of a climatic oscillation of interstadial magnitude came from Allerød in Denmark, where organic lake muds containing plant remains were found between upper- and lower-clay deposits. Both clays were rich in the remains of the mountain avens (*Dryas octopetala*), a plant which indicates a severe climate. This plant gave its name to the Older Dryas and Younger Dryas periods. The intervening Allerød lake muds showed a cool-temperate flora with the remains of tree birches and plants, which are more warmth-demanding than the Arctic Alpine group, so there was evidence for an interstadial period. Later, when pollen analysis was carried out on similar Late Glacial profiles, a pollen

(a)   (b)

Figure 30.29 Coring from the southern basin of Lake Windermere (English Lake District): (a) the Mackereth corer used from a boat; (b) core from the last few thousand years of the Late Pleistocene and Holocene

Approximate Dates bc	Zones	Periods	Characteristic Vegetation
	VIII	Sub-Atlantic	Alder, Oak, Birch
500			
	VIIb	Sub-Boreal	Alder, Oak, Lime 'Elm decline'
3,000			
	VIIa	Atlantic	Alder, Oak, Elm, Lime ('Mixed Oak forest')
5,500			
	VI	Later Boreal	Hazel, Pine c. Oak, Elm, Lime / b. Oak, Elm / a. Elm, Hazel
7,000			
	V	Early Boreal	Hazel, Birch, Pine
7,600			
	IV	Pre-Boreal	Birch, Pine forest
8,300			
	III	Younger Dryas	Park tundra
8,800			
	II	Allerød Interstadial	Birch with Park tundra
10000			
	I	Older Dryas	Park tundra
12000			

Figure 30.30 Late Glacial and postglacial vegetation zones in southern Britain based on pollen and radiocarbon dating. (Source: after Walker, 2003)

zonation was established (Figure 30.30). The different periods became known as:

- **Zone I:** the Older Dryas, a pre-Allerød cold phase.

- **Zone II:** the Allerød cool-temperate oscillation.

- **Zone III:** the Younger Dryas cold phase. The first evidence for this type of subdivision came from Ballybetagh bog in the Wicklow Mountains, Ireland.

The typical vegetation association of this period was the so called 'park tundra' and in the cold phases and possibly at all times in more northern latitudes and at high altitudes there was tundra vegetation, characterized by a lack of trees and open communities. The four common Dryas plants were dwarf birch (*Betula nana*), Arctic or least willow (*Salix herbacea*), a prostrate variety of juniper and the mountain avens (Figure 30.31). Many species are now found in open habitats and species known as colonizers are found in new habitats on mountains, such as the fir club moss (*Lycopodium selago*).

In the Allerød period, tree birches, juniper and some pine were common, either as continuous woodlands or as copses in the herbaceous tundra vegetation (see Figure 30.32) for a modern counterpart. In Denmark, Germany and Holland there is some evidence for a synchronous climatic oscillation before the Allerød in Zone I. This oscillation marked the first migration of tree birches into north-west Europe, dated to around 12 750 BP. It was first noted at Bølling in Denmark and therefore Pollen Zone I was subdivided into PZIa, b and c. Evidence can be found for the Bølling Oscillation in Britain too, but only in the east of the country, for example at Tadcaster in Yorkshire, where tree birches became relatively much more abundant at a point in PZI, followed by a recession before the opening of the Allerød (PZII).

However, in western Britain there have been great difficulties in zoning the pollen diagrams for the Late Glacial into PZI–III and it has been found to be much better to correlate this period by C^{14} dates. For example, at Blelham Bog in the English Lake District the cored sediments showed two organic layers separated

(a)

(b)

(c)

(d)

(e)

Figure 30.31 Common Dryas-stage plants: (a) dwarf birch (*Betula nana*) from the Mesters Vig region of eastern Greenland; (b) Arctic or least willow (*Salix herbacea*) from a mountain summit on Arran, western Scotland; (c) prostrate juniper from Hverfjell, northern Iceland; (d) mountain avens (*Dryas octopetala*) from Morsadalur, Iceland; (e) fir club moss (*Lycopodium selago*) from east Greenland, near Mesters Vig

Figure 30.32 Birch and willow woodland, Kjos (southern Iceland), a similar environment to the Allerød cool-temperate oscillation vegetation

by 5 cm of silt of low organic content, originally suggesting a sequence comparable with that at Bølling (Figure 30.33), but C[14] dates from the lower organic layer proved a much older date than at the type location: 14 330 ± 220 BP. This has been confirmed by duplicate dates, by consistent dates above and by dates at other sites, like Glanllynau in north-west Wales (14 468 ± 300 BP). What we see here is that sites in western Britain were accumulating organic material and pollen before the date chosen as the opening of the classic Late Glacial in north-western Europe. Hence the pollen sequences in western Britain differ from the classic sites of Scandinavia, but it nevertheless seems possible to correlate them by dating. The differences just indicate a regional

differentiation of climate and vegetational response, as you would expect since the amelioration from the full glacial climate began much earlier in the western margins of northern Europe than at sites further east at the same latitudes.

An aid to the interpretation of the Late Glacial sediments and environments comes from an analysis of the chemistry of the sediments. The argument goes that the maximum accumulation of biogenic sediments takes place under ecologically-stable interglacial climates. Minimum concentrations of organic matter occur under glacial conditions when all the soils have been destroyed by glacial erosion and the deposits consist of rock flour and minerogenic sediments. This argument can be taken a step further in that in periods of high erosion rate the soil is rich in potassium, sodium and magnesium and is rapidly transferred to lake basins. In periods of low erosion rate, by comparison, these elements are rapidly removed by leaching from the catchment surface. Hence the material that is transferred to the lake basins is relatively poor in K, Na and Mg. For Late Glacial profiles, analyses show relatively high concentrations of the erosion indicators in both pre-interstadial and post-interstadial deposits. Minimum concentration occurs in the interstadial deposits, where the maximum values for total carbon show maximum humus accumulation in the maturing soil profiles (Figure 30.34).

30.5.3 What was happening in western Britain during the Late Glacial?

It is possible to characterize several stages and time periods with typical vegetation associations from this part of Britain:

1. **Pre-13 000 BP:** The pollen spectra from sites within the area of the last glaciation suggest snow-bed communities

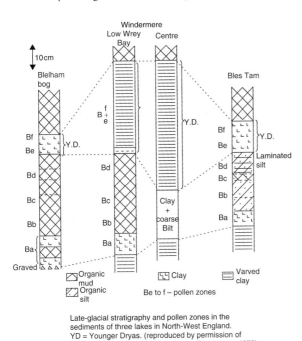

Figure 30.33 Blelham Bog (English Lake District) stratigraphy and pollen zonation

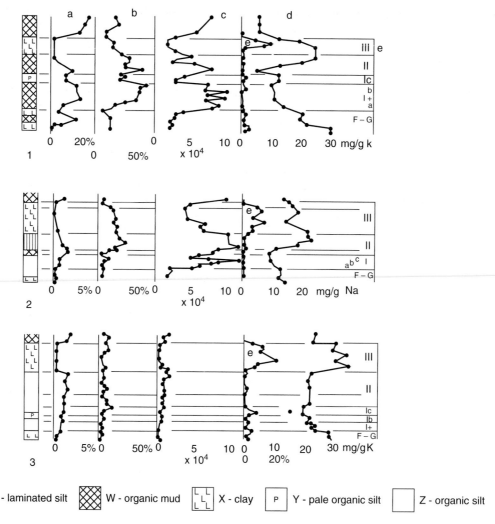

The Late-Weichaelian climatic oscillation; pollen

Figure 30.34 Pollen and chemical stratigraphy from Blelham Bog in the Late Weichselian climatic oscillation (after Pennington, 1975). (a) carbon as % dry weight, (b) Birch (Betula) pollen as % of total pollen, (c) total pollen concentration per cc., (d) intensity of erosion as measured by proportion of potassium or sodium, (e) Artemisia pollen as % of total.

dominated by *Salix herbacea* at the higher latitudes and altitudes, for example at Blea Tarn and Easedale in the Lake District and at Cam Loch in Sutherland. At the more lowland sites there was a grass-sedge tundra with much sorrel and low percentages of juniper (interpreted as prostrate shrubs depending on a winter snow cover), as at Blelham Bog and Lake Windermere in the Lake District. At Blelham we are in PZIIa (see Figure 30.43), the Rumex-Graminae PAZ, with low pollen concentrations, tundra deposition rates of pollen at under 200 grains per cm^2 per year. At this time there was polar desert in Holland.

2. **The Windermere Interstadial, ~13 000–11 000 BP:** At Blelham Bog the lowest boundary of the main organic layer falls at ~13 000 BP and corresponds with a general climatic amelioration in north-western Europe. Above this horizon there is a rapid rise in annual deposition rates of all major

pollen taxa and in particular a large increase in juniper, which seems to have been an immediate response to the improving climate of juniper already present in the area. Through the Juniper PAZ there is a progressive increase in birch, which indicates a steady dispersal of tree birches towards the site. The overlying Betula PAZ (IIc) in which the maximum Late Glacial percentages of birch pollen are found can be interpreted as the time of maximum extent of birch woodland. The juniper was suppressed by the superior competitive ability of the birches. There is no evidence of a vegetation regression between the juniper pollen zone and the birch maximum, but rather there was a progressive unidirectional vegetation succession. All this took place during the continental Bølling period (13 000–12 000 BP). This was the period of maximum temperature, as evidenced by the percentage of tree pollen and the higher rates of pollen deposition (1000–3000 grains per cm^2 per year).

3. **The Older Dryas, PZIc, 12 000 – 11 800 BP:** At Blelham there was a change to a zone of lower deposition rates of birch and juniper. This small absolute decline was due to a fall in temperatures sufficient to affect pollen production by thermophilous plants but not to increase soil movement by freeze–thaw processes. This minor recession ended the warmest period in the interstadial and from then on there were fluctuating, but on the whole declining, temperatures in the Allerød equivalent (IId).

4. **The Birch–Juniper Pollen Zone (IId), 11 800 – 11 000 BP:** There was an interplay between birch and juniper, illustrative of a fluctuating climate, and at the same time a rise of pollen deposition of open-environment herbs.

There is supportive evidence for this pollen record in the interpretation of beetle remains from sites like Glanllynau, Windermere and St Bees. At the first site the dating shows that deposition occurred in the kettlehole from ~14 000 – 10 000 BP and the beetles indicate that at around 13 000 BP there was an intensely cold, continental climate, which gave rise suddenly to a period with summer temperatures at least as warm as those of today. However, the landscape was entirely devoid of trees. From 12 000 to about 10 000 BP there was a progressive deterioration of the temperature curve and the period of birch forest is shown to have a less thermophilous fauna than that of the previous pollen zone. So there is evidence in western Britain of major, synchronous climatic amelioration at around 13 000 BP from a range of evidence. However, the Bølling Oscillation seems to have been metachronous and represents the local arrival of birch woodland at each site as it took time for the dispersal or migration of tree birches from their glacial refuges to occur. We can see this because there is a significant upward shift in annual birch productivity at 12 500 BP at Blelham, at 12 300 BP at Bølling and at 12 650 BP at Brøndmyra (southern Norway). At Blelham the maximum Late Glacial percentage of birch occurred in the Bølling equivalent, whereas at Tadcaster, as at the type site of Bølling, the highest values occurred in the Allerød. In western Britain there was an irregular temperature fall in the Allerød, with a break-up of the ecological equilibrium of the birchwoods, an increased representation of herbs and increased soil erosion. All suggest lower temperatures than in the continental time period. This was probably caused by cooling from the west as the amount of ice in the North Atlantic increased.

A case study of one location in East Yorkshire, Gransmoor is presented in Box 30.14 (see the companion website).

30.5.4 The Older Dryas and its vegetation

The Older Dryas was a somewhat variable cold, dry period in Northern Europe, roughly equivalent to Pollen Zone Ic. It was preceded by the Bølling and followed by the Allerød periods. It may or may not appear in the climatological evidence for different regions. If it does not appear, then the Bølling and Allerød phases are considered one interstadial period. The strength of the Older

Dryas depends to some degree on latitude and it seems to have been strongest in northern Eurasia. In the Greenland oxygen-isotope record, the Older Dryas appears as a downward peak establishing a small, low-intensity gap between the Bølling and the Allerød.

In other parts of the world it may be indicated in pollen. In the island of Hokkaidō in Japan it is indicated by the records of a *Larix* pollen peak and matching sphagnum decline at 12 400 – 11 800 BP uncal., 14 600 – 13 700 BP cal. In the White Sea a cooling occurred at 14 700 – 13 400/13 000 BP, which resulted in a re-advance of glaciers in the initial Allerød. In Canada, the Shulie Lake phase (a re-advance) is dated to 14 000 – 13 500 BP. On the other hand, varve chronology in southern Sweden indicates a range of 14 050 – 13 900 BP. Capturing the Older Dryas through high-resolution dating continues to be of concern to researchers in palaeoclimatology.

Description of the vegetation

Northern Europe had an alternation of steppe and tundra environments depending on the permafrost line and the latitude. In moister regions around lakes and streams were thickets of dwarf birch, willow, sea buckthorn and juniper. In the river valleys and uplands to the south were open birch forests.

The first trees, birch and pine, had spread into Northern Europe 500 years previously. During the Older Dryas, the glaciers advanced again and the trees retreated southward, to be replaced by a mixture of grassland and cool-weather Alpine species. This type of biome has been called 'park tundra' or 'Arctic tundra' and exists today in the transition between taiga and tundra in Siberia. Then it stretched from Siberia to Britain in a more or less unbroken expanse.

To the north-west was the Baltic Ice Lake, which was truncated by the edge of the ice sheet. Species had access to Denmark and southern Sweden, but most of Finland and the Baltic countries were under the ice or the lake for most of the period. Northern Scandinavia was glaciated. Between Britain and today's European mainland were rolling hills prolifically populated with animals, and thousands of bone specimens have been recovered by dredging and fishing from the current North Sea, an area known as Doggerland. See the comanion website for Box 30.14 Late–Glacial Palaeoenvironmental palaeoenvironmental reconstructions from Gransmoor, East Yorkshire.

30.5.5 The Younger Dryas

The Younger Dryas was the most significant rapid climate change event to occur during the last deglaciation of the North Atlantic region. It occurred about 12 900 BP and is the canonical example of abrupt climate change. It is best seen in the Greenland ice cores, although it had very marked consequences over Europe, North America and as far as New Zealand.

The Younger Dryas is an invaluable case study: it occurred recently enough that records of it are well-preserved and it seems to have left traces all over the world. In Ireland, the period is

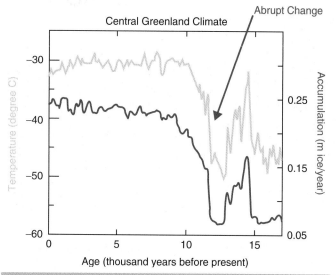

Figure 30.39 Temperature curve from central Greenland for the last 18 000 years and the phases of climate

dramatic example is provided by the calcium series covering the last 10 000–18 000 years. Prominent periods of increased dustiness have been observed in the record, peaking approximately every 500 years during the early PB at ~11 400 BP; throughout the Younger Dryas at ~11 810, 12 220 and 12 640 BP; during the Bølling–Allerød (B/A) at ~13 180, 13 650 and 14 020 BP; and during much of the glacial. Such events have been attributed by Mayewski *et al.* (1993) to changes in the size of the polar atmospheric cell and in source regions (e.g. growth and decay of continental biogenic and terrestrial source regions).

The climate change that accompanied the Younger Dryas was not restricted to Greenland. The record of variations in the CH^4 concentration of trapped gases in the GRIP ice core (Chappellaz *et al.*, 1993) shows that tropical and subtropical climates were colder and drier during the Younger Dryas and earlier cold events. The major natural source region of CH^4 is low-latitude wetlands and higher atmospheric concentrations are presumably caused by the greater areal extent of tropical and subtropical wetlands.

The ammonium flux record from GISP2 ice provides an estimate of continental biogenic source strength (Mayewski *et al.*, 1993) during the Younger Dryas. Although at the onset of the Bølling–Allerød ammonium flux levels and outliers rose dramatically, during the Younger Dryas ammonium flux levels dropped only minimally and the number of ammonium outliers decreased slightly. Since ammonium concentrations are highest near continents and decrease with transport as a consequence of deposition, it appears that continental sources close to Greenland (North America and Europe) were not as dramatically affected during the Younger Dryas as were low-latitude wetland regions, as evidenced by the CH^4 record. This may indicate the continued importance of ice sheets and permafrost in limiting the growth of vegetation at higher latitudes until the end of the Younger Dryas. Both low-latitude-source CH^4 and ammonium rose at the end of the Younger Dryas (Chappellaz *et al.*, 1993; Mayewski *et al.*, 1993). Thermally-fractionated nitrogen- and argon-isotope data from GISP2 ice indicate that the summit of Greenland was ~15 °C colder during the Younger Dryas than today. In the UK, coleopteran (fossil beetle) evidence suggests mean annual temperature dropped to approximately 5 °C and periglacial conditions prevailed in lowland areas, while ice fields and glaciers formed in upland areas. Nothing of the size, extent or rapidity of this period of abrupt climate change has been experienced since.

known as the Nahanagan Stadial, while in the United Kingdom it is known as the Loch Lomond Stadial, and most recently as Greenland Stadial 1 (GS1).

The Younger Dryas is also a Blytt–Sernander climate period (PZIII) detected from layers in Northern European bog peat. It lasted approximately from 12 900 to 11 500 BP cal., or 11 000 to 10 000 BP uncal.

When the temperature curve from central Greenland for the last 18 000 years is inspected (Figure 30.39), several phases can be seen. Previous ice-core studies have focused on the abrupt termination of this event (Dansgaard *et al.*, 1989) because this transition marks the end of the last major climate reorganization during the deglaciation. More recently the Younger Dryas has been re-dated, using precision, subannually-resolved, multivariate measurements from the GISP2 ice core, as an event of 1300 ± 70 years' duration that terminated abruptly, as evidenced by a ~7–10 °C rise in temperature and a twofold increase in accumulation rate, at ~11 640 BP. Fluctuations in the electrical conductivity of GISP2 ice on the scale of <5–20 years have been used to reveal rapid changes in the dust content of the atmosphere during the same periods and throughout the last glacial. These rapid changes appear to reflect a type of 'flickering' between preferred states of the atmosphere, which provides a new view of climate change. Holocene climates are by comparison stable and warm.

High-resolution, continuous measurements of major anions (chloride, sulphate and nitrate) and cations (sodium, magnesium, potassium, calcium and ammonium) of GISP2 ice have been used to reconstruct the palaeoenvironment during the Younger Dryas, since these series record the history of the major soluble constituents transported in the atmosphere and deposited over central Greenland (Mayewski *et al.*, 1993). These multivariate glaciochemical records provide a robust indication of changes in the characteristics of the sources of these soluble components, or changes in their transport paths, in response to climate change. A

30.5.6 The Younger Dryas (PZIII) cold phase in the British uplands

Sediments in lake cores indicate a return to glacial conditions in the British upland areas for a 400–500-year period (10 800–10 300 BP). For example, varved clays are found in Windermere, there are glacial moraines demarcating corrie glaciers in Wales (Figure 30.40(a)), the Lake District (Figure 30.40(b)) and Scotland, and protalus ramparts associated with snow banks are found in these upland areas too. At Blelham, four pollen assemblages coincide with this period (Figure 30.33).

(a)

(b)

Figure 30.40 Glacial moraines marking the Younger Dryas corrie glacier limits at Cader Idris (mid-Wales) and Langdale (English Lake District)

The Artemisia zones (Be and Bf) coincide with a solifluction clay in the core and there is a fall in birch and juniper pollen rate to under 100 grains per cm² per year, which indicates the disappearance of the local woodland. Pollen is scarce. Sometimes one finds the leaf macrofossils of the dwarf willow, which is only seen on the mountain summits of the Lake District today.

The two later zones, the Graminae-herbs and Empetrum (Bg) and the Juniper (Bh) coincide with the second half of the period, when there was a rapid temperature rise, a stabilization of the land surface by vegetation cover and mud of increasing organic content. This becomes transitional to the Flandrian (Holocene) when birch percentages expand. Sedges, grasses and herbs expand at the expense of the more open community plants. See the companion website for Box 30.15 Other explanations for the Younger Dryas climatic event.

30.5.7 Why did the Younger Dryas take place?

This is currently the focus of much research. One explanation involves a thermohaline-circulation (THC) shutdown triggered by a catastrophic discharge of fresh water from Lake Agassiz in North America, which resulted in a rapid reduction in northward ocean heat transports, leading to an abrupt cooling over Northern Europe and North America. That is why so much attention is focused on the behaviour of the North Atlantic Ocean circulation.

A problem with this hypothesis is the timing of the melt-water pulse that was supposed to have triggered the THC shutdown, as it was found that a second melt-water pulse, albeit slightly smaller than the first, occurred at the end of the Younger Dryas. Why did this not trigger a similar chain of consequences in the climate system? Broecker (2006) discusses the possible routes for the flood water, including a subglacial path, and although he favours flood waters from Lake Agassiz, the inability to identify the routes is disconcerting.

An alternative explanation (Clement et al., 2001) invokes the abrupt cessation in the El Niño – Southern Oscillation in response to changes in the orbital parameters of the Earth, although how such a change would impact regions away from the tropics remains to be explained.

The respective merits of both hypotheses have discussed by Broecker (2003) but the issue is far from being settled, as we can see from the two suggested explanations in Box 30.15, see the companion website.

30.5.8 Is there evidence that the Younger Dryas was a global event?

Any attempt to answer this question is hampered by the lack of a precise definition of 'Younger Dryas' in all the records. In Western Europe and Greenland, the Younger Dryas is a well-defined synchronous cool period. But cooling in the tropical North Atlantic may have preceded this by a few hundred years, while in South America there is a less well defined initiation but a sharp termination. The Antarctic Cold Reversal appears to have started a thousand years before the Younger Dryas and has no clearly defined start or end. It has been argued that there is some confidence in the absence of the Younger Dryas in Antarctica, New Zealand and parts of Oceania. Similarly, the southern-hemisphere cooling known as the Deglaciation Climate Reversal (DCR) began approximately 1000 years before the Younger Dryas, between 14 000 and 11 500 years ago, as noted in the Sajama ice core. The Andean climate returned to Last Glacial Maximum conditions, with colder temperatures coupled with higher precipitation (high lake stands in the Altiplano).

In western North America it is likely that the effects of the Younger Dryas were less intense than in Europe, but there is evidence of glacial re-advance, which indicates Younger Dryas cooling in the Pacific Northwest.

However, in the Younger Dryas there was:

1. Replacement of forest with glacial tundra in Scandinavia.

2. Glaciation or increased snow in mountain ranges around the world.

3. Formation of solifluction layers and loess deposits in Northern Europe.

4. More dust in the atmosphere, originating from deserts in Asia.

5. Drought in the Levant, perhaps motivating the Natufian culture to invent agriculture. It is argued that the cold and dry Younger Dryas lowered the carrying capacity of the area and forced the sedentary Early Natufian population into a more mobile subsistence pattern. Further climatic deterioration is thought to have brought about cereal cultivation. While there is relative consensus regarding the role of the Younger Dryas in the changing subsistence patterns during the Natufian, its connection to the beginning of agriculture at the end of the period is still debated (Munro, 2003).

The Huelmo/Mascardi Cold Reversal in the southern hemisphere began slightly before the Younger Dryas and ended at the same time.

30.5.9 What happened at the end of the Younger Dryas?

Measurements of oxygen isotopes from the GISP2 ice core suggest the ending of the Younger Dryas took place over just 40–50 years in three discrete steps, each lasting five years. Other proxy data, such as dust concentration and snow accumulation, suggest an even more rapid transition, requiring a $\sim 7\,^{\circ}$C warming in just a few years; the total warming was $10 \pm 4\,^{\circ}$C.

The end of the Younger Dryas has been dated to around 9620 BC (11 550 calendar years BP, or 10 000 radiocarbon years BP – a 'radiocarbon plateau') by a variety of methods, with mostly consistent results:

*   11 530 ± 50 BP – GRIP ice core, Greenland.

*   11 530 BP – Kråkenes Lake, western Norway.

*   11 570 BP – Cariaco Basin core, Venezuela.

*   11 570 BP – German oak/pine dendrochronology.

*   11 640 ± 280 BP – GISP2 ice core, Greenland.

30.6 The Medieval Warm Period or Medieval Climate Optimum and the Little Ice Age

30.6.1 Introduction

This period had an unusually warm climate in the North Atlantic region. It occurred around 800–1300 AD, during the European middle ages. The Medieval Warm Period (MWP) is often invoked in contentious discussions of global warming. Some scientists refer to the event as the Medieval Climatic Anomaly as this term emphasizes that effects other than temperature were important.

Initial research on the MWP and the following Little Ice Age (LIA) was largely completed in Europe, where the phenomenon was most obvious and most clearly documented.

It was at first believed that the temperature changes were global. However, this view has been questioned: the National Oceanic and Atmospheric Administration (NOAA) says that the 'idea of a global or hemispheric "Medieval Warm Period" that was warmer than today however, has turned out to be incorrect' and that what those 'records that do exist show is that there was no multi-century periods when global or hemispheric temperatures were the same or warmer than in the 20th century.' Indeed, global temperature records taken from ice cores, tree rings and lake deposits have shown that the Earth was actually slightly cooler (by 0.03 °C) during the 'Medieval Warm Period' than in the early and mid-twentieth century.

Palaeoclimatologists developing region-specific climate reconstructions of past centuries conventionally label their coldest interval as 'LIA' and their warmest as 'MWP'. Others follow this convention and, when a significant climate event is found in the 'LIA' or 'MWP' time frame, associate it to that period. Some 'MWP' events are thus wet or cold events rather than strictly warm events, particularly in central Antarctica, where climate patterns opposite to those of the North Atlantic area have been noticed.

30.6.2 North Atlantic and North American regions

The Vikings took advantage of ice-free seas to colonize Greenland and other outlying lands of the far north. The MWP was followed by the LIA, a period of cooling that lasted until the nineteenth century. In Chesapeake Bay, researchers found large temperature excursions during the MWP ($\sim$800–1300 AD) and the LIA ($\sim$1400–1850), possibly related to changes in the strength of North Atlantic THC. Sediments in the Piermont Marsh of the lower Hudson Valley show a dry MWP from 800 to 1300 AD (see Box 30.16, see the companion website). Prolonged droughts affected many parts of the western United States, especially eastern California and the western Great Basin. Alaska experienced three time intervals of comparable warmth: 1–300, 850–1200 and post-1800 AD.

A radiocarbon-dated box core in the Sargasso Sea (Keigwin, 1996) shows that the sea-surface temperature was $\sim 1\,^{\circ}$C cooler than today approximately 400 years ago (during the LIA) and 1700 years ago, and approximately 1 °C warmer 1000 years ago (during the MWP). During the MWP wine grapes were grown in Europe as far north as southern Britain, as they are today.

30.6.3 Other regions on Earth

The climate in equatorial east Africa has alternated between drier than today and relatively wet. The drier climate took place during the MWP ($\sim$1000–1270 AD).

An ice core from the eastern Bransfield Basin, Antarctic Peninsula, clearly identifies events of the LIA and MWP (Khim *et al.*, 2002). The core shows a distinctly cold period about 1000–1100 AD, neatly illustrating the fact that 'Medieval Warm Period' is a moveable term, and that during the 'warm' period there were, regionally, periods of both warmth and cold.

Corals in the tropical Pacific Ocean suggest that relatively cool, dry conditions may have persisted early in the millennium, consistent with a La Niña-like configuration of the El Niño–Southern Oscillation patterns.

Although there is an extreme scarcity of data from Australia (for both the MWP and LIA), evidence from wave-built shingle terraces for a permanently full Lake Eyre during the ninth and tenth centuries is consistent with this La Niña-like configuration, though of itself inadequate to show how lake levels varied from year to year or what climatic conditions elsewhere in Australia were like.

Adhikari and Kumon (2001), whilst investigating sediments in Lake Nakatsuna in central Japan, verified the existence there of both the MWP and the LIA.

30.6.4 The Little Ice Age

Introduction

The Little Ice Age was a period of cooling after a warmer medieval climate optimum. Climatologists and historians find it difficult to agree on either the start or the end dates of this period. Some confine the LIA to approximately the sixteenth century until the mid-nineteenth century, but it is generally agreed that there were three minima, beginning about 1650, about 1770 and 1850, each separated by slight warming intervals.

It was initially believed that the LIA was a global phenomenon but it is now less clear that this is the case. The Intergovernmental Panel on Climate Change (IPCC) describes the LIA as 'a modest cooling of the Northern Hemisphere during this period of less than 1 °C,' and says, 'current evidence does not support globally synchronous periods of anomalous cold or warmth over this timeframe, and the conventional terms of "Little Ice Age" and "Medieval Warm Period" appear to have limited utility in describing trends in hemispheric or global mean temperature changes in past centuries.' There is evidence, however, that the LIA did affect the southern hemisphere.

Dating of the Little Ice Age

There is no agreed beginning year to the LIA, although there is a frequently-referenced series of events preceding the known climatic minima. Starting in the thirteenth century, pack ice began advancing southwards in the North Atlantic, as did glaciers in Greenland. The three years of torrential rain beginning in 1315 ushered in an era of unpredictable weather in Northern Europe, which did not lift until the nineteenth century. There is also anecdotal evidence of expanding glaciers almost worldwide. In contrast, a climate reconstruction based on glacial length shows no great variation from 1600 to 1850, though it shows strong retreat thereafter.

For this reason, any of several dates ranging over 400 years may indicate the beginning of the LIA:

- 1250, for when Atlantic pack ice began to grow.
- 1300, for when warm summers stopped being dependable in Northern Europe.
- 1315, for the rains and Great Famine of 1315–1317.
- 1550, for the theorized beginning of worldwide glacial expansion.
- 1650, for the first climatic minimum.

In contrast to its uncertain beginning, there is a consensus that the LIA ended in the mid-nineteenth century.

The Little Ice Age in the northern hemisphere

The LIA brought bitterly cold winters to many parts of the world but is most thoroughly documented in Europe and North America. In the mid-seventeenth century, glaciers in the Swiss Alps advanced, gradually engulfing farms and crushing entire villages. The River Thames and the canals and rivers of the Netherlands often froze over during the winter, and people skated and even held frost fairs on the ice. The first Thames frost fair was in 1607 and the last in 1814 (although it should be noted that changes to the bridges and the addition of an embankment also affected the river flow and depth, and hence the possibility of freezes). The freeze of the Golden Horn and the southern section of the Bosphorus took place in 1622. In 1658 a Swedish army marched across the Great Belt to Denmark and invaded Copenhagen. The winter of 1794/95 was particularly harsh, allowing the French invasion army under Pichegru to march on the frozen rivers of the Netherlands whilst the Dutch fleet was fixed in the ice in Den Helder harbour. In the winter of 1780, New York Harbor froze, allowing people to walk from Manhattan to Staten Island. Sea ice surrounding Iceland extended for kilometres in every direction, closing that island's harbours to shipping.

The severe winters affected human life in ways large and small. The population of Iceland fell by half, but this was perhaps also due to fluorosis caused by the eruption of the volcano Laki in 1783. The Viking colonies in Greenland died out (in the fifteenth century) because they could no longer grow enough food there. In North America, American Indians formed leagues in response to food shortages.

Lamb (1995) noted that, in many years, 'snowfall was much heavier than recorded before or since, and the snow lay on the ground for many months longer than it does today.' Many springs and summers were extremely cold and wet, although there was great variability between years and groups of years. Crop practices throughout Europe had to be altered to adapt to the shortened, less-reliable growing season, and there were many years of death and famine (such as the Great Famine of 1315–1317, although this may have been before the LIA proper). Wine growing entirely disappeared from some northern regions. Violent storms caused massive flooding and loss of life, and some resulted in permanent losses of large tracts of land from the Danish, German and Dutch coasts.

Impact on agriculture

Lamb (1966) points out that in the warmest times of the last 1000 years, southern England had the climate that northern France has now. For example, the difference between the northernmost vineyard in England in the past and the northernmost present-day vineyard locations in France is about 350 miles: that means the growing season changed by 15–20% between the warmest and coldest times of the last millennium. That is enough to affect almost any type of food production, especially crops highly adapted to use the full-season warm climatic periods. During the coldest times of the LIA, England's growing season was shortened by one to two months compared to present day values. Therefore, climate changes had a much greater impact on agricultural output in the past.

Figures 30.44 and 30.45 show the price of wheat and rye, respectively, in various European countries during the LIA. Each of the peaks in prices corresponds to a particularly poor harvest, mostly due to unfavourable climates, with the most notable peak in the year 1816 – 'the year without a summer.' One of the worst famines in the seventeenth century occurred in France due to the failed harvest of 1693. Millions of people in France and surrounding countries were killed.

The effect of the LIA on Swiss farms was also severe. Due to the cooler climate, snow covered the ground deep into spring. A parasite known as *Fusarium nivale*, which thrives under snow cover, devastated crops. Additionally, due to the increased number of days of snow cover, the stocks of hay for the animals ran out, so the livestock was fed on straw and pine branches. Many cows had to be slaughtered.

In Norway, many farms located at higher latitudes were abandoned for better land in the valleys. By 1387, production and tax yields were between 12 and 70% of what they had been around 1300. In the 1460s it was being recognized that this change was permanent. As late as the year 1665, the total Norwegian grain harvest is reported to have been only 67–70% of what it had been about the year 1300 (Lamb, 1995).

Figure 30.46 shows a chronology of dearth and famine in Scotland during the LIA. Broken lines are years with reported dearth and full lines are years with reported famine.

Impact on wine production

People keep records of their most important crops, and the grapes used for wine-making are no exception. There were many 'bad years' for wine during the LIA in France and surrounding countries due to very late harvests and very wet summers.

The cultivation of grapes was extensive throughout the southern portion of England from about 1100 to 1300. Grapes were also grown in northern France and Germany at that time: areas

which even today do not sustain commercial vineyards. At the time of the compilation of the Doomsday Survey in the late eleventh century, vineyards were recorded in 46 places in southern England, from East Anglia through to modern-day Somerset. By the time King Henry VIII ascended to the throne there were 139 sizeable vineyards in England and Wales, 11 of them owned by the Crown, 67 by noble families and 52 by the Church. In fact, Lamb (1995) suggests that during that period the amount of wine produced in England was substantial enough to provide significant economic competition to the producers in France. With the coming cooler climate in the 1400s, temperatures became too cold for grape production and the vineyards in southern England gradually declined.

German wine production also declined during the cooling experienced after the MWP and during the LIA. Between 1400 and 1700, German wine production was never above 53% of that before 1300, and at times it was as low as 20% (Lamb, 1995).

The extent of glaciers and evidence for cold conditions

The extent of mountain glaciers had been mapped by the late nineteenth century. In both the north and the south temperate zones of our planet, snowlines (the boundaries separating zones of net accumulation from those of net ablation) were about 100 m lower then than they were in 1975, and the LIA marked the recent glacial advances in mountainous areas like the European Alps and Norway (Figure 30.47). In Glacier National Park the last episode of glacier advance came in the late eighteenth and early nineteenth centuries. In Chesapeake Bay, Maryland, large temperature excursions during the LIA (~1400–1900) and the MWP (~800–1300) may have been related to changes in the strength of North Atlantic THC (Cronin *et al.*, 2003).

In Ethiopia and Mauritania, permanent snow was reported on mountain peaks at levels where it does not occur today. Timbuktu, an important city on the trans-Saharan caravan route, was flooded at least 13 times by the Niger River; there are no records of similar

Figure 30.47 LIA moraines, Engadine Valley (Swiss Alps), showing lateral and terminal moraines. (Source: Wikimedia Commons)

flooding before or since. In China, warm-weather crops, such as oranges, were abandoned in Jiangxi Province, where they had been grown for centuries. In North America, the early European settlers also reported exceptionally severe winters. For example, in 1607/08 ice persisted on Lake Superior until June (Lamb, 1995). Fagan (2001) tells of the plight of European peasants during the 1300–1850 chill: famines, hypothermia, bread riots and the rise of despotic leaders brutalizing an increasingly dispirited population. In the late seventeenth century agriculture had dropped off so dramatically that 'Alpine villagers lived on bread made from ground nutshells mixed with barley and oat flour' (Fagan, 2001). Finland lost perhaps a third of its population to starvation and disease.

Depictions of winter in European painting

Burroughs (1981) analyses the depiction of winter in paintings. He claims (quite wrongly) that this occurred almost entirely from 1565 to 1665, and was associated with the climatic decline from 1550 onwards. He hypothesizes that the unusually harsh winter of 1565 inspired great artists to depict highly original images, and that the decline in such paintings was a combination of the 'theme' having been fully explored and mild winters interrupting the flow of painting.

The famous winter paintings of Pieter Bruegel the Elder (for example, see Figure 30.48) all appear to have been painted in 1565. Snow also dominates many villagescapes by Pieter Bruegel the Younger, who lived from 1564 to 1638. Burroughs states that Pieter Bruegel the Younger 'slavishly copied his father's designs. The derivative nature of so much of this work makes it difficult to draw any definite conclusions about the influence of the winters between 1570 and 1600...'

Dutch painting of the theme appears to begin with Hendrick Avercamp after the winter of 1608. There is then an interruption between 1627 and 1640, with a sudden return thereafter, which may suggest a milder interlude in the 1630s. The 1640s to the 1660s cover the major period of Dutch winter painting, which fits with the known proportion of cold winters then. However, the

Figure 30.48 Winter Landscape with a Bird Trap, 1565, Pieter Bruegel the Elder

final decline in winter painting, around 1660, does not coincide with an amelioration of climate and so caution must be used when trying to read climatic events into artistic output as fashion also plays a part. Winter painting recurs around the 1780s and 1810s, which again marks a colder period.

Scottish painting and contemporary records demonstrate that curling and skating were formerly popular outdoor winter sports, but it is now seldom possible to curl outdoors in Scotland due to unreliable conditions. The revival of interest in painting such scenes as Raeburn's *Skating Minister* may owe as much to the Romantic movement, which favoured depictions of dramatic landscapes, as to any meaningful observation on climate, so again care has to be taken when using this type of evidence for inferences related to climate.

The Little Ice Age in the southern hemisphere

An ocean-sediment core from the eastern Bransfield Basin in the Antarctic Peninsula shows centennial events which Khim *et al.* (2002) link to the LIA and MWP. The LIA is also easily distinguished in the Quelccaya ice cap (Peruvian Andes, South America). The Siple Dome (SD) had a climate event with an onset time that is coincident with that of the LIA in the North Atlantic based on a correlation with the GISP2 record. This event is the most dramatic climate seen in the SD Holocene glaciochemical record. The SD ice core also contained its highest rate of melt layers (up to 8%) between 1550 and 1700, most likely due to warm summers during the LIA.

There is limited evidence for conditions in Australia, though lake records in Victoria suggest that conditions at least in the south of the state were wet and/or unusually cool. In the north of the continent the evidence suggests fairly dry conditions, whilst coral cores from the Great Barrier Reef show similar rainfall to today but with less variability.

Tropical Pacific coral records indicate the most frequent and intense El Niño–Southern Oscillation activity occurred in the mid-seventeenth century, during the LIA.

Law Dome ice cores show lower CO_2 mixing ratios during 1550–1800, probably as a result of a colder global climate (see Box 30.17, see the companion website).

Climatic patterns and their causes

In the North Atlantic, sediments accumulated since the end of the last ice age, nearly 12 000 years ago, show regular increases in the amount of coarse sediment grains deposited from icebergs melting in the now open ocean, indicating a series of 1–2 °C cooling events recurring every 1500 years or so. The most recent of these cooling events was the LIA. These same cooling events are detected in sediments accumulating off Africa, but the events there appear to be larger, ranging 3–8 °C.

Scientists have identified two possible causes of the LIA from outside the ocean/atmosphere/land systems: decreased solar activity and increased volcanic activity. Research is ongoing on more ambiguous influences, such as the internal variability of the climate system and anthropogenic influences. For example, it has been speculated that the depopulation of Europe, East Asia and the Middle East during the Black Death, with the resulting decrease in agricultural output and reforestation taking up more carbon from the atmosphere, may have prolonged the LIA. Further speculation suggests that massive depopulation in the Americas following the European contact in the early 1500s had similar effects.

One of the difficulties in identifying the causes of the LIA is the lack of consensus on what constitutes 'normal' climate. While some scientists regard the LIA as an unusual period caused by a combination of global and regional changes, others see glaciation as the norm for the Earth and the MWP (as well as the Holocene interglacial period) as the anomaly requiring explanation (Fagan, 2001).

Solar activity

During 1645–1715, in the middle of the LIA, there was a period of low solar activity known as the Maunder Minimum. The physical link between low sunspot activity and cooling temperatures has not been established, but the coincidence of the Maunder Minimum with the deepest trough of the LIA is suggestive of such a connection. The Spörer Minimum has also been identified, with a significant cooling period near the beginning of the LIA. Other indicators of low solar activity during this period are levels of the isotopes carbon-14 and beryllium-10.

Volcanic activity

Throughout the LIA, the world also experienced heightened volcanic activity. When a volcano erupts, its ash reaches high into the atmosphere and can spread to cover the whole of the Earth. This ash cloud blocks out some of the incoming solar radiation, leading to worldwide cooling that can last up to two years after an eruption. Also emitted by eruptions is sulphur in the form of SO_2 gas. When this gas reaches the stratosphere it turns into sulphuric acid particles, which reflect the Sun's rays, further reducing the amount of radiation reaching the Earth's surface.

The 1815 eruption of Tambora in Indonesia blanketed the atmosphere with ash, and the following year, 1816, came to be known as the Year without a Summer, when frost and snow were reported in June and July in both New England and Northern Europe.

Ocean conveyor shutdown

Another possibility is that there was a shutdown or slowing of the THC, also known as the 'great ocean conveyer' or 'meridional overturning circulation'. The Gulf Stream could have been interrupted by the introduction of a large amount of fresh water to the North Atlantic, possibly caused by a period of warming before the LIA. There is some concern that shutdown of THC could happen again as a result of global warming.

The end of the Little Ice Age

Beginning around 1850, the climate began warming and the LIA ended. Some global warming critics believe that the Earth's climate is still recovering from the LIA and that human activity is not the decisive factor in present temperature trends, but this idea is not widely accepted. Instead, mainstream scientific opinion on climate change is that warming over the last 50 years has been caused primarily by the increased proportion of CO_2 in the atmosphere caused by human activity. There is less agreement over the warming from 1850 to 1950. See the companion website for Box 30.18 Europe's chill linked to disease.

Exercises

1. What is the evidence for Snowball Earth in the Proterozoic?

2. How could life possibly survive Snowball Earth conditions?

3. How have the super warm climates of the Late Cretaceous been explained?

4. What types of evidence have been used to document the change from Eocene greenhouse climates to the Oligocene glaciation?

5. What explanations have there been for the Palaeoecene supergreenhouse climate?

6. How does the isotope record in ocean and ice cores help in reconstructing the climate of the time of deposition of the ice or sediments?

7. Explain how the shutdown or 'slowdown' of the thermohaline circulation is a possible effect of global warming.

8. How can the Milankovitch cycles help explain the origins of ice ages?

9. Why did the Younger Dryas cold phase occur?

10. What types of evidence have been used to explain the European Little Ice Age?

References

Adhikari, D.P. and Kumon, F. 2001. Climate changes during the past 1300 years as deduced from the sediments of Lake Nakatsuna, central Japan. *Limnology*. **2**, 157–168.

Allen, P.A. and Etienne, J.L. 2008. Sedimentary challenge to Snowball Earth. *Nature Geoscience*. **1**, 817–825.

Berner, R.A. and Kothavala, Z. 2001. GEOCAB111: A revised model of atmospheric CO_2 over Phanerozoic time. *American Journal of Science*. **301**, 182–204.

Bowen, G.J. 2007. When the world turned cold. *Nature*. **445**, 607–608.

Broecker, W.S. 2003. Does the trigger for abrupt climate change reside in the ocean or in the atmosphere? *Science*. **300**, 1519–1522.

Broecker, W.S. 2006. Was the Younger Dryas triggered by a flood? *Science*. **312**, 1146–1148.

Burroughs, W.J. 1981. Winter landscapes and climatic change. *Weather*. **36**, 352–357.

Came, R.E., Eiler, J.M., Veizer, J., Azmy, K., Brand, U. and Weidman, C.R. 2007. Coupling of surface temperatures and atmospheric CO_2 concentrations during the Paleozoic era. *Nature*. **449**, 198–201.

Chappellaz, J., Blumer, T., Raynaud, D., Barndla, J.A., Schwander, J. and Stauffer, B. 1993. Synchronous changes in atmospheric CH_4 and climate between 40 and 8 ky b.p.

Clement, A.C., Cane, M.A. and Seager, R, 2001 An prbitally driven tropical source for abrupt climate change. *Journal of Climate* **14**, 2369–2.

Crawford, K.R. 1997. The Late Cenozoic Sedimentary Record of the Antarctic Continental Shelf. Unpublished PhD Thesis, Liverpool John Moores University.

Dansgaard, W. *et al.* 1989 The abrupt termination of the Younger Dryas climate event. *Nature*. **339**, 532–534.

Dupont-Nivet, G., Krijgsman, W., Langereis, C.G., Abels, H.A., Dai, S. and Fang, X. 2007. Tibetan Plateau aridification linked to global cooling at the Eocene–Oligocene transition. *Nature*. **445**, 635–638.

Eyles, N. and Januszczak, N. 2004. 'Zipper-rift': A tectonic model for Neoproterozoic glaciations during the breakup of Rodinia after 750 Ma. Earth Science Reviews. **65**, 1–73.

Fagan, B.M. 2001. The Little Ice Age: How Climate Made History, 1300–1850. Basic Books.

Frakes, L.A., Francis, J.E. and Syktus, J.I. 1992. Climate Modes of the Phanerozoic: The History of the Earth's Climate over the Past 600 Million Years. Cambridge, Cambridge University Press.

Hambrey, M.J., Ehrmann, W.U. and Larsen, B. 1991. Cenozoic glacial record of the Prydz Bay continental shelf, East Antarctica. In: Barron, J., Larsen J.B. *et al.* (Eds). Proceedings of the Offshore Drilling Program Scientific Results 119B, College Station, Texas (Ocean Drilling Program). pp. 77–132.

Hays, J.D., Imbrie, J. and Shackleton, N.J. 1976. Variations in the Earth's orbit: Pacemaker of the ice ages. *Science*. **194**, 1121–1132.

Huybers, P. and Wunsch, C. 2005. Obliquity pacing of the late Pleistocene glacial terminations. *Nature*. **434**, 491–494.

Ivany, L.C., Van Simaeys, S., Domack, E.W. and Samson, S.D. 2006. Evidence for an earliest Oligocene ice sheet on the Antarctic Peninsula. *Geology*. **34**, 377–380.

Keigwin, L.D. 1996. The Little Ice Age and Medieval Warm Period in the Sargasso Sea. *Science*. **264**, 1503–1508.

Khim, B.-K., Kang, C.Y. and Bahk, J.J. 2002. Unstable climate oscillations during the Late Holocene in the Eastern Bransfield Basin, Antarctic Peninsula. *Quaternary Research*. **58**, 234–245.

Kopp, R.E., Kirschvink, J.L., Hilburn, I.A. and Nash, C.Z. 2005. The Paleoproterozoic Snowball Earth: A climate disaster triggered by the evolution of oxygen photosynthesis. *Proceedings of the National Association of Science*. **102**, 11131–11136.

Lamb, H.H. 1966. The Changing Climate, Methuen, London. pp. 236.

Lamb, H.H. 1995. The Little Ice Age. In: Climate, History and the Modern World. London, Routledge. pp. 211–241.

Mayewski, P.A., Meeker, L.D., Whitlow, S., Twickler, M.S., Morrison, M.C., Alley, R.B., Bloomfield, P. and Taylor, K. 1993. The atmosphere during the Younger Dryas. *Science*. **261**, 195–197.

Moran, K. *et al.* 2006. The Cenozoic palaeoenvironment of the Arctic Ocean. *Nature*. **441**, 601–605.

Munro, N.D. 2003. Small game, the Younger Dryas, and the transition to agriculture in the Southern Levant. *Mitteilungen der Gesellschaft für Urgeschichte*. **12**, 47–64.

Pearson, P.N., McMillan, I.K., Wade, B.S., Dunkley Jones, T., Coxall, H.K., Brown, P.R. and Lear, C.H. 2008. Extinction and environmental change across the Eocene–Oligocene boundary in Tanzania. *Geology*. **36**, 179–182.

Rieu, R., Allen, P.A., Plötze, M. and Pettke, T. 2007. Climatic cycles during a Neoproterozoic 'snowball' glacial epoch. *Geology*. **35**, 299–302.

Ruddiman, W.F. and Kutzbach, J.E. 1991. Plateau uplift and climate change. *Scientific American*. **264**, 66–74.

Sohl, L. and Chandler, M. 2006. EdGCM: Climate modeling for research and education: Simulating a Snowball Earth. A brief look at a Snowball Earth: The glaciation of the Neoproterozoic Era (750Ma). http://edgcm.columbia.edu/outreach/showcase/snowball_earth_750ma.html.

Stanley, S.M. 2005. Earth System History. 2nd edition. New York, Freeman.

Veizer, J., Godderis, Y. and François, L.M. 2000. Evidence for decoupling of atmospheric CO_2 and global climate during the Phanerozoic Eon. *Nature*. **408**, 698–701.

Walker, G. 2003. Snowball Earth. London, Bloomsbury Publishing.

Zachos, J., Pagani, M., Sloan, L., Thomas, E. and Billups, K. 2001. Trends, rhythms and aberrations in global climate 65 Ma to present. *Science*. **292**, 686–693.

Zanazzi, A., Kohn, M.J., MacFadden, B.J. and Terry, D.O. 2007. Large temperature drop across the Eocene–Oligocene transition in central North America. *Nature.* **445**, 639–642.

Further reading

Barratt, P.J. 2003. Cooling a continent. *Nature.* **421**, 221–223.

Broecker, W.S. 2000. Was a change in thermohaline circulation responsible for the Little Ice Age? *Proceedings of the National Academy of Sciences.* **97**, 1139–1342.

Crowley, T.J. 2000. Causes of climate change over the past 1000 years. *Science.* **209**, 270–277.

Crowley, T.J. and Kim, K. 1995. Comparison of longterm greenhouse projections with the geologic record. *Geophysical Research Letters.* **22**, 933–936.

Hoffman, P.F. and Schrag, D.P. 2000. Snowball Earth. *Scientific American.* **282**, 68–75.

Milankovitch, M. 1920. Theorie Mathematique des Phenomenes Thermiques produits par la Radiation Solaire. Paris, Gauthier-Villars. Trans. 1998. Canon of Insolation and the Ice Age Problem. Alven Global.

Prothero, D.R, Ivany, L.C. and Nesbitt, E.A. (Eds). 2002. From Greenhouse to Icehouse. New York, Columbia University Press.

Sankaran, A.V. 2003. Neoproterozoic 'Snowball Earth' and the 'cap' controversy. *Current Science.* **84**. http://www.ias.ac.in/currsci/apr102003/871.pdf.

The Snowball Earth Web site is the definitive online resource, although it represents the views of pro-Snowball Earth scientists: http://www.snowballearth.org/index.html.

31 Global Environmental Change in the Future

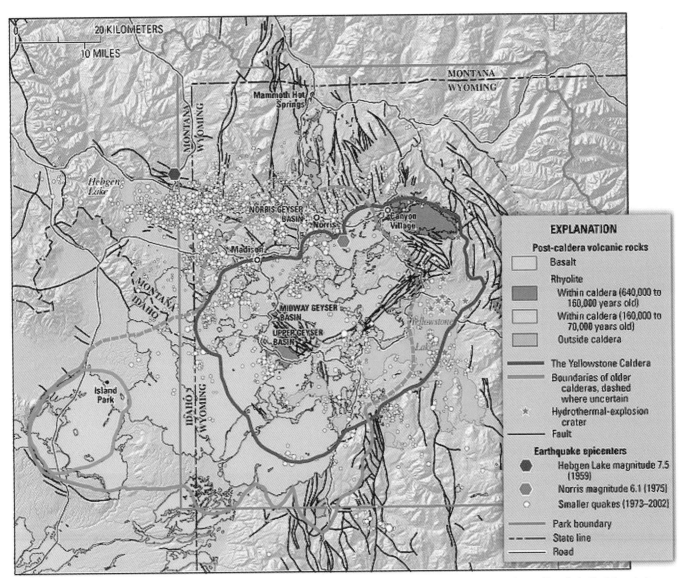

Yellowstone National Park, showing the recent caldera and the boundaries of older calderas. This is a likely supervolcanic eruption site in the future, but when? (Source: Wikipedia Commons from mapping by USGS, Http://pubs.usgs/gs/2005/3024/press-images/fig_03_yellowstone_map.jpg)

Earth Environments   David Huddart and Tim Stott
© 2010 John Wiley & Sons, Ltd

Learning Outcomes

After reading this chapter and completing the exercises you will be able to:

➤ Understand what is meant by dangerous climate change.

➤ Understand the role and main findings of the Intergovernmental Panel for Climate Change (IPCC) and the need for international agreements on carbon emissions.

➤ Describe the potential effects of climate change on the geosphere and the links with volcanic and earthquake hazards.

➤ Understand the likely future changes associated with plate tectonic movements and the future changes associated with various geomorphic, astronomical and geological processes.

➤ Outline the potential effects of climate change on sea level and ocean circulation.

➤ Describe the likely effects of dangerous climate change on plant and animal extinction, food production and human migration.

➤ Outline the likely events over the next two centuries if greenhouse gas emissions remain unchecked.

31.1 Introduction

This concluding chapter aims to draw together the likely future changes to the global Earth environments that have been mentioned in earlier chapters. We have seen that the Earth is an ever-changing, dynamic system with both exogenic and endogenic processes operating to generate change. Here we can only indicate some of the more important changes, both short- and long-term, and some examples of climate change, geomorphic change (including changes to the cryosphere), ecological change and finally geological change.

In climate change, future predictions are made by complex atmospheric and ocean modelling, with expected climate parameter changes added to indicate the change in prediction. The assumptions used are based on trends in global warming, which is caused by the concentration of greenhouse gases (GHGs) in the atmosphere: future climate change is likely to depend on how rapidly we add extra GHGs.

In terms of geomorphic change, processes like the rise in sea level through the melting of ice sheets and glaciers will cause problems in coastal areas that are already experiencing crustal subsidence, such as the North Sea coasts of the UK. River discharges are likely to change as rainfall regimes are altered; snow melt will occur in different amounts and in different periods of the year; and there will be more frequent floods. Fresh water

may become scarce in some parts of the world as supplies decline and demand increases due to a greater population and a higher demand per person. There will also be a general decrease in the area and volume of ice sheets and glaciers, and some increase in fast-moving glaciers may occur where there is abundant subglacial melt water. Permafrost melting in northern Arctic regions will cause implications for the land surface in terms of drainage, vegetation and methane release.

All the above types of change will have ecological impacts. Soils will be affected both directly and indirectly through climate and vegetation change, and warmer temperatures will increase the rate of organic matter decomposition in soils. These warmer temperatures and the higher carbon dioxide levels should create more net primary productivity and this will affect the pattern of biomes and the range of individual species. The changes in precipitation in terms of amounts and regimes will also have an ecological effect, but the predictions related to precipitation are more variable and therefore the ecological consequences are not as certain. In high mountains there will be major changes in the vegetational zonation and some zones and species will become extinct. Currently we see that the Earth is moving into a period of accelerated environmental change, through global warming and a wide range of human-induced stresses (Steffen *et al.*, 2004). Yet the spatial scale of abrupt, geologically-induced change can vary from the continental and regional levels to local, individual landscapes.

The time frame is also important in change. The impacts of a geological or geomorphic event may be immediate, as when a volcano erupts or a major landslide occurs, or they may take years, decades or centuries to reveal themselves. It is easiest for humans to visualize change over a period of up to several decades; any longer and it is difficult for us. It is clear too that a single change can have different results depending on the peoples in a region. For example, Viking settlements failed in south-west Greenland during the cooling of the Little Ice Age, whilst the Inuit nearby carried on as usual (Diamond, 2005)

In this chapter we will deal first with future climate change, drawing upon the numerous models used by the IPCC Fourth Assessment Report (IPCC 4AR, 2007), then look at the likely changes in the lithosphere, hydrosphere and biosphere, before concluding with a tentative time line for the next two centuries. Note that the further into the future we try to gaze, the more uncertain our forecasts become.

31.2 Future climate change

The scientific evidence is now overwhelming: global climate change presents very serious social, environmental and economic risks and demands an urgent global response. This was the message sent by leaders of over 150 global business organizations to the United Nations Climate Change Conference in Bali in December 2007.

The IPCC had just published its Synthesis Report, in conclusion of the IPCC 4AR (2007), on the science of climate change. It issued a warning that, with current climate change mitigation policies and related sustainable development practices, global

GHG emissions will continue to grow, and that, without urgent action, anthropogenic global warming may lead to impacts that are abrupt or irreversible.

The IPCC has sent the clear and unequivocal message that we are not doing enough quickly enough to avoid dangerous climate change: time is of the essence. James Hansen, one of the world's leading authorities on climate change, has warned that the Earth's climate is nearing a point of no return, beyond which it will be impossible to avoid climate change with far-ranging, undesirable consequences.

We must have a global agreement on emissions control that is sufficient to solve the problem faster than we are creating it. Unless we do, sustainable development will be impossible. Concentration and emissions reduction targets must be embodied in an international agreement framed to meet the objective of the United Nations Framework Convention on Climate Change (UNFCCC) if the markets and new technology are to become the mainspring of the new low-carbon economy. In the absence of this agreement, we will continue to struggle under the 'greatest market failure ever seen', as diagnosed in the Stern Review of 2006.

The future of the Earth's climate is an issue which, over the past two to three years, has risen up political agendas around the world. The publication of the Stern Report on the economics of climate change brought this issue firmly into the political arena. Barely a news broadcast goes by without some mention of global warming and its effects or some discussion of how we should mitigate against it.

The most credible future predictions are probably those contained within the IPCC 4AR (2007). IPCC 4AR states that a 50–80% cut in GHGs is required by 2050 if the Earth is to avert dangerous climate change. In effect, this requires stabilization of current GHG production by 2015.

Dangerous climate change is popularly defined as an increase of $2\,^{\circ}C$ above pre-industrial temperatures. The global mean temperature is currently $0.74\,^{\circ}C$ higher than pre-industrial levels, is rising at a rate of $0.2\,^{\circ}C$ per decade and has an estimated $0.6\,^{\circ}C$ still to rise due to GHG concentrations already in the atmosphere. This means that the Earth is already only $0.7\,^{\circ}C$ below that dangerous climate change threshold. Instrument records show that 11 of the hottest years on record have occurred in the last 13 years and the Earth's global mean temperature is within $1\,^{\circ}C$ of the highest it has been for the past 1 million years.

The polar regions are warming much faster than other areas. Greenland temperatures have been on average $3\,^{\circ}C$ higher in the past 20 years, while the Antarctic peninsular is warming at five times the global average ($2.5\,^{\circ}C$ in the last 50 years). Worldwide, global mean temperature has varied by $2–3\,^{\circ}C$ in the last 10 000 years. There is a threat of wholesale melting of the Greenland ice sheet and fringing ice shelves in Antarctica. The likely consequences of this are discussed in Section 31.4.

Another definition for dangerous climate change is when global atmospheric CO_2 levels exceed 450 ppm. In the 1960s, the global atmospheric CO_2 level was 279 ppm; in 1970 it was 325 ppm; and in 2007 it was 387 ppm. In 2005 the atmospheric CO_2 level rose by 2.6 ppm, one of the highest annual increases ever recorded. Currently CO_2 levels are rising four times faster

than in the 1990s. Having stated these facts, it is generally agreed that there is only a 20% chance that holding CO_2 levels below 450 ppm will prevent a $2\,^{\circ}C$ rise in global mean temperature. For CO_2 levels of 550 ppm, the expected warming is $2–4.5\,^{\circ}C$.

In a warmer Earth weather hazards could account for 90% of natural disasters and 60% of deaths. There would be more extreme weather events of all types. Dry parts of the Earth would become drier while wet places would get wetter. The monsoon may become more extreme, with prolonged severe drought in Africa, Australia, central USA and Southern Europe commonplace and desertification spreading in North Africa and into Southern Europe.

In the UK the annual number of major floods has increased from ~100 in the early 1990s to ~250 in 2007. Intense winter rainfall has become increasingly common since 1900, with an accelerating trend. Annual flood costs may reach £22 billion by 2050, exacerbated by the 350 000 properties built on flood plains in the last 20 years.

Around 40% of the recent rise in Atlantic hurricane activity has been linked to increased sea-surface temperatures. The Atlantic hurricane season is longer than in the past and the area affected is greater. Worldwide, twice as many intense tropical cyclones are now monitored than 30 years ago. The UK is twice as stormy as 50 years ago and more intense European storms are predicted for a warmer Earth.

In 2005 some 97% of Portugal was under severe drought conditions and parts of Spain required emergency boreholes for water. The 427 million people who live around the Mediterranean will increase by a further 100 million by 2030. Some 300 000 km^2 of Southern Europe (home to 16 million people) will be threatened by deserts, an area larger than the UK. Similar problems will be faced in the US, Australia, China and elsewhere.

However, some scientists argue that the IPCC and the Stern Report have not taken all factors into full consideration, such as: tipping points or thresholds, including the physical collapse of the Greenland ice sheet and rapid melting in Antarctica (e.g. the Larsen B ice shelf, an area the size of Luxembourg, collapsed in 2002); the release of CO_2 from the soil, melting permafrost and sea-floor sediments; and other potential feedback effects, including the geosphere response.

An average temperature rise of $7\,^{\circ}C$ has been predicted if all the world's resources of conventional hydrocarbons are burnt. If unconventional sources (tar sands, clathrates) are included, the rise could be as great as $13\,^{\circ}C$ – hotter than the Earth has been for 50 million years. Augmented by positive feedback effects, the oceans, soils and plants could become carbon sources rather than sinks (as they currently are). If CO_2 levels were maintained above 350 ppm for a few centuries, this would be sufficient to melt all the polar ice.

31.3 Change in the geosphere

Rapid climate change involves the geosphere as well as the atmosphere and hydrosphere. Potential mechanisms include: ocean-loading seismicity; submarine landslides; tsunamis; gas hydrate destabilization; ocean-loading volcanism; ice-unloading volcanism; and earthquake activity.

In the Grimsvotn area of Iceland, volcanic eruptions are modulated by crater-lake draining. Changes in the mass balance and melt rate of the Icelandic glaciers could trigger further volcanic eruptions. Melting of the Vatnajokull ice cap in Iceland is predicted to increase mantle melt production, leading to increased eruptive activity and more earthquakes. In south-west Alaska the level of seismicity in recent decades has been modulated by glacial mass fluctuations. This suggests that in some areas, melting of glaciers could increase the earthquake risk. Likewise, ice-sheet unloading in Greenland and western Antarctica is predicted to increase seismicity.

Studies have compared the ages of Mediterranean explosive eruptions with the rate of sea-level rise and identified three periods of good correlation. This correlation has also been identified in Greenland ice cores, associated with a 200% increase in volcanic activity. Pavlof is the most active volcano in Alaska, having erupted 40 times since the late 1700s. Like many Alaskan volcanoes, it is known to geologists as a 'wet' volcano because its summit is covered with snow and ice. Further melting could lead to violent, steam-rich, hydromagmatic explosions, with rivers of hot water laden with broken rock and ash (mudflows' or lahars), which would clog river channels and cause flooding on the flanks of the volcano. Tidal stresses have been proposed to modulate eruptions at some volcanoes, and changes in sea level may have the same effect.

Of course, plate tectonic processes will continue into the foreseeable geological future. Figure 31.1 illustrates the current configuration of the plates and continents and makes predictions, based on known rates of spreading, for 50 million and 100 million years into the future. If the present-day plate motions continue, the Atlantic will widen; the African plate will collide with the Eurasian plate, closing the Mediterranean Sea and leading to increased seismicity and volcanicity; there will be the continued compression of India and other plates northward into Asia; and Australia will collide with South East Asia. In North America, California will slide northwards up the coast towards Alaska. However, we know from the geological record that ocean basins only last approximately 200 million years so plate configurations at this time scale will be very different.

On a much shorter time scale, Dutch (2006) provides a stimulating article titled 'The Earth Has a Future', in which he suggests that events which are rare or unknown in recorded history will become almost inevitable and frequent in the near geological future. He reviews changes over four time scales: 1000, 10 000, 100 000 and 1 million years and gives examples of tectonic and volcanic processes, geomorphic processes, astronomical processes and human processes. We will look initially at the San Andreas and associated fault systems in California.

Palaeoseismic studies have shown that there is an average recurrence interval of 150–200 years for M = 8 earthquakes on the most active segments of the San Andreas Fault (Grant and Lettis, 2002). Therefore, in the next 1000 years it is likely that there will be between five and seven such events, with a total fault motion of around 25 m. By 10 000 years into the future there will

be slippage of 250 m and between 50 and 70 M = 8 earthquakes. In 100 000 years, at the current slip rates, there will have been 2.5 km movement on the San Andreas Fault and around 700 such earthquakes. That amount of movement will be enough to bring the coastal hills west of the fault nearly to the Golden Gate, but not enough to close it. In 1 million years' time the fault will have moved 25 km and experienced around 7000 M = 8 earthquakes. The Golden Gate will be blocked by the hills of the present San Mateo County, although erosion will probably maintain a valley northward along the fault trace as slip takes place.

It is obvious that seismic risk is large along the length of this fault and other faults in California. At http://geosphere. gsapubs/org/content/2/3/113.figures-only and http://dx.doi.org/ 10.1130/GES00012.S2 there is an animation illustrating the motion of the fault near San Francisco over this time period, but this region can be seen to have moved a considerable distance after 1 million years.

The Holocene volcanic record in Simkin and Siebert (1994) shows around 80 eruptions for Vesuvius in the past 2000 years, suggesting that we can expect around 40 in the next 1000, and most likely an eruption in the Campi Flegrei volcanic field west of Naples too. Over the next 10 000 years there will probably be several hundred eruptions of Vesuvius, enough to fill the Monte Somma caldera completely. Mount Fuji in Japan will probably erupt 5–10 times and in the Cascades (USA) there will be around 12 eruptions. After 10 000 years there will probably have been 100 eruptions of Fuji and a similar number in the Cascades. New eruptions centres will have appeared in many major volcanic chains as well as in places with Holocene volcanism but little or no historic activity, such as Australia, Syria, the Arabian peninsula, Turkey, Iran and Manchuria. After 1 million years several of the present day Cascade volcanoes will have become extinct and others will have collapsed due to caldera collapse or major debris avalanches. However, new eruptive centres will have formed mountains similar in size to the current peaks in the Cascades, whilst given the Quaternary record of Long Valley, Yellowstone and Toba amongst others, there will be several eruptions causing the collapse of major magma chambers (see frontispiece for Yellowstone calderas, past and present). Lake Taupo in the North Island of New Zealand has had many rhyolitic, destructive volcanic episodes and the volcano is now considered dormant, rather than extinct. It was created about 26 500 years ago and has erupted 28 times in the last 27 000 years.

In terms of geomorphic processes, studies of erosion, crustal uplift and exhumation generally find denudation rates up to an average of 39 m per million years in the Sierra Nevada (USA), or up to 2100–2900 m per million years in the Himalayas (Riebe et al., 2001; Galy and France, 2001). Valley incision rates can be an order of magnitude greater, commonly limited by the rate of uplift; for example, 200 m per million years in the southern Sierra Nevada (Stock et al., 2004) and 4000–15 000 m per million years in Nepal (Lavé and Avouac, 2001). Mountains could not exist without uplift exceeding erosion and uplift rates of kilometres per million years in active orogenic belts are not uncommon, for example 16 000–34 000 m per million years in the Alps (Rubatto and Hermann, 2001). In some areas therefore there will be

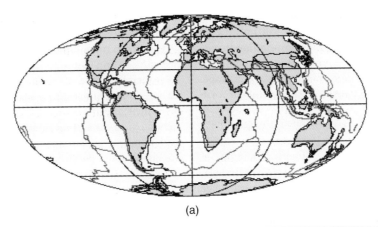

(a)

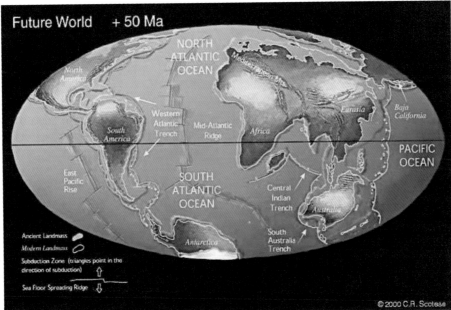

(b)

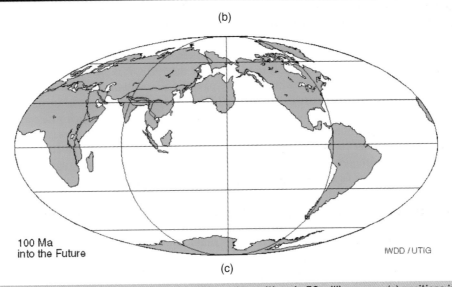

(c)

Figure 31.1 Predicted plate tectonic movements: (a) current positions; (b) positions in 50 million years; (c) positions in 100 million years. The 100-million-year prediction is based on Dalziel (1999) and plate rotations are based on the PLATES Project plate model and the Nuvel-1a poles (DeMets *et al.*, 1994). (Sources: http://www.ig.utexas./research/projects/plates/index.htm and the Paleomap project: http://www.scotese.com). Reprinted by kind permissions of Christopher Scotese from the PALEO Map project www.scotese.com

1 km of uplift in 100 000 years, and if erosion takes place at rates comparable to uplift, the overall effect will only be to expose deeper rocks without much changing the overall landscape. However, in areas where one of the processes is dominant we will expect to see significant uplift or lowering of the landscape. This amount of time is also sufficient for laterite soils to form (Gunnell, 2003).

In the Chinese karst regions Zhang *et al.* (2000) and Liu (2000) found corrosion rates of 1 m or more per 1000 years, so in 100 000 years these areas would undergo 100+ m of corrosion. Some of the famous Chinese tower karst, such as that in the Guilin area, will disappear in that time. After 1 million years the most active mountain belts could see 10 km of uplift and the erosion will be sufficient to remove all presently-exposed rocks. By then the Earth will have had 10 major glaciations and each will have been followed by catastrophic outburst floods (as in the Altay mountains in Siberia, described by Rudoy and Baker (1993)) and major drainage reorganization (as in the Green River, Kentucky, as suggested by Granger *et al.* (2001)).

The Mississippi river is likely to change its course soon and engineers already allow it to do so in a controlled fashion by diverting flood waters down the Atchafalaya spillway. In 1000 years the river will have changed its course, and given the post-glacial delta growth it seems likely that the Atchafalaya delta will be outgrown and the river will have a new delta. The corollary to this is that new floodplains and deltas will result in sediment starvation and coastal subsidence in former delta areas. There is an animation illustrating the possible future delta evolution at http://geosphere.gsapubs/org/content/2/3/113.figures-only. and http://dx.doi.org/10.1130/GES00012.S2. It assumes ice advance and sea-level lowering will begin 20 0000–30 000 years in the future, reach a maximum after 80 000 years and return to interglacial conditions after around 100 000 years. After 10 000 years the delta will probably have shifted location between 10 and 20 times.

There will be astronomical changes too and there is a significant chance of a meteor impact, with a rating of 8 or above on the Torino scale. This would cause significant damage and excavate a crater 100 m or more in diameter. After 10 000 years there will have been several impacts of 8 or more on this scale, with a good chance of at least one creating a crater 1 km or more in diameter. There is a significant possibility of a level 9 event, capable of causing regional destruction on a scale of hundreds of kilometres, caused by blast, ejecta and tsunami. After 100 000 years there will probably have been tens of impacts rating 8 or above and several rating 9, and there is a significant chance of a level 10 event, which would have global effects. After 1 million years there will have been hundreds of meteor impacts above 8, tens above 9 and a high probability of one or more 10 event with major global effects.

It is likely too that asteroids and comets will have a role to play in the longer term. In fact, the 100 m objects of the kind that made the 20 megaton Tunguska impact over Siberia in 1908 are likely every few decades. Bigger objects, on a scale of 1 km – like the 1997 XF11, which came close enough to Earth to cause media concern – would be on a scale of 1000 megatons and would cause widespread devastation over a continental scale. However,

artificial geoengineering will perhaps shield the Earth from such astronomical events.

It is much more difficult to predict human processes because it is almost impossible to imagine what society will be like in 1000 years. The present anthropogenic increase in atmospheric carbon dioxide should have waned because of the exhaustion of fossil fuels and the switch to alternative energy, or else due to technological collapse as a result of climate change and resource depletion. But it is difficult to predict how long changes from human climate modification will last or what they will be.

We cannot even hazard a guess at what human society will be like in 100 000 years' time. This is a significant fraction of the time since *Homo sapiens sapiens* first appeared. As species have lifetimes of several million years or more (Smith and Peterson, 2002) it might appear likely that we will still be around, but if there is a significant collapse of society into geographically-isolated gene pools, it is possible that these might undergo speciation. Even if we do not cause a mass extinction, it is likely that many species will have become naturally extinct and new ones will have evolved.

31.4 Change in the oceans and hydrosphere

31.4.1 Sea-level change

Melting of the Greenland ice sheet has been estimated to have been 96 km^3 in 1996 and 240 km^3 in 2005. Some glaciers have been moving up to 15 km per year, with some surging up to 5 km in 90 minutes. Some estimates suggest the break-up of the Greenland ice sheet could happen in less than 300 years. IPCC AR4 suggests an 18–59 cm sea-level rise by 2100, but there is a growing consensus that it will be far more rapid than this: if the Greenland ice sheet totally melted it would account for a rise of 7 m, and if the western Antarctic ice sheet totally melted it would account for a rise of 12 m.

Perennial Arctic sea ice has been vanishing at ~9% per decade since the 1970s. It is estimated to have thinned by 40% over the past 35 years, with the north pole ice-free in the summer of 2000. There was record summer melting in 2007, with a 40% reduction, and in August 2008 the melt rate was 78 000 km^2 per day. The north-east and north-west passages were open in 2007 and some suggest that summer sea ice may be gone by 2013. This will lead to reduced summer albedo, with consequences for oil and gas exploitation and trade.

The IPCC 4AR suggests that there is a 50% probability that major ice sheet loss cannot be avoided, leading to a 1–2 m rise this century and several more in the next. This clearly has major implications for all coastal towns and cities, and for low-lying coastal nuclear installations. A 1 m rise threatens ~1 billion people and one third of the world's agricultural land. A 1 mm rise could lead to an average of 1.5 m coastline retreat.

During the penultimate interglacial (130 000 BP) global mean temperature was ~1 °C higher than now and sea level was ~4 ± 2 m higher. During the Pliocene (~5–2 million years ago) global mean temperature was ~3 °C higher than now and sea level was 25 ± 10 m higher. This suggests that if global mean temperature is already >1 °C above pre-industrial levels, sea level must have a considerable way still to rise, and that perhaps a lag-time is involved here. Long-term maintenance of atmospheric CO_2 above 350 ppm over one or more centuries could result in a 70+ m rise in sea level.

31.4.2 Change in ocean circulation systems

We saw in Chapter 28 how melting ice sheets can cause the ocean's circulation to weaken in response to global warming. For example, the Gulf Stream is part of the Atlantic thermohaline circulation, which in turn is part of a global ocean circulation system that helps keep the northern North Atlantic and North West Europe about 9 °C warmer than comparable latitudes. IPCC AR4 states that it is very likely that the Atlantic Conveyor will slow (up to 50%) by 2100. Although abrupt shutdown is thought very unlikely over such a period, uncertainties still remain. Some predictions indicate that the north of Norway could experience cooling of up to 12 °C, while the UK and Scandinavia could be 1–3 °C cooler.

During the 1990s the first generation of coupled climate models predicted that the ocean's overturning circulation would weaken markedly over the next 100–200 years in response to global warming and a stronger hydrological cycle, which make the ocean surface waters less dense and less able to sink in relation to the water below. These models therefore suggested that circulation would be less vigorous in a warmer climate, but there is no firm evidence for weakening of the overturning and more stratification of the polar oceans. However, the climate models used in the last round of assessments, IPCC AR4, predict that the westerlies will shift polewards and become stronger in the twenty-first century. They still suggest that the overturning of the Atlantic Ocean will weaken, though not nearly as much as predicted by the first generation of climate models in the 1990s.

31.5 Change in the biosphere

The predicted climate change is likely to bring about one of the greatest mass extinctions for 65 million years. Extinction rates could be 1000 times greater than normal. By 2050 a quarter of all land plants and animals could be extinct. Some 15 million acres of prime forest are lost every year – along with the life therein. A 2 °C rise will see the demise of polar bears, the loss of 97% of coral reefs and the loss of 50% of the Mediterranean wetlands.

Malaria kills 1.5 million a year, with up to 340 million more at risk due to climate change. Suitable breeding conditions are predicted to arrive in Europe (including parts of the UK) by 2050, with the likely northward spread of malaria, dengue fever, yellow fever, West Nile virus and others. The natural spread of these disease vectors will be accelerated by air traffic. In 1999, mosquitoes carrying West Nile virus arrived in New York, and since then 21 000 people have been infected and 800 have died.

In terms of human migration, there are predicted to be up to 250 million environmental refugees by 2050, with up to 4 billion in water-stressed areas. The United Nations predicts a rapid increase in 'water wars'. The loss of Himalayan glaciers will result in insufficient water for South Asia and China.

Food prices will be affected. Between 1975 and 2005 food staple prices fell by 75% but between 2005 and 2008 they have risen by 75%.

31.6 A timeline for future Earth

While putting specific dates on potential future Earth events is challenging, the following structure outlines the Earth's future based on several recent studies and the IPCC 4AR:

- **2009:** Currently more of the world's population lives in cities than in rural areas, which has changed land-use patterns. The world population is 6.7 billion. Global oil production peaks sometime between 2006 and 2018, according to a model by one Swedish scientist; others say this turning point, known as 'Hubbert's Peak', won't be reached until 2020. However, once reached, global oil production will begin an irreversible decline, which could well trigger a global recession, food shortages, and wars and conflicts over the remaining oil supplies.

- **2020:** The IPCC predicts an increase in the likelihood of flash floods across all parts of Europe, while less rainfall in many parts of the world could reduce agricultural yields by up to 50%. World population will have reached 7.6 billion.

- **2030:** Diseases related to diarrhoea are likely to have increased by 5% in low-income parts of the world. World population will have reached 8.3 billion. Up to 18% of the world's coral reefs (perhaps up to 30% in Asian waters) will likely have been lost as a result of climate change and other environmental stresses. Warming temperatures will cause temperate glaciers on equatorial mountains in Africa to disappear (Figure 31.2).

  The urban population in developing countries will have more than doubled to around 4 billion people, resulting in overcrowding in many cities. The urban population of developed countries could also have increased by 20%.

- **2040:** In summer the Arctic Sea could be ice-free and in winter ice depth will have shrunk drastically. Other predictions suggest that the Arctic will continue to have summer ice up until 2060, or even 2105.

- **2050:** Small Alpine glaciers will very likely have disappeared completely, while large ones will have shrunk by 30–70%. Austrian scientist Roland Psenner at the University of Innsbruck says that this is a conservative estimate, and that small Alpine glaciers could have melted completely as early as 2037.

  World population will have reached 9.4 billion.

  There are likely to be an extra 3200–5200 heat-related deaths each year in Australia, with a greater proportion of these deaths being among people over 65. An extra 500–1000

(a)

(b)

Figure 31.2 Disappearing glaciers on Kilimanjaro's summit crater, August 2007

people in New York City will die of heat-related deaths each year. In the United Kingdom the opposite will occur, with cold-related deaths outpacing heat-related ones.

Crop yields could increase by up to 20% in East and South East Asia, and decrease by up to 30% in Central and South Asia. Similar shifts in crop yields can be expected on other continents. As biodiversity hotspots are more threatened, a quarter of the world's plant and vertebrate animal species could face extinction.

- **2070:** Electricity production by the world's existing hydropower stations is likely to decrease as glaciers disappear and areas affected by drought spread. In Europe, hydropower potential is expected to decline by 6% on average, but around the Mediterranean region it could be as much as 50%. Warmer, drier conditions will result in more frequent and longer droughts, increased risk of forest fires and a longer fire-season.

- **2080:** Scientists predict that up to 20% of the world's population currently lives in low-lying areas or near rivers where

there will be an increased risk of flood hazards. Up to 100 million people could experience coastal flooding each year, with densely-populated and low-lying areas that are least able to adapt to rising sea levels suffering most, particularly in the wake of tropical storms. Sea levels could rise around New York City by more than 1 m, potentially flooding much of southern Brooklyn and Queens and lower Manhattan.

Between 1.1 and 3.2 billion people could be experiencing water shortages, with up to 600 million suffering food shortages. The risk of dengue fever from climate change is estimated to increase enough to affect 3.5 billon people.

- **2100:** Many ecosystems will be pushed to the limit by a combination of global warming and other factors, forcing them to exceed their natural ability to adapt to climate change. Levels of carbon dioxide in the atmosphere will be higher than at any time in the past 650 000 years. Ocean pH levels will likely decrease by as much as 0.5, reaching the lowest they have been for 20 million years, compromising the ability of marine organisms such as corals, crabs and oysters to form shells or exoskeletons.

  Thawing permafrost and other factors will make the Earth's land surface a net source of carbon emissions, meaning it will emit more carbon dioxide into the atmosphere than it will absorb from it. Around 20–30% of species could be extinct by 2100 if global mean temperatures rise 2–3 °C in excess of pre-industrial levels. New climate zones will appear on up to 40% of the Earth's land surface, radically transforming the planet. Increased droughts could significantly reduce moisture levels in the American south-west, northern Mexico and parts of Southern Europe, Africa and the Middle East, more or less recreating the 'dust bowl' environments of the 1930s in the Midwest of the United States.

- **2200:** Rising ocean temperatures will cause oceans to expand. Most of this expansion will take place in the North Atlantic, near the north pole. Since the poles are closer to the Earth's axis of rotation, the greater mass around them will increase the planet's speed of rotation enough to make an Earth day 0.12 milliseconds shorter.

31.7 Causes for future optimism?

31.7.1 Approaching the point of no return

The famous 1990 Kyoto agreement may slow temperature rise by 0.02–0.28 °C by 2050. The UK production of CO_2 has increased by 19% compared to 1990 levels, with aviation and shipping emissions increasing faster than ever. US emissions nearly doubled in the last 14 years and the increasing demand for power could increase global emissions by 52% by 2030, or 30% if renewables play a key role in our energy production. In short, at least a 90% cut in GHG emissions is needed if we are to have any chance of averting dangerous climate change.

Little less than an all-inclusive international agreement is needed to make massive GHG reductions. Massive investment in renewable energy systems is required. For example, micro-generation could provide 30–40% of the UK's total electricity by 2050. Energy efficiency must improve; technology transfer needs to be facilitated. A huge worldwide reforestation drive must be agreed and major lifestyle changes must be made, particularly in high-energy-user societies like the USA and Europe. Carbon capture and storage techniques and systems for scrubbing atmospheric CO_2 need rapid development and widespread implementation.

The good news is that climate is now at the top of the world agenda. 10 000 participants, including representatives from over 180 countries, attended the December 2007 Climate Change Conference in Bali. Parties to the United Nations Framework Convention on Climate Change (UNFCCC) decided to launch formal negotiations on a strengthened international deal on climate change. The conference culminated in the adoption of the Bali Road Map, which consists of a number of forward-looking decisions that represent the various tracks that are essential to reaching a secure climate future. UNFCCC Executive Secretary Yvo de Boer stated that:

> This is a real breakthrough, a real opportunity for the international community to successfully fight climate change. Parties have recognized the urgency of action on climate change and have now provided the political response to what scientists have been telling us is needed.

The Bali Road Map includes the Bali Action Plan, which charts the course for a new negotiating process designed to tackle climate change, and aims to complete this by 2009. These negotiations are set to be concluded by the end of 2009 at the Climate Change Conference in Copenhagen.

At the Bali conference the US recognized the need for 'deep emission cuts'. US cities with >80 million people aim to achieve the United States of America Kyoto obligation. The European Union is seeking a 20–30% cut in GHGs by 2020, while China is aiming for 15% renewable energy production by 2020 and is closing 1000 of its most inefficient power plants.

31.7.2 Renewable technologies

Investment in renewable technologies globally is now more than US$100 billion (18% of all new investment in the energy sector), and by 2020 this could be as much as US$2 trillion. In 2005 the US environment sector had US$340 billion in sales and 5.3 million jobs. By 2030, 40 million Americans are likely to be working in the renewables and energy-efficiency sectors. However, on a more sombre note, only 0.8% of US energy currently comes from renewable sources, with close to half coming from coal.

In the UK the government is planning major road and airport expansion and currently just 2% of the UK's energy needs come from renewables. China is opening a new coal-fired power station every 4 days (550 by 2030). Carbon capture and storage technologies are proven but are still seen as 'too expensive' for widespread adoption.

The Confederation of British Industry members have agreed to cut GHG emissions by 1 million tonnes over the next three years, but with emissions totalling around 370 million tonnes this doesn't look to be a very significant cut. The number of motor vehicles is set to rise to 1 billion within 20 years. Aircraft emissions are rising 3–4% year on year, and China plans to increase its aircraft fleet from 800 to 3000 in 15 years.

Sweden has committed to becoming oil-free by 2020, while New Zealand and other small states have committed to being carbon-free by 2050. Germany has undertaken a 40% emissions cut by 2020. In the UK the Liberal Democrats promise a zero-carbon Britain by 2050, with hydrocarbon vehicles phased out by 2040, but at present they are not in power.

Clean power is available from the Sun, wind and waves. Ground vehicles can be powered by hydrogen, third-generation biofuels, renewable electricity or even air. Aircraft can be fuelled by algal-based biofuels.

31.7.3 Peak oil and coal production

British Petroleum has reported that oil reserves will support current consumption levels for a further 40 years. An independent Energy Watch report, however, claims that peak oil was reached in 2006 and that oil production will be just 39 million barrels per day by 2030, when demand would be expected to be 113 million barrels per day. A London-based Oil Depletion Analysis Centre predicts peak oil will be in 2011, but currently oil production is falling in two-thirds of oil-producing states. Depending on which predictions about oil supplies are correct, one report suggests that peak coal production could be as early as 2025.

Prof. David Rutledge of the California Institute of Technology has led a vigorous debate about fossil-fuel production and whether it will be sufficient in the future. At the same time, great efforts are being made to predict the contribution to future climate change that will result from consuming this fuel. He reports that trends for future fossil-fuel production are less than any of the 40 UN scenarios considered in climate-change assessments. The implication is that producer limitations could provide useful constraints in climate modelling.

Time constants for fossil-fuel exhaustion are about an order of magnitude smaller than the time constant for temperature change, which means that to minimize the effects of climate change associated with future fossil-fuel use, reducing ultimate production is more important than slowing it down. Peak hydrocarbons may limit atmospheric carbon dioxide levels to 470 ppm but as we have seen, this firmly puts the planet in the 'dangerous climate change' scenario.

31.7.4 Contraction and convergence: The last hope?

Supported by China, Germany, the European Parliament, Stern and many others, this concept is based on the idea that everyone on planet Earth has the right to emit the same quantity of GHG.

At present a US citizen emits 20 tonnes of CO_2 every year, a UK citizen emits 11 tonnes, while a Nigerian only emits 0.09 tonnes.

Contraction and Convergence (C&C) is the Global Commons Institute's proposed UNFCCC-compliant climate mitigation strategy for an equitable solution to cutting carbon emissions through global collective action. The ultimate objective of the UN climate treaty is to move to a safe and stable GHG concentration in the atmosphere and C&C starts with this.

C&C recognizes that subject to this limit, we all have an equal entitlement to emit GHGs to the global atmosphere, since continuing unequal use will make it impossible to get the global agreement needed for success. The Kyoto protocol cannot be the basis of this success because it is not science-based and, due to divergent national interests, it does not include all countries.

Scientists have advised on the safe concentration of CO_2 in the atmosphere and on the global cap on emissions necessary to achieve it. A level of 450 ppm has until recently been regarded as the upper limit for keeping under the maximum global temperature increase of 2 °C above the pre-industrial average.

From the inception of a global agreement, C&C schedules the mandatory annual global contraction (reduction of emissions) that will keep CO_2 concentrations from rising beyond the agreed safe level. This rate of contraction must be periodically adjusted to take account of the increasing release of GHGs caused by climate warming.

C&C proposes emission entitlements to every country. While starting with current emissions, it proposes a scheduled convergence to equal per-person entitlements for everyone on the planet by an agreed date (Figure 31.3). That way, convergence will reduce the carbon shares of the developed over-emitting countries sharply until they converge with the (temporarily rising) shares of developing under-emitting countries. The latter will be able to sell their surplus carbon shares to wealthier nations. Emissions trading will be subject to rapid investment in renewable energy.

The 14th session of the Conference of the Parties to the Climate Change Convention (COP 14) will be held in conjunction with the 4th Conference of the Parties Serving as the Meeting of the Parties to the Kyoto Protocol (CMP 4) in Poznań, Poland, from 1 to 12 December 2008.

In 2012 the Kyoto Protocol expires. To keep the process going there is an urgent need for a new climate protocol. In 2012 the Kyoto Protocol to prevent climate changes and global warming runs out. It is to be hoped that discussions at the Climate Conference in Copenhagen in 2009 and subsequent agreements lead to a Copenhagen Protocol to prevent global warming and climate change.

31.8 Concluding remarks

The 2001 Amsterdam Declaration on Global Change stated that the Earth behaves as a single, self-regulating system comprising physical, chemical, biological and human components. Earth system dynamics are characterized by critical thresholds and abrupt changes. Global change cannot be understood in terms of a simple cause–effect paradigm. For example, many local changes cannot easily be related to simple and identifiable causes.

There is a major problem in separating human-induced from natural environmental change, which is challenging. The two sets of change-inducers are interlinked. But as Berger (2006) concluded: Earth scientists have an important role to play by decoding the past and establishing the record of environmental change throughout human history. Recognizing more clearly the role of non-human inputs to abrupt environmental change could make a difference to the way in which humans think about the Earth and the kinds of policies that might be adopted in the search for sustainability.

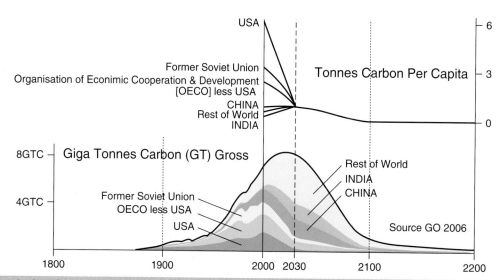

Figure 31.3 Regional rates of contraction and convergence. The same contraction budget converges on shares equal to population by 2030. (Source: Global Commons Institute: http://www.gci.org.uk/kite/Carbon_Countdown.pdf)

One final question should be posed here: Have humans created a new geological epoch? The discussion on the *New Scientist* Web page stimulated by Brahic (2008) suggests that this is possible, although of course this epoch is not official. According to the Stratigraphy Commission of the Geological Society of London a new geological epoch needs:

1. Change in the atmosphere's composition, modifying plants.

2. Change in the distribution and diversity of species, changing the future fossil record.

3. Acidification of the oceans, modifying mineral deposits on the ocean floor.

This suggests that the takeover of the Earth by humans has started a new geological age: the Anthropocene, the age dominated by human activities. But to get formal ratification of this epoch scientists would have to show that 10, 100 or 500 million years in the future there would be a clear identification of a distinct marker layer that resulted from our reign on Earth. This may be the case: for example, ocean acidification could bring an end to corals, changing the nature of marine rocks in some parts of the world, while human activities like agricultural soil erosion and tree felling are creating large amounts of erosion and sediments, generating a new sediment layer on the ocean floor. Animal species are becoming extinct at a great rate too, and the result is that the fossil record of our time will be distinctive from the pre-Anthropocene record. Only time will tell, but it is likely that such a stratigraphic layer will contain heavy metals, toxic waste, organic matter, glass, plastics and polystyrene.

It is hoped that this book has provided the reader with a knowledge of the complex history of the Earth and its life and that it has made a contribution towards a greater understanding of Earth sciences for society. Written during the International Year of Planet Earth, it may help towards the main aim of that year, which was to demonstrate the great potential of the Earth sciences to lay the foundations of a safer, healthier and wealthier society. It also clearly demonstrates that we are unlikely to survive far into the geological future without experiencing significant natural changes brought about by meteor impact, supervolcanic events, continuing major earthquakes in some parts of the world, major landslides and tsunami. Humans can be seen as a very small part of the Earth system and we are likely to remain relatively insignificant in the longer term.

Exercises

1. What do we mean by 'dangerous climate change'?

2. What are the likely effects of plate tectonic processes over the next 100 million years?

3. What is likely to happen in California over the next 100 million years due to movement on the San Andreas Fault?

4. Over the next 10 000 years, what changes are likely to occur in volcanic activity throughout the world?

5. In the next 1 million years, what is likely to happen to the Mississippi delta in terms of geomorphic change?

6. What are the likely effects on sea level of global warming in the next century?

7. Is there any cause for optimism that humans will counteract global warming in the next 50 years?

8. What do we mean by 'contraction and convergence' with regard to greenhouse gases?

References

Berger, A. 2006. Abrupt geological changes: Causes, effects and public issues. Quaternary International. **151**, 3–9.

Brahic, C. 2008. http://newscientist.com/blog/environment/2008_01_01_archive.html

Dalziel, I.W.D. 1999. Vestiges of a beginning and the prospects of an end. In: Craig, G.Y. and Hull, J.H. (Eds). James Hutton: Present and Future. Geological Society of London Special Publication 150. pp. 119–155.

DeMets, C., Gordon, R.G., Argus, D.F. and Stein, S. 1994. Effect of recent revisions to the geomagnetic reversal time scale on estimates of current plate motions. *Geophysical Research Letters*. **21**, 2191–2194.

Diamond, J. 2005. Collapse: How Societies Choose to Fail or Succeed. New York, Viking.

Dutch, S.I. 2006. The Earth has a future. *Geosphere*. **2**(3), 113–124.

Galy, A. and France, L.C. 2001. Higher erosion rates in the Himalaya: Geochemical constraints on riverine fluxes. *Geology*. **20**, 23–26.

Granger, D.E., Fasbel, D. and Palmer, A.N. 2001. Pliocene–Pleistocene incision of the Green River, Kentucky, determined from radioactive decay of cosmogenic Al^{26} and Be^{10} in Mammoth Cave sediments. *Bulletin of the Geological Society of America*. **113**, 825–836.

Grant, D. and Lettis, W.R. 2002. Paleoseismology of the San Andreas Fault system. *Bulletin of the Seismological Society of America*. **92**, 2551–2877.

Gunnell, Y. 2003. Radiometric ages of laterites and constraints on long term denudation rates in West Africa. *Geology*. **31**, 131–134.

IPCC. 2007. Climate Change 2007: The Physical Science Basis, Contribution of Working Group to the Fourth Assessment Report of the Intergovernmental Panel on Climate Change. *Fourth Assessment Report of the Intergovernmental Panel on Climate Change.* Cambridge University Press. Available at http://www.ipcc.ch/[accessed 21-10-07].

Lavé, J. and Avouac, J.P. 2001. Fluvial incision and tectonic uplift across the Himalayas of central Nepal. *Journal of Geophysical Research.* **106**, 26 561–26 592.

Liu, Z. 2000. Field experimental research on the corrosion kinetics of limestone and dolomite in allogenic water: Case from Yaoshan Mt., Guilin, China. *Carsologica Sinica.* **19**, 1–4.

Riebe, C.S., Kirchner, J.W., Granger, D.E. and Finkel, R.C. 2001. Minimal climatic control on erosion rates in the Sierra Nevada, California. *Geology.* **29**, 447–450.

Rubatto, D. and Hermann, J. 2001. Exhumation as fast as subduction? *Geology.* **29**, 3–6.

Rudoy, A.N. and Baker, V.R. 1993. Sedimentary effects of cataclysmic late Pleistocene glacial outburst flooding, Altay Mountains, Siberia. In: Fielding, C.R. (Ed.). Current research in fluvial sedimentology: Papers from the 5th International Conference on Fluvial Sedimentology. *Sedimentary Geology.* **85**, 53–62.

Simkin, T. and Siebert, L. 1994. Volcanoes of the World. Tucson, AZ, Geoscience Press.

Smith, A.B. and Peterson, K.J. 2002. Dating the time of origin of major clades: Molecular clocks and the fossil record. *Annual Review of Earth and Planetary Sciences.* **30**, 65–88.

Steffen, W., Sanderson, A., Tyson, P.D., Jäger, J., Matson, P.A., Moore, B., III, Oldfield, F., Richardson, K., Schellnhuber, H.J., Turner, B.L. and Wasson, R.J. 2004. Global Change and the Earth System: A Planet under Pressure. Berlin, Springer.

Stern, N. 2006. Stern Review on The Economics of Climate Change (pre-publication edition). Executive Summary. HM Treasury, London. http://www.hm-treasury.gov.uk/sternreview_index.htm.

Stock, G.M., Anderson, R.S. and Finkel, R.C. 2004. Pace of landscape evolution in the Sierra Nevada, California, revealed by cosmogenic dating of cave sediments. *Geology.* **32**, 193–196.

Zhang, C, Pei, J, Xie, Y. and Weng, J. 2000. Karst development and karst water resources exploitation in typical solutional hill regions: A case study in Sangzi Town, Xinhua County, Hunan, China. *Carsologica Sinica.* **19**, 58–64.

Further reading

Gore, A. 2006. An Inconvenient Truth: The Planetary Emergency of Global Warming and What We Can Do About It. London, Bloomsbury Publishing.

Houghton, J. 2004. Global Warming: The Complete Briefing. 3rd edition. Cambridge, Cambridge University Press.

Kennett, J.P., Cannariato, K.G., Hendy, I.L. and Behl, R.J. 2003. Methane Hydrates in Quaternary Climate Change: The Clathrate Gun Hypothesis. American Geophysical Union Special Publications Series Volume 54. Washington, DC.

McGuire, B. 2008. Seven Years to Save the Planet: The Questions . . . and Answers. Weidenfeld.

Pittock, A.B. 2007. Climate Change: Turning Up the Heat. Collingwood, Victoria, CSIRO Publishing.

Index

Aa types of lava, 191
Abiotic mode, carbonate sedimentary
    environments, 381
Ablation, 504
Abrasion, 453
Absolute humidity, 40–41
Absolute temperature, 27
Absolute zero, 26
Accidental dispersal hypothesis, 645
Accretionary prism, 176
Accretionary wedge, 220
Acid brown earth soil, 707
Acid deposition, 22
ACID or Felsic igneous rocks, 153
Acid rain, 21–22
    anthropogenic sources, 22
    natural sources, 22
Acid sand and micropodzols, 729
Acidification, 22
Acidity, 22
Acrisols, 721
Active continental margin, 181
Active rifting, 184
Advection cooling, 41
Aeolian (wind) processes, 611–637
    aeolian bedforms, 616–629
    current controls on, 613
    intertropical convergence zone, 613
    reptation, 615
    ripples, 617–618
    saltation, 614–615
    sand movement process, 613–614
    sediment entrainment, 613–614
    spiral vortices, 620
    suspension, 615–616
Aeolian dunes, classification, 621
Aeolian sediments, 629–631
Aeolian silt or loess, 599
'Aerodynamic ripples', 617
Aerosol production, global, 21
Aerosols, 129–131
Afro trailing-edge, 490
Air-fall deposits, 204
Air masses, 62
Air movement on a non-rotating planet, 56
Air stability and instability, 42

Aklé dunes, 623
Alaskan tide-water glaciers, 509
Alaskan-type fjords, lithofacies in, 534
    diamicton, 534
    gravel (poorly sorted), 534
    gravel (well sorted), 534
    mud, 534
    rhythmites (laminated sand and mud),
        534
    sand (poorly sorted), 534
    sand (well sorted), 534
    mud with dispersed clasts, 534
Albedo, 34, 80, 245
Albeluvisols, 721
Alfisols, 715, 753
Alisols, 721
Alkali basalt xenoliths, 771
Allochthonous, 220
Alluvial fan sediments, 365–371
    bajadas, 365
    braided fluvial fan, 366
    debris flow dominated fan, 366
    fan area, 365–369
    fan slope, 369
    landform assemblage as part of fluvial
        piedmont, 365–371
    low sinuosity/meandering fluvial fan,
        366
    morphology, 365
    playas, 365
    processes of formation, 369–370
        debris flow, 369
        mudflow, 369
        stream flows, 369
Alluvial fans in periglacial region, 602–603
Alluvial gley soils, 718
Alpine permafrost, 567
Altitude, 30, 44
Amero trailing-edge, 490
Amphibole groups, 146–147
Amphibolite, 777
Anabatic wind, 73
Anastomosing, 343
Andesite, 154
Andesitic magma, 188
Andosols, 720

Animal life, 750
Antarctic Bottom Water (AABW), 109
Antarctic ice sheets, 843–844
Antarctic Intermediate Water (AAIW), 109
Antarctic winter, 30
Antecedence, 315–318
Antecedent drainage development, 372
Antecedent morphology, 476
Anthropogenic sources acid rain, 22
Anthrosols, 720
Anticline, 229
Anticyclones, 55, 71
Antidunes, 335
Antiforms, 229
Apatite, 150
Applied stress, 229
Apron reef, 392
Aragonite, 387
Arboreal hypothesis, 815
Arc–forearc, 767
Arc–interarc, 767
Archaean granite-gneiss terranes, 767
Archaean greenstone belts, 766
    collisional graben, 767
    compressional foreland, 767
    cratonic extensional, 767
    stable shelf, 767
    volcanic basin, 767
Archaean rocks, 764
    granite-gneiss terranes, 764
    greenstone belts, 764
    late Archaean sedimentary basins, 764
Archaeological indicators of climate change,
    120
Arctic barrens, 743
Arctic Ocean for Cenozoic climate change, 844
Arctic summer, 30
Arctic, tundra soils, 721–723
    platy or lenticular structure, 722
    reorientation of skeletal grains, 722
    skeletal grain coatings, 722
    vesicular voids, 722
Areal scouring, 537
Arenosols, 721
Argillic brown Earths, 717
Argillic gley soils, 718

Argillic pelosols, 718
Argon (Ar), 19
Aridisols, 715
Arterial drainage systems, 507
Asthenosphere, 166, 177
Atmos (meaning vapour), 18
Atmosphere, 5, 17–23
  acid rain, 21–22
  aerosol production, global, 21
    anthropogenic, 21
    natural, 21
  carbon dioxide, 20–21
  layers of, 18
    mesosphere, 18
    stratosphere, 18
    thermosphere, 18
    troposphere, 18
  methane, 20–21
  ozone in upper atmosphere, 19–20
    chlorofluorocarbons (CFCs), 20
    ozone holes, 20
  particulate matter, 21
  vertical structure of, 19
  water vapour, 21
Atmospheric circulation, global pattern, 55–58
  Rossby waves and jet streams, 55–58
  tricellular model, 55, 57
Atmospheric evolution, *See* Evolution of
  Earth's atmosphere
Atmospheric motion, 49–58
  atmospheric pressure, 50–51
  basic principle, 50
  pocket-size aneroid barometer, 51
  winds and pressure gradients, 51–55
Atmospheric oxygen, rise of, 782
Atmospheric pressure, 50–51
Atoll reef, 392
Atoms and their structure, 140–141
Atterberg limits, 275
Atterberg Scale, 696–697
Attrition, 335
Avialae, 812
Avulsion, 356
Azonal raw sands and sand pararendzinas or
  sandy regosols, 729

Back-arc basin, 176
'Bacteria', 806
Back-scattering, 34
Backwash (downrush), 439
Bajadas, 365
Ballistic hypothesis, 617
Ballot's law, 54
Banded ironstone formations (BIF), 831
Bank reef, 392
Bar development, 357–358
Barbuilt estuaries, 476
Barchan dunes, 622–623
Barometric pressure, 68
Barrier beaches, 464
Barrier islands, 468

Barrier reef, 392
Barriers, 460–471
Basal debris, 519
Basal sliding, 511
Basalt, 154, 198
Basaltic magma, 185, 188
'Basket of eggs' topography, 538
Base surges, pyroclastic materials, 203–204
BASIC or Mafic igneous rocks, 153
Basin and range, 233–234
Beaches, 456, 460–471
  changes, in swell and storm conditions,
    458
  classification, 471
  cusps, 463
  dominant morphology and terms, 459
  profile variation, 462
  sea-level rise on, 496
Beaufort Wind Scale, 83
Bedding, 370
Bedforms, 337
Bergeron–Findeisen theory, 45
Bevelled cliff, 448
Big Bang explosion, 4
Bioerosion, 384, 456
Biogenic sedimentary rocks, 158
Biogeochemical cycling, 660–665
Biogeography, 643–683
  analytical biogeography, 644
Biological era, 112–114
Biological indicators of climate change, 120
Biological weathering, 240
Biosphere evolution and biosphere change,
  797–825
  chemo-autotrophy, 808
  earliest life in fossil record, 805–806
  evolutionary change modes, 799
  evolving metazoans, 807–808
  life from nonliving chemicals, 801
    catalysis, 801
    concentration, 801
    energy, 801
    protection, 801
  life on land, challenges to, 811
    gas exchange, 811
    physical support, 811
    reproduction, 811
    retention of water, 811
  mechanisms, 798–801
  origins of life, 801–803
  outline history, 803–817
  photo-autotrophy, 808
  Pre-Cambrian–Cambrian transition,
    808
Biosphere, 6
  change in, 873
Biota, 688
Biotic evolution, 216–217
Biotic subclimax, 679
Biotically-controlled mode, carbonate
  sedimentary environments, 381

Biotically-induced mode, carbonate
  sedimentary environments, 381
Biotite, 149
Bioturbated mudstone, 546
Bird's foot delta, 375
Birds, evolution, 814–815
  behaviour, 814
  bone characteristics, 814
  feathers, 814
  microstructure of the eggs, 814
    display and fighting hypothesis, 815
  sleeping posture, 814
Black body, 27
Blanket bog, 677
Block-and-ash flows, 205
Block entrainment, or removal, 516
Blue-green algae, 113
Blytt–Sernander climate period (PZIII), 858
Bögli's mixture corrosion hypothesis, 410
Bogs, 675
  blanket bog, 677
  flat bog, 677
  and marsh, 743
  raised bog, 677
Bonding in atoms, 141
  covalent bonding, 141
Boulders, 595
Braided river landform–sediment assemblage,
  357–362
  bar development, 357–358
  causes of braiding, 358–359
Breaking wave, 437
Breccias, 155–156
Brekknafjöll-Jarlhettur linear ridge system, 561
British classification, soils, 714–719
Brown and red Mediterranean soils, 723–724
Brown Earth soil, 707
Brule formation, slopes on, 307
Building, 357
  longitudinal bars, 357
  transverse bars, 357
Bulge, 439
Bulk-interaction steam explosivity, 196
Buoyancy-dominated river mouths, 482
Buys Ballot's law, 64

Calcareous pelosols, 718
Calcareous soils, 711
Calcification, 694
Calcisols, 721
Calcite ($CaCO_3$), 145, 150, 387
Calcium carbonate content in soil, 702
Calvin cycle, 653
Cambering
  and associated structures, 596–597
  and valley bulging, 599
Cambic gley soils, 718
Cambisols, 721
Cambrian explosion of life, 808–809
Cap carbonate rocks, 831
Capillary tension, 274

Caprocks, 597
Carbon cycle, 661
Carbon dioxide, 20–21, 131, 256, 402
    of equilibrium, 402
Carbonate coastal landforms and sediments,
    484–489
Carbonate facies models, 395–401
    facies patterns, 398–399
    platform interior
        brackish, 401
        evaporitic, 401
        normal marine, 400
        restricted, 400
    from ramp to rimmed platform, 398–399
    siliciclastic beach–open shelf model, 398
    siliciclastic beach–shelf model, 398
    siliciclastic shelf models, 398
    standard facies belts, 399
        cratonic deep-water basins, 399
        deep sea, 399
        deep shelf, 400
        reefs of platform margin, 400
        sand shoals of platform margins, 400
        slope, 400
        toe-of-slope apron, 400
    Wilson model, 399
Carbonate minerals, 147
Carbonate platforms, 488
Carbonate ramps, 389, 488
Carbonate sedimentary environments,
    379–427
    bioerosion, 384
    carbonate factory, types of, 381
        cool-water carbonate factory, 383
        mud-mound factory, 383
        tropical factory, 381
    carbonate material production, 381
        abiotic mode, 381
        biotically-controlled mode, 381
        biotically-induced mode, 381
    carbonate rock characteristics, 380–394
    carbonate rocks, 387
        chemistry, 387
        classification of, 387
        Dunham's classification, 388
        Folk's classification, 388
        grainstone, 388
        minerology, 387
        packestone, 388
        textures of, 387
        wackestone, 388
    chemical dissolution, 383
    chemical erosion, 384
    cyanobacteria, 392
    erosion in carbonate factories, 383
    Folk's classification, 390
    limestones classification based on
        depositional texture, 390
    mechanical erosion, 383
    mounds, 394
    salinity effect, 385

shallow platforms, 389
    carbonate ramps, 389
    rimmed carbonate platforms, 389
    siliciclastics, 389
shoal-water carbonate accumulations,
    geometry of, 386–387
shoal-water carbonate systems, growth
    potential, 384
        carbonate sedimentation rates, 384
        siliciclastic depositional systems, 384
Carbonate slopes, 394
Cataclastite, 230
Cation-exchange capacity, 700–702
Cave development, 412
    factors influencing, 416
    stages of, 421
Cave speleothems, 423
Cave systems, calcite deposits by degassing,
    423
Cavernous weathering, 265–267
Cementation, 274
Cenozoic climate change, Arctic Ocean for,
    844
Chadron formation, slopes on, 307
Chalk, 157
Channel network, 8
Channels in planform, 343–348
    braid bar development, 347
    braided streams, 345
    straight channels, flow in, 347
Cheluviation, 693
Chemical dissolution, 383
Chemical erosion, 384
Chemical sedimentary rocks, 158
Chemical weathering, 240, 251–262
    hydration, 256
    intensity, controls on, 260–262
        discontinuities, 260
        extrinsic factors, 260–262
        intrinsic factors, 260
    isomorphous replacement, 257
    oxidation, 257
    products of, 257–260
        better drainage with faster water flow,
            257
        poor drainage with slow water flow, 257
        rapid drainage, with very fast water
            flow, 257
    reaction-limited, 254–255
    reduction, 257
    rock-forming minerals, 251–252
    solution, 252–257
    transport-limited, 254–255
Cheniers, 460–471
Chernozems (USDA mollisol, suborder
    boroll), 720, 723
Chert, 157
Chestnut-brown soils (USDA mollisol,
    suborder xeroll), 723
Chèzy formula, 325
'Chimney', 47

Chlorites, 147, 149, 701
Chlorofluorocarbons (CFCs), 20, 131–132
Chute cutoff, 353
Circulation
    ocean, 105–109
    types of, 476
Classic Late Glacial, 852
Clast velocity, 515
Clastic sedimentary rocks, 155
Clay dunes, 628–629
Clay smear, 231
Clay translocation or lessivage, 693
Clays, 147
Cleavage, minerals, 142
Cliff erosion, 448
Cliffs, 447
Climate change, 111–114, 117–136, 781–788,
    See also Internal forcing
    causes of, 125–136
    climate models, 783
    emissions scenarios used by the IPCC,
        783–784
    evidence for, 118–125
    external factors in, 126–128
        changes in the Earth's orbit, 126–128
        variations in solar output, 126
    feedback effects, 133–134
        positive, 133
    future climate change, 783–787
        predictions, 783
        projections, 783
        scenarios, 783
    geological evidence, 118–119
    geomorphological evidence, 118–119
    glacial periods, 122–123
    global average surface warming, 786
    human-induced global warming, mitigating,
        787–788
        be a catalyst for change, 787–788
        consume less and conserve more, 787
        get around on less, 787
        save energy at home, 787
    Little Ice Age, 123–124
    Milankovitch-type orbital oscillations,
        783
    oceans in, 135
    physical processes that control, 128
    present climate, 124–125
    principles, 117–136
    proxy indicators for, 118–122
        archaeological, 120
        biological, 120
        glaciological, 120
Climate, 89–100, 313–315, 687–688, See also
    World climates
    classification, 90–100
Climaxes, 665–681
Clints, 410
Closed flow systems, 8–9
Closed-system pingo formation, mechanism
    for, 581

Clouds, 42–44
    altitude, 44
    interference with atmosphere, 134
    and precipitation, 80
    reflection, 34
    shape, 44
    structure, 44
    vertical extent, 44
Coal, 157
Coalescence theory, 45
Coarse-grained deltas, 487
Coarse-grained texture in igneous rocks, 153
Coastal dunes, 471, 476
    free dunes, 476
    impeded dunes, 476
    primary dunes, 476
    remnant dunes, 476
    secondary dunes, 476
    transgressive dunes, 476
Coastal erosion landforms, 450
Coastal landsystems, 445
    climatic change impact on, 492
    dune coasts, 492
    estuaries, 492
    lagoons, 492
    rising sea level impact on, 492
    salt marshes, 492
Coastal landsystems, distribution, 489
    collision coasts or convergent-margin coasts, 489
    marginal sea coasts, 490
    trailing-edge coasts or passive-margin coasts, 490
Coastal microclimates, 79
Coastal processes, 431–498, See also Wave-dominated coastal landform assemblages
    boundaries, 433
    longshore circulation, 440
    longshore currents, 439
    near-shore circulation, 439–440
    rip cell circulation and bar formation, 441
    terminology, 433
    tide patterns, formation, 442
Coastal slopes, 447
Coelophysis, 812
Coelurosaurs, 813
Cohesionless soils, 275
Cold-based glaciers, 511
Collapse dolines, 405
Collision, 45
Collisional graben, 767
Colloids, 257, 700–702
Colonization of land, 809–811
Colour, minerals, 141–142
Competent velocity, 325
Composition and fluid pressure, 229
Compressional foreland, 767
Condensation, 39, 41–42
    advection cooling, 41
    convective cooling, 42
    orographic and frontal uplift, 41
    radiation (contact) cooling, 41
Conditional instability, 42
Conglomerates, 155–156, 546
Conrad discontinuity, 169
Consumers, 651
Contact force, 515
Contact metamorphism, 159
Continental and collage collisions, 222
Continental collision, 773
Continental crust, 218
    evolution, 776
Continental deposits, 155
Continental margin orogenesis, 221–222
Continental margin topography, 102–104
Continental or desert loess, 632
Continental rift zone magmatism, 182–186
Continental rifting, 182, 219, 220
Continental shelf break, 105
Continental shelves
    depositional glaciomarine sequences on, 534
    ice shelf in, 535
Continental subarctic climates with extremely severe winters, 98
Continental subarctic or boreal (taiga) climates, 98
Continental/microthermal climates (Group D), 98
    hot summer continental climates, 98
    warm summer continental or hemiboreal climates, 98
Continuous chains, 148
Continuous creep, 284
Continuous permafrost, 567
Continuous sheets, 148
Contraction and Convergence (C&C), 875–876
Convection, 26
Convective cooling, 42
Convective or adiabatic cooling, 41
Convergent plate boundaries
    geomorphology, 176–177
    volcanism at, 179–180
Cool-water carbonate factory, 383
Coral reefs, 392
    bank reef, 392
Cord grass, 678
Core, 166
'Corestones', 589
Coriolis effect, 53–55, 105
Corridor caves, 417
Corrie or cirque glaciers, 502
Corrosion or solution, 334, 453
Cosmic collisions, 129–131
Covalent bonding in atoms, 141–143
Crassulacean acid metabolism (CAM), 654
Cratonic deep-water basins, 399
Cratonic extensional, 767
Cratons, 764
Craven Faults, 232
Cretaceous–Tertiary boundary, extinctions at, 820–823
Cretaceous–Tertiary geological boundary, 129
Crevasse-fill ridges, 557
Crevasse splays, 355
Crevasses, 507
Critical shear stress, 326
Cross-grading, 331
Crustal uplift, 232
Cryoplanation or altiplanation processes, 585–588
    small-scale cryoplanation processes, 588
Cryoplanation terraces, development, 589
Cryosols, 720
Cryoturbation, 721
Cryptofauna, 393
Crystallization pressure, 249
Cursorial hypothesis, 815
Cuspate forelands, 463
Cusps, 461
Cutoff, 353
    chute cutoff, 353
    neck cutoff, 353
    plugging, 353
Cyanobacteria, 392, 805
Cyanophyta, 113
Cycles, 8
    of variation, 457
Cyclones, 55

Dansgaard-Oeschger events, 128
Darcy's law, 305
Dartmoor tors, 592
Dating, 119
Davis' model of slope decline, 305
De Geer moraines, 530
    formation, 525
Debris avalanches, 207–210
    dynamic emplacement models for, 211
    hummocky topography, 207
    phreatic explosion, 207
    submarine debris avalanches, 207
Debris flow, 297, 369–370
Decomposers, 651
Deep-ocean currents, 107–109
Deep-sea sediments, 105
Deflation, 599
Deformation structures, 229–234, See also Faults
    anticline, 229
    antiforms, 229
    basin and range, 233–234
    folds, 229–230
    joints, 233
    neutral folds, 229
    syncline, 229
    synforms, 229
Deglaciation, 510
    Deglaciation Climate Reversal (DCR), 859
    proportion of maritime ice margin, 510

varying air temperature, 510
varying precipitation, 510
Deinonychus, 814
Delamere forest, woodland soils in, development, 730–731
Deltaic processes and environments, 372–375
climate, 373
coastal processes, 373
estuarine delta, 375
river regime, 372
simple delta types, 373–375
Bird's foot delta, 375
classical arcuate delta, 375
Gilbert delta, 373
structure of the region, 373
Deltas, 482, 524
buoyancy-dominated river mouths, 482
friction-dominated rivermouths, 482
hyperpycnal-dominated river mouths, 482
inertia-dominated river mouths, 482
Density, minerals, 142–144
Deposition processes, 515–517, 711
Depositional environment, 155
continental, 155
marginal, 155
marine, 155
Depressions, 63–67
open-wave depression, 63
Destructive plate boundaries, 176, 180–181
crustal growth at, 777
Detached beaches, 462
Detrital sedimentary rocks, 155–156, 158
breccia, 156
chalk, 157
chert, 157
coal, 157
conglomerate, 156
gypsum, 157
limestone, 157
rock salt, 157
sandstone, 156
shale, 156
siltstone, 156
Detritus food chain, 659
Devonian period, 809–810
Diamond, 144, 771
Diatomaceous mudstone, 546
Diatomaceous ooze/diatomite, 546
Diffuse radiation, 30, 34
Dinosaur ecology and ethology, 813–814
Dinosaurs, 812–816
Diorite, 154
Direct solar radiation, 30
Discontinuities, 243, 260
Discontinuous permafrost, 567
Dissipative beaches, 471
Distributary mouths, 483
density conditions, 484
morphology, 484
offshore slope, 484
river-current velocity, 484

Diurnal cycle, 11
Divergent plate boundaries, geomorphology, 175–176
Dokuchaeiv approach, 712–714, 721
Dokuchaeiv soil classification, 714
Dolines, classification, 408
Dolomite, 365, 387
Domes, 190
Drag forces, 326
Drainage basins, 295, 328–332
Dropstones, 328–332, 532
Drowned river valleys, 476
Drumlins, 538
Dry (arid and semiarid) climates (Group B), 95–97
Dryas-stage plants, 854
Dune seral relationships, 672
Dune soils, development, 729–730
acid sand and micropodzols, 729
azonal raw sands and sand pararendzinas or sandy regosols, 729
peaty gleys, 729
Sefton coast dune system, 730
Dune system, 474
Dunefields, 629
Dunes, 336, 618
Aklé dune, 623
associated with obstacles, 627–628
clay dunes, 628–629
linear dunes, 623–625
star dunes, 626–627
zibar, 627
Dungeness cuspate foreland, 466
Dunham's classification, 388
Durisols, 720
Dust, 631–633
Dwarf-planet, 5
Dynamic equilibrium, 10
Dynamic threshold, 614

Early Earth, 764
Early geological record, 764–768
Early glaciations in Proterozoic phase of Pre-Cambrian, 828–833
ice age and glacials, 828
interglacial, 828
interstadial, 828
stadial, 828
Earth
early Earth, 764
Earth's core form, 768–769
first Earth system, 768
history, 217–222
materials, 139–162
oldest rocks on, 767–768
as a planet, 764–765
Earth spheres, 5–6
atmosphere, 5
biosphere, 6
components of, 6

hydrosphere, 5
lithosphere, 5
water on Earth, 6
Earth's atmosphere, changes in, 847
Earth's formation, 4–5
Earth's mantle, evolution, 769–777
Earth's orbit, variations in, 848–849, 851
Earth's short-wave radiation cascade, 34
Earth–Sun relationship, 6
Earthquake seismic waves as Earth probes, 168–171
Earthy peat soils, 719
East-Greenland or open-system pingos, 579
Eccentricity, 848–849
Ecesis, 127, 668
Ecology, 643–683
Ecosystems
components of, 647–650
energy flow in, 651–657
Ecozone concept, 736, See also Tundra ecozone
Ecozones, global warming effects on, 757–759
Edaphic subclimax, 679
Edge waves, 437
Ediacaran fauna, 807–808
Effusive volcanism, 188–190
effusive basalts, 188–190
physical attributes of, 189
composition, 189
cooling, velocity, 189
dimension, 189
discharge rate, 189
temperature, 189
texture, 189
volume, 189
'Egg box' topography, 406
El Niño–Southern oscillation (ENSO), 212–214
Elastic limit, 242
Electrical tension, 274
Electromagnetic radiation, 26
Electromagnetic spectrum, 28
atmospheric opacity in relation to, 29
Elongated depressions, 219
Eluviation, 693
Embryonic and foredunes, 471, 475
Endosymbiosis, 806
End-Pleistocene extinctions, 823–824
End-Triassic period, extinctions in, 820
Energy flow in ecosystems, 651–657
consumers, 651
decomposers, 651
gross primary production, 653
primary productivity, 653
producers, 651
Energy in atmosphere and the earth heat budget, 25–36
'internal energy', 26
'Enhanced F Scale', 70

Entisols, 715
Environmental lapse rate (ELR), 18, 42
Enzyme, 801
Eocene, 817, 838–841
Eoraptor, 812
'Epigenetic' ice wedges, 578
Epsilon cross stratification, 352
Equilibrium, 10
    dynamic, 10–11
    static, 10–11
    steady-state, 10
    in systems, 9–11
Equinoxes, 849–850
Erosion, 155, 334
    attrition, 335
    corrosion or solution, 334
    in carbonate factories, 383
    hydraulic action, 335
    on rocky coasts, 455
        bloerosion, 455
        mass movements, 455
        mechanical wave erosion, 455
        weathering, 455
Erosional notches, 453
Eruptions
    basalt eruptions, 198
    subaerial processes, 199
        hydrovolcanic, 199
        magmatic eruptions, 199
        petrographic features of volcaniclastic
            sediments, 200
        subaqueous deposits from subaerial
            volcanoes, 199
        tephra from subaqueous eruptions, 200
    subaqueous eruptions, 198
    subglacial eruptions, 198
Esker formation, 539
    in proglacial lake, 540
Estuaries, 476, 492
    controls on, 474
    fully-mixed or macrotidal estuaries, 476
    partially-mixed or mesotidal estuaries, 476
    salt-wedge or microtidal estuaries, 476
Estuarine delta, 375
Estuary fill, types of, 476
Eucarya, 806
Eucytes, 806
Eukaryotes and diversification of life, 806–807
European painting, winter depictions in,
    862–863
European Project for Ice Coring in Antarctica
    (EPICA), 119
Evaporation, 39–42
    determining factors, 39
    radiation, 39
    saturation vapour pressure, 39
    temperature, 39
    wind speed, 39
Evaporites, 394
    depositional environment of, 397
        'clinoform' platform, 398

'undaform' platform, 398
Evolution of Earth's atmosphere, 112–114
    biological era, 112–114
    and climate change, 781–788
    cyanophyta or blue-green algae, 113
    early atmosphere and effect of volcanic
        eruptions, 112
    effect of humans on, 114
    rise of oxygen, 112–114
Exfoliation of granite, 244
Exogenous dome, 193–194
Explosive volcanic activity, 215
    individual volcanic eruption climatic
        disturbances, 215–216
        Laki fissure eruption of 1783, 215
    primary cause, 215
    trigger mechanism, 215
Explosive volcanism, 194–200
    andesitic, 194
    bulk-interaction steam explosivity, 196
    cooling-contraction granulation, 196
    hydromagmatic processes, 196
    Plinian eruptions, 197
    rhyolitic, 194
Extrusive vulcanicity, 152

Factor of Safety (FS), 272
Fan slope, 369
Fault gouge, 231
Faulting, 161
Faults, 230–233
    cataclastite, 230
    clay smear, 231
    craven faults, 232
    faultline scarp, 232
    mylonite, 231
    pseudotachylite, 231
    San Andreas fault system, 232
    San Jacinto fault zone, 232
    strike-slip faults, 232–233
    tectonic or fault breccia, 231
Feedback effects, 7–8, 133–134
'Feldspar', 147
Fellfields, 256, 743
Ferrallitization process, 695–696
Ferralsols, 720
Ferrugination, 695
Fetch, 433
Fill, 352
Fine-and coarse-grained point bars,
    362
Fine-grained deltas, 487
Fine-grained point bars, 349–353
Fine-grained texture in igneous
    rocks, 153
Fjords, 476
    sedimentation in, 529
        depositional landforms in, 531
Flat bog, 677
Floating ice, 503, 529
Flocculation, 480

Flood basalts, 216–217
Flood basin, 355–356
Floodplain construction, 356–357
Flow regime concept, 336–339
Flow systems, 8
    closed, 8–9
    inputs, 8
    open, 8–9
    outputs, 8
Fluctuations in oxygen, 114
Fluidization, 203
Fluted ground moraine, 552
Fluvial channel geomorphology, 332–375
    antidunes, 335
    base level, changes in, 340–344
        Pleistocene period, 340
        rejuvenation, 340–341
        river long profile, 341
        river terrace sequences, 340
        terrace aggradations, 340
    bedforms and cross-stratification types,
        337
    downstream size and sorting changes,
        348–349
    drainage patterns, 333
    dunes, 336
    fluvial landform, sediment assemblages, 349
    large-scale current ripples or dunes, 335
    large-scale lineations, 336
    lower flow regime, 338
    meandering river landform–sediment
        assemblages, 349–357
    megaripples, 336
    morphological activities of rivers, 335–339
        sand-bed rivers, 335–336
    radial drainage, 333
    ripples, 336
    river channel adjustments in cross-section,
        339–340
    river erosion processes, 334
    sand waves, 336
    transitional forms of bed roughness, 335
    upper flow regime, 338
    upper-phase plane beds, 335
Fluvial erosion landscapes in karstarst areas,
    426
Fluvial landform characteristics in rock
    channels, 371–372
Fluvial processes, 321–375
    channels and drainage basins, origin,
        331–332
    cross-grading, 331
    drainage basin, 328
    energy of a river and its ability to do work,
        323–325
    flow types in, 323
    fluvial channel geomorphology, 332–375
    Hjülstrom and Sundborg diagrams, 327
    internal distortion resistance, 325
    laminar flow, 322
    loose boundary hydraulics, 322–323

open river channels, types of flow in, 322–323
in periglacial environment, 600–602
rainfall–runoff responses, 328–329
rill systems, 331
river hydrology, 328–329
rolling, 325
sediment load, transport of, 325–327
sediment load, types, 327–328
stream hydrograph, 328
suspension, 325
turbulent flow, 322–323
Fluvisols, 720
Fog frequency, 77
Föhn winds, 73–74
Folds, 229–230
Fold Belt, 235
terminology of, 229
Foliated or layered metamorphic rock, 160
Foliation, 161
Folk's classification, 388
of carbonates, 390
Food and Agriculture Organization soil reference groups (1998), 720
Food chains, 657–660
carbon cycle, 661
detritus food chain, 659
grazing food chain, 659
nitrogen cycle, 661–662
parasite chains, 659
phosphorus cycle, 663–665
potassium cycle, 665–666
predator chains, 659
saprophyte chains, 659
sulphur cycle, 662–663
'Fool's gold', 148, 150
Fore-arc basin, 176
Forecasting, 82–86
Forest, 79
Fossilization, 158
giant fossil ammonite, 158
trilobite fossil, 158
Fossils, 157–159
trace fossils, 157
Fracture propagation, 516, 517
Free dunes, 476
'Free energy', 302
Freely drained soil, 705
Friction, 290, 322
Friction-dominated rivermouths, 482
Fringing reef, 392
Frontal or cyclonic rainfall, 47
Fronts, 63
standard notation of, 64
Frost cracking, 574–578
Frost heaving, 571–573
'Frost hollow', 73
Frost sorting, landforms associated with, 573–574
Frost weathering, 247

Frost wedging (frostrost shattering or splitting), 569–570
Frost thrusting, 571–573
Froude number, 323
Fully-mixed estuaries, 479
Fully-mixed or macrotidal estuaries, 476
Future climate change, 868–869

Gabbro, 154
Galena, 150
Gaping Gill system, fault and joint control in, 419
Garnet group of minerals, 145–146
Gelifluction, 587
landforms produced by, 594–595
single terraces, 595
soil tongues, 595
terraces on slight slopes, 595
sediments created by, 595–596
Geochemical system, mantle, 775–776
Geology, vegetation and soils, relationship, 728
Geomorphic processes affecting carbonates, 403
Geosphere, 869–872
Geosynclines, 174
Geotectonics, 227–235, See also Deformation structures
landforms, 227–235
processes structures, 227–235
rock deformation, controls on, 229
composition and fluid pressure, 229
heat, 229
lithostatic pressure, 229
Geozonal ecosystem, 736
Giant fossil ammonite, 158
Gilbert deltas, 373, 482
Glacial climate
Birch–Juniper Pollen Zone (IId), 11 800–11 000 BP, 855
classic Late Glacial, 852
Older Dryas and its vegetation, 857
Older Dryas, PZIc, 12 000–11 800 BP, 857
'park tundra', 853
Pre-13 000 BP, 855
vegetation description, 857
western Britain during, 855–856
Windermere Interstadial, 13 000–11000 BP, 856
Younger Dryas, 856–858
Glacial erosion, 515–517
clast velocity, 515
contact force, 515
debris concentration within the basal ice layer, 515
debris renewal, 515
hardness of the clasts relative to bed, 515
Glacial lakes, 521
deposition from melt-water flows, 522
direct deposition of supraglacial debris from the ice front, 522

iceberg rainout, 523
resedimentation by sediment gravity flows, 523
suspension, 523
thermal stratification in, 523
Glacial landsystems, 536
preservational, cold-based continental ice sheets, 536
Glacial melt water, 507
ice melt, 507
'pressure melting point', 507
surface melting, 507
Glacial periods, 122–123, 846–847
mitigating processes, 846
negative feedbacks in, 846, 851–852
positive feedbacks in, 846–848, 850–851
postglacial period, 123
Glacial processes and landsystems, 501–562, See also individual entries
Alaskan tide-water glaciers, 509
effects on mass balance, 509
firn to glacier ice, 505
glaciers formation, 504
iceberg calving, 508
mass balance and glacier formation, 504
Medvezhii glacier, 512
newly-fallen snow crystals to granular or old snow, 504
old snow to firn or névé, 505
Otto glacier, 512
particle lodgement beneath glacier, 516
Steele glacier, 512
sublimation, 509
surging or galloping glaciers, 512
Variegated glacier, 513
Glacial quarrying or plucking, 516
Glacial sediments, 517
ice marginal, 521
proglacial, 521
subglacial, 517, 521
supraglacial, 519
Glacier cap or lowland ice cap, 503
Glacigenic trough mouth fan systems, 106
Glaciofluvial channels, 528
Glaciofluvial deposition, 520
Glaciological indicators of climate-related variables, 120
Glaciomarine environments, 529
deposition from the glacier, 529
floating ice, 529
glaciomarine sediments, 531
controlling, 529
recognition, 529
process systems in, 532
carbonate sedimentation, 531
delivery of sediment to glaciomarine environment, 531
grounding-line fans, 532
ice-contact deltas, 532
ice-ocean interactions, 531
processes of reworking, 531

Glaciomarine systems, 548
Glaciovolcanic landsystems, 558
Gley soils, 718
  alluvial gley soils, 718
  argillic gley soils, 708, 718
  cambic gley soils, 718
  humic gley soils (sensu stricto), 719
  humic-alluvial gley soils, 719
  humic-sandy gley soils, 719
  sandy gley soils, 718
  stagnogley soils, 718
  stagnohumic gley soils, 719
Gleying, 693
Gley-podzols, 717
Gleysols, 720
Global atmospheric effects of volcanic
    processes, 210–215
Global dimming, 131
Global environmental change in the future,
    867–877
  biosphere, change in, 873
  change in geosphere, 869–872
  Contraction and Convergence (C&C),
    875–876
  future optimism, 874–876
    peak oil and coal production, 875
    point of no return, approaching,
    874–875
    renewable technologies, 875
  ocean circulation systems, change in,
    873
  oceans and hydrosphere, change in, 872–873
  predicted plate tectonic movements,
    871
  sea-level change, 872–873
  timeline for future Earth, 873–874
    2009, 873
    2020, 873
    2030, 873
    2040, 873
    2050, 873
    2070, 874
    2080, 874
    2100, 874
    2200, 874
Global glaciation, 830–831
Global hydrological cycle, 38–42
Global long-wave radiation cascade, 34
Global Ocean Conveyor, 109
Global precipitation regimes, 92
Global radiation, 29–32
  altitude of the Sun in the sky, 30
  distance from the Sun, 30
  insolation through the year and by latitude,
    32
  length of day and night, 30–32
  solar constant, 29–30
Global seismicity, 175
Global tectonics, 174
Global temperature, 90–91, 792–793
Global warming

impacts of, 784–787
implications of, 784–787
Gneiss, 162
Gondwana-Pangea, 773
Grabens, 231–232
Grading, 464
Grainstone, 388
Granite, 151, 154
Granite-gneiss terranes, 764, 767
Granular disintegration, 285–295
Granule ripples, 618
Graphite, 144
Grass sedge tundra, 743
Gravel, 155
Gravity, 322
Grazing food chain, 659
Greenhouse climate in the Middle and Late
    Cretaceous, 834–838
Greenhouse effect, 34–35
'Greenhouse Earth', 828
Greenhouse gas concentrations, variations in,
    131–133
Greenhouse to icehouse climates, changes
    from, 833–844
Greenhouse warming, 782
Greenland Stadial 1 (GS1), 858
Greenstone belts, 764
Grey brown podzolic soil, 707–708
Gross primary production, 653
Ground ice, growth, 578
Ground water flow, 302–305
Gullies, 300
Gypsisols, 720
Gypsum, 150, 157

Halides, 146
Halobacteria, 806
Halophytes, 677–678
  cord grass, 678
  red fescue, 678
Haloseres, 677–679
  in the United Kingdom, 680
Hammada, 635
Hand-texturing of soils, 706
Hardness, minerals, 142
  Mohs scale of, 145
Harzburgite-type mantle, 771
Hawaiian–Emperor volcanic chain,
    181
Hawaiian volcano, 183
Head scarp retreat, 303
Heat conduction, 26
Heat, 229
Heaving, frost, 571–573
  mechanisms of, 573
Helictites, 423, 425
Hematite, 150
Henry's Law, 402
Hillslope hydrology, 297–305
  aquifer/ground water basin, 304
  ground water flow, 302–305

gullies, 300
head scarp retreat, 303
overland flow, 300–302
rainsplash, 298–299
rills, 300–301
slope angle, 300
subsurface flow and piping, 302
Hillslopes, 270
Histosols, 720
Hjülstrom and Sundborg diagrams, 327
Holocene floodplain development, 357
Holocene volcanic record, 870
Holocene, 468, 644
'Honeycomb' or 'alveolar' weathering,
    264–266
Horizontally-stratified sands, 352
Hornblende, 149
Horsts, 231–232
Hot spots, 181–182
Hot summer continental climates, 98
Huelmo/Mascardi Cold Reversal, 860
Human-induced global warming, mitigating,
    787–788
Humans effect on Earth's atmosphere
    evolution, 114
Humic gley soils (sensu stricto), 719
Humic-alluvial gley soils, 719
Humic-sandy gley soils, 719
Humid subtropical climates (Cfa, Cwa), 98
Humid tropics, climate, 746
Humidity, 40–41, 80–81
  relative humidity (RH), 41
  specific humidity, 40
Hummocky topography, 207
Humus, 699
  Moder, 699
  Mor, 699
  Mull, 698
Hurricanes, 68
  Saffir–Simpson scale for, 69
HWMNT (high water mark neap tides), 677
Hyaloclastites, 194
Hydration, 256
Hydraulic action, 335
'Hydraulic geometry', 339
Hydrogen sulphide (H_2S), sulphates from, 21
Hydrological cycle, 8, 38–42
Hydrological methods, 507
Hydrological regime, 338
Hydromagmatic fragmentation, 196
Hydromagmatic processes, 196
Hydroseres, 668, 673–679
  cyperacae, 675
  development, 673
  fens, 675
  mires, 75
Hydrosphere, 5, 789–795
Hydroturbation, 753
Hydrovolcanic subaerial process, 199
Hydroxides, 146
Hyperpycnal-dominated river mouths, 482

Ice ages, 790, 792, 847–851
Ice bubbles, 845
Ice cap climate, 99, 828
Ice crystal mechanism, 45
Ice masses, 502
   corrie or cirque glaciers, 502
   niche, cliff or wall-sided glaciers, 502
   transection glaciers, 502
   valley glacier of the alpine type,
     502
Ice melt, 507
Ice segregation, 247
Ice sheets, 503
Ice shelf advanced to edge of continental shelf,
   535
Ice shelf in recessed state, 535
Ice wedges
   development, 574–578
   growth of, 576
Iceberg calving, 508
Iceberg rainout, 523
Ice-contact glaciofluvial sediments, 557
Ice-core isotopic record, 845
Igneous rocks, 151–153
   andesite, 154
   basalt, 154
   classification, 153
     ACID or Felsic, 153
     BASIC or Mafic, 153
     Intermediate, 153
   diorite, 154
   gabbro, 154
   granite, 154
   obsidian, 154
   pumice, 154
   rate of cooling affecting, 153
     coarse-grained texture, 153
     fine-grained texture, 153
     rapid cooling, 153
     slow cooling, 153
   rhyolite, 154
   tuff, 154
Impact threshold, 614
Impeded dunes, 476
Imperfectly drained soil, 705
In situ weathering, 711
Inceptisols, 715
Inertia-dominated river mouths, 482
Inferred cave evolution, 422
Infragravity waves, 437
Injected-ice stage, 579
Insolation losses in atmosphere, 32–34
   solar energy cascade, 32–34
Insolation weathering, 245, 745
Interglacial, 828
Intergovernmental Panel for Climate Change
   (IPCC), 124, 783, 861
Intermediate igneous rocks, 153
'Internal energy', 26
Internal distortion resistance, 325
Internal forcing, 128–133

atmospheric composition, changes in, 129
   aerosols, 129–131
   cosmic collisions, 129–131
   volcanoes, 129–131
clearance of rainforest, 130
global dimming, 131
greenhouse gas concentrations, variations in,
   131–133
surface changes, 128–129
Internal structure of earth, 165–171
   asthenosphere, 166
   core, 166
   Earth's composition from drilling, 166–167
   Earth's composition from meteorites,
     167–168
   Earth's composition from volcanoes, 167
   earthquake seismic waves as Earth probes,
     168–171
   lithosphere, 166
   mesosphere, 166
Interstadial, 828
Intertropical convergence zone (ITCZ), 55
Intrazonal soils caused by topographic change,
   731
Intrazonal upland soils in Britain, 724–728
   biotic factor, 725
   climate, 725
   parent material, 725
   time factor, 725–728
   topography or the relief factor, 725
Intrusive dome, 193–194
Intrusive vulcanicity, 152
Isobars, 52
Isolated tetrahedra, 148
Isomorphous replacement, 257

Jet streams, 55–58
Jigsaw cracks, 209
Joint plane caves, 417
Joints, 233

Kame hillocks, conical kames or moulin
   kames, 543
Kames, formation, 543
   kame formation, models for, 544
   kame hillocks, conical kames or moulin
     kames, 543
   kame plateaux or flat-topped hills, 543
   Kame terraces, 543
   Kettled sandar, 544
Kaolinite, 147, 149, 257, 701
Karren, 403
   Kluftkarren, 403
   Rillenkarren, 403
   Rundkarren, 403
   Trittkarren, 403
Karst processes, 401
   aggressive carbon dioxide, 402
   carbon dioxide pressure, 402
   complex controls on, 402
   and landforms, 379–427

solution of carbonates, 401
Karst streets and tower karst, development of,
   427
Karst areas, fluvial erosion landscapes in,
   426
Kastanozems, 720
Katabatic winds, 73
Kelvin temperature scale, 26
Kimberlites, 186, 771
King's model of parallel retreat, 305
Kluftkarren, 403
Kolka–Karmadon rock, 290
Köppen climate classification, 92–99
   continental/microthermal climates (Group
     D), 98
   dry (arid and semiarid) climates (Group B),
     95–97
   polar climates (Group E), 98–99
     ice cap climate, 99
     tundra climate, 98–99
   temperate/mesothermal climates (Group C),
     97–98
     humid subtropical climates (Cfa, Cwa),
       98
     maritime subarctic climates or
       subpolar oceanic climates (Cfc), 98
     maritime temperate climates or oceanic
       climates (Cfb, Cwb), 98
     mediterranean climates (Csa, Csb), 98
   tropical/megathermal climates (Group A),
     95
K-species, 681
Kuiper belt, 5

Lagoons, 492
Lahars, 205–207
Lake plain deep-water silts and clays, 391, 525
Lakes, 79
Laki fissure eruption of 1783, 215
Laminar flow, 322
Land-based weather station, measuring
   weather elements at, 80–82
   humidity, 80–81
   temperature, 80–81
   thermometer shelter or Stevenson screen,
     81
   wind speed and direction, 81
Land breeze, 53, 71–73
Landforms, 431–498
   associated with frost sorting, 573–574
   of glacial erosion, 554
   weathering, 262–264
Landform–sediment assemblages, 321–375,
   570–607, 611–637
Landsliding, 271
Large Hadron Collider (LHC), 4
Large-scale current ripples or dunes, 335
Large-scale detached beach, 463
Large-scale lineations, 336
Late Archaean sedimentary basins, 764,
   766–767

Late Cenozoic ice ages, 845, 847, 849, 851
　　rapid climate change in Quaternary,
　　　　845–851
　　　　chemical evidence, 845
　　　　geological/geomorphological, 845
　　　　palaeontological evidence, 845
Late Devonian period, 819
Late glacial climates and evidence for rapid
　　change, 852–860
Late Ordovician period, 819
Late-Glacial climatic oscillations, 824
Latent heat, 26
Lateral accretion, 356
Lava flows, 190–194
　　Aa types, 191
　　blocky surfaces, 190
　　derived from rhyolitic magmas, 190
　　lava dome classifications, 193
　　　　exogenous dome, 193–194
　　　　intrusive dome, 193–194
　　　　peléan dome, 193–194
　　　　spine, 193–194
　　　　upheaved plug, 193–194
　　pahoehoe type lava, 191
　　shield volcano forms, 191
　　strato-volcano forms, 192
　　subaqueous basaltic lavas, 190
　　subaqueous silica-rich, 190
Lava, physical properties, 187–188
　　lava viscosity, controls on, 189
　　　　bubbles, 189
　　　　chemical composition, 189
　　　　crystals, 189
　　　　pressure, 189
　　　　temperature, 189
　　　　volatiles, 189
　　Pahoehoe lava, 190
Lava-lamp model for Earth's mantle,
　　776
Leaching, 693
'Leap year', 30
Leptosols, 720
Lesothosaurus, 813
Levees, 353–355
Lherzolites, 771
Lichen weathering, 247
　　chemical mechanisms, 247
　　physical mechanisms, 247
Lift, 326
Limestone, 157, 410
　　classification based on depositional texture,
　　　　390
Linear dunes, 623–625
Linear volcanic chains, hot-spot model for,
　　182
Linton's deep rock rotting hypothesis for tor
　　formation, 592
Lithology, 311–313, 448
Lithomorphic (A/C) soils, 716
Lithoseres or psammoseres, 669
Lithosphere, 5, 166, 745

Lithostatic pressure, 229
Little Ice Age, 123–124, 603, 860–864
　　climatic patterns and their causes, 863–864
　　　　ocean conveyor shutdown, 864
　　　　volcanic activity, 864
　　dating of, 861
　　end of, 864
　　European painting, winter depictions in,
　　　　862–863
　　in the northern hemisphere, 861–863
　　　　extent of glaciers and evidence for cold
　　　　　　conditions, 862
　　　　impact on agriculture, 861–862
　　　　impact on wine production, 862
　　in southern hemisphere, 863
Littoral limestone zonation, 456
Lixisols, 721
Local winds, 71–74
Loch Lomond Stadial, 858
Loess deposition, 631–633
　　continental or desert loess, 632
　　periglacial loess, 632
Longitudinal bars, 357
Longitudinal compression, 519
Longshore bars, 461
Longshore currents, 433–439
Long-spreading ridges, 179
Lower flow regime, 338
Lunettes, 628–629
'The lungs of the Earth', 128
Lustre, minerals, 142
Luvisols, 721

Maars, 202
Macroevolution, 798
Macroscale synoptic systems, 62–71
　　air masses, 62
　　depressions, 63–67
　　tornados, 70–71
　　weather fronts, formation, 62–63
Magma eruption, 185–186
Magma generation, 177–178
Magma ocean scenarios, 770
Magmatism, 180
Magnitude and rate of applied stress, 229
Mammals, origin of, 816–817
Mangrove swamps, 481
Maniraptora, 812
Man-made soils, 719
Manning formula, 325
Mantle
　　convection, 772
　　Earth's earliest mantle, 776
　　evolution, 769–777
　　　　models for, 776–777
　　lava-lamp model for, 776
　　mantle rocks on the Earth's surface, 771–772
　　　　harzburgite-type, 771
　　　　lherzolite-type, 771
　　melting, 772
　　plumes, 772–775

　　plum-pudding model for, 776
Mantle plume, 773
MAPSS model, 757
Maquis, 723
Marble, 162
Marginal deposits, 155
Marginal sea coasts, 490
Marine deposits, 155
Maritime temperate climates or oceanic
　　climates (Cfb, Cwb), 98
Mass balance
　　and glacier flow, 510
　　　　basal sliding, 511
　　　　internal deformation or ice creep,
　　　　　　510
　　　　subglacial deformation, 511
　　and glacier formation, 504
　　measuring, 505
　　　　direct methods, 505
　　　　hydrological methods, 507
　　　　photogrammetric methods,
　　　　　　507
Mass extinctions, 216–217, 817–819
Mass movement, types, 282–297, 455
　　debris flow, 297
　　Earth flow, 297
　　granular disintegration, 285–295
　　mud flows, 297
　　rockfalls, 285–295
　　rockslides/avalanches, 285–295
　　slab failures, 285–295
　　slides, 295–296
　　soil creep, 284–285
　　soil falls, 295–296
　　topples, 295–296
Massive diamictite, 546, 548
Massive mudstone, 546, 548
Meandering, 344
　　and braided, types in between, 362–364
　　river landform–sediment assemblages,
　　　　349–356
　　　　fill, 352
　　　　fine-grained point bars, 349–353
　　　　flood basin, 355–356
　　　　floodplain construction, 356–357
　　　　horizontally-stratified sands, 352
　　　　natural levees, 353–355
　　　　small-scale cross stratification, 352
　　　　swale fills, 352
Mechanical erosion, 383, 455
Mechanical weathering, 240
　　controls on the intensity, 250–251
　　mechanisms of, 243–250
　　　　volumetric change of a rock mass,
　　　　　　243–246
Medieval Warm Period or Medieval Climate
　　Optimum, 860–864
Mediterranean climates (Csa, Csb), 97–98
Medvezhii glacier, 512
Megabarchans, 622
Megaripples, 336

Mega-scale glacial lineations, 539
Mercury barometer, 51
  pocket-size aneroid barometer, 51
Meridional overturning circulation,
    109
Mesoproterozoic Eon, 828
Meso-scale, 71–74
  Föhn winds, 73–74
  land breeze, 71–73
  local winds, 71–74
  mountain and valley winds, 73
  sea breeze, 71–73
Mesosphere, 18, 166
Metallic meteorites, 168
Metamorphic rocks, 151, 159–162
  contact metamorphism, 159
  foliated or layered, 160
  gneiss, 162
  marble, 162
  nonfoliated or recrystallized, 161
  quartzite, 162
  regional metamorphism, 159
  schist, 162
  slate, 162
Metasomatism, 161
Metazoans, 807–808
Meteorites, 4
  Earth's composition from, 167–168
  metallic meteorites, 168
  stony meteorites, 168
Methane (CH$_4$), 20–21, 131
Methanobacteria, 806
Mica (illite), 147, 701
Microclimates, 74–79
  of forest, lake and coast, 80
  urban microclimate, 74–79
Microevolution, 798
Microfauna, 456
Microflora, 456
Microjointing, 290
Micrometre or μm, 27
Microtidal regime, 451
Mid-Atlantic depression, 64
Middle and Late Cretaceous, greenhouse
    climate in, 834–838
Mid-latitude ice sheet, 547
Mid-ocean ridge systems (MORs), 178–179
  long-spreading ridges, 179
  slow-spreading ridges, 179
Milankovitch hypothesis
  on future climate change, 127, 847–851
  problems associated, 851
  'traditional' Milankovitch theory, 851
Milankovitch-type orbital oscillations, 783
Mineral matter, pathways of, 660–665
Mineralogy, 139–162
  amphibole groups, 146–147
  carbonate minerals, 147
  chlorites, 147
  clays, 147
  common groups of minerals, 144–148

garnet group, 145–146
  halides, 145–146
  Mohs scale of hardness for minerals,
      145
  nonsilicates, 144
  olivine group, 145
  silicates, 144
  sulphides, 145
  environments in which minerals form, 148
  'feldspar', 147
  identifying minerals, 141–144
    cleavage, 142
    colour, 141–142
    density, 142–144
    diamond, 144
    hardness, 142
    lustre, 142
    streak on a streak plate, 141
  micas, 147
  mineral, description, 140
  ore minerals, 147–148
  phosphate minerals, 147
  pyroxenes, 146–147
  quartz, 147
  sulphate minerals, 147
Moder, 699
Mohorovičić discontinuity, 169
Mohs scale of hardness for minerals, 145
Moisture in the atmosphere, 37–47
  global hydrological cycle, 38–42
Mollisols, 715
Molten magma ocean, 769
Montmorillinite, 701
Mor, 699
Moraine formation, 545
Moulins, 507
Mounds, 391, 394
Mount Huarascaran rock avalanche, 290
Mountain ice caps, 503
Mountain-building scenario, 221–222
Mud flows, 297, 369–370
Mudflats, 479, 481
  sea-level rise on, 496
Mud-mound factory, 383
Mudrocks, 155
Mudstone breccia, 546
Mudstone, 548
Mull, 698
Muscovite, 149
Mylonite, 231

National Oceanic and Atmospheric
    Administration (NOAA), 860
Natural levees, 353–355
Natural sources of acid rain, 22
Near-shore bar morphology, 461
Neck cutoff, 353
Needle ice development, 574
Negative feedback, 7
Neo trailing-edge coasts, 490
Neoproterozoic Eon, 828

Net radiation and energy balance, 35–36
Networks, 148
Neutral folds, 229
Nevado de Toluca volcano, 193
New England and Wales soil classification of
    soil, 716
  argillic brown earths, 717
  brown podzolic soils, 717
  gley soils, 718
  gley-podzols, 717
  lithomorphic (A/C) soils, 716
  man-made soils, 719
  paleo-argillic brown earths, 717
  pararendzinas, 717
  peat soils, 719
    earthy peat soils, 719
    raw peat soils, 719
  pelosols, 718
  podzols (sensu stricto), 717
  rankers, 716
  rendzinas, 716
  sand-pararendzinas, 717
  sand-rankers, 716
  stagnopodzols, 717
Niche or wall-sided glacier, 502–503
Nitisols, 721
Nitrogen cycle, 661–664
Nitrogen gas, 18
Nitrous oxide (N$_2$O), 131–132
Nival stream hydrographs, 601
Nivation processes, 583–585
Non-calcareous pelosols,
    718
Nonfoliated or recrystallized metamorphic
    rock, 161
Nonseasonal or random creep, 284
Nonsilicates, 144–146
  carbonates, 146
  halides, 146
  hydroxides, 146
  native elements, 146
  oxides, 146
  sulphates, 146
  sulphides, 146
North Atlantic and North American regions,
    860
North Atlantic Deep Water (NADW),
    109
Nuclear fusion, 28
Nuclear winter, 129, 822
Nutrients, soil, 703
Nye-channels, 507

Oblique bars, 461
Obliquity or axial tilt, 849
Obsidian, 154
Obstacles, dunes associated with, 627–628
Ocean circulation, change in, 789–795, 873
  global temperature variation, 793
  global temperature, 792–794
  ice ages, 792–794

Ocean circulation, change in (*continued*)
 sea-level change, 790–794
 supercontinental cycle, 790–794
 in warming climate, 794–795
 Wilson cycle, 791
Ocean structure and circulation patterns,
   101–110
 composition, 102
 continental margin topography, 102–104
 continental shelf break, 105
 deep-sea sediments, 105
 ocean circulation, 105–109
  deep-ocean currents, 107–109
  salinityy profile, 107
  surface ocean currents, 105–107
  vertical temperature, 107
 ophiolite suite layers, 103
 physical structure, 102–105
 sea-floor, 102–104
 sea-level change, 109–110
  meridional overturning circulation,
    109
  thermohaline circulation, 109
 temperature structure, 105
 thermocline, 105
 turbidity currents, 105
Ocean tides, 443
Oceans role in climate changes, 135
Offshore topography, 435
Oligocene Antarctic glaciation and its potential
   impacts, 841–843
Oligocene, 817
Olivine group of minerals, 145
Olivine, 149, 169, 769
One newton per square metre, 51
Open community, 665
Open flow system, 8–9
Open river channels, types of flow in, 322–323
Open-wave depression, 63
Ophiolite suite layers, 103
Ore minerals, 147–148
Organic matter in soil, 705, 707
Organic soils or peats, 708–711
Organic wastes and their decomposition,
   745
Organisms, soil, 704
Ornithischians, 813
Orographic and frontal uplift, 41
Orographic or relief rainfall, 46
Orthopyroxene, 769
Otto glacier, 512
Overconsolidation, 274
Overland flow, 300–302
Oviraptors, 813
Oxidation, 257
Oxides, 146
Oxisols, 716, 753
Oxygen, 18, 112–114
 fluctuations in, 114
Ozone holes, 20
Ozone in upper atmosphere, 19–20

Pachycephalosaurus, 813
Pacific and Indian Ocean Common Water
   (PICW), 109
Packestone, 388
Pahoehoe lava, 190
Pahoehoe types of lava, 191
Palaeocene supergreenhouse and changes in
   Eocene, 838–841
Palaeoclimate, 118
Palaeomagnetic data, 175
Palaeomagnetic measurements, 831
Palaeontology, 157, 803
Palaeosols, 731–732
Palaeozoic era, 833–834
Paleo-argillic brown Earths, 717
Paleocene, 817
Paleoproterozoic Eon, 828
Palsas, 582–583
Pans, 634
Parabolic dunes anchored by vegetation, 630
Pararendzinas, 717
Parasaurolophus, 813
Parasite chains, 659
Partially-mixed ormesotidal estuaries, 476
Particulate matter, 21
Passive margins, evolution, 219
Passive rifting, 184
Patch reef, 392
Patterned ground, 575
 Washburn's genetic classification of,
   575
Peat soils, 719
 earthy peat soils, 719
 raw peat soils, 719
Peaty gleys, 729
Pedospere, 708, 745
Peléan dome, 193–194
Pelean eruptions, 200
Pelosols, 718
 argillic pelosols, 718
 calcareous pelosols, 718
 non-calcareous pelosols, 718
Penck's model, 305
Periglacial alluviation, 600
Periglacial cambering, 601
Periglacial climate, wind action in, 597–600
Periglacial environment, fluvial processes in,
   600–602
Periglacial loess, 632
Periglacial processes, 448
 and landform–sediment assemblages,
   565–607
  characteristics, 566
  permafrost, 566
 lowland Britain, 607
 in upland Britain landscape shaping,
   603–607
Periglacial region, alluvial fans in, 602–603
Periodic table of the elements, 142
Permafrost, 566, 601
 alpine permafrost, 567

 continuous, 567
 discontinuous, 567
 sporadic permafrost, 567
 subsea permafrost, 567
 types, 567
Permo-Triassic period, 819–820
pH of soil, 702
Phaeozems, 720
Phanerozoic life story, 805
Phosphate minerals, 147
Phosphorus cycle, 663–665
Photogrammetric methods, 507
Phreas, 412
Phreatic cave passages, 413
Phyletic gradualism, 798
Phyllosilicates, 257
Physical or mechanical weathering, 242–251
 discontinuities, 243
 elastic limit, 242
 forces, 242–243
 rock strength, 242–243
Physical structure of oceans, 102–105
Physiographic subclimax, 679
Piedmont glaciers, 503
Piedmont lobes, 510
Pingos, 578–583
 Adventdalen, Spitsbergen, 580
 closed-system pingo formation, mechanism
   for, 581
 injected-ice stage, 579
 Mackenzie-Delta type, 580
 pore-ice stage, 579
 segregated-ice stage, 579
 Skeldal, East Greenland type, 580
Pits, weathering, 264–265
Plagioclase feldspar, 150
Plagioclimax conditions, 679–681
Planar slide, 290
Plane beds, 618
Planetary accretion, 166
Planosols, 720
Plastic deformation of ice under its own
   weight, 505
Plastic limit, 274
Plate area, 176
Plate motion mechanisms, 177
Plate tectonics, 173–175, 217–222, 491, 870
 back-arc basin, 176
 convergent plate boundaries,
   geomorphology, 176–177
 divergent plate boundaries, geomorphology,
   175–176
 fore-arc basin, 176
 global seismicity, 175
 global tectonics, 174
 late motion mechanisms, 177
 palaeomagnetic data, 175
 transform margins, geomorphology of,
   177
Platea, 426
Platy or lenticular structure, 722

Playas, 365
Pleistocene period, 340
Pleistocene, 644
Plinian eruptions, 197
Plinthosols, 720
Ploughing block movement, 596
Ploughing blocks or boulders, 595
Plugging, 353
Plume tectonics, 217–219
Plumes, mantle, 772–775
Plum-pudding model for Earth's mantle, 776
Plunging breakers, 435
Podzol, 708
Podzolic soils, 717
Podzolization, 710
    deposition, 711
    inorganic theory, 711
    organic theory, 711–713
    in situ weathering, 711
    translocation, 711
Podzols (sensu stricto), 717, 720
Point bars, 362
    coarse-grained point bars, 362
    fine-grained point bars, 362
    morphological zones, 362
Polar climates, 98–99
Poljes, 406
Polygonal ground, 576
Polygonal karst, 406
Poorly drained soil, 705
Pore-ice stage, 579
Porosity, 243, 304
    primary, 243, 304
    secondary, 243, 304
Postglacial period, 123
Potassium cycle, 665–666
Potential evapotranspiration (PET), 95
Powered flight in birds, 815–816
Precipitation, 44–47, 81–82
    clouds making, 44–46
    frontal or cyclonic rainfall, 47
    types of, 46–47
Predator chains, 659
Present atmospheric level (PAL), 782
Pressure gradient force (PFG), 55
Pressure gradients, 51–55
    coriolis effect, 53–55
    land and sea breezes, 53
    pressure change with altitude, 52
    surface winds, 53–55
'Pressure melting point', 507
Primary dunes, 476
Primary porosity, 243, 304
Primary productivity, 653
Priseres, 666
Producers, 651
Proglacial outwash plain (sandur),
    359–362
Protalus ramparts in upland Britain,
    formation, 584–585
Proterozoic Eon, divisions of, 828

mesoproterozoic, 828
neoproterozoic, 828
paleoproterozoic, 828
Proxy indicators for climate change,
    118–122
Psammoseres, 670–673
Pseudotachylite, 231
Pumice, 154
P-waves, 168
Pyramid of biomass, 660
Pyramid of numbers, 660
Pyrites, 148
Pyroclastic flows, 204–205
Pyroclastic materials, 200–210
    air-fall deposits, 204
    base surges, 203–204
    block-and-ash flows, 205
    cold base surges, 204
    convecting eruption column, 202
    deposition of, 200–210
    directly related to eruptions, 206
    hyperconcentrated flows, 205, 207
    indirectly related to eruptions, 206
    lahars, 205–207
    maars, 202
    pyroclastic density flows, 200–210
    pyroclastic fall or tephra deposits, 200
    pyroclastic flow deposits, 205
        associated materials, 205
        components and chemical
            composition, 205
        flow units, 205
        texture, 205
        welding and the effects of high
            temperaturess, 205
    transport of, 200–210
    tuff cones, 202
    tuff rings, 202
Pyroxenes, 146–147, 149, 169
Pyrrhotites, 148

Qattara depression, 637
Quarrying, 452
Quartz, 147, 150
Quartzite, 162, 365
Quaternary geological period, 122

Radiant energy, 8
Radiation (contact) cooling, 41
Radiation, 26, 39
    radiation laws, 27–29
        Stefan–Boltzmann law, 27
Radiative effects, 211–212
Radioactive isotopes, 140
Radiocarbon dating, 119
Rainfall–runoff responses, 328–329
Rainfall seasonality, 258
Rainsplash, 298–299
Raised bog, 677
Ramps, 488
Ranker-like alluvial soils, 716

Rankers, 716
Rapid cooling in igneous rocks, 153
6–7 Rapid flow, 335
Rapid Quaternary climate change, 845–846
    ice-core isotopic record, 845
    oxygen-isotope record in ocean, 845–846
Rapids, development, 371
Rare single-celled fossils, 806
Raw peat soils, 719
Reaction-limited chemical weathering, 254
Red fescue, 678
Reduction, 257
Reefs, 391–394
    evolution on the pacific plate, 489
Reflected waves, 437
Reflective beaches, 471
Reg, 635
Regional metamorphism, 159
Regional scale, 7
Regosols, 721
Regression, 792
Rejuvenation, 340
Relative humidity (RH), 41, 80
Relief, 688
    altitude affects, 688
    aspect, 688
    morphology, 688
Remnant dunes, 476
Rendzinas, 716
Renewable technologies, 875
Reorientation of skeletal grains, 722
Reptation, 615
Reptilian radiation, 811–812
    coelurosaurs, 813
    Deinonychus, 814
    dinosaurs and their predecessors, 812–816
        avialae, 812
        coelophysis, 812
        eoraptor, 812
        maniraptora, 812
        Meilong, 813
        theropoda, 812
        Unenlagia, 813
    Lesothosaurus, 813
    mammal-like reptiles, 811
    ostrich dinosaurs, 813
    oviraptors, 813
    Pachycephalosaurusa, 813
    parasaurolophus, 813
    Tenontosaurus, 814
    Triassicsynapsid group, 811
Re-radiation, 34–35
Retreat, 448
Reynolds number, 323
Rhyolite, 154
Rhyolitic magma, 188
Rhythmic beach landforms, 460
Rhythmite, 546, 548
Ribbon reef, 392
Ridges, 460–471
Rifting, 220

Rill systems, 331
Rillenkarren, 403
Rills, 300–301
Rimmed carbonate platforms, 389
Rimstone pools, 426
Rinds, weathering, 264
Rip-currents, 461
Ripple marks, 159
Ripples, 336, 617–618
 'aerodynamic ripples', 617
 granule ripples, 618
River channel adjustments, 339–340
River hydrology, 328–329
River, energy of, 323–325
River-dominated coastal landforms, 481
 gently-angled bottom, 482
 inclined foresets, 482
 topsets 482
Rocks, 148–151
 'building blocks' of, 148
  'fool's gold', 150
  apatite, 150
  augite crystal, 149
  biotite, 149
  calcite, 150
  chlorite, 149
  galena, 150
  gypsum, 150
  hematite, 150
  hornblende, 149
  kaolinite, 149
  muscovite, 149
  olivine, 149
  plagioclase feldspar, 150
  pyroxene, 149
  quartz, 150
  uraninite, 150
 classification, 151
 fossils, 153–159
 hydraulic properties of, 243
 igneous rocks, 151–153, See also individual
  entry
 metamorphic rocks, 151, 159–162, See also
  individual entry
 rock cycle, 148–151
 rock deformation, controls on, 229
 rockfalls, 285–295
 rock-forming minerals, 251–252
 rock glide, 290
 rock mass, stages, 241
 rock salt, 157
 rock slide, 290
 rock topples, 289
 rock weathering, 455
 sedimentary rocks, 153–159, See also
  individual entry
 sedimentary structures, 153–159
Rockslides/avalanches, 285–295
Rocky coasts, 449
Rodinia, 773
Rolling, 325

Rossby waves, 55–58
Rotating tidal system, development,
 444
Rotational slumps, 295
Röthlisberger-channels, 507
Rundkarren, 403
Running raptor hypothesis, 815

Saffir–Simpson scale for hurricanes, 69
Salinity effect on modern animals and plants in
 the marine environment, 385
Salinization, 695
Salt marshes, sea-level rise on, 498
Salt-marsh successions, 677–679
Salt weathering, 249
Saltation, 614–615
Saltmarshes, 479
Salt-wedge estuaries, 476
San Andreas fault, 177, 232
San Jacinto fault zone, 232
Sand-bed rivers, 335–336
 small-scale current ripples, 335
 tranquil flow, 335
 rapid flow, 335
Sand movement, 613–614
Sand-pararendzinas, 717
Sand-rankers, 716
Sand seas, 629
Sand shoals of platform margins, 400
Sand streaks, 629
Sand transport, 621
Sand waves, 336
Sand wedges, 575
Sandstone, 155, 156
Sandur, 359–362
Sandy beaches, 471
Sandy gley soils, 718
Saprophyte chains, 659
Saturation vapour pressure, 39
Savanna ecozone, 750–757
 animals, 754–755
 climate of the seasonal tropics, 752–753
 environmental relationships in, 752
 savanna fires, 755
 vegetation characteristics, 753–754
 weathering regime, 753
Scales in space and time, 6–7
 regional scale, 7
 time scales, 7
Scallops, 415
Schist, 162
Sea breeze, 53, 71–73
Sea cliff erosion, 452
Sea-floor, 102–104
Sea-level change, 109–110, 790–794, 872–873
 responses to, 488
Sea waves, 433–439
 and shore-normal and longshore currents,
  433–439
Seasonal creep, 284
Seasonal tropics, 750–757

Secondary porosity, 243
Sedan nuclear test, 203
Sediment entrained basally, 517
Sediment entrainment, 613–614
Sediment load
 transport of, 325–327
 types, 327–328
Sediment transport, 473
Sedimentary basins
 closure of, 219–221
 opening of, 219
Sedimentary rocks, 151, 153–159
 biogenic sedimentary rocks, 158
 chemical sedimentary rocks, 155, 158
 classifying, 155–157
 clastic sedimentary rocks, 155
 detrital sedimentary rocks, 155, 158
Sedimentary structures, 153–159
Sedimentation in fjords, 529
Sediments, 431–498
Sefton coast dune system, 730
 dune ridges, 730
 dune slacks, 730
 pinewoods, 730
Segregated-ice stage, 579
Seismic waves as Earth probes, 168–171
 P-waves, 168–170
 S-waves, 168–170
 velocity variation, 170
Sensible heat, 26
Sepkoski's three great faunas, 804
Serir, 635
Settling velocity, 326
Shale, 156
Shear forces, 322
Shear velocity, 621
Shield volcano forms, 191
Shoal-water carbonate accumulations,
 geometry of, 386–387
Shoal-water carbonate systems, growth
 potential, 384
Shooting, 323
Shore platforms, 447
Shore-normal beach profiles, 457
Shore-normal currents, 433–439
Silica tetrahedra, 148
 continuous chains, 148
 continuous sheets, 148
 isolated tetrahedra, 148
 networks, 148
Silicates, 144–145
Siliciclastic beach–open shelf model, 398
Siliciclastic beach–shelf model, 398
Siliciclastic shelf models, 398
Siliciclastics, 389
 mounds, 391
 reefs, 391
 shore-to-slope profiles of, 391
  carbonate ramp, 391
  rimmed platform, 391
  siliciclastic shelf, 391

tidal flat and lagoonal environments, 391
Siltstone, 156
Single climax community, 667
Skeletal grain coatings, 722
Slab failures, 285–295
Slate, 162
Sleet, 46
Slides, 295–296
Slope morphology
    measuring, 305–307
        annotated slope profile, 306
    and its evolution, 305–318
Slope processes associated with short summer
    melt season, 593–596
Slopes, 269–318
    antecedence, 315–318
    on Brule formation, 307
    on Chadron formation, 307
    continuous and ubiquitous, 272
    continuous but localized, 272
    controls on process and slope form, 311–318
        climate, 313–315
        lithology, 311–313
    discontinuous and ubiquitous, 272
    discontinuous but localized, 272
    evolution of, 317
        high-magnitude or catastrophic events,
            317
    Factor of Safety (FS), 272
    mass movement, 270–297
    process and form, 307–311
    slope strength, 272–277
        capillary tension, 274
        cementation, 274
        continuity of partings, 272
        electrical tension, 274
        high shear stress, contributing factors,
            276
        infill along partings, 274
        low shear strength, contributing
            factors, 277
        orientation of partings, 272
        outflow of water, 272
        overconsolidation, 274
        quaternary sediments, 274
        rock mass strength classification and
            ratings, 273
        scale of mass weathering grades, 273
        spacing of partings, 272
        width of partings, 272
    slope stress, 277–282
        lateral pressure, 278
        lateral support removal, 277
        slope loading, 278
        underlying support removal, 278
Slow cooling in igneous rocks, 153
Slow-spreading ridges, 179
Small-scale cross stratification, 352
Small-scale cryoplanation processes, 588
Small-scale current ripples, 335
Snowball Earth, 828–833

    arguments against, 832–833
    evidence for, 782, 828–829
    initiation, 830
    life survival, 833
Snowbanks, processes associated with,
    583–585
Snowflakes, 46
Soil chemistry, 700–703
    calcium carbonate content, 702
    cation-exchange capacity, 700–702
    chlorite (vermiculite), 701
    colloids, 700–702
    degree of humification, 701
        medium humified peat, 701
        slightly humified peat, 701
        strongly humified peat, 701
    description in the field, 704–707
        dig the soil pit, 704
        pit size, 704
        site characteristics, 704
    Kaolinite, 701
    Mica (illite), 701
    Montmorillinite, 701
    pH, 702
    soil nutrients, 703
    soil organisms, 704
    soil water, 703–704
Soil classification, 712–720
    genetic method or Dokuchaeiv approach,
        712–714
        azonal, 713
        intrazonal, 712
        zonal, 712
Soil creep, 284–285
    continuous creep, 284
    movement in, 286
    nonseasonal or random creep, 284
    qualitative indices for, 287
    seasonal creep, 284
Soil falls, 295–296
Soil flows, 297
Soil organic matter, 698–700
Soil profile description, 704–707
    hand-texturing of soils, 706
    profile description, 705–707
        colour, 705
        depth of horizon and thickness in cm,
            705
        moisture, 705
        organic matter, 705
        stoniness, 705–707
        texture, 705
    profile drainage, 704–705
        freely drained, 705
        imperfectly drained, 705
        poorly drained, 705
Soil profile development, 690–696
    horizons in soils derived from peat, 691
    horizons, 690–693
    mineral sub-surface horizons, 691
    nomenclature, 690–693

    organic and organic–mineral surface
Soil properties, 696–704
    Atterberg Scale, 696–697
    grain size, 696–698
        moderate, 698
        strong, 698
        structureless or massive, 698
        weak, 698
    United States Department of Agriculture
        'American' Scale, 697–698
Soil, types
    acid brown earth, 707
    based on texture and structure, 698
        clayey, 698
        loamy, 698
        sandy, 698
    blocky, 700
    crumb, 700
    gley, 708
    granular, 700
    grey brown podzolic, 707–708
    peaty gleyed podzol, 708
    platy, 700
    podzol, 708
    prismatic, 700
Soil-forming processes and products, 685–732
    biological processes, 695–696
    British classification, 714–719
    Brown and red Mediterranean soils,
        723–724
    calcareous soils, 711
    calcification, 694
    Chernozems (USDA mollisol, suborder
        boroll), 723
    Chestnut-brown soils (USDA mollisol,
        suborder xeroll), 723
    clay loam, 687
    clay, 687
    controls on, 685–689
    ferrallitization, 695
    ferrugination, 695
    Food and Agriculture Organization soil
        reference groups (1998), 720
    gleying, 693
    intrazonal upland soils in Britain, 724–728
    loamy sand, 687
    New England and Wales soil classification,
        716, See also individual entry
    parent materials, 688–689
    podzolization, 710
    regional and local soil distribution, 720–729
        arctic, tundra soils, 721–723
        zonal soils, 720–724
    salinization, 695
    sand, 687
    sandy clay loam, 687
    sandy clay, 687
    sandy loam, 687
    sandy silt loam, 687
    silt loam, 687
    silty clay loam, 687

Soil-forming processes and products
        (*continued*)
    silty clay, 687
    soils as systems, 689–690
    time, 689
    USDA (7th Approximation) Classification,
        714
Soils, 750, 753
    alfisols, 753
    ultisols, 753
Solar activity, 863
Solar constant, 28–30
Solar energy cascade, 32–34
'Solar forcing', 29
Solar radiation, 26
Solonchaks, 720
Solonetz, 720
Solutional depressions, 404
    solutional dolines, 404
    uvulas, 404
Solutional dolines, 404
Solutional microforms, classification, 405
Special Report on Emissions Scenarios (SRES),
        783–784
Specific humidity, 40
Speleothems, 423
Sphaira (meaning sphere), 18
Spilling breakers, 435
Spine, 193–194
Spiral vortices, development, 620
Spits, 460–471
Spitsbergen catchments, 602
Spodosols, 715
Sporadic permafrost, 567
Spring tide, 443
Stable shelf, 767
Stadial, 828
Stagnogley soils, 718
Stagnohumic gley soils, 719
Stagnopodzols, 717
Stalagmites, 434
Star dunes, 626–627
Static equilibrium, 10
Steady-state equilibrium, 10
Steele glacier, 512
Steep beaches, 439, 457
Steep rock slopes, 234
Stefan's law, 28
Stefan–Boltzmann law, 27
Stevenson screen, 81
Stokes' Law of Settling, 326
Stone pavements, 634–637
Stoniness in soil, 705–707
Stony meteorites, 168
Stratification, 336
Stratified sandstone, 548
Stratosphere, 18
Stratospheric dynamics, volcanic effects on,
        211
Strato-volcano forms, 192
Streak on a streak plate, 141

Streaming, 323
Stress, 229
Strike-slip faults, 232
Strike-slip geomorphology, 233
Subaerial erosion, 243
Subaerial processes, 199
Subaqueous basaltic lavas, 190
Subaqueous eruptions, 198
Subaqueous silica-rich, 190
Subclimax conditions, 679–681
    biotic, 679
    edaphic, 679
    physiographic, 679
    sub-climax vegetation, 681
Subduction, 180
Subglacial basaltic volcanism, 559
Subglacial bedforms, 538
Subglacial channels, 507
Subglacial deformation, 511
Subglacial eruptions, 198
Subglacial rhyolite, 198
Subglacial sediments, 517
Sublimation, 509
Submarine debris avalanches, 207
Subsea permafrost, 567
Sub-sere, 679
Subsurface flow and piping, 302
Subsystems in the Earth, 12
Subtropical jet stream, 58
Sullivan, 645
Sulphate minerals, 147
Sulphates, 146
Sulphides, 146
Sulphur cycle, 662–664
Sulphur dioxide (SO_2), 210
Sun, 6
    sun's energy output, variations in, 852
Supercontinents, 217–219
    formation of, 783
    supercontinent cycle, 790–794
Superplumes, 217–219
Supraglacial sediments, 519–520
Surf zone, 437
Surface ocean currents, 105–107
Surface solutional landforms, 403–410
    cave formation processes and landforms, 410
    collapse dolines, 405
    Karren, 403
    limestone pavements, 410
        clints, 410
    shakehole development, 406
    solutional depressions, 404
    stream resurgences and surface valleys,
        407–410
Surface uplift, 232
Surface winds, 53–55
Surging glaciers, 512–513, 555
    active phase, 512
    inactive phase, 512
    theories to explain, 514
Surging or galloping glaciers, 512

Suspension, 325, 523, 615–616
Swale fills, 352
Swallow holes, 407
Swash (uprush), 439
S-waves, 168
Syncline, 229
Synforms, 229
Systems and feedback, 7–8
Systems Theory, 11

Table reef, 392
'Tafoni', 264–265
Tectonic basins within plate interiors, 222
Tectonic denudation, 243
Tectonic or fault breccia, 231
Tectonic structures as lines of weakness in
        landscape evolution, 234
Temperature, 26, 39, 80–81
    inversion, 18
    and rainfall seasonality, relationship, 258
    regimes, global, 90–91
    structure of oceans, 105
    temperate/mesothermal climates, 97–98
Tenontosaurus, 814
Tephra from subaqueous eruptions, 200
Terranes, 220
Terrestrial planets, 5
Textures of carbonate rocks, 387
The Origin of Continents and Oceans, 175
Thermal plumes, 181–182
Thermal stratification in glacial lakes, 523
Thermistors, 80
Thermocline, 105
Thermohaline circulation, 109
Thermohaline-circulation (THC) shutdown,
        859
Thermometer shelter, 81
Thermosphere, 18
Theropoda, 812
Tholeiitic flood basalts, 182
Thufurs, 581–582
Tidal bulge, 443
Tidal flat and lagoonal environments, 391
Tide-dominated coastal landform assemblages,
        473
Tides, 433–445, 451
    bulge, 439
Time cycles in systems, 11–13
Time scales, 7
Toba supervolcano, 215–216
Tombolo, 463–464
Tonalite-trondhjemite granodiorite (TTG),
        767
Topples, 295–296
Toppling failure, 285–295
Tornados, 70–71
Tors
    Dartmoor tors, 592
    development, 588–593
    Linton's hypothesis, 592
Tortula, 672

Trace fossils, 157
Trade winds, 90
Transection glaciers, 502
Transform faults, 177
Transform margins, geomorphology of, 177
Transgressive dunes, 476
Transitional forms of bed roughness, 335
Translational slides, 295
Translocation processes, 693, 711
cheluviation, 693
clay translocation or lessivage, 693
leaching, 693
Transport-limited chemical weathering, 254
Transpression, 222
Transverse bars, 357
Transverse dunes, 620–622
Transverse sand bars, 461
Tranquil flow, 335
Trewartha's system, 99
Tricellular model, 55, 57
Trilobite fossil, 158
Tripartite glacial sediment development,553
Trittkarren, 403
Tropical (equatorial) rain forest, 744–750
Tropical cyclones, 68–70
Tropical factory, 381–382
Tropical rain forest, 746–751
animal life, 750
ecosystem model of, 749
functional organization, 748–750
soils, 750
structural organization, 747–748
vulnerability, 750
Tropical/megathermal climates, 92–95
tropical rainforest climates (Af), 95
tropical monsoon climates (Am), 95
Troposphere, 18
Tropospheric cooling, 212
Tsunamis, 433–439, 445
Tuff, 154
tuff cones, 202
tuff rings, 202
Tundra climate, 98–99
Tundra ecozone, 737–744
animal life, 745
climate of, 739–741
climatic environmental conditions, 745
desert, 738
environmental conditions and vegetation,
interactions between, 741
forest, 738
grassland, 738
insolation, 745
land use, 745
lithosphere, 745
organic wastes and their decomposition, 745
pedospere, 745
savanna, 738
terrestrial biomes, 738
tundra diversity, 742–743
tundra ecosystem structure, 742

tundra plant communities, 743–744
arctic barrens, 743
bogs andmarsh, 743
fellfields, 743
grass sedge tundra, 743
tundra scrub, 743
tundra soil system, 742
tundra, 738
vegetation, 745
Tundra heath, 743
Tundra scrub, 743
Turbidity currents, 105
Turbulent flow, 322–323
types of, 323
shooting, 323
streaming, 323
Turloughs (dry lakes), 407
Tyrannosaurs, 813
Tyrannosaurus rex, 814

Ultisols, 716, 753
Umbrisols, 721
Unenlagia, 813
Uniformitarianism, 217–222
United Nations Environment Programme
(UNEP), 783
United Nations Framework Convention on
Climate Change (UNFCCC), 869
United States Department of Agriculture
'American' Scale, 697–698
Upheaved plug, 193–194
Upland soils in Britain, 724–728
Uplift, 232
crustal uplift, 232
surface uplift, 232
Upper flow regime, 338
Upper-phase plane beds, 335
Uraninite, 150
Urban microclimate, 74–79
Urban surfaces, wind at, 77–78
USDA (7th Approximation) Classification,
soil, 714–716
Uvulas, 404

Vadose cave passages, 412, 414–417
Valley asymmetry, 602
Valley bulging, 599
Valley glacier of the alpine type, 502
Valley systems (glaciers constrained by
topography), 550
Variegated glacier, 513
Vegetation succession, 665–681
bracken, 672
building stage, 667
calluna, 672
dune scrub, 672
dune seral relationships, 672
ecesis, 668
forest, 672
grassland, 672
mature (climax) stage, 667

pioneer or invasion stage, 666
psammoseres, 670–673
reaction, 668
stabilization, 668
Vegetation types in hypothetical continent, 737
Ventifacts, 597, 634
Vermillion Cliffs area, 159
Vernier scale, 40
Vertical accretion, 356
Vertisols, 715, 720, 753
Very poorly drained soil, 705
Vesicular voids, 722
Volcanic aerosols, 210–211
Volcanic basin, 767
Volcanic eruptions and biotic evolution,
relationships, 216–217
flood basalts, 216–217
mass extinctions, 216–217
Volcanic processes, 177–186, See also Effusive
volcanism; Explosive volcanism
active continental margin, 181
andesitic magma, 188
asaltic magma, 188
basaltic magma eruption, 185–186
at convergent plate boundaries, 179–180
destructive plate boundaries where one plate
is continental, 180–181
global tectonic model, 177–186
Hawaiian volcano, 183
Hawaiian–Emperor volcanic chain, 181
lava, physical properties, 187–188
linear volcanic chains, hot-spot model for,
182
magma generation, 177–178
mid-ocean ridge systems (MORs), 178–179
passive rifting, 184
rhyolitic magma, 188
subduction, 180
thermal plumes or hot spots, 181–182
tholeiitic flood basalts, 182
volcanic arc, 180
Wadati–Benioff zone, 180
Volcanic processes and the Earth's atmosphere
and climate, 210–216
dynamic emplacement models for debris
avalanches, 211
El Niño–Southern oscillation and volcanic
events, 212–214
global atmospheric effects of volcanic
processes, 210–215
volcanic aerosols, 210–211
Handler's model linking aerosol production,
214
tropospheric cooling, 212
volcanic activity and climatic change, links
between, 214–215
'Volcanic winter', 215
Volcaniclastic sediments, petrographic
features, 200
Volcanism, 173–222, 852
Volcanoes, 129–131

Volcanoes (*continued*)
  Earth's composition from, 167
Volumetric change of a rock mass, 243–246
  exfoliation of granite, 244
  subaerial erosion, 243
  tectonic denudation, 243
Vulcanicity, 151–153
  extrusive, 152
  intrusive, 152

Wackestone, 388
Wadati–Benioff zone, 175, 176, 180
Warm summer continental or hemiboreal
  climates, 98
Warm-arm-based continental ice sheets, 537
  landscapes of areal scouring, 537
  subglacial bedforms, 538
Warming climate, ocean circulation in,
  794–795
Washburn's genetic classification of patterned
  ground, 575
Watchtower, 592
Water on Earth, 6
Water turbulence, 338
Water vapour, 21
Water, soil, 703–704
Waterfalls, 371
'Waterspout', 70
Water-table caves, 420
Wave dispersion, 434
Wave-dominated beach, 472
Wave-dominated coastal landform
  assemblages, 447
  bevelled cliff, 448
  cliffs, 447
  coastal slopes, 447
  shore platforms, 447
Wave heights, 491
Wave-induced cliff erosion, 451
Wave motion from deep to shallow water, 435
Wave processes, 438
  deep water, 438

  intermediate and shallow water, 438
  surf zone, 438
  swash zone, 438
Wave refraction, 436
Wave setup, 438
Weakly stratified diamictite, 546, 548
Weakly stratified mudstone, 546
Weakly stratified sandstone, 546
Weather fronts, formation, 62–63
Weather maps, 82–86
Weather observation and forecasting,
  79–86
Weather systems, 61–86, *See also* Macroscale
  synoptic systems
  forecasting, 79–86
  weather observation, 79–86
Weathering regime, 746
Weathering, 239–267, 688, *See also* Physical or
  mechanical weathering
  advanced weathering stages, 688
  biological, 240
  cavernous, 265–267
  chemical, 240
  early weathering stages, 688
  frost weathering, 247
  fully-developed weathering profile on
    granitic rock, 242
  insolation weathering, 245
  intermediate weathering stages, 688
  landforms, 262–264
  Lichen weathering, 247
  mechanical, 240
  pits, 264–265
  rates, measuring, 262
  rinds, 264
  rock mass, stages, 241
  salt weathering, 249
  traditional weathering, 240
  volumetric change within pores, voids and
    fissures, 246–250
Webs, 657–660
Wein's law, 27–28

Well stratified diamictite, 546, 548
Well-stratified mudstone, 546
Well-stratified sandstone, 546
Westerlies, 90
Wet-bulb thermometer, 83
Wet volcano, 870
Wilson Cycle, 219, 791
Wilson model, 399
Wind erosion landforms, 633–637
  pans, 634
  stone or desert pavements, 635–637
  ventifacts, 634
  yardangs, 634
  Zeugen, 635
Wind speed, 39
Wind-eroded depressions and plains, 637
Wind transport, process, 614–616
Wind speed and direction, 80
Winds, 51–55
'Whole Earth' tectonic model, 218
World climates, 89–100
  global precipitation regimes, 92
  global temperature regimes, 90–91
World ecosystems, 735–739
World Meteorological Organization (WMO),
  783

Xeroseres, 668

Yardangs, 634
'Year without a Summer', 129
Younger Dryas, 510, 856–858
  cold phase in the British uplands, 858
  as global event, 859
  happenings at the end of, 860
  occurrence, 857

Zeugen, 635
Zibar, 627
'Zipper rift' hypothesis, 832
Zonal soils, 720–724